ENCYCLOPEDIA OF APPLIED PHYSICS

SPONSORS

AMERICAN INSTITUTE OF PHYSICS
DEUTSCHE PHYSIKALISCHE GESELLSCHAFT
JAPAN SOCIETY OF APPLIED PHYSICS
PHYSICAL SOCIETY OF JAPAN

DEUTSCHE PHYSIKALISCHE GESELLSCHAFT

JAPAN SOCIETY OF APPLIED PHYSICS

PHYSICAL SOCIETY OF JAPAN

ENCYCLOPEDIA OF APPLIED PHYSICS

VOLUME 10
Mechanics, Classical to Monte-Carlo Methods

Edited by

GEORGE L. TRIGG

Associate Editors

EDUARDO S. VERA

WALTER GREULICH

Managing Editor

EDMUND H. IMMERGUT

Assistant Managing Editor

LINDEN BYASS

Editorial Assistant

CHRISTOPHER THOMAS MORAN

George L. Trigg
275 Beaver Dam Road
Brookhaven, New York 11719

Walter Greulich
Ziegeleiweg 30
D-69488 Birkenau
Federal Republic of Germany

Eduardo S. Vera
NTT Advanced Technology Corporation
NTT Interdisciplinary Research Laboratories
Musashino-shi, Tokyo 180, Japan.

Edmund H. Immergut
Linden Byass
Christopher Thomas Moran
2 Sidney Place
Brooklyn, New York 11201

Library of Congress Cataloging-in-Publication Data

Encyclopedia of applied physics.
(Revised for Vol. 10)

"Sponsors, American Institute of Physics . . . [et al]."
Includes bibliographical references and index.
1. Physics—Encyclopedias. 2. Engineering—Encyclopedias. I. Trigg, George L. II. Vera, Eduardo S. III. Greulich, Walter. IV. American Institute of Physics.
QC5.E543 1991 530′.05 91-8738
ISBN 1-56081-069-6 (VCH Publishers)

British Library Cataloguing in Publication Data

Encyclopedia of applied physics.
Vol. 10
I. Trigg, George L. (George Lockwood) II. Vera, Eduardo S. III. Greulich, Walter
621
ISBN 1-56081-058-0 (set)
ISBN 1-56081-069-6 (vol. 10)

Printed in the United States of America.

ISBN 3-527-28132-0 (Volume 10) VCH Verlagsgesellschaft mbH
ISBN 3-527-26841-3 (set) VCH Verlagsgesellschaft mbH

Printing History:
10 9 8 7 6 5 4 3 2 1

Published jointly by:

VCH Publishers, Inc.
220 East 23rd Street
New York, NY 10010

VCH Verlagsgesellschaft mbH
P.O. Box 10 11 61
69451 Weinheim
Federal Republic of Germany

VCH Publishers (UK) Ltd.
8 Wellington Court
Cambridge CB1 1HZ
United Kingdom

ADVISORY BOARD

EDITORIAL CONSULTANTS

MAIN ENTRIES

The subject matter in the *Encyclopedia of Applied Physics* is presented in approximately 500 individual articles, arranged alphabetically. The topics can be classified into 20 sections, similar to the AIP Physics and Astronomy Classification Scheme (PACS):

01	General Aspects: Mathematical, Computational, and Information Techniques
02	Measurement Science, General Devices and/or Methods
03	Nuclear and Elementary Particle Physics
04	Atomic and Molecular Physics
05	Electricity and Magnetism
06	Optics (classical and quantum)
07	Acoustics
08	Thermodynamics and Properties of Gases
09	Fluids and Plasma Physics
10	Condensed Matter A: Structure and Mechanical Properties
11	Condensed Matter B: Thermal, Acoustic, and Quantum Properties
12	Condensed Matter C: Electronic Properties
13	Condensed Matter D: Magnetic Properties
14	Condensed Matter E: Dielectrical and Optical Properties
15	Condensed Matter F: Surfaces and Interfaces
16	Materials Science
17	Physical Chemistry
18	Energy Research and Environmental Physics
19	Biophysics and Medical Physics
20	Geophysics, Meteorology, Space Physics, and Aeronautics

Each article has been assigned a code number consisting of two digits which denotes the section, and a letter which gives the type of article. There are six types: A = Devices, Equipment; B = Materials; C = Methods, Processes; D = Phenomena, Effects; E = Scientific or Technological Fields; F = Institutions, Companies, Societies and other organizations.

CONTRIBUTORS

J. Bass, Department of Physics and Astronomy, Michigan State University, East Lansing, Michigan, U.S.A., and Max-Planck-Institut für Festkörperforschung, Hochfeld Magnetlabor, Grenoble, France
Metals and Alloys, Conductivity in

K. Binder, Institut für Physik, Johannes-Gutenberg-Universität Mainz, Mainz, Germany
Monte-Carlo Methods

Reinhard Bruch, Department of Physics, University of Nevada, Reno, Nevada, U.S.A.
Molecular and Atomic Collision Processes with Ion Beams

Robert W. Cahn, Department of Materials Science and Metallurgy, Cambridge University, Cambridge, United Kingdom
Metallurgy, Physical

A. Y. Cho, AT&T Bell Laboratories, Murray Hill, New Jersey, U.S.A.
Molecular Beam Epitaxy

Julian L. Davis, Cranbury, New Jersey, U.S.A.
Mechanics, Classical

Fereshteh Ebrahimi, Materials Science and Engineering Department, University of Florida, Gainesville, Florida, U.S.A.
Metals and Alloys, Structure of

T. Egami, Department of Materials Science and Engineering, University of Pennsylvania, Philadelphia, Pennsylvania, U.S.A.
Metallic Glasses

Paul Engelking, Department of Chemistry, University of Oregon, Eugene, Oregon, U.S.A.
Molecules

R. J. Fischer, AT&T Bell Laboratories, Murray Hill, New Jersey, U.S.A.
Molecular Beam Epitaxy

Stephan Fülling, Department of Physics, University of Nevada, Reno, Nevada, U.S.A.
Molecular and Atomic Collision Processes with Ion Beams

Sigmar German, Physikalisch-Technische Bundesanstalt, Braunschweig, Germany
Metrology

Kevin Hamilton, Geophysical Fluid Dynamics Laboratory/NOAA, Princeton University, Princeton, New Jersey, U.S.A.
Meteorology and Climatology

Kohji Hohkawa, Faculty of Engineering, Kanagawa Institute of Technology, Atsugi, Kanagawa, Japan
Modulators and Demodulators, Electrical

Joanna M. Hunter, Department of Chemistry, Northwestern University, Evanston, Illinois, U.S.A.
Molecular and Atomic Clusters

Irving Itzkan, G.R. Harrison Spectroscopy Laboratory, Massachusetts Institute of Technology, Cambridge, Massachusetts, U.S.A.
Medical Use of Lasers

Joseph A. Izatt, Research Laboratory of Electronics, Department of Electrical Engineering and Computer Science, Massachusetts Institute of Technology, Cambridge, Massachusetts, U.S.A.
Medical Use of Lasers

Martin F. Jarrold, Department of Chemistry, Northwestern University, Evanston, Illinois, U.S.A.
Molecular and Atomic Clusters

Michael J. Kaufman, Materials Science and Engineering Department, University of Florida, Gainesville, Florida, U.S.A.
Metals and Alloys, Structure of

Mats Larsson, Manne Siegbahn Institute of Physics, Stockholm, Sweden
Molecular and Atomic Collision Processes with Ion Beams

Sven Mannervik, Manne Siegbahn Institute of Physics, Stockholm, Sweden
Molecular and Atomic Collision Processes with Ion Beams

Giorgio Margaritondo, École Polytechnique Fédérale de Lausanne, Lausanne, Switzerland
Monochromators

Simon Middelhoek, Department of Electrical Engineering, Delft University of Technology, Delft, The Netherlands
Microsensors

A. Morino, NEC Corporation, Kawasaki, Kanagawa, Japan
Microelectronics

Yoshimasa Murayama, Advanced Re-

search Laboratory, Hitachi, Ltd., Hatoyama, Saitama, Japan
Mesoscopic Systems

D. A. Ramsay, Herzberg Institute of Astrophysics, National Research Council of Canada, Ottawa, Ontario, Canada
Molecular Spectroscopy

Elizabeth Rauscher, Department of Physics, University of Nevada, Reno, Nevada, U.S.A.
Molecular and Atomic Collision Processes with Ion Beams

T. Maurice Rice, Eidgenössische Technische Hochschule, Zürich, Switzerland
Metal-Insulator Transitions

P. L. Rossiter, Department of Materials Engineering, Monash University, Clayton, Victoria, Australia
Metals and Alloys, Conductivity in

Dieter Schneider, Lawrence Livermore National Laboratory, Livermore, California, U.S.A.
Molecular and Atomic Collision Processes with Ion Beams

Henry I. Smith, Department of Electrical Engineering and Computer Science, Massachusetts Institute of Technology, Cambridge, Massachusetts, U.S.A.
Microlithography

Peter Staecker, M/A-COM, Inc., Corporate Research and Development Center, Lowell, Massachusetts, U.S.A.
Microwave Circuits

Jean-Louis Staehli, École Polytechnique Fédérale de Lausanne, Lausanne, Switzerland
Monochromators

T. Sugano, RIKEN, Wako, Saitama, Japan
Microelectronics

Nick Tredennick, Tredennick, Inc., Los Gatos, California, U.S.A.
Microprocessors

Wolfgang Wöger, Physikalisch-Technische Bundesanstalt, Braunschweig, Germany
Metrology

S. K. Yao, Optech Laboratory, Santa Ana, California, U.S.A.
Modulators and Demodulators, Optical

Celia E. Yeack-Scranton,* IBM Corporation, San Jose, California, U.S.A.
Memories

E. H. Young, NEOS Technologies, Inc., Melbourne, Florida, U.S.A.
Modulators and Demodulators, Optical

---

*Deceased

MECHANICS, CLASSICAL

JULIAN L. DAVIS, *Cranbury, New Jersey, U.S.A.*

> Science is built on facts, as a house is with stones. But a collection of facts is no more a science than a heap of stones a house. . . . Scientists believe that there is a hierarchy of facts and that a judicious selection can be made. One has only to open one's eyes to see that the triumphs of industry, which have enriched so many practical men, would never have seen the light if only these practical men had existed, and if they had not been preceded by disinterested fools who died poor, who never thought of the useful, and yet had a guide that was not their own caprice. What fools did, as Mach has said, was to save their successors the trouble of thinking. . . . It is necessary to think for those who do not like thinking, and as they are many, each one of our thoughts must be useful in as many circumstances as possible. For this reason, the more general a law is, the greater its value.
>
> Henri Poincaré (1902)

INTRODUCTION

This article is an advanced treatment of classical mechanics, sometimes called *analytical dynamics*. The material presupposes an elementary knowledge of mechanics such as that given in an undergraduate course. Some of the references in the list of Further Read-

3-527-28132-0/94/$5.00 + .50

ing give a full discussion of these aspects of mechanics. Particularly recommended is Synge and Griffith (1959).

Analytical dynamics is described by Whittaker (1974) as "that branch of knowledge in which the motions of material bodies, considered as due to the mutual interactions of the bodies, are discussed by the aid of mathematical analysis." In this spirit, mathematical analysis is used and developed within the context of the material. Mathematics is not used for its own sake, but as a powerful tool to codify and explain complex physical phenomena arising in mechanics. The material to be covered is "classical" in the sense that it will not include the special and general theories of relativity. Neither will it discuss quantum mechanics. These topics are adequately discussed in other articles in this Encyclopedia (RELATIVITY, SPECIAL; GRAVITATION AND GENERAL RELATIVITY; QUANTUM MECHANICS).

1. PARTICLE MECHANICS—A SYSTEM OF PARTICLES

Consider a system of n particles acted on by external forces and internal forces. Let $\mathbf{F}_i$ be the external force acting on the ith particle and $\mathbf{f}_{ij}$ be the internal force acting on the ith particle due to the jth particle. (Clearly, $\mathbf{f}_{ii}$ is zero.) Let $\mathbf{p}_i = m_i d\mathbf{r}_i/dt$, where m_i is the mass of the ith particle, be the linear momentum of the ith particle. Then Newton's second law yields the following equations of motion, one for each of the n particles:

$$\sum_j (\mathbf{f}_{ij} + \mathbf{F}_i) = \frac{d\mathbf{p}_i}{dt} = \dot{\mathbf{p}}_i, \tag{1}$$

where i and j range from 1 to n. Note that there are n equations, in each of which we sum all the interactions due to the other $n - 1$ particles. All the forces are assumed to obey Newton's third law.

1.1 Newton's Laws of Motion

We now obtain equations of motion in terms of the position of each particle. Consider an inertial reference system, defined effectively as one relative to which a particle's motion is determined only by the real forces exerted on it. A frame of reference fixed relative to the sun and having a fixed orientation relative to the fixed stars forms an excellent approximation to an inertial system; a frame of reference fixed on the earth's surface does not, as there are apparent forces (centrifugal and Coriolis) that affect the motion of a particle relative to such a frame. Let $\mathbf{r}_i$ be the radius vector of the ith particle. Then the linear velocity of the ith particle is $\mathbf{v}_i = d\mathbf{r}_i/dt$. We now sum Eqs. (1) over i (over all the particles). The equations of motion (1) become

$$\frac{d^2}{dt^2}\sum_i m_i\mathbf{r}_i = \sum_i \mathbf{F}_i + \sum_{i,j(i\neq j)} \mathbf{f}_{ij}. \tag{2}$$

The first sum on the right is the total external force acting on the system of n particles, while the second sum vanishes—the terms cancel by pairs—because we have assumed Newton's third law, $\mathbf{f}_{ij} = -\mathbf{f}_{ji}$. We may simplify these equations of motion by introducing a vector $\mathbf{R}$ which is the mass-weighted average of the radii vectores of the particles. This means

$$\mathbf{R} = \sum m_i \frac{\mathbf{r}_i}{M}, \quad \text{where} \quad M = \sum m_i. \tag{3}$$

The vector $\mathbf{R}$ defines the *center of mass* or *center of gravity* of the system. Then Eq. (2) becomes

$$M\frac{d^2\mathbf{R}}{dt^2} = \sum_i \mathbf{F}_i \equiv \mathbf{F}, \tag{4}$$

where $\mathbf{F}$ is the sum (resultant) of all the external forces acting on the system. We see that the internal forces acting on each particle have no effect on the system. It is clear that the center of mass moves as if the resultant external force were acting on the entire mass M concentrated at the center of gravity. Examples occur in rocket and jet propulsion as well as in many other situations. For example, the motion of the fragments of an exploding shell is such that the center of mass of the fragments travels as if the shell were still in a single piece. Note that we must neglect air resistance, since the drag force due to air resistance depends on the shape and speed of the object. In this analysis we assume that the mass is independent of time.

Let **P** be the total linear momentum of the object. We then obtain

$$\mathbf{P} = \sum m_i \frac{d\mathbf{r}_i}{dt} = M \frac{d\mathbf{R}}{dt}. \tag{5}$$

Taking the time derivative of Eq. (5), we obtain the vector equation of motion for our system of particles:

$$\frac{d\mathbf{P}}{dt} \equiv \dot{\mathbf{P}} = \mathbf{F}. \tag{6}$$

1.2 Angular Momentum

We define the *angular momentum* of a particle as $\mathbf{L}_i = \mathbf{r}_i \times \mathbf{p}_i$. The total angular momentum of the system of particles is then

$$\mathbf{L} = \sum_i \mathbf{L}_i = \sum_i \mathbf{r}_i \times \mathbf{p}_i = \sum_i \mathbf{r}_i \times m_i \frac{d\mathbf{r}_i}{dt}. \tag{7}$$

Now set $\mathbf{r}_i = \mathbf{R} + \boldsymbol{\rho}_i$; it is seen that $\boldsymbol{\rho}_i$ is the radius vector of the ith particle relative to the center of mass. If we substitute this into Eq. (7), we obtain

$$\begin{aligned}\mathbf{L} &= \sum_i (\mathbf{R} + \boldsymbol{\rho}_i) \times m_i \frac{d}{dt}(\mathbf{R} + \boldsymbol{\rho}_i) \\ &= \sum_i (\mathbf{R} \times m_i\dot{\mathbf{R}} + \mathbf{R} \times m_i\dot{\boldsymbol{\rho}}_i \\ &\quad + \boldsymbol{\rho}_i \times m_i\dot{\mathbf{R}} + \boldsymbol{\rho}_i \times m_i\dot{\boldsymbol{\rho}}).\end{aligned} \tag{8}$$

But since $\boldsymbol{\rho}_i = \mathbf{R} - \mathbf{r}_i$, then

$$\begin{aligned}\sum_i m_i\boldsymbol{\rho}_i &= \sum_i m_i\mathbf{R} - \sum_i m_i\mathbf{r}_i \\ &= M\mathbf{R} - M\mathbf{R} \equiv 0,\end{aligned} \tag{9}$$

and so the second and third terms in the last form of Eq. (8) vanish. If we examine the forms of the remaining two terms, we have the important result that *the angular momentum of a system of particles about an arbitrary origin is the sum of the angular momentum about their center of mass and the angular momentum of the system, considered as a single particle concentrated at the center of mass, about the arbitrary origin.*

Next consider a force **F** acting at a point whose radius vector from some arbitrary origin is **R**. We define the moment of the force about the origin, or the *torque* about the origin, as $\mathbf{R} \times \mathbf{F}$. Returning to our system of particles, we can find the total torque **N** acting on it as

$$\begin{aligned}\mathbf{N} &= \sum_i \mathbf{r}_i \times \left(\mathbf{F}_i + \sum_j \mathbf{f}_{ij}\right) \\ &= \sum_i \mathbf{r}_i \times \mathbf{F}_i + \sum_{i,j} \mathbf{r}_i \times \mathbf{f}_{ij}.\end{aligned} \tag{10}$$

Now for every term $\mathbf{r}_i \times \mathbf{f}_{ij}$ in the double sum, there is a term $\mathbf{r}_j \times \mathbf{f}_{ji}$. We assume, as before, that the internal forces are equal and opposite, so that $\mathbf{f}_{ij} = -\mathbf{f}_{ji}$, and combine the two terms as $(\mathbf{r}_i - \mathbf{r}_j) \times \mathbf{f}_{ij}$. We now further assume that the internal forces are not only equal and opposite, but *central* forces, that is, that the forces between any two particles act along the line joining them. Then the terms in the double sum cancel by pairs, and we are left with

$$\mathbf{N} = \sum_i \mathbf{r}_i \times \mathbf{F}_i: \tag{11}$$

the total torque on a system whose internal forces are central forces obeying the law of action and reaction is due only to the external forces.

Let us now find the rate of change of angular momentum. We have

$$\begin{aligned}\frac{d\mathbf{L}}{dt} &= \frac{d}{dt}\sum_i \mathbf{r}_i \times \mathbf{p}_i \\ &= \sum_i (\mathbf{r}_i \times \dot{\mathbf{p}}_i + \dot{\mathbf{r}}_i \times \mathbf{p}_i).\end{aligned} \tag{12}$$

The last term vanishes, since $\mathbf{p}_i = m_i\mathbf{v}_i$ is parallel to $\dot{\mathbf{r}}_i$. In the remaining sum, we recognize that $d\mathbf{p}_i/dt$ is just the total force acting on the ith particle, and so the ith term in the sum is the torque on the ith particle, and the sum is the total torque on the system. We are left with

$$\frac{d\mathbf{L}}{dt} = \mathbf{N}: \tag{13}$$

the rate of change of the angular momentum of a system of particles is equal to the total torque exerted on it—which, as we have seen, is the torque due to the external forces alone.

1.3 Work and Energy

Consider a constant force **F** acting on an object that moves along a straight path segment described by the vector **S**. We define the *work* done by the force as $W = \mathbf{F} \cdot \mathbf{S}$. We can easily generalize this to a force varying in magnitude. We take $\mathbf{n}$ as the unit vector along $\mathbf{S}$, so that $\mathbf{S} = S\mathbf{n}$. We let ds be an element of length along $\mathbf{S}$. Then we write $W = \int \mathbf{F} \cdot \mathbf{n} ds$. Note that in either case, it is only the component of $\mathbf{F}$ along $\mathbf{S}$ that is significant. Note also that there is no consideration of time in the discussion. No matter how long the motion takes, the work done is the same. Moreover, as long as we can relate a time variation in $\mathbf{F}$ to a position variation along $\mathbf{S}$, the time variation of $\mathbf{F}$ is also immaterial. The further generalization to a curvilinear path is straightforward, involving only the concept of a line integral. The expression is

$$W = \oint \mathbf{F} \cdot d\mathbf{r}. \tag{14}$$

In all of these cases, it is possible for the work done to be negative. This is interpreted as (positive) work being done *by* the system on the agent exerting the force.

The concept of energy involves some arbitrariness. The energy of a system is always defined in terms of some reference state. More precisely, we define the *difference* in energy between two states of a system as the work that is done to change the system from the initial state to the final (that is, if positive work is done to change from the initial state to the final, the system has greater energy in the final state than in the initial).

Let us calculate the work done in giving a body of mass m a velocity $\mathbf{v}$. We have

$$\begin{aligned} W &= \int \mathbf{F} \cdot d\mathbf{r} = \int m \frac{d\mathbf{v}}{dt} \cdot d\mathbf{r} \\ &= \int m d\mathbf{v} \cdot \frac{d\mathbf{r}}{dt} = m \int \mathbf{v} \cdot d\mathbf{v} = \tfrac{1}{2} m v^2. \end{aligned} \tag{15}$$

Since this is the energy due to motion, it is called *kinetic energy*.

1.4 Conservation Laws

We have seen [Eq. (5)] that the rate of change of the total momentum of a system is equal to the sum of the external forces acting on it, provided that the internal forces obey the law of action and reaction. (It must be pointed out that the forces between moving charged particles do not always obey that law.) We immediately deduce the *law of conservation of linear momentum*: If the external forces on a system vanish, the linear momentum is constant.

Likewise, we have seen [Eq. (13)] that the rate of change of angular momentum is equal to the external torque exerted on the system, provided the internal forces are central and obey the law of action and reaction. This leads to the *law of conservation of angular momentum*: If the external torques on a system are zero, the angular momentum is constant.

The case of energy is a bit more complicated. Equation (15) tells us that the work done on a body is equal to the increase of kinetic energy. If we denote kinetic energy by T, we have

$$W_{12} = T_2 - T_1. \tag{16}$$

Now suppose that the force varies with position in such a way that the work done is independent of the shape of the path between the two end points. This means that W_{12} must be expressible in terms of a function that depends only on the end points. If we reduce the path to a differential, then we must have

$$dW = \mathbf{F} \cdot d\mathbf{r} = -dV, \tag{17}$$

where the minus sign is chosen for future convenience. Such a force field is called *conservative*. From the last two forms of this relationship, we find that the component of $\mathbf{F}$ along $\mathbf{s}$, F_s, is given by

$$F_s = -\frac{\partial V}{\partial s}. \tag{18}$$

This means that

$$\mathbf{F} = -\nabla V. \tag{19}$$

Clearly, the function V is arbitrary to within an additional constant, which is in harmony with our discussion of energy as being determined only as a difference. The function V is called the *potential energy*.

Combining Eqs. (16) and (19) gives, after some rearrangement,

$$T_1 + V_1 = T_2 + V_2: \tag{20}$$

In a conservative force field, the sum of the

kinetic and potential energy is constant. This is what is usually meant by the conservation of energy.

There are many interesting problems in particle mechanics: the two-body problem in a gravitational field, the theory of vibration of a system of coupled particles, the oscillations of a pendulum, etc. We leave it to the reader to supplement the foregoing concepts with books on elementary mechanics such as Synge and Griffith (1959).

1.5 Generalized Coordinates

We take as our mathematical model of a dynamical system a discrete system of particles characterized by a finite number of *degrees of freedom*. The number of degrees of freedom is defined as the number of independent coordinates necessary to describe the positions of all the particles of the system. A system composed of N particles has $3N$ degrees of freedom in three-dimensional space. A rigid body has six degrees of freedom: three to define the position of its center of mass and three angles to define the orientation of the body (see Sec. 2.1.1). These independent coordinates are called *generalized coordinates* and are expressed as a set $\mathbf{q} = (q_1, q_2, \ldots, q_n)$, where $n = 3N$ for our set of N particles. If a system of N particles has k constraints (see Sec. 1.6), then the number of generalized coordinates is $3N - k$. It is important to note that these generalized coordinates need not have the significance of Cartesian coordinates, nor even of space coordinates of any kind. We discuss later situations in which some of the generalized coordinates are momenta or even energies.

Corresponding to the generalized coordinates are generalized velocities. The set of generalized velocities is $\dot{\mathbf{q}} = (\dot{q}_1, \dot{q}_2, \ldots, \dot{q}_n)$, and the generalized momentum is given by the set $\mathbf{p} = (\dot{p}_1, \dot{p}_2, \ldots, \dot{p}_n)$, $p_i = m_i \dot{q}_i$, $i = 1, 2, \ldots$

Suppose a system of particles is described by n generalized coordinates. Then the instantaneous position or configuration of the system is described by these n coordinates in an n-dimensional hyperspace called *configuration space*. This means that at each instant of time, the system is represented by a point in configuration space. As the time parameter increases, the position of the entire system changes so that the point traces a curve called the path of motion of the system. If, as previously assumed, the system consists of N particles, and there are $n = 3N$ degrees of freedom, the configuration space has $3N$ dimensions. A trajectory in configuration space does not have any resemblance to the path in actual space—not necessarily even for a single particle moving in three dimensions so that its configuration space also has three dimensions, except when the three generalized coordinates happen to coincide with the three Cartesian coordinates.

A more useful concept is *phase space*. This is a hyperspace consisting of n generalized coordinates and n generalized momenta for a system with n degrees of freedom. A given set of initial conditions (given by prescribing the initial generalized positions and momenta of all the particles) defines the configuration of the dynamical system at $t = 0$. As time increases, the motion or trajectory of the system unfolds so that it is completely described by its generalized coordinates and momenta. Phase space is essential in Hamilton's description of a dynamical system (see Sec. 4).

1.6 Constraints

A *constraint*, roughly speaking, is a restriction on the motion of the particles constituting the system. It is important to understand the nature of constraints. If a system has n coordinates, then the number k of constraints must satisfy $0 < k < n$; for if $k = 0$ then the system is free, and if $k = n$ then the system is not dynamic. We shall now classify constraints and give examples.

For a single particle in three-dimensional space moving on a surface, there is one constraint, since the equation of the surface on which the particle moves ties up one degree of freedom. This is most easily seen if the surface is taken to be one of the coordinate planes, since then the equation of the surface fixes one coordinate. In either this or the more general case, the constraint is called *holonomic*, which means that it is defined by an equation connecting the coordinates. A rigid body consisting of N particles is another simple example of a system having holonomic coordinates. Since there is no relative motion between the particles, the constraints are given by

$$(\mathbf{r}_i - \mathbf{r}_j)^2 = C_{ij}, \quad i \neq j, \quad i, j = 1, 2, \ldots, n, \tag{21}$$

where the C_{ij} are constants.

In general, constraints are holonomic if they can be expressed as equations connecting the coordinates of the particles (and possibly time). The general form of the equations for N particles is

$$f(r_1, r_2, \ldots, r_N) = 0. \tag{22}$$

A *nonholonomic* constraint is one that cannot be expressed by an algebraic equation. In some cases, it can be expressed as a relationship between the derivatives (or differentials) of the generalized coordinates, and possibly time (if it is *rheonomic*—see below). For example, a vertical disk rolls on an inclined plane with a frictional force that prevents slipping (see Fig. 1). The constraint is given by a simple relationship

$$ds = r d\theta \tag{23}$$

between the differential of the arc length and the differential angle of rotation. A nonholonomic constraint may also be expressed by an inequality. For example, a particle is placed on top of a large sphere of radius R and given an initial velocity. The particle and sphere are assumed to be in a gravitational field, so that the particle will slide down the surface of the sphere for a time and then fall off. The condition for falling off is that $r > R$ where r is the magnitude of the radius vector of the particle as measured from the center of the sphere. The constraint is that the radius vector must satisfy the inequality $r \geq R$.

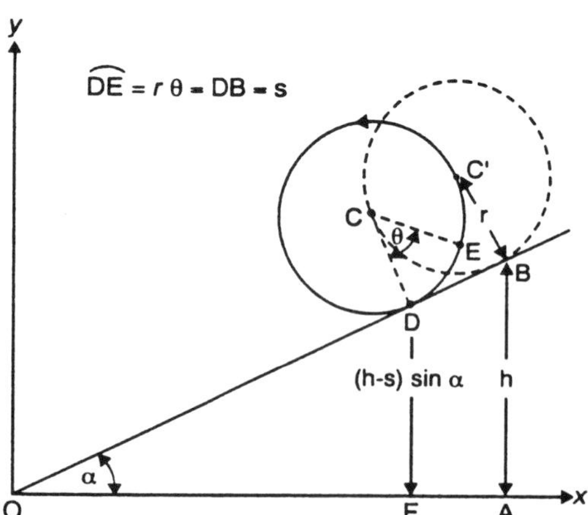

FIG. 1. Hoop rolling down an inclined plane.

Constraints are also classified as *scleronomic*, or time independent, and *rheonomic*, or time dependent. The examples given so far involve scleronomic constraints since surfaces are assumed to be stationary. If a bead slides along a moving wire, then the constraint is rheonomic.

Suppose we have a holonomic rheonomic constraint whose equation is

$$f(q_1, q_2, \ldots, q_n, t) = 0 \tag{24}$$

for a system with n degrees of freedom. This means that the generalized coordinate set $\mathbf{q}$ has n independent components. Since $df = 0$, this is equivalent to the differential equation

$$\sum_j \frac{\partial f}{\partial q_j} dq_j + \frac{\partial f}{\partial t} dt = 0. \tag{25}$$

If we set

$$\frac{\partial f}{\partial q_j} = a_{ij}, \quad \frac{\partial f}{\partial t} = a_{it}, \tag{26}$$

we get

$$\sum_j a_{ij} dq_j + a_{it} dt = 0 \tag{27}$$

for the ith constraint. If we have k constraints then there are k equations, for $i = 1, 2, \ldots, k$. These constraint equations give relations connecting the differentials dq and dt. They may also apply to nonholonomic constraints. If the time is constant during the displacements, then $dt = 0$. Therefore virtual displacements (see Sec. 1.7) must satisfy constraints of the form

$$\sum_j a_{ij} \delta q_j = 0, \quad i = 1, 2, \ldots, k, \tag{28}$$

where the symbol δ means a differential change keeping time constant.

Regardless of the nature of the constraints, if our system of n degrees of freedom has k constraints then the number of generalized coordinates is $n - k$.

1.7 The Principle of Virtual Work and D'Alembert's Principle

We next introduce the concept of *virtual displacements*. A virtual displacement is an arbitrary infinitesimal displacement of the configuration of a system of particles from its original state, consistent with the forces and constraints imposed on the system at a given instant. The important point is that *time does not change during a virtual displacement*. This is in contrast to an actual displacement, where time changes by an amount dt. Let $\delta \mathbf{r}_i$ be a virtual displacement of the ith particle. Note that the radius vector $\mathbf{r}$ is not a set of generalized coordinates, since it is subject to constraints. The actual displacement of the ith particle is $d\mathbf{r}_i$. Figure 2 shows plots of virtual and actual displacements of n particles in the (x,y) plane. The particles in the equilibrium configuration at t_0 are on the (fictitious) curve C_E. They are transformed to the curve C_V under a virtual displacement, and to the curve C_A under an actual displacement. The position of the ith particle in equilibrium is given by $\mathbf{P}_i$. Upon undergoing a virtual displacement $\delta \mathbf{r}_i$, it is transformed to the point $\mathbf{Q}_i$. If the particle at $\mathbf{P}_i$ undergoes an actual displacement $d\mathbf{r}_i$, it is transformed to the point $\mathbf{R}_i$. Clearly, these transformations hold for all the N particles. The particles on C_V are at t_0 and are consistent with the constraints acting on them. The particles on C_A are at time $t_0 + dt$.

Virtual work is defined as the work done on a dynamical system due to the external

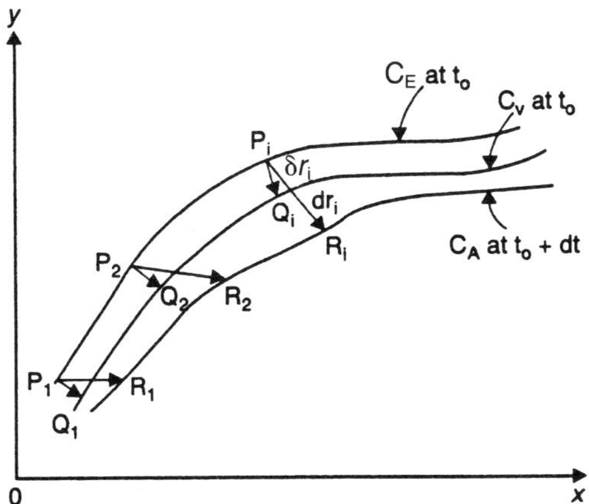

FIG. 2. Virtual and actual displacements of n particles in the (x,y) plane.

forces acting on it when the system undergoes a virtual displacement (time independent) compatible with the constraints. We again consider a system of N particles in equilibrium. Then the resultant force on each particle vanishes so that $\mathbf{F}_i = 0$, $i = 1, 2, \ldots, N$. Let δW_i be the virtual work done on the ith particle by the force $\mathbf{F}_i$ producing a virtual displacement $\delta \mathbf{r}_i$. Then $\delta W_i = \mathbf{F}_i \cdot \delta \mathbf{r}_i$, and the condition that the system be in equilibrium is that the virtual work due to the resultant force $\mathbf{F}$ on the system will vanish:

$$\sum_{i=1}^{N} \mathbf{F}_i \cdot \delta \mathbf{r}_i = 0. \tag{29}$$

Let the resultant force on the ith particle be given by

$$\mathbf{F}_i = \mathbf{F}'_i + \mathbf{f}_i, \tag{30}$$

where $\mathbf{F}'_i$ is the applied force and $\mathbf{f}_i$ is the constraining force on the ith particle. If we restrict ourselves to a system where the virtual work done by the constraining forces is zero, then Eq. (30) reduces to the condition

$$\sum_{i} \mathbf{F}'_i \cdot \delta \mathbf{r}_i = 0. \tag{31}$$

This states that the condition for a system of n particles to be in equilibrium is that the virtual work of the applied forces on the system must vanish. This will not be true for a system subject to frictional or other dissipative forces. Virtual work must be distinguished from the *actual* work dW done by the forces acting during the time dt.

Since the principle of virtual work [Eq. (30)] assumes that the system is in equilibrium, we do not have a dynamical system. The question then arises: How can the principle of virtual work be manipulated to include dynamical systems? The answer was given by D'Alembert. He ingeniously wrote the equations of motion in the form

$$\mathbf{F}_i - \dot{\mathbf{p}}_i = \mathbf{0}, \quad i = 1, 2, \ldots, N. \tag{32}$$

This is a vector equation of motion for the ith particle. Note that the right-hand side of this equation is the zero force vector. The observation that the difference between the external and the inertial force on the ith particle is equal to the zero force has the advantage of treating the inertial force $\dot{\mathbf{p}}_i$ in the same manner as the external force $\mathbf{F}_i$. By this technique, D'Alembert essentially reduced a problem in dynamics into one in which the principle of virtual work can be applied.

Another way of stating the principle of virtual work is the following: A system of N particles in equilibrium under the action of external forces $\mathbf{F}$ and the "reversed inertial force" $-\dot{\mathbf{p}}_i$ has zero virtual work done on it. This is expressed by the equation

$$\sum_{i=1}^{N} (\mathbf{F}_i - \dot{\mathbf{p}}_i) \cdot \delta\mathbf{r}_i = 0. \tag{33}$$

Since we restrict ourselves to conservative systems, $\mathbf{F}$ is the applied force. Equation (33) is called D'Alembert's principle.

1.7.1 Transformation to Generalized Coordinates In carrying out our program of transforming the coordinate system from $\mathbf{r}$ to $\mathbf{q}$, we seek a coordinate transformation of the form

$$\mathbf{r}_i = \mathbf{r}_i(q_1, q_2, \ldots, q_{3N-k}), \quad i = 1, 2, \ldots, n \tag{34}$$

for a system of N particles subject to k constraints. Each vector $\mathbf{r}_i$ is to be obtained as a function of all the scalar q_i's. We now express the velocity of the ith particle $\mathbf{v}_i = d\mathbf{r}_i/dt \equiv \dot{\mathbf{r}}$ as a linear combination of the generalized velocity components $\dot{q}$. Expanding by the chain rule gives

$$\mathbf{v}_i = \sum_{j=1}^{3N-k} \frac{\partial \mathbf{r}_i}{\partial q_j} \dot{q}_j + \frac{\partial \mathbf{r}_i}{\partial t}, \quad i = 1, 2, \ldots, n. \tag{35}$$

Note that $\mathbf{v}$ is a vector while $\dot{q}$ is the jth component of the set $\dot{\mathbf{q}}$. It is clear that in each of the n equations of (35), we sum over all the scalar values of q. In a similar manner, the virtual displacement (vector) of the ith particle is expressed as a linear combination of the generalized virtual displacements (scalars):

$$\delta\mathbf{r}_i = \sum_{j=1}^{3N-k} \frac{\partial \mathbf{r}_i}{\partial q_j} \delta q_j. \tag{36}$$

In terms of the generalized virtual coordinates, the virtual work of the system becomes

$$\begin{aligned} \delta W &= \sum_i \mathbf{F}_i \cdot \delta\mathbf{r}_i = \sum_{i,j} \mathbf{F}_i \cdot \frac{\partial r_i}{\partial q_j} \delta q_j \\ &= \sum_i Q_i \delta q_i. \end{aligned} \tag{37}$$

Here Q_j is the *generalized force* on the jth particle, defined as

$$Q_j = \sum_i \mathbf{F}_i \cdot \frac{\partial \mathbf{r}_i}{\partial q_j}. \tag{38}$$

1.8 Lagrange's Form of the Equations of Motion

Having obtained the virtual work, generalized forces, etc., in terms of generalized coordinates, we are now in a position to derive Lagrange's form of the equations of motion, one of the cornerstones of analytical dynamics. We derive them from D'Alembert's principle, expressed in the form

$$\sum_i \dot{\mathbf{p}}_i \cdot \delta\mathbf{r}_i = \sum_j Q_j \delta q_j. \tag{39}$$

We perform the following series of calculations:

$$\begin{aligned} \sum_i \dot{\mathbf{p}}_i \cdot \delta\mathbf{r}_i &= \sum_i m_i \ddot{\mathbf{r}}_i \cdot \delta\mathbf{r}_i \\ &= \sum_i m_i \ddot{\mathbf{r}}_i \cdot \frac{\partial \mathbf{r}_i}{\partial q_j} \delta q_j \\ &= \sum_i \left[\frac{d}{dt} \left(m_i \dot{\mathbf{r}}_i \cdot \frac{\partial \mathbf{r}_i}{\partial q_i} \right) \right. \\ &\qquad \left. - m_i \dot{\mathbf{r}}_i \cdot \frac{d}{dt} \left(\frac{\partial \dot{\mathbf{r}}_i}{\partial q_i} \right) \right] \delta q_j. \end{aligned} \tag{40}$$

By interchanging the order of the operators d/dt and $\partial/\partial q_i$, we obtain the following two relationships:

$$\frac{d}{dt} \frac{\partial \mathbf{r}_i}{\partial q_j} = \frac{\partial \mathbf{v}_i}{\partial q_j}, \quad \frac{\partial \mathbf{r}_i}{\partial q_j} = \frac{\partial \mathbf{v}_i}{\partial \dot{q}_j}. \tag{41}$$

Using these expressions and the definition of the kinetic energy of the system, we obtain

$$\begin{aligned}\sum_i \dot{\mathbf{p}}_i \delta\mathbf{r}_i &= \sum_i \left[\frac{d}{dt}\left(m_i\mathbf{v}_i\cdot\frac{\partial\mathbf{v}_i}{\partial\dot{q}_j}\right) - m_i\mathbf{v}_i\cdot\frac{\partial\mathbf{v}_i}{\partial q_j}\right] \\ &= \sum_j \left[\frac{d}{dt}\left(\frac{\partial}{\partial\dot{q}_j}\right)\left(\sum_i \tfrac{1}{2}m_iv_i^2\right)\right. \\ &\quad \left. - \frac{\partial}{\partial q_j}\sum_i \tfrac{1}{2}m_iv_i^2\right]\delta q_j \\ &= \sum_j \left[\frac{d}{dt}\left(\frac{\partial T}{\partial\dot{q}_j}\right) - \frac{\partial T}{\partial q_j}\right]\delta q_j. \end{aligned} \tag{42}$$

From this expression we get

$$\sum_j \left[\frac{d}{dt}\left(\frac{\partial T}{\partial\dot{q}_j}\right) - \frac{\partial T}{\partial q_j} - Q_j\right]\delta q_j = 0. \tag{43}$$

Since the virtual displacements are independent, the coefficient of each one must vanish. This yields

$$\frac{d}{dt}\left(\frac{\partial T}{\partial\dot{q}_j}\right) - \frac{\partial T}{\partial q_j} = Q_j, \quad j = 1, 2, \ldots, n. \tag{44}$$

These are Lagrange's equations for T and the Q_j's in terms of the generalized coordinates and velocities.

In the case of a conservative system, the resultant force $\mathbf{F}_i$ on each particle is derivable from the potential energy V, so that

$$\mathbf{F}_i = -\nabla_i V, \tag{45}$$

where ∇_i is the gradient operator on the ith particle. The generalized force on the jth particle becomes

$$Q_j = \sum_i \mathbf{F}_i\cdot\frac{\partial\mathbf{r}_i}{\partial q_j} = -\sum_i \nabla_i V\cdot\frac{\partial\mathbf{r}_i}{\partial q_j} = -\frac{\partial V}{\partial q_j}. \tag{46}$$

Substituting this result into Eq. (44) and rearranging slightly, we have

$$\frac{d}{dt}\left(\frac{\partial T}{\partial\dot{q}_j}\right) - \frac{\partial}{\partial q_j}(T - V) = 0, \quad j = 1, 2, \ldots, n. \tag{47}$$

Now define the Lagrangian L as $T - V$. Since V is independent of the $\dot{q}_j$, it is true that $\partial T/\partial q_j = \partial L/\partial q_j$, and Eq. (47) becomes

$$\frac{d}{dt}\left(\frac{\partial L}{\partial\dot{q}_j}\right) - \frac{\partial L}{\partial q_j} = 0, \quad j = 1, 2, \ldots, n. \tag{48}$$

These are Lagrange's form of the equations of motion. They can be seen as generalizations of Newton's equations as given by his second law, if we make the following interpretation: The term $\partial L/\partial\dot{q}_i$ represents the generalized momentum of the ith particle, and $\partial L/\partial q_i$ represents the "generalized force" derivable from the potential V.

2. RIGID-BODY MECHANICS

We define a *rigid body* as a continuous and bounded distribution of particles having the property that the distance between any two of the particles is invariant with time. The most general motion of a rigid body is a combination of a pure rotation about an arbitrary axis and the pure translational motion of the center of mass of the body with respect to an inertial coordinate system. This is known as *Chasles' theorem*. Rather than a detailed proof, we present the following intuitive reasoning: Under rotation, one point in the body is fixed, and the rotational motion is defined in terms of the three degrees of freedom, usually described in terms of the Euler angles (see Sec. 2.1.1). Allowing the one point to move (removing the constraint) introduces translational motion represented by three more degrees of freedom. This gives the most general motion of the rigid body.

Since the motion of the center of mass of a rigid body is the same as if all the mass were concentrated at that point and all the external forces applied there, the translational motion of the center of mass of a rigid body is determined by the same methods used in earlier sections to discuss particle motion; we will therefore drop it from consideration and treat only the rotation of a body having one point fixed.

2.1 Rotation of a Body about a Fixed Point

To describe the general motion of a rigid body we need (as reference) a space-fixed coordinate system. In general, this will be a three-dimensional inertial system. We also need a coordinate system fixed in the body.

Figure 3 shows these two coordinate systems in terms of Cartesian coordinates; for convenience, we take them to have a common origin fixed in the body. We use a symmetric notation to relate to matrix manipulation. We denote the space-fixed system by $\mathbf{x} = (x_1,x_2,x_3)$ and the body-fixed system by $\mathbf{x}' = (x'_1, x'_2,x'_3)$.

We further take the common origin of the two systems to be the fixed point of the body. We then determine the rotation of the body by rotating the body-fixed coordinates through an arbitrary angle θ with respect to the space-fixed coordinates. Let $(\mathbf{e}_1,\mathbf{e}_2,\mathbf{e}_3)$ be unit vectors along the (x_1,x_2,x_3) axes, and let $(\mathbf{e}'_1,\mathbf{e}'_2,\mathbf{e}'_3)$ be the unit vectors in the body-fixed system. Figure 3 shows the situation. It is convenient to introduce direction cosines describing the primed coordinate system with respect to the unprimed coordinate systems. There are nine direction cosines of the form l_{ij} defined by

$$l_{ij} = \cos(\mathbf{e}'_i,\mathbf{e}_j) = \mathbf{e}'_i \cdot \mathbf{e}_j, \quad i,j = 1, 2, 3. \tag{49}$$

l_{ij} is the direction cosine of the ith body-fixed axis with respect to the jth space-fixed axis. Let $\mathbf{r}$ be the radius vector from the common origin to an arbitrary point P in the body with respect to the unprimed axes, and let $\mathbf{r}'$ be the radius vector to the same point with respect to the primed axes. A rotation of the body is represented by the rotation of the vector $\mathbf{r}$ into $\mathbf{r}'$ by the matrix transformation

$$\mathbf{r}' = \mathbf{A}\mathbf{r}, \tag{50}$$

where $\mathbf{r}$ and $\mathbf{r}'$ are now regarded as column vectors and the rotation matrix $\mathbf{A}$ is given by

$$\mathbf{A} = \begin{pmatrix} l_{11} & l_{12} & l_{13} \\ l_{21} & l_{22} & l_{23} \\ l_{31} & l_{32} & l_{33} \end{pmatrix}. \tag{51}$$

Equation (50) is thus three equations, of which the ith ($i = 1,2,3$) is

$$x'_i = \sum_{j=1}^{3} l_{ij}x_j. \tag{52}$$

The matrix $\mathbf{A}$ has the property that the scalar product of any column, considered as a column vector, and any other column is zero, and the scalar product of any column with itself is 1; similar statements hold for the rows. This is expressed mathematically as

$$\sum_k l_{ik}l_{jk} = \delta_{ij}, \quad \sum_k l_{ki}l_{kj} = \delta_{ij}, \tag{53}$$

where the symbol δ_{ij}, known as the *Kronecker delta*, equals 0 if $i \neq j$ and 1 if $i = j$. Such a matrix is called *orthonormal*. From these properties of $\mathbf{A}$, it is easily seen that

$$\mathbf{A}^{-1} = \mathbf{A}^{\mathrm{T}}, \tag{54}$$

where $\mathbf{A}^{-1}$ is the inverse matrix and $\mathbf{A}^{\mathrm{T}}$ is the transpose matrix. Premultiplying both sides of Eq. (50) by $\mathbf{A}^{-1}$, we obtain the inverse mapping $\mathbf{r}' \rightarrow \mathbf{r}$, given by

$$\mathbf{A}^{-1}\mathbf{r}' = \mathbf{A}^{\mathrm{T}}\mathbf{r}' = \mathbf{r}. \tag{55}$$

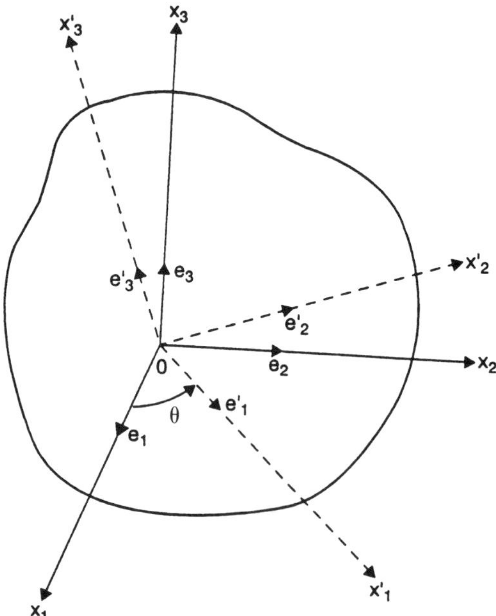

FIG. 3. Body-fixed coordinates rotated about a fixed point relative to space-fixed coordinates.

2.1.1 Eulerian Angles Since the direction cosines are not linearly independent, we cannot use $\mathbf{A}$ to define the generalized coordinates associated with a rotating body. We need a set of three independent angles to define an arbitrary rotation. Euler attacked this problem by defining such a set of angles, which are aptly called the *Eulerian angles of rotation* and which represent three independent modes of rotation. Figure 4 shows the space-fixed axes (x, y, z) with z vertical and (x, y) in the horizontal plane. The body-fixed axes (x', y', z') are also shown. Here z' makes the angle θ with z. (x', y') lie in the tilted plane that cuts the horizontal plane at the line ON, called the *line of nodes*. Let us assume the body to be a symmetrical top spinning about the z' axis, its axis of symmetry. Varying only ψ means that the body spins about this axis, and so ψ is called the angle of spin. Varying only θ rotates the body about the line of nodes; this mode of rotation is called *nutation*, and θ is called the angle of inclination. Varying only ϕ causes the body to rotate about the z axis, the mode called *precession*; ϕ is called the azimuth angle. The Eulerian angles are clearly independent, since each angle describes a rotation about an axis independent of the others. They are used in Sec. 2.4 as generalized coordinates in Lagrange's equations to develop Euler's equations of motion.

2.2 Moment of Inertia

Let us return to consideration of the rigid body as a system of discrete particles. Suppose that the body moves with one point, taken as the origin, fixed so that it rotates. We have defined the angular momentum as

$$\mathbf{L} = \sum_i m_i \mathbf{r}_i \times \mathbf{v}_i, \tag{56}$$

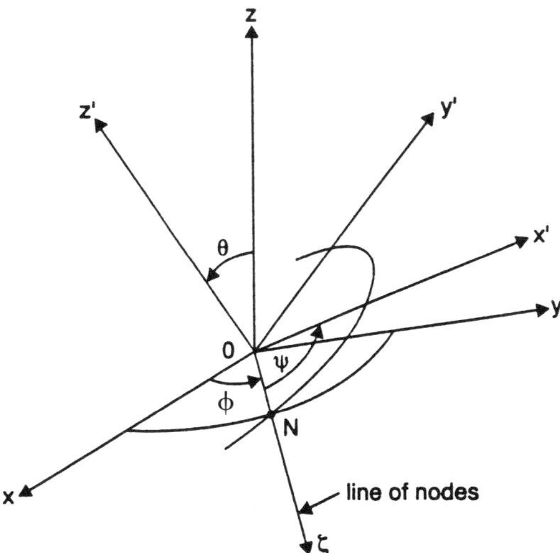

FIG. 4. Euler's angles for a symmetric body.

where $\mathbf{r}_i$ and $\mathbf{v}_i$ are the radius vector and velocity relative to the origin. In rotational motion, the body possesses *angular* velocity $\boldsymbol{\omega}$, which is related to the linear velocity of each particle by

$$\mathbf{v}_i = \boldsymbol{\omega} \times \mathbf{r}_i. \tag{57}$$

If we substitute this into Eq. (56) and expand the vector triple product, we have

$$\mathbf{L} = \sum_i m_i[\boldsymbol{\omega} r_i^2 - \mathbf{r}_i(\mathbf{r}_i \cdot \boldsymbol{\omega})]. \tag{58}$$

This shows that $\mathbf{L}$ is related to $\boldsymbol{\omega}$ by a linear transformation. We can see this more clearly by expanding Eq. (58). The x component of $\mathbf{L}$ is

$$\begin{aligned} L_x &= \omega_x \sum_i m_i(y_i^2 + z_i^2) + \omega_y \sum_i m_i x_i y_i \\ &\quad + \omega_z \sum_i m_i x_i z_i \\ &= I_{xx}\omega_x + I_{xy}\omega_y + I_{xz}\omega_z. \end{aligned} \tag{59}$$

The other two components of $\mathbf{L}$ can be obtained by cyclic permutation. The matrix equation showing $\mathbf{L}$ as a linear combination of the components of $\boldsymbol{\omega}$ is

$$\begin{pmatrix} L_x \\ L_y \\ L_z \end{pmatrix} = \begin{pmatrix} I_{xx} & I_{xy} & I_{xz} \\ I_{yx} & I_{yy} & I_{yz} \\ I_{zx} & I_{zy} & I_{zz} \end{pmatrix} \begin{pmatrix} \omega_x \\ \omega_y \\ \omega_z \end{pmatrix} = \mathbf{L} = \mathbf{I}\boldsymbol{\omega}. \tag{60}$$

The matrix $\mathbf{I}$ represents the *inertia tensor*. The diagonal elements are known as the *moments of inertia* about the x, y, and z axes, respectively; the off-diagonal elements are the *products of inertia*. $\mathbf{I}$ is a second-rank symmetric tensor.

We now consider a rigid body composed of a continuous distribution of mass. Let $dm = \rho d\tau$ be the element of mass in a differential element of volume $d\tau$, where ρ is the (continuous) density of the body. We now generalize the sums in Eq. (58) to integrals over the volume occupied by the body to obtain the elements of the inertial tensor for a continuous mass distribution. The result can be expressed concisely as

$$I_{ij} = \int_V (r^2\delta_{ij} - x_i x_j)\rho d\tau. \tag{61}$$

2.3 Rotational Kinetic Energy

We now show that the kinetic energy of rotation of a rigid body depends on the inertia tensor **I** and is a quadratic form in ω. For simplicity, the body is considered to be a system of discrete particles. We use the following facts: The velocity of the ith particle is $\mathbf{v}_i = \boldsymbol{\omega} \times \mathbf{r}_i$, and $\mathbf{L} = \mathbf{I}\boldsymbol{\omega}$. The kinetic energy T becomes

$$\begin{aligned} T &= \tfrac{1}{2}\Sigma m_i \mathbf{v}_i \cdot (\boldsymbol{\omega} \times \mathbf{r}_i) = \tfrac{1}{2}\boldsymbol{\omega} \cdot \Sigma m_i(\mathbf{r}_i \times \mathbf{v}_i) \\ &= \tfrac{1}{2}\boldsymbol{\omega} \cdot \mathbf{L} = \tfrac{1}{2}\boldsymbol{\omega} \cdot \mathbf{I}\boldsymbol{\omega}. \end{aligned} \tag{62}$$

It is clear that T is a quadratic form in $(\omega_x,\omega_y,\omega_z)$ since $\boldsymbol{\omega} \cdot \mathbf{I}\boldsymbol{\omega}$ is the scalar product of two vectors. We show this quadratic form by expanding the last form of Eq. (62) to get

$$\begin{aligned} 2T = {} & I_{xx}\omega_x^2 + I_{yy}\omega_y^2 + I_{zz}\omega_x^2 + 2I_{xy}\omega_x\omega_y \\ & + 2I_{yz}\omega_y\omega_z + 2I_{zx}\omega_z\omega_x. \end{aligned} \tag{63}$$

A standard result of linear algebra (Birkhoff and Maclane, 1977; Noble and Daniels, 1977) assures us that since **I** is symmetric, a coordinate system can always be found in which it is diagonal; that is, $I_{xy} = I_{yz} = I_{zx} = 0$. The axes of this coordinate system are called *principal axes* of the body, and the transformation to this system from the original one is a principal-axis transformation.

2.4 Euler's Equations of Motion for Rigid-Body Rotation

For a rigid body undergoing rotation with one point fixed, we perform a principal-axis transformation. The moments of inertia along the principal axes are (I_1,I_2,I_3), and the angular velocity has components $(\omega_1,\omega_2,\omega_3)$ along these axes. The fundamental vector equation of motion for a rotating body is

$$\frac{d\mathbf{L}}{dt} = \mathbf{N}, \tag{64}$$

This equation only holds for an inertial system of coordinates. To obtain an equation valid for a rotating system, we use the following relationship between the time derivative operator in an inertial system and that in a rotating system:

$$\left.\frac{d}{dt}\right|_{\text{space}} = \left.\frac{d}{dt}\right|_{\text{body}} + \boldsymbol{\omega}\times. \tag{65}$$

Operating on **L** with this operator, we get

$$\left.\frac{d\mathbf{L}}{dt}\right|_{\text{space}} = \left.\frac{d\mathbf{L}}{dt}\right|_{\text{body}} + \boldsymbol{\omega} \times \mathbf{L}. \tag{66}$$

The x component of this equation (along the principal axis) is

$$\frac{dL_1}{dt} + \omega_2 L_3 - \omega_3 L_2 = N_1. \tag{67}$$

However, the components of **L** along the principal axes are proportional to the corresponding components of $\boldsymbol{\omega}$. We therefore get

$$I_1\dot{\omega}_1 - \omega_2\omega_3(I_2 - I_3) = N_1. \tag{68}$$

This is the component of Euler's equations of motion along the principal x axis. The other two equations are obtained by cyclic permutation of the indices. Note the coupling between the components of $\boldsymbol{\omega}$. This has the consequence that even if the motion starts with ω_x, say, equal to zero, it will not remain so if $I_2 \neq I_3$.

2.4.1 Torque-Free Motion of the Earth

As an example of the use of Euler's equations, we discuss the rotation of the earth about its axis of symmetry, neglecting torques due to the sun or moon. The origin is chosen at the center of mass. Because the earth is an oblate spheroid, the principal moments of inertia are (I_1,I_1,I_3), where I_3 is taken along the axis of symmetry (north–south poles). Euler's equations become

$$\begin{aligned} & I_1\dot{\omega}_1 + \omega_2\omega_3(I_3 - I_1) = 0, \\ & I_1\dot{\omega}_2 + \omega_3\omega_1(I_1 - I_3) = 0, \\ & I_3\dot{\omega}_3 = 0. \end{aligned} \tag{69}$$

Integrating the last of these equations tells us that the component ω_3 of angular velocity along the axis of symmetry is constant. If we let $\alpha = \omega_3(I_3 - I_1)/I_1$, differentiate the first equation, and substitute from the second for $\dot{\omega}$, the solution becomes $\omega_1 = a\cos(\alpha t + \phi)$ where a is the amplitude and ϕ the phase angle. These constants are arbitrary since they depend on the earth's position and velocity at the arbitrary time from which t is counted. The resultant angular velocity becomes $\omega = (a^2 + \omega_3^2)^{1/2}$, which is constant. Figure 5 shows a circle of radius a centered at the origin O in the equatorial plane of the earth. ω_3 points

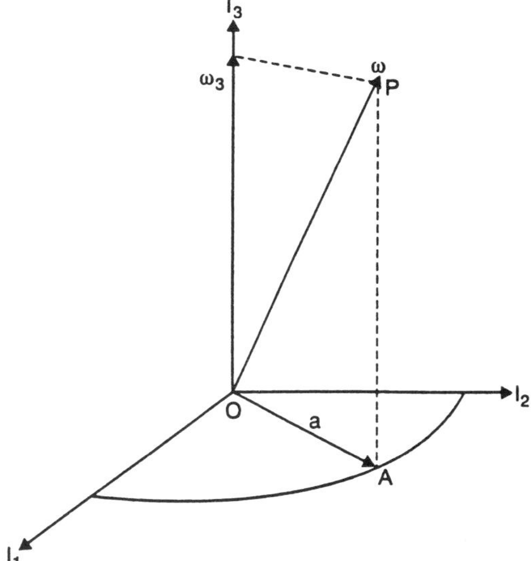

FIG. 5. Precession of the angular velocity about the axis of symmetry in force-free motion of the earth as a rotating rigid body.

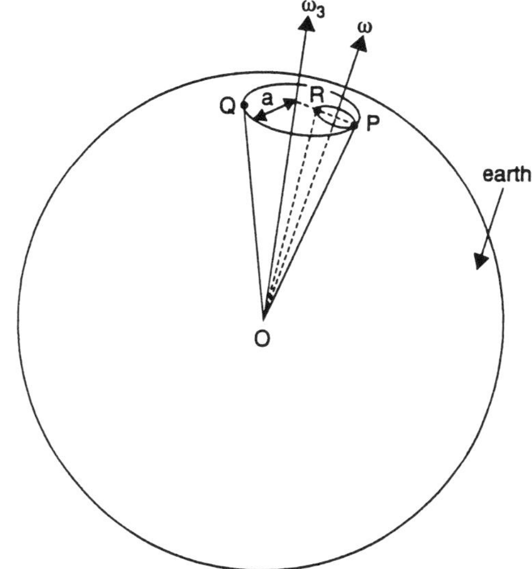

FIG. 6. Space- and body-fixed cones for torque-free rotation of the earth. The cone *OPQ* fixed in the earth rolls on the cone *OPR* whose instantaneous axis of rotation ω rotates about ω_3.

from the origin to the north pole (the axis of symmetry). From any point *A* on the circle, we draw a vertical vector *AP* of length equal to ω_3. The vector *OP* is then the vector $\boldsymbol{\omega}$ at a particular time. Clearly, *P* rotates about the axis of symmetry in a circular path. Figure 6 shows the earth as an oblate spheroid. A cone *OPQ* is drawn in the earth with the earth's axis of symmetry as its axis and a circle of radius *a* on the surface centered at the north pole. This cone is clearly body fixed. The figure also shows a cone *OPR* whose axis is $\boldsymbol{\omega}$ rotating with respect to the body-fixed cone as $\boldsymbol{\omega}$ rotates about the origin. In addition, *P* rotates on the circle of radius *a*. For this reason, $\boldsymbol{\omega}$ is called the axis of rotation.

Since $\omega_3 = 2\pi/\text{day}$, the period of rotation of $\boldsymbol{\omega}$ is given by $\tau = 2\pi/\alpha = (2\pi/\omega_3)I_1/(I_3 - I_1)$, which turns out to be about 300 days. It is called the Euler period, but it is not observed. However, a deeper analysis, taking into account the earth's imperfect rigidity, gives a period of 427 days called the Chandler period.

2.5 Motion of a Top

An analysis of the spinning top is important not only for its academic importance in the study of complicated rotational motion, but in its technical applications to gyroscopic motion used in navigational instruments. Top theory was also very important in contributing to military research on spinning shells during World War II.

We now discuss the motion of a heavy symmetrical top with one point (the origin) fixed, from an elementary viewpoint. Frictional forces are not considered. Referring to Fig. 7, z' is the axis of the top, which will *spin* about its axis (varying ψ, angle of spin), *nutate* (varying θ, angle of inclination), and *precess* (varying ϕ, the azimuth angle) about the fixed point *O*. *O'* is the center of mass of the top, a distance *l* from the origin. The weight of the top is $M\mathbf{g}$, acting at *O'* to produce the torque $\mathbf{N}$, which acts along the line of nodes *ON*: $\mathbf{N} = \boldsymbol{\rho} \times M\mathbf{g}$, where $\boldsymbol{\rho}$ is the vector *AB* normal to the top axis. There is no component of $\mathbf{N}$ along the vertical axis z or the axis of the top z'. Therefore, the components of angular momentum along these axes must be constant in time.

We now derive Lagrange's equations for the rotating top. I_3 is the longitudinal component of the moment of inertia (along the axis of symmetry of the top), and $I_2 = I_1$ are the equal transverse moments of inertia. In terms of Euler's angles, the kinetic energy

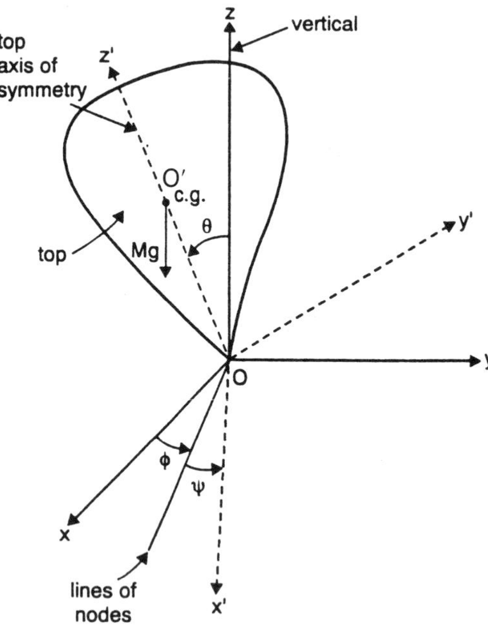

FIG. 7. Symmetrical top showing Euler's angles.

becomes

$$T = \tfrac{1}{2}I_1(\dot{\theta}^2 + \dot{\phi}^2 \sin^2 \theta) + \tfrac{1}{2}I_3(\dot{\psi} + \dot{\phi}\cos\theta)^2. \tag{70}$$

The potential energy is $V = Mgl\cos\theta$. The Lagrangian becomes

$$L = \tfrac{1}{2}I_1(\dot{\theta}^2 + \dot{\phi}^2 \sin^2\theta) + \tfrac{1}{2}I_3(\dot{\psi} + \dot{\phi}\cos\theta)^2 - Mgl\cos\theta. \tag{71}$$

Since ϕ and ψ do not appear in the Lagrangian, they are called *cyclic* coordinates, and the corresponding generalized momenta p_ϕ and p_ψ are constant in time, as already shown by more elementary considerations. The two first integrals of the motion are then obtained as

$$p_\phi = \frac{\partial L}{\partial \dot{\phi}} = (I_1 \sin^2\theta + I_3 \cos^2\theta)\dot{\phi} + I_3\dot{\psi}\cos\theta \equiv I_1 a, \tag{72}$$

and

$$p_\psi = \frac{\partial L}{\partial \dot{\psi}} = I_3(\dot{\psi} + \dot{\phi}\cos\theta) = I_3\omega_3 \equiv I_1 b. \tag{73}$$

(We have expressed the two constants as multiples a and b of I_1.) An additional integral is furnished by invoking the conservation of energy:

$$E = T + V = \tfrac{1}{2}I_1(\dot{\theta}^2 + \dot{\phi}^2 \sin^2\theta) + \tfrac{1}{2}I_3\omega_{z'}^2 + Mgl\cos\theta. \tag{74}$$

By manipulating Eqs. (72) and (73), we eliminate $\dot{\psi}$ and obtain

$$\dot{\phi} = \frac{a - b\cos\theta}{\sin^2\theta}. \tag{75}$$

Substituting Eq. (75) into Eq. (74) results in

$$\dot{\psi} = \frac{I_1 b}{I_3} - \cos\theta\,\frac{a - b\cos\theta}{\sin^2\theta}. \tag{76}$$

Equations (75) and (76) are then used, along with the energy equation, to obtain a differential equation involving θ. We now introduce the constants

$$E' = E - \tfrac{1}{2}I_3\omega_{z'}^2, \quad \alpha = \frac{2E'}{I_1}, \quad \beta = \frac{2Mgl}{I_1}. \tag{77}$$

If we now make the change of variable $x = \cos\theta$ and use Eqs. (79), after a little manipulation the energy equation (74) becomes

$$\dot{x}^2 = (1 - x^2)(\alpha - \beta x) - (a - bx)^2. \tag{78}$$

Solving for t, we obtain the following quadrature in the form of an elliptic integral:

$$t = \int_{x(0)}^{x(t)} \frac{dx}{[(1 - x^2)(\alpha - \beta x) - (a - bx)^2]^{1/2}}. \tag{79}$$

Using Eqs. (74) and (75), we may then also obtain ϕ and ψ in terms of quadratures.

However, it is not necessary to evaluate these integrals to obtain the nature of the nutational motion. We study the right-hand side of Eq. (78), which we denote by $f(x)$. This is a cubic polynomial in $x = \cos\theta$. The roots of this polynomial furnish the turning angles, those values of θ for which $\dot{\theta}$ changes sign. Figure 8 shows a rough plot of $f(x)$ vs x. The roots are (x_1,x_2,x_3), where x_3 does not correspond to a real angle. The nutational motion

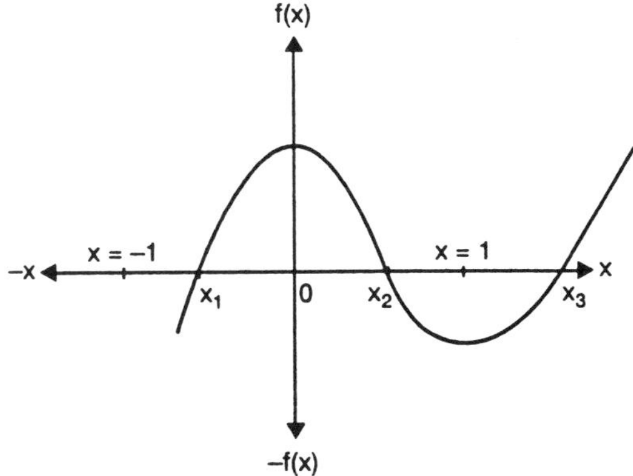

FIG. 8. Plot of $f(x)$ vs x showing location of turning angles of θ for nutational motion of top.

of the top is expressed in Fig. 9, where the motion of the top axis (called the *locus*) is traced on a unit sphere whose center is the fixed point of the top. The curves lie between the latitudes θ_1 and θ_2. On these latitudes, $\dot{\theta} = 0$. The shapes of the curves are determined by the initial conditions, which can be expressed in terms of the value $y = a/b$ (the root of $a - bx$). Figure 9(a) shows the locus curve tangent to the latitudes; Fig. 9(b) shows the curve with loops; and Fig. 9(c) shows the transition curve between Figs. 9(a) and 9(b).

If the initial conditions are such that $\dot{\phi}$ always has the same sign when $\theta_1 \leq \theta \leq \theta_2$, then there are no loops, and Fig. 9(a) obtains. According to Eq. (75), these initial conditions must satisfy the condition that $y > x_2$. The axis of the top then precesses about the vertical axis since ϕ increases in one direction or the other. If $x_1 < y < x_2$, the direction of the precessional motion will be different on the two latitudes, and the locus curve will have loops as shown in Fig. 9(b). This means that there are two values of $\dot{\phi}$, giving two different directions of precession, for each value of θ in a range $\theta_1 < \theta < \theta_2$. The average rate of precession will not vanish, however, so that there is always a net precession in one direction or the other.

Finally, for the case where a/b coincides with one of the roots of $f(x)$, then $\dot{\theta} = \dot{\phi} = 0$ on the latitude representing that root. This is shown in Fig. 9(c), where cusps touch the latitude $\theta = \theta_2$. One set of initial conditions yielding this case is $\theta = \theta_0$, $\dot{\theta} = \dot{\phi} = 0$ at $t = 0$. The initial angle θ_0 is the root that corresponds to the upper latitude in Fig. 9(c). Note that the initial value of E' is $Mgl \cos\theta_0$, and that the terms involving $\dot{\theta}$ and $\dot{\phi}$ are always positive. As $\dot{\theta}$ and $\dot{\phi}$ begin to differ from their initially zero values, the potential energy must decrease in order to conserve energy. This means that θ must increase from its minimum value at $\theta_0 = \theta_2$. When the top is released in this manner, it continues to precess, and it falls until $\theta = \theta_1$, the other latitude. Then it rises until $\theta = \theta_2$, and the process repeats, as shown in Fig. 9(c).

2.5.1 Stability of a Sleeping Top A top is said to be "sleeping" if its spin axis coincides with its axis of symmetry and is in the vertical direction. Let $s = \dot{\psi}$. The variable s is called the spin. If s is large enough, then the spin axis becomes vertical for a small disturbance (the top will perform small oscillations about the vertical). If s is smaller than a critical value, then the oscillations

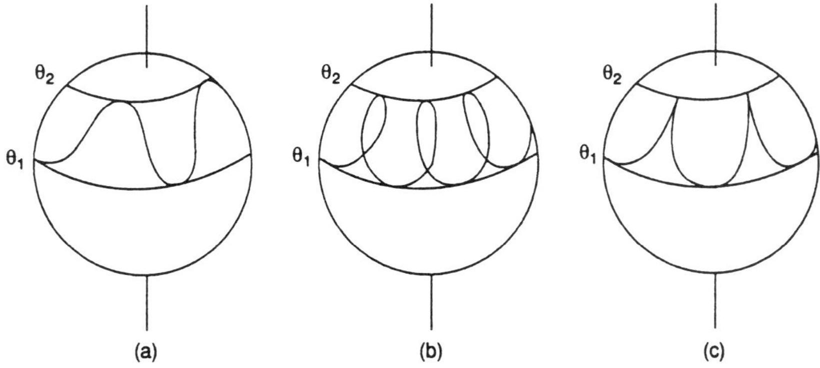

FIG. 9. Possible nutational modes of top, shown by locus curves traced by axis of top on a unit sphere: (a) tangent to latitudes, (b) loops, (c) transition.

about the spin axis will increase so that the top will wobble and eventually fall down. What is this critical value of s? The answer is obtained by study of the cubic $f(x)$. For a sleeping top, we have $\theta = \dot{\theta} = 0$. The cubic becomes

$$f(x) = (1 - x)^2\left[\frac{2Mgl}{I_1}(1 + x) - \left(\frac{I_3}{I_1}\right)^2 s^2\right], \tag{80}$$

since $a = b$ and $\alpha = \beta$. Here $x = 1$ is a double root of $f(x)$. The other root,

$$x = I_3^2 s^2/2MglI_1 - 1, \tag{81}$$

is either greater than unity (x_3) or less than unity (x_1). The case of a root greater than unity is physically unattainable, since x is the cosine of an angle. In that case, then, the only possible motion is that which makes $x = 1$, which corresponds to $\theta = 0$, or the axis vertical. The condition for this is that s be large enough to make the root greater than 1. The critical value is the solution of

$$4Mgl/I_1 = (I_3/I_1)^2 s^2, \tag{82}$$

or

$$|s| = (4I_1 Mgl)^{1/2}/I_3. \tag{83}$$

The condition for stability of a sleeping top is that $|s|$ equal or exceed this value. If the condition is not fulfilled, the top if disturbed from the vertical position will fall to the inclination of θ_1 where $\cos\theta_1 = x_1$.

3. VARIATIONAL PRINCIPLES AND THE CALCULUS OF VARIATIONS

Many important results in physics are obtained from what are known as variational principles. Such principles state, in general, that the actual course of physical events is such as to render some quantity an extremum—either a minimum or a maximum. Perhaps the best known of these is Fermat's principle, which applies to optics. But there are equally important principles in mechanics. To appreciate them, we must first review some results in the calculus of variations.

We consider an elementary situation in the differential calculus where we have a differentiable function $I = I(y_i(x), y_i'(x))$, where the prime denotes derivative with respect to x, that we want to *extremize*, that is, to find the values of the y_i that yield an extremum—either a maximum or a minimum—of the function. In the calculus of variations, this concept of extremizing a function is generalized to that of extremizing a *functional*, which is roughly defined as a "function of a function." Specifically, a functional is a definite integral whose integrand is one of a class of functions each of which produces a particular value of the integral. The basic problem in the calculus of variations is to find the function that extremizes the integral (for fixed limits of integration, for simplicity). In mechanics, as well as in optics, the extremum is usually a minimum rather than a maximum. It turns out that each of the admissible functions defines a path joining the end points of the integral. The techniques of the calculus of variations allow us to find that path that minimizes the integral. We set

$$I = \int_{x_1}^{x_2} f(y_i(x), y_i'(x))dx. \tag{84}$$

We now perform variations δy_i and $\delta y_i'$, subject to the conditions that all the variations vanish at the end points of the integral. The condition that I be an extremum is that the resulting $\delta I = 0$. This means

$$\delta I = 0 = \int_{x_1}^{x_2} \sum_i \left(\frac{\partial I}{\partial y_i}\delta y_i + \frac{\partial I}{\partial y_i'}\delta y_i'\right)dx. \tag{85}$$

We present a geometric or physical interpretation of the variations of a function and functional, which relates to the principle of virtual work (Sec. 1.7). Figure 10 shows a minimizing curve $y = y(x)$ in the (x,y) plane, passing through the fixed end points A and B. In mechanics, y represents the position of a particle at a time equal to x. The minimizing curve is the trajectory of the particle, or the time history of its motion. Let $y = \bar{y}(x)$ be a neighboring curve passing through the same end points. δy is the change in y from the minimizing curve *for a fixed* x. [Note that this infinitesimal change in y is different from $dy(x)$, which represents an infinitesimal change in y *along* the minimizing curve.] Let $\delta I(y)$ be the corresponding variation in the

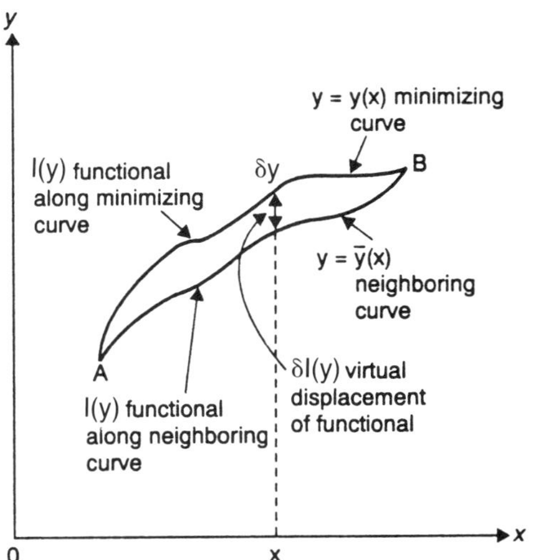

FIG. 10. Minimizing and neighboring curves in (x,y) plane showing virtual displacement of coordinates.

functional $I(y)$. Clearly, δI is the variation in the functional from the minimizing to the neighboring curve for a fixed time.

Given this interpretation, it is easy to show that

$$\delta y_i' = \frac{d}{dx}\,\delta y_i. \tag{86}$$

We use this in the last term of the integrand in Eq. (85) and integrate by parts:

$$\begin{aligned}\int_{x_1}^{x_2} \sum_i \frac{\partial I}{\partial y_i'}\,\delta y_i' dx &= \int_{x_1}^{x_2} \sum_i \frac{\partial I}{\partial y_i'}\frac{d\delta y_i}{dx}\,dx \\ &= \sum_i \frac{\partial I}{\partial y_i'}\,\delta y_i\bigg|_{x_1}^{x_2} \\ &\quad - \int_{x_1}^{x_2} \sum_i \frac{d}{dx}\frac{\partial I}{\partial y_i'}\,\delta y_i dx.\end{aligned} \tag{87}$$

Since the variations were specified to vanish at the end points, the first sum in this result vanishes, and Eq. (85) becomes

$$\delta I = \int_{x_1}^{x_2} \sum_i \left(\frac{\partial I}{\partial y_i} - \frac{d}{dx}\frac{\partial I}{\partial y_i'}\right)\delta y_i dx. \tag{88}$$

But since the variations δy_i are arbitrary (except for vanishing at the end points), the only way the integral can be guaranteed to vanish is for

$$\frac{\partial I}{\partial y_i} - \frac{d}{dx}\frac{\partial I}{\partial y_i'} = 0. \tag{89}$$

This set of equations, known as *Euler's equations*, is our key result. They constitute a necessary condition for the existence of an extremum of I.

Note that we say *necessary* condition. This means that if an extremum exists, then Euler's equations must be satisfied. But that does not guarantee an extremum. We give an example that illustrates Fermat's principle but shows that an extremum does not exist. This principle, enunciated by the French mathematician Pierre Fermat (1601–1655), deals with light rays in a medium of variable refractive index so that the paths of the rays are not necessarily straight lines. The statement of the principle is this: Of all the possible paths taken by a light ray in going from A to B (fixed points), the actual path taken by the ray is the one that takes the least time. Consider Fig. 11, which shows an optically transparent (x,y) plane embedded in a medium of constant index of refraction. Let A be a fixed point on the x axis such that the length of the line segment OA is L. Suppose a light beam starts at point O, travels along the y

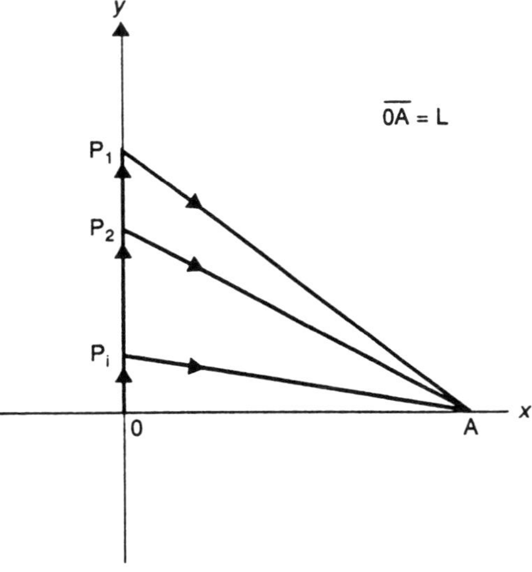

FIG. 11. Example of Fermat's principle showing the nonexistence of an extremum in a two-dimensional optically transparent medium of constant refractive index.

axis a variable distance to the point P_1, then goes along the straight line from P_1 to A. Each trajectory is a curve given by $y = y_i(x)$ for $i = 1, 2, \ldots$. Every admissible path is composed of the broken line segments $y_i = OP_i + P_iA$. According to Fermat's principle, the time taken to traverse each path is a minimum for that path. Therefore, the functional $I(y_i)$ represents the time taken to traverse each broken line. Clearly, the length of each path is greater than L by the amount OP_i. In the limit, $P_i \rightarrow P_\infty = 0$. The minimum distance is the line segment $OA = L$. But OA is *not an admissible path*, since the light beam must start at O *in the direction of the y axis* and then travel on a line of sight to A. As P_i tends to O, the length of each admissible path gets closer to L. However, *there is no minimal path*, since we obtained an infinite number of admissible paths as $P_i \rightarrow O$. We have thus demonstrated that an extremum does not exist for at least one particular functional.

This brief treatment of the calculus of variations will enhance our understanding of Hamilton's variational principle, which supplies the foundation for Hamilton's equations of motion and an alternative (and generalizable) basis for Lagrange's equations of motion, the two cornerstones of classical mechanics. Again, we emphasize that the key equation is Euler's equation. For a system of N particles in one dimension, there are N Euler equations, one for each particle.

3.1 Lagrange's Equations from Hamilton's Variational Principle

Both Lagrange's and Hamilton's formulations of mechanics involve energy considerations. In these formulations, the mechanical properties of our system are completely described by its kinetic energy T and its potential energy V. In general, we assume that T is a quadratic function of the generalized velocities, possibly also dependent on the generalized coordinates, with the following form:

$$T = \sum_{i,j} P_{ij}(q_1,q_2,\ldots,q_n,t)\dot{q}_i\dot{q}_j, \tag{90}$$

where the P_{ij} are prescribed functions of the set $\mathbf{q}$. Equation (90) allows T to depend arbitrarily on $\mathbf{q}$. For the time being, we consider only conservative systems so that $T + V = E$, where the total energy E is constant.

The solution of Lagrange's equations of motion yields the generalized coordinates and velocities as functions of time, which completely describes the trajectory of each particle of the system in phase space. On the other hand, as we will see, the solution of Hamilton's equations of motion yields the generalized coordinates *and momenta* as functions of time, which describes the system. However, there is a conceptual difference between the two phase spaces. Lagrange's equations are second-order differential equations, thus requiring two sets of initial conditions (generalized coordinates and velocities) to describe the trajectory of the system. On the other hand, Hamilton's equations consist of pairs of first-order differential equations, each pair requiring a knowledge of initial generalized coordinates and momenta. This is reflected in the meaning of the trajectories in the two spaces.

Hamilton applied the variational principle to the Lagrangian $L(\mathbf{q}(t),\dot{\mathbf{q}}(t),t)$. His principle can be stated as follows: Between any two instants of time, t_A and t_B, the actual motion of a dynamical system of particles, defined by their trajectories $q = q(t)$ in configuration space, proceeds in such a way as to make the functional

$$I(\mathbf{q},\dot{\mathbf{q}}) = \int_{t_A}^{t_B} L(\mathbf{q}(t),\dot{\mathbf{q}}(t),t)dt \tag{91}$$

a minimum.

We relate Eq. (91) to the functional defined by Eq. (84) and the corresponding Euler equations given by Eq. (89) as follows: the term x corresponds to t, y corresponds to q, dy/dx corresponds to dq/dt, and F corresponds to L. Therefore, L must satisfy Euler's equation and we obtain

$$\frac{d}{dt}\frac{\partial L}{\partial \dot{q}_i} - \frac{\partial L}{\partial q_i} = 0, \quad i = 1, 2, \ldots, n. \tag{92}$$

As seen by reference to Sec. 1.8, these are Lagrange's equations of motion. Because of the way they have been derived here, they are sometimes called the Euler-Lagrange equations.

As an example of the use of Lagrange's equations, we consider a body of mass m in a central force field. The potential is $V = -k/r$, where k is proportional to the gravitational constant. The kinetic energy is $T = \frac{1}{2}m\dot{r}^2$. This gives

$$L = \left(\frac{m}{2}\right)\dot{r}^2 + \frac{k}{r}, \quad \frac{\partial L}{\partial \dot{r}} = m\dot{r} = p, \quad \frac{\partial L}{\partial r} = -\frac{k}{r^2}. \tag{93}$$

Using Lagrange's equation (88), we get

$$m\ddot{r} = -k/r^2 = f, \tag{94}$$

where f is the inverse-square force.

As another example, we take a system of N uncoupled harmonic oscillators in three-space. The ith particle has mass m_i, and all the particles have the spring constant k. Since the particles are noninteracting, there are no constraints, so that we may use Cartesian coordinates for the generalized coordinates. Therefore

$$q = (x_1,y_1,z_1,\ldots,x_N,y_N,z_N),$$
$$\dot{q} = (u_1,v_1,w_1,\ldots,u_N,v_N,w_N),$$

where

$$\dot{x}_i = u_i, \quad \dot{y}_i = v_i, \quad \dot{z}_i = w_i, \quad i = 1, 2, \ldots, N.$$

The Lagrangian becomes

$$L = \tfrac{1}{2}\sum_{i=1}^{i=N} m_i[u_i^2 + v_i^2 + w_i^2] - \tfrac{1}{2}k\sum_{i=1}^{i=N} [x_i^2 + y_i^2 + z_i^2]. \tag{95}$$

Differentiating L with respect to the velocity components yields the momentum components for the ith particle:

$$\frac{\partial L}{\partial u_i} = m_i u_i = (p_i)_x, \quad \frac{\partial L}{\partial v_i} = m_i v_i = (p_i)_y,$$
$$\frac{\partial L}{\partial w_i} = m_i w_i = (p_i)_z. \tag{96}$$

The spring force is given by

$$\frac{\partial L}{\partial x_i} = -kx_i, \quad \frac{\partial L}{\partial y_i} = -ky_i, \quad \frac{\partial L}{\partial z_i} = -kz_i, \tag{97}$$

which are the equations of motion for our system of uncoupled harmonic oscillators.

This has been an integral approach to Lagrange's equations, in the sense that it uses energy considerations in the variational technique to compare admissible paths with the actual path in configuration space. This is to be contrasted with the differential approach used in deriving Lagrange's equations from D'Alembert's principle, where we consider the instantaneous state of the system and superimpose small virtual displacements from this state.

3.1.1 Dealing with Constraints As we noted in Sec. 1.6, the presence of constraints means that the virtual displacements are not independent. In that case, the derivation of Lagrange's equations from Hamilton's principle fails, since that derivation depended on the independence of the virtual displacements (variations of the generalized coordinates). Lagrange found a way to deal with this problem. Note that each of the constraint conditions given by Eq. (28) equates a certain sum to zero. We can therefore add or subtract any multiple of each equation from any other quantity without effect. Specifically, we multiply each of the equations (28) by a corresponding constant λ_i, giving

$$\lambda_i \sum_j a_{ij}\delta q_j = 0, \quad i = 1, 2, \ldots, k. \tag{98}$$

We subtract the sum of these terms from the integrand in Eq. (88):

$$\delta I = 0 = \int_{x_1}^{x_2} \sum_i \left(\frac{\partial I}{\partial y_i} - \frac{d}{dx}\frac{\partial I}{\partial y_i'} - \sum_{j=1}^{k} \lambda_j q_{ji}\right) \times \delta q_i dx. \tag{99}$$

Of course, this does not affect the fact that the δq_i's are not independent. Only $n - k$ of them may be chosen independently. However, the values of the λ_i's have not yet been determined. We may therefore define them by setting the integrand in Eq. (99) equal to zero for the last k δq_i's. That is,

$$\frac{d}{dt}\frac{\partial L}{\partial \dot{q}_i} - \frac{\partial L}{\partial q_i} - \sum_{j=1}^{k} \lambda_j a_{ji} = 0,$$
$$i = n - k + 1, n - k + 2, \ldots, n. \tag{100}$$

These are the equations of motion for those q_i's that are related by the constraints. They form a set of k linear equations for the λ_j's. Having determined them from these equations, we can now set the integrand of Eq. (99) equal to zero for the remaining $n - k$ independent δq_i's. Combining these results yields the complete set of n equations for a nonholonomic system,

$$\frac{d}{dt}\frac{\partial L}{\partial \dot{q}_j} - \frac{\partial L}{\partial q_j} = \sum_{i=1}^{k} \lambda_i a_{ij}, \quad j = 1, 2, \ldots, n. \tag{101}$$

The n equations (101) combined with the k equations (100) yield a complete solution in that they allow us to calculate the n q_j's and the k λ_i's.

The right-hand side of Eq. (101) is the force applied to the jth particle by the constraints and is called a *generalized force*. If we removed the k constraints on the jth particle and applied external forces equal to the constraining forces, the system would not change.

As an easy example, we take the case of a hoop rolling without slipping down an inclined ramp. Figure 1 shows a profile of the hoop, whose mass is m and whose radius is r, rolling in the (x,y) plane down the ramp OB inclined at an angle α to the horizontal. The hoop starts at the top of the ramp tangent to the incline at B, at height h, and rolls down the incline. At some time t, the hoop is tangent to the incline at D, having rotated about its horizontal axis through the angle θ. Its center has translated a distance s parallel to the incline, having dropped through a vertical distance $(h - s)\sin\alpha$. Let the generalized coordinates be $q_1 = s$, $q_2 = \theta$. There is one constraint equation; it is given by Eq. (23), which we repeat:

$$ds - r d\theta = 0. \tag{102}$$

This is a specific case of the constraints given in differential form [Eq. (27)] with $i = 1$, $j = 1, 2$, $a_{11} = 1$, and $a_{12} = r$. The Lagrangian is

$$L = T - V = \tfrac{1}{2}m\dot{s}^2 + \tfrac{1}{2}r^2\dot{\theta}^2 - mg(h - s)\sin\alpha. \tag{103}$$

The first term represents the kinetic energy of the pure translational motion of the center of mass; the second term represents the kinetic energy of the pure rotational motion about the horizontal axis through the center of mass. Lagrange's equations for the two degrees of freedom are

$$\begin{aligned} &m\ddot{s} - mg\sin\alpha = \lambda, \\ &mr^2\ddot{\theta} - r\lambda. \end{aligned} \tag{104}$$

The constraint equation is

$$r\dot{\theta} = \dot{s}. \tag{105}$$

Differentiating with respect to t gives

$$r\ddot{\theta} = \ddot{s}. \tag{106}$$

From these equations we obtain

$$\begin{aligned} \lambda &= -\tfrac{1}{2}mg\sin\alpha, \\ \ddot{\theta} &= -\frac{1}{2r} g\sin\alpha. \end{aligned} \tag{107}$$

The equation for λ is the friction force of constraint. It is directed up the inclined plane and is $\frac{1}{2}$ the projection of the weight of the hoop normal to the incline.

3.2 Nonconservative Systems

We now consider a system of N particles with n degrees of freedom subject to nonconservative forces such as those due to friction and viscosity. These forces dissipate heat, so that the law of conservation of energy is not valid. This means that $\mathbf{F}$ is not derivable from a potential V that depends only on $\mathbf{q}$. Then $\mathbf{Q}$ cannot be obtained from Eq. (46). But all is not lost! Viscous forces depend on the velocity as well as position. As examples, the slow motion of a rough sphere in a viscous fluid is proportional to its velocity; the motion of an airplane or projectile in air depends on the square of its velocity (because of air drag). Therefore, we replace V by a velocity-dependent potential U that depends on the generalized velocities as well as coordinates, so that $U = U(\dot{q},q)$. The potential U is sometimes called the *generalized potential*. The Lagrangian is now given by

$$L = T - U. \tag{108}$$

We now set

$$T(\dot{q},q) = L(\dot{q},q) + U(\dot{q},q), \tag{109}$$

and insert this expression in Eq. (44) to obtain

$$\frac{d}{dt}\left(\frac{\partial L}{\partial \dot{q}_i}\right) - \frac{\partial L}{\partial q_i} = Q_i + \frac{\partial U}{\partial q_i} - \frac{d}{dt}\left(\frac{\partial U}{\partial \dot{q}_i}\right), \quad i = 1, 2, \ldots, n. \tag{110}$$

If we set the right-hand side of this equation equal to zero, we obtain Lagrange's equations. Then the generalized force for the ith particle becomes

$$Q_i = -\frac{\partial U}{\partial q_i} + \frac{d}{dt}\left(\frac{\partial U}{\partial \dot{q}_i}\right). \tag{111}$$

For the special case of a conservative force system, this reduces to Eq. (46).

3.2.1 Rayleigh's Dissipation Function For our system of particles in a nonconservative force field, we restrict ourselves to frictional forces obeying the simple law

$$\mathbf{f}_i = -\mathbf{k} \cdot \mathbf{v}_i, \tag{112}$$

where $\mathbf{f}_i$ is the friction force and $\mathbf{v}_i$ is the velocity of the ith particle. The friction coefficient $\mathbf{k}$ is a tensor, which we take to be diagonal with nonzero components k_x, k_y, k_z, and to be the same for each particle. Lord Rayleigh (1878) introduced a scalar quadratic function of the velocity called the *Rayleigh dissipation function* $\mathcal{F}$, which is defined by

$$\mathcal{F} = \tfrac{1}{2}\sum_{i=1}^{n} (k_x u_i^2 + k_y v_i^2 + k_z w_i^2). \tag{113}$$

From this definition we obtain

$$\mathbf{f} = -\nabla_{\mathbf{v}}\mathcal{F}, \quad \text{where } \nabla_{\mathbf{v}} = \mathbf{i}\frac{\partial}{\partial u} + \mathbf{j}\frac{\partial}{\partial v} + \mathbf{k}\frac{\partial}{\partial w}, \tag{114}$$

where $(\mathbf{i},\mathbf{j},\mathbf{k})$ are the unit vectors in the (x,y,z) directions. Note that the gradient operator is with respect to the velocity. We give a physical interpretation of $\mathcal{F}$ by considering a differential amount of work $dW_{\mathbf{f}}$ done by the system against the friction force $\mathbf{f}$. This is given by

$$\begin{aligned} dW_{\mathbf{f}} &= -\mathbf{f} \cdot d\mathbf{r} = -\mathbf{f} \cdot \mathbf{v}dt \\ &= (k_x u^2 + k_y v^2 + k_z w^2)dt. \end{aligned} \tag{115}$$

This means that the Rayleigh dissipation function is *half* the rate of energy dissipated by friction. The jth component of the generalized force due to friction is

$$Q_j = \sum_i \mathbf{f}_i \cdot \frac{\partial \mathbf{r}}{\partial q_j} = -\sum_i \nabla_{\mathbf{v}}\mathcal{F} \cdot \frac{\partial \dot{\mathbf{r}}}{\partial \dot{q}_j} = -\frac{\partial \mathcal{F}}{\partial \dot{q}}. \tag{116}$$

Lagrange's equations now become

$$\frac{d}{dt}\frac{\partial L}{\partial \dot{q}_j} - \frac{\partial L}{\partial q_j} = -\frac{\partial \mathcal{F}}{\partial \dot{q}_j}, \quad j = 1, 2, \ldots, n. \tag{117}$$

We see from these equations that the two scalars L and $\mathcal{F}$ must be prescribed in order to solve for the motion of a system subject to a viscous force.

4. HAMILTON'S EQUATIONS OF MOTION

Recall that Lagrange's formulation of the equations of motion is a description of a dynamical situation in terms of trajectories defined by the generalized coordinates and velocities as functions of time. Hamilton's alternative formulation of the equations of motion is based on introducing a generalized p_i called the *canonical* or *conjugate momentum* (each p_i is conjugate to the corresponding q_i). Recall that the term $\partial L/\partial \dot{q}_i$ is the momentum of the ith particle. It is only natural that Hamilton would define the conjugate momentum as

$$p_i = \frac{\partial L}{\partial \dot{q}_i}. \tag{118}$$

Again, considering a conservative system, Lagrange's equation for the ith particle is transformed into

$$\frac{d}{dt}p_i \equiv \dot{p}_i = -\frac{\partial V}{\partial q_i}, \tag{119}$$

which is a first-order system of equations for p_i and V. But this form of Lagrange's equations was not used by Hamilton, since he wanted a system of equations of motion where the independent variables are $(\mathbf{p},\mathbf{q})$. To this end, he defined a function, aptly called the Hamiltonian, as follows:

$$H = H(\mathbf{p},\mathbf{q}) = T + V, \tag{120}$$

where H is the total energy of a conservative system and is a function of $(\mathbf{p},\mathbf{q})$. Expanding the time derivative of L gives

$$\frac{dL}{dt} = \sum_i \frac{\partial L}{\partial \dot{q}_i}\frac{d\dot{q}_i}{dt} + \sum_i \left(\frac{d}{dt}\frac{\partial L}{\partial \dot{q}_i}\right)\dot{q}_i = \sum_i \frac{d}{dt}\left(\dot{q}_i \frac{\partial L}{\partial \dot{q}_i}\right). \tag{121}$$

Integrating this equation gives

$$\sum_i \dot{q}_i \frac{\partial L}{\partial \dot{q}_i} - L = \text{const.} = H, \tag{122}$$

where we provisionally set the constant of integration equal to the Hamiltonian. From this we obtain

$$H = \sum_i \dot{q}_i p_i - L, \tag{123}$$

which is an important equation, provided we can show that this constant of integration is indeed the total energy of a conservative system. We show this by first using the definition of L and recognizing that V depends only on $\mathbf{q}$. This gives

$$p_i = \frac{\partial T}{\partial \dot{q}_i}. \tag{124}$$

We also have

$$\sum_i \dot{q}_i p_i = \sum_i \dot{q}_i \frac{\partial T}{\partial \dot{q}_i} = 2T, \tag{125}$$

since T is a homogeneous quadratic function of $\dot{q}_i$. From these expressions, we obtain

$$H = 2T - L = 2T - (T - V) = T + V = E. \tag{126}$$

We now derive Hamilton's equations, which consist of a pair of first-order partial differential equations in $(\mathbf{p},\mathbf{q},t)$. To this end, we expand dH and use the above relations to obtain

$$dH = \sum_i \frac{\partial H}{\partial p_i} dp_i + \sum_i \frac{\partial H}{\partial q_i} dq_i + \frac{\partial H}{\partial t} dt. \tag{127}$$

From Eq. (123) we obtain another expansion for dH:

$$dH = \sum_i \dot{q}_i dp_i + \sum_i p_i d\dot{q}_i - \sum_i \frac{\partial L}{\partial \dot{q}_i} d\dot{q}_i - \sum_i \frac{\partial L}{\partial q_i} dq_i - \frac{\partial L}{\partial t} dt. \tag{128}$$

The coefficients of the $d\dot{q}_i$'s are zero because of the definition of the p_i's. From Lagrange's equations we obtain

$$\frac{\partial L}{\partial q_i} = \dot{p}_i. \tag{129}$$

Equation (128) then reduces to the simple form

$$dH = \sum_i \dot{q}_i dp_i - \sum_i \dot{p}_i dq_i - \frac{\partial L}{\partial t} dt. \tag{130}$$

Equating coefficients of like terms in Eqs. (127) and (130) yields

$$\dot{q}_i = \frac{\partial H}{\partial p_i}, \quad \dot{p}_i = -\frac{\partial H}{\partial q_i}, \quad i = 1, 2, \ldots, n; \tag{131}$$

$$-\frac{\partial L}{\partial t} = \frac{\partial H}{\partial t}. \tag{132}$$

Equations (131) are *Hamilton's canonical equations of motion*. They constitute a system of $2n$ first-order partial differential equations for $H(\mathbf{p},\mathbf{q},t)$ that replaces the Lagrange system of n second-order equations for $L(\mathbf{q},\dot{\mathbf{q}},t)$. Equation (132) relates L to H for the time-varying case.

As an example, we again take the system of n uncoupled simple harmonic oscillators, so that $\mathbf{q}$ is interpreted as a set of normal coordinates. Let the spring constant k be the same for each particle. The Hamiltonian becomes

$$H = T + V = \frac{1}{2m}\sum_i p_i^2 + \frac{k}{2}\sum_i q_i^2 = \tfrac{1}{2}\sum \mathbf{p}\cdot\frac{\mathbf{p}}{m_i} + \tfrac{1}{2}k\sum \mathbf{q}\cdot\mathbf{q} = H(\mathbf{p},\mathbf{q},t). \tag{133}$$

From this we obtain

$$\frac{\partial H}{\partial q_i} = kq_i, \quad \frac{\partial H}{\partial p_i} = \frac{p_i}{m_i}. \tag{134}$$

Applying Hamilton's equations, we get

$$\dot{p}_i = -kq_i, \quad \dot{q}_i = p_i/m_i. \tag{135}$$

The second equation is merely a rearrangement of the definition of the momentum of the ith particle.

4.1 Cyclic Coordinates

If the Lagrangian of a system does not contain a particular generalized coordinate q_i (although it may contain the corresponding generalized velocity), then that q_i is said to be a *cyclic coordinate*. Clearly, we have

$$\frac{\partial L}{\partial q_i} = 0, \tag{136}$$

so that the Lagrange equation of motion for the ith particle becomes

$$\frac{d}{dt}\frac{\partial L}{\partial \dot{q}_i} = \dot{p}_i = 0. \tag{137}$$

This expresses the important result

$$p_i = \text{const.} \quad \text{for } q_i \text{ cyclic,} \tag{138}$$

which can be put in the form of the following conservation theorem: *The generalized momentum conjugate to a cyclic coordinate is conserved.*

Since the momentum conjugate to a cyclic coordinate is constant, it follows from one of Hamilton's equations that H does not contain the coordinate. This is expressed in the following principle: *A cyclic coordinate will not appear in the Hamiltonian. Conversely, if a generalized coordinate does not occur in the Hamiltonian, then the conjugate momentum is conserved.*

5. CANONICAL EQUATIONS OF TRANSFORMATION

The *canonical transformations* that we shall discuss are those transformations of coordinates and momenta that conserve Hamilton's canonical equations of motion (preserve their form). Recall that $(\mathbf{q},\mathbf{p})$ are the generalized position and momentum vectors in phase space (which was introduced in Sec. 1.5). Let $(\mathbf{Q},\mathbf{P})$ be the corresponding transformed vectors. The transformation equations from the old to the new vectors have the following form:

$$\begin{aligned}&\mathbf{Q} = \mathbf{Q}(\mathbf{q},\mathbf{p}),\\ &\quad \text{or } Q_i = Q_i(q_1,q_2,\ldots,q_n,p_1,p_2,\ldots,p_n),\\ &\mathbf{P} = \mathbf{P}(\mathbf{q},\mathbf{p}),\\ &\quad \text{or } P_i = P_i(q_1,q_2,\ldots,q_n,p_1,p_2,\ldots,p_n).\end{aligned} \tag{139}$$

This is the general form of the transformation equations from original to transformed phase space. We seek the properties of such a transformation that makes Hamilton's equations invariant.

Let $H = H(\mathbf{p},\mathbf{q},t)$ be a general Hamiltonian. Let the transformed Hamiltonian be designated as K. Then $K = K(\mathbf{P},\mathbf{Q},t)$. ($t$ transforms into itself.) Hamilton's canonical equations of motion in the transformed space are assumed to be given by

$$\dot{Q}_i = \frac{\partial \mathbf{K}}{\partial P_i}, \quad \dot{P}_i = -\frac{\partial K}{\partial Q_i}. \tag{140}$$

Those transformation equations (139) that satisfy Eqs. (140) are said to be *canonical* transformations. Some authors call these *contact* transformations, because neighboring curves are transformed into neighboring curves.

The key to finding the appropriate set of canonical transformations is to start with Hamilton's variational principle, where the integrand in the functional $I = \int_{t_1}^{t_2} L\,dt$ is given explicitly in terms of the Hamiltonian. This is easily done since

$$L = \sum_i p_i\dot{q}_i - H. \tag{141}$$

Therefore, Hamilton's principle may be put in the form

$$\delta \int_{t_1}^{t_2} \left[\sum_i p_i\dot{q}_i - H(\mathbf{q},\mathbf{p},t)\right] dt = 0. \tag{142}$$

Under a canonical transformation, Hamilton's principle becomes

$$\delta \int_{t_1}^{t_2} \left[\sum_i P_i \dot{Q}_i - K(\mathbf{Q},\mathbf{P},t) \right] dt = 0. \qquad (143)$$

If we set the difference between the two integrands equal to the derivative of an arbitrary function of time $F(t)$, we get

$$\delta \int_{t_1}^{t_2} \left\{ \left[\sum p_i \dot{q}_i - H \right] - \left[\sum P_i \dot{Q}_i - K \right] - \frac{dF(t)}{dt} \right\} dt = 0. \qquad (144)$$

(Obviously, dF is an exact differential.) From this equation, it is clear that $F(t)$ is an extremum with respect to the fixed end points (t_1,t_2), so that

$$\delta \int_{t_1}^{t_2} \frac{dF(t)}{dt} dt = \delta F(t_2) - \delta F(t_1) = 0. \qquad (145)$$

Knowing F as a function of one of the following sets of coordinates—$(\mathbf{q},\mathbf{Q})$, $(\mathbf{q},\mathbf{P})$, $(\mathbf{p},\mathbf{Q})$, or $(\mathbf{p},\mathbf{P})$—allows us to calculate the transformation equations given by Eq. (139). For this reason, F is called a *generating function*, because it "generates" the transformation equations. In addition to time, for a system with n degrees of freedom there are a total of $2n$ generalized coordinates and $2n$ generalized momenta (old and new together). The transformation equations (139) consist of $2n$ equations, which yield $2n$ independent relationships between these variables. Here F is usually taken to have one of the following forms: $F_1(\mathbf{q},\mathbf{Q},t)$, $F_2(\mathbf{q},\mathbf{P},t)$, $F_3(\mathbf{p},\mathbf{Q},t)$, or $F_4(\mathbf{p},\mathbf{P},t)$; but many other forms are possible.

Suppose $F = F_1(\mathbf{q},\mathbf{Q},t)$. Recall that the difference of the integrands in Eqs. (142) and (143) must be a perfect differential. Therefore, we set

$$\begin{aligned} &\sum_i p_i dq_i - H dt - \sum_i P_i dQ_i + K dt \\ &= dF_1(q,Q,t) \\ &= \sum \frac{\partial F_1}{\partial q_i} dq_i + \sum \frac{\partial F_1}{\partial Q_i} dQ_i + \frac{\partial F_1}{\partial t} dt. \end{aligned} \qquad (146)$$

Since $(\mathbf{q},\mathbf{Q},t)$ are independent variables, the coefficients of each dq_i, each dQ_i, and dt must vanish. This yields

$$p_i = \frac{\partial F_1}{\partial q_i}, \quad P_i = \frac{\partial F_1}{\partial Q_i}, \quad K - H = \frac{\partial F_1}{\partial t}. \qquad (147)$$

The first two sets of these equations can be solved for $\mathbf{Q} = \mathbf{Q}(\mathbf{q},\mathbf{p})$ and $\mathbf{P} = \mathbf{P}(\mathbf{q},\mathbf{p})$. The third equation allows us to calculate the new Hamiltonian $K = K(\mathbf{Q},\mathbf{P},t)$ in terms of the old $H = H(\mathbf{q},\mathbf{p},t)$ and the transformation equations.

5.1 Hamilton-Jacobi Theory

There is a general procedure that allows us to calculate the appropriate generating function without using this *ad hoc* method. It involves manipulating the Hamiltonian into the form of a partial differential equation and thereby deriving the generating function (specifically, its spatial derivatives) as a solution. This technique is called the Hamilton-Jacobi theory.

5.1.1 Hamilton-Jacobi Equations The first point to make in investigating this theory is that we need a transformation from $(\mathbf{q},\mathbf{p})$ space to $(\mathbf{Q},\mathbf{P})$ space that will simplify Hamilton's equations in the transformed coordinates. The key is the use of cyclic coordinates. We saw (Sec. 4.1) that if the Hamiltonian does not contain a particular q_i, then the conjugate momentum is conserved. In particular, if the transformed Hamiltonian is independent of $\mathbf{Q}$, then all the transformed momenta are conserved, and we can set the set of momenta $\mathbf{P} = \boldsymbol{\beta} = \mathbf{P}(0)$, the initial value of the set. We can go even further by making the set $\mathbf{Q} = \boldsymbol{\alpha}$ where $\boldsymbol{\alpha}$ may be taken as the initial value also. Then we consider the transformation

$$\begin{aligned} \mathbf{q} &= \mathbf{q}(\boldsymbol{\alpha},\boldsymbol{\beta},t), \\ \mathbf{p} &= \mathbf{p}(\boldsymbol{\alpha},\boldsymbol{\beta},t), \end{aligned} \qquad (148)$$

where the original system is transformed into the system of initial coordinates and momenta (the initial conditions). The transformed Hamiltonian K is constant.

To derive the Hamilton-Jacobi equations, we refer to Eq. (144). Let $F = F(\mathbf{q},\mathbf{P},t)$. We again consider the integrand for a system of n degrees of freedom and add the exact differential $\Sigma P_i dQ_i + \Sigma Q_i dP_i$ to the differential dF, which is expanded in terms of $d\mathbf{q}$, $d\mathbf{P}$, $d\mathbf{Q}$, and dt. (At this stage, we consider $\mathbf{P}$ and $\mathbf{Q}$ as variable sets; later, we shall set them equal to the initial position and momentum sets.) We now use the condition that K = const., and we get

$$\sum p_i dq_i - \sum P_i dQ_i - H dt - \sum \frac{\partial F}{\partial q_i} dq_i - \sum \frac{\partial F}{\partial P_i} dP_i - \frac{\partial F}{\partial t} dt + \sum P_i dQ_i + \sum Q_i dP_i = 0. \tag{149}$$

Equating the coefficients of $d\mathbf{q}$, $d\mathbf{Q}$, $d\mathbf{P}$, and dt to zero yields

$$p_i = \frac{\partial F}{\partial q_i}, \quad Q_i = \frac{\partial F}{\partial P_i}, \tag{150}$$

$$H + \frac{\partial F}{\partial t} = 0. \tag{151}$$

Recall that $H = H(\mathbf{q},\mathbf{p},t)$. Equation (151) then becomes a partial differential equation for F when the expression for the components of $\mathbf{p}$ in the first equation of (150) is substituted into H. We get (in expanded form)

$$H\left(q_1,\ldots,q_n,\frac{\partial F}{\partial q_1},\ldots,\frac{\partial F}{\partial q_n},t\right) + \frac{\partial F}{\partial t} = 0. \tag{152}$$

Equation (152) is called the *Hamilton-Jacobi* (H-J) equation. It is a first-order partial differential equation in the n variables $(q_1,\ldots,q_n,t)$ for the solution F. For a given dynamical problem, it is clear that H is a prescribed function of the arguments shown. It is customary to denote the generating function F by S, which is called *Hamilton's principal function*. Note that S itself does not appear explicitly in the H-J equation, but only its partial derivatives with respect to the components of $\mathbf{q}$. Therefore, S plus an additive constant is also a solution of the H-J equation. There must be $n + 1$ constants of integration. We may select n of those to be the components of the initial momentum set.

The additive constant may then be taken as the total energy E. The complete solution of the H-J equation is given by

$$S = S\,(q_1,\ldots,q_n,\beta_1,\ldots,\beta_n,t), \quad \beta_i = P_i. \tag{153}$$

Suppose we have a complete solution for a given dynamical problem. We rewrite Eq. (150) as

$$p_i = \frac{\partial S(q_1,\ldots,q_n,\beta_1,\ldots,\beta_n,t)}{\partial q_i}. \tag{154}$$

Note that S is also a known function of the initial momentum set. Therefore,

$$Q_i = \alpha_i = \left.\frac{\partial S(q_1,\ldots,q_n,\beta_1,\ldots,\beta_n,t)}{\partial \beta_i}\right|_{t=0}, \quad \alpha_i = q_i(0), \quad \beta_i = p_i(0). \tag{155}$$

We can perform an inverse transformation on Eq. (139) and obtain our final results:

$$q_i = q_i(\alpha_1,\ldots,\alpha_n,\beta_1,\ldots,\beta_n,t), \quad i = 1,\ldots,n; \tag{156}$$

$$p_i = p_i(\alpha_1,\ldots,\alpha_n,\beta_1,\ldots,\beta_n,t). \tag{157}$$

Combining these equations and using t as a parameter, we get the trajectory of the solution in phase space. This is a $2n$-dimensional space where the trajectory starts at (α,β) and unwinds as t increases from $t = 0$. The trajectory thus gives the complete time history of the dynamical system for a given set of initial conditions. Thus, Hamilton's principal function S yields a solution to a dynamical problem. In addition, S is the generating function that generates a canonical transformation to constant coordinates and momenta. Instead of constructing *ad hoc* generating functions, the problem is transformed into that of solving a first-order partial differential equation for $\mathbf{q} = \mathbf{q}(t)$.

5.1.2 Hamilton's Characteristic Function

If H is not an explicit function of t, then the H-J equation is

$$\frac{\partial S}{\partial t} + H\left(\mathbf{q}, \frac{\partial S}{\partial \mathbf{q}}\right) = 0 \tag{158}$$

(where $\partial S/\partial \mathbf{q}$ stands for the n spatial derivatives of S). We may separate the t-dependent part of S by setting

$$S(\mathbf{q},t) = W(\mathbf{q},\boldsymbol{\beta}) - Et, \tag{159}$$

where E is the total energy. The t-independent function is called *Hamilton's character-*

istic function. It satisfies the H-J equation

$$H\left(\mathbf{q}, \frac{\partial W}{\partial \mathbf{q}}\right) = E. \tag{160}$$

Here W generates a canonical transformation (independent of t) in which all the transformed coordinates are cyclic (the initial coordinates and momenta).

If W can be put in the form

$$W = \sum_i W_i(q_i,\beta_1,\ldots,\beta_n), \tag{161}$$

then the H-J equation can be split into n ordinary differential equations of the form

$$H_i\left(q_i, \frac{\partial W_i}{\partial q_i}, \beta_1,\ldots,\beta_n\right) = e_i, \quad i = 1, 2, \ldots, n, \tag{162}$$

where the e_i are several constants. (For a discussion of separation of variables, see Landau and Lifshitz, 1976.) Each of these n equations is a first-order ordinary differential equation and can clearly be reduced to quadratures. We need only solve for $\partial W_i/\partial q_i$ and then integrate with respect to q_i.

5.1.3 Keplerian Motion of an Electron about a Nucleus As an example of the use of the H-J equation, we consider the motion of an electron of charge $-e$ and mass m about an atomic nucleus of charge Ze and mass M. Here Z is the atomic number (the number of electrons in the neutral atom). We use the approximation that the nucleus remains stationary since $m \ll M$. In polar coordinates (r,θ), the kinetic energy of the electron is

$$T = \tfrac{1}{2}m(\dot{r}^2 + r^2\dot{\theta}^2), \tag{163}$$

and the potential energy is

$$V = -e^2Z/r. \tag{164}$$

From Eq. (163) we obtain the components of the generalized momentum in polar coordinates:

$$p_r = m\dot{r}, \quad p_\theta = mr^2\dot{\theta}^2. \tag{165}$$

Therefore, the Hamiltonian is

$$H = \frac{1}{2m}\left(p_r^2 + \frac{1}{r^2}p_\theta^2\right) - \frac{e^2Z}{r}. \tag{166}$$

We then obtain the H-J equation

$$\frac{1}{2m}\left[\left(\frac{\partial S}{\partial r}\right)^2 + \frac{1}{r^2}\left(\frac{\partial S}{\partial \theta}\right)^2\right] - \frac{e^2Z}{r} = E, \tag{167}$$

where E is the total energy. This is a case where the variables can be separated:

$$\frac{\partial S}{\partial \theta} = \left[2mEr^2 + 2me^2Zr - r^2\left(\frac{\partial S}{\partial r}\right)^2\right]^{1/2} = \beta, \tag{168}$$

where β is a constant. From this we obtain the two subsidiary equations

$$\frac{\partial S}{\partial \theta} = \beta \tag{169}$$

and

$$\frac{\partial S}{\partial r} = \left(2mE + 2\frac{me^2Z}{r} - \frac{\beta^2}{r^2}\right)^{1/2}. \tag{170}$$

The first equation tells us that the angular momentum p_θ is constant. Indeed, θ is a cyclic coordinate, since it does not appear explicitly in the H-J equation; therefore the conjugate momentum p_θ is conserved. Since

$$dS = \frac{\partial S}{\partial \theta}d\theta + \frac{\partial S}{\partial r}dr,$$

we can determine S by integrating Eqs. (169) and (170) and adding. This gives

$$S = \beta\theta + \int\left(2mE + 2me^2\frac{Z}{r} - \frac{\beta^2}{r^2}\right)^{1/2} dr. \tag{171}$$

To calculate the orbit, we differentiate Eq. (171) with respect to β and obtain

$$\theta - \beta\int\frac{dr}{r(-\beta^2 + 2me^2Zr + 2mEr^2)^{1/2}} = \gamma. \tag{172}$$

Using any table of integrals, we evaluate the integral and solve for the reciprocal of the radius vector, thus obtaining

$$\frac{1}{r} = \frac{me^2Z}{\beta^2} + \left(\frac{m^2e^4Z^2}{\beta^4} + \frac{2mE}{\beta^2}\right)^{1/2}\cos(\theta - \phi), \tag{173}$$

where ϕ is a constant phase angle. Equation (173) is the equation of a conic section. It is shown in any text on analytic geometry that the eccentricity ϵ is

$$\epsilon = \left(1 + \frac{2E\beta^2}{me^4Z^2}\right)^{1/2}. \tag{174}$$

This shows at once that the orbit is elliptic, parabolic, or hyperbolic according to whether E is negative, zero, or positive. An electron rotating about a nucleus is only stable if E is negative, for only in this case can the orbit be closed. If we set $E = -W$, then the semimajor axis of the ellipse is

$$a = e^2Z/W. \tag{175}$$

This tells us that all elliptic orbits having the same energy have the same major axis irrespective of their eccentricities.

6. PRINCIPLE OF LEAST ACTION

In Hamilton's variational principle (for a conservative system), the neighboring paths of an extremum might not conserve energy, meaning that $T + V =$ const. only along the dynamical path (extremum). There exists another variational principle, the *principle of least action*. Its purpose is to rectify this weakness in Hamilton's principle.

The principle of least action is like Hamilton's principle in that it compares a dynamical path with a neighboring path that has the same fixed end points. Unlike Hamilton's principle, the principle of least action allows the virtual displacements to depend on time. This allows us to impose the constraint that if H is conserved on the actual path, it is also conserved on the varied path. Because of this constraint, corresponding points on the dynamical path and a neighboring path are not chosen as points reached at the same time, so that a varied path is not necessarily traversed in the same time as the dynamic path. We show that the functional that is made stationary and fulfills the criterion of energy conservation on neighboring paths is the *action integral*, which we now describe.

The action integral A is defined by

$$A = \int_{t_1}^{t_2} \sum_j p_j\dot{q}_j dt. \tag{176}$$

Now, a virtual displacement does not correspond to an actual physical displacement, as was pointed out in Sec. 1.7. In the latter type of displacement, H is conserved, while in a virtual displacement it is not conserved. If we wish to consider a new type of displacement, which moves to a neighboring trajectory but still conserves H, it is clear that it must involve a change in time. Thus, the transit times of points along varying paths obtained with this new variation need not be constant, and the end points may differ in time also although the variations of the q_i at the end point remain zero.

We define this new type of variation Δq_i by tagging each trajectory with a parameter α. Since each curve has the parameter t associated with each point, t must be a function of α. Thus, we must have $q_j = q_j(\alpha,t)$. We now expand δq_j and obtain

$$\Delta q_j \rightarrow d\alpha\left(\frac{dq_j}{d\alpha}\right) = d\alpha\left(\frac{\partial q_j}{\partial\alpha}\right) + d\alpha\left(\frac{\partial q_j}{\partial t}\right)\left(\frac{dt}{d\alpha}\right). \tag{177}$$

The first term is just δq_j, while the last is Δt since t depends only on itself, and thus the change in t is just a Δ variation. Thus Eq. (177) becomes

$$\Delta q_j = \delta q_j + \dot{q}_j\Delta t. \tag{178}$$

Since this relation between the two difference operators holds for any function $f(\mathbf{q},t)$, we have the more general relation

$$\Delta f(\mathbf{q},t) = \delta f + \dot{f}\Delta t = \sum_j \left(\frac{\partial f}{\partial q_j}\delta q_j\right) + \left[\sum_j \left(\frac{\partial f}{\partial \dot{q}_j}\delta\dot{q}_j + \frac{\partial f}{\partial t}\right)\right]\Delta t. \tag{179}$$

The first sum represents the expansion of the change in f due to a virtual displacement; the second, the expansion of the time derivative of f.

With this mathematical apparatus at our disposal, we can now prove that $\Delta A = 0$, which tells us that the action integral is an extremum along a dynamical path. Applying the operator Δ to the action integral gives

$$\Delta A = \Delta \int_{t_1}^{t_2} (L + H)dt$$
$$= \Delta \int_{t_1}^{t_2} Ldt + H(\Delta t_2 - \Delta t_1). \quad (180)$$

If we now represent the integral $\int Ldt$ as $I(t)$, then the definite integral in Eq. (180) is $I(t_2) - I(t_1)$, and its variation is $\Delta I(t_2) - \Delta I(t_2)$, and we have

$$\Delta \int_{t_1}^{t_2} Ldt = \Delta I(t_2) - \Delta I(t_1)$$
$$= \delta I(t_2) - \delta I(t_1) + \dot{I}(t_2)\Delta t_2 - \dot{I}(t_1)\Delta t_1$$
$$= \delta I + L\Delta t|_{t_1}^{t_2}. \quad (181)$$

Since we have not required the δq_j to vanish at the end points of the integral, we cannot use Hamilton's principle to assert that δI vanishes. Instead, we must calculate δI by expanding δL and then using Lagrange's equations and Eq. (178). We have

$$\delta \int_{t_1}^{t_2} Ldt = \int \delta L dt$$
$$= \int \sum_j \left(\frac{\partial L}{\partial q_j} \delta q_j + \frac{\partial L}{\partial \dot{q}_j} \frac{d}{dt} \delta q_j \right) dt$$
$$= \sum \int \frac{d}{dt} \left(\frac{\partial L}{\partial \dot{q}_j} \delta q_j \right) dt$$
$$= \sum \int \frac{d}{dt} \left(\frac{\partial L}{\partial \dot{q}_j} \Delta q_j - \frac{\partial L}{\partial \dot{q}_j} \dot{q}_j \Delta t \right) dt$$
$$= \sum \left(\frac{\partial L}{\partial \dot{q}_j} \Delta q_j - \frac{\partial L}{\partial \dot{q}_j} \dot{q}_j \Delta t \right) \Bigg|_{t_1}^{t_2}$$
$$= -\sum p_j \dot{q}_j \Delta t|_{t_1}^{t_2}, \quad (182)$$

since $\partial L / \partial \dot{q}_j = p_j$. (Note that the Δq_j's vanish at the end points, but Δt does not.) Combining the above results yields

$$\Delta A = \left(-\sum_j p_j \dot{q}_j + L + H \right) \Delta t|_{t_1}^{t_2} = 0. \quad (183)$$

This completes the proof that the action integral is an extremum.

Example: Consider a single particle of mass m, total energy E, and kinetic energy T moving through a force field with potential energy $V(x,y,z)$. Clearly, the force field is conservative. The action integral becomes

$$A = \int_{t_1}^{t_2} 2T dt = \int_A^B [2m(E - V)]^{1/2} ds$$
$$= \int_A^B p(E,x,y,z) ds, \quad (184)$$

where ds is the element of arc length along a path in three-dimensional space whose end points are (A,B), and p is the momentum of the particle, which clearly depends on space and the total energy. The principle of least action tells us that

$$\Delta \int_A^B p(E,x,y,z) ds = 0 \quad (185)$$

on a dynamical path.

The action integral is most important in H-J theory. Consider Hamilton's characteristic function $W(\mathbf{q},\boldsymbol{\beta})$, which we recall is the steady-state part of Hamilton's principal function $S(\mathbf{q},\beta,t)$. For the steady-state case, Eq. (154) becomes

$$\frac{\partial W(q_1,\ldots,q_n,\beta_1,\ldots,\beta_n)}{\partial q_i} = p_i. \quad (186)$$

The total time derivative of W is

$$\frac{dW}{dt} = \sum_j \frac{\partial W}{\partial q_j} \dot{q}_j = \sum_j p_j \dot{q}_j. \quad (187)$$

Integrating this from t_1 to t_2 yields

$$W = \int_{t_1}^{t_2} \sum_j p_j \dot{q}_j dt. \quad (188)$$

Comparing this equation with Eq. (176), we see that Hamilton's characteristic function W is indeed the action integral. Therefore, with respect to the H-J theory, the principle of least action tells us that W is an extremum on a dynamical path. This demonstrates the importance of the principle of least action in H-J theory. H-J theory invokes the more realistic condition that all paths neighboring the dynamical path satisfy the conservation of energy, rather than Hamilton's principle where energy is not conserved along neighboring paths.

7. ACTION-ANGLE VARIABLES AND PHASE INTEGRALS

Dynamical systems undergoing periodic motion are most important in classical mechanics. Action-angle variables and phase integrals are introduced into H-J theory to handle such problems. Before treating these topics, we first briefly discuss periodic motion.

Essentially, two types of periodic motion occur in dynamics. We illustrate them by one-dimensional models in phase space.

1. A harmonic oscillator. Both q and p are periodic functions of time with the same frequency. Since q and p return to their starting values after one period, it is clear that the trajectory in phase space is closed.
2. Rotation. Here q is not periodic, but when it is increased by some constant value the configuration of the system remains unchanged. Suppose q is the angle of rotation about the axis of rotation of a rigid body. Increasing q by 2π radians produces no change in the system configuration. On the other hand, p is periodic. The trajectory in phase space is not closed.

To illustrate both types of periodic motion in a one-dimensional model, we consider the simple pendulum oscillating with no friction in the vertical plane. Let q be the angle θ of deflection from the vertical and p_θ be the conjugate momentum. The conservation of energy tells us that $E = p_\theta^2/2mL^2 - mg\cos\theta$, where L is the length of the pendulum suspension, m its mass, E the total energy, and g the acceleration of gravity. Solving this equation for p_θ gives

$$p_\theta = [2mL^2(E + mgL\cos\theta)]^{1/2}. \qquad (189)$$

If the motion is periodic, then one period extends between two consecutive zeros of p_θ. For a zero, we must have

$$\cos\bar{\theta} = -E/mgL. \qquad (190)$$

There are three possibilities:

1. If $E < mgL$, then the motion is of the vibration type where the pendulum oscillates between $\bar{\theta}$ and $-\bar{\theta}$.
2. If $E > mgL$, all values of θ are possible, so that the angle can increase without limit, producing periodic motion of the rotation type. In this case, the total energy is great enough to swing the pendulum through $\theta = \pi$ so that it will continue to rotate.
3. If $E = mgL$, we have a transitional case between **1** and **2**.

Figure 12 shows the three types of trajectories in phase space for the simple pendulum. Curve (1) is periodic of the vibration type. Curve (2) is periodic of the rotation type. Curve (3) is the transitional curve.

We now turn to the more general case of systems with more than one degree of freedom. We restrict the discussion to those periodic problems where Hamilton's characteristic function W is separable in at least one set of canonical variables. We repeat Eq. (186):

$$p_j = \frac{\partial W_j(q_j,\beta_1,\ldots,\beta_n)}{\partial q_j}. \qquad (191)$$

The solution of this equation is

$$p_j = p_j(q_j,\beta_1,\ldots,\beta_n). \qquad (192)$$

We see that Eq. (192) is the projected trajectory in phase space of the system point in the (q_j,p_j) plane. Periodic motion will occur if Eq. (192) describes either a closed orbit or a periodic function of q_j. Note that it is not necessary for each coordinate and momentum pair to have the same frequency. For the example of a three-dimensional oscillator with different components of the spring force con-

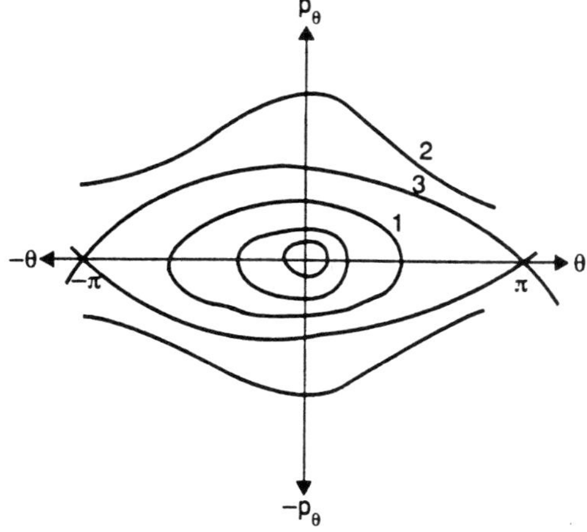

FIG. 12. Types of trajectories in phase space for a simple pendulum.

stant (different frequencies), the complete motion is not necessarily simply periodic. Indeed, if the separate frequencies are not rational fractions of each other, the phase trajectory will be an open Lissajous figure. Such a trajectory is called *conditionally periodic*.

We now discuss the *action variables* and the corresponding *phase integrals*. For a system with n degrees of freedom, we have n action variables $J_1, J_2, \ldots, J_n$ defined by

$$J_i = \oint p_i dq_i. \tag{193}$$

J_i is the ith action variable, and the contour integral on the right-hand side is the corresponding phase integral. The integration is carried out over a complete period of oscillation or rotation of q_i, as the case may be. Note the resemblance of the phase integral to the action integral given by Eq. (176) for the ith component. Using Eq. (186) we get

$$J_i = \oint \frac{\partial W(q_i,\beta_1,\ldots,\beta_n)}{\partial q_i} dq_i. \tag{194}$$

It is clear that each action variable J_i is a function only of the n constants of integration appearing in the solution of the H-J equation. Note that these constants are the n components of the constant momentum, which may be taken as the initial value of the momentum. Since the generalized coordinates and momenta are linearly independent, it follows that the n action variables form n independent functions of the β_i's and hence are suitable as a new set of constant momentum components. This means that we can express the constant momentum components as functions of the components of the action variable. We may thus write W as

$$W = W(q_1,\ldots,q_n;J_1,\ldots,J_n), \tag{195}$$

while the Hamiltonian appears only as a function of the action variables.

The *angle variable* w_i conjugate to J_i is a generalized coordinate defined by

$$w_i = \frac{\partial W_i}{\partial J_i} \tag{196}$$

If we now turn to the first equation in (131), we see that q_i is analogous to w_i (since it is a generalized coordinate), and p_i is analogous to J_i. Therefore, the equations of motion in terms of the action and angle variables are

$$\dot{w}_i = \frac{\partial H(J_1,\ldots,J_n)}{\partial J_i} = \nu_i(J_i,\ldots,J_n), \tag{197}$$

where the ν_i's are constant functions of the action variables. We immediately solve Eq. (197) and obtain

$$w_i = \nu_i t + \alpha_i, \tag{198}$$

where α_i is the constant of integration. We can always calculate q_i as a function of t, ν_i, and α_i from the above equations. But the real value of this analysis lies in the physical interpretation of the ν_i's. We show that each ν_i is the frequency associated with the periodic motion of the corresponding q_i.

To show this, we consider the change Δw_i in the angle variable w_i when the coordinate q_i goes through a complete cycle of vibration or rotation. Let δw_i be the infinitesimal change as a result of an increment in an arbitrary q_j; we write

$$\Delta w_i = \oint \delta w_i, \quad \text{where } \delta w_i = \frac{\partial w_i}{\partial q_j} dq_j. \tag{199}$$

We now use Eq. (196) and perform the following calculation:

$$\begin{aligned}\Delta w_i &= \oint \delta w_i = \oint \frac{\partial w_i}{\partial q_j} dq_j = \oint \frac{\partial^2 W}{\partial q_j \partial J_i} dq_j \\ &= \frac{\partial}{\partial J_i} \oint \frac{\partial W}{\partial q_j} dq_j = \frac{\partial}{\partial J_i} \oint p_j dq_j = \frac{\partial J_j}{\partial J_i} = \delta_{ij}.\end{aligned} \tag{200}$$

This tells us that w_i changes by a unit amount when q_i goes through a complete cycle, but is unaffected by a similar change in q_j. Let τ_i be the period associated with q_i. It follows from the above that $\nu_i = 1/\tau_i$. This is what we set out to prove: that ν_i is the period associated with the periodic variation of q_i.

To illustrate the use of the action-angle variables and the phase integrals, we discuss in detail the one-dimensional simple harmonic oscillator. There is only one action variable and one phase integral. They are obtained by constructing the Hamiltonian and the H-J equation and solving for W. The Hamiltonian is

$$H = p^2/2m + kq^2/2, \tag{201}$$

where k is the spring constant. The H-J equation is

$$\frac{1}{2m}\left(\frac{\partial W}{\partial q}\right)^2 + \frac{kq^2}{2} = E. \tag{202}$$

Solving for W, we get

$$W = (mk)^{1/2}\int(2E/k - q^2)^{1/2}dq. \tag{203}$$

The action integral or phase variable is

$$\begin{aligned} J &= \oint p dq = \oint \frac{\partial W(q,E)}{\partial q} dq \\ &= (2m)^{1/2} \oint \left(\frac{2E}{k} - q^2\right)^{1/2} dq \\ &= 2E\left(\frac{m}{k}\right)^{1/2} \int_0^{2\pi} \cos^2\theta d\theta = 2\pi E\left(\frac{m}{k}\right)^{1/2}, \end{aligned} \tag{204}$$

where we used the substitution $q = (2E/k)^{1/2} \sin\theta$. (Note that we integrated over a complete cycle of q.) From this we get

$$E = H = \frac{J}{2\pi}\left(\frac{k}{m}\right)^{1/2}, \tag{205}$$

and the frequency of oscillation of the oscillator is

$$\frac{\partial H}{\partial J} = \nu = \frac{1}{2\pi}\left(\frac{k}{m}\right) = \frac{\omega}{2\pi}, \tag{206}$$

where ω is the angular frequency.

GLOSSARY

Angle of Inclination: Angle between fixed vertical axis and axis of symmetry of top.

Angle of Spin: Angle between x' axis of top and line of nodes.

Azimuth Angle: Angle between fixed x axis and line of nodes.

Configuration Space: $2n$-dimensional space whose axes are n generalized coordinates and n generalized velocities.

Constraints: Conditions imposed on coordinates, making them not independent of one another.

Cyclic Coordinate: A generalized coordinate not appearing in the Lagrangian. Its conjugate momentum is conserved.

D'Alembert's Principle: A method of describing equations of motion by treating the negative of the time derivative of momentum as a force, so that the method of virtual work may be applied.

Eulerian Angles: Three independent angles that define the orientation of a rotating body with one point fixed. See **Angle of Inclination, Angle of Spin**, and **Azimuth Angle**.

Euler's Equation: A partial differential equation obtained as a necessary condition for existence of an extremum of a functional. When used in Hamilton's principle, Euler's equation becomes Lagrange's equation of motion.

Functional: A definite integral involving a class of integrands; a function whose value depends on which of a set of functions is chosen as its argument.

Generalized Coordinates: A set of independent variables that define the position of a dynamical system.

Generalized Momentum: A dynamical variable associated with a particular generalized coordinate in the same fashion as a component of linear momentum is associated with the corresponding Cartesian coordinate.

Generalized Velocity: The time derivative of a generalized coordinate.

Hamiltonian: A function of the generalized coordinates and momenta, defined as the sum of the products of the generalized velocities by the corresponding generalized momenta, minus the Lagrangian. For a conservative system, it is equal to the sum of the kinetic and potential energies.

Hamilton's Principle: The statement that the actual path of a dynamical system between two fixed points in configuration space is that which minimizes the time integral of the Lagrangian between the two points.

Lagrangian: A function of the generalized coordinates and velocities equal to the difference between the kinetic and potential energies; sometimes called the *kinetic potential.*

Line of Nodes: In the motion of a top, the intersection of the horizontal fixed plane and the plane normal to the axis of symmetry of the top. Also, in treating the motion of a planet, the line of intersection of the plane of the orbit with the equatorial plane.

Nutation: Motion of a top characterized by time variation of the angle of inclination.

Phase Space: $2n$-dimensional space whose axes are the n generalized coordinates and the n generalized momenta.

Precession: Motion of a top characterized by time variation of the azimuth angle.

Principal Axes of Inertia: A set of mutually perpendicular directions in a rigid body with respect to which the products of inertia vanish so that the inertia tensor is diagonal.

Virtual Displacement: An infinitesimal variation of coordinates compatible with any constraints that may be acting, at constant time.

Virtual Work: Work done on a dynamical system in the course of a virtual displacement.

Works Cited

Birkhoff, G., Maclane, S. (1977), *A Survey of Modern Algebra*, 4th ed., New York: Macmillan.

Landau, L. D., Lifshitz, E. M. (1976), *Course of Theoretical Physics*, Vol. 1, *Mechanics*, 3rd ed., London: Pergamon.

Noble, B., Daniels, J. W. (1977), *Applied Linear Algebra*, 2nd ed., Englewood Cliffs, NJ: Prentice-Hall.

Poincaré, Henri (1902), *Science et l'Hypothèse*, Paris: Flammarion; translation (1952), New York: Dover, Chaps. 6 and 7.

Lord Rayleigh (1878), *The Theory of Sound*, New York: Macmillan; reprint (1945), New York: Dover, Vol. 1, Chap. 4.

Synge, John L., Griffith, Byron A. (1959), *Principles of Mechanics*, 3rd ed., New York: McGraw-Hill.

Whittaker, E. T. (1937), *A Treatise on the Analytical Dynamics of Particles and Rigid Bodies*, 4th ed., London: Macmillan; reprint (1974), New York: Dover.

Further Reading

Corben, H. C., Stehle, Philip (1974), *Classical Mechanics*, 3rd ed., New York: Wiley.

Davis, Julian L. (1990), *Wave Propagation in Electromagnetic Media*, Berlin: Springer-Verlag. Appendix to Chap. 3 on similarity and orthogonal transformations; Chap. 5 on canonical transformations and Hamilton-Jacobi theory.

Davis, Julian L. (1988), *Wave Propagation in Solids and Fluids*, Berlin: Springer-Verlag, Chap. 9. Variational methods, Hamilton's principle, Lagrange's and Hamilton's equations, etc.

Feynman, R. P., Leighton, R. B., Sands, M. (1963), *The Feynman Lectures on Physics*, Vol. 1, Reading, MA: Addison-Wesley.

Goldstein, Herbert (1950), *Classical Mechanics*, 1st ed., (1980), 2nd ed., Reading, MA: Addison-Wesley.

Klein, Felix (1897), *The Mathematical Theory of the Top*, New York: Scribners; reprint (1967) in: *Congruences of Sets and Other Monographs*, New York: Chelsea.

Lanczos, Cornelius (1970), *The Variational Principles of Mechanics*, 4th ed., Toronto: University of Toronto Press.

Landau, L. D., Lifshitz, E. M. (1976), *Course of Theoretical Physics*, Vol. 1, *Mechanics*, 3rd ed., London: Pergamon.

Meirovitch, Leonard (1970), *Methods of Analytical Dynamics*, New York: McGraw-Hill.

Synge, John L., Griffith, Byron A. (1959), *Principles of Mechanics*, 3rd ed., New York: McGraw-Hill.

Whittaker, E. T. (1937), *A Treatise on the Analytical Dynamics of Particles and Rigid Bodies*, 4th ed., London: Macmillan; reprint (1974), New York: Dover.

MEDICAL USE OF LASERS

IRVING ITZKAN, *G.R. Harrison Spectroscopy Laboratory, Massachusetts Institute of Technology, Cambridge, Massachusetts, U.S.A.*

JOSEPH A. IZATT, *Research Laboratory of Electronics, Department of Electrical Engineering and Computer Science, Massachusetts Institute of Technology, Cambridge, Massachusetts, U.S.A.*

INTRODUCTION

Lasers are now used routinely in medicine to cut, shape, remove, heat, and examine biological tissue. Largely in the last decade, lasers have made the transformation out of the research laboratory and into the medical clinic, garnering an impressive penetration into many medical specialties, as illustrated in Table 1. Commercialization of lasers and delivery systems specifically designed for particular applications has kept pace with this transformation from research tool to surgical necessity (see Fig. 1).

The growth of interest in recent decades in the applications of light, optics, and spectroscopy in medicine is related to several key technological advances. The principal advance is the development of lasers to the state where they can be used in medical environments. The special qualities of lasers that make the light they generate especially attractive for medical applications include high spatial coherence, tunability, the generation of high peak and average powers, and the lack of significant rf interference. Spatial coherence is important for beam handling, shaping, and focusing, and for efficient coupling

3-527-28132-0/94/$5.00 + .50

Table 1. Estimates of laser utilization in surgical procedures, 1990–1995.

Specialty	Active surgeons	Potential laser procedures 1990	Potential laser procedures 1995	Laser use, 1990 Procedures per year	Laser use, 1990 %	Laser use, 1995 (Forecast) Procedures per year	Laser use, 1995 (Forecast) %
Otolaryngology	7 400	950 000	1 045 000	95 000	10	209 000	20
Gastroenterology	5 700	650 000	715 000	65 000	10	143 000	20
Obstetrics/gynecology	30 300	1 100 000	1 210 000	110 000	10	242 000	20
Urology	8 800	1 100 000	1 210 000	55 000	5	242 000	20
General surgery	35 500	1 500 000	1 650 000	45 000	3	330 000	20
Orthopedics	17 200	1 500 000	1 650 000	15 000	1	412 500	25
Potential laser procedures		6 800 000	7 480 000	385 000	6	1 578 500	21
Minimally invasive procedures		282 000	2 500 000	141 000	50	875 000	35
All U.S. surgeries		25 000 000	27 500 000	500 000	2	1 100 000	4

Source: I. Arons, Arthur D. Little, Inc., 1993 (adapted from Needham & Co. analysis, 1991).

into optical fibers. Tunability provides the capability of selecting optimal wavelengths for specific applications, matching the absorption and scattering properties to the problem at hand. High power accompanying spatial coherence allows for the precise delivery of high energy densities to selected tissue sites. The absence of rf interference assures compatibility with operating room equipment and pacemakers.

The second major technical advance that has been critical to laser-tissue interactions research and applications is the development of optical fibers, which has been driven primarily by the communications industry. The spatial coherence of laser light enables it to be coupled efficiently into low-loss optical fibers, which can deliver the light to remote sites in the body, minimizing trauma to adjacent tissues. Where normal body openings exist, treatment can be performed through endoscopes. Otherwise, the fiber can be passed through a catheter, a laparoscope, or even a hypodermic needle in a minimally invasive manner. In the past, tissue would have had to be cut away or removed to reveal the site of interest.

The development of intensified detector arrays, particularly optical multichannel analyzers with high quantum efficiencies and large dynamic ranges, has brought real-time spectroscopic analysis of tissue to the medical clinic and the operating room. In many cases, the use of slower, conventional tools to accomplish the same task would not be practical in the clinical setting.

Finally, the development of high-speed, large-capacity computers for analysis and

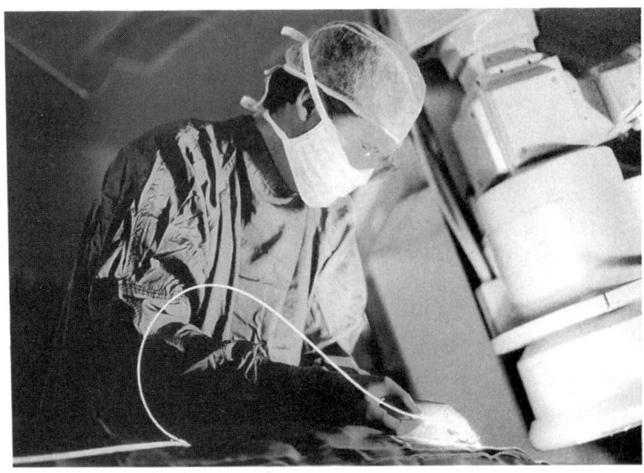

FIG. 1. An example of a commercial medical laser system for rapid thrombus (blood clot) removal. After the physician has inserted a thin, fluid-core catheter into the affected artery, laser energy is fired through the catheter to vaporize the clot. Photo courtesy of Palomar Medical Technologies, Inc., Beverly, Massachusetts.

real-time control of medical instrumentation has permitted the use of elaborate algorithms for therapy, imaging, and diagnostics, and has amplified the impact of the other technical advances already mentioned.

Current research in laser applications in medicine may be roughly divided into four areas: biomedical optics, therapeutic applications, diagnostic applications, and biomedical instrumentation. Biomedical optics is the study of light transport in tissue and the optical properties of tissue. This includes theory, such as analytical and numerical modeling of radiation transport, and experiment, such as techniques for measuring tissue optical properties. The primary difficulty in modeling light propagation in tissue is that, at most laser wavelengths, scattering dominates absorption in most biological tissues (Cheong *et al.*, 1990). While many approximate or numerical models have been suggested, no universally acceptable analytical models have been developed.

Recent developments in biomedical optics include the study of time-resolved photon migration in tissue (Wilson *et al.*, 1992). Using fast electronics and operating in either the time or frequency domain, this technique reduces the light-scattering problem, since the transit time of photons (and, hence, their approximate path length) is measured. Another advantage is that it is a particularly direct way to obtain absorption coefficients of tissues. Applications of time-resolved radiation transport have included *in situ* optical ranging in biological tissue, and initial steps toward optical tomography (Muller, 1993).

Lasers are being used in a wide range of therapeutic applications. The most widely known and commercially successful is laser cutting or ablation of tissue, the "laser scalpel." Although laser cutting is a capital-intensive technology, its advantages are still compelling under the proper circumstances. Since it requires no physical contact between the light-delivery system and the target tissue, the possibility for infection is greatly reduced. In contrast to a sharp scalpel blade, which cuts by exerting pressure in excess of the local tissue tensile strength, laser ablation exerts only minimal pressure on the tissue generated by the reaction forces. (Unfortunately, tactile feedback to the surgeon is also absent.) Avoiding the traction caused by the laser scalpel is particularly important in nerve and brain tissue surgery. The use of the laser for excision of brain tumors decreases the likelihood of initiating convulsions, and requires a smaller orifice through the skull and intervening brain tissue. With the use of optical fibers inserted through endoscopes, catheters, and needles, surgery is possible in regions of the body that are inaccessible to the scalpel blade without the collateral damage that results from conventional surgery.

Other applications of lasers in medical therapy involve tissue heating only. Nonablative laser treatments have been developed in dermatology, for example, which involve coagulation of subcutaneous blood vessels or destruction of pigment cells for treatment of skin disorders (Tan *et al.*, 1989). Lasers are used as distributed heat sources for hyperthermia treatment (Svaasand *et al.*, 1985). Laser welding of tissue is used in vascular surgery and wound repair (Schober *et al.*, 1986). The activation with laser light of phototoxic exogenous chromophores, primarily hematoporphyrin derivatives, continues to receive attention as a treatment for cancer and certain immune disorders (Berns, 1984).

Lasers are also used for medical diagnosis and monitoring. Laser-induced fluorescence in tissue can serve as a probe for architectural and compositional changes signifying the onset and development of disease. Work in this area has demonstrated the successful identification and discrimination of various stages of atherosclerotic plaque, and neoplastic conversion in urinary bladder and gastrointestinal tissues. Related techniques using infrared wavelengths, which yield molecular level information, hold out the promise of gaining functional as well as morphological information.

Finally, several novel applications of lasers in biomedical instrumentation deserve comment. Laser-induced fluorescence microscopes are commonly used in pathology research, contributing to the study of disease. Systems have been developed for rapid cell sorting by laser-induced fluorescence-based flow cytometry (Melamed *et al.*, 1990). A novel technique for suspending transparent microscopic objects at the focus of a laser beam, or "optical tweezers," is capable of holding and manipulating individual, living biological cells (Ashkin *et al.*, 1987). This technique has been used to study the mechanics of the biochemical motors contained in cells.

1. PHYSICS OF LASER-TISSUE INTERACTION

1.1 Laser-Tissue Interaction Regimes

The nature of the laser-tissue interaction process may be divided into several regimes, determined primarily by the intensity of the laser beam and its interaction time with the tissue (Fig. 2). The physical basis of most surgical applications of lasers is tissue cutting and removal at relatively high laser-beam intensities with exposure times of milliseconds to seconds, resulting in the rapid deposition of heat and subsequent vaporization or decomposition. These are the processes characteristic of thermal or thermoacoustic ablation. For nanosecond pulses at relatively high photon energies (ultraviolet wavelengths), photons can directly break specific chemical bonds, resulting in strong absorption and particularly clean cuts. Picosecond and shorter pulses can induce nonlinear absorption in biological tissue, providing the capability for localized absorption of laser energy in otherwise transparent tissue such as the eye. Finally, some interactions have been observed at very low laser-intensity levels delivered over minutes or hours. These interactions are strongly coupled to the host tissue response to photochemistry, and remain as yet poorly understood.

Following a review of the physical properties of biological tissue that are of concern for laser-tissue interactions, these mechanisms are discussed in some detail.

1.2 Tissue Physical Properties

1.2.1 Tissue Optical Properties The ability to deliver therapeutic laser light to or collect diagnostic light from a specific location within the body depends upon an accurate understanding of how light is distributed and absorbed in biological tissues. The distribution of laser light in tissue is, in general, a very complicated function of its wavelength-dependent intrinsic optical properties. Light propagation in tissue can be considered as a special case of radiation transport in random media (Ishimaru, 1978). Conventional problems in electromagnetics usually deal with deterministic wave propagation, such as is encountered in antenna, waveguide, and diffraction theory. In contrast, problems of wave propagation in random media may involve unpredictable variations in wave amplitude and phase. Optically, tissue is a continuous medium with a spatially varying complex index of refraction. For modeling purposes, it is usefully approximated as a distribution of discrete absorbers and scatterers, each of which has intrinsic scattering properties. Under the assumption of noninteracting particles, the bulk absorption (μ_a) and scattering (μ_s) coefficients (units of cm^{-1}), are connected to the cross sections for absorption (σ_a) and scattering (σ_s) of individual particles by their number density n (particles/mm^3):

$$\mu_a = n\sigma_a, \tag{1}$$

$$\mu_s = n\sigma_s. \tag{2}$$

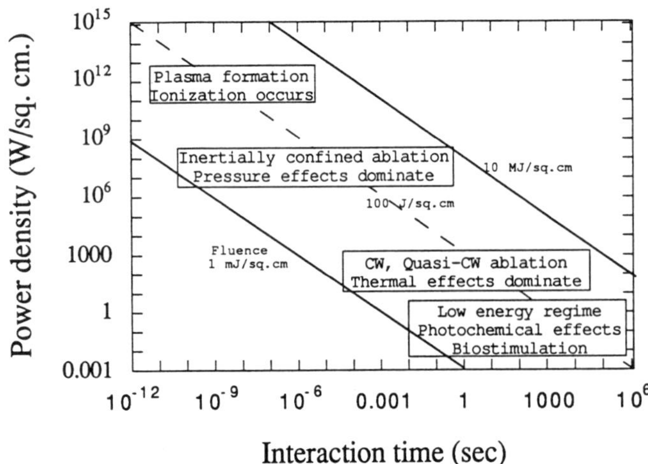

FIG. 2. Laser-tissue interaction map, indicating primary laser-tissue interaction regimes. After Boulnois (1986).

Light is absorbed in tissue through the action of a variety of chromophores, including water, various proteins, hemoglobin in blood, and tissue pigments such as melanin. A graph of the wavelength dependence of absorption for several prominent tissue chromophores appears in Fig. 3. The low absorption of almost all tissue constituents in the 600–1300-nm region has resulted in that wavelength range being referred to as the "therapeutic window" because of the resulting accessibility of deep tissues to light in that range. Light is very strongly scattered in tissue from index gradients within organs and cells, as well as from discrete structures such as collagen fibrils and red blood cells. A significant characteristic of tissue optics is that in almost all tissues except for portions of the eye, the scattering coefficient is much larger than the absorption coefficient at most wavelengths. A summary of the relevant parameters for radiative transport calculations in biological tissues and typical values appear in Table 2.

Several methods are available for modeling the interaction of light with large collections of absorbers and scatterers. Multiple-scattering theory starts with a description of the interaction of waves with individual particles consistent with Maxwell's equations, and then introduces the interaction effects of many particles with statistical averaging. This method is very difficult and has not been applied to light propagation in tissue. Another method for describing light distribution within a large collection of absorbers and scatterers is transport theory. Transport theory describes the propagation of intensities, based on phenomenological observations of transport characteristics. It does not include diffraction and interference between fields; in this sense it is less rigorous than multiple-scattering theory.

1.2.1.1 Transport Theory. Transport theory describes light propagation in a random collection of scatterers and absorption by the radiative transport equation:

$$\mathbf{s}\cdot\nabla L(\mathbf{r},\mathbf{s}) = -(\mu_a + \mu_s)L(\mathbf{r},\mathbf{s}) + \mu_s\int_{4\pi} p(\mathbf{s},\mathbf{s}')L(\mathbf{r},\mathbf{s}')d\omega'. \quad (3)$$

Here $L(\mathbf{r},\mathbf{s})$ is the radiance (W/mm^2-sr) of light at position $\mathbf{r}$ traveling in the direction of the unit vector $\mathbf{s}$, and ω' is the differential solid angle in the direction of $\mathbf{s}'$. The first term on the right represents the decrease in the radiance resulting from absorption and scattering out of the direction $\mathbf{s}$, while the second term describes the increase in the radiance resulting from scattering from other directions $\mathbf{s}'$ into the direction $\mathbf{s}$. Absorption and scattering are described by the coefficients μ_a and μ_s and the phase function $p(\mathbf{s},\mathbf{s}')$. The phase function $p(\mathbf{s},\mathbf{s}')$ represents the normalized probability that light scattered out of the direction $\mathbf{s}'$ is scattered into the direction $\mathbf{s}$. Often, the effect of the phase function can be adequately characterized by the average value of the cosine of the angle between $\mathbf{s}$ and $\mathbf{s}'$, denoted by g:

$$g = \int_{4\pi} p(\mathbf{s},\mathbf{s}')(\mathbf{s}\cdot\mathbf{s}')d\omega'. \quad (4)$$

The three quantities g, μ_a, and μ_s represent the fundamental tissue optical properties used in most treatments of laser light distribution in tissue.

The exact form of the phase function $p(\mathbf{s},\mathbf{s}')$ is often unknown. In many problems in tissue optics, the Heyney-Greenstein phase function, first used for scattering problems in astrophysics, has been found to be useful:

$$p(\mathbf{s},\mathbf{s}') = \frac{1 - g^2}{[1 + g^2 - 2g(\mathbf{s}\cdot\mathbf{s}')]^{3/2}}. \quad (5)$$

When defined in this way, the parameter g in the Heyney-Greenstein phase function has the same meaning as in the defining equation, Eq. (4).

Heat deposition in tissue occurs by light absorption from all directions. The integral of the radiance over all solid angles is the fluence rate F (W/mm^2, also called the space irradiance). The rate of energy deposition in the tissue per unit volume, the absorbed power density U (W/mm^3), is given by the product of the local absorption coefficient and the fluence rate,

$$U = \mu_a F. \quad (6)$$

Analytical solutions for the radiative transport equation are very difficult to obtain. Two approaches are generally taken to obtain approximate solutions. The first is numerical simulation, in which individual photon trajectories through a model tissue are

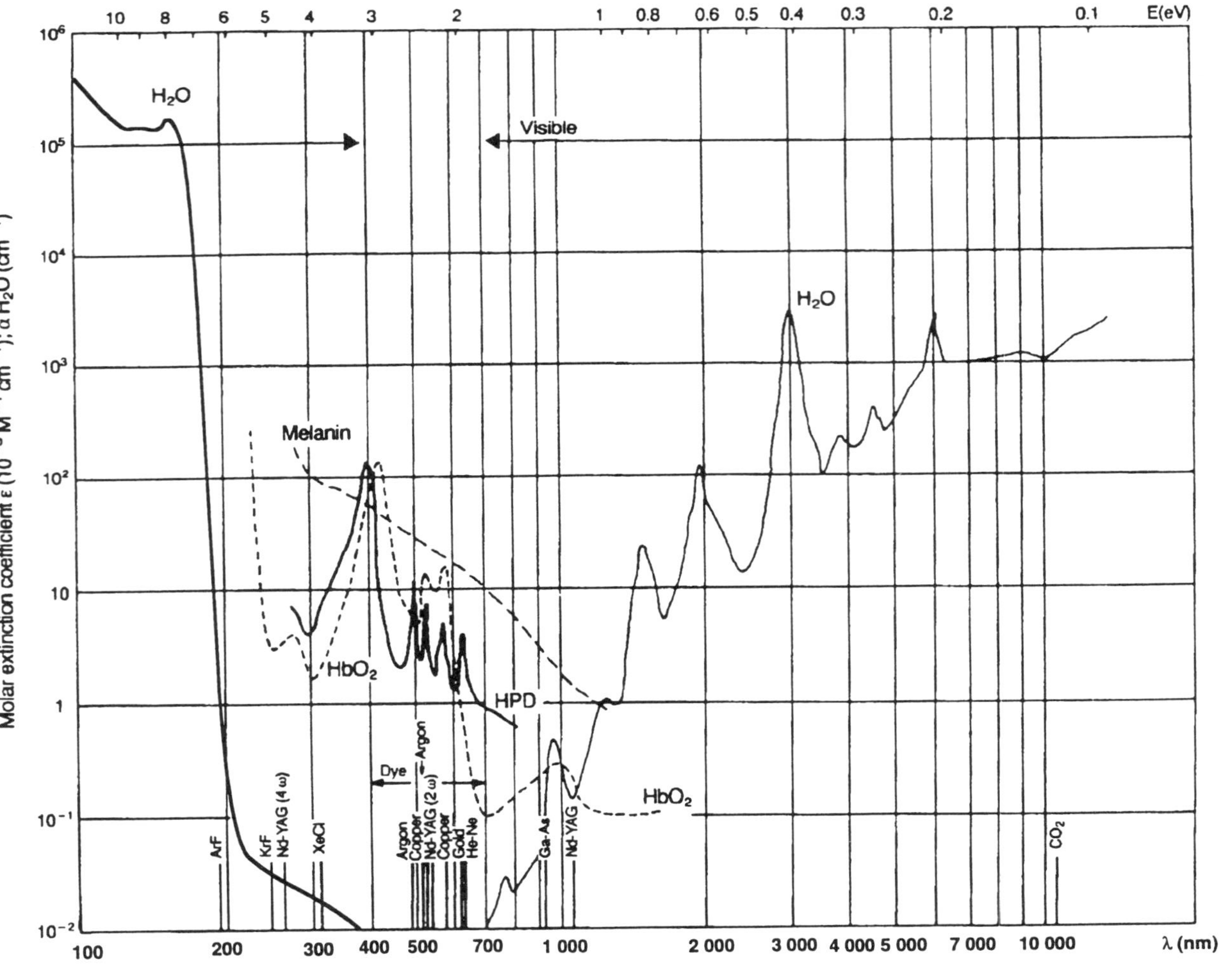

FIG. 3. Absorption spectrum of several important tissue chromophores, together with the wavelength positions of the most widely used medical lasers. HbO_2 is a commonly used symbol for oxygenated hemoglobin, and HPD stands for hematoporphyrin derivative, the dye used in photoradiation therapy. After Boulnois (1986). The ultraviolet portion of the H_2O spectrum is adapted from Hale and Querry (1973).

Table 2. Summary of tissue physical properties.

Parameter	Symbol	Optical Properties: Typical values in therapeutic window ($\lambda \sim$ 600–1300 nm)	Optical Properties: Approximate value in breast (λ = 635 nm) (Key et al., 1991)
Absorption coefficient	μ_a	0.1–10 cm^{-1}	0.2 cm^{-1}
Scattering coefficient	μ_s	10–1000 cm^{-1}	200 cm^{-1}
Total attenuation coefficient	$\mu_t = \mu_a + \mu_s$	10–1000 cm^{-1}	200 cm^{-1}
Albedo	$\mu_s/(\mu_a + \mu_s)$	0.5–0.99	0.99
Anisotropy	g	0.7–0.95	0.95
Reduced scattering coefficient	$\mu_s' = \mu_s(1 - g)$		10 cm^{-1}
Forward attenuation coefficient	$\mu_{\text{forward}} = \mu_s' + \mu_a$		10 cm^{-1}
Forward-scattering mean free path	$(\mu_s' + \mu_a)^{-1}$		1 mm
Diffusion coefficient	$D = (3[\mu_a + \mu_s(1 - g)])^{-1}$		300 μm
Effective attenuation coefficient	$\mu_{\text{eff}} = \sqrt{\mu_a/D}$		2.5 cm^{-1}

Parameter	Symbol	Thermal properties
Specific heat/unit volume	ρC	3.96×10^{-3} (J/°C-mm^3)
Heat diffusivity	κ	0.106 mm^2/sec

Parameter	Symbol	Mechanical properties: Soft tissue	Mechanical properties: Bone
Speed of sound	c	1500 m/s	2800 m/s
Grüneisen coefficient	Γ	0.3	1–2
Maximum elongation		25–100%	2.5%
Maximum stress	σ	600 bars	500–2500 bars

calculated with Monte-Carlo methods, and the resulting fluence rate is summed over a large number of trajectories. A second method which gives useful, although approximate, analytical results is the diffusion approximation.

1.2.1.2 Diffusion Approximation. One approach to solving the radiative transport equation is the diffusion approximation. Here the radiance in Eq. (3) is separated into scattered and unscattered components:

$$L(\mathbf{r},\mathbf{s}) = L_c(\mathbf{r},\mathbf{s}) + L_d(\mathbf{r},\mathbf{s}). \quad (7)$$

The unscattered component is attenuated exponentially according to Beer's law. The scattered component then approximately satisfies the diffusion-like equation

$$-D\nabla^2\Phi(\mathbf{r}) + \mu_a\Phi(\mathbf{r}) = S(\mathbf{r}), \quad (8)$$

where

$$\Phi(\mathbf{r}) = \int_{4\pi} L_d(\mathbf{r},\mathbf{s}')d\omega' \quad (9)$$

is the diffuse flux,

$$D = 1/3[\mu_a + \mu_s(1 - g)] \quad (10)$$

is the diffusion coefficient, and $S(\mathbf{r})$ is a source term.

Solutions to the diffusion equation have been obtained for the fluence rate, total reflectance, and transmittance through a tissue slab, including time dependence (Patterson *et al.*, 1989). The diffusion approximation is accurate for optically dense and highly scattering media, far from tissue boundaries or internal sources of light. A particularly useful result of the diffusion approximation is that the diffuse flux attenuates exponentially according to an effective attenuation coefficient

$$\mu_{\text{eff}} = \sqrt{\mu_a/D} = \sqrt{3\mu_a[\mu_a + \mu_s(1 - g)]}. \quad (11)$$

A problem with the diffusion approximation is that the assumptions are poorest for forward scattering, i.e., where g is near unity. Unfortunately, biological tissue is predominantly forward scattering.

1.2.2 Tissue Thermal Properties Absorption of visible and near-ultraviolet light in tissue occurs by exciting electronic transitions in tissue molecules, primarily proteins and lipids. Near- and mid-infrared light are absorbed via vibrational-rotational excitations in biomolecules, as well as in water. The lifetimes of both the electronic and vibrational-rotational states are much less than laser pulse durations of hundreds of nanoseconds or longer. As soon as the initial excited state distributions thermalize, the energy that is deposited by the laser pulse may be considered as heat, and its dissipation is described by thermal diffusion.

The thermal properties of tissue relevant to heating (without phase change) and heat diffusion are the density (ρ), specific heat (C), and thermal conductivity (σ). For thermal diffusion calculations, the important quantity is the thermal diffusivity, obtained from these quantities via

$$\rho C = \text{specific heat per unit volume,} \tag{12}$$

and

$$\kappa = \sigma/\rho C = \text{thermal diffusivity.} \tag{13}$$

Estimates for ρC and κ in biological media are listed in Table 2. These values are very close to the thermal properties of pure water, which comprises ~60–90% of all biological tissues except bone, teeth, hair, and pathological calcifications.

The attenuation depth μ_{eff}^{-1} is inversely proportional to the concentration of the absorbing chromophores times their molar extinction coefficient. Figure 3 shows the spectral variation of the molar extinction coefficients of the principal tissue chromophores and provides a qualitative sense of why the body is most translucent in the range of 600 nm to 1.3 μm. Measured attenuation depths in most tissues of clinical interest range from fractions of a micron in the deep ultraviolet and mid-infrared regions up to several millimeters in the near infrared. Practical laser spot diameters on tissue, on the other hand, are rarely less than a few hundred microns. For many situations of clinical interest, it is thus sufficient to consider the limit in which the illuminated region of the tissue corresponds to a relatively thin disk. Most of the heat loss from the disk will occur out of the bottom, and so a semi-infinite one-dimensional slab geometry may be used to describe heat diffusion.

The time-dependent equation of heat conduction in this geometry reduces to

$$\frac{\partial T(z,t)}{\partial t} = \kappa \frac{\partial^2 T(z,t)}{\partial z^2}. \tag{14}$$

Here $T(z,t)$ is the temperature, a function of both distance into the tissue (z) and time (t) after the (assumed short) laser pulse is over. A solution to this equation for an exponential initial temperature distribution

$$T(z',t'=0) = T_0 e^{-z'} \tag{15}$$

is

$$T(z',t') = \tfrac{1}{2}T_0 \left\{ e^{(t'-z')}\left[1 + \operatorname{erf}\left(\frac{z'}{2\sqrt{t'}} - \sqrt{t'}\right)\right] + e^{(t'+z')}\left[\operatorname{erfc}\left(\frac{z'}{2\sqrt{t'}} + \sqrt{t'}\right)\right]\right\}, \tag{16}$$

where

$$z' = \mu_{\text{eff}}\, z \quad \text{and} \quad t' = \mu_{\text{eff}}^2 \kappa t \tag{17}$$

are dimensionless intermediate variables.

It is common to refer to

$$\tau = 1/4\kappa\mu_{\text{eff}}^2 \tag{18}$$

as an approximate characteristic time for thermal diffusion in tissue. This value is used both for determining how short laser pulse durations must be in order to ablate without causing thermal damage in peripheral tissue, and also to determine the maximum repetition rate for avoiding a cumulative rise in local temperature with successive laser exposures.

1.2.3 Tissue Structural Properties Human tissue can be conveniently divided into two classes based on its response to laser ablation. Soft tissue includes muscle, fat, skin, blood vessels and the various internal organs. Hard tissue includes bone, teeth, and calcified plaque, which is often the result of a disease process. Soft tissue consists primarily of proteins, with a large amount of water and many trace constituents, such as electrolytes and complex organic molecules, which frequently determine the optical-absorption coefficient. Hard tissue is a composite material consisting of calcium phosphate salts embedded in a soft tissue matrix.

An important point applicable to most biological tissue is that the optical scattering coefficient is usually much larger than the absorption coefficient. Depending on wavelength, it can be ten to one hundred times larger; i.e., biological tissue is a translucent material. In modeling the effects of lasers on tissue, it is often appropriate to assume that the material has the optical properties of the particular type of tissue, the thermodynamic properties of water, and the mechanical properties of structural proteins.

Almost all calcified tissues of the body are composite materials, in which microcrystallites of a calcium salt are embedded in a soft connective tissue matrix. The matrix provides tensile strength, while the salt provides compressive strength. In compact bone, the inorganic fraction is mostly made up of hydroxyapatite salts, densely deposited in a highly organized collagen matrix, with a small percentage of residual water (Table 3). Individual salt crystals in bone are needle-shaped, with diameters on the order of nanometers; however, their deposition into the tissue matrix organizes them into granules with dimensions on the order of a few microns. Cardiovascular calcified deposits are not organized, but consist mostly of mixed apatite salts embedded in a soft component matrix of lipids and cross-linked proteins. The hardest calcified tissues have little or no soft component and are very brittle. Tooth enamel consists of almost all hydroxyapatite, with a small amount of keratin as the soft component. Urinary calculi (kidney stones) have no soft component at all, and so do not constitute a composite substance. For this reason, they are very brittle and can be broken up by ultrasound. Laser ablation of these stones proceeds by depositing energy in a surface plasma rather than in the stone, and using the shock wave generated in the plasma to disintegrate the stone.

1.3 Nonablative Tissue Effects

Significant tissue alterations may be induced by the cumulative effects of relatively slight laser heating. A moderate rise in temperature on the molecular level effects the breakage of hydrogen and other van der Waals bonds that stabilize the conformation of the molecules. Such breakage can lead to denaturation of proteins and enzymes, resulting in inactivation of their function. It has been shown that denaturation reactions are fairly well described by a rate-process formalism from first-order chemical kinetics, the Arrhenius damage integral:

$$\Omega = P \int_0^t \exp\left(\frac{-E_a}{RT}\right) dt'. \tag{19}$$

Here Ω is a measure of the cumulative damage, P is a pre-exponential constant, E_a is an activation energy for the reaction, R is the universal gas constant, T is the temperature, and t is the time of exposure. For denaturation reactions, P and E_a are determined experimentally, with typical values in the range of $P \sim 10^{70}\ s^{-1}$ and $E_a \sim 4 \times 10^5$ J/mol for liver tissue. The cumulative damage is thus a highly nonlinear function of the tissue temperature and heating history.

1.4 Laser Tissue Ablation

A review of the experimental evidence shows, for a wide range of parameters, that the energy density in tissue during ablation

Table 3. The composite nature of calcified tissues. mp = melting point; bp = boiling point.

Tissue	Hard component	Soft component
Cardiovascular calcified deposits	75% mixed apatite salts (mp ~1500 °C), granules 1 to 10 μm in size	20% lipids, cross-linked proteins (bp ~300 °C) 5% water (bp 100 °C)
Bone	75% hydroxyapatite (mp ~1500 °C), granules 1 to 10 μm in size	20% collagen (bp ~300 °C) 5% water (bp ~100 °C)
Tooth enamel	Hydroxyapatite	Keratin
Urinary calculi	Calcium oxalate	None

is an order of magnitude less than that required for vaporization of tissue water alone. Consideration of the thermodynamics of water shows that, under conditions of inertial confinement, when there is not enough time for material to move, these energy densities can lead to the generation of tremendous pressure. This pressure exceeds a threshold that depends on the structural properties of the tissue and leads to ablation. The relationship among fluence threshold, penetration depth, and pulse duration for a host of experiments reported in the literature is consistent with this model.

Since soft biological tissue contains up to 70% water, the extremely well-known thermodynamic properties of water can be used to model its thermal behavior. These properties, including the density, pressure, internal energy, entropy, enthalpy, and temperature, have been tabulated over a wide range of values. Figure 4 shows the most important parameters required for an understanding of the ablation process. On this plot of pressure vs internal energy, contours of constant temperature and density are displayed, and the boundary of the phase-change region is marked with a bold line.

The value of 2500 J/cm^3 required to vaporize pure water applies to a constant-pressure process. This process is illustrated in Fig. 4 by the horizontal path A–C. This path assumes that the conditions of the heating are such that the pressure remains constant. However, if the heating is rapid compared to the time of pressure relaxation, the pressure will increase rapidly, and the system will traverse a different thermodynamic path. In the extreme case of instantaneous heating, the volume (or density) of the system remains constant. To illustrate the difference between heating at constant pressure and heating at constant volume, consider paths A–B and A–D in Fig. 4. In path A–B, 300 J/cm^3 are slowly added to the water. The pressure remains constant at 1 bar while the temperature increases to 92 °C, and the density decreases to 0.964 g/cm^3. In path A–D, 300 J/cm^3 are added to the water faster than the pressure can relax. While the density remains constant at 1.0 g/cm^3 and the temperature increases to 96 °C, the pressure rises to 900 bars.

A concept of inertially confined ablation may be defined on the basis of the generation of pressure that occurs while heating a substance at constant volume. In pulsed laser ablation of tissue, the energy can be supplied so rapidly that the material is inertially confined during the length of the laser pulse. This high pressure, which corresponds to large compressive stresses in tissue, leads to dislocations between the layers of tissue and the nucleation and growth of microcracks, which can be modeled by solving the kinetic equation for the total volume of cracks. Ablation occurs when laser-induced pressure exceeds a threshold value (which is dependent on the structural properties of the tissue) and proceeds through a disassembly of the tissue at the microcracks and subsequent acceleration away from the crater as a result of the stress gradient. Although a tremendous pressure is

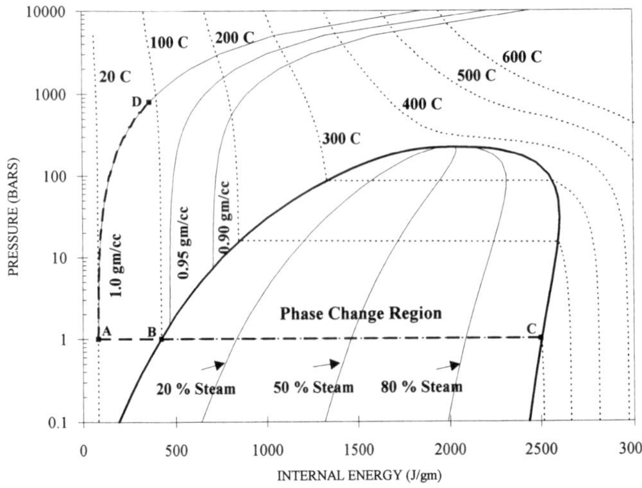

FIG. 4. Thermodynamic phase diagram for pure water, illustrating the concept of inertially confined ablation. 2500 J/g of internal energy are required to vaporize water in a constant-pressure process (path A-C). However, the addition of only 300 J/g of internal energy to water at STP in a constant volume process increases the pressure to 900 bars (path A-D). This pressure rise, when achieved in tissue water by the rapid deposition of energy by a pulsed laser, is sufficient to exceed the tensile strength of tissue and lead to tissue fracture and removal. Reprinted with permission from Itzkan *et al.* (1993).

generated, the small temperature increase is about the same as that predicted by the assumption of heating at constant pressure (the lines of constant temperature on the left portion of Fig. 4 are nearly vertical).

Ablation of hard tissue can be described in terms of the inertial confinement model for soft tissue discussed above. Under these conditions, inertial confinement causes the same large pressures and pressure gradients as in soft tissue. The soft-component bonds fail in tension, and the material is accelerated, ejected, and leaves behind a crater in the tissue. The salt crystals remain intact; however, the soft component rapidly expands and exerts a force on the salt granules through momentum transfer and viscous drag. At sufficiently high irradiance, the salt granules become dislodged and are accelerated out of the crater. Experimentally, ablation of calcified tissue is achieved at all wavelengths (ultraviolet, visible, and infrared) at irradiances above 10^6 W/cm^2. This is consistent with calculations that indicate that vapor velocities of a few hundred meters per second are needed to drag hydroxyapatite particles 1–10 μm in size out of the crater. Gas-dynamic calculations show that these vapor velocities are obtained with incident laser irradiances of about 10^6 W/cm^2.

1.5 Nonlinear Laser-Tissue Interactions

The sharpness of a laser scalpel cut and the efficiency of tissue ablation are both limited by the depth over which laser light is absorbed in the target tissue. This depth is a function of the laser wavelength being used and the optical properties of the tissue. In some applications, where a sufficiently short absorption depth cannot be obtained through the choice of an optimum wavelength or the use of a dye additive, it is possible to use the high intensities available from short pulse laser sources to induce nonlinear absorption. The most successful application of nonlinear absorption to achieve a desired clinical end point is in intraocular surgery, since the eye has negligible absorption for visible wavelengths.

The process through which nonlinear absorption is used in laser medicine is termed laser-induced optical breakdown, and it involves the interplay of several complex physical mechanisms. When a short laser pulse is brought to a tight focus within a target tissue, nonlinear absorption can occur through the processes of multiphoton absorption and/or avalanche ionization. For sufficiently high pulse intensity, the resulting dielectric breakdown of the medium results in a microexplosion accompanied by plasma formation and cavitation. The rapidly expanding plasma generates strong acoustic transients, including shock waves that propagate outward from the breakdown site. When the pulse is over, the extinguished plasma leaves behind a cavity which may continue to expand and, depending upon the tissue mechanical properties, may contract and reexpand repeatedly. These explosive actions lead to the desired therapeutic cutting effect but also may produce undesirable collateral damage. Studies conducted using pulses from nanoseconds to femtoseconds have shown that, in general, the volume of the collateral damage varies as the pulse energy, while the threshold of the laser breakdown process is intensity dependent. Thus, shorter pulses are capable of initiating breakdown resulting in less collateral damage and finer cuts. This rule applies until the pulses are so short (femtoseconds) that nonlinear propagation effects such as dispersive pulse broadening and/or self-focusing begin to inhibit delivery of pulses to the target tissue.

Laser-induced breakdown using nanosecond pulses is used in intraocular surgery for cutting of transparent structures within the eye, such as piercing the lens posterior capsule if it opacifies in the months or years following cataract surgery. Picosecond photodisruption is used where collateral damage is an important consideration, such as in the cutting of vitreous membranes, which can grow very near the retina as the result of a disease process.

1.6 Laser Micromanipulation

A novel use of lasers in medicine is to use the pressure exerted by light itself to exert controlled forces on cells or microscopic particles without physical contact. A laser beam of a few watts intensity focused to a few microns spot size exerts sufficient light pressure to nudge or grasp objects of up to tens of microns in dimension. The principle of operation of such an optical "tractor beam" can be understood with simple ray optics, as illustrated in Fig. 5. Light propagating through a transparent particle positioned at the focus of a laser beam is refracted in a pattern of directions such that the forces exerted by the light on the particle tend to keep the particle near the focus. The particle must be relatively transparent at the laser wavelength used so that it will not be destroyed or killed in the process. This ingenious device was conceived by Ashkin *et al.* (1987).

Laser traps have been used for holding and manipulating single biological cells and cell components such as chromosomes. Laser traps are used in research to measure the strength of intracellular biochemical motors, and may be useful in practice to select and position individual cells for operations such as selection of viable sperm cells in fertility enhancement, and in genetic engineering.

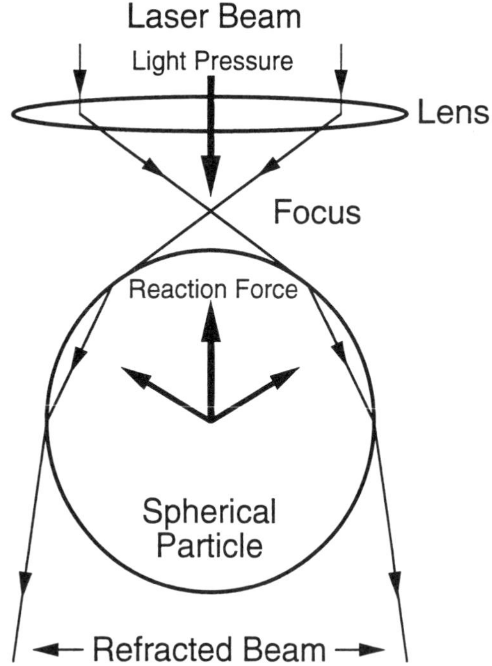

FIG. 5. Schematic of a gradient force optical trap, illustrating how a spherical particle is captured near the focus of a strongly focused laser beam. The particle is held in balance between the force of light pressure from one direction, and the sum of reaction forces resulting from refraction at the particle boundaries. After Ashkin *et al.*, (1987).

1.7 Sub-Diffraction-Limit Microscopy and Ablation

The resolution of a conventional optical microscope depends upon the quality of the lenses and/or mirrors of which it is constructed, but is ultimately limited by diffraction to approximately the wavelength of light. The same limit applies to the minimum spot size achievable when focusing a laser beam for ablation. In many applications, however, it is desirable to have the benefits of optical techniques, such as noninvasive pump/probe analysis of spectroscopic and polarization properties, available at nanometer resolution. It was realized early in this century that the diffraction limit could be overcome if a sample were illuminated through an aperture much smaller than the wavelength of light, with the sample positioned very close to the aperture, i.e., in the near field. In this case, the resolution depends on the size of the aperture rather than on the light wavelength. It was not until the scanning tunneling microscope was developed in 1982 that the technology for creating such small apertures could be matched with the technology for positioning them precisely enough with respect to the sample to create the scanning near-field optical microscope (SNOM). An optical waveguide with a nanometer-scale aperture can be made by stretching a heat-softened pipette tip out to a very long length, after which the inside is coated with a reflective metal (to enhance the pipette's intrinsic waveguide properties). The tip is then held in very close proximity to a sample with feedback techniques developed in scanning tunneling microscopy. This tip can then be used to obtain nanometer-resolution optical images by raster-pattern scanning or, alternatively, can be used to deliver laser energy to vaporize specific cellular components with the same spatial resolution.

The ablative technique has been used recently to make a hole in the outer coating of a mouse ovum, providing just enough space for the passage of a sperm cell, thus rendering fertile a group of ova that were previously infertile because of a defect in the outer coating.

2. THERAPEUTIC APPLICATIONS

2.1 Ophthalmology

The use of lasers in ophthalmology is more mature than in any other medical specialty, and with good reason. Because of the transparent nature of the eye to optical wavelengths from 400 to 1200 nm, lasers can deliver the energy required to probe, heat, and cut ocular tissue in a far less invasive manner than any other surgical technique. Figure 6 is a diagram of the essential parts of the eye. Lasers are currently in clinical use in a variety of procedures addressing diseases and defects in many different parts of the eye. The therapeutic uses of lasers in ophthalmology can be divided into three classes: photocoagulation, photodisruption, and corneal surface remodeling.

Laser wavelengths that are transmitted freely through the optical pathway of the eye, but absorbed in the surrounding tissues, can be used at relatively low intensities to coagulate structures and thus alter or disable their function. Diabetic retinopathy, which is characterized by retinal hemorrhages, microaneurisms, and retinal neovascularization, has become a leading cause of blindness in the United States. If caught early, the progress of the disease can be halted by the technique of panretinal photocoagulation, in which short bursts of cw argon laser light are used to create an array of thermal damage spots in the portion of the retina surrounding the macula. The macula is the central portion of the retina, which contains most of the color receptors and is the region of highest visual acuity. The laser light, which is absorbed in the pigment epithelium of the retina, creates a form of scar tissue, which reduces the rate of severe visual loss in diabetic patients by blocking neovascularization. Thus,

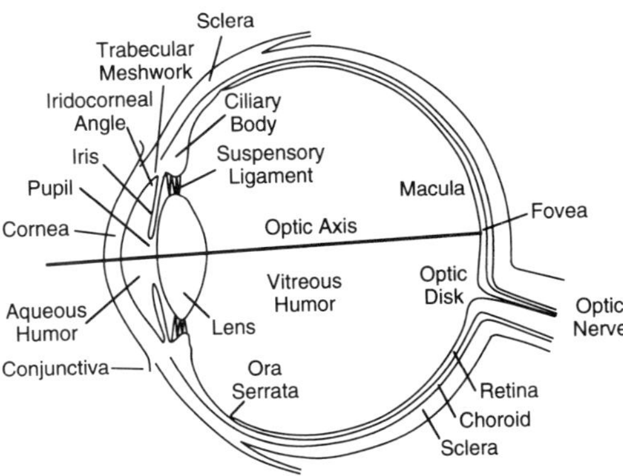

FIG. 6. Schematic of the anatomy of the human eye, illustrating the primary structures relevant to ophthalmic laser therapy and diagnostics.

the macula is saved at the expense of the loss of some peripheral vision.

Glaucoma is also a leading cause of blindness in the western world. It is characterized by elevated intraocular pressure, which leads to visual field loss. Under normal conditions, aqueous humor, the fluid that fills the space between the lens and the cornea, is continually produced in the ciliary body, an organ located next to the suspensory ligaments. This fluid enters the space between the lens and the iris, flows through the pupil into the space between the iris and the cornea, and is drained to the venous system through a porous network located at the iridocorneal angle called the trabecular meshwork. Glaucoma can occur when the aqueous outflow is reduced by either

1. contact between the iris and the lens, which leads to bulging of the iris and blocking of the anterior angle (angle-closure glaucoma), or
2. excessive resistance to aqueous flow in the trabecular meshwork (open-angle glaucoma).

In argon laser trabeculoplasty, laser light is used to coagulate the trabecular meshwork, allowing increased aqueous outflow in open-angle glaucoma.

Finally, the cw argon laser is also used as a welding tool to "tack" the retina to the underlying choroid in regions where it has become detached as a result of head trauma or pathological conditions that place traction on the retina.

Pulsed laser light can be used to cut transparent structures of the eye through the process of photodisruption. In photodisruption, high-intensity laser light is strongly focused at the desired incision location, leading to the creation of a plasma. The plasma, consisting of ionized material at very high temperature, creates a local damage zone whose volume scales with the laser pulse energy. The plasma also absorbs the laser light very strongly, effectively shielding underlying ocular structures from laser damage. Cataracts are a disease in which the lens of the eye becomes cloudy, usually caused by aging, and exacerbated by excess exposure to ultraviolet or infrared light. The treatment is to surgically remove the lens and replace it with a plastic implant. During surgery, it is customary to leave the posterior capsule (the sac which encloses the lens) in place, to serve as a barrier against infection, to retain the vitreous humor, and to provide a bed for the implant. In approximately 30% of the cases, the posterior capsule, which is initially clear, will become cloudy within the first year after surgery. Posterior capsulotomy is an outpatient procedure during which this opaque capsule is opened using a Q-switched Nd:YAG laser. It replaces the older, invasive technique in which the posterior capsule was simply torn open with a needle. Posterior capsulotomy is the most commonly performed photodisruption surgery. Photodisruption is also used in a procedure known as iridectomy to puncture holes in the iris for the treatment of closed-angle glaucoma.

The major portion of the refractive power of the eye is provided by the interface between the cornea and the air. Since the curvature of the cornea does not change under normal conditions, the refractive power of the cornea-air interface is fixed. The lens, whose shape can be varied by its associated musculature, provides the fine focal adjustment for the distance of the viewed object. Excess curvature of the cornea causes myopia, or nearsightedness, and insufficient curvature of the cornea causes hyperopia, or farsightedness. Laser wavelengths that are strongly absorbed by the cornea can be used to shave very small amounts of tissue from the cornea, changing its shape and thus correcting the refractive power of the eye. Light from ultraviolet excimer lasers, which is absorbed in the first few microns of corneal tissue, can be used to decrease the corneal curvature and thus correct myopia by ablating more material near the optic axis than at the periphery. In a newer development, aimed at correcting hyperopia, infrared lasers are being investigated to coagulate corneal tissue in a ring-shaped pattern around the optic axis, shrinking the irradiated tissue and increasing the curvature of the central cornea.

2.2 Surgery

2.2.1 Urology Unlike bone, many of the various types of urinary calculi (kidney stones) are brittle and are thus susceptible to fragmentation by ultrasound. They can also be fragmented by the pressure wave generated by laser-induced plasmas created on their surface. The smaller fragments can then be passed out through the urinary tract. The laser procedure is useful in removing stones lodged in the portion of the ureter (the tube which connects the kidney to the bladder) that is shielded by the pelvic bone and is therefore inaccessible to external ultrasound. This is accomplished by inserting an optical fiber via the bladder into the ureter until it is in contact with the stone, then applying pulsed dye laser pulses until the stone is fragmented.

Bladder tumors on the inner wall of the bladder are now commonly destroyed with a laser. This procedure has the advantage of leaving the bladder wall intact and watertight, a major advantage for any form of bladder treatment.

Hypertrophy of the prostate gland, a common problem among older men, causes difficulty in urination. This condition is usually treated surgically, often through the urethra. Laser treatment of prostate hypertrophy is currently being actively studied. It has the major advantage of minimizing blood loss, since avoiding the need for blood transfusion has become a high priority in medical practice.

2.2.2 Gynecology Cervical dysplasia is a common precancerous condition of the surface of the cervix. Surgical treatments may create a problem for women wishing to bear children by compromising the cervix's ability to retain a fetus to full term. Laser removal of the diseased tissue can preserve the mechanical strength of the cervix, retaining its ability to perform its function during pregnancy. Here, too, decreased blood loss is a major advantage.

2.2.3 Otolaryngology A frequent cause of hearing loss is calcification of, or damage to, the tiny bones of the middle ear. Today, this problem can often be corrected by surgically removing one or more of these bones and replacing them with a prosthesis. Because of the confined space and the small amount of tissue to be removed, fiber optic delivery of laser energy is an ideal way to perform this operation.

One of the earliest uses of laser surgery was in the larynx, to remove growths on the vocal cords. Because of the oxygen being supplied to the breathing patient, some cases of airway fires occurred. Great care is now taken to avoid such accidents in all laser surgery performed near airways.

2.3 Dermatology

Anderson and Parrish (1983) invented the term selective photothermolysis to describe the process in which specific cell types are targeted for destruction by a laser while leaving intact adjacent cells of differing types. The technique involves selecting the laser wavelength to coincide with the absorption spectrum of a particular chromophore present in the targeted cells, but absent in the adjacent cells. The concentration of laser energy deposited in the target cells causes their temperature to rise to the point where cell death is induced by photocoagulation or hyperthermia. An additional requirement is selecting the laser pulse duration such that the diffusion of heat from the targeted cells does not injure the adjacent cells.

A successful application of selective photothermolysis has been the treatment of vascular lesions. By targeting the hemoglobin present in the unwanted blood vessels, these vessels are selectively destroyed (the absorption spectrum of hemoglobin is shown in Fig. 3). Port wine stain (PWS), for example, is a congenital birthmark resulting from an assembly of abnormal blood vessels located in the dermis, often appearing on the face or neck. Normal blood vessels in the skin are usually about 90% empty (except when flushed as a result of exercise, cold, spicy food, or embarrassment). In contrast, PWS vessels are normally filled. By irradiating the area containing PWS vessels, the abnormal vessels are selectively destroyed while sparing the remainder of the dermis. The thermal diffusivity of biological tissue (see Table 2) leads to an approximate thermal diffusion time of about 1 μs for abnormal vessels, which are typically 40 to 100 μm in diameter. Many different lasers have been used to treat PWS. Considerable success has been achieved with a pulsed dye laser with a 300-μs pulsewidth, tuned to a wavelength near the 577-nm absorption peak of oxyhemoglobin (Tan *et al.*, 1989). Usually, multiple treatments are required to obtain a satisfactory result.

Other successful applications of selective photothermolysis include pigmented lesions, where the pigmentation can be natural, due to melanin (see spectrum, Fig. 3), or artificial, as in tattoos. The basal layer, which is the boundary between the dermis and epidermis, contains cells that continually manufacture new epidermal skin cells. These new cells dehydrate as they migrate to the surface, where the outermost layer of dead cells is continually being sloughed off. In the basal layer, there is also a sprinkling of melanocyte cells, whose function is to generate the pigment melanin and inject it into the newly forming skin cells. Benign pigmented lesions containing an excess of melanin are successfully treated with lasers, which selectively destroy the melanocytes. Malignant lesions usually call for complete excision.

Tattoos consist of artificial pigments that have been deposited in the skin. The pigments are insoluble, and the granules are too large to be removed by the normal body processes, such as phagocytosis. By using short-pulse, high-intensity lasers, the pigments can be broken into smaller fragments that are then cleared by the body, causing minimal damage to the surrounding tissue. Lasers that have been successfully used to remove tattoos include alexandrite, 1-μs 510-nm pulsed dye, Q-switched ruby, and Q-switched Nd:YAG (both the fundamental and the second harmonic). The choice of laser depends upon the pigment being attacked.

2.4 Cardiology

Although the subject of intense study for the past ten years, the utility of lasers in cardiology is not yet clearly established. Initially, it was conjectured that a fiber optic passed through a catheter to the site of a coronary artery blockage could be used to deliver laser energy to ablate and remove the diseased tissue causing the blockage, and thus successfully recanalize the artery. This procedure would avoid the need for costly and painful bypass operations involving open heart surgery. Balloon angioplasty, in which a deflated balloon is inserted into the partially blocked opening and then inflated to force an enlargement of the opening, has achieved partial success toward this goal, but is only about 50% successful. This is principally because restenosis can occur as soon as during the first six months after treatment. Restenosis is usually not a recurrence of the original disease, but is caused by the growth of scar tissue called intimal hyperplasia, which closes down the balloon-inflated opening. The balloon is also contraindicated for use in many of the more difficult lesions.

Initially, the argon-ion laser was used for laser angioplasty. Although this laser could ablate soft plaque, it was unable to penetrate calcified plaque, and would occasionally perforate vessel walls. The use of pulsed ultraviolet lasers solved the problem of ablating calcified tissue, and using various forms of guidance, such as guidewires, reduced the danger of perforation. However, the restenosis rate for laser angioplasty has proven to be no better than that of the balloon. There are a limited number of special lesions for which the laser may have a useful and unique role, such as enlarging the opening of a lesion so that a balloon may be inserted, or recanalizing partially occluded bypass grafts. Laser systems are also being studied to determine their utility in opening partially occluded vessels in the legs.

In mammals, the oxygenated blood supply to the heart muscle is via the coronary arteries, and it is the blockage of these vessels that is responsible for much heart disease. In reptiles, the heart muscle is supplied with oxygenated blood directly from the heart chamber. A repair method which mimics the reptilian heart has been pioneered by Mihroseini (Mihroseini *et al.*, 1988). In this procedure, a CO_2 laser is used to drill a hole directly through the heart muscle into the left ventricle. Oxygenated blood then perfuses into the heart muscle directly via this new passage. The outside of the hole is sealed by blocking the opening until a clot forms.

Acute heart attacks are very often the result of a thrombus (blood clot) flowing downstream and blocking the opening of an already partially occluded coronary artery. Emergency treatment involves dissolving this clot quickly, before serious damage can occur to oxygen-starved tissues. There are several effective drugs available for this purpose; however, in many cases, these drugs are contraindicated or unsuccessful. The clot is very soft compared to artery wall or plaque and is easily ablated using laser energies well below levels which could damage adjacent vessel walls. At these levels, the energy can be delivered using a fluid-filled catheter, in which the fluid guides the light in much the same manner as an optical fiber. Figure 1 is a photograph of a system developed by Palomar Medical Technologies, Inc., which uses a 480-nm pulsed dye laser and a fiber optic inside a fluid-core catheter, and which is undergoing clinical trials.

2.5 Orthopedics

Initial attempts to use lasers in arthroscopic surgery and, in particular, surgery involving the knee began with the use of the CO_2 laser. This laser had two serious drawbacks. First, this wavelength is strongly absorbed in water, and, therefore, the knee had to be flushed with CO_2 gas to provide a dry field. This added complexity to the procedure and occasionally led to embolisms. Second, acceptable optical fibers at this wavelength are not yet available, thus requiring the use of articulated arms or hand-held units in which the surgeon manipulates the entire laser. Both were cumbersome approaches.

More recently, surgeons have investigated using the Ho:YAG laser, taking advantage of the fact that its 2.1-μm light can propagate through conventional fibers. Although it can be used in a wet field, light at this laser wavelength is partially absorbed by water, and, thus, the distance from the fiber tip to the tissue must be precisely controlled. A substantial amount of the energy is dissipated in the saline field.

An examination of the spectroscopic characteristics of meniscus tissue in the human knee indicates that the near ultraviolet may be a useful wavelength for this application. Pyridinoline, the collagen cross-linking agent in meniscus tissue, has an absorption peak at 340 nm. Preliminary studies indicate that the third harmonic of Nd:YAG laser light can be used to create cuts that are clean and smooth without charring. The combination of good propagation through fibers, good propagation through saline, and very clean cutting makes the use of the long-pulse tripled Nd:YAG laser in arthroscopy an exciting prospect.

2.6 Dentistry

In recent years, the laser has been suggested for use in preventive and reconstructive dentistry. Among the potential applications that have been considered in the literature are removal of caries, sealing fissures in enamel, inhibition of demineralization, etching of enamel, preparation of retention pin holes, and cleaning and preparation of root canals. The principal laser studied, the CO_2 laser, has two serious drawbacks. First, although useful for cutting and removing soft tissues, such as during gum surgery or frenectomies, the CO_2 laser does not ablate hard tissue without leaving considerable char. Second, in order to be effective, a large amount of heat must be deposited. This can raise the temperature of the pulp, causing pulpal damage. The maximum permissible temperature rise for pulp is 5.5 °C, after which irreversible damage occurs.

Since near-ultraviolet and 2.9-μm lasers have proven effective with other hard tissues, (see Sec. 1.4), these sources have been considered for ablation of tooth enamel. This application is not immediately obvious, since the ablation mechanism appears to involve absorption in the soft component of the tissue, which acts as the principal chromophore. In bone, the soft component is collagen; in calcified plaque, it is lipids and proteins; and, in tooth enamel, it is keratin. In all three materials, the hard component is hydroxyapatite. Recent results show that a Q-switched near-ultraviolet laser produces clean, well-defined cuts in tooth enamel, with sharp edges and no char, demonstrating that keratin may be an adequate absorber in this wavelength region.

The near-ultraviolet laser overcomes the thermal heating problem in two ways. Because of the rapid ablation resulting from the short pulses, most of the energy is carried off in the ablated material, and very little is deposited in the underlying substrate. Thus, residual char is avoided, and excessive heating of the pulp is prevented. Also, since the beam propagates through water unattenuated, the tooth can be kept cool, if necessary, using a liquid spray. Although several preliminary laboratory results have appeared in the literature, no clear-cut clinical applications for the use of the laser on teeth have been demonstrated.

2.7 Photoradiation Therapy

The two principal methods now used to destroy malignant cells in the treatment of cancer *in situ* are radiation and chemotherapy. In both cases, the treatments also damage healthy cells. This damage is minimized by careful aiming during radiation treatment and by precise dose control in chemotherapy; however, the fine line between destroying all of the tumor but not the healthy tissue is a difficult one to maintain. Oncologists have long wished for a "magic bullet" that would selectively destroy only tumor cells and not normal cells. Photoradiation therapy, sometimes called photodynamic therapy (PDT), offers the prospect of such a magic bullet.

Photoradiation therapy relies on the fact that a great many organic compounds possess the property of phototoxicity; i.e., in the absence of light they are benign, but, when irradiated with certain specific wavelengths, they become toxic. Some well-known drugs, such as tetracycline, possess this property, and users are cautioned to avoid exposure to sunlight. Another essential property of a good PDT drug, often called a sensitizer dye, is that it is preferentially absorbed by tumor tissue, or else more rapidly cleared from normal tissue and therefore selectively retained in the tumor. Nile Blue A, a benzothiazine, is a dye that shows a large differential absorption.

Figure 7 is a schematic of photoradiation therapy of cancer. Prior to irradiation, the patient is injected with the phototoxic dye. To date, the preponderance of clinical trials have used a commercial dye, Photofrin®, which is a form of hematoporphyrin derivative (HPD). This dye depends on selective clearing, requiring 2 to 3 days for the dye concentration in the normal tissue surrounding the tumor to be reduced to a safe value. Phototoxic dyes are chosen to have their excitation wavelength in the wavelength region in which tissue is translucent, from approximately 600 to 1200 nm (Fig. 3). The excitation wavelength for HPD is 630 nm (the absorption spectrum for HPD is also shown in Fig. 3)

Although incoherent light can be used for surface tumors, the light source for PDT is usually a laser, so that the light may be efficiently coupled into optical fibers. This enables the light to be directed to the tumor through endoscopes, laparoscopes, or even hypodermic needles.

The mechanism for cell destruction is thought to be via the generation of singlet oxygen ($^1O_2$), a metastable excited state of molecular oxygen, which is known to be toxic to cells (Fig. 7). The incoming photon excites

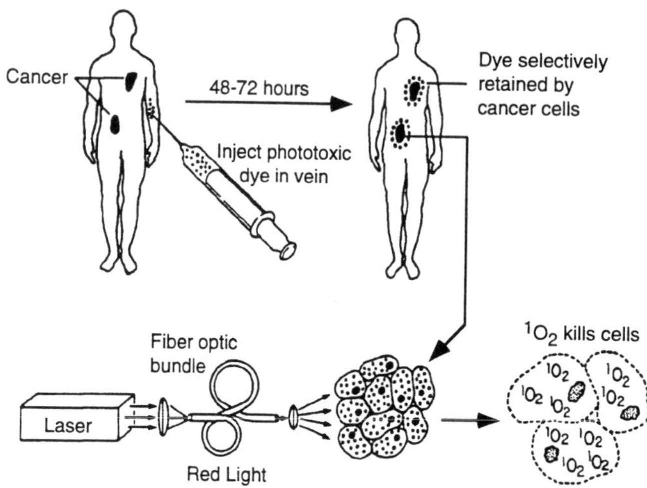

FIG. 7. Schematic of photoradiation therapy of cancer. The patient is first injected with a phototoxic dye. After dye localization in tumor tissue has occurred, the drug is activated by illumination with laser light delivered through a catheter, endoscope, or laparoscope. Tumor tissue containing the drug is selectively destroyed by singlet oxygen generated by the activated drug. After Berns (1984).

the dye to a singlet state. Good phototoxic dyes have a high probability for intersystem crossing to the triplet manifold, and a large population of metastable triplets is formed. When one of these triplets collides with a ground state oxygen molecule, excited-state singlet oxygen is created, which can then destroy membranes or vital proteins.

At present, a number of dyes are under active study, although none has yet proven to be the sought-after magic bullet. There are several major classes of principal interest, including the phthalocyanines (of which the nonsulfonated zinc analog is about to undergo clinical trials), the benzoporphyrin derivatives, various cationic dyes, which include benzothiazines and rhodamines, and the chlorins, which are chlorophyll derivatives. Results to date show great promise. The most likely result will be that specific dyes prove to be effective for specific types of tumors.

2.8 Biostimulation

An area of great controversy has been variously termed biostimulation or low–light-level laser therapy. There have been a large number of reports in the literature of the observation of enhanced wound healing, accelerated tissue growth, and a measurable decrease in pain induced by the application of doses of low-intensity laser light. Many of these reports have come from reputable and careful workers. Unfortunately, much of this work proves to be irreproducible, and there are a host of reports in which no effect is observed.

There is some physiological basis for the possibility of such effects. Even though the light intensities used are too low to cause measurable temperature changes, significant changes in the rate of cell growth *in vitro* have been observed by Karu (1987). These changes are highly dependent on the wavelength, fluence, and cell type. Itzkan and Bourgelais (1988) have postulated a mechanism involving the photodissociation of oxyhemoglobin, in which physiologically significant levels of additional O_2 could be liberated. Nevertheless, until this field can be placed on a sounder scientific basis, biostimulation will remain controversial.

2.9 Laser Tissue Welding

The first attempts at laser welding were used to perform blood vessel anastomoses, i.e., reconnecting the severed ends of two blood vessels. This was accomplished by heating the ends of the vessels with a laser, causing the tissue proteins to become sticky, and adhere on cooling. A successful result requires that the joined vessels remain patent (have an adequate internal opening), withstand the blood pressure, and not leak. Initially, "stay" sutures were used to keep the ends of the vessels together during the procedure. This vitiated some of the advantage over conventional suturing of avoiding the scar tissue caused by the sutures. The results were no better than those obtained by a good vascular surgeon.

Recent improvements include the use of protein materials, such as fibrin, to act as "solders," providing more strength and better sealing, and the use of soluble mandrels to hold the ends in apposition while the laser heat is being applied. The mandrel then dissolves, leaving the vessel patent. It also insures that tissues of the same type align with one another, which speeds healing.

These techniques are especially useful in reconnecting nerves and bowel, where the absence of leakage is more important. Blood vessels have the ability to repair small leaks by forming clots. On the other hand, a micron-sized hole in a repaired nerve sheath will allow the extremely fine axons to grow out of the sheath, impairing the result, while leakage in a bowel can cause peritonitis.

2.10 Laser Energy-Delivery Systems

Optical fibers capable of transmitting substantial amounts of energy are available for wavelengths from about 300 nm to 2.5 μm. When the emerging energy is to be used directly from the fiber, various focusing and diffusing tips may be used, the choice depending on the application. For example, the Nd:YAG laser can conveniently supply large amounts of average power, but its 1.06-μm wavelength penetrates too deeply for some applications. For surgery, it may be coupled to a shaped sapphire tip, which is heated by the laser and then serves as a combination mechanical knife, heated tip, and laser light deliverer.

For external work, the practitioner may use a handpiece that provides a focus at a fixed point in space. The handpiece is usually optically coupled to the laser via a fiber for flexibility and convenience. For wavelengths longer than about 2.5 μm, the laser is usually optically coupled to the handpiece via an articulated arm, containing folding mirrors, similar in mechanical operation to the kind used for dentist's drills. Some optical fibers and hollow flexible waveguides are available for these IR wavelengths, but they are, as yet, not entirely satisfactory. Small lasers are sometimes mounted directly on an articulated arm. Handpieces have been designed incorporating a computer-controlled scanning function to provide more precise control of the delivered intensity and spatial uniformity of the dose. Surgical laser systems are usually fitted with a low-power aiming beam to permit precise placement of the high-power beam on the tissue.

2.11 Surgical Laser Safety

Safety issues that arise during laser surgery are eye protection, fire, and infection. Protocols requiring the use of safety glasses or goggles for all personnel when the laser is operating are mandatory in most medical facilities. Because a stray laser beam can easily ignite flammable material, all surgical drapes in use during laser surgery must be fire resistant and must be kept wet at all times when the laser is in use, particularly if oxygen is also being used. Special care is taken in or near the patient's airway. Also, live organisms and viable DNA have been found in the laser plume. Cases of practitioners developing warts in their nostrils after removing warts from patients have been reported. The laser plume must be considered potentially infectious, and appropriate ventilation and masking systems must be in use at all times.

3 DIAGNOSTIC APPLICATIONS

One of the most active areas of research in laser medicine in recent years has been the application of optics and photonics technology for tissue imaging and diagnosis of disease. This area has included the conversion of several state-of-the-art technologies to medical applications, such as ultrafast optics, holography, confocal microscopy, and various forms of high-resolution spectroscopy. In comparison to other advanced medical diagnostics developed in recent decades, such as magnetic resonance imaging (MRI) and computed tomography (CT), optical techniques for imaging and diagnosis have the advantages of being simultaneously nonionizing, potentially inexpensive, and often noninvasive. In this section, highlights of recent advances in this active field are reviewed.

3.1 Laser-Based Optical Imaging in Medicine and Biology

3.1.1 Laser Scanning Confocal Microscopy An important application of the laser in medicine is its use as a high-brightness light source for confocal microscopy. In the decades since the first demonstrations of the laser and of confocal microscopy, the combination of these two inventions has revolutionized high-resolution optical imaging in biology and medicine. In a confocal microscope, illumination light emanating from a point source is focused with a condenser lens to a minimal spot size in the sample, and only that light is collected that passes through a pinhole placed in an image plane of the point source. Two-dimensional images are built up by scanning the incident beam or the sample in two dimensions, while synchronously recording the transmitted light-intensity level. The brightness of the image depends upon the intensity and the degree of spatial coherence of the illumination light, making the laser an ideal source. The confocal microscope provides resolution superior to that of the conventional microscope in both lateral and axial directions. In the lateral direction, the overlap of the illuminating light-intensity distribution and the objective point spread function is slightly smaller than the resolution that could be obtained by the objective lens alone. The primary benefit is in the axial direction, however; almost all of the light scattered from planes in the sample either above or below the object plane (the plane of dual focus of both the illumination light and the objective lens) does not pass through the pinhole aperture, and is rejected from the image. By translating the focal plane axially through a relatively thick sample, one can thus optically "section" planes with micron resolution without having to slice the sample physically. This feature is particularly useful in living biological samples, permitting detailed examination of living organisms in their natural environments.

3.1.2 Laser Imaging for Ophthalmic Diagnosis An application of laser scanning confocal microscopy that has made the leap into current medical practice is in imaging of the eye and retina. Confocal microscopic images of the living cornea can be obtained and reconstructed in three dimensions with sufficient resolution to discern cellular structure. The scanning laser ophthalmoscope is a confocal microscope optimized for imaging of the retina (Webb *et al.*, 1987). It provides substantial advantages over the conventional hand-held indirect ophthalmoscope for retinal viewing, in that the images are crisper and cleaner (because of confocal rejection of scattered light), and, at the same time, it is much more comfortable for the patient because of more efficient usage of the illumination intensity.

High-resolution imaging of the living retina is necessary for diagnosis and treatment of almost all retinal diseases. These include emergency conditions, such as torn retinas and dislocations, as well as degenerative diseases, such as retinopathy. Of particular current research interest is imaging of the retina and optic disk for early diagnosis of glaucoma (see Sec. 2.1). The retinal nerve fiber layer (RNFL) is the uppermost layer of the retina, consisting of nerve axons carrying visual stimuli that converge from all parts of the retina toward the optic disk and out of the eye to the brain. In glaucoma, significant degeneration of the RNFL can occur before conventional tests of intraocular pressure and visual field can detect the presence of the disease. This layer is less than 50 μm thick in most regions of the eye. Unfortunately, this is too thin to be sectioned by confocal microscopy, since the axial resolution obtainable at the back of the eye is limited by the pupil and ocular aberrations to a few hundred microns.

Two new laser-based retinal imaging devices have recently been developed that have the capability to measure directly the thickness of the RNFL, in the hope that such a direct measurement will provide for early diagnosis of glaucoma before vision is threatened. Scanning laser tomography is an enhanced form of scanning laser ophthalmology, in which the thickness of the RNFL is estimated by measuring the optical retardation of backscattered light (Bille *et al.*, 1990). This technique depends upon the fact that the RNFL is a birefringent structure, composed of thin, directionally oriented nerve fibers. The birefringence is measured with Fourier ellipsometry, in which the polarization states of the incident and scattered light are modulated and analyzed, respectively, in such a way as to provide a fast, accurate estimation of the layer's birefringence.

Optical coherence tomography (OCT) is a new scanning laser confocal imaging technique for micron-resolution imaging at arbitrary working distance, which has the potential to perform direct measurements of

RNFL thickness (Huang *et al.*, 1991). Optical coherence tomography is based on low-coherence interferometry, in which light from the sample arm of a Michelson interferometer is directed onto the tissue to be imaged. The interferometer is illuminated with a broad bandwidth light source, having a coherence length of only a few microns. Measurements of reflectivity as a function of depth into the tissue are obtained by scanning the reference arm of the interferometer while synchronously monitoring the signal at the interferometer output. Interference occurs only when the distance to a reflectance in the sample matches the reference arm length to within the coherence length of the light source. Micron-resolution cross-sectional images of tissue microstructure are built up by transverse scanning of the sample arm beam. Optical coherence tomography is particularly useful for imaging structures in the back of the eye, since the axial resolution depends only on the source coherence length and is independent of the available pupil aperture and the optical quality of the eye. This technique has produced the highest-resolution images of the living retina to date (Plate 1).

3.1.3 Optical Tomography The ultimate goal in optical imaging of tissue would be to develop a general-purpose tomographic imaging technology, which could obtain images of optical scattering or tissue spectroscopic properties from deep within the body. A practical goal for such investigations has been to develop a method for optical transillumination imaging of the breast with sufficient sensitivity and resolution to visualize the early stages of breast cancer, which could serve as an alternative to x-ray mammography. The primary difficulty in imaging deep into tissues is that almost all tissues, except for the eye, are highly scattering, and image information is severely degraded after just a few scattering mean free paths into the tissue, typically ~1 mm in breast tissue (see Table 2).

Several approaches are under investigation for resolving the problem of optical imaging through highly scattering media. One is to attempt to reconstruct an image of the interior of an illuminated diffusing medium by recording the position and direction of all of the photons that emerge from it, and then to trace them backward in time computationally to derive the positions of the internal absorbers and scatterers. This "inverse problem" is tremendously complex at best and has not yet been shown to be soluble. Another approach is to use time-resolved methods to detect only those photons that have not lost image information in passing through the tissue. A pulse of light propagating through a tissue sample will be severely lengthened by multiple scattering inside the sample. The photons in the transmitted pulse can be classified into so-called ballistic, snake, or diffuse components, depending upon their arrival times (Fig. 8). Ballistic light is the unscattered component, which has taken a straight-line path through the sample. Snake light is that component that has been scattered but has remained close to the straight-line path, and diffuse light is the remainder. A shadowgram of hidden objects within the medium may be obtained by time-gating the earliest-arriving light, either the ballistic or snake light components.

Several techniques are available for distinguishing between the different components of light transmitted through a turbid medium. Conceptually simplest is time gating, in which ultrashort pulses are used to illuminate a tissue sample, and the transmitted light is classified according to its arrival

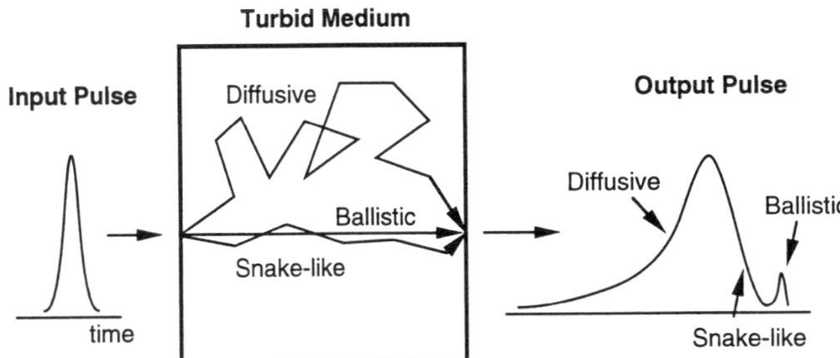

FIG. 8. Schematic of time-resolved optical transillumination imaging in a turbid medium, illustrating the ballistic, snake, and diffuse components of a transmitted ultrashort pulse of light. In the pulse diagrams time increases to the left. After Wang *et al.* (1991).

time through the use of fast electronic detectors (such as streak cameras), or even faster nonlinear optical detection techniques (such as the ultrafast Kerr shutter). This approach has resulted in shadowgraph images of small animals (Benaron and Stevenson, 1993) and in tissue phantoms up to a few cm thickness (Wang *et al.*, 1991), but has not yet resulted in demonstrable clinical utility. Two additional techniques have been introduced for time-resolved imaging that do not use explicit time gating. The first is frequency-domain imaging, in which the amplitude and phase characteristics of the transmitted portion of light from a radio-frequency–modulated continuous light source are used to distinguish photon arrival time. The second is coherence-domain imaging, in which the coherence length of a broad-bandwidth light source is used in place of a time gate in order to distinguish early-arriving photons. The longitudinal ranging capability of optical coherence tomography (Sec. 3.1.2) is an example of the use of this technique in one dimension, but it is also possible to image in two or three dimensions using holography (Leith *et al.*, 1992). These latter techniques have the advantage that ultrashort pulse sources and fast time gates are not required. More detailed information on time-resolved medical imaging techniques may be found in the list of suggestions for further reading.

A few theoretical results have been obtained that are useful for evaluating different transmitted light components for optical tomography. Since ballistic photons are unscattered in traversing the tissue, diffraction-limited imaging theoretically is possible. However, the ballistic component is severely attenuated in the scattering medium. The maximum tissue thickness for ballistic light imaging may be written as

$$L_{\max} = \frac{1}{\mu_t} \ln\left(\frac{1}{2}\frac{E}{h\nu}\right), \tag{20}$$

where μ_t is the total attenuation coefficient, $h\nu$ is the photon energy, E is the light energy deposited per resolution pixel, and a signal-to-noise ratio of 1 has been assumed as the detectivity limit. Tissue optical damage thresholds limit the maximum energy that can be deposited in the tissue in a reasonable image-acquisition time. This limits the maximum ballistic imaging thickness to approximately $\mu_t L_{\max}$ ~40 mean free paths, corresponding to ≲4 mm of breast tissue (Hee *et al.*, 1993).

The attenuation of snake light is experimentally observed to be about a factor of 10 less than ballistic light, so that imaging through several centimeters of tissue is possible. However, widening of the time gate to accept snake photons inevitably degrades the resolution of the resulting image. An expression for the full-width–half-maximum spatial image resolution Δx as a function of the tissue thickness has been obtained from diffusion theory, which is independent of the tissue-scattering properties as long as the conditions for the diffusion approximation are satisfied:

$$\Delta x \sim 0.2d. \tag{21}$$

Here d is the tissue thickness, and a detectivity of 1 part in 10^{10} of the incident optical signal is assumed (Moon *et al.*, 1993). These two theoretical results demonstrate that this approach for imaging through thick tissue using time-resolved methods, i.e., accepting snake photons, comes at a considerable cost in image resolution.

3.2 Spectroscopic Monitoring and Diagnosis

An exciting new application for lasers in medicine is the prospect of locating, identifying, and classifying diseased tissue inside the living patient in a minimally invasive manner. Among current clinical methods available to study disease, pathology is the gold standard. Pathology requires biopsy samples, which can be difficult and painful to obtain and can involve considerable delay before the results are available, and requires a highly skilled pathologist to interpret the results. Even highly skilled pathologists often differ on interpretation and conclusions. Clinical techniques capable of evaluating the nature of diseased tissue *in situ*, in real time, and in a safe, valid, and reproducible manner are therefore of critical importance for diagnosis, for selecting and evaluating the effects of various therapies, and for epidemiological and clinical research. Other technologies that can provide some of this information include magnetic resonance imaging (MRI), x-ray computed tomography (CT), radionuclear imaging, and ultrasound. All of these techniques have limitations. It is therefore desirable to develop complementary techniques for probing and quantitatively monitoring disease.

Using flexible optical fibers to excite the tissue and collect spectroscopic signals, optical systems provide access to many organs in the body through endoscopes, laparoscopes, catheters, and hypodermic needles, and high-performance diagnoses have been demonstrated. Optical spectroscopy using natural fluorescence emission from tissue has been demonstrated to yield valuable information about the state of the health of many tissues. For example, fluorescence spectroscopy can distinguish between oxygenated and ischemic tissue. The principal source of energy for cells is chemical processes localized in the mitochondria, which employ a cyclical process called the Krebs cycle, followed by oxidative phosphorylation. A simplified view of this process is that an energy-rich compound called adenosine triphosphate (ATP) is produced, and molecular oxygen and various carbohydrates are consumed. Two important routes for transporting electrons from the foodstuffs to the oxygen involve the equilibria of the reactions

$$\mathrm{FAD} + 2\mathrm{H}^+ + 2e^- \leftrightarrow \mathrm{FADH_2}, \qquad (22)$$

$$\mathrm{NAD}^+ + \mathrm{H}^+ + 2e^- \leftrightarrow \mathrm{NADH}. \qquad (23)$$

The compounds FAD and $FADH_2$ are the oxidized and reduced forms, respectively, of flavin adenine dinucleotide, and NAD^+ and NADH are the corresponding forms of nicotinamide adenine dinucleotide. The compound NADH has a fluorescence peak in the blue, FAD has a fluorescence peak in the green, and $FADH_2$ and NAD^+ have no appreciable fluorescence. Consequently, when there is insufficient O_2 to accept the electrons from NADH and $FADH_2$, the equilibrium shifts to the right, and the ratio of blue to green fluorescence increases. Conversely, insufficient nutrient shifts the equilibrium to the left, and the ratio of blue to green decreases. By this means, laser-induced fluorescence becomes a sensitive measure of the metabolic state of cellular activity.

It has been demonstrated that, by analyzing the shape of the fluorescence emission spectrum using a suitable algorithm, it is possible to distinguish between the various forms of arterial plaque (fibrous, fatty, calcified) and normal arterial wall. It has also been demonstrated that it is possible to distinguish between malignant and normal tissue in many organs, including breast and colon. A comparison of fluorescence spectra obtained from normal and adenomatous colon tissue is shown in Fig. 9. The capacity to differentiate is clearly evident.

Fluorescence techniques, however, are not capable of providing detailed biochemical information suitable for histochemical analysis of the tissue. For this purpose, the use of Raman spectroscopy as an *in situ* optical probe of disease holds the promise of providing information unobtainable by other *in situ* methods. The Raman spectrum contains information about the quantitative biochemical composition of the tissue. The signals are inherently weak, but a number of recent technological advances have combined to permit data to be obtained on clinically useful time scales. Figures 10 and 11 illustrate what can be accomplished. Atheromatous plaque is the fatty deposit that occurs within the blood vessels of a patient with coronary artery disease, and, as it progresses, it narrows the vessel lumen, resulting in a decrease in the blood supply to tissue downstream. Figure 10 compares the Raman spectrum obtained from such a deposit with that of pure cholesterol. The detailed similarity shows that this form of plaque contains a high concentration of cholesterol and its esters, and that such a spectrum can serve

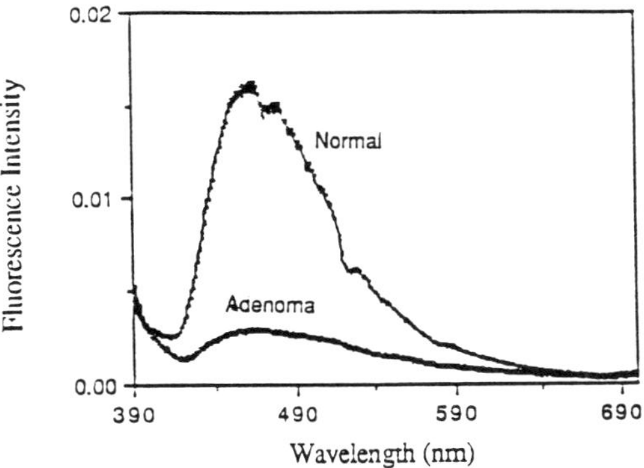

FIG. 9. Laser-induced fluorescence spectra of normal and adenomatous colon tissue, illustrating the diagnostic capability of this technique. The excitation wavelength was 369.9 nm. Reprinted with permission from Richards-Kortum (1990).

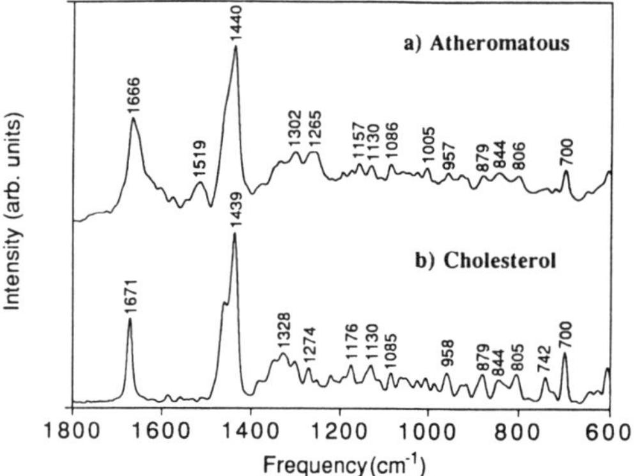

FIG. 10. Near-infrared Fourier transform Raman spectra, illustrating the correspondence between (a) typical atheromatous plaque and (b) pure, anhydrous cholesterol. The well-characterized vibrational spectrum of cholesterol is clearly identified in the atheromatous data. Reprinted with permission from Baraga (1992).

as a unique detector for these kinds of deposits. In the more advanced stages of this disease, calcified deposits form within the plaque. These deposits contain phosphates that have a unique Raman peak at 960 cm^{-1}. Figure 11 shows that the peak from a subsurface calcification can be clearly observed even when it is covered by a fibrous cap about 1.5-mm thick. The signal increases by about a factor of 8 when the cap is removed. Thus,

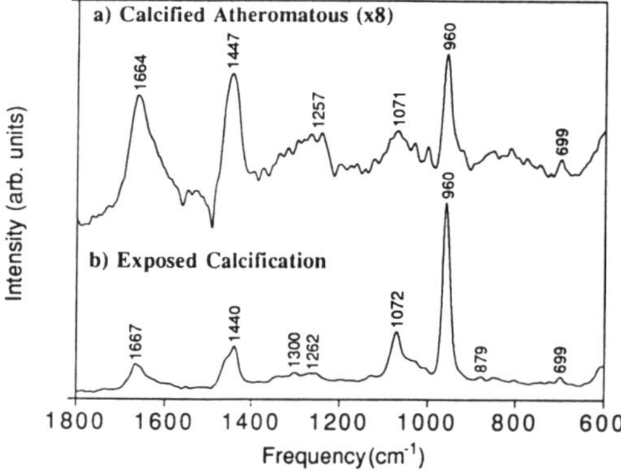

FIG. 11. Near-infrared Fourier transform Raman spectrum of calcified atheromatous plaques: (a) obtained through an intact fibrous cap approximately 1.5-mm thick, and (b) with the calcification exposed. The 960-cm^{-1} peak of the PO_3^{--} ion, which is unique to calcification, is clearly observed through the fibrous cap. Spectrum (a) is expanded by a factor of 8 relative to spectrum (b). Reprinted with permission from Baraga (1992).

calcified tissue can be detected even when hidden by intervening tissue.

The techniques described above rely on the naturally occuring, or endogenous, fluorophores present in the tissues. As was discussed in the section on photoradiation therapy (Sec. 2.7), fluorescent dyes exist that, when introduced into the body, selectively deposit in specific tissues, such as tumors. The presence of these exogenous dyes can be detected by their characteristic fluorescent spectra, and the information can be used to localize the diseased tissue.

3.3 Laser-Based Biomedical Instrumentation

Lasers are playing an increasingly important role in the biomedical instrumentation that underlies much of modern-day medicine. Examples of laser-based biomedical instrumentation include pulsed laser oximetry, laser-induced fluorescence microscopy and cytometry, and laser scanning confocal microscopy.

The pulsed laser oximeter uses infrared laser light to monitor a patient's heart rate and blood oxygenation in clinics. This is especially important in newborn and premature infants. The operation of the oximeter depends on the fact that oxygenated and deoxygenated hemoglobin possess different absorption spectra (see Fig. 3). This information is used to determine the degree of oxygen saturation in the blood by monitoring the differential transmission of laser light between two different wavelengths through the patient's finger or earlobe.

The monochromaticity of laser light is useful in creating fluorescence images in microscopy samples. With a spectrally narrow laser illumination source, the laser light can be blocked from saturating the image with a line-blocking filter. Laser-induced fluorescence of endogenous (native) tissue chromophores provides information supplemental to pathological analysis for discrimination between normal and diseased excised tissues during microscopic analysis (Plate 2). In the last few years, several commercial laser scanning confocal microscopes using fluorescence have been introduced to the market. These devices are capable of generating bright, clear images of optically thin sections of living cells and embryos. These instruments are of great interest in pathology, as well as in general biomedical research.

Cells tagged with exogenous fluorescent chromophores can be detected and manipulated in special flow cytometers (Melamed *et al.*, 1990). In these instruments, the cells are injected into a stream of droplets and directed to fall through an excitation beam. When a tagged cell passes through the beam, the fluorescence signal is detected, and that droplet is deflected by a pair of electrostatic plates into a collecting cup.

4. CONCLUSION

In this article, we have endeavored to illustrate the increasing importance of the use of lasers in medicine. Lasers are now routinely used in many medical applications, both diagnostic and therapeutic. We have briefly discussed the physics of the various types of laser-tissue interaction processes. We have surveyed a number of the more interesting therapeutic laser procedures, and described highlights of recent work on laser-based imaging of tissue and spectral diagnosis of disease. These discussions have included both those procedures and diagnostics in actual clinical practice, as well as those which are still in the research phase or in clinical trials.

GLOSSARY

Adenoma: An ordinarily benign tumor of epithelial tissue.

Atheroma: The lesion of atherosclerosis.

Atherosclerosis: Reduction of a blood vessel opening caused by lipid (fatty) deposits.

Axon: The nerve cell filament that conducts nerve impulses.

Beer's Law: The exponential attenuation of a propagating beam.

Cervix: The neck of the uterus.

Chromophore: The molecule upon which the color of a substance depends, and which therefore determines its absorption properties.

Coherence: The property of light waves that describes the degree to which different wave components separated in space or time maintain a fixed phase relationship to each other.

Collagen: The major structural protein of the body. It is a primary component of connective tissue, such as bone and cartilage.

Dysplasia: Abnormal tissue development.

Embolism: The obstruction of a vessel by transported foreign material, such as blood clots, large air bubbles, or bacterial clumps.

Endogenous: Originating within an organism.

Endoscope: An instrument for examining the inside of a canal.

Exogenous: Originating from outside an organism.

Fibrin: The elastic, filamentary protein that gives blood clots strength.

Frenectomy: Cutting the membrane beneath the tongue to allow freer movement of the tongue.

Hematoporphyrin: A dark red biochemical resulting from the decomposition of hemoglobin. It is the starting material in the production of hematoporphyrin derivative (HPD).

Hydroxyapatite: The primary mineral component of bone, teeth, and other calcified tissue. The chemical formula is $Ca_{10}(PO_4)_6(OH)_2$.

Hyperthermia: Raising tissue temperature above a normally safe limit, sometimes used therapeutically in order to selectively kill pathological tissue.

In situ: In place, without having to be excised or removed.

Ischemic: Oxygen-starved because of diminished blood supply resulting from a mechanical obstruction of the vasculature.

Keratin: A hard, fibrous protein; the principal component of hair, nails, horns. Also found in tooth enamel.

Laparoscope: An instrument for seeing and operating inside the peritoneal cavity through a small incision.

Laser Ablation: Removal of material by vaporization with an intense laser beam.

Mitochondria: Cell organelles which provide the cell's principal energy supply.

Neoplasia: The pathologic process that results in the growth of a tumor.

Neovascularization: The growth of new blood vessels; normal in wound repair, abnormal when associated with a tumor.

Peritonitis: An inflammation or infection of the sac that lines the abdominal cavity.

Phagocytosis: The action by which bacteria, dead tissue, and other debris are ingested and digested by specialized cells.

Photodissociation: The process by which a molecule absorbs a photon and splits into smaller fragments.

Phototoxic: The property in which a normally safe substance turns poisonous when exposed to light.

Plaque: A small differentiated area on a body surface.

Q-Switch: A means of extracting energy from a laser that generates a short, very intense pulse.

Tomography: A means of obtaining sectional views of an object using x rays or other radiation.

Transcutaneous: Through unbroken skin.

Works Cited

Anderson, R. R., Parrish, J. A. (1983), *Science* **220**, 524–527.

Ashkin, A., Dziedzic, J. M., Yamane, T. M. (1987), *Nature* **330**, 769–771.

Benaron, D. A., Stevenson, D. K. (1993), *Science* **259**, 1463–1466.

Baraga, J. J. (1992), *In Situ Chemical Analysis of Biological Tissue: Vibrational Raman Spectroscopy of Human Atherosclerosis*, Ph.D. Thesis, MIT Archives, p. 80.

Berns, M. W. (1984), *Laser Surg. Med.* **4** (1–4); Vol. 4, No. 1 is a special issue on Hematoporphyrin Derivative Photoradiation Therapy.

Bille, J. F, Dreher, A. W., Zinser, G. (1990), in: B. R. Masters (Ed.), *Noninvasive Diagnostic Techniques in Ophthalmology*, New York: Springer-Verlag, p. 528.

Boulnois, J.-L. (1986), *Laser Med. Sci.* **1**, 47–66.

Cheong, W. F., Prahl, S. A., Welch, A. J. (1990), *IEEE J. Quantum Electron.* **QE-26**, 2166–2185.

Hale, G. M., Querry, M. R. (1973), *Appl. Opt.* **12**, 555–565.

Huang, D., Swanson, E. A., Lin, C. P., Schuman, J. S., Stinson, W. G., Chang, W., Hee, M. R., Flotte, T., Gregory, K., Puliafito, C. A., Fujimoto, J. G. (1991), *Science* **254**, 1178–1181.

Hee, M. R., Izatt, J. A., Jacobson, J. M., Fujimoto, J. G., Swanson, E. A. (1993), *Opt. Lett.* **18**, 950–952.

Ishimaru, A. (1978), *Wave Propagation and Scattering in Random Media*, New York: Academic Press.

Itzkan, I., Bourgelais, D. B. C. (1988), *Laser Life Sci.* **2**, 249–255.

Itzkan, I., Albagli, D., Banish, B. J., Dark, M., von Rosenberg, C., Perelman, L. T., Janes, G. S., Feld, M. S. (1993), in: J. C. Miller, D. B. Geohegan (Eds.), *Laser Ablation: Mechanisms and Applications-II*, AIP Conference Proceedings No. 288: New York: American Institute of Physics, p. 491.

Karu, T. (1987), *IEEE J. Quantum Electron.* **QE-23**, 1703–1717.

Key, H., Davies, E. R., Jackson, P. C., Wells, P. N. T. (1991), *Phys. Med. Biol.* **36**, 579–590.

Leith, E., Chen, C., Chen, H., Chen, Y., Dilworth, D., Lopez, J., Rudd, J., Sun, P.-C., Valdmanis, J., Vossler, G. (1992), *J. Opt. Soc. Am. A* **9**, 1148–1153.

Melamed, M. R., Lindmo, T., Mendelsohn, M. L. (Eds.) (1990), *Flow Cytometry and Sorting*, New York: Wiley-Liss.

Mihroseini, M., Shelgikar, S., Cayton, M. M. (1988), *Ann. Thorac. Surg.* **45**, 415–420.

Moon, J. A., Mahon, R., Duncan, M. D., Reintjes, J. (1993), *Opt. Lett.* **18**, 1591–1593.

Muller, G. (Ed.) (1993), *Medical Optical Tomography: Functional Imaging and Monitoring*, Bellingham, WA: SPIE Optical Engineering Press.

Patterson, M. S., Chance, B., Wilson, B. C. (1989), *Appl. Opt.* **28**, 2331–2336.

Richards-Kortum, R. R. (1990), *Fluorescence Spectroscopy as a Technique for Diagnosis in Human Arterial, Urinary Bladder, and Gastrointestinal Disease*, Ph.D. Thesis, MIT Archives, p. 200.

Schober, R., Ulrich, F., Sander, T., Durselen, H., Hessel, S. (1986), *Science* **232**, 1421.

Svaasand, L. O., Boerslid, T., Oeveraasen, M. (1985), *Laser Surg. Med.* **5**, 589–602.

Tan, O. T., Sherwood, K., Gilchrest, B. A. (1989), *New Engl. J. Med.* **320**, 416–421.

Wang, L., Ho, P. P., Liu, C., Zhang, G., Alfano, R. R. (1991), *Science* **253**, 769–771.

Webb, R. H., Hughes, G. W., Delori, F. C. (1987), *Appl. Opt.* **26**, 1492–1499.

Wilson, B. C., Sevick, E. M., Patterson, M. S., Chance, B. (1992), *Proc. IEEE* **80**, 918–930.

Further Reading

Major Journals Specific to Laser Medicine

Lasers in Surgery and Medicine, C.A. Puliafito, Editor-in-Chief, Wiley-Liss, New York, published 6 times per annum.

Lasers in the Life Sciences, M. Wolbarsht, Editor-in-Chief, Harwood Academic Publishers, New York, published 4 times per annum.

Recent Journal Issues Devoted to Laser Medicine

Applied Optics, Special Issue on Photon Migration in Tissue and Biomedical Applications of Lasers, M. Motamedi, Guest Editor, Vol. 32, No. 4, 1993.

IEEE Journal of Quantum Electronics, Special Issue on Lasers in Biology and Medicine, R. Birngruber, Guest Editor, Vol. 26, No. 12, 1990.

IEEE Transactions on Biomedical Engineering, Special Issue on Laser-Tissue Interaction, A. J. Welch, Guest Editor, Vol. 36, No. 12, 1989.

Optics and Photonics News, Special Issue: Time-Resolved Imaging & Diagnostics in Medicine, J. G. Fujimoto, Guest Editor, Vol. 4, No. 10, 1993.

Books

Chance, B. (Ed.) (1990), *Photon Migration in Tissues*, New York: Plenum Press.

Chance, B., Alfano, R. R. (Eds.) (1993), *Photon Migration and Imaging in Random Media and Tissues*, SPIE Proceedings Vol. 1888, Bellingham, WA: Society of Photo-Instrumentation Engineers.

Dixon, J. A. (1987), *Surgical Applications of Lasers*, Chicago: Year Book Medical Publishers.

Goldman, L. (Ed.) (1981), *The Biomedical Laser: Technology and Clinical Applications*, New York: Springer-Verlag.

Hillenkamp, F. (Ed.) (1980), *Lasers in Biology and Medicine*, New York: Plenum Press.

Muller, G. (Ed.) (1993), *Medical Optical Tomography: Functional Imaging and Monitoring*, Bellingham, WA: SPIE Optical Engineering Press.

Steinert, R. F., Puliafito, C. A. (1985), *The Nd:YAG Laser in Opthalmology: Principles and Clinical Applications of Photodisruption*, Philadelphia: W.B. Saunders.

White, R. A., Grundfest, W.S. (1987), *Lasers in Cardiovascular Disease*, Chicago: Year Book Medical Publishers.

MEMBRANE MECHANICS

See PHYSICS OF FLEXIBLE MEMBRANES

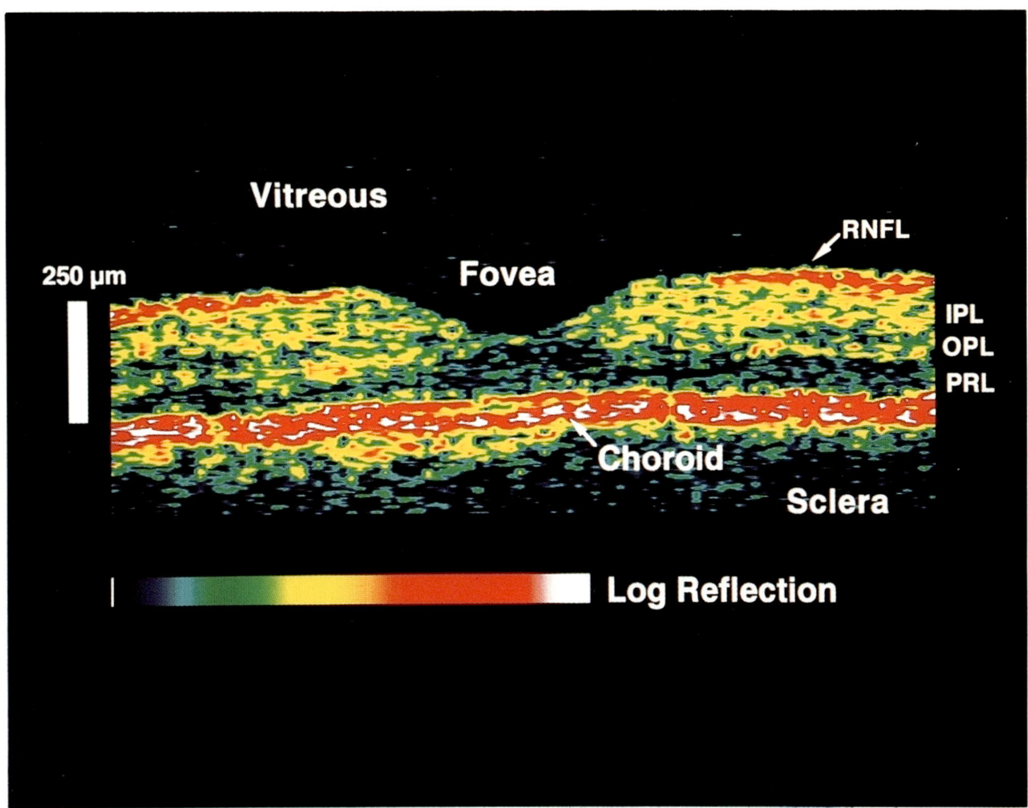

PLATE 1. Optical coherence tomography image of the retina taken in a living human subject. The fovea (the region of highest visual acuity) is centrally located; lateral to the fovea, identifiable retinal layers include RNFL = retinal nerve fiber layer; IPL = inner plexiform layer; OPL = outer plexiform layer; PRL = photoreceptor layer. The false color scale represents the logarithm of the power reflectivity of the retina, ranging 35 dB from a minimum detectable reflectivity of approximately 1 part in 10^9. Photo courtesy of Prof. J. G. Fujimoto, Massachusetts Institute of Technology, and Prof. C. A. Puliafito, Tufts University School of Medicine.

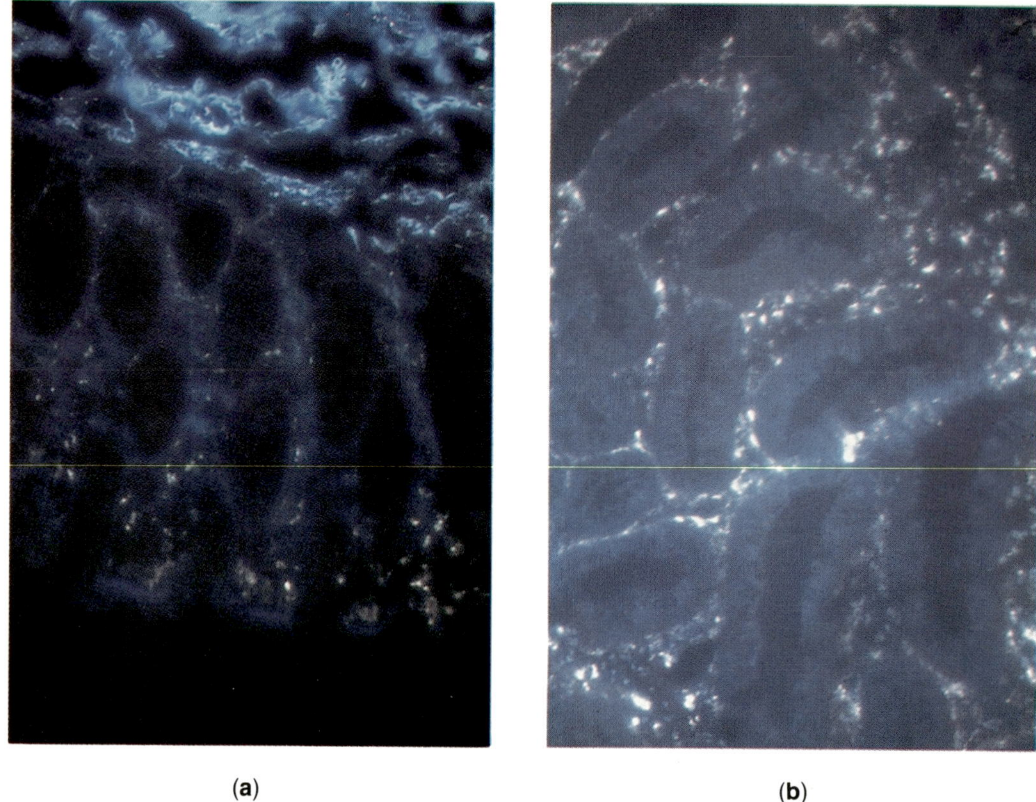

(**a**) (**b**)

PLATE 2. Laser-induced fluorescence microphotographs revealing differentiation between (a) normal and (b) dysplastic colon crypt cells. The crypt cells line the crypts (the oval structures) and do not fluoresce in normal tissue, and therefore are not discerned. In abnormal tissue, the crypt cells emit a fluorescence which appears light blue. The bright white spots are eosinophylls in the spaces between the crypts. Photo courtesy of Prof. M. Fitzmaurice, Case Western Reserve University School of Medicine.

MEMORIES

CELIA E. YEACK-SCRANTON*, *IBM Corporation, San Jose, California, U.S.A.*

INTRODUCTION

Elements of modern computing were envisioned by Charles Babbage in his Analytical Engine of 1868, but it was the invention and practical implementation of punched-card data-processing machines by Hollerith and the use of these machines in the 1890 U.S. census that began a steady progression of invention and development toward modern computer memories.

A key challenge in computing technology is in data memory. Memory must provide accurate, long-term retrieval of data at access speeds compatible with the computing device. As the speed, accessibility, and uses of computing have increased, so have the types, uses, and sheer volume of memory. The economic significance has also mushroomed, amounting to a $52 billion industry in 1990 in magnetic storage, the largest single component. A brief overview of the general features of memory and its evolution as a synergy between invention and commercial development is given in Sec. 1.

*Deceased.

Memories continue to become faster, denser, less expensive, lower in power consumption, and more rugged. Yet the underlying physical phenomena used in today's practical memory and storage devices are remarkably the same as in the 1950s and 1960s, when the first modern electronic computers with program storage and significant reliance on memory devices were introduced. Magnetic and semiconductor technologies remain dominant, with emerging devices based on optics, magneto-optics, and ferroelectrics. These phenomena are reviewed in Sec. 2.

The economic success of computing has supported a rapid pace of invention and development of many memory technologies. Historically, each new device has solved a particular technical problem and/or made possible a new application area in computing. Continued utility of the device has depended on the technological extendability and the rate of improvement of cost and function relative to that of other technologies. The design, function, and operation of a selection of devices are described in Sec. 3. Magnetic core memories are included for historical signifi-

3-527-28132-0/94/$5.00 + .50

cance. Tape, disk, and dynamic random-access memory (DRAM) are economically the most significant. Magneto-optic devices, electrically erasable programmable read-only memory (EEPROM), and ferroelectric memories offer potentially significant opportunities for the future. Static random-access memory (SRAM), read-only memory (ROM), programmable read-only memory (PROM), and bubble devices fill other markets.

The ideal memory device would simultaneously be the fastest, least expensive, most dense, lowest in power consumption, longest lived, and most robust. In practice, however, optimization is achieved at the system level via a computer storage hierarchy. Here, small, fast, expensive memories are used close to the (fast) processor. Additional levels of memory are configured in tiers away from the processor, each level having an increasing number of bytes at decreasing cost and performance. Those in which access time to any bit is independent of its physical location are termed "random access," and are generally referred to as "memory." Those that store data in serial blocks, requiring much longer access times, are generally termed "storage." Typical hierarchies and new applications made possible by the current trends in cost, size, and power reduction will be discussed in Sec. 4.

1. GENERAL PRINCIPLES AND BRIEF HISTORY

Tabulation for large businesses and government was the dominant use of the early electromechanical computers used in the first half of the 20th century. In these systems, a few numbers were physically stored in mechanical relays and, later, dials. Storage and retrieval times were limited by the motion of these large mechanical structures, and this rudimentary memory was used primarily for looking up constants, for instance, in insurance tables as the computation proceeded.

By the late 1940s, a recognized need in the fledgling computer industry was for large-capacity data storage. Focused projects, notably to provide a "defense calculator," fostered a series of inventions and developments leading to the introduction of the first modern computers, such as the IBM type 701 system, in 1952. Both the memory types and the mix of memory devices in this computer bear similarities to those in use today. Electrostatic devices, later supplanted by magnetic cores and then by semiconductor memories, provided fast-access memory that worked in synchrony with the computing unit. The capacity of 2048 words of 36 bits each was accomplished by use of 72 cathode ray tubes. Since the tubes could be operated in parallel, all the bits from a word were read or written simultaneously, giving a word every few microseconds. This random-access memory was "dynamic," as it needed to be rewritten, or "refreshed," every 12 ms since the charge stored in the tubes dissipated in that time. Thus, when the power was turned off, information stored in the electrostatic storage unit was lost and, hence, was termed "volatile."

The IBM 650 magnetic drum memory in the 701 computer provided a slower but more cost-effective and nonvolatile storage. Here, a specially designed electromagnetic "head" (described further in Secs. 2.1 and 3.1) was used to write individually magnetized "bits" or regions onto a magnetic thin-film surface coated onto a rotating cylindrical drum. By the inverse process, the head could be used to sense the presence of bits, or read information, since a voltage was produced as the bit pattern passed under the head. The head was held at a specific spacing above the drum, since abrasive wear from contact would quickly destroy both the head and drum. Rotation allowed writing around one circumference, creating a "track" of data. Arranging many heads side by side parallel to the drum axis allowed the creation of multiple tracks, and five were read or written in parallel to increase the word speed. The 10-cm (4-in.) diameter, 40-cm (16-in.) long drum was rotated at 12 500 rpm to provide an average access (half-rotation) time of 2.4 ms to the first bit of a desired set of data. Four drums storing a total of 8192 36-bit words were arranged in a unit. They worked in synchrony with the main electrostatic computer memory, with some clever designs to minimize the effect of the long access time.

The function of the IBM 701 computer was augmented greatly with the introduction of the RAMAC, a magnetic disk file, in 1956. Here, the rotating medium was a stack of disks, and a single head pair was moved from track to track and from disk to disk to read and write all of the data surfaces, one head for the upward-facing disk surfaces and the other head for downward-facing surfaces. This allowed up to 5 megabytes (MB) of information to be stored, but with a maximum retrieval time of 0.9 s due to the mechanical motion required to insert the head into the disk stack at the desired disk and to move to the correct track. The use of a single head pair minimized cost, but because of the enormous performance difference, drum memory was reserved for frequently used information while larger volumes of data were stored on the RAMAC.

The memory in Remington Rand's UNIVAC computer, announced in 1952, was based on a technology in which a sequence of electrical pulses was converted to a sequence of ultrasonic pulses, launched down a delay line consisting of a cylinder filled with mercury, and reconverted to the electrical waveform at the other end. Since sound travels about 10 000 times more slowly than electrons, much computation was accomplished in the processor during the acoustic propagation time, and the results from that computation were used to update the reconverted electrical pulse stream. After the update, the revised electrical pulse sequence was converted to ultrasonic pulses again, and the process continued indefinitely. While attractive in its simplicity, it was difficult to scale this memory type to larger capacities, and this memory was supplanted within a few years by ones using magnetic ferrite cores.

Magnetic core memories (described in Sec. 3.1.1) were initially introduced as a buffer between the input/output devices (magnetic tape and punch cards) and the main computer. However, since their random-access capability eliminated the access delay associated with drum memory, since their reliability exceeded that of the electrostatic storage tubes, and since they were more scalable to smaller dimensions (and thus larger storage capacities) than the acoustic delay-line memories, they quickly supplanted all three and persisted as a dominant choice for main memory until the mid-1970s. At that time, further shrinking of the individual cores to increase device capacity became less cost effective than the fabricating of integrated-circuit semiconductor memories.

Magnetic thin-film memories were developed in the 1960s, using the same principles as core memories, but the structures were created in a thin-film device, facilitating miniaturization by eliminating discrete wires and cores. While used in some products, they were supplanted quickly by the rapidly evolving semiconductor memory devices.

Semiconductors emerged as early as the 1960s as a basis for memories, but early devices relied upon assembling many discrete transistors and components in arrays and were less reliable than other choices. In 1959, however, integrated circuits were first fabricated. In integrated circuits, many transistors and other components are built on the surface of a silicon wafer and interconnected with thin films of metal. This integration made possible an ever decreasing cost per bit by steadily decreasing the size of each cell with improvements in design and processing. Higher reliability and speeds accompanied this, and semiconductor memory began displacing magnetic core memories in the early 1970s. Static random-access memory, which requires a number of transistors to store each bit, was later replaced with DRAM, which uses only a single transistor per storage bit but requires constant refreshing of the information. Semiconductors remain the highest performance memories available, with a clear scalability to much finer dimensions, faster speeds, and lower costs for at least one more decade.

In addition to storage and memory devices in which data are both read and written by the user, a class of memory devices is designed for reading fixed data. Read-only memory (ROM), first proposed in the 1960s, was initially a device in which data were fixed at the time of manufacturing by the presence or absence of transistors at each bit cell. Later devices allowed the user to "burn in" connections, creating a desired bit pattern. More modern devices store charge in such a way that it is very hard to remove, allowing some rewriting (up to 10^6 cycles). Of these, Flash EEPROMs are gaining rapid acceptance, although slow write and erase times and limited cyclability leave room for improvement.

Flash EEPROM memories offer nonvolatile random access storage with read/write capability (although writing is limited). This avoids the time delays associated with most magnetic storage but offers nonvolatility, unlike conventional DRAM and SRAM. Other technologies, such as ferroelectric random-access memory (RAM), achieve nonvolatility

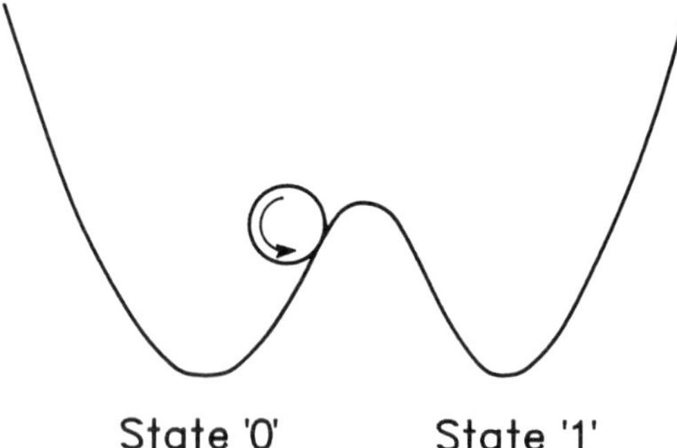

FIG. 1. Simple illustration of the principles of memories using a ball rolling between two wells.

by use of different materials, and lower-power conventional DRAM designs facilitate the use of batteries to achieve nonvolatility. Cost, performance, and application determine which nonvolatile technology is used, and increased use of portable, high-performance computers makes this a direction of growth.

2. PHYSICAL PHENOMENA ASSOCIATED WITH MEMORY DEVICES

This section describes the phenomena that are key to the different memory technologies. Each fulfills basic requirements for useful devices. In particular, the system must exhibit bistability, or two stable states, so that 1's and 0's can be represented. A simple example, Fig. 1, can be used to illustrate this and other attributes. Here, a ball can roll into either of two potential wells. To move from one well to the next requires sufficient energy to roll over the hill in the middle. A smaller hill will require less switching energy, but it must not be so low that a ball may accidentally roll from one state to the other. Switching must be as fast as possible, and it is desirable that neither the hill nor the balls wear out with time or use. The state of the ball should be read without moving it and with minimal expended energy, although in some memory devices, the read process can be likened to plucking the ball from the hole, then putting it back.

2.1 Magnetism

Magnetic storage is easily visualized by imagining a string of permanent magnets in which individual magnets can be flipped to give a desired pattern of north and south poles, as shown in Fig. 2. The orientation of each magnet can be viewed as the "1" or "0" state described above. Controlling and sensing the orientation of these magnets are the analogs of writing and reading.

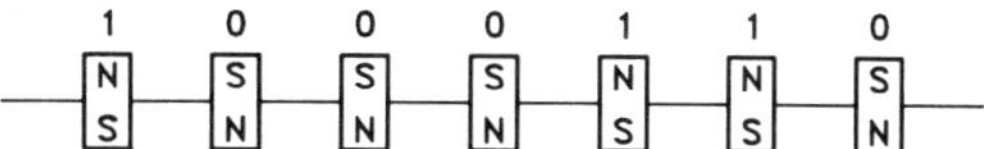

FIG. 2. A string of magnets oriented to represent a string of 1's and 0's, analogous to data storage on magnetic media.

Practical magnetic devices use continuous materials in which the magnetization (direction of the internal magnetic flux) can be locally flipped within small regions of the material. The most common materials used are ferromagnetic or ferrimagnetic, whose properties are illustrated in the hysteresis loop in Fig. 3. Initially, without an applied field H, the spontaneous magnetization direction, M, is comprised of tiny domains or particles whose directions tend to cancel each other, giving rise to a net $M = 0$ (*a* in Fig. 3). However, with a large enough applied field, these domains line up in the direction of the field (*b* in Fig. 3). Then, even after the field is removed, a remanent magnetic flux density M_r remains (*c* in Fig. 3). This remanence is analogous to one orientation of the magnets of Fig. 2 and determines the strength of the read-back signal for magnetic recording. The

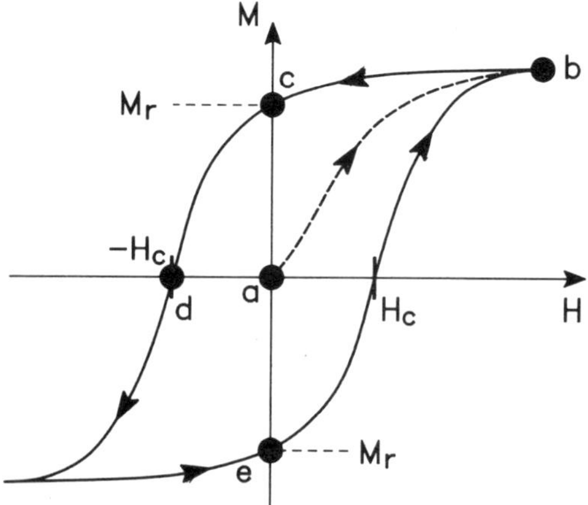

FIG. 3. Hysteresis loop showing the relationship between the applied field H and the resulting internal field M for a ferromagnetic or ferrimagnetic material.

magnetization can be reduced to zero by applying a sufficiently large field in the opposite direction, $-H_c$ (d in Fig. 3). An even larger opposing field causes a reversal of this magnetization and an opposite sense of remanence, $-M_r$ (e in Fig. 3), analogous to the oppositely oriented magnets of Fig. 2. H_c is called the coercive force or coercivity.

A key advantage of magnetic storage is that it is nonvolatile. That is, the magnetization direction is permanent until reset, even in the absence of power. Further, the magnetization direction can be flipped as many times as desired without wearing out the material, giving it infinite cyclability. Magnetic switching times are fundamentally limited only by the gyromagnetic resonance, which is in the gigahertz range. Current device designs are still an order of magnitude away from this. The minimum stable domain size, determined by the product of media magnetization and coercive force, has continued to decrease with better materials. Thus, given continued engineering advances, continued magnetic recording density improvements are possible for at least ten years.

2.2 Magneto-optic and Optical Phenomena

Magneto-optic storage relies on creating tiny magnetized regions, in a thin film, which interact with an optical beam to alter its reflection. The magneto-optic Kerr effect is commonly utilized, and the physics of this phenomenon is illustrated in Fig. 4. A thin film of magneto-optic material is used as the storage medium, with the magnetization oriented vertically up or down with respect to the surface of the film. To write information onto the thin film, a laser pulse is focused onto a tiny spot in the presence of a weak external magnetic field opposing the magnetization direction of the overall film. The laser locally heats the material very quickly above a critical temperature (T_c) at which the external magnetic field aligns the magnetization to this reverse direction, creating a stored bit. As in magnetic recording, the recording surface is moved under the head to write or read a sequence of bits. Reading is accomplished by sensing a change in polarization of a reflected low-powered laser beam as the stored bit passes under it. Erasing is done in exactly the same way as writing except with a reversed external field. The areal recording density for magneto-optic storage is very high since the optical beam used to create it can be focused to about a micrometer.

Another important optical storage mechanism for read-only applications is to use optical contrast between featured and unfeatured regions of a smooth disk. The pattern of features represents the bits. The simplest way of creating these features is to stamp pits in a plastic disk with a master. This mechanism is the basis for consumer compact-disk (CD) players and fixed data storage in CD-ROM (CD read-only memory). Another mechanism relies upon using the laser heat pulse to create spots on the disk with material structures different than those of the surrounding material, for instance, a different crystal phase. This is the basis for "write-once" optical storage.

For all optical storage phenomena, the potential number of stored bits per area increases as the wavelength is decreased, since this tightens the beam focus. Several generations of this fundamental type of improvement are expected before radical changes are necessary.

2.3 Charge Storage

Charge storage is easily visualized by imagining a planar grid of wires with a box (cell) comprised of a charge bucket and a valve

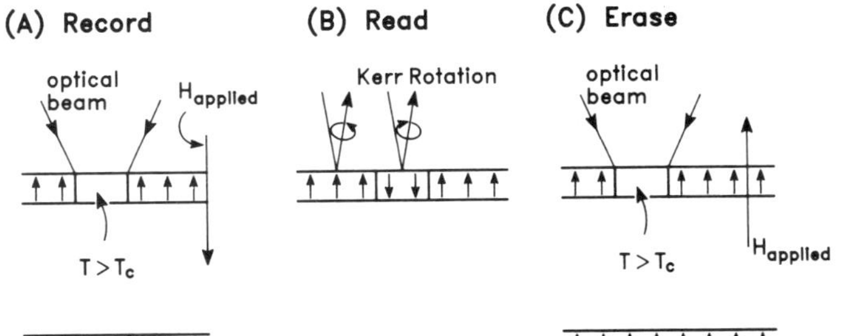

FIG. 4. Magneto-optic recording processes.

at each intersection, as sketched in Fig. 5. The valves for all the buckets in a row are turned on in unison through a single control (the word line), and charge can then be added to the desired buckets through the appropriate vertical (bit) lines, creating the desired charge storage pattern. Sensing is done by switching on the row, then measuring the presence or absence of charge in each charge bucket as a 1 or a 0 at the respective vertical line. This matrix scheme allows minimal space to be used for addressing, decoding, and routing data.

Early charge storage was accomplished by means of cathode ray tubes, but these devices were supplanted by semiconductors, as silicon-integrated circuits. While many attributes of today's devices vary, the underlying principles remain the same. Bit cells are arranged in a grid as described above, where the rows are word lines and the columns are bits lines. In each cell, transistors are used as valves, and charge can be put on a capacitor as a voltage. Sensing is done by measuring the stored charge as a voltage or as a current when the voltage is drained off.

Most semiconductor memory cells are based on metal-oxide-semiconductor field-effect transistors (MOSFETs), in which a conducting channel of electrons or holes is switched on or off by an electric field applied from the gate. In the *n*-channel MOSFET sketched in Fig. 6, a *p*-type semiconductor is implanted to form *n*-doped regions. The conductivity of a semiconductor depends on the doping and the applied field, exhibiting a very nonlinear increase in conductivity above a well-defined "threshold" voltage. When a voltage above the threshold is applied to the gate, a current of conducting electrons flows through this channel, between the source and the drain. This provides an electrically controlled switch. Most FET-based memory devices use a complementary metal-oxide-semiconductor (CMOS) in which *n*-channel and *p*-channel devices are fabricated on the same silicon substrate.

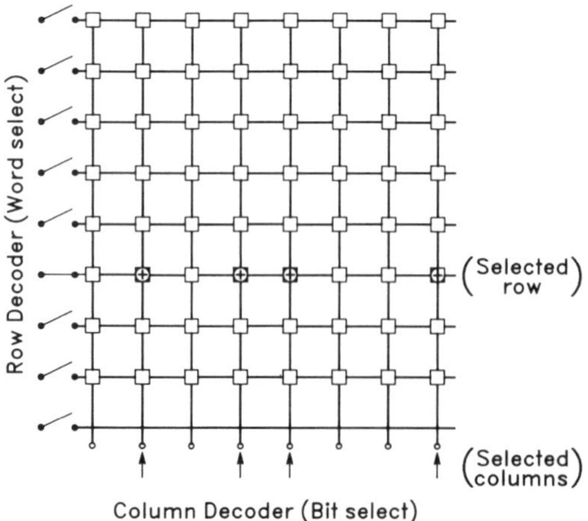

FIG. 5. Charge bucket array analogy illustrating the principles of random access semiconductor memory devices.

The number and arrangement of the transistors, choice of capacitor, and wiring determine the differences between the memory cells discussed in Sec. 3. Important functional differences are described briefly here as an introduction. "Static memory" is that in which the state of each cell is permanent until it is reset. "Dynamic memory" requires "refreshing" or rewriting the state of the cell periodically. Generally, both static and dynamic memory are "volatile"; that is, the state of each cell is lost when the power is turned off. Thus, "nonvolatile" memory is a type of memory in which the state of each cell is retained even in the absence of power. "Destructive readout" is a method of sensing the state of the cell in which the state is lost as a consequence. Hence, the cell is generally "restored" or rewritten after the bit is sensed. For all of these memory cells, important metrics are cell area (since this translates directly to cost), speed, and power consumed per bit cell.

2.4 Ferroelectricity

The use of a ferroelectric material as the insulator in a thin-film capacitor of a semiconductor memory cell provides the valuable attribute of nonvolatility in semiconductor memory. This arises from the unique properties of ferroelectrics, which exhibit a nonlinear dielectric behavior and retain a residual polarization when an external electric field is applied and then removed. When the field is reversed, spontaneous polarization in the opposite direction occurs. These two stable polarization states allow creation of an electrically controlled bistable charge state on the capacitor that is retained without any power for at least a year, hence providing nonvolatile memory.

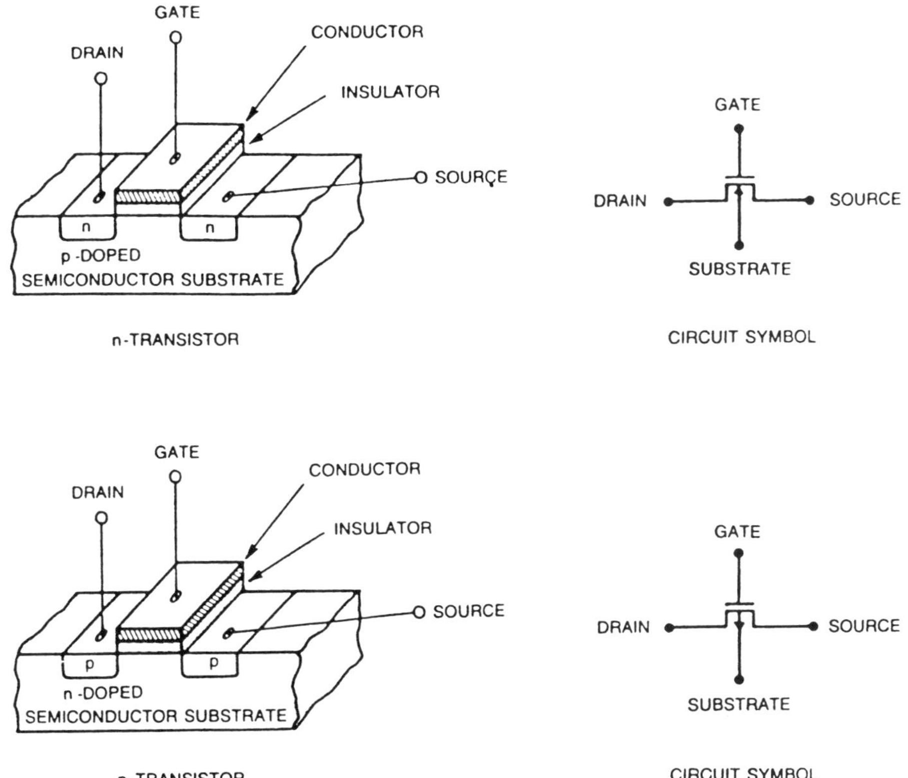

FIG. 6. MOSFET transistor and its circuit representation. (Weste and Eshraghian, 1985, © 1985 by AT&T Bell Laboratories, Inc. Reprinted by permission of Addison-Wesley Publishing Company.)

The intrinsic bistability is seen in a plot of polarization vs applied field, which gives a hysteresis loop analogous to those of ferromagnets (see Fig. 3). This similarity is the basis for the use of "ferro" in the name "ferroelectric." However, in ferromagnets, opposite magnetic polarizations are produced by flipping electron spin states, while in ferroelectrics, opposite electric polarization states are produced by altering the crystal structure. One important consequence of this is that the number of times the electric polarization can be flipped (cyclability) is much less than that for magnetic spins.

The readout of a ferroelectric material can be accomplished by applying a field that exceeds the coercive electric field. The two different polarization states give different responses, and the states can thus be distinguished. However, the memory cell then has to be rewritten to restore the original information. The limited cyclability (about a million cycles) and the somewhat slower switching speed compared to semiconductors are both serious constraints in some applications.

3. DESIGN, FUNCTION, AND OPERATION

3.1 Magnetic Devices

3.1.1 Core Memories Magnetic core memories are comprised of arrays of magnetic toroids, or cores. A magnetic core is a ring of ferrimagnetic material that can be magnetically switched from one circumferential magnetization direction to the opposite one. Switching is accomplished by applying a field in the desired direction to oppose the internal magnetization, as described in Sec. 2.1. The field is produced by applying a current to a wire passing through the core, as sketched in Fig. 7. Opposite currents give opposite fields.

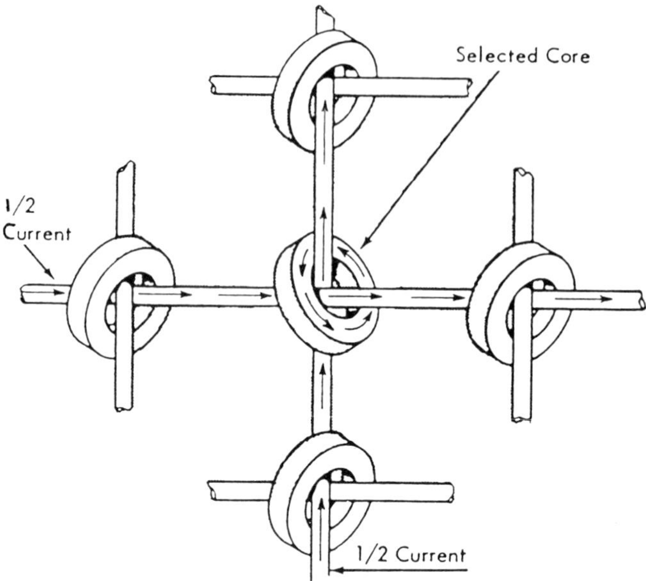

FIG. 7. A small section of a magnetic core memory, illustrating the principles of coincident-current selection. (IBM News, 10 November 1967)

Devices use the abrupt threshold for switching apparent in the hysteresis loop shown in Sec. 2.1. In particular, one-half of the coercive field is not sufficient to switch the core. Thus, to make a device, cores are arranged in a matrix so that there is a core at each intersection of two wires, one from the horizontal (i) and one from the vertical (j) direction. To selectively write a core at the (i,j) position, half of the required current is applied to each of the ith horizontal and jth vertical wires. Only at the (i,j)th core is there sufficient current to switch the core to a new state.

The matrix arrangement of cores allows random access to any desired bit (core). The switching speed of each bit is limited mainly by the electrical resonance of the core in the read/write circuit, typically in the megahertz range, giving about 1-μs cycle time per bit. Many bits are accessed in parallel for faster overall memory speeds. Readout is destructive, requiring a rewrite operation for continued storage.

Ferrite core memories were cost competitive in the 1960s with cycle times of about 1 μs and a cost of about $10 000/MB. However, assembling core memories required individually wiring each magnetic core onto a wire matrix, one core per stored bit, and was practically accomplished when memory sizes were only up to a few thousand bytes. The difficulties of fabricating these ceramic cores smaller than about 1 mm, materials frequency limits, and the challenge in wiring the tiny cores all limited continued cost-performance improvements relative to integrated semiconductor memories. The latter became the dominant random access memory choice by the mid 1970s.

3.1.2 Rotating-Disk Storage In rotating-disk storage, writing and reading information is accomplished with a magnetic recording head, which is conceptually described as a wire-wound toroid with a small slice removed to create a gap in the magnetic material, as shown in Fig. 8. Head operation is as follows. A current representing the input signal is applied to the coil. The magnetic field thus created goes across the gap but also induces a fringing field in the region below the head gap. This field magnetizes small regions of the moving magnetic medium, creating the desired bit pattern. These tiny magnetic poles in the material in turn cause magnetic flux to fringe out of the disk, which can be sensed by the same head as an induced voltage.

Early magnetic recording heads used machined magnetic cores similar to the one shown in Fig. 8. These were laminated onto a slider designed to float above the disk on a tiny cushion of air to prevent wear from rubbing against the disk. In 1979, thin-film heads were introduced in which the core was re-

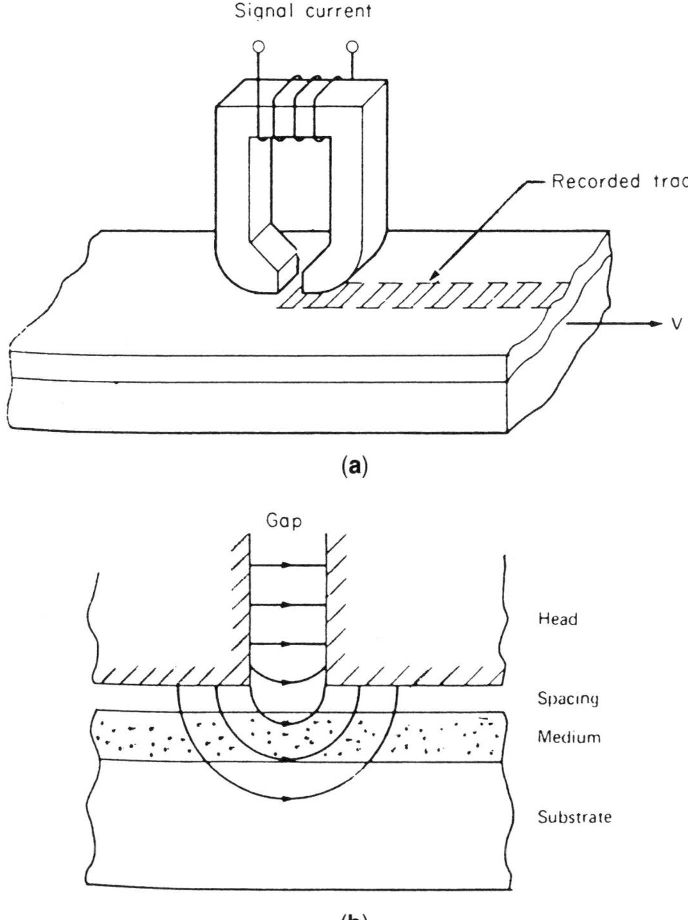

FIG. 8. Illustration of the recording process using a ring recording head. (a) Three-dimensional view; (b) cross section showing the magnetic field from the gap. (Mee and Daniel, 1987, © 1987 by McGraw-Hill, Inc. Reprinted by permission of C. D. Mee.)

placed with thin-film plated and patterned magnetic layers, the wires were replaced with patterned thin-film conductors, and the gap was replaced with a thin-film deposited structure. Many thin-film head sliders are created at once on the same wafer by means of multiple patterning and deposition steps in a manner analogous to semiconductor processing.

Further head improvements have been achieved by using heads separately optimized for reading and writing, fabricated in the same thin-film structure. The most sensitive read head used in the industry today is based on a magnetoresistive film, which has a different resistance for different directions of the magnetic field from the disk. This allows scaling to finer dimensions and higher-frequency operation, and hence higher bit densities.

An important attribute of rotating storage and tape applications is that information is stored on an unfeatured medium, within which tiny regions of material are individually switched. This means that while it may be very complex and costly to create the head that writes and reads the information, the head can be used to write many bits onto a relatively simple structure, generally lowering the overall device cost. Using this head, one track of information is created or read at a time, and the head is physically moved to write or read another track. In hard-disk storage, there are generally tens of thousands of bits per centimeter along the track direction and thousands of tracks per centimeter. A typical disk drive is sketched in Fig. 9. Note that much of the volume and power dissipation comes from the mechanical components required for disk rotation and track accessing.

Fundamental to the continued density in-

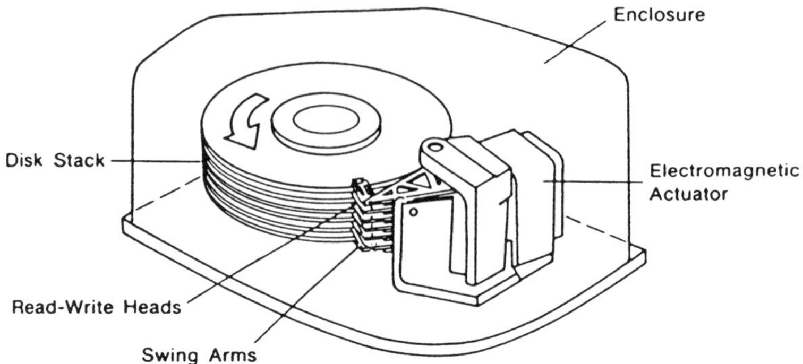

FIG. 9. Rigid disk drive. (Mee and Daniel, 1988, © 1988 by McGraw-Hill, Inc. Reprinted by permission of C. D. Mee.)

crease in magnetic storage is the ability to scale to finer dimensions. The near-field (fringe field) operation of the heads requires that the head-to-disk separation is decreased in proportion to the density increase, which in turn requires smoother, more durable surfaces to withstand the inevitable occasional contacts. Thin-film head dimensions must also be reduced. Materials and circuits must be improved to reduce magnetic and electronic noise. Finally, mechanical tracking accuracy must be increased. Nonetheless, continued extendability is expected for magnetic-recording hard-disk drives. Densities up to 300 Mb (megabits) per square centimeter, or 2 Gb (gigabits) per square inch, have been demonstrated. Product densities are at about 30 Mb/cm^2, increasing at 60% per year with no fundamental limits anticipated through the year 2000. Approximate cost and performance figures are $1/MB, access times of 12 ms, and data rates of 9 MB/s.

Disks for magnetic recording are created by coating a thin magnetic layer onto a very smooth substrate. Substrate materials are predominantly a highly polished aluminum-magnesium alloy, although in recent years glass and ceramics have been used. The first disk coatings were a mixture of iron oxide and alumina particles coated as a slurry onto the substrate, then hardened by baking to a layer several micrometers thick. The alumina particles provided durability so that the recording head slider could land onto the disk without damaging the soft iron oxide. Requirements for higher recording densities include smoother surfaces (for closer head media separation and lower noise) and thinner magnetic films (for better resolution) with higher magnetization and lower coercivities (to maintain high signal-to-noise ratios). This led to the development of thin-film deposited magnetic layers, ranging from 25 nm to 200 nm thick. Generally, high wear resistance is achieved by coating the magnetic film with a thin (25 nm) layer of carbon. Most of today's disk files use thin-film disks.

A small proportion of disk files are floppy drives, so-called because the disk itself is a flexible material coated with a thin magnetic layer. Most of the same principles apply, except that these disks are designed to be removable and interchanged with any other drive. Mechanical alignment of the disk hub to the spindle, the warping of the softer disk material with temperature and humidity, and the differences from one head to another all make the densities for these devices much less than those for hard-disk drives. Capacities of a few megabytes per disk are typical, compared with up to hundreds of megabytes for a similar-sized rigid disk, but they represent an important option for transporting programs and data.

3.1.3 Tape Devices Tape devices, like most magnetic recording devices, operate by using a magnetic head to read and write a magnetic medium that is moving with respect to the head. Unlike disk drives, the recording medium is a flexible magnetically coated tape wound in a removable cartridge. Further, the heads are generally run in contact with this flexible medium, which is open to the environment, requiring robust heads tolerant of debris and wear at the magnetic pole tips. However, because the cartridges are removable, the heads can be cleaned periodically and tapes can be copied if signs of degradation are detected. Since the cost of tape cartridges is very low and many cartridges are used with one read/write unit, tape devices offer a very inexpensive storage option, as low as $0.01/MB.

Two basic head types shown in Fig. 10 are widely used today. The simplest structure is a linear, stationary head array in which a string of head elements creates an equal number of tracks along a tape simply by moving the tape across it. In rotary-head recording, the tape is moved and the head is rotated at an angle on a cylinder to produce diagonal tracks. The latter allows a much higher head-to-tape velocity (about 1000 cm/s), higher track densities (about 1000 tracks/cm), and a very rapid search mode for desired data blocks. High velocity allows high data rates (up to 20 MB/s), and high track densities allow greater tape capacities (typically greater than 1 GB for a small cassette and up to 200 GB in larger ones).

The magnetic tape is critical to performance and has been improved substantially in the last decade. Early types consisted of iron oxide particles coated onto a polyethylene tape, giving coercivities of about 300 Oe. Cobalt surface treatments and, most recently, metal iron particle (MP) tapes offer improvements of up to 1450 Oe coercivity. Smoother surface finishes and greater durability have also contributed to steady recording improvements.

The chief disadvantage of tape devices is that access is serial. In the worst case, the entire length of the tape would have to be searched in order to get to a desired data block, requiring at least several seconds. High

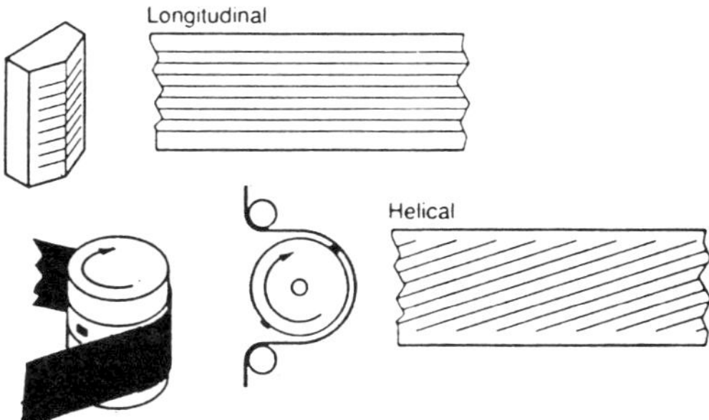

FIG. 10. Two modes of tape recording: fixed-head longitudinal and rotary-head helical (Mallinson, 1990, © 1990 IEEE).

winding speed is required for short access times. Improved tape transport design is important for tracking accuracy at high tape velocities and for thinner tapes (from 25 μm in 1980 to as low as 7 μm in small rotary cassettes). Better mechanical design as well as improved wear characteristics due to carbon coatings on the back of the tape have facilitated continued tape volumetric density increases and faster data searches.

3.1.4 Bubbles A magnetic bubble is a self-contained, cylindrical magnetic domain that is polarized in a direction opposite to the magnetization of the surrounding thin magnetic film, typically a rare-earth iron garnet deposited on a nonmagnetic substrate. A peculiar property of this film is that any domain that you put into this material spontaneously relaxes to a circular domain of a standard size that can be easily moved around, counted, or destroyed. Further, if you cut a bubble in half you get two full-size bubbles. Utilizing these attributes, sequences of bubbles (in which 1's and 0's are represented by the presence or absence of bubbles) are used to store information in the film, providing a sequential block storage device without the limitation of using mechanically moving parts.

Physically, the device, shown in Fig. 11, consists of a substrate coated with a thin garnet film (which contains the bubbles), a patterned conductor layer, and a patterned magnetic Permalloy layer, all separated by layers of silicon dioxide to provide electrical and physical isolation and optimum magnetic coupling (between the garnet and Permalloy layers). Mechanical and environmental protection is achieved with an overcoat of silicon dioxide, and electrical contact pads allow electrical control and input and output of the bit patterns. The final device, like a silicon chip, is a multipin package mounted onto a circuit board.

The bulk of the area of a magnetic bubble memory is dedicated to physically storing strings of bubbles (each about a micron in diameter), but an important fraction of the chip is dedicated to operating and controlling the device. Four basic functions are required in order to operate a magnetic bubble memory: propagation (access), generation (write), detection (read), and annihilation (read and erase). For some architectures, a method of

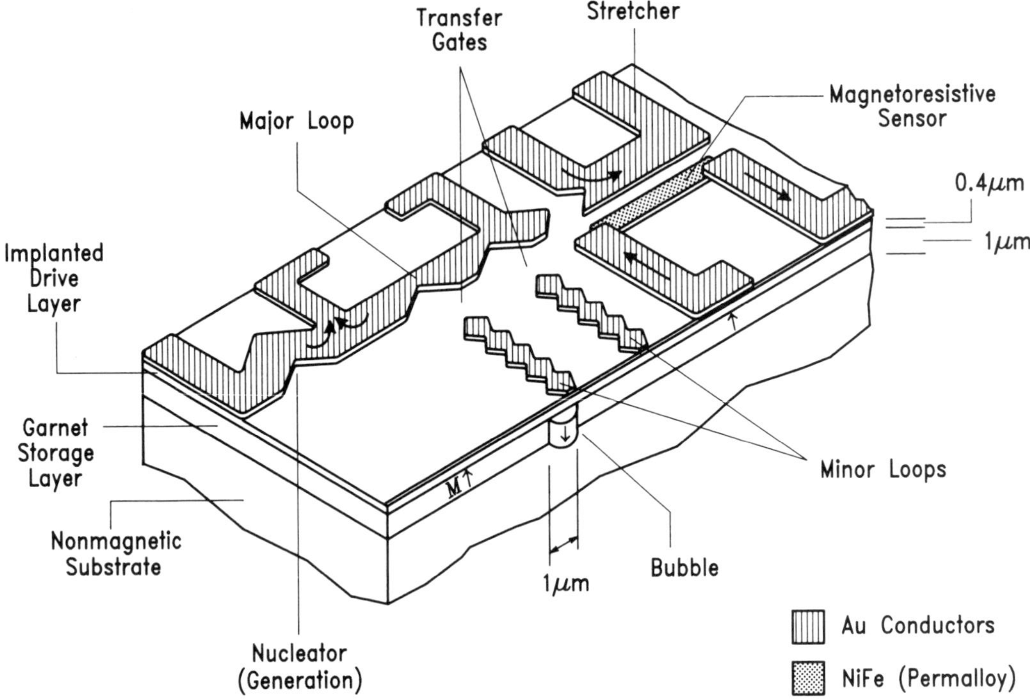

FIG. 11. Schematic diagram of one type of magnetic bubble memory device, showing major and minor loops for bubble storage and access, a nucleator for bubble generation, and a bubble stretcher and magnetoresistive sensor for bubble detection. (After Lin *et al.*, 1979.)

replicating bubble patterns is also necessary. These functions are accomplished through electrical control, using special patterning of the layers and materials properties (such as magnetoresistance, a change in resistance of a magnetic metal film depending on the magnetization).

Bubble memories are organized to provide many small sequences (minor loops) of bubbles accessed and controlled by one master sequence (major loop). Thus, when a desired piece of data is required; only the minor loop containing that data needs to be accessed. Bubble memories are used in calculators and other small-volume, nonvolatile memory applications. Difficulties in scaling bubbles to sizes below about 1 μm, materials challenges, bubble propagation speed limits, the rapid development of low-power and nonvolatile semiconductor devices, and the continued development of very small, high-performance magnetic storage devices all have limited the application area for magnetic bubble devices.

3.2 Magneto-optic and Optical Storage

Optical storage devices are in many ways similar to magnetic-disk files in that they have a specially designed head to write bits onto a rotating medium. Further, rotation provides circumferential tracks, and moving the head from track to track is accomplished with an electromechanical actuator. Common important attributes are linear (circumferential) bit density, data rate, access times to the first bit of information in the string of data, and the number of tracks per disk. Thus, the engineering requirements and progress described in Sec. 3.1.2 generally apply to this technology.

There are several key distinctions between magnetic and optical storage. First, the information is stored in a magneto-optic thin film. Second, transparent substrates are generally used, such as plastic and glass. Another distinction is that a focused laser beam is used to read and write information. This gives circular bits and thus comparable linear and track densities, requiring much greater tracking accuracy than disk drives. Also, very small, low-cost, reliable lasers and small, integrated focusing optics are essential. Today's optical storage heads generally require somewhat longer access times (due to the higher mass components), operate at a slightly lower data rate (constrained by laser power), and have larger head structures (influencing the number of disks possible in a stack) than magnetic-disk files. However, continued mass reduction and integration are expected.

An advantage of using a focused beam system, rather than operating in the near field as in magnetic recording, is that scaling to higher areal densities can be accomplished without bringing the optical head physically closer to the disk surface. Instead, the head-to-disk spacing is set by the optical focal length, generally many microns and much greater than the bit cell size. This substantially reduces concerns about head-disk interface wear and obviates the need for an air bearing to maintain accurately a small spacing between the head and disk. The larger spacing also allows the use of stamped grooves in the disk for simple tracking schemes and makes it possible to embed the information layer for optical storage into the disk below protective layers for robustness against handling damage and debris. These are among the properties that make magneto-optic and optical storage suited for very high-density removable-media storage devices.

Magneto-optic storage offers rewrite capability, infinite cyclability, and very fast reading and writing. However, erasing requires some time to achieve thermal equilibrium. This means that rewrite is done in two steps by first erasing and then writing the new information, requiring at least 2 disk revolutions. Also, a low-resolution magnet must be near the bit cell. Because this magnet takes space, it increases the height of the storage device relative to magnetic drives.

CD-ROM type storage allows low-cost media replication since the physical deformations that give rise to the optical contrast can be created by stamping with a master. However, like its (more costly) semiconductor analog described in Sec. 3.3.3, it is useful only for fixed data and program storage since it cannot be altered.

Phase-change storage has a limited cyclability and hence is used for write-once storage applications, analogous to (but much lower in cost than) PROM described in Sec. 3.3.4. Devices introduced in the mid-1980s offered 2 GB of storage per disk at 400 000 bits per square centimeter, 0.4-MB/s data rate, and 100-ms access time.

Continued extendability of optical and magneto-optical storage relies on reducing the bit size by focusing the laser beam to a tighter spot, generally by reducing the optical wavelength. In addition, direct overwrite techniques currently under research, which allow simultaneous erase and rewrite, will increase rewrite speed. Several generations of this type of improvement are expected before significant breakthroughs are required.

3.3 Charge Storage

Random-access memory is memory in which the access time is independent of the physical location of the data. It is a characteristic of all the device classes described below and offers significantly reduced access times compared to serial-access memory such as disk files, tape, or magnetic bubbles. Further, charge storage bits are accessed electrically, avoiding delays associated with the mechanical motion in disk files and tape and the physical motion in bubbles.

Fundamental to these devices is the use of a matrix of identical memory cells addressed by word and bit lines. Each cell contains one (or more) transistors arranged in a circuit, as discussed in Sec. 2.3. The CMOS offers the smallest area and power per bit and is the basis for most of today's devices.

Also fundamental to these devices is wafer fabrication in which many identical memory chips are created at the same time on a single silicon wafer. The silicon wafer (usually 20-cm diameter) is processed by superimposing several layers of conducting, insulating, and semiconducting materials. Each layer is patterned by means of masks and photolithographic processes to create the desired features. The wafer is then physically sliced up into several hundred identical chips. Each chip holds several million identical storage bit cells, connected together with thin-film wiring in an addressable matrix. Each bit cell is comprised of one to several transistors and possibly other integrated components.

Smaller bit cell sizes mean that more bits can be stored on a chip, generally lowering cost per bit. Thus, enormous effort is aimed at the physics and engineering of reducing transistor and cell areas. For instance, as cell sizes shrink, mask dimensions must be reduced to include linewidths well below a micron. Linewidths as low as 0.25 μm appear feasible with use of short-wavelength light and clever mask design. Mask dimensions substantially below 1 μm can be envisioned using x rays for patterning, and research is aimed at this approach. Also, the cell itself can be made smaller by orienting large-area capacitors vertically with respect to the rest of the films.

In addition to reducing bit cell area, desired directions of device improvement include lower power dissipation per bit, increasing speeds (faster bit rates, shorter access times, and shorter cycle times), and improving reliability. Nondestructive readout (NDRO) and nonvolatility are also important goals. No one device excels in all these parameters, giving rise to the variety of structures described below.

3.3.1 Static Random-Access Memory Static RAM exhibits the property that the state of the cell is permanent unless rewritten, as long as the device power remains on. Hence it is static, but volatile. Static RAM is high speed but requires six transistors and resistors per cell, resulting in a large cell area and, thus, more cost per bit than many other RAM options.

A schematic of an SRAM cell is shown in Fig. 12. In this flip-flop circuit, when one cross-coupled MOSFET transistor is on, the other is off. Appropriate voltage inputs to the word and bit lines determine which transistor is turned on, making this bistable circuit into an addressable memory cell.

Bipolar SRAM is the oldest integrated memory circuit technology and dominated the 1960s and early 1970s, but was outpaced first by NMOS (*n*-channel MOS) and then by CMOS as the bipolar cell area and power consumption limited chip capacities and kept the cost high. The fastest RAM memories today are CMOS SRAM, with an 8-ns access time. However, the large cell size relative to DRAM (typically 4×) limits the use of CMOS SRAM to applications in which the faster cycle times [typically (5 to 10)×] are so important that the cost premium is justified.

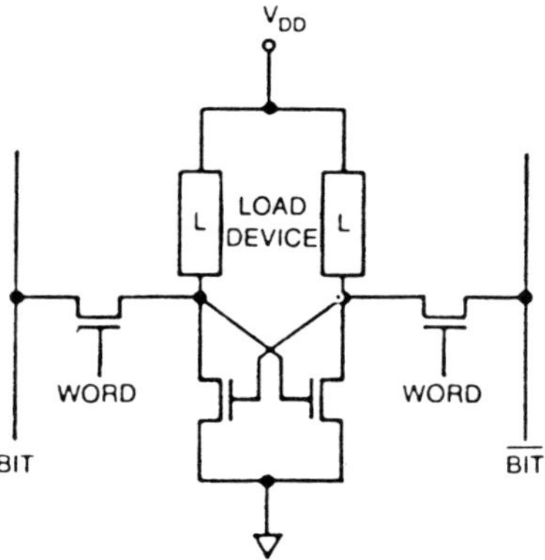

FIG. 12. Static RAM cell, based on a flip-flop circuit. The load device may be resistors or transistors and is usually transistors to minimize cell area. (Weste and Eshraghian, 1985, © 1985 by AT&T Bell Laboratories, Inc. Reprinted by permission of Addison-Wesley Publishing Company.)

3.3.2 Dynamic Random-Access Memory Dynamic RAM exhibits the property that the memory cell must be refreshed, or rewritten, every few milliseconds to restore the charge that has leaked from the cell's capacitor. While this requires added complexity in the chip's control circuitry and adds to the access and cycle times, it also makes possible the use of very compact bit cells. This in turn gives larger capacity chips at a lower cost per bit.

A DRAM cell schematic is shown in Fig. 13. Information is stored as a charge across a single capacitor. The storage capacitance must be kept low enough to minimize the area occupied on the wafer. However, the smaller the capacitance, the more quickly the cell discharges, requiring refreshing. Thus, de-

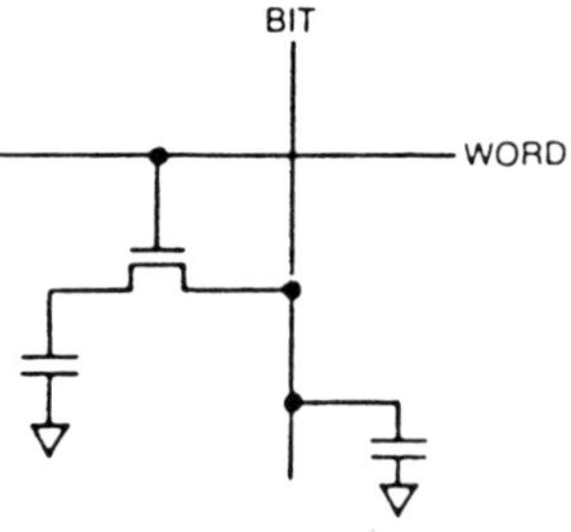

FIG. 13. Dynamic RAM cell using a single transistor. The separate capacitor shown represents the parasitic capacitances that must be included for proper device design. (Weste and Eshraghian, 1985, © 1985 by AT&T Bell Laboratories, Inc. Reprinted by permission of Addison-Wesley Publishing Company.)

sign efforts are focused on maximizing the storage capacitance within the area limits of the cell. Current designs require rewriting every 64 ms, giving a 98 to 99% availability of the cell for reading and writing.

The very small cell size, low cost, and good performance make DRAM the dominant random-access memory choice today. For instance, typical 4-Mb chips have access speeds of about 60 ns, and 16-Mb chips have access speeds of about 50 ns, at prices ranging from $15 to $50 per MB. Technology demonstrations give a high level of confidence that the historical improvement of on average 4× density increase every three years since 1965 (with associated cost reduction and speed improvement) will continue for several generations, although there is some slowdown in the rate of improvement.

3.3.3 Read-Only Memory Read-only memory differs fundamentally from other semiconductor memories in that its stored information is fixed, unalterable, and nonvolatile. A simple form of ROM is shown in Fig. 14. The state of each (i,j) position is fixed at the time of fabrication by the presence or absence of a transistor at that position. Practically, this means that ROM cells may be fabricated with only one transistor per bit of storage, although any updates to the content require another unique chip.

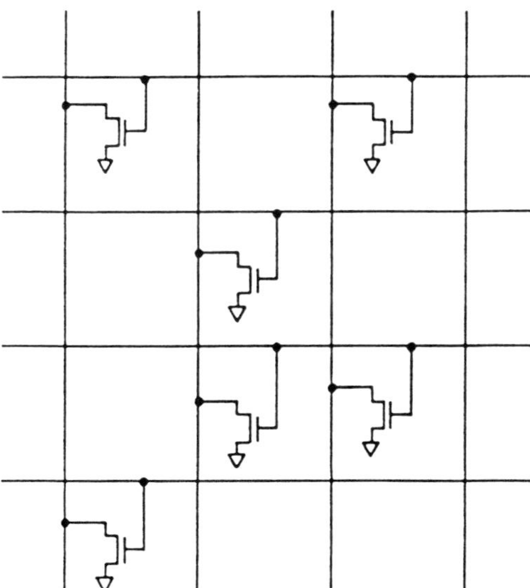

FIG. 14. ROM layout showing 1's and 0's represented by the presence or absence of an active transistor at matrix positions, requiring unique patterns for each distinct memory device. (Weste and Eshraghian, 1985, © 1985 by AT&T Bell Laboratories, Inc. Reprinted by permission of Addison-Wesley Publishing Company.)

ROM devices are fabricated like most semiconductor memories with standard materials and processing technology, except that unique personalization is required in one or more masks during fabrication to create the desired unique bit pattern. Techniques used to program the masks include contact programming, presence or absence of a transistor, or a special doping to turn the desired transistors permanently on or off.

ROM is used for fixed-program storage and offers a convenient way to "boot up" or initialize computer systems. It is very rugged, since the program is a part of the device itself. Its primary disadvantage is that any change, however small, requires an entirely new ROM to be designed and produced.

3.3.4 Programmable Read-Only Memory Programmable read-only memory offers fixed, nonvolatile storage that is set by the user as a one-time operation. This eliminates the need for personalization as the device is fabricated but requires special programming processes at the user level. This also requires that each (i,j) memory will be potentially either a 1 or a 0, making the area per unit of storage somewhat larger.

Programming the PROM is called "burning" since physically the process is to open fuse links at locations where a binary 1 is desired, leaving all 0's intact. Burning the fuses is accomplished by applying a high-energy pulse to the desired addresses of the 1's. This high-energy pulse alters the properties of a transistor memory cell so as to create the 1 state. PROMs are somewhat more expensive than ROMs but give useful added flexibility for the user.

Erasable PROM (EPROM) overcomes the primary disadvantage associated with ROM and PROM; that is, it can be erased and rewritten up to thousands of times by the user. To program an EPROM, the user selects the desired address, places the desired bit pattern of 1's and 0's on the data line, and pulses the PROGRAM pin. Charge is physically stored on the floating gate of a MOSFET transistor, described in more detail in Sec. 3.3.5. To erase an EPROM, the trapped charges are removed by exposing the chip to ultraviolet light through a transparent quartz window, exciting holes and electrons. This allows the trapped charge to escape over the potential barrier, erasing all stored data on the chip at once. While the erasability is a large advantage over ROM and PROM, an important limitation is that the chip itself must be physically removed from the system for erasing and reprogramming.

3.3.5 Electrically Erasable PROM An important class of EPROMs are those that are electrically erasable. The low cost and ease of erasability of this nonvolatile RAM offer significant advantages for program and data storage.

The floating gate memory cell, the basis for EEPROMs and EPROMs, contains an FET whose gate is electrically floating. Charge on the floating gate modulates the charge in the channel below the gate. The state of the cell is sensed by measuring the channel conductivity, typically in about 80 ns, and inferring from this the state of the charge on the floating gate.

A typical structure is shown in Fig. 15. To write, charges are injected from the silicon through the insulator onto the floating gate. This is generally done by putting a large bias voltage, V_D, between the source and drain, then switching the transistor on by use of the control gate (labeled V_G in Fig. 15). The large bias causes a flow of very energetic electrons between the source and drain in the channel underneath the floating gate. About 1 of every 1000 of these "hot" electrons is injected onto the floating gate and is trapped there, taking about 10 to 100 μs to establish enough charge for device use. This net charge on the floating gate, with proper design, can be retained for many years.

The charge on the floating gate, once established, can be removed via a tunneling mechanism, thus erasing the data. A large electric field is applied between the source and control gate, causing the electrons trapped on the floating gate to "tunnel" back into the silicon. The thinner the oxide, the less is the voltage required for this process, since voltage is the product of electric field and thickness. Typical oxide thicknesses are 10 nm. Enriching the oxide with silicon or using nitride layers also helps reduce the required erase field. Erase times are typically 5 to 20 ms, and thus it is a very attractive feature of this technology that an entire block or an entire chip may be erased simultaneously. EEPROMs with this block-erase characteristic are called Flash EEPROMs.

The large fields and energetic electrons required for the write and erase operations in Flash cause, over repeated cycles, a change in the insulator properties. Charges and traps build up in the insulators and at surfaces, resulting in inefficient injection and tunneling. Details of the cell design, materials, and operation mode all affect the endurance, but typical devices last only up to 10^6 cycles. EEPROMs are configured in RAMs in the same manner as described in Sec. 3.3.

The cell size of Flash EEPROMs can be very small, since there is no need for a separate capacitor. With a 0.7-μm minimum feature size, 16-Mb chips costing about $20/MB have been shipped. The random-access capability allows access times of 15 μs, 1000 times faster than magnetic direct-access storage device and more than 100 times slower than DRAM, giving Flash EEPROMs an intermediate performance. The electrical erasability makes possible field updates of chips used in pro-

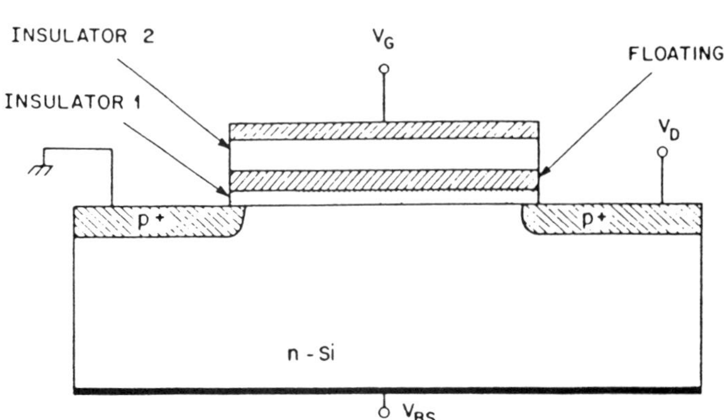

FIG. 15. EEPROM structure illustrating the floating gate that controls injection and removal of trapped charge for nonvolatile storage. (Sze, 1981, copyright © 1981 by John Wiley and Sons, Inc. Reprinted by permission of John Wiley and Sons, Inc.)

gram storage applications. These attributes make Flash EEPROMs very useful for small–form-factor, very robust, modest storage-volume applications such as laptop and palmtop portable computers.

3.4 Ferroelectric Storage

Ferroelectric RAM is fabricated with a material such as lead zirconate titanate (PZT) as the dielectric in the capacitor of an otherwise conventional DRAM-integrated memory cell, as shown in Fig. 16. Since the polarization state of a ferroelectric is permanent until reset, even in the absence of power, this provides nonvolatile random-access storage in a compact (single transistor) cell structure.

Physically, the device is created by standard semiconductor processes to create CMOS transistors. The ferroelectric thin-film capacitors are built on top of each transistor. The polarization state of this capacitor is used as nonvolatile storage in a manner analogous to the charge state of conventional capacitors in DRAM.

Commercial devices became available around 1990, although the concepts of ferroelectric storage have been known since the 1950s. Significant engineering and materials challenges were overcome to make practical devices, and densities are still very low compared to DRAM. For instance, only recently was it possible to fabricate ferroelectric thin films of sufficient quality for use in these devices, allowing lower-voltage operation and the high level of integration necessary for acceptable costs. Film material uniformity, process temperatures, thickness, cyclability, and electrode material compatibility are some of the specific challenges. The cyclability of about 10^6 still limits performance and con-

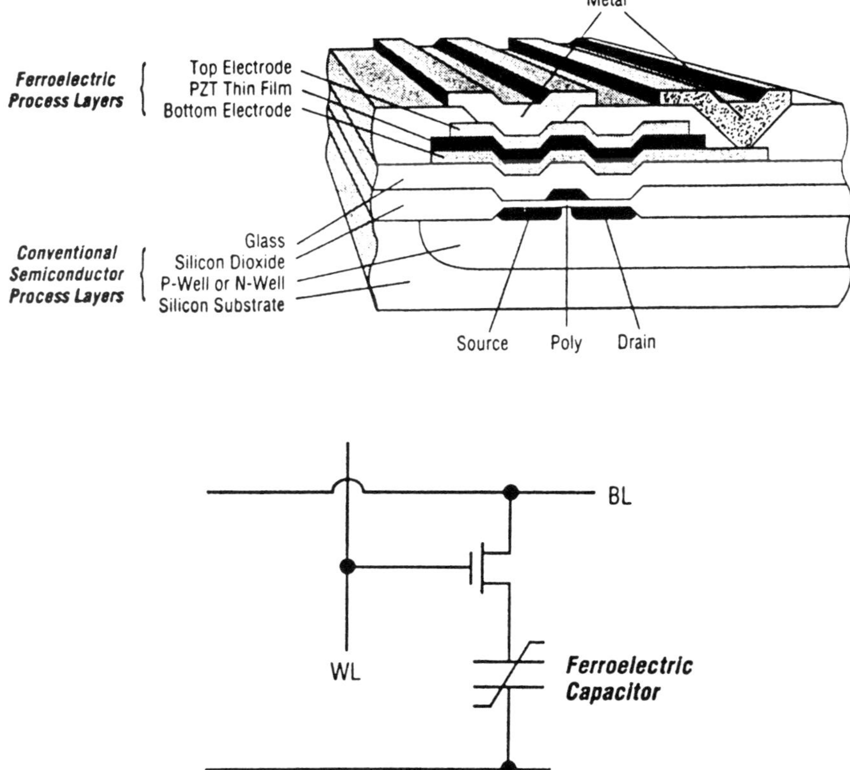

FIG. 16. Ferroelectric memory cell showing (a) the ferroelectric capacitor on conventional DRAM to provide nonvolatile storage and (b) the circuit representation of this structure. (Illustration © Ramtron International Corporation, 1993.)

strains the applications, but the demand for truly nonvolatile storage continues to motivate further development.

4. APPLICATIONS

In today's computers, many memory and storage technologies are utilized, organized into a hierarchy as shown in Fig. 17. The goal of using a hierarchy is to optimize system-level cost performance. This is largely because the fastest technologies are typically the costliest, with slower alternatives available at orders of magnitude less cost per bit. Judicious use of high-performance, high-cost memory fed by increasing amounts of lower-cost, lower-performance technologies can give optimum cost performance. In addition, nonvolatility and low power per bit become increasingly important for large volumes of data, and these have usually been characteristics of lower-cost storage devices.

In particular, at the top level in the hierarchy are the processor's internal registers. These are very fast since they reside on the processor chip itself, but chip area severely limits their number. At the second level is a memory running in synchrony with the processor, called cache. The speed requirement necessitates SRAM, and, thus, the size is constrained by the high cost of this memory. Information most likely to be used by the processor is kept in cache, while the rest is stored in lower levels of the hierarchy. The third level is main memory, generally comprised of DRAM, which is about ten times slower than SRAM but five to ten times less costly. Programs and data are kept in this memory. Through the third level in the hierarchy, memory, that is, random-access technologies, are used.

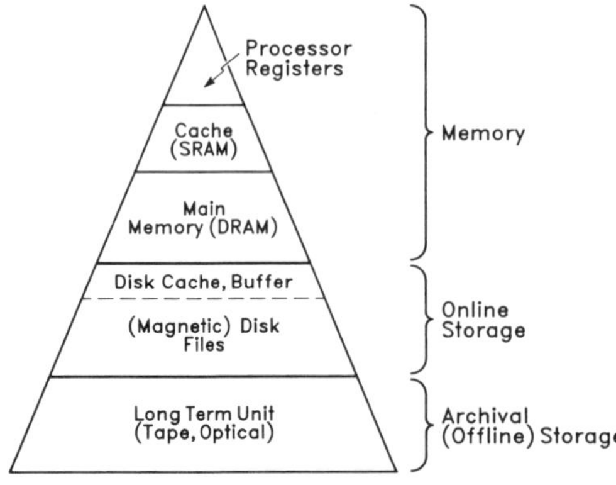

FIG. 17. Computer memory hierarchy.

The fourth level is mass storage, generally disk files. These are serial block-storage devices, connected to the RAM by special channels that optimize availability of the data in the presence of mechanical access delays. Typically, tens of milliseconds are required for access, but this provides "on-line" nonvolatile storage at a cost of \$1 to \$5 per MB. Even larger volumes of data, system backup information, and archival files are stored "off-line," on tapes (at costs as low as \$1/MB) or on optical storage disks that have to be physically mounted onto the drive when information is requested. This can amount to petabytes in some applications and is often physically stored elsewhere as insurance against disaster.

The very large performance gap between semiconductor and magnetic memory requires a sophisticated link between the two to hold blocks of data temporarily and schedule requests for information flow between main memory and magnetic storage. Special memories are used to facilitate this process, known as buffers and cache. Because it is important to retain this information if power is lost, this is an important application for battery-backed DRAM and the emerging nonvolatile RAM technologies discussed in Sec. 3.3.

New solutions are also in sight for the performance gap between on-line and off-line storage. An area of growth is "near-line" storage, that is, storage of large volumes of seldom used data in "juke box" type tape or optical files for access in a few seconds. This is likely to be of increasing significance as high-volume data storage applications of video, high-definition television, and computer-aided design continue to grow.

Steady cost and size reductions of memory technologies have had a profound influence on the numbers and types of memories available. No longer is computer memory confined to specially protected facilities servicing large mainframe computing devices, so-called "glass houses." In fact, the fastest growth areas are in small computers and the mobile computer market, necessitating compact, robust, low-power, nonvolatile storage. Flash EEPROM and small, highly shock-tolerant disk files are particularly attractive here. While the cost per bit of semiconductor storage has historically been orders of magnitude greater than that of magnetic storage, that gap has steadily narrowed, and for the small volumes of storage (a few megabytes) required for many applications, the cost of Flash EEPROM is acceptable.

Trade-offs in cost and performance and new uses of computing have played a key role in the diversification of today's memory technologies. With future technology improvements and as the volumes of memory continue to grow, a rich mix of alternatives, complementary and competitive, will continue to evolve.

GLOSSARY

Areal Density: The number of data bits that can be stored per unit area on a thin-film storage medium.

Coercive Field, Coercivity: The reverse applied magnetic or electric field at which the material's magnetization or polarization, respectively, can be reduced to zero.

Crystal Phase: An organization of atoms into a specific geometric structure. Often the same atoms organize in a different structure (phase) at different temperatures or under different process conditions.

Cyclability: The number of times a memory can be erased and rewritten before it can no longer be used.

Dielectric: An electrically nonconducting material.

Domains: Regions within a material in which electric polarization or magnetic spins align in the same direction.

Ferrite Cores: Rings of ceramic magnetic material.

Ferroelectric: A material exhibiting an electric polarization even in the absence of an applied electric field.

Ferromagnetic: A material exhibiting a magnetic moment even in the absence of an applied magnetic field.

Holes: Vacant states in an otherwise filled electron band in a crystal, which can be represented as a positive charge.

Hysteresis: A property of a material that exhibits different behavior for increasing and decreasing applied fields.

Magnetic Bubble: A cylindrical domain in a thin magnetic film that is magnetized oppositely and vertically with respect to the overall film.

Magneto-optic: A coupling between magnetic and optical properties of a material.

Magnetoresistance: A change in electrical resistance of a material depending on the applied magnetic field.

Masks: Patterned glass or metal plates used to create desired features on a thin-film fabricated structure. Light travels through transparent sections and is absorbed or reflected by opaque sections.

Media: The material in which data bits are physically stored.

Near Field: A region of a wave very near the source compared to a wavelength.

Photolithography: The process by which images are transferred onto a structure. A photosensitive layer called photoresist is coated onto the surface. Masks are used to expose selected regions of the resist and then developed so that the desired pattern of resist remains. Layers are deposited or removed using this photoresist template.

Remanence: The residual magnetization of a material after the applied field is removed.

Substrate: A base material onto which thin films are deposited and patterned appropriately to make devices.

Works Cited

Lin, Y. S., Almasi, G. S., Keefe, G. E., Pugh, E. W. (1979), "Self-Aligned Contiguous-Disk Chip Using 1 μm Bubbles and Charged Wall Functions," *IEEE Trans. Magn.* **MAG 15**, 1642–1647.

Mallinson, J. C. (1990), "Achievements in Rotary Head Magnetic Recording," *Proc. IEEE* **78**, 1004–1016.

Mee, C. D., Daniel, E. D. (1987), *Magnetic Recording, Volume I: Technology*, New York: McGraw-Hill, p. 5.

Mee, C. D., Daniel, E. D. (1988), *Magnetic Recording, Volume II: Computer Data Storage*, New York: McGraw-Hill, p. 6.

Sze, S. M. (1981), *Physics of Semiconductor Devices*, 2nd ed., New York: Wiley, p. 496.

Weste, N., Eshraghian, K. (1985), *Principles of CMOS VLSI Design*, Reading, MA: Addison-Wesley, pp. 6, 349, 353, 355.

Further Reading

Basche, C. J., Lyle, R. J., Palmer, J. H., Pugh, E. W. (1986), *IBM's Early Computers*, Cambridge, MA: The MIT Press.

Bobeck, A. H., Scovil, H. E. D. (1971), *Sci. Am.*, **224** (6), p. 78–90.

Luecke, G., Mize, J. P., Carr, W. N. (1973), *Semiconductor Memory Design and Application*, London: McGraw-Hill.

Mallinson, J. C. (1987), *The Foundations of Magnetic Recording*, New York: Academic.

Mee, C. D., Daniel, E. D. (1987), *Magnetic Recording*, New York: McGraw-Hill, Vols. I, II, and III.

Sze, S. M. (1981), *Physics of Semiconductor Devices*, 2nd ed., New York: Wiley.

Watson, J. K. (1980), *Applications of Magnetism*, New York: Wiley.

Weste, N., Eshraghian, K. (1985), *Principles of CMOS VLSI Design*, Reading, MA: Addison-Wesley.

MESOSCOPIC SYSTEMS

YOSHIMASA MURAYAMA, *Advanced Research Laboratory, Hitachi, Ltd., Hatoyama, Saitama, Japan*

INTRODUCTION

Throughout the history of electronics, we have pursued smaller cellular devices in, e.g., integrated circuits, which means better performance per cost. Most people on the engineering side vaguely conjectured that these tiny devices should be ruled by quantum physics. However, they hardly recognized that these devices form a bridge between a bulk system governed by classical physics and a system governed by quantum physics, and consequently, they open up a new front of physics. This new physical world, neither classical nor quantum, is named *mesoscopic* physics, after the Greek word *μεσος* meaning "middle." The mesoscopic world has many partially coherent phenomena, which seem to be successfully described by introducing *decoherence* mechanisms into quantum mechanics.

3-527-28132-0/94/$5.00 + .50

1. BACKGROUND

1.1 Quantum Effects

During the revolutionary two decades of the discovery of quantum mechanics, de Broglie proposed the ingenious idea that massive particles are also waves (see QUANTUM MECHANICS). This idea was soon experimentally confirmed using electron beams that can really interfere. As far as an electron in a vacuum is concerned, such a free particle behaving like a free wave can be simulated as an actual wave like a light wave, i.e., $e^{-i\omega t+ikx}$, if its frequency ω and wave vector $\mathbf{k}$ are identified with $p^2/2\hbar m$ and $p/\hbar$ ($\hbar$ is the Planck constant/2π; p, momentum). This is true when the *coherence length* of the electron beam is long enough. Otherwise, it is common to describe the beam as a *wave packet* with a specific coherence length, in other words, a span in real space.

On the other hand, in crystals, the energy-momentum relation is not as simple as in a vacuum, since a series of periodic potentials strongly modulates the plane-wave-like nature. In particular, electron wave states specified by energies around $\hbar^2\pi^2/2ma^2$ (a is the lattice constant) interfere destructively and, consequently, become extinguishing waves. This phenomenon is essentially the same as Bragg reflection, and the energies for which waves extinguish are known to be within a certain energy gap. In ordinary solids the separation between successive bands does not tend to zero at the Brillouin-zone (BZ) boundary, and the minimum separation is called the energy gap (see ELECTRON STRUCTURE OF SOLIDS).

For this kind of complicated dispersion between energy and momentum, a renormalized mass can be introduced, which is ordinarily called the *effective mass m*. Electronics engineers, especially semiconductor device engineers, frequently view electrons as billiard balls with this effective mass. This is because the concept of effective mass has been very successfully applied to analyzing the properties of solid states in those devices. Here, the momentum $\mathbf{p} = m\mathbf{v}$ should be replaced by $\hbar\mathbf{k}$, since $\mathbf{k}$ describes a state better than $\mathbf{v}$.

If we use the effective mass m instead of the bare mass m_0 of the elementary particle, most properties look as if electrons are classical particles with their modified mass. In fact, the Molecular Dynamics simulation of devices was based on this transinterpretation.

However, it has been well recognized that elementary quantum processes, such as photon absorption and electron-phonon scattering, should be treated by calculating the scattering cross section σ or the scattering time τ, which means that each process occurs not *deterministically* but *probabilistically* following quantum mechanical rules. Engineers used to describe electronic properties in terms of classical parameters such as effective mass and σ or τ, to avoid recourse to quantum mechanics. Thus, in engineering, quantum effects have been laid aside so far, at least apparently. However, in the world of electronics, there began to appear quantum devices such as *superlattices* (SLs), *quantum wells* (QWs), and resonant tunneling devices (RTD), since the beginning of the 1970s. These are recognized as explicitly showing quantum effects.

In optics, to our surprise, the most quantumlike device, the laser, can be described mostly in terms of a classical monochromatic light source. A pure quantum state behaves like a classical wave. This is because a laser has extremely long coherence length and is thus similar to a classical plane wave. In this sense, quantum effects in optics seem to appear only in nonlinear effects or fluctuations in a vacuum. A similar situation occurs for macroscopic quantum effects such as in superfluids and superconductors. Inside superconductors, the coherence length covers the whole sample so that their states are conventionally described by a macrowave. The easiest example to illustrate quantum effects is size quantization; i.e., intrinsic Bloch states must be requantized according to additionally imposed boundary conditions, because the size of concern is small enough to allow *ad hoc* quantization in ordinary mesoscopic systems. Thus, there appear new quantum effects in addition to the above-mentioned implicit quantum effects. These new quantum effects are concerned in the present mesoscopic problems.

1.2 History

When discussing quantum effects in electronic devices, we might use the term "quantum electronics." However, for historical reasons, quantum electronics has usually been used to mean electronics using lasers and related optical devices. Thus, quantum electronics can no longer be used for purely electronic quantum devices.

To avoid confusion, the present author proposed the name "quantum microelectronics" (Murayama, 1984). However, a better name, "mesoscopic physics," became popular after it was proposed by van Kampen (1976). This terminology describes a broader scope of physics.

Currently, mesoscopic physics has attracted much attention, since such small electronic devices have been widely fabricated, opening a new physical world, not only in electronics and optics but in other fields where the importance of novel physical phenomena has been more and more recognized.

2. QUANTUM VS CLASSICAL PHYSICS

2.1 Wave-Particle Duality

Let us first consider what scattering process means in quantum mechanics. Consider a wide-spread plane wave of light before the physical process of scattering takes place. Then, imagine an atom of dimensions on the order of several angstrom that can absorb the light wave with a wavelength on the order of 5000 Å. A quantum absorption process occurs, as is easily imagined, only on the portion of the amplitude of the light wave that has the same spatial extension as the atom. At the moment of absorption, the light wave must behave like a particle, say, a photon, since the absorption process occurs on an all-or-nothing basis. This nature is a manifestation of the *particle-wave duality* of a quantum mechanical entity. The process occurs *probabilistically*, depending on the amplitude of the light wave distributed over the atomic scale. The cross section is proportional to the square of the product of these terms: the extension of the atom and the amplitude of the original light wave. When this probability is sufficiently small, most photons pass through without being absorbed.

This fact should be interpreted as follows. The entity to be absorbed is *not* a part of the light wave, but an entire photon. In other words, absorption occurs only when the photon behaves as if it were atom-sized. Here, the notion of a wave is nothing but a wave representing a distribution of the probability amplitude of the existence of photons. The probability distribution is wavelike, as it can be specified by its amplitude and phase.

From an early stage in the discovery of quantum mechanics, it has been recognized that the rules governing quantum mechanics, like the Schrödinger equation, are linear and can be *superposed*. Superposition means that two quantum mechanical states (or waves), defined by ψ_1 and ψ_2, must have a physical, not mathematical, meaning superposed as $\psi_1 + \psi_2$. Actually, there should exist a superposer, which can exert a physical action on both states so as to superpose them. For example, in electron interferometry with an Aharonov–Bohm geometry, two particle waves of an electron hit a fluorescence screen, where they are superposed by interacting with atoms inside the screen, which works simultaneously as a detector.

According to the experiment cited in Fig. 1, the interference pattern is the pileup of the records of individual fluorescence spots. Each spot shows an elementary interaction between a superposed electron and a fluorescent atom. Thus, as long as the quantum states of electrons emitted from a source are well-defined, i.e., they have a similar energy within a very narrow energy spectrum and the source is pointlike, then the quantum behavior of all the electrons is also well-defined, although it is never defined where on the screen each electron jumps to, but only the probability of spot distribution is determined by the squared wave function. This eventually causes the well-known sinusoidal interference pattern. The same sort of experiment was also performed using photons (Tsuchiya *et al.*, 1985).

Waves interfere only when both the waves are *coherent*. If both are *incoherent*, their superposition can never give rise to interference, even if they reach a small region simultaneously, where some superposition mechanism should work.

Next, we will see that it is not *size* that differentiates a macroscopic system from a microscopic one. According to the correspondence principle by Niels Bohr, any microscopic state has a corresponding entity in classical physics, especially states with large quantum numbers (*correspondence principle*). Take a hydrogen atom, for instance; the states with sufficiently large principal quantum

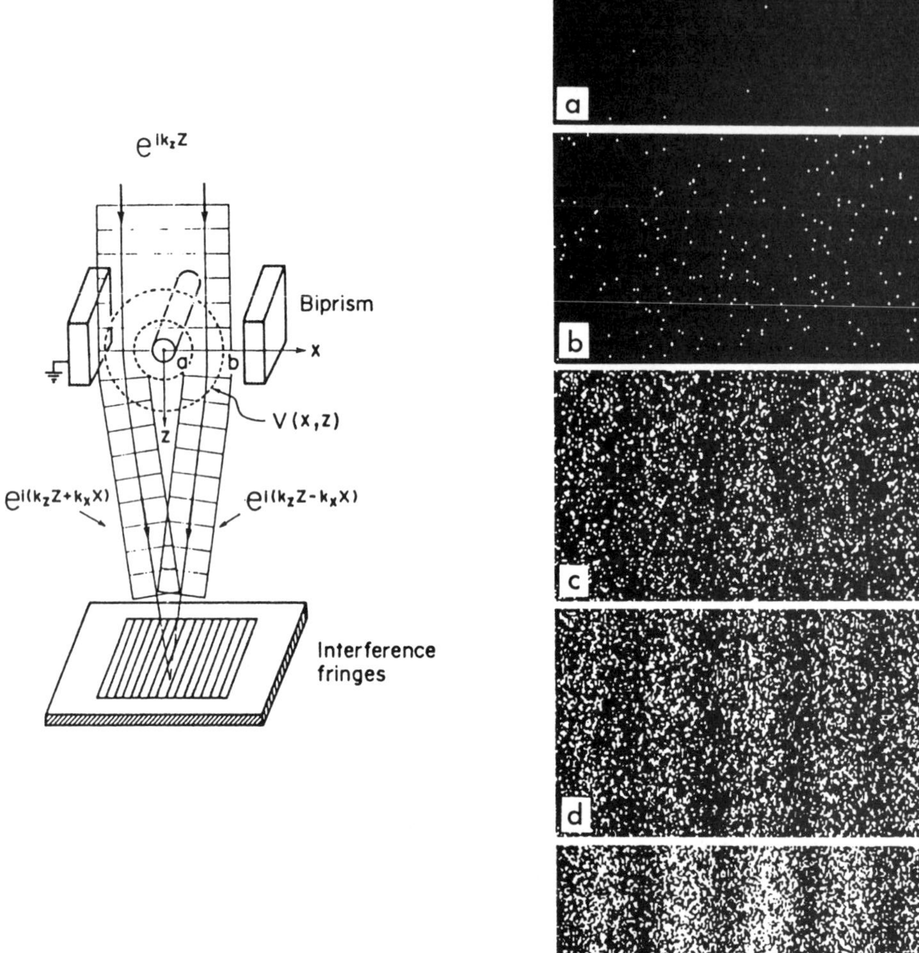

FIG. 1. Superposition of divided electron plane waves on a fluorescence plate, where an interference fringe is formed. Building of the interference pattern is shown progressively for number of electrons equal to (a) 10, (b) 100, (c) 3000, (d) 20 000, and (e) 70 000 (after Tonomura *et al.*, 1989).

numbers n are close to each other and look as if they were continuously distributed. For example, it is known that so-called Rydberg states in hydrogen atoms have orbitals larger than $\sim 1\ \mu m$ for $n \gtrsim 100$ (orbital radius $\propto n^2$).

In this context, it can be said that the boundary between macroscopic and microscopic world is not concerned with *size*. It is *coherence*. It is well established by now that the classical system is the limiting case of a quantum system when the number of degrees of freedom and/or the quantum number $\rightarrow \infty$. The former is intrinsically included in the latter but appears to be a different world because of, e.g., innumerable degrees of freedom. Then, what brings a microscopic system into a macroscopic one? It is not system size, as we have shown. It is what makes a *coherent* state *incoherent*. Any *decohering* mechanisms are thus the subjects of interest.

2.2 Coherence

Figure 2 shows the concept of a localized state in solids, after Bergmann (1983). This kind of localization is known as Anderson localization (see ELECTRON STATES, LOCALIZED), named after the first paper on the topic by Anderson (1958). In solids with disorders, an electron is scattered *elastically* many times by these disorders and eventually returns to its original position. When a wave is superposed with itself after an itinerary, it must have a peak intensity at a specific site on the path where the phase relation is *constructive* between itself and the adjoint after itinerary. This resembles an interference experiment in an electron microscope (Fig. 1). This state is well defined in the sense that it is stable and long lasting, unless it obtains enough energy to be activated from a localized to a delocalized state. The activation is only possible through *inelastic scattering* by optic phonons, photons, plasmons and so on.

Historically, it had been *a priori* presumed that, even in very small devices, we cannot directly observe coherent phenomena, since there are lots of elastic scatterers in solids and the product $k_F l_{el}$ rarely exceeds unity (l_{el} is the mean free path under elastic scatterings; k_F, the Fermi momentum/$\hbar$). However, this

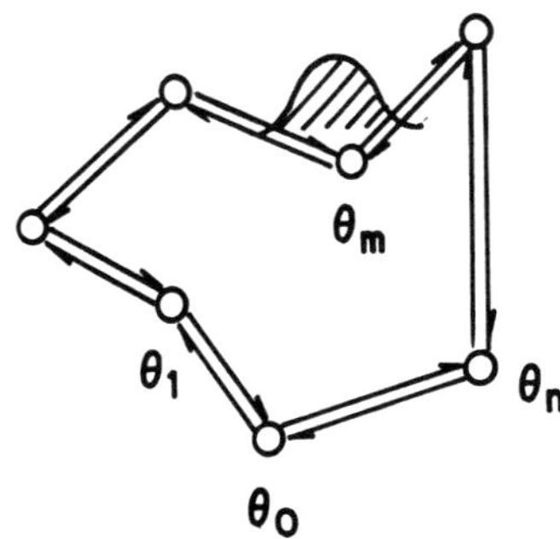

clockwise and anti-clockwise waves interfere constructively at site m such that

$$\sum_{i=1}^{m} \theta_i \approx \sum_{i=m}^{n} \theta_i$$

FIG. 2. Conceptual description of the Anderson localized state. Each node represents a defect, causing an elastic scattering (after Bergmann, 1983).

presumption was proved not to be the case. As is suggested in Fig. 2, elastic scatterings do not reduce coherence. Elastic scattering causes only a phase change.

On the other hand, *inelastic* scatterings change the energy from the initial value, which reduces the coherence. This is *decoherence*. After physicists realized this fact, many phenomena investigated so far in small electronic devices were recognized to show at least a small fraction of coherence, especially at lower temperatures with few inelastic scattering events. Thus, the above-mentioned criterion should be replaced by $k_F l_{in} \gtrsim 1$, although $k_F l_{el} \ll 1$, where l_{in} means the mean free path under inelastic scatterings, which ordinarily equals the phase-breaking mean free path l_ϕ. In the time domain, the scattering time for inelastic scatterings, τ_{in} or τ_ϕ, is ordinarily used, the latter of which is called the phase-breaking time.

Understanding this is very important from the viewpoint of quantum measurement theory. It is *a priori* assumed that, in order to acquire information from a detection, the detection must be an inelastic process. This is also wrong. If we can somehow determine a phase shift using a detector, then this is a kind of measurement. To detect a phase change, we need lots of similar events, not a single event, because of the uncertainty between phase and the number of events of concern. In fact, we cannot deduce anything from a single event on the fluorescence screen shown in Fig. 1, because a small number of spots can never make an interference pattern, and consequently, no information about phase is obtained from such a sparsely spotted pattern. Elastic scattering is actually a limiting case of inelastic scattering with an infinitesimal energy transfer.

A well-known difficulty in quantum measurement theory is in how a microscopic system evolves into a macroscopic one at the moment of detection. Thus, unless an inelastic process exists, the detection itself is elastic and the change caused by the process is merely a phase shift, and hence, the whole system remains in a quantum mechanical microscopic state, as was discussed by Wigner (1963). Even in this case, we can detect the change in phase by means of an ensemble average of similar events (Machida and Namiki, 1980; Murayama, 1990).

2.3 Quantization

In a periodic lattice, an extended wave is quantized so that the wave vector **k** is a good quantum number according to the Bloch theorem. Here, the minimum unit of periodicity is the lattice constant **a**. In order to form an extended wave over the whole crystal, every atomic orbital should be coherently superposed. The disturbance that hinders this superposition is, as was discussed before, inelastic scatterings. Inelastic and quasielastic scatterings such as optic- and acoustic-phonon interaction reduce the periodicity and, hence, invalidate the quantization condition claimed by the Bloch theorem. Ordinarily, a well-defined **k** value is thought to smear out according to this disturbance on the lattice points. Let us express the displacement of lattice points by $x_j = x_{j0} + \delta x_j$ (x_{j0}: regular lattice points at the jth site); then

$$\langle e^{ikx_j}\rangle_{av} \approx \exp(ikx_{j0})\exp(-\tfrac{1}{2}k^2\langle\delta x_j\rangle^2),$$

the second exponential of which is known as the Debye–Waller factor. In order that a state is well-quantized within a certain space (e.g., a box), it is necessary that $\frac{1}{2}k^2\langle\delta x_j\rangle^2 \ll 1$, if the decoherence mechanism is assumed to be electron-phonon interaction.

Again consider a box with waves inside. To obtain well-quantized states in the box, the waves on the two sides must be coherent; in other words, the information on the left extreme must be conveyed accurately to the right. This is a necessary condition for *size quantization*. The state in a solid is composed of Bloch waves with the lattice periodicity, the envelope of which is further modified according to the size quantization specified by the boundary condition of the box of concern. The latter modification according to the boundary condition of the box is called an *envelope function*. Ordinarily, the de Broglie wavelength of the envelope function is much longer than that of the Bloch wave on the order of lattice spacing. Accordingly, the two quantizations are dealt with separately; one is for the Bloch quantization, and the other is for the size quantization.

3. MESOSCOPIC SYSTEMS

In the following sections, we discuss n-dimensional quantization ($n = 0,1,2,3$). Since real space has three dimensions, we have three degrees of freedom to quantize.

3.1 3D Quantum Systems

In an ideal 3D system, all Bloch waves are extended all over the system. This means that it is still a microscopic system, however large it is, so long as there are no decoherence mechanisms.

Besides these ordinary systems, it is worth mentioning that localized states due to disorder in three dimensions behave similarly to mesoscopic systems. Take a hypercube of d dimensions with a volume L^d. As shown in Fig. 2, localization is expected as a result of the disorder. Disorder tends to decrease the extension of electron waves from delocalized to localized states. This occurs because of the change in the boundary conditions or, in other words, because the translational symmetry is broken.

According to Thouless (1977), each localized state has a scale length ξ, over which the state extends. If $L \lesssim \xi$, the system seems ideally microscopic, where all states are extended all over the hypercube.

The following discussion makes use of the scaling theory due to Abrahams, Anderson, Licciardello, and Ramakrishnan (AALR: Abrahams *et al.*, 1979). The size dependence of conductance is illustrated in Fig. 3. When we discuss the conductance G in this system, it is natural to assume that the normalized conductance

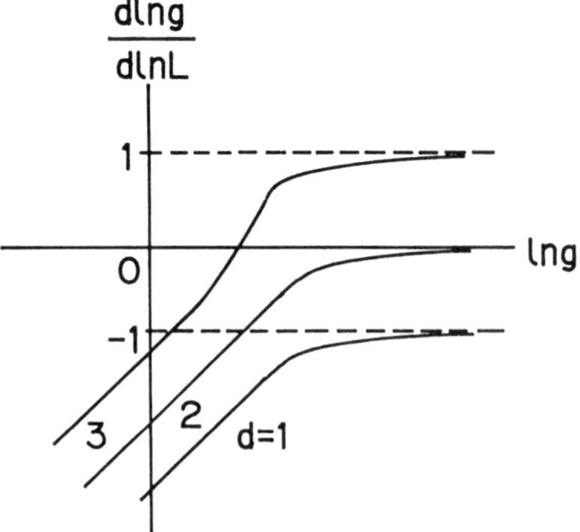

FIG. 3. Schematic diagram of $d\ln g/d\ln L$ vs $\ln g$, where $g = (2\hbar/e^2)G$, for d = 0, 1, 2 (after Abrahams *et al.*, 1979).

$$g = (2\hbar/e^2)G = (2\hbar/e^2)\sigma L^{d-2} \quad (1)$$

is proportional to $e^{-L/\xi}$ (σ is the conductivity); i.e., $g(L) = g_a e^{-L/\xi}$, or $d\ln g/d\ln L \approx \ln(g/g_a)$ for small g. Here, g_a is the conductivity that an extended wave shows. For large g, $g(L)$ is asymptotically given by the solution of the equation

$$\frac{d\ln g}{d\ln L} \sim d - 2 - \pi^2/g + \ldots \quad (2)$$

Thus, the asymptotes for $d = 1$, 2, and 3 are -1, 0, and 1, respectively. In three dimensions, this scaling implies $G(L) = \sigma L$, which increases with the size of the system L. The case for two dimensions is typical; $d\ln g/d\ln L$ is always negative and $g(L) \to 0$ for $L \to \infty$; $g(L)$ shows a crossover from logarithmic (for large g: *weakly localized regime*) to exponential localization (for small g: *strongly localized regime*). In the weakly localized regime, $g(L) \sim \text{const} - \pi^{-2} \ln L$, where L is very often replaced by l_ϕ or $\sqrt{D\tau_\phi}$ (D is the diffusion constant). Since τ_ϕ is proportional to T^{-p} (T, temperature; p, a number of order unity) according to the perturbational calculation of the electron-phonon and electron-electron interaction, it is seen that conductance has a logarithmic dependence on T. This reflects the fact that localized states can diffuse when inelastic scattering supplies them with the necessary amount of energy to delocalize.

The strongly localized regime is for $l_\phi \gg \xi$. In the opposite case, i.e., $l_\phi \lesssim \xi$, the conduction process is classified as follows:

1. for $L > l_\phi$, classical hopping conduction occurs; and
2. for $L \lesssim l_\phi$, electron waves are extended and coherent.

In order to discuss conductance in this system, Thouless (1977) introduced a dimensionless number called the "Thouless number," $g(L) = v(L)/w(L)$, where $v(L)$ is a typical energy, which may be expressed in terms of $\hbar/\tau_D$ (τ_D is the specific time for diffusion, defined by $D = L^2/\tau_D$), and $w(L)$ is the characteristic separation of levels at the Fermi energy, given by $1/N(\epsilon_F)L^d$ for a density of states $N(\epsilon_F)$. The equivalence of this Thouless number $g(L)$ to the above-mentioned $g(L) = (2\hbar/e^2)\sigma L^{d-2}$ can be easily proved from the generalized Einstein relation $\sigma = e^2 D N(\epsilon_F)$. It is interesting to note that localized states can convey an electric current via hopping through the most probable percolation path. Hopping itself is possible by the assistance of, e.g., phonon absorption. Following Mott (1974), dc conductivity is given by $\sigma \sim \exp[-(T_0/T)^{d+1}]$, where d is the number of dimensions in the system of concern (variable-range hopping).

Lastly, note that, in ordinary band theory in 3D, $N(\epsilon)L^3$ is given by $(2m/\hbar^2)^{3/2}$ $(L^3/2\pi^2) \times \sqrt{\epsilon}$; i.e., $N(\epsilon) \propto \sqrt{\epsilon}$ is typical for three dimensions.

3.2 2D Quantum Systems

In a very thin slab of material, it is easily imagined that the Bloch waves are confined inside it with extinguishing wave functions outward from both the surfaces. This is an analog of the Fabry–Pérot resonator. In condensed matter, the waves can be confined, not by a vacuum but by an energy gap. Within a QW with barriers of height V_0 and width w, bound energy levels are created at $E_n = \hbar^2 k_n^2/2m$, where k_n is the solution of

$$k_n w = n\pi - 2\sin^{-1}\frac{\hbar k_n}{\sqrt{2mV_0}}.$$

Except in the case of an infinite V_0, the number of bound levels within the QW is finite.

In the following, the most well-known example of GaAs-AlAs heterostructures is numerically solved. The difference in electron affinity in the GaAs and $Al_{0.3}Ga_{0.7}As$ system is about 240 meV, and this causes a step-like change in the conduction-band bottom at the interface. Likewise, if the sum of the electron affinity and the energy gap is different in the well from that in the barrier, the system has a QW for holes. When the conduction bands of the well and barrier materials are represented by specific effective masses of 0.0665 and 0.5 times m_0, respectively, the solution of the Schrödinger equation is as shown in Fig. 4 for a well width of 50 Å. The lowest bound level within the QW is at -199.5 meV, whereas the second level is at -2.1 meV, very near but below the barrier conduction band of AlGaAs. The probabilities of these waves $|\Psi(x)|^2$ are also plotted.

In such 2D systems, the degrees of freedom parallel to the well still survive, and consequently, electrons move almost freely

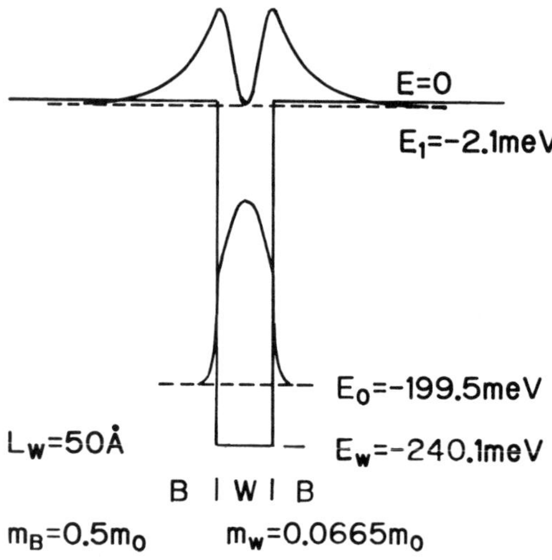

FIG. 4. Calculated $|\Psi(x)|^2$ for the ground and the excited level within a QW. The QW assumes the parameters of GaAs and $Al_{0.3}Ga_{0.7}As$ for the well and the barrier, respectively.

in that direction. Thus, 2D quantum systems can be realized by 1D confinement. If several *single-quantum wells* (SQW) are almost independent because of spaces between them, they are called *multiple-quantum wells* (MQW). Whether they are coupled or not is determined by simple quantum mechanics (see ELECTRON STATES IN ZERO-, ONE-, AND TWO-DIMENSIONAL STRUCTURES). Any textbook teaches that the wave function within a QW leaks out to the neighbors separated by barriers of height V_0 and width $d - w$ with a probability of $\exp\{-2\kappa(d - w)\}$, where $\kappa = \sqrt{2m(V_0 - E)/\hbar^2}$.

When all QW's are closely coupled to each other, especially when they are ordered periodically with a period d, they are called an SL. An SL gives doubly periodic Bloch-like quantized states with two periods of a and d ($d \gg a$; d is ordinarily 15–100 Å; a is the lattice constant). The overlaid periodicity gives the system an additional BZ boundary in one dimension and minigaps on it. If an appropriate strength of electric field is applied perpendicular to the well, SLs can either completely reflect according to the Bragg reflection at the minigaps, or conduct incident electrons across the minibands via interminiband tunneling.

The previous section stated that the 2D systems are more likely to localize electrons, following AALR theory, than 3D systems. This is because any possible percolative paths are more strongly inhibited in 2D or, in other words, there are fewer redundancies in reaching a specific point from some starting point via a 2D percolative path.

In 2D, the density of states (DOS) is given by $N(\epsilon_F)L^2 = mL^2/\pi\hbar^2$, which is constant and independent of energy. If there is a finite number of quantized levels, the overall DOS is a sum of steps with offsets corresponding to quantized energies E_n.

3.3 1D Quantum Systems

One-dimensional quantized systems are more easily localized under the existence of disorder. They are called *quantum wires*. In them, two degrees of freedom are frozen, and only one survives, parallel to the wire.

The DOS in 1D systems is given by $N(\epsilon)L = \pi^{-1}(m/2\hbar^2)^{1/2}\, L/\sqrt{\epsilon - E_n}$, where E_n is the quantized energy. This DOS resembles that of 3D electrons under a magnetic field. In 3D, the degree of freedom parallel to the magnetic field is never affected by applying a field, whereas those perpendicular to the field are Landau quantized, and motions in 2D are reduced to the center-of-mass motion and the motion of relative coordinate. Although the center-of-mass coordinate gives a quantum number other than the Landau quantum number n, the quantized energy is incidentally independent of the center-of-coordinate quantum number or, in other words, degenerate with respect to this quantum number, so long as an electric field does not coexist. In fact, the DOS is proportional to $1/\sqrt{\epsilon - \hbar\omega_c(n + \frac{1}{2})}$, with $\omega_c = eB/m$.

In the case of intrinsically 2D systems, a magnetic field perpendicular to the system quantizes the energy into a zero-dimensional, dispersionless scheme. However, by applying an electric field within the 2D system, it is easy to lift the degeneracy with respect to the center-of-mass motion, which makes the system look like a 1D system within the 2D plane. Thus, it is easy to understand in the Hall bar geometry that the edge currents along both the sample edges parallel to the fed-in current still play an important role in experiments such as those on the quantized Hall effect (QHE), as will be discussed later.

3.4 0D Quantum Systems

These systems are called *quantum boxes* or *quantum dots*. The energy scheme is a series of levels, just like in clusters in solid states. Every state is confined within the box. In disks larger than ~100 Å, for example, the states may be approximately described by an envelope function and highly oscillatory Bloch-like states with modulation on the order of the lattice constant a. The envelope function is specified by the boundary conditions and is thus disk-shaped.

The DOS in the quantum boxes is obtained by simply counting those discrete levels existing between ϵ and $\epsilon + d\epsilon$. Ordinarily, these disks are with difficulty accessed electrically through leads. Accordingly, high-frequency properties as in optics and, sometimes, diamagnetic properties due to Landau motion are more suited for observing a zero-dimensional quantized system.

4. PHYSICAL PROPERTIES

4.1 Optical Properties

4.1.1 Single/Multiple-Quantum Wells Within semiconductors, shallow donors and acceptors, as well as excitons, can be described by hydrogenic wave functions. In order to discuss excitons, the mass should be taken to be a reduced value μ of the effective masses of an electron and a hole. Hence, the exciton binding energy E_n is $-\Re/n^2$, $n = 1, 2, \ldots$, where $\Re = m_0e^4/2\epsilon_0^2\hbar^2$ is the effective Rydberg energy, $13.6(\mu/m_0)/\epsilon_0^2$ eV. The term ϵ_0 is the static dielectric constant. This is still the case within a QW, if the width is large enough. For a narrow QW, one degree of freedom of the spherical orbital is squeezed. An extremely thin QW will look like a 2D hydrogen atom. The binding energy in this limit is known to be given by $E_n = -\Re/(n - \frac{1}{2})^2$, $n = 1, 2, \ldots$. If both the lowest binding energies in 3D and 2D are compared, it is obvious that $4E_1(3D) = E_1(2D)$. Correspondingly, the radial extension of the s orbital of a 2D exciton is half of that in 3D.

The fact that the binding energy of an exciton in a QW is larger than that in 3D has an important meaning for applications. For example, when a QW is formed as an active layer in a semiconductor laser, the emission efficiency is improved, partly because the binding energy can be larger than thermal excitation energy at room temperature (RT), and the exciton is not easily excited into a pair of a free electron and a hole that can more easily recombine without emitting luminescence. The other reason is that the DOS in 2D is constant, as was stated before, and consequently, the emissive transition probability is higher than in 3D, where the DOS starts from zero and is in proportion to $\sqrt{E}$.

Figure 5 shows the dependence of the binding energy of an exciton against the well width for the $GaAs\text{-}Al_{0.4}Ga_{0.6}As$ system. For narrow widths, the energy first increases and then decreases. The reason is that for a vanishing well width, the bound energy level of an electron comes very close to the conduction-band bottom of the barrier material, and consequently, the wave function is no longer confined within the well, just as was shown for the second excited level E_1 in Fig. 4. This is also the case for holes (see EXCITONS). The

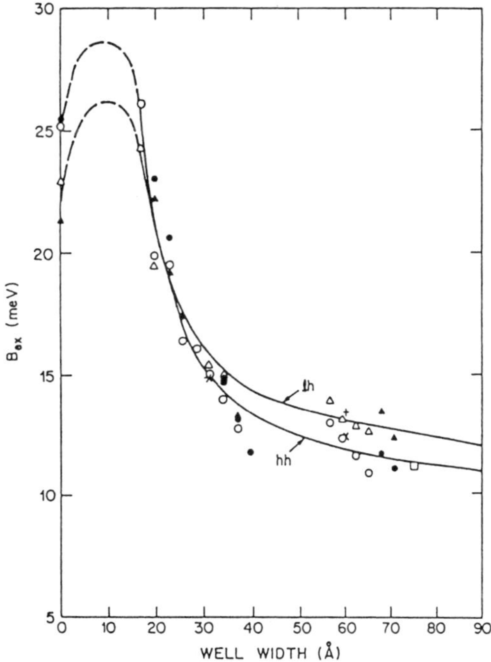

FIG. 5. Exciton binding energy within QW's vs well width. The labels *lh* and *hh* designate the exciton binding energy formed between, respectively, an electron and a light hole, and an electron and a heavy hole. Since the exciton belongs to the bound state within the QW, which has an energy inversely proportional to the effective mass, an *lh* exciton has a larger binding energy (after Nelson *et al.*, 1987).

excitonic state consisting of these extended orbitals is never confined within the well, and therefore, the binding energy of the exciton approaches again the value for 3D, lower than the value for 2D.

Figure 6 shows the photoluminescence (PL) as a function of the well width (see PHOTOLUMINESCENCE). The PL photon energy corresponds to the difference between the lowest bound-state energies for electrons and holes minus the exciton binding energy. The binding energy is given in Fig. 5. The limiting case for a vanishing well width in this figure must be equal to the energy gap E_g minus the exciton binding energy, both in 3D bulk.

Another interesting phenomenon is seen in the efficiency of PL. When MQW samples are fabricated with barriers too thick to couple with each other, it is easy to observe a number of PL spectra, under a single-shot excitation, from wells with various widths. From a well of a certain thickness, PL of a specific wavelength comes out, and its intensity is proportional to the mth power of the excitation intensity, $I_{PL} \propto I_{ex}^m$. This power index depends on the well width, as is shown in Fig. 7. The measurement was done by Mishima *et al.* (1986). As already stated, when the well width w increases from a very small value, the lowest bound level lowers from the conduction-band bottom of the barrier material. Around $w \approx 4.7$ nm for $Al_{0.3}Ga_{0.7}As/GaAs$ MQW, a second bound level appears. This must look as shown in Fig. 4. Note that PL always comes from the recombination of an exciton belonging to the lowest-electron and highest-hole bound states. When the highest electron level comes near the barrier conduction band, i.e., the *resonance condition* holds, photoexcited electrons can easily be freed from within the well. On the other hand, when there is a considerable energy depth between the conduction-band bottom of the barrier and the highest level within the well, i.e., the *off-resonance condition* holds, photoexcited carriers are tightly trapped within the well and contribute significantly to the PL. This situation reflects on the PL efficiency, which depends on the well width in the manner shown in Fig. 7. For $m \sim 2$, the efficiency is low, whereas for $m \sim 1$ it is high. This periodicity was accurately interpreted on the basis of the argument above (Murayama, 1986).

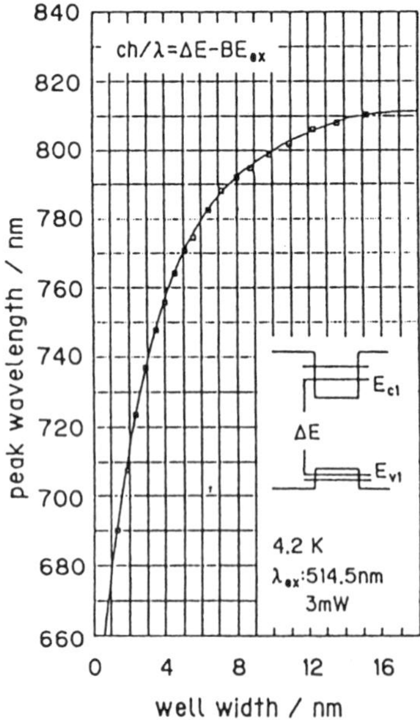

FIG. 6. PL from QWs vs well width. The PL energy is equal to the separation between the lowest electron and the highest hole level minus the exciton binding energy (after Murayama, 1992).

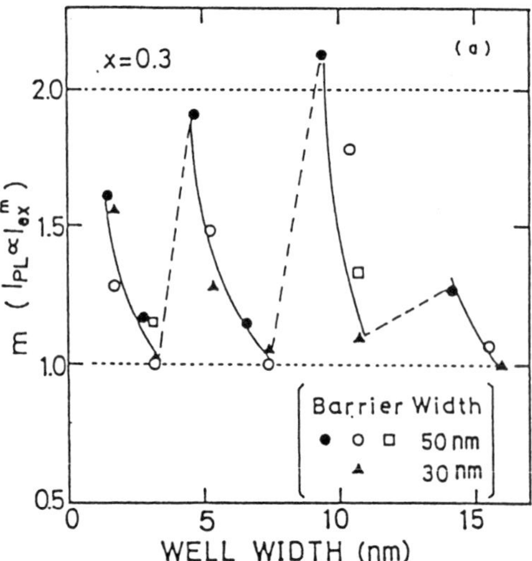

FIG. 7. Power index m: $I_{PL} \propto I_{ex}^m$ vs well width. m = 2 and m = 1 correspond to low-efficiency resonance and high-efficiency off-resonance, respectively (after Mishima *et al.*, 1986).

4.1.2 Superlattice In an SL, there are minibands and minigaps (see HETEROSTRUCTURES AND SUPERLATTICES, SEMICONDUCTOR), as shown in Fig. 8(a). When an electric field

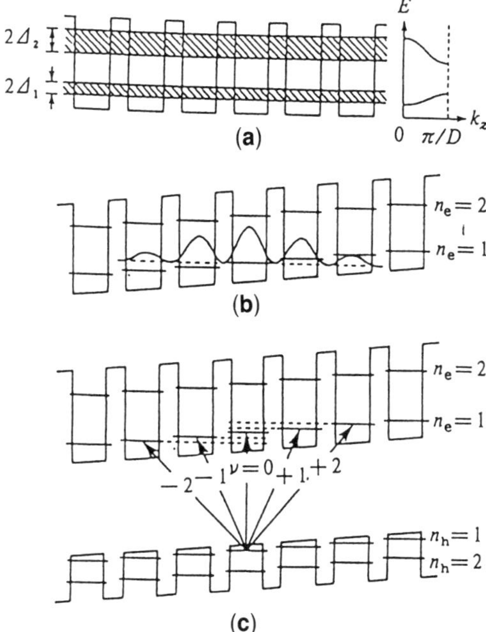

FIG. 8. (a) Minibands in an SL. (b) An electron state that has the largest amplitude at the center of the SL, when an electric field is applied. (c) Suggested optical excitation of an electron in the valence band at the center of the SL to several electron states. The largest transition probability occurs between the electron and hole within the same well (after Nakayama *et al.*, (1991).

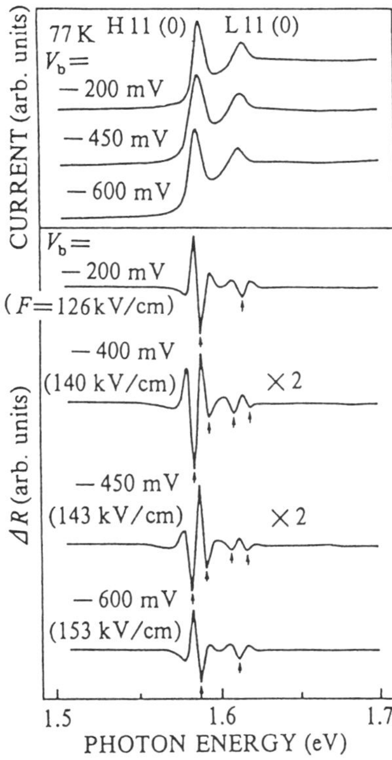

FIG. 9. Spectra of photocurrent and reflectance change vs photon energy as a function of applied electric potential V_b (after Nakayama *et al.*, 1991).

is applied perpendicularly to this device, the potential of each QW varies as a function of the well position, as schematically shown in Fig. 8(b). So, we can expect a series of optical transitions between a hole level, e.g., in the SL center, and electron levels with offset energies according to the Stark effect. Figure 9 plots the photocurrent spectrum and the variation in reflectance spectrum with electric field against the photon energy for various fields. The strongest signal is obtained for both the electron and hole within the same well, and weaker ones are for the transitions to the electron levels in neighboring wells.

It is seen in Fig. 9 that resonant tunneling occurs around $V_b \approx -(400\text{–}450)$ mV between the $n_e = 1$ miniband for $\nu = 0$ and the $n_e = 2$ miniband for $\nu = -2$ QW, causing very weak reflectance changes ΔR. When the resonance condition holds, both the $n_e = 1$ and 2 minibands are strongly mixed so that they lose specific spike structures according to the QW features.

4.1.3 Selection Rules Silicon is the best material to manufacture inexpensive transport devices in mass fabrication at the present time. However, it is never applied to light-emitting devices, because it is known to perform only indirect optical transition, and hence, the emitting efficiency is extremely low. Using SLs with an overlaid periodicity of n times the original lattice constant, it is recognized that the BZ in the bulk state can be folded into $1/n$ of the original reciprocal lattice vector in one dimension. For example, by making an SL by alternately growing Si and SiGe alloy, the conduction-band minimum of Si may come close to the Γ point in the BZ. Does such Si with a folded BZ show a higher light-emitting efficiency? This is not necessarily so, because the transition matrix element remains the same after the folding, although the k-selection rule can change from indirect to direct transition.

Regarding the optical properties of Si, recent PL studies in porous Si (Canham, 1990) are worth noting from the viewpoint of application of mesoscopic systems. When Si wafers are immersed in HF acid and anodized, the surface is mesoscopically processed and optical transition is no longer indirect as in bulk. So far, cluster and quantum wire models have been proposed, where Si dangling bonds on the surface of the mesoscopic systems must be terminated by H and/or O atoms.

4.2 Transport Properties

4.2.1 Transport Parallel/Perpendicular to QW As described in the preceding sections, single/multiple QWs are specified by their size-quantized energy levels in the direction perpendicular to the wells. Transport properties are first of all classified into (1) transport parallel to the wells and (2) transport perpendicular to the wells.

4.2.1.1 Parallel Conduction—Metal-Oxide Semiconductor (MOS). Conduction within the well planes is the same as that within thin slabs. Ordinarily, these are nearly 2D systems. Sometimes there is a heterostructure between the well and the barrier material, which causes a sharp potential dip and makes the electronic system perfectly 2D. This case resembles the Si-SiO_2 interface in an MOS device. The 2D electron gas (2DEG) sheet is formed in an MOS by the built-in electric field generated between SiO_2 and Si so as to equalize the chemical potentials. (see SI-SIO_2 INTERFACE, ELECTRONIC PROPERTIES OF; TRANSISTORS, FIELD EFFECT).

Let us explain the band structure at the interface in terms of the built-in field F. Bulk n-type Si is known to have four degenerate valleys, each situated around one of the minimum points: $(\pm 0.85,0,0)\pi/a$, and the equivalents. Let us assume that an MOS is formed on the (001) plane (Fig. 10). Then the energy band in the [001] valley is given by

$$E_n(\mathbf{k})^{[100]} = E_n^l + \hbar^2 k_y^2/2m_t + \hbar^2 k_z^2/2m_t,$$

with

$$E_n^l = (3\pi\hbar eF/2\sqrt{2m_l})^{2/3}(n + 3/4)^{2/3}.$$

This analytical result is obtained in the triangular electric potential, which is good for a sufficiently strong field (Ando *et al.*, 1982). For the valleys in the directions of [100] and [010], the quantized energies E_n^t are given by

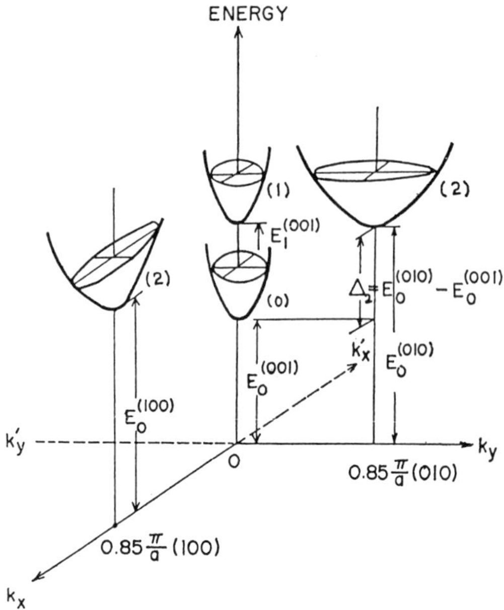

FIG. 10. Suggested valley structure quantized by a built-in electric field at the (100)-Si-SiO_2 interface (after Murayama *et al.*, 1972).

replacing m_l by m_t in the above equation; m_l and m_t are the effective masses in the longitudinal and transverse directions in the [001] valley, respectively. The term $m_l \approx 1.58m_0$ and $m_t \approx 0.081m_0$, and therefore, $E_n^t \gg E_n^l$. Thus, in a (001)-MOS device, all four valleys [$\pm$100] and [0 $\pm$10] are degenerate, but the other two valleys, [00 $\pm$1], are not and have a lower energy. For the [100] valley, the parabolic band mass is m_t, whereas, in the other valleys, the parabolic masses are m_t and m_l. Consequently, higher bands have a larger mass and the lower band has a smaller mass. This difference in masses can cause a phenomenon similar to that called the Gunn effect in bulk GaAs, i.e., a differential negative resistance (DNR) device. Both the experimental work (Katayama *et al.*) and the theoretical analysis (Murayama *et al.*) were done in 1972.

4.2.1.2 Perpendicular Conduction—Resonant Tunneling Device. A QW is formed between two barrier materials and is specified by a size-quantized energy level. If electrons flow perpendicular to the well, the requirements of energy and momentum conservation limit the allowable tunneling between the energy levels outside the barrier and those inside the well. This relation is schematically described in Fig. 11(a). When an electric field is applied across the QW, most of the potential drop occurs across the barrier materials because of their high resistivity. First, for an electric field, tunneling is allowed at the Fermi level satisfying the energy-momentum conservation condition (see RESONANCE TUNNELING). For a higher field, tunneling may no longer be allowed. This increase in electric field sometimes means a decrease in current, i.e., a DNR (Tsu and Esaki, 1973). An example of observations is shown in Fig. 11(b). This DNR characteristic is specific to RTDs.

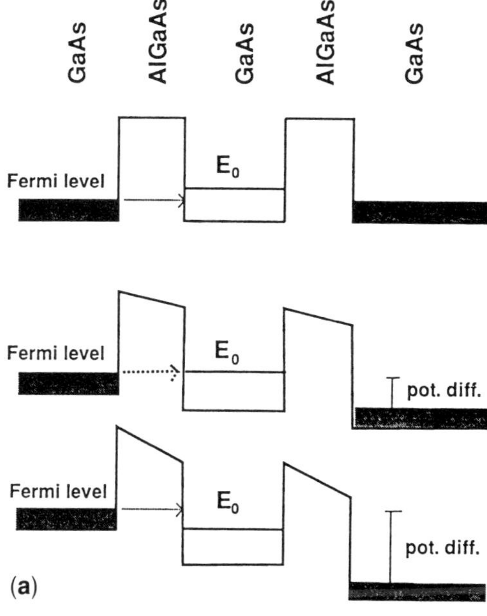

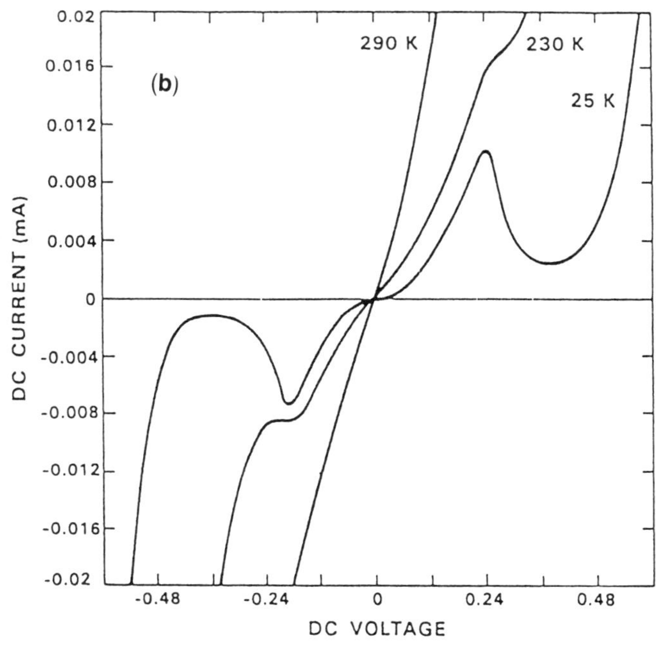

FIG. 11. (a) Suggested mechanism of an RTD. (b) Observed dc current vs dc voltage showing a negative differential resistance (after Sollner *et al.*, 1983, reprinted with the permission of Lincoln Laboratory, Massachusetts Institute of Technology, Lexington, MA).

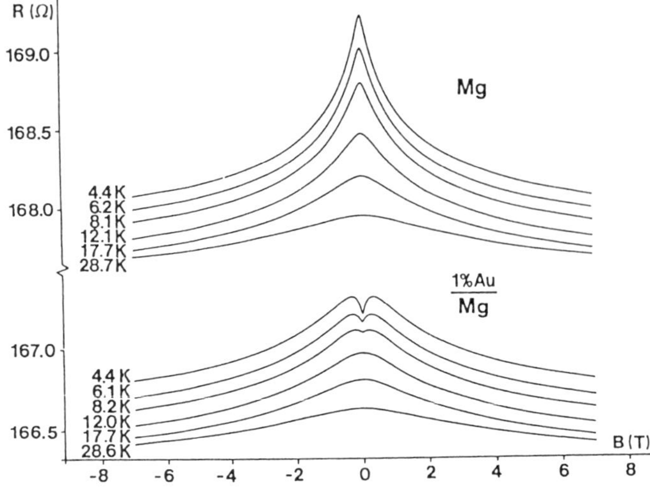

FIG. 12. Observed negative magnetoresistance of Mg thin films uncovered (upper data) and covered (lower data) with Au, as a function of temperature (after Bergmann, 1983).

4.2.2 Negative Magnetoresistance Ordinary nonmagnetic materials show a slight increase in transverse resistance under an applied magnetic field. The increase is proportional to B^2. This is natural because of symmetry; i.e., the system should behave symmetrically for B and $-B$. Since no classical transport theory explains this dependence on B (the van Leeuwen theorem), the magnetoresistance must be attributed to a quantum effect.

In a 2DEG in a very thin Mg film, magnetoresistance was measured by Bergmann (1983), as shown in Fig. 12. The resistance decreases with increasing magnetic field B. This sample seems to be in a weakly localized regime with slight disorder. In Anderson localization, coherence causes an electron to localize around a specific site on the quasi-1D path shown in Fig. 2. Let us simplify Fig. 2 by assuming a homogeneous phase distribution on the ring in Fig. 13. Here, the total phase change Θ is composed of the accumulated phase shift on each scattering site, $\Sigma_i \theta_i$, and of kL, where k is the wave vector and L the perimeter. Now, Θ is divided into $c\Theta$ and $(1 - c)\Theta$. The wave is assumed to start from point O and the superposition of the waves is considered at point A. The analytical expression is shown in Fig. 13. The Aharonov–Bohm phase is denoted by $\chi = 2\pi BS/\Phi_0$ with the enclosed area S and the quantum of flux $\Phi_0 = h/e$.

When a magnetic field B is applied to this system, the most probable existence of the wave is at a site different from that in the absence of B, since

$$|\Psi(\mathrm{A})|^2 = |\psi_c + \psi_{1-c}|^2 \propto \{1 + \cos[\chi - \Theta(1 - 2c)]\}. \tag{3}$$

It is obvious that the most probable constructive interference occurs on the antipodal point A, i.e., $c = \frac{1}{2}$, in such a simplified situation as $B = 0$.

Another effect of a magnetic field on mesoscopic systems is the change in the scale length under a strong field; i.e., the system size L should be replaced by a magnetic length $L_B = \sqrt{\hbar/eB}$, when $L \gtrsim L_B$. Thus, $g(L_B)$ in Eq.

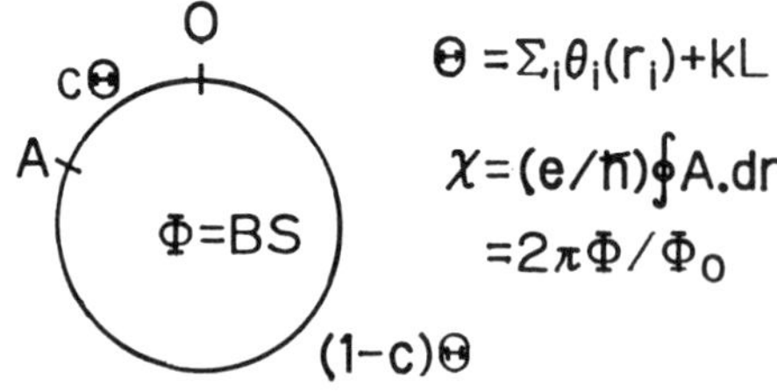

$$\Psi(A) = \psi_c + \psi_{1-c} = \psi_0 e^{ic\chi} e^{i(\Theta-\chi)/2}$$

$$\times [e^{ic\Theta} e^{-i(\Theta-\chi)/2} + \mathrm{C.C.}]$$

FIG. 13. Simplified version of Fig. 2, explaining the superposed wave function after a series of elastic scatterings. The wave function divided at O is eventually superposed at A, giving the amplitude shown. Θ is the total phase along the path and χ is the Aharonov–Bohm phase.

(2) gives the dependence on the field in the form $\Delta g(L_B) \sim -\pi^{-2} \ln L_B$, which increases with B (see Section 3.1).

The situation differs when there is strong spin-orbit coupling. In Fig. 12, the lower part shows the data for Mg film covered with Au. There is strong spin-orbit coupling between Mg and Au, and the characteristic time is the spin-orbit scattering time τ_{SO} rather than the ordinary inelastic scattering time, especially at lower temperatures. This τ_{SO} is of an elastic nature and is almost independent of temperature. It affects the magnetoresistance positively because the spin-orbit interaction breaks the time-reversal symmetry. All these experiments are very accurately interpreted by the phenomenological theory by Hikami *et al.* (1980) in terms of a few parameters τ_{el}, τ_{in}, τ_{SO}, and B.

The same kind of negative magnetoresistance is also observed in a pure 2DEG system in the channel of MOS devices, as shown by Kawaguchi *et al.* (1978).

4.2.3 Aharonov–Bohm Effect Figures 2 and 13 showed a conceptual description of Anderson localization, where the point was that after multiple elastic scatterings the wave comes back to itself, and the original and the returned waves can be superposed constructively at a certain site or destructively at other sites. When the superposition is constructive, the wave can have a larger amplitude than the average.

In 1981, Sharvin and Sharvin (1981) prepared a quartz fiber with a diameter of ~2 μm covered with vacuum-evaporated Mg thin film and measured the transport properties of the sample with two leads separated from each other in the direction of the fiber under a parallel magnetic field (Fig. 14). This experiment was suggested by a preceding theory of Al'tshuler *et al.* (1981). This sample simulates the geometry shown in Fig. 13, although scattering occurs against defects, not within a connected but within a disconnected region. On the equation shown in the figure, assume $c = 0$ and $c = 1$, which mean clockwise and anticlockwise travel of the wave, respectively. Since for $c = 0$,

$$\Psi(c = 0) = \psi_0(1 + e^{i(\Theta-\chi)}), \quad (4)$$

while for $c = 1$,

$$\Psi(c = 1) = \psi_0(e^{i(\Theta+\chi)} + 1), \quad (5)$$

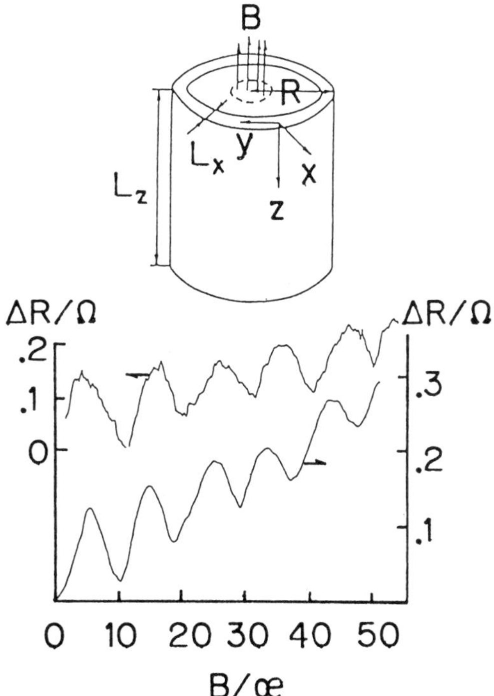

FIG. 14. Observed Aharonov–Bohm effect on disordered thin Mg hollow cylinder under a parallel magnetic field. The data on two samples are shown, scaled on the left and right ordinates as shown by the arrows (after Sharvin and Sharvin, 1981).

the superposed wave at point O for both directions gives its probability of existence to be proportional to $1 + 2\cos\Theta \cos\chi + \cos^2\chi$. This is peculiar in that a periodicity is expected for a magnetic field change of $S\Delta B = \Phi_0/2$, as actually happened in the experiment (Sharvin and Sharvin, 1981).

Another conventional Aharonov–Bohm effect in solid state was observed using a small metallic ring (Webb *et al.*, 1985, 1987; Washburn *et al.*, 1985) (Fig. 15). In this case, the effect is understood analogously to the electron-beam interference experiment. The period is, in contrast to the Sharvins' experiment, $\Delta B = \Phi_0/S$. The difference comes from whether the travel of the waves occurs along the fiber axis or does so in the plane merely in one turn, when the orbital is localized within the plane perpendicular to the direction of B. After the first observation of periodic variation of conductance through a small ring, other experiments improved the periodic behavior by using samples with a higher aspect ratio (the diameter versus the width

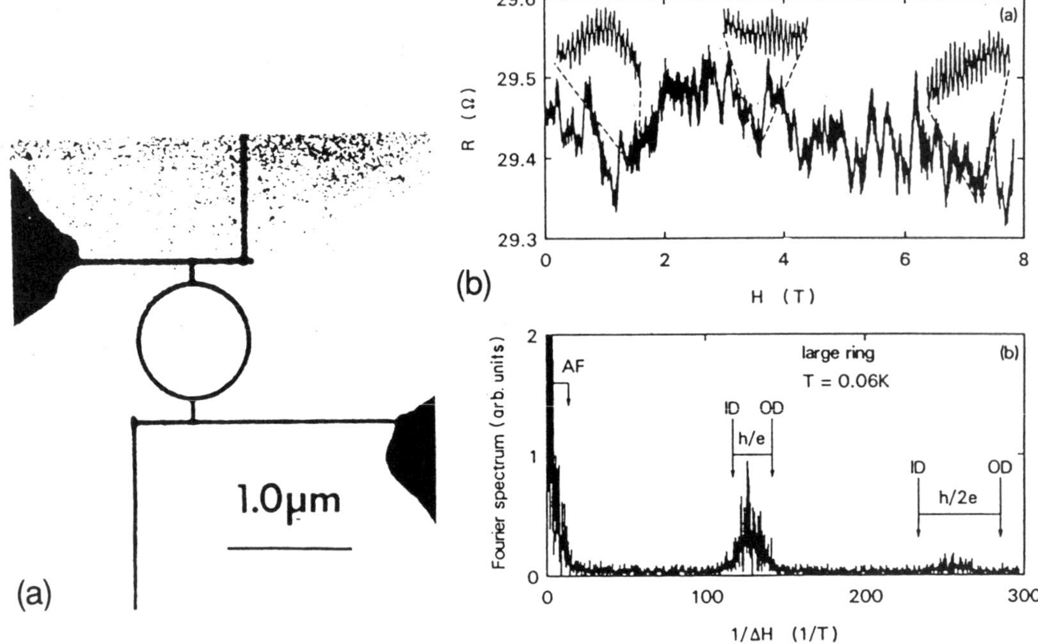

FIG. 15. (a) Photo of a small Au ring sample. (b) Measured variation in resistance vs magnetic field H, and its Fourier-transformed spectrum vs $1/H$ (after Washburn *et al.*, 1985).

of the ring). This improvement results from the fact that the magnetic field can penetrate the arm of the ring itself, when the arm is very wide. An enhanced aspect ratio gives fewer channels to be averaged over. If there are many paths inside the ring arm, the Aharonov–Bohm effect occurs for each path, and the conductance must eventually be averaged over all the paths.

The latter Aharonov–Bohm effect with a period Φ_0/S is sensitive to the ensemble average. This fact is obvious from the strong dependence on the aspect ratio of the sample. In contrast, the former effect, with the period of $\Phi_0/2S$, is not sensitive to the ensemble average.

4.2.4 Landauer–Büttiker Formula of Conductance Before discussing the universal conductance fluctuation, it seems helpful to derive first an intuitional formula for conductance that is specifically applicable to mesoscopic systems. This is called the Landauer formula (Landauer, 1957), after the inventor's name, and sometimes Landauer–Büttiker formula (Büttiker *et al.*, 1985), attaching the name of the researcher who extended the formula to a more general case (see ELECTRON STATES IN ZERO-, ONE-, AND TWO-DIMENSIONAL STRUCTURES).

Let us follow Imry's discussion (1986). As shown in Fig. 16, the electron incident from the left channel i tunnels into the right channel j with a probability $\Sigma_j f_L T_{ij}$. There are numerous channels in an ordinary case. The $f_{L,R}$ is the Fermi distribution function in the left (L) or right (R) lead acting as a reservoir. Similarly, the electron incident from the right channel i is reflected into the right channel j with a probability $\Sigma_j f_R R_{ij}$. The sum of these probabilities minus the equilibrium distribution gives the current transported on

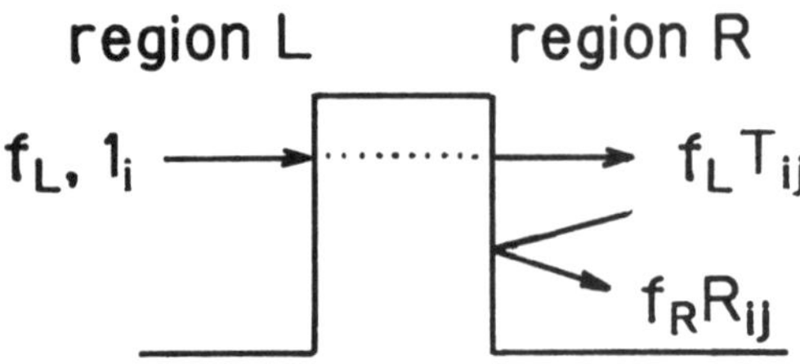

FIG. 16. Definition of channel i and its probability of occupation f_i around regions R and L with a tiny barrier in between. T and R are transmission and reflection, respectively.

channel i from left to right, i.e.,

$$I_i = \int dEev_iN_i(E) \sum_j (f_LT_{ij} + f_RR_{ij} - f_R) \quad (6)$$

$$\sim [e(\mu_L - \mu_R)/\pi\hbar] \int dE(-df/dE) \times \Sigma_jT_{ij}(E), \quad (7)$$

where the density of states in 1D is $N_i(E) = 1/\pi\hbar v_i$ (v_i is velocity) and $R + T = 1$. μ_L (μ_R) is the chemical potential in region L (R). Since the conductance G is given by $\Sigma_iI_i/[(\mu_L - \mu_R)/e]$, it is easy to obtain the Landauer formula $G = (e^2/\pi\hbar)\Sigma_{ij}T_{ij}$. Very often $\Sigma_{ij}T_{ij}$ is written as $\mathrm{tr}tt^\dagger$, using the transmission probability amplitude t: $T_{ij} = t_{ij}t_{ji}^\dagger$. This formula holds when electrons are fed directly from both the "classical" reservoirs. Classical means that electrons are completely incoherent within the reservoir because of frequent inelastic scattering. On the other hand, with two "ideal" leads between the sample and the reservoirs, the formula is given by Landauer's original expression: $G = (e^2/\pi\hbar)T/R$.

4.2.5 Universal Conductance Fluctuation In Fig. 13, it was shown that a localized state is defined on an idealized quasi-1D ring with a distributed phase on the perimeter. It is important that the state thus defined is localized around α; i.e., Ψ_α is very sensitive to parameters such as magnetic field B and chemical potential μ: $\Psi_\alpha(B,\mu, \ldots)$. Thus, in what manner the state is localized appears in its dependence on B, μ, and so on. This phenomenon was called a magnetofingerprint by Stone (1987), since the system behaves always specifically as a function of B.

The system of concern intrinsically has disorder. It is characterized by the mode in which such disorder is distributed. In mesoscopic systems, these disorder-induced fluctuations appear in, e.g., conductance, and are never averaged out. This is called *conductance fluctuation* specific to the system of concern. When the system is made larger and larger, the fluctuation disappears, as can be imagined easily, and approaches a macroscopic limit.

In contrast, when the size is reduced, the fluctuation is enhanced. Let us consider the conductance G; G is proportional to $(2e^2/h) \times \int dt \langle\Psi_\alpha|v(t)|\Psi_\beta\rangle\langle\Psi_\beta|v(0)|\Psi_\alpha\rangle$, which is the sum over all possible pairs $\alpha\beta$. The term v is the velocity operator. The deviation in G from the mean value as a function of B, μ, and so on should be given by a higher-order term: $\langle\delta G(B,\mu, \ldots)\delta G(B + \Delta B,\mu + \Delta\mu, \ldots)\rangle$ which is the correlation among four localized states. As suggested above, this fluctuation approaches a *universal conductance fluctuation* (UCF) given by

$$\Delta G = \frac{e^2}{h} \times \begin{pmatrix} 0.729 \\ 0.862 \\ 1.088 \end{pmatrix} \quad \text{for} \quad \begin{pmatrix} 1\text{D} \\ 2\text{D} \\ 3\text{D} \end{pmatrix} \quad (8)$$

after Lee *et al.* (1987). These three fractional numbers are only numerically calculated.

Experimentally, Webb *et al.* (1987) (Fig. 17) measured the voltage fluctuation as a function of the length of thin gold wires. As shown in the inset, the variance decreases in proportion to $L^{1/2}$ and approaches a constant value equal to the UCF. For a sufficiently small system $L \ll \xi$ (ξ is coherence length), the voltage fluctuation should disappear as suggested in the inset, because coherent conduction means "lossless" conduction.

4.2.6 Quantized Conductance In the arrangement shown on the left or right half in Fig. 18(a), electrons are emitted through an orifice formed by potential barriers on both sides and a through-channel with a lower po-

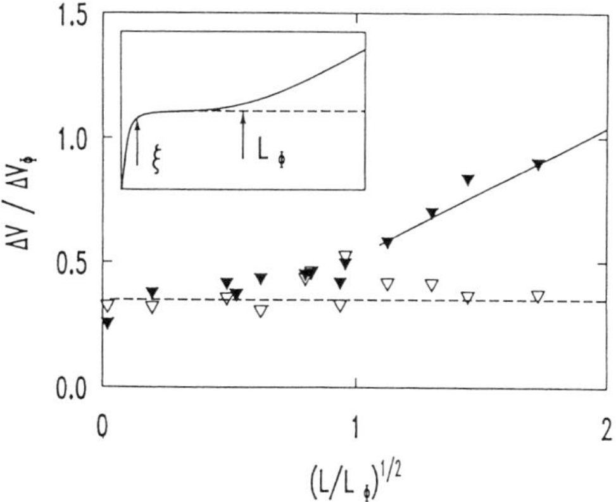

FIG. 17. Variance in voltage measured on short thin Au quantum wires. For a vanishing length a UCF appears. The inset explains schematically a theoretical variation of V vs L. The solid and open symbols represent, respectively, symmetric and antisymmetric contributions against the exchange of current and voltage leads. Only the symmetric data should be compared with the theory (after Webb *et al.*, 1987).

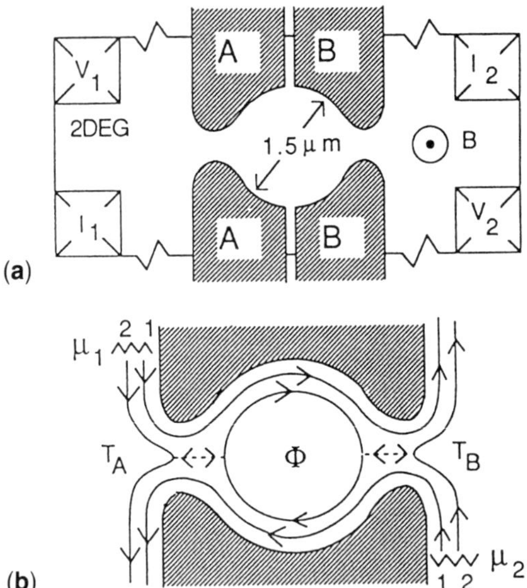

FIG. 18. Circuit within which coherent current channels are controlled by applying a voltage to the gates. With both gates A and B switched, closed-current channels are formed, and for either A or B switched, an orifice is formed (after van Wees *et al.*, 1989).

tential in between. This kind of orifice is called a *quantum point contact* (QPC). Figure 18(b) as a whole shows a circuit with two QPC's, between which there is the room to accomodate intereference between edge currents. When there is a single QPC, conductance is known to show a series of quantized values with plateaus on each step, as shown in Fig. 19.

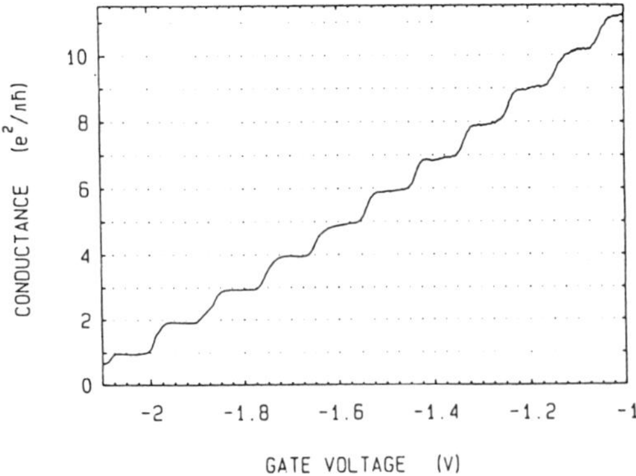

FIG. 19. Conductance quantized in units of $e^2/\pi\hbar$ observed on an orifice, which allows a number of channels determined by the gate voltage to pass (after van Wees *et al.*, 1988).

This behavior is understandable if the transport is attributed to a bundle of 1D channels, as the Landauer formula says. Let the orifice be sufficiently wide and low in its potential well. Then most channels run through the orifice without being blocked by the barrier potential. Otherwise, some may be blocked. One can count the number of channels running through the orifice. Since each channel carries a conductance of $e^2/\pi\hbar$, the total aggregate conductance may manifest a pileup of integral multiples of $e^2/\pi\hbar$, the number of which increases from unity to some integer, when the voltage increases. This is because the applied voltage lowers or widens the potential at the orifice.

In the limit where no potential channel remains wide or shallow enough to accomodate a single 1D channel, current is only transported by thermal excitation over or quantum tunneling through the potential barrier at the orifice.

4.2.7 Interference between Edge Currents Figure 20, curve A, shows observed conductance as a function of magnetic field

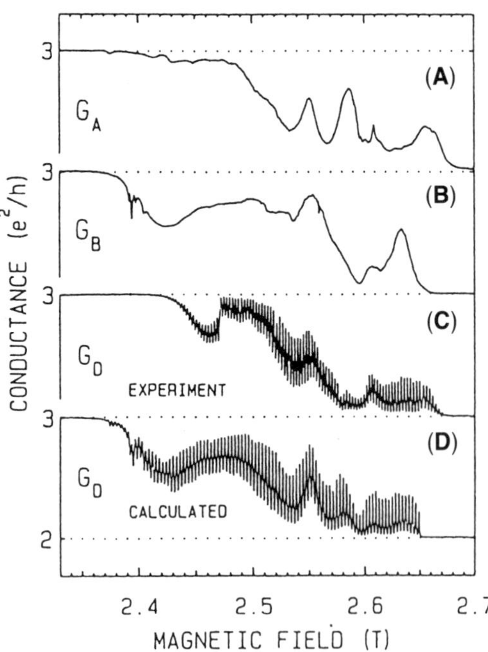

FIG. 20. Observed conductance as a function of magnetic field, when a single orifice (A) at *A* or (B) at *B* is formed and (C) when a closed-channel circuit is formed. (D) A simulation result (after van Wees *et al.*, 1989).

when the barrier potential B does not exist, and hence, the circuit appears to have a single QPC A. Figure 20, curve B, is for a similar case with a single QPC B. If the barrier potentials are switched on, a circular quantum dot is formed by the two QPC's and the circumferences of the barriers. Figure 20, curve C, depicts the observed oscillation of conductance, with a period in B given by $2\pi S\Delta B/\Phi_0 = 2\pi$, where S is the area of the circle shown in Fig. 18(b).

Here it is instructive to compare classical and quantum conditions for geometrical resonance of electrons under a magnetic field. It is known that, by application of the correspondence principle, a cyclotron motion is specified by the condition $E = \frac{1}{2} mr_c^2 \omega_c^2 = (n + \frac{1}{2})\hbar\omega_c$. Thus, the classical cyclotron orbit radius is known to be $r_c = \sqrt{\hbar/eB}$ according to the correspondence principle. On the other hand, quantum mechanics tells us that the phase of an electron circulating under a magnetic field is $\pi l^2 2\pi B/\Phi_0$, which results in $l = \sqrt{2}r_c$ if the acquired phase equals 2π. Thus, the motion of a classical particle under a magnetic field can be exactly reinterpreted quantum mechanically as a change in the Aharonov–Bohm phase, resulting from the vector potential generating the field.

4.2.8 Integral Quantum Hall Effect As stated before, conductance in 2D can be independent of the system geometry as $G \propto L^0$, and accordingly, conductance is equal to conductivity. This means that G is given by constants like e, $\hbar$, and m, independently of the sample size. This condition actually occurs in the Hall geometry in the 2DEG in a MOS under a magnetic field.

Let the width and length of a 2D channel be W and L in the y and x directions, respectively. Then

$$I_x/W = \sigma_{xx}V_x/L + \sigma_{xy}V_y/W,$$
$$I_y/L = \sigma_{yx}V_x/L + \sigma_{yy}V_y/W. \tag{9}$$

For a vanishing I_y and $V_{\rm H} = V_y$ ($\sigma_{xy} = -\sigma_{yx}$) (ordinarily this is the case for the Hall effect),

$$BR_{\rm H} = \frac{V_{\rm H}}{I_x} = \frac{\sigma_{xy}}{\sigma_{xx}\sigma_{yy} + \sigma_{xy}^2}. \tag{10}$$

This means that for the observed voltage $V_{\rm H}$ and current I_x, the ratio is given by a quantity including universal constants and, at least, τ. When $\omega_c\tau \gg 1$, $\sigma_{xx} = \sigma_0/[1 + (\omega_c\tau)^2] \to 0$ and $\sigma_{xy} = -ne/B$ (Fig. 21). The density of Landau states in 2D is shown in Fig. 22(a) as a function of energy. Without scattering, they must be sharp levels with a degeneracy of $1/2\pi l^2$. Here, l is the magnetic length defined by $l = \sqrt{\hbar/eB}$. If the Landau sub-bands are filled up to i, the density of electrons $n = i/2\pi l^2$ gives $BR_{\rm H} \sim 1/\sigma_{xy} = -h/ie^2$. Thus, the observed ratio $V_{\rm H}/I_x$ must be equal to a universal constant. This is why the effect is called the integral QHE, because the Hall constant times B is quantized to a universal constant divided by an integer. This only interprets that the observed $BR_{\rm H}$ value should reach $-h/ie^2$ at the point where $n = i/2\pi l^2$. To determine the integer n experimentally, we need the plateaus around the n on the observed curves.

Every experiment is performed as a function of the gate voltage V_G, or the density of carriers n. When the DOS is replotted as a function of n, the variation should read as in Fig. 22(b). However, it is known that electrons are very easily localized in 2D. Thus, each Landau subband has localized states besides the central delocalized narrow band.

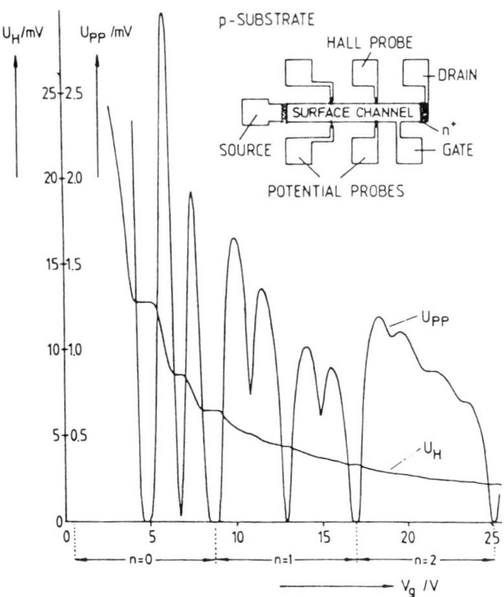

FIG. 21. Observed Hall voltage $U_{\rm H}$ ($V_{\rm H}$ in text) between the Hall probes (see the inset) and voltage U_{PP} (V_x) between the source and drain under a constant applied current as functions of gate voltage V_g on n-channel MOS Hall bar sample. $U_{\rm H}$ shows a number of plateaus, i.e., integral QHE (after von Klitzing *et al.*, 1980).

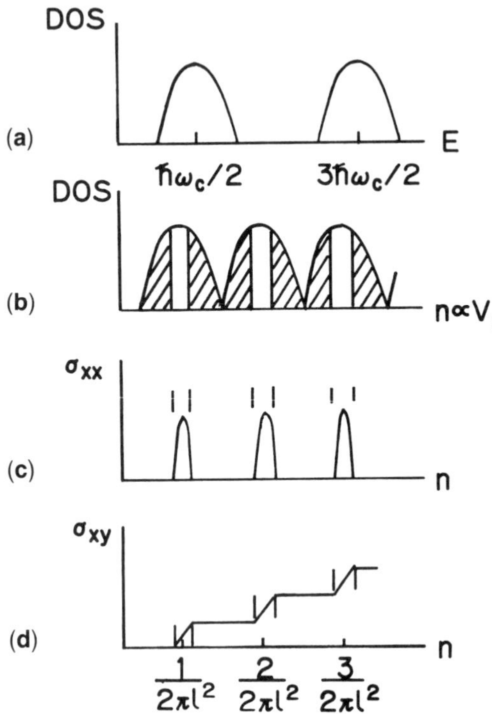

FIG. 22. Conceptual sketch of 2D Landau levels (a) as a function of energy and (b) as a function of carrier density n. When localization occurs on both sides of 2D Landau levels, σ_{xy} is quantized as shown in (d), where σ_{xx} vanishes (c).

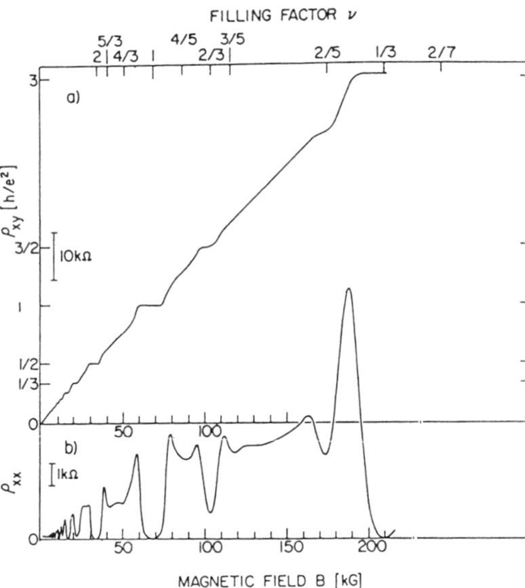

FIG. 23. Observed ρ_{xy} and ρ_{xx} as a function of magnetic field B. A number of plateaus appear for fractional filling factors shown on the upper axis, i.e., fractional QHE (after Störmer *et al.*, 1983).

Unless there is localization, the Hall-effect experiment would never have plateaus in the Hall coefficient as a function of n or V_G. The existence of localization was never thought of beforehand in experiments, but was only deduced from research on the integral QHE.

4.2.9 Fractional Quantum Hall Effect In the integral QHE, plateaus appear for parameters n that let the Fermi energy stand between neighboring Landau subbands. That is, all the subbands below the quantum number N are occupied, but those above are unoccupied. Historically, the integral QHE was discovered in a 2DEG at the Si-SiO_2 interface.

On the other hand, it is well established that the heterostructure consisting of GaAs and AlGaAs gives an excellent quality 2DEG. It shows a very high mobility up to $\sim$1000 m^2/V s and a mean free path of several microns. This kind of sample often shows very flat plateaus at magnetic fields, as seen in Fig. 23 (Störmer *et al.*, 1983), for which the carrier densities are such that a Landau subband is occupied fractionally to $\nu = p/q$ of the total electrons of $1/2\pi l^2$ per subband and unoccupied beyond. The variable p ($\neq q$) is an integer and $q = 3, 5, 7, \ldots$, an odd number, because of the Fermi statistical requirement.

This effect is called "fractional" QHE. Experimentally, integral QHE is known to occur in dirty samples, which seem to ease localization. On the other hand, fractional QHE is only observed in clean samples with a high mobility. Fractional QHE shows plateaus within the same Landau subband as in integral QHE. To simplify the explanation, let us limit ourselves to the $q = 3$ case. To interpret the effect, Laughlin (1983) introduced the possibility of correlated three-particle states.

The 2D Landau electron gas is degenerate, as already mentioned, with a degeneracy of $1/2\pi l^2$, which comes from the fact that the eigenenergy is not explicitly dependent on the quantum number, the center coordinate of cyclotron motion. However, because of the correlation energy between Landau electrons, they are mutually repulsive and are likely to form, e.g., a triangular lattice, just as in the Wigner crystal. The ground three-body state has threefold symmetry due to its correlation energy, and its occupation is $\nu = \frac{1}{3} - \delta$ (δ: infinitesimal). The Laughlin state is a trial function to perform variation on and is constructed so as to maximize mutual repulsion. More accurate calculations show that the excited state with an occupation of $\frac{1}{3} + \delta$ has a tiny Coulomb gap over the ground state energy with an occupation of $\frac{1}{3} - \delta$. This Coulomb gap plays the same role as the gap between Landau subbands in the integral case. Thus, plateaus should be observed at around $\frac{1}{3}$ occupation as a function of B. The plateau at $\nu = \frac{2}{3}$ is easily understood from the electron-hole symmetry within a single Landau subband.

4.2.10 Ballistic Transport Through a sufficiently short path, e.g., in a GaAs-AlGaAs heterostructure with high mobility, electrons can be transported without being scattered. In old vacuum tubes, this was the case since there is ordinarily no scattering in a vacuum. In this case, electrons are accelerated up to a velocity limited by the accumulated space charge, as given by the Richardson formula. In the solid state, a similar situation can occur, especially in high-quality interfaces such as in GaAs-AlGaAs (see HETEROSTRUCTURES AND SUPERLATTICES, SEMICONDUCTOR). Figure 24 shows the variation in resistance as a function of the magnetic field when a current flows from lead k to lead l while the voltage is measured between leads m and n. The measured resistance is denoted by $R_{kl,mn}$.

Even if the current path is bent, there is no resistance due to the bend, so long as the generated channels consist of eigenmodes of the potential problem. However, the path might be narrowed around the bends, and consequently, there may be considerable blocking and mixing between channels. Or, there may also be a different disturbance at the corners, which must cause a resistance on the electrons curving around the corners. This additional resistance may be called the *bend resistance*. In the absence of a magnetic field, a straight path gives a lower resistance, whereas under a magnetic field paths are more natural with bends, and hence, the difference in resistance is reduced.

4.2.11 Coulomb Blockade In the band theory of condensed matter physics, it is known that we can neglect most of the correlation energy between electrons, so long as their orbitals are extended sufficiently. The correlation energy is nothing but Coulomb repulsion energy. However, if electron orbitals are localized, this is not the case. They feel mutual repulsion. Throughout this section, we follow the line described in Averin and Likharev (1991).

In quantum dots, orbitals are limited by the boundary, so that Coulomb repulsion is not negligible in these dots. Let us consider a tunnel junction [Fig. 25(a)]. One electron tun-

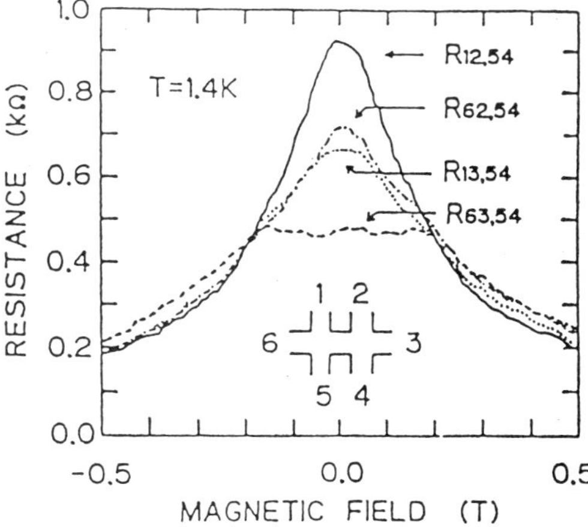

FIG. 24. Observed magnetoresistance $R_{kl,mn}$ on a sample with a current fed between leads k and l and a voltage measured between leads m and n. For a straight current path, the zero-field resistance is minimum. For a higher magnetic field, the resistance is almost independent of the path, since the current path tends to be curved (after Takagaki *et al.*, 1989).

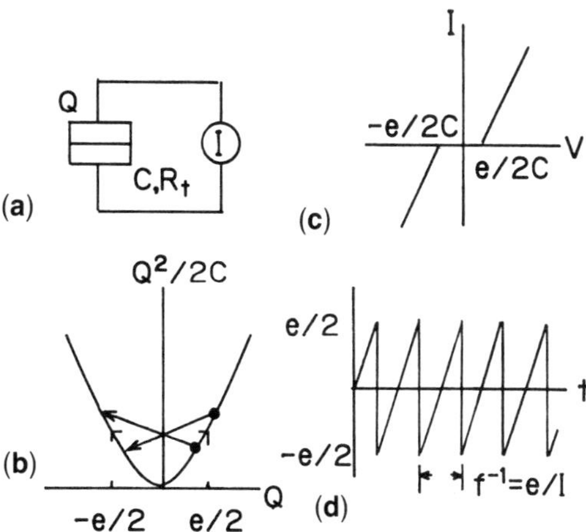

FIG. 25. (a) Diagram of the Coulomb blockade circuit. (b) Schematic interpretation of Coulomb blockade. (c) Expected I-V characteristic. (d) Single-electron tunneling oscillation with time.

nels into the opposite electrode. Then another electron may or may not tunnel in the same way because of the repulsive force. Figure 25(b) shows this concept. The parabolic curve is the electrostatic energy, $\mathscr{E} = Q^2/2C$, where Q is the charge of concern and C is the capacity of the junction. If there is a charge larger than $e/2$ and one electron tunnels out, then $\mathscr{E}$ decreases. Otherwise, $\mathscr{E}$ increases. Thus, even under an application of a voltage V where $-e/2C < V < e/2C$, the junction does not allow one electron to tunnel. Since the current I is given by dQ/dt, I is suppressed for the voltage mentioned above. This phenomenon is called *Coulomb blockade* [Fig. 25(c)].

In a circuit with a constant current I, Q is expected first to increase from 0 to $e/2$. Then Q must be suddenly reduced to $Q - e/2$ when it reaches $Q + e/2$, and starts to increase again. This is expressed by the equations $(Q_g - ne)^2/2(C + C_g)$ for the electrostatic potential energy and $V \propto (Q_g - ne)$ for the voltage between the capacitor and the tunnel junction. Thus, the charge Q oscillates as shown in Fig. 25(d). This is called *single-electron tunneling* (SET) oscillation.

Figure 26(a) shows a circuit consisting of a SET junction and an ordinary capacitor. From the discussion above, it is obvious that the electrostatic energy $\mathscr{E}$ varies as in Fig. 26(b) as a function of the number of tunneling electrons, and the inner potential drop V varies as $Q_g - ne$ when increasing $Q_g = C_gU$, where U is the external potential.

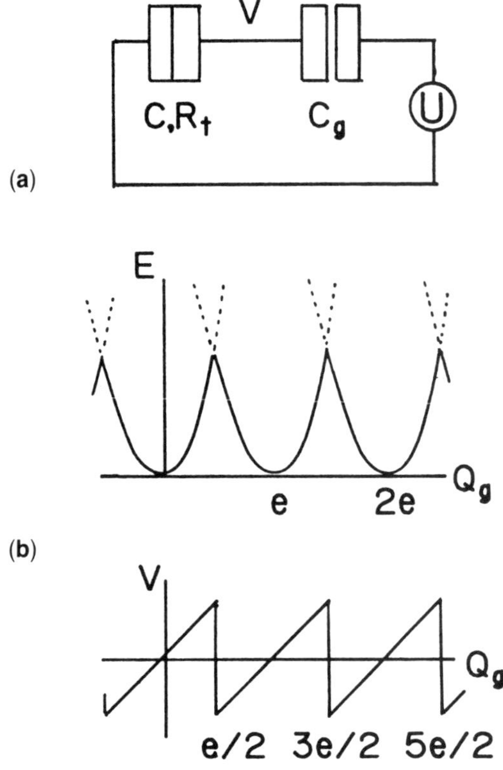

FIG. 26. (a) Circuit consisting of a conventional capacitor C and a single-electron tunnel junction C_g. (b) Electrostatic potential on the tunnel junction as a function of tunneling charge Q_g. The inner potential V varies in proportion to $Q_g - ne$, as shown below.

One example of experimental SET oscillation is shown in Fig. 27 (Lafarge *et al.*, 1991). A tiny tunnel junction is formed using Al, and a tiny electrometer was also formed on the same chip to detect the oscillating current I (Fig. 28).

4.3 Magnetic Properties

4.3.1 Fine Particles Among mesoscopic systems, 0D systems such as quantum dots and clusters are peculiar in that it is hard to measure the dc transport properties, and the only detectable properties are ac effects, i.e., responses to high frequencies including light beams. Besides optical responses, magnetic properties are also detectable.

In 1962, Kubo discussed the thermodynamic properties of metallic fine particles such as their heat capacity and paramagnetic susceptibility. Since, in fine particles, the typical energy difference between neighboring levels can be larger than ambient temperature, thermal excitation is sometimes suppressed at an ambient temperature much more than that in the bulk, continuously excitable system. This reduces the heat capacity and increases the paramagnetic susceptibility. According to this theory, the number of spins in the particles affects the magnetic properties strongly, depending on whether it is even or odd. When it is odd, the particles must behave like magnetic particles with uncompensated spin (see MAGNETIC ORDERING IN SOLIDS).

Almost at the same time, Néel (1961) considered a similar problem. Let fine particles be composed of an odd number of antiferromagnetic sublattices. The magnetic moment on the surface sublattice rotates to parallel to the external magnetic field and the

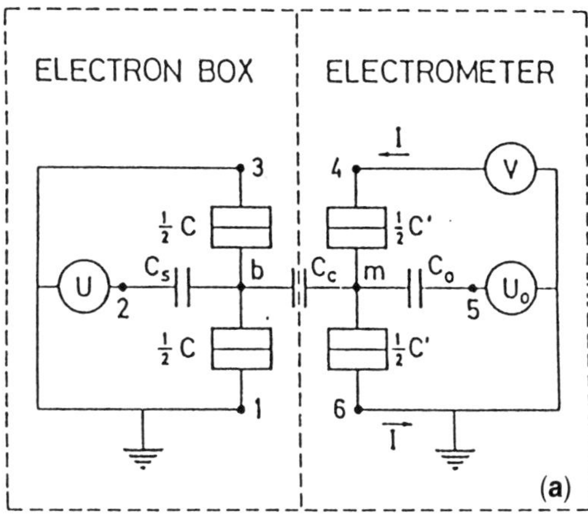

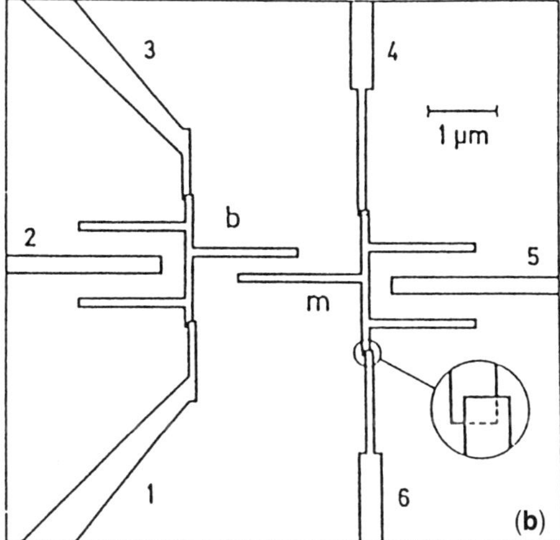

FIG. 27. Circuit used for SET oscillation experiment and its physical layout (after Lafarge *et al.*, 1991).

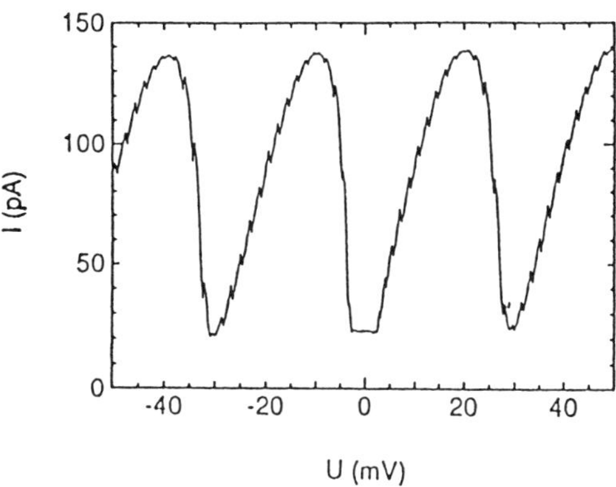

FIG. 28. Observed oscillatory current vs externally applied potential U (after Lafarge *et al.*, 1991).

moments inside are sequentially affected by the neighboring spin orientation. In this manner, antiferromagnetic particles show higher magnetic susceptibility, because they behave like a spiral spin state with a remanent magnetic moment. He called this particle state *superantiferromagnetism.*

When particles have an uncompensated magnetic moment, e.g., in single-domain ferromagnetic particles, they look like particles with a giant magnetic moment. Since the particles are not ordered when they are in a nonmagnetic medium, the whole system does not ordinarily show ferromagnetism. If the particles are small enough, ~100 Å, then they behave as a paramagnetic medium under thermal agitation. This state is called *superparamagnetism.*

It is well known that a γ ray is emitted without recoil in bulk radioactive materials, a phenomenon called the *Mössbauer effect.* However, such is not the case in a mesoscopically small-sized particle. The recoil energy must be compensated for by the motion of the particle as a whole. So, when it is sufficiently small and behaves without coupling to its environment, the recoil energy must be appreciable and manifest itself eventually as a shift of the emitted γ ray. This fact was pointed out by Murayama (1966a).

4.3.2 Magnetic Thin Films Two-dimensional magnetic systems show very interesting phenomena. The first example is the stripe domain in Permalloy thin films (see PLATES AND FILMS, MAGNETIC). The domain structure is schematically shown in Fig. 29 and was discovered by Spain (1963) and by Saito *et al.*

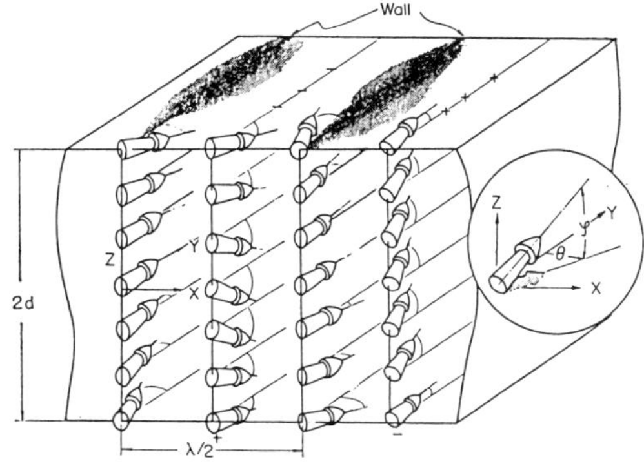

FIG. 29. Micromagnetic configuration of spins inside stripe domains (after Murayama, 1966b).

(1964), independently. Theoretically, Kaczér *et al.* (1963) proposed the simplest model, where the spin tilting from the plane varies sinusoidally as a function of the coordinate in the direction perpendicular to the spin. Murayama (1966b) showed there is a *micromagnetic* state with a lower energy, as shown in Fig. 29, where spins form a circulating Landau-type closure domain in the cross section of the film perpendicular to the mean direction of spins. In this domain structure, the thickness and stripe width are on a comparable scale, a few times 1000 Å.

In 1967, Bobeck discovered that a serpentine stripe domain in the absence of a magnetic field changes into magnetic cylindrical (bubble) domains in the presence of a perpendicular dc field in some orthoferrites. These bubbles can be driven by applying an ac magnetic field, from beneath one magnetized segment on the surface to beneath another. The domains have been developed as a BORAM (block-oriented random access memory) just like a CCD (charge-coupled device) in an MOS structure.

4.3.3 Superlattice of Magnetic Thin Films Magnetic thin films sometimes show unusual effects when they are formed in an SL. Figure 30 shows *giant magnetoresistance* on a [(Fe 30 Å)/(Cr 9 Å)]$_{60}$ SL at 4.2 K (Baibich *et al.*, 1988). Ordinary bulk material has a magnetoresistance of at most several percent, whereas in this SL it amounts to several tens of percent. From the application point of view, this effect is worth development.

4.4 Properties of Macroscopic Quantum States

Three typical examples of macroscopic quantum states are lasers, superfluids, and superconductors. Here, we are most interested in superconductivity (SC) (see SUPERCONDUCTIVITY, LOW-TEMPERATURE).

It is well established that the SC state can be described in terms of a macrowave, which is a coherent wave extending from one edge of the sample to the other. This is almost always the case even if the sample is macro-

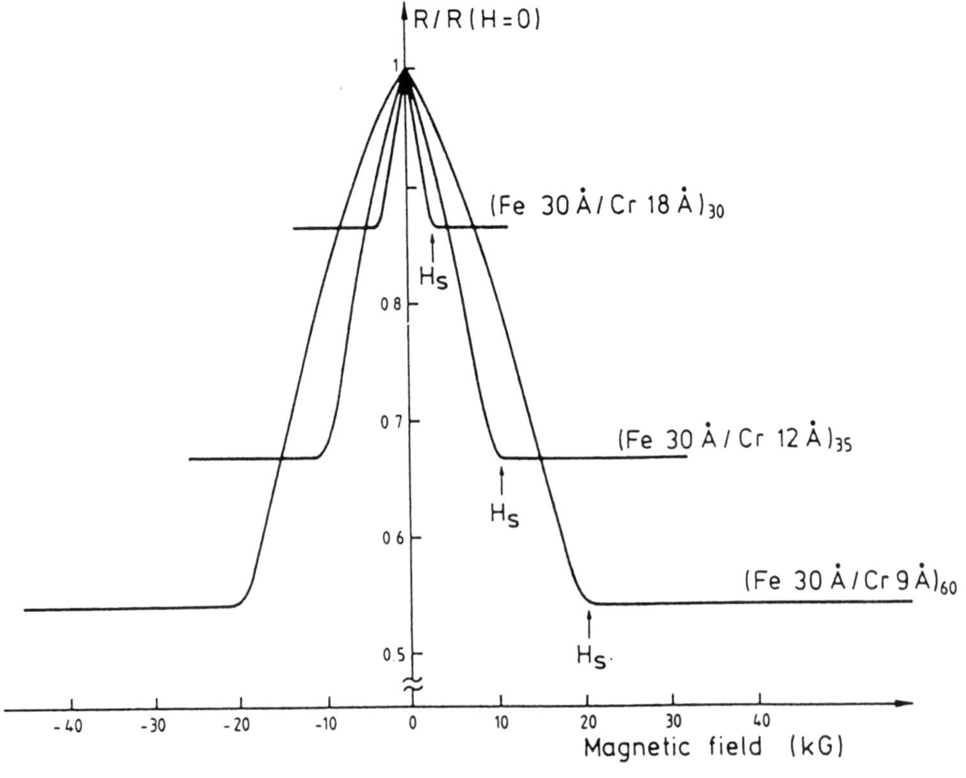

FIG. 30. Observed giant magnetoresistance on superstructured samples with Fe and Cr (after Baibich *et al.*, 1988).

scopically large, so long as it is superconducting. This means that the large sample behaves just like a tiny microscopic system from the point of view of its physical properties. The mesoscopic systems we are discussing are limited to those where the physical properties are between the microscopic and macroscopic scales. In SC, this kind of degraded coherence can occur, e.g., when the system is thin enough.

An ideal SC is made in bulk samples where there is no decay in the current between two leads or in a persistent ring current. This kind of current can flow persistently only without damping. However, there is always a slight damping mechanism even in the SC state. In high-T_c superconductors, SC is specified by 2D features. In 2D systems, it is known that thermally excited fluxon-antifluxon pairs interact with the current fed to measure the voltage drop through the sample. Thus, the *I-V* characteristic is subject to a power law, according to the Kosterlitz-Thouless (KT) (1973) modes intrinsic to 2D systems. This means that a voltage drop of exactly zero is expected only for a vanishing measuring current.

Besides the damping owing to the KT mechanism, superconducting diamagnetism caused by a Meissner current can change as a result of thermal excitation beyond barriers that separate a potential minimum with N from neighboring minima with $N \pm 1$ fluxons (Fig. 31). The case of persistent current through a ring circuit is schematically depicted in the figure. This quantized state of fluxons is known as the London quantization. This quantization is nothing other than the case for the Aharonov–Bohm phase equal to 2π, i.e., $2\pi\Phi/\Phi_0^* = 2n\pi$, or $\Phi = n\Phi_0^*$ with $\Phi_0^* = h/2e$ for Cooper pairs. This is because the macrowave must continue to itself without any phase shift (mod 2π).

4.4.1 Superconducting Wire Let us consider a 1D wire system. Figure 32 shows the variation in the phase of the macrowave along the wire when a bias current I is applied, according to the fact that the energy $E(\phi)$ is given by $E_J \cos\phi - I\Phi$. The terms ϕ and Φ are related to each other by the gauge $\phi = 2\pi\Phi/\Phi_0^*$. The variable E_J is known as the energy stored in a Josephson junction. The horizontal axis of the figure is phase ϕ and incidentally corresponds to the coordinate along the wire, since the phase is continuously rounded in one sense through the wire in regular, homogeneous states of the sample. When one fluxon jumps from one potential minimum to the neighboring minimum thermally or via resonant tunneling, the total winding number through the wire decreases. This means that a fluxon crosses the wire perpendicularly.

Thus, in wires so thin that fluxons can cross anywhere (the width $w \lesssim \lambda_L$, the London penetration length), the wires always show a voltage drop or, in other words, are resistive. Ordinarily, the resistance depends on the temperature, but, at low enough temperatures, thermally activated phase changes can no longer occur so that only tunneling takes place. The crossover is at a temperature specific to the sample (Fig. 33: Giordano, 1988). The theoretical justification of the crossover

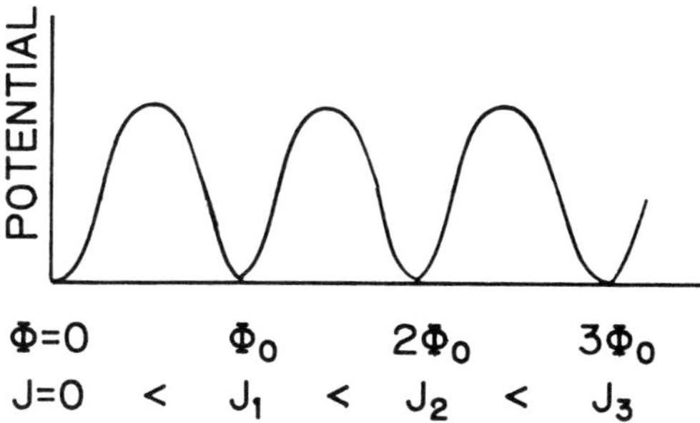

FIG. 31. Schematic interpretation of potential variation on a superconductive persistent-current-carrying ring circuit. The horizontal axis denotes a phase corresponding to a hypothetical superconducting state with a fractional number of fluxons. The fluxon-carrying state is equal to a state with a persistent current.

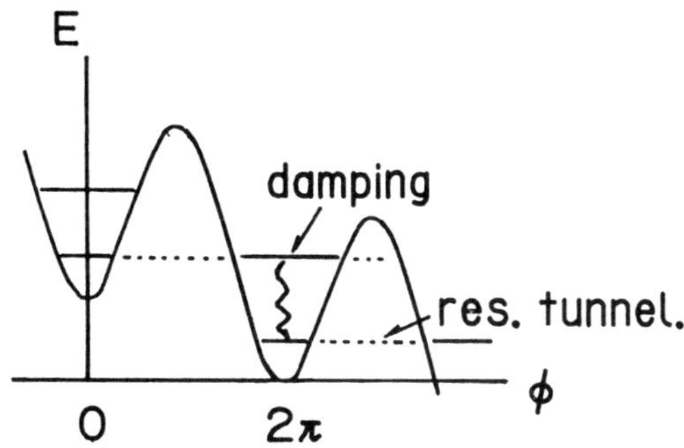

FIG. 32. "Washboard" potential of a current-biased superconducting single wire vs phase of the macrowave. Phase slippage occurs via thermal excitation and quantum tunneling.

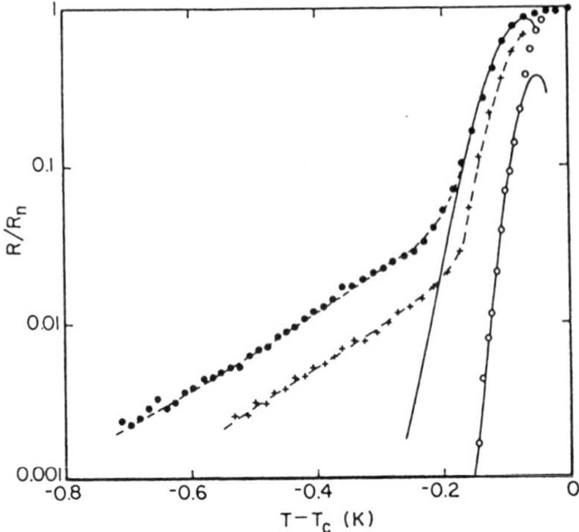

FIG. 33. Temperature vs dc resistance of thin superconducting wires with diameters 410 Å (•), 505 Å (+), and 720 Å (○) (after Giordano, 1988).

from thermal activation to tunneling was given by Saito and Murayama (1989).

It is also interesting to observe the same phenonemon in a superconductor as in an SET in its normal state. This time it is not the tunneling of *single normal electrons,* but *quasiparticles* in the superconductor. Cooper pairs may show a similar phenomenon when the Josephson junction has a small enough area.

4.4.2 Andreev Reflection Superconductivity is known to result from pairs of electrons that behave like bosons. According to the BCS theory, the most probable pair is an electron with momentum $\hbar\mathbf{k}$ and a hole with $\hbar(-\mathbf{k})$. Figure 34(a) shows schematically the case where an electron incident from the normal-metal side has an energy below the SC gap. Since the energy is not high enough for the electron to become a quasiparticle in the superconductor, it must be reflected but as a hole with an energy $\epsilon_h = 2\epsilon_F - \epsilon_{el}$. This kind of reflection was first discussed by Andreev (1964). Note that the generated hole has a momentum of $-\mathbf{k}$ and, hence, traces the same path as the electron but in the opposite direction. This is shown in Fig. 34(c), in comparison with the ordinary normal reflector [Fig. 34(b)].

In an experiment on a mesoscopic scale, Nishino (1990) observed a geometrical resonance between an Nb injector and an Nb reflector through Si. If the distance d between them is such that $\sqrt{\Delta^2 + (nhv_F/2d)^2} \sim eV$ (V

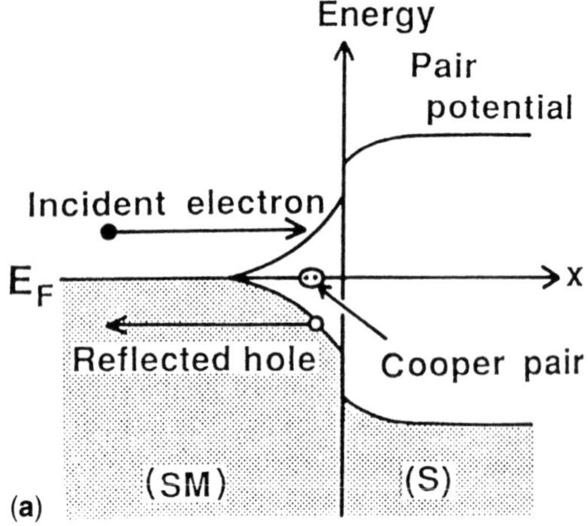

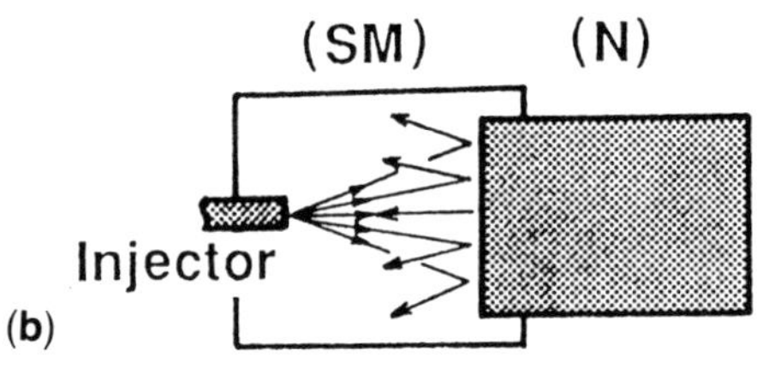

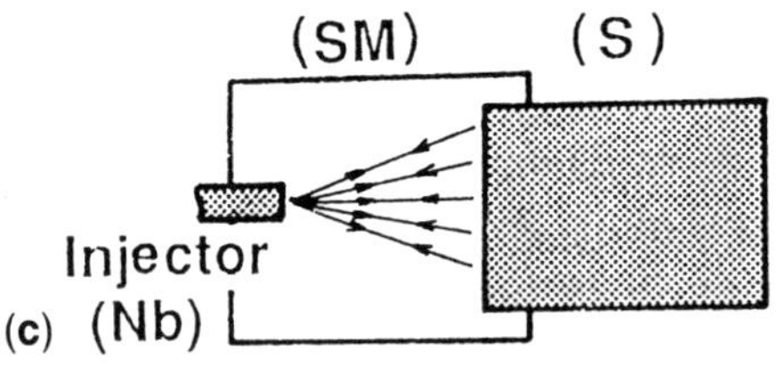

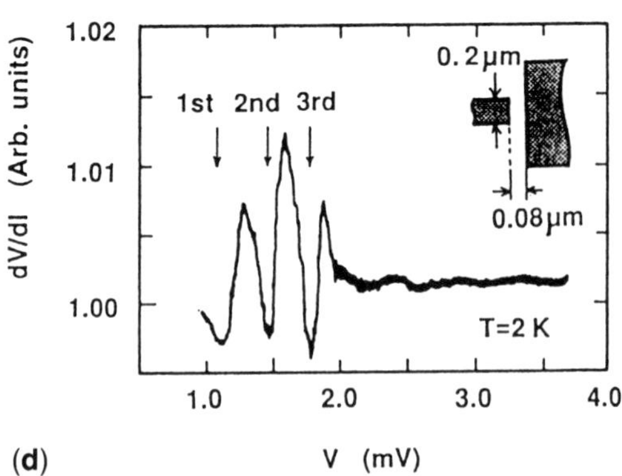

FIG. 34. (a) Conceptual explanation of the Andreev reflection. Expected paths on reflection for specular scattering (b) in the normal state and (c) for Andreev reflection. (d) Observed differential resistance vs voltage showing a few geometrical resonances (after Nishino *et al.*, 1990).

is applied voltage) with an integer n, a resistance minimum appears. These data are plotted in Fig. 34(d), which clearly shows first-, second-, and third-order geometric resonances.

5. FUTURE PROSPECTS

5.1 Physics

All the phenomena described so far are related to physics. Future prospects with respect to applications are given below. The most interesting matter in mesoscopic physics is that an appreciable degree of coherence is lost there. Regarding microscopic systems such as a hydrogen atom in its excited state, it may decay to the ground state with a specific decay time. The ground state itself can be excited by interacting with its environment such as photons, phonons, and other particles. These all reduce coherence. In fact, hydrogen atom beams do not show an interference pattern in a bad vacuum where they suffer from frequent scatterings. Thus, even an atomic-level microscopic entity may not be called "microscopic" in the exact sense in a certain environment.

On the other hand, a macrowave within superconductors may show long-lasting macroscopic coherence, even if it is macroscopically large. However, the environment always has pair-breaking (dephasing) mechanisms causing SC to be broken. There, the collectiveness of the Bose condensate and the stiffness of the macrophase prevent such dephasing. Whether we should consider a system to be microscopic or macroscopic depends on the time scale over which we observe it. The time scale typical for a mesoscopic system ranges from a very tiny scale to a considerably large one, depending what we are discussing. The only clear point is that the temporal and spatial variations are respectively specific to the system; in this sense, it becomes rather difficult to expect *a priori* the validity of ergodicity in such mesoscopic systems (Stone, 1987). The specificity in the sample of concern plays an important role in mesoscopic systems (e.g., *magnetofingerprint* in Section 4.2.5) and, accordingly, the time-averaged physical properties can deviate from the average over an ensemble of similar replicas.

Another interesting matter, seen in the examples listed so far, is that many quantities are observed in terms of universal constants, such as resistance in $\pi\hbar/e^2$, current in e times frequency f, and voltage $fh/2e$ in the ac Josephson effect. Actually, it was possible to determine the fine-structure constant, $\alpha = (\mu_0 c/2)(e^2/2h)$, where μ_0 is the permeability of the vacuum, in a mesoscopic experiment, e.g., the integral QHE.

The new physics in mesoscopic systems is Anderson localization owing to disorder. Localized states stand at the opposite position to band states. In tiny systems, localized states often appear, especially at low temperatures. It is known that low-dimensional systems ($d \leq 2$) localize electrons easily.

When electrons are localized, not in the sense of Anderson but according to the mesoscopic boundary conditions, the correlation energy between the two outweighs the difference in one-particle energy. This effect appears as the Coulomb blockade. When the correlation is large, the condensed-matter physical world can be very different from the conventional one based on band energy states.

Localization is likely to occur for a narrow band. A narrow band has a larger DOS. In condensed-matter physics, the DOS is an important concept. For a larger DOS, every physical property is enhanced. For SC to occur, a large DOS is necessary. This is the case also in a high-T_c SC (Murayama and Nakajima, 1991). Sometimes, SC competes against localization; localized electrons can obviously never show SC (Haviland *et al.*, 1989). Typical experimental evidence of competition between localization and SC in Nb thin films is shown in Fig. 35 (Nishida *et al.*, 1991).

Again, it should be emphasized that for elastic scattering any electronic state only undergoes a phase shift, which has no relation to damping or dephasing. It only shifts its original state to another (unitarily transformed) state. Physically, the difference is trivial. On the other hand, for inelastic scattering, the energy of the electron is lost to other degrees of freedom, which means damping on the electron. The degrees of freedom work frequently as a heat reservoir. This causes dephasing, and the coherence cannot survive this scattering. This degradation in coherence is a crucial concept in mesoscopic physics.

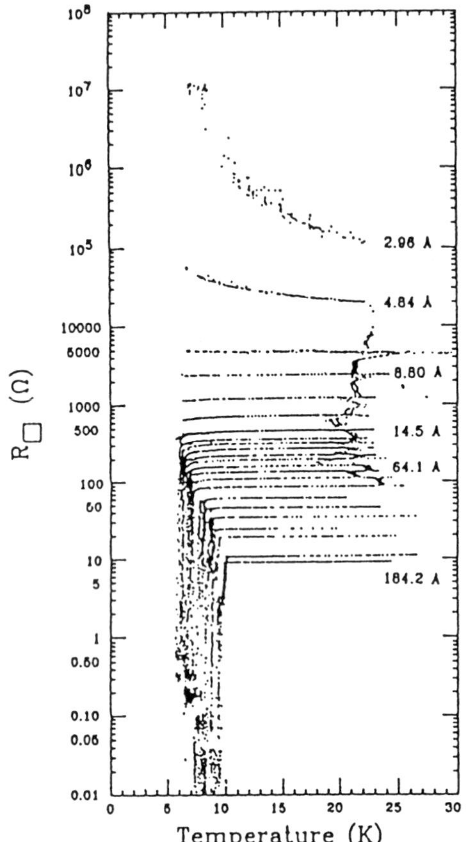

FIG. 35. Observed variation in resistance vs temperature as a function of sample thickness. For thinner samples, localization occurs, and for thicker ones, they show a transition to SC (after Nishida *et al.*, 1991).

5.2 Applications

The use of 2D structures is already conventional in semiconductor devices such as semiconductor lasers, light-emitting diodes, photo diodes/transistors, solar cells, and photoconducting membranes in image pickup tubes. In transport devices, the 2D confinement technology is applied to MOS transistors, high-electron mobility transistors (HEMT's), and RTD. All these devices aim at confining carriers, which works well for radiative recombination in optical applications. That is, radiative recombination must be over before irradiative recombination mechanisms work in order to improve the efficiency of luminescence. For semiconductor lasers, this confinement of the states is so effective in double heterostructures that continuous oscillation has been achieved at RT. Even in this device, the other two degrees of freedom in the active layer are still free. These degrees of freedom can also be frozen by implementing disorder within the layer so as to cause Anderson localization (Murayama *et al.*, 1988).

One of the most remarkable recent achievements in optical-device applications of mesoscopic systems is seen in microcavity lasers. There, an electromagnetic field is confined inside a mesoscopically small cavity, where the boundary condition affects the quantum mechanical features of the field. Even the vacuum state is controlled in such a microcavity. This topic is discussed separately in the article QUANTUM OPTICS. Several related studies are found in Ezawa and Murayama (1993).

The idea of photon equivalents of various electronic states in solids was proposed by Yablonovitch (1987). He described how, first of all, a photonic band can be formed by preparing a periodic structure composed of different electromagnetic potential layers, i.e., of different refractive indexes, for photons. Theoretically, forming such a superstructure yields all kinds of specific electronic states with photonic equivalents: the band energy states (*E-k* curves) with appropriate band gaps as well as the Anderson localization, the quantum well, the superlattice and so. The spontaneous emission rate in bulk materials is now controlled by giving the appropriate boundary conditions to the electromagnetic field. This is also a quantum optical issue.

For magnetic applications, we began with magnetic thin-film memory devices. However, this type of memory device was short-lived and was soon replaced by the magnetic core memory. Then the bubble domain memory appeared, which was actually utilized for a short while. The recent discovery of giant magnetoresistance looks promising for application, since magnetic recording heads are becoming smaller and smaller down to a thin-film structure. This is also on the mesoscopic scale. In magnetic applications, metallic superlattices are formed nonepitaxially, in a different way from in semiconductor mesoscopic systems. Nonepitaxial growth, if it produces good devices, is highly preferable for industrial applications.

For device applications, it is crucial that the effects described so far work well at RT, or at least at 77 K. The above-mentioned optical devices are used at RT. SC devices such as Josephson devices and SQUIDs will be limited to low temperatures, even if high-T_c SC's can be developed. In order to apply physical phenomena to actual devices, the energy difference between adjacent energy levels must exceed the ambient thermal energy. This means that the higher the difference, the easier the application. In this respect, a large difference in the correlation between two electrons in the phenomenon of the Coulomb blockade is highly useful. For example, a 10-Å distance means an energy difference of 1.5 eV, unless any screening exists.

Localized orbitals cause a larger correlation energy. In the future, devices on the other side of the edge opposite to the conventional band energy states may be developed. In a high-T_c SC substance like $Bi_2Sr_2Ca_{1-x}Y_xCu_2O_{8+\delta}$, holes are known to be in a localized state for $x \gtrsim 0.4$ at RT. If we inject excess carriers into this substance, they are expected to delocalize holes and change it to an SC phase. This results from the easily localizable 2D layer of CuO_2 and the fact that the band responsible for SC is sufficiently narrow. Thus, on one hand, the narrowness of bands allows many interesting phenomena due to the high DOS, and, on the other hand, localization, charge-density waves, and so on compete against the SC. This competition must be investigated in more detail for applications of mesoscopic physics to materialize.

As the last example of mesoscopically sized devices, we may mention atomic-scaled devices, which can be produced only by scanning tunneling microscope (STM) based manipulation of atoms on a certain substrate. This technology has very recently been introduced and is promising, in principle; however, it involves a number of basic problems to be resolved before becoming practical.

ACKNOWLEDGMENTS

The author is very grateful to Dr. K. Nakazato, Dr. B. J. van Wees, Dr. M. Büttiker, Dr. A. Tonomura, Dr. Y. A. Ono, and Professor T. Ando for their valuable suggestions in preparing this article. Above all, encouragement from Professor M. Tanaka in undertaking this job is highly appreciated.

GLOSSARY

Coherence: A necessary condition for two waves to interfere. The property is seen in an interferometry experiment. Coherence only occurs for two split waves that originally belonged to a single state of a fermion. Two *incoherent* waves cannot interfere. Under inelastic scattering, two waves partially lose their energy, which causes originally coherent states to degrade into incoherent ones. This process is called *decoherence*.

Localization: A concept opposite to the extended state described by the conventional band theory. In disordered solids, the Bloch theorem does not apply, and electronic states are localized on a scale much larger than the atomic size but smaller than the macroscopic sample size. Often it is called Anderson localization after the first author of the investigation.

Macrowave: When a single state is occupied by many identical waves, the ensemble of the waves acquires a definite phase, since the number-phase uncertainty allows it. This phenomenon is known to be a macroscopic quantum state. There are three typical examples: one is superconductivity, where fermions are condensed into a single ground state (Bose–Einstein condensation) by forming bosonlike Cooper pairs; the other two are superfluidity and lasers. Here, an enormous number of bosons occupy the ground state. All these states are described by macrowaves.

Quantum Well: When two materials with different electron affinities χ are epitaxially grown in series as a barrier material with a smaller χ_B, a well material with a larger χ_W, and then again a barrier material, the intermediate material forms a potential well where electrons are confined. For holes, the quantity χ plus the band gap E_g plays the same role as χ for electron confinement. A quantum well is a 2D quantized system.

Quantum Dot: A 0D quantized system, where all the degrees of freedom are lost by confining electrons into a small particle.

Quantum Wire: A 1D quantized system, where two degrees of freedom are lost by confining electrons into a thin wire.

Superlattice: Periodically repeated quantum wells form a man-made periodic system, which has a miniband structure and minigaps, following the Bloch theorem. In order to have a significant coupling between quantum wells, the barrier materials in between must be thin enough for electrons to tunnel.

Wave Packet: An analog of a classical particle with quantum mechanical properties. It is not a rigid but a soft particle in the sense that it is a superposition of eigenstates and obeys a time-dependent Schrödinger equation. The inverse of half its extension in k space is called the *coherence length*.

Works Cited

Abrahams, E., Anderson, P. W., Licciardello, D. C., Ramakrishnan, T. V. (1979), *Phys. Rev. Lett.* **42**, 673–676.

Al'tshuler, B. L., Aronov, A. G., Spivak, B. Z. (1981), *JETP Lett.* **33**, 94–97.

Anderson, P. W. (1958), *Phys. Rev.* **109**, 1492–1505.

Ando, T., Fowler, A. B., Stern, F. (1982), *Rev. Mod. Phys.* **54**, 437–672.

Andreev, A. F. (1964), *J. Exptl. Theoret. Phys.* **46**, 1823–1828.

Averin, D. V., Likharev, K. K. (1991), in: B. L. Al'tshuler, P. A. Lee, R. A. Webb (Eds.), *Mesoscopic Phenomena in Solids*, Amsterdam: North-Holland, pp. 173–272.

Baibich, M. N., Broto, J. M., Fert, A., Nguyen Van Dau, F., Petroff, F. (1988), *Phys. Rev. Lett.* **61**, 2472–2475.

Bergmann, G. (1983), *Phys. Rev. B* **28**, 2914–2920.

Bobeck, A. H. (1967), *Bell Syst. Tech. J.* **46**, 1901–1925.

Büttiker, M., Imry, Y., Landauer, R., Pinhas, S. (1985), *Phys. Rev. B* **31**, 6207–6215.

Canham, L. T. (1990), *Appl. Phys. Lett.* **57**, 1046–1048.

Giordano, N. (1988), *Phys. Rev. Lett.* **61**, 2137–2140.

Haviland, D. B., Liu, Y., Goldman, A. M. (1989), *Phys. Rev. Lett.* **62**, 2180–2183.

Hikami, S., Larkin, A. I., Nagaoka, Y. (1980), *Prog. Theor. Phys.* **63**, 707–710.

Imry, Y. (1986), in: G. Grinstein, G. Mazenko (Eds.), *Directions in Condensed Matter Physics*, Singapore: World Scientific, pp. 101–163.

Kaczér, J., Zelený, M., Sůda, P. (1963), *Czech. J. Phys.* **13**, 579–585.

Katayama, Y., Yoshida, I., Kotera, N., Komatsubara, K. F. (1972), *Appl. Phys. Lett.* **20**, 31–33.

Kawaguchi, Y., Kitahara, H., Kawaji, S. (1978), *Surf. Sci.* **73**, 520–527.

Kosterlitz, J. M., Thouless, D. J. (1973), *J. Phys. C: Solid State Phys.* **6**, 1181–1203.

Kubo, R. (1962), *J. Phys. Soc. Jpn.* **17**, 975–986.

Lafarge, P., Pothier, H., Williams, E. R., Esteve, D., Urbina, C., Devoret, M. H. (1991), *Z. Phys. B Cond. Matter* **85**, 327–332.

Landauer, R. (1957), *IBM J. Res. Dev.* **1**, 223–231.

Laughlin, R. B. (1983), *Phys. Rev. Lett.* **50**, 1395–1398.

Lee, P. A., Stone, A. D., Fukuyama, H. (1987), *Phys. Rev. B* **35**, 1039–1070.

Machida, S., Namiki, M. (1980), *Prog. Theor. Phys.* **63**, 1457–1473, 1833–1847.

Mishima, T., Kasai, J., Morioka, M., Sawada, T., Murayama, Y., Katayama, Y., Shiraki, Y. (1986), in: M. Fujimoto (Ed.), *Gallium Arsenide and Related Compounds*, Institute of Physics Conference Proceedings No. 79, Bristol: Institute of Physics, Chap. 8, pp. 445–450.

Mott, N. F. (1974), *Metal-Insulator Transition*, London: Taylor & Francis.

Murayama, Y. (1966a), *Phys. Lett.* **23**, 332–334.

Murayama, Y. (1966b), *J. Phys. Soc. Jpn.* **11**, 2253–2266.

Murayama, Y., Kamigaki, Y., Yamada, E. (1972), in: Proceedings of the 3rd Conference on Solid State Devices, Tokyo, 1971, *Suppl. Oyo Buturi*, **41**, 133–140.

Murayama, Y. (1984), *Buturi*, **39**, 485 (in Japanese).

Murayama, Y. (1986), *Phys. Rev. B* **34**, 2500–2507.

Murayama, Y., Takeda, Y., Nakamura, M., Shiraki, Y., Katayama, Y., Chinone, N. (1988), "Semiconductor Laser," U. S. Patent No. 4,759,030 (19 July 1988).

Murayama, Y. (1990), *Found. Phys. Lett.* **3**, 103–127.

Murayama, Y., Nakajima, S. (1991), *J. Phys. Soc. Jpn.* **60**, 4265–4279.

Murayama, Y. (1992), *Surf. Sci.* **263**, 604–608.

Nakayama, M., Tanaka, I., Nishimura, H., Kawashima, K., Fujiwara, K. (1991), *Phys. Rev. B* **44**, 5935–5938.

Néel, L. (1961), *Comptes Rendus*, Séance du 26 Juin, 4075–4080; Séance 3 Juillet, 9–12; Séance 10 Juillet, 203–208.

Nelson, D. F., Miller, R. C., Tu, C. W., Sputz, S. K. (1987), *Phys. Rev. B* **36**, 8063–8070.

Nishida, N., Okuma, S., Asamitsu, A. (1991), *Physica* **B169**, 487–488.

Nishino, T., Hatano, M., Hasegawa, H., Murai, F., Kure, T., Yamada, E., Kawabe, U. (1990), in: S. Kobayashi *et al.* (Eds.), *Proc. 3rd Int. Symp. Foundations of Quantum Mechanics*, Tokyo: The Physical Society of Japan, pp. 263–269.

Saito, N., Fujiwara, H., Sugita, Y. (1964), *J. Phys. Soc. Jpn.* **19**, 421–424, 1116–1125.

Saito, S., Murayama, Y. (1989), *Physica* **139**, 85–90.

Sharvin, D. Yu., Sharvin, Yu. V. (1981), *JETP Lett.* **34**, 272–275.

Sollner, T. C. L. G., Goodhue, W. D., Tannenwald, P. E., Parker, C. D., Peck, D. D. (1983), *Appl. Phys. Lett.* **43**, 588–590.

Spain, R. J. (1963), *Appl. Phys. Lett.* **3**, 208–209.

Stone, A. D. (1987), in: M. Namiki *et al.* (Eds.) *Proc. 2nd Int. Symp. Foundations of Quantum Mechanics*, Tokyo: The Physical Society of Japan, pp. 207–217.

Störmer, H. L., Chang, A., Tsui, D. C., Hwang, J. C. M., Gossard, A. C., Wiegmann, W. (1983), *Phys. Rev. Lett.* **50**, 1953–1956.

Takagaki, Y., Wataka, F., Takaoka, S., Gamo, K., Murase, K., Namba, S. (1989), *Jpn. J. Appl. Phys.* **28**, 2188–2192.

Thouless, D. J. (1977), *Phys. Rev. Lett.* **39**, 1167–1169.

Tonomura, A., Endo, J., Matsuda, T., Kawasaki, T. (1989), *Am. J. Phys.* **57**, 117–120.

Tsu, R., Esaki, L. (1973), *Appl. Phys. Lett.* **22**, 562–564.

Tsuchiya, Y., Inuzuka, E., Kurono, T., Hosoda, M. (1985), in: B. L. Morgan (Ed.), *Advances in Electronics and Electron Physics*, Vol. 64A, New York: Academic, pp. 21–31.

van Kampen, N. G. (1976), in: L. Pál, P. Szépfalus, (Eds.), *Statistical Physics, Proc. of Int. Conf.*, 25–29 August 1975, Budapest, Amsterdam: North-Holland.

van Wees, B. J., van Houten, H., Beenakker, C. W. J., Williamson, J. G., Kouwenhoven, L. P., van der Marel, D., Foxon, C. T. (1988), *Phys. Rev. Lett.* **60**, 848–850.

van Wees, B. J., Kouwenhoven, L. P., Harmans, C. J. P. M., Williamson, J. G., Timmering, C. E., Broekaart, M. E. I., Foxon, C. T., Harris, J. J. (1989), *Phys. Rev. Lett.* **62**, 2523–2526.

von Klitzing, K., Dorda, G., Pepper, M. (1980), *Phys. Rev. Lett.* **45**, 494–497.

Washburn, S., Umbach, C. P., Laibowitz, R. B., Webb, R. A. (1985) *Phys. Rev. B* **32**, 4789–4792.

Webb, R. A., Washburn, S., Umbach, C. P., Laibowitz, R. B. (1985), *Phys. Rev. Lett.* **54**, 2696–2699.

Webb, R. A., Washburn, S., Benoit, A. D., Umbach, C. P., Laibowitz, R. B. (1987), in: M. Namiki *et al.* (Eds.), *Proc. 2nd Int. Symp. Foundations of Quantum Mechanics*, Tokyo: The Physical Society of Japan, pp. 193–206.

Wigner, E. P. (1963), *Am. J. Phys.* **31**, 6–15.

Yablonovitch, E. (1987), *Phys. Rev. Lett.* **58**, 2059–2062.

Further Reading

Al'tshuler, B. L., Lee, P. A., Webb, R. A. (Eds.) (1991), *Mesoscopic Phenomena in Solids*, Amsterdam: North-Holland.

Bergmann, G. (1984), *Phys. Rep.* **107**, 1–58.

Ezawa, H., Murayama, Y. (Eds.) (1993), *Quantum Control and Measurement*, Amsterdam: North-Holland.

Grinstein, G., Mazenko, G. (Eds.) (1986), *Directions in Condensed Matter Physics*, Singapore: World Scientific.

James, T. W. (Ed.) (1985), *Characterization and Behavior of Materials with Submicron Dimension*, Singapore: World Scientific.

Kamefuchi, S., Ezawa, H., Murayama, Y., Namiki, M., Nomura, S., Ohnuki, Y., Yajima, T. (Eds.) (1984), *Proc. Int. Symp. Foundations of Quantum Mechanics*, Tokyo: The Physical Society of Japan.

Kobayashi, S., Ezawa, H., Murayama, Y., Nomura, S. (Eds.) (1990), *Proc. 3rd Int. Symp. Foundations of Quantum Mechanics*, Tokyo: The Physical Society of Japan.

Nagaoka, Y. (Ed.) (1985), in: *Anderson Localization, Prog. Theor. Phys.* Suppl. No. 84, Kyoto: Research Institute for Fundamental Physics and The Physical Society of Japan.

Namiki, M., Ohnuki, Y., Murayama, Y., Nomura, S. (Eds.) (1987), *Proc. 2nd Int. Symp. Foundations of Quantum Mechanics*, Tokyo: The Physical Society of Japan.

Ovchinnikov, A. A., Ukrainskii, I. I. (Eds.) (1991), *Electron Correlation Effects in Low-Dimensional Conductors and Superconductors*, Berlin: Springer.

Prange, R. E., Girvin, S. M. (Eds.) (1987), *The Quantum Hall Effect*, New York: Springer.

Reed, M. A., Kirk, W. P. (Eds.) (1989), *Proc. Int. Symp. Nanostructure Physics and Fabrication*, New York: Academic Press.

Thouless, D. J. (1978), *Les Houches, Ecole d'Eté de Physique Théorique, Session XXXI*, Amsterdam: North-Holland.

METAL-INSULATOR TRANSITIONS

T. MAURICE RICE, *Eidgenössische Technische Hochschule, Zürich, Switzerland*

INTRODUCTION

Most solids can be classified simply as metallic or insulating depending on the size of their electrical conductivity. This difference becomes more pronounced as the temperature is lowered, and in this chapter a metal will be defined as a material whose conductivity extrapolates to a finite value at zero temperature and an insulator as a material whose conductivity vanishes when extrapolated to zero temperature. In a small number of cases metal-insulator transitions can be triggered by changes in temperature or by the application of pressure. More generally, an insulator can be made metallic by changing the stoichiometry or by doping, i.e., by adding a small concentration of foreign elements. Such metal-insulator transitions lead to dramatic changes of many orders of magnitude in the electrical conductivity.

The explanation of the difference between metallic and insulating solids was one of the earliest successes of the quantum theory of solids. Bloch energy bands, which are the energy levels of a single electron moving in the periodic potential of the crystal, can be exactly filled in a stoichiometric material with an even number of electrons per unit cell. This leads to a finite energy gap between the ground state and all excited states. Such a material will have insulating behavior. On the other hand, if the Bloch bands are partially filled, there is no energy gap and a current-carrying state can have an energy arbitrarily close to the ground state. This can happen with an even number of electrons per unit cell if Bloch bands overlap at the Fermi (i.e., the highest occupied) energy. It must be the case if the number of electrons per unit cell is odd, and then the single-electron theory predicts metallic behavior.

It was soon recognized that exceptions to this rule occurred. In particular de Boer and Verwey (1937) raised the case of NiO, which is an insulator despite the fact that the Ni-derived 3*d* bands are partially filled. The explanation was quickly provided in a remark by Peierls, who pointed out that the Coulomb interaction between electrons could destroy metallic behavior. This idea was extended and formalized by Mott (1949, 1956, 1961). In Mott insulators, electrons are localized to individual atoms by the Coulomb interaction, and a finite energy is required to move an electron from one atom to another. The localized electrons have magnetic moments from spin, and

3-527-28132-0/94/$5.00 + .50

possibly orbital, degrees of freedom, and these moments undergo an ordering transition as the temperature is lowered. This magnetic behavior distinguishes Mott insulators from band insulators. There is no (or at least only a very small) energy gap in the spin degrees of freedom, but there is a large energy gap for the charge degrees of freedom, and it is this charge gap that is responsible for the insulating behavior.

It is a curious, but poorly understood, fact that a transition between metallic and insulating behavior is rarely seen for stoichiometric materials even under the application of pressure. This is true although on general grounds at high enough pressure the broadening of the energy bands will inevitably lead to overlapping of bands and so to metallic behavior for all materials. However, the energy gaps in Mott insulators are large, and consequently the pressures required to reach the metallic state are large.

The nature of the metal-insulator transition was discussed in early articles by Landau and Zel'dovich (1943) and Mott (1949), who stressed the important difference in the effective Coulomb interaction between charge carriers in insulators, where it remains long range, and in metals, where it is short range because of screening by the other carriers. Metallic screening, however, requires a finite density of carriers, and so they argued that a continuous metal-to-insulator transition with arbitrarily small values of the conductivity at zero temperature is impossible, at least in an ordered material.

Changes in the stoichiometry through doping immediately alter the conductivity of an insulator and can lead to a transition to metallic behavior. The randomness that is an inevitable by-product of the doping procedure changes the character of the transition, so that it is necessary to distinguish between the relatively rare cases of metal-insulator transitions in ordered stoichiometric crystals and the much more common cases of transitions caused by doping away from stoichiometry with the inevitable randomness that is introduced in the doping process.

1. METAL-INSULATOR TRANSITIONS IN ORDERED SYSTEMS

1.1 Band-Crossing Transitions

Conceptually the simplest kind of metal-insulator transition is that due to band crossing leading to an overlap at the Fermi energy. The width of a Bloch band in a solid is a sensitive function of the atomic separation and therefore of the pressure. If the energy gap between the occupied and unoccupied energy bands can be closed under pressure, then a transition from insulator to metal or, more precisely, from semiconductor to semimetal will ensue. Assuming this transition were to be continuous, then as the bands cross a small number of electrons would be transferred from the occupied valence bands to the unoccupied conduction bands; but there is a Coulombic attraction between the electrons in the conduction band and the holes left behind in the valence band. A single electron-hole pair in a semiconductor will always bind to form a neutral exciton as a result of the long-range nature of the Coulombic attraction. So to create a conducting state, a finite density of electrons and holes is required that are able to screen the Coulomb force and so form an electron-hole plasma. In practice a suitable system that displays such a semiconductor-semimetal transition under pressure has not been found. Instead, if pressure is applied to a typical semiconductor such as Si or Ge, the insulator-to-metal transition coincides with a crystallographic transition between two crystalline phases with different symmetries and is not in principle different from a standard crystallographic phase transition. Of interest are the cases where the metal-insulator transition takes place without a change of lattice symmetry or between phases with closely related symmetries.

The best example of such a band-crossing transition within a single crystallographic phase is provided by the divalent fcc metals Ca, Sr, and Yb. Actually in this case the transition is from metal to insulator under pressure due to an increase of the direct band gap even as the overall energy bands broaden under pressure. The best studied example is Yb, which has the lowest transition pressure, ~10 kbar (McWhan *et al.*, 1969). The metal-insulator transition is found to be continuous rather than discontinuous or first order as discussed above. It is believed that this difference arises because of a very small energy scale and therefore a great sensitivity to impurities, crystalline defects, etc., which mask the true nature of the transition in a pure system.

Nonetheless there is an elegant way to explore the band-crossing transition by examining the phase diagram of an electron-hole fluid. By optical pumping of an indirect semiconductor such as Si or Ge, an electron-hole fluid with a long lifetime can be created and its thermodynamic behavior studied (for reviews see Hensel *et al.*, 1977; Rice, 1977; and Jeffries and Keldysh, 1984). A metallic self-bound electron-hole liquid is found to be the lowest energy state, lower than exciton states. The relationship to the semiconductor-semimetal transition is clear when one considers that this transition occurs by varying the energy to create an electron-hole pair, i.e., the chemical potential of the electron-hole fluid. The electron-hole fluid has a first-order liquid-gas transition below a critical temperature. Assuming a linear relationship between the chemical potential and an external variable such as pressure, then this implies that a first-order metal-insulator transition below the critical temperature (see Fig. 1) occurs in a band-crossing transition. Note that the energy scale is typically small ($\sim 10^{-3}$ eV) because of dielectric screening of the Coulomb interaction due to higher-energy interband transitions.

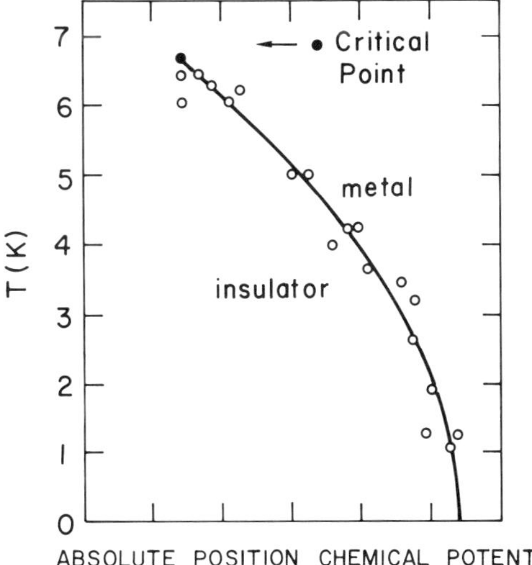

FIG. 1. The phase diagram of the electron-hole fluid in optically pumped Ge as determined by Thomas *et al.* (1974). A metallic phase is stable for values of chemical potential larger than a critical value below a critical temperature. In the analogous band-crossing transition this transition to a metallic phase corresponds to a discontinuous transition between an insulating phase, with a finite energy gap between conduction and valence energy bands, and a metallic phase, with finite overlap between these bands.

1.2 Mott Transitions

In the simplest view, a Mott transition arises because of the competition between the intra-atomic Coulomb interaction U, which opposes charge fluctuation, and the bandwidth W, which represents the kinetic energy that can be gained by allowing charged excitations to propagate through the crystal. The criterion separating metallic from insulating behavior then is determined by the comparison $U \lesssim W$ or $U \gtrsim W$. Since U is an intra-atomic Coulomb interaction, it should not be sensitive to pressure, unlike bandwidth, W. Therefore, in principle it should be possible to convert all Mott insulators into metals by applying enough pressure. In practice this is not so easy since Mott insulators generally have large energy gaps for charge excitations; e.g., NiO, the classic Mott insulator, has an energy gap ≈ 5 eV. Thus very large pressures are required, which, in turn, prevents detailed experimental study.

Vanadium sesquioxide, V_2O_3, is the best known exception. At atmospheric pressure, it displays a first-order metal-insulator transition (Föex, 1946) (Fig. 2) upon cooling at a temperature $T_{MI} \approx 150$ K with a resistivity increase by a factor $\sim 10^7$. The low-temperature phase is insulating and antiferromag-

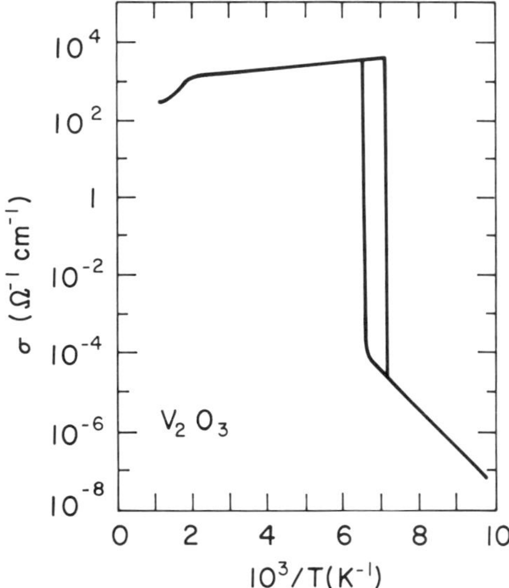

FIG. 2. The conductivity of V_2O_3 showing the metal-insulator transition found by Föex (1946) at 150 K. In addition, there is an anomaly in the conductivity at temperatures ≈ 600 K as discussed in the text.

netically ordered with an ordered moment $\sim 1.2\mu_B/(\mathrm{V\ atom})$. In addition at high temperature, ~600 K, a broad anomalous increase in the resistivity is observed.

McWhan and coworkers (1971), by a judicious use of doping and pressure, established the full phase diagram of the V_2O_3 system. Substitution of Cr for V was found to act as a negative pressure that can be counteracted by applying real pressure, and this allows the complete phase diagram to be established (see Fig. 3). The key feature is an almost vertical line separating two phases with the same corundum crystal structure, a higher-temperature insulating phase and a lower-temperature metallic phase. In the insulating phases the electrons are localized to form V^{III} oxidation states, whereas the oxidation state fluctuates through charge fluctuations in the metal. This change does not require a change of symmetry so that the metal-insulator transition line ends at a critical point, which at atmospheric pressure is ≈400 K. The anomalous increase in the resistivity in pure V_2O_3 at ~600 K (see Fig. 2) can now be seen as a crossover at temperatures above the critical point. A key feature of the localization of the electrons is that the associated spin degrees of freedom remain free, leading to an increase of entropy, so that now the transition is from metal to insulator with increasing temperature. Only the charge degrees of freedom are frozen out by localization. The spins of the V^{III} oxidation state are coupled by superexchange processes and order antiferromagnetically as the temperature is lowered. The existence of this magnetic ordering transition is an important distinction from the band-crossing case discussed above. It leads to a distortion of the metal-insulator phase boundary due to the additional gain in free energy of the insulator caused by magnetic ordering. The metal-insulator transition remains first order, in agreement with the general considerations discussed above, although the actual size of the discontinuity may well be influenced by coupling with the lattice.

The metallic phase of V_2O_3 shows clear signs of the nearby localization transition. The effective degeneracy or Fermi temperature of the metal is greatly reduced near the metal-insulator boundary. This shows up in an enhancement by roughly an order of magnitude in the Pauli spin susceptibility and linear specific-heat coefficient and very clearly in the resistivity versus temperature. The latter (Fig. 4) displays a T^2 law at low temperatures, characteristic of Fermi liquids. The term Fermi liquid is used to describe an electronic state that is only minimally modified by the Coulomb interaction between the electrons

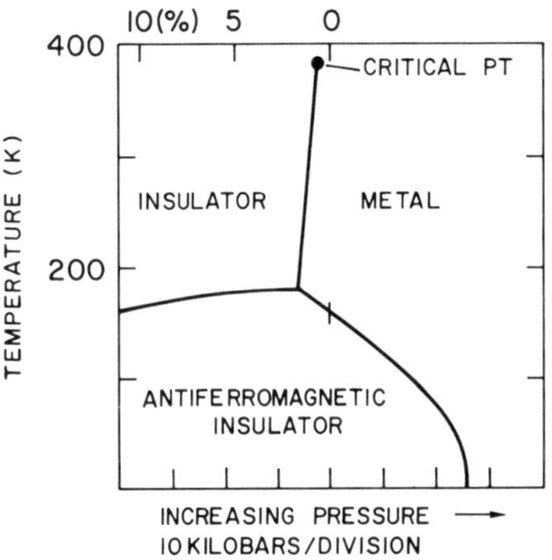

FIG. 3. Phase diagram of the metal-insulator transition in V_2O_3 and $(V_{1-x}Cr_x)_2O_3$ as a function of temperature T and pressure P as determined by McWhan *et al.* (1971).

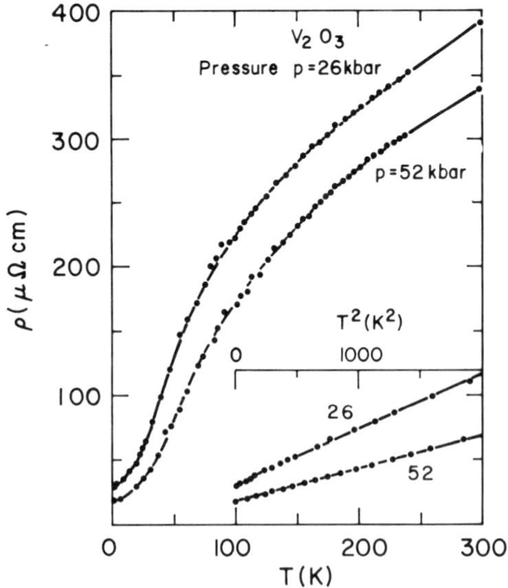

FIG. 4. The resistivity of metallic V_2O_3 under pressure as a function of temperature showing the enhanced T^2 dependence as the critical boundary to the insulating phase is approached (McWhan and Rice, 1969).

from the single-electron description in terms of partially filled energy bands. The coefficient of the T^2 term is unusually large, however, and it is a strong function of pressure and increases as the metal-insulator boundary is approached. This in turn leads to a reduction in the temperature that characterizes the crossover to coherent motion of the electrons through the periodic lattice that is a key property of a Fermi liquid (see Brinkman and Rice, 1970). The behavior of the metallic state near the Mott transition contrasts with the metallic state near a band-crossing transition that has a small overlap of the valence and conduction bands. In this latter case characteristic metallic properties such as the Pauli spin susceptibility and the linear specific-heat coefficient at low temperatures vanish as the metal-insulator transition is approached, whereas in the approach to a Mott transition they are enhanced. This difference reflects the difference in the insulating state—a band insulator has an energy gap for both spin and charge excitations, whereas in a Mott insulator the gap is only for charge excitations, and the spin excitations develop a low energy scale.

1.3 Metal-Insulator Transitions due to Lattice Distortion

In vanadium dioxide, VO_2, a strong first-order metal-insulator transition with a rise of roughly a factor of 10^5 in the resistivity at a temperature of 340 K was found by Morin (1959) (see Fig. 5). In contrast to the case of V_2O_3, the crystal structure changes at the transition from the tetragonal rutile phase, characterized by uniform chains of V atoms, in the metallic phase, to a monoclinic insulating phase with close V-V pairs. These have a separation of only 2.65 Å compared to a regular 2.88-Å separation in the high-temperature phase. The oxidation state V^{IV} has an electronic configuration $3d^1$ and a spin $s = \frac{1}{2}$ and so favors the formation of singlet pairs and a nonmagnetic insulating phase.

The large displacements of the V atoms that accompany the metal-insulator transition show that in this system the electron-lattice interaction plays an important role in stabilizing the insulating phase, and, in contrast to the V_2O_3 system, the phase transition is not sensitive to pressure. Nonetheless there are indications that electron localization plays a role here, too. In this regard there is the curious fact that an intermediate insulating phase can be stabilized by substitution of Cr or, more clearly, by applying uniaxial pressure (see Fig. 6). In this intermediate phase one-half of the V chains in the rutile structure pair, but the other half undergo a zigzag distortion, remaining unpaired. They form instead a chain of localized V^{IV} ions with spins $s = \frac{1}{2}$.

There are several Ti oxides that also have metal-insulator transitions due to the formation of singlet pairs of Ti^{III} ions. In the case of titanium sesquioxide, Ti_2O_3, the corundum structure contains Ti-Ti pairs, so that a continuous transition without a change of lattice symmetry is possible between a low-temperature insulating phase with close Ti-Ti pairs and a high-temperature metallic phase with an increased separation of the Ti-Ti pairs. In Ti_2O_3, this gradual transition with a change of a factor of 10 in the resistivity takes place around 500 K (Abrahams, 1963).

1.4 Metal-Insulator Transition due to Charge Ordering

The previous discussion was limited to cases where the average valence is an integer and all sites are equivalent. However, metal-insulator transitions also occur in cases when the average valence is a fractional number, i.e., in materials containing a mixture of at least two different oxidation states. In such systems, insulating behavior occurs when the intersite Coulomb repulsion, and possibly also the electron-lattice interaction, stabilizes a charge-ordering transition. The insulating phase has an ordered superstructure formed from ions with different oxidation states.

The Magneli phases, which are intermediate between the dioxides and sesquioxides with a chemical formulae T_nO_{2n-1} (T = Ti or V), generally undergo charge-ordering transitions. Particularly interesting is the case of Ti_4O_7, which has equal concentrations of Ti^{III} and Ti^{IV} oxidation states with $3d^1$ and $3d^0$ electronic configurations, respectively. The charge-ordering transition occurs in two stages at T = 150 and 130 K with a narrow intermediate range (~20 K wide) of reduced conductivity that is interpreted as a fluid state of singlet Ti^{III}-Ti^{III} pairs (Lakkis *et al.*, 1976). At lower temperatures ($T < 130$ K) charge or-

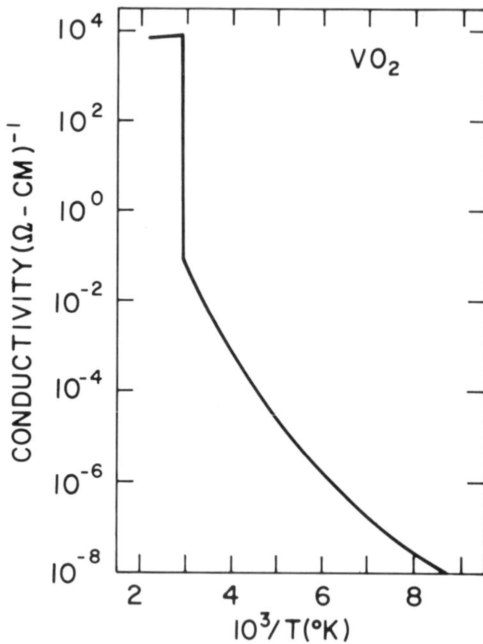

FIG. 5. The resistivity of VO_2 as a function of the temperature showing the metal-insulator transition at 340 K (from Ladd and Paul, 1969).

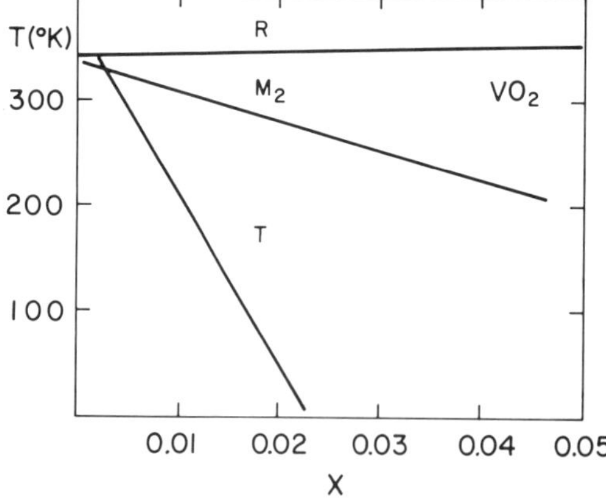

FIG. 6. The phase diagram of the $V_{1-x}Cr_xO_2$ system (Pouget *et al.*, 1974). Only the high-temperature rutile (*R*) phase is metallic. The insulating monoclinic phases M_1 and M_2 have either all V-V pairs (M_1) or one-half the V sites in pairs and the other half in localized V^{IV} chains (M_2). *T* is a triclinic insulating phase intermediate between M_1 and M_2.

dering of the Ti^{III}-Ti^{III} pairs occurs and the conductivity is further reduced.

The Verwey (1939) transition in magnetite (Fe_3O_4) is another example of a metal-insulator transition driven by charge ordering. In this case the rise in resistivity is modest (a factor of 10^2) although the transition temperature is relatively low (119 K), and the charge ordering is between Fe^{II} and Fe^{III} oxidation states on one of the sublattices in the structure.

1.5 Metal-Insulator Transitions in Lower-Dimensional Systems

The case of a one-dimensional metal is special. One-dimensional metals occur when electrons are constrained to move along chains of atoms or molecules. Peierls (1954) showed that they are inherently unstable toward the formation of charge-density-wave (CDW) states, which are characterized by a periodic charge modulation with a periodicity related to the electronic density on the chains. Such a CDW introduces a gap in the energy spectrum for charge and spin excitations. A motion of the CDW (i.e., the modulated charge) relative to the background lattice, which would contribute to the electrical conductivity, is generally not possible at small electric fields because of various effects that lock the two periodic systems to each other. As a result the CDW state is insulating and the Peierls transition is also a metal-insulator transition.

Several examples of Peierls transitions are known, and in particular the cases of the organic conductor TTF-TCNQ (tetrathiafulvalene tetracyanoquinodimethane) and the inorganic chain compound $K_{0.3}MoO_3$ are well studied. In both cases the metal-insulator transition is continuous at temperatures of 60 and 200 K, respectively.

2. METAL-INSULATOR TRANSITIONS IN DISORDERED SYSTEMS

2.1 Uncompensated Doped Semiconductors

When dopants with a different nuclear charge from that of the host are introduced into a semiconductor, shallow donor or acceptor levels may result. Such systems can be well described by a hydrogenic model.

Well-known examples are the donor systems Si:P and Ge:Sb and the acceptor systems Si:B and Ge:Ga. In the hydrogenic model the Coulomb interaction between the charges is reduced by the dielectric constant κ of the semiconductor. The electron or hole effective mass m^*, determined by the curvature of the conduction or valence bands, enters the kinetic energy. As a result, the hydrogenic donor or acceptor orbits can be very large and have characteristic radii that are many lattice constants, justifying the hydrogenic model.

In the dilute limit, the uncompensated donor (acceptor) system is a form of Mott insulator with the electrons (holes) localized on individual donors (acceptors) to form an array of well-separated hydrogenic atoms. As the dopant density n increases, so does the overlap of the hydrogenic atoms, and a Mott transition to a metallic state ensues. In Fig. 7 a series of curves for the resistivity ρ versus $1/T$ at various acceptor concentrations for the system Ge:Ga illustrate the extreme sensitivity of ρ to the concentration. For acceptor concentrations $n \lesssim 10^{17}\ \text{cm}^{-3}$, $\rho(T)$ continues to rise down to the lowest temperatures, ≈ 1 K in the figure. But for values of $n \gtrsim 10^{17}\ \text{cm}^{-3}$, $\rho(T)$ extrapolates to a finite value as $T \to 0$ K. In the clearly insulating regime with $n \ll 10^{17}\ \text{cm}^{-3}$ the resistivity has an exponential temperature dependence with an activation energy ϵ_i ($i = 1, 2$, or 3) depending on the concentration and temperature regime. This difference in activation energy is believed to reflect different processes that control the resistivity, but we will not discuss this further. Rather we wish to stress the point that an activation energy in the resistivity reflects an energy gap for charge excitations. The charge carriers (electrons or holes) freeze out at low temperatures on to hydrogenic donor or acceptor orbits. This process shows up also in an exponential rise in the Hall coefficient R_H, which is inversely proportional to the effective number of carriers n^* ($R_H \propto 1/n^*$). Thermal excitation across the energy gap between localized orbits and ex-

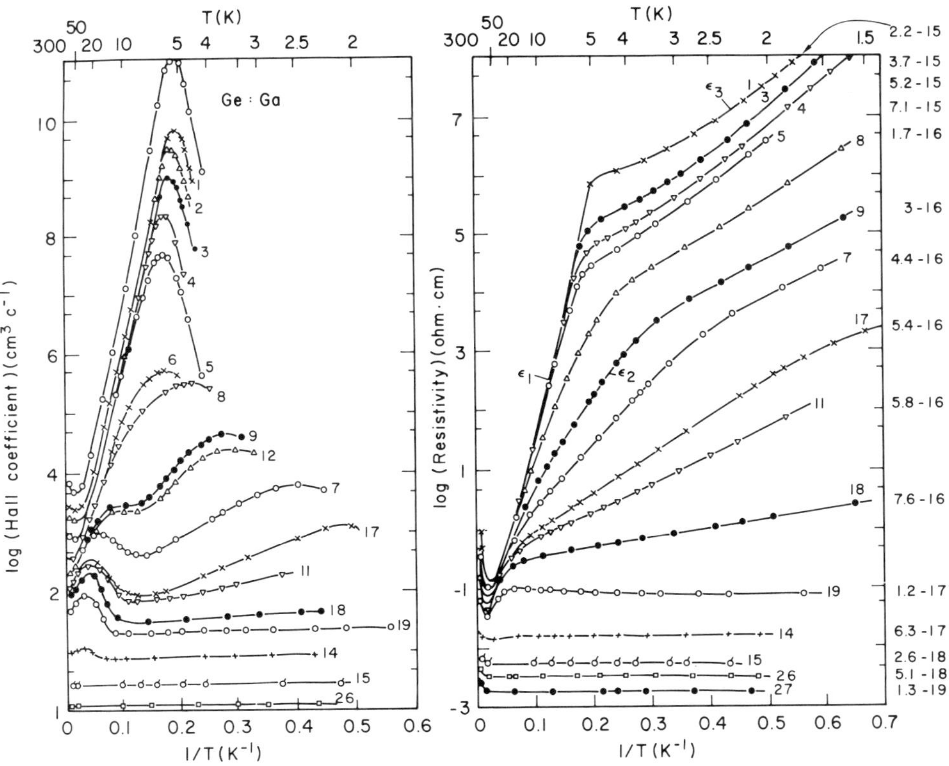

FIG. 7. The Hall coefficient and resistivity of Ga-doped Ge samples as functions of temperature for different Ga concentrations, as determined by Fritzsche (1955, 1978). The concentrations of the Ga acceptors are shown using the notation $X\text{-}Y \Rightarrow X \times 10^Y\ \text{cm}^{-3}$.

tended states is then the source of the exponential form for n^*.

Using scaled hydrogenic values for the intrasite Coulomb repulsion and the band-width for charged excitations, Mott (1949, 1961) proposed a formula for the critical density n_c separating metallic and insulating behavior:

$$n_c^{1/3}\, a_H^* \simeq 0.25, \tag{1}$$

where a_H^* is a scaled Bohr radius ($a_H^* = \hbar^2\kappa/m^*e^2$). This critical density depends on the material parameters κ and m^*. In a wide variety of semiconductors one finds that the critical dopant density agrees well with this criterion (see Fig. 8).

In contrast to the ordered systems, the metal-insulator transition here appears as a true critical point only at zero temperature. At finite temperatures a smooth crossover is observed with arbitrarily small values of the conductivity (see Fig. 9), rather than the discontinuous transition with a jump in conductivity discussed in Sec. 2. An important physical difference here is the strong disorder present in the doped semiconductors. Since the dopants generally are introduced

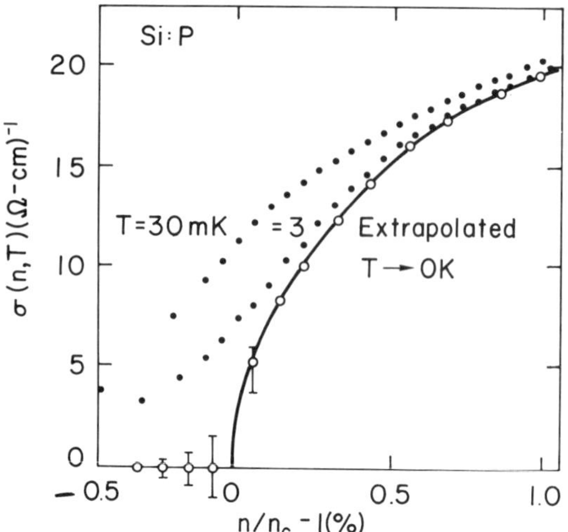

FIG. 9. The dc conductivity of Si:P as it is tuned through the metal-insulator transition by applying an external stress to vary n_c. The solid circles are data at temperatures of 30 and 3 mK. The open circles are data extrapolated to T = 0 K (Thomas *et al.*, 1983).

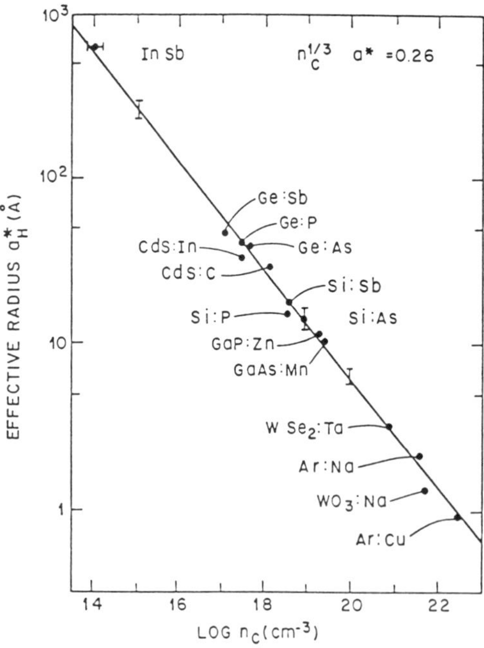

FIG. 8. The Mott criterion, Eq. (1), for the critical concentration dividing metallic ($n > n_c$) from insulating ($n < n_c$) behavior in a variety of systems. Data compiled by Edwards and Sienko (1978).

at high temperatures, the spatial correlations between the dopant positions are on the scale of a lattice constant. This scale is much smaller than the size of the scaled hydrogenic orbits of the donor or acceptor wave functions. As a result, the disorder in the doped semiconductors is very strong and represents an important physical difference from the ordered systems discussed earlier.

The nature of the metal-insulator transition in doped semiconductors has been the subject of many experimental investigations over the years. A particularly detailed set of measurements has been reported recently by Thomas *et al.* (1983) on the classic donor system Si:P. They extended the temperature range of the measurements down to 3 mK. Further they used an applied stress S to tune continuously across the critical region. In this way they could obtain much more precise measurements of the critical behavior. On the metallic side they extrapolated their results down to zero temperature using a form $\sigma(T) = \sigma(0)(1 + aT^{1/2})$ and obtained the critical behavior of $\sigma(0,n,S)$.

The extrapolated values fit well to a critical form

$$\sigma(0,n,S) = \sigma_0\{[n/n_c(S)] - 1\}^{\mu}. \tag{2}$$

The prefactor σ_0 and, especially, the critical

density n_c are functions of the external stress S, and it is this sensitivity that allows one to tune through the metal-insulator transition. Theoretical interest has focused on the fact that the conductivity varies continuously to zero and on the value of the exponent μ, which characterizes the critical region. Paalenen *et al.* (1983) report a value of $\mu = 0.5$ for experiments on Si:P (see Fig. 10).

On the insulating side for values of the density n/n_c, the approach to the metallic state is characterized by increasing values of the low-frequency limit of the dielectric polarizability χ. Note that low frequency in this context means that the corresponding energy is below the characteristic energies of the electronic transitions associated with the donors or acceptors. These characteristic energies decrease as the metallic state is approached and the value of χ increases. In fact, Herzfeld (1927), in a very early work on the criterion for metallic behavior, used this idea that a breakdown of an insulator occurs through a divergence of the low-frequency dielectric polarizability. The Herzfeld criterion separating metallic and insulating, when applied to a system of hydrogenic atoms, is very similar to Mott criterion, Eq. (1)—actually the difference is only that the constant on the right-hand side is 50% higher. In a metal the value of the low-frequency and long-wavelength polarizability is not well defined and depends on the ratio of the frequency to the wave vector as both go to zero in this limit. In an insulator the limiting value is well defined but can become arbitrarily large as $n \to n_c$. A power-law divergence in $\chi(n)$ with exponent ζ is observed,

$$\chi = \chi_0[(n_c/n) - 1]^{-\zeta}. \tag{3}$$

Experimentally, a value $\zeta = 2\mu = 1.0$ is found for the exponent of the dielectric constant (see Fig. 10).

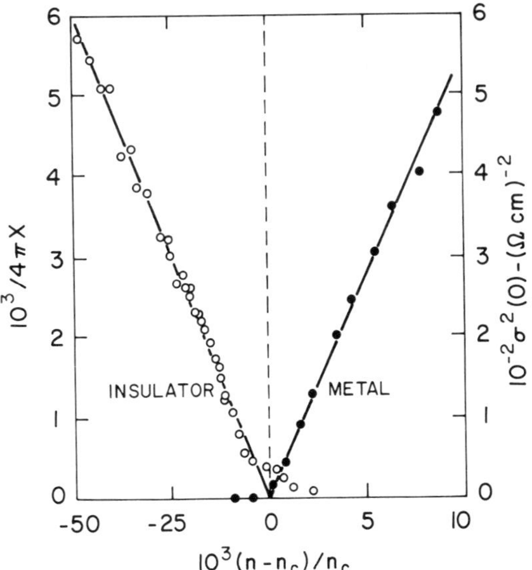

FIG. 10. The inverse donor polarizability and the square of the conductivity plotted against the deviation of the concentration from the critical value n_c, which is tuned by applying an external stress. The linear variation gives a value $\mu = \frac{1}{2}$ for critical exponent (Paalenen *et al.*, 1983).

As discussed previously, at a Mott transition in an ordered system, a change in the magnetic properties from paramagnetic metal to antiferromagnetic insulator is observed. Similarly on the metallic side, i.e., at large doping $n \gg n_c$, there is a paramagnetic Pauli spin susceptibility χ_s that is temperature independent at low temperature. On the dilute or insulating side the susceptibility is found to be strongly temperature dependent with a power-law divergence of $\chi_s(T) \propto c/T^a$. Typical values are $a \sim 1$. Such a behavior was explained by Bhatt and Lee (1982) as a successive singlet formation by donor pairs as the temperature is lowered. The question of the detailed evolution of the magnetic properties between these two limiting behaviors and the relationship to the evolution of the electrical conductivity remains open.

There are two important ingredients in the theoretical description of uncompensated doped semiconductors. One is the role of the intrasite and intersite Coulomb interactions, and the other that of the randomness inherent in the doping process. As discussed above, the former alone would cause a finite-temperature critical point and a discontinuous transition at lower temperature. However, experimentally only a zero-temperature transition is observed. This points to an essential role for the random disorder in the system.

Indeed, it was the properties of doped semiconductors that first stimulated Anderson (1958) to study the quantum-mechanical problem of a particle moving in a random potential. In a classic article he showed that a transition occurs from extended to localized states as the strength of the disorder increases. This in turn implies a zero-temperature transition from a metallic state with finite conductivity and with extended states at the Fermi energy to an insulating state with zero conductivity and localized states at the Fermi energy. The next important development was the introduction of a scaling hypothesis for the conductance as a function of the characteristic length by Abrahams *et al.* (1979). These authors calculated the scaling function perturbatively in the weak-disorder limit and showed that it was a sensitive function of the dimension of the system under consideration and that two is the critical dimension. In a lower dimension, such as a one-dimensional chain of atoms, the conductance decreases exponentially with length and reflects the fact that in a random chain all states are localized. In the special case of planar motion or two dimensions, the initial corrections are logarithmic and give rise to a set of effects known as weak localization effects, e.g., a reduction in the conductivity that is logarithmic in the temperature at low temperatures and that is very sensitive to an applied magnetic field. These effects have been observed in thin films, and Bergmann (1984) has given an elegant physical interpretation of weak localization as an interference effect between time-reversal-related pairs of paths that an electron can take in a two-dimensional film.

Returning to three-dimensional samples, in this case a transition is predicted from metallic and extended to insulating and localized with increasing disorder. The characteristic length ξ that enters the scaling theory determines the dc conductivity σ, through a relationship of the form

$$\sigma = c^1 e^2/\hbar\xi, \tag{4}$$

where c^1 is a numerical constant. This characteristic length ξ has a power-law divergence as the localization or Anderson transition is approached, $\xi \propto (n/n_c - 1)^{-\mu}$. This divergence causes critical behavior in the conductivity, of the form (2). On the insulating side, the characteristic length ξ is the localization length, which determines the extent of the localized wave functions, and this diverges also at the critical point. Assuming that on shorter length scales the static wavevector–dependent polarizability has a metallic form, the divergence in ξ leads to a divergence in χ of the form

$$\chi = \chi_0(n_c/n - 1)^{-2\mu}. \tag{5}$$

Such effects are indeed observed as discussed previously.

The one-component-scaling theory of Abrahams *et al.* (1979) predicts a value $\mu = 1$, in contrast to experiment on uncompensated Si:P quoted earlier, which give a value of $\mu = \frac{1}{2}$. It is plausible to ascribe this discrepancy to the Coulomb interaction, which may cause the transition region to be more first-order-like, or in other words to have a smaller value of the critical exponent, μ. A number of attempts have been made to extend the scaling theory to include electron-electron interactions. The initial corrections in the weak-disorder limit, e.g., to weak-localization effects, are well understood, but it has proved difficult to extend the theory to describe the region of the critical point. In particular, key questions concerning the role of possible magnetic instabilities remain to be resolved. At present there is no generally accepted theory for the transition region and the value of the critical exponent (see Lee and Ramakrishnan, 1985).

A key idea taken over from the study of other phase transitions is that the critical exponents such as μ and ζ should have universal values that should be independent of the material system. This universality is now under question as a result of a recent series of elegant experiments by Katsumoto (1988) on the system $Al_{0.3}Ga_{8.7}As$ doped with Si. He used a special feature of this system, that Si donors can form either deep (i.e., strongly bound) states or shallow (i.e., hydrogenic) states, and that the deep states can be converted to long-lived shallow states by shining light on the material. Katsumoto used this technique to tune continuously through the critical region, and under the reasonable assumption that deep states are inert, then the metal-insulator transition takes place among the shallow states. By varying the total light flux, the effective donor concentration can be continuously changed and tuned through the metal-insulator transition. However, Katsumoto finds values $\mu = 1$ and $\zeta \sim 2$ in agreement with the theoretical estimates in the Anderson limit (see Fig. 11). At present the origin of the discrepancy between the experiments on the two systems remains unclear.

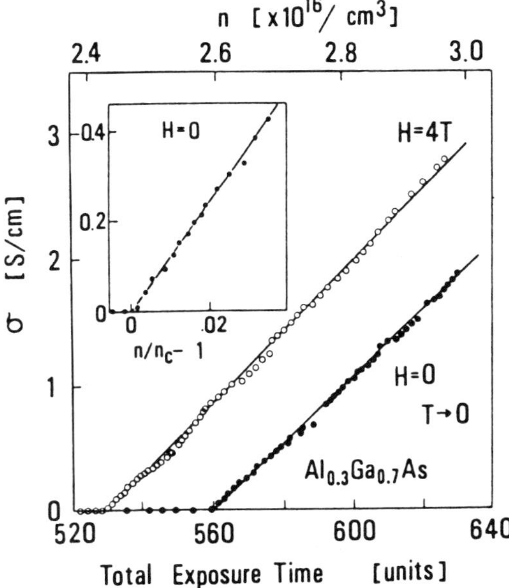

FIG. 11. The extrapolated zero-temperature conductivity as a function of the concentration of shallow Si donors, n, in the photoexcited $Al_{0.3}Ga_{0.7}As$:Si system (Katsumoto, 1988). Inset: the linear dependence of σ on $n/n_c - 1$ close to the critical concentration n_c. This linear dependence remains unaltered in a magnetic field $H = 4$ T.

2.2 Compensated Doped Semiconductors

A doped semiconductor is said to be compensated if both donors and acceptors are added. In this case the majority dopants of one valence are compensated by minority dopants so that average valence change is reduced. Consider the case of Si:P discussed previously. The material without acceptors is said to be uncompensated and has one electron carrier added per P donor. If now B acceptors are added, the average valence change is reduced and electrons are trapped at the B sites. As a result, the number of electron carriers is now less than one per P dopant. In this case an insulating state in the dilute limit occurs as a consequence not of the intrasite Coulomb interaction but of the randomness and possibly of the intersite Coulomb interactions. Note that the random character of the potential is strengthened by the trapped charges introduced at compensating impurity sites. As a result, compensated doped semiconductors are closer to the Anderson limit, and the description in terms of a single particle in a random potential should work better. Experimentally, values $\mu = 1$ are generally found, in agreement with this hypothesis.

Metal-semiconductor transitions can also be obtained by forming random alloys of metals and semiconductors. Examples of such systems are the amorphous alloys Nb_xSi_{1-x} (Hertel *et al.* 1983) and Au_xSi_{1-x}. In both cases the metal-insulator transition occurs as a zero-temperature critical point, and critical exponents $\mu \approx 1$ are measured. In these systems the difference between the electronic structure of the two components is very large, and presumably this determines the character of the metal-insulator transition and leads to randomness as the dominant effect.

2.3 Doping of Mott Insulators

Generally it is considerably more difficult to achieve a transition to a metallic state by doping a Mott insulator. For example, attempts to dope NiO, the best studied Mott insulator, by substituting Li atoms do not lead to a conducting metallic state. Rather the carriers seem to polarize the lattice strongly and thereby form polaron states with a small radius that have a low mobility. Such small polarons are easily localized and insulating behavior follows.

The system $La_{1-x}Sr_xVO_3$ is an exception. The parent compound $LaVO_3$ is an antiferromagnetically ordered insulator with a Néel temperature $T_N \approx 120$ K and an ordered moment of 1.2 μ_B/(V atom). Substitution of Sr for La atoms decreases the resistivity and also the magnetic susceptibility. At concentrations $x \gtrsim 0.22$, the resistivity is temperature independent and indicates that a transition to a metallic state has taken place (Dougier and Hagenmuller, 1975). Detailed studies of the low-temperature behavior are not available.

The high-T_c superconducting state is obtained by doping the parent antiferromagnetic cuprates, e.g., La_2CuO_4, and this will be discussed below.

2.4 Metal-Insulator Transitions in Expanded Liquid Metals

In one of the earliest articles to consider the nature of the metal-insulator transition, Landau and Zel'dovich (1943) speculated that it could appear as a separate critical point from the critical point of the liquid-gas phase transition in metals. In general the study of the liquid-gas critical points of metals is technically difficult, but in the case of metals with low melting points such as Hg or Cs the critical point is more easily accessible. A number of studies of the conductivity and other properties of fluid Hg and Cs have been made [for a recent review see Freyland and Hensel (1985)]. Only one critical point of the liquid-gas type is observed, and the conductivity evolves continuously. In the case of Hg, which is divalent, the basic metal-insulator transition should be of the band-crossing type, whereas in Cs, which is monovalent, it is of the Mott type. Apparently in both cases the metal-insulator critical temperature lies below the liquid-gas phase boundary and in the physically inaccessible two-phase region. In the case of mercury the evolution from metallic to nonmetallic on the liquid-gas phase boundary occurs continuously around a density ~7.5 g cm^{-3}, which is above the density of the critical point (4.9 g cm^{-3}).

3. SUPERCONDUCTOR-INSULATOR TRANSITIONS

3.1 The $BaBi_{1-x}Pb_xO_3$ and $Ba_{1-y}K_yBiO_3$ Systems

Interest in direct transitions from insulating to superconducting metallic states has risen since the recent discoveries of high-temperature superconductivity in several such cases. One of these systems starts with the parent compound $BaBiO_3$, which would be a metal in the cubic perovskite structure due to a half-filled band derived from antibonding Bi(6*s*)-O(2*p*) hybridized states. In fact it is an insulator with a band gap ≈2 eV due to a lattice distortion in which the Bi^{IV} oxidation states disproportionate into Bi^{III} and Bi^{V} oxidation states coupled to breathing-mode distortions of the O lattice (Cox and Sleight, 1979). This distortion can be thought of as a commensurate charge-density wave (CDW) that splits the band and causes the energy gap and insulating character.

The metallic and superconducting states are reached by substituting either Pb atoms on the Bi site or K atoms on the Ba site. In both cases, the insulating character continues to be observed up to a substantial critical concentration. In the former case $x_c \sim 0.65$ while the latter case $y_c \sim 0.35$. X-ray studies in both cases do not show coherent CDW distortions in these concentration ranges, but in both cases optical studies show that the CDW-derived energy gap has not closed. Therefore, it is plausible to ascribe the insulating character to local CDW distortions that persist up to the critical concentration, at least. The transition to the metallic state occurs as the energy gap closes, although in the normal phase the conductivity is quite low. Nonetheless these poor metals are good superconductors with relatively high values of T_c [13 and 30 K respectively; see Sleight *et al.* (1975) and Mattheiss *et al.* (1988)].

Detailed studies of the insulator-superconductor transition have been made in the case of the earlier compound, $BaBi_{1-x}Pb_xO_3$, with values of $x \sim x_c$. At the optimum concentration for superconductivity ($x \approx 0.75$) the metallic state is only poorly conducting with large (~1 mΩ cm) and temperature-independent resistivity. But at smaller values of x, as shown in Fig. 12, a spectacular insulator-superconductor transition is observed. The normal phase shows a resistance that rises rapidly as the temperature is lowered until the superconductivity is reached, where it drops precipitously. This system is the clearest example of a direct insulator-superconductor transition.

3.2 High-T_c Cuprate Superconductors

In the case of the high-T_c cuprate superconductors, again a direct transition from insulating to superconducting is observed as the oxidation state of the Cu ions is changed away from Cu^{II} without any intermediate normal metallic phase. In this case the parent insulating phases are antiferromagnetic with $s = \frac{1}{2}$ moments on the Cu^{II} sites and so fall in the general class of Mott insulators. The materials La_2CuO_4 and $YBa_2Cu_3O_6$ are well-studied examples with optical energy gaps ≈2 eV. The oxidation state of the Cu is increased by substituting Sr or Ba for La in the former system or by adding O in the latter case. This introduces low-spin ($s = 0$) Cu^{III} oxidation states, which are mobile in the background of Cu^{II} sites through the exchange of an electron and very effectively destroy the long-range magnetic order. In Fig. 13, the evolution of the resistivity in the $La_{2-x}Sr_xCuO_4$

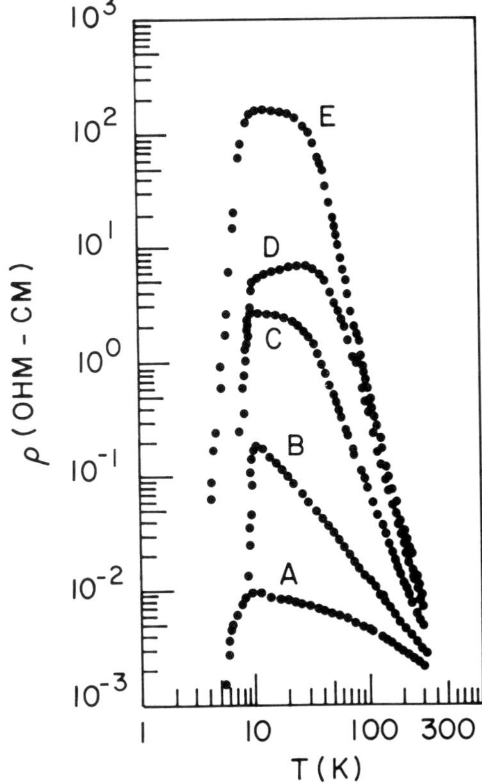

FIG. 12. The variation of the resistivity through the insulator-superconductor transitions in $BaBi_{1-x}Pb_xO_3$ samples with Pb concentrations $x = 0.65$ (*A*), 0.62 (*B*), 0.61 (*C*), and 0.59 (*D*, *E*) measured by Takagi *et al.* (1985).

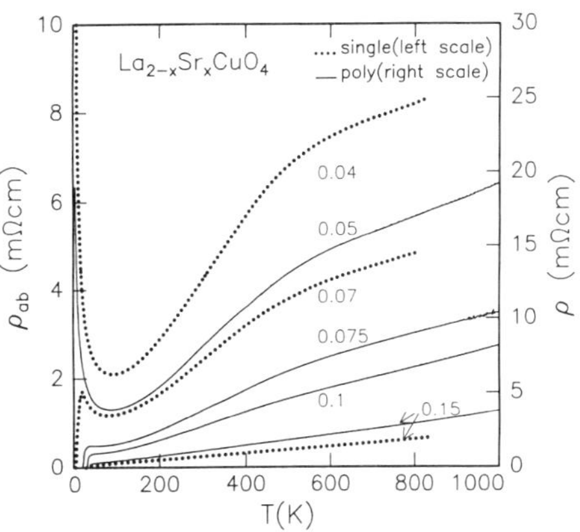

FIG. 13. The resistivity vs temperature for the high-temperature superconductor $La_{2-x}Sr_xCuO_4$ in a series of polycrystal and single-crystal samples with varying Sr concentration (Takagi *et al.*, 1992). Note that no samples show normal metallic behavior (i.e., finite resistivity) at low temperatures so that the zero-temperature transition is directly from an insulator to a superconductor with increasing values of *x*.

system is displayed as recently reported by Takagi *et al.* (1992). Again we see a direct evolution from insulating behavior with a divergent resistivity at the lowest temperatures to a superconducting behavior with zero resistivity below a critical temperature. No normal metallic state, characterized by a finite value of the zero-temperature resistivity, is observed as the concentration of Cu^{III} states is increased. The magnetic properties evolve from antiferromagnetic to paramagnetic, but the detailed relationship to the insulator-superconductor transition continues to be studied and to be sensitive to the system.

At present, the nature of the normal phase at temperatures above the transition to superconductivity and the microscopic cause of the superconductivity continue to be actively debated.

4. APPLICATIONS

The obvious potential applications for metal-insulator transitions lies in their use as switches. In particular, the vanadium dioxide system (VO_2) with a critical temperature of 340 K or 67°C could be interesting for a number of uses. However, devices built on these principles have not been popular. The relatively small change in resistance (only a factor of 10^5), the slow switching time, and the difficulty of repeated cycling of a sample through the transition are the most important drawbacks.

GLOSSARY

Acceptor: A dopant (or foreign) element with a smaller ionic charge than that of the host, which accepts (or binds) an electron from the conduction band into a state in the lower part of the energy gap.

Band-Crossing Transition: A transition between electronic states without overlap between the conduction and valence energy bands to a state in which these bands overlap in energy.

Band Insulator: An insulator caused by a gap in energy between the conduction and valence energy bands.

Charge-Density Wave (CDW): A static modulation of the charge density at atomic sites accompanied by a periodic lattice distortion.

Charge Ordering: An ordered configuration of oxidation states of different valences.

Conduction Band: The lowest unoccupied electron energy band.

Critical Concentration (n_c): The concentration of dopants that separates metallic ($n > n_c$) from insulating ($n < n_c$) behavior.

Donor: A dopant (or foreign) element with a larger ionic charge than that of the host, which donates (or liberates) an electron into a state near to the lowest energy in the conduction band.

Electron-Hole Fluids: A fluid of electrons in the conduction band and holes (i.e., unoccupied states) in the valence band that is analogous to a fluid of electrons and positrons.

Energy Bands: The quantum-mechanical státes of a single electron moving in the periodic potential of the crystal have energies in allowed bands separated by gaps or forbidden zones.

Extended States: Quantum-mechanical states of a single electron in a random potential that extend throughout the whole sample.

Fermi Liquid: A quantum-mechanical state which is only minimally modified by Coulomb interaction from the single-electron description in terms of a partially occupied energy band.

Insulator: A material whose electrical conductivity, extrapolated to zero temperature, vanishes.

Localized States: Quantum-mechanical states of a single electron in a strongly random potential that are localized in space so that their amplitude is concentrated near a single point in the sample.

Metal: A material whose electrical conductivity, extrapolated to zero temperature, does not vanish.

Mott Insulator: A material that in the band description should be a metal but nonetheless is an insulator because of localization of electrons on individual atoms by the Coulomb interaction.

Mott Transition: The transition between the Mott insulating and metallic states.

Oxidation State: The formal valence of an ion in an oxide obtained by assigning formal charge of -2 to the oxygen ions. It determines the symmetry properties of the highest occupied electronic states.

Semiconductor: A band insulator with a relatively small energy gap between valence and conduction bands.

Semimetal: A metal due to small overlap between valence and conduction bands.

Superconductor: A metal whose conductivity is infinite below a critical temperature.

Weak Localization: A set of phenomena occurring in thin films, characterized by weak (logarithmic) decrease of the conductivity to zero at low temperatures.

Works Cited

Abrahams, E., Anderson, P. W., Licciardello, D. W., Ramakrishnan, T. V. (1979), *Phys. Rev. Lett.* **42,** 673–676.

Abrahams, S. C. (1963), *Phys. Rev.* **130,** 2230–2237.

Anderson, P. W. (1958), *Phys. Rev.* **109,** 1492–1505.

Bergmann, G. (1984), *Phys. Rep.* **107,** 1–58.

Bhatt, R. N., Lee, P. A. (1982), *Phys. Rev. Lett.* **48,** 344–347.

Brinkman, W. F., Rice, T. M. (1970), *Phys. Rev. B* **2,** 4302–4304.

Cox, D. E., Sleight, A. W. (1979), *Acta Crystallogr.* **B35,** 1–10.

de Boer, J. H., Verwey, E. J. W. (1937), *Proc. Phys. Soc.* **A49,** 59–71.

Dougier, P., Hagenmuller, P. (1975), *J. Solid State Chem.* **15,** 158–166.

Edwards, P. P., Sienko, M. J. (1978), *Phys. Rev. B* **17,** 2575–2579.

Föex, M. (1946), *C. R. Acad. Sci. (Paris)* **223,** 1126–1128.

Freyland, W., Hensel, F. (1985), in: P. P. Edwards, C. N. R. Rao (Eds.), *The Metallic and Non-Metallic States of Matter,* London: Taylor & Francis, p. 93.

Fritzsche, H. (1955), *Phys. Res.* **99,** 406–419.

Fritzsche, H. (1978), in: L. R. Friedman, D. P. Tunstall (Eds.), *The Metal-Nonmetal Transition in Disordered Systems,* Scottish Universities Summer School in Physics, Edinburgh: SUSSP Publications, p. 193.

Jeffries, C. D., Keldysh, L. V. (1984) (Ed.), *Electron-Hole Droplets in Semiconductors,* Amsterdam: North-Holland.

Hensel, J. C., Phillips, T. G., Thomas, G. A. (1977), in: F. Seitz, D. Turnbull (Eds.), *Solid State Physics,* Vol. 32, New York: Academic, pp. 87–314.

Hertel, G. *et al.* (1983), *Phys. Rev. Lett.* **50,** 743–746.

Herzfeld, K. F. (1927), *Phys. Rev.* **29,** 701–705.

Katsumato, S. (1988), in: T. Ando, H. Fukuyama (Eds.), *Anderson Localization*, Springer Proceedings in Physics Vol. 28, New York: Springer, pp. 45–52.

Ladd, L. A., Paul, W. (1969) *Solid State Commun.* **7**, 425–428.

Lakkis, S., Schlenker, C., Chakraverty, B. K., Buder, R., Marezio, M. (1976), *Phys. Rev. B* **14**, 1429–1440.

Landau, L. D., Zel'dovich, G. (1943), *Acta Phys. Chem. (USSR)* **18**, 194. [See D. ter Haar (Ed.), *Collected Papers of L. D. Landau*, New York: Gordon and Breach.]

Lee, P. A., Ramakrishnan, T. V. (1985), *Rev. Mod. Phys.* **57**, 287–337.

Mattheiss, L. F., Gyorgy, E. M., Johnson, S. W. (1988), *Phys. Rev. B* **37**, 3745–3746.

McWhan, D. B., Rice, T. M. (1969), *Phys. Rev. Lett.* **22**, 887–890.

McWhan, D. B., Rice, T. M., Schmidt, P. H. (1969), *Phys. Rev.* **177**, 1063–1071.

McWhan, D. B. *et al.* (1971), *Phys. Rev. Lett.* **27**, 941–943.

Morin, F. J. (1959), *Phys. Rev. Lett.* **3**, 34–36.

Mott, N. F. (1949), *Proc. Phys. Soc.* **A62**, 416–422.

Mott, N. F. (1956), *Can. J. Phys.* **34**, 1356–1368.

Mott, N. F. (1961), *Philos. Mag.* **6**, 287–309.

Paalenen, M. A., Rosenbaum, T. F., Thomas, G. A., Bhatt, R. N. (1983), *Phys. Rev. Lett.* **51**, 1896–1899.

Peierls, R. E. (1954), *Quantum Theory of Solids*, New York: Oxford University.

Pouget, J. P. *et al.*, (1974), *Phys. Rev. B* **10**, 1801–1815.

Rice, T. M. (1977), in: F. Seitz, D. Turnbull (Eds.), *Solid State Physics*, Vol. 32, New York: Academic, pp. 1–86.

Sleight, A. W., Gillson, J. L., Bierstedt, P. E. (1975), *Solid State Commun.* **17**, 27–28.

Takagi, H., Naito, M., Uchida, S., Kitazawa, K., Tanaka, S., Katsui, A. (1985), *Solid State Commun.* **55**, 1019–1022.

Takagi, H. *et al.* (1992), *Phys. Rev. Lett.* **69**, 2975–2978.

Thomas, G. A., Rice, T. M., Hensel, J. C. (1974), in: M. H. Pilkuhn (Ed.), *Proceedings of the 12th International Conference on Physics and Semiconductors*, Stuttgart: Teubner, p. 105–109.

Thomas, G. A., Paalenen, M. A., Rosenbaum, T. F. (1983), *Phys. Rev. B* **27**, 3897–3900.

Verwey, E. J. W. (1939), *Nature* **144**, 327–328.

Further Reading

The most comprehensive account of the field is in Mott, N. F. (1990), *Metal-Insulator Transitions*, 2nd ed., London: Taylor & Francis; further details on most of the topics mentioned here can be found there. Other good sources are Edwards, P. P., Rao, C. N. R. (Eds.) (1985), *The Metallic and Non-Metallic States of Matter*, London: Taylor & Francis; Efros, A. L., Pollak, M. (Eds.) (1985), *Electron-Electron Interactions in Disordered Systems*, Amsterdam: North-Holland.

Conferences and summer schools devoted to this field, and especially to the study of disordered semiconductors, are organized fairly often, although there is no regular conference series devoted to this subject. The proceedings of these meetings also give valuable overviews of the subject. Recent examples include Friedman, L. R., Tunstall, D. P. (Eds.) (1978), *The Metal-Nonmetal Transitions in Disordered Systems*, Scottish Universities Summer School in Physics, St. Andrews, Edinburgh: SUSSP Publications; "Impurity Bands in Semiconductors," *Philos. Mag.* **B42**, 723–1003; Landsberg, P. T. (Ed.) (1985), "Transition in Semiconductors," *Solid State Electron.* **28**, 3–216; Kramer, B., Bergmann, G., Bruynseraede, Y. (Eds.) (1985), *Localization, Interaction and Transport Phenomena*, Springer Series in Solid State Science, Berlin: Springer; Ando, T., Fukuyama, H. (Eds.) (1988), *Anderson Localization*, Springer Proceedings in Physics Vol. 28, Berlin: Springer; Benedict, K. A., Chalkov, J. T. (Eds.) (1990), *Localization 1990*, Institute of Physics Conference Series No. 108: Bristol: The Institute of Physics.

A recent review of the metal-insulator transition in Si:P has been written by Löhneysen, H. v. (1990), *Festkörperprobleme* **30**, 95–111.

METALLIC GLASSES

T. EGAMI, *Department of Materials Science and Engineering, University of Pennsylvania, Philadelphia, Pennsylvania, U.S.A.*

INTRODUCTION

In our daily life, the word glasses implies oxide glasses used as window panes and kitchenware. Scientifically, however, the glassy state is defined as an amorphous or vitreous solidlike state of matter having an aperiodic and disordered atomic structure (see GLASSES), in contrast to the crystalline state, which is characterized by a periodic atomic structure (see CRYSTALLINE STATE). Thus, glasses are not limited to oxide glasses, and include polymers, other chalcogenides than oxides, and the most recent addition to the family of glasses, metallic glasses. Some of the characteristics associated with oxide glasses are not necessarily found in other glasses. For instance, oxide glasses are usually transparent and brittle, but metallic glasses show a shiny metallic appearance and are ductile as produced. But they share certain properties intrinsic to glasses, such as the presence of a glass transition and structural relaxation behavior. The nature and properties of metallic glasses are better understood by recognizing which properties depend more strongly upon chemical composition, and which properties reflect more sensitively the glassy nature of the atomic structure. This combination of properties found in crystalline metallic solids and those seen in oxide glasses makes metallic glasses scientifically interesting and uniquely useful for various applications. For instance, iron- or cobalt-based metallic glasses are beginning to be widely used as soft magnetic materials in power transformers, transducers, and sensors, because of their high magnetic permeability and resistance to mechanical deformation. High specific mechanical strength coupled with appreciable ductility of aluminum-based metallic glasses makes them attractive for application in aviation. Some chromium-containing glasses show remarkable corrosion resistance and are considered for various applications in hostile environments.

Metallic glasses became known only recently, because they do not form naturally, and it usually takes deliberate measures, such as rapid cooling of the melt or vapor depo-

3-527-28132-0/94/$5.00 + .50

sition, to obtain them. Because of the need for rapid extraction of heat during cooling, the dimensions of metallic glasses have been limited for some time to thin films, ribbons, and sheets. However, as a result of the recent discovery of certain compositions that yield metallic glasses even with the cooling rate commonly used for bulk casting, bulk metallic glasses are beginning to become available. Metallic glasses became commercially available in the early 1970s, and intense research activities ensued in the following ten years or so. In the last ten years, there has been marked progress in commercial application of metallic glasses.

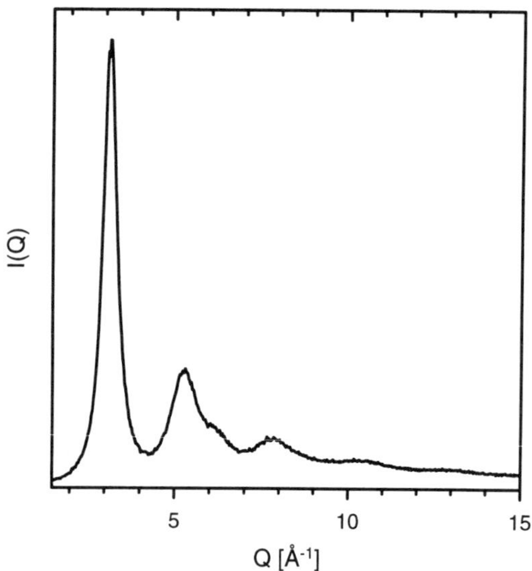

FIG. 1. Intensity of x rays scattered from a metallic glass, as a function of diffraction vector **q**.

1. ATOMIC STRUCTURE OF METALLIC GLASSES

1.1 Method of Structural Characterization

Diffraction measurements of x rays, electrons, and neutrons are commonly used in characterizing the atomic structure of crystalline solids. These methods are also basically effective in characterizing metallic glasses; however, there are some complications, among them the fact that it is not easy to differentiate glasses from microcrystalline solids by diffraction methods alone. When a glasslike substance was first obtained by vapor deposition in the 1930s, researchers thought it was likely to be microcrystalline. Interest in vapor-deposited films resumed after World War II, in the 1950s, and investigators started to explore their physical properties. Since then, there have been long, and often heated, discussions regarding whether metallic glasses are truly glassy or merely microcrystalline. The glassy nature of the metallic glass was confirmed only after many detailed structural studies involving atomic structural modeling and corroboration by thermodynamic evidence.

The intensity of x rays diffracted from a polycrystalline solid as a function of diffraction vector, **q** (= $4\pi \sin\theta/\lambda$, where θ is the diffraction angle and λ is the wavelength of the x ray), shows a number of sharp Bragg peaks. On the other hand, the x-ray intensity diffracted from a glass, including a metallic glass, is continuous with several broad peaks, as shown in Fig. 1. The electron-diffraction pattern from a crystal consists of many Bragg spots, but the electron-diffraction pattern from a glass has several weak diffuse rings. The problem is that, as the size of the crystalline grain becomes small, the Bragg diffraction peak width increases, and it becomes virtually impossible to distinguish true glasses from polycrystalline aggregate with the grain size below 15 Å by diffraction, without a very detailed analysis. The diffracted intensity from a metallic glass is similar to that from a liquid metal, except that the peaks are slightly sharper. This is very strong evidence that the atomic structure of a metallic glass is similar to that of a liquid. As we will discuss below, the structure of a liquid is basically modeled by a dense random packed structure, and the metallic glasses are also described by this model.

1.2 Atomic Pair-Distribution Function

The continuous diffraction intensity, such as the one shown in Fig. 1, is difficult to interpret as is, in contrast to the diffraction pattern from a crystal, which can readily be interpreted in terms of crystal lattice symmetry and lattice parameters. Therefore, it is customary to process the data to extract the atomic interference function, apply the Fourier transformation, and obtain the atomic pair-density, or distribution, function (PDF), as shown in Fig. 2. The PDF, $\rho(r)$, describes

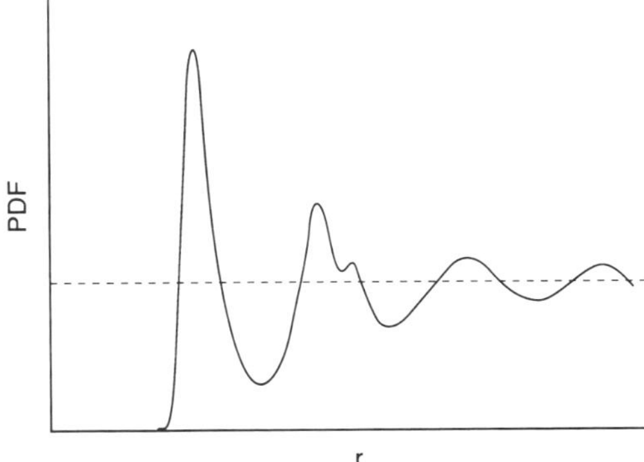

FIG. 2. Atomic pair-distribution functions of a metallic glass, obtained by Fourier transforming the normalized x-ray or neutron-scattering intensity.

the probability of finding an atom at a distance r, centering on another atom. The PDF of a metallic glass is characterized by a relatively sharp first peak, which corresponds to the nearest-neighbor atoms, and oscillations that become weaker as the distance increases. By integrating the PDF times $4\pi r^2$ over the first peak, one obtains the number of nearest neighbors, which is usually about 13.

1.3 Structural Models

Since the PDF gives only one-dimensional information, it is necessary to build a three-dimensional model to interpret the PDF further. The first successful models of the liquid structure were proposed simultaneously by Bernal and Scott in 1960, and were named the dense random packing model. The physical model that Bernal made consisted of many steel balls randomly packed in a bag, glued together with wax. The structure they created was clearly distinct from a microcrystalline aggregate, with very few well ordered regions. By analyzing this model, Bernal noticed that local clusters rarely found in crystals, such as an icosahedron (Fig. 3), are abundantly found in this model. Since an icosahedron is characterized by fivefold symmetry, it is not compatible with crystalline periodicity, except in complex multielement compounds. Earlier, in 1952, Frank had already pointed out that, in isolation, an icosahedron is the most stable form of a cluster made of 13 atoms, and that the presence of such clusters stabilizes the liquid structure.

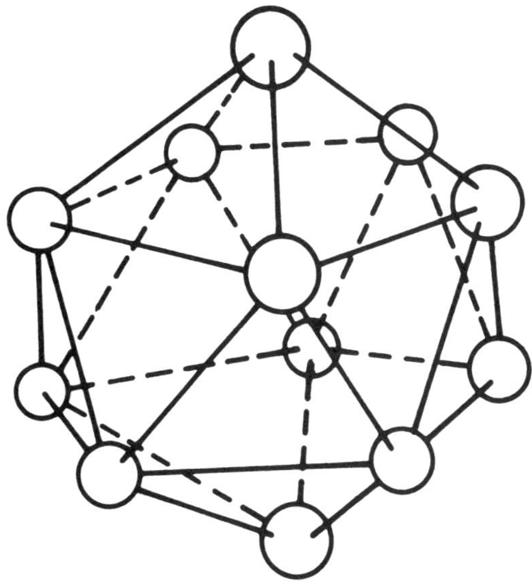

FIG. 3. Icosahedral cluster of atoms.

A solid with the symmetry of the icosahedron was later discovered (1984), and was named a quasicrystal (see QUASICRYSTALS). The atomic structure of a quasicrystal is completely ordered,with two periodicities that are incommensurate to each other, while the structure in a glassy state is highly disordered, with no periodicity. In liquids and glasses, with the constraint of periodicity removed, local structures that are usually not allowed in a crystal but are locally more stable can be found. A glass is formed with the principle of minimizing the local energy when the kinetic constraint does not allow a wide phase space (in statistical mechanics) to be probed by the system, while a crystal is formed by globally minimizing the total energy when the atomic mobility is high, and the system is allowed to look for the free-energy minimum in the entire phase space.

2. THERMODYNAMICS

2.1 Structure of Liquids and the Glass Transition

When a crystalline solid is heated up, the amplitude of lattice vibration is increased, but the average structure itself is unchanged and remains periodic except for slight thermal expansion. In contrast, the topology of the structure of liquid defined by the connectivity of the nearest-neighbor atoms depends upon temperature, so that a change in temperature results in rearrangements of atomic bonds, involving cutting and reforming of bonds. These rearrangements cannot be made instantaneously, so that a finite relaxation time is required for a liquid to assume the equilibrium structure when the temperature is changed abruptly. This structural relaxation time is proportional to the viscosity, and, thus, it is inversely related to the diffusivity. It is a strong function of temperature, and it rapidly increases when a liquid is deeply supercooled below the melting temperature. At a particular temperature, usually 40% to 70% of the melting temperature, the viscosity becomes so high that the system starts to behave like a solid. This temperature is the glass transition temperature T_g, and is usually defined by the viscosity reaching 10^{13} P. The glass transition, however, is not a phase transition, but is merely a thermal arrest phenomenon. If a liquid is cooled at a constant cooling rate, initially the structure of the liquid continuously keeps changing to the equilibrium structure at each temperature. However, the relaxation time increases quickly as the temperature is lowered, and eventually the structure becomes unable to follow the changing temperature, resulting in the thermal arrest phenomenon.

2.2 Kinetics of the Liquid-to-Crystal Transition

It appears that below the melting temperature, the crystalline state is always the state with the lowest energy, and the glassy state is metastable, although a theorem stating this has not been proven. Thus, when a liquid is slowly cooled, it transforms to a crystalline state. However, the boundary between liquid and crystal is well defined, having the thickness of a few atomic distances, and has an energy higher than either liquid or solid. Therefore, in order for a crystalline particle to form in a liquid, this interfacial energy barrier has to be overcome; consequently, it takes a finite amount of time before a crystalline particle is formed, even below the freezing point. An example of the time-temperature-transformation (T-T-T) diagram describing this transformation kinetics is shown in Fig. 4. Here the alloy is in the liquid state

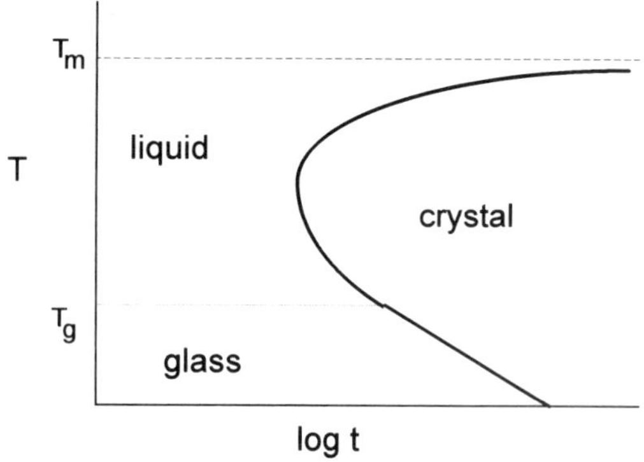

FIG. 4. T-T-T (Time-Temperature-Transformation) diagram for liquid-to-crystal transformation. The horizontal axis is the time a liquid takes to transform itself into a crystal, on a logarithmic scale. By circumventing the nose of the *C* curve by rapid cooling, glass can be obtained.

at $t = 0$, after the melt is cooled to temperature T infinitely fast. The *C* curve indicates the boundary between the liquid and crystalline states. On the left of the *C* curve, the material remains a supercooled liquid. If cooling is fast enough, it is possible to suppress the nucleation of crystalline particles altogether, through reducing atomic motion by bringing the liquid well below its glass transition temperature, and to form a solid-like state without a liquid-to-crystal transition. The solid thus produced is a glass. A metallic glass of Au-Si was first obtained by rapid quenching of the melt by Duwez and his students in 1960 (Klement *et al.*, 1960). They then discovered that various alloys of transition metals and metalloids can be quenched into a glass. It is interesting to note that the principle of rapid cooling discussed above applies in general. In other words, if one could cool the liquid fast enough, any material could be made glass. Turnbull (1969) was the first to recognize this intriguing fact, and established the theoretical basis for the existence of a metallic glass.

2.3 Glass-Forming Composition

While, in principle, any material can be made into a glass, the critical cooling rate necessary to form a glass varies greatly from one system to another. For a pure metal, the critical rate is estimated to be over 10^{12} K/s, a rate virtually impossible to achieve except by vapor deposition. Also, a glassy pure metal is so unstable that it crystallizes well below room temperature. Thus, only alloys over certain composition ranges can form a glass that is stable at room temperature by rapid cooling from the melt with the cooling rate of the order of up to 10^6 K/s. Various factors have been considered to influence the glass-forming ability, but the size difference among the constituent atoms seems to be the most dominant factor. The glass-forming composition range is also characterized by the eutectic phase diagram. Near the eutectic composition, liquids are relatively more stable, and thus glasses are more stable as well.

The relative stability of glass can best be understood by considering the so-called T_0 line in the phase diagram (Fig. 5). The T_0 line defines the composition range within which a liquid is the phase with the minimum free energy, if the phase separation is suppressed by rapid cooling. Figure 6 shows the free-energy curves for two solid phases and a liquid, defining the T_0 composition. Note that the free energy is the lowest for a liquid in the composition range from x_1 to x_2, provided the material remains a single phase. If phase separation is allowed by diffusion, the mixture of α and β phases has a lower energy than a liquid. By bringing the liquid within the T_0 lines below the glass transition temperature, a glass can be obtained with a cooling rate high enough to suppress diffusion and phase separation. Typical glass forming compositions are transition-metal–metalloid alloys, such as $TM_{100-x}M_x$, where TM is an alloy of Fe, Ni, Co, and Cr, M is a metalloid element, such as B, Si, C, and P, or their mixture, and $15 < x < 25$; transition-metal alloys, such as $Zr_{100-x}Cu_x$, where $40 < x < 75$; and aluminum–transition-metal alloys, such as $Al_{100-x}Fe_yCe_{x-y}$, where $9 < x < 15$. The exact range of glass-forming composition depends on the details of the production method, but, since the critical cooling rate very strongly depends upon composition, the glass-forming ranges determined by various researchers are remarkably similar. Figure 7 shows one of these composition ranges for the Fe-B-Si alloy system. Glass forming is easiest at the center of the range shown. The glass transition temperature for these compositions weakly depends on the composition, and is about 300–450 °C.

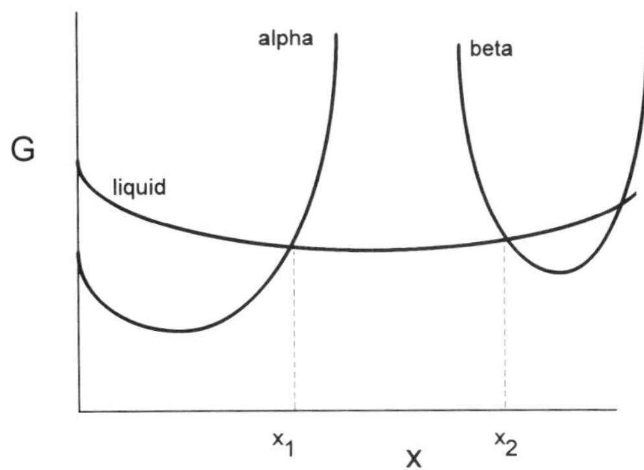

FIG. 6. Free-energy diagram of a binary system. The figure shows the free energy G of various phases (α, liquid, β) vs composition x at temperature T. Between x_1 and x_2, liquid (glass) is the most stable phase, except for the mixture of the α and β phases. The loci of x_1 and x_2 at various temperatures define the T_0^α and T_0^β lines.

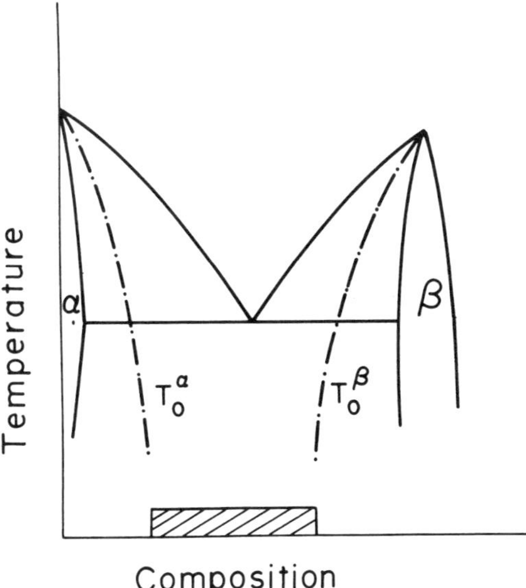

FIG. 5. A binary phase diagram showing the T_0 line (chained curves). The α phase is a solid solution of B atoms in the A phase, and the β phase is a compound of A and B. Hatched range indicates glass-forming composition.

2.4 Crystallization and Structural Relaxation

Once a glass is formed, it is stable well below T_g. However, if the temperature is raised up to the vicinity of T_g, the atomic mobility becomes high enough to nucleate crystalline particles, and the glass crystallizes. Since this

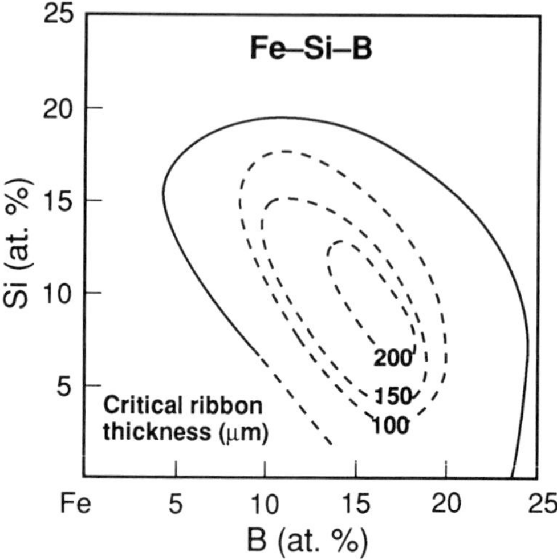

FIG. 7. Glass-forming range and crystallization temperature for Fe-B-Si alloys (inside the solid curve, after Waseda *et al.*, 1990). Numbers denote the maximum thickness (in microns) of the ribbon that can be cast as metallic glass.

is just a liquid-to-crystal transition taking place at a low temperature, the kinetics is basically the same as that for the crystallization near the freezing (solidification) temperature. The crystallization temperature T_x is defined usually by the volume fraction of the crystalline portion reaching 0.5 for a constant heating rate. However, since crystallization is a kinetic phenomenon, T_x depends upon the heating rate. The temperature dependence of T_x is Arrhenian, with the apparent activation energy ranging from 2 to 5 eV. The value of T_x shows moderate composition dependence.

A glass produced by rapid cooling retains the structure that it had at a temperature, called the fictive temperature, where thermal arrest has happened during cooling. Therefore, the higher the cooling rate, the higher the fictive temperature. When a glass with a high fictive temperature is annealed, or held at an elevated temperature, the structure gradually reverts to a more stable structure, and its fictive temperature approaches the annealing temperature. This structural relaxation occurs even well below the glass transition temperature, and is accompanied by various changes in the properties, including magnetic as well as mechanical properties. Incidentally, the structural relaxation is not an incipient crystallization. It results in a more stable glass that is structurally totally distinct from the crystalline state.

It is useful to recognize the distinction between topological short-range ordering and compositional short-range ordering during structural relaxation. The topology of the structure described by its fictive temperature cannot attain equilibrium, except for the immediate vicinity of T_g. Thus, additional annealing treatments are always additive. However, the compositional short-range order among similar atoms, for instance, among the transition-metal elements, can reach an equilibrium in a relatively short time, and can exhibit reversible behavior.

3. METHODS OF PRODUCING METALLIC GLASSES

3.1 Rapid Quenching Methods

For most glass-forming compositions, a cooling rate of the order of 10^6 K/s is required to form metallic glasses from the melt, suppressing crystalline nucleation. In order to achieve such a high cooling rate, various methods have been developed to extract heat from the melt by contact to a chill block. In each case, since heat has to be extracted in a very short time, the dimension of the solid is limited to a small one, typically below 50 μm. The method most widely used and most successful is the melt-extraction technique shown in Fig. 8. In this method, the molten liquid metal alloy is pushed out of a nozzle by gas

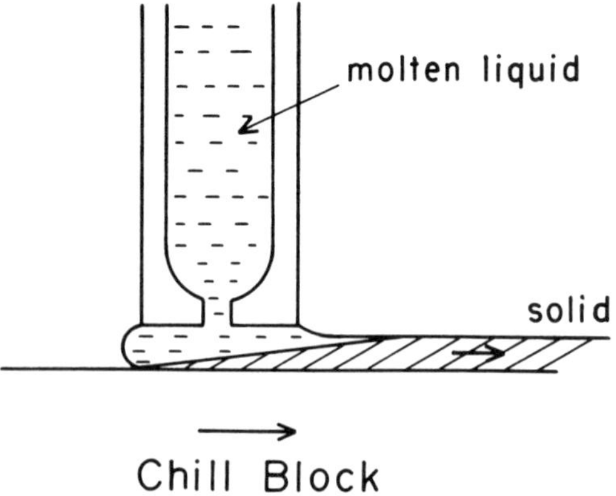

FIG. 8. Melt-extraction method. The chill block is usually a copper wheel rotating with a surface speed over 30 m/s.

pressure, forms a layer between the nozzle and the chill block, a rotating metallic disk, and is continuously extracted in the form of a ribbon by the rotating disk. A ribbon about 1 cm wide, 30 μm thick can be easily produced by this method in the laboratory. By fabricating a wide, flat nozzle, much wider sheets, up to half a foot wide, can be produced in industrial plants. Recently, certain ranges of composition, such as $(Ti,Zr)_{55}(Ni,Cu)_{25}Be_{20}$, were found to have a very low critical cooling rate comparable to those of oxide glasses, and, therefore, not to require special rapid quenching to form a glass (Peker and Johnson, 1993). They can be cooled into a bulk glass by regular casting.

3.2 Solid-State Amorphization

Metallic glasses can also be obtained without going through a liquid phase. For instance, alloys subjected to radiation damage by particles such as electrons and ions can be amorphized. Most interestingly, a two-phase crystalline material annealed at a moderate temperature can achieve amorphization through interdiffusion. An example is the multilayered thin film of Ni-Zr, which becomes amorphous when annealed at 350 °C for several hours. Furthermore, mechanical alloying can also achieve amorphization. By mixing small particles of two kinds of crystalline phase using the ball-milling machine for a long time, it is possible to force the two components to become atomistically mixed, and, if the composition is right, to become vitreous. In each case, the purpose of solid-state amorphization is to achieve a single-phase material with the composition between the T_0 compositions shown in Fig. 5. Since the most stable phase in this composition range is a glass, amorphization occurs. By this technique, bulk metallic glasses that cannot be obtained by rapid cooling methods can be produced.

3.3 Physical and Chemical Deposition

The method first used in producing metallic glasses was vapor deposition in vacuum at low temperatures. By depositing atoms one by one onto a cold substrate, a very high quenching rate can be achieved, because an atom is cooled in a time scale comparable to that of the lattice vibration (10^{-12} s). Vapor of metallic alloys can be produced by various methods, such as evaporation, ion sputtering, and electron-beam heating. Since the deposition rate is usually of the order of 10^{-2}–$\sim10^{+1}$ nm/s, only thin films ranging from 10 nm to 10 μm can be easily produced by this method, but thicker materials can be deposited by special high-rate deposition systems.

Electrodeposition (plating) or electroless chemical deposition sometimes can produce amorphous alloys, for instance, of transition metal and phosphorus. However, it is difficult to control the homogeneity of the product, and the films produced often contain hydrogen. Thus, electroplating is not the most preferred method of producing metallic glasses.

4. PROPERTIES

4.1 Mechanical Properties

What makes the metallic glasses so distinct from oxide glasses is their ductility. When a stress is applied, metallic glasses can yield and plastically deform. Oxide glasses have no ductility, so that small cracks grow fast when a stress is applied, resulting in brittle behavior. Metallic glasses are ductile, so that the stress concentration at the tip of the crack is diffused by local deformation, stopping the propagation of the crack. Thus, they show tensile strength comparable to the strongest steel (piano wire), ranging from 3000 to over 4000 MPa for transition-metal–metalloid glasses. Al-based alloys, such as $Al_{90}Ce_3Fe_7$, have a lower strength, about 1000 MPa, but show the highest specific strength (strength divided by the density) of all known metallic materials, reaching 30 $(kg/mm^2)/(g/cm^3)$ and above (He *et al.*, 1988). However, particularly in the case of iron-based alloys, much of the ductility is often lost when they are annealed and structural relaxation takes place. Also, unlike crystalline alloys, they show no work hardening, so that, as soon as they start to yield, they fail, often catastrophically in the case when tensile stress is applied. Therefore, special care has to be exercised in using these glasses in applications requiring strength. Deformation occurs uniformly when the temperature is close to the glass transition temperature; however, at lower temperatures, deformation is localized to form shear bands.

Metallic glasses show a fair amount of internal friction and sound attenuation above about 200 °C. However, the room-temperature sound attenuation is surprisingly small, making them attractive for certain applications involving ultrasonic propagation.

4.2 Magnetic Properties

Metallic glasses containing a substantial amount of Fe and Co are usually ferromagnetic. Most of the basic magnetic properties, such as the saturation magnetization and the Curie temperature, depend primarily upon composition and, thus, are similar in the crystalline and amorphous states. For instance, the magnetic moment of transition metals and their alloys is determined by the difference in the spin density in the majority-spin (for instance, up-spin) electron band and that in the minority-spin (for instance, down-spin) electron band. Since the majority-spin band of cobalt is full, and the magnetic moment depends only upon the number of holes in the minority-spin band (this case is called strong magnetism), the magnetic moment of cobalt depends only on composition, and is thus independent of the structure. On the other hand, both the majority- and minority-spin bands are not full in iron (this case is called weak magnetism), so that the magnetic moment of iron does depend somewhat on the structure, varying when the structure changes from crystalline to amorphous. For the transition-metal–metalloid (TM-M) glasses, if the composition of TM is changed while keeping the composition of M fixed, the magnetic moment is a linear function of the *d*-electron density, as shown in Fig. 9. Such a curve for TM crystalline alloys, shown by the dotted curve in the figure, is known as the Slater-Pauling curve. The curve for the TM-M glasses is slightly shifted because of the effect of metalloid elements.

Secondary magnetic properties, such as magnetic anisotropy and magnetostriction (change of dimension due to magnetization), reflect the atomic structure a little more sensitively. A glass in an ideal state should be isotropic, but real glasses are not quite isotropic because of directional solidification or mechanical deformation. Thus, real ferromagnetic metallic glasses usually have magnetic anisotropy. In particular, sputter-deposited amorphous-iron–rare-earth alloys, such as $Fe_{75}Tb_{25}$, show strong anisotropy as deposited. If the deposition condition is right, the anisotropy can be strong enough to keep the magnetization perpendicular to the film, even though magnetostatic energy prefers it to be parallel to the plane of the film. This anisotropy is critical in the use of these alloys as magnetic recording media. Also, magnetic anisotropy can be induced by annealing either in a magnetic field or under stress. The microscopic origin of this anisotropy has been

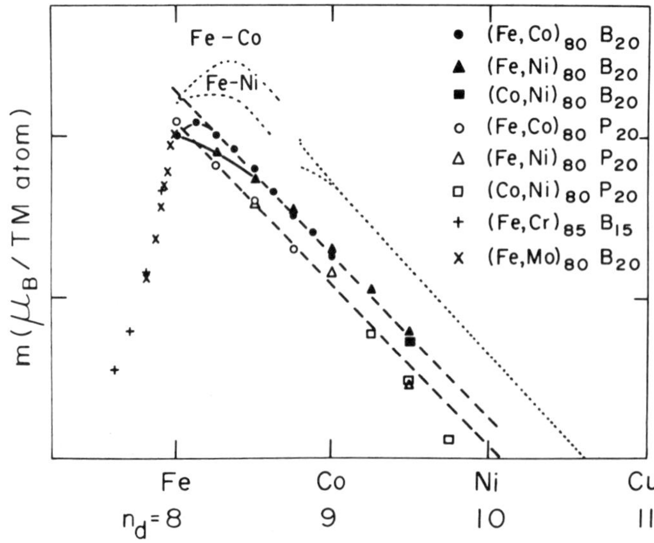

FIG. 9. Dependence of the saturation magnetic moment on the nominal *d*-electron concentration in the transition-metal–metalloid metallic glasses (after Egami, 1984).

the subject of numerous studies. The topology of the structure and compositional anisotropic short range ordering are considered to be responsible.

Low-field magnetic properties, such as permeability, depend more on the microstructure at the length scale equal to the domain-wall thickness than on the atomic structure. Metallic glasses are random and inhomogeneous at an atomic scale, but at the scale of domain-wall thickness (~100 nm), they are much more homogeneous than crystalline alloys, since they do not have any grain boundaries or lattice dislocations. Thus, ferromagnetic metallic glasses usually show high magnetic permeability and are excellent soft magnetic materials. In particular, magnetostriction is zero in certain cobalt-based alloys, making them ideal for soft magnetic applications. An example of the composition dependence of magnetic permeability around the zero-magnetostriction composition is shown in Fig. 10.

However, effects of structural relaxation have to be taken into account in using metallic glasses. Metallic glasses as cast are unstable, and their properties, particularly the low-field properties, are strongly affected by subsequent heat treatments. Thus, it is customary to anneal them beforehand, often in a magnetic field, in order to stabilize their properties. In designing the annealing treatment, a detailed knowledge of the structural relaxation phenomena is required. Many of the magnetic properties, such as the Curie temperature and magnetic anisotropy, are affected by the compositional short range ordering. Thus, their changes during annealing are fast, and they can attain equilibrium during annealing.

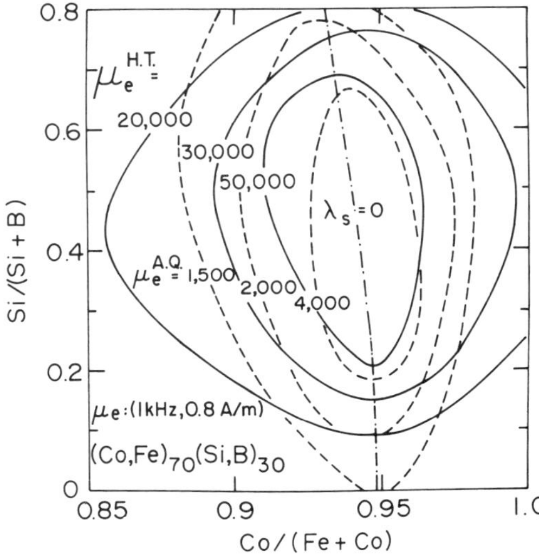

FIG. 10. Composition dependence of magnetic permeability μ (= dB/dH, where B is the magnetic flux density and H is the magnetic field) of $(Co,Fe)_{70}(B,Si)_{30}$ alloys, as quenched (μ_e^{AQ}, dashed curve) or heat treated at 450 °C and water quenched (μ_e^{HT}, solid curve) (after Masumoto and Egami, 1981). The dot-dashed line indicates the zero-magnetostriction composition.

Not all amorphous films show high magnetic permeability. Sputter-deposited thin films of transition-metal–rare-earth alloys, such as $(Fe,Co)_{75}Tb_{25}$, show relatively high coercivity, of the order of 1 kOe, because of high local anisotropy of Tb and the surface effect. Together with the high magnetic anisotropy mentioned above, which keeps the magnetization perpendicular to the film, this high coercivity stabilizes small circular ferromagnetic domains called bubble domains. These bubble domains are used in the magneto-optical data storage application of these films.

4.3 Electrical Properties

In crystalline solids, because of periodicity in the atomic structure, only the electrons with specific wave vectors are scattered (Bragg's law). Consequently, at T = 0 K, the electrical resistivity of an ideal metal goes to zero, and only the deviations from the perfect periodicity, lattice vibrations, produce resistance. In metallic glasses, Bragg's law does not apply because of their disordered atomic structure. Thus, all electrons are scattered, resulting in a very short mean free path of only a few atomic distances. Consequently, the resistivity is high even at T = 0 K, and is usually of the order of 100 μΩ-cm. Since the resistivity is already high at 0 K, temperature has little effect on resistivity. It can either weakly increase or weakly decrease, depending upon the conduction electron concentration.

Metallic glasses containing 4d transition metals such as Zr, Mo, and Nb are superconducting at low temperatures. Their superconducting transition temperature is usually slightly lower than the crystalline counterpart, but the opposite can be true when the transition temperature of the crystalline solid is low. On the other hand, the critical magnetic field is usually higher for superconducting metallic glasses than for crystals. Superconducting metallic glasses provide an interesting proof that periodicity, which is assumed in the conventional theories of superconductivity, is not a prerequisite for superconductivity.

4.4 Chemical Properties

The homogeneity of the structure of metallic glasses at a mesoscopic level, mentioned above, is also important for surface chemical properties. Metallic glasses containing significant amounts (usually more than 5%) of chromium form very uniform surface oxide layers that inhibit corrosion. The presence of phosphorus, in addition to chromium, is also known to be helpful. Thus, Cr-containing metallic glasses show excellent corrosion resistance. They also show interesting catalytic activities and are promising as catalysts, as well.

5. APPLICATIONS

5.1 Magnetic Applications

As has been mentioned several times above, metallic glasses containing Fe or Co are used in several applications, including uses in electrical power-distribution transformers, in high-frequency electrical power supplies, in sensors (for instance, in antitheft devices), in magneto-optical data storage media, and as transient materials for permanent magnets.

Wide ribbons of rapidly cooled Fe-B-Si alloys, such as $Fe_{78}B_{13}Si_9$, are now beginning to be used as core materials in many electrical power-distribution transformers. Since their magnetic permeability is high, their magnetic core loss is small, resulting in considerable energy saving. Currently, the power loss due to the distribution transformer with crystalline silicon-iron core amounts to about 1% of the total power generated, or about 4×10^9 dollars a year in the United States. Replacing these conventional distribution transformers with amorphous-core transformers will reduce this loss to a half, resulting in enormous savings. In order to choose the best alloy composition for this application, several factors have to be considered. One wants to maximize the magnetic flux density and magnetic permeability, as well as thermal stability, but these requirements often conflict with each other. For instance, the magnetization increases as the iron content is increased. However, if the iron content is too high, magnetic permeability and thermal stability start to degrade. Furthermore, economics also interfere in the decision. When the cost of energy is high, as it was right after the start of the Iran-Iraq war in 1983, it makes sense to place priority on reducing the core loss rather than the flux density. However, if the energy cost is not so high, it is better to choose a material with a higher flux density and minimize the size of the transformer to reduce the cost of the transformer. Currently, the preferred alloy system is Fe-B-Si alloys with Fe 78–80 at. %. Magnetic metallic glasses are useful not only in transformers, but also in motors, for the same reason.

High magnetic permeability of cobalt-based zero-magnetostrictive metallic glasses makes them attractive for applications as magnetic shielding materials and magnetic sensors, such as the recording heads for tape recorders. Since the electrical resistivity of metallic glasses is relatively high, and it is easy to produce thin foils of metallic glasses, cobalt-based metallic glasses retain high magnetic permeability up to high frequencies (100 kHz or more), and they are excellent for application in high frequency power supplies. They are used in switching-mode transformers in which ac or dc voltage is changed into high-frequency ac voltage, transformed, and switched back to the original frequency. They are also used in antitheft devices in retail stores, to prevent shoplifting. Their high-frequency response, particularly at very high harmonics because of their square *B-H* loop, makes them excellent tagging materials. High magnetic permeability coupled with reasonable magnetostriction makes some metallic glasses attractive as transducer materials. They are used in various force sensors, for instance, in engine and power trains in automobiles.

Sputter-deposited transition-metal–rare-earth amorphous films are used as magneto-optical data storage media. Small bubble domains of about 1 μm in diameter are produced by locally heating the film by a semiconductor laser. Domains are read by the rotation of the polarization of light reflected from the film (Kerr effect). By this method, a recording density as high as the compact disk can be attained, and yet, unlike the conventional aluminum-based compact disks, the recording can be erased and overwritten. Already several manufacturers are commercially producing optical data storage systems (CD-RAM) using amorphous films.

Metallic glasses are also used in fabricating a crystalline permanent magnet, $Fe_{14}Nd_2B$. Because of a complex phase diagram, it is not easy to produce this compound, which shows excellent magnetic induction and coercivity. Thus, in a technique widely used now, the alloy is first cast as a ribbon in an amorphous state and then annealed into a crystalline permanent magnet.

5.2 Other Applications

The possibility of using metallic glasses in mechanical applications as structural materials was recognized already in the 1960s. While magnetic applications increased rather quickly, mechanical applications remained merely a possibility for some time, mainly because of the price, dimensions, and structural relaxation. Functional materials, such as magnetic materials, can be reasonably costly, but structural materials need to be inexpensive. Also, the thickness of the sheet, usually of the order of 30 μm, severely limited its use. An additional problem is the loss of ductility upon structural relaxation in iron-based glasses.

The recent discovery of glass compositions with very low critical cooling rates for vitrification can change this situation dramatically. With these compositions, a bulk glass can be cast easily and inexpensively. By machining these bulk glasses, complex tools, such as gears, can be produced. It is most likely that they will find major markets in the near future.

Metallic glasses are found in other applications, for instance, as brazing alloys. Brazing alloys, usually in the form of foils, are used in joining two metallic blocks upon heating that melts the brazing alloy. Since metallic glasses are ductile and can be cut into various forms, they are convenient in joining pieces with complex shapes. When the joining is done, the glasses are crystallized. Mainly nickel-based alloys are used for this purpose.

6. SUMMARY

While metallic glasses belong to the family of glasses, their properties are only in limited cases similar to those of oxide glasses; in some other cases, they are closer to those of crystalline alloys, and in other cases, they resemble neither. This combination of properties makes the metallic glass an interesting scientific subject of study, and offers unique opportunities for industrial and commercial applications. While the structure of metallic glasses is disordered at an atomistic scale, at a scale slightly larger than the atomistic scale, the structure is more homogeneous than crystalline solids because of the absence of grain boundaries and lattice dislocations. This gives rise to interesting mechanical as well as low-field magnetic properties, which are attractive to applications. Metallic glasses are beginning to be used widely as soft-magnetic materials, and their use as structural materials is likely to grow. They are clearly one of the advanced materials that are important in technology today and tomorrow.

Works Cited

Egami, T. (1984), *Rep. Prog. Phys.* **47**, 1601–1725.

Frank, F. C. (1952), *Proc. R. Soc. London, Sec. A* **215**, 43–46.

He, Y., Poon, S. J., Shiflet, G. J. (1988), *Science* **241**, 1640–1642.

Klement, K., Willens, R. H., Duwez, P. (1960), *Nature* **187**, 869–870.

Masumoto, T., Egami, T. (1981), *Mater. Sci. Eng.* **48**, 147–165.

Peker, A., Johnson, W. L. (1993), *Appl. Phys. Lett.* **63**, 2342–2344.

Turnbull, D. (1969), *Contemporary Phys.* **10**, 471–488.

Waseda, Y., Ueno, S., Hagiwara, M., Aust, K. T. (1990), *Prog. Mater. Sci.* **34**, 149–280.

Further Reading

Beck, H., Güntherodt, H.-J. (Eds.) (1981, 1983), *Glassy Metals I, II*, Berlin: Springer-Verlag.

Liebermann, H. H. (Ed.) (1993), *Rapidly Solidified Alloys*, New York: Marcel Dekker.

Luborsky, F. E. (Ed.) (1983), *Amorphous Metallic Alloys*, London: Butterworths.

METALLURGY, PHYSICAL

ROBERT W. CAHN, *Department of Materials Science and Metallurgy, Cambridge University, Cambridge, United Kingdom*

INTRODUCTION

"Physical metallurgy" is a term coined in 1914 by Walter Rosenhain, of the National Physical Laboratory near London, to denote the application of physical principles and experimental techniques to the understanding, processing, and improvement of alloys. Among the physical experimental techniques applied to metals and alloys, microscopy is crucial. Metallography, an older discipline concerned largely with the microscopic examination of alloys in various stages of heat treatment, is now part of physical metallurgy. The methods and approach of physical metallurgy have in recent years been extended to ceramics also.

The four basic activities of physical metallurgists are

1. the study of equilibrium between phases as a function of composition and temperature;
2. the study of the geometry and development of microstructures;
3. the relation of useful properties, particularly mechanical properties, to the phases present and to their proportions and microstructural disposition; and
4. the rate and the mechanism by which a microstructure and its constituent phases change when the temperature of an alloy is changed.

The central concept of physical metallurgy is that of equilibrium among phases in an alloy; this is expressed graphically by means of equilibrium diagrams (see also PHASE EQUILIBRIA). The mechanism and kinetics of the transition of a phase into one or more other phases, with different crystal structures, when the temperature is changed, is a central study of physical metallurgy. However, alloys are frequently trapped in nonequilibrium, "metastable" states, for instance by rapid cooling from a high temperature. Examination of the transition from a metastable to a stable (equilibrium) state is also an important part of physical metallurgy (see also PHASE EQUILIBRIA).

Metals and alloys are always full of metastable defects that govern their behavior un-

3-527-28132-0/94/$5.00 + .50

der load, aggressive environments, or irradiation. The interpretation of such behavior, as well as the examination of the geometry and dynamics of defects as a topic in its own right, forms a third constituent of physical metallurgy.

Finally, but by no means least, physical metallurgists accord a place of honor to the study of microstructure, that is to say, the geometrical and crystallographic disposition of the individual crystals, or grains, of a single phase, and of the distinct phases in a multiphase alloy. The physical and mechanical properties of an aggregate of grains and phases depend in subtle ways on the scale and morphology of the constituents of microstructure; to pick just one variable, grain size affects the strength, ductility, and toughness of any metal or alloy (see also MECHANICAL PROPERTIES OF SOLIDS).

Since World War II, the variety and subtlety of the physical techniques used to examine metals and alloys have been vastly extended. As a result, physical metallurgy has been revolutionized.

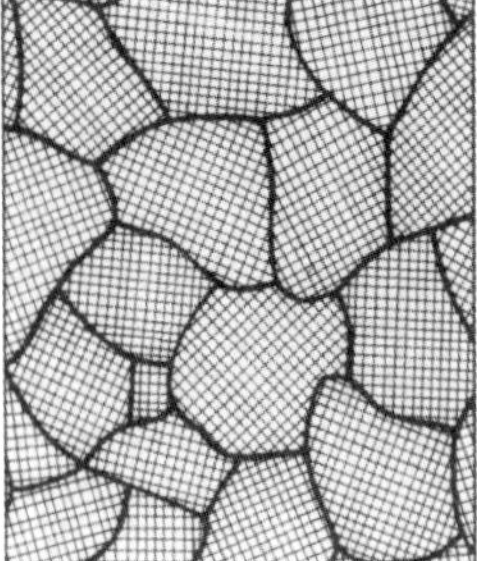
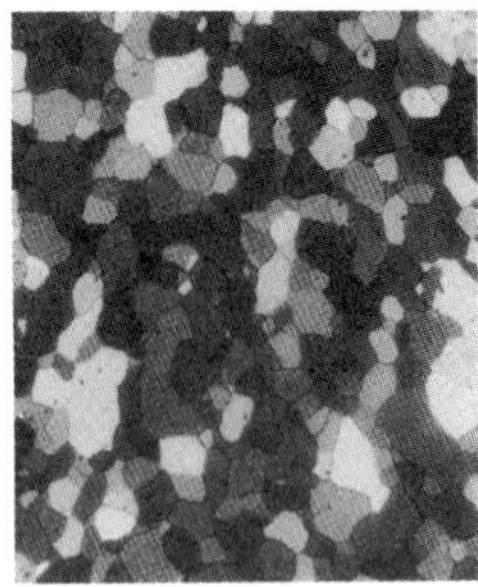

FIG. 1. Polycrystalline metal, consisting of a jumble of tiny crystal grains (typically ≈0.1 mm across). Left, diagrammatic; right, an optical micrograph of a polished and oxidized copper surface, viewed by polarized light (courtesy R. M. J. Cotterill).

1. PHASE STRUCTURE OF METALS AND ALLOYS

Metals and alloys are generally crystalline, though they usually lack the crystalline faces familiar on minerals. The crystal structure of metals and alloys determines their properties, in particular that of anisotropy: the variation of properties along different directions in a crystal. Thus, properties such as elastic stiffness, magnetic permeability, and thermal expansion can differ along different crystal axes. Anisotropy is often disguised in a metallic object because such an object consists of many grains (see Fig. 1) whose crystal axes, randomly oriented, tend to average out the anisotropy in the object as a whole. If, however, the crystal axes have a tendency to line up in a preferred direction, the object is said to have a preferred orientation, or texture, and many properties then differ, for instance, along the length or width of a rolled sheet.

1.1 Lattice Structure

Most elementary metals crystallize with very simple structures, such as cubic or hexagonal close packing, or the body-centered cubic structure in which atoms reside at the corners and center of a cubic unit cell (Fig. 2). Tin, gallium, and uranium are examples of the few metals that have more complex crystal structures with tetragonal or orthorhombic unit cells (see also CRYSTALLOGRAPHY).

When one metal is mixed intimately with another on the atomic level, e.g., by melting together or by mechanical alloying, either the second metal dissolves in the first or the metals form an intermetallic compound. In the first case, a solid solution is created in which the atoms are distributed at random and the mean spacing between them varies, but the crystal structure does not change. An intermetallic compound usually has a crystal structure distinct from that of either constituent metal, and its atoms are usually distributed in an ordered manner.

Intermetallic compounds mostly have simple formulas such as CuZn, Ni_3Al, or

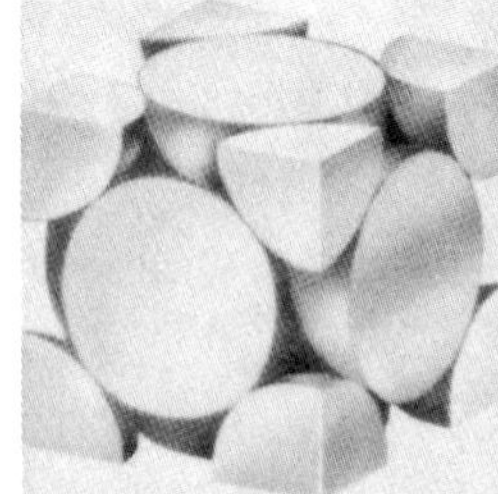

FIG. 2. Simple metallic crystal structures: left, a face-centered cubic, or cubic close-packed structure (representing Al, Cu, Au, etc.); right, a body-centered cubic structure (representing Na, Mo, the low-temperature form of iron, etc.) (courtesy R. M. J. Cotterill).

Cu_3Sn. Some are rigorously limited to their ideal compositions; many others, however, have a considerable range of possible compositions, or homogeneity range. In the early days of physical metallurgy, when the nature of chemical bonding was as yet mysterious, the two types of intermetallic compound were known as daltonides and berthollides, terms referring back to two early chemists: John Dalton, the 18th-century proponent of the atomic basis of chemical combination, and Claude Berthollet, who was skeptical of this concept. The bonding of metals with each other is now much better understood, and the terms are falling into desuetude. (The International Union of Pure and Applied Chemistry has now recommended that these antiquarian terms no longer be used!) A berthollide of nonideal composition is in effect a solid solution based on a compound of ideal composition; a solid solution based on a pure metal is known as a terminal solid solution.

1.2 Phases

An alloy may contain regions, in juxtaposition, of two or more crystal structures, each of a different composition. Typically, one region will consist of a terminal solid solution, and others will contain one or more intermetallic compounds. The occupants of such regions are called *phases*. A particular phase has its own characteristic crystal structure, which defines it. The composition of a phase can vary according to the composition of the alloy as a whole, but in equilibrium the composition, in a particular alloy, will be uniform throughout that phase. In a partly or wholly molten alloy, the liquid region is also a distinct phase, and so is the gas above the alloy itself. A phase can be regarded as the central entity in physical metallurgy.

The number, lattice structures, composition ranges, and relative proportions of the phases of an alloy, studied as functions of the temperature and overall makeup of the alloy, are the basis of phase or equilibrium diagrams. To these we turn next.

2. EQUILIBRIUM DIAGRAMS

The notion of a stable combination of phases for a material of specified composition at a given temperature and pressure goes back to the work of the American physicist Josiah Willard Gibbs (1839–1903). Gibbs recognized that, in principle, even a pure element might adopt a number of different crystal structures. Of these, the structure with the lowest free energy would be that adopted in equilibrium. The free energy, or Gibbs free energy, G, is based upon two subsidiary concepts: internal energy and entropy.

The internal energy H of a gram of metal can be thought of as the sum of the energies of all the bonds between neighboring pairs of atoms in that gram. The internal energy depends upon crystal structure because both the number of atom pairs and the distances between neighboring atoms will vary from structure to structure. (By convention, the most firmly bonded collection of atoms has the lowest internal energy.)

However, the structure with the lowest internal energy will not necessarily be adopted by the metal at all temperatures. This is because the thermal vibrations of the atoms will affect the strength of bonding: the higher the temperature, the greater the vibration and the weaker the bonding. The magnitude of these vibrations is expressed by a quantity termed the entropy S, most conveniently regarded as a measure of disorder. The free energy G is given by the relation $G = H - TS$, where T is the temperature on the absolute or Kelvin scale. In a solid solution (see above), the entropy is affected not only by the thermal vibrations of the atoms but also by the number of different ways in which dissimilar atoms can be distributed in the crystal structure. This number, which is an aspect of disorder, becomes greater as the solid solution becomes more concentrated.

The relation $G = H - TS$, combined with Gibbs's conclusion that the lowest possible value of G determines which crystal structure is stable, indicates that internal energy and entropy are in permanent competition in physical metallurgy. An equilibrium diagram indicates how this competition works out for two or more metals in various proportions at various temperatures. The principles underlying this idea will be made clear by considering the equilibrium diagram for each of three very different alloy systems.

A metastable phase (a concept introduced in the Introduction) is one that has a free energy higher than the phase or phase combination that would be stable for a specific alloy composition and temperature but has been created by some process of rapid or forceful change such as quenching from a high temperature: The metastable phase is then trapped because at ambient temperature, the atoms are incapable of changing places as would be necessary to change the crystal structure and local composition to that (or those) of the state(s) that would be in equilibrium. Metastability is a central concept in understanding the genesis of commonly found alloy structures.

The concepts briefly touched on here are dealt with in more detail in the articles on THERMODYNAMICS, EQUILIBRIUM and THERMODYNAMICS, NONEQUILIBRIUM, and particularly in PHASE EQUILIBRIA.

2.1 The Silver-Magnesium System

The equilibrium diagram for this combination of metals is shown in the lower part of Fig. 3. The vertical scale is temperature, and the horizontal scale is atom fraction of magnesium in the silver-magnesium mixture. The diagram shows five single-phase fields: at high temperatures, liquid; at lower temperatures, the α field, a range of terminal solid solutions based on the cubic close-packed silver crystal structure; the γ field, based on the hexagonal close-packed magnesium crystal structure; the β' field, a [berthollide] intermetallic compound centered on the formula AgMg with a simple cubic structure of Ag atoms at cube corners and Mg atoms at cube centers; and the ϵ phase, a narrower berthollide field based on $AgMg_3$, the structure of which has a more complicated hexagonal unit cell. These single-phase fields are separated by two-phase fields in which the adjacent phases coexist, at compositions defined by the boundaries on either side and varying with temperature, and in relative proportions depending on the point within the field that defines the temperature and the overall composition. [Thus, point P, in the β' + ϵ field, defines an alloy with 0.6 atom fraction of Mg, equilibrated at 600 K (327 °C).] An alloy such as this illustrates the fact that overall alloy composition is quite distinct from the composition of the equilibrium phases where two or more of these exist.

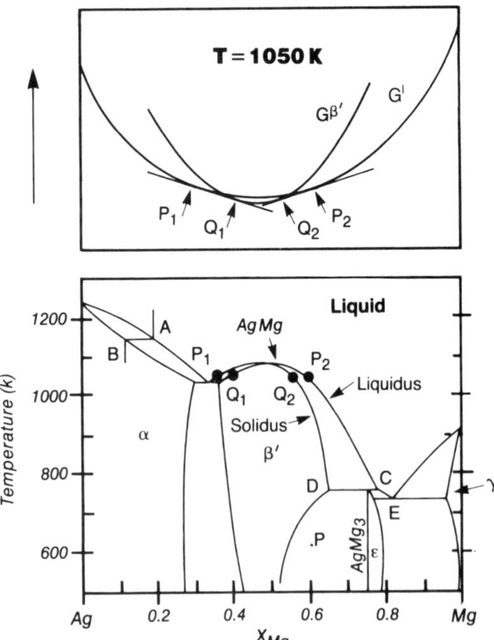

FIG. 3. Silver-magnesium equilibrium diagram, with corresponding free-energy diagram for 1050 K (777 °C) above. (G represents the Gibbs free energy.)

An alloy melt usually cannot freeze without a change in composition. Thus, when a melt of composition A reaches the liquidus, it will initially freeze as a solid of composition B, defined by the solidus. This process alters the composition of the remaining melt, enriching it in magnesium, so that it freezes at successively lower temperatures, producing a solid of steadily changing composition, with an onionlike configuration. Such an alloy is said to be "cored," or "microsegregated"; it cannot be in true equilibrium until migration of the atoms in the hot solid has evened out the compositional gradient so that the entire solid returns to the uniform composition A. If, however, the frozen solid is promptly cooled to ambient temperature, the atoms are unable to move, and the structure remains permanently cored. This is a simple example of a metastable structure, in the sense that the material is not in equilibrium and is yet, for purely kinetic reasons, unable to reach that state.

The first intermetallic compound, at its ideal composition, AgMg, freezes "congruently," without change of composition—that is to say, the liquidus and solidus coincide at this alloy composition. This often serves to define the ideal composition of a berthollide. The ϵ phase, however, freezes out by a more complex process termed peritectic solidification, which involves a reaction between a liquid of composition C and a solid of composition D.

Finally, at the right-hand side of the diagram, we have a "eutectic" solidification process. In this, an alloy of overall composition E freezes to form an intimate two-phase mixture of ϵ phase and Mg-rich terminal solid solution, a so-called eutectic mixture, solidifying at a constant eutectic temperature. If the alloy as a whole contains a little more or less Mg than does composition E, then one or the other of the two phases mentioned above freezes out first until the residual liquid achieves composition E, and the eutectic mixture then forms. The result is a curious two-phase mixture consisting (in the case of magnesium excess) of regions of coarse magnesium-rich solid solution interspersed with regions of the fine eutectic mixture, so that the same phase exists in two different microstructural forms.

The upper diagram of Fig. 3 shows the estimated variation of the Gibbs free energy G, at a constant temperature of 1050 K (777 °C), with phase composition for the liquid and β' phases. For a given overall alloy composition, the stable phase structure is always that corresponding to the lower value of G, except in those composition ranges where a common tangent to the two curves lies below both curves, such as the regions P_1Q_1 and P_2Q_2. Here, a two-phase mixture will have a lower total Gibbs free energy than if the alloy consisted exclusively of one or the other phase. All other portions of the equilibrium diagram can be likewise interpreted.

2.2 The Sodium-Bismuth System

This system (Fig. 4) exemplifies well the circumstances that lead to the appearance of extreme daltonide phases, of vanishingly small homogeneity ranges. Two such phases exist in the sodium-bismuth system: Na_3Bi and NaBi. Very narrow (daltonide) phase fields are often referred to as line compounds, because the single-phase fields have to be drawn as single lines in the equilibrium diagram. There is also a eutectic, in which both constituent phases, NaBi and Bi, have vanishingly small mutual solubilities.

The upper diagram shows the variation of the Gibbs free energy for a line compound at

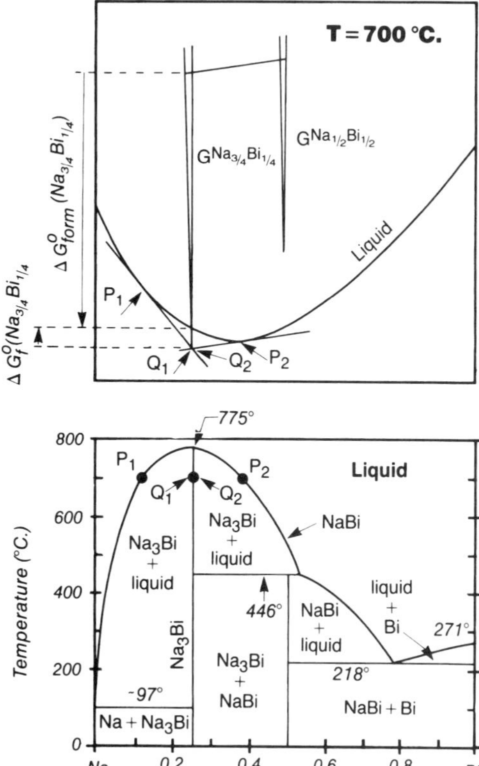

FIG. 4. Sodium-bismuth equilibrium diagram, with corresponding free-energy diagram for 700 °C above. (*G* represents the Gibbs free energy.)

a fairly high temperature. Very small deviations from the ideal composition lead to very sharp increases in the Gibbs free energy. This happens because such phases have extremely strong bonds between the neighboring Na and Bi atoms but weak bonds between pairs of identical atoms. The structure of compounds of the sodium-bismuth type is such that there are no, or very few, atoms of the same kind as next neighbors. When there is a small excess of either Na or Bi, then additional Na-Na or Bi-Bi neighbor pairs necessarily exist. This gives rise to a sharp increase of the internal energy H, with but a small change in entropy S (because there are so few "wrong" atoms present); hence, at a fixed temperature, the free energy G rises sharply. In the upper diagram, the contact points of the tangents referring to the phase Na_3Bi at the points Q_1 and Q_2 are so close that the homogeneity range of the phase is of negligible width.

2.3 The Copper-Gold System

This system, often regarded as a metallurgical classic, illustrates a further consequence of the arguments presented in the preceding paragraph.

Figure 5 shows a section of the Cu-Au equilibrium diagram in simplified form. Cu_3Au is a berthollide compound, the crystal structure of which is shown in Fig. 6(a). When this compound in its equilibrium state is slowly heated, it undergoes (at a critical temperature T_c = 390 °C) an abrupt transformation to a state in which the copper and gold atoms are distributed at random on the sites of a face-centered cubic structure. The transformation is accordingly called an order-disorder transition. On further heating, the disordered form remains in equilibrium until it reaches the melting temperature. If the compound is of a nonstoichiometric or nonideal composition, for example $Cu_{78}Au_{22}$, then T_c falls to a lower value, ≈350 °C, and a narrow region exists in which the ordered and disordered forms of the compound, of slightly different compositions, coexist in equilibrium.

An ordered form such as that of Cu_3Au exists because it has as many unlike (Cu-Au) nearest neighbors in its crystal structure as are compatible with the formula. Such neighbors have relatively stronger bonds,

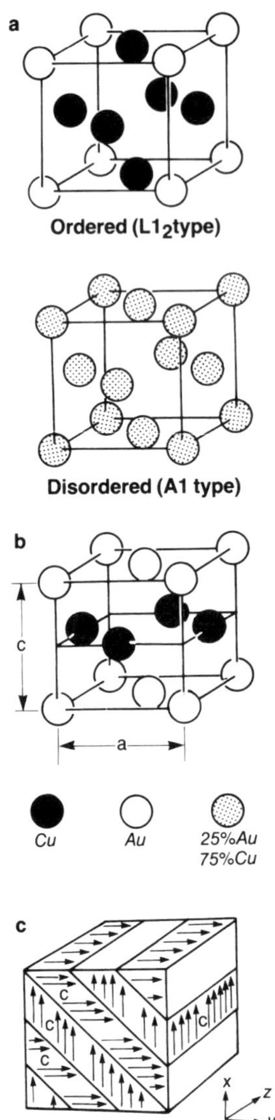

FIG. 6. Crystal structures of (a) ordered and disordered Cu_3Au and (b) ordered CuAu. (c) Schematic of an array of CuAu domains (two out of three possible orientations) in what was originally a single disordered crystal.

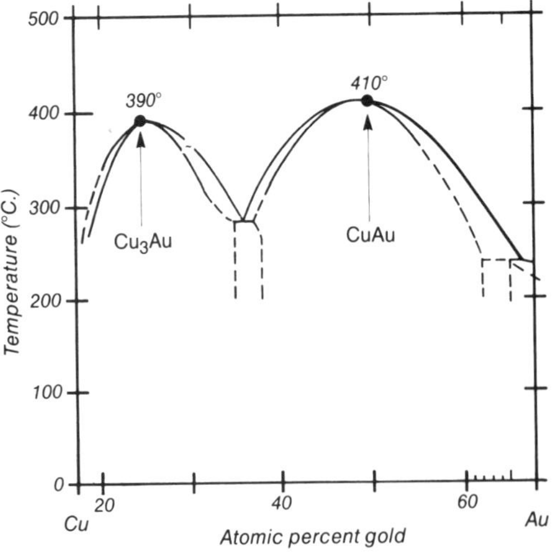

FIG. 5. Copper-gold equilibrium diagram (partial, simplified). The dashed lines represent phase boundaries that have not been located with certainty.

implying a lower overall internal energy H of the ordered form.

An order-disorder transition occurs because the TS in the expression $G = H - TS$ rises with rising temperature until it becomes advantageous—in terms of free energy—for the alloy to assume disorder, which entails a greatly enhanced entropy. In this disorder, a large number of "like" atom pairs (Cu-Cu and Au-Au) come into existence. These pairs have relatively weak bonds, and *H* rises. This rise is more than compensated by the sharp rise in configurational entropy resulting from the random distribution of the Cu and Au atoms.

The disappearance of ferromagnetism when a piece of iron is heated above its Curie temperature is an analogous phenomenon. The individual atomic magnets in the iron behave much as do the Cu and Au atom pairs in Cu_3Au (see also MAGNETIC ORDERING IN SOLIDS).

To explain why the transition can be abrupt (as is the case for Cu_3Au) requires the techniques of statistical mechanics, the theory of the collective behavior of large aggregates of atoms (see STATISTICAL MECHANICS, CLASSICAL). When the critical temperature is reached, a kind of avalanche takes place: The more atoms disorder, the easier it becomes for the remainder to follow. Statistical mechanics can make sense of structural transitions, given information about relative bond strengths; it cannot interpret the fact that Cu-Au bonds are stronger than Cu-Cu and Au-Au bonds. To do that, it is necessary to apply the theory of quantum mechanics. Both classical statistical mechanics and quantum mechanics (*q.v.*) are important in the study of modern physical metallurgy.

When more gold is added to the alloy to reach the composition CuAu, another order-disorder transition is observed. The ordered form of this berthollide compound is shown in Fig. 6(b). Ordered CuAu has a tetragonal structure consisting of successive sheets of Cu only and Au only. Counting shows that this structure has more unlike nearest neighbors than does the disordered (face-centered cubic) structure. This kind of order-disorder transition, associated with a change of symmetry and unit cell dimensions, leads to interesting complications. The disordered cubic form has three crystallographically equivalent cube axes at right angles to each other. Any of these axes can turn into the unique tetragonal *c* axis, so that a single disordered crystal can turn into an array of so-called domains, each of which has a *c* axis oriented differently from its immediate neighbors; this is shown schematically in Fig. 6(c). Since the *c* axis of an ordered domain is shorter than the cube edge of a disordered unit cell, ordering will impose localized strains on the alloy. One consequence is that the configuration of domains [Fig. 6(c)] is optimized so as to minimize the total elastic strain energy, because differently oriented domains relieve each other's elastic strains to some degree. The other consequence is that an external stress will modify the preferred pattern of domains and in the extreme case will ensure that an entire crystal consists of a single domain. The role of stress in determining structure has increasingly been recognized in recent research in physical metallurgy.

Alloys near the CuAu composition, and containing small amounts of other noble metals, are used for dental crowns and bridges. When quenched from a high temperature, such an alloy is metastably disordered and in this state is soft. When heated at 200 to 300 °C, the alloy orders and assumes a complex domain structure that is much harder. In this way, dental crowns can be strengthened.

Further discussion of the categories of binary phase diagrams and their relationship to the compositional and temperature dependence of the Gibbs free energy will be found in the article PHASE EQUILIBRIA.

2.4 The Iron-Carbon System

The iron-carbon system is of crucial industrial importance. Figure 7 shows the accepted version of the phase diagram for this system. The diagram, in which the metastable phase Fe_3C, cementite, appears, is appropriate to the study of steels. In cast irons, which contain large carbon concentrations, carbon forms as the most stable phase, that is, elemental carbon (graphite). In steels, carbon forms so sluggishly that cementite, which forms easily, is found instead. Figure 7 is thus an example of a metastable phase diagram.

The essential point to grasp concerning the heat treatment of simple (that is, iron-carbon) steels is that they have two different crystal structures, depending on temperature. Just below the freezing temperature, iron and dilute alloys are body-centered cubic (δ phase). At lower temperatures, the steel transforms to austenite (γ phase), which is face-centered cubic. At still lower temperatures, the form reverts to body-centered cubic, here called α phase. (The transition from body-centered cubic to face-centered cubic to body-centered cubic again is unique to iron.) Beta (β) iron disappeared from the scientific

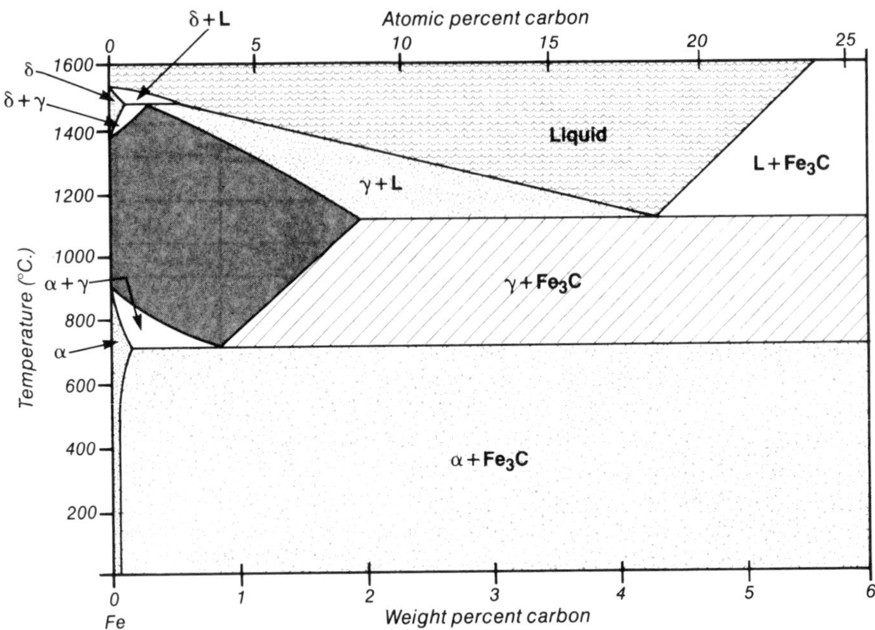

FIG. 7. Iron-carbon equilibrium diagram (omitting equilibria involving graphite).

vocabulary when it was recognized that the phase it was intended to describe entails no crystallographic but only a magnetic transition. At low temperatures, α is ferromagnetic, but it becomes paramagnetic at higher temperatures.

When a plain carbon steel containing 3.6 atoms of carbon per 100 atoms of alloy (3.6%) is slowly cooled, it undergoes a eutectoid reaction. This reaction is akin to a eutectic reaction (see above) but takes place in the solid state. The product is a mixture called pearlite, consisting of nearly pure body-centered iron (called ferrite) and cementite, Fe_3C. Figure 8(a) shows a polished and etched section of pearlite under the microscope. If, however, the austenite is quenched rapidly in water, pearlite has no time to form, and a product called martensite forms instead. This transformation does not appear in the phase diagram, because martensite is metastable in relation to pearlite and forms by a shear process requiring no migration of carbon atoms, in contrast to pearlite. (Indeed, it cannot be displayed in a conventional phase diagram, because the fraction of the alloy transformed into martensite varies continuously with temperature below a critical value, called M_S.)

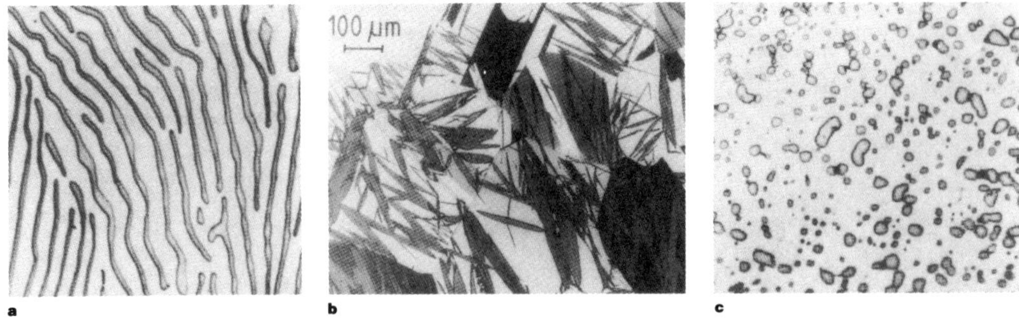

FIG. 8. Optical micrographs of (a) pearlite, (b) martensite, and (c) a distribution of cementite particles in ferrite (courtesy E. Hombogen).

Figure 8(b) is a micrograph of a steel that has been partly transformed into martensite. Each "needle" is the product of a single shear process.

Martensite is tetragonal, with a severely distorted form of body-centered cubic crystal structure, making it extremely hard and brittle. When martensite is heated to moderate temperature, it begins to decompose into an intimate mixture of iron and cementite, becoming progressively softer and more deformable [Fig. 8(c)]. This is the age-old process of tempering. A high-quality knife or sword is shaped while the steel is soft, then heat treated to quench the cutting edge very fast and the thicker part more slowly, and finally mildly tempered. The result is an edge that is very hard and a backing that is less martensitic and thus less brittle. Many actual steels are the result of far more complex heat treatment than the simple Fe-C steel just discussed and also contain other alloying elements such as chromium, vanadium, niobium, and molybdenum. These additions can enhance the tendency to form martensite, so that a thicker section can still be effectively hardened. Each steel has characteristic time-temperature transformation curves (Fig. 9 shows an example) so that an engineer knows how thick a section of that steel can be effectively hardened.

Until recently, rolled steel sheet, as used for the manufacture of automobile bodies, could not be heat treated because the huge coils of sheet had far too much thermal inertia to permit rapid changes of temperature. In continuous annealing, the sheet is cold rolled and then passed through a furnace and a forced cooling device before coiling. The temperature-time schedule is regulated by computer control of every stage of the process. In this way, lightly alloyed steel sheet

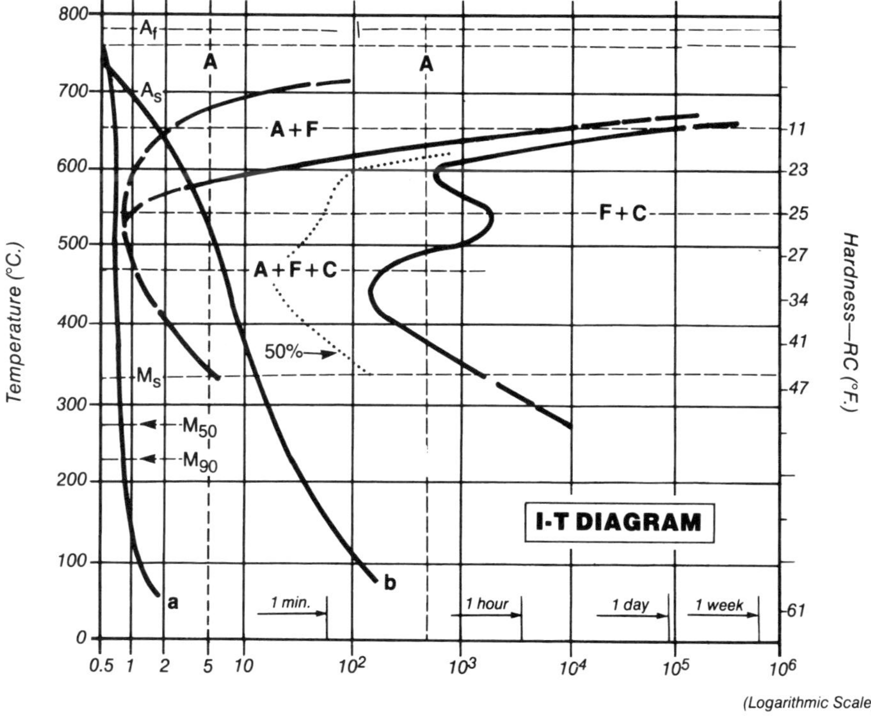

FIG. 9. Time-temperature transformation curves for isothermal heat treatment of an initially quenched steel containing 0.4 wt% C and 1.0 wt% Mn. A = austenite (solution of carbon in face-centered cubic iron); F = ferrite (solid solution of carbon in body-centered cubic iron); C = cementite, Fe_3C. M_s is the temperature at which martensite starts to form on cooling; M_{90} indicates 90% completion of the transformation. Curve **a** represents the temperature-vs-time curve characteristic of a water quench, which generates martensite only; curve **b** represents the cooling rate for an oil quench (slower cooling), which generates a mixture of martensite, ferrite, and cementite. The other curves represent the start, 50% stage, and termination of the isothermal transformation of austenite to yield ferrite and cementite.

can be manufactured to standards of strength and uniformity that were previously unattainable. The result is an automobile body that is light and strong. Heat treatment of steels, and indeed that of nonferrous alloys, is further discussed in MATERIALS TREATMENT; see also STEEL.

2.5 Ternary Alloys

Many important phase diagrams refer to ternary systems—systems comprising three elements. The Fe-Ni-Cr alloy system is a prominent instance; it is the key to understanding stainless steels. Ternary diagrams involve three-phase, two-phase, and single-phase fields. They are thus quite complex and require three-dimensional graphical representation by means of a prism on an equilateral triangular base. The axis of the prism represents temperature, and each point on the base represents a unique ternary composition. Phase boundaries are displayed in two-dimensional sections. Quaternary systems are even more complex and are not often examined. They are represented by means of a regular tetrahedron in which each point represents a unique composition. All such points refer to a single temperature.

2.6 Calculation of Phase Diagrams

The time and effort involved in the accurate determination of equilibrium diagrams, especially ternary systems, have been greatly reduced by the development of theoretical methods such as CALPHAD (CALculation of PHAse Diagrams). Thermochemical quantities such as heats of formation and activities of solid solutions can be measured experimentally and then used to predict the location of phase boundaries. In particular, from measurements for three binary systems, phase boundaries in the corresponding ternary system can be calculated. This activity is a flourishing subspecialty in physical metallurgy.

3. INSTRUMENTATION FOR CHARACTERIZATION OF ALLOYS

Nothing has yet been said about the experimental methods used to obtain the information discussed in the preceding pages. We now turn to the role of metallurgical instrumentation in determining the nature of crystal structure, the position of phase boundaries in equilibrium diagrams (which involves both temperature and composition measurement), the particulars of order-disorder transitions, and the kinetics of phase transformation in alloy steels at various temperatures. For convenience, the methods of attack can be divided into three categories:

1. means of determining crystal structure;
2. means of observing microstructure; and
3. means of determining local variations in composition.

3.1 X-Ray Diffraction

Until 1912, no method existed for the determination of even the simplest crystal structures. In that year, Max von Laue (German), who was interested in establishing the nature of the recently discovered x rays, sought to diffract x rays from crystals, which he estimated from indirect evidence to have interatomic dimensions of the same order of magnitude as the postulated wavelength of x rays. He argued that a periodic three-dimensional structure would scatter (diffract) a beam of x rays strongly in particular directions, much as a diffraction grating scatters a beam of light. There is a simple mathematical relationship between the directions of incident and diffracted beams and the periodic spacing of the diffraction grating or crystal lattice planes (Fig. 10). Von Laue's experiment was successful, and the wave nature of x rays was proved. (At the time, he could not know that, soon after, quantum wave mechanics and the experiments of Arthur Compton would show that the particle model for x rays is just as valid as the wave model; both models are correct.)

Von Laue's lead was followed up at once by the English physicists William Henry and William Lawrence Bragg, father and son, who determined the simple structures of materials such as rock salt and copper. The Braggs were the first to realize that it is possible in principle to locate the geometric positions of atoms in a crystal from measurements of the directions and intensities of the scattered beams. Soon after, the technique was used by metallurgists to determine the structures not only of elementary metals and solid solu-

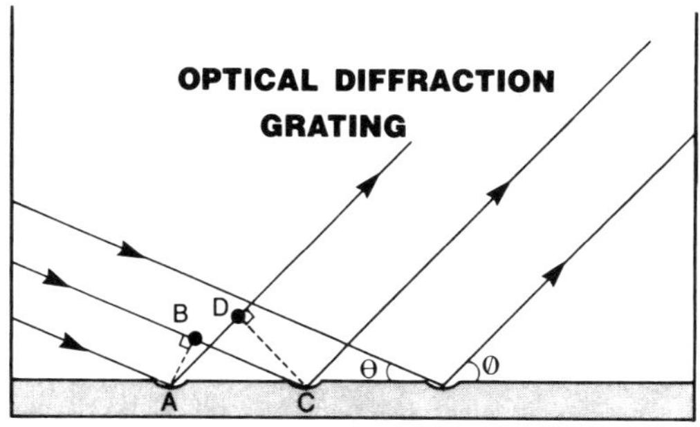

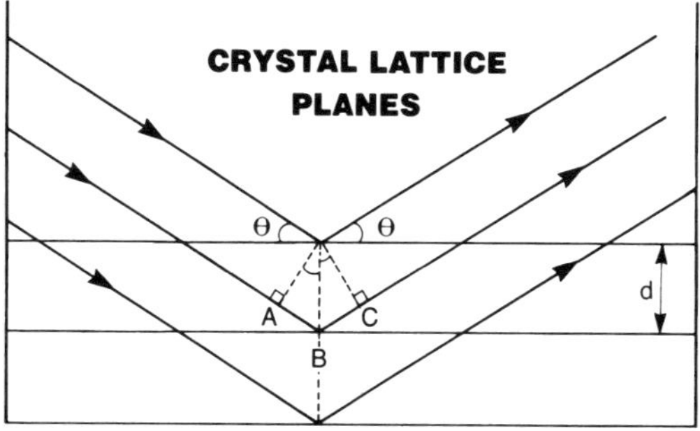

FIG. 10. X-ray diffraction from an optical grating and from a family of lattice planes in a crystal. For the lattice planes but not for the uniperiodic grating, the angles of incidence and diffraction must be equal. In both instances, successive diffracted rays are an exact integral (*n*) number of wavelengths (λ) out of step.

tions, but also of numerous intermetallic compounds. By now, thousands and thousands of such individual structures are known, many of extreme complexity. X-ray diffraction was the most important single experimental innovation of the 20th century in the field of physical metallurgy (see X-RAY DIFFRACTION).

Two types of x-ray technique developed. In the powder method, an assembly of millions of minute, randomly oriented crystallites is exposed to a beam of single wavelength. In the single-crystal method, a small monocrystal is rotated or oscillated while being bathed by a single-wavelength x-ray beam or is exposed motionless to a multiwavelength beam. In either case, diffracted beams flash out in specific preferred directions. Single-crystal methods are used to determine crystal structure and also to identify the orientations of single metal crystals used for research. The powder method is used for such purposes as identifying unknown compounds, determining phase boundaries in phase diagrams, and searching for nonrandom (preferred) orientation of the crystal grains in metal sheets. It is also used to determine internal strains within the grains of a polycrystalline metal object by using the spacing of lattice planes as a built-in gauge. One of the more sophisticated applications of the method is to the measurement of the relationship in crystal orientation between a parent phase in an alloy and a phase formed from it by partial transformation during heat treatment. (Solid-state chemists call this sort of relationship topotaxy.)

Diffraction experiments were originally done with special cameras, using photographic film. Diffractometers, in which the diffracted beams are located and measured by means of devices such as Geiger counters or photomultipliers, are preferred nowadays.

When progressively increasing amounts of a solute are incorporated in solid solution in a metal (or in a berthollide intermetallic compound), the spacing of the lattice planes progressively alters according to the relative sizes of the two atom species. When the solubility limit is reached, the change of spacing ceases, and the position of the phase boundary of, say, a terminal solid solution at a particular temperature can then be accurately determined. X-ray diffraction is widely used in physical metallurgy for this rather simple type of measurement.

3.2 Optical Microscopy

The light microscope was first applied systematically to the study of metals and alloys by Henry C. Sorby, working between 1847 and 1887. Ever since, seeing has been believing for the physical metallurgist, and a substantial subsidiary science of metallography has grown from the use of the light microscope.

To examine an alloy under the light microscope, a section is cut and mechanically or chemically polished to a very fine finish. The sample is then etched in acid or other reagent, so that different phases can be distinguished. Extensive training and skill are required to produce a section of optimum appearance. Figure 8 illustrates alternative microstructures of a simple carbon steel. Figure 8(a) shows pearlite; Figure 8(b) the shear-generated structure of martensite; and Fig. 8(c) the fine distribution of cementite in tempered martensite. These micrographs show clearly that very distinct microstructures can be created in one and the same alloy by different heat treatments.

A major use of the light microscope is to define phase fields in equilibrium diagrams. An alloy is water quenched from a high temperature, and a section is then polished, etched, and examined to identify the phase or phases in the microstructure. This identification may require the use of other instruments as well, such as the electron probe microanalyzer described below.

3.3 Electron Microscopy

Over the last several decades, the light microscope has been supplemented by a pair of instruments that permit much finer detail to be seen. The resolving power of a light microscope is limited by the wavelength of visible light, so that features separated by less than 10^{-3} mm (1 μm) cannot be properly resolved. The transmission electron microscope can improve this resolution by more than a thousandfold, allowing, in the best instruments, features about 0.2 nm (2×10^{-7} mm) apart to be resolved. The image is either recorded on a photographic plate or projected onto a fluorescent screen for immediate viewing.

The specimen is in the form of a foil thin enough (about 0.1 μm) to allow the electrons to pass through. The resolving power of the microscope should ideally approximate the electron wavelength, $\approx$0.004 nm. If this were possible, fine detail within individual atoms ($\approx$0.2 nm diameter) should be resolvable. In practice, however, the resolving power is limited by aberrations intrinsic to the construction of the instrument. Accordingly, the resolution achieved is less than the ideal by a factor of about 50.

The formation of the image entails diffraction of the electrons from crystal planes in the specimen. Crystallographic information can be obtained by manipulating the diffracted beam in various ways. Thus, it is possible to check the crystal structure, and the spatial orientation of the crystal axes, of the phases in the foil, as well as the spatial disposition of the phases. An electron microscope combines the best features of a light microscope and an x-ray diffraction camera and in addition has a greatly superior ability to resolve fine detail.

Figure 11 shows two microstructures as seen in the transmission electron microscope. Figure 11(a) shows the structure of an "age-hardened" aluminum-copper alloy. The individual "zones" consist of regions in which copper atoms are locally enriched along specific crystal planes of the aluminum-rich solid solution that has been abruptly cooled from

a

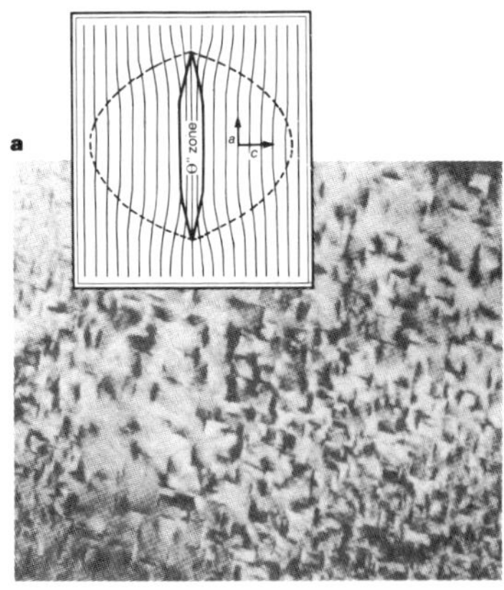

b

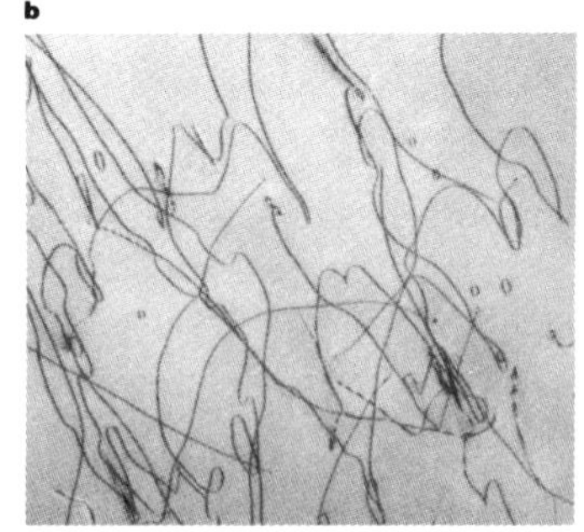

c

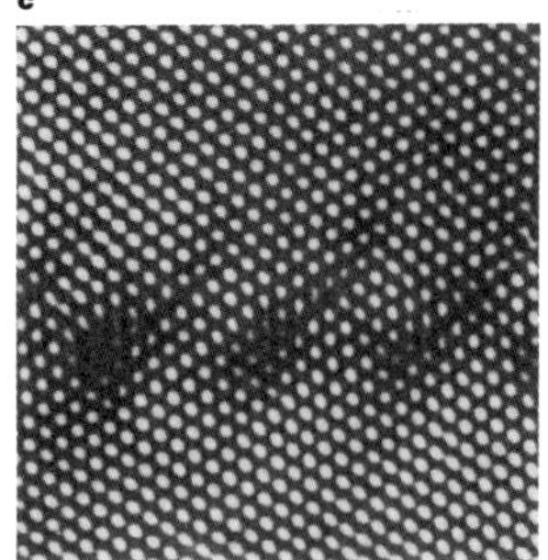

FIG. 11. Transmission electron micrographs: (a) Localized zones enriched in copper in a supersaturated (quenched) and subsequently aged aluminum alloy containing 4 wt% Cu; the inset shows the distortion of lattice planes near a copper-enriched zone (courtesy J. Nutting). (b) A network of dislocations in a magnesium oxide crystal (courtesy Y. Moriyoshi). (c) A high-resolution micrograph of two adjacent grains, lightly misoriented, after computer processing to enhance the contrast; the magnification is approximately ten million times (courtesy W. Krakow).

a high temperature at which all the copper is in solid solution. The solid solution is unstable because the compound $CuAl_2$ should be present in equilibrium at room temperature. But migration of atoms is too slow for $CuAl_2$ to be deposited, and the micrograph shows instead the attainment of a halfway stage: The diffuse areas appearing around the zones represent regions of intense strain, occasioned by the fact that copper atoms are smaller than aluminum atoms. Strain alters the spacing of the lattice planes, and local differences in spacing are revealed by the electron beam as local variations in diffraction intensity. The localized strain, in turn, hardens the alloy; this is the basis of age-hardening, used to strengthen an aircraft structure after it has been shaped and assembled.

Figure 11(b) shows a thin ceramic section containing dislocations. These defects, which are present in all metals and alloys as well as in ceramics, move through a crystal under applied stress and cause the crystal to change shape. The capacity of an alloy to be shaped and to resist mechanical shock depends entirely on these crystal defects, and physical metallurgists spend much time in examining the geometry of dislocations. (More about these entities below.) There is no space in this article to describe the geometry of dislocations; particulars can be found in the article POINT DEFECTS AND OTHER IMPERFECTIONS IN CRYSTALS.

The multiplication and interaction of dislocations causes work-hardening, that is, the increase of resistance to plastic deformation when a metal is extensively deformed. The lines of dislocations show up in the electron microscope because the lattice is distorted in the immediate vicinity of the defect, and the electron beam is scattered less strongly when it passes close to such lines.

Figure 11(c) shows a high-resolution image of an ultrathin crystal foil, which was made by an electron microscope of particularly precise construction. The image shows the positions of stacks of atoms, viewed parallel to the incident electron beam. To a first approximation, one can say that such an image shows the positions of individual atoms in the crystal structure of the specimen. Images of this kind are widely used in examining the fine structure of the boundaries between crystal grains [one such is seen in Fig. 11(c)] and the interfaces between two phases. Grain boundaries in polycrystalline assemblies (the normal condition of industrial metals and alloys) have become particularly important in understanding mechanical and electrical properties and are nowadays the subject of intense study: Electron microscopy has made it possible to demonstrate that the fine structure, and hence the properties, of grain boundaries depend in a complex way on the exact orientation relationship between two adjacent grains. For instance, for certain orientation relationships, dislocations transmit with particular readiness from one grain to its neighbor, and the prevalence of such grain boundaries enhances plastic deformability. The statistical control of grain-boundary crystallography by appropriate deformation and heat treatment has been termed grain-boundary engineering.

The scanning electron microscope is a type of electron microscope used to examine the surface of a solid. Here, an electron beam is made to scan a surface continuously, and scattered electrons are picked up by a particle counter. The reading of the counter is continuously amplified and fed into a television tube that forms an image in synchrony with the beam. It is this image that is observed. Figure 12 shows such an image for a directionally solidified eutectic. This kind of alloy consists of rods of one phase regularly disposed in a matrix of another and is useful

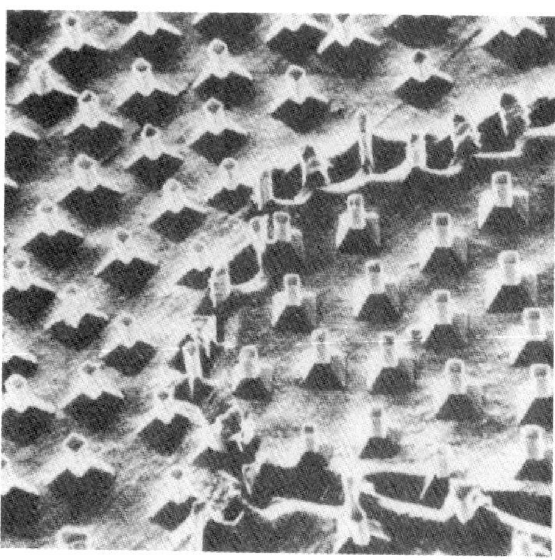

FIG. 12. Directionally frozen eutectic microstructure consisting of tantalum carbide needles in a nickel-chromium matrix. The micrograph was taken using a scanning electron microscope. (Courtesy M. J. Henry.)

for resisting stress at high temperatures. Scanning electron micrographs are particularly useful for revealing perspective detail, at a resolution and depth of focus considerably higher than is possible with a light microscope, which is also used to examine the surface of a solid.

For extensive further details concerning both types of electron microscopy, see ELECTRON MICROSCOPY.

3.4 Determination of Local Composition

Undergraduate students of metallurgy once studied chemical analysis, and most metallurgical research laboratories employed professional analytical chemists. Today, physical metallurgists have a range of instruments that not only analyze an alloy automatically but also can reveal variations in composition from point to point in a microstructure.

All such instruments use the principles of physics to obtain chemical information. They include secondary ion mass spectrometers, x-ray photoelectron spectrometers, Auger spectrometers, and electron microprobe analyzers (EMPAs). The EMPA has become an indispensable tool of every research metallurgist. It is briefly described here. Further details can be found in ELECTRON MICROSCOPY.

In an EMPA, an electron beam is focused on a point, on the order of 1 μm diameter, on the surface of a polished specimen. Two kinds of information result. The backscattered electrons form an image as in a scanning electron microscope, allowing different phases to be distinguished. Also, when an electron beam is slowed by the alloy it generates a burst of x rays, of wavelengths characteristic of the particular metals present. The x-ray wavelengths are measured. The electron beam can be scanned over the surface, and the x rays emitted at a particular electron wavelength can be fed into a picture tube, so that an image of the point-to-point variation of the concentration of the relevant metal can be displayed on the screen.

The EMPA has become an indispensable research tool because of its ability to localize, measure, and identify impurities and phases within a microstructure. Indeed, the EMPA, the x-ray diffractometer, the transmission electron microscope, and the scanning electron microscope have, together, transformed the study of alloys since World War II.

The techniques outlined in Secs. 3.1–3.4 are further discussed in the context of characterization in general in MATERIALS ANALYSIS AND CHARACTERIZATION.

4. MICROSTRUCTURE AND PROPERTIES

An equilibrium diagram specifies only the compositions, crystal structures, and atomic or mass fractions of two or more phases existing in mutual equilibrium. The shapes, sizes, and mutual separation of the regions occupied by the distinct phases are not specified by the Gibbs free energy equation or by the equilibrium diagram depending on this equation. The study of these variables constitutes the study of microstructures.

The quantitative study of microstructures entails the measurement of variables such as the mean diameter of a dispersed phase (for example, the rods in the eutectic of Fig. 12); the form of such a phase (spheres, cylinders, laths, sheets); the mean separation between units of a dispersed phase; the volume fractions of two phases; and the mean free path—the average distance a point moving in a straight line must go between successive intersections with a dispersed phase. All such measurements must be treated statistically and a mean value and a variance determined, since the variables are not constant for any particular microstructure. The science of deducing this information from statistical sample measurements on two-dimensional sections is termed quantitative metallography, or stereology. Computerized image analyzers are used to assemble the statistical array of measurements and to calculate the numerical information from the raw data. Thus, an image analyzer can rapidly determine the distribution of mean diameters of the crystal grains shown in Fig. 1, or the mean spacing and thickness of the cementite plates in the pearlite structure of Fig. 8(a).

4.1 Development of Microstructures

No microstructure is ever wholly stable. When a polycrystalline assembly of grains such as that of Fig. 1 is heated, some grains grow larger by consuming others. Again, a distribution of dispersed particles such as that seen in Fig. 8(c) will progressively coarsen. The control of coarsening is especially important for the optimization of mechanical behavior; thus, the dispersion of phases in a superalloy used in an aero engine must remain fine after long service at high temperatures. This kind of coarsening is also known as Ostwald ripening, after the German physical chemist Wilhelm Ostwald (1853–1932) who first examined it. An example will help to make the point clear. Suppose we have two spheres of phase *A* disposed in a matrix of phase *B*. Suppose further that atoms of the predominant constituent *x* of phase *A* can diffuse (migrate) through the *B* matrix. The local solubility of *x* in *B* will be greater near the smaller sphere, because its sharper curvature exposes the *x* atoms at the interface more freely to the matrix. On balance, therefore, the smaller sphere will dissolve, and the *x* atoms will diffuse along a composition gradient through the matrix and deposit on the more gradually curved interface of the larger sphere. (This is essentially what happens when a large raindrop in a cloud grows at the expense of a smaller one—except that in this case, water molecules evaporate and migrate singly through the air.) The mathematical laws of the process are well understood. To slow down or even eliminate the process, *x* should have a very low solubility in *B*, and the rate of migration of *x* atoms should be low. Furthermore, the atomic fit between the two phases *A* and *B* at their interface should be as nearly perfect as possible. This last feature stabilizes superalloys, as used in the construction of aero engines, against Ostwald ripening.

Ostwald ripening, as discussed above, is an important illustration of the kind of microstructural study that physical metallurgists engage in. In this particular case, the proportions of the two phases *A* and *B* do not change during the process. A more common case is a phase transformation. Thus, Fig. 9 expresses the kinetics of the decomposition of an austenitic steel structure into other phases, and the physical metallurgist's task is to measure these kinetics experimentally and to interpret them in terms of atomic diffusion and events at the interface between the old and the new phases. Indeed, as remarked at the outset, the study of phase transformations is a central concern of physical metallurgy.

Another form of microstructural feature, which originally provided early evidence of the existence of dislocation lines in crystals, is the growth spiral on a crystal surface. Such spirals were first observed in the early 1950s; an example, an optical micrograph of silicon carbide, is seen in Fig. 13. When a particular type of dislocation named a screw dislocation (see POINT DEFECTS AND OTHER IMPERFECTIONS IN CRYSTALS) emerges at a free crystal surface, then an "inextinguishable" step, typically one unit cell in height, is formed at the surface. Such steps are essential for a crystal face to grow, because itinerant atoms are easily captured by a growing crystal at such a step (see CRYSTAL GROWTH). Under certain circumstances, the step height can be larger, as in Fig. 13, and it is then easily captured even in an optical micrograph with its limited resolution. This kind of micrograph caused enormous interest and provoked much

FIG. 13. A growth spiral on a silicon carbide crystal, originating from the point of emergence of a screw dislocation (courtesy S. Amelinckx).

research when dislocations were new and disputed entities.

4.2 Mechanical Properties of Metals and Alloys

The mechanical behavior of solid materials in general, including metals, is discussed in MECHANICAL PROPERTIES OF SOLIDS. The physical metallurgist needs to make sense of the ways in which changes of microstructure modify properties such as ductility and strength. One way in which this has been done is to turn a wire or bar of an alloy into a single large crystal grain—by slow solidification from one end, for example. If the crystal is then deformed, it is found that plastic deformation is concentrated in separate packets, or glide planes, as seen in Fig. 14, so that the crystal looks like a pack of cards being shuffled. In general, concentration of deformation along a single plane as seen in the figure, instead of dispersion among a large number of parallel planes close together (one thick "card" instead of many thin cards, as it were) makes for limited plastic deformability.

The ductility of a metal or alloy has been found to be intimately linked with the number of alternative, differently oriented lattice planes on which such deformation takes place. Thus, metals with high (cubic) crystal symmetry, where many crystallographically equivalent but differently oriented planes exist, are more likely to be highly ductile. The rate at which the crystal hardens as it is progressively deformed is determined by the specific ways in which the dislocation lines [see Fig. 11(b)] multiply and interact with each other. To make sense of this, the investigator combines experiments with single crystals under tension with examination of thin slices of the deformed crystal in a transmission electron microscope.

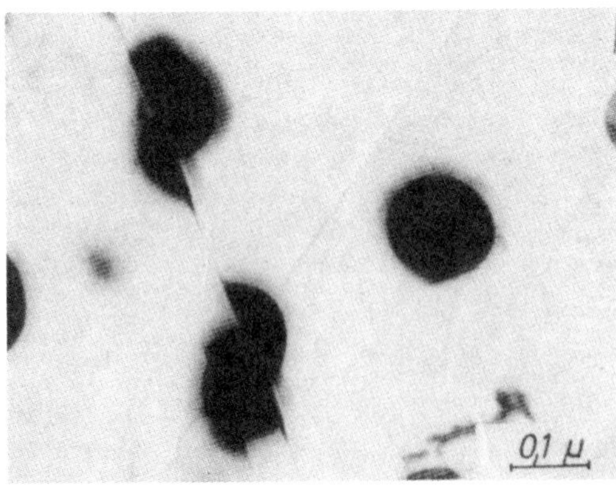

FIG. 14. Precipitated particles in an alloy containing iron, nickel, and titanium. A slip plane arising from plastic deformation is seen shearing the particles. (Transmission electron micrograph, courtesy E. Hombogen.)

Figure 14 shows particles being sheared by the passage of a succession of dislocations on the same plane (though the dislocations themselves are not seen imaged in this micrograph). Depending on the composition and strength of bonding of such particles, such shearing action may entail too large an energy penalty and thus does not happen at all readily. Then particles are not sheared at all but repel dislocation lines; especially if the particle spacing has a critical value, this leads to extensive strengthening of the alloy. Dislocation lines being repelled from fine particles in a high-strength aluminum alloy are seen in Fig. 15. Smaller particles can "pin" a dislocation line; to unpin it, the dislocation line has to cut right through the pinning particles, and that requires much energy and therefore a high stress. An example of dislocation lines pinned at several points by fine particles is seen in Fig. 16, again in an aluminum alloy.

When a metal has been intensely deformed, it becomes filled with a tangle of dislocation lines and loses its capacity to accept further deformation; that is, its ductility becomes exhausted. To restore the ductility, it is necessary to anneal the metal; that is, it must be heated so that a new population of crystal grains consumes the damaged set. This process is known as recrystallization, and its study is a major branch of physical metallurgy (see RECRYSTALLIZATION, DYNAMIC; RECRYSTALLIZATION, STATIC). One aspect of importance is the rate at which the new grain population replaces the old. Another is the statistical tendency for the new grains to line up with their crystal axes in particular orientations in space; the result is termed a preferred orientation, or texture, and can be characterized by x-ray diffraction. Certain preferred orientations are desirable because they enhance the ability of the annealed metal to accept subsequent deformation. Thus, steels to be pressed into shapes for automobile bod-

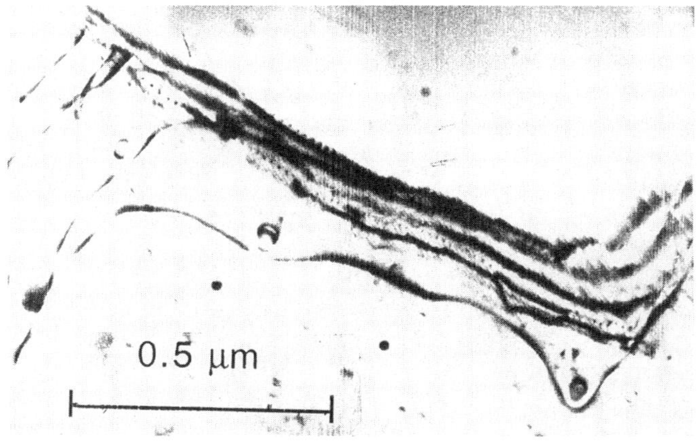

FIG. 15. Dislocations held up by precipitated particles in an aluminum alloy (transmission electron micrograph, courtesy R. A. Ricks).

ies, or steel sheets for the magnetic core of a transformer, are required to have a pronounced preferred orientation.

Another deformation mechanism that is important in some metals, for instance zinc, titanium, and uranium, is deformation twinning. Here, successive crystal planes shear through equal, crystallographically determinate vectors, generating thereby a new lamella of the same phase but in a distinct orientation.

4.3 Mechanically Induced Phase Transformations

Intense mechanical deformation generates a pronounced preferred orientation or deformation texture, and indeed this is a precondition of the formation of annealing textures as just mentioned. But quite apart from this, the application of an external stress can have strange consequences.

It can happen that a phase can be impelled to change into a crystallographically quite distinct form by mechanical stress alone, without necessarily any change in temperature. This has become industrially important in the family of shape-memory alloys, one of which is nearly equiatomic NiTi. These alloys are unstable against martensitic transformations induced by stress. Since martensitic transformation entails a mechanical shear, such alloys can change shape substantially by phase transformation alone, though they are entirely brittle in the sense that dislocations cannot move in them. A shape-memory

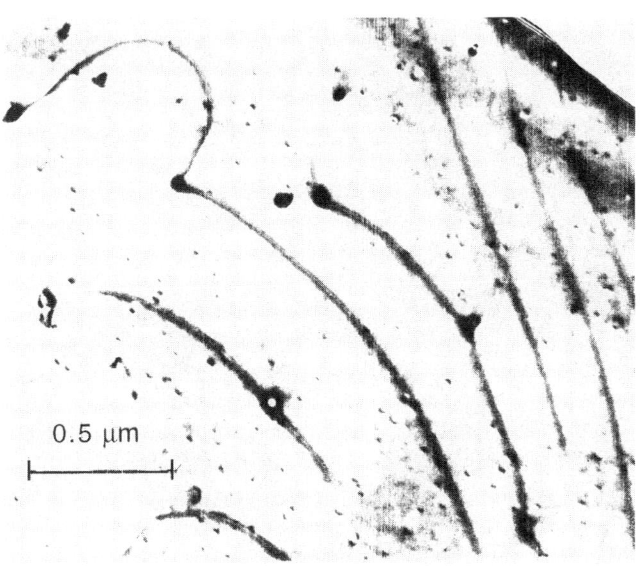

FIG. 16. Dislocations pinned by particles in an aluminum alloy (transmission electron micrograph, courtesy R. A. Ricks).

alloy has the striking property that if it is distorted by an applied stress from its stable shape, it remains metastably in the distorted shape, but on gentle heating, the martensitically generated phase changes back to the original phase, and in the process, the material returns to its pristine shape. Such alloys find extensive use, both in mass applications such as thermostat valves in automobile radiators and in more exotic uses such as wire aerials for satellites, crumpled into a small space before takeoff and deployed into the proper shape by passage of a heating current once the satellite is in orbit. A large number of physical metallurgists have recently been studying this type of alloy to understand just what determines the details of the memory and the force against which the shape-memory effect can successfully operate.

Some very surprising examples of stress-induced phase changes have been discovered recently. Thus, silicon, the most studied of any material, an element of cubic symmetry, has been found to be unstable against a concentrated stress in a limited temperature range and found to transform into a crystal form of hexagonal symmetry when loaded with a sharp diamond. Germanium, an element similar to silicon, behaves in the same way. The hexagonal form is produced where two deformation twins intersect (Fig. 17). Such a twin is a thin lamella produced by shear in a phase *without* change of phase; they are apt to form at explosive speed. In some metals, twinning offers an alternative mode of plastic deformation to slip of the kind shown in Fig. 14. What is seen in Fig. 17, however, is a phase change from cubic to hexagonal symmetry, in a specimen held at about 350 °C, in the intersecting zone where two twins, parallel to two different lattice planes, intersect and produce a high instantaneous stress at this site because of the speed with which the second twin lamella crashes into the first-formed one. This high stress generates the unexpected hexagonal phase.

This recently prepared (1992) micrograph indicates that physical metallurgists can still expect surprises even in their relatively mature science! The fact that Fig. 17 refers to germanium and Fig. 13 referred to silicon carbide, neither of which is a metal, also shows that no branch of materials science can nowadays be hermetically sealed off from its neighboring fields. Most of the journals carrying papers on physical metallurgy now also

FIG. 17. Stress-induced phase transformation in germanium held at ≈350 °C and deformed under a diamond indentation. M = matrix; T_1, T_4 = primary deformation twins; H = hexagonal polymorph generated where the primary twins intersect. Lattice planes are clearly visible. (High-resolution transmission electron micrograph.) (From Xiao and Pirouz, 1992, courtesy P. Pirouz.)

invite studies of nonmetallic materials that can cast light on the behavior of metallic systems.

5. COMPUTER MODELING

In the last few years, the understanding of the various phenomena that together make up the science of physical metallurgy has been much enhanced by modeling, or simulation, of the processes by means of high-speed computers. This is particularly beneficial when the analytic theoretical expressions that describe processes such as recrystallization, Ostwald ripening, texture formation, or melting, to give just a few examples, are too complex to solve readily, or to solve at all, or alternatively when analytical expressions are not attainable and one is restricted to numerical computations. When it further becomes desirable to carry out simulations for many different values of one or more parameters that enter into the equations, then computers become indispensable.

A typical example of a series of simulated micrographs, here used to map the structural evolution of a coherent particle (i.e., one whose interface with the matrix is of particularly low energy) in the presence of an internal stress in the system, is shown in Fig. 18. This kind of splitting is an alternative response of a microstructure, distinct from the shape-memory response outlined above. The application of computer modeling to problems thrown up by physical metallurgy has recently reached the point where journals entirely dedicated to this technique have been launched (see also PHYSICS APPLICATIONS OF COMPUTERS—MODELING).

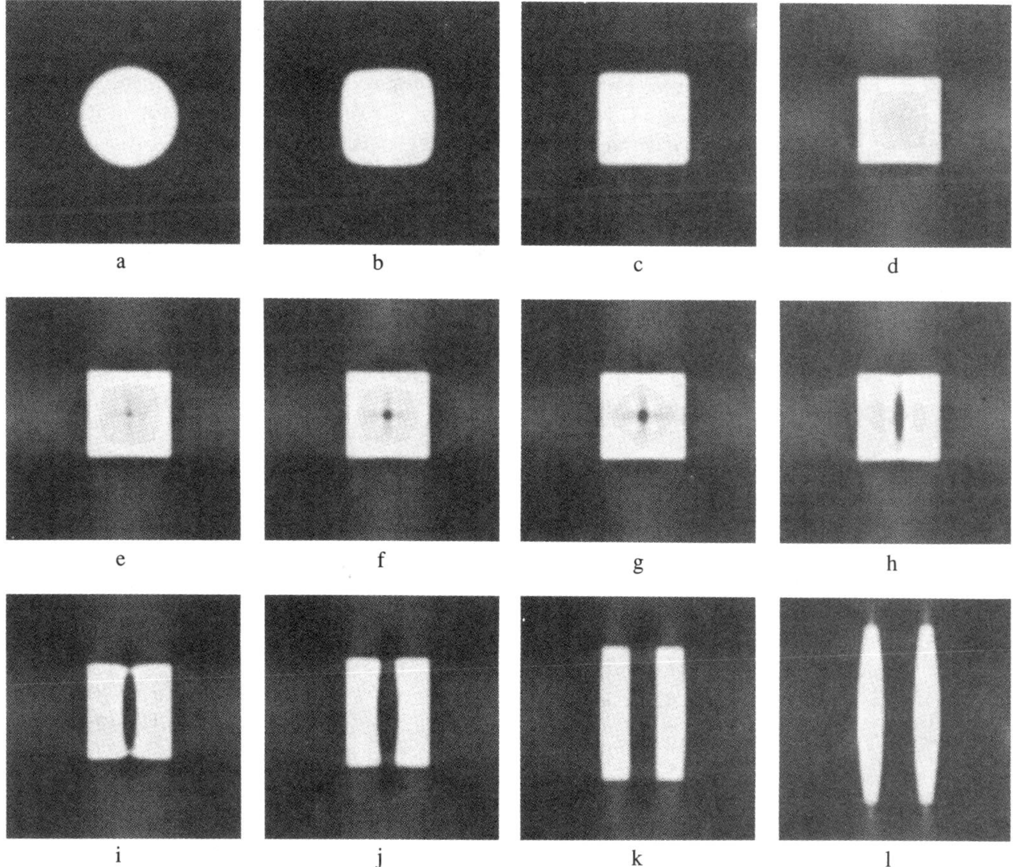

FIG. 18. Computer-simulated sequence of shape relaxation and splitting of a coherently precipitated particle during stress-induced coarsening (from Y. Wang *et al.* 1993, courtesy A. G. Khachaturyan).

ACKNOWLEDGMENTS

A shorter version of this article was published in *Collier's Encyclopedia*, and I am most grateful to Dr. Quigley of Macmillan in New York for his permission to print the expanded version here and to use the versions of some of the illustrations as first printed in *Collier's Encyclopedia*.

I am also deeply indebted to the following colleagues for the provision of original micrographs: Dr. M. J. Henry, Prof. E. Hornbogen, Prof. A. G. Khachaturyan, Dr. W. Krakow, Dr. Y. Moriyoshi, Prof. J. Nutting, Dr. P. Pirouz, and Dr. R. A. Ricks.

GLOSSARY

Cubic Crystal Structure: A structure that is assembled by stacking unit cells in the form of cubes. Such crystals have a high symmetry, which implies that many properties are isotropic.

Diffusion: The process that makes possible atomic transport within a solid. The most common form of this process involves the migration of vacancies (sites where an atom is missing), which entails a counterflow of atoms. Phase transformations requiring a local change of composition can only take place if the temperature is high enough for diffusion to take place at an appreciable rate.

Dislocation: A crystal defect in the form of a line, at which the stacking regularity of a crystal is interrupted. Plastic deformation of crystals requires that dislocations be free to move, under the action of a stress, along a lattice plane, termed the "glide plane" or "slip plane." The direction of displacement of one-half of the crystal relative to the other is termed the "slip direction" and is parallel to a unique vector that characterises any dislocation, the "Burgers vector." During extensive plastic deformation, the density of dislocations is increased by several orders of magnitude by a process of geometrical multiplication.

Eutectic Transformation: A form of solidification in which a homogeneous liquid phase solidifies, at constant temperature, into an intimate mixture of two phases of different compositions (and usually of different crystal structures).

Eutectoid Transformation: Analogous to an eutectic transformation, but here the phase of origin is solid.

Equilibrium Diagram: A graphical representation, with composition as abscissa and temperature as ordinate, of the domains of stability of different phases in an alloy system. "Phase diagram" is an alternative name. Binary diagrams refer to alloys with two components, ternary diagrams to those with three components.

Hexagonal Crystal Structure: A structure that is assembled by stacking unit cells in the form of right regular hexagonal prisms. Such crystals have a lower symmetry than cubic ones and accordingly are more anisotropic in their properties.

Martensitic Transformation: A phase transformation involving the coordinated shear displacement of successive planes of atoms, without any diffusional change of composition. Because no diffusion is involved, the process can be very rapid and can take place at low (even subambient) temperatures.

Metastability: A state (of an alloy, in the present context) in which one or more phases have crystal structures or compositions, or both, that are not in accordance with the equilibrium diagram. To qualify as metastable, such a state must be "frozen in," in the sense that diffusional processes are too slow for the system to evolve at a measurable rate in the direction of the equilibrium configuration. (Term coined by Ostwald in 1893.)

Microstructure: A broad term, denoting the geometrical disposition of the constituent phases of an alloy, whether in equilibrium or in a metastable state. The concept is distinct from that of "crystal structure," which refers to the disposition of atoms within the unit cell of a phase.

Ostwald Ripening: A process during which a population of particles in a biphase alloy progressively coarsens, through the preferential growth of the larger particles at the expense of the smaller. The overall volume fraction of the particles and their crystal structure do not alter during Ostwald ripening.

Pearlite: The product of a eutectoid transformation in a steel, produced by relatively slow cooling. The phase of origin is austenite (a face-centered cubic structure), and the product is an intimate lamellar mixture of almost pure body-centered iron and cementite (Fe_3C), with a complex orthorhombic crystal structure.

Phase: A region in microstructure that is uniform with respect to crystal structure (or can be amorphous, liquid, or gaseous) and, when in equilibrium, is also uniform in composition.

Phase Diagram: See **Equilibrium Diagram**.

Preferred Orientation: The condition of a polycrystalline metal or alloy in which the individual crystal grains tend statistically toward an ideal orientation (which would be equivalent to a single crystal). Such a condition can be produced either by intense plastic deformation or by annealing following such deformation. An alternative name is "texture."

Recrystallization: The process that leads to softening of a work-hardened metal or alloy during high-temperature annealing. The original population of crystal grains, full of dislocations, is replaced by a new population of undamaged grains.

Solid Solution, Terminal: A crystal with two or more components, generated by dissolving solute(s) in a pure metal. Such a solid solution necessarily has the same crystal structure as the pure "parent" element.

Texture: See **Preferred Orientation**.

Work Hardening: The progressive increase in hardness, or strength, of a deformable metal or alloy, resulting from plastic deformation.

Works Cited

Wang, Y. *et al.* (1993), *Acta Metall. Mater.* **41**, 279.

Xiao, S.-Q., Pirouz, P. (1992), *J. Mater. Res.* **7**, 1406.

Further Reading

Numerous books on physical metallurgy and its broader modern variant, materials science, are available, some suitable for the general reader, others more appropriate for the specialist. The first on the list is a general reference work very suitable for the nonspecialist.

Cotterill, Rodney (1985), *Cambridge Guide to the Material World*, Cambridge, U.K.: Cambridge University Press. A new edition is in preparation.

The next two titles assume that the reader has some background in the physical sciences.

Hume-Rothery, W., Smallman, R. E., Haworth, C. W. (1969), *The Structure of Metals and Alloys*, London: Institute of Metals.

Weaire, D. L. W., Windsor, C. G. (Eds.) (1987), *Solid State Science*, Bristol, U.K.: Hilger.

Next is a well-tried exposition that has been used for many years as an introductory text leading on to an intermediate level. A recent edition is cited.

Reed-Hill, Robert E., Abbaschian, R. (1992), *Physical Metallurgy Principles*, Boston: PWS-Kent Publishers.

The last three books are for more advanced readers, and each includes a good deal of theory.

Cahn, Robert, Haasen, Peter (1983), *Physical Metallurgy*, 3rd ed., Amsterdam: Elsevier-North-Holland; (1995), 4th ed.

Haasen, Peter (1986), *Physical Metallurgy*, 2nd ed., Cambridge, U.K.: Cambridge University Press.

Porter, D.A., Easterling, K.E. (1981), *Phase Transformations in Metals and Alloys*, New York: Van Nostrand Reinhold.

METALS

See ALUMINUM; COPPER; PRECIOUS METALS; STEEL

METALS AND ALLOYS, CONDUCTIVITY IN

P. L. Rossiter, *Department of Materials Engineering, Monash University, Clayton, Victoria, Australia*

J. Bass, *Department of Physics and Astronomy, Michigan State University, East Lansing, Michigan, U.S.A., and Max-Planck-Institut für Festkörperforschung, Hochfeld Magnetlabor, Grenoble, France*

3-527-28132-0/94/$5.00 + .50

INTRODUCTION

The electrical transport properties of metals and alloys have been extensively studied for over 100 years, both in order to understand the basic phenomena and to provide a tool for monitoring other effects such as phase transformations, defect levels, and electron states and mobilities. The conductivities of solids cover a range of about 10^{30} from the generally highly conducting metals and alloys at one extreme to insulators at the other. Applications based upon specific conductivity characteristics include electrical power distribution systems, stable electrical resistances, heating elements, and transducers to measure temperature and strain, for example. Similarly, Hall-effect devices find wide application in magnetic field measurement and various rotation and displacement transducers, while the thermopower determines the operating characteristics of thermocouples, and thermoelectric cooling is quite commonplace.

All of the properties of interest here—the conductivity (or resistivity), thermopower, magnetoresistance, and Hall coefficient—are determined by the rate of scattering of conduction electrons from initial to final states on the Fermi surface. This is described in an approximate way by the Boltzmann transport equation and the simplified idea of a relaxation time. Scattering mechanisms are briefly discussed leading to a statement of the scattering problem in terms of the nature of the atomic potentials (form factor) and their distribution in space (structure factor).

Specific results for the resistivity, thermopower, magnetoresistance, and Hall coefficient are given for a range of conducting materials under various temperature and field conditions.

1. ELECTRONIC TRANSPORT PROPERTIES

1.1 Background

In 1900, Drude attempted to explain the transport properties of metals by assuming that the electrons behaved as classical particles moving freely through the metal but occasionally bouncing off positive ions, with a mean time between collision (or relaxation time) of τ. While this model led to a seemingly plausible explanation of the electrical resistivity, it failed in other areas such as accounting for the relatively small electronic specific heat observed in metals at room temperature. Following the realization that electrons must obey the Pauli exclusion principle, Sommerfeld replaced the classical Maxwell–Boltzmann distribution of electron energies in the Drude model with the Fermi–Dirac distribution. This theory, known as the "free-electron theory," accounted for a wide variety of metallic properties but could still shed no light on the actual scattering process. Nor could it account for many other observations such as positive Hall constants, the temperature dependence of conductivity, or, indeed, why not all elements are good conductors.

These matters were resolved by properly taking into account the effects of the potential field generated by the ions. In a crystalline solid, this produces diffraction effects leading to the notion of energy bands, with scattering of the conduction electrons (and hence finite conductivity) produced by deviations from perfect periodicity of the lattice potential. It turns out that in many cases the potential seen by the electrons is only weak, and so much of the subsequent work falls under the title "nearly-free-electron theory." This theory forms the basis for the discussion that follows.

We wish to examine how the electrons in metals transport charge and heat under the influence of an applied electric field **E** and/or a temperature gradient $\nabla_r T = \Delta T/\Delta x$. Here, ΔT is a temperature difference applied over a small distance Δx, and boldface indicates a vector.

In zero magnetic field, the quantities of main interest are the conductivity σ (or the electrical resistivity, $\rho = 1/\sigma$) and the thermopower S. When a magnetic field **B** is applied, we shall focus upon the magnetoresistance $MR = [\rho(\mathbf{B}) - \rho(\mathbf{B} = 0)]/\rho(\mathbf{B} = 0)$, and the Hall coefficient R_H. The third of the (generally) independent transport coefficients, the thermal conductivity κ, is discussed only briefly, in terms of the Wiedemann-Franz Law, which relates κ to ρ at low and high temperatures. The reader interested in further information about κ is referred to the articles HEAT TRANSFER and THERMAL CONDUCTION IN SOLIDS.

In this section, we first briefly describe how each of these four quantities is measured, and then discuss the general linear transport equations relating the current density of charge $\mathbf{j}$ and the current density of heat $\mathbf{j}_q$ to $\mathbf{E}$ and $\nabla_r T$. We then outline the microscopic theory of electronic transport in metals and subsequently examine a variety of experimental data.

1.2 Electrical Resistance *R* and Resistivity *ρ*

The electrical resistance R of a metallic sample is measured by injecting a known current I_x through a wire [Fig. 1(a)] or foil [Fig. 1(b)] having cross-sectional area $A = \pi r^2$ or Wt, respectively, where r is the wire radius and W and t the foil width and thickness. From Ohm's law,

$$R = V_x/I_x, \tag{1}$$

where V_x is the emf (or voltage) generated over length L, and R is an extensive quantity that doubles if L is doubled or A is halved. Arrangements for liquid metals and alloys generally follow these configurations, with the additional requirement of some form of containment. The intrinsic property of the metal

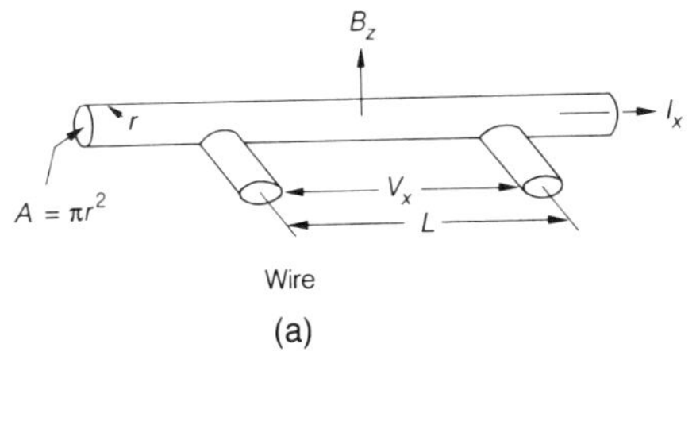

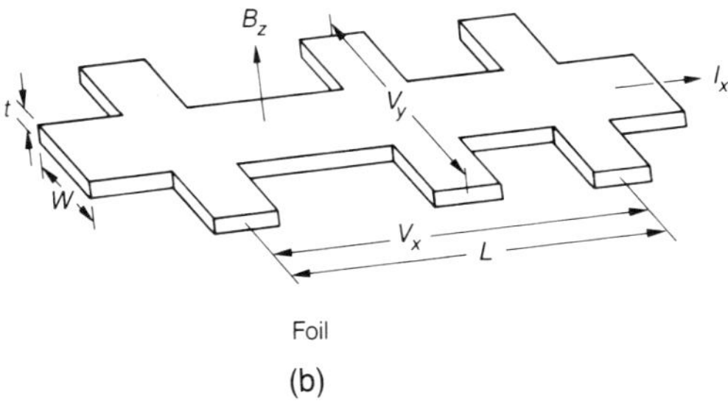

FIG. 1. Specimen configuration for measuring the electrical resistance of (a) a wire and (b) a foil. Arrangement shown in (b) also allows determination of the Hall voltage V_y.

is ρ, related to R by

$$\rho = R(A/L), \tag{2}$$

provided that the material is homogeneous and that $\mathbf{j}_x = I_x/A$ is uniform across the sample area. The resistivity ρ is discussed in Sec. 5.

1.3 Thermopower *S*

There are three experimental thermoelectric phenomena with three corresponding coefficients: the Thomson effect (coefficient μ), the Peltier effect (coefficient Π), and the Seebeck effect (thermopower coefficient S). Only the Thomson effect can be measured on a single metal. It was discovered in 1854 by William Thomson (Lord Kelvin) and involves the reversible generation of heat when current flows in a conductor while a temperature gradient is present (i.e., the production of heat changes to absorption if the direction of either the current or the temperature—but not both at once—is reversed). In the Peltier effect, discovered by French watchmaker Charles Peltier in 1834, heat is reversibly generated at the junction between two different metals when a current flows through the junction. The most widely known of the thermoelectric effects is the Seebeck effect, discovered by a Dutchman, Thomas John Seebeck. In 1821, he observed that a magnetic needle deflected when the two junctions of a nearby closed electrical circuit consisting of two different solid or liquid conductors were held at different temperatures; we now know that this deflection was caused by an electrical current that the temperature difference caused to flow in the circuit. The fundamental thermoelectric property of the conductors comprising the circuit is the voltage (or electromotive force) produced by the circuit when one junction is electrically disconnected (open circuit), since the current in the fully connected circuit depends upon the electrical resistivities of the circuit elements. Such an open-circuit configuration is called a thermocouple (Fig. 2). Thermocouples find wide use in temperature measurement and regulation. Connecting a number of thermocouples in series, and heating one common junction (e.g., with a small piece of radioactive material) of the resulting "thermopile," produces thermoelectric "bat-

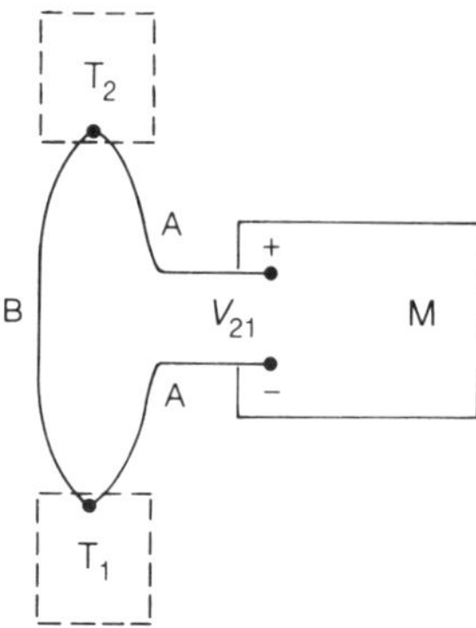

FIG. 2. Diagram of apparatus usually used for measuring thermoelectric (Seebeck) voltage V_{21}. M is an instrument for measuring potential.

teries" or "generators" that find use in space probes and remote places where long life, small size, absence of moving parts, and insensitivity to environmental changes are needed.

The three thermoelectric coefficients are connected by the Kelvin relations:

$$S_1 = \int_0^T (\mu_1/T')dT', \tag{3a}$$

$$\mu_1 = T(\partial S_1/\partial T), \tag{3b}$$

$$\Pi_{12} = T(S_2 - S_1), \tag{3c}$$

so that complete knowledge of any one yields complete knowledge of all three; S is usually the easiest to measure.

When a temperature difference $\Delta T = T_H - T_L$ is applied across a thermocouple, the resulting thermoelectric voltage is

$$\begin{aligned} V_{21} &= \int_{T_L}^{T_H} [S_2(T) - S_1(T)]dT \\ &= \int_{T_L}^{T_H} S_2(T)dT - \int_{T_L}^{T_H} S_1(T)dT, \end{aligned} \tag{4}$$

where S_1 and S_2 are the thermopowers of metals 1 and 2, respectively. If both metals are homogeneous, the integrals depend only upon the materials and the end temperatures T_H and T_L, and not upon how the temperature varies along each arm of the couple. If ΔT is small, then

$$S_2 - S_1 = V_{21}/\Delta T. \tag{5}$$

Equation 5 shows that S_1 for a sample with unknown properties can only be determined if S_2, the thermopower of a reference material, is known. At very low temperatures, the reference material can be a superconductor, for which $S = 0$ in its superconducting state. "High-temperature" ceramic superconductors can be used up to around 90 K. At higher temperatures, Pb has been calibrated as a reference standard up to 350 K. Above 350 K, there is no universal standard, although Au, Cu, and Pt have been used.

The thermopower S consists of two very different parts: S_d, the electron diffusion component, involves a differential flow of electrons from one end of the metal to the other under the influence of a temperature gradient, with the phonons in the metal in thermal equilibrium; S_g, the phonon-drag component, results from the "dragging along" of electrons by the nonequilibrium phonon flow from the hot end of the metal to the cool end that is caused by the temperature gradient. In pure metals, S_g often dominates S from a few kelvins to above 100 K. The quantities S_d and S_g are discussed further in Sec. 8.

1.4 Magnetoresistance *MR* and Hall Coefficient R_H

In 1879, while studying the force exerted on a current carrying metal by an applied magnetic field, and looking for a change in resistance with increasing field (*MR*), the Englishman E. H. Hall discovered that applying such a field perpendicular to the direction of current flow gave rise to a voltage in the direction perpendicular to both the current and the field. This "Hall effect" is characterized by a coefficient R_H; *MR* and R_H can be measured together on the foil sample shown in Fig. 1(b). A current I_x is passed through the sample from left to right, a field $\mathbf{B}_z$ is applied normal to the sample, and the voltages V_x and V_y are measured. The direction of $\mathbf{B}$ is then reversed, and the two voltages measured again. The *MR* is defined to be an even function of $\mathbf{B}$ and is determined from the average of the values of V_x for the two field directions. The term R_H is defined to be an odd function of **B**, and it is determined from the difference between the two values of V_y. This reversing procedure is one of the steps necessary to minimize systematic errors in MR and R_H. Errors can result if care is not taken to ensure spatial uniformity of **B**, good alignment of the sample in the field, and a uniform current flow through the sample.

Since $V_x = E_x L$ and $I_x = j_x Wt$, we have $\rho(\mathbf{B}) = V_x Wt/I_x L$ for each of the two terms to be averaged in determining the MR. Since $V_y = E_y W$, each term for R_H must have the form $R_H = V_y t/BI_x$; R_H is defined with B divided out, because in many cases V_y is proportional to B.

The terms MR and R_H are discussed in Secs. 6 and 7, respectively.

1.5 General Transport Equations and Coefficients

The formal transport equations for metals provide linear relations among four vector quantities: **E**, $\nabla_r T$, the electrical current density **j**, and the heat current density $\mathbf{j}_q = \mathbf{I}_q/A$, where $\mathbf{I}_q$ is the total heat current. These equations can be written in two different forms.

The first form defines the directly measured quantities. The vectors **E** and $\mathbf{j}_q$ are related to the vectors **j** and $\nabla_r T$ by tensor coefficients (also designated by bold symbols):

$$\mathbf{E} = \boldsymbol{\rho}(\mathbf{B}) \cdot \mathbf{j} + \mathbf{S}(\mathbf{B}) \cdot \nabla_r T, \tag{6a}$$

$$\mathbf{j}_q = \boldsymbol{\Pi}(\mathbf{B}) \cdot \mathbf{j} - \boldsymbol{\kappa}(\mathbf{B}) \cdot \nabla_r T, \tag{6b}$$

where the coefficients are generally functions of **B**. The thermal conductivity $\boldsymbol{\kappa}$ is found by applying $\mathbf{j}_q$ and measuring $\nabla_r T$ while holding $\mathbf{j} = 0$. Similarly, $\boldsymbol{\rho}$ is determined by applying **j** and measuring **E** while holding $\nabla_r T = 0$, and **S** is determined by applying $\nabla_r T$ and measuring **E** while holding $\mathbf{j} = 0$. From the Onsager relation (Ziman 1960, 1972), $\boldsymbol{\Pi}(\mathbf{B}) = T\mathbf{S}(-\mathbf{B})$, and there are only three independent tensors in Eqs. (6): $\boldsymbol{\rho}$, $\boldsymbol{\kappa}$ and **S**.

The second pair of equations defines the calculated quantities. The terms **E** and $\nabla_r T$ are treated as "forces" and **j** and $\mathbf{j}_q$ as derived "flows." These equations have the advantage of being directly comparable with the Boltzmann transport equation which we discuss in Sec. 2:

$$\mathbf{j} = \boldsymbol{\sigma}(\mathbf{B}) \cdot \mathbf{E} + \boldsymbol{\gamma}(\mathbf{B}) \cdot \nabla_r T, \tag{7a}$$

$$\mathbf{j}_q = \boldsymbol{\gamma}(-\mathbf{B})T \cdot \mathbf{E} - \boldsymbol{\kappa}'(\mathbf{B}) \cdot \nabla_r T, \tag{7b}$$

For $\mathbf{B} = 0$, Eqs. (6) and (7) reduce to scalars in cubic metals. Comparing Eq. (7) and Eq. (6) in scalar form shows that $\kappa' = \kappa - S^2/\sigma$ and $\gamma = -S\sigma$. In metals, S^2/σ is nearly always negligibly small compared to κ. The differences between Eqs. (6) and (7) are of little importance when the coefficients are simply scalars, since converting from, e.g., ρ to σ simply involves scalar inversion. When the full tensors must be retained, however, the conversion from one form to another involves matrix inversion. Writing $\boldsymbol{\rho}$ as a tensor with elements ρ_{ij} where $i,j = x,y,z$, the definitions of the MR and R_H given above yield

$$MR = \frac{\frac{1}{2}[\rho_{xx}(B) + \rho_{xx}(-B)] - \rho(B = 0)}{\rho(B = 0)}, \tag{8a}$$

$$R_H = [\rho_{yx}(B) - \rho_{xy}(-B)]/2B. \tag{8b}$$

2. THE BOLTZMANN TRANSPORT EQUATION

2.1 Background

If an electric field is applied to a conductor, all electrons not in completely filled bands are displaced at a uniform rate in **k** space. Scattering by defects tends to restore the electrons to their equilibrium distribution. The steady-state distribution is thus determined by a dynamical balance between the acceleration of the electrons in the applied field and their scattering by the lattice. A distribution function $f(\mathbf{k},\mathbf{r},t)$ may be used to describe the location of the state of an electron in six-dimensional $(\mathbf{k},\mathbf{r})$ phase space. It is defined so that $(4\pi^3)^{-1}f(\mathbf{k},\mathbf{r},t)\, d^3k d^3r$ is the number of electrons that lie in an element d^3r of real space and d^3k of wave-vector space at time t, such that $f = 1$ if all states are filled or $f = 0$ if all are empty. If the electron gas is at equilibrium at some temperature, f is simply the Fermi–Dirac distribution function, designated f_0.

In the presence of applied electric and magnetic fields **E** and **B**, but without scattering, any particular state will move through phase space according to semiclassical equations, but its occupancy will not change (i.e., if it was initially filled it will remain so). However, if scattering occurs, an electron can discontinuously change its momentum from some initial state $\mathbf{k}$ to a final state $\mathbf{k}'$, and the time derivative of f is given by

$$\frac{df}{dt} = \left.\frac{\partial f}{\partial t}\right|_{\text{scatt}}, \tag{9}$$

where the last term is the change in f due to that scattering process.

Expanding the left-hand side in terms of its partial derivatives gives

$$\frac{\partial f}{\partial t} + \frac{\partial f}{\partial \mathbf{r}} \cdot \frac{d\mathbf{r}}{dt} + \frac{\partial f}{\partial \mathbf{k}} \cdot \frac{d\mathbf{k}}{dt} = \left.\frac{\partial f}{\partial t}\right|_{\text{scatt}} \tag{10}$$

Since

$$\hbar \frac{d\mathbf{k}}{dt} = \mathbf{F}, \tag{11}$$

where $\mathbf{F}$ is any applied force, Eq. (10) may be written as

$$\frac{\partial f}{\partial t} + \mathbf{v}(\mathbf{k}) \cdot \frac{\partial f}{\partial \mathbf{r}} + \frac{e}{\hbar}[\mathbf{E} + \mathbf{v}(\mathbf{k}) \times \mathbf{B}] \cdot \frac{\partial f}{\partial \mathbf{k}} = \left.\frac{\partial f}{\partial t}\right|_{\text{scatt}} \tag{12}$$

where $\mathbf{v}(\mathbf{k})$ is the velocity of electrons in state $\mathbf{k}$ and e the electronic charge, which is one form of the celebrated Boltzmann equation. The solution of this equation for a specific system will give the distribution function in the presence of external fields or temperature gradients. Such a function determines the electrical and thermal conductivities and all thermoelectric effects, including their dependence upon magnetic fields. However, finding this function is not trivial. The difficulty lies in the complexity of the scattering term, which involves the rates of transition from all states to some particular state, $\mathbf{k}'$, and these in turn depend upon the occupation numbers of these states. This term will thus generally involve an integral over all values of $\mathbf{k}'$ with the distribution function itself appearing in the integrand, leading to computational difficulties.

2.2 Linearized Boltzmann Equation

Under steady-state conditions, the deviation in f from the equilibrium electron distribution is usually only small. By expressing f explicitly in terms of this deviation $g(\mathbf{k},\mathbf{r},t)$,

$$f(\mathbf{k},\mathbf{r},t) = f_0(\mathbf{k},\mathbf{r}) + g(\mathbf{k},\mathbf{r},t), \tag{13}$$

one may substitute f_0 for f on the left-hand side of Eq. (12) and retain only the lowest power of $f - f_0$ that does not vanish in the scattering term. This procedure leads to the linearized Boltzmann equation:

$$\mathbf{v}(\mathbf{k}) \frac{\partial f_0}{\partial t} \cdot \boldsymbol{\nabla}_r T + e \frac{\partial f_0}{\partial \epsilon_{\mathbf{k}}} \mathbf{v}(\mathbf{k}) \cdot \mathbf{E} = \left.\frac{\partial f_{\mathbf{k}}}{\partial t}\right|_{\text{scatt}} - \mathbf{v}(\mathbf{k}) \cdot \frac{\partial g_{\mathbf{k}}}{\partial \mathbf{r}} - \frac{e}{\hbar}[\mathbf{v}(\mathbf{k}) \times \mathbf{B}] \cdot \frac{\partial g_{\mathbf{k}}}{\partial \mathbf{k}}, \tag{14}$$

where $\epsilon_{\mathbf{k}}$ is the energy of an electron of wave vector $\mathbf{k}$ (see e.g., Ziman, 1972, p. 213), and the subscript $\mathbf{k}$ has been added to f and g to indicate their explicit dependence upon $\mathbf{k}$.

This linearized form yields the observed linear response of the electron and thermal fluxes to electric and magnetic fields and thermal gradients. Equation (14) is the starting point for many derivations of the electrical and thermal conductivities. However, its solution still requires evaluation of the scattering term. If $P_{\mathbf{kk}'}$ is defined as the probability of transition from some state $\mathbf{k}$ (assumed to be filled) to some other state $\mathbf{k}'$ (assumed to be empty), the scattering term may be written as

$$\left.\frac{\partial f_{\mathbf{k}}}{\partial t}\right|_{\text{scatt}} = \frac{\Omega}{2\pi^3} \int (g_{\mathbf{k}'} - g_{\mathbf{k}}) P_{\mathbf{kk}'} d^3k, \tag{15}$$

where Ω is the specimen volume.

An evaluation of Eq. (15) then requires determination of the transition probability, often with the assistance of first-order perturbation theory.

2.3 Relaxation-Time Approximation

Instead of evaluating the transition probability $P_{\mathbf{kk}'}$ directly, one may make the phenomenological assumption that the distribution will relax exponentially in time if the field is switched off. Equation (15) then simply becomes

$$\left.\frac{\partial f_{\mathbf{k}}}{\partial t}\right|_{\text{scatt}} = -\frac{f_{\mathbf{k}} - f_0}{\tau_{\mathbf{k}}} = -\frac{g_{\mathbf{k}}}{\tau_{\mathbf{k}}}, \tag{16}$$

where $\tau_{\mathbf{k}}$ is again the relaxation time, which may now vary with the magnitude of $\mathbf{k}$ (but not its direction). This approximation is rigorous if the scattering is purely elastic (i.e., $|\mathbf{k}| = |\mathbf{k}'|$) and the Fermi surface (FS) is spherical (Ashcroft and Mermin, 1976). In most situations, the assumption of isotropic scattering over a spherical FS is not valid, and more general solutions must be found. Nevertheless, the approximation leads to considerable simplification and serves to illustrate the evaluation of the transport coefficients from the linearized Boltzmann Eq. (14).

2.3.1 Electrical Conductivity In the presence of a uniform electric field, but no magnetic field or thermal gradient, Eqs. (14) and (16) give

$$e\frac{\partial f_0}{\partial \epsilon_{\mathbf{k}}}\mathbf{v}(\mathbf{k}) \cdot \mathbf{E} = \frac{-g_{\mathbf{k}}}{\tau_{\mathbf{k}}}. \tag{17}$$

The current density $\mathbf{j}(\mathbf{r},t)$ is obtained by summing the contributions $e\mathbf{v}(\mathbf{k})$ for all electrons:

$$\begin{aligned}\mathbf{j}(\mathbf{r},t) &= \frac{e}{4\pi^3}\int \mathbf{v}(\mathbf{k})f(\mathbf{k},\mathbf{r},t)d^3k \\ &= \frac{e}{4\pi^3}\int \mathbf{v}(\mathbf{k})g_{\mathbf{k}}d^3k\end{aligned} \tag{18}$$

[from Eq. (13)]. Substitution of $g_{\mathbf{k}}$ from Eq. (17) ultimately leads to

$$\mathbf{j}(\mathbf{r},t) = \frac{e^2}{4\pi^3\hbar}\int \tau_{\mathbf{k}}\mathbf{v}(\mathbf{k})[\mathbf{v}(\mathbf{k}) \cdot \mathbf{E}]\frac{dS}{|\mathbf{v}(\mathbf{k})|}, \tag{19}$$

where the integration over a volume $d\mathbf{k}$ of $\mathbf{k}$ space has been transformed into an integration over an element of area of the FS dS.

A comparison of Eq. (19) with Eq. (7a) ($\nabla_r T = 0$) leads to an expression for components of the conductivity tensor:

$$\sigma_{ij} = \frac{e^2\tau_{\mathbf{k}}}{4\pi^3\hbar}\int \frac{v_i(\mathbf{k})v_j(\mathbf{k})dS}{|\mathbf{v}(\mathbf{k})|}. \tag{20}$$

The tensor is symmetric and can be reduced to diagonal form giving the three principal conducting coefficients parallel to the principal axes of the crystal. In cubic crystals, the tensor becomes a scalar, and, for free electrons, Eq. (20) becomes the simple Drude expression

$$\sigma = 1/\rho = n_e e^2\tau/m, \tag{21}$$

where n_e is the electron density and m the electron mass.

2.3.2 Thermoelectric Effects If the temperature gradient is nonzero, Eqs. (14) and (16) give

$$\mathbf{v}(\mathbf{k}) \cdot \frac{\partial f_0}{\partial T}\nabla_r T + e\frac{\partial f_0}{\partial \epsilon_{\mathbf{k}}}\mathbf{v}(\mathbf{k}) \cdot \mathbf{E} = \frac{-g_{\mathbf{k}}}{\tau_{\mathbf{k}}}. \tag{22}$$

This may be substituted into Eq. (18) to give the electric current, which is now Eq. (19) plus an additional term

$$\frac{e^2}{4\pi^3\hbar}\int \tau_{\mathbf{k}}\mathbf{v}(\mathbf{k})\left[\mathbf{v}(\mathbf{k})\frac{\partial f_0}{\partial T} \cdot \nabla_r T\right]d^3k, \tag{23}$$

i.e., a temperature gradient $\nabla_r T$ alone produces an electric current through the thermoelectric effect. Comparison with Eq. (7a) (with $\mathbf{E} = 0$) allows determination of γ.

2.3.3 Electronic Thermal Conductivity An additional effect of a temperature gradient is to cause a flow of heat. The heat current density $\mathbf{j}_q$ is then

$$\mathbf{j}_q = \frac{1}{4\pi^3}\int (\epsilon_{\mathbf{k}} - \mu)\mathbf{v}(\mathbf{k})g_{\mathbf{k}}d^3k, \tag{24}$$

since the heat current is just the electron flux times the difference between the kinetic energy per electron $\epsilon_{\mathbf{k}}$ and the chemical potential μ of the electron system.

By analogy with the previous derivations, one obtains the electronic thermal conductivity (see e.g., Ziman, 1972). In cubic materials,

$$\kappa = \frac{\pi^2}{3}\frac{n_e\tau}{m}k_B^2 T, \tag{25}$$

where k_B is the Boltzmann constant.

Comparison with Eq. (21) allows definition of the Wiedemann-Franz ratio

$$\frac{\kappa}{\sigma} = \frac{\pi^2}{3}\frac{k_B^2 T}{e^2}. \quad (26)$$

Because it was derived from the relaxation-time approximation, Eq. (26) is valid only at very low or very high temperatures (see e.g., Ziman, 1972).

2.3.4 Hall Effect In the presence of both electric and magnetic fields, Eq. (14) becomes

$$e\frac{\partial f_0}{\partial \epsilon_{\mathbf{k}}}\mathbf{v}(\mathbf{k})\cdot\mathbf{E} = \frac{-g_{\mathbf{k}}}{\tau_{\mathbf{k}}} - \frac{e}{\hbar}[\mathbf{v}(\mathbf{k})\times\mathbf{B}]\cdot\frac{\partial g_{\mathbf{k}}}{\partial \mathbf{k}}. \quad (27)$$

This equation cannot be solved directly for $g_{\mathbf{k}}$. However, by using a trial function of the same form as Eq. (17),

$$g_{\mathbf{k}} = -e\tau\frac{\partial f_0}{\partial \epsilon_{\mathbf{k}}}\mathbf{v}(\mathbf{k})\cdot\mathbf{A}, \quad (28)$$

equation (27) becomes

$$\mathbf{v}(\mathbf{k})\times\mathbf{E} = \mathbf{v}(\mathbf{k})\times\mathbf{A} + \frac{e\tau}{m}[\mathbf{v}(\mathbf{k})\times\mathbf{B}]\cdot\mathbf{A}, \quad (29)$$

which has a solution

$$\mathbf{E} = \mathbf{A} + (e\tau/m)(\mathbf{B}\times\mathbf{A}). \quad (30)$$

By analogy with Eq. (7a)

$$\mathbf{j} = \sigma\mathbf{A}, \quad (31)$$

and Eq. (30) becomes

$$\mathbf{E} = \frac{1}{\sigma}\mathbf{j} + \frac{e\tau}{m\sigma}\mathbf{B}\times\mathbf{j}. \quad (32)$$

This equation shows that there are two components to the field: the normal longitudinal field parallel to **j** and, if **B** is perpendicular to **j**, a transverse field known as the Hall field, which may be written as

$$E_H = \frac{1}{n_e e}|\mathbf{B}||\mathbf{j}| = R_H|\mathbf{B}||\mathbf{j}|. \quad (33)$$

In the relaxation-time approximation, it has the same sign as the charge carrier:

$$R_H = 1/n_e e. \quad (34)$$

A more general analysis of R_H is given in Sec. 7.

2.3.5 Magnetoresistance The relaxation-time approximation does not predict any dependence of conductivity upon the magnetic field since the longitudinal electric field (i.e., that component of **E** parallel to **j**) is given simply by $\mathbf{E}_L = \sigma^{-1}\mathbf{j}$ [see Eq. (32)]. However, if there is more than one band of carriers, or if the scattering is not isotropic, there does appear a component of the longitudinal electric field that depends upon **B**. This is just the magnetoresistance introduced in Sec. 1 and discussed further in Sec. 6.

The detailed behavior of the conduction electrons in a magnetic field is further complicated by the fact that the Lorentz force $\mathbf{v}(\mathbf{k}) \times \mathbf{B}$ causes them to orbit around the FS as shown in Fig. 3. Different FS's have very different topologies. These topologies are most important at high fields, where they lead to

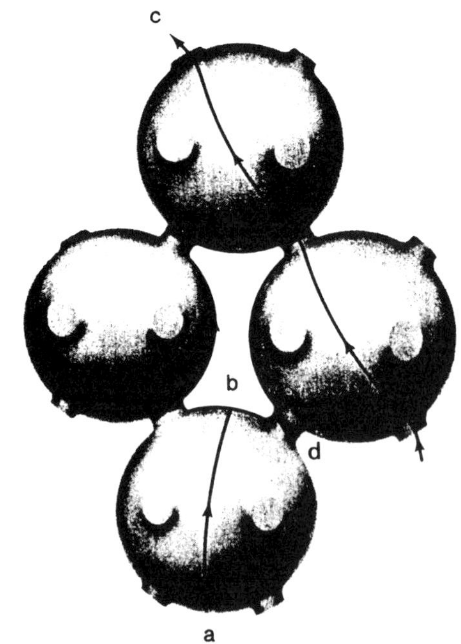

FIG. 3. The FS of Cu. (c) Open orbits, (a) and (d) closed electron-like orbits, and (b) holelike orbits for electrons in a uniform magnetic field (from Ashcroft and Mermin, 1976).

different field dependences of both R_H and MR as discussed in later sections.

3. SCATTERING MECHANISMS

A perfect, infinite lattice would produce no obstacle to the passage of an electric current. Only deviations from a perfect periodicity of the lattice potential cause the scattering of electrons from initial to final states, leading to a reduced conductivity. These deviations may be separated into those that do not have any intrinsic dependence upon thermally induced atomic displacements ("static" atomic or magnetic disorder) and those directly resulting from the thermal motion of atoms (thermally induced disorder).

One form of static disorder in solids results from atomic displacements such as occur around point and extended defects, from interstitials, or from atomic size effects in alloys containing atoms with different atomic radii. Such disturbances are characterized by the total atomic displacements produced.

Liquid metals and amorphous solid metals may be regarded similarly, with the static disruption to the atomic arrangement so severe that there is no long-range translational periodicity, although there will still be some local sense of atomic positional order ranging over the first few neighbors (see AMORPHOUS MATERIALS, STRUCTURE OF; LIQUIDS, SIMPLE, STRUCTURE OF; METALS AND ALLOYS, STRUCTURE OF). In alloys there is an additional disruption to the periodicity of the lattice potential as a result of imperfect spatial ordering of the different atomic potentials. (A structure with perfect long-range atomic order would again have no resistivity.)

Grain boundaries, free surfaces, and interphase and antidomain boundaries also disrupt the periodicity of the potential and so will cause electron scattering. Furthermore, if the scale of a decomposition (e.g., grain size, precipitation, spinodal decomposition, and antiphase domains), exceeds the conduction-electron mean free path $l = v_F\tau$ (where v_F is the velocity of the electron at the FS), it is necessary to regard each region and the boundary as separate elements and to develop models consisting of arrays of these elements to represent the microstructure adequately.

The sensitivity of the scattering process to the scale of the structure or microstructure presents a significant complication. In principle, the Boltzmann equation should be solved within each characteristic region to obtain a "local" mean free path $l_{(loc)}$. However, the scattering term in the Boltzmann equation depends upon the scattering matrix, which in turn depends upon the size of region $\sim l_{(loc)}$ over which the potential is averaged. Thus, a complete solution to the problem strictly requires a self-consistent solution to the equation.

At any realistic temperature, thermal excitations will cause the atoms to be displaced from their equilibrium positions. As far as the conduction electrons are concerned, the adiabatic approximation holds, and the atoms may be considered to be fixed or "frozen" into this distorted configuration. The electron scattering is thus determined by the size of the displacements and hence by the temperature.

Conduction electrons are also scattered by magnetic moments so that any deviation from perfect periodicity in an array of these moments will also contribute to the resistivity. The magnetic structure may be described in terms of spin-spin correlation parameters, in direct analogy to the atomic order parameters described above, or alternatively by the use of spin waves.

4. ELECTRONS IN A MAGNETIC FIELD

4.1 Cyclotron Orbits and $\omega_c\tau$

A fundamental quantity for characterizing electronic transport in the presence of **B** is the dimensionless parameter $l/r_c = \omega_c\tau$, which specifies the average number of times an electron traverses a given cyclotron orbit between scattering events. Here r_c is the cyclotron radius, and ω_c is the cyclotron frequency given by (Abrikosov, 1972; Ashcroft and Mermin, 1976)

$$\omega_c = \frac{h}{2\pi}\frac{\partial A}{\partial \epsilon}, \tag{35}$$

where h is Planck's constant and $\partial A/\partial\epsilon$ is the energy derivative of the cross-sectional area of the cyclotron orbit in **k** space. We must distinguish four different regimes:

1. Low magnetic fields, $\omega_c\tau \ll 1$. In this limit, electrons traverse only a small fraction of a cyclotron orbit before being scattered. Their behavior is determined primarily by the scattering processes and by the properties of the FS in the local vicinity of their wave vectors before and after scattering.
2. Intermediate magnetic fields, $\omega_c\tau \approx 1$. Here, electrons traverse about one full orbit between scattering events. Both scattering processes and surface topology are important, making this usually the most difficult region of **B** in which to analyze data.
3. High magnetic fields, $\omega_c\tau \gg 1$. In this limit, electrons traverse their cyclotron orbits many times between scattering events. The field dependences of both MR and R_H (i.e., whether MR is proportional to B^0, B^1, or B^2, and whether R_H is proportional to B^0 or B^{-2}) are determined solely by the topological structure of the FS in the plane in **k** space perpendicular to **B**, with details of scattering affecting only the magnitudes of the transport properties.

At low temperatures and high magnetic fields, quantum oscillations appear in transport properties because of Landau quantization of the electronic energy levels. These occur in a fourth limit:

4. The Landau level-quantization limit: $\omega_c\tau > k_BT$, where T is the absolute temperature.

4.2 High-Field Forms of *MR* and R_H

We start by considering metals that in the free-electron approximation have odd numbers of conduction electrons per atom (i.e., one or three), and thus odd numbers per primitive unit cell (which for these metals contains only one atom). Since each Brillouin zone (BZ) can accommodate two electrons/atom, whatever changes in the FS occur when the number of conduction electrons per unit cell N_e increases, the numbers of electrons and holes per atom in such metals must remain unequal; e.g., for Al, $N_h - N_e = 1$. Metals for which $N_e \neq N_h$ are called "uncompensated." In contrast, metals or alloys that in the free-electron approximation have an even number of conduction electrons per primitive unit cell must have $N_e = N_h$. Such materials are called "compensated." The alternative high-field forms of MR and R_H for different FS topologies are listed in Table 1 and illustrated in Fig. 4. Note the generally different behaviors for compensated and uncompensated metals. For further details, see Abrikosov (1972, 1988).

5. ELECTRICAL RESISTIVITY

All of the scattering mechanisms discussed previously produce one or both of two effects:

1. a change in the nature of the total scattering potential $W(\mathbf{r})$ at the "defect" (whether a foreign atom or vacancy) and
2. a disruption to the spatial periodicity of the lattice potential.

If the disturbances are not too large, first-order perturbation theory allows a fairly straightforward evaluation of the scattering associated with the defect. Equation (20) then leads to the expression (see e.g., Rossiter, 1987)

$$\rho = \frac{3\pi m^2 \Omega_0 N}{4\hbar^3 e^2 k_F^6} \int_0^{2k_F} |\langle \mathbf{k}+\mathbf{q}|W(\mathbf{r})|\mathbf{k}\rangle|^2 q^3 dq, \tag{36}$$

where N and Ω_0 are the number and volume of unit cells, k_F the magnitude of the Fermi wave vector (see Electron Structure of Solids), the integrand $\langle \mathbf{k}+\mathbf{q}|W(\mathbf{r})|\mathbf{k}\rangle$ represents the matrix elements of the total scattering potential $W(\mathbf{r})$, and the integration is over the magnitude of the scattering wave vector $\mathbf{q}$ defined by

$$\mathbf{q} = \mathbf{k} - \mathbf{k}'. \tag{37}$$

The scattering potential may be factorized into two terms:

$$\langle \mathbf{k}+\mathbf{q}|W(\mathbf{r})|\mathbf{k}\rangle = \mathfrak{S}(\mathbf{q})\langle \mathbf{k}+\mathbf{q}|w(\mathbf{r})|\mathbf{k}\rangle, \tag{38}$$

where $\mathfrak{S}(\mathbf{q})$ is the structure factor

$$\mathfrak{S}(\mathbf{q}) = \frac{1}{N}\sum_{\mathbf{r}} \exp(-i\mathbf{q}\cdot\mathbf{r}) \tag{39}$$

and depends only upon atom positions, and $\langle \mathbf{k}+\mathbf{q}|w(\mathbf{r})|\mathbf{k}\rangle$ [often written simply as $w(\mathbf{q})$] is the form factor that is determined entirely by the atomic potential $w(\mathbf{r})$ associated with site $\mathbf{r}$.

Table 1. Types of behavior of the Hall and transverse magnetoresistance effects in metals (from Hurd, 1972).

	Behavior in the high-field condition		Nature of orbits in planes normal to the applied field direction; state of compensation
Type	Magnetoresistance[a]	Hall coefficient	
Single-crystal sample			
1	$\Delta\rho/\rho(0)$ saturates with H	$R_{\mathrm{H}} \propto (n_e - n_h)^{-1}$	All closed; $N_e \neq N_h$
2	$\Delta\rho/\rho(0)$ tends to H^2 variation (transverse) $\Delta\rho/\rho(0)$ saturates with H (longitudinal)	R_{H} tends to a constant[b]	All closed; $N_e = N_h$
3	$\Delta\rho/\rho(0)$ saturates with H	$R_{\mathrm{H}} \propto (n_e - n_h)^{-1}$	Negligible number of open orbits; $N_e \neq N_h$
4	$\Delta\rho/\rho(0)$ tends to $H^2 \sin^2\theta$[c] variation	R_{H} tends to a constant[b]	Open orbits in one direction only
5	$\Delta\rho/\rho(0)$ saturates with H	$R_{\mathrm{H}} \propto H^{-2}$	Open orbits in more than one direction
Polycrystalline sample			
6	$\Delta\rho/\rho(0) \propto H$	R_{H} tends to a constant[b]	All types if the cyrstallites have random orientations
	Behavior in the low-field condition		
Single-crystal or polycrystalline sample			
7	$\Delta\rho/\rho(0) \propto H^2$	R_{H} tends to a constant[d]	It is irrelevant what type of orbit predominates

[a]Transverse magnetoresistance except as indicated for type 2.
[b]The Hall coefficient is not related to the effective number of carriers in any simple manner.
[c]Note that $\Delta\rho/\rho(0)$ tends to saturation when θ, the angle between the applied electric field and the axis of the open orbit in real space, is zero.
[d]The Hall coefficient depends upon the anistropy of the dominant electron scattering process and the electron's velocity and effective mass at each point on the FS.
[e]N_e and N_h are, respectively, the number of electrons and holes per unit cell of the Bravais lattice; n_e and n_h are, respectively, the density of electrons and holes in real space.

It must be emphasized that Eq. (36) and the simple separation of Eq. (38) are only strictly valid for metals with a simple electronic structure (i.e., nearly-free-electron-like bands) and isotropic scattering over a spherical FS. These requirements are not satisfied in many metals and alloys, particularly those involving transition metals, where more complex methods are needed.

If the scattering mechanisms are independent and the scattering isotropic, the contributions to the total resistivity are simply additive. In an alloy, this leads to Matthiessen's rule

$$\rho_{\mathrm{TOT}} = \rho_0 + \rho_p(T), \tag{40}$$

where all of the static mechanisms are collected into the "residual" resistivity ρ_0, and $\rho_p(T)$ is the temperature-dependent phonon-scattering term. The assumption of independent scattering is only partly true and there is a large body of work devoted to studying deviations from this rule (see e.g., Bass, 1972).

5.1 Pure Metals

5.1.1 Static Displacements A vacancy at some site $\mathbf{r}'$ removes the average lattice potential $\bar{w}(\mathbf{r}')$ from that site (neglecting the effect of charge redistribution) and causes lattice distortions around that site leading to a change in the structure factor $\mathfrak{S}(\mathbf{q})$. The former produces a positive contribution to the resistivity that is reduced somewhat by lattice relaxation. Values for the resistivity due to 1 at. % of vacancies in some pure metals are given in Table 2. Since a typical equilibrium vacancy density is only about $1 \times 10^{13}/\mathrm{cm}^3$ ($\sim 10^{-9}$ at. %) at 300 °C and $2.7 \times 10^{16}/\mathrm{cm}^3$ ($\sim 10^{-6}$ at. %) at 600 °C, the total contribution of the equilibrium vacancies to ρ_0 is very small.

The resistivity contribution due to dislocations may be considered as due to the potential discontinuity at the core and the effects of atomic displacements in the surrounding strain field, with the former probably dominant (see e.g., Brown, 1977). Some

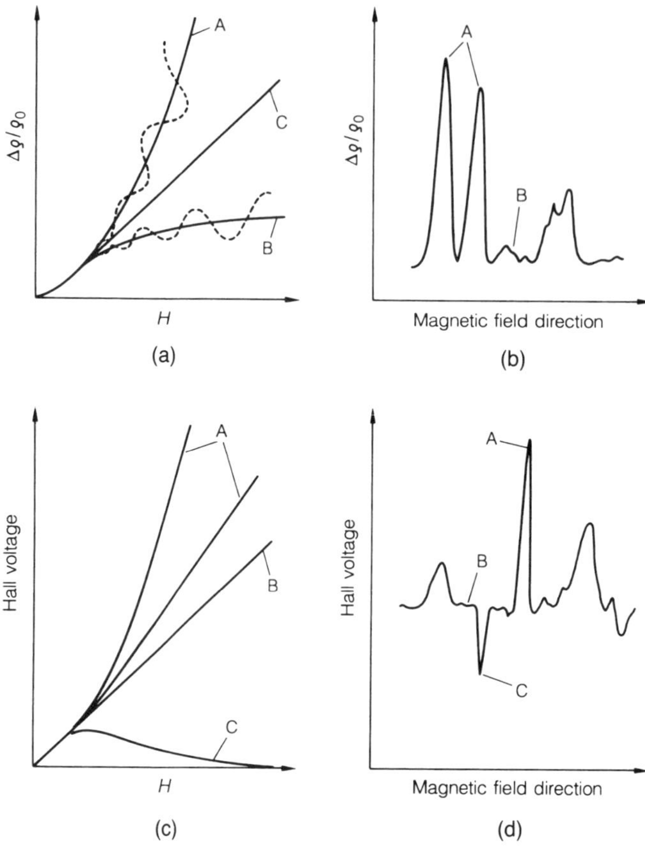

FIG. 4. Types of field dependence observed experimentally for the transverse magnetoresistance and the Hall voltage (from Hurd, 1972). (a), (b) *A* illustrates a quadratic (i.e., B^2) high-field variation in $\Delta\rho$, *B* illustrates saturation in $\Delta\rho$, and *C* illustrates the linear (i.e., *B*) variation in $\Delta\rho$, which is generally observed in polycrystalline samples. Quantum oscillations [dashed curves in (a)] may be superimposed upon behaviors *A* and *C* for single-crystal samples. (c), (d) *A* indicates high-field linear variations of different magnitudes; *B* illustrates a linear variation extending all the way down to $B = 0$; and *C* illustrates a B^{-1} high-field variation (from Hurd, 1972).

data are compared with the prediction of a resonant scattering model due to Brown in Table 3. Since the dislocation density in a heavily cold-worked metal is of order $10^{12}/\text{cm}^2$, the maximum dislocation resistivity is $\sim 0.1\ \mu\Omega$ cm.

The resistivity due to grain boundaries has been considered using a number of models, including an array of scattering potentials or an array of dislocations. Some results for a number of metals are given in Table 4, along with calculated values based simply upon the assumption that the grain boundary is represented by an array of dislocations and that these act as independent scatterers.

Table 2. Electrical resistivity due to 1 at. % of vacancies in the metals indicated.

Metal	ρ_0 (μΩ cm)/at. %	References
K	1.9–2.1	Benedek and Baratoff (1971)
Sn	4.4	Sun and Ohring (1976)
Al	1.0–3.3	Benedek and Baratoff (1971)
Cu	1.0–3.0	Simmons and Balluffi (1960)
Ag	1.3	Dugdale (1977)
Au	1.5	Dugdale (1977)

Mechanisms proposed for the scattering of conduction electrons at free surfaces, which become important in thin foils or fine wires, include microscopic surface roughness and localized surface charges (see e.g., Ziman, 1960; Greene, 1964). Such approaches often lead to the notion of a specularity parameter *p* that describes the fraction of electrons that are scattered in a specular (i.e., nondiffuse) manner. The resistivity or conductivity of a foil or wire can then be written in the form

$$\rho_0(d,T)/\rho_\infty = 1 + A(p,l_\infty,d)\, l_\infty/d, \tag{41}$$

where the parameter $A(p,l_\infty,d)$ depends upon the model used, *d* is the foil or wire thickness, l_∞ is the bulk mean free path, and ρ_∞ and σ_∞ are the corresponding bulk values of the

Table 3. Dislocation specific resistivities per unit dislocation density N (measured in lines/cm^2). These figures represent averages over all dislocation orientations. (From Brown, 1977, 1982.)

Metal	Temp. (K)	ρ_0/N 10^{-13} $\mu\Omega$ cm^3	Theory (Brown, 1977)
K	4.2	4	8
Cu	4.2	1.6 ± 0.2	1.3
Ag	4.2	1.9	1.9
Au	4.2	2.6	1.9
Be	80	34	28
Cd	80	24	25
Al	4.2	1.7 ± 0.3	1.8
Zr	80	100	40
Ti	4.2	100	29
Pb	80	1.1	4.2
Bi	1.3	2×10^5	1.7×10^5
Mo	4.2	5.8	3.7
W	4.2	7.5	7.4
Pt	80	9	4
Fe	80	10 ± 4	1.9
Ni	80	10	1.1
Rh	80	32	1

resistivity and conductivity. The dependence of A upon d/l_∞ for some different models is discussed in other literature (Rossiter, 1987). Note that this size-effect scattering. Eq. (41), represents an explicit deviation from Matthiessen's rule since it depends explicitly upon T.

A similar form of analysis may also be applied to grain boundaries where the grain-boundary resistivity may be expressed in the form of Eq. (41) where d is now the mean grain diameter. The corresponding variation of $A(p,l_\infty,d)$ with d/l_∞ is shown in Fig. 5. This size effect is demonstrated in Fig. 6, which shows the normalized grain-boundary resistivity as

Table 4. Grain-boundary specific resistivities in terms of area of boundary per unit volume S (measured in cm^2/cm^3) (from Brown, 1982).

Metal	ρ_0/S 10^{-12} Ω cm^2	Temp. (K)	Theory (Brown, 1977) (10^{-12} Ω cm^2)
Cu	2.4	4.2	2.7
Au	3.5	300	3.9
Cd	17	4.2	32
Al	1.1–2.4	4.2	2.6
Pd	0.3, 1.3	5–20	1.4
Bi	6.9×10^4	77	17×10^4
W	20	77	22
Fe	80–160	4.2	33
Ni	140	77	17

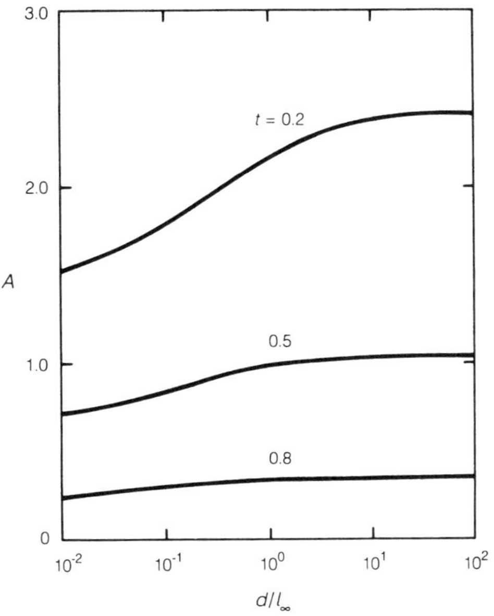

FIG. 5. The parameter A as a function of d/l_∞ from the grain-boundary scattering model for various values of the transmission coefficient t (from Rossiter, 1987).

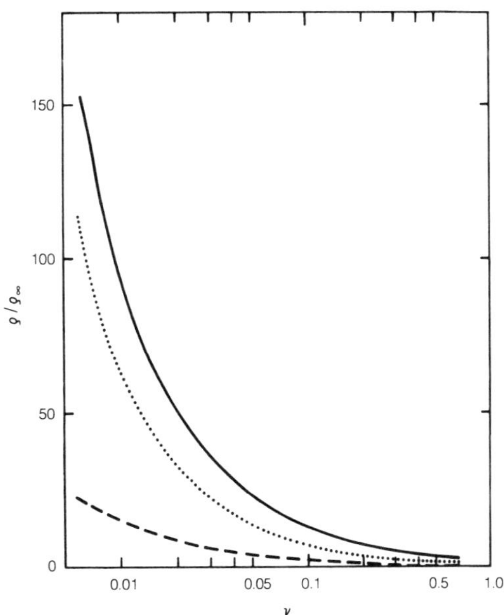

FIG. 6. The normalized grain-boundary resistivities as functions of the grain-size–dependent parameter ν. Solid line, $\rho_g(\nu)/\rho_\infty$; dotted line, $\rho_\perp(\nu)/\rho_\infty$; dashed line, $\rho_\parallel(\nu)/\rho_\infty$ where $\rho_\perp$ and $\rho_\parallel$ are the resistivities perpendicular and parallel to the current flow; ρ_g is the average resistivity for a three-dimensional array of boundaries, and ρ_∞ is the resistivity for infinite boundary spacing (from Rossiter, 1987).

a function of the grain-size–dependent parameter $\nu = d/l_\infty ln(1/t)$, where t is an empirical specular transmission coefficient.

5.1.2 Thermal Effects The effects of thermal scattering may be incorporated into the structure factor by allowing the atom to be displaced away from the lattice site $\mathbf{r}_i$ by a distance δ_i, obtained as a sum over all phonon wave vectors and modes. Such information is available either from a solution of the dynamical matrix based upon known force constants or from experimentally determined phonon dispersion curves (see PHONONS IN CRYSTAL LATTICES).

An alternative approach is to write the structure factor directly in terms of the dynamical matrix, allowing $\rho_p(T)$ to be determined directly from known force constants. Once again, for all but the simple metals, it is necessary to go beyond the first-order perturbation-theory approach.

Some experimental results for Cu are shown in Fig. 7 and indicate that reasonable agreement between theory and experiment is now possible. A compilation of results for a number of pure metals is given in Bass (1983).

A simpler though less satisfactory approach is to use the Debye (or Einstein) model to evaluate the thermal displacements, leading to the general prediction of $\rho_p(T) \propto T^5$ at low temperatures ($T < \theta_D$) and $\rho_p(T) \propto T$ at higher temperatures, in general agreement with experiment for free-electron-type metals, although the predicted magnitudes of the resistivity are usually not correct. Nevertheless, this approach leads to a useful phenomenological expression due to Bloch (1930), Grüneisen (1933), and Wilson (1937):

$$\frac{\rho_p(T)}{\rho(\theta_D)} = 4.225\left(\frac{T}{\theta_D}\right)^5 \int_0^{\theta_D/T} \frac{z^5 dz}{(e^z - 1)(1 - e^{-z})}, \tag{42}$$

where $\rho(\theta_D)$ is the resistivity at the Debye temperature. In practice, a better fit between experiment and theory is obtained by allowing θ_D in Eq. (42) to vary slightly from the actual Debye temperature (as obtained from specific-heat measurements, for example). The corrected characteristic "resistivity" temperatures, usually denoted θ_R, for some metals are shown in Table 5 and the results plotted in Fig. 8. Some other results are given in Meaden (1965).

At room temperature, $\rho_p(T) \sim 1\ \mu\Omega$ cm which is at least one order of magnitude larger than the contribution from lattice defects, even in severely cold-worked materials.

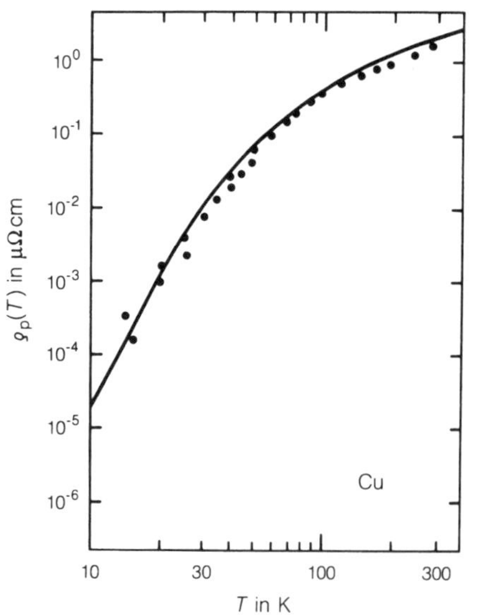

FIG. 7. The temperature dependence of the resistivity of Cu derived from the form factors of Borchi and de Gennaro (1970) (solid line). The solid circles represent a collection of experimental data (from Rossiter, 1987).

5.2 Alloys

Substitutional and interstitial impurities lead to changes in the form factor as well as the structure factor through the local lattice

Table 5. Some representative high-temperature values for θ_R (valid for $\theta_R/3$ to θ_R) compared with corresponding values for θ_D (from Meaden, 1965).

Metal	θ_D (K)	θ_R (K)
Be	1000	1240
Na	160	195
Mg	325	340
Al	385	395
K	100	110
Ca	225	~145
Ti	355	342
Cr	450	485
Cu	320	320
Zn	245	175
Pd	300	270
Ag	220	200
Cd	165	130
W	315	333
Pt	225	240
Au	185	200
Pb	88	~100

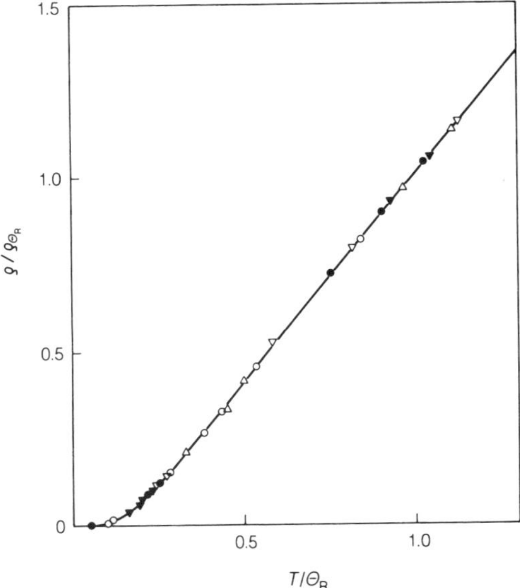

FIG. 8. The temperature dependence of the resistivity ρ/ρ_θ plotted against T/θ_R. ●, Li, θ_R = 363 K; ○, Na, θ_R = 202 K; ▽ Cu, θ_R = 333 K; △ Au, θ_R = 175 K; ▼ Pb, θ_R = 86 K (from Olsen, 1962). These values of θ_R differ from those shown in Table 5 as the data have been fitted by Eq. (42) over a range of lower temperatures.

distortions. In dilute alloys, one finds that the residual resistivity contribution varies linearly with impurity concentration,

$$\rho_0 \propto C_i, \tag{43}$$

where C_i is the concentration of the impurity atoms. This expression may be obtained directly from Eq. (36) since, if the effects of lattice strain are neglected,

$$\rho_0 = \frac{3\pi m^2 \Omega_0}{4\hbar^3 e^2 k_F^6} C_i \int_0^{2k_F} |w_i(\mathbf{q}) - \bar{w}(\mathbf{q})|^2 q^3 dq \tag{44}$$

where $w_i(\mathbf{q})$ and $\bar{w}(\mathbf{q})$ are the form factors for the impurity and host, respectively. This linear relationship is well obeyed up to a few atomic percent; see, e.g., Bass (1983, 1985).

From Eq. (44), one expects a variation of ρ_0 with the square of the magnitude of the scattering potential. Substitution of form factors for some simple model potentials (such as screened Coulomb or empty core potentials) leads to the Linde-Norbury rule, i.e., the impurity resistivity should depend upon the square of the difference in valency Δz between the solute and host,

$$\rho_0 \propto \Delta z^2. \tag{45}$$

Such simple behavior (Fig. 9) is generally observed only in free-electron-like metals.

In more concentrated binary alloys, the form factor of interest turns out to be the difference in the form factors of the alloy components, in which case

$$\rho_0 = \frac{3\pi m^2 \Omega_0}{4\hbar^3 e^2 k_F^6} N \int_0^{2k_F} |\mathfrak{S}^d(\mathbf{q})|^2 \times |w_A(\mathbf{q}) - w_B(\mathbf{q})|^2 q^3 dq, \tag{46}$$

where $\mathfrak{S}^d(\mathbf{q})$ is the structure factor that describes the spatial distribution of the A and B atoms. Since the form factor is squared in Eq. (46), a random alloy containing 1 at. % of A in B should lead to the same resistivity as 1% B in A. This is known as Mott's rule.

Short- and long-range atomic correlations in concentrated alloys directly affect the structure factor, leading to expressions of the form

$$\rho_0 = \frac{3\pi m^2 \Omega_0}{4\hbar^3 e^2 k_F^6} C_A C_B \int_0^{2k_F} \sum_i c_i \alpha_i \frac{\sin(qr_i)}{qr_i} \times |w_A(\mathbf{q}) - w_B(\mathbf{q})|^2 q^3 dq, \tag{47}$$

where c_i is the number of atoms in the ith

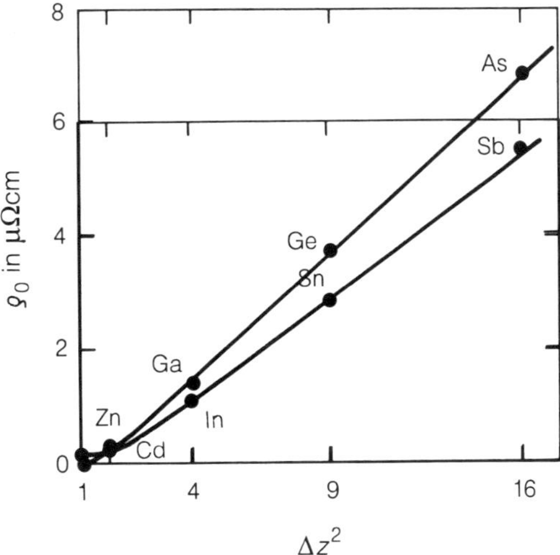

FIG. 9. The change in resistivity for 1 at. % impurity in Cu as a function of $|\Delta z|^2$ (from Rossiter, 1987; using the data of Linde, 1932).

coordination shell of radius r_i, and α_i is the Warren-Cowley short-range order parameter (Warren, 1969). In a random alloy, there is no order; $\alpha_0 = 1$ and $\alpha_i = 0$ for all $i > 0$. Equation (47) then gives Nordheim's rule:

$$\rho_0 \propto C_A C_B. \tag{48}$$

In concentrated alloys containing transition metals, it is necessary to take into account the composition dependence of the band structure and also to allow for the fact that the charge carriers may occupy (and scatter between) a number of energy bands. Examples of data for Pd-Au and Pd-Ag alloys are given in Fig. 10.

The presence of short-range atomic ordering or clustering may lead to either an increase or a decrease in the resistivity from the random-alloy value. However, when the size of the ordered or clustered region becomes comparable with l, any further increase in the degree of atomic correlation (whether clustering or ordering) will cause the resistivity to decrease since this represents a decrease in the degree of disorder on a scale of l. This form of behavior is typified in the changes in resistivity that occur during Guinier-Preston zone formation and growth (see e.g., Porter and Easterling, 1981), as shown in Fig. 11. As the zones evolve, scattering can become quite anisotropic over the Fermi surface (see Sec. 2.3), necessitating a more detailed calculation (Rossiter, 1987).

In the case of long-range order, Eq. (47) leads to

$$\rho_0(s) = \rho_0(0)(1 - s^2), \tag{49}$$

where $\rho_0(s)$ is the residual resistivity of an alloy with Bragg-Williams long-range order parameter s (Warren, 1969). Since complete long-range ordering ($s = 1$) is only possible at stoichiometric compositions, one obtains a composition dependence of the residual resistivity of fully ordered specimens as shown in Fig. 12.

If a material separates into two phases, it is necessary to assume some model of the microstructure in order to calculate the resistivity. Some results for a two phase mixture of Mg_2Pb–Pb are shown in Fig. 13.

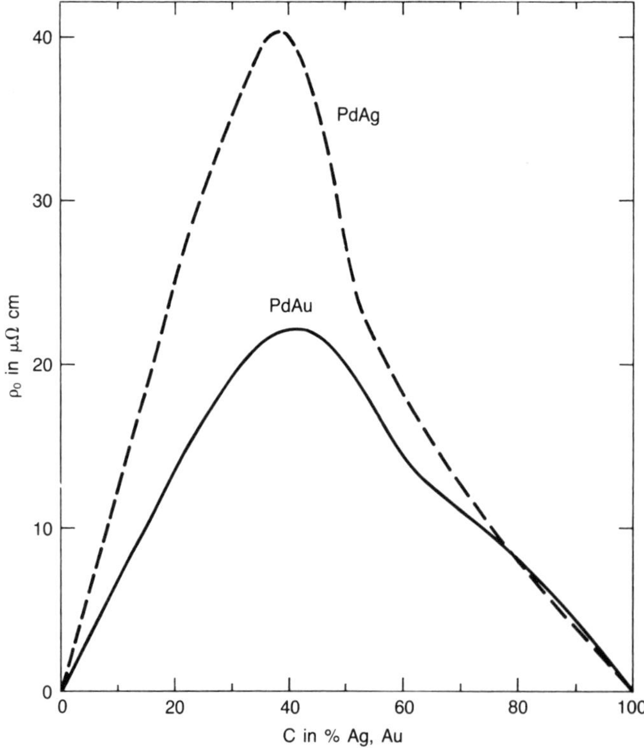

FIG. 10. Composition dependence of the residual resistivity of Pd-Au (solid line) and Pd-Ag (dashed line) (from data of Coles and Taylor, 1962).

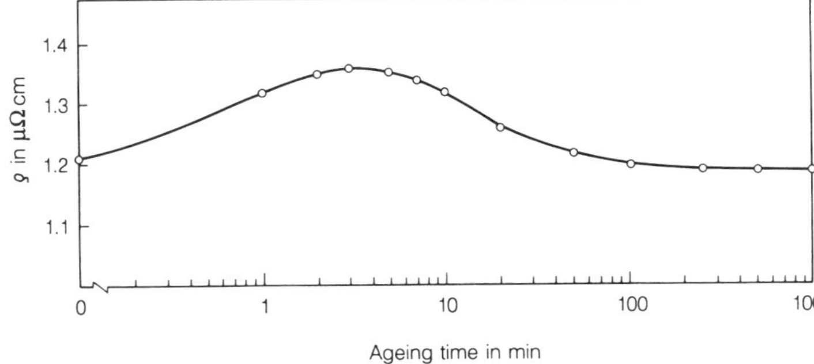

FIG. 11. Electrical resistivity measured at 4.2 K for Al–4.9 at. % Zn as a function of aging time (aging temperature 20 °C) (from Osamura et al., 1973).

5.3 Magnetic and Nearly Magnetic Metals and Alloys

Scattering from disordered magnetic moments in the transition metals is complicated by their decidedly non-free-electron band structure and the associated interband scattering effects. The conduction electrons may be classified into two groups having spin directions either parallel (spin "up," ↑) or antiparallel (spin "down," ↓) to the local magnetization. Elastic scattering of the electrons will not change the population of the subbands, but inelastic (spin-flip) scattering will result in a transfer of momentum between the two groups of electrons.

In the simplest approximation, it is assumed that the conduction band contains nearly free electrons. The current is then carried by the spin-up and spin-down bands in parallel. Solution of the Boltzman equation

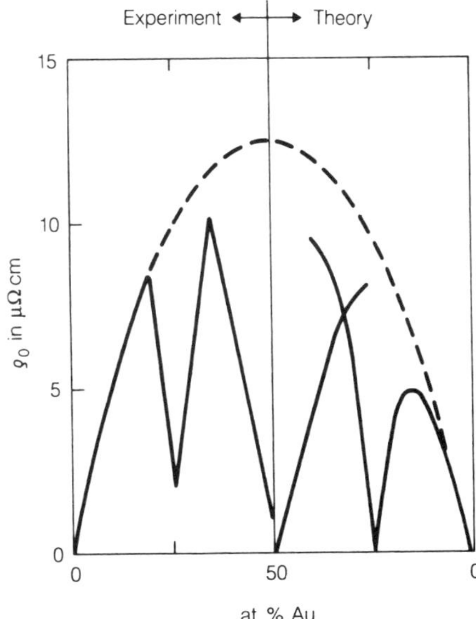

FIG. 12. Variation of resistivity with composition for long-range–ordered Cu-Au alloys. The experimental results of Johansson and Linde (1936) are shown in the full curve on the left, whereas the theoretical curve is shown on the right-hand side. The resistivity of a disordered alloy is shown as the dashed curve. The theoretical curve is scaled so that the disordered resistivities are the same at Cu-50%Au. The experimental data have been reduced by 1.7 $\mu\Omega$ cm to give $\rho = 0$ for pure Cu (from Rossiter, 1987).

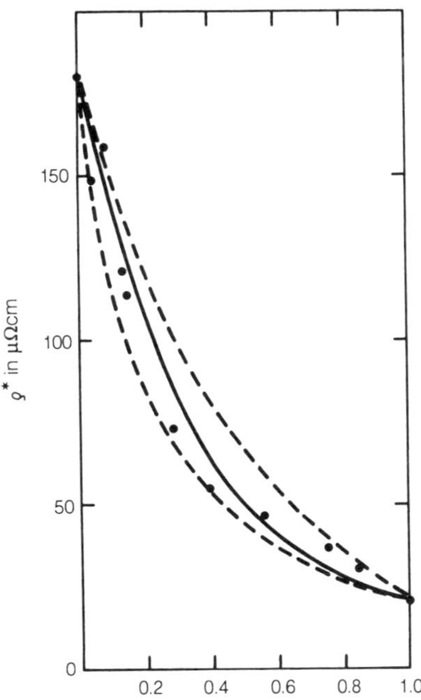

FIG. 13. Resistivity of two-phase Mg_2Pb–Pb mixture as a function of volume fraction of Pb: dashed lines, upper and lower bounds; solid line, effective-medium theory for random spheres; points, experimental data (from Rossiter, 1987, using the data of Hashin and Shtrikman, 1962).

for each band then leads to

$$\rho_{\mathrm{TOT}} = \frac{\rho_{\uparrow}\rho_{\downarrow} + \rho_{\uparrow\downarrow}(\rho_{\uparrow} + \rho_{\downarrow})}{\rho_{\uparrow} + \rho_{\downarrow} + 4\rho_{\uparrow\downarrow}}, \tag{50}$$

where $\rho_{\uparrow\downarrow}$ describes the contributions of the spin-flip processes and each resistivity $\rho_{\uparrow}, \rho_{\downarrow}$ is given by the sum of residual and phonon-scattering contributions:

$$\rho_{\uparrow} = \rho_{0\uparrow} + \rho_{p\uparrow}(T), \tag{51}$$

and similarly for $\rho_{\downarrow}$ (i.e., Matthiessen's rule is assumed valid within each sub-band). Note that the spin-disorder resistivity is contained in $\rho_{0\uparrow}$ and $\rho_{0\downarrow}$ and that different sub-band resistivities will lead to a dependence of ρ_{TOT} upon the direction of the current in relation to the direction of magnetization.

The phonon contribution to the resistivities is complicated by band-structure effects. At low temperatures, the excitations within the spin system are spin waves (magnons), which lead to inelastic scattering. This mechanism introduces a temperature-dependent resistivity $\rho_{p\uparrow\downarrow}(T)$ that must freeze out as $T \rightarrow 0$ K. Other possible contributions to $\rho_{\uparrow\downarrow}$ include up-electron–down-electron interactions and spin-orbit interactions, although these are expected to be negligible in all but very dilute alloys. Thus, as $T \rightarrow 0$, all temperature-dependent parts also go to zero and

$$\rho_0 = \frac{\rho_{0\uparrow}\rho_{0\downarrow}}{\rho_{0\uparrow} + \rho_{0\downarrow}}. \tag{52}$$

Note that even though each of the sub-bands is assumed to obey Matthiessen's rule, the total resistivity does not.

At high temperatures, there will be complete spin mixing and

$$\rho_{\mathrm{TOT}} = 1/4[\rho_{0\uparrow} + \rho_{0\downarrow} + \rho_{p\uparrow}(T) + \rho_{p\downarrow}(T)]. \tag{53}$$

By varying temperature and composition, it is possible to separate out the different contributions. Some results for Ni are shown in Table 6.

As in the case of long-range atomic ordering, long-range antiferromagnetic ordering may introduce new energy gaps at superlattice BZ boundaries that will modify this behavior and possibly result in an increase in ρ_{TOT} just below the Néel temperature. Experimental data given in Meaden (1965) and Shröder (1983) indicate a rich variation in the types of behaviors observed.

Competing ferromagnetic and antiferromagnetic interactions can give rise to an amorphous spin structure—a spin glass. At low temperatures, a $T^{3/2}$ behavior has been attributed to scattering from long-wavelength magnon modes, damped ferromagnetic modes, and discrete excitations. However, a T^2 behavior is also sometimes observed, supposedly as a result of random spin reversals at higher temperatures. In many such systems, spin fluctuation effects are also important, leading to low-temperature resistivity minima and maxima.

5.4 Liquid and Amorphous Alloys

Liquid and amorphous alloys exhibit a wide range of behaviors (see, e.g., Mizutani, 1983; Naugle, 1984).

In many nearly-free-electron materials, the resistivity may be evaluated by substituting appropriate static or dynamical structure factors (obtained from diffraction experiments or by calculation) into Eq. (38) and then using Eq. (36). However, this diffraction approach is not valid if $l \sim$ lattice spacing, since the electron wave packets and associated wave vectors $\mathbf{k}$ are no longer clearly defined, and one would expect a localized (rather than extended) state model to be more realistic. The transition should occur where $\rho_{\mathrm{TOT}} \approx 100$–$160$ $\mu\Omega$ cm. (The same statement applies to high-

Table 6. Values of the parameters used to interpret the sub-band resistivities of Ni-based alloys (after Fert and Campbell, 1976).

Temperature (K)	$\rho_{\uparrow\downarrow}(T)$ ($\mu\Omega$ cm)	$\rho_0(T)$ ($\mu\Omega$ cm)	$\rho_{p\uparrow}(T)$ ($\mu\Omega$ cm)	$\rho_{p\downarrow}(T)$ ($\mu\Omega$ cm)
77	0.9 ± 0.3	0.32	0.38	1.9
200	5 ± 2	2.5	3	15
300	11 ± 4	5.4	6.7	27

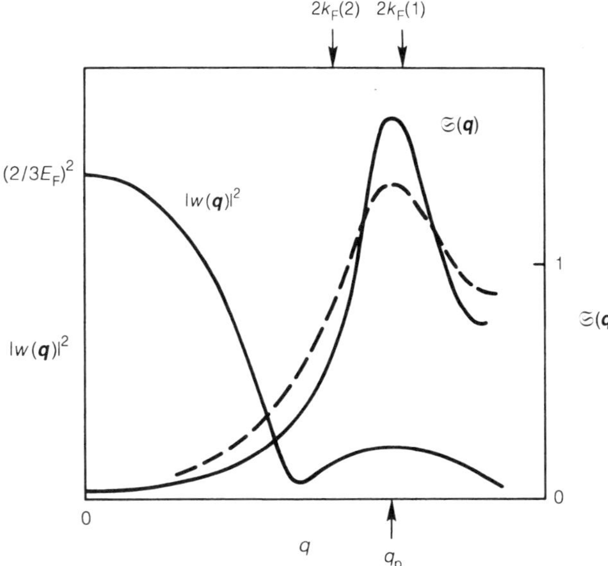

FIG. 14. Schematic representation of the factors $|w(q)|^2$ and $\mathfrak{S}(q)$ that appear in the integrand of the resistivity integral. The structure factor $\mathfrak{S}(q)$ is represented by the full line at the temperature T_1, and the dashed line at $T_2 > T_1$ (from Rossiter, 1987).

resistance crystalline materials—see, e.g., Rossiter, 1987.) Nevertheless, where the model may be applied it can explain the observed positive and negative temperature coefficients of resistivity. In particular, Fig. 14 gives a schematic representation of the structure factor at two different temperatures. The resistivity given by Eq. (36) is heavily weighted by the q^3 term in favor of the $q = 2k_F$ region. Thus, for systems with $2k_F$ indicated by $2k_F(1)$ on the diagram, the resistivity will increase with increasing temperature, whereas with $2k_F(2)$, a negative $d\rho/dT$ will result. From this simple argument, it is expected that the behavior is determined by the position of $2k_F$ in relation to the peak in the structure factor at q_p. Such a correlation is in fact observed, as illustrated in Fig. 15. For the same reason, divalent metals exhibit negative $d\rho/dT$ (at constant volume) whereas others have positive $d\rho/dT$.

In amorphous metals, two competing effects appear at temperatures well below θ_D. The phonon scattering goes as $+T^2$, but the static structure factor is modified by a Debye-Waller factor $\exp(-2M)$, the importance of which depends upon the size of ρ_0. At low temperatures, $M \propto T^2$, and so there is a direct

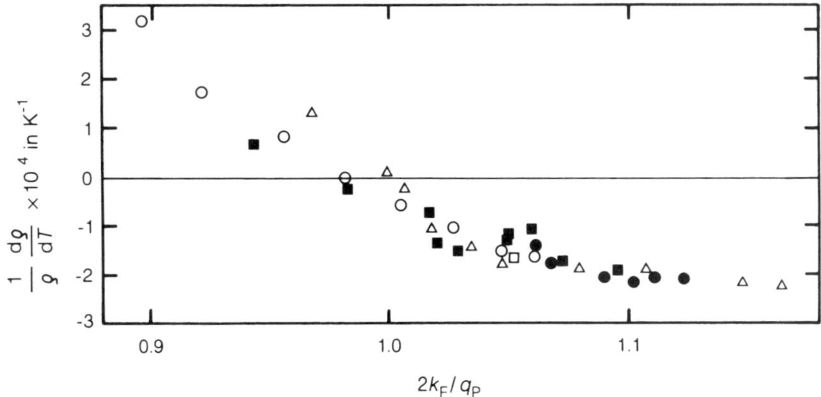

FIG. 15. $d\rho/dT$ as a function of $2k_F/q_p$ for a number of simple metallic glasses: ○: Ag-Cu-Mg; △: Ag-Cu-Al; ■: Ag-Cu-Ge; ●: Mg-Zn; □: Mg-Cu (from Rossiter, 1987, using the data of Mizutani, 1983).

competition between the $+T^2$ phonon term and $-T^2$ Debye-Waller term. If ρ_0 is small, the phonon term is dominant and gives an overall $+T^2$ dependence, but if ρ_0 is large, a $-T^2$ dependence will prevail.

The resistivity of Fe- and Ni-based ferromagnetic amorphous alloys tends to vary as $T^{3/2}$ at temperatures less than $\sim T_C/2$, where T_C is the Curie temperature, and then varies as T^2 up to T_C. The former results from spin-wave scattering, which breaks down into random spin reversals at higher temperatures leading to the T^2 variation.

Many amorphous metals exhibit resistivity minima at low temperatures. This behavior does not appear to depend upon whether the material is paramagnetic or ferromagnetic, and behavior is insensitive to composition, suggesting that it is a fundamental property of the disordered lattice rather than some specific fluctuation effect.

Because the scattering due to thermal motion of the solid lattice is already quite large at temperatures near the melting point, the change in ρ upon melting is not very large, usually less than a factor of 2 (see, e.g., Fig. 24). Liquid alloys tend to show a similar dependence upon composition to that in their disordered solid alloy counterparts.

5.5 ac Effects

The equations described previously are not significantly changed for time-varying (i.e., ac) fields for frequencies up to megahertz, except that the field inside a conductor will now decay exponentially from the surface into the conductor. A measure of the depth of penetration is the classical skin depth δ_0 given by

$$\delta_0 = c(2\pi\sigma\omega)^{-1/2} \tag{54}$$

and is valid for $\delta_0 \gg l$. In this expression, ω is the frequency of the applied electric field, and c is the speed of light. If δ_0 is less than the specimen size, the cross section of specimen-carrying current will differ from the specimen cross section (see Fig. 1), significantly complicating the determination of resistivity.

At frequencies in the microwave range, the field penetration will become dependent upon features of the FS ("anomalous" skin-depth effect), and cyclotron resonance will occur when the electric field equals the cyclotron frequency ω_c [Eq. (35)]. At still higher frequencies ($\omega\tau \sim 1$), one should replace the conductivity σ by a frequency-dependent (ac) conductivity

$$\sigma(\omega) = \frac{\sigma_0}{1 - i\omega\tau}. \tag{55}$$

Finally if $\omega\tau \gg 1$, oscillating excitation of the conduction electrons (charge density waves, or "plasmons") may occur at the plasma frequency ω_p given by

$$\omega_p^2 = 4\pi\sigma_0/\tau. \tag{56}$$

The material will become transparent to electromagnetic radiation at frequencies above ω_p.

6. MAGNETORESISTANCE (*MR*), TRANSVERSE (T-) AND LONGITUDINAL (L-)

There are two separate magnetoresistances (*MR*): transverse (T-*MR*), where the magnetic field is applied perpendicular to the current direction; and longitudinal (L-*MR*), where the field is applied parallel to the current.

6.1 Kohler's Rule

If the electrons on the spherical FS of a free-electron metal are all scattered with a single relaxation time τ, then both the T-*MR* and L-*MR* are zero as noted in Sec. 2.3. This occurs because the average force eE_y produced on an electron by the Hall voltage V_y exactly cancels the Lorenz force $e\langle v_x\rangle B_z$ acting upon the same electron. Here, B_z is the applied magnetic field, pointing in the z direction, and $\langle v_x\rangle$ is the electron's average drift velocity in the x direction. If the electrons occupy a more complex FS, with different values of the effective mass m^* and different relaxation times for electrons on different parts of the FS, then the single Hall field cannot simultaneously cancel the different Lorenz forces felt by different electrons, and nonzero T-*MR* and L-*MR* both appear.

If the electron scattering on a real FS can still be approximated by a single relaxation time, the *MR* should be a function only of the unitless quantity $\omega_c\tau$ defined in Sec. 4, and all data for a given metal should fall on a single curve when the *MR* is plotted as a function of $\omega_c\tau$. Such behavior, called Kohler's rule, is found to be experimentally valid under somewhat more general conditions than just described. Figure 16 shows how Kohler's rule applies to the T-*MR* of polycrystalline samples of In at temperatures ranging from 2 to 20 K. Similar plots of the T-*MR* and the L-*MR* of a variety of polycrystalline metals are shown in Figs. 17 and 18, respectively. The abscissa scale in these figures is proportional to $\omega_c\tau$.

6.2 Low Magnetic Fields ($\omega_c\tau \ll 1$)

At very low fields, the general theory of *MR* predicts a B^2 increase of the *MR* independent of the FS topology and whether the sample is a single crystal or polycrystalline. This form automatically satisfies the Onsager-relation requirement that the *MR* must be an even function of **B**. Examples of approximately B^2 variations are shown in the low-field regions of Figs. 16 and 17.

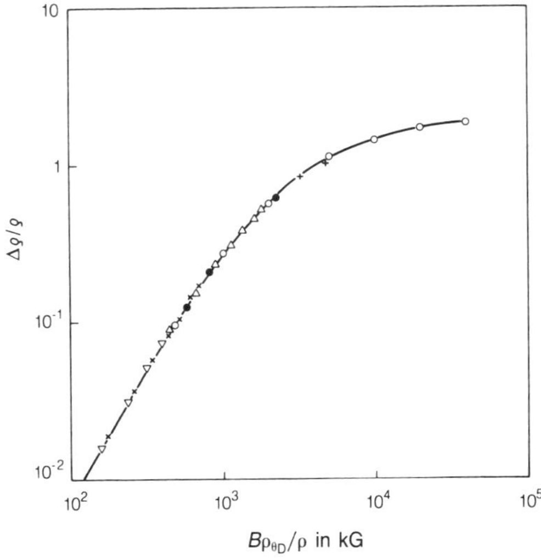

FIG. 16. Magnetoresistance in polycrystalline indium in a transverse magnetic field showing agreement with Kohler's rule. Some points affected by boundary scattering have been suppressed. ▽: In (specimen 1), 14 K, ρ/ρ_{273} = 0.024; X: In (specimen 6), 20 K, ρ/ρ_{273} = 0.023; △: In (specimen 6), 14 K, ρ/ρ_{273} = 0.0086; +: In (specimen 6), 4.2 K, ρ/ρ_{273} = 0.0012; ○: In (specimen 2), 4.2 K, ρ/ρ_{273} = 0.00007; ●: In (specimen 2), 2 K, ρ/ρ_{273} = 0.00003 (from Olsen, 1962).

6.3 High Magnetic Fields ($\omega_c\tau \gg 1$)

In single-crystal samples, the high-field L-*MR* becomes independent of B ("saturates") for all FS topologies and for both compensated ($N_e = N_h$) and uncompensated ($N_e \neq N_h$) metals. The T-*MR*, in contrast, can either become independent of B or vary as B^2, depending upon the topology of the FS and the orientation of **B** with respect to the crystallographic axes; the permitted alternatives were listed in Table 1.

For an uncompensated metal with no open orbits (see Sec. 4.3), the T-*MR* "saturates" for all directions of **B** when $\omega_c\tau \gg 1$. If the FS contains open orbits, then the T-*MR* saturates when there are no open orbits in the plane perpendicular to **B**, increases as B^2 when there is one band of open orbits in the plane perpendicular to **B**, and saturates again when there are two perpendicular open orbits in the plane perpendicular to **B**. Saturation and B^2 variations for different field directions are illustrated in Fig. 19 for the uncompensated intermetallic compound $AuGa_2$, which can be produced in a highly ordered state that permits the high-field condition to be reached.

For a compensated metal with no open orbits or one band of open orbits, the T-*MR* is proportional to B^2 independent of field orientation, except that the T-*MR* saturates when the open orbit direction in **k** space is exactly the direction of current flow in the sample. The B^2 behavior for compensated metals is illustrated in Fig. 20 for Pb.

Finally, in polycrystalline samples that contain open orbits, an approximately linear T-*MR* may result from averaging of B^0 and B^2 behaviors on different orbits.

6.4 Additional Examples

6.4.1 Deviations from the Ideal The analyses just given are rigorous for a completely homogeneous sample in which the current density is uniform across the sample area and the magnetic field is uniform along the sample length. If the current density is perturbed, for example by defects in the sample or on its surface, or the magnetic field is not uniform, the T-*MR* can become linear in B.

6.4.2 Quantum Oscillations At low temperatures and high magnetic fields, quantum

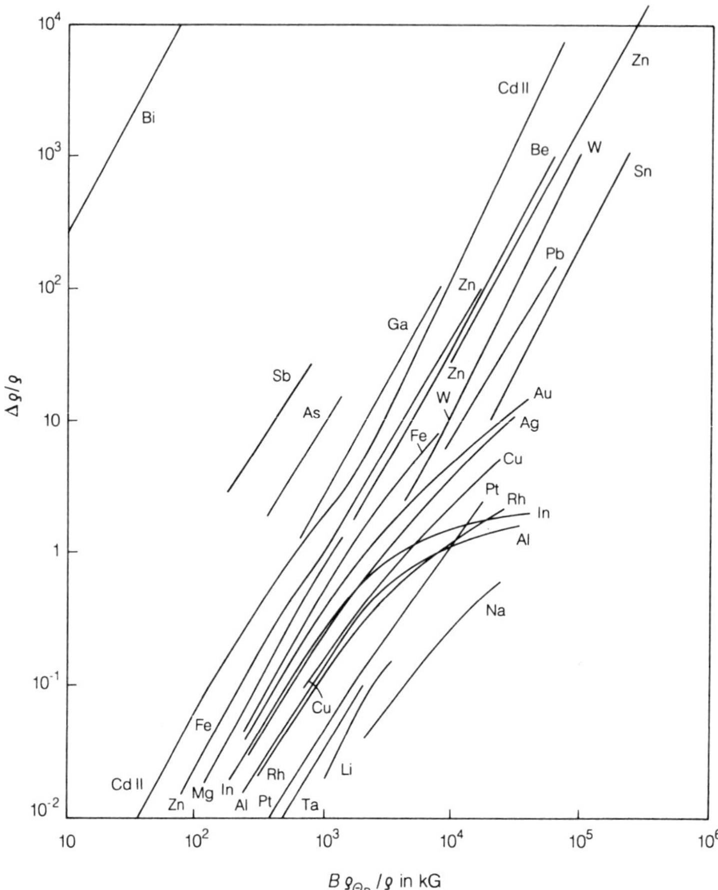

FIG. 17. The transverse magnetoresistance of polycrystalline metals in the reduced Kohler diagram. $\Delta\rho/\rho$ is plotted against $B\rho_{\theta_D}/\rho$, where ρ is the resistivity in zero field, and ρ_{θ_D} that at $T = \theta_D$ in zero field (from Olsen, 1962).

oscillations (called Shubnikov–de Haas oscillations) are seen in single crystals of high-purity metals, as illustrated in Fig. 21. The periods of these oscillations are determined by the cross-sectional areas of the FS perpendicular to the direction of **B**.

6.4.3 Size Effects If a transverse **B** is directed in the plane of a thin, high-purity foil, Sondheimer oscillations (Ziman, 1972) appear in the T-*MR*. The period of these oscillations is determined by the average Fermi momentum of the electrons. If **B** points along the axis of a thin wire, the resulting L-*MR* first increases slightly with increasing B and then decreases, the decrease beginning when the electron's cyclotron orbit becomes comparable to the wire radius; thereafter, increasing the field reduces the number of electrons that reach the wire surface, and thus reduces the contribution of surface scattering to ρ.

6.4.4 Ferrogmagnets The *MR* of ferromagnets can be complex and anomalous. For example, Fig. 22 shows that the initial values of the T-*MR* for Ni are negative, and those of the L-*MR* are positive. At fields high enough to align all of the magnetic domains parallel to the field, both the T-*MR* and L-*MR* decrease with increasing field, probably as a result primarily of magnetostrictive effects (see e.g., Chikazumi, 1964).

7. HALL EFFECT, R_H

7.1 Free Electrons

The free-electron model of a metal provides the simplest picture of the physical phenomenon underlying the Hall effect. The current density j_x in a foil of width W is given by $n_e e\langle v_x\rangle$. When a magnetic field B_z is applied in the negative z direction, it produces

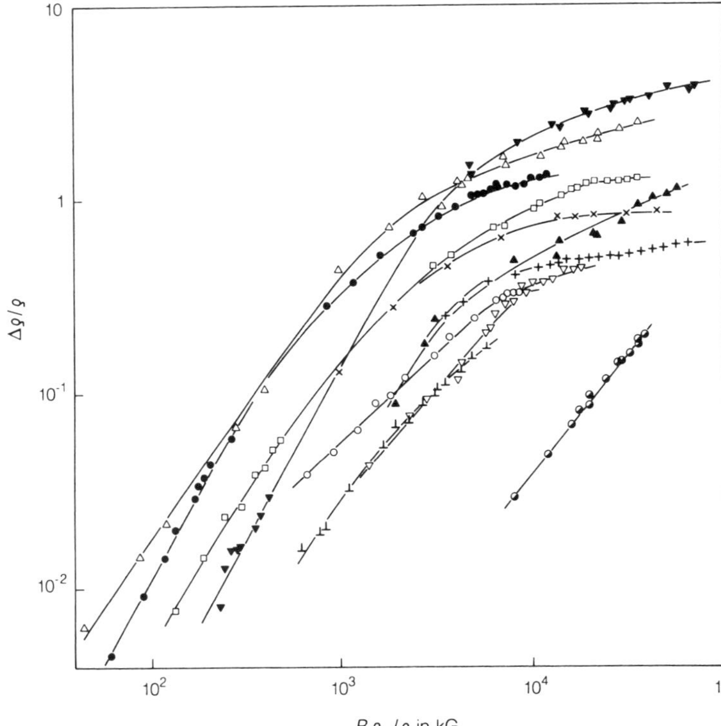

FIG. 18. The magnetoresistance of polycrystalline metals in longitudinal fields. ρ is the zero-field resistivity at the temperature of observation, and $\rho_{\theta D}$ that at $T = \theta_D$. The measurements were made in fields up to 200 000 gauss. ◑, Li, $\theta_D = 430$ K; +, Al, $\theta_D = 410$ K; ▼ Sn, $\theta_D = 160$ K, ○ Ni, $\theta_D = 410$ K, □ Ag, $\theta_D = 220$ K, × In, $\theta_D = 100$ K; ▲ Pb, $\theta_D = 90$ K; ▽ Pt, $\theta_D = 240$ K; ● Au, $\theta_D = 320$ K; △ Zn, $\theta_D = 240$ K; ⊥ Fe, $\theta_D = 355$ K (from Olsen, 1962).

on the electron stream a force $e\langle v_x \rangle B_z$, which is directed in the y direction and causes the electrons to try to flow also in this direction. Since the sample has finite width, the electrons can only pile up on the edge of the foil, thereby generating a charge separation in the foil that leads to an electric field E_y. The electronic system comes into dynamic equilibrium when this field produces a force just large enough to balance the magnetic force,

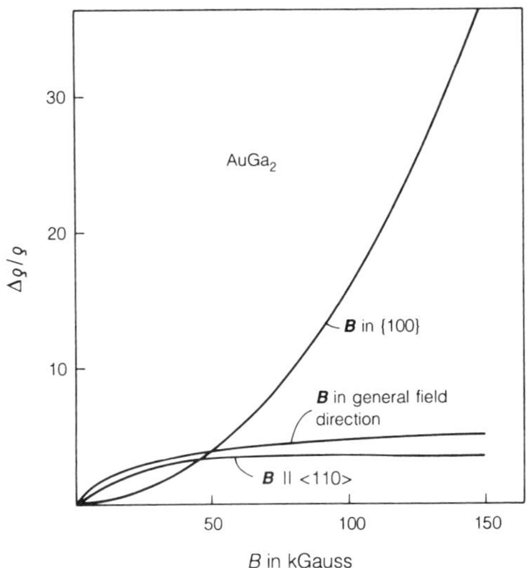

FIG. 19. $\Delta\rho/\rho$ of $AuGa_2$ versus B for various directions of **B** (from Longo et al., 1969).

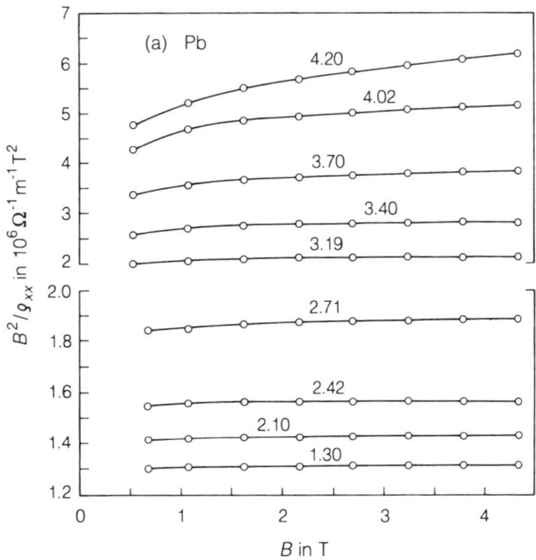

FIG. 20. The quantity B^2/ρ_{xx} plotted as a function of magnetic field B for Pb (after Fletcher, 1977).

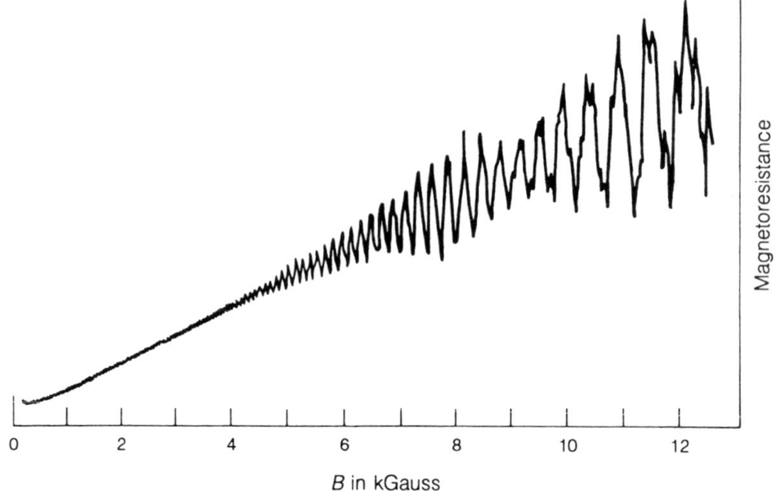

FIG. 21. Shubnikov–de Haas oscillations in the magnetoresistance of gallium versus field at 1.3 K (from Ashcroft and Mermin, 1976).

i.e.,

$$eE_y = e\langle v_x\rangle B_z. \tag{57}$$

Substituting $j_x = n_e e\langle v_x\rangle$ into the right-hand side of Eq. 57 and defining $R_H = E_y/J_xB_z$ yields

$$R_H = 1/n_e e, \tag{58}$$

as already derived under similar assumptions with the Boltzmann transport equation in Sec. 2.3. Equation (58) predicts R_H to be directly proportional to the electron charge (i.e., strictly negative) and inversely proportional to the electron density.

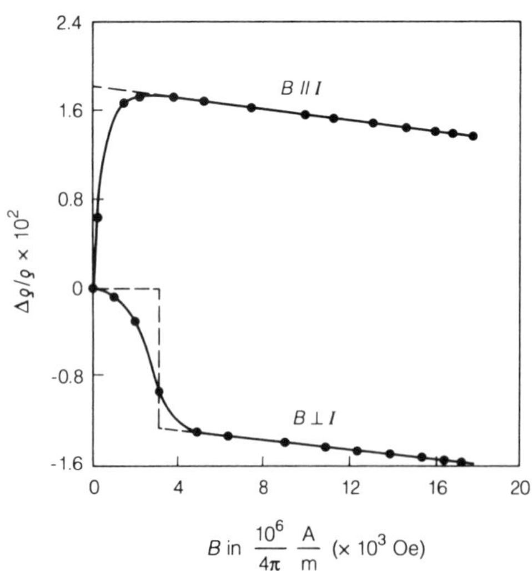

FIG. 22. Variation of Ni resistivity as a function of a magnetic field (from Chikazumi, 1964, using the data of Englert, 1932).

7.2 Beyond Free Electrons

Table 7 lists approximate normalized values of R_H at low temperatures and high magnetic fields for several uncompensated metals (i.e., $N_e \neq N_h$). Equation (58) gives the correct order of magnitude of R_H for all of these metals, but rarely gives precisely the correct value, and sometimes gives even the wrong sign. These deviations from Eq. (58)

Table 7. Hall coefficients of selected elements in moderately high fields (from Ashcroft and Mermin, 1976).

Metal	Valence	$-(R_H n_e e)^{-1}$
Li	1	0.8
Na	1	1.2
K	1	1.1
Rb	1	1.0
Cs	1	0.9
Cu	1	1.5
Ag	1	1.3
Au	1	1.5
Be	2	−0.2
Mg	2	−0.4
In	3	−0.3
Al	3	−0.3

arise because different sets of electrons in a metal have different effective masses and are scattered in very different ways, neither of which was taken into account in the free-electron model.

The Boltzmann-equation analysis given in Sec. 2 can be generalized to allow for different effective masses on different parts of the FS. It is useful to consider this expression in two extreme limits: the low-field limit ($\omega_c\tau \ll 1$) and the high-field limit ($\omega_c\tau \gg 1$).

7.2.1 Low Magnetic Fields The simplest generalization is a two-band model, with relaxation times τ_1 and τ_2, conductivities σ_1 and σ_2, and effective masses m_1^* and m_2^*, where τ and σ are positive quantities but m^* can be positive (electron-like) or negative (holelike). This model yields (Olsen, 1962)

$$R_H = e\frac{\tau_1\sigma_1/m_1^* + \tau_2\sigma_2/m_2^*}{\sigma_1^2 + \sigma_2^2}. \tag{59}$$

For a free-electron FS with single values of τ, σ, and m^*, Eq. (59) reduces to Eq. (58), since $\sigma = n_e e^2\tau/m^*$. In general, however, the sign of R_H depends on a balance between the two bands.

7.2.2 High Magnetic Fields Three different cases occur in this limit, as were listed in Table 1.

1. For an uncompensated metal ($N_e \neq N_h$) with no open orbits, R_H takes the value

$$R_H = 1/(n_e - n_h)e, \tag{60}$$

which is simply a generalization of Eq. (34) or (58) to take into account that the FS's of real metals can contain both electron-like and holelike portions. For one-electron metals such as the alkali and noble metals, where $N_e = 1$ and $N_h = 0$, Eq. (60) reduces to Eq. (58), and the numbers in Table 7 are nearly unity. For Al, in contrast, the second-zone surface contains $\approx$2 holes/atom, and the third-zone surface contains $\approx$1 electron/atom. In high magnetic fields, Eq. (60) thus predicts R_H to be $-\frac{1}{3}$ of the free-electron value. In low fields, however, electrons execute only a small fraction of a cyclotron orbit before being scattered, and thus, most of them never collide with a BZ boundary. The BZ boundaries can then be neglected, and Eq. (60) would reduce to Eq. (58) if the details of scattering were unimportant. The high-field value of R_H in Al should be well defined, but the low-field value changes as the dominant scatterer changes. This behavior is illustrated in Fig. 23.

For uncompensated metals, and compensated metals with open orbits, the situation is more complex.

2. In most cases, R_H still becomes constant at high fields, but the value of this constant depends upon details of both scattering and the FS.
3. When open orbits are present in more than one direction, R_H decreases like $1/B$ as B increases.

7.3 Additional Examples Including Magnetic Metals

The temperature dependences of R_H for metals and alloys can be quite complex. Both ρ and R_H change upon melting. For simple metals, the change in R_H (and ρ) is usually not large, as illustrated in Fig. 24 for Li.

As illustrated in Fig. 25 for a $Mg_{0.7}Zn_{0.3}$ alloy, R_H can exhibit different behaviors in crystalline, amorphous, and liquid forms of the same metal or alloy.

Although quantum oscillations of all kinds occur in R_H under the same conditions that they occur in *MR*, they are usually very small (an exception is quantum size effects in the semimetal Bi).

An additional phenomenon, called the anomalous Hall effect, has made Hall-effect measurements important for studies of metals containing large, localized magnetic moments (e.g., ferromagnets), because of an additional contribution, called the anomalous Hall effect. Figure 26 shows schematically what is expected for the Hall resistivity $\rho_H = R_H B$ as a function of B in such a metal.

8. THERMOPOWER *S*

8.1 The Temperature Dependence of *S* in Pure Metals

Figure 27 shows the thermopower S for several pure metals (for more complete data see Foiles, 1985). For each metal, the data can be divided into two components (see Sec. 1.2).

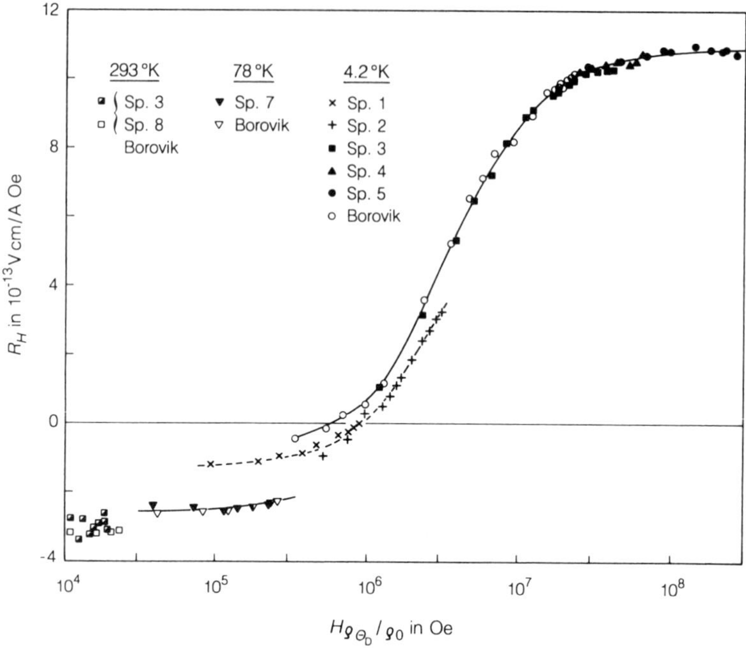

FIG. 23. Hall coefficient of Al as a function of the reduced magnetic field measured through the high-field/low-field transition. Sp1 and Sp2 refer, respectively, to Al + 0.4 at. % Zn and Al + 0.1 at. % Zn. The other specimen numbers refer to pure Al specimens of various thicknesses (from Hurd, 1972, using the data of Forsvoll and Holwech, 1964).

The first, S_d, is approximately linear in T, as indicated by the dashed lines. The second, S_g, passes through a maximum value well below room temperature.

8.2 Rough Derivations of S_d and S_g

The presence of a temperature gradient $\nabla_r T$ causes both electrons and phonons to flow from the hot end of the metal to the cold end. In each case, we define an effective "force per unit volume" associated with this flow that acts on electrons:

$$F_{\text{flow}} = \frac{dU}{dx} = \frac{\partial U}{\partial T}\frac{dT}{dx} = C\,\frac{dT}{dx} \tag{61}$$

where U and C are the energy density and heat capacity per unit volume, respectively.

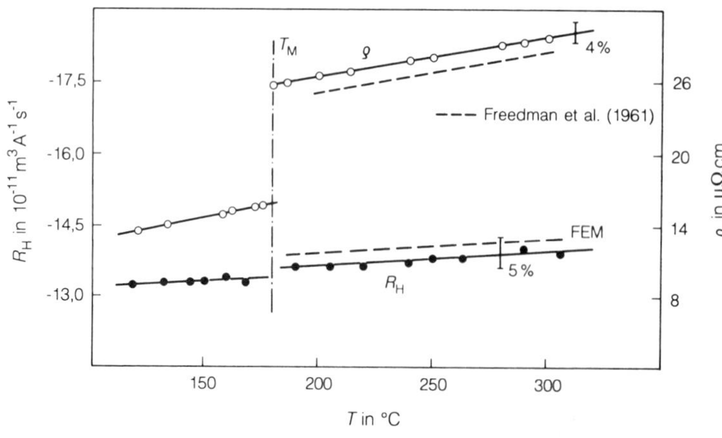

FIG. 24. Hall coefficient and electrical resistivity of solid and liquid Li. T_M = melting point, and FEM = free-electron value (from Kunzi and Guntherodt, 1980).

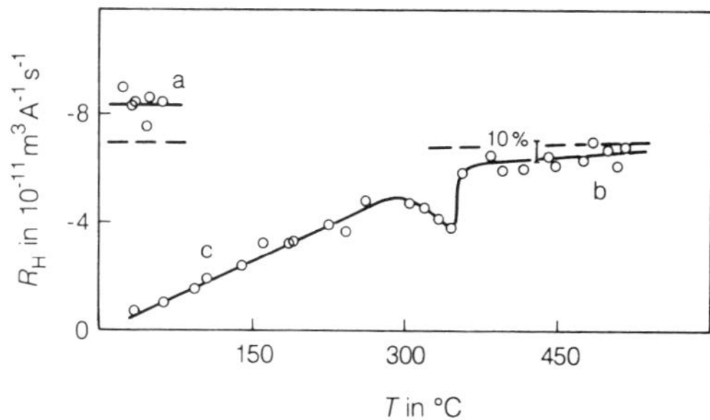

FIG. 25. Hall coefficient of (a) an amorphous, (b) a liquid, and (c) a crystalline $Mg_{70}Zn_{30}$ alloy; dashed line indicates the free-electron values for the Hall coefficient (from Kunzi and Guntherodt, 1980).

In Sec. 1, we noted that S is measured under conditions of no net current flow. This means that the electrons being "pushed" by the force of Eq. (61) must pile up at one end of the sample to produce an electric field E that opposes further electron flow. The force per unit volume due to this field acting on the electron system must be

$$F_E = n_e eE. \quad (62)$$

For electron flow alone, writing $C = C_e$ and equating the forces of Eqs. (61) and (62) gives

$$S_d = \frac{E}{dT/dx} \sim \frac{C_e}{n_e e} = \pi^2 \frac{k_B}{e} \frac{k_B T}{\epsilon_F}, \quad (63)$$

where ϵ_F is the Fermi energy. Equation (63) predicts that S_d should increase linearly with T. For a typical $\epsilon_F \approx 1$ eV, the coefficient in Eq. (63) is $\sim 10^{-8}$ V/K^2, leading to room-temperature values of $S_d \sim 1\ \mu$V/K^2.

For phonon flow, we initially assume that the phonons collide only with electrons, thereby driving them along with their own motion. As above, we then find

$$S_g = E/(dT/dx) \sim C_g/n_e e. \quad (64)$$

Aside from a factor of $\frac{1}{3}$, this is the correct term for free electrons interacting with Debye phonons via normal scattering processes alone.

Since phonons also scatter off other phonons, impurities, etc., Eq. (64) must be multiplied by a factor α, the ratio of the probability for phonon-electron scattering (the only scattering that drives the electrons along with the phonons) to the probability for all scattering (including electron-phonon scattering). Writing the probability for phonon-electron scattering as inversely proportional to one relaxation time τ_{pe}, and the probability for all other types of scattering as inversely proportional to another, τ_{po}, gives

$$S_g \sim (C_g/n_e e)\alpha \sim (C_g/n_e e)\tau_{po}/(\tau_{po} + \tau_{pe}). \quad (65)$$

At low temperatures compared to the Debye temperature θ_D (i.e., $T \ll \theta_D$), phonon-electron scattering dominates, τ_{pe} is short compared to τ_{po}, and the ratio of the τ's cancels out. This leaves $S_g = C_g/n_e e$, so that $S_g = (12\pi^4/5)(k_B/e)(T/\theta_D)^3$. For $T > \theta_D$, C_g becomes constant at the classical value $3nk_B$, where n is the num-

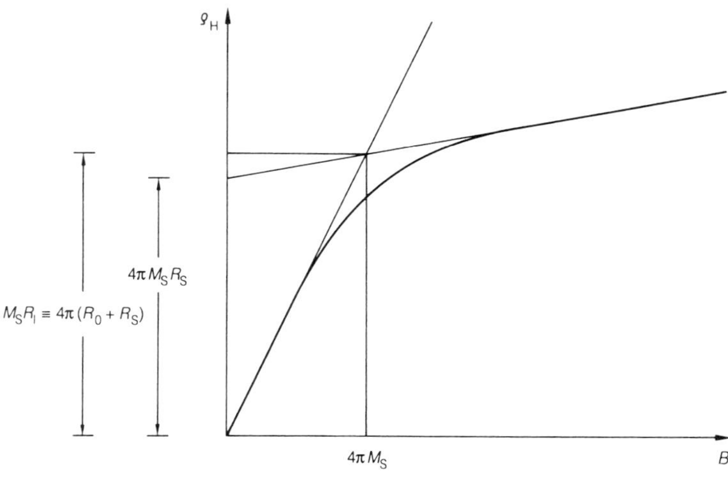

FIG. 26. Schematic behavior of the Hall resistivity ρ_H as a function of magnetic induction B in a metal showing appreciable magnetization (from Hurd, 1972). M_S is the spontaneous magnetization; R_o and R_S are constants called, respectively, the ordinary and spontaneous Hall coefficients.

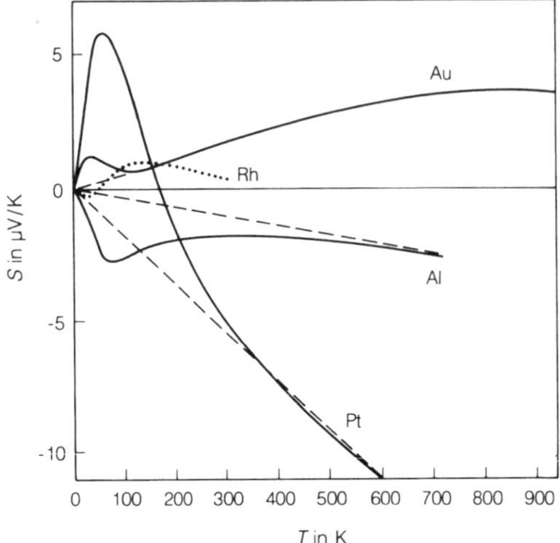

FIG. 27. The thermopowers S of the metals Au, Al, Pt (solid curves), and Rh (dotted curve) as a function of temperature. For Pt, Al, and Au, the differences between the solid curves and the dashed lines indicate the magnitude of the phonon-drag components S_g (from Bass, 1982).

ber of atoms per unit volume in the metal. τ_{pe} becomes temperature independent, whereas τ_{po} is dominated by phonon-phonon scattering which varies as T^{-1}; S_g thus decays as T^{-1}. Inserting appropriate values for τ_{pe} and τ_{po} reveals that as T increases from $T = 0$ K, the S_g of Eq. (65) first increases as T^3, then passes through a maximum value of order $\sim\mu V/K^2$ at $T \approx \theta_D/5$, and finally decreases as T^{-1}.

Combining S_d from Eq. (63) with S_g from Eq. (65) yields an S containing one term linear in T with room-temperature values of order μV/K, and another which peaks at $\sim\theta_D/5$ with maximum values also of order μV/K. This combination correctly describes both the forms and magnitudes of the data of Fig. 27. The model predicts, however, that both terms will be strictly negative, whereas Fig. 27 shows that each can be either negative or positive.

This simple model has neglected

1. details of the scattering of electrons on real FSs by realistic phonons and by other scatterers;
2. the presence of "Umklapp" scattering events, which can cause the scattered electron to end up moving oppositely to the direction of the incoming phonon; and
3. contributions to S_d from higher-order scattering events and from many-body effects.

The first two items provide mechanisms for changes in the sign of S_g, and all three provide mechanisms for changes in the sign and magnitude of S_d.

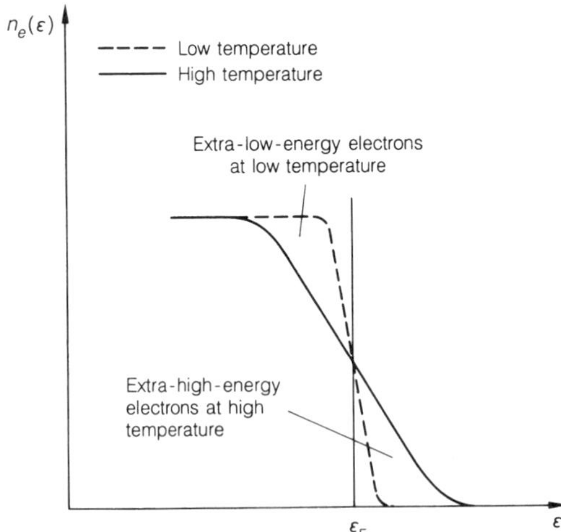

FIG. 28. The variation with energy ϵ of the number of conduction electrons n_e in a metal in the vicinity of the Fermi energy ϵ_F for two different temperatures. In the interests of simplicity, a small variation of ϵ_F with temperature has been neglected in drawing this figure (from Bass, 1982).

8.3 More Detailed Analysis of S_d

Direct application of the Boltzmann transport equation described in Sec. 2 yields for S_d (e.g., Barnard, 1972)

$$S_d = \frac{\pi^2 k_B^2 T}{3e\epsilon_F}\left[\frac{d\ln\sigma(\epsilon)}{d\ln\epsilon}\right]_{\epsilon_F} = \frac{\pi^2 k_B^2 T}{3e\epsilon_F}\left[\frac{d\ln n_e}{d\ln\epsilon} + \frac{d\ln\tau}{d\ln\epsilon}\right]_{\epsilon_F} \tag{66}$$

where ϵ is the electron's energy, $\sigma(\epsilon)$ is the energy-dependent conductivity, and the derivatives are to be evaluated at ϵ_F. The second expression involves the single-relaxation-time approximation. Since for a spherical FS $n_e \propto \epsilon^{3/2}$, the first term in Eq. (66) is simply the generalization of Eq. (63). The second term in Eq. (66) contains the effects of scattering alluded to at the end of Sec. 8.2. The simplest way to understand the physical sources of the two terms in Eq. (66) is to examine the Fermi distributions at the

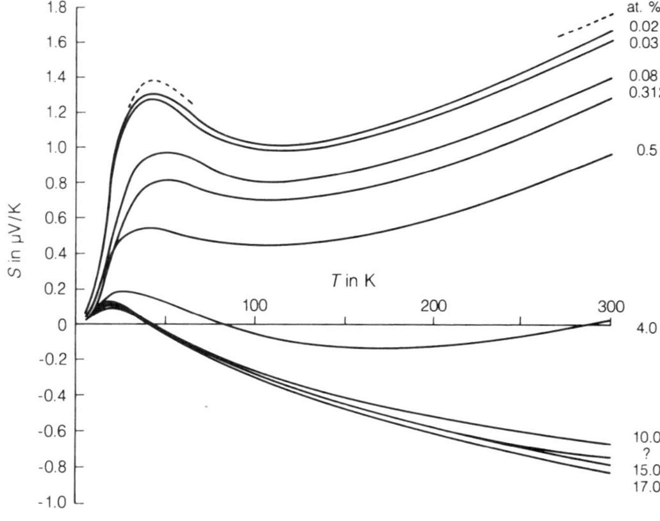

FIG. 29. Measured thermopowers as functions of temperature for the Ag-Hg system (from Craig and Crisp, 1978). The concentrations of Hg in at. % are shown adjacent to each curve.

two ends of the metallic sample of interest. If T_H and T_L are the temperatures at the high- and low-temperature ends of a bar, the resultant Fermi distributions are as shown in exaggerated fashion in Fig. 28. This diagram shows that there are more "high-energy" electrons (i.e., energy greater than ϵ_F) at T_H than at T_L, and more "low-energy" electrons (energy less than ϵ_F) at T_L than at T_H. Thus, if both n_e and τ are larger at higher electron energies, the energy derivative of each will cause electrons to flow more easily from the hot end of the metal to the cold end, thereby building up an excess of electrons at the cold end. These electrons will produce an electric field, which will grow until it just counteracts the flow due to the two "forces" just described. This is exactly what happens on a spherical FS with simple scattering and yields the normal negative S_d. If, however, the energy dependences of n_e or τ are opposite to those assumed in the foregoing, then electrons will tend to pile up at the hot end of the metal, yielding a positive contribution to S_d. This can occur, for example, for n_e, in the case where the FS contacts the BZ boundary.

Even in the free-electron model, the magnitude of the linear coefficient in S_d changes with temperature. The simplest models yield $\tau \propto \epsilon^{3/2}$ for electron-phonon scattering (which is dominant at high temperatures), but they yield $\tau \propto \epsilon^{-1/2}$ for electron-impurity scatter-

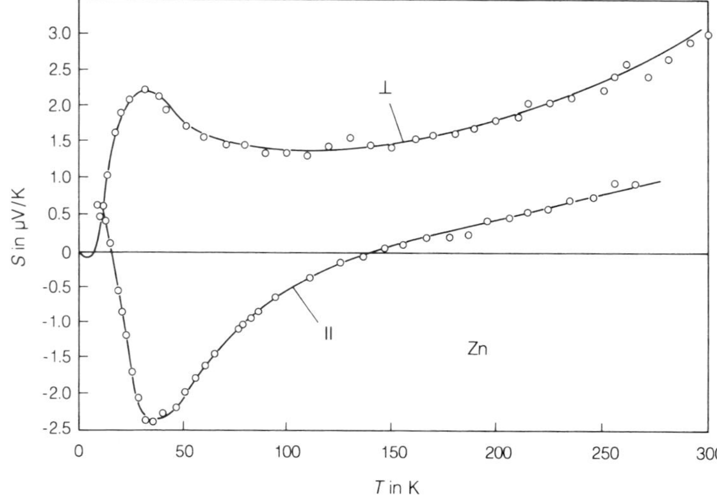

FIG. 30. The thermopower of Zn parallel (||) and perpendicular (⊥) to the hexagonal axis (from Rowe and Schroeder, 1970).

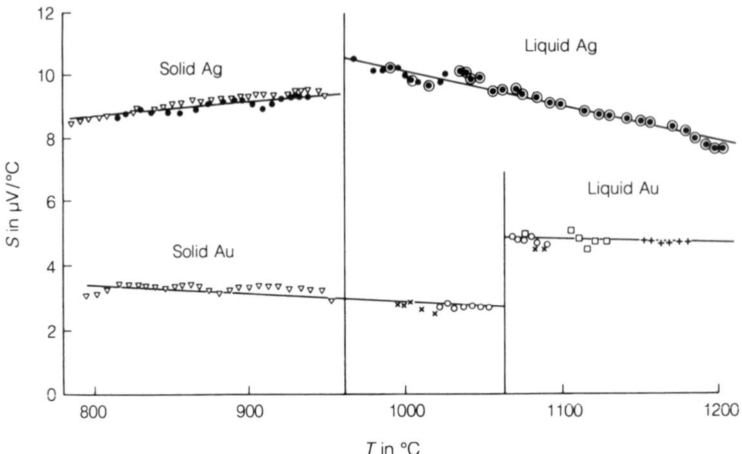

FIG. 31. The changes in the thermopowers of Au and Ag upon melting (from Howe and Enderby, 1967).

ing (which is dominant at low temperatures). Combining each of these terms with $n_e \propto \epsilon^{3/2}$, we find that the coefficient of S_d decreases by $\frac{1}{3}$ from high to low temperatures.

More generally, $\sigma(\epsilon)$ is a complex integral over the FS that depends upon how $\tau_{\mathbf{k}}$ varies from place to place, and it is not possible to separate out terms related simply to n_e and τ. A complete calculation of S_d must also include contributions from higher-order scattering events (i.e., events involving virtual scattering contributions) and corrections due to many-body effects.

Adding impurities to a metal generally reduces S_g and changes the slope of S_d, as illustrated in Fig. 29 (see also Foiles, 1985). The

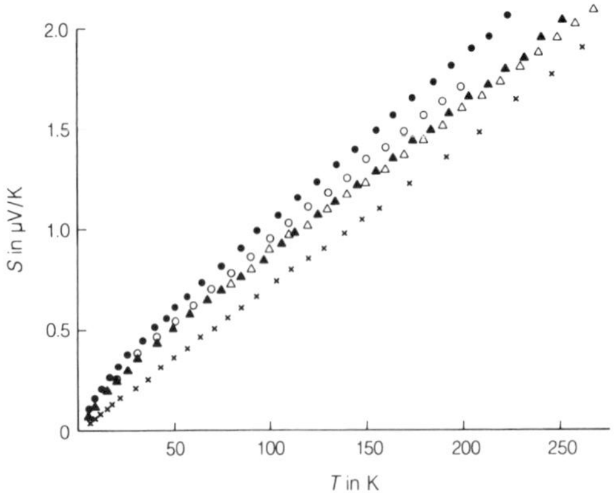

FIG. 32. Measured thermopowers for three Cu-Zr systems: ●, $Cu_{45}Zr_{55}$; ○, $Cu_{27.5}$ $Zr_{72.5}$; △, ▲, $Cu_{60}Zr_{40}$; and × $Cu_{50}Zr_{40}Fe_{10}$ (from Gallagher and Greig, 1982).

change in slope of S_d can be understood in terms of the Nordheim-Gorter rule:

$$\begin{aligned} S_{\mathrm{TOT}} &= (\rho_p S_p + \rho_0 S_0)/\rho_{\mathrm{TOT}} \\ &= (\rho_p/\rho_{\mathrm{TOT}})(S_p - S_0) + S_0, \end{aligned} \tag{67}$$

which is derived from Eq. (66) by assuming that the impurities represent an additional scatterer for which ρ can be treated by Matthiessen's rule (Eq. 40). In this equation, S_{TOT} is the total thermopower, S_p is the temperature-dependent thermopower of the ideally pure metal, and S_0 is the impurity thermopower. The far right-hand side of Eq. (67) predicts that a plot of S_{TOT} versus $1/\rho_{\mathrm{TOT}}$ should yield a straight line with slope $\rho(S_p - S_0)$ and ordinate-axis intercept S_0. The reductions in S_g shown in Fig. 29 are usually attributed to quenching of phonon drag due to impurity scattering, which drives the phonons back toward thermal equilibrium before they can "drag" the electrons along with them.

8.4 Additional Examples

The thermopower of a single crystal of a noncubic metal can be very different along different crystallographic axes, as illustrated in Fig. 30.

Straining or thinning a metal can also produce significant changes in S.

For a liquid metal, S generally has a magnitude similar to that for the solid; the tem-

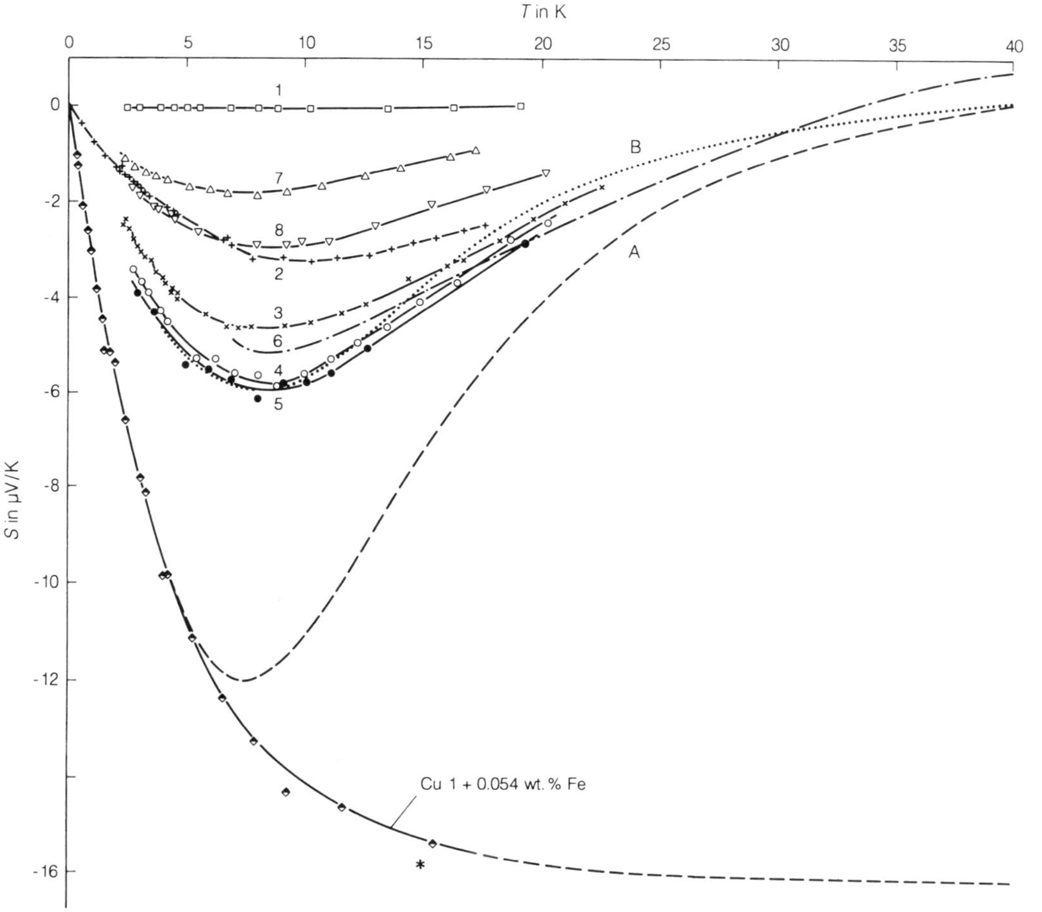

FIG. 33. The low-temperature thermopowers of various samples of copper containing very small concentrations of iron. Sample 1 is most representative of pure Cu, because the Fe is present as oxide (from Gold et al., 1960).

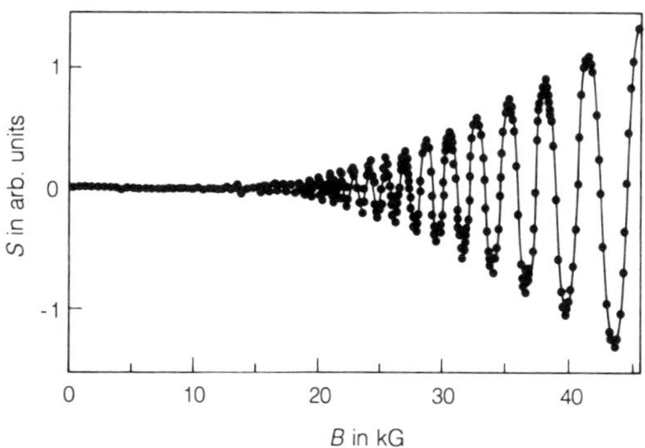

FIG. 34. Magnetic-field dependence of the thermoelectric power of Al for **B** 2° off a ⟨001⟩ direction. One unit in the vertical scale represents about 5 μV per degree (from Kesternich and Papastaikouidis, 1974).

perature-dependent component of S normally varies approximately linearly with T, and the coefficient of this linear term can have the opposite sign from that in the solid. These behaviors are illustrated in Fig. 31.

Like the resistivity (see Sec. 5.2), the thermopower undergoes characteristic changes when a ferromagnet becomes paramagnetic.

The scattering of electrons by defects and disorder in nonmagnetic amorphous metals is so large that phonon drag is completely quenched. Examples of S for several nonmagnetic amorphous metals are given in Fig. 32. The change in slope of S with increasing temperature is attributed to the disappearance with increasing temperature of many-body electron-phonon mass enhancement. (Howson and Gallagher, 1988).

When a magnetic impurity, such as Fe, is dissolved in a nonmagnetic noble metal, such as Au, a "giant" low-temperature anomaly (also called a Kondo anomaly; cf. Sec. 5.3) appears in *S*. Compare, for example, the curve of *S* for 0.054% Fe in Cu in Fig. 33 with that for nominally pure Cu (curve 1).

8.5 Magnetic-Field Effects

Application of a magnetic field to a pure metal can produce significant changes in *S*, in both the quasiclassical and quantum regimes defined in Sec. 4. Unusually large oscillations in *S* occur in single crystals of Al in the quantum regime, because of magnetic breakdown. Fig. 34 shows that at high fields these oscillations become much larger than the nonoscillatory component.

In Sec. 7, it was shown that the Hall coefficient of Al changes sign when a magnetic field is applied at low temperatures, a phenomenon attributed to a change from electronlike to holelike behavior with increasing *B*. The same behavior is displayed in *S* of dilute Al-based alloys.

Application of a magnetic field also leads to more complex thermomagnetic effects. For example, when a field **B** is present in the *z* direction, application of a temperature gradient in the *x* direction gives rise to a voltage in the *y* direction. If the sample is thermally isolated, this phenomenon is called the adiabatic Nernst-Ettingshausen (NE) effect. If the sample is immersed in a constant-temperature bath, the voltage is more nearly representative of the isothermal NE effect. Such measurements have been used to show that thermoelectric coefficients are enhanced by many-body effects.

GLOSSARY

Alloy: A general term used to describe a metal that contains two or more different components. An alloy is characterized by its composition, which defines the relative amounts of each component (usually as either an atomic or weight percentage, or as an atomic ratio). A solid alloy may be *crystalline* (with long-range spatial order), or it may be *glassy* or *amorphous* (no long-range spatial order). The term *liquid alloy* is often used to describe an alloy above its melting point.

Amorphous (or Glassy) Structures: Solid structures that have no long-range spatial periodicity (cf. **Liquids**).

Atomic Order: A description of the extent to which the distribution of atoms over a lattice is periodic or repetitive in space. Ordering may occur over a distance that is large compared to scale of interest (or coherence length of the probe used to investigate the state of order) in which case it is termed *long-range* atomic ordering, or it may occur over a shorter distance in which case it is termed local, or short-range, atomic order. The term ordering is usually reserved to describe structures with unlike nearest neighbors (e.g. ABABAB . . .) whereas clustering or precipitation is used to describe like nearest-neighbor structures (e.g. AAAABBBB . . .). Ordering is a cooperative effect that vanishes above a critical order-disorder temperature.

Brillouin Zone (BZ): Geometrical construction in wave-vector (or reciprocal) space, such that the BZ about a given reciprocal-lattice point is the region of reciprocal space closer to that point than to any other reciprocal-lattice point.

Conductance: An extrinsic property of a body that measures the rate at which charge or heat is conducted through that body. Conductance is related to conductivity by the size and shape of the body. The inverse of conductance is *resistance*.

Conductivity: An intrinsic property of a material that describes the ease with which electrical charge or heat is conducted through it. The inverse of conductivity is *resistivity*.

Electron Orbit: Path followed by an electron on the Fermi surface. Orbits traversed under the influence of a magnetic field may be "open" if the path is open-ended, or "closed" if the path traces out a closed loop.

Electron State: A state of given energy and wave vector that is available for occupation by an electron. An occupied state is a "filled state"; an empty state is a "vacant" or "unfilled" state.

Energy Band: A continuum of electron energy levels in a solid derived from the atomic energy levels of the isolated atoms. A band may overlap with an adjacent band or may be separated from it by an energy gap (also known as a band gap).

Fermi Surface (FS): A surface of constant energy in wave-vector space which, at 0 K, separates occupied electron states from empty ones.

Fermi–Dirac Distribution: A function that describes how the probability of occupation of an electron state varies with energy and temperature.

Free Electrons: Electrons that behave totally independently of any potential fields resulting from the atoms or ions in a solid, except that they are constrained to remain within the solid.

Hall Effect: Applying a magnetic field transverse to the direction of current flow in a conductor produces a "Hall" electric field transverse to both the current and the magnetic field.

Hole: A fictional "quasiparticle" concept, used to describe the motion of a vacant state that results from the collective motion of all of the filled electron states in an energy band.

Lattice: A periodic array of points in space.

Lattice Defects: Disruption of the perfect spatial periodicity of an atomic lattice caused by a missing atom (vacancy), an extra atom between lattice sites (interstitial), a missing line of atoms (dislocation), a plane of atomic misfit (grain boundary), or a free surface. Displacements of atoms away from perfectly periodic positions (due to local strain or thermal excitation) may also be classed as defects.

Magnetic Order: Same as for atomic order, except that ordering of magnetic spin directions is of concern rather than the type of atom. Parallel nearest-neighbor spin alignment is termed ferromagnetic ordering, whereas antiparallel spin alignment is termed antiferromagnetic or ferrimagnetic ordering.

Magnetoresistance: An additional contribution to the electrical resistance that results from the application of a magnetic field.

Mean Free Path: A measure of the average distance traveled by a conduction electron between scattering events.

Nearly Free Electrons: Electrons whose motion suffers a moderately small perturbation from the atomic or ionic potential fields in a solid.

Peltier Effect: The reversible absorption or emission of thermal energy at the junctions of a thermocouple when a current passes.

Relaxation Time: A measure of the average time between scattering events of a conduction electron.

Scattering Process: A process that causes an electron (or phonon) to undergo a transition from some initial state into a final state, such as interaction with another electron or with deviations in the periodicity of the lattice potential. Such deviations may result from lattice defects, from the presence of different atomic species, or from displacement of atoms away from periodic lattice sites (caused by local strain or thermal motion). The process may be *elastic* if there is no difference in energy between the initial and final states, or *inelastic* if a change in energy occurs.

Seebeck Effect: The generation of voltage by a thermocouple.

Thermocouple: Two different conducting materials connected together with one junction at a different temperature from the other.

Thermoelectric Effects: Effects involving the direct conversion of heat into electrical energy, or vice versa (see **Thermopower, Seebeck Effect, Peltier Effect, Thomson Effect**).

Thermopower (Thermoelectric Power): The voltage developed in a thermocouple per unit increase in temperature difference between the thermocouple junctions.

Thomson Effect: The reversible absorption or emission of heat produced by the passage of an electrical current through a conductor along which a temperature gradient has been imposed.

Transport Properties: Properties of a material that depend upon the transport of electrical charge or heat under the influence of external fields or temperature gradients.

Wave Vector: A vector that describes the crystal momentum of a particular mode of vibration of a lattice (i.e., a phonon) or of a particular electron state.

Works Cited

Abrikosov, A. A. (1988), *Fundamentals of the Theory of Metals*, Amsterdam: North-Holland.

Abrikosov, A. A. (1972), "Introduction to the Theory of Normal Metals," in: F. Seitz, D. Turnbull, H. Ehrenreich (Eds.), *Solid State Physics*, Supplement 12, New York: Academic Press.

Ashcroft, N., Mermin, N. D. (1976), *Solid State Physics*, New York: Holt, Rinehart and Winston.

Barnard, R. D. (1972), *Thermoelectricity in Metals and Alloys*, London: Taylor and Francis.

Bass, J. (1972), *Adv. Phys.* **21**, 431–604.

Bass, J. (1982), "Thermoelectricity," in: S. P.

Parker (Ed.), *McGraw-Hill Encyclopedia of Science and Technology*, Vol. 13, p. 295.

Bass, J. (1983), in: J. Bass, K. H. Fischer (Eds.), *Landolt-Börnstein: Numerical Data and Functional Relationships in Science and Technology*, New Series, Group 3, Vol. 15a, *Electronic Transport Phenomena*, Berlin: Springer.

Bass, J. (1985), in: J. Bass, J. S. Dugdale, C. L. Foiles, A. Myers (Eds.), *Landolt-Börnstein: Numerical Data and Functional Relationships in Science and Technology*, New Series, Group 3, Vol. 15b, *Electronic Transport Phenomena*, Berlin: Springer.

Benedek, R., Baratoff, A. (1971), *J. Phys. Chem. Solids* **32**, 1015–1024.

Bloch, F. (1930), *Z. Phys.* **59**, 208–214.

Borchi, G., de Gennaro, S. (1970), *Phys. Lett.* **32A**, 301–302.

Brown, R. A. (1977), *J. Phys. F. Metal Phys.* **7**, 1477–1488.

Brown, R. A. (1982), *Can. J. Phys.* **60**, 766–778.

Chikazumi, S. (1964), *Physics of Magnetism*, New York: John Wiley & Sons.

Coles, B. R. Taylor, J. C. (1962), *Proc. R. Soc. A* **267**, 139–145.

Craig, R., Crisp, S. (1978), in F. J. Blatt, P. A. Schroeder (Eds.), *Thermoelectricity in Metallic Conductors*, New York: Plenum Press. p. 51.

Dugdale, J. S. (1977), *The Electrical Properties of Metals and Alloys*, London: Arnold.

Englert, E. (1932), *Ann. Phys.* **14**, 589–613.

Fert, A., Campbell, I. A. (1976), *J. Phys. F. Met. Phys.* **6**, 849–871.

Fletcher, R. (1977), *Solid State Commun.* **21**, 1139.

Foiles, C. L. (1985), in: J. Bass, J. S. Dugdale, C. L. Foiles, A. Myers (Eds.), *Landolt-Börnstein: Numerical Data and Functional Relationships in Science and Technology*, New Series, Group 3, Vol. 15b, *Electronic Transport Phenomena*, Berlin: Springer.

Forsvoll, K., Holwech, I. (1964), *Phil. Mag.* **9**, 435–450.

Gallagher, B. L., Greig, D. (1982), *J. Phys. F.* **12**, 1721–1741.

Gold, A. V., MacDonald, D. K. C., Pearson, W. B., Templeton, I. M. (1960), *Phil. Mag.* **5**, 765–786.

Greene, R. F. (1964), *Surface Sci.* **2**, 101–113.

Grüneisen, E. (1933), *Ann. Phys.* **16**, 530–540.

Hashin, Z., Shtrikman, S. (1962), *J. Appl. Phys.* **33**, 3125–3131.

Howe, R. A., Enderby, J. E. (1967), *Philos. Mag.* **16**, 467–476.

Howson, M. A., Gallagher, B. L. (1988), *Phys. Rep.* **170**, 265–324.

Hurd, C. M. (1972), *The Hall Effect in Metals and Alloys*, New York: Plenum Press.

Johansson, C. H., Linde, J. O. (1936), *Ann. Phys.* **25**, 1–48.

Kesternich, W., Papastaikouidis, C. (1974), *Phys. Stat. Sol.* **B64**, K41–K43.

Kunzi, H. U., Guntherodt, H.-J. (1980). in C. J. Chien, C. R. Westgate (Eds.), *The Hall Effect and Its Applications*, New York: Plenum Press.

Linde, J. O. (1932), *Ann. Phys. (Leipzig)* **15**, 219–248.

Longo, J. T., Schroeder, P. A., Sellmyer, D. J. (1969), *Phys. Rev.* **182**, 658–670.

Meaden, G. T. (1965), *Electrical Resistance of Metals*, London: Heywood Books.

Mizutani, U. (1983), *Prog. Mater. Sci.* **28**, 98–228.

Naugle, D. G. (1984), *J. Phys. Chem. Solids* **45**, 367–388.

Olsen, J. L. (1962), *Electronic Transport in Metals*, New York: John Wiley & Sons.

Osamura, K. O., Hiraoka, Y., Murakami, V. (1973), *Philos. Mag.* **28**, 809–825.

Porter, D. A., Easterling, K. E. (1981), *Phase Transformations in Metals and Alloys*, New York: Van Nostrand Reinhold.

Rossiter, P. L. (1987), *The Electrical Resistivity of Metals and Alloys*, Cambridge, U.K.: Cambridge University Press.

Rowe, V. A., Schroeder, P. A. (1970), *J. Phys. Chem. Solids* **31**, 1–8.

Schröder, K. (1983), *CRC Handbook of Electrical Resistivity of Binary Metallic Alloys*, Boca Raton, FL: CRC Press, p. 1.

Simmons, R. O., Balluffi, R. W. (1960), *Phys. Rev.* **117**, 62–68.

Sun, P. H., Ohring, M. (1976), *J. Appl. Phys.* **47**, 478–485.

Warren, B. E. (1969), *X-Ray Diffraction*, Reading, MA: Addison-Wesley.

Wilson, A. H. (1937), *Proc. Cambridge Philos. Soc.* **33**, 371–379.

Ziman, J. M. (1960), *Electrons and Phonons*, Cambridge: Cambridge University Press.

Ziman, J. M. (1972), *Electrons and Phonons*, 2nd ed., Oxford: Clarendon Press.

Further Reading

Abrikosov, A. A. (1988), *Fundamentals of the Theory of Metals*, Amsterdam: North-Holland Physics Publishing.

Ashcroft, N., Mermin, N. D. (1976), *Solid State Physics*, Philadelphia: Holt, Rinehart, and Winston.

Barnard, R. D. (1976), *Thermoelectricity in Metals and Alloys*, London: Taylor and Francis.

Blatt, F. J. (1968), *Physics of Electronic Conduction in Solids*, New York: McGraw-Hill.

Blatt, F. J., Schroeder, P. A. Foiles, C. L., Greig, D. (1976), *Thermoelectric Power of Metals*, New York: Plenum Press.

Chien, C. J., Westgate, C. R. (Eds.) (1980), *The Hall Effect and Its Applications*, New York: Plenum Press.

Hurd, C. M. (1972), *The Hall Effect in Metals and Alloys*, New York: Plenum Press.

Kittel, C. (1986), *Introduction to Solid State Physics*, 6th ed. New York: John Wiley & Sons.

Mott, N. F., Jones, H. (1936), *The Theory of the Properties of Metals and Alloys*, Oxford: Clarendon Press.

Rossiter, P. L. (1987), *The Electrical Resistivity of Metals and Alloys*, Cambridge: Cambridge University Press.

Ziman, J. M. (1965), *Principles of the Theory of Solids*, Cambridge: Cambridge University Press.

Ziman, J. M. (1972), *Electrons and Phonons*, 2nd ed. Cambridge: Cambridge University Press.

METALS AND ALLOYS, STRUCTURE OF

FERESHTEH EBRAHIMI AND MICHAEL J. KAUFMAN, *Materials Science and Engineering Department, University of Florida, Gainesville, Florida, U.S.A.*

INTRODUCTION

Metals are generally distinguished from nonmetals by their high electrical conductivity, which increases with decreasing temperature. Most of the elements in the periodic table are metals, with nonmetals occupying the right upper corner of the table (see Fig. 1). Metals, like other elements, may exist in gas, liquid, or solid state; this article is concerned mainly with the last state.

When metallic atoms are gathered together, the most loosely bound electrons (valence electrons) become separated from their parent atoms and wander almost freely through the entire body of the material. These free electrons are responsible for the electrical, optical, and magnetic properties and partially for the thermal conductivity of metals. The cohesion between metallic atoms is controlled mainly by the electrostatic attraction between the ionic (positively charged) cores and the free electrons, and, therefore, the energy of a group of metallic atoms gathered together is not very sensitive to the positions of ionic cores, and there are no directional atomic bonds. Consequently, metallic atoms tend to gather in close-packed structures, making it much easier to shear than to compress them. Hence, metals tend to be ductile and may be formed into complicated shapes without failure. Normally, in solid metals, the atoms arrange themselves in a periodic array known as the crystal structure. However, under nonequilibrium conditions, such as those encountered in fast cooling from the liquid state, condensation from the vapor state, or bombardment with high-energy particles, the atoms may be arranged in an irregular manner (amorphous structure) that resembles the structure of liquids.

Metals are mixed with other elements (metals and nonmetals) to form alloys, in order to improve properties such as strength and environmental stability. A system composed of two or more elements may consist of a variety of phases—a phase being defined as a structurally homogeneous arrangement of atoms. Most metals mix homogeneously in the gaseous and liquid states, in contrast to the solid state, where complete mixing is achieved at the atomic level only in a few binary (two-component) systems.

In this article, we describe the structure of metals and alloys in terms of the crystal structure—the arrangement of atoms in crystals—and the microstructure—the ar-

3-527-28132-0/94/$5.00 + .50

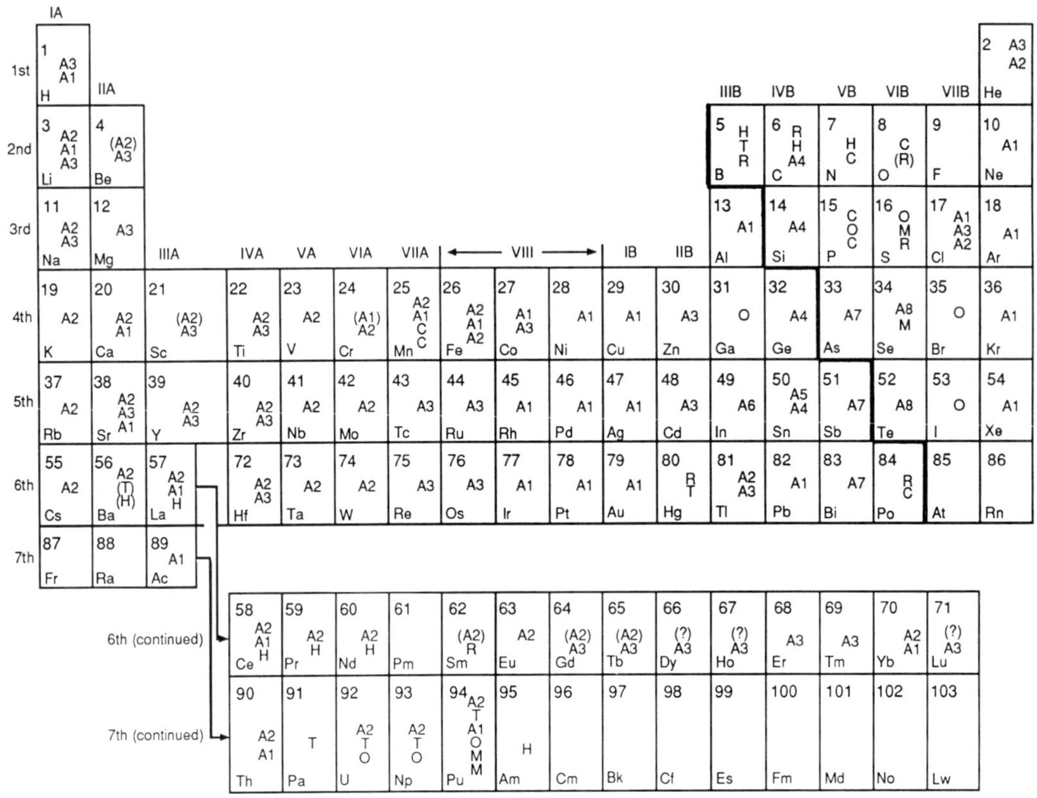

	IA	IIA	IIIA	IVA	VA	VIA	VIIA	VIII	VIII	VIII	IB	IIB	IIIB	IVB	VB	VIB	VIIB	
1st	1 H A3 A1																	2 He A3 A2
2nd	3 Li A2 A1 A3	4 Be (A2) A3											5 B H T R	6 C R H A4	7 N H C	8 O C (R)	9 F	10 Ne A1
3rd	11 Na A2 A3	12 Mg A3											13 Al A1	14 Si A4	15 P C O C	16 S O M R	17 Cl A1 A3 A2	18 Ar A1
4th	19 K A2	20 Ca A2 A1	21 Sc (A2) A3	22 Ti A2 A3	23 V A2	24 Cr (A1) A2	25 Mn A2 A1 C C	26 Fe A2 A1 A2	27 Co A1 A3	28 Ni A1	29 Cu A1	30 Zn A3	31 Ga O	32 Ge A4	33 As A7	34 Se A8 M	35 Br O	36 Kr A1
5th	37 Rb A2	38 Sr A2 A3 A1	39 Y A2 A3	40 Zr A2 A3	41 Nb A2	42 Mo A2	43 Tc A3	44 Ru A3	45 Rh A1	46 Pd A1	47 Ag A1	48 Cd A3	49 In A6	50 Sn A5 A4	51 Sb A7	52 Te A8	53 I O	54 Xe A1
6th	55 Cs A2	56 Ba A2 (T) (H)	57 La A2 A1 H	72 Hf A2 A3	73 Ta A2	74 W A2	75 Re A3	76 Os A3	77 Ir A1	78 Pt A1	79 Au A1	80 Hg R T	81 Tl A2 A3	82 Pb A1	83 Bi A7	84 Po R C	85 At	86 Rn
7th	87 Fr	88 Ra	89 Ac A1															

6th (continued)	58 Ce A2 A1 H	59 Pr A2 H	60 Nd A2 H	61 Pm	62 Sm (A2) R	63 Eu A2	64 Gd (A2) A3	65 Tb (A2) A3	66 Dy (?) A3	67 Ho (?) A3	68 Er A3	69 Tm A3	70 Yb A2 A1	71 Lu (?) A3
7th (continued)	90 Th A2 A1	91 Pa T	92 U A2 T O	93 Np A2 T O	94 Pu A2 T A1 O M M	95 Am H	96 Cm	97 Bk	98 Cf	99 Es	100 Fm	101 Md	102 No	103 Lw

A1 = f.c.c.; A2 = b.c.c.; A3 = c.p.h.; A4 = diamond cubic; A5 = b.c.t.; A6 = f.c.t.; A7 = rhombohedral; A8 = trigonal; H = hexagonal (usually ABAC · · · close-packed); R = rhombohedral; O = orthorhombic; C = complex cubic; T = Tetragonal; M = monoclinic; ( ) uncertain.

FIG. 1. Periodic table of the elements (Barrett and Massalski, 1980).

rangement of crystals and phases in polycrystals; we conclude with a brief description of both amorphous and quasicrystalline structures.

1. CRYSTAL STRUCTURE

The crystal structure of metals and alloys can be identified and characterized using various diffraction techniques (x-ray, neutron, and electron). As mentioned above, atoms of elemental metals tend to be packed closely, since there is no significant interaction between their ionic cores. However, additions of alloying elements may modify the electron distribution significantly and add covalent and/or ionic characters to the forces holding the atoms together. This modification of the metallic binding may result in the ordering of the atoms and/or the formation of complex crystal structures as discussed in Secs. 1.2–1.4.

1.1 Elemental Metals

The crystal structure of metals is given in the periodic table presented in Figure 1, which clearly shows that the majority of metals crystallize in relatively close-packed structures, namely face-centered cubic (fcc), hexagonal close-packed (hcp), and the less closely packed body-centered cubic (bcc) structures. Elements of the same group tend to have similar structures; for example, the alkali metals (Group IA), Li, Na, K, Rb, and Cs, are bcc, and Group IB metals, Cu, Ag, and Au, are fcc. In the group of metals known as transition metals (Groups IV, V, VI, VII, and VIII elements), the *s* energy states are filled with electrons before the *d* energy states are filled. At the usual interatomic spacing of these metals, the atomic *d* orbitals do not overlap significantly, and a tight-binding approach based on the density-of-states distribution seems to explain the types of crystal structures along the three transition-metal rows (Cottrell, 1991).

In the case of the hcp crystals, the c/a ratio (c and a are the atomic lattice parameters) depends on the electronic structure. For example, both Group IVA, Ti, Zr, and Hf, and Group IIB, Zn and Cd, have the hcp structure; in the latter group, the outer electron shells are nearly filled, and the structure is looser and, hence, has a larger c/a ratio. The large c/a ratio of Zn and Cd limits the extent of plastic deformation to slip in the basal planes, and, consequently, through a lack of a large enough number of slip systems, polycrystalline forms of these metals are relatively brittle.

Since the cohesion of atoms in metals is not very sensitive to the positions of the ionic cores, metals often exist in more than one crystal structure. For example, the Group IVA elements are hcp at low temperatures and transform to the more loosely packed bcc structure, which has higher vibrational entropy, at elevated temperatures. This solid-state phase change is known as an allotropic transformation, and it may also be induced by a change in pressure. In the case of iron, the volume decrease due to a change from the ferromagnetic to the paramagnetic state overrides the temperature effect, and iron transforms from bcc to fcc with increasing temperature. On further heating, however, it transforms back to the bcc structure.

1.2 Solid Solutions and Superlattices

If the crystal structure of a metal (solvent) remains unchanged by the addition of another element (solute), then the mixture is called a solid solution. There are two types of solid solutions: substitutional and interstitial. Figure 2 shows the positions of interstitial and substitutional atoms in a lattice.

The interstitial solute atoms, since they are squeezed in between the regular lattice atoms, need to be considerably smaller than the solvent atoms and are limited to a few elements, including H, B, C, N, and O. Since interstitials strain close-packed crystals greatly, their solubility in metals is limited and depends on the crystal structure. In general, the interstitial solubility is much higher in transition metals than in other metals. Although the bcc structure is a less closely packed structure than fcc, the free volume is distributed as smaller spaces in the former structure, and, hence, bcc phases tend to have a lower solubility for interstitial atoms. The higher solubility of carbon in fcc iron compared with bcc iron is the basis for achieving a wide variety of microstructures and properties in steels (see STEEL).

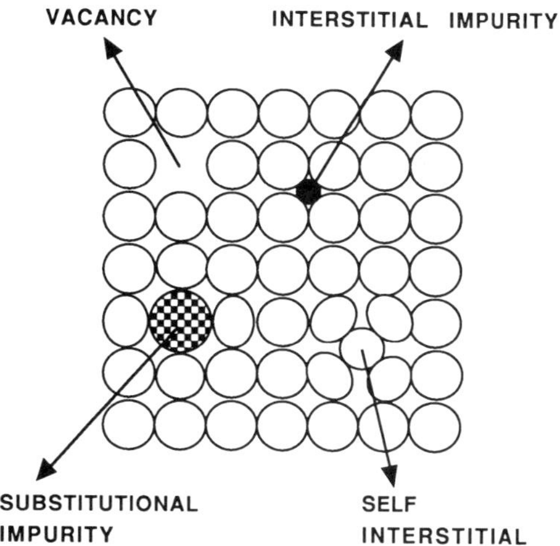

FIG. 2. Schematic showing various types of point defects in disordered solid solutions.

Metals may dissolve extensively in each other by forming substitutional solid solutions. The extent of the solubility is governed by three empirical rules, known as Hume-Rothery rules:

1. the atomic size difference must be less than 15% for extensive solubility;
2. the difference in electronegativity (or electropositivity) of the atoms must be small (otherwise they prefer to form compounds); and
3. a metal of lower valency tends to dissolve less in a metal of higher valency than vice versa.

The solubility range also depends on temperature; for example, Nb-Ta, Nb-Ti, and Cu-Au systems show complete solubility in the solid state at high temperatures. While Nb and Ta remain completely soluble at low temperatures, the bcc structure of Nb is not favored by Ti because of the polymorphism of the latter element, and a new hcp solid solution evolves at lower temperatures. In the case of the Cu-Au system, because of the attraction between copper and gold, as the temperature drops, the atoms tend to order, thereby increasing the number of copper-gold bonds and creating ordered solid solutions (superlattices). The lattice sites chosen by the Cu and Au atoms depend on the composition. For example, as shown in Fig. 3(a), at about 25% gold, Cu atoms occupy the face centers of the fcc lattice, and Au atoms go to the cube corners, creating an ordered structure known as the Cu_3Au-I or $L1_2$ (*Strukturbericht* designation) superlattice. In this structure, the copper atoms have eight nearest neighbors of similar kind and four of the opposite kind, whereas the gold atoms have twelve nearest neighbors of the opposite kind. Because of the ordering, this structure is no longer fcc, but it can be generated by repeating the small four-atom unit marked in Fig. 3(a) on a simple cubic lattice. When the composition is close to 50 at. % gold, the copper and gold atoms occupy alternate planes, as shown in Fig. 3(b). Because of the difference in the ionic radii of copper and gold, and a modification of the cohesion forces, the resulting structure is slightly tetragonal and is known as $L1_0$ or CuAu-I–type superlattice.

Since the majority of metals are fcc, bcc, or hcp, the common superlattices are derived from these structures. Table 1 presents a list of common superlattices and typical examples of each kind.

Many ordered solid solutions go through an order-disorder transition with temperature. The ordering is considered to be a first-order reaction whenever it occurs by the nucleation of highly ordered regions that grow into a disordered matrix. When the ordered regions touch each other, the atomic ordering may be in or out of phase; when it is out-of-phase, it creates a boundary, known as an antiphase boundary (APB), across which the crystal structure and crystal orientation remain unchanged, but atoms have the "wrong" nearest neighbors. Figure 4 presents schematics of the ordered B2 lattice structure containing an APB, along with actual boundaries as revealed by transmission electron microscopy (TEM).

Table 1. A list of common superlattices that occur in ordered solid solutions and intermetallics.

Structure Designation	Examples
$L1_2$	Cu_3Au, Zr_3Al, Co_3V, Ni_3Fe, Ni_3Al, Zn_3Ti
DO_3	Fe_3Al, Li_3Bi, Cu_3Sb, Al_3Cu
DO_{19}	Mg_3Cd, Fe_3Sn, Co_3Mo, Ti_3Al, Co_3V
B2	CuZn, NiAl, AuCd, NiZn, LiTl
$L1_0$	CuAu, TiAl, CoPt, CuTi, FePt, NiZn

1.3 Short-Range Ordering and Clustering

The formation of superlattices as explained above is known as long-range ordering and occurs below some critical temper-

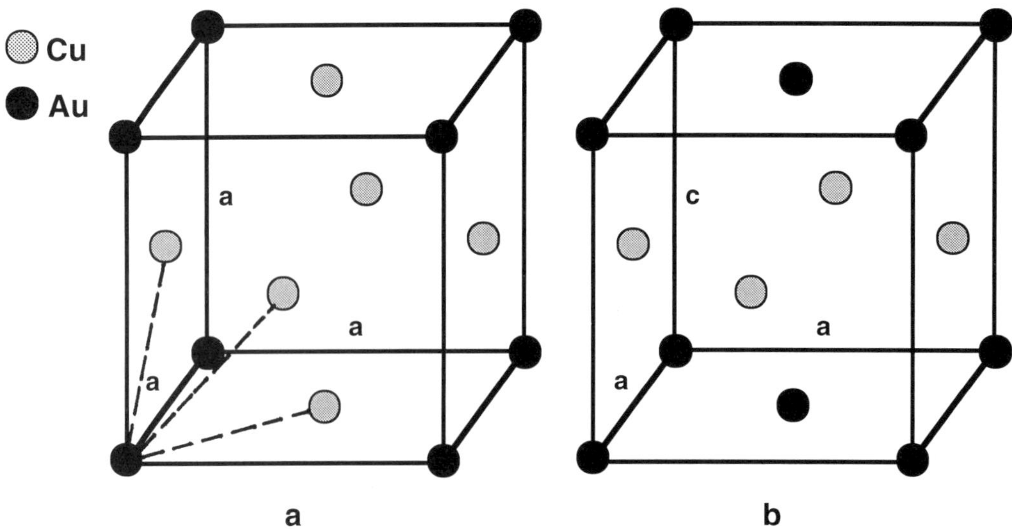

FIG. 3. The structures of (a) Cu_3Au and (b) CuAu superlattices.

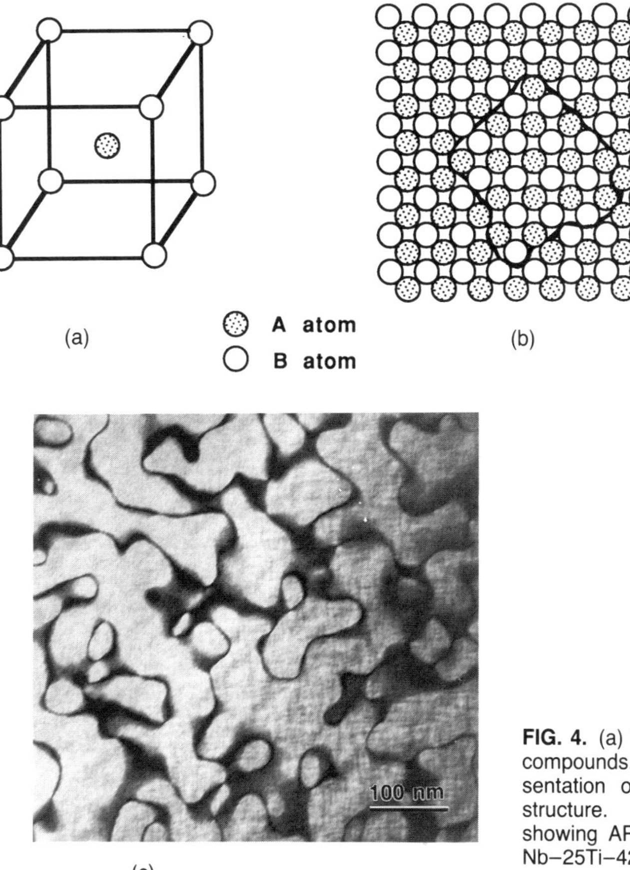

FIG. 4. (a) The B2 structure of AB compounds. (b) Schematic representation of an APB in the B2 structure. (c) TEM micrograph showing APB's (black areas) in a Nb–25Ti–42Al alloy with B2 structure.

ature. The degree of ordering in an alloy is measured as the proportion of atoms occupying the correct lattice site. Short-range ordering, in contrast to long-range ordering, is measured by the extent to which an atom is surrounded by unlike neighbors, with no reference to the lattice site. Short-range ordering has been observed in solid solutions just above the critical (long-range) order-disorder temperature and in alloys quenched from above this temperature. The opposite of short-range ordering is clustering, which occurs in solid solutions that prefer like atoms to unlike atoms as nearest neighbors. Clustering is usually observed in systems with limited solubility and as a precursor to precipitation of a second phase. Both short-range ordering and clustering can cause tremendous strengthening in alloys. For example, the enhanced clustering of copper and phosphorus in iron caused by neutron irradiation is responsible for radiation embrittlement in water reactor pressure vessels (Ebrahimi *et al.*, 1988). On account of the small size of clusters and short-range–ordered regions, field-ion microscopy (FIM) and small-angle neutron and x-ray scattering (SANS and SAXS) techniques are conventionally employed to study these phenomena. More recently, high-resolution transmission electron microscopy techniques have been successfully used to reveal such fine structures.

1.4 Metallic Intermediate Phases

Intermediate phases form at compositions beyond the solid solubility limits in systems that do not show complete mixing, even at high temperatures. These phases can be disordered and similar to solid solutions, they may become ordered upon cooling, or they may have such a strong ordering tendency that they remain ordered up to their melting point or to temperatures at which they transform to other phases. Metallic intermediate phases (intermetallics) are phases in which the atomic bonding is intermediate between metallic and covalent or metallic and ionic. In cases where the ionic or covalent component of the atomic bonding is significant, the intermediate phases behave like chemical compounds; i.e., they have a highly ordered structure with a limited or negligible solubility range. At the outer limits are "ionic compounds," such as NaCl and CaF_2, which have mainly ionic bonding and low conductivities, and "covalent compounds," such as GaAs and InP, which have mainly covalent bonding and are semiconductors.

When there is a large difference between the atomic sizes of the elements, the ordering and the limited solubility may be induced by a "size-factor" effect; examples include interstitial compounds such as carbides, nitrides, borides, and hydrides, and substitutional compounds such as Laves and sigma phases. If the size effect is the dominant factor governing the structure, then the intermediate phases composed of atoms of similar atomic-size ratios should have similar crystal structures and show significant mutual solid solubilities. Examples include complete solid solubility between the carbides of V, Ti, Nb, Ta, and Zr, and the formation of carbonitrides when both carbon and nitrogen are mixed with these elements (Barrett and Massalski, 1980).

Another group of intermediate phases involves the so-called "electron compounds," which occur at specific electron-to-atom ratios (Barrett and Massalski, 1980). Examples of this group include the α, β, γ, and ϵ phases that form in alloys based on Cu, Ag, and Au mixed with Group B metals. These phases have a relatively large solubility range and may become ordered at low temperatures.

There is a multitude of crystal structures that are formed by combining various elements, and it is beyond the scope of this article to cover them all. In general, the interplay of atomic size, electronic structure, and temperature must be considered, and easily categorized compounds are merely those in which one of these factors is dominant. Many intermetallics behave like metals i.e., they are good conductors and reflect light, while the bond directionality results in high melting points and large elastic moduli, which make them possible candidates for high-temperature structural applications. For example, aluminides such as Ni aluminides (Ni_3Al, NiAl), Ti aluminides (Ti_3Al, TiAl, $TiAl_3$), and Nb aluminides (Nb_3Al, Nb_2Al, $NbAl_3$) are being considered for high-temperature structural applications such as turbine blades and vanes in next-generation aircraft engines (Liu and Stiegler, 1984).

2. DEFECTS IN CRYSTALLINE STRUCTURES

Metal crystals, like other crystals, contain defects. Crystal defects are usually characterized as point, line, surface, and volume defects (see POINT DEFECTS AND OTHER IMPERFECTIONS IN CRYSTALS).

The primary point defects are vacancies, self-interstitial atoms, and foreign atoms (see Fig. 2). Point defects are the only thermodynamically stable defects, and may sometimes further reduce the energy of the system by coming together and forming extended defect complexes. Point defects, particularly impurities, affect the properties of metals considerably. For example, the shear yield stress of copper is increased by 45% when the purity is decreased from 99.999 to 99.98% (McLean, 1962). Likewise, the high concentration of vacancies produced by quenching aluminum from high temperatures can result in a fivefold increase of its shear yield strength (Smallman, 1985).

In ordered structures, the substitution on a lattice site of a wrong type of atom creates an "antisite" defect or substitutional disorder. In strongly ionic (or covalent) compounds, point defects should ionically (or electronically) compensate each other. For example, the absence of a positively charged ion from a lattice site should be compensated by the absence of a negatively charged ion (Schottky defect) or the existence of a positive ion in an interstitial site (Frenkel defect).

Dislocations are line defects, which are usually produced in metallic crystals by plastic deformation. Ductile polycrystalline metals and alloys are used extensively in the wrought form and may contain a dislocation density of about 10^8 cm/cm^3 in a completely annealed state and a dislocation density of up to 10^{12} cm/cm^3 in a heavily cold-worked condition. Cast structures tend to have lower dislocation densities than wrought products.

Undeformed single crystals may contain a much lower density of dislocations; in the extreme case of metallic whiskers, the dislocation density can be as low as 10^3 cm/cm^3. Usually, an increase in the dislocation density results in strengthening of metals, except when the dislocation density is initially very low, in which case the contribution of the dislocation nucleation process to deformation—a harder process than dislocation motion—results in an inverse relationship. The most widely used techniques for the observation of dislocations are TEM and etch-pitting, examples of which are shown in Fig. 5.

Surface and volume defects, similar to dislocations, are produced during the solidification and thermomechanical treatment of metals (see MATERIALS TREATMENT). Examples of surface defects are grain boundaries, interfaces between phases, twin boundaries, antiphase boundaries, and stacking faults. All surface defects, except for highly disordered grain and interfacial boundaries, can be constructed by networks of dislocations. Indeed, twin boundaries, antiphase boundaries, and stacking faults can be generated by the motion of dislocations during plastic deformation. Cold working produces an increase in the number of grain and interfacial boundaries, which may act as both sources and sinks for dislocations. Voids are the most common type of volume defects. They may form as solidification defects, or develop during plastic deformation or by diffusion processes during high-temperature deformation.

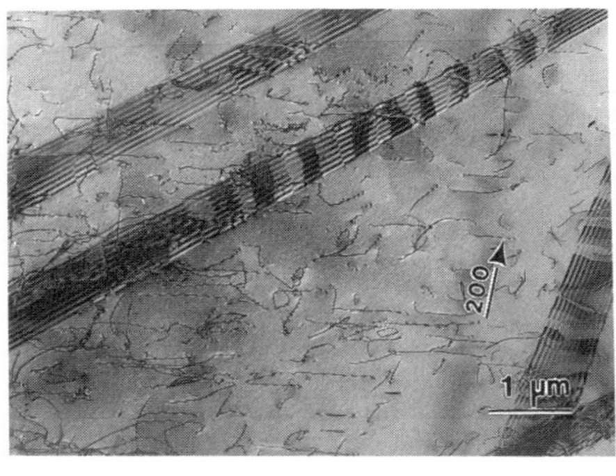

(a)

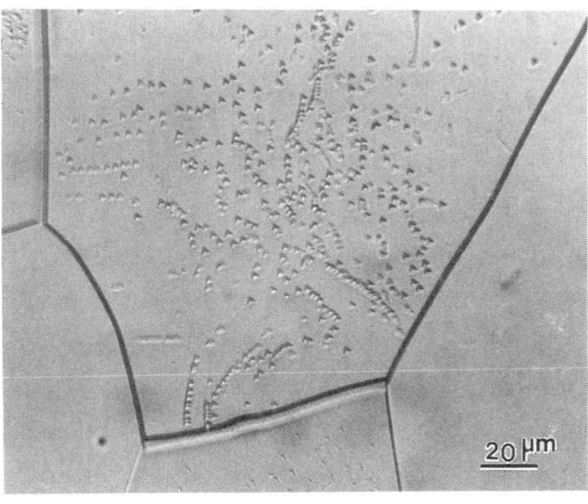

(b)

FIG. 5. Dislocations as observed with TEM and etch-pitting techniques. (a) TEM micrograph showing dislocations in a TiAl alloy. (b) Optical micrograph showing dislocation etch pits in pure niobium.

3. MICROSTRUCTURE

Microstructure is the term that materials scientists use to describe the distribution, shape, and size of phases, grains, etc. in a given material. It is important in view of the fact that the mechanical properties of any material depend strongly on microstructure. Consequently, it is necessary to understand how microstructures evolve during processing and/or application and how to use this understanding to "tailor" the microstructure and thereby achieve desirable properties.

There are numerous metallurgical processes used to manipulate the microstructure of metals and alloys. The most common processes, and those that are discussed here, include solidification (liquid to solid) and various solid-state processing schemes such as heat treating, mechanical working, and combinations of these (see MATERIALS TREATMENT). Consequently, it is important to understand the structure of liquids and solids, their similarities and differences, and the factors that influence their transformations during both types of processing.

One of the most useful tools available to the materials scientist is the phase diagram (see PHASE EQUILIBRIA). Phase diagrams are graphical representations of the phases that exist as a function of composition and temperature (most are published for pressures of 1 atmos) when the system is in equilibrium. The phase diagram for the two-component (or binary) Ag-Cu alloy system is shown in Fig. 6 and will be referred to in the following sections in order to describe the microstructural features of interest and how they might be

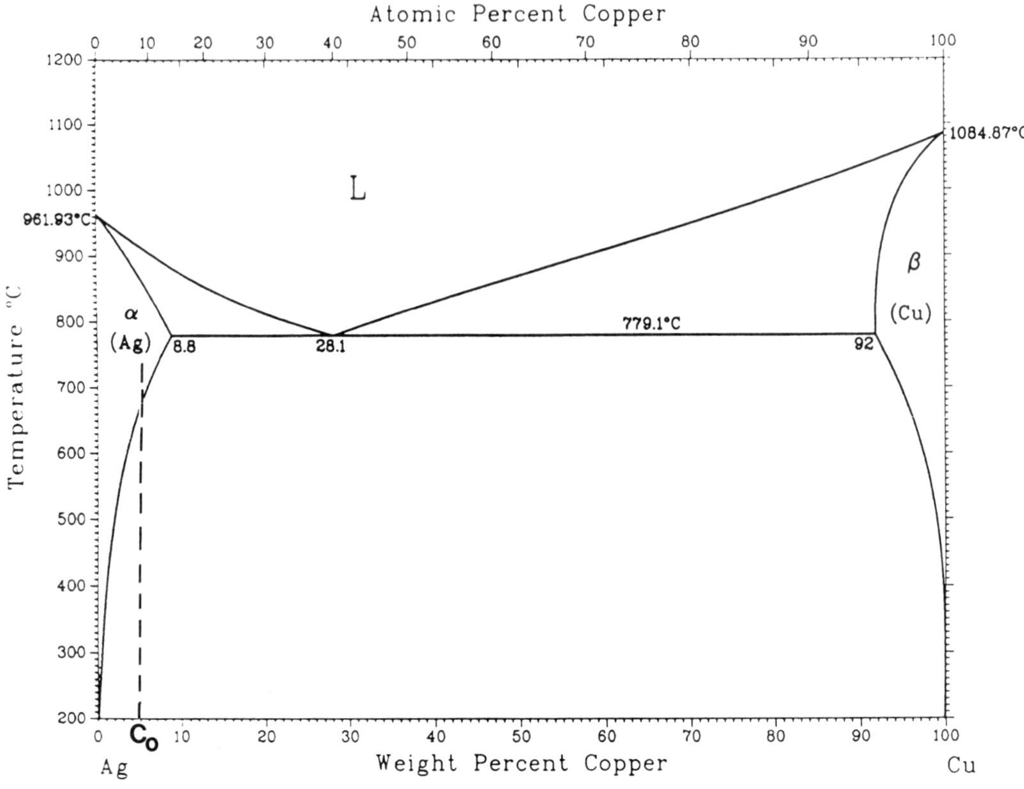

FIG. 6. Ag-Cu phase diagram (ASM Handbook, 1992).

controlled through composition and temperature variations. For simplicity, only binary systems are considered in the following discussion, although it should be recognized that more complex, multicomponent phase diagrams and alloys are frequently encountered in practice.

3.1 Liquid Alloys

Since most commercial metallic objects are formed from a liquid phase at some point during their evolution, it is important to discuss the structure of the liquid state in some detail (see LIQUIDS, SIMPLE, STRUCTURE OF). First of all, liquid alloys of practical importance occur as single homogeneous phases. Unlike solid crystalline phases, where the arrays of atoms have long-range periodicity, or gases, where the atoms exhibit a complete lack of periodicity, liquid alloys tend to have "intermediate structures," in which the atoms display some degree of local short-range periodicity. This implies that the atoms, over short distances, are arranged in a fashion similar to that characteristic of the solid phase. However, because of the presence of a large number of structural defects, the long-range periodicity characteristic of the crystalline state is not achieved. The volume of the liquid phase is typically within a few percent of that of the solid crystalline phase: it may be larger, as in the case of close-packed solids with high coordination numbers, or smaller, as in the case of less-close-packed solids with lower coordination numbers.

Although the exact nature of the structural defects in liquids remains unknown, it is believed that these defects are responsible for many of the properties of liquid metals. For example, the diffusivity of atoms in the liquid state just above the melting point is typically several orders of magnitude larger than that in the solid state just below the melting point. This implies that the liquid is constantly changing such that the local order existing at any one position in space changes continuously with time. Because of this feature, liquids cannot support shear stresses even when they are very small in magnitude. Finally, it should be mentioned that most liquid metals tend to have similar properties, indicating that the properties are controlled primarily by the structural defects rather than the interatomic bond energies. This is in direct contrast to the properties of metallic solids, which depend strongly on the bonding and crystal structures that exist and less on structural defects.

3.2 Solid Alloys

3.2.1 Cast Structures The microstructures that develop when a metal or alloy is cooled from above to below its melting point depend on such things as composition, cooling rate, diffusivity of the various species in the liquid and solid phases, and the nature of the phase boundaries for the particular alloy under investigation. For pure metals, the structure will typically consist of columnar grains that are oriented in a direction parallel to the direction in which the heat is extracted. For alloys, the situation becomes considerably more complex. Typically, alloys will solidify in either a cellular or dendritic (treelike) fashion, with compositional variations from the center to the periphery of the cells or dendrites. These compositional differences are often undesirable and have to be removed (by some sort of homogenization treatment) prior to application. Such homogenization treatments are frequently achieved by reheating the alloys to elevated temperatures sufficient for the atoms to move (diffuse) from regions of higher concentration to regions of lower concentration. Whenever possible, this type of treatment is performed by heating the alloy of interest into a single-phase field (e.g., the α field in Fig. 6). If the elements involved are "slow diffusers," and/or the scale of the segregation is large, the times required for homogenization may become impractical. In such cases, the segregation is typically removed by mechanically working the alloy. Although there are a variety of methods for achieving this goal, these will not be discussed here, as this topic is beyond the scope of this article.

In addition to the cellular and dendritic structures described above, some alloys can be solidified into two-phase structures. The most common two-phase structures are known as eutectics and occur when two solid phases having different compositions grow "cooperatively" from a parent liquid (liq $\rightarrow \alpha + \beta$ in Fig. 6). The thicknesses and amounts of the two phases are strong functions of their compositions relative to the initial liquid and the imposed cooling rate. In some alloys, the mechanical properties can be enhanced by aligning the eutectic phases; this is achieved by cooling the liquid directionally and is known as directional solidification. An example of a directionally solidified eutectic is shown in Fig. 7.

3.2.2 Wrought Structures The microstructural features of the wrought alloys that are most frequently referred to include grain size, shape, texture, second-phase distribution, and dislocation density (or amount of

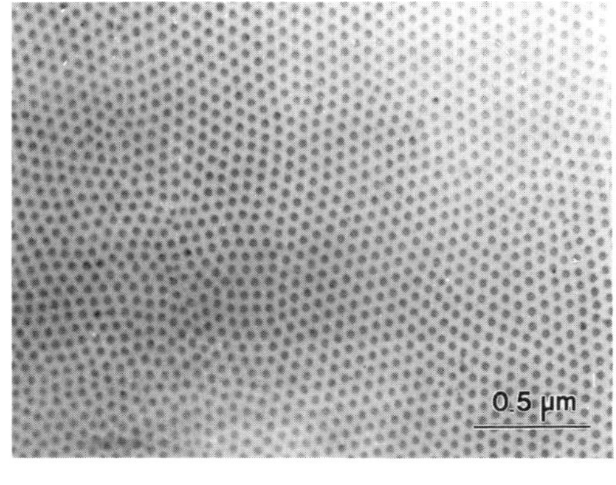

(a)

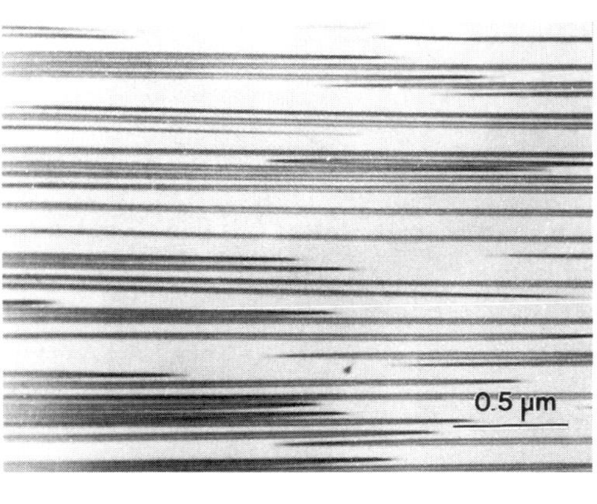

(b)

FIG. 7. Optical micrographs showing the unidirectionally solidified structure of MnSb-Sb eutectic: (a) transverse section, (b) longitudinal section (Cole, 1971).

cold work). Many of the properties of metals and alloys depend strongly on these features. Figure 8 depicts two examples of grain structures—equiaxed and elongated. As might be expected, the mechanical properties of the alloys are a strong function of the grain distribution. In addition, the properties of the structures with elongated grains are anisotropic; i.e., they vary with test direction relative to the grain boundaries. Another microstructural feature that is commonly observed in the wrought structure of fcc metals is annealing twins, which can be seen in the microstructures shown in Fig. 8.

In some instances, cold working of ductile two-phase cast structures results in very strongly elongated structures, which resemble those found in conventional fibrous or lamellar composites. Since very small (submicron) interlamellar or interfibrous spacing can be produced in heavily deformed alloys, they are frequently referred to as "nanocomposites" (see, e.g., Fig. 9).

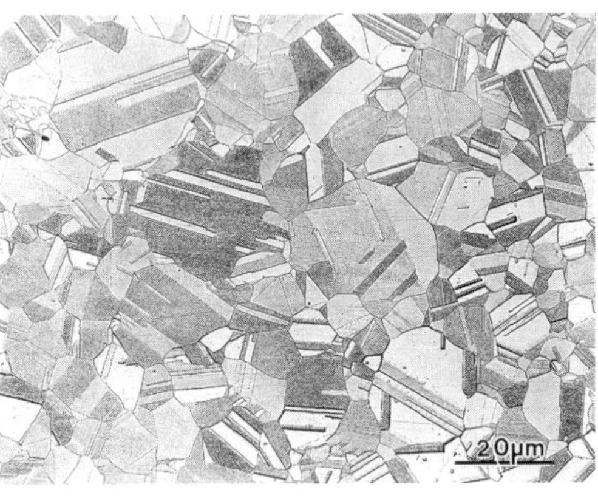

(a)

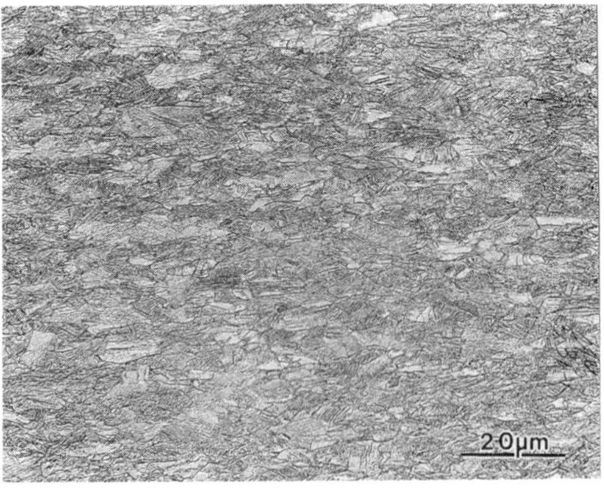

(b)

FIG. 8. Optical micrographs showing (a) equiaxed and (b) elongated grain structures in recrystallized and cold-rolled red-brass (Cu–15% Zn alloy). The features in (a) with straight boundaries are annealing twins.

3.2.3 Solid-State Phase Transformations

3.2.3.1 Precipitation. Many alloys are strengthened by second phases, which may be introduced by a variety of methods. Probably the most commonly employed method is precipitation hardening, in which the alloy of interest is heat treated in a manner that produces a fine distribution of submicron second-phase particles that are effective in inhibiting dislocation movement (see, e.g., Fig. 10). Typically, this is achieved by homogenizing the alloy of interest in a single-phase field prior to cooling and then reheating somewhere in the two-phase field in order to cause formation (i.e., precipitation) of a second phase. Referring to the phase diagram in Fig. 6, an alloy of composition C_0 might be heated into the single-phase α field, quenched to room temperature, and then reheated to an intermediate temperature in the two-phase $\alpha + \beta$ field. During this treatment, the fine β precipitates can form in the α matrix, giving rise to considerable strengthening.

Alternatively, some alloys are strengthened with insoluble dispersoids, which may be introduced by rapid solidification processing (see the following) or mechanical alloying, a process in which insoluble phases

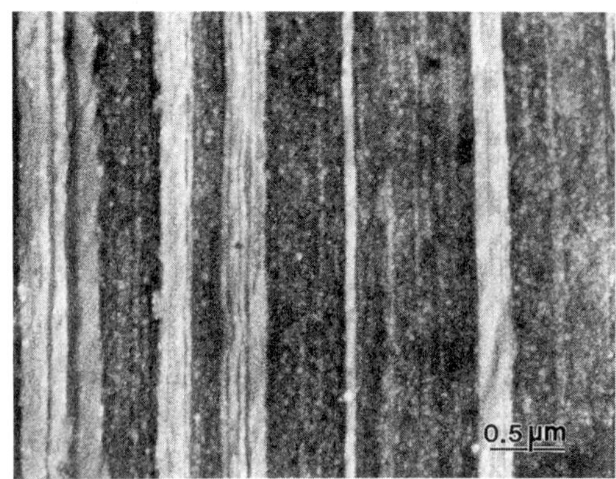

FIG. 9. SEM micrograph of a heavily worked Cu-Ag alloy. Note the fibrous structure of the Ag (light phase) in the Cu matrix. The drawing direction is vertical in the micrograph.

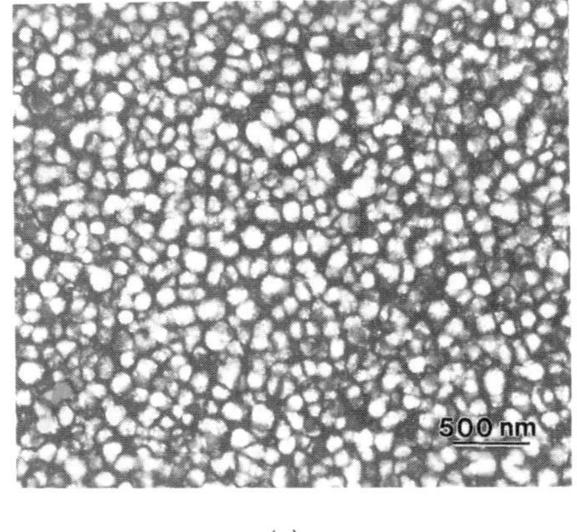

(a)

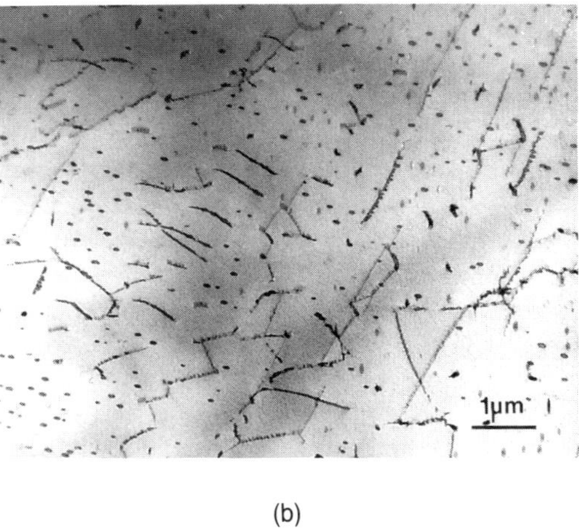

(b)

FIG. 10. (a) Dark-field transmission electron micrograph of Al_3Li precipitates (white) in an aluminum matrix (dark). (b) Bright-field transmission electron micrograph of nitride precipitates in an Fe–0.3Cu–0.7Ni alloy. Note the uniform distribution of the precipitates in (a) in contrast to the preferential precipitation on dislocations in (b).

are mixed intimately using mechanical means rather than thermal means.

3.2.3.2 Eutectoid Reaction. The equivalent of the eutectic reaction also occurs in the solid state, in which one phase transforms into two other phases of different compositions ($\alpha \rightarrow \beta + \gamma$). The classical and most common example of the eutectoid reaction occurs in steels (Fe-C–based alloys), in which the high-temperature austenite (fcc) phase transforms into the carbon-lean ferrite (bcc) phase and the carbon-rich cementite (Fe_3C) phase. As with eutectic structures, the scale of the eutectoid reaction product is a strong function of cooling rate and is finer at the higher cooling rates (Fig. 11).

3.2.3.3 Martensitic Transformation. In addition to precipitate formation during heat treatment, there are other transformations of technological significance. One of the most important is the martensitic transformation, where the transformation product is known as martensite (see MATERIALS TREATMENT). Martensitic transformations usually occur when an alloy is cooled quickly from a single-phase field at elevated temperatures into a two-phase field at lower temperatures (Fig. 12). Unlike precipitation, which requires long-range diffusion and rearrangement of the atoms in the material, martensite is said to occur in a "diffusionless" manner. Specifically, these reactions occur by movement of the at-

(a)

(b)

FIG. 11. Optical micrographs of a eutectoid steel after (a) air cooling and (b) furnace cooling from the high-temperature single-phase (austenite) field. Note the coarser microstructure after furnace cooling in comparison with that obtained by the faster air cooling.

FIG. 12. TEM micrograph of martensite in a Ti–24Al–11Nb alloy after cooling from elevated temperatures. The overall complexity of this type of structure should be noted.

oms in a cooperative or regimented fashion. Furthermore, since the atoms move only short distances, martensitic transformations occur at speeds approaching the speed of sound in the material.

There are many alloys which undergo martensitic transformations—for example, Cu-Zn, Cu-Mn, Ti-Mo, Fe-Ni, and many more. The most common example of martensite occurs in steels. Specifically, when certain iron-carbon alloys (steels) are quenched rapidly from high temperatures, where the γ phase is stable, to room temperature, a body-centered tetragonal phase is formed in which the distortion from the normal α-bcc structure increases with increasing carbon concentration. The morphology of the martensite in the Fe-C system depends on the carbon content and varies from a lath to a plate shape with increasing carbon content (Marder and Krauss, 1969).

3.3 Rapid Solidification Processing

As mentioned above, there has been a considerable amount of work in recent years in the area of rapid quenching of metals and alloys from the liquid and vapor states. This has led to a wide range of metastability-related phenomena such as

1. microstructural refinement in terms of the distribution of phases and segregation during solidification;
2. extended metastable solid solubility of solute atoms in the metallic matrix phase;
3. formation of metastable crystalline phases that have crystal structures considerably different from those predicted from the equilibrium phase diagram;
4. formation of quasicrystalline phases; and
5. formation of metallic glasses.

In order to account for these phenomena, it is important to describe the structures of the metals in their various equilibrium and nonequilibrium states. This section will describe the structure of metallic glasses, nanocrystals, and quasicrystals.

3.3.1 Metallic Glasses When certain alloys are cooled rapidly below their melting points, the formation of the stable crystalline phase may not occur, leading to the formation of what is referred to as a metallic glass. This occurs primarily in those systems where the viscosity of the liquid increases rapidly with decreasing temperature. In these instances, the liquid simply reaches a sufficiently high viscosity that the atomic mobility becomes insufficient for the atoms to come together to form crystalline nuclei, and, consequently, the liquid transforms to a glassy solid. Because of their low atomic mobility, the atoms are "frozen" in the positions they occupied in the liquid, i.e., they have similar short-range periodicity.

The physical and mechanical properties of metallic glasses tend to be considerably different from those of regular crystalline materials. For example, metallic glasses sometimes exhibit high strengths coupled with high ductilities and unusual electrical and magnetic properties. Also, since metallic glasses are single phase with essentially no segregation, they tend to have considerably better corrosion resistances than their crystalline counterparts, which are typically multiphase with some compositional segregation within the microstructure.

Finally, the mechanical properties of metallic glasses vary depending on whether deformation is being performed above or below the glass transition temperature (the temperature at which the glass transforms to a supercooled liquid). For deformation above the glass transition temperature, metallic glasses tend to deform homogeneously, provided they do not crystallize, since they are essentially supercooled liquids. Below this temperature, metallic glasses tend to deform in a very inhomogeneous manner, leading to the formation of highly localized shear bands. Normally, metallic glasses exhibit little if any work hardening, so that their tensile strengths are comparable to their yield strengths.

3.3.2 Nanocrystals A relatively new class of materials receiving considerable current attention is that referred to as nanocrystals. As the name implies, nanocrystals consist of nanometer-size crystals, which tend to be oriented randomly relative to each other as a result of the manner in which they are produced. Because of their fine structures, nanocrystals are comprised of about 50% grains that exhibit the long-range order characteristic of bulk material and 50% grain-boundary regions.

Unlike the glassy structures just described, in which there is a considerable degree of short-range order between the atoms, the grain-boundary regions in nanocrystals have essentially no order; this lack of order between the atoms gives rise to completely different physical properties (heat capacities, conductivities, etc.) compared with their crystalline and amorphous counterparts. These differences have spurred considerable interest and research into this new class of materials.

Since nanocrystals tend to have unusual properties, they are being considered for unique applications. However, it should be noted that structures with such fine grains are metastable with respect to forming larger grains, and this "coarsening" will occur, given sufficient thermal energy, leading to structures with more bulklike properties. Thus, it becomes important to realize that there are probable temperature limitations on the application of these materials. Furthermore, while descriptions of the production methods are beyond the scope of this article, it is

(a)

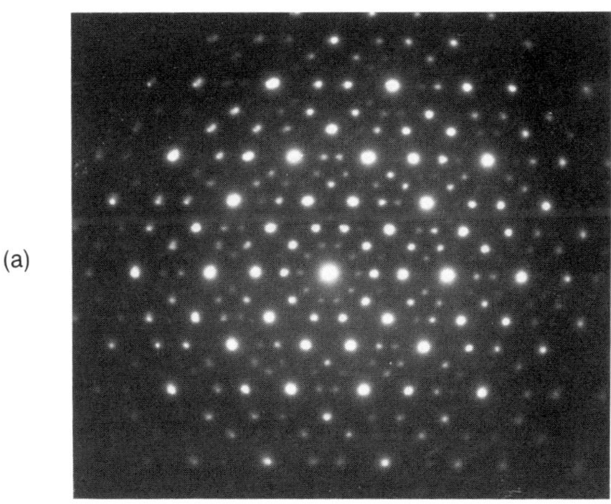

(b)

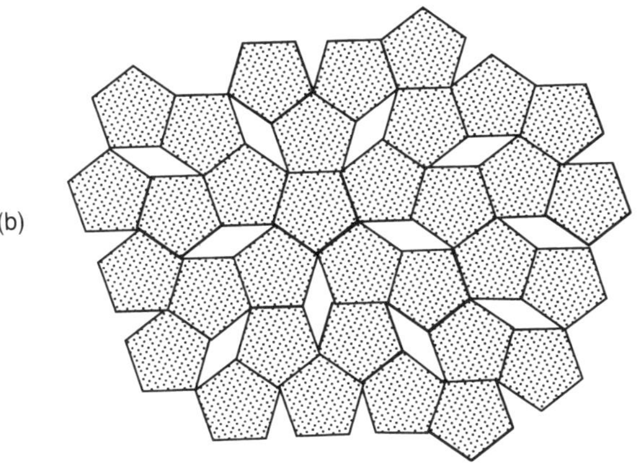

FIG. 13. (a) Electron diffraction pattern demonstrating the fivefold symmetry of the icosahedral phase that forms in Al-Mn alloys when rapidly solidified. (b) Two-dimensional illustration showing that it is impossible to fill space using objects with five-fold symmetry.

worth noting that most of the methods used to date are capable of producing very small quantities of powder only; these small quantities are then consolidated into sections of sufficient size for various physical property measurements.

3.3.3 Quasicrystals In 1984, Shechtman and co-workers from the National Institute of Science and Technology (NIST) reported the formation of a rather unique structure in binary Al-Mn alloys after rapid solidification. Specifically, electron diffraction patterns obtained from this phase (Fig. 13) appeared to display fivefold symmetry, which is unexpected because of the impossibility of filling space with objects of such symmetry [the reader can appreciate this in two dimensions by drawing pentagons and trying to arrange them side by side in a space-filling manner; see Fig. 13(b)]. Consequently, it is difficult to discuss the long-range order of this phase, as discussed above for crystalline phases; for this reason, other structural models have been developed to account for this anomaly.

Since the initial report of the icosahedral phase in Al-Mn alloys, quasicrystalline phases with icosahedral, decagonal (tenfold), and other structures have been reported in numerous other systems. While many of these can only be produced by rapid solidification processing, there are exceptions. For example, the stable T2 phase in the Al-Li-Cu ternary system has the icosahedral structure, and relatively large "crystals" of it have been reported in the shrinkage cavities of slowly cooled alloys; these have been useful for achieving a better understanding of the atomic arrangements in this phase.

As with nanocrystals, the unusual structures of quasicrystals also lead to some rather unique properties. Unfortunately, the property differences between the quasicrystals studied to date and their crystalline counterparts have been insufficient to warrant the application of these materials, and they remain a laboratory curiosity at this time. In fact, the formation of this phase in commercial Al-Li alloys is undesirable from a mechanical properties standpoint and, therefore, is usually avoided if possible. Even so, the complex structures characteristic of quasicrystals continue to generate considerable interest and curiosity among the scientific community.

GLOSSARY

Cohesive Properties: The forces and energies that hold atoms together.

Covalent Bonding: The bonding that occurs when adjacent atoms "share" electrons. Such a bond precludes the possibility of these elements being good conductors, since the electrons are not sufficiently free to conduct a current.

Dendrites: When, during the growth of crystals from liquids, the temperature falls in the liquid in advance of the liquid-solid interface, the solid may grow by branching out, with the appearance of a miniature tree. The word "dendrites" is a Greek word meaning "of a tree."

Diffusion: Thermally activated motion of atoms in matter. Diffusion may occur as a result of the existence of either a temperature or a concentration gradient.

Ductility: Ductility is a measure of the amount of plastic deformation that can be applied to a material before it breaks. Ductility is conventionally measured by tensile testing of metals.

Grain Boundary: The boundary between two grains in a polycrystalline material. For example, during casting, metals solidify by nucleation and growth of many independent grains or crystals that are oriented differently in the space. Obviously, when two crystals meet each other, their atomic arrangements do not match, and, consequently, the intersection plane, i.e., the grain boundary, will be disordered and hence distinguishable from the two crystals.

Ionic Bonding: The bonding that occurs when one element (positive ion) "transfers" electrons to another element (negative ion). The principal bonding force in ionic solids arises from the electrostatic attraction between opposite-sign ions.

Stacking Fault: The atomic arrangement of close-packed structures of metals can be defined by stacking of close-packed planes. In a fcc crystal, the stacking sequence is given by ABC ABC ABC, etc., and, in hcp structure, by AB AB AB, etc. Errors, or faults, in the stacking sequence may develop during solidification or by plastic deformation; these are known as stacking faults.

***Strukturbericht* Designation:** Books that codify all the structural work that is recorded in the literature. They cover the period from 1913, just after discovery of the use of x rays for identifying crystal structures, to the present. Initially, they were published in German, but, since World War II, their publication has been in English.

Texture: A metal that has been deformed extensively—for example, by rolling or wire drawing—may develop a preferred crystallographic orientation or texture. Texturing may also develop during solidification. In a textured structure, the grains of the metal tend to orient themselves in a particular crystallographic orientation.

Twinning: Twinning is a process by which a portion of the crystal takes up an orientation that is related to the rest of the untwinned lattice in a symmetrical fashion. The twinned portion of the crystal is a mirror image of the parent crystal. Twins may be produced by mechanical deformation (mechanical twins) or as a result of annealing following plastic deformation (annealing twins).

Wrought Structure: The microstructure of metals and alloys that have been manufactured by processes that involve plastic deformation.

Yield Stress: Yield stress is conventionally measured by tensile testing and defines the stress at which permanent deformation, i.e., plastic deformation, becomes apparent.

Works Cited

ASM Handbook (1992), *Alloy Phase Diagrams*, Vol. 3, Metals Park, OH: ASM International.

Barrett, C., Massalski, T. B. (1980), *Structure of Metals*, 3rd ed., New York: Pergamon Press, pp. 227, 261, 247.

Cole, G. S. (1971), in: T. J. Hughel and G. F. Bolling (Eds.), *Solidification*, Metals Park, OH: ASM International, p. 290.

Cottrell, A. (1991), *Introduction to the Modern Theory of Metals*, London: The Institute of Metals, p. 94.

Ebrahimi, F., Hoelzer, D. T., Venables, D., Krishnamoorthy, V. (1988), *Development of a Mechanistic Understanding of Radiation Embrittlement in Reactor Pressure Vessel Steels*, NUREG/CR-5063, U.S. Nuclear Regulatory Commission.

Liu, C. T., Stiegler, J. O. (1984), *Science* **226**, 636–642.

Marder, A. R., Krauss, G. (1969), *Trans. ASM* **62**, 2343–2357.

McLean, D. (1962), *Mechanical Properties of Metals*, New York: Wiley, p. 75.

Schechtman, D., Belch, I., Gratias, D., Cahn, J. W. (1984), *Phys. Rev. Lett.* **53**, 1951–1954.

Smallman, R. E. (1985), *Modern Physical Metallurgy*, 4th Ed., Boston: Butterworths, p. 297.

Further Reading

Barrett, C., Massalski, T. B. (1980), *Structure of Metals*, 3rd Ed., New York: Pergamon Press.

Reed-Hill, R. E., Abbaschian, R. (1992), *Physical Metallurgy Principles*, Boston: PWS-Kent Publishing Company.

METALS, LIQUID

See LIQUID METALS, STRUCTURE OF

METEOROLOGY AND CLIMATOLOGY

KEVIN HAMILTON, *Geophysical Fluid Dynamics Laboratory/NOAA, Princeton University, Princeton, New Jersey, U.S.A.*

INTRODUCTION

Mankind has no doubt always had a fascination with and a practical interest in the weather. Meteorological phenomena were a major subject of speculation in the philosophical works of classical antiquity, but scientific study of the weather is generally dated from the invention of the thermometer and barometer in the 17th century. There were sporadic attempts to plot weather maps from surface observations in the 18th century. The invention of the telegraph in the 19th century opened the prospect of producing and disseminating real-time forecasts using data gathered over a large geographical area. Government-sponsored observing networks were begun in several countries in the mid- and late 19th century. The 19th century also saw important developments in basic fluid dynamics and thermodynamics, which put the study of the atmosphere on a firm basis as a problem in applied physics. In recent decades, spectacular advances have been made in both observational and theoretical studies of the atmosphere. This recent progress has been greatly facilitated by the availability of satellite platforms for atmospheric observing systems and the development of digital computers allowing approximate numerical solution of the nonlinear governing equations.

Historically, the study of the atmosphere has been divided into the disciplines of *meteorology* and *climatology*. Climatology could be defined as the study of those processes that determine the time-mean state of the atmosphere—where the mean is defined as an average over some substantial period (a year or perhaps a number of years). Meteorology then deals with the physics of the higher-frequency components of atmospheric variability. It has become increasingly obvious that this distinction is quite arbitrary and not particularly useful. The circulation in the atmosphere displays variability at all time scales, and there are important interactions among the various frequency components.

Much recent work on atmospheric moni-

3-527-28132-0/94/$5.00 + .50

toring and modeling has been motivated by an awareness that mankind has the potential to alter significantly (if inadvertently) the global climate. Particular concerns have been raised about the increasing atmospheric levels of so-called *greenhouse gases* such as CO_2 caused by industrial and agricultural activities. Reliable predictions of the sensitivity of climate to such anthropogenic influences would be enormously valuable in formulating strategies to mitigate the social and economic consequences of global environmental change.

1. GOVERNING EQUATIONS AND SOME BASIC CONSIDERATIONS

1.1 The Primitive Equations

Below very high altitudes (say ~100 km), the molecular nature of the atmosphere can be neglected, and the usual continuum approximation of fluid mechanics may be used. The air in this region is sufficiently well mixed that species with very long chemical lifetimes (notably O_2, N_2, and Ar) always occur in the same proportions. Those species with shorter atmospheric lifetimes (notably water vapor) are observed to have concentrations that vary significantly in space and time. The chemical composition of the atmosphere is discussed in more detail in the article ATMOSPHERIC STRUCTURE.

The motions of the atmosphere can be modeled quite well by the Navier-Stokes equations of fluid mechanics formulated for a gas in a gravitational field. Thus, three scalar momentum equations, the first law of thermodynamics, an equation of state, and a continuity equation allow the determination of the three components of the wind and the temperature, pressure and density of the air. In addition, any realistic model of the atmosphere must include a continuity equation for the water vapor. To a very good approximation, the ideal gas law may be used as the equation of state for the atmosphere (noting that the exact value of the gas constant depends on the local concentration of water vapor). The usual approach in meteorology is to formulate the equations so that air motions are defined relative to the earth's surface. The wind components used are u in the zonal (east-west) direction, v in the meridional (north-south) direction, and w in the local vertical (see Fig. 1). The use of the rotating reference frame of the earth necessitates the inclusion of Coriolis and centrifugal forces in the governing equations. The importance of the Coriolis force on motions in the atmosphere (and ocean) is perhaps the most significant feature that distinguishes meteorology (and oceanography) from most other branches of applied fluid mechanics.

The relevant governing equations can be written as

$$\frac{D\mathbf{v}}{Dt} = -\frac{1}{\rho}\nabla p + \mathbf{g} - 2\mathbf{\Omega} \times \mathbf{v} + \Omega^2\mathbf{r}, \tag{1}$$

$$C_p\frac{DT}{Dt} = \frac{1}{\rho}\frac{Dp}{Dt} + Q, \tag{2}$$

$$\frac{D\rho}{Dt} = -\rho\nabla \cdot \mathbf{v}, \tag{3}$$

$$p = \rho RT, \tag{4}$$

where $\mathbf{v}$ is the three-dimensional wind vector defined relative to the rotating frame of the earth, p is the pressure, ρ is the density, $\mathbf{\Omega}$ is

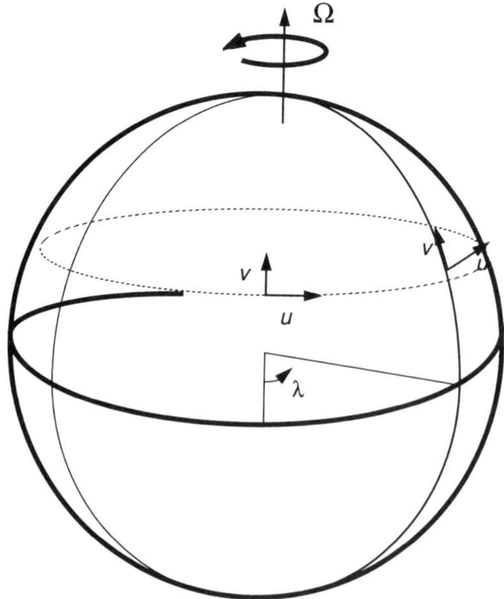

FIG. 1. An illustration of the usual notation used by meteorologists to describe the horizontal wind field. In addition to the horizontal components u and v, a component w is defined as parallel to the local vertical.

the angular velocity vector, Ω is the magnitude of the earth's angular velocity, **g** is the vector acceleration of gravity, T is the temperature, **r** is the position vector normal to (and pointing outward from) the earth's rotation axis, C_p is the specific heat of air at constant pressure, Q is the diabatic heating, and R is the gas constant. The operator D/Dt is the material (or substantial) derivative reflecting the rate of change following air parcel motions. The effects of molecular viscosity and heat conduction have been omitted in writing Eqs. (1)–(4) but can easily be included.

One very important simplification of the full Navier-Stokes equations results from the smallness of large-scale vertical air velocities in the atmosphere. Generally, w is of the order of 0.1 m s^{-1} or less, and vertical air-parcel accelerations are many orders of magnitude smaller than the gravitational acceleration g. Thus to an extremely good approximation, the vertical momentum equation reduces to hydrostatic balance:

$$\frac{\partial p}{\partial z} = -g\rho = -\frac{gp}{RT}, \tag{5}$$

where z is the local vertical (usually defined relative to sea level). Some minor simplifications to the other governing equations (based on the facts that the atmospheric depths of interest are much less than the earth's radius and that typical vertical velocities are small) are also generally made. Notable among these is the neglect of the part of the Coriolis force associated with the component of the earth's angular velocity vector that lies in the local tangent plane to the earth's surface. The Navier-Stokes equations on a rotating sphere with hydrostatic balance (and these other shallow-atmosphere approximations) are referred to as the *primitive* equations. The primitive equations are generally regarded as the appropriate starting point for even the most comprehensive models of the large-scale atmospheric circulation. Nonhydrostatic equations are normally required only to understand quite small-scale features in the atmosphere (such as intense thunderstorms).

One useful feature of the hydrostatic assumption is that it implies a one-to-one relation between pressure and geometric height. It is generally convenient to work with pressure as the independent vertical coordinate and height as the dependent variable to be computed. Thus, most weather maps are plotted on constant-pressure or *isobaric* surfaces. It is easy to show that the horizontal gradients of height along an isobaric surface are linearly related to the pressure gradients on a constant-height surface. Thus, high- (low-) pressure centers appear on isobaric maps as areas of high (low) height.

The hydrostatic equation can be integrated to relate the pressure at any height to the pressure at some reference height and the temperature in the intervening layer, i.e.,

$$p(z) = p(z^*)\exp\left(-\int_{z^*}^{z} \frac{1}{H}\,dz\right), \tag{6}$$

where the pressure scale height $H(z) = RT(z)/g$. For typical values of atmospheric temperature, H is about 7 km. The global-mean atmospheric pressure at sea level is about 1.013 $\times 10^5$ Pa (= 1.013 bar). Day-to-day fluctuations of surface pressure are typically 1–2% of the mean. Thus, it is very convenient in meteorology to measure pressure in terms of millibars (mb). The 500-mb level (which marks the height below which roughly half the atmospheric mass resides) is typically found at an altitude between 5 and 6 km.

1.2 Geostrophy

An approximation to the horizontal momentum equations can sometimes be useful in describing atmospheric circulation. In particular, the dominant components of the momentum balance are usually the pressure-gradient term and the Coriolis force. When these terms are exactly equal, the winds and pressure fields are said to be in *geostrophic balance*, i.e.,

$$2\Omega \sin\phi u = \frac{-1}{a\rho}\left(\frac{\partial p}{\partial \phi}\right)_z = -\frac{g}{a}\left(\frac{\partial z}{\partial \phi}\right)_p \tag{7}$$

in the meridional direction and

$$\begin{aligned} 2\Omega \sin\phi v &= \frac{1}{\rho a \cos\phi}\left(\frac{\partial p}{\partial \lambda}\right)_z \\ &= \frac{g}{a \cos\phi}\left(\frac{\partial z}{\partial \lambda}\right)_p \end{aligned} \tag{8}$$

in the zonal direction. Here ϕ is the latitude, λ is the longitude, and a is the radius of the earth. Only the component of the Coriolis force associated with the projection of the earth's angular velocity vector in the local vertical is considered (hence the factor $2\Omega \sin\phi$). The pressure gradient at constant height has been related to the height gradient at constant pressure by use of the hydrostatic relation. Under geostrophic balance, the horizontal wind and horizontal pressure gradients are perpendicular. This accounts for the tendency apparent on any weather map for the winds to blow around isolated high- and low-pressure regions. Of course, geostrophic balance is only a fairly rough approximation to the full momentum equations (errors typically ~10%). This relation breaks down at low latitudes where $2\Omega \sin\phi$ is small. There are also strong deviations from geostrophic balance in the lowest kilometer or so of the atmosphere where stresses from turbulence affect the large-scale horizontal momentum balance. As a result of these turbulent stresses, the wind near the ground generally has a significant component from high pressure toward low pressure. Thus, the surface level winds around a low- (high-) pressure center tend to spiral inward (outward).

The geostrophic and hydrostatic equations may be combined to produce the *thermal wind relation,* which connects the vertical derivative of horizontal wind to the horizontal temperature gradient. In particular, for the zonal wind component,

$$\frac{\partial u}{\partial \ln p} = \frac{R}{2\Omega a \sin\phi} \frac{\partial T}{\partial \phi}. \tag{9}$$

When we note that geometric height is very roughly proportional to $-\ln p$, the left-hand side is a measure of the strength of the vertical shear in the zonal wind. Since in the lower atmosphere the temperature normally decreases toward the pole (i.e., $\partial T/\partial\phi < 0$ in the Northern Hemisphere), one usually finds that the westerly wind intensifies with height.

The geostrophic approximation to the horizontal momentum equations eliminates the time derivative term, and so this relation has no predictive power. There are approximations to the primitive equations that use geostrophic balance to first order but consider the effects of deviations from geostrophy on the time tendency of the pressure and wind fields. These simplified prognostic equations are known as the *quasigeostrophic* system. The quasigeostrophic equations were the basis of early attempts at numerical weather forecasting, and they are still widely used in theoretical studies of atmospheric behavior in the extratropics (Phillips, 1963).

1.3 Potential Temperature, Stability, and Available Potential Energy

The potential temperature θ of air is defined as the temperature it would attain if compressed (or expanded) adiabatically to some standard pressure p^* (usually taken to be 1000 mb). Using the first law of thermodynamics and the equation of state, it is easy to show that $\theta = T(p^*/p)^{R/C_p}$. The potential temperature is obviously conserved during an adiabatic process.

Consider a horizontally homogeneous column of the atmosphere in which $\partial\theta/\partial z > 0$. If a small parcel of air in this column is displaced upward, it will adjust rapidly to the lower ambient pressure at its new level. If the process is rapid enough, it will be adiabatic, and the potential temperature of the parcel will be unchanged (and hence lower than the ambient θ). This means that the parcel temperature will be lower (and the parcel density will be higher) than in the ambient atmosphere. Buoyancy forces will then push the parcel down toward its initial position. Thus, this atmospheric column is regarded as stable to vertical displacements (a situation called *convective stability* by meteorologists). The stronger the positive stratification of θ, the more resistant the atmosphere is to vertical mixing of air parcels. Similar arguments can be made to demonstrate that if $\partial\theta/\partial z < 0$, adiabatic vertical air displacements are reinforced by buoyancy forces (i.e., *convective instability*) and that this should lead to strong vertical mixing within the air column. The large-scale mean temperature structure of the atmosphere is stably stratified, but local regions of convective instability appear each day scattered over the globe.

Figure 2(a) shows a schematic meridional section through a hemisphere of the atmosphere marked with lines of equal θ. Note that $\theta_1 < \theta_2 < \theta_3 < \theta_4 < \theta_5$ so that each atmospheric column is convectively stable. The

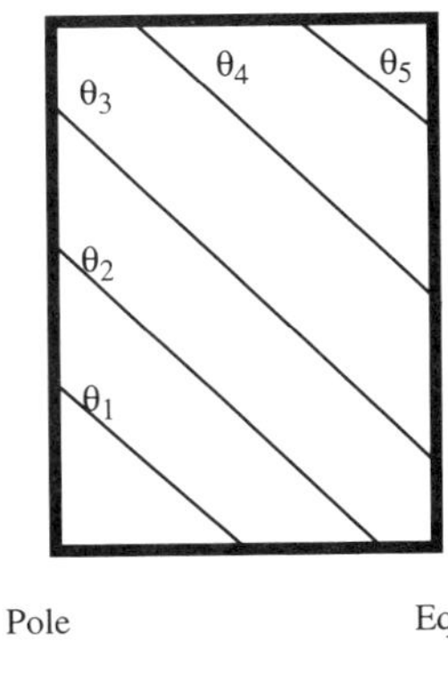

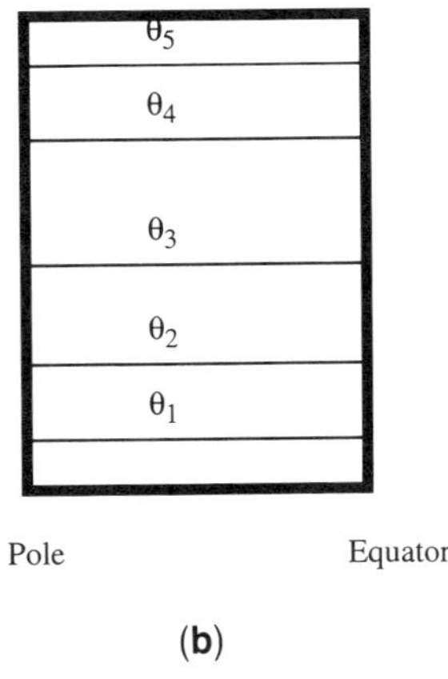

FIG. 2. Schematic meridional section through the atmosphere in one hemisphere. Contours of constant potential temperature are shown. (a) A realistic state for the atmosphere; (b) the lowest energy reference state.

meridional gradient of θ shown in this figure simply reflects the fact that on any horizontal pressure surface, the temperatures are lower at the pole than at the equator. Figure 2(b) shows a hypothetical situation in which the air has been rearranged adiabatically so that the lowest-θ air is at the bottom. This represents the lowest gravitational energy state (consistent with hydrostatic balance) that can be obtained from 2(a) through adiabatic air motions. The difference between the gravitational potential energy in 2(a) and 2(b) is referred to as the *available potential energy* (APE) of the atmosphere in state (a). The APE obviously increases with increasing horizontal temperature gradient. Atmospheric wind systems are ultimately driven by the equator-to-pole gradient in radiative heating that produces APE. Atmospheric motions then grow by converting the APE to kinetic energy of the winds. The strength of the observed atmospheric flow reflects the resulting balance between radiative generation of APE and its release by motions that bring low-θ air downward and equatorward and high-θ air upward and poleward.

This simple picture needs to be modified when the effects of moisture and latent-heat release are included. In particular, the upward displacement of an air parcel that is saturated with water vapor will not conserve θ. As the parcel rises, it will cool less rapidly than a dry (or unsaturated) air parcel because of the release of latent heat as water condenses out. Discussions of the thermodynamics of moist air can be found in any standard meteorological textbook (e.g., Wallace and Hobbs, 1977).

2. OBSERVED ASPECTS OF LARGE-SCALE CIRCULATION

2.1 Observational Network and Satellites

Our detailed knowledge of the atmospheric circulation and its variations comes largely from analysis of the routine measurements taken every day for the purposes of operational weather forecasting. One component of this observational data set consists of measurements of wind speed and direction, barometric pressure, precipitation, air temperature, and humidity at three-hour intervals at several thousand surface stations (and on commercial ships) throughout the world. Another very important component of the data archive is the collected measurements from *radiosondes* borne aloft by unmanned balloons. These radiosondes are instrument packages that transmit values of the pressure, temperature, and humidity at specified intervals during the balloon ascent. By tracking the balloon visually or with radar, the horizontal wind speed and direction can also be determined as a function of height. These radiosonde ascents are typically made once or twice a day at several hundred stations worldwide. Figure 3 shows the distribution of the 1093 stations where upper-air balloon measurements were performed for some period during 1963–73. On any given day, roughly 600 stations actually reported observations. Weather balloon measurements are normally limited to heights below about 30 km.

Balloon measurements are critical for exploring the three-dimensional structure of the

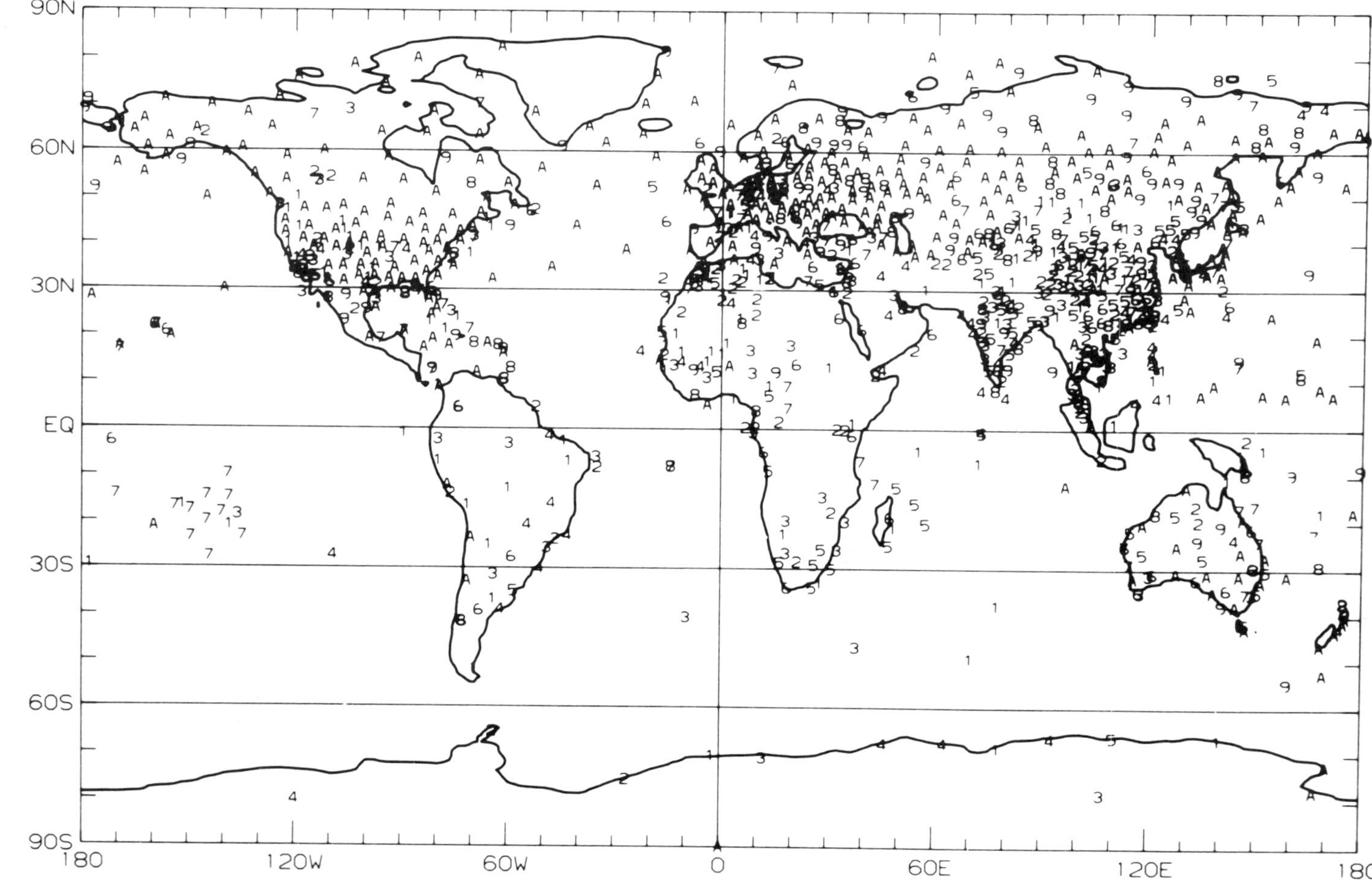

FIG. 3. The 1093 stations for which balloon-borne radiosonde data were available at 500 mb for at least one January between 1964 and 1973. The number at each station location shows the number of years for which fairly complete January data records exist ("A" indicates that data is available from all ten years). From Peixoto and Oort (1992).

atmospheric circulation, but they have obvious limitations in terms of geographical coverage (particularly over the oceans). The gaps in station coverage can be filled in to some extent by meteorological satellites. *Geostationary satellites* provide fairly continuous observations of the cloud field over the tropics and midlatitudes. Figure 4 shows a typical image from a geostationary satellite. *Polar orbiting satellites* in fairly low orbits carry radiometers that measure outgoing radiation at the top of the atmosphere. By performing these measurements at several wavelengths, information concerning the vertical structure of temperature and humidity can be extracted (see ATMOSPHERIC PHYSICS for more details). These satellites provide information along a sun-synchronous satellite trajectory as the earth rotates beneath it. Thus, global coverage is achieved roughly twice each day.

2.2 Surface Climate

The observed long-term mean of the surface air temperature in January is shown in Fig. 5. This figure displays a feature common to many climatological statistics, namely that the zonal gradients are generally weaker than the meridional gradients. This tendency reflects the fact that the daily-average solar radiative flux incident at the top of the atmosphere is symmetric about the rotation axis. The deviations from axial symmetry in long-term mean climate data such as that shown in Fig. 5 are caused by the inhomogeneity of the earth's surface (i.e., land-sea contrast and topography).

FIG. 4. A visible-light image of the earth taken on 18 May 1986 from a satellite in geostationary orbit above 160°E. The continental outlines and latitude-longitude grid have been superposed on the actual image. Reproduced from Philander (1991), courtesy of Taiwan Central Weather Bureau.

Figure 6 shows the long-term mean of the horizontal wind obtained from surface observations for December–February. Noteworthy here is the predominance of westerly winds (i.e., winds blowing from the west) in the midlatitudes and easterly winds in the tropics (the so-called *trade winds*). Beyond this some significant deviations from zonal symmetry are evident. In particular, there is a net tendency for surface flow off the continent onto the oceans in the winter hemisphere and a reverse flow in summer. In midlatitudes, this effect generally makes a fairly small perturbation to the dominant zonal flows, but in the tropics this *monsoon* circulation can dominate the wind field. In particular, the flow over the Indian Ocean is oriented outward from the Asian land mass in winter (winter monsoon) and completely reversed in summer (summer monsoon). The wind field in the 40°S–60°S band is particularly zonal, presumably reflecting the virtual absence of east-west inhomogeneity in the lower boundary at these latitudes.

Also shown in Fig. 6 are contours of the height of the 1000-mb surface (very nearly proportional to the sea-level pressure field) averaged for the same period. The tendency for the winds to blow parallel to the constant-height surfaces is quite evident in the extratropical regions. In the tropics, the component of flow from high to low pressure is more pronounced.

2.3 Zonal-Mean Climate

Given the tendency for climate elements to display a very approximate axial symmetry, it is useful to examine the zonally averaged circulation to obtain a rough picture of the state of the atmosphere. In the usual terminology, the circulation can be divided into the *zonal-mean* (i.e., the average around each latitude circle) and the *eddy* (the deviation from the zonal-mean) components. Figure 7 shows the height-latitude section of the long-term average of the zonal-mean tem-

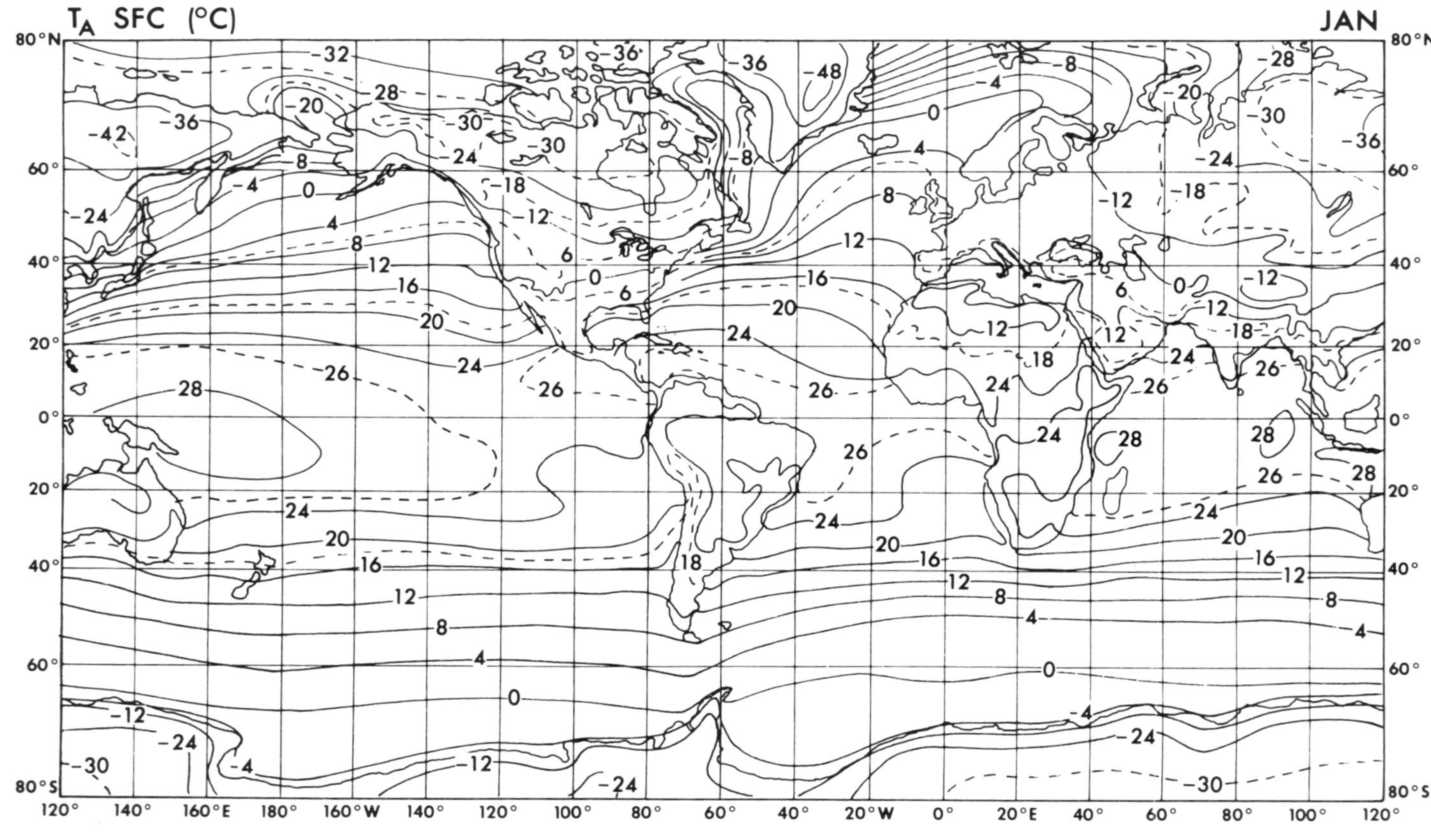

FIG. 5. The long-term mean (°C) of the observed surface air temperature in January. From Peixoto and Oort (1992).

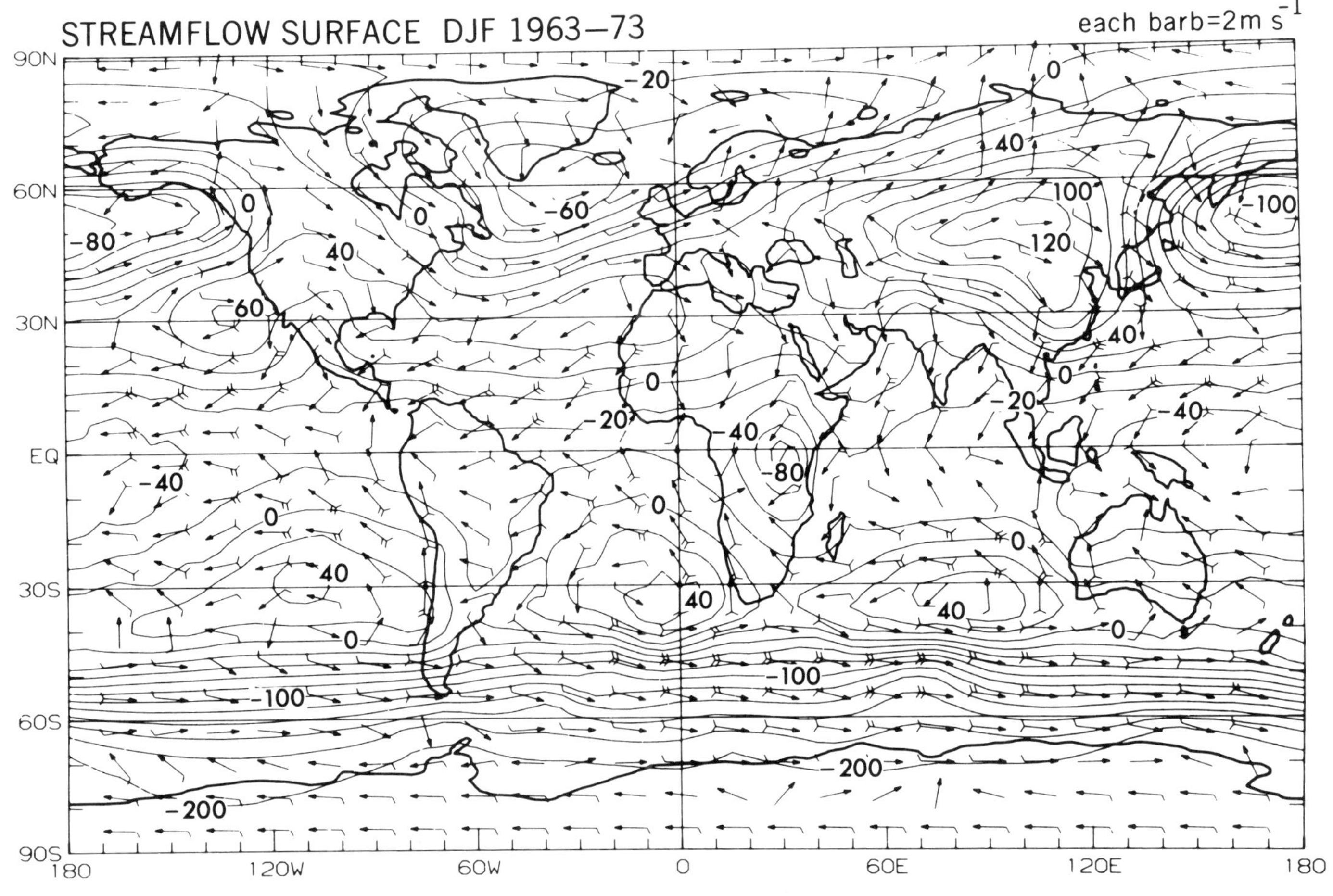

FIG. 6. Long-term mean surface-flow climatology for the December–February period. The arrows show the direction of the horizontal wind measured about 2 m above the surface. The wind speed is represented by the number of barbs on each arrow (each barb represents 2 m s^{-1}). The contours show the height above sea level of the 1000-mb surface relative to the global mean value of 113 m. From Peixoto and Oort (1992).

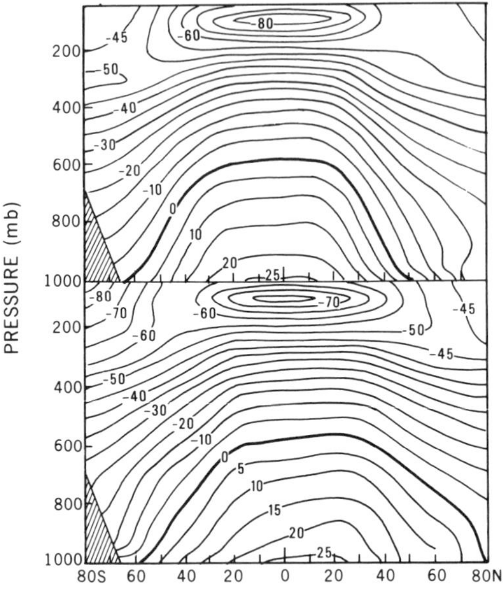

FIG. 7. Long-term mean of the zonally averaged temperature in the December–February (top) and June–August (bottom) periods. Contour labels in °C. Adapted from Peixoto and Oort (1992).

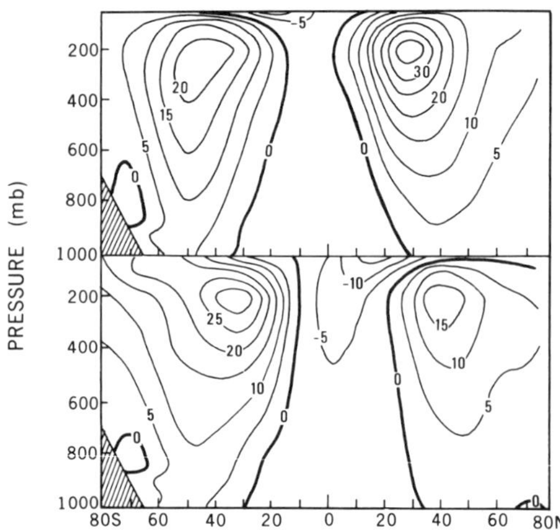

FIG. 8. Long-term mean of the zonally averaged zonal wind in the December–February (top) and June–August (bottom) periods. Contour labels in m s^{-1}. Positive values denote westerly winds (i.e., winds blowing from the west). Adapted from Peixoto and Oort (1992).

perature up to 100 mb (about 18 km altitude) in December–February (top) and June–August (bottom). This figure clearly shows two regions with different vertical temperature structures. The *troposphere,* where the temperature decreases with height, extends from the ground to the minimum-temperature level called the *tropopause.* The tropopause is at about 300 mb (~8 km) near the poles and reaches to about 120 mb (~17 km) at the equator. Above this region is the *stratosphere* where temperature increases with height. The stratosphere extends up to about 0.5 mbar (~50 km). See ATMOSPHERIC STRUCTURE for more details on the temperature profile of the upper atmosphere. In the troposphere, the temperatures at any level are generally warmer near the equator than in the polar regions. This meridional temperature gradient is significantly stronger in the winter hemisphere than in the summer. This difference reflects the stronger pole-to-equator contrast in incident solar radiation in the winter. Remarkably, the meridional temperature gradient is reversed (so that the pole is actually warmer than the equator) in the winter lower stratosphere (e.g., near 150 mb in Fig. 7).

Figure 8 shows the comparable height-latitude sections for the long-term mean zonally-averaged zonal wind. In both seasons and both hemispheres, the most prominent feature in this figure is the core of strong westerlies peaking in the midlatitudes near 200 mb. Each of these wind maxima is the time-averaged and zonally-averaged manifestation of a *subtropical jet stream.* Even in this average sense, the winds in these jets are quite strong, and at a particular time and place the jet stream level winds may exceed 100 m s^{-1}. The fact that there is westerly shear (i.e., u increases with height) up to 200 mb is consistent with thermal wind balance between the zonal wind and the meridional gradient of temperature field in Fig. 7.

2.4 Time-Mean Circulation

The eddy component of the atmospheric circulation has particularly strong day-to-day variability associated with the development, decay, and propagation of weather disturbances. However (as seen above in the surface fields shown in Figs. 5 and 6), there are important eddy components in the long-term mean of the circulation as well. The deviations from the zonal average seen in long-term mean fields are referred to as the *stationary eddies* or *stationary waves.*

The contours in Fig. 9 show the long-term mean November–March 500-mb height field

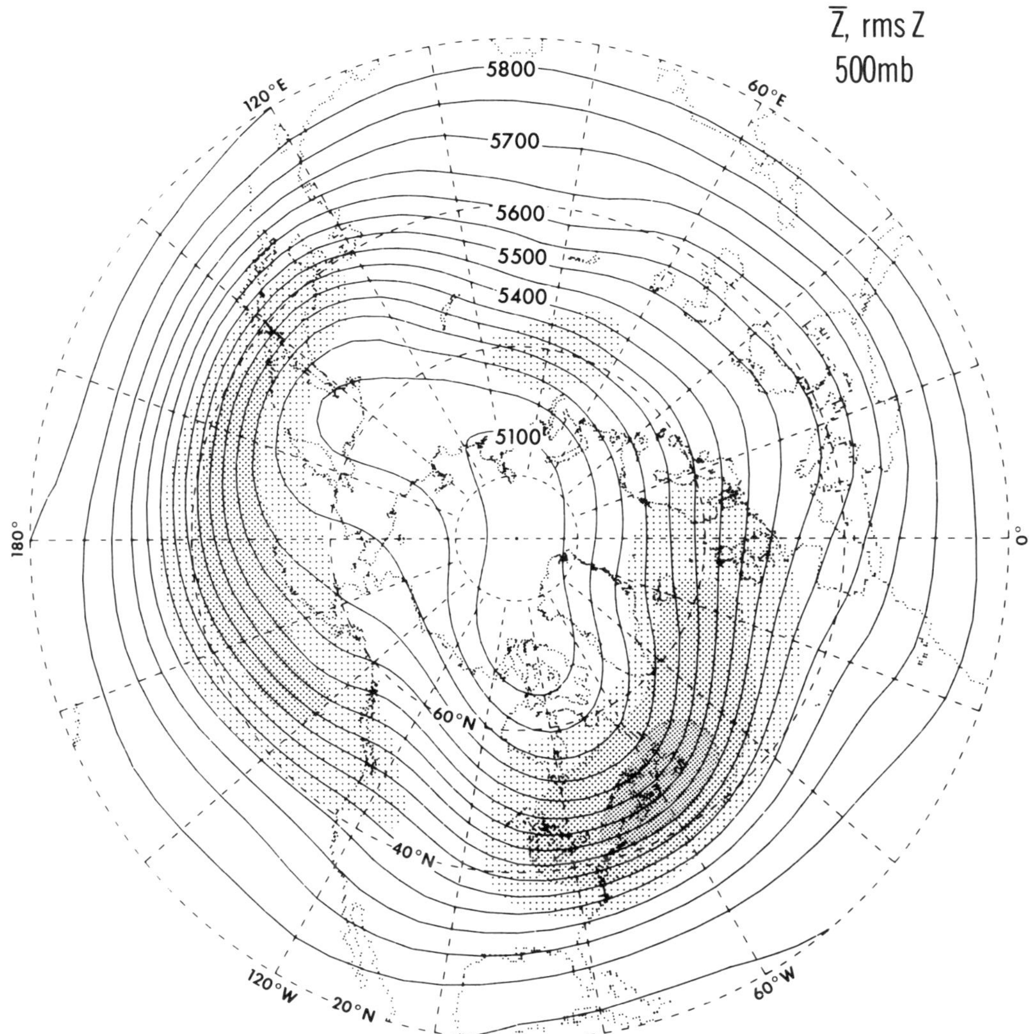

FIG. 9. Long-term mean for the November–March period of the height of the 500 mb surface over the Northern Hemisphere. Contour labels in m. The shading denotes regions of high day-to-day transient fluctuations in the 500-mb height. In particular, the light, medium, and heavy shading correspond to regions where the standard deviation of the 500-mb height time series, filtered to retain only periods between 2.5 and 6 days, exceeds 50, 60, and 70 m, respectively. From Lau (1988), courtesy of the American Meteorological Society.

over the Northern Hemisphere. The most evident feature is the decrease of height with latitude. This is geostrophically consistent with the strong mean westerlies at this height (Fig. 8). Superimposed on this zonally symmetric variation is a longitude-dependent component that seems to be dominated by a wave-number 2 pattern at high latitudes and a wave-number 3 pattern in midlatitudes. In particular, the meridional gradient of the geopotential (and hence the zonal wind) is enhanced locally over the eastern North Pacific and the eastern North Atlantic regions. This is the mid-tropospheric manifestation of the stationary wave pattern seen in the surface pressure field in Fig. 6. The stationary waves play an important role in controlling the day-to-day weather. In particular, regional-scale transient weather systems tend to be strongest in particular geographical regions called *storm tracks*. The intensity of the transient disturbances is shown by the shading in Fig. 9. The time series of the 500-mb height at each location was bandpass filtered to retain only fluctuations with periods between 2.5 and 6 days. Then the standard deviation of the filtered time series was computed. The three levels of shading in Fig. 9 show where this standard deviation exceeds 50, 60, or 70 m. It is apparent that the two principal Northern Hemisphere storm tracks are oriented along a rough west-southwest/east-northeast axis in the North Atlantic and in the North Pacific. These correspond closely to the regions of enhanced meridional height gradients seen in the time-mean pattern in Fig. 9.

3. DYNAMICS OF CLIMATE

3.1 Radiative Forcing

On an annual-mean, global-mean basis, 30% of the solar radiation incident at the top of the earth's atmosphere is reflected elastically back into space, 19% is absorbed by various atmospheric constituents, and 51% is absorbed by the earth's surface. The direct atmospheric absorption of solar radiation (particularly by O_2 and O_3) is crucial for maintaining the temperature structure of the stratosphere and higher regions. This is discussed in more detail in the articles ATMOSPHERIC PHYSICS and ATMOSPHERIC STRUCTURE.

For the troposphere, the dominant radiative effects are the solar heating of the ground and the exchange of infrared (IR) radiation between the ground and the overlying atmosphere. The IR opacity of the atmosphere is provided primarily by water vapor, clouds, CO_2, and O_3. The downward flux of IR radiation from the atmosphere helps to heat the surface in a manner described in the article ATMOSPHERIC PHYSICS. This phenomenon is sometimes called the *greenhouse effect,* and it will be enhanced (and hence surface temperatures are expected to rise) if the IR opacity of the atmosphere increases (e.g., by anthropogenic CO_2 emissions). The detailed mathematical treatment of both solar and IR radiative transfer in the atmosphere is reviewed in Liou (1980).

3.2 Radiative Equilibrium and Radiative-Convective Models of Climate

The simplest model of climate is one in which the solar heating at each point in the atmosphere is balanced by the IR cooling, a situation called *radiative equilibrium.* Actually to implement such a model for a single column of the atmosphere, one needs to specify the distribution of all radiatively active constituents as well as the reflectivity of the surface and cloud layers. A detailed computer code that solves a finite-difference version of the governing radiative transfer equation can be used to calculate the radiative heating rates for any given vertical temperature structure. Then an iterative algorithm is applied to determine the atmospheric $T(z)$ (as well as the earth's surface temperature) that leads to exactly zero heating rates. The solid curve in Fig. 10 shows the result of such a calculation using a very detailed radiative model and employing absorber distributions, solar zenith angle, etc. appropriate for 35° latitude at equinox. The result is quite unrealistic with temperatures of more than 60 °C predicted for the ground and a drop-off of more than 100 °C in the lowest kilometer of the atmosphere. In reality, the large-scale mean temperature is observed to fall off at a fairly constant rate of about $-6.5\ °C\ km^{-1}$ in the troposphere (see Fig. 7), and actual midlatitude surface temperatures are closer to 20 °C.

The failure of the purely radiative model to simulate a realistic structure for the troposphere suggests that the effects of air motions are of critical importance in determining atmospheric temperature. The radiative equilibrium model can be modified in a simple way to take into account the first-order effects of dynamics. The pure radiative-model prediction has temperature decreasing so fast that the result is convectively unstable. The real tropospheric $T(z)$ is determined by a balance between the effects of radiation trying to heat the near-surface layer, and the effects of vertical mixing produced by convective instability in transporting warm (i.e., high-θ) air upward and cold (low-θ) air downward. Convective instability can homogenize θ in the vertical in a few hours, while the time scale for radiative processes to establish temperature gradients in the lower atmosphere is of the order of weeks. Thus, the troposphere might be expected to be in a condition of nearly uniform θ (which can be shown to be equivalent to a uniform temperature decrease $\partial T/\partial z = -9.8\ °C\ km^{-1}$). This simple prediction must be modified to account for the fact that most vertical mixing in the

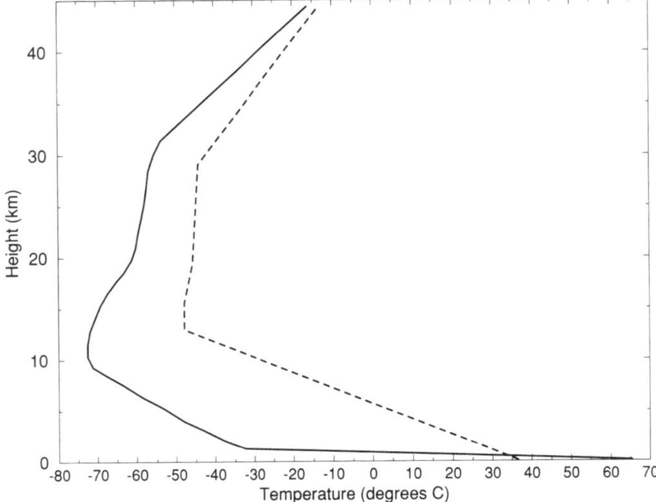

FIG. 10. The predicted vertical temperature structure $T(z)$ for a radiative equilibrium model (solid curve) and a radiative-convective equilibrium model (dashed curve). Results appropriate for cloudless skies at 35° latitude at equinox. Adapted from Manabe and Wetherald (1967).

troposphere occurs in clouds, and the associated latent-heat release actually maintains a positive stratification in the $\theta(z)$ profile. In *radiative-convective equilibrium* (RCE) climate models, the details of the maintenance of the tropospheric stratification are ignored, and a simple constraint is imposed so that enough heat is mixed upward to ensure that the predicted $\partial T/\partial z$ is never steeper than the observed value of -6.5 °C km^{-1}. The dashed curve in Fig. 10 shows the result of adding this fix to the radiative model. The result is very encouraging in that more reasonable surface temperatures are predicted. The temperatures are still too warm in this particular simulation because the reflectivity of clouds was not considered (when this effect is included, a very realistic surface temperature is obtained; see Manabe and Wetherald, 1967). The application of RCE models to climate studies is reviewed in Ramanathan and Coakley (1978).

3.3 Effects of Horizontal Heat Transport

While the RCE model can produce reasonable predictions of the global-mean vertical temperature structure, it fails in simulating the full range of seasonal and latitudinal variation of climate. As an extreme example, in the polar night region a straightforward application of the RCE model would result in a predicted temperature of 0 K . The actual temperature structure in the atmosphere is very strongly influenced by the dynamical transport of heat from low latitudes to the polar regions. Figure 11 summarizes the terrestrial heat budget using the climatological radiative fluxes at the top of the atmosphere (measured from satellite platforms). The top panel gives the annual mean of the absorbed solar flux (i.e., the incident downward flux minus upward reflected shortwave flux) as a function of latitude. Naturally, this quantity peaks in the tropics and drops rapidly with latitude. The middle panel shows the annual mean outgoing IR radiation at the top of the atmosphere. This also peaks in low latitudes, but the equator-to-pole contrast is much weaker than for the absorbed solar component. The bottom panel gives the difference between the curves in the upper panels, thus representing the net radiative heating of the atmosphere/ocean column (the area-weighted global mean of this curve is zero). This shows that the tropical region is heated by radiation, while the regions poleward of about 40° latitude are cooled. This net heating and cooling must be balanced by dynamical transports in the atmosphere and ocean that act to move heat poleward. More detailed analysis of atmospheric data suggests that in the extratropics this transport is accomplished largely by the day-to-day weather disturbances, which act to mix cold polar air equatorward and warm air poleward. The dynamics of this process will be examined in Sec. 5.1. By contrast, in low latitudes most of the atmospheric heat transport is in the form of latent heat of condensation (i.e., moist

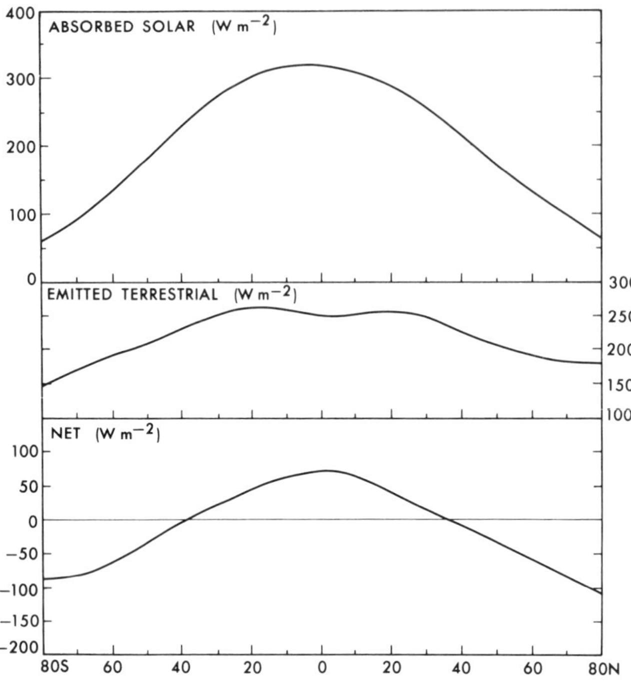

FIG. 11. Components of the long-term mean radiation budget as measured at the top of the atmosphere. Top panel, the net incident downward solar radiative flux. Middle panel, the outgoing terrestrial IR flux to space. Bottom panel, the difference between these quantities. Adapted from Peixoto and Oort (1992).

air is transported poleward where it condenses and releases latent heat), and the poleward transport of heat within the ocean is significant. The prediction of realistic meridional heat transports is a critical test of the adequacy of a climate model.

3.4 General Circulation Models

The most comprehensive numerical models for the climate system are referred to as general circulation models (GCM's). At the minimum, a GCM involves a time-dependent simulation of the full three-dimensional wind and temperature field in the global atmosphere. The model climate is then found as a mean of a very long numerical integration (e.g., several years of model time) with the numerical code.

GCM's now exist with varying degrees of sophistication. All involve some numerical approximation to solve the basic primitive equations (and the continuity equation for atmospheric water vapor). This is normally done with centered second-order finite differences in time. In *grid point* models, the horizontal derivatives are approximated by finite differences on a specified three-dimensional grid. In *spectral* models, the dependent variables are written as linear combinations of some finite set of spatial basis functions, and so the original differential equations are replaced by a large coupled set of first-order ordinary differential equations (in time) for the coefficients of each of the basis functions. In practice, spectral GCM's have used finite differences in the vertical and a spectral representation in terms of spherical harmonics in the horizontal (e.g., Gordon and Stern, 1982).

At present, the typical resolution of a model employed in climate studies (i.e., one that can be readily integrated for many years of model time) might be ~400 km in the horizontal and ~2 km in the vertical (with a total of ~15 levels). Such a GCM would be integrated with a time step that would be in the range of ~5–30 minutes (depending on the details of the model formulation).

In addition to the dynamical core of the model designed to solve explicitly for those aspects of the atmospheric flow resolvable on the numerical grid, a GCM needs to include other processes. In particular, the effects of motions with scales smaller than those resolvable by the model must be incorporated. These effects have to be parametrized; i.e., the contributions of the subgrid-scale flow must be expressed somehow in terms of the resolved variables. The usual approach to this problem is to assume that unresolved scales act as a dissipation of the resolved-scale flow. A special case of such parametrization is the treatment of clouds (which in the real atmosphere have scales much smaller than resolved in current GCM's). The degree of cloudiness, the vertical mixing, and the latent-heat release in any vertical column of model grid points are generally specified as functions of the convective stability as well as the humidity field on the resolved grid. The model precipitation is determined as part of this subgrid-scale vertical-mixing parametrization. The effects of radiative transfer are also included by solving the radiative transfer equations in each column of grid points (using the instantaneous temperatures as well as the distribution of water vapor and cloudiness predicted by the model).

A GCM also requires the effects of interactions between the atmosphere and the underlying surface to be included. Heat, moisture, and momentum are all transferred between the atmosphere and the underlying ocean or land. The atmospheric model needs to know the surface temperature and, over land, the soil wetness (as well as a measure of the roughness of the surface). These quantities can be either specified (usually using climatological data) or calculated using very simple models of the land surface and ocean. A more ambitious approach is actually to couple the atmospheric GCM to an analogous three-dimensional, time-dependent simulation model for the global ocean. In such a coupled model, the ocean provides surface temperature to the atmospheric model, which in turn provides the heat flux, water flux (through precipitation and evaporation), and surface wind stresses that drive the ocean circulation.

A fairly complete description of a typical finite-difference GCM is found in Holloway and Manabe (1971), and a description of a typical spectral GCM is given in Gordon and Stern (1982). While the development of more sophisticated GCM's is an area of active research, these models have already produced impressively realistic results. The tropospheric zonal-mean temperature and wind structure, the strength and position of the stationary waves and storm tracks, and even the climatological distribution of precipitation can now be reasonably well simulated.

4. CLIMATE VARIABILITY

The interannual variability of seasonal-mean temperature and precipitation at a particular location is a matter of common experience. To some extent, this simply reflects the sampling of day-to-day weather variability. This sampling effect can be reduced by considering area-averaged quantities (and not surprisingly, the variance in the time series of an area-averaged variable decreases as the area included in the mean is enlarged). However, even at global scales there is clear evidence for systematic interannual variability in climate. Detailed understanding of the nature of this variability would have great practical value. In particular, recent developments in modeling have raised the possibility of producing useful forecasts of seasonal-mean climate anomalies several months in the future. Detection and modeling of possible anthropogenic effects on climate also require an understanding of the natural "background" climate variability.

The current understanding of the basic nature of interannual climate variability is closely connected to the concept of *atmospheric predictability*. If two parallel runs of a sophisticated three-dimensional GCM (with fixed surface boundary conditions) are conducted, each starting with slightly different initial conditions, the results are always found to diverge very strongly after a period of a few weeks. The period it takes for the two integrations to diverge is called the *theoretical limit of predictability*. After this time limit, the model has essentially "lost" memory of its initial conditions (behavior which has analogs in many other kinds of turbulent fluid systems). This somewhat vague definition can be made more precise, and other means have also been used to estimate the predictability limit (Lorenz, 1969). For the present purposes, the limited persistence of memory in the atmosphere suggests that interannual fluctuations must critically depend on the interactions of the atmosphere with the ocean or land surfaces (which act as the long-term memory for the coupled system). While the land surfaces can play a significant role (through storage of soil moisture and changes in vegetation), it is the oceans with their enormous capacity to store and transport heat that are thought to be most important in the generation of interannual variability in the climate system.

4.1 The Southern Oscillation and El Niño

One quite systematic component of large-scale interannual variability in the atmosphere has its roots in the tropical Pacific Ocean. The *Southern Oscillation* (SO) involves the shifting of atmospheric mass (and hence surface pressure) between the eastern and western tropical Pacific regions. Figure 6 shows the long-term mean pattern with high surface pressure in the equatorial eastern Pacific and low pressure in the equatorial western Pacific. The mean winds near the equator then blow from east to west. At one extreme of the Southern Oscillation, this climatological pattern is intensified, while at the other extreme both the zonal pressure gradient and wind direction in the equatorial Pacific reverse. The period of oscillation varies between about 2 and 6 years. The phase of the oscillation has traditionally been measured by the Southern Oscillation Index (SOI) defined as the surface pressure measured at Tahiti (18°S, 149°W) minus that measured at Darwin (12°S, 131°E). The SOI varies by about 4 mb between extremes of the SO.

The variations in the atmosphere during the SO are accompanied by profound changes in the ocean. Figure 12 shows the ocean surface temperature in November 1982 (near the low SOI extreme) and in November 1983 (near the high SOI extreme). The eastern and central Pacific surface waters are as much as 4 °C warmer at the low SOI extreme. The warm (low SOI) phase of the Southern Oscillation has received particular study. The warming along the west coast of Southern America (and related changes in ocean currents and fish populations) have been known to fishermen for centuries. The local people called this phenomenon *El Niño*, and in the last decade the term *El Niño/Southern Oscillation* (*ENSO*) *event* has become standard in the scientific literature. The ENSO events have a strong phase locking with the annual cycle, so that the peak warming in the eastern Pacific normally occurs in the December–February period. A detailed description of the typical evolution of oceanic and atmospheric con-

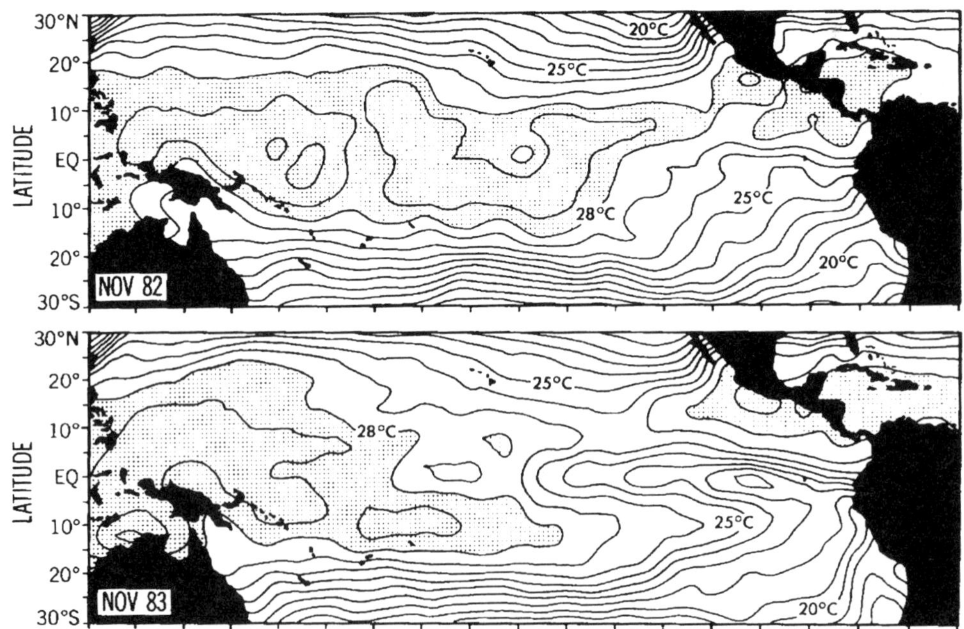

FIG. 12. Ocean surface temperature in the tropical Pacific measured at two extremes of the Southern Oscillation: (top) November 1982 and (bottom) November 1983. Regions where the ocean temperature exceeds 28 °C are shaded. From Philander (1991).

ditions during an ENSO event is given by Rasmusson and Carpenter (1982).

Some aspects of the dynamics of the SO are now well understood, notably the manner in which the atmosphere and ocean interact to excite the extreme warm and cold phases. The warming of the ocean surface in the eastern Pacific during ENSO heats the overlying atmosphere, generating upward motion and consequent low-level convergence of air. This convergence is accomplished by the anomalous westerly winds along the equatorial Pacific seen during El Niño (and a strengthening of the low-level easterly winds over the South American continent). The wind changes affect the stresses acting on the ocean surface. It turns out that westerly wind stresses near the equator inhibit the upwelling of cold deep ocean water, thus reinforcing the ocean surface warming. Analogous arguments can account for the excitation of the cold extreme of the SO. The mechanism by which the atmosphere-ocean system switches from one phase to the other is still not completely clear, however. This presumably involves the internal dynamics of the ocean in the Pacific basin (and this oceanic circulation must largely determine the dominant 2–6-yr time scale for the SO). The SO and the ENSO phenomenon are reviewed by Philander (1991).

While primarily a tropical phenomenon, the SO has detectable effects on the climate in the extratropics as well (e.g., Horel and Wallace, 1981). The clearest signal is found in Northern Hemisphere winter over the eastern North Pacific and North America. Figure 13 presents a schematic picture of the perturbation to the seasonal mean circulation of the troposphere induced at the peak warming of an ENSO event in the tropical Pacific. The effect is a wave train of alternating high- and low-pressure regions propagating along a nearly great-circle trajectory. This basically reinforces the long-term mean standing-wave pattern, most notably in the intensification of the low-pressure region seen in winter over the eastern North Pacific (see Fig. 6). SO effects can account for as much as 25% of the variance seen in the interannual time series of winter-mean surface pressure at some midlatitude locations.

4.2 Interdecadal Variability

Figure 14 shows observational estimates of the time series of yearly-mean surface air temperature averaged over the Northern and

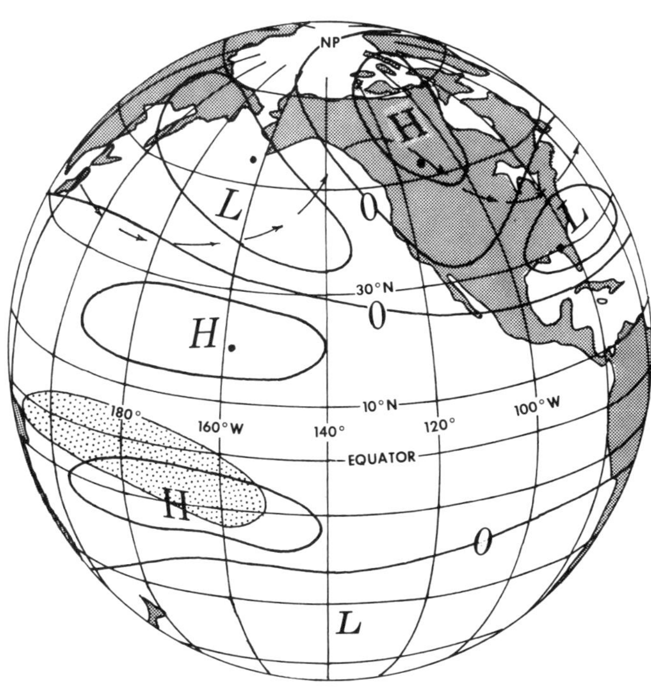

FIG. 13. Schematic diagram showing the perturbation to the usual seasonal mean pressure patterns forced by the ocean warming associated with an ENSO event. The shaded region in the equatorial central Pacific shows the region of most anomalous precipitation during ENSO. The alternating pattern of high- and low-pressure regions extending into the Northern Hemisphere extratropics describes the observed effect at all levels in the troposphere. From Horel and Wallace (1981), courtesy of the American Meteorological Society.

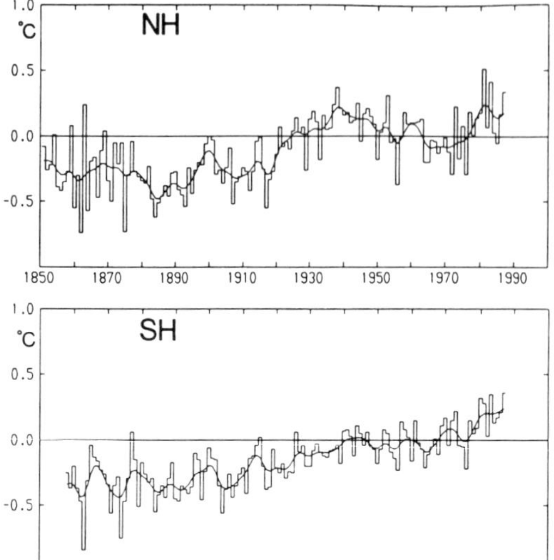

FIG. 14. Time series of yearly mean surface air temperature averaged over (top) the Northern Hemisphere and (bottom) the Southern Hemisphere. In each case, only departures from the long-term mean are shown. The smooth curves are Gaussian-weighted ten-year running means. From Peixoto and Oort (1992).

Southern Hemispheres (the long-term mean in each case has been removed). See Jones *et al.* (1986a, b) for a discussion of the issues involved in the construction of such a time series. This figure reveals a great deal of noisy year-to-year variability (some may be connected with the Southern Oscillation and some with volcanic activity, and some may simply represent sampling error). In addition to this, however, there seems to be significant longer-period variability. The long-term warming trend seen in both hemispheres may be due to anthropogenic modification of atmospheric composition (see Sec. 4.4 below), but the variations on scales of a few decades are thought to be natural fluctuations of the coupled atmosphere-ocean system. Recent research suggests that this may be largely connected with the rate of overturning of the waters in the North Atlantic Ocean. Periods of high overturning rate are times when more cold deep ocean water rises to the surface, leading to colder global air temperature. During periods of weak oceanic overturning, the atmosphere is more insulated from the cold deep ocean, resulting in warmer atmospheric conditions. Since much of the water in the global deep ocean originates from sinking surface waters north of Iceland, this variability appears most strongly in the high-latitude North Atlantic. Levitus (1989) presents detailed observations of substantial variability in oceanic conditions in the North Atlantic during two periods separated by 15 years.

In a quite recent development, rather realistic interdecadal variability has been found in a very long simulation performed with a coupled atmosphere-ocean GCM (Delworth *et al.*, 1993). The modeling of interdecadal climate variations is now an area of extremely active research.

4.3 Anthropogenic Effects

Anthropogenic increases in the IR opacity of the atmosphere will result in increased radiative forcing of the surface. Modeling the effects of this increased radiative driving on the actual global and regional climate is now a major activity in the environmental sciences. The most important driver for this anthropogenic climate change is the increase of CO_2 concentration resulting from burning of fossil fuels and deforestation (particularly in the tropical rain forests). Atmospheric CO_2 concentration has been measured since the 19th century and has been monitored very carefully since the 1950s. Preindustrial values are estimated to be ~280 ppmv (parts per million by volume), while present-day values are close to 360 ppmv and are increasing at almost 2 ppmv per year. Without significant governmental regulation, the CO_2 increase is likely to proceed at an accelerating rate, and doubling of the preindustrial concentrations could well occur before the end of the 21st century. This likely scenario has led modelers to consider a standard benchmark experiment which involves simulating climate with something close to the preindustrial value and then repeating the simulation after doubling the CO_2 concentration. Recent research has shown that trends in other trace constituents (e.g., CH_4, N_2O, and chlorofluorocarbons) may also significantly affect the IR opacity of the atmosphere. It has been estimated that anthropogenic changes in these trace constituents have contributed about half as much as the CO_2 rise itself to the increase in radiative forcing of surface climate since preindustrial times.

The earliest useful predictions of anthropogenic climate change were made with a simple one-dimensional radiative-convective model by Manabe and Wetherald (1967). They found that the global-mean surface temperature should increase by about 3 °C in response to a doubling of CO_2. Later work has focused on using comprehensive GCM's for such studies. A number of different general circulation models have been applied to the problem of calculating the "equilibrium" response of climate to doubling CO_2. In these experiments, the upper layer of the ocean is included solely as a heat reservoir, and the preindustrial and double CO_2 cases are both integrated for a period sufficiently long to obtain stable climate statistics. The report IPCC (1990) summarizes 22 such calculations from groups all over the world. All predict a global-mean warming of the surface of between 2 and 5.5 °C (with most predictions between 3 and 4.5 °C). There is also a consistent prediction that the warming will be most intense in the high-latitude regions. This is largely the result of the feedback of snow cover (and sea ice) with temperature. As the climate warms, the snow line (or ice pack) is pushed poleward, reducing the reflectivity of the surface. Beyond this basic feature, however, there remains considerable disagreement among models concerning the predicted regional patterns of climate change.

Some recent studies have used coupled atmosphere–ocean GCM's in an attempt to model the transient evolution of climate expected over the next century. Such studies typically begin from preindustrial conditions and then slowly increase the CO_2 in a ~100–200-yr integration. The results from these experiments differ considerably from those obtained in the equilibrium experiments. When the full effect of the ocean is included, the predicted surface climate warming is considerably slower, since the ocean acts as a huge heat reservoir. Changes in ocean circulation induced by the changes in atmospheric conditions also play an important role in the coupled-model simulations. In particular, the ocean response tends to reduce greatly the warming over the next century predicted for the high-latitude Southern Hemisphere (IPCC, 1990).

The warming found in even the least sensitive doubled-CO_2 experiments would be easily detectable. More troublesome is the question of whether the long-term global warming trends actually observed over the last century (see Fig. 14) can be attributed to anthropogenic effects. At least one reputable research group now regards the observed temperature record as confirming model predictions of anthropogenic warming (Hansen *et al.*, 1991), but at present there is no consensus on this issue.

5. DYNAMICS OF WEATHER

5.1 Baroclinic Instability and the Formation of Extratropical Cyclones

The day-to-day weather fluctuations in midlatitudes are largely associated with the west-to-east passage of cyclones (low-pressure centers) and anticyclones (high-pressure centers). The scale and intensity of these systems vary considerably, but as a general rule one can identify a train of roughly six cyclones alternating with anticyclones circling the midlatitude zone in each hemisphere, resulting in a dominant scale for a cyclone/anticyclone pair of roughly 5000 km. The mean speed of propagation is ~10–15 m s^{-1}, suggesting a period of ~5 days for the passage of the cyclone and anticyclone. Power spectra computed from observed tropospheric pressure in midlatitudes generally show a broad peak centered near periods of 5 days. The actual day-to-day evolution of the weather is complicated by the simultaneous propagation, development, or decay of the cyclones and anticyclones. Growth and decay of individual systems generally occur on time scales of a few days.

From a practical standpoint, the low-pressure systems merit more attention than the high-pressure systems, since cyclones generally develop more intensely and are associated with more severe weather than anticyclones. A schematic picture of the structure of a developing cyclone is shown in Fig. 15. At midtropospheric levels, the cyclone is manifested as an equatorward dip (sometimes called a *trough*) in the prevailing westerly wind. Near the ground, the perturbation to the zonal mean is greater, and closed height contours around the central low pressure are observed. There is also a westward tilt of the cyclone with altitude; i.e., the midtropospheric trough is located west of the surface low. This phase tilt is a very char-

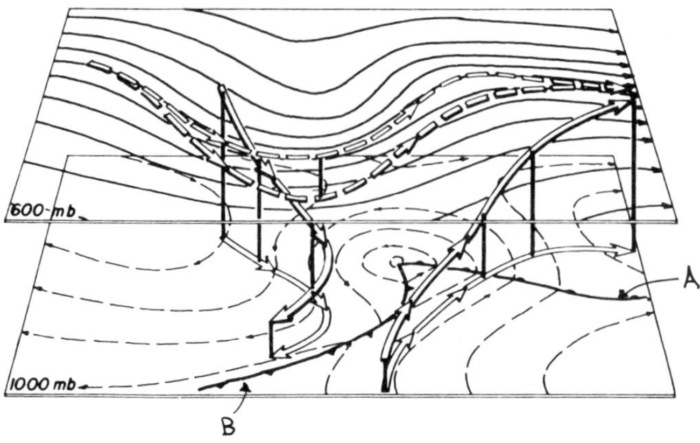

FIG. 15. Schematic depiction of a developing midlatitude cyclone. In the Northern Hemisphere, the storm is being viewed from the south (so that east is to the viewer's right). The dashed curves on the lower plane represent the 1000-mb heights (very nearly proportional to sea-level pressure), while the solid curves on the upper plane are contours of the heights of the 600-mb surface. The continuous heavy arrows represent the three-dimensional trajectories of air parcels, one originating at midtroposphere north of the storm center and one originating near the ground south of the storm center (the same heavy arrows are also used to depict the projection of the trajectories onto the 1000-mb surface). The dashed heavy arrows show trajectories of parcels originating at a midtroposphere level west of the storm center. The solid curves on the lower plane marked "A" and "B" are the warm and cold fronts associated with the cyclone. From Palmen and Newton (1969).

acteristic feature of intensifying weather systems in midlatitudes, and it is easy to show that the westward shift with increasing height implies that the net heat transport by the geostrophic wind will be poleward. The air parcel motions depicted in Fig. 15 illustrate the actual details of heat transport. Air parcels in the cold air north and west of the cyclone center are brought downward and southward, while air parcels in the warm southern sector travel upward and northward. Since small-scale moist convection and precipitation are normally set off by rising motion (leading to adiabatic cooling and then condensation of water vapor), the worst weather normally occurs during the passage of the warm sector (followed by fair weather in the cold sector).

The dynamics of the day-to-day weather systems in the midlatitudes can be largely understood as a manifestation of the phenomenon of *baroclinic instability*. Looking once more at Fig. 2(a), one might imagine that the release of APE would be a straightforward matter. In a small-scale laboratory experiment, one would simply expect the cold (low-θ) air to flood down from the left and displace the high-θ air near the bottom in a simple sloshing motion. On the rotating earth, however, the pressure gradients associated with the meridional temperature gradients in Fig. 2(a) can simply be balanced by the Coriolis force on the zonal component of the wind, and the situation depicted in this figure can be a stationary state of the system. However, for realistic values of the temperature gradient, the situation in Fig. 2(a) is indeed stationary but also unstable to perturbations with wavelike structures in the zonal direction. Thus, the release of APE (and poleward net heat transport) occurs in the cyclones and anticyclones that develop.

The mathematical theory of baroclinic instability was first presented by Charney (1947). The usual approach is to linearize the governing equations about a zonal-mean state. Then the normal-mode solutions (each with a single zonal wave number) are computed. Some of these normal modes will represent neutral traveling-wave solutions, while others will be associated with exponential growth (as well as zonal propagation). For realistic meridional temperature gradients (and hence realistic mean zonal flow shear), the linear theory predicts that the most unstable modes should have zonal wavelengths of several thousand kilometers and have propagation speeds ~ 10 m s^{-1}. The *e*-folding times for growth are typically ~ 2 days. The theoretical results have other realistic features, such as the westward phase tilt with height and the intensification of the pressure perturbations near the ground. The predicted growth rates increase with increasing meridional temperature gradient, and so it is easy to explain why large-scale cyclones and anticyclones are more intense in winter than summer, and why the storm tracks (see Sec. 2.4) tend to occur in specific geographic areas where the shear in the time-mean zonal wind is strongest.

The linear theory of baroclinic instability provides a direct explanation only for the initial growth phase of midlatitude weather disturbances. More complicated nonlinear theories have been advanced to model the entire life cycle of cyclones and anticyclones (see Pedlosky, 1979, for a review).

5.2 Fronts

One very characteristic aspect of the weather in the extratropics is the concentration of gradients in very narrow regions called *fronts*. The passage of a very intense front over some locality can result in temperature changes of more than 20 °C and a complete shift in wind direction in less than an hour. The typical situation is depicted in Fig. 15. The developing cyclone produces a poleward penetration of a wedge of warm air. The temperature gradients at the edges of this wedge intensify into the two fronts. The zone marked A in Fig. 15 leads the warm air as the cyclone moves eastward and is called a warm front, while the one marked B is a cold front. Typically, the cold front will move faster than the warm front. If the cold front actually overtakes the warm front, an *occlusion* is said to occur. An occluded cyclone obviously has little APE to feed on, and thus occlusion is a sign that the winds in a cyclone are about to diminish.

The basic reason for the prominence of fronts in the atmospheric flow may be understood as follows. A blob of passive tracer in any complicated flow tends to get stretched into long filaments by the action of the deformation component of the flow field. Dynamically active quantities such as the temperature may be more resistant to this process. However, in a rotating system like the atmosphere, it is easy to sustain a long narrow zone of temperature gradient. The temperature gradients produce pressure gradients normal to the front, but these pressure forces are balanced by the Coriolis force acting on the wind component along the front. Thus, the pressure gradient provides no resistance to further stretching in the along-front direction. The result is a narrow zone of strong cross-front temperature gradient accompanied by strong vertical shear in the along-front component of the wind. The classical mathematical theory for the formation of atmospheric fronts was presented by Hoskins and Bretherton (1972).

The concentration of horizontal temperature gradients in frontal regions makes them potential sites for development of small-scale baroclinically unstable waves. A theory for the formation of frontal cyclones is discussed by Moore and Peltier (1987).

5.3 Tropical Cyclones and Hurricanes

Meteorology in the tropics differs from that in the extratropics for three principal reasons. First, the factor $2\Omega \sin\phi$ becomes small, thus rendering the geostrophic approximation less accurate. The meridional gradient of radiative forcing decreases rapidly near the equator (see Fig. 11), and thus baroclinic instability is a much less important excitation for the eddy motions in the tropics than in the extratropics. Since the saturation concentration of water vapor rises very rapidly with temperature, the importance of latent-heat release associated with phase changes of atmospheric water is a much more important aspect of the dynamics of the tropics than of the colder, higher-latitude atmosphere.

These differences in the underlying meteorology affect the day-to-day weather systems seen in the tropics. The most frequently observed organized weather systems at low latitudes are the *tropical easterly waves*, weak (~1–2 mb) low-pressure systems with zonal scales ~2000–3000 km, which propagate in the direction of the easterly flow that prevails at low levels (see Fig. 6). In the Atlantic, such waves typically originate over Africa and intensify as they propagate across the ocean in ~10 days. These waves are invariably accompanied by regions of enhanced rainfall. The main energy source for tropical easterly waves during their oceanic propagation is the latent-heat release associated with the precipitation. Detailed observations of the structure of Atlantic tropical easterly waves is given by Reed *et al.* (1977). There are analogous disturbances over the Pacific Ocean.

Some waves in the tropical ocean regions

develop further into tropical cyclones. These are storms in which the low-pressure center is strong enough to form an isolated region of closed contours in surface isobaric maps. When the surface winds in a tropical cyclone exceed 33 m s^{-1}, it is referred to as a *hurricane* (sometimes called a *typhoon* if it occurs in the Pacific region). Hurricanes differ greatly in size and intensity (radii from 100 to 1000 km and maximum pressure perturbations from 20 to 100 mb), but all display a rough circular symmetry.

The development of tropical cyclones and hurricanes can be thought of as a kind of instability. The surface winds spiraling into the low pressure bring moist air, which rises near the center, releasing the latent heat that ultimately drives this storm-scale convection cell. The hurricane is then a heat engine with a source of fuel that increases as the engine works faster. The explosive growth phase of a hurricane can last from 12 hours to a couple of days. The hurricane then generally persists for a few days until it reaches land or heads northward over colder water. Either of these contingencies leads to a reduction in the latent-heat energy source for the storm, and the hurricane will then generally dissipate in one or two days.

The instability process that leads to hurricane formation depends on having sufficiently warm ocean surface waters. This imposes both geographical and seasonal constraints on hurricane formation. Almost all North Atlantic hurricanes and strong tropical cyclones form in the late summer or early fall, when ocean surface temperatures north of the equator are highest. In addition, almost all Atlantic hurricane formation is restricted to the western half of the Atlantic basin where the ocean is warmest. Over the Pacific Ocean, there are also specific seasons and geographical regions where hurricane formation is favored. Recently, there has been some speculation that one consequence of enhanced greenhouse warming of climate will be an increase in the number or intensity of severe tropical cyclones.

GLOSSARY

Anthropogenic: Induced by the activities of man.

Convection: Used by meteorologists to denote overturning fluid motions (this use differs from some other branches of fluid mechanics).

Greenhouse Gases: Trace constituents that contribute to the IR opacity of the atmosphere.

Greenhouse Effect: Blanketing effect of the downward emission of IR radiation from the atmosphere, leading to a warming of the earth's surface climate.

Radiative Equilibrium: A hypothetical state in which the net radiative heating at each point in the atmosphere (and the earth's surface) is identically zero.

Spectral Method: A numerical technique for the approximate solution of time-dependent differential equations in which the dependent variables are written as finite sums over a set of spatial basis functions (e.g., spherical harmonics in a global model).

Stratosphere: The layer of the atmosphere above the troposphere, characterized by increasing temperature with height.

Trace Constituent: Chemical species present in the atmosphere in very small concentrations.

Tropopause: The level between the troposphere and stratosphere.

Troposphere: The lowest layer of the atmosphere, characterized by decreasing temperatures with height.

Wind Shear: Spatial gradient of the wind.

Works Cited

Charney, J. G. (1947), *J. Meteorol.* **4**, 135–162.

Delworth, T., Manabe, S., Stouffer, R. (1993), *J. Climate* **6**, 1993–2011.

Gordon, C. T., Stern, W. F. (1982), *Mon. Weath. Rev.* **110**, 625–644.

Hansen, J., Johnson, D., Lacis, A., Lebedeff, A., Rind, D., Russel, G. (1991), *Science* **213**, 957–966.

Holloway, J. L., Manabe, S. (1971), *Mon. Weath. Rev.* **99**, 335–370.

Horel, J. D., Wallace, J. M. (1981), *Mon. Weath. Rev.* **109**, 813–829.

Hoskins, B. J., Bretherton, F. P. (1972), *J. Atmos. Sci.* **29**, 11–37.

IPCC (1990), *Climate Change—The IPCC Scientific Assessment*, Intergovernmental Panel on Climate Change, Cambridge, U.K.: Cambridge University Press.

Jones, P. D., Raper, S. C. B., Bradley, R. S., Diaz, H. F., Kelly, P. M., Wigley, T. M. L. (1986a), *J. Climate Appl. Meteorol.* **25**, 161–179.

Jones, P. D., Raper, S. C. B., Bradley, R. S., Diaz, H. F., Kelly, P. M., Wigley, T. M. L. (1986b), *J. Climate Appl. Meteorol.* **25**, 1213–1230.

Lau, N.-C. (1988), *J. Atmos. Sci.* **45**, 2718–2743.

Levitus, S. (1989), *J. Geophys. Res.* **94**, 16125–16131.

Liou, K. N. (1980), *Atmospheric Radiation*, New York: Academic Press.

Lorenz, E. N. (1969), *J. Atmos. Sci.* **26**, 636–646.

Manabe, S., Wetherald, R. T. (1967), *J. Atmos. Sci.* **24**, 241–259.

Moore, G. W. K., Peltier, W. R. (1987), *J. Atmos. Sci.* **44**, 384–409.

Palmen, E., Newton, C. W. (1969), *Atmospheric Circulation Systems*, New York: Academic Press.

Pedlosky, J. (1979), *Geophysical Fluid Dynamics*, New York: Springer-Verlag.

Peixoto, J. P., Oort, A. H. (1992), *Physics of Climate*, New York: American Institute of Physics.

Philander, S. G. H. (1991), *El Niño, La Niña, and the Southern Oscillation*, New York: Academic Press.

Phillips, N. A. (1963), *Rev. Geophys.* **1**, 123–176.

Ramanathan, V., Coakley, J. D. (1978), *Rev. Geophys. Space Phys.* **16**, 465–489.

Rasmusson, E. M., Carpenter, T. (1982), *Mon. Wealth. Rev.* **110**, 354–384.

Reed, R. J., Norquist, D. C., Recker, E. E. (1977), *Mon. Weath. Rev.* **105**, 317–333.

Wallace, J. M., Hobbs, P. V. (1977), *Atmospheric Science*, New York: Academic Press.

Further Reading

Andrews, D., Holton, J. R., Leovy, C. B. (1987), *Middle Atmosphere Dynamics*, New York: Academic Press.

Houghton, J. T. (1986), *Physics of Atmospheres*, Cambridge, U.K.: Cambridge University Press.

Riehl, H. (1979), *Climate and Weather in the Tropics*, New York: Academic Press.

METROLOGY

WOLFGANG WÖGER AND SIGMAR GERMAN, *Physikalisch-Technische Bundesanstalt, Braunschweig, Germany*

INTRODUCTION

Metrology usually is described as the field of knowledge concerned with measurement. However, this rather general description can be slightly misleading, as it does not sufficiently emphasize that any metrological consideration and operation ultimately focus on the notions of the accuracy and uniformity of measurement. In fact, as metrology concentrates on these two aspects of measurement, it does not encompass the bulk of experimental physics, but rather can be viewed as a branch of applied physics. As such, it provides the scientific basis of measurement with special regard to the accuracy of measurement as well as to methods and means to ensure uniformity of measurements. Accuracy and uniformity of measurement contribute to the growth of scientific knowledge, bring about reliability and confidence, support the quality of products, protect the consumer, and add to the capability of economy to successfully compete in the markets. Therefore, metrology is important to all fields of science, technology, and commercial or social transactions.

In practice, metrological tools and operations can be expensive. This is particularly true for measurements on the highest possible level of accuracy such as measurements of the fundamental constants of physics. They usually are performed in national metrology laboratories or academic institutions. But even at somewhat lower levels of accuracy, such as in manufacturing industries, adequate metrology may still cause significant cost. Nonetheless, appropriate on-line measurements of critical parameters of a manufacturing process, acknowledged by many as introducing "non-value-added" expense, can often be shown to be ultimately less expensive than the application of only poor production metrology (see e.g. Kudva and Potter, 1992). In fact, the present level of development of some technologies, such as microelectronics, would not have been possible without paralleling advances in production metrology.

Metrology may be divided into three partially overlapping fields:

1. Scientific metrology, dealing with the theory of systems of units; the theory of measurement; methods of evaluation of results of measurement and uncertainty of measurement (numerical expression of the accuracy of measurement); terminology; realization of the units on the highest possible accuracy level; and very accurate measurements of, for instance, fundamental constants, or measurements to test or compare physical theories.
2. Applied scientific metrology, dealing with all components of practical importance that influence the accuracy and uniformity of measurements. Such components are the closeness between the actual ref-

3-527-28132-0/94/$5.00 + .50

erence quantity used in measurement and the defined unit the reference quantity is supposed to represent; the properties of the measuring instruments; and the choice of the measurement procedure. Thus, applied metrology includes the representation of the units or of their multiples and submultiples by standards arranged in steps of uncertainty, the maintenance of the standards, the dissemination of the units by comparison of standards; and calibration of measuring instruments.

3. Legal metrology, dealing with the inspection examination of properties of measuring instruments intended to be used in the public interest. The properties must meet legally specified requirements (mostly in the form of specified maximum permissible errors of the instrument). Regulations are set up, for instance, for instruments used in consumer protection or in medicine, or for instruments monitoring environmental pollution. Legal metrology is not intimately connected to scientific metrology and its direct applications.

Historically, up to the last few decades, metrology was perceived by much of the scientific community as a science of auxiliary and custodial character, not contributing essential elements to scientific progress. This view simply disregards important scientific discoveries by metrologists such as the establishment of the laws of blackbody radiation in 1900. After development of a method to realize accurately the concept of the blackbody, it was the subsequent accurate measurement of its radiation's spectral energy distribution that enabled Planck to initiate quantum physics.

But undoubtedly the tasks of metrology for a long time were predominantly understood as to provide, maintain, and disseminate a set of units—in particular, of length and mass—in order to support uniformity of measurements in commercial and social transactions. But the units of length and mass in former times were only locally defined by arbitrary man-made artifacts, and these were reproduced as accurately as possible to yield reliable secondary material measures valid in the geographical region for which the units were defined. In Europe, for example, this local restriction often made trade difficult even between different cities of the same country, as the cities usually insisted on having their own units of length and mass.

It was only in 1875 that a first step to national and international uniformity was made. In this year, because of the apparent growing need for common standards and accurate measurements in the emerging industrial society, the Convention du Mètre (agreement on the metric system) was signed by 17 countries, including the U.S.A. A decimal system of weights and measures was established and the agreed-on units, the prototypes, of length and mass were decided to be kept in Paris. This event can be viewed as the birth of modern metrology. Additionally, to propagate and further to develop the metric system, the Conférence Générale des Poids et Mesures (CGPM), the Bureau International des Poids et Mesures (BIPM), and the Comité International des Poids et Mesures (CIPM) were founded at that time (for details see UNITS).

In the sequel, national metrological institutions were installed that, on the basis of the above-mentioned agreement, were in charge of scientific metrological research and of the national uniformity of measurement in technology and industry by realizing, maintaining, and disseminating national standards in concordance with the internationally agreed-on units. The first of such institutes were the Physikalisch-Technische Reichsanstalt (now Physikalisch-Technische Bundesanstalt, PTB), founded in 1887 in Germany, the National Physical Laboratory, NPL, founded in 1900 in England, and the National Bureau of Standards (now National Institute of Standards and Technology, NIST), founded in 1901 in the U.S.A.

In the course of time, progress in measurement showed that particularly the accuracy of a length measurement was nearly exclusively determined by the accuracy with which the actual reference quantity in measurement could be compared with the prototype in Paris. For example, whereas the lattice parameter of a crystal could very accurately be measured in terms of a specific x-ray wavelength, on expressing the lattice parameter in the length unit meter the experimental accuracy immediately was lost.

It was only when improved interferometric measurement techniques suggested redefining the meter as a multiple of the wavelength of a characteristic electromagnetic radiation of krypton atoms (^{86}Kr) that the accuracy in length measurements was increased. The multiple was chosen in such a way that for all practical purposes there was no change to the original prototype definition of the meter (concept of the "continuity of the units").

The unit of length with its above new definition was one member of the set of seven base units of the coherent International System of Units (SI) that in 1960 replaced all of the various earlier unit systems (see UNITS). The definitions of these base units, however, should not be viewed as fixed once and for all within the SI. For example, according to the needs of even higher accuracy and taking into account the principle of continuity, a redefinition of the length unit was agreed on in 1983, tracing it to the definition of the unit of time, the second, which can be realized with a higher accuracy than any other of the base units. Also, in 1979, the 1967 definition of the unit of luminous intensity, the candela, was changed (for details, see UNITS).

The least satisfying definition in the SI is that of the unit of mass, the kilogram, which since 1889 is given by an artifact maintained in Paris at the BIPM. Any change with time of the mass of the prototype results in the change of the unit. Great efforts are being made in modern metrology to replace this anachronism by a very accurate determination of the Avogadro constant (see CONSTANTS, FUNDAMENTAL).

However, whereas the increasing sophistication of measuring instruments and techniques allow for improvement of the utmost accuracy of measurement by appropriate definition of the SI units, the corresponding primary or even secondary standards as well as the methods of unit transfer may differ markedly from their prototype predecessors. An example is the standard of the meter, represented by the wavelength of a specific visible line of the He-Ne laser. The line is linked experimentally to the frequency (in the infrared) of the radiation corresponding to a well-defined transition of the cesium atom, i.e., it is derived from the radiation used in defining the second. This standard meets scientific needs and is known to be of a high accuracy, but it is far from the much more practical line scale, which allows for meter comparisons by a simple physical operation. Elaborate interferometry must be used to compare the wavelength with a line scale accurately.

Therefore, quite a large part of metrological work in the national institutions consists of providing a hierarchy of standards of decreasing accuracy. The chain starts from the first transfers from the unit and, as in the case of the line scale, ends in an appropriate physical embodiment of the unit to be used in practice. The hierarchy of standards is of special importance in quality management of measuring equipment and standards. Here, for calibration purposes measurement standards must have the property of "traceability," which means that the accuracy of the standard in use must be traceable to that of a corresponding national or international standard through a documented unbroken chain of comparison (see CALIBRATION AND MAINTENANCE OF TEST AND MEASURING EQUIPMENT).

The basic idea underlying the SI is to define, whenever possible, a unit not by an arbitrary artifact, but by a property of a prescribed specific natural phenomenon that can be realized very accurately at any place and time. The impact of this concept on metrology resulted in a change of its role and scope in science. Modern metrology is acknowledged to be a universal discipline intimately linked to nearly all frontier developments of the physical sciences and of technology.

To treat so vast a topic as metrology and its wide range of practical applications in detail is impossible in a brief space. This article thus concentrates on basic and general metrological aspects that are related to applied physics. For additional information on, for instance, methods and algorithms to evaluate measurements, on the system of fundamental constants, or on calibration and maintenance of standards and measuring instruments, the reader is referred to the corresponding articles in the Encyclopedia.

In the first section, basic terms in metrology are discussed. The important concept of a system of units completely based on fundamental constants of nature is introduced. The second section is devoted to the result and accuracy of measurement. Quantitative measures of the accuracy such as the uncertainty of measurement, as well as other related terms, are described. The article closes with a list of selected basic and general metrological characteristics of measuring instruments.

1. BASIC CONCEPTS IN METROLOGY

In the first two subsections, a series of definitions and statements is presented concerning some of the most important terms related to measurement. Section 1.3 deals with the problem of the value of a measurand defined prior to measurement. In Section 1.4, the concept of completely basing a system of units on the fundamental constants of nature is introduced as a major topic of scientific metrology.

1.1 Quantity, Value, and Unit

The qualitative and quantitative description of an attribute of a physical phenomenon (body, process, state) is accomplished by the introduction of a (physical) "quantity." The magnitude of a quantity is its "value."

To establish an appropriate physical phenomenon under prescribed physical conditions means to establish a definite value of a quantity. Obviously, there are arbitrarily many ways to establish the same value of a quantity. Thus, a quantity cannot be specified by a value but only by a complete description of the physical context fixing the value of the quantity of interest. In reality, a description of a quantity in most cases only approximates the ideal complete description, the approximate description being regarded as sufficient for the purposes under consideration. For example, to talk of the resistance of a given specimen of wire at a specified temperature is only an approximate description of the quantity, as its value is known to depend also, at least, on the existing magnetic field. In metrology, the definition of a quantity plays a role when evaluating experimental information.

A quantity can be compared quantitatively with an arbitrary reference quantity of the same kind in order to express its value relative to that of the reference quantity.

The value of a quantity usually is written as the value of the considered reference quantity multiplied by a number called "numerical value." The value of the reference quantity is the "unit" considered in the comparison. Its numerical value conventionally is accepted to be equal to 1. Depending on the chosen unit, the value of a quantity may be expressed in more than one way.

1.2 Measurement

The set of operations having the object of comparing quantitatively a quantity with a chosen unit (of measurement) is called "measurement." In metrology, a measurement always includes a set of experimental operations but, in general, also consists of theoretical considerations. (There are more general interpretations of the term measurement, where the comparison need not be performed experimentally, and the result may consist of a symbol.)

A quantity subject to measurement is the "measurand." It need not be directly measured. Its value may be derived from those of directly measured quantities by known physical or mathematically defined relationships. The measurement of a measurand always is based on a chosen principle of measurement and a method of measurement, both influencing accuracy and precision of the measurement.

The aim of measurement need not be the value of the measurand expressed by the chosen unit of measurement, but may also be just a statement about whether the value is larger or smaller than some multiple of the unit.

Any result of a comparison need not be observed immediately. It may be automatically processed further in a measuring system.

The evaluation of information obtained from measurement, whether performed by hand or by an integrated computer, is considered to be the final part of measurement.

When the same measurand is measured repeatedly, the measurements should be performed under fixed conditions in such a way that the observed data can be regarded as independent of one another. Otherwise, the evaluation of the obtained data to infer the value of the measurand will prove to be difficult. If additionally the conditions chosen to be the same during repetition allow for the assumption of constant systematic errors (see Sec. 2), the measurement commonly is said to be performed under "repeatability conditions." In this case, the fluctuation of the observed data stems solely from uncontrollable influences of the measuring instrument. Repeatability conditions consist at least of the same observer, the same measuring procedure, the same measuring instrument, and the same influence quantities, which are quantities other than the measurand that affect the result of measurement.

1.3 Value of the Measurand

The measurand is the quantity to be measured. In the simplest case, the measurand may be defined to be the quantity that is directly recorded by the measuring instrument under the actual physical conditions existing at the time and place of measurement. Although these conditions never can be known completely, they uniquely determine the value of the actual measurand. The instrument provides direct but incomplete information about that value.

However, a comparison process always is characterized by an interaction energy between the observed actual measurand and the measuring instrument. Therefore, the value of the measurand observed is not identical with that of the corresponding undisturbed measurand. The latter value must be inferred from knowledge of the disturbing process. It is an essential part of the evaluation of given experimental information to set up a model for the relationship between the value of the actual measurand and that of the undisturbed measurand if, as usual, the latter is the value of interest.

In practice, often, the measurand is specified in advance of measurement by a list of prescribed physical states and conditions. But this description of the measurand may not be complete and, thus, may not fix the measurand's value uniquely. Rather, the given conditions defining the measurand will allow for a set of values consistent with the conditions, where each value out of the set is considered to fit the purpose of measurement. It is meaningless in case of a so-defined measurand to talk of "the" value of the measurand before measurement.

But if the quantity realized for measurement is known to be fully consistent with the (incomplete) definition of the measurand, its value is defined to be the value of the measurand. For example, if the measurand is the resistance of a given wire for any temperature between 20 and 30 °C, then realizing a temperature within this interval will yield a value of the resistance not in contradiction to the defined measurand. The value of the realized quantity can be inferred from the information gained by measurement.

If the definition of the measurand is regarded as complete, i.e., is regarded as fixing the value of the measurand, then, as a rule, the quantity realized for measurement only approximates the measurand. It is the realized quantity that will be directly subject to measurement. To infer the value of the measurand from the gained information, one not only needs a model of measurement of the realized quantity, but also a model of the relationship between measurand and realized quantity.

1.4 Units and Fundamental Constants

Although the choice of the unit of measurement (see Sec. 1.1) in principle is quite arbitrary, to achieve uniformity in measurement, the same unit for all measurements of quantities of the same kind should be used.

A convenient definition of a unit consists of an appropriately chosen phenomenon under prescribed physical conditions such that

1. the value of the quantity of interest, i.e., the unit, does not change in space and time; and
2. any experimental reproduction of the phenomenon and the conditions is possible very accurately at any place and time (also see UNITS).

If the result of a measurement is to be stated in terms of a well-defined unit, the measurement cannot be more accurate than the definition of the unit was realized. Thus, it is a primary metrological task to realize such a definition as accurately as possible and to find ways to disseminate the unit for practical use.

Fundamental constants are of particular interest to modern metrology because of the idea of establishing a system of units entirely based on fundamental constants of nature. The constants to our best knowledge satisfy requirement **1**. Additionally, it is well known that there are some constants like the magnetic flux quantum $h/2e$ (h is Planck constant, e, the elementary charge) that are connected to phenomena satisfying requirement **2**.

In particular, the magnetic flux quantum

in the ac Josephson effect (see CONSTANTS, FUNDAMENTAL) relates a voltage to a frequency. Whereas the voltage, and thus the value of the flux quantum, can be reproduced with a relative uncertainty of a few parts in 10^{13}, the value of the flux quantum in SI units is only known with a relative uncertainty that is six orders of magnitude larger. The reason is the poorly known relationship of the as-maintained unit of voltage to the SI volt.

The general idea to improve this situation is to fix the numerical value of the flux quantum to introduce via the ac Josephson effect a new unit of voltage without the deficiencies of the existing SI unit. The procedure could be repeated with other phenomena such as the quantum Hall effect where the von Klitzing constant h/e^2 can be used to establish a new ohm (see CONSTANTS, FUNDAMENTAL). Similarly, the recently discovered single-electron tunneling effects (SET) (Geerligs *et al.*, 1990) could lead via the elementary charge e to a new unit of electric current closing the "electrical triangle" between fundamental constants and the electrical units.

However, because of the demand of continuity of units and the fact that most of the constants are interrelated, the choice of units based on phenomena the theory of which contains fundamental constants cannot be arbitrary. Such a system of units can only be found on the basis of a very accurately known consistent set of experimentally determined values of fundamental constants expressed in the adopted SI units. At present, a set satisfying these needs does not exist. It is a major goal of scientific metrology to improve this situation.

For more details on units and fundamental constants, see CONSTANTS, FUNDAMENTAL and UNITS. The first of these articles to some extent treats the problem of the unit of mass to be based on the Avogadro constant.

2. RESULTS OF MEASUREMENT

As stressed in the Introduction, the accuracy of measurement is an important element of all metrological considerations. In this section, an overview is given of the metrological concepts related to accuracy.

A measuring instrument used to determine the value of a measurand yields an output from which the "measured value" can be extracted (often the output directly is the measured value). The measured value is equal to the product of some numerical value and the unit of measurement. It must be assumed to deviate unpredictably from the unknown value of the measurand, but it is associated with the measurand as it contains information about the measurand's value.

The same is true for the "uncorrected result of measurement" (see below). It is a function of the measured values obtained from repeated measurements of the measurand under the same conditions. The uncorrected result of measurement serves as a starting point for stating a best estimate of the unknown value of the measurand. It is this best estimate that finally is stated as the "result of measurement."

In establishing the result of measurement, apart from measured values and the uncorrected result of measurement, additional knowledge on the properties of influence quantities and measuring instruments must be taken into consideration. Moreover, if the measurand is to be determined indirectly from measurement of directly measured quantities, known physical relationships between the quantities must be used to find the result of measurement.

In the context of the evaluation of experimental data, the unknown value of the measurand commonly is called the "true value" of the measurand. For a given purpose, it often is convenient to replace the unknown true value of the measurand by a known "conventional true value," regarded as sufficiently close to the true value for the difference to be insignificant for the given purpose. For instance, for the purpose of calibration, an appropriate experimentally determined value of a measurand may be substituted for the true value by convention (see CALIBRATION AND MAINTENANCE OF TEST AND MEASURING EQUIPMENT).

2.1 Error of Measurement

The result of measurement is an estimate of the true value of the measurand and presumably will deviate from it (if it happens not to deviate from the true value, the occurrence of this event is unknown). The deviation of the result x_{res} of measurement from the true value x_{tr} is called "error of (the result of) measurement."

The error of measurement

$$e = x_{res} - x_{tr} \quad (1)$$

originates from at least one of the following causes:

1. The actually measured quantity is not exactly equal to the measurand. In principle, this always happens when the measuring instrument is coupled to the measurand. Also, when the measurand is prescribed and has to be realized experimentally, the realization does not exactly agree with the prescribed measurand.
2. The actual unit of measurement is not exactly equal to the unit intended to be used (usually the SI unit).
3. The comparison of the value of the actually measured quantity with the actual unit of measurement is not exact.
4. The values of the actual influence quantities do not agree completely with those the influence quantities are intended to have.
5. The result of measurement is calculated from directly measured quantities on the basis of a relationship that does not correspond to the correct physical relationship between the quantities.

The estimation of the true value from the known data of measurement most often is accomplished by application of the methods of estimation developed in statistics. These methods can be applied to a series of observed indicated values, obtained independently of one another from repeated measurements under the same conditions. This, in particular, means that a new measurement must only be made if the experiment as a whole is in the same state as it was immediately before taking the first measurement of the series.

The basic ingredient for construction of an estimate x_{res} for the unknown true value x_{tr} of a measurand is the measured value. Taking n measurements, it will be denoted by x_i $(i = 1, \ldots, n)$.

Under repeatability conditions, the unknown error $x_i - x_{tr}$ of the measured value can be regarded as the sum of two components, which differ drastically in their behavior when the measurement is repeated. Whereas the so-called "systematic error" e_s for any i can be assumed to be the same, the "random error" e_{ri} fluctuates at random from measured value to measured value;

$$x_i - x_{tr} = e_i = e_{ri} + e_s \quad (2)$$

The systematic error mainly is caused by items **1**, **2**, **4**, and **5** listed above, whereas the random error may be traced to causes like the uncontrollable influences of the measuring instruments, uncontrollable changes of the values of the influence quantities, unpredictable changes of the value of the measurand, or varying influences of the observer (e.g., in reading the measured values).

The systematic error is constant for all measured values of a measurement series of size n, provided the chosen set of experimental conditions can be assumed to be constant during repetition. It is only in this sense that the error is "systematic," because this constant error occurs at random immediately before the first measured value is observed. The systematic error is not observable and cannot be extracted from the measured values of the measurement series. If error source **5** above has to be taken into account, it introduces an unobservable systematic error, which is not of a random character. Usually two limits enclosing that error can be stated. Then, on the basis of this knowledge, each of the values between the limits is assumed to have the same probability to be equal to the actual systematic error.

In metrology, the random error is understood to satisfy the demand that it averages out, when one takes the arithmetic mean

$$\bar{x} = \frac{1}{n} \sum_{i=1}^{n} x_i \quad (3)$$

of a very large number (ideally, an infinite number) of measured values obtained under the same conditions. This demand is an essential part of the model of the measured value [Eq. (2)].

The scatter of the observed measured values within a measurement series of size n usually is characterized by the standard deviation

$$s = \left[\frac{1}{n-1} \sum_{i=1}^{n} (x_i - \bar{x})^2 \right]^{1/2}, \quad (4)$$

which is independent of the systematic error in a series of measurements under repeatability conditions.

The standard deviation of the measured values is a numerical measure of the closeness of the agreement between the measured values and, thus, of the "precision" of the measurement performed. The use of the qualitative term precision for "accuracy" should be avoided, as the latter term is understood as the closeness of agreement between the result of measurement and the true value of the measurand.

The constant value approached by $\bar{x}$ in the limit $n \to \infty$ is the expectation value of the measured values, denoted by μ. (The probability of $\bar{x}$ deviating from μ vanishes for $n \to \infty$). The expectation value in principle is observable but in practice always is unknown. It can be estimated by $\bar{x}$ with finite n. The arithmetic mean $\bar{x}$ is the uncorrected result of measurement. It is an estimate of the expectation value.

The expectation value of the measured value is equal to the true value if there is no systematic error. For, inserting Eq. (2) into Eq. (3), letting $n \to \infty$, and using the fact that the average random error in this limit approaches zero, i.e., the expectation value of the random error is zero, yields

$$\mu = x_{\text{tr}} + e_s. \tag{5}$$

In Eq. (5), it is assumed that for all measured values the systematic error is the same, i.e., it is possible to completely control the set of constant experimental conditions even in an infinite measurement series.

Expressed in terms of μ, it follows that

$$e_s = \mu - x_{\text{tr}}. \tag{6}$$

Using this in Eq. (2), one finds the random error

$$e_{ri} = x_i - \mu. \tag{7}$$

The measured values x_i $(i = 1, \ldots, n)$ are grouped around their expectation value in such a way that the average deviation from μ approaches zero for a large number of measured values.

The random error e_{ri} $(i = 1, \ldots, n)$ is unknown. If something is known at all about the error e_i [Eq. (2)], this can only refer to e_s. Apart from exceptional cases, the systematic error is not known completely. Therefore, it commonly is decomposed into an estimate of e_s, called the "known systematic error" e_{sk}, and the "unknown systematic error" e_{su};

$$e_s = e_{\text{sk}} + e_{\text{su}}. \tag{8}$$

In practice, the knowledge of the systematic error often merely consists of a set of conceivable systematic errors enclosed between two limits a and b $(a < b)$. It is assumed that a value outside this interval will not occur in the measurement. As an estimate of e_s, the known systematic error e_{sk} is chosen as the average of all conceivable systematic errors for the measurement under consideration. Therefore, it is an element of that set and mostly is situated in the middle of the interval. The unknown systematic error e_{su} is the unknown deviation of the actual systematic error e_s from e_{sk}. As e_s can be any of the conceivable errors, the average of e_{su} vanishes.

The expression

$$x_i^c = x_i - e_{\text{sk}} \tag{9}$$

is the "corrected measured value." The value $K = -e_{\text{sk}}$, which compensates for the known systematic error when algebraically added to the measured value, is called the "correction."

Figure 1 schematically shows the connection of a measured value to the various kinds of errors introduced above.

The corrected measured values are the basis of the statement of the result of measurement. In fact, under repeatability conditions, the result of measurement is

$$x_{\text{res}} = \overline{x^c} = \frac{1}{n} \sum_{i=1}^{n} x_i^c. \tag{10}$$

It can be taken as the best estimate of x_{tr} on the basis of the knowledge about the measurements performed; x_{res} is a statement about one particular measurement series under the same conditions.

If, for the systematic error, limits a and b were stated inappropriately, e_{sk} is not a valid estimate of e_s. This unknown fact can only be revealed by comparison of x_{res} with one or more other results of measurement for the same measurand.

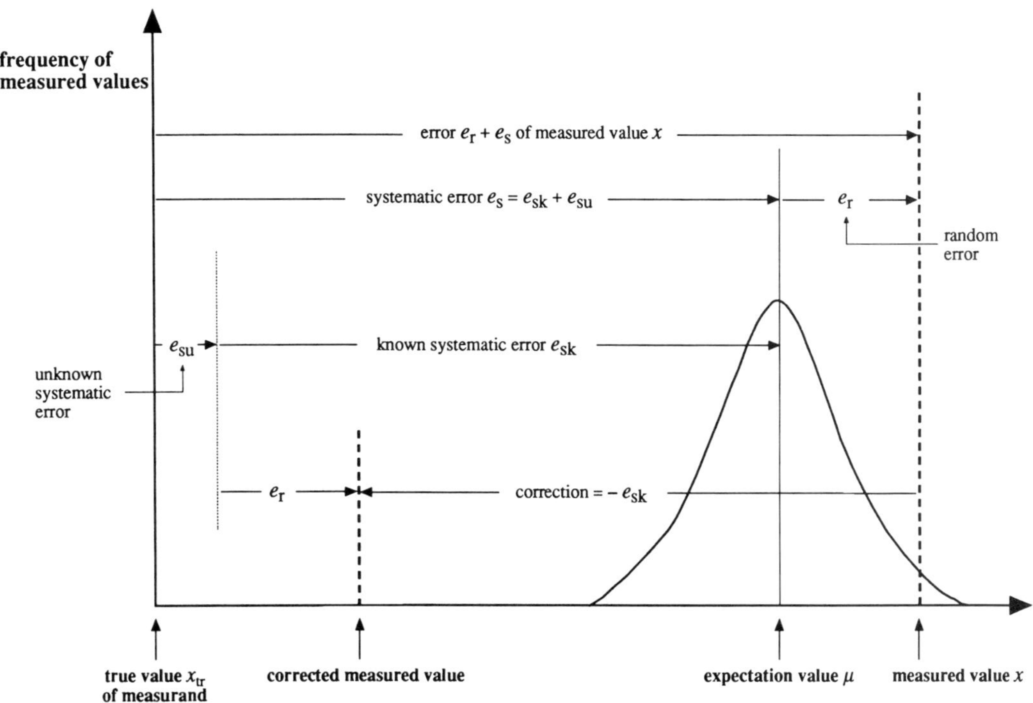

FIG. 1. Schematic representation of the relations between measured value, corrected measured value, expectation value, and true value of the measurand for measurements under repeatability conditions. According to a distribution, the measured values group around the expectation value, which deviates by the systematic error from the true value of the measurand. The corrected measured value results from addition of the correction to the measured value and deviates from the true value by the sum of random and unknown systematic error. The representation is valid also when one substitutes the uncorrected result of measurement for the measured value and the result of measurement for the corrected measured value (the distribution of uncorrected results of measurement then replaces that of the measured values).

Inserting Eqs. (2) and (8) into the right-hand side of Eq. (10), one has

$$x_{\mathrm{res}} - x_{\mathrm{tr}} = \overline{e}_r + e_{\mathrm{su}} \tag{11}$$

The error of measurement can be reduced by increasing n, the number of measurements under the same conditions, because then the average random error gets smaller and smaller. But e_{su}, which is constant during repetition of measurement, can only be reduced by reducing $b - a$, the width of the interval containing all conceivable systematic errors.

2.2 Uncertainty of Measurement

After measurement, the knowledge of the value of the measurand has increased compared to the situation before measurement. But this knowledge still is incomplete. There always remains a set of values, consisting of reasonable estimates of the true value of the measurand, so that any of these values can be attributed to the measurand without contradicting the (incomplete) information obtained from the experiment. Thus, the set represents uncertainty in the knowledge of the measurand's value after measurement.

The actual result of measurement is an element of this set. But its statement alone obviously cannot be regarded as a complete description of what one has learned from the measurement. It must be taken into consideration that any of the other values of the set could also have occurred as the result of measurement. The complete description consists of the result of measurement and a numerical characterization of the uncertainty in the knowledge of the true value.

From all that one knows from and about the measurement (directly measured values, influence quantities, properties of measuring instruments, relations between quantities, etc.), a typical such set or interval I_{est} of estimates can be found out of the set of all possible estimates of the true value of the measurand (for methods see STATISTICAL DATA AND ERROR ANALYSIS). It is centered around the actual result x_{res} of measurement and, to repeat, contains those values considered to be reasonably attributable to the measurand on the basis of the measurement.

The spread of the estimates in I_{est} on either side of x_{res} is given by the numerical value of an appropriately chosen parameter. This numerical parameter value is called the "uncertainty of measurement," the "uncertainty of measurand," or simply "uncertainty," and will here be denoted by u.

In stating the uncertainty, one takes account not only of the typical spread of the observations but also of the spread of values of the total systematic error that could reasonably occur in the measurement performed. Both spreads influence the spread of the reasonable estimates. Thus, the uncertainty of measurement decreases for increasing information about the true value.

As the uncertainty of measurement is determined by the knowledge of the properties of the measurement and nothing else, it is to be distinguished from the spread of values characteristic of a confidence interval with given confidence level (see below).

It is meaningless to talk about the "uncertainty of the result of measurement," as this result is one of the few things certainly known after measurement. Additionally, this wording supports the misleading identification of the error of the result of measurement with the uncertainty of measurement. The two are distinct concepts. In fact, although there always is a nonvanishing uncertainty of measurement, the error of the result of measurement unknowingly can be zero.

The interval I_{est} with limits $x_{res} - u$ and $x_{res} + u$ cannot be regarded as necessarily containing the true value of the measurand. The degree of confidence or the probability attached to the statement to be correct that the actually quoted I_{est} contains the true value depends on the uncertainty u calculated on the basis of the chosen parameter that measures the spread of estimates around the result of measurement. Consequently, if two such parameters are chosen, one of which on the basis of the same experimental information consistently yields larger uncertainty than the other, the corresponding interval of estimates will be preferred in situations where a high degree of confidence is asked for. Parameters of this kind are recommended for commercial, industrial, regulatory, and other applications (e.g., concerning health and safety).

But in scientific metrology and most of applied scientific metrology the parameter is chosen to yield the uncertainty as the standard deviation of the result of measurement [this standard deviation is not equal to s in Eq. (4)], in which case the uncertainty is called "standard uncertainty." The standard uncertainty does not satisfy the demand of an interval with a high level of confidence.

The main reasons for this choice are that in science one frequently compares different measurements of the same measurand or often uses a previous result of a measurement of some measurand to evaluate the measurement of a different measurand. But if there are differences in two or more measurements of the same quantity, they should not be masked by stating too large an uncertainty. Only if such differences are clearly detectable can one learn how to remove them by improved measurements. Additionally, it is only the standard uncertainty of measurement of a measurand for which an algorithm exists to propagate the uncertainty consistently to compute the final standard uncertainty of measurement of a related measurand.

The algorithm, along with least-squares adjustment, is dealt with in the article STATISTICAL DATA AND ERROR ANALYSIS (also see Martin, 1971; Press *et al.*, 1989). In its most general form, it allows us, for instance, to calculate several measurands of interest and their uncertainties on the basis of their known relationships to a set of other quantities simultaneously measured. This kind of evaluation task gets more and more important because of the increasing trend to complex measuring systems, where integrated computers facilitate the calculations.

2.3 Accuracy of Measurement

The "accuracy of measurement" is the closeness of the agreement between the result of a measurement and the true value of the measurand. It should be distinguished from the precision of measurement, as the accuracy implicitly contains statements on systematic errors. Therefore, a precise measurement can be very far from being accurate.

The uncertainty u is a quantitative measure of the accuracy of the measurement in the sense that smaller uncertainty means higher accuracy. Another measure for the accuracy of measurement is given by the statement of a confidence interval with accompanying confidence level. Here the confidence level $1 - \alpha$ is chosen in advance and numerically expresses the confidence that an interval of estimates to be determined contains the true value. On the basis of the measurement, one then finds a corresponding interval using an assumed distribution of estimates (mostly a normal distribution). For a symmetric distribution, this confidence interval with confidence level $1 - \alpha$ is centered around the corrected mean (result of measurement) of a measurement series of size n under repeatability conditions. Half of its width w, when stated together with confidence level $1 - \alpha$, can be used as a numerical measure of the accuracy of the measurement. For nonsymmetric distributions, the total width of the resulting interval characterizes the accuracy of measurement. For a normal distribution and if the unknown systematic error in the measurement series is negligible, w is given by the product of the standard deviation $s_{res} = s/\sqrt{n}$ [for s see Eq. (4)] of the result of measurement and the Student factor $t(n - 1, 1 - \alpha)$, which depends on n and the arbitrarily chosen α (see, e.g., Martin, 1971).

It must be stressed that w cannot be stated as the uncertainty of measurement, which, by definition, is constructed independent of the confidence in the trueness of the assumption that the true value of the measurand is contained in the stated interval I_{est}. Moreover, there is no rule to propagate w.

The complete final result of a measurement consists of the result of measurement and any of the above-mentioned quantitative statements characterizing the accuracy of measurement.

3. GENERAL CHARACTERISTICS OF MEASURING INSTRUMENTS

The properties of a measuring instrument not only influence the result of measurement, but also the uncertainty of measurement. To describe numerically the properties of a measuring instrument, various metrological characteristics are of interest (ISO/IEC/OIML/BIPM, 1993). Their values depend on the conditions under which the measuring instrument is used. In practice, the properties of instruments often are investigated under rated operating conditions or reference conditions.

This section contains a list of characteristics of measuring instruments that are of general importance to metrology. The list is not complete. In particular, characteristics specific to the various types of instruments such as mass spectrometers may be found in the corresponding articles of the Encyclopedia.

The "response characteristic" is the relationship between an input signal (also called stimulus) to a measuring instrument and the corresponding output signal (also called response) under defined conditions.

The stationary-state properties of a measuring instrument are described by a response characteristic obtained by finding the relationship between different fixed values of the input quantity and the corresponding final steady values of the output quantity. The relationship usually is represented by the characteristic curve.

The dynamical behavior of a measuring instrument is characterized by varying the input signal in time in a specific way (e.g., stepwise variation, sinusoidal variation) and observing the response. The response characteristic, then, often is represented in form of a transfer function (Laplace transform of the response divided by that of the stimulus).

The "sensitivity (coefficient)" of a measuring instrument is understood as the change in its response divided by the corresponding causing change in the stimulus. The value of the coefficient may depend on the actual value of the stimulus.

The slow variation with time of a metrological characteristic of a measuring instrument is called "drift."

The "response time" of a measuring instrument is the time interval between the instant when a stimulus is subjected to a specified abrupt change and the instant when the response reaches and remains within specified limits around its final steady value.

The "error of a measuring instrument" is that contribution to the error of measurement that is caused by the measuring instrument. The two errors, of course, need not be the same. The error of a measuring instrument consists of a random and a systematic part.

In practice, often the "bias error (of a measuring instrument)" is of particular interest. It is an estimated contribution of an indicating measuring instrument to the systematic error of a measurement. The bias error A_s usually is determined as the difference between the uncorrected result of measurement (which may consist of a single value) and the conventional true value x_r of the measurand:

$$A_s = \bar{x} - x_r.$$

This estimate sufficiently describes the contribution of the instrument to the systematic error of measurement, if the absolute value of the random error of the instrument is negligible compared to the absolute value of its systematic error, and if the absolute value of the deviation of x_r from the true value of the measurand is much smaller than $|A_s|$.

The "measuring (or working) range" of a measuring instrument is that set of values of measurands for which the error of the measuring instrument is intended to lie within specified limits.

The "maximum permissible errors of a measuring instrument" are the extreme values of an error permitted by specifications, regulations, etc. for a given measuring instrument.

A class of measuring instruments meeting certain metrological requirements such that errors of the instruments are kept within specified limits is called "accuracy class." The accuracy class characterizes the properties of the instruments with respect to the error of a measuring instrument. But it must be stressed that, whereas the error of measurement obtained by an instrument of a given accuracy class does contain the error of the instrument, it is not characterized by the accuracy class.

GLOSSARY

Calibration: Set of operations that establish, under specified conditions, the relationship between values of quantities indicated by a measuring instrument, or values represented by a material measure, and the corresponding values realized by standards.

Influence Quantity: Quantity that is not the measurand but that affects the result of measurement.

Measuring Instrument: Device intended to be used to make measurements, alone or in conjunction with supplementary device(s).

Measurement Signal: Quantity that represents the measurand and that is functionally related to it.

Method of Measurement: Set of operations in measurement that is independent of the principle of measurement and appropriate to yield the measurand's value (examples: the null method, the substitutional method).

Principle of Measurement: Scientific basis of measurement consisting of a physical phenomenon or effect that uniquely relates the measurand of interest to a quantity to be directly measured. The relationship is known from theory or established empirically (example: the thermoelectric effect applied to measurement of temperature).

Rated Operating Conditions: Conditions of use for which specified metrological characteristics of a measuring instrument are intended to lie within given limits.

Reference Conditions: Conditions of use prescribed for testing the performance of a measuring instrument or for intercomparison of results of measurements.

Relative Uncertainty: Uncertainty divided by the modulus of the result of measurement.

(Measurement) Standard: Measuring instrument or material measure intended to define, realize, conserve, or reproduce a unit or one or more values of a quantity to serve as a reference.

Works Cited

Geerligs, L. J., Anderegg, V. F., Holweg, P. A. M., Mooij, J. E., Pothier, H., Esteve, D., Urbina, C., Devoret, M. H. (1990), *Phys. Rev. Lett.* **64**, 2691–2694.

ISO/IEC/OIML/BIPM (1993), *International Vocabulary of Basic and General Terms in Metrology (VIM)*, Geneva: International Organization for Standardization.

Kudva, S. M., Potter, R. W. (1992), in: M. T. Postek (Ed.), *6th Annual Conference on Integrated Circuit Metrology, Inspection and Process Control*, SPIE Proceedings Vol. 1673, Bellingham, WA: SPIE.

Martin, B. R. (1971), *Statistics for Physicists*, New York: Academic Press.

Press, W. H., Flannery, B. P., Teukolsky, S. A.,

Vetterling, W. T. (1989), *Numerical Recipes (The Art of Scientific Computing)*, Cambridge, U.K.: Cambridge University Press.

Further Reading

Petley, B. W. (1985), *The Fundamental Physical Constants and the Frontier of Measurement*, Bristol: Adam Hilger.

Two different points of view concerning the expression of the uncertainty of measurement are taken in the following two papers:

Colclough, A. R. (1987), *J. Res. Nat. Bur. Stand.* **92**, 167–185.

Weise, K., Wöger, W. (1993), *Meas. Sci. Technol.* **4**, 1–11.

MICROELECTRONICS

T. SUGANO, *RIKEN, Wako, Saitama, Japan*

A. MORINO, *NEC Corporation, Kawasaki, Kanagawa, Japan*

3-527-28132-0/94/$5.00 + .50

INTRODUCTION

The need for and importance of miniaturized circuits were recognized long ago, for applications to a variety of fields, including space development and military systems. Microminiaturization, that is, the development of electronics systems with very small-dimensioned circuits, was carried out in several ways. A typical example was the assembling of microminiaturized components in a very small area on a printed wiring board or a ceramic substrate.

The term "microelectronics" usually refers to the technology of design, manufacture, and use of electronics devices with very small dimensions. An epoch-making event in the history of microelectronics is the invention of the transistor in 1947, by J. Bardeen, W. H. Brattain, and W. Shockley, all with Bell Telephone Laboratories, AT&T. The invention of the integrated circuit (IC) in 1959 by J. Kilby of Texas Instruments, and R. Noyce of Fairchild, independently, is another monumental landmark, leading to the development of a completely new world of microelectronics. Since then, the number of components integrated on a silicon chip has shown a constant increase, as shown in Fig. 1 (*IEEE ISSCC Digest*, 1970–1993; *Digest of Symposium on VLSI Circuits*, 1987–1993). Usually, the terms LSI (Large-Scale Integrated Circuits), VLSI (Very Large-Scale Integrated Circuits), and ULSI (Ultralarge-Scale Integrated Circuits) refer to integrated circuits with numbers of components equal to or larger than 10^3, 10^5, and 10^7, respectively. Based on these situations, microelectronics, in recent years, usually refers to a field of electronics that includes the integrated circuits and the area of applications that has been developed by them.

The introduction of microelectronics has had a tremendous impact on industry and human society. The first impact is the realization of electronic systems with smaller size and lighter weight. As is shown in Fig. 1, the sizes of memory and digital functions have been reduced roughly by 1/100 and 1/30, respectively, for every ten years. The corre-

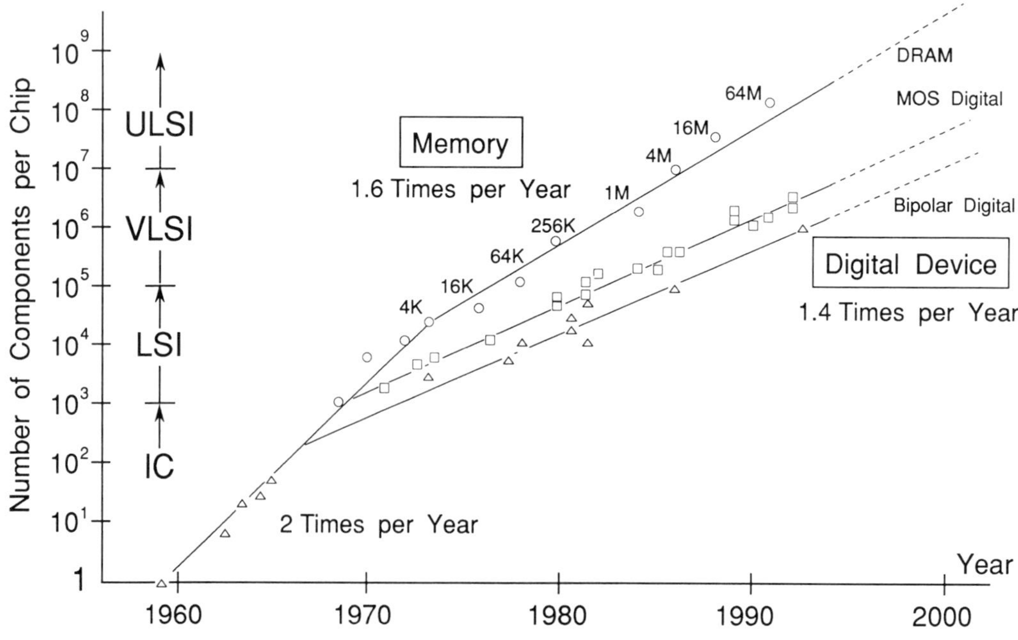

FIG. 1. Trend of the number of components integrated on a chip.

sponding weight reduction is easily understood.

The second impact is cost reduction. It results from both component and interconnections sides, as follows. As is well known, ICs are fabricated by batch processing of wafers containing large numbers of chips. The cost of each component integrated on a chip is reduced as the number of components integrated on a chip increases. Furthermore, the production cost of each chip is reduced as the production volume increases. An experimental rule, called a learning curve, says that the production cost is reduced by roughly 30% every time the accumulated production volume becomes double. Another cost reduction results from making large numbers of interconnections between components on a chip, by batch processing. On the other hand, making interconnections outside chips, on a printed wiring board, for example, requires much more time, and so leads to much higher cost.

Increase in reliability is the third impact of microelectronics. Transistors, as compared with electron tubes, have essentially higher reliability because of not employing a heated electrode. Furthermore, the drastic reduction in the number of soldered contacts leads to the excellent reliability of the electronic systems. Generally speaking, soldered contacts are the least reliable portions in electronic systems. Interconnections between components on a chip, on the other hand, are made by using thin-film metal layers, without using soldered contacts.

The final, but not the least, impact is the performance improvement. Important factors to determine the operational speed of electronic systems are the time required to charge/discharge parasitic capacitances and the time required for the signal to propagate through the signal lines. Microelectronics consisting of components with smaller size and shorter wiring length contributes to the realization of high-speed operation in terms of both of these mentioned factors.

1. COMPONENTS OF INTEGRATED CIRCUITS

1.1 Design Targets of Integrated Circuits

Typical targets of integrated circuits products include the following items:

1. functions;
2. performance, including operational speed;
3. power consumption;
4. price; and
5. timing of introduction into the market.

The importance of item **3** is now increasing, because of the increase in power consumption resulting from the increase in the number of components integrated on a chip, and also because of the growing market of portable/personal terminals.

The design of IC products starts with selecting the most appropriate basic circuit type, such as CMOS (Complementary symmetry Metal-Oxide-Semiconductor), BiCMOS (Bipolar-CMOS), and so on. The CMOS circuit is characteristically of low power consumption, whereas the BiCMOS circuit affords higher switching speed, but at larger power consumption, compared with CMOS (see CIRCUITS, DIGITAL).

Design styles of LSIs can be divided roughly into three types. The first is standard products, such as memory LSIs and microprocessors. The second refers to so-called ASIC (Application-Specific Integrated Circuit), such as gate arrays and cell-based ICs. These devices are used to integrate such functions as are specific to particular systems. Between these two types is the third one, which is called ASSP (Application-Specific Standard Product). This is the device widely used in specific fields, such as telecommunications, networks, computer terminals, and so on. Examples include the CODEC (COder+DECoder; see Sec. 4.1), and the floppy-disk controller. Generally speaking, production volume is the largest for the first type, and the smallest for the second one.

1.2 Basic Device Structures of Integrated Circuits

1.2.1 Transistors Typical examples of the basic device structures used in integrated circuits are briefly described in this section.

Fig. 2(a) shows a typical cross-sectional view of a CMOS configuration, a basic structure of CMOS integrated circuits. Here, the right-hand side corresponds to an *n*-channel MOS (Metal-Oxide-Semiconductor) transistor. Three parts, that is, the metal, oxide, and semiconductor, correspond to the polycrystalline silicon, silicon dioxide, and *p* well, or

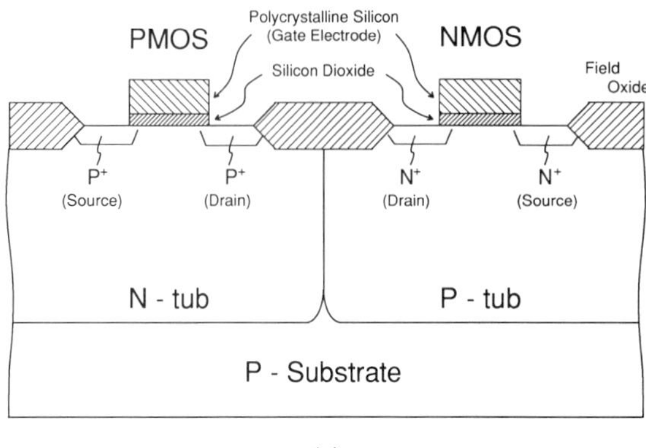

(a)

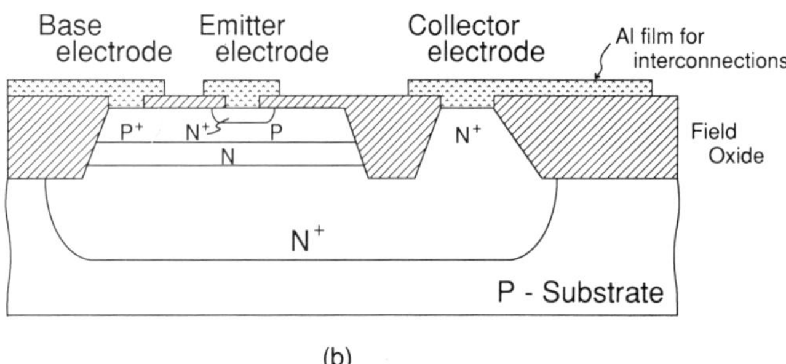

(b)

FIG. 2. Cross-sectional view of transistors. (a) CMOS device, (b) bipolar device.

"tub," respectively. The polycrystalline silicon performs as the gate electrode of a MOS transistor. Two $N+$ regions correspond to the source and drain of a transistor. For the left-hand side of the figure, which corresponds to a p-channel MOS transistor, a similar structure is formed. Each tub is prepared to provide an isolated island for the respective transistor. Field oxide is formed to assure electrical isolation between adjacent transistors, by eliminating the leakage current through the surface.

Shown in Fig. 2(b) is a typical cross-sectional view of a bipolar transistor. Here, the three electrodes, that is, the emitter, base, and collector, are shown. Three regions beneath the emitter electrode, $N+$, P, and N, correspond to the emitter, base, and collector regions, respectively, of an npn bipolar transistor. The $N+$ region under the collector region and through the collector electrode is formed to reduce the collector series resistance. Similarly, the $P+$ region beneath the base electrode is formed for the reduction of the base series resistance. These $N+$ and $P+$ regions contribute to the realization of high-speed operation of the transistor.

The most favorable feature of the bipolar transistor is its high-speed operation. The circuit type that affords the highest switching speed is ECL (Emitter Coupled Logic; see CIRCUITS, DIGITAL.) Key parameters to determine the operational speed of ECL circuits are cutoff frequency f_T, base resistance r_b, collector parasitic capacitance C_c, and collector series resistance R_c. Approaches to realization of high-performance ECL include the reduction of lateral and vertical dimensions, employment of trench isolation, and/or self-aligned transistor structures. The cutoff frequency of leading-edge transistors in practical applications is around 30 GHz.

Usually, the switching speed of a digital circuit is expressed by the propagation delay time T_{pd}, which is defined as the time interval between the output and the input signals. The smallest values of T_{pd}, for the typical digital circuits under production in large quantity at present, are as follows: ECL, 30 ps/gate; CMOS circuit, 350 ps/gate; BiCMOS circuit, 300 ps/gate.

1.2.2 Memory Cells A memory cell is one of the most significant components of memory integrated circuits. Figure 3(a) shows an equivalent circuit of a memory cell of Dy-

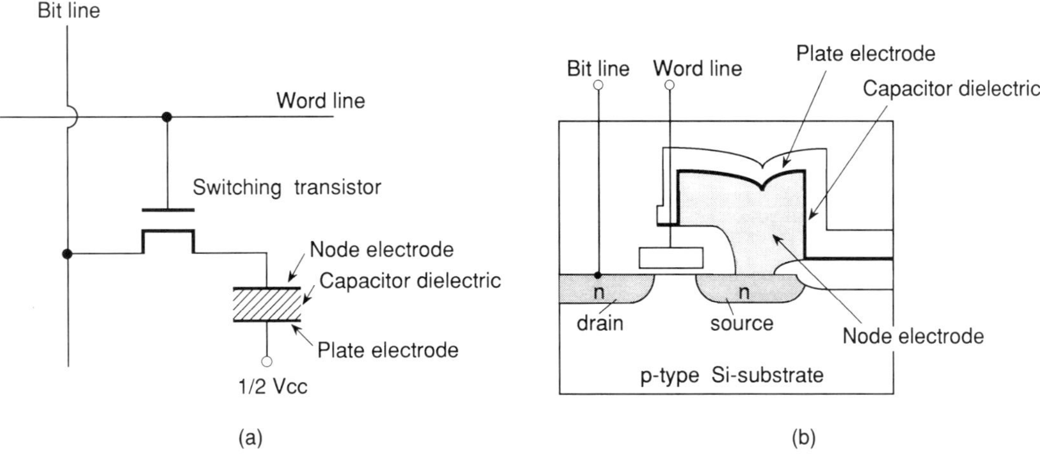

FIG. 3. (a) Equivalent circuit of a DRAM cell. (b) Cross-sectional view of a DRAM cell (stacked capacitor cell).

namic Random-Access Memory (DRAM; see CIRCUITS, DIGITAL). The cell consists of a capacitor for charge storage and a switching transistor for read/write control. The voltage setup on the bit line is written into the capacitor when the word line is made high. In the read phase, the word line is made high, and the signal charge is derived from the capacitor to the bit line.

The key point of a memory-cell design is to realize a cell area as small as possible, in order to develop a memory device with a chip area as small as possible. In fact, the cell area per bit has continuously been reduced, corresponding to the continuous increase in the bit capacity per chip. It is required, on the other hand, that the charge stored in a cell is no smaller than a critical value that is determined from the consideration of (a) immunity to soft errors (see Sec. 1.3.2.6) and (b) the minimum value that can stably be amplified by the sense amplifier. The corresponding critical capacitance is roughly 30 fF. As the reduction of the cell area per bit proceeds, keeping pace with the increase in the bit density, realization of 30-fF capacitance is becoming increasingly difficult. For high-density (4 Mbit and beyond) DRAMs, the following approaches are employed to avoid the decrease in capacitance:

1. employment of three-dimensional cell structures in order to increase the effective area of the electrodes; and
2. the use of materials with higher dielectric constant.

The first example of approach **1** described above is a stacked capacitor cell, as shown in Fig. 3(b). This is one of the most typical cell structures used in high-density DRAMs. In the cell, effective cell area is increased through the use of both the area over the switching transistors and the sidewall of the capacitor as electrodes for the capacitor. Here, the height of the pad is optimized by taking into account the target capacitance and the process complexities. Typical value of the pad height is from 0.2 to 0.5 μm.

Another typical example of approach **1** is a memory cell that employs a trench capacitor. A silicon substrate is etched to form a narrow, deep groove in the silicon, and then specific films are formed. The trenched groove is typically 3–4 μm in depth. Although this structure can realize a relatively large capacitance, the issue is how to decrease the junction leakage current due to the crystal defects induced by the digging of the silicon substrate.

As for approach **2**, silicon nitride, with the dielectric constant of roughly 7.5, is used as a capacitor material. (The dielectric constant of silicon dioxide is around 3.9.) For future DRAMs with higher bit density, applications of other materials with larger dielectric constant, including Ta_2O_5 (the dielectric constant is around 25) and $BaSrTiO_3$ (the dielectric constant is over 100), are being developed.

1.2.3 Isolation Isolation is one of the most important structures of integrated circuits. First, all of the components on a chip are required to be electrically isolated from each other. Leakage current between them should be small enough to meet specifications. Second, the parasitic capacitance associated with the isolation structure should be as small as possible. Third, the area of the isolated region is required to be as small as possible, because it has a big impact on chip size. Here, three typical examples of isolation structures will be reviewed.

The first one is the LOCOS (LOCal Oxidation of Silicon) structure, which is extensively used in almost all MOS integrated circuits. It employs a thick oxide selectively formed in the field regions, as shown in Fig. 2(a). The term field regions means those parts outside the islands in which transistors and various components are formed. The thick oxide is effective in reducing the leakage current between components, thus avoiding the electrical coupling between those components. Selective oxidation of the silicon is carried out by using a silicon nitride layer as an oxidation mask. The field oxide is partially formed under the silicon surface, resulting in a decreased step height of the field oxide. This relatively small step is important to avoid open failures of interconnects.

In order to reduce the leakage current more completely, a channel stopper, a region with high impurity concentration, is formed near or under the field oxide. The high impurity concentration avoids channel formation, thus keeping the electrical isolation more thorough.

On the other hand, as the reduction of the device feature size proceeds, reduction of the area for the isolated region is also necessary. The bird's beak, a characteristic structure formed at the boundary of the field oxide and silicon, causes difficulty for area reduction of the LOCOS structure. A typical approach to eliminate the bird's beak is the employment of a polysilicon-buffered LOCOS structure. By inserting the polysilicon layer under the silicon nitride layer, field oxide with a reduced step height can be formed.

The second approach is trench isolation. The structure is formed by etching a narrow groove in silicon and filling it with oxide and/or polysilicon and the like. Figure 4 shows an example of a cross-sectional view of trench isolation. The trench is filled with BPSG (Boro-Phospho-Silicate Glass). It leads to

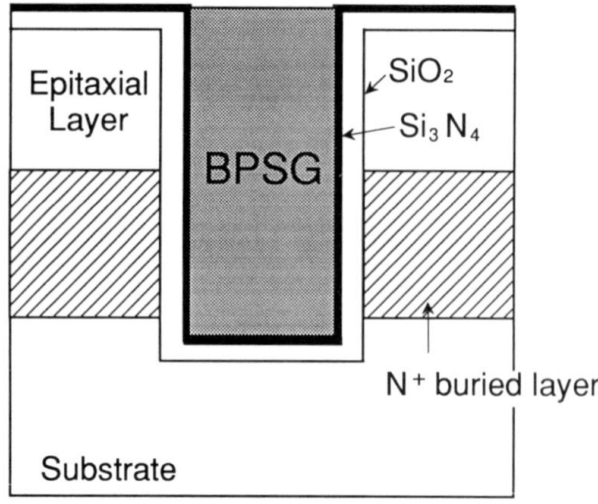

FIG. 4. An example of a cross-sectional view of a trench isolation, filled with BPSG.

compact isolation with a width of 2 μm or smaller. Generally, the requirements for the isolation include low leakage current and low parasitic capacitance. From this viewpoint, it is found that the trench structure filled with BPSG is superior to those filled with polysilicon. The low glass-transition characteristics of BPSG induce less mechanical stress, thus resulting in low leakage current of the isolated region.

The third approach is the use of SOI (Silicon On Insulator) substrate. Starting with the silicon substrate, silicon oxide and silicon layers are formed successively. The thickness of the silicon layer ranges from 50 to 500 nm, depending on the application. Isolation is made by means of either oxidation of, or etching off, the silicon layer, up to the underlying silicon oxide layer. Because of the isolation by silicon oxide or air, the leakage current between the components can be made quite small. Furthermore, latchup-free CMOS devices can be realized, because parasitic *pnpn* switches are not formed. SOI substrates may be fabricated by SIMOX (Separation by IMplanted OXygen) or BESOI (Bonding and Etchback SOI) technology.

1.3 Chip Yield and Reliability

In the actual applications of integrated circuits to electronic systems, price and stable operations are important factors, as well as chip functions, performance, and so on. Here, chip yield and reliability issues will be reviewed.

1.3.1 Chip Yield The first factor that determines chip yield is the point defects on a chip. One of the most common causes of point defects is dust or other particles present in the environment. These particles, when they have fallen on wafers, may induce imperfect patterns, leakage current through insulating layers, and so on. This may lead to chip yield loss. For this reason, wafer processing is done in clean rooms that contain particles within a specified number. A "class 1 room," for example, refers to one that contains no more than one particle with a size of 0.15 μm in diameter (see AEROSOLS). In response to the continuous decrease in device feature size, clean rooms with tighter cleanness control will be required.

The second factor to determine chip yield is the fluctuation of process parameters, such as the thickness of oxide layers, or impurity concentration. These fluctuations are due to the variations from the specified values of the process conditions, including temperature, and gas flow in production equipment. They will lead to unevenness of the device parameters and, hence, may result in chip yield loss. The third factor is a circuit design-related issue. When the circuit performance changes sensitively in response to fluctuations of the device and process parameters, high chip yield is not expected. A circuit design that affords circuit performance insensitive to parameter changes, that is, low-sensitivity circuit design, is highly desired.

For the sake of quantitative understanding of the chip yield issue, yield modeling and cost predictions have been made, taking various related factors into account. The results are used also to get insight into the possibilities of improvements in process facilities, design of process and circuit, and so on.

The chip yield is also an important factor that determines the maximum chip size. That is, larger chip size leads to lower yield and, hence, higher cost. Given the cost target of a chip, the maximum chip size is determined from the chip yield considerations based on the process technology level at the time. In fact, the maximum chip size has increased at the rate of roughly 1.14 times per year, or a factor of 15 during the 20 years from 1970 to 1990 (*IEEE ISSCC Digest*, 1970–1993; *Digest of Symposium on VLSI Circuits*, 1987–1993).

1.3.2 Reliability Fabricated devices should exhibit their functions and performance throughout their intended life. Generally, the term *reliability* is defined as the probability that a device will afford required functions under specified conditions for a stated period of time.

Failure rates for typical integrated circuits are classified into three regions, as shown in Fig. 5. After introduction into practical usage, the devices will encounter a relatively high failure rate, so-called early failures. Generally, this type of failure results from manufacturing defects. In the steady-state region, device failure occurs accidentally. The final stage is called the wearout region. In most ICs, there are no wearout mechanisms observed, as opposed to electron tubes. However, some time-dependent failure modes, such as electromigration or oxide breakdown, are

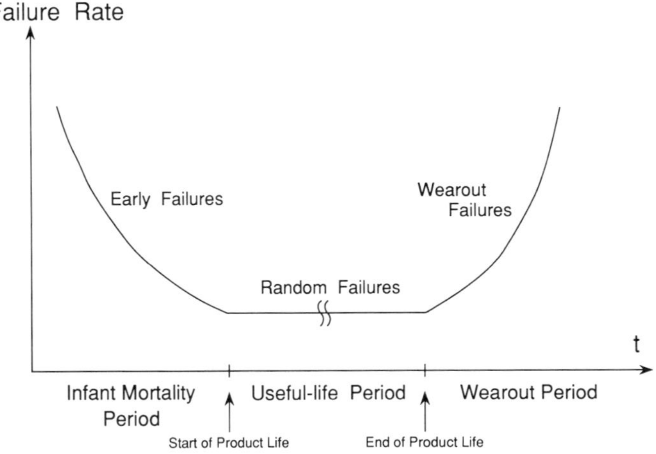

FIG. 5. Failure rate vs time of typical integrated circuits.

observed, especially in devices with poor reliability.

Many types of effort have been devoted to avoiding device failures after shipping the products. A typical approach to reject early failures is a screening, which is carried out by the burn-in process. In this process, the products are operated under accelerated conditions, such as increased temperature or increased voltage, for some period of time. During this process, most of the devices that are subject to early failure actually fail.

With regard to the steady-state failure shown in Fig. 5, a common approach is to estimate the failure rate of a device lot itself. The actual procedure is to pick out a specified number of devices from each lot, and subject them to accelerated conditions for some period of time. These conditions are set up to simulate those the devices will actually encounter in the market. Based on comparison of the measured failure rate with the target value, the decision is made whether each lot is able to be shipped out or not. As for the wearout failure shown in Fig. 5, identification of the possible failure mechanisms, and modifications of the process parameters or design, are carried out.

Several measures of failure rate have been used, including these:

1. ppm (parts per million); one ppm corresponds to one device in failure per one million devices. This measure is frequently used to express the reliability level of shipped devices, especially in the early failure region. The present market requires less than 50-ppm level for usual applications, and less than 10-ppm level for severe environment, such as automobile applications and so on.
2. FIT (Failure Unit); one FIT refers to one failure per 10^9 device hours of use of the products.
3. MTTF (Mean Time To Failure); this measure expresses the average lifetime of the device before it goes wrong.

From the viewpoint of reliability improvement, identification and elimination of failure mechanisms are quite important. It is especially true for devices with ever-increasing density and performance. For CMOS LSIs, the following typical failure mechanisms have been identified.

1.3.2.1 Gate-Oxide–Related Failures. The voltage applied to the gate oxide, or the current flow through it, is known to induce time-dependent breakdown of the oxide. This mode is TDDB (Time-Dependent Dielectric Breakdown), or simply wearout. A higher temperature leads to a higher failure rate. The mechanism of oxide breakdown is considered to be destruction of a Si-O bond by either a locally increased electric field or thermal runaway triggered by an increased current flow. The possible candidates for their origins are sodium contamination, electron traps, and hole traps in the oxide. Sodium ions segregated at the interface induce a lowering of the Si–SiO_2 barrier height, resulting in an increase in the localized gate current. The sodium-ion model gives a good explanation of the temperature dependence of TDDB. The electron-trap or hole-trap model explains the increased internal electric field by charge traps. These models are attempts to explain the mechanisms of intrinsic breakdown. The TDDB test is a tool to estimate the quality of the oxidation process and the silicon substrate, and also to screen a weak spot of gate oxide. Gate-oxide failure will be decreased by making the electric field applied to the oxide film no higher than the critical value.

1.3.2.2 Hot-Carrier–Induced Failures. Hot-carrier–induced failure refers to a change of the electrical characteristics of metal-oxide-semiconductor field-effect transistors (MOSFETs), such as threshold voltage or ON current, due to the injection of high-energy (hot) carriers into the gate oxide. Generally, hot electrons, as compared with hot holes, give more serious impacts on the device failure. So, the following descriptions will be limited to hot-electron–induced failure only.

With the continuous decrease in the transistor feature size, the lateral electric field between the drain and source electrodes has constantly increased, under the constant power-supply voltage of 5 V. The resulting high electric field accelerates electrons in the channel region, and also secondary generated electrons. As a result, part of the accelerated carriers gain the energy to flow into the gate oxide over the Si–SiO_2 barrier. Injected electrons introduce charge traps and interface states, through interactions between the injected electrons and Si-OH or Si-Si bonding. The major part of the degradation mechanisms in the gate oxide, however, is not fully identified.

Several models have been proposed to describe the population of hot electrons. Two typical models are a "lucky-electron" model and an electron-gas model. The lucky-electron model treats the electrons on the basis of the probability concept: Each electron has a specific probability to be accelerated to an energy higher than the barrier energy ϕ. The probability is $\exp(-\phi/\lambda E)$, where λ is the mean free path of electrons and E is the local electric-field strength. The electron-gas model, on the other hand, assumes a nondisplaced Maxwellian for the mass of the electron. The ratio of the number of electrons that have energy higher than ϕ is given by $\exp(-\phi/kT_e)$, where k is Boltzmann's constant and T_e is the electron temperature.

The common approach for this mode of failure is to employ a device structure that avoids hot-electron effects by lowering the electric field near the drain region. Examples include LDD (Lightly Doped Drain) and DDD (Double Diffused Drain), where some region is introduced with reduced electric-field strength. By use of these structures, devices with submicron feature size can operate under 5-V supply. The structure, on the other hand, requires precise design of transistor parameters, to avoid both the parasitic series resistance and degradation of the transistor drivability.

When the device feature size is around 0.5 μm or smaller, however, a 5-V power supply cannot be employed, to avoid hot-electron–induced failure. For devices with 0.5-μm feature size, for example, a power-supply voltage around 3 V is employed. Further reduction in the power supply and/or the introduction of new transistor structures will be necessary, depending on further reduction of the feature size of devices.

1.3.2.3 Metallization- and Interconnection-Related Failures. There are two typical modes for metallization- and interconnection-related failures. The first is electromigration. It is a failure resulting from material transport in metals, under enhanced and directional mobility of atoms caused by the electric field. The electric field is produced either by the applied voltage or by the current flow itself. The most typical example is electromigration of aluminum. Aluminum atoms are accumulated in the direction of electron flow, resulting in a discontinuity of the interconnect, which leads to cessation of the device operation. This type of failure is one of the most serious failure modes in integrated circuits.

The aluminum atoms that are upstream in terms of electron flow from a vacancy have a higher probability of occupying a vacancy site, and thus inducing electromigration. In polycrystalline aluminum films, the electromigration occurs mainly along grain boundaries, where aluminum atoms are bound less tightly than in lattice positions. An equation for MTTF of aluminum interconnects, called Black's formula, has been empirically derived as follows:

$$t_{50} = AJ^{-n} \exp(Q_0/kT),$$

where t_{50} is median time to failure, A is a process- and material-dependent constant, J is the current density, n is a current-dependent constant, Q_0 is the activation energy for the diffusion of aluminum atoms, k is Boltzmann's constant, and T is the absolute temperature. Typical approaches to suppress electromigration are design of the integrated circuits so as to limit the value of J within the critical value, addition of Cu to Al, and formation of structured Al layers by employing transition metals.

The second failure mode is stress-induced migration. For aluminum lines with a linewidth of roughly 1 μm or smaller, the failure becomes more serious. When Al interconnects covered with passivation layers are subject to IC fabrication process (400–500 °C), or high-temperature (150–250 °C) storage, voids are induced. This results in an open-circuit failure. Possible mechanisms include compressive intrinsic stress in passivation films, tensile stress in the Al interconnects due to the difference in the thermal-expansion coefficients between Al and CVD films, and deformation of passivation films. Countermeasures for this failure include addition of Cu or Ti to Al, or employment of a layer of Al laminated with Ti, W, or TiN.

1.3.2.4 Electrostatic Discharge (ESD). In the handling of devices, voltages higher than the breakdown voltage of the gate oxide may be applied on the circuits, thereby inducing device failure. Typical examples of induction of very high voltages include a person walking near ICs, and removal of an IC from its plastic packaging material. To avoid this type of failure, input and output circuits are designed so as to provide paths for the discharge current, thus preventing induction of excessive voltage across the gate oxide of the device.

1.3.2.5 Latchup. In CMOS devices, there are many source and drain regions close to adjacent tubs. These structures may be causes of parasitic bipolar transistor action, and, thus, may lead to parasitic *pnpn* switches. When the firing condition is met, parasitic *pnpn* switches are turned on, and the chip turns into a very low-impedance state. In this state, the device is not controlled by the input signals. The phenomenon is reversible. That is, the device will come back to its old state if the power supply is turned off. This failure mode will be avoided by the design of the distance between the impurity regions comprising the parasitic bipolar transistors large enough to avoid the firing of the *pnpn* switches, employment of a trench isolation structure, and/or employment of epitaxial wafers. For the final case, the substrate for the epitaxial layer plays a role as a stable equipotential plane, thus avoiding the parasitic bipolar actions.

1.3.2.6 Alpha-Particle–Induced Soft Errors. Alpha particles emitted by radioactive elements contained in IC packaging materials cause soft errors in the ICs. The penetration of alpha particles into the silicon causes the generation of electron-hole pairs. The generated carriers may lead to the loss of charge stored in dynamic circuits, including the memory cells of a dynamic memory, or dynamic shift registers. The term soft error refers to a failure that does not lead to a physical defect of the device. Approaches to avoid this type of failure include making the capacitance value larger, and the coating of chips with material with a very low density of radioactive elements.

With the continuous decrease in device feature size, increase in the number of components on a chip, and also the increase in process complexities, conventional approaches for reliability assurance are becoming unsatisfactory. One novel concept for more reasonable reliability assurance is the building-in of reliability, or design for reliability (DFR). This is a typical example showing the shift of approaches from measurement and analysis, to design and synthesis, of device reliability. Deeper understanding of the physical phenomena, and development of better modeling and simulation systems, are highly required (Sze, 1988; Takeda *et al.*, 1993).

1.4 Circuit Configurations of Integrated Circuits

1.4.1 Circuit Configurations for VLSIs and Beyond The circuits to be applied to VLSIs and beyond are required to have better figures of merit, that is, higher-speed operation, lower power consumption, and smaller chip area. From this viewpoint, several circuit configurations of CMOS or BiCMOS circuits have been proposed and applied to practical usage.

The first circuit type that has the above-mentioned features is a dynamic circuit. One of the most typical examples that employ this circuit type is DRAM (Dynamic Random-Access Memory), which has always realized the largest number of components on a chip. Through the development of the DRAM chip, a number of novel circuits have been designed (Itoh, 1990). Varieties of other circuit types for better figures of merit have been proposed, including clocked CMOS and Domino Logic circuits. These circuits require high-level design techniques. In order to utilize fully the apparent advantages of these circuit types, judicious studies on the circuit operations and the design strategies are indispensable.

For further power reduction, circuit operation under low supply voltage is extensively studied. In this case, simultaneous realization of low power consumption and high-speed operation is the requirement (Chandrakasan, *et al.*, 1992).

The BiCMOS circuit has the advantage of higher operational speed compared with the CMOS circuit. This is because the bipolar devices have higher driving capability than MOS devices. The power consumption and chip area of BiCMOS circuits, on the other hand, are larger than for CMOS. In the selection of circuit type, comparison of pros and cons of the candidate circuits should be carried out (Santo, 1989).

High-speed interface circuits are becoming increasingly important. This is because the high-speed signal transmission between chips on a printed wiring board is one of the key factors to realize high-speed digital systems, as well as high-speed chips (Gunning *et al.*, 1992).

For the basic circuit types, see CIRCUITS, DIGITAL and CIRCUITS, ANALOG.

2. SELECTION OF PROCESS TECHNOLOGIES FOR INTEGRATED CIRCUITS

In chip development, selection of the best possible process technologies and fabrication facilities is of utmost importance. Generally, candidates for the selection have their own strong and weak points. Therefore, detailed comparison of the pros and cons of the candidates is indispensable for the reasonable selection of process technologies.

Important factors to be taken into account in the selection of the process technologies will be grouped into three. They are (1) high performance, (2) high productivity, and (3) high reliability. Higher productivity is important to realize lower production cost and also shorter turnaround time in chip development and production. Higher reliability is important for long and stable operation of the electronic systems, which is an important condition for customer satisfaction. Each of the above three groups consists of multiple factors. The starting point of the selection of process technologies is to list those factors in the order of importance. Shown below are some examples of discriminating criteria used in the selection of alternative process technologies. There may be, of course, different criteria, depending on the situation.

Here, factors to be considered in the selection will be reviewed, with some specific examples. As for the element process technology, see VLSI TECHNOLOGY.

2.1 Lithography

There are three approaches for the lithography process: optical, electron-beam, and x-ray systems. Factors to be taken into account in the selection of the technology are listed as follows, in the order of importance.

1. Resolution. The resolution, that is, the minimum dimension that can be defined by the technology, is the most important factor in lithography systems. At present, optical systems with various wavelengths are employed in the actual fabrication or development: *g* line (wavelength equal to 436 nm) or *i* line (365 nm) from a mercury lamp, KrF (248 nm) and ArF (193 nm) lasers. Electron-beam systems that employ an electron beam with the wavelength of 0.02–0.05 nm, depending on the energy, afford much higher resolution compared with the optical systems. As will be described later, issues of electron-beam systems include low throughput and expensive production facility. On the other hand, lithography systems using x-rays with the wavelength of 0.3–5 nm have also much higher resolution capability compared with optical systems. The issues for this type of system include difficulty in the fabrication of masks and the expensive production facility. Efforts are now being devoted to overcome the above-mentioned issues.
2. Depth of focus (DOF). In order to define patterns on silicon surfaces with unevenness of the silicon oxide generated during the fabrication processes, good focus should be assured within some range of depth. DOF refers to this focus margin. Generally, electron-beam and x-ray systems afford better DOF, compared with optical systems.
3. Field size. In order to make a pattern transfer for every mask layer at once, the field size of the system should be large enough to cover the chip size. To keep pace with the increase in chip size, increasing field size is required. The maximum field size for present optical systems is around 22 mm $\times$ 22 mm.
4. Uniformity.
5. Throughput. From the viewpoint of production cost, the throughput of the process is quite important. At present, electron-beam systems have a throughput much lower than that of optical systems.
6. Reliability. From the reliability standpoint, the radiation damage to the transistors induced by the incident electron beams is an issue for electron-beam systems.

2.2 Etching

In regard to the etching process, there are two types of approaches: wet and dry etching systems. Dry etching is further divided into several technologies: ion milling, chemical dry etching, and reactive ion etching (RIE). Factors to be taken into account in the selection of the technologies are listed as follows, in the order of importance.

1. CD (critical dimension) loss. CD loss refers to the deviation of the dimension of the most important pattern (usually the minimum dimension), after processing, from the designed one. Zero CD loss means the formation of patterns just the same as the designed ones. There are several sources leading to CD loss in the mask-making, lithography, and etching processes. Dry etching is superior to wet etching in this respect, thanks to its anisotropic nature.
2. Gate-oxide destruction. Destruction of the gate-oxide film by charge-up damage during the etching process can be a fatal problem.
3. Selectivity. One common approach to enhance the selectivity is the use of a protection layer. For example, selective etching of vias can successfully be carried out with the addition of CO gas, which forms the protection layer.
4. Uniformity.
5. Throughput. Etching rate is important to realize high throughput.
6. Particle contamination. Etching by ion milling technology, for example, has many favorable features. However, it is not employed in volume production, because of the particle contamination.

2.3 Metallization

The metallization process refers to the formation of the electrical contacts and interconnection layer(s). There are three approaches for that; evaporation, sputtering, and chemical-vapor deposition (CVD). Factors are listed as follows in a similar way.

1. Resistance. Low resistance of the interconnection layers and, also, low contact resistance are of utmost importance. The former requirement is the issue of materials selection. As for the latter requirement, employment of several metals in layers is being tried to realize good adhesion between a metal and silicon.
2. Coverage. Especially, coverage of the vias is important to realize a low contact resistance and high reliability. From this viewpoint, sputtering is better than evaporation. Evaporation is unable to realize full coverage of vias with a large aspect ratio, because of the incident-angle issue. Furthermore, the chemical-vapor deposition of W affords better coverage of vias with both a large aspect ratio and small size.
3. Adherence. Whereas the adherence of Al to silicon oxide is enough, that of W is not so good. Addition of TiW between W and the silicon oxide is very effective to improve the adherence to the oxide.
4. Immunity to electromigration. See Sec. 1.3.2.3.
5. Reproducibility. As an example, the selective chemical-vapor deposition of W has not been applied to practical usage, because of its poor reproducibility.

2.4 Thin-Film Formation

For the formation of thin films for gate dielectric and capacitor applications, oxidation and chemical-vapor deposition are the two types of approaches employed. The related factors are as follows.

1. Breakdown voltage and leakage current. From the applications point of view, high breakdown voltage and low leakage current are the most important requirements.
2. Step coverage.
3. Uniformity.
4. Particle contamination. Avoidance of particle contamination is highly desired.
5. Throughput. The rate of the oxidation or deposition is important.

2.5 Common Issues

In addition to the above-mentioned factors, there are other important factors that are common to each process technology. First, typical requirements for the production facilities include

1. price;
2. stable operation;
3. volume;
4. maintainability; and
5. extensibility.

With regard to item **3**, the recent trend is the growing introduction of single-wafer systems. It has attractive features from the viewpoint of flexibility and facility cost. Regarding item **5**, efforts are being devoted to the utilization of fabrication equipment over several device generations, after necessary modifications. This is an example of a countermeasure to the rapid increase in investment on fabrication equipment due to the decrease in device dimensions and the increase in process complexities.

Second, requirements for the process-flow-related issues include the following:

1. The number of processing steps should be as small as possible;
2. the process can integrate any type of function, such as memory, digital, analog/digital mixed circuits, and so on; and
3. there must be high controllability and high reproducibility.

Regarding item **1**, one example is the CMP (Chemical Mechanical Polishing) technology, which has a smaller number of processing steps than the etchback approach.

3. DESIGN METHODOLOGY AND COMPUTER-AIDED DESIGN OF INTEGRATED CIRCUITS

With the continuing increase in the number of components integrated on a chip, increase in the complexities of transistor structures, and decrease in the feature size of transistors, design of integrated circuits (ICs) is becoming increasingly sophisticated. The demand for shorter design turn-around time (TAT), on the other hand, is increasing, because of the time-to-market competition of the new product development. From these points of view, development and introduction of both advanced design methodology and computer-aided design (CAD) technology are greatly needed.

3.1 Design Methodology of Integrated Circuits

In the typical design of integrated circuits, product specifications are successively transformed to more detailed descriptions (top-down design) (see COMPUTER-AIDED DESIGN IN ELECTRONICS). Here, each design stage makes full use of the library. The library stores the generic basic circuits such as gates or flip-flops, which are called "primitive cells," and function units, called "function macros." This library is prepared from process, device, and circuit designs, which are carried out by using simulations and refined on the basis of the results of measurements. This approach is effective for realizing shorter design TAT.

So-called design methodology refers to any strategy for improving the design efficiency to obtain shorter TAT using less engineering resources. The first is top-down design. It is also called hierarchical design, because of the transformation from abstract description to a more detailed one. The second strategy is the so-called divide and conquer design. It refers to the design of a chip after dividing it into several portions with data sizes that can easily be processed. Each portion will be designed in a short design TAT. Also, a design can be carried out in parallel, by multiple designers.

The third strategy is concurrent engineering (CE) (Rosenblatt *et al.*, 1991). In CE, many jobs that are related to the chip development, that is, marketing, design, and manufacturing, for example, are carried out simultaneously, keeping good communication among groups. By working together from the start of the device development, each group can anticipate problems and eliminate them at an early stage of the development. Compared with the so-called serial approach, where each job is carried out in cascade, this parallel approach leads to shorter development time and cheaper development cost. In this case, computer-aided design, engineering, and manufacturing technology is one of the key factors for success, in addition to a good framework.

For the CAD tools for high-level design (that is, behavior, register-transfer, and structural level design), and layout design, see COMPUTER-AIDED DESIGN IN ELECTRONICS.

3.2 Test CAD

Technology for chip testing is quite important in the production of integrated circuits. There are two serious issues associated with the increase in the number of components integrated on a chip: the increase in the time required for testing fabricated chips and the decrease in the test coverage. The term "test coverage" refers to the percentage of cases that chips in failure are exactly detected. Design style that provides means for easy chip testing is generally called "design for testability (DFT)." A typical example of DFT is a so-called scan design. It converts flip-flops to combinational circuits by using additional control circuits, thus drastically reducing the number of test patterns required for testing. As a result, a reduction in testing time and an increase in test coverage can be obtained. The technology has been widely applied to the testing of ASIC products. This approach, however, has not been applied to the custom design of chips in volume production, including microprocessors. This is because the increase in the chip area caused by the application of the scan technology is too large from the chip cost point of view. DFT technology that affords a chip area increase much less compared with the existing approach, while keeping high test coverage, is greatly needed. Such approaches, including partial scan and built-in self-test (BIST) systems, are being developed for that purpose.

Furthermore, automatic logic synthesis that takes into account the test design, called "synthesis for testability (SFT)," is a promising approach for high-quality design and short design time, by avoiding repetitive rework cycles in the chip design.

Regarding the testing of integrated circuits after assembly on the printed circuit board, boundary scan technology is widely used (Levitt, 1992).

3.3 CAD for Process, Device, and Circuit Design

Conventionally, the development of a new process is carried out by a trial-and-error approach. Test chips are fabricated under varieties of process parameters and then evaluated. By the comparison of the measured data with the target values, the optimal process conditions are determined. This approach, however, is becoming increasingly inefficient. It is because reduction in the device feature size and the employment of more complex device structures lead to an increase in the process complexities and the number of process steps, and coupling between process conditions, device behavior, and circuit performance in an increasingly intimate fashion. Hence, the conventional approach requires a great deal of development time and cost, which is becoming impractical from the time-to-market point of view in new product development.

On the other hand, thanks to both the progress of simulation technology and related CAD tools, device structure, doping profile, device behavior, and circuit performance can be simulated on computers. By employing these technologies, conventional test fabrication and performance evaluation in the process technology development can be replaced by simulations on computers. It is expected that the full use of so-called computer experiments will lead to a drastic saving of time and cost of process technology development.

The sequence of the process, device, and circuit simulation is roughly as follows. Process simulation covers the element process (see VLSI TECHNOLOGY), including lithography, ion implantation, oxidation, and so on. The results of the process simulation, such as the impurity profile, are forwarded to the device simulation. Starting with the data supplied from the process simulation, and the given device structure, the device simulator then simulates the device behavior by calculating the carrier transport and the electrostatic potentials induced by the external bias voltages. Parasitic parameters of the interconnect are also extracted by calculating the electrostatic and vector potentials. Then the circuit performance is estimated through the circuit simulation, by using the device behavior supplied by the device simulation, together with the circuit configuration.

The above-mentioned simulation consists of the modeling of each phenomenon or device, and the subsequent numerical calculations based on the model. Here, employment of a model with reasonable accuracy is of utmost importance. This is because an excessively accurate model does not necessarily give results leading to good insight into the phenomena, and a complex model generally requires too lengthy computing time, which may be impractical from the viewpoint of the time frame of the integrated circuit development.

3.3.1 Process Simulation The history of general purpose process simulators effectively started in 1977, with SUPREM and SAMPLE programs developed by Stanford University and the University of California, Berkeley, respectively. Such processes as lithography, etching, deposition, oxidation, ion implantation, diffusion, and epitaxial growth are covered by both or one of the above simulators. Since then, improvements have been carried out in many respects (Antoniadis *et al.*, 1977; Sze, 1988).

3.3.1.1 Lithography. For the lithography process, optical imaging and exposure of resist materials are successfully modeled by calculating irradiation distributions and light propagation within the layer, respectively, and are used in practice in the optimization of optical projection systems. On the other hand, development of resist and subsequent etching simulation are not well established yet, because of the lack of knowledge about the chemical and physical phenomena at the wafer surface combined with the wide variety of process and equipment variations.

3.3.1.2 Ion Implantation. Two types of modeling methods are well established:

1. Monte Carlo simulation that calculates the motion of dopants through their successive collisions with target particles, and the resulting defect distributions;
2. impurity distribution functions that are analytically characterized by their moments such as projection range and standard deviation.

These two models complement each other. Monte Carlo simulation is accurate but very CPU intensive, while the analytical distribution-functions model is less expensive, but requires the moment-parameter extraction. Both models cover the various doping species and the wide range of dose and energy conditions, including high-energy implantations, often used in modern processes. Figure 6 shows an example of an analytically simulated impurity profile, as compared with the measured one.

3.3.1.3 Thermal Diffusion. In full generality, the impurity diffusion process is simulated by calculating the time evolution of dopant concentrations governed by the continuity equations for all charged particles. Here, species can be donor, acceptor, vacancy, or interstitial. By determining the diffusion coefficient for each species, the impurity distribution is obtained with good accuracy. To take interstitial diffusion into consideration is important to simulate profiles for modern processes accurately, such as

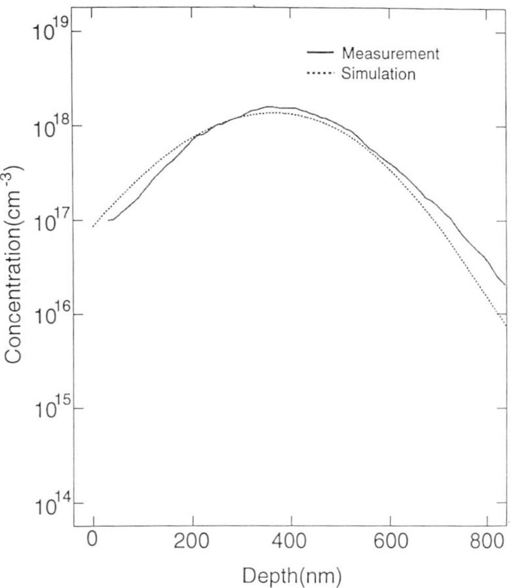

FIG. 6. An example of an analytically simulated impurity profile, as compared with the measured one. (Implanted boron profile under the condition of 120 keV, 5×10^{13} cm^{-2}, followed by annealing at 900 °C for 30 min.)

low-temperature and short-time annealing processes.

3.3.1.4 Oxidation. The growth of SiO_2 is successively simulated by integrating the oxygen flux diffusing through the existing SiO_2 layer. A compact equation representing the growth of the layer was derived by Deal and Grove (Sze, 1988).

3.3.2 Device Simulation The following two simulation models are used:

1. A drift-diffusion model, which assumes that carrier distribution is in thermal equilibrium with the semiconductor lattice and that the current is a sum of the drift and diffusion components.
2. A hydrodynamic model, which further considers the carrier energy-conservation equations (Engl, 1986; Selberherr, 1984).

Model **1** predicts the electrostatic potentials and the carrier distributions fairly well. The drain current characteristics down to 0.1-μm MOS channel-length regions are predicted. Model **2** is useful, for example, to predict hot-carrier–related characteristics.

Figure 7 shows an example of simulated electrostatic potential for a MOS device.

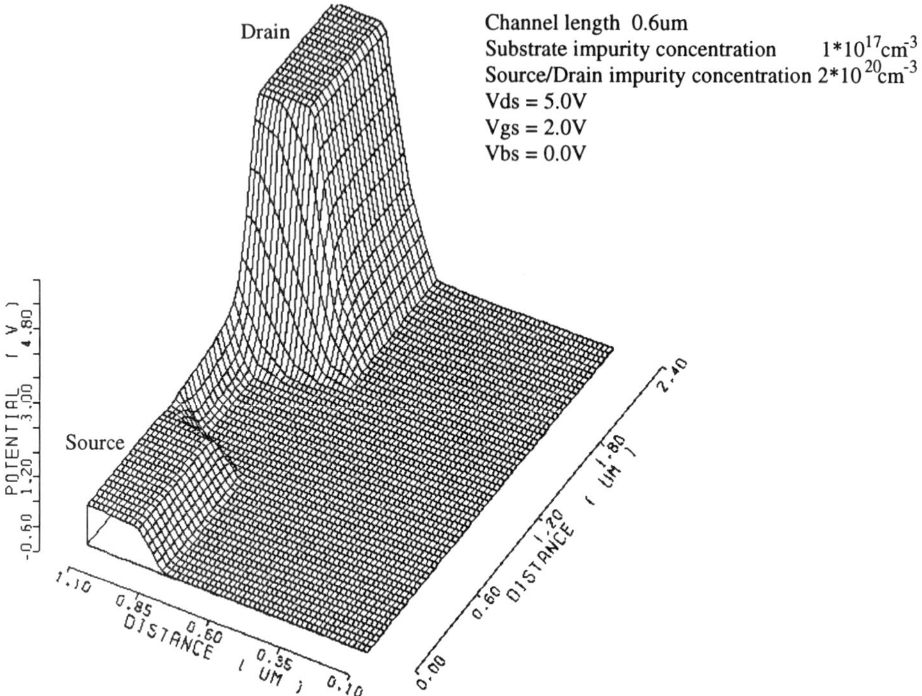

FIG. 7. An example of simulated electrostatic potential of a MOS device.

3.3.3 Circuit Simulation A set of linear equations of the entire circuit is solved to obtain circuit voltages and currents. Device characteristics are formulated in closed forms after the Poisson and the drift-diffusion equations are integrated under the appropriate assumptions. Typical models are the Gummel–Poon model for bipolar devices, and the BSIM (Berkeley SIMulation) model for MOS devices. The performance of a LSI with device technology down to 0.35-μm MOS channel length can be simulated (Nagel, 1975; Sangiovanni-Vincentelli *et al.*, 1980).

3.4 Design Environment and Technology Standardization

At present, varieties of CAD tools are supplied from many tool vendors. From the viewpoint of chip design at maximum design efficiency, designers want to choose the best possible CAD tools and combine them into an integrated design system. In this case, each CAD tool is operated on various platforms, such as engineering workstations (EWSs) or personal computers (PCs) (Comerfold and Mogal, 1993). On the other hand, ASIC vendors are delivering databases for their ASIC devices, such as gate arrays and cell-based ICs. Under these multivendor situations, in the areas of both CAD tools and IC databases, establishment of common bases for promoting mutual data exchange and smooth combination of tools is gaining increasing importance. From this viewpoint, world-wide activities on technical standardization are on the way for the realization of data interchangeability and interoperability. Described below are some examples.

3.4.1 CFI (CAD Framework Initiative) The purpose of the CFI's activity is the establishment of a common interface between individual CAD tools and the framework into which individual CAD tools should be incorporated. When the common interface is set up, individual CAD tools developed by any tool vendor can be incorporated into the framework. This opens a way to the realization of an optimal CAD system, comprised of the combination of the best possible individual CAD tools.

3.4.2 EDIF (Electronic Design Interchange Format) EDIF is one of the most typical standardized formats, for design data exchange.

3.4.3 HDL (Hardware Description Language) HDL is widely employed in describing electronic circuits, in place of the schematic expression. At present, VHDL (VHSIC Hardware Description Language) and Verilog-HDL are typical ones (Waxman *et al.*, 1989).

3.5 Future Trends

In order to realize shorter TAT for the design of chips with increasingly larger numbers of components, the following approaches will be indispensable:

1. Development of CAD tools to support higher-level design, including automatic synthesis of register-transfer–level description.
2. Establishment of an integrated design system. The final goal of the design system is a silicon compiler, which automatically translates chip specifications to mask-making data, the final level of the description.
3. Development of layout technology to produce high component-packing density. Realizing a chip size as small as possible is one of the most important requirements, from the viewpoint of the cost competitiveness of the chip.
4. Introduction of powerful platforms. Thanks to the rapid increase in the performance of microprocessors, EWSs and PCs with increasingly higher performance will be developed. Furthermore, CAD engines, dedicated hardware systems for specific design, will be developed. Timely introduction of these platforms will be more and more important for the increase in design efficiency.

4. APPLICATIONS OF INTEGRATED CIRCUITS TO ELECTRONICS SYSTEMS

Integrated circuits have been applied to almost all electronics systems, and have contributed to the creation of new functions and services with drastically reduced cost. Functions of the electronics systems are integrated in a form selected out of the following three types: standard products, ASIC, and ASSP, depending on the target specifications (see Sec. 1). This section briefly reviews such a trend and shows some typical examples.

4.1 Communication Systems

Steady progress has been made in communication systems to get closer to the goal, "Anytime, Anywhere, Access to Anyone."

The telephone system is one of the most typical systems in a communications network. At the center of the system is an exchange system, in which incoming signals are connected to a specified telephone number, via switching subsystems. Following the initial mechanical exchange systems employing relays or crossbar switches, electronic exchange systems employing electron tubes or transistors were developed to realize lower system cost. Then, based on the progress of integrated circuits and digital transmission technologies, a digital exchange system was developed in the 1960s and put into practical service in the 1970s. A digital exchange system has many favorable features, including higher transmission quality, lower system cost, and versatility in system implementations. It is a key system component of the ISDN (Integrated Services Digital Network).

One of the most important functions in digital exchange systems is the conversion of the speaker's voice signal to a Pulse-Code Modulated (PCM) signal, which is called "coding," and the reverse function, which is called "decoding." The device to perform both functions, which is called a CODEC (COder+DECoder), was first realized as a combination of transistors and passive components on a printed wiring board, then as a combination of small-scale integrated circuits and passive components on a ceramic board (that is, a hybrid integrated circuit). In accordance with the progress of integrated-circuit technology, the CODEC was first integrated in a single silicon chip in early 1980s, by use of CMOS technology. CMOS devices have the significant features of low power consumption and high packing density, which are very effective to realize a single-chip CODEC. By employing a single-chip CODEC, the size, power, and production cost of digital exchange systems were drastically reduced, which made a big contribution to the widespread penetration of the telephone network.

Regarding "Anytime, Anywhere, Access" service, the contribution of personal communications, that is, the mobile and portable telephone service, is remarkable. Especially, in cellular telephone systems, communications between handheld transceivers are made through networked base stations or "cells" (Fischetti, 1993). They are in wide use all over the world. The majority of the handsets for cellular telephones employ CMOS devices, for their features of low power consumption and operation under low supply voltages, as well as high packing density of components on a chip. The former two features are of utmost importance in battery-drive operation of portable terminals. The present battery operating time of the telephone handset is typically one to two hours. Realization of much longer service times between battery recharging is a target for the moment. Reduction of the power consumption of the chips, as well as increase of the capacity of the battery itself, is necessary for that purpose. Lowering the supply voltage and reducing the device feature size will be practical approaches for the power reduction of chips, for the time being.

Future personal communications systems are now under development, where a single phone number is assigned to each individual, no matter where the subscriber is. The system will afford "Anytime, Anywhere, Access" service in a more advanced way. One of the most important issues for that is the development of low-power, small, and light-weight personal communicators, for which the role of integrated circuits is quite significant (Heilmeier, 1992; Kobb, 1993).

Image signals, including still pictures and motion video, play quite important roles in communication systems. In just the same way as for voice signals, image signals are converted to pulse code modulated signals by a coder at the transmitting terminals, and converted back to the original signals by the decoders at the receiving ends. The amount of information inherent to the image signals, however, is quite enormous, say a thousand times as much as that for the case of voice signals. Therefore, the clock frequency required for image processing is high, roughly over 10 MHz, for example. In earlier days, these frequency ranges could be covered only by bipolar devices. Bipolar devices, however, are not appropriate to integrate on a chip enough functions for image-processing applications, because of the large chip area occupied by transistors, and large power consumption. On the other hand, MOS devices, which have the features of small chip area and low power consumption, were not fast enough to meet the speed requirements for image processing.

Thanks to the continuous reduction of the feature size of MOS transistors and the improvement in circuit technology, the clock frequency of MOS LSIs has made a steady increase. In the mid-1980s, the clock frequency of CMOS LSIs exceeded 30 MHz, which opened a way to image processing by use of CMOS LSIs. Since that time, varieties of CMOS LSIs for image-processing applications have been developed, thanks to the constant increase in the clock frequencies of CMOS LSIs. So-called multimedia, that is, image signals in addition to voice and data signals, have the possibilities to create new and broader areas of services. In this sense, the mid-1980s are considered to be the starting point of the so-called multimedia era. From this viewpoint, the goal of communication services may be expressed as "Anytime, Anywhere, Access to Anyone with Anymedia."

One of the most typical examples of multimedia services is a video teleconference, where multiple attendees in remote places join the mutual discussions, information exchange, and so on, by using image signals, voice, and data. As an extended version of the video teleconference system, desk-top video teleconference systems, operating on workstations or personal computers, are now being developed by many vendors. The systems will make possible video teleconferences among attendees sitting at their own desks, which will be a powerful tool for the promotion of a type of CSCW (Computer-Supported Cooperative Work). In the future, handheld terminals will be able to handle multimedia information, as a result of the drastic progress of CMOS technology, in terms of the large number of components on a chip and also very low power consumption.

Establishment of several technical standards for compression/decompression of images and voices, and the development of CMOS LSIs following the standards, are in progress (Yurgan, 1992).

One of the most urgent issues to realize

multimedia services is the establishment of a broadband network that can transmit broadband signals in a reasonable time. A broadband integrated services digital network (abbreviated as B-ISDN) is an important infrastructure for future communication network systems (Kaplan, *et al.*, 1991). It has a transmission speed of 150 Mb per second or higher. Several devices for B-ISDN are now under development in many organizations.

In the United States, an "Information Super Highway," a next-generation high-speed information network, is under establishment.

4.2 Computer Systems

Computer systems have always provided the largest markets for semiconductor devices and also stimulated the development of leading-edge devices to realize the most advanced computer systems. Transistors were first introduced into computer systems in the 1960s, and then integrated circuits, in the 1970s. This contributed to the realization of a system with much smaller size, higher reliability, and much lower cost, compared with the system comprising electron tubes. Integrated circuits widely used at the initial stage were bipolar small-scale integrated circuits, such as TTL (Transistor-Transistor Logic), ECL gates, and flip-flops. Here, TTL had the features of low power consumption with fairly high speed at that time, and ECL showed the highest operational speed (see CIRCUITS, DIGITAL). Since then, ECL integrated circuits with an increasing number of components on a chip have been developed, making contributions to the realization of so-called main-frame computers which have leading-edge performances. Here, gate array or cell-based ICs have been mainly employed to integrate digital system functions.

On the other hand, MOS integrated circuits have many advantages, including smaller chip area and lower power consumption, compared with ECL devices, although with lower operational speed. In the early 1970s, two MOS LSIs were first disclosed, a 1-kbit DRAM in 1970 and a 4-bit microprocessor in 1971. Microprocessors are devices that integrate the central processing unit (CPU) of the computers on a chip (see COMPUTERS; COMPUTER HARDWARE; MICROPROCESSORS). After that, the number of components integrated on a chip for memory and digital devices has steadily increased, as shown in Fig. 1. These MOS LSIs have been widely applied to minicomputers and desk-top machines (that is, workstations and personal computers).

The performance of microprocessors has increased at the rate of roughly 2.7 times per every three years. Recently, RISC (Reduced Instruction Set Computer) microprocessors have given big contributions to the rapid increase in microprocessor performance. Compared with the conventional CISC (Complex Instruction Set Computer) microprocessor, the RISC microprocessor has an essential advantage of realizing a short cycle time. As a result, the performance of the desk-top machines has been approaching that of main-frame computers. As a consequence, the percentage of computing jobs that are processed on small-sized machines is increasing year by year. This trend is called "downsizing." The term refers to the reduction in both the size and price of the machines. This is a big and typical change generated by widespread introduction of microelectronics (Gelsinger *et al.*, 1989; Comerfold and Mogal, 1993).

The rapid increase in the performance of MOS LSIs is stimulating a new trend in the computer field. So-called massively parallel processors (MPP), composed of large numbers of processors connected in parallel, can afford high throughput performance as a whole. This type of machine can be applied to heavy computing jobs, including two- or three-dimensional simulations of semiconductor devices, or the simulation of the fabrication processes of integrated circuits. As a result, MPP is becoming one of the most promising approaches for recent most advanced supercomputers.

LSIs for controlling peripheral functions, such as controllers for floppy and hard disks, and display devices, are also indispensable to realize high-performance computers without overloading microprocessors. In addition, recently, displaying a large amount of data in a visual form is becoming increasingly important, for easier understanding of the data. This field of technology is called "scientific visualization." For this purpose, a high-speed graphic display controller (GDC) chip is one of the most important devices for workstations and personal computers (see COMPUTER GRAPHICS).

Historically, because of the rapid increase in the performance of workstations, the price-to-speed ratio of the computers for computer graphics (CG) applications, for example, has been reduced by a factor of roughly 1/100 during the past five years. This is one of the most important factors to accelerate the applications of CG technology to many fields, including the production of movies or pictures.

4.3 Consumer Electronics

One of the most typical products in consumer electronics is a desk-top calculator. The first model was developed in the late 1960s using bipolar ICs, including TTL. In the 1970s, 1-chip CMOS LSI with low power consumption and battery-drive capability was developed. A low-price palm-top calculator was put to practical applications by the combination of this chip and the liquid crystal display device. Since then, the palm-top calculators have always created new markets, based on the increasingly enhanced functions/performances and cost reduction.

The automobile system is another example in which the introduction of microelectronics to electronic systems has been making big impacts. Employment of semiconductor devices in automobile systems was first made in 1960, with a rectifying semiconductor diode. Following the first introduction of an *n*-type metal-oxide semiconductor (NMOS) microprocessor in the mid-1970s, CMOS microprocessors and LSIs have extensively been introduced since the mid-1980s. Microprocessors are widely used to control engine systems, the sequence of door or seat operations, and information systems such as a car navigation system. Contributions of microelectronics will further include safety and amenity improvement, and information assistance, as well as optimal engine operation control from the viewpoint of environmental protection.

A television system is also a typical consumer product to which varieties of integrated circuits have been applied, resulting in cost reduction and reliability increase. Recently, a HDTV (High-Definition Television) is under development all over the world. The system has a resolution of roughly two times as large as the present television system, in both horizontal and vertical directions. This high-resolution presentation of video signals is expected to develop new services in consumer electronics, and also other applications, including medical and industrial fields. At present, experimental broadcasting service has started in Japan. In the United States and Europe, examinations on the system specifications are being carried out. A big commercial market for HDTV systems is expected to build up late this century or in the beginning of the next century. Important requisites for promoting the buildup of HDTV service will be preparation of both low-price receiver sets and abundant software. One of the key factors for success is the development of LSI chip sets with a chip count as small as possible.

When multimedia information is expressed in a digital format, the signal can be treated seamlessly in television, computer, and communication, which leads to the creation of varieties of new services. For example, CD-ROM, a type of disk, can store 540 Mbytes of data. On the other hand, video information, when compressed according to the MPEG1 standard, has a data transfer rate of roughly 1.5 Mbit/s. Therefore, roughly 48 minutes of compressed video information is stored in the CD-ROM [$(540 \times 8/1.5)/60 =$ 48 minutes] (Yurgan, 1992; Adam and Cole, 1993). This capacity, however, is not enough to store varieties of types of information, from the viewpoint of recording time and signal quality. The development of a disk with much larger capacity is greatly needed.

Figures 8(a) and 8(b) provide an insight into the storage capacity required for the typical services. Figure 8(a) shows some examples of services employing still pictures. The horizontal axis shows to what extent the original signal is compressed, in order to limit the signal bandwidth. The vertical axis corresponds to the number of sheets that can be recorded on a memory chip. The figure shows that the development of a memory chip with 64 Mbit capacity or larger will find varieties of services. Figure 8(b), on the other hand, shows typical examples of video-signal–related services and required memory capacity. Here, it is assumed that the video signal has the usual National Television System Committee (NTSC) format. In this case, to record video information on a single chip is a very tough issue, even if a high compression ratio is assumed. These applications are

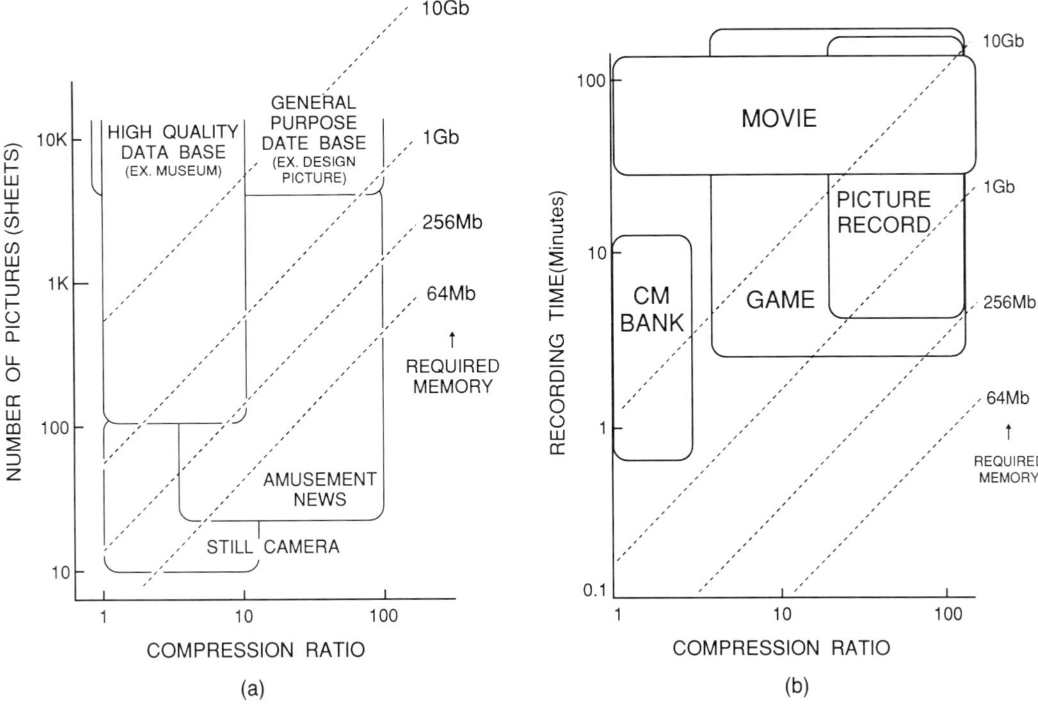

FIG. 8. Typical examples of image-signal–related services and necessary memory size: (a) still pictures, (b) motion video.

waiting for the development both of the advanced semiconductor technology and efficient signal-compression algorithms. As for the recording of the HDTV signals, the required memory capacity is several times as large as that for signals with NTSC format, corresponding to the broader signal bandwidth of the original signal. This application will be a target of big technical challenge.

4.4 Human Interface

One of the most familiar forms of Human Interface (HI) with electronics systems is the user interface with computers. The development of GUI (Graphical User Interface) has contributed to better user interface (see COMPUTER GRAPHICS).

For portable computers, on the other hand, making an input of a name, address, phone number, and so on, through handwriting using a pen may be convenient compared with the use of keyboards. In order for such a system to be practical, any person's handwriting will have to be recognized accurately and speedily. This target is much harder to meet than OCR (Optical Character Recognition) systems, which recognize printed characters. At present, several recognition algorithms are being developed on 32-bit low-power microprocessors.

With regard to voice recognition, the present system can recognize very limited numbers of words spoken by persons whose spoken words are recorded beforehand. The main target of present development is the realization of a system that can recognize a large number of words spoken by anyone, accurately and speedily. Varieties of algorithms are being developed on digital signal processors, aiming at practical applications.

A voice synthesis system integrated on a chip has already been in practical application. The present and future target of chip development is the synthesis of a comfortable voice.

4.5 Future Trends

It can naturally be anticipated that boundaries among consumers, computers, and communications markets will disappear in the future, and new types of services will be generated. One typical example is an interactive television, which will afford video-on-demand service, home shopping, and so on. Factors accelerating such movement include the incorporation of digital technology and integrated circuit technology. Deregulation will favor the early introduction of those services.

Based upon the increase in the number of components on a chip, handheld terminals with varieties of multimedia functions will be available at a reasonable price. Integration of all or part of a personal computer, a cellular telephone, a television, and so on, will provide users with attractive services on portable terminals.

5. FUTURE PROSPECTS AND LIMITATIONS OF MICROMINIATURIZATION

5.1 Future Prospects

5.1.1 Expectations of Features Features of LSIs in the year 2000 will be anticipated as follows, if the trends up to now are simply extrapolated:

Number of components on a chip	
DRAM	4 Gbit
Digital LSI	50 M components
Operational frequency	1 GHz
Performance of a microprocessor	1000 MIPS
Chip size	25×25 mm^2

A complete desk-top computer, with additional functions, will be integrated on a single chip. Moreover, multimedia handheld terminals with attractive functions and performances are expected to be available, at an attractive price.

5.1.2 Technical Issues and Solutions Reduction in device feature size and increase in the number of components integrated on a chip raise the following issues:

1. Degradation of the device reliability, due to the reduction of the device feature size. Lowering the supply voltages will be effective.
2. Rapid increase in the power consumption, due to the increase in the number of components integrated on a chip. Low-power design is one of the most significant issues for the realization of high-performance, portable terminals. For this purpose, reduction of the device feature size and supply voltages, and also circuit and system designs, will be indispensable.
3. Increase in the design time, corresponding to the increase in both the complexities of the device structures and the number of components integrated on a chip. Employment of top-down design style and powerful CAD systems will be the solution. Development of the most powerful CAD system, through the selection of the CAD tools, should be carried out strategically.
4. Rapid increase of manufacturing cost. Decrease in the device feature size and accompanying increase in the complexities of the fabrication process are giving rise to exponential cost increases for equipment and facilities. It is likely that this may slow down the rate of progress of semiconductors (Moore, 1991). Furthermore, an advanced system-on-silicon chip requires the integration of varieties of functions, including digital, memory [SRAM (static RAM), EEPROM (electrically erasable PROM), or DRAM], and analog-digital mixed circuits, which tends to cause further cost increase. Efforts should be directed to the development of process technology with a reduced number of process steps, based on the overall design strategy.
5. Development of new markets and application technology.

Throughout the above activities, the necessity of effective cooperation among engineers from the system, circuit, and device/process sides is increasing.

5.2 Limitations of Microminiaturization

The great progress in Si VLSIs has been realized by MOSFET miniaturization based on scaling theory. High speed, low power, and high packing density have resulted from scaling down MOSFET dimensions by a factor of k, as well as supply voltage. The scaling theory is still considered to be a good guideline for designing future MOSFETs.

However, as a result of several physical and practical limitations, it will not be a perfect rule for very small MOSFETs. Approaches for pushing the limits are becoming more important when the design rule is less than a quarter micron.

Since the 1970s and 1980s, there have been many discussions on the limits of digital electronics (Fukuma, 1988; Meindl *et al.*, 1993). In the 1970s, "0.25 μm" seemed to be a minimum design rule, and, in the 1980s, "0.1 μm" was believed to be a limit for room-temperature operation. However, now we have an experimental 256-Mbit DRAM using 0.25-μm MOSFETs (Sugibayashi *et al.*, 1993) and a room-temperature test vehicle using ~0.1-μm MOSFETs (*IEEE IEDM Digest*, 1993). In this section, the future prospects of MOS VLSIs are reviewed, emphasizing several limitations based on modern knowledge.

5.2.1 Physical and Practical Limitations

Although scaling theory provides a guideline for reducing MOSFET size, it does not give optimum MOSFET parameters. Hence, some parameters that optimize VLSIs may be inconsistent with scaling theory. This means that, in addition to purely physical limitations, practical considerations also seriously hamper device miniaturization. Here, several limiting factors and countermeasures are discussed.

5.2.1.1 Tunneling Effects. When the gate-oxide thickness is reduced to 2–3 nm, direct tunneling between the gate and the substrate occurs. Similarly, when the channel length approaches 10 nm, tunneling between source and drain occurs. MOSFETs with leakage current due to such tunneling phenomena will not be acceptable for VLSI circuits where MOSFETs are treated as switches. As long as SiO_2 and Si are used for the gate insulator and the substrate, respectively, the minimum sizes mentioned above are the absolute limits in MOSFET miniaturization.

5.2.1.2 Hot-Carrier Effects. In order to increase the operational speed of VLSI circuits, higher supply voltage is preferable. Increasing the supply voltage, however, induces many hot carriers near the drain. These hot carriers generate states at the Si–SiO_2 interface and/or may be trapped in the gate oxide. Interface states and trapped carriers lead to long-term characteristic degradation in MOSFETs, so that the hot-carrier effect is one of the most serious limiting factors for maximum supply voltage. Note that MOSFET characteristic degradation is observed even when the supply voltage is much lower than the Si–SiO_2 barrier height or threshold voltage (energy) for interface state generation. This is because several modes of carrier scattering provide some high-energy carriers. Since electron mobility is larger than hole mobility and since the Si–SiO_2 barrier height for an electron is lower, characteristic degradation is more severe for n-channel MOSFETs.

According to scaling theory, the supply voltage should be scaled down by the same scaling factor, k, as that for MOSFET dimensions, to keep the inner electric field constant. However, because of nonstationary carrier transport, the carrier energy is much lower than that predicted from the local electric field. As a result, the maximum supply voltage allowed, $V_{DD,\max}$, is almost proportional to $k^{-0.5}$ even if conventional single-drain structure is used. Recent experiments have shown a maximum supply voltage of 1.5 V for 0.1-μm n-channel MOSFETs; therefore, 0.5 V may be expected for 0.01-μm MOSFETs.

5.2.1.3 Subthreshold Leakage Current (Threshold Voltage Scaling). Drain current in the subthreshold region is proportional to $\exp(qV_G/nkT)$. Here, V_G is gate voltage, and other symbols have conventional meanings. Since we need sufficiently low leakage current at $V_G = 0$ V for dynamic circuit operation and low standby power dissipation, the threshold voltage V_T has to be greater than 0.5 V at room temperature; i.e., V_T cannot be scaled down. In this case, however, the gate delay τ_{pd} increases when the design rule is less than 0.1 μm. Because the gate delay is proportional to $V_{DD}/(V_{DD} - V_T)^m$, the lower supply voltage required by hot-carrier reliability considerations cancels out the MOSFET miniaturization merits with respect to speed. To overcome this problem, we have to accept relatively large off-state current due to the lower threshold voltage. Although we will not be able to employ dynamic-type circuits, the standby power issue will be solved by introducing a sophisticated power-management system.

If low-temperature operation can be employed, threshold voltage can be reduced without any leakage current increase. For instance, at liquid nitrogen temperature (77 K), a threshold voltage of 0.1–0.2 V is possible. A compact cooling system is a breakthrough for realizing high-performance sub-0.1-μm CMOS VLSIs.

5.2.1.4 Punchthrough Phenomenon (Short Channel Effect). Generally the "punchthrough" of a MOS transistor refers to the state in which the depletion regions on the drain and source sides touch each other. In this case, a large current will flow through the transistor, and the device ceases normal operation. Punchthrough occurs for a large applied voltage and/or small device dimensions.

Hence, the width of the depletion-layer spread from source and drain should be scaled down to suppress the punchthrough phenomenon. Ideally, we need to increase the substrate impurity concentration by a factor of k and to decrease the effective substrate bias $V_{bi} + V_{sub}$, where V_{bi} is built-in potential and V_{sub} is the actual substrate bias. However, V_{bi} related to the Si band gap cannot be scaled down, and, therefore, the punchthrough phenomenon is extremely serious when the design rule is less than 0.1 μm. To reduce the effective substrate bias, forward biasing on the substrate with reducing operation temperature is attractive. This corresponds to virtual band-gap scaling. Another approach is to use extremely thin Si film on an insulator (SOI: Silicon On Insulator). If the Si film thickness is comparable with the inversion-layer thickness, the potential in the whole Si film is fully controlled by the gate electrode. Then, the punchthrough phenomenon is well suppressed.

5.2.1.5 Parasitic Elements. Resistances and capacitances associated with wiring are becoming dominant factors in determining the limits of VLSI operational speed. Both (a) wiring resistance per unit length and (b) contact resistance are proportional to k^2, while (c) MOSFET resistance in the ON state is constant. Thus, the sum of terms (a) and (b) will be equal to the value of (c) for the design rule around 0.1 μm. Finer design rule will make the former terms larger. New materials such as Cu and Ag are expected to reduce wiring resistance. To reduce the contact resistance, increasing effective contact area by silicidation may be more practical than reducing specific contact resistance by new materials.

On the other hand, wiring capacitance is not simply scaled down because of coupling with neighboring wires and because of longer wirings necessary in a large chip. New interlayer dielectrics with low permittivity such as organic materials will help this situation. However, chip architectures that avoid long wiring are more important.

5.2.1.6 Power Dissipation. Power dissipation for CMOS circuits is expressed by $P = V_{DD}^2\,\Sigma fC$, where f is frequency and C is node capacitance. The ideal scaling does not change the power dissipation per unit area. However, in the practical case, power dissipation density increases with smaller design rules, because possibly high supply voltage is adopted for high-speed operation, and wiring capacitance is not ideally scaled down.

The maximum power dissipation of a chip, determined mainly by packaging technology, is around 30 W for an air-cooling system. Present high-performance processors have almost reached this limit. Therefore, the power dissipation is an important issue for increasing operational speed, even at present.

Several approaches may be used to overcome this difficulty, e.g., reducing wiring capacitance by sophisticated layout tools and/or interlayer dielectric material with low dielectric constant; reducing supply voltage and clock frequency with massively parallel architecture; using memory-oriented architecture; using dynamic power and clock control; and improving package.

5.2.1.7 Fluctuations. Fluctuation in device structures caused by poor process controllability or statistical randomness is one of the most serious practical limiting factors for miniaturization. For example, when the design rule is 0.1 μm (10^7–10^8 MOSFETs can be integrated), the standard deviation allowed for the effective channel length is estimated to be about 3 nm in order to realize reasonable production yield. Note that the effective channel length is determined by the controllability of various process steps, including the lithography for the polycrystalline silicon as a gate electrode, etching, source/drain implantation, annealing, and so on. At present, the standard deviation of 3 nm seems to be a very difficult target. Another example is the threshold voltage fluctuation caused by statistical fluctuation of impurity distribution. Since the total number of substrate impurities involved in each MOSFET

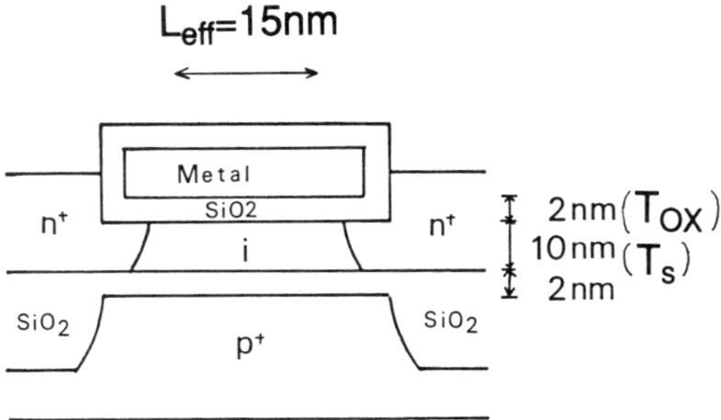

FIG. 9. Cross-sectional view of an ultimately scaled-down *n*-MOSFET.

is scaled down by k^{-1}–k^{-2}, it rapidly decreases with the decrease in the design rule. Then the distribution becomes Poisson type, and the fluctuation is proportional to the square root of the total impurity number, which results in large fluctuation in the threshold voltage. This statistical fluctuation cannot be completely controlled.

To overcome these fluctuation problems, it is essential to develop new processes with high controllability, and new devices with low process sensitivity. New circuits and architectures including fault-tolerant technique will be appreciated as well.

5.2.2 Ultimately Scaled-Down MOSFET and Future Prospect of VLSIs Among the limitations discussed in Sec. 5.2.1, the tunneling effect is the final one in determining the minimum physical size. The smallest possible value of the oxide thickness is ~2 nm, and the effective channel length is ~10 nm.

Figure 9 illustrates an ultimately small *n*-channel MOSFET. The MOSFET is based on SOI structure, and the thickness of the active Si layer is designed to be equal to that of the inversion layer. To avoid the impurity fluctuation problem, this Si layer is intrinsic type. The threshold voltage adjustment must be done by designing Si and SiO_2 thicknesses and/or by controlling a bias on the p^+ substrate. The p^+ substrate is also effective to prevent punchthrough. The maximum supply voltage will be ~0.5 V from the hot-carrier reliability consideration. This value is reasonably high even at room temperature for avoiding thermal noise problems. Figures 10(a) and 10(b) show the simulated electrical characteristics of an *n*-channel MOSFET with L_{eff} = 15 nm (L_{gate} = 20 nm), at 77 K. Sufficiently large drain current and high on/off ratio are predicted. Estimated propagation delay time is 1–2 ps.

To realize VLSIs using the MOSFETs shown in Fig. 9, many problems mentioned in Sec. 5.2.1 must be solved. Some of these problems might be overcome individually by countermeasures described in that section. However, difficulties concerned with power dissipation and impurity fluctuations are is-

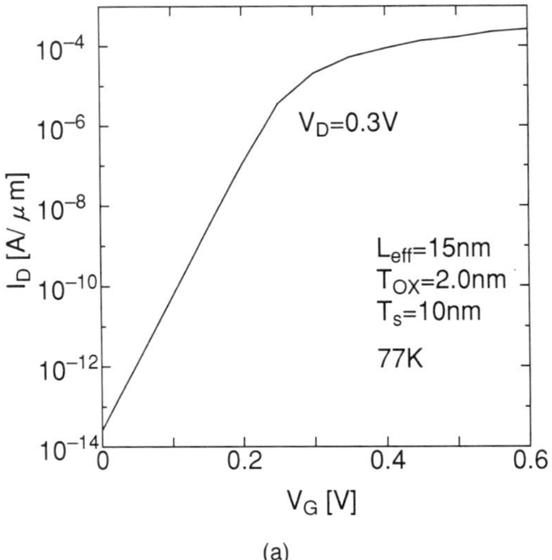

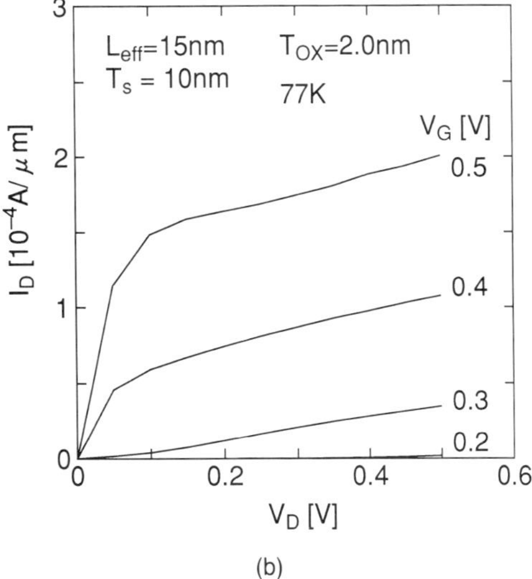

FIG. 10. Simulated electrical characteristics of the *n*-MOSFET shown in Fig. 9, at 77 K: (a) drain current vs gate–source voltage, (b) drain current vs drain–source voltage.

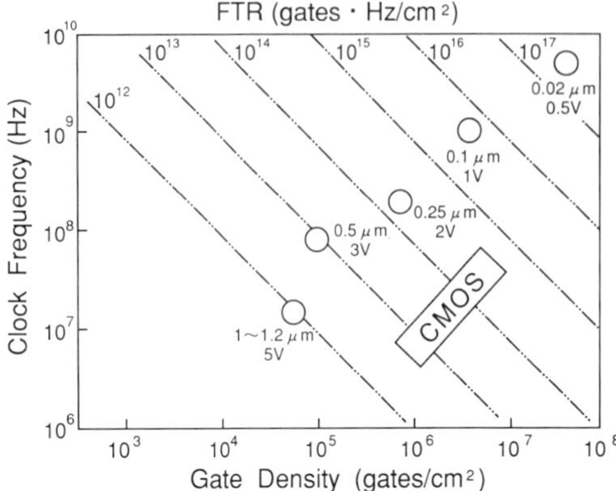

FIG. 11. Estimated clock frequency vs gate density for CMOS microprocessors, consisting of the transistors shown in Fig. 9.

sues of trade-off between performance and the practical situation, including economics. This means the necessity of comprehensive optimization from the system viewpoint. In particular, parallel processing and fault-tolerant technique will be important breakthroughs. Once the optimization is successful, progress similar to that shown in Fig. 11 might be expected. Here, a CMOS microprocessor has been taken as an example. A functional throughput rate (FTR) of 10^{17}–10^{18} may be expected with these ultimately scaled-down MOSFETs, where

$$\text{FTR} = (\text{gate density}) \times (\text{clock frequency}).$$

GLOSSARY

ASIC (Application-Specific Integrated Circuit): An IC designed for application to some specific systems or markets, as opposed to standard products, such as memories or microprocessors.

ASSP (Application-Specific Standard Product): An IC that is widely used in some specific market. Examples include a floppy-disk controller.

B-ISDN (Broadband Integrated Services Digital Network): A public end-to-end digital communications network that supports a wide range of services, such as audio, voice, video, and data, over standard interfaces.

CD-ROM: Compact-Disc ROM (Read-Only Memory), that is, a laser-encoded disc that in a typical format stores 540 Mbytes (540 × 8 Mbit) of randomly accessible video, audio, and/or text data.

Cellular Telephone: A mobile and portable radio telephone service that uses networked stations or "cells."

CISC (Complex Instruction Set Computer): A computer in which individual operations carry out elaborate data processing. It is much slower than RISC.

CMOS (Complementary symmetry MOS): A circuit type that employs both *n*-channel and *p*-channel MOS transistors. It has a favorable feature of very low power consumption.

Depth of Focus: An amount of defocus tolerance within which the image shows a quality to satisfy the specifications.

Design for Reliability (DFR): Chip design so as to realize high reliability.

Design for Testability (DFT): Chip design so as to make chip testing easier.

ESD (Electrostatic Discharge): Potentially damaging discharge into the pins of an IC from an outside source, including a person's hands.

MPEG: Moving Picture coding Experts Group. A technical group investigating international standards on the compression and decompression of video and audio signals.

Multimedia: Delivery of information that combines different content formats (motion video, still pictures, graphics, animation, audio, text, and so on).

RISC (Reduced Instruction Set Computer): A computer based on a set of simplified instructions. Instructions' simplicity affords higher throughput, compared with CISC.

Virtual Reality (VR): A virtual environment generated by computers around a person, based on information sensed by a person himself.

Works Cited

Adam, J. A., Cole, B. (1993), "Special Report on Interactive Multimedia," *IEEE Spectrum* **30** (3), 22–39.

Antoniadis, D. A., Hansen, S. E., Dutton, R. W., Gonzalez, A. G. (1977), *SUPREM 1—A program for IC process modeling and simulation*, Stanford University Stanford Electronics Labs. Technical Report No. 5019-1.

Chandrakasan, A. P., Sheng, S., Broderson, R. W. (1992), "Low-Power CMOS Digital Design," *IEEE J. Solid-State Circuits*, **SC-27**, 473–484.

Comerfold, R., Mogal, J. (1993), "Focus Report

and Guide to Engineering Work stations and PCs," *IEEE Spectrum* **30** (5), 37–79.

Engl, W. L. (Ed.) (1986), *Process and Device Modeling*, Amsterdam: North-Holland, Elsevier Science Publishers B.V.

Fischetti, M. (1993), "The Cellular Phone Scare," *IEEE Spectrum* **30** (6), 43–47.

Fukuma, M. (1988), "Limitations on MOS ULSIs," *Technical Digest of Symposium on VLSI Technology*, 7–8.

Gelsinger, P. P., Gargini, P. A., Parker, G. H., Yu, A. Y. C. (1989), "Microprocessors circa 2000," *IEEE Spectrum* **26** (10), 43–47.

Gunning, B., Yuan, L., Nguyen, T., Wong, T. (1992), "A CMOS Low-Voltage-Swing Transmission-Line Transceiver," *ISSCC Digest of Technical Papers*, 58–59.

Heilmeier, G. H. (1992), "Personal Communications: Quo Vadis," *ISSCC Digest of Technical Papers (Invited)*, 24–25.

Kobb, B. Z. (1993), "Personal Wireless," *IEEE Spectrum* **30** (6), 20–25.

IEEE ISSCC Digest of Technical Papers, from 1970 to 1993, and the *Digest of Technical Papers of the Symposium on VLSI Circuits*, from 1987 to 1993.

Itoh, K. (1990), "Trends in Megabit DRAM Circuit Design," *IEEE J. Solid-State Circuits*, **SC-25**, 778–789.

Kaplan, G., Gerla, M., Sharma, R. L., Delisle, D., Pelamourgues, L. (1991), "Special Guide to Data Communications: High-Speed Networks and Interconnections," *IEEE Spectrum* **28** (8), 21–44.

Levitt, M. E. (1992), "ASIC Testing Upgraded," *IEEE Spectrum* **29** (5), 26–29.

Meindl, J. D., De, V. K., Agrawal, B. (1993), "Prospects for Gigascale Integration (GSI) Beyond 2003," *ISSCC Digest of Technical Papers* 124–125.

Moore, G. E. (1991), "Semiconductor Technology Reaches Middle Age," *IEEE Comput. Mag.*, 105.

Nagel, L. (1975), "SPICE2: A Computer Program to Simulate Semiconductor Circuits," University of California, Berkeley, ERL Memo ERLM75/520.

Rosenblatt, A., Watson, G. F., Shina, S. G., Hall, D., Reddy, R., Wood, R. T., Cleetus, K. J., Turino, J., Wheller, R., Burnett, R. W. (1991), "Special Report on Concurrent Engineering," *IEEE Spectrum* **28** (7), 22–37.

Sangiovanni-Vincentelli, A. L. (1980), "Circuit Simulation," in: P. Antognetti, D. O., Pederson, H. De Man (Eds.), *Computer Design Aids for VLSI Circuits*, Alphen aan den Rijn, The Netherlands: Sijthoff & Noordhoff, pp. 19–112.

Santo, B. (1989), "BiCMOS Circuitry: the Best of Both Worlds," *IEEE Spectrum* **26** (5), 50–53.

Selberherr, S. (1984), *Analysis and Simulation of Semiconductor Devices*, New York: Springer-Verlag.

Sugibayashi, T., *et al.* (1993) , "A 30ns 256Mb DRAM with Multi Divided Array Structure," *ISSCC Digest of Technical Papers*, 50–51.

Sze, S. M. (1988), *VLSI Technology*, 2nd Ed., New York: McGraw-Hill.

Takeda, E., *et al.* (1993), "VLSI Reliability Challenges: From Device Physics to Wafer Scale Sysems," *Proc. IEEE*, **81** (5), 653–674.

Waxman, R., Saunders, L., Carter, H. (1989), "VHDL Links Design, Test, and Maintenance," *IEEE Spectrum* **26** (5), 40–44.

Yürgan, R. K. (1992), "Digital Video," *IEEE Spectrum* **29** (3), 24–30.

Further Reading

Hennessy, J. L., Jouppi, N. P. (1991), "Computer Technology and Architecture: An Evolving Interaction," *IEEE Comput. Mag.*, 18.

Mavor, J., Jack, M., Denyer, P. (1983), *Introduction to MOS LSI Design*, London: Addison-Wesley.

Mead, C., Conway, L. (1980). *Introduction to VLSI Systems*, Reading, MA: Addison-Wesley.

Sze, S. M. (1981), *Physics of Semiconductor Devices*, 2nd Ed., New York: Wiley.

Weste, N. H. E., Eshraghian, K. (1985), *Principles of CMOS VLSI Design; A Systems Perspective*, Reading, MA: Addison-Wesley.

MICROLITHOGRAPHY

HENRY I. SMITH, *Department of Electrical Engineering and Computer Science, Massachusetts Institute of Technology, Cambridge, Massachusetts, U.S.A.*

INTRODUCTION

This brief review of microlithography focuses on the basic principles of the technology, and the methods which have proven most effective in forefront research and commercial applications, most notably the fabrication of microelectronic and optoelectronic devices. These methods are photolithographic shadow printing, optical-projection lithography, scanning-electron-beam lithography, and x-ray lithography.

1. BASIC PRINCIPLES OF MICROLITHOGRAPHY AND RESISTS

1.1 The Planar Process

Figure 1 illustrates a process, known as the planar fabrication process, whereby a complex patterned structure can be created in or on the surface of a substrate. As depicted, it proceeds in four stages. A substrate to be patterned is first coated with a film, commonly called a "resist," and this resist is exposed to a pattern of radiation that can be visible or ultraviolet photons, x-ray photons, electrons, ions, or even neutral atoms if they have sufficient energy. As discussed below, the chemical alteration of the resist produced by the irradiation is used to produce a relief structure or "relief image" in the resist layer. Thereafter, this relief image can be used as a mask for subsequent modification of the substrate. As depicted in Fig. 1, this modification can be

1. an etching of the substrate in unprotected areas by a chemical solution, a reactive gaseous plasma, or energetic atomic or ion bombardment;
2. a growth of material up through the interstices in the resist by a process such as electroplating;
3. a patterned doping via diffusion of impurity atoms or by ion implantation into areas unprotected by the resist coating; or
4. a patterning of deposited material through a process known as liftoff.

In the latter case, material is deposited over

3-527-28132-0/94/$5.00 + .50

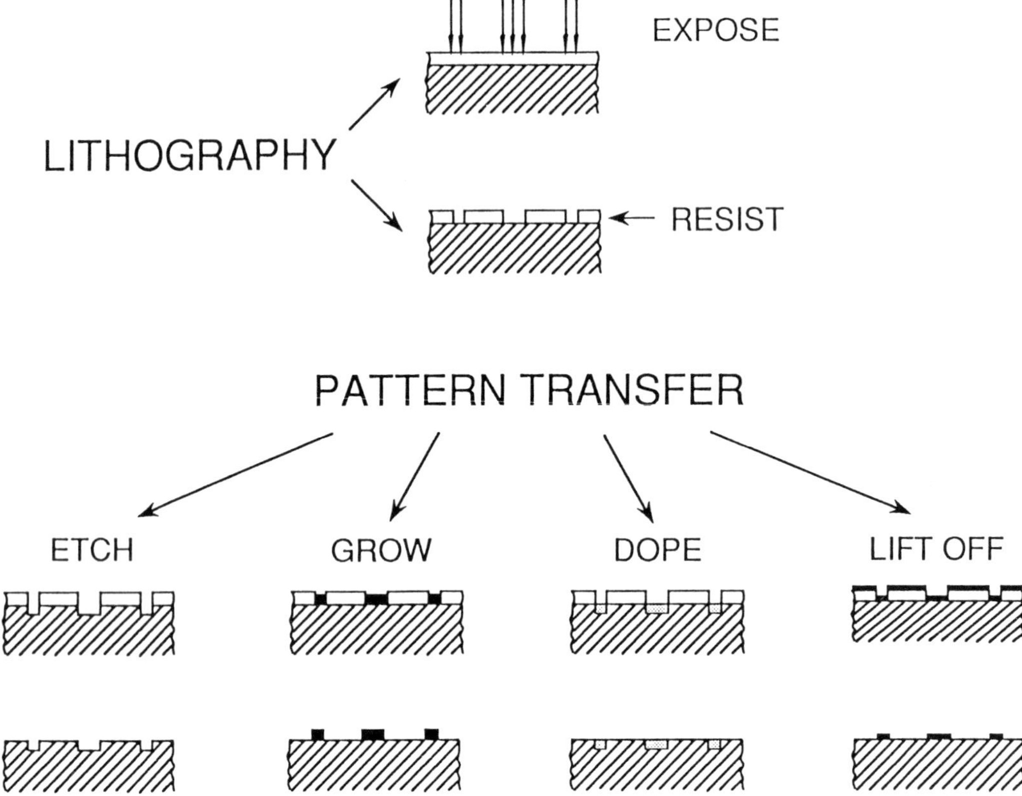

FIG. 1. Schematic of the planar fabrication process.

the entire relief image, and then, after removal by dissolution or etching of the resist, a patterned deposit is left behind.

The basic concept inherent in the planar process was actually invented well over 1000 years ago by neolithic people (Haury, 1967). The Hohokam Indians of the American Southwest had a thriving agricultural civilization which lasted from about 300 B.C. to 1400 A.D. As a resist, the Hohokam used a pitch obtained from the mesquite plant to coat the substrates, which were commonly seashells obtained in trade with native Americans living along the seacoast. The Hohokam were not able to image photons or charged particles, of course, but instead produced the relief image in the resist by mechanical means, such as scratching. A weak acidic acid solution produced from the blossoms of the saguaro cactus enabled them to etch relief images into the seashell substrates. This process, which is usually called lithography, from the Greek *lithos,* meaning stone, and *graphein,* meaning to write, was later reinvented by the Europeans and used for such things as decorating armor and, many years later, for producing plates used in the printing of graphics. It is a curious irony of technological history that the very first photographic image, by Nicephore Niepce in 1826, was actually produced in a photosensitive polymer (bitumen of Judea) and made permanent by chemical etching into a substrate, effectively duplicating the planar process of Fig. 1, not the silver-halide-based process that we now call photography. The planar process of Fig. 1 is used in a wide range of industries from printing and the production of mechanical parts to the production of the microelectronic "chips" that are the basic memory and logic components of computers.

From the point of view of etymology, the term *lithography* should refer to the entire planar process, as depicted in Fig. 1. However, today it is more common that the terms lithography, microlithography, and nanolithography refer only to the process of producing the relief image in the resist film; the subsequent stages are referred to as "pattern transfer." We will follow this usage in the rest of this article.

The printing industry also uses the term lithography for the process in which a pattern is formed in an organic material that repels water-based inks on a substrate to which the inks adhere. Pressing the substrate onto a paper sheet transfers the ink pattern onto the paper.

1.2 Lithography, A Binary Process

The term "microlithography" is used to designate the creation of structures in resist with minimum feature sizes from a few micrometers to 0.1 μm (100 nm). The terminology "nanolithography" is widely used in reference to sub-100-nm features, and we will follow this custom here.

In this article, we refer frequently to "resolution," a term used rather loosely to designate the minimum size of features (lines, or openings) or the minimum "spatial period" or "pitch" achievable with a given lithographic technique. The minimum-pitch criterion is preferred because it is directly relatable to resolution criteria, such as the Rayleigh criterion, used with optical imaging systems (i.e., telescopes, camera, microscopes). Although not all applications require the highest resolution, the following discussions will be framed around the systems and methods that yield the highest practical resolution.

It is important to emphasize that lithography is a binary process and hence is quite distinct from photography. By this we mean that the relief image in the resist is either "on" or "off"; i.e., the substrate is either protected by the resist mask or not. In photography, on the other hand, the task is to record scenes from the world around us, which involves a wide range of light intensities. As a result, a photographic process must represent "shades of gray" and, in some cases, colors. In lithography, there are no shades of gray, only "black and white," that is, "on" and "off."

1.3 Positive-Tone and Negative-Tone Resists

A positive resist is one in which the areas irradiated by photons, ions, or electrons are removed, and the unirradiated areas remain as the relief image in resist. With a negative-tone resist, the relief image corresponds to the irradiated areas. Figure 1 depicts the operation of a positive resist. A wide range of resists is in common use, and an even wider range of resists has been the subject of experimentation and research. We will discuss only a few, those which have found wide application.

Perhaps the simplest resist to discuss is polymethyl methacrylate (PMMA), a common polymer, known under various trade names as Plexiglas, Lucite, etc. The polymer PMMA, as it is commonly called, is sensitive to ionizing radiation, that is, photons with wavelengths below about 200 nm (6.2 eV), electrons, and ions. For lithography purposes, the PMMA molecular weight is usually between about 400 000 and 950 000. The effect of the ionizing radiation is to break bonds, leading to chain scission and reduction of molecular weight in irradiated areas. Thereafter, a weak solvent can be used to remove irradiated areas selectively. Some results (Hawryluk *et al.*, 1975) are shown in Fig. 2,

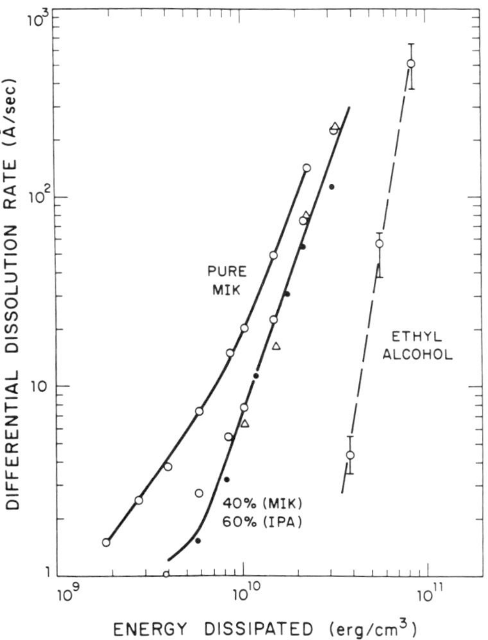

FIG. 2. Plot of the development rate of polymethyl methacrylate resist as a function of absorbed energy. For a developer solution consisting of a 40/60 mixture of methyl isobutyl ketone (MIK) and isopropyl alcohol (IPA), there is a cubic dependence, which implies a high-contrast resist.

which is a logarithmic plot. Note that beyond a certain threshold dose, the development rate (i.e., the rate of dissolution) in the 40/60 mixture has a cubic dependence. That is, a doubling of the irradiation dose leads to an increase in dissolution rate of a factor of 8. Such a strong nonlinearity is referred to as "high-contrast resist."

High-contrast resist response is critically important to the lithography process. Figure 1 depicts abrupt discontinuity in the irradiance distribution between exposed and unexposed regions. This is almost never the case in practice. Rather, the irradiance might have a distribution as depicted in Fig. 3. Because the ratio of I_{max} to I_{min} is ~3 to 2, rather than a more desirable 5 to 1 or 10 to 1, this distribution is said to have a "low contrast." The task of the resist response then is to convert this low-contrast irradiance distribution into a binary relief structure in resist. The "clipping" level depicted in Fig. 3 implies that an irradiance above this level leads to full development of the resist, whereas no resist development occurs for irradiance below this level. This implies a highly nonlinear or "high-contrast" resist response. This is certainly the case for PMMA, as illustrated in Fig. 2.

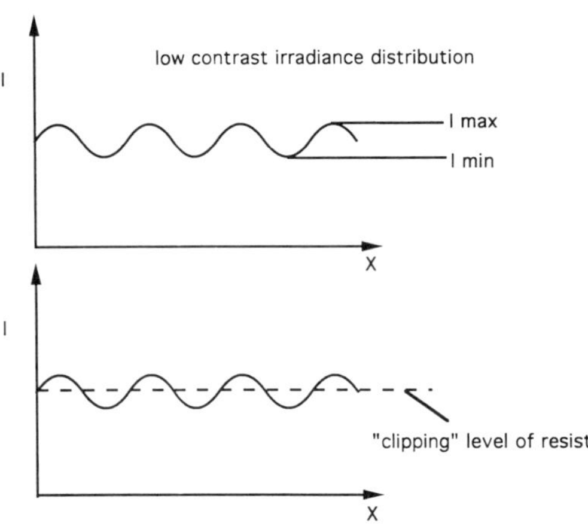

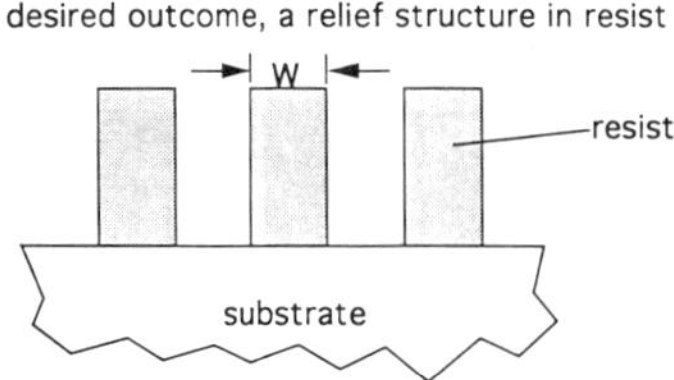

FIG. 3. Schematic illustrating the binary nature of the lithographic process. A low-contrast irradiance distribution can be converted into a binary resist structure with vertical profiles if the resist has a highly nonlinear response equivalent to "clipping" (i.e., development occurs only for irradiance above a threshold or clipping level).

An added virtue of PMMA is its resolution: Relief features as fine as 18 nm on a 36-nm pitch (centerline-to-centerline spacing) have been produced. As indicated above, PMMA is sensitive only in the deep ultraviolet (UV) and beyond. Most photolithography is done in the violet or near UV, and, for such work, a totally different resist chemistry is required.

The most common photoresists, useful in the blue and near UV, are based on a combination of the acidic polymer novolak and the photosensitive material napthaquinone diazide (known as the inhibitor). The combination is relatively insoluble in basic solutions because the inhibitor protects the novolak from attack by the base. However, with irradiation the inhibitor is transformed, allowing rapid dissolution or etching of the novolak. This family of photoresists also has high contrast, i.e., highly nonlinear response.

There are many other types of resists, the most novel being the so-called chemically amplified resists (CAR). With these, the sensitivity and resolution are controlled by a post-exposure baking which determines how far an acid group, released upon exposure, diffuses beyond the point of exposure, creating additional reactions along the way. We will not have space to discuss resists further. Suffice it to say that a resist must have a high contrast and be capable of resolving fine linewidth features.

1.4 Approaches to Lithography with Photons, Electrons, and Ions

A number of approaches can be taken to expose resist. We distinguish between two generic categories: pattern generation and pattern replication. Pattern replication is done using a reticle or mask made, in advance, by a pattern-generation instrument. Examples of pattern-generation schemes include the controlled scanning, usually under computer control, of a laser beam, an electron beam, or an ion beam over a resist-coated substrate. Holographic exposure of resist should also be considered pattern generation. Pattern replication can be done by shadow printing the reticle or mask, or by projecting its image through a lensing system, often with demagnification. Any one of the radiations—UV, x rays, electrons, or ions—can be employed to replicate a reticle or a mask. However, the structure of the mask must be quite different in each case. The following types of radiation have been used, or proposed for use, in microlithography and nanolithography instruments: UV photons (300 nm $< \lambda <$ 400 nm), deep UV (150 nm $< \lambda <$ 300 nm), very soft x rays (5 nm $< \lambda <$ 15 nm), soft x rays (0.4 nm $< \lambda <$ 5 nm), hard x rays ($\lambda <$ 0.4 nm), electrons (200 eV $< E <$ 400 keV), and ions (20 keV $< E <$ 500 keV). In most cases, these forms of radiation can be used in either the pattern-generation or the pattern-replication mode. (A notable exception: No x-ray-based pattern *generation* scheme has been proposed since x-ray beams cannot be readily focused and scanned.) Hence, one could consider perhaps 50 different types of microlithographic instruments that either exist, are under development, or have been proposed. However, most would prove to be impractical or ineffective. We will confine this exposition to only those few that have proven to be effective: photolithographic shadow printing, optical-projection lithography, scanning-electron-beam lithography, and x-ray lithography.

2. PHOTOLITHOGRAPHIC SHADOW PRINTING

Photolithographic shadow printing is the simplest type of microlithography, the most economical, and, in many cases, highly effective. The scheme is depicted in Fig. 4. When required, alignment of the mask to a pattern preexisting on the substrate is done by simultaneously viewing, with an optical microscope, complementary alignment marks on the mask and substrate, and mechanically shifting one or the other until the marks superimpose. Typically, the mask consists of the desired pattern in a metal such as chromium, or absorber such as iron oxide, on a glass or quartz plate. The mask must be made in advance by another lithographic technique, one capable of pattern generation. In shadow printing, the quality of the areal image at the resist layer depends on the magnitude of the mask-sample gap, because it is diffraction in the gap that degrades the image. If the gap is zero, i.e., "contact," the quality and resolution of the image can be extremely high since the resist would then be recording the transmitted irradiance distribution prior to any significant diffraction. In addition, the diffraction that takes place in traversing the resist is, in effect, suppressed somewhat because the resist index of refraction is typically about 1.6. As depicted in Fig. 4, mask and substrate are perfectly flat and parallel, but this is almost never the case in practice. Real substrates are not generally very flat relative to UV wavelengths. Moreover, buildup of resist occurs at substrate edges, and dust particles or other surface asperities are often present and prevent one from achieving a zero gap. Nevertheless, even with commercially available masks (which are mechanically rigid and optically flat) and typical, non-flat substrates, one can get good contact over limited areas of the substrate and thus achieve high-resolution photolithography. Many high-performance binary optical and microelectronic devices, especially microwave devices in *III-V* materials, are made in this way. However, contact photolithography is not considered a mass production technique because of damage to mask and substrate that can occur when the two are brought into hard contact.

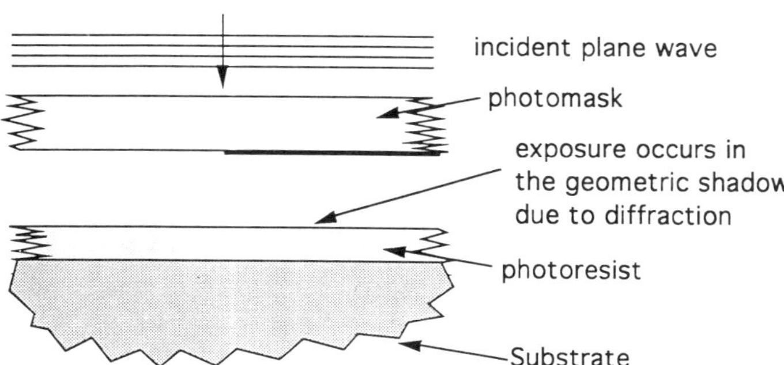

FIG. 4. Schematic of photolithographic shadow printing.

The problem of damage can be avoided through the use of flexible or conformable photomasks. These are made on thin glass, similar to that used in microscope cover glass. Damage is avoided, and contact can be achieved over a much larger fraction of the substrate. Special provisions must be made to avoid loss of contact by the nitrogen evolution that occurs during exposure of novolak-diazide types of photoresist. Extremely high-resolution photolithography is achievable with conformable photomask lithography (Smith, 1974; Everett *et al.*, 1991).

3. OPTICAL-PROJECTION LITHOGRAPHY

Figure 5 is a schematic of an optical-projection lithography (OPL) system. An image of the mask or "reticle" is cast onto a resist-coated substrate. For reasons that will become clear below, the mask pattern is depicted as a simple grating. Figure 5 depicts in-line imaging with a single refractive lens.

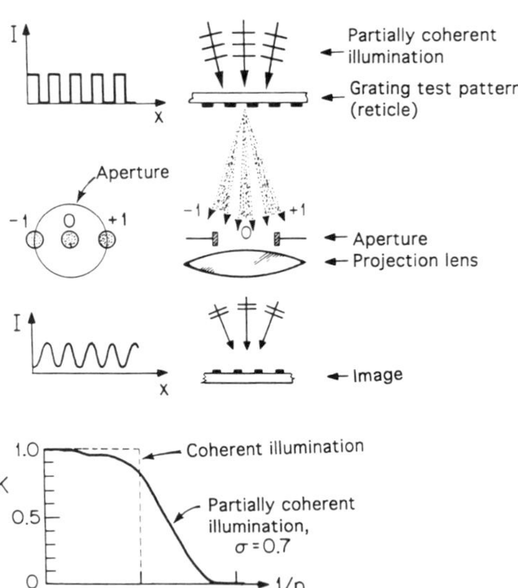

FIG. 5. Schematic of an optical-projection lithography system. The reticle is depicted as a binary transmission grating. The illumination is monochromatic, but the cone of illumination partially fills the entrance aperture, so-called partially coherent illumination. The image, having pitch or spatial period p, has a contrast K that depends on the value of p and the type of illumination, as given in the plot of K vs $1/p$. The contrast K is defined as $(I_{max} - I_{min})/(I_{max} + I_{min})$.

Actually, many types of projection systems are in use. Some project the reticle image at the same scale (1-to-1 systems), others provide demagnification (typically 5× or 10×). Some use entirely refractive optics, others primarily reflective optics. Reduction refractive systems are probably the most widely used. We limit our discussion to such systems.

In principle, a reduction projection system that uses refractive optics is similar to an optical microscope operated in reverse (i.e., image and object positions are reversed). In fact, optical microscopes can be, and have been, used for lithography, and with them the highest-resolution optical projection lithography has been obtained (Kawata *et al.*, 1989). It has been known since the work of Ernst Abbé in the mid-19th century that the resolution of a lens is related to its numerical aperture (NA), defined as the sine of the half angle of collection (or emission) multiplied by the local index of refraction. Specifically, the minimum resolvable pitch p is given by

$$p_{min} = \lambda/2\text{NA}, \tag{1}$$

where λ is the wavelength used. Optical microscopy gets very close to this resolution limit. During the last century, driven by the need to achieve the finest microscope resolution for medical and biological research, lenses with NAs as high as 0.95 in air and 1.4 in oil were developed. When used for lithography in a research setting, such lenses have produced minimum features of 140 nm, finer than $\frac{1}{3}$ the wavelength of the blue light used (Kawata *et al.*, 1989). Microscope lenses are not suitable for commercial lithography because of their relatively small fields of view, and distortion within the field.

The lenses used in modern optical projection systems are enormous relative to their microscope and camera forebears. Their fields of view, as measured by the number of resolvable features across them, are 10 to 100 times larger, and distortion is significantly lower. Projection lenses must be telecentric; i.e., demagnification should not change with defocus. The wavelengths used include the *G* and *I* lines of Hg (436 and 365 nm, respectively), the KrF excimer laser (248 nm), and, perhaps in the future, the ArF excimer laser (λ = 193 nm). Numerical apertures range from 0.3 to 0.6 and fields of view from 10 × 10 to

20 × 20 mm² (Okazaki, 1991). In order to achieve such enormous fields of view with low distortion and minimal aberrations, between 20 and 30 separate lens elements are required. They are assembled in a precision housing, up to 40 cm in diameter, with a total weight of over 200 kg.

The substrate "wafers" used in the production of Si integrated circuits range from 10 to 20 cm in diameter, and so, to expose an entire substrate, multiple exposures are required. Hence, the wafer must be "stepped" and realigned multiple times. For this reason, OPL systems are often called "steppers." A variety of techniques are used for aligning or positioning the reticle image on the substrate, which generally must be done to a tolerance finer than the minimum lithographic feature. The most effective of these techniques image a mask feature onto a substrate mark (which might consist of a single line, or a grating) and converge to alignment by interpreting the pattern of scattered light. The x-y position of the table upon which the wafer is held is monitored by laser interferometry, which can detect a change in stage position as small as 1 nm, although perhaps not in a production environment. Because of the high precision of table positioning, direct image alignment need be done at only a few sites on the substrate, other sites being positioned for exposure under control of the laser interferometer alone.

Because of the widespread use of OPL in Si microelectronics production, and the enormous costs and dislocations that would be caused by any major change in production technology, there is an urgent imperative to continue using OPL as the minimum feature sizes in microelectronic circuits follow their historic trend of continuous shrinkage.

3.1 The Push to Higher Resolution

The resolution of OPL can be expressed in a manner similar to Eq. (1):

$$p_{\min} = 2W_{\min} = 2k_1(\lambda/\mathrm{NA}), \qquad (2)$$

where $W_{\min}$ is the minimum feature size. The value of the parameter k_1 depends on a number of factors, including the angular distribution of the reticle illumination, the type of reticle, the image contrast one is willing to tolerate, and the required depth of focus.

The angular distribution of reticle illumination affects the image contrast at a given pitch, as illustrated in Figure 5. Traditionally, the illumination of the reticle has been uniform over about half the range of angles that could be accepted by the lens entrance pupil. This is referred to as "partially coherent illumination" with "filling factor $\sigma \simeq 0.5$." Such illumination tends to boost the image contrast for spatial periods larger than $0.8\lambda/\mathrm{NA}$ relative to the contrast one would have with incoherent illumination, which generally is used in optical microscopy. The connection between the angular distribution of illumination and contrast at a given pitch has been known since the last century. Off-axis and annular illumination was used in microscopy to boost the contrast of a specific range of spatial periods. Recently, this strategy has been investigated in OPL, where it is particularly attractive for Si integrated circuit manufacturing, which uses only a limited repertoire of shapes and sizes (Tamechika *et al.*, 1992).

A contrast K of 0.5 or higher is considered essential for production lithography. For partially coherent illumination, this corresponds to a k_1 value of about 0.4. To approach this value in production, every aspect of the OPL technology is being scrutinized, and trade-offs are made between minimum feature size and the latitude of the processing. In this engineering optimization of OPL, the reticle can also be optimized. Figure 5 assumes that only the irradiance transmission is altered. But one could also alter the relative phase of the transmitted light. In the limit of a "pure phase" mask of a grating, k_1 could be effectively reduced by the factor of 2! However, for patterns of general interest, the gain cannot be so large. The "attenuating phase-shift mask" is applicable to all pattern geometries (Lin, 1992). It provides about 10-to-1 contrast and a π phase shift at the reticle, which gives some gain in image contrast.

One can also introduce phase and amplitude filtering in the pupil plane of the projection lens (Inoue *et al.*, 1992), something that is closely related to phase-contrast microscopy introduced by Zernike in the 1930s.

In summary, OPL is a highly developed production technology and represents the forefront of optical systems engineering. It is currently in a state of intensive development in an effort to provide cost-effective commercial production of integrated circuits as minimum feature sizes are shrunk from 0.35 μm to 0.25 and 0.13 μm. Whether OPL will succeed in this before competitive technologies such as x-ray lithography is a major question of the day.

4. SCANNING-ELECTRON-BEAM LITHOGRAPHY

Electron-beam lithography (EBL) systems can take many forms, from those that scan a beam of circular cross section and sequentially fill in a pattern shape to those that project an entire pattern using a reticle mask. The most successful and widely used are the "round beam" and "shaped beam" scanning systems. The latter generally project a rectangular shape and scan it to fill in a pattern more efficiently than a round beam can. Figure 6 illustrates schematically the main features of a scanning-electron-beam lithography (SEBL) system. The similarity to a scanning electron microscope (SEM) is evident. Many research groups have, in fact, modified SEMs to serve as SEBL systems, but these have serious limitations in accuracy and speed, which make them inadequate for all commercial and many research applications.

Because a SEBL system can also function as a SEM, alignment of patterns relative to marks on the substrate (or a stage) is relatively straightforward. One merely views the alignment mark and adjusts the position of the scan field, either electronically or mechanically, so that the mark has the desired location in the scan field. Thereafter, a laser-interferometer-controlled stage moves the substrate to desired x and y locations. This viewing of alignment marks and establishment of a coordinate system are done under computer control.

The electron gun must provide a stable emission, and as high a brightness as possible, as this directly determines current density in the focal spot on the sample, which in turn determines exposure time. Thermal-field-emission sources (see ELECTRON MICROSCOPY), which achieve brightnesses approaching 10^8 A/cm^2 sr, are preferred. The lenses demagnify the source onto the sample. The objective lens usually has a relatively large bore to enable a large field to be scanned. The design of the deflection coils also influences how large a field can be scanned with minimal distortion. The role of the beam blanker is to turn the beam on and off as needed to create the desired pattern. A laser interferometer can determine stage position to better than 10 nm. If a stage position error occurs, a correction signal is fed to the electron deflector.

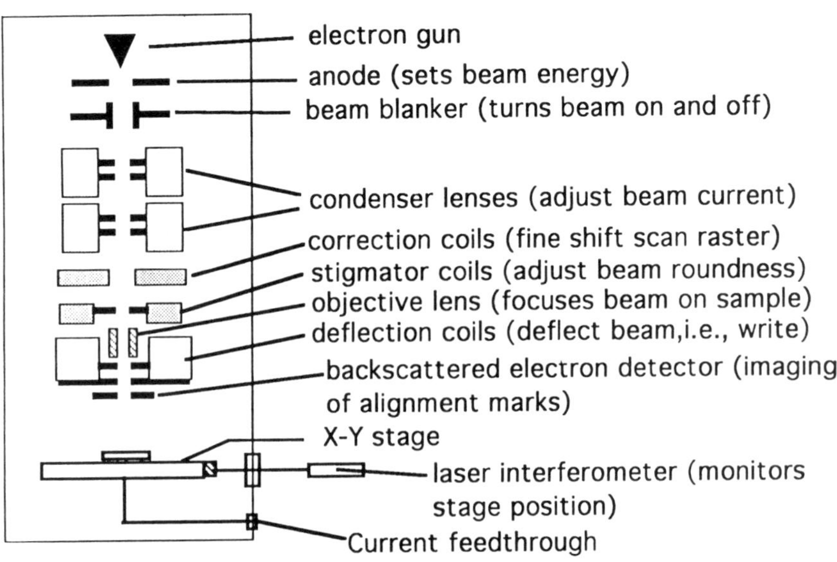

FIG. 6. Schematic of a scanning-electron-beam lithography system.

4.1 Writing Strategies

It is convenient to subdivide any pattern that is written into square "pixels" with edge dimension $\frac{1}{5}$ the minimum feature size of interest W, as shown in Fig. 7, and to assume that the pixel size ϵ is equal to the minimum beam diameter d_o. The writing of a pattern is done within a scan field that is almost always much smaller than the substrate that is being written on. The scanning beam addresses pixels in sequence under the control of a digital computer.

It is appropriate to describe the size of a scan field in terms of the number of pixels it encompasses. If we then assume that the written pattern elements cannot be allowed to deviate from their assigned positions by more than one pixel, for example, this fixes the maximum field size. In other words, the maximum "distortion free" field size is dictated by the requirement that the pixel grid not deviate from a perfect Cartesian grid by more than ϵ.

Two factors determine how many pixels span the maximum allowable field: the electron optics and the digital electronics. It is difficult to design an objective lens and deflector that can meet the above field size criterion at a field size that is much more than 10^4 pixels across. We might then consider a 14- to 16-bit digitally specified field to be the limit imposed by the electron optics, although this is not a truly fundamental limit. Similarly, although probably not a fundamental limit, it is difficult to construct a digital control system that can scan a 16-bit field with 1-bit accuracy at high speed. Nevertheless, we will take 65 536 pixels as the limit of the scan field size, realizing that current SEBL scan fields are generally $\frac{1}{2}$ to $\frac{1}{4}$ this size.

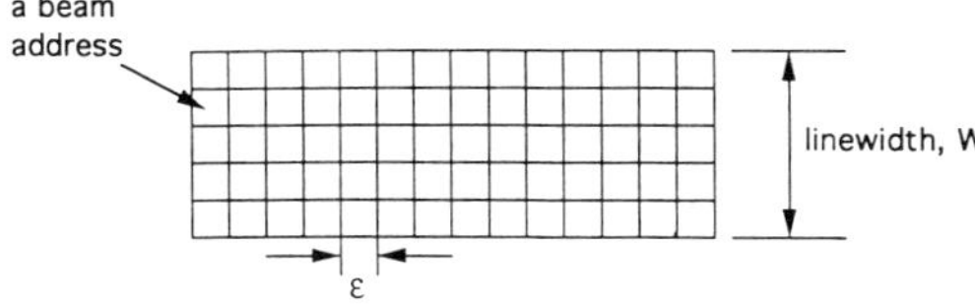

FIG. 7. The "pixel model," useful in discussing SEBL. The minimum linewidth W is subdivided into square pixels, each approximately equal in edge dimension ϵ to the beam diameter d_o. Each pixel is separately addressed by the scanning beam.

For a minimum beam diameter of 10 nm, with a corresponding minimum feature size of 50 nm, the maximum scan field size is 0.65 mm.

Within a scan field, writing can be done in a raster fashion, as in a TV, or by vector scanning. In raster scanning, all points are addressed, and the beam is blanked or unblanked as needed. In vector scanning, only the parts that make up a desired pattern are addressed. Vector scanning is more efficient from an exposure-time viewpoint but is more difficult from a control point of view, and its exposure-time advantage tends to disappear for very dense patterns.

A variant on the raster scanning strategy, which has been highly successful in SEBL systems designed for mask making, is raster scanning with the stage continually moving under laser interferometer control (Abboud *et al.*, 1992). The scanning pattern on the sample is thus serpentine. Because of the dependence on a laser interferometer for pattern integrity, the scan field (which is just a line scan along one axis) need not be very large. It is typically only 10^3 to 10^4 pixels across, which implies simpler electron optics and control electronics.

4.2 Writing Time vs Linewidth

When the pixels that constitute a pattern are addressed in a serial fashion, writing time can quickly become excessive as sample area and pattern complexity are increased, which is the trend in modern applications of microlithography. The dwell time per pixel, t_P, is given by $t_P = S/J$, where S is the resist sensitivity expressed in charge per unit area and J is current density in the beam. With thermal-field-emission sources, current densities of 1000 A/cm^2 and somewhat higher are realizable. Resist sensitivities (i.e., the amount of charge input per unit area required to achieve full resist development) of 10^{-6} C/cm^2 are realizable, giving $t_P = 10^{-9}$ s. However, this ignores the stochastic nature of the current, i.e., shot noise in the beam (Smith, 1986, 1988). If we assume that the mean number of electrons per pixel is some number (e.g., 100) in order that the probability of error in pixel exposure be below some given value, then we can determine the minimum possible writing time for a given current density. This is given in Fig. 8. The writing time per unit area

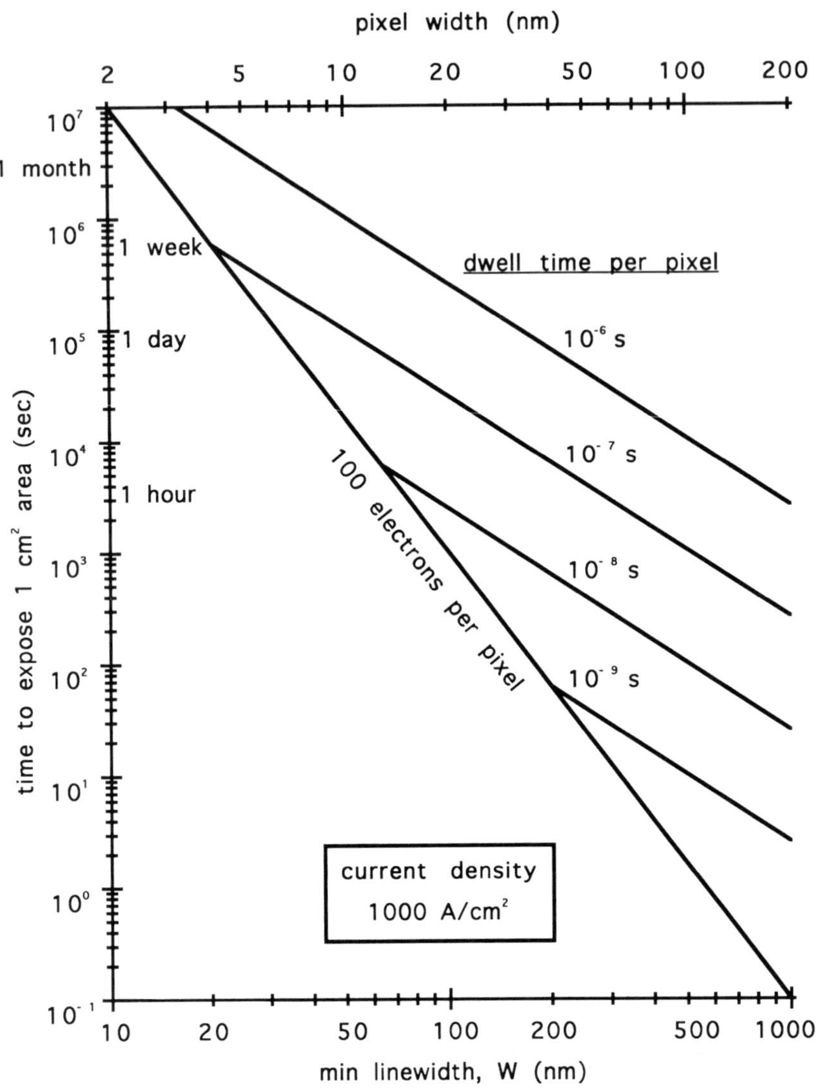

FIG. 8. Plot of *e*-beam writing time for a 1-cm^2 area as a function of minimum linewidth for various dwell times per pixel, assuming a current density of 1000 A/cm^2, and that every pixel is addressed. The line marked 100 electrons per pixel represents an assumed shot-noise limit.

is limited either by the speed with which the pixels can be addressed or by fundamental shot-noise considerations. One can, of course, make different assumptions with regard to current density or minimum numbers of electrons per pixel, but the general conclusion will still be the same: Writing time can easily get out of hand for complex patterns, and so, for industrial applications, SEBL should be considered primarily a mask-making tool.

4.3 Field Stitching

Because field size in SEBL is small compared to most sample areas of interest, fields must be stitched together. Generally, this is done via laser interferometer control of stage position. To achieve distortion free patterns, the scan field axes, sample axes, and stage axes have to be made collinear to within less than one pixel over the scan field, or other suitable corrections made.

A fundamental problem in all forms of SEBL is that they are run "open loop." That is, the electron beam can drift with respect to the stage coordinate system as a result of thermal expansion, charging, and a variety of other causes. Thus, although the laser interferometer's computer can know exactly where the stage is at any time, the position of the beam is not known precisely unless some reference is made to a fiducial mark on the stage or sample. A proposal for "closing the loop" in SEBL, known as "spatial-phase-locked SEBL," would use a fiducial grid patterned directly on the substrate in such a way as to not disturb the electron beam writing (Smith *et al.*, 1991; Ferrera *et al.*, 1993). By giving rise to a modulation of secondary (or backscattered) electrons, the grid would enable a computer to keep track of the beam position on the sample at all times, independent of influences that cause drift. To ensure accuracy and freedom from distortion, the fiducial grid is generated by holographic lithography.

4.4 Reducing Exposure Time

The most successful approach to reducing the exposure time in SEBL has been the so-called shaped-beam systems (Pfeiffer *et al.*, 1993). Although such systems allow many pixels to be projected in parallel, the equivalent pixel transfer rate does not generally go up in proportion because the current density is significantly reduced over that achievable in round-beam systems with thermal-field-emitter sources. The mutual repulsion of electrons, which causes shape distortion, limits the current density. In addition, the throughput gain with shaped-beam systems depends very much on the pattern type. For complex patterns of fine lines, the gain goes down in proportion. When all these factors are considered, the gain in throughput with shaped beams is generally about an order of magnitude, seldom two orders.

4.5 Electron Scattering

When a high-energy electron enters a material, it interacts with the resident electrons, losing energy more or less continuously through the excitation of secondary electrons, most of which have energies of only several eV. The primary electron undergoes multiple small-angle scattering events. By this mechanism, a well-defined incident beam spreads out as it passes through material. Depending on the substrate atomic number, between 10% and 40% of the incident electrons backscatter and emerge from the substrate at points that can be quite remote from the original entry point. As a result, the exposure dose at a given location depends not only on the local incident dose but also on the dose incident at all points within the electron range. This is known as cooperative exposure or "proximity effect." In principle, by knowing the electron scattering characteristics, which can be determined either experimentally or by Monte-Carlo methods (*q.v.*) (Hawryluk *et al.*, 1975), one can correct for the proximity effect, but with a lowering of the image contrast (Owen, 1990).

5. X-RAY LITHOGRAPHY

The basic principle of proximity x-ray lithography is illustrated schematically in Fig. 9. A membrane, which is x-ray transparent, carries a patterned x-ray absorber whose shadow is cast on a substrate held in close proximity. An x-ray wavelength around 1 nm, or slightly longer, provides adequate image contrast and resolution down to feature sizes of 20 nm, which is about the practical limit of the lithographic process using polymeric resists. A unique virtue of x-ray lithography is that scattering is negligible. In addition, the deleterious effects of photoelectrons released when a photon is absorbed are negligible for wavelengths longer than about 0.8 nm (Early *et al.*, 1990; Ocola and Cerrina, 1993). Hence, proximity effects are essentially absent, and high-quality lithography is possible, irrespective of pattern geometry, substrate composition, or topography.

5.1 Sources

Electron-bombardment x-ray sources are used in research but are too inefficient for commercial x-ray lithography. Both synchrotrons (Grobman, 1981) and dense plasma sources are used instead. In a synchrotron, an electron beam with an energy of 600–1000 MeV is forced by strong magnetic fields to

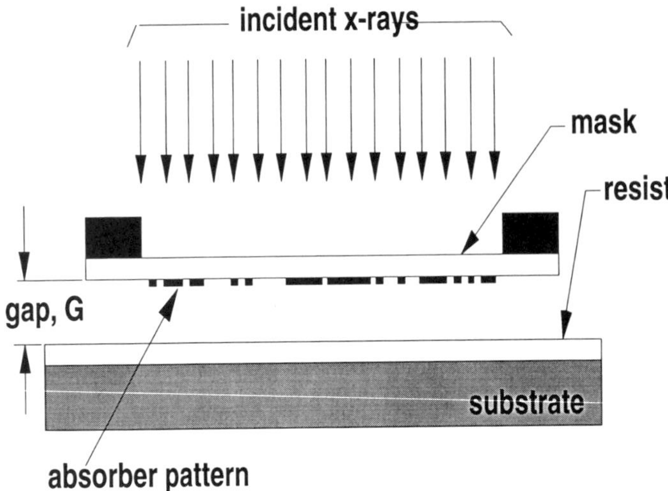

FIG. 9. Schematic of the x-ray lithography technique. A pattern, in the form of a high–atomic-number x-ray absorber, resident on a thin, x-ray–transparent membrane, is replicated into a resist film on a substrate separated from the membrane by a gap *G*.

follow a circular or racetrack orbit. The radiation emitted when the electron trajectory is altered is most intense in the soft x-ray regime around 1 nm and is quite adequate for mass-production lithography. A synchrotron is a fixed facility that provides 10 to 20 exposure stations or "beam lines" that radiate out tangent to the electron orbit. A superconducting synchrotron can be somewhat compact, measuring only a few meters in its largest dimension.

At the present time, laser plasma sources lag behind synchrotrons in output flux per exposure station but make up in part for this by operating at somewhat longer wavelengths (e.g., 1.4 nm) where more of the incident radiation is absorbed in the resist (Hector *et al.*, 1993). Other plasma sources, such as the "plasma focus" (Prasad, 1994), offer the possibility of significantly higher flux, comparable to a synchrotron. Such sources require additional development.

5.2 Masks

A mask structure suitable for use with a wavelength of 1.3 nm is shown in Fig. 10. One's intuition, based on everyday experience, would say that such a structure is highly fragile and subject to distortion. However,

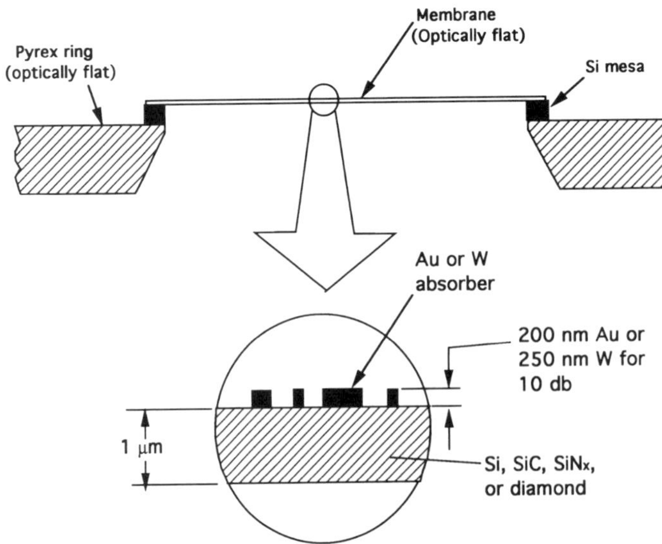

FIG. 10. Configuration of x-ray lithography mask suitable for use at the 1.3-nm wavelength. Typically, the membrane is flatter than 0.25 μm and over 30 mm in diameter. The glass ring provides rigidity.

FIG. 11. Scanning electron micrograph of grating pattern exposed in PMMA resist using a 4.4-nm-wavelength x-ray and a mask-substrate gap less than 1 μm (Smith and Craighead, 1990).

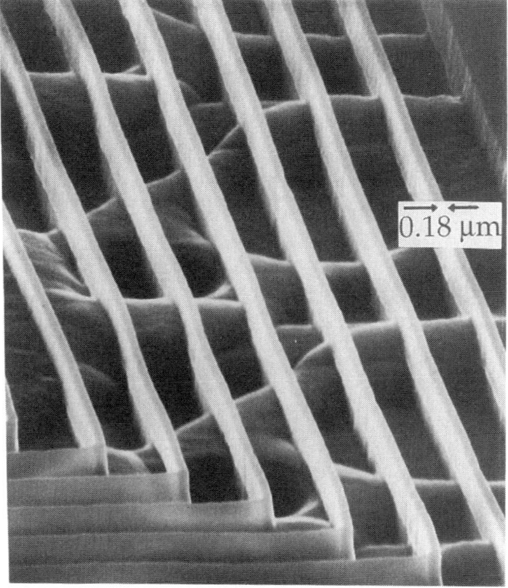

FIG. 12. Scanning electron micrograph of pattern exposed in a chemically amplified resist (PN 114) using a laser-plasma x-ray source (1.4 nm average wavelength) and a mask-substrate gap of 20 μm (courtesy of I. Plotnik).

membranes of Si, SiN_x, and SiC, 1 to 2 μm thick, have proven to be unusually robust. They are readily handled in research and manufacturing environments and are even used as vacuum windows.

Pattern distortion can arise from compressive or tensile stress in the absorber. By keeping such stress below 10 MPa, pattern distortion is well below 1 part per million for the types of patterns encountered in microelectronics.

5.3 Results

Figures 11 and 12 illustrate some results. The former was exposed at a near-zero gap between mask and substrate, and illustrates high-resolution nanolithography in thick resist. The latter was exposed at a 20-μm gap and illustrates exposure over topography. The relationship between gap G and minimum feature size W is given by

$$G = \alpha W^2/\lambda, \tag{3}$$

where λ is the x-ray wavelength and α is a parameter. Experiment and numerical simulation of diffraction effects have shown that one can readily operate at quite large gaps, corresponding to α in the range 1 to 1.5, provided the effective source size (i.e., its spatial incoherence) is sufficiently large to smooth out edge ripples that occur with spatially coherent illumination (Hector *et al.*, 1992).

5.4 Alignment

In x-ray lithography, because the mask and substrate are in close proximity, alignment of a mark on the mask to a mark on the substrate is technologically simpler than in other forms of lithography. It can be done with 10-nm precision using a variety of methods, the most sensitive of which employ optical interferometry (Ishihara *et al.*, 1989). However, overlayers grown on top of alignment marks can confuse the interferometric alignment. To avoid such confusion, it is preferable to use a range of optical wavelengths, which is not feasible with traditional interferometric methods. Recently, an on-axis interferometric scheme that uses white light was described. In conjunction with modern digital image-processing technology, high precision, together with greater insensitivity to overlayer growth, should be achievable (Moel *et al.*, 1993).

6. CONCLUSIONS

Each of the several forms of lithography has distinct advantages, disadvantages, and a domain of applicability. The commercial manufacture of high-density integrated circuits is the most publicized application of microlithography, and the one that provides the greatest financial motivation. In this sphere, many nontechnical factors, such as existing infrastructure and costs, affect the choice of lithography technique. Such considerations are largely beyond the scope of this analysis, which has focused on engineering and physical aspects.

GLOSSARY

Irradiance: A measure of the power per unit area of radiation incident on a substrate.

Microlithography: The creation of structures in resist with minimum feature sizes from a few micrometers to 100 nm.

Nanolithography: The creation of structures in resist with minimum feature sizes below 100 nm.

Photoresist: A resist film sensitive to visible or ultraviolet radiation.

Pitch: The center-to-center distance between two adjacent lines.

Resist: A film coating onto a substrate, which, upon exposure to a pattern of radiation, can be converted, via a development process, into a relief image.

Resolution: A term used rather loosely to designate the minimum size of relief structures achievable with a given lithographic technique. In microscopy, resolution usually refers to the minimum center-to-center distance between two objects where one can distinguish that there are two objects rather than one.

Spatial Period: The center-to-center distance between two adjacent lines.

Works Cited

Abboud, F., Gesley, M., Colby, D., Comendant, K., Dean, R., Eckes, W., McClure, D., Pearce-Percy, H., Prior, R., Watson, S. (1992), "Electron Beam Lithography using MEBES IV," *J. Vac. Sci. Technol. B* **6**, 2734–2742.

Early, K., Schattenburg, M. L., Smith, H. I. (1990), "Absence of Resolution Degradation in X-ray Lithography for λ from 4.5 nm to 0.83 nm," *Microelectron. Eng.* **11**, 317–321.

Everett, P. N., Delaney, W. F., Griswold, M. P. (1991), "Flexible and Rigid Masks and Alignment in Binary Optics," in: *Proceedings of the Lasers and Electro-Optics Society (Institute of Electrical and Electronics Engineers) Meeting, July 31–August 2, 1991, Newport Beach, California*, New York: IEEE, Sec. 2, pp. 19–20.

Ferrera, J., Wong, V. V., Rishton, S., Boegli, V., Anderson, E. H., Kern, D. P., Smith, H. I. (1993), "Spatial-Phase-Locked Electron-Beam Lithography: Initial Test Results," *J. Vac. Sci. Technol. B* **11**, 2342–2345.

Grobman, W. D. (1981), "Synchrotron Radiation X-ray Lithography," in: E. E. Koch, D. E. Eastman, P. Farge (Eds.), *Handbook on Synchrotron Radiation*, Vol. 1, Amsterdam: North Holland.

Haury, E. W. (1967), "The Hohokam," *Natl. Geogr. Mag.* **131** (May), 670–695.

Hawryluk, R. J., Smith, H. I., Soares, A., Hawryluk, A. M. (1975), "Energy Dissipation in a Thin Polymer Film by Electron Beam Scattering—Experiment," *J. Appl. Phys.* **46**, 2528–2537.

Hector, S. D., Schattenburg, M. L., Anderson, E. H., Chu, W., Wong, V. V., Smith, H. I. (1992), "Modeling and Experimental Verification of Illumination and Diffraction Effects on Image Quality in X-ray Lithography," *J. Vac. Sci. Technol. B* **10**, 3164–3168.

Hector, S. D., Smith, H. I., Schattenburg, M. L. (1993), "Simultaneous Optimization of Spectrum, Spatial Coherence, Gap, Feature Bias, and Absorber Thickness in Synchrotron-Based X-ray Lithography," *J. Vac. Sci. Technol. B* **11**, 2981–2985.

Inoue, S., Fujisawa, T., Tamaushi, S., Ogawa, Y., Nakase, M. (1992), "Optimization of Partially Coherent Optical System for Optical Lithography," *J. Vac. Sci. Technol. B* **6**, 3004–3007.

Ishihara, S., Kanai, M., Une, A., Suzuki, M. (1989), "A Vertical Stepper for Synchrotron X-ray Lithography," *J. Vac. Sci. Technol. B* **7**, 1652–1656.

Kawata, H., Carter, J. M., Yen, A., Smith, H. I. (1989), "Optical Projection Lithography using Lenses with Numerical Apertures Greater than Unity," *Microelectron. Eng.* **9**, 31–36.

Lin, B. J. (1992), "The Attenuated Phase-Shifting Mask," *Solid State Technol.* (January), 43–47.

Moel, A., Moon, E. E., Frankel, R., Smith, H. I. (1993), "Novel On-axis Interferometric Alignment Method with Sub-10 nm Precision," *J. Vac. Sci. Technol. B* **11**, 2191–2194.

Ocola, L. E., Cerrina, F. (1993), "Parametric Modelling Photoelectron Effects in X-ray Lithography," *J. Vac. Sci. Technol. B* **11**, 2839–2844.

Okazaki, S. (1991), "Resolution Limits of Optical Lithography," *J. Vac. Sci. Technol. B* **9**, 2829–2833.

Owen, G. (1990), "Methods for Proximity Effect Correction in Electron Lithography," *J. Vac. Sci. Technol. B* **8**, 1889–1892.

Pfeiffer, H., Davis, D. E., Enichen, W. A., Gordon, M. S., Groves, T. R., Hartley, J. G., Quickle, R. J., Rockrohr, J. D., Stickel, W., Weber, E. V. (1993), "EL-4, A New Generation Electron-Beam Lithography System," *J. Vac. Sci. Technol. B* **11**, 2332–2341.

Prasad, R. R., Krishnan, M., Mangano, J., Greene, P. A., Qi, N. (1994), "Neon Dense Plasma Focus Point X-Ray Source for ≤0.25 μm Lithography," in: D. Patterson (Ed.), *Electron-Beam, X-Ray, and Ion-Beam Submicrometer Lithographies for Manufacturing IV*, SPIE Proceedings Vol. 2194, Bellingham, WA: SPIE.

Smith, H. I. (1974), "Fabrication Techniques for Surface-Acoustic Wave and Thin-Film Optical Devices," *Proc. IEEE* **62**, 1361–1387.

Smith, H. I. (1986), "A Statistical Analysis of UV, X-ray and Charged-Particle Lithographies," *J. Vac. Sci. Technol. B* **4**, 148–153.

Smith, H. I. (1988), "A Model for Comparing Process Latitude in UV, Deep-UV and X-ray Lithography," *J. Vac. Sci. Technol. B* **6**, 346–349.

Smith, H. I., Craighead, H. G. (1990), "Nanofabrication," *Phys. Today* **43** (2), 24–30.

Smith, H. I., Hector, S. D., Schattenburg, M. L., Anderson, E. H. (1991), "A New Approach to High Fidelity E-Beam Lithography Based on an In-Situ, Global Fiducial Grid," *J. Vac. Sci. Technol. B* **9**, 2992–2995.

Tamechika, E., Matsuo, S., Komatsu, K., Takeuchi, Y., Mimura, Y., Harada, K. (1992), "Investigation of Single Sideband Optical Lithography using Oblique Incidence Illumination," *J. Vac. Sci. Technol. B* **10**, 3027–3031.

Further Reading

Brodie, I., Muray, J. J., (1992), *The Physics of Micro/Nano-Fabrication*, New York: Plenum.

Moreau, W. M., (1988), *Semiconductor Lithography*, New York: Plenum.

Thompson, L. F., Willson, C. G., Bowden, M. J., (1983), *Introduction to Microlithography*, ACS Symposium Series Vol. 219, Washington, D.C.: American Chemical Society.

MICROMECHANICAL DEVICES

See SYSTEMS TECHNOLOGY, MICROMECHANICAL

MICROPROCESSORS

NICK TREDENNICK, *Tredennick, Inc., Los Gatos, California, U.S.A.*

INTRODUCTION

Figure 1 is the block diagram of a simple computer (see COMPUTERS). The simple computer consists of a central processing unit (CPU), a memory, and peripheral-interface logic connected to a common bus. The CPU is the engine for the computer. It runs the programs and controls the system. The memory holds the programs, the input data, and the results. The peripheral-interface logic is the means for getting the data and programs into the computer (from a keyboard, mouse, or disk, for example) and the means for getting the results from the computer (to a printer or display screen, for example).

The first computers, built in the early 1940s, were room sized and were built from vacuum tubes. The transistor was invented in 1947 by Bardeen, Brattain, and Shockley (Davies and Wharton, 1983). Later computer generations used transistors instead of vacuum tubes and became smaller and more reliable. After the integrated circuit, which allowed numerous transistors on a single semiconductor chip, was invented in 1958, computers got even smaller. Today, large integrated circuits sometimes contain several million transistors.

There is very little conceptual difference between this simple computer system, first built with vacuum tubes, and a computer built with a microprocessor. (There are enormous differences in physical implementation, however.) Figure 2 shows a simple microprocessor system. A microprocessor is a CPU on a single integrated circuit (sometimes called a "chip"). In the illustration of a simple microprocessor system. I have broken the memory into the random-access memory (RAM) and the read-only memory (ROM). RAM is useful because it allows rapid access to any location for reading or writing and is generally faster than ROM, but it loses any stored information when power is lost. ROM does not change or lose its contents when power is lost or when the system is restarted, and so it is useful for holding initialization programs and other programs often referenced, but not changed.

Figure 3 shows how a computer program, residing in the memory, looks to the programmer and to the computer. The programmer thinks of the computer program as being a sequential series of instructions such as add and subtract, punctuated by branches. The computer sees the program a little differently; it sees only strings of ones and zeros.

3-527-28132-0/94/$5.00 + .50

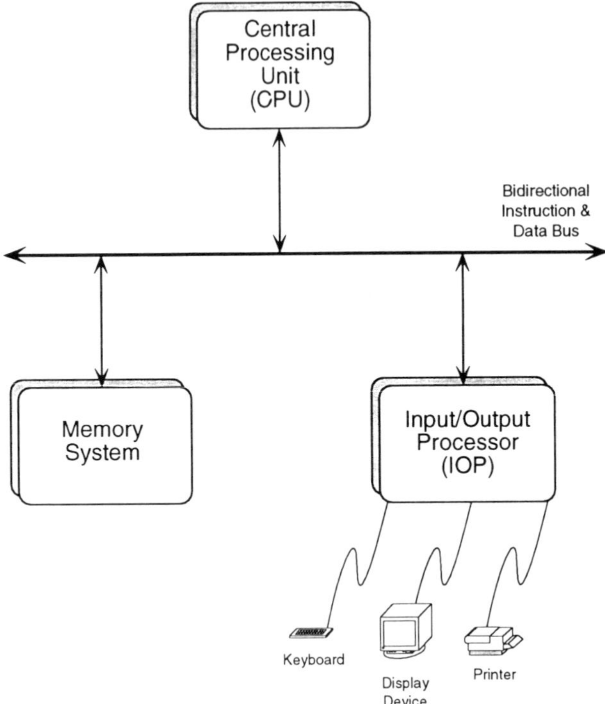

FIG. 1. Simple computer system. The simplest computer is a central processing unit (CPU) connected on a common bus to a memory and an input/output processor (IOP). (Figure components courtesy of Joe Higham.)

Programmers work with symbols, numbers, and mnemonics so that the programs are easier to understand. The computer works with binary ones and zeros because that is all it understands. A look at the contents of a location in the memory system would reveal only ones and zeros: the symbols, numbers, and mnemonics are constructed for the convenience of the programmer. Programmers create programs of symbols, numbers, and mnemonics to be translated (by another computer program called an assembler or compiler) to ones and zeros for use by the computer.

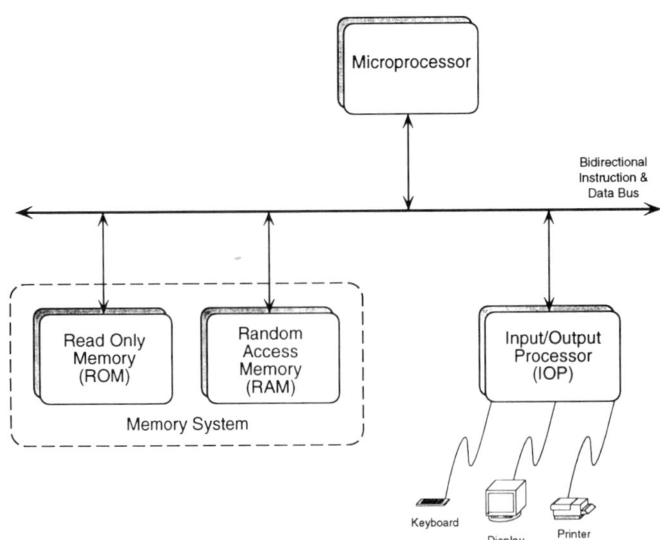

FIG. 2. A simple microprocessor system has a microprocessor, ROM, RAM, and an IOP connected to a common instruction and data bus.

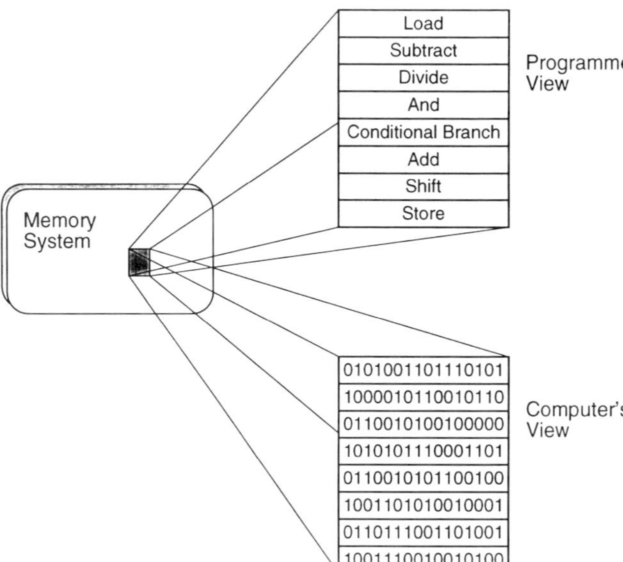

FIG. 3. Simple computer program. To the programmer, the computer program looks like a sequence of actions. To the computer, the computer program looks like a sequence of ones and zeros.

1. HISTORY

Even though the block diagrams of the simple computer system and the simple microprocessor system are strikingly similar, the microprocessor did not start out as part of a general-purpose computer. The microprocessor started as an alternative way to implement digital circuits built from standard logic modules. Before the invention of the integrated circuit by Jack Kilby at Texas Instruments in 1958 (Davies, 1983), logic circuits were built on boards (called circuit boards or logic boards) using discrete components: transistors, diodes, resistors, capacitors, and inductors (Bell *et al.*, 1978). The integrated circuit permitted multiple transistors on a single chip. Transistors, diodes, capacitors, and resistors could be built on an integrated circuit. After the invention of the integrated circuit, standard logic families were invented—the most notable being the 74xxx series. For example, the Texas Instruments SN7400 integrated circuit contained four two-input NAND gates (the output of each two-input gate is the inverse of the Boolean AND of the two inputs) in a fourteen-pin package. The logic family began with basic logic gates (NAND, NOR, AND, OR, and NOT) and grew to include more complex functions [such as register files, arithmetic-logic units (ALUs), and multiplexers]. These standard logic families placed modular logic functions in the hands of designers and led to increased complexity and variety in circuit board design. As integrated circuit densities improved, complexity and variety of the logic modules increased correspondingly. Soon, manufacturers were supporting catalogs of hundreds of different standard logic modules (see, for example, Texas Instruments, 1973).

1.1 Microprocessor as a Logic Device

The first implementation of a microprocessor was the result of a special design request from Busicom, a Japanese company, contracting with Intel for the custom design of calculator logic in 1969 (Faggin, 1992). Busicom was not content to build their calculator product using available standard logic modules and felt that they could get a more efficient implementation with custom integrated circuits; they wanted to specify the logic modules and have Intel build them. The innovative logic structure proposed by Dr. Marcian "Ted" Hoff of Intel led to the development of the MCS-4 chip set. One of the MCS-4 chips, the 4004 (see Fig. 4), was the first microprocessor. The MCS-4, with its 4-bit 4004 microprocessor, was introduced to the market by Intel in 1971. While the first microprocessors were expensive, they still found their way into some applications. Eventually, prices began to decrease and industry came to view the microprocessor as a

FIG. 4. Intel's 4004 introduced in 1971 was probably the world's first commercial microprocessor. (Courtesy Intel Corporation.)

good substitute for circuits built from standard logic modules. The microprocessor found application in calculators, washing machines, microwave ovens, instruments, and even computers. Rather than have a large number of standard logic modules on a circuit board, the designer could use a microprocessor with some RAM and ROM and program the logic functions. This could reduce component count, system cost, and the variety of components stocked by the circuit board manufacturer. For this to work, the microprocessor had to provide equivalent function, performance, and reliability at a lower cost. Simple bus protocols (the way the microprocessor "talks" to other chips in the circuit) allowed the microprocessor to be connected to a common bus with a variety of memory and peripheral-interface components. Market emphasis on the use of microprocessors to replace standard logic modules led to the following characteristics:

1. lowest cost design with adequate performance (efficient, optimized designs lead to smaller, cheaper chips);
2. simple, flexible bus protocols (to allow the chips to work with a variety of memory and peripheral components with differing speed requirements);
3. few package pins so that the microprocessor would fit in commonly available integrated circuit packages and, therefore, use standard sockets (most of the standard logic-module integrated circuit packages were 14 or 16 pins; an integrated circuit with 28 to 40 pins would be considered large).

These microprocessors were designed to be manufactured in very high volumes. Even today, many of these microprocessors are shipped in quantities of millions of units per year. Almost all of these microprocessors are used as alternatives to standard logic modules. Low-end microprocessors of all kinds are shipped at more than a billion units per year.

1.2 Microprocessor as a CPU

Use of the microprocessor as the central processing unit in a computer system began with the invention of the personal computer in the mid 1970s (Smith, 1981). The market for personal computers grew rapidly after IBM introduced their version of the personal computer in 1981. Just 10 years later, in 1991, about 21 million IBM-compatible personal computers were shipped worldwide. Intel Corporation makes the microprocessors used as the CPU in IBM-compatible personal computers. In 1991 Intel (and other manufacturers of 80386-compatible microprocessors) shipped over 2 million 80486 microprocessors and more than 15 million 80386 microprocessors, almost all of which ended up as CPUs in personal computer systems (Slater, 1992). During the same year, Intel also shipped more than 10 million 80286 microprocessors (also used in versions of the IBM-compatible personal computer). Probably most of these became the CPU's in personal computer systems. Motorola makes the 68000 family of microprocessors used as the CPU in Apple's popular Macintosh computers, as well as in the Amiga and Atari computer systems. In 1991, Motorola shipped over 3 million 680x0 microprocessors for use as the CPUs in computer systems (McWilliams, 1992). Motorola and Intel make the most popular microprocessors used in personal computers, but several other manufacturers share the much smaller market for microprocessors used as the CPUs in engineering work stations (a computer for engineering design that is more powerful and more expensive than a personal computer). So the market for microprocessors as a CPU may be 25 to 30 million units per year (Tredennick, 1991). It is not a large part of the total market for microprocessors, which is about 1.5 billion units per year—just about 2%. But a microprocessor used as a CPU tends to be more expensive than a microprocessor for other applications, and so the dollar value of this market is high.

The problem with the microprocessor-as-a-CPU market was its requirements: the CPU market demands performance. The CPU market is performance driven, not cost driven. The requirements of the market for microprocessors replacing standard logic modules (low cost and adequate performance) and the CPU market (maximum performance at any cost) were in direct conflict. From a marketing viewpoint, there is no contest—the logic replacement market is about 20 times the size of the CPU market. Performance of the early microprocessors was limited by the simple, leisurely bus protocols, multiplexed instruction and data buses, packages with few pins, and slow, immature technology. The instruction sets and the implementations were an excellent fit for applications where the microprocessor replaced standard logic modules.

The market in logic applications, being 98% of the market volume, was too large to be ignored, and so microprocessor manufacturers concentrated their efforts on devices to support its requirements: low cost and adequate performance. This meant that the market for microprocessors used as CPUs was not well served by available microprocessors. In the early 1980s, researchers began to pro pose microprocessor architectures specifically suited to the market for central processors (Hennessy *et al.*, 1982; Patterson and Séquin, 1980). These microprocessors increased the number of available package pins and improved the speed of memory transactions. Implementers provided separate 32-bit buses for addresses and instructions or data. Conventional microprocessors had few pins, and so pins were shared for addresses, instructions, and data. Bus protocols were sped up at the cost of requiring tightly coupled special memory systems called caches and not allowing direct connection to peripherals. These designs considered performance first. Today we have two markets for microprocessors: one market for the microprocessor replacing standard logic modules on a circuit board (98% of the volume) and one market for the microprocessor as a CPU (2% of the volume). The market for the microprocessor replacing standard logic modules (called embedded control) is characterized by high volumes and extreme price competition. The market for the microprocessor as a CPU is characterized by low volumes and competition for performance.

2. ARCHITECTURE

The "architecture" of a microprocessor refers to the view of the microprocessor as seen by a programmer (Stallings, 1990). Elements of a microprocessor's architecture are usually described in a user's manual or programmer's reference manual (much like the owner's manual for an automobile). The architecture description includes the instruction set (what the microprocessor can do), the instruction formats (what commands to the microprocessor look like), and the internal register set (storage registers available to the programmer).

Instructions are the commands to the microprocessor. Figure 5 is an example of a 32-bit instruction format for a microprocessor with 64 general-purpose internal registers (2^6, the length of the source and destination specifiers). The instruction format consists of an operation field, two source operand fields, a mode field, and a destination field. The operation field tells the microprocessor the operation (e.g., add, subtract, multiply, shift). In a modern microprocessor architecture, the source operand fields point to register locations on the microprocessor. The destination field tells the microprocessor where to place any result from the operation.

I divide instructions into five categories: arithmetic, logical, control, transfer, and branching. Arithmetic instructions include add, subtract, multiply, and divide. Logic instructions include AND, OR, and EOR (exclusive OR). Control instructions help the programmer set the internal and external context for the execution of the other instructions. Transfer instructions include the load and store instructions and more operands between the memory system and the CPU. Branching instructions change the sequential execution of the program, often in conjunction with the testing of condition codes resulting from an arithmetic or logical operation.

3. OPERATION

The microprocessor is controlled by a program stored in the memory. A program might be written in a high-level language (e.g., FORTRAN, COBOL, C^{++}, BASIC, ADA; see COMPUTER PROGRAMMING LANGUAGES). The high-level language representation is translated by a compiler into instructions that the microprocessor understands (the ones and zeros of Fig. 3). The program might also be written in an

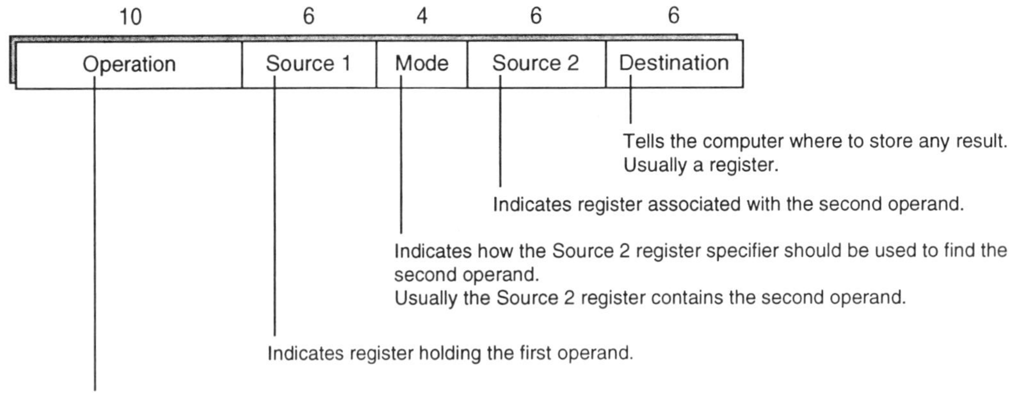

FIG. 5. Example format for a microprocessor instruction. The instruction tells the microprocessor where the source operands are, what operation to perform, and where to store any result.

assembly language, which is a mnemonic and symbolic representation of the microprocessor instruction set. An instruction-set assembler is then used to translate the assembly-language program into the ones and zeros that the microprocessor understands. The program and data are stored on a floppy disk or hard disk that can be accessed by the microprocessor through the peripheral-interface logic. Tiny magnetic domains on the floppy disk or hard disk are flipped in one direction to store a one and in the opposite direction to store a zero. The direction of the domain remains stable even when power is no longer applied to the device. Since the floppy disk and hard disk retain their data even when the power is turned off, they are nonvolatile storage. Floppy disks, hard disks, tapes, and other magnetic media provide nonvolatile storage for programs and data. ROM is also nonvolatile storage. RAM and the register file in the microprocessor lose their information when power is removed. Storage devices that lose information when power is removed are called volatile.

Now that the microprocessor, computer system block diagram, and instructions have been introduced, it is possible to describe operation of a simple microprocessor. (From the description of the simple microprocessor, the explanation will progress to more complex implementations.) To run a program (see Fig. 3), the microprocessor fetches an instruction from the memory system by communicating with the memory system over the bidirectional instruction and data bus (see Fig. 2). The microprocessor decodes the instruction to determine the required operation and the location of source operands. Then the microprocessor "executes" the instruction by obtaining the necessary operands, sending them to the ALU, and storing the result. For the example instruction shown in Fig. 5, the microprocessor would fetch the instruction from the memory system. Decoding the instruction would tell the microprocessor to set up the ALU for addition (as opposed to subtraction or some other operation) and that the source operands are in registers R3 and R2. The microprocessor executes the instruction by reading the contents of R3 and R2 from the register file, sending them to the ALU (which is set for addition), and storing the result in R48. The simple microprocessor updates the program counter (which is an internal pointer to the next instruction in the program) and fetches the next sequential instruction from the memory system.

3.1 Reset

When power is turned off, the microprocessor and the RAM lose their internal state information (including the contents of the program counter). When power is applied, voltage and current transients occur throughout the chip and may cause unpredictable states inside the microprocessor. A reset pin allows the user to set the microprocessor to a known internal state. The microprocessor might have an additional reset pin that it can drive to set the system in which it resides to a known internal state. Once the microprocessor is in a known internal state, it should be able to bring the rest of the computer system to a known operational state. But the RAM, where the program and data for execution normally reside, will not contain useful information (since information in the RAM was lost the last time the power was removed). The ROM attached to the microprocessor as a part of the memory system (see Fig. 2) does not lose its information content when power is lost; so, when power is applied to the system, the microprocessor goes to the ROM to get the instructions for system initialization.

3.2 Initialization

The microprocessor will reset the system and then move programs and data from the nonvolatile magnetic storage devices (hard disk, floppy disk, or tape drive) into the RAM in the memory system for execution. One of the first programs brought into the memory system for execution will be the operating system. The operating system is the program that sets the policies and procedures for executing application programs on the microprocessor. Why bother with the RAM? Why not just run the programs from the magnetic storage devices? The reason is speed. Typical RAMs have access times measured in nanoseconds (billionths of a second), while typical magnetic storage devices have access times measured in milliseconds (Rao, 1978). Even though the access time for a magnetic storage device is about a million times slower than the access time of the RAM in the memory system, transfer rates only differ by a factor of 10 or 100. A typical memory system might transfer 4 bytes every 40 ns, for example, for a transfer rate of 100 MB/s. A hard disk attached to the peripheral-interface logic in a microprocessor system usually has a transfer rate of between 1 and 10 MB/s. The efficiency of the computer system is greatly increased, therefore, by transferring blocks of programs and data from the magnetic storage device to the memory system and running the programs from the memory system.

3.3 Cache

Cache is a special form of memory. A cache memory resides between the main memory system and the CPU. In most modern microprocessors, the cache is located on the chip with the CPU and might be between 4 kB (in computer terms, k represents 1024, so that this is really 4096 bytes) and 32 kB (really 32 768 bytes). Main memory might be about 1 to 32 MB. The cache memory is much faster than the main memory system because it is smaller, located closer to the CPU, and built of more expensive technology. Cache memory exploits a characteristic of programs known as locality of reference. Instructions and data used by a computer program tend to be used again within a short time. The first time an instruction or memory operand is referenced, it is brought into the cache. Subsequent references to the same instructions or operands produce a "hit" in the cache, and the access is returned to the CPU much quicker than if the reference had to go to the main memory system. Even though the main memory system might be a thousand times the size of the cache, instructions and data might "hit" in the cache 90% of the time because of locality of reference. The cache can have a dramatic effect on the performance of the microprocessor. Suppose, for example, that it takes two clock cycles to access the cache on the microprocessor chip and four clock cycles to access the main memory system. If there is no cache, the time to execute a program is the number of accesses n in the program times the number of clock cycles, 4, in a memory system access:

$$t = 4n. \quad (1)$$

If there is a cache with a 90% "hit" ratio, then the time to execute the same program becomes

$$t = 1 \times 0.9n + 4[(1 - 0.9)n] = 1.3n. \quad (2)$$

Locality of reference helps a cache reduce execution time from $4n$ to $1.3n$. In this example, the small cache makes the CPU think the memory system is three times faster than it really is.

3.4 Interrupts and Exceptions

The microprocessor is the CPU running the system, but it cannot conveniently do everything (Money, 1990). It is simpler, for example, to have a separate processor in the keyboard watching for keystrokes. When you strike a key on your keyboard, the keyboard processor sends a message to the CPU requesting processing time for the keyboard. The form of this message is called an interrupt. The interrupt allows the CPU to process program instructions that are its primary responsibility and not have to waste time explicitly checking for numerous infrequent external activities. Interrupts are the means for devices outside the CPU to request attention and get service for infrequent (relative to the speed of CPU processing) events. Interrupts, since they are requests from outside the microprocessor, are usually asynchronous, meaning that their timing is not synchronized with the microprocessor's clock. An interrupt request may occur at any time during the processing of any instruction. Interrupt requests must, therefore, be synchronized with the microprocessor clock. This can be accomplished by use of synchronizing circuit to capture the request in a register. The microprocessor will then recognize and process the request at the next convenient interval (usually at the end of the current instruction and before beginning the following instruction).

An exception occurs when the microprocessor cannot complete the current instruction. If, for example, the microprocessor attempts to decode an instruction with an illegal bit pattern (i.e., a bit pattern not defined for the instruction set), normal processing cannot continue since no operation is defined for the bit pattern. The microprocessor will execute instead an exception sequence in an ef-

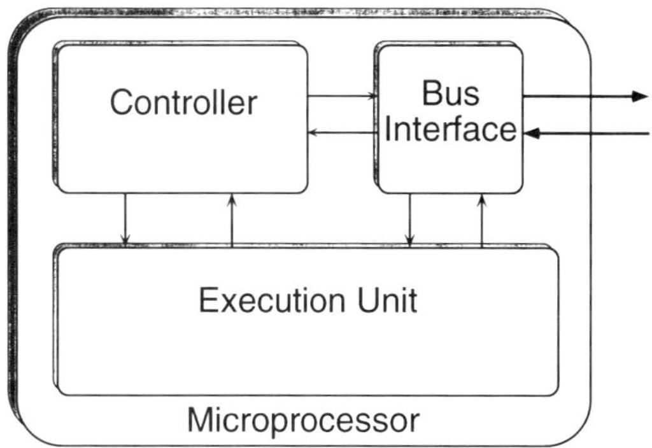

FIG. 6. The major blocks inside a simple microprocessor are the controller, the execution unit, and the bus interface.

fort to identify the problem and continue processing. Exceptions can also occur if the microprocessor cannot complete execution of an instruction because it is unable to obtain data from memory. Arithmetic exceptions such as overflow and divide by zero may also be defined.

4. DESIGN

Figure 6 shows the inside of a microprocessor. The three necessary components of the microprocessor are the bus interface, the controller, and the execution unit. The execution unit (or data path) contains the resources for computing such as the register file, program counter, and arithmetic units. The controller runs the execution unit and issues commands to the bus interface. The bus interface is responsible for transactions between the external bus and the execution unit. This block diagram is good for describing functions on the first microprocessor and it is good for describing modern microprocessors.

4.1 Execution Unit

Figure 7 shows an execution unit with a three-bus structure. The execution unit contains the resources seen by the programmer such as the programmer's register set, the program counter, and the condition code register. In addition, the execution unit contains the instruction register, the ALU, data and address temporary storage registers, an address output register, and input and output registers for data.

The internal *A* and *B* buses in the execution unit transmit operands from the register file or other source to the ALU (or other destination). The internal *C* bus transmits the result to the register file or other destination. Since the execution unit shown in Fig. 7 has two source buses and a destination bus, it is able to return the result of the previous operation while executing the current operation. The execution unit could be made cheaper by leaving out one bus (so that the registers would only have two access ports, rather than three), but then the execution of the current instruction could not be overlapped with storing the result of the previous operation.

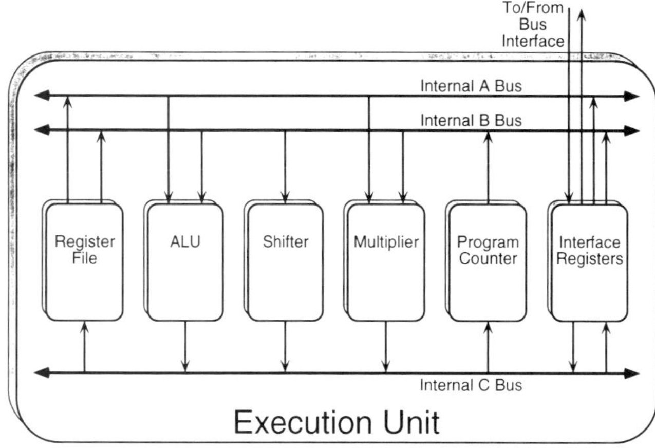

FIG. 7. The execution unit contains the resources (such as the registers and arithmetic units) that the controller manipulates to execute a program.

4.2 Bus Interface Unit

The data in and out registers and the address out register are shared with the bus interface (see Interface Registers in Fig. 7). The bus interface communicates with the external bus using commands from the controller and data and addresses from the execution unit. For a store-to-memory operation, the bus interface will send the address from the address out register and the data value from the data out register to the memory system via the external bus. To load a value from memory to a register in the microprocessor, the bus interface will send the address from the address out register to the memory system and place the value returned on the external bus in the data in register in the execution unit.

4.3 Controller

Figure 8 is the controller of a microprocessor. The controller is built of an instruction decoder, a state sequencer, and a control decoder. The instruction decoder looks at the ones and zeros in the instruction and directs the initial action of the state sequencer. The state sequencer steps the microprocessor through the states necessary to execute an instruction. The control decoder mixes cycle-by-cycle dynamic control signals from the state sequencer with static decode signals from the instruction decoder. The output of the control decoder helps the state sequencer determine the next control state and also directly controls actions in the execution unit. Input to the instruction decoder might be 16 to 32 bits. Output of the instruction decoder might be 9 to 14 bits. Output of the state sequencer could be 30 to 150 bits. Output of the control decoder might be 180 to 400 bits with 9 to 14 of that being the feedback to the state sequencer to help the state sequencer determine what needs to be done next.

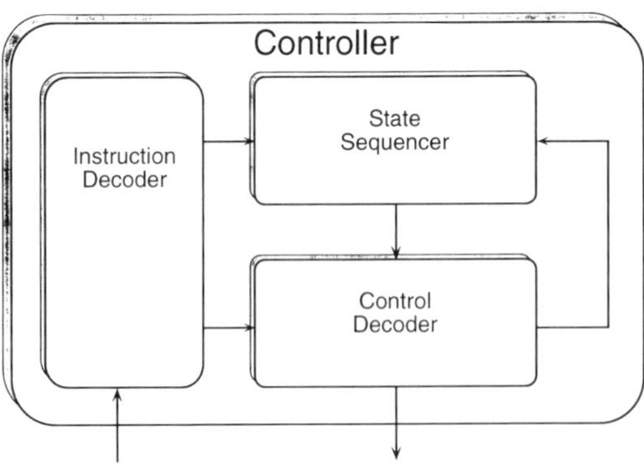

FIG. 8. The controller is the engine that runs the microprocessor. The controller decodes the instructions and sequences the microprocessor through the states necessary to execute the instruction.

The instruction decoder uses the operation field to determine what the initial action of the state sequencer should be. Values of the register fields are sent to the control decoder. The state sequencer needs to know which operation is to be accomplished, but the operations performed on a cycle-by-cycle basis do not depend on which registers hold the source operands or store the results. This is a good example of the difference between static and dynamic control information. Pointers to the source and destination registers are static control information. Dynamic control information tells the execution unit when to do what and static control information tells which resources to use.

The state sequencer is the heart of the controller. The state sequencer controls the resources of the execution unit (through the control decoder). The state sequencer determines when to fetch the next instruction, access operands, perform calculations, store results, or process interrupts. The state sequencer holds the repertoire of commands for the control decoder and options for next state actions. If this information is held in a memory array, it is called microcode. If the same information is held in a collection of logic gates [such as a programmed logic array (PLA)], it is often called random logic. In a random logic implementation, the boundary between the state sequencer and the control decoder may blur. In either case (microprogrammed or random logic), the information in the state sequencer is likely to be encoded to save space.

The control decoder translates commands from the state sequencer into actions in the execution unit. The control decoder mixes control information from the static decoder (e.g., pointers to the source and destination registers) with the dynamic information (e.g., when to read the operands and when to store a result) from the state sequencer to control the cycle-by-cycle action of the resources in the execution unit.

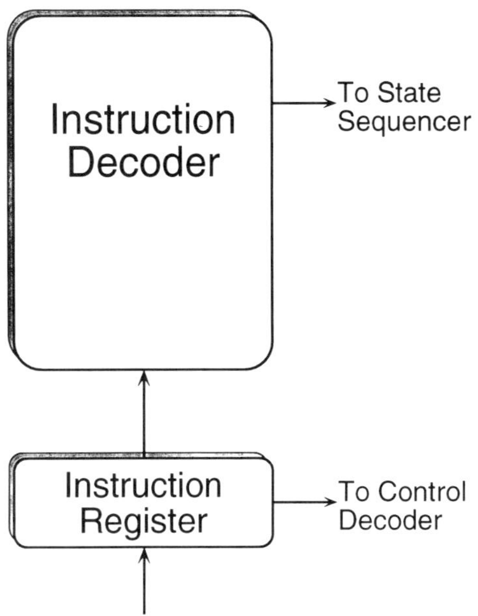

FIG. 9. Instruction register and instruction decoder in a simple microprocessor. A single register drives the instruction decoder and the static values for the control decoders.

4.4 Pipelining

The first microprocessors used only a single register to hold the instruction being executed, as is shown in Fig. 9. This is the simplest hardware to support instruction decoding, but also leads to a limited execution model. The processor first fetches the instruction into the instruction register. Once the instruction is fetched, it is decoded to determine initial state for the state sequencer. The state sequencer cycles through the proper states to execute the instruction. The state sequencer provides the dynamic control information to the control decoder and the instruction register provides the static control information (e.g., pointers to the operand source and destination register), and so it cannot be changed until after the last use of any of the static control information.

Figure 10 is the execution model for a microprocessor using the simple instruction decoder shown in Fig. 9. In Fig. 10, each row represents a single instruction and time progresses from left to right. Each instruction has to go through fetch, decode, and execute stages before the next instruction can be allowed to start. The microprocessor fetches the first instruction, then decodes the first instruction, and then executes the first instruction. When the first instruction is completed, the microprocessor fetches the second instruction, then decodes the second instruction, and so on. In the simplest microprocessors, these stages (fetch, decode, and execute) were not overlapped because the instruction decoder had only a single register. If a new instruction is fetched into the instruction register before the previous instruction completes execution, the pointer to the result register for the previous instruction would be overwritten by the new instruction. The performance bottleneck in early microprocessors is the controller (since the external bus, decoder, and execution unit are busy only every third cycle).

Figure 11 is an instruction decoder to support an instruction pipeline. This instruction decoder has separate registers for fetch, decode, and execute. This instruction decoder configuration makes the execution model of

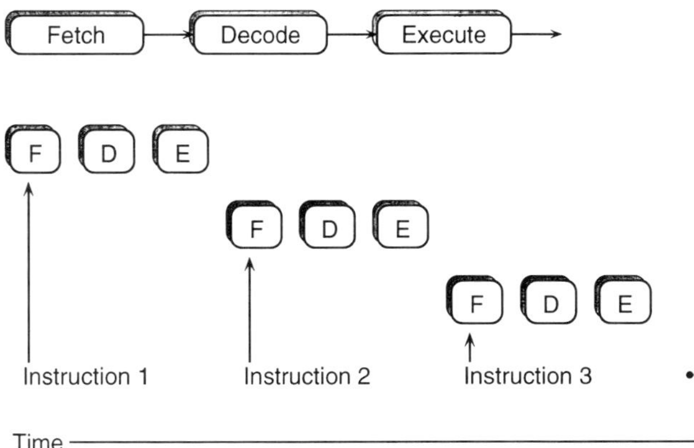

FIG. 10. In the simplest microprocessor implementation, the CPU steps sequentially through fetch, decode, and execute cycles.

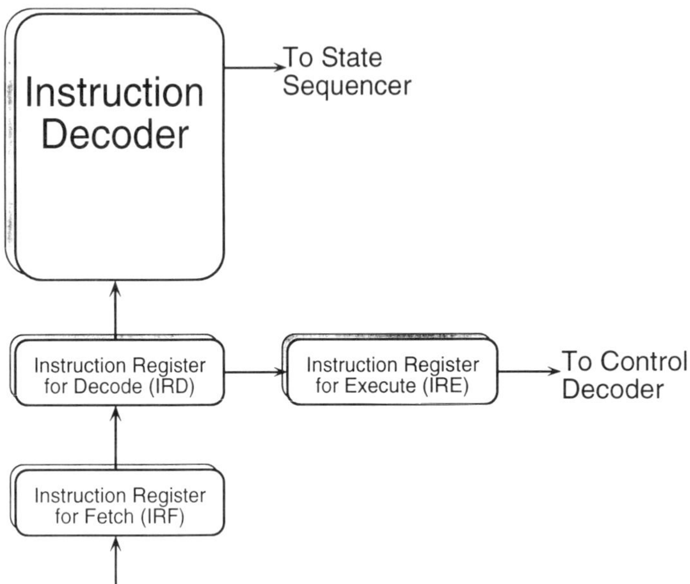

FIG. 11. The instruction decoder for overlapped execution has separate instruction registers for fetch, decode, and execute.

Fig. 12 possible. The IRE holds the instruction ready for execution and supplies static control information (e.g., pointers to the registers) to the control decoder. At the same time, the IRD holds the next instruction for the instruction decoder. The IRF is available to capture the next instruction. All three registers can be active at the same time.

Figure 12 shows the execution model for the microprocessor with the instruction decoder of Fig. 11. Since there are separate registers to preserve the instruction values for the decode and execute phases, fetch, decode, and execute can be overlapped. The bottleneck in the microprocessor is the memory system. One instruction is completed every cycle, but the external bus is busy on every cycle. If one of the instructions is a load or store, additional cycles will be introduced and subsequent instructions will be correspondingly delayed. In Fig. 12 each row represents an instruction, and time flows from the left of the diagram to the right. The first instruction (at the top of the diagram) is fetched. As the second instruction is fetched, the first instruction is decoded. As the third instruction is fetched, the second is decoded, and the first is executed. To determine resources required

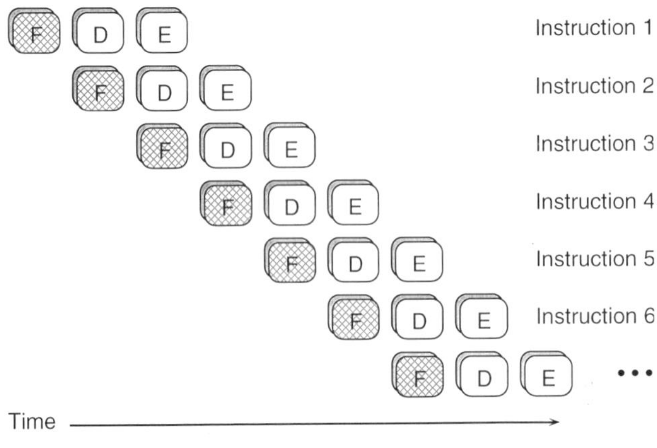

FIG. 12. Execution model for a simple instruction pipeline. Each instruction is represented by a fetch, decode, execute row. A new instruction enters the pipeline on each cycle.

to build this microprocessor, the designer need only look at a vertical slice in the execution model (this gives a complete sample of activity during a single cycle).

The execution model of Fig. 12 shows an instruction completing every cycle and it shows the external bus being used on every cycle. But it does not show any loads and stores. What happens if the microprocessor executes loads and stores? A load instruction will have to read an operand from the memory system. During the execute cycle for the load instruction, the operand is read from the memory system into the microprocessor. The only external bus available to the microprocessor will be used for the operand read transaction, so that the instruction fetch, which normally would occur simultaneously with the execute cycle (for the instruction after the one currently being decoded and two instructions after the one being executed) will have to be delayed. If an instruction fetch is delayed, there will be no instruction to decode in the next cycle, and so on. Bubbles are introduced in the pipeline to share the external bus between the instruction fetches and data reads and writes. If this happens very often, it will not be possible to complete an instruction each cycle. Suppose that for those same load and store instructions, execution unit resources (such as the ALU) are needed to calculate the operand address. During the cycles when the execution unit resources are being used to calculate the operand address, they are not available for the execute phase of another instruction. If the execute phase of an instruction is blocked, the decode phase and the fetch for the two instructions behind it are prevented from advancing. Another bubble is introduced in the pipeline.

Figure 13 is an early microprocessor for overlapped instruction execution. There is only a single external bus for instructions and data. The execution unit contains a dual-port register file. Overlapping fetch, decode, and execute gave good performance improvement over earlier microprocessors using sequential execution of fetch, decode, and execute, while not significantly increasing required resources.

Figure 14 shows the instruction pipeline extended to prevent delays caused by operand reads, writes, and address calculation. I have extended the execution model from fetch, decode, and execute (F, D, E, respectively) to fetch, decode, address calculate, operand read, execute, and operand write (F, D, A, R, E, W, respectively). Not every instruction uses all of these execution phases, but every instruction must pass through all of the phases.

A vertical slice through this execution model determines the resources required to support it. Supporting requirements for a vertical slice in this execution model will ensure that delays are not introduced because of a lack of resources. The fetch stage will need a single read port to the external memory system. The address calculate stage may require two ports to the register file (the contents of two registers may be added to compute an operand address). The operand read stage may require a read port to the external memory and a write port to the register file (some address calculations update the value

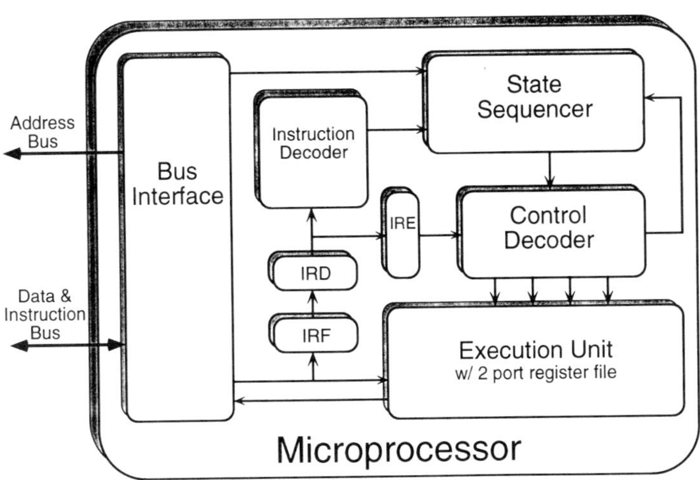

FIG. 13. Block diagram of a microprocessor with overlapped instruction execution. The instruction decoder contains three instruction registers, the register file has two read/write ports, and there is a single external bus for instructions and data.

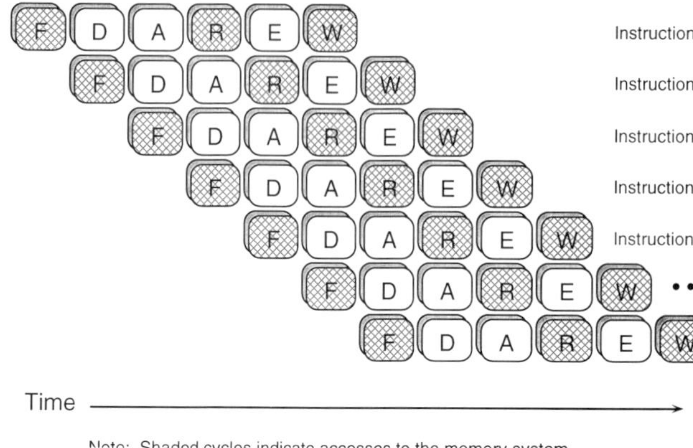

FIG. 14. The extended instruction pipeline has fetch, decode, address calculate, operand read, execute, and write stages. Pipelining is increased to reduce conflicts that delay instruction execution.

of a register used to calculate an operand address). The execute stage may require two read ports to the register file. The operand write stage may require a write port to the external memory system and a write port to the register file. To support the extended instruction pipeline, the microprocessor will need two read ports and one write port to the memory system. It will also need two ALUs, one for address calculation and one for instruction operation, and a six-port register file. Many of the recently introduced high-end microprocessors, such as the 80486 (see Fig. 15) and the Motorola 68040 (see Fig. 16), implement an execution model similar to the extended instruction pipeline. Early microprocessors contained a few thousand transistors. The 80486 and the Motorola 68040, which were both announced in 1989 (the 80486 shipped in 1989 and the 68040 shipped in 1991), contain about 1.2 million transistors each on a chip of approximately 1.5 cm^2 area.

In the execution model of Fig. 14, it looks as if the microprocessor has finally reached the goal of executing one instruction every cycle, even if there are loads and stores. This is not so. Instructions such as multiply and divide might tie up the execution unit for several cycles. If they do, they block everything behind them for a few cycles, too. How does the microprocessor prevent bottlenecks in the execution unit from delaying the whole pipeline? One solution advocates throwing out all the instructions taking more than a single execute cycle. One might also provide enough hardware to execute any instruction in a single cycle. A third solution, becoming more popular in the industry, is to provide multiple execution units. This solution is based on the premise that instructions taking more than one cycle will not usually occur next to each other. In other words, instructions will be distributed over the available execution units to improve performance reasonably.

With multiple execution units, it becomes possible to initiate or complete the execution of more than a single instruction each processor cycle (Diefendorff and Allen, 1992). Implementations attempting to issue multiple instructions per cycle are called superscalar. Superscalar implementations attempt to break through the barrier of one instruction per cycle as the theoretical maximum execution rate. Figure 17 is a die photo of the Motorola MC88110, which is a modern superscalar microprocessor. It employs multiple integer execution units, multiple floating-point execution units, a graphics execution unit, a load/store execution unit, and separate on-chip instruction and data caches.

The return on invested hardware is decreasing. The microprocessor improved performance by a factor of 3 initially by adding two registers to the instruction decoder to overlap instruction fetch, decode, and execute. The next improvement, the extended instruction pipeline, added considerable hardware to the execution unit and memory system and lots of pins to the microprocessor, and the improvement was only seen when operand accesses would have conflicted with instruction fetches. Since operand accesses

FIG. 15. The Intel i486™, containing 1.2 million transistors, is a good example of a modern microprocessor with an extended instruction pipeline. The Intel i486™ has an 8-kB combined instruction and data cache. (Courtesy Intel Corporation.)

are less than half the frequency of instruction accesses, performance is now only one and a half times the original performance. This is still a substantial improvement, but the design has added substantial complexity. Adding multiple function units and allowing the decoder to issue multiple instruction per cycle has the potential to improve performance substantially, but there is a corresponding increase in complexity. The decoder must decide which instructions are eligible to execute concurrently. If out-of-order completion of instructions is allowed, cleaning up the state of the processor after an exception may be difficult.

Hardware can also be added to compensate for delays introduced by operand or address conflicts. An operand conflict occurs when the result of the current instruction is needed by some following instruction. The pipeline is stalled until the result is available. This problem is worse if the result of the current instruction is needed to complete the address calculation for the following instruction. Adding hardware for these cases can improve performance. New microprocessors,

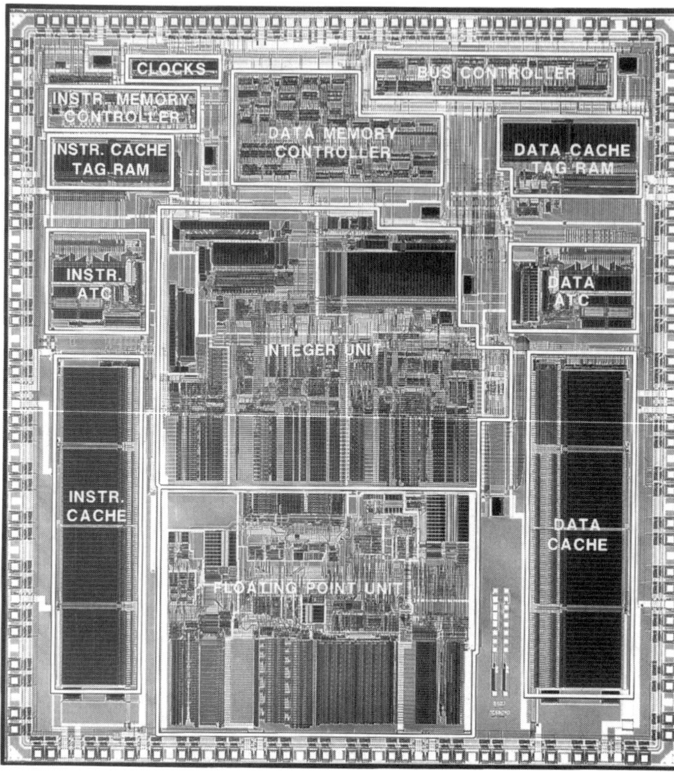

FIG. 16. Motorola's MC68040 is a good example of a modern microprocessor with an extended instruction pipeline. The MC68040 has separate buses to its internal 4-kB instruction cache and internal 4-kB data cache. (Reprinted with permission of Motorola.)

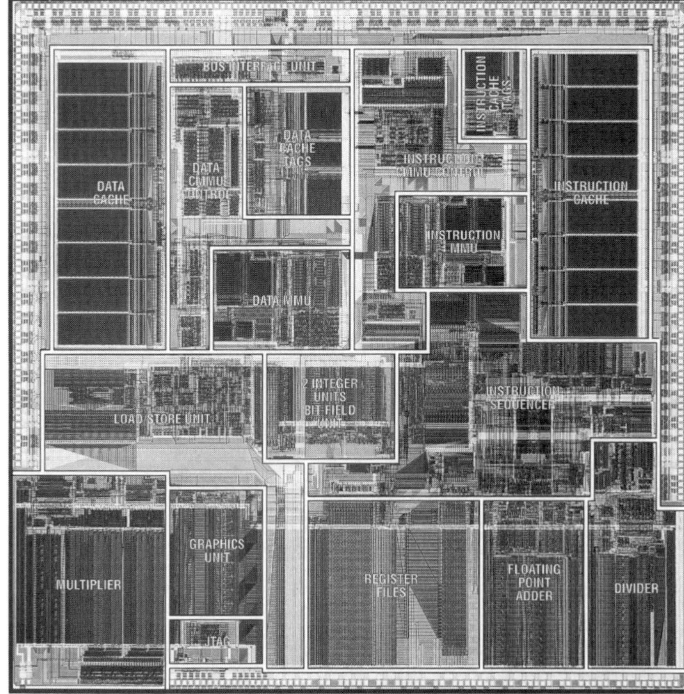

FIG. 17. Motorola's MC88110, containing 1.2 million transistors, is one of the first superscalar microprocessors. A superscalar microprocessor is capable of issuing and completing more than a single instruction each clock cycle. (Reprinted with permission of Motorola.)

however, are still adding more and more hardware for less and less return in performance.

4.5 Branch Execution Model

The description to this point tells how a microprocessor works and reviews possible hardware improvements from a pipeline point of view. These improvements consider throughput but have not considered the effect of branches. Figure 18 is an execution model showing the effect of a branch instruction with the branch condition satisfied. A branch taken causes a significant penalty because the pipeline is blocked until the outcome of the branch condition is known. Since the outcome of a branch may not be known until the end of the execute cycle, the fetch to the branch destination instruction may not begin until the write cycle of the current instruction. A five-cycle bubble is introduced in the pipeline. If branches occur frequently, performance could suffer considerably.

But branches are frequent. Branches constitute about 20% of instructions by dynamic frequency of occurrence. Figure 19 illustrates the effect of branches on processor performance. The vertical axis is processor performance normalized so that performance is one in the absence of branch instructions. The horizontal axis shows branch frequency of occurrence. The family of curves represents the likelihood that a branch is successful. The calculations assume that instructions other than branches complete every 1.5 clock cycles on average. Branches are 5 clock cycles if taken and 1 clock cycle if not taken. The example lines show that for a branch frequency of occurrence of 20% (a branch occurs every fifth instruction) and a branch success rate of only 40%, the performance of the processor is only about 70% of its performance in the absence of branches. If the branch success rate is 80%, the performance is reduced to only 58% of performance in the absence of branches. Plenty of effort in microprocessor design goes into reducing the cost of branches.

One suggestion for reducing the cost of branches is to incorporate a delayed branch instruction. The instruction immediately after the delayed branch instruction is always

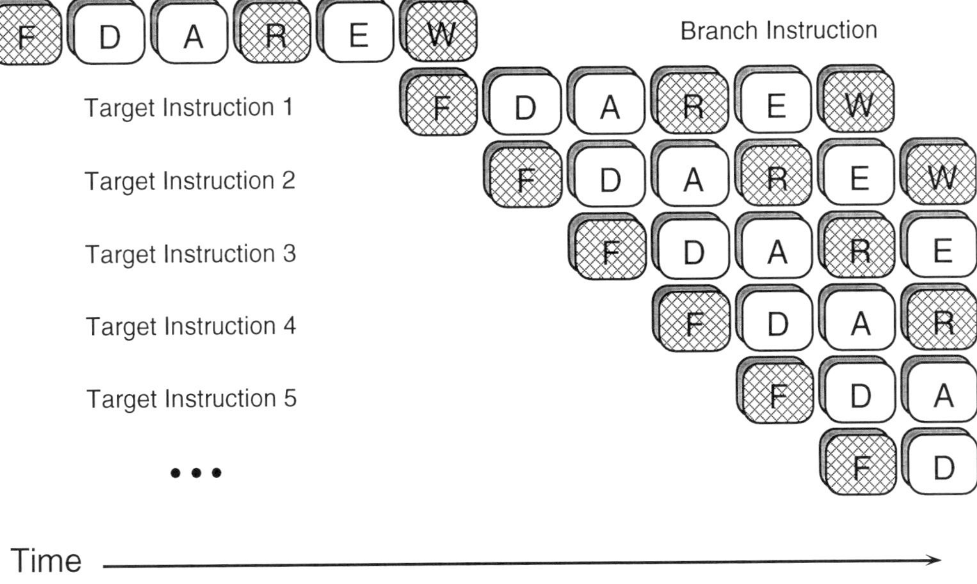

FIG. 18. Execution model showing the effect of a branch instruction with the branch condition satisfied. The top row is the branch instruction. The second row is the target of the branch. The pipeline is blocked until the outcome of the conditional branch is known.

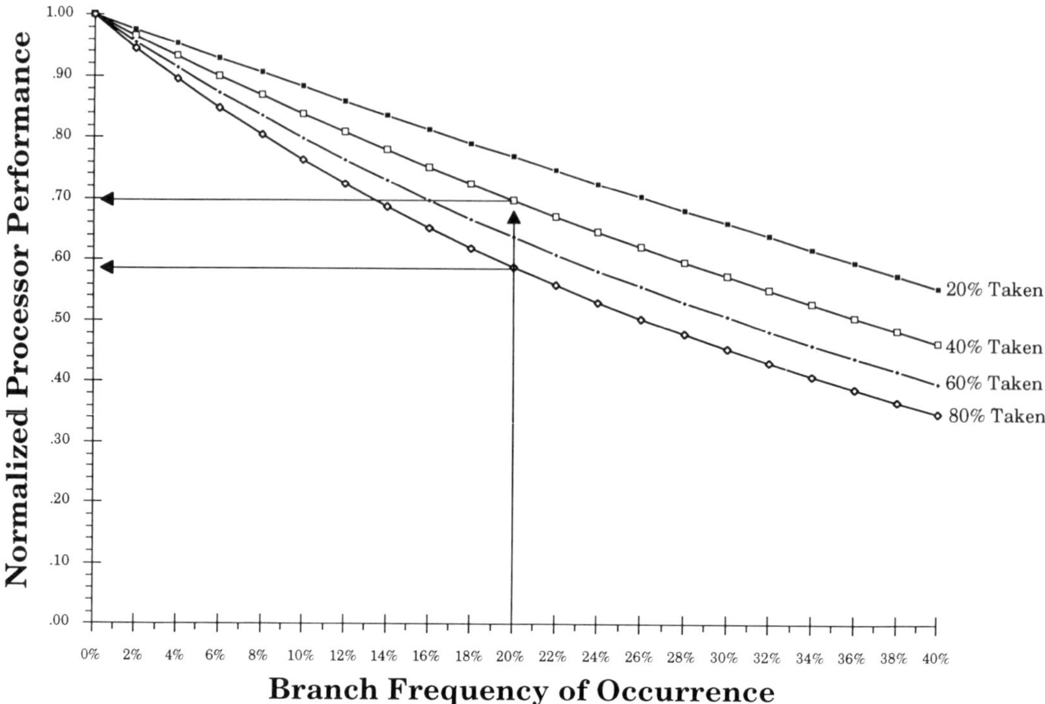

FIG. 19. Branch penalty significantly reduces processor performance. If branches occur 20% of the time and are successful (taken) 40% of the time, the processor is only 70% as efficient as it would be if there were no branches.

executed. This recovers one of the cycles lost when a branch is successful and is based on the premise that the instruction after the branch was already mostly through the pipeline when the outcome of the branch was decided, so its execution might as well finish.

The branch prediction cache, shown in Fig. 20, tracks the program counter and maintains a record of branches recently encountered. When the hardware sees a branch instruction, it checks the branch prediction cache. The branch prediction cache may have recorded the success rate for this branch instruction. If the instruction is expected to branch, the branch prediction cache provides the target address to the memory system. With enough hardware and early recognition, the entire branch penalty may be eliminated. Hardware is added to detect the branch instructions, record the branch history and branch target addresses, and recover in the event the branch does not meet its prediction.

The branch target cache is a variation of the branch prediction cache. The branch target cache is much like the branch prediction cache except that it stores the target instructions rather than the address of the target instructions. This further helps reduce branch penalty because the branch target cache can provide the target instructions to the decoder and help cover the latency in accessing the external memory system.

4.6 External Bus Protocol

Figure 21 shows how the microprocessor communicates with the memory system to obtain the program instructions (and data). The microprocessor, using the common bus connection, sends the memory system a request saying "I want to read an instruction from the following memory location," or, perhaps, "I want to write a result to the following memory location." What the microprocessor is really placing on the bus is a bunch of ones and zeros (the address) representing the location of the instruction in the memory and some tags indicating what kind of transaction (read or write, for example) the microprocessor is requesting. For an instruction read request, the memory system will

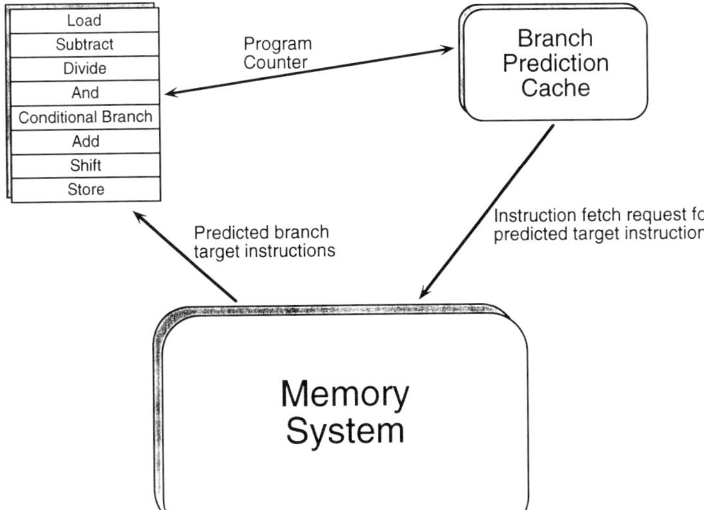

FIG. 20. The branch prediction cache keeps track of program branching history and attempts to fetch the target of branches expected to succeed before the pipeline stalls.

retrieve the contents of the memory location and place the value on the data and instruction bus lines along with a tag saying "Here's the instruction." Once the microprocessor has taken the instruction, it will signal the memory system "Thank you" to tell the memory system it is okay to end the current transaction. If the microprocessor initiates the transaction and waits an undetermined number of cycles for the response from the memory system, the transaction is an asynchronous handshake protocol. If the microprocessor initiates a transaction and expects a response after a fixed number of clock cycles, the transaction is a synchronous protocol. Fast microprocessors in 1992 can do about 20 million transactions with an external memory system every second. Today's computer system built with a microprocessor is significantly faster and more powerful than the giant computers of only a few years ago.

5. TECHNOLOGY AND MANUFACTURING

The original Intel 4004 microprocessor in 1971 contained about 2300 transistors (Smith, 1992), was designed with about 8-μm feature sizes (Hennessy and Jouppi, 1991), operated at 108 kHz, and came in a 16-pin package. The early packages for microprocessors were

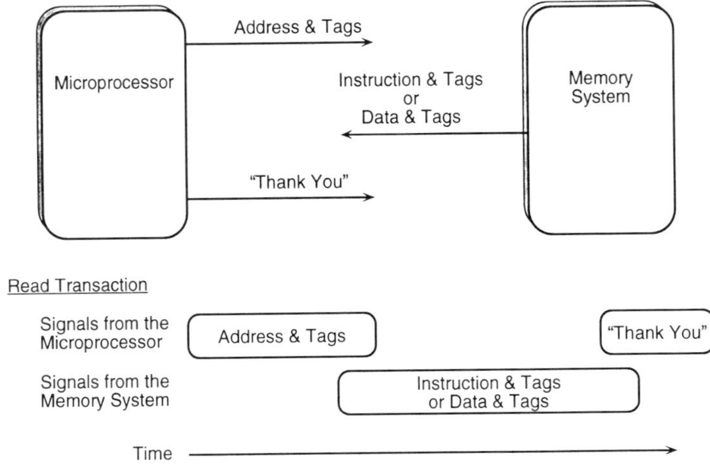

FIG. 21. A transaction between the microprocessor and the memory system using a handshake protocol.

designed to be compatible with the contemporary transistor-transistor-logic packages, so that they could be more easily designed into standard circuit boards. The most popular package for 8-bit microprocessors such as the Zilog Z80 was the 40-pin plastic package. The Motorola MC68000, announced in 1979, is a 32-bit processor with a 16-bit external interface and was introduced in a 68-pin ceramic package. In 1992, Intel introduced a version of the 80486 containing 1 200 000 transistors, designed with 0.8-μm feature sizes, operating internally at 66 MHz, in a 168-pin package. Figure 22 shows the progression of transistors on Intel microprocessors from the 4004 through the 80486 to the latest-generation microprocessor, now on the market as Pentium (data from Smith, 1992). Data points labeled with the name of the microprocessor indicate number of transistor for each microprocessor generation. The plain line is the theoretical curve for a doubling of the transistor count every two years. This curve shows the combined effects of the reduction in linewidth (allowing more transistors per unit area) and increase in manufacturable chip sizes (allowing more transistors per chip). Linewidths have decreased from about 8 μm in 1971 to between 0.8 and 0.5 μm in 1992 (Hennessy and Jouppi, 1991). As the linewidth decreases by a factor of 2, the number of transistors in a fixed area goes up by a factor of 4. Doubling the size of the chip doubles the number of transistors on the chip.

The first microprocessors were small chips about 4 mm or less on a side and were produced on small wafers about 50 mm in diameter. Recent microprocessors may be 15 mm on a side and are commonly produced on wafers 150 mm in diameter and sometimes on wafers 200 mm in diameter. Today's highest performance microprocessors may contain 3 000 000 transistors, have about 500 pins, and operate internally at frequencies in the range of 100 MHz. At these frequencies, external chips connected to the microprocessor will not be able to keep up with the speed of the microprocessor. One technique being used today to help solve this problem allows the microprocessor to operate at some multiple (2×, 3×, 4×) of the external system clock rate. In the near future, microprocessors will contain more than 10 000 000 transistors, be designed with 0.2-μm feature sizes, and operate internally above 200 MHz (5-ns cycle time).

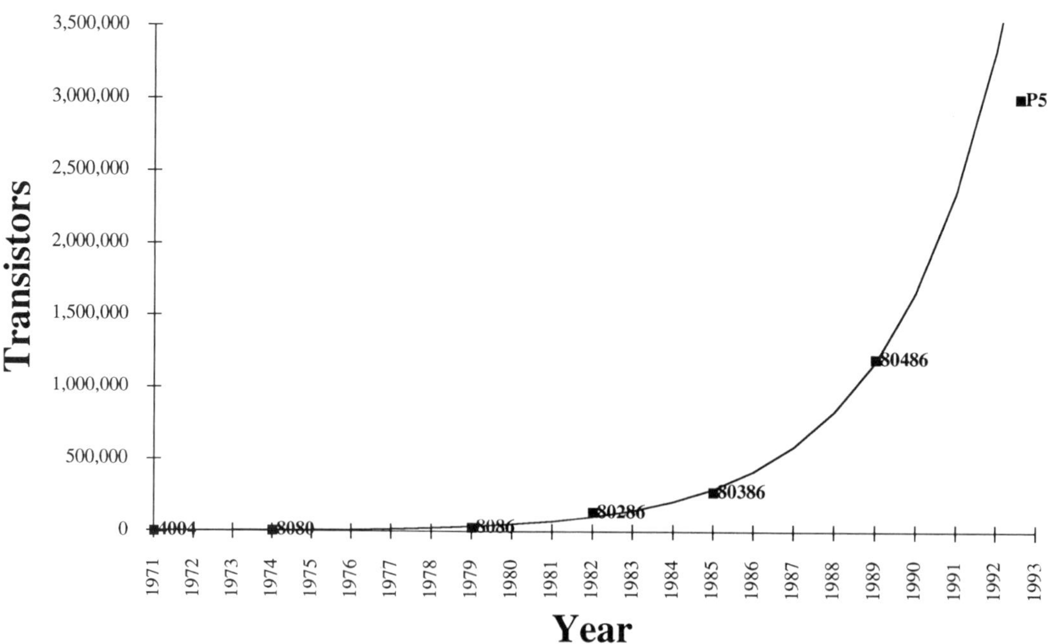

FIG. 22. The number of transistors in some Intel microprocessors from 1971 to 1993. The point labeled P5 refers to the microprocessor that has now reached the market as Pentium.

6. APPLICATIONS

6.1 The Microprocessor as a CPU

More than 21 million IBM-compatible personal computers were shipped in 1991. All of those computers used microprocessors based on the architecture of the Intel 80x86. Personal computers from Apple, Atari, and Commodore and based on the Motorola 680x0 microprocessor shipped probably between 2 and 3 million systems in 1991 (McWilliams, 1992). In addition to the personal computer market, there is a smaller market for engineering work stations. Engineering work stations are more expensive than personal computers and are based on chips from many different companies such as Intel, Motorola, MIPS, IBM, Cypress, DEC, Intergraph, Texas Instruments, and Hewlett-Packard. The total shipment of engineering work stations from all companies in 1991 was below 5×10^5 (Slater, 1992). There are also other computer systems using the microprocessor as the CPU, but their shipping volumes do not account for a significant number of microprocessor sales. The total market for the microprocessor used as a CPU in a computer system is probably below 30 million units.

Microprocessors designed for use as the CPU in a computer system may cost from below \$50 to well over \$1000, depending on the development cost, technology of the processor, and performance features of the design. Computer applications for microprocessors fall into two general categories: desktop computers (designed to be plugged into a permanent power source) and portable computers (designed with an internal battery for some operation independent of a permanent power source). Today, the same microprocessors serve for both applications. In the future, microprocessors designed for desktop computers will diverge from microprocessors designed for portable computers. Microprocessors designed for desktop computers will emphasize performance. The transistor budget for these microprocessors will be spent on special circuits for multiply and divide, additional arithmetic units for address calculation, on-chip instruction and data caches, and instruction and execution pipelining. Microprocessors designed for portable computers will emphasize low power and maximum function (to reduce the number of additional chips required to build a system) because portable computers often run using batteries and battery life is an important measure of perceived market value for the user. Performance is still important, but the first consideration is conserving power. The transistor budget for these microprocessors will be used to integrate as many of the computer system peripheral functions as possible, such as the serial and parallel communication ports, keyboard scanner, and memory controller. In addition, microprocessors for portable computers will have sophisticated on-chip power management circuits. Microprocessors for portable computers will become more like microprocessors for embedded control.

6.2 Embedded Control

An embedded control application is any use of a microprocessor other than as the CPU in a computer system. By contrast with the relatively small number of microprocessors used as CPUs in computer systems (fewer than 30 million per year), a relatively large number of microprocessors are used in embedded control applications. Microprocessors are used in microwave ovens, automobiles, printers, computer keyboards, calculators, games, gasoline pumps, and toys. They are everywhere. In 1991, about 1.5×10^9 microprocessors were shipped worldwide for use in embedded control applications.

Requirements of embedded-control applications make them significantly different from applications for microprocessors used as the CPU in a computer system. Most computer systems demand maximum performance from the CPU, and so the design of microprocessors for computers is driven primarily by performance. High-volume applications in embedded control are driven by cost. In these applications, the microprocessor competes to displace standard logic modules, and so the microprocessor must provide adequate performance (equal to the standard logic) at a lower cost. The requirements of these two applications (CPUs and embedded control) have split the design of microprocessors along two diverging lines. As the underlying technology improves, microprocessors designed for use as CPUs in computer systems use the additional transistors, speed, and chip area to improve the performance of the microprocessor. Microprocessors designed for embedded-control applications take advantage of improvements in chip technology to increase chip function, to reduce power consumption, and to reduce chip size. They try to squeeze onto the chip as much of the target application's system as possible. These chips, called microcontrollers or microcomputers, may incorporate most of the functions of the simple computer system from Fig. 2, including ROM, RAM, peripheral-interface logic, and even the peripheral functions themselves. In high-volume embedded-control applications, the average selling price of the microprocessor may range from under $1 to about $3.

GLOSSARY

Address Mode: Information in an instruction that says how to calculate the address of the operand(s).

ALU (Arithmetic and Logic Unit): The ALU is used for all general-purpose arithmetic.

Architecture: The view of the computer as seen by the user. Generally includes the instruction set, address modes, instruction formats, and registers visible to the computer user.

Asynchronous (Hardware): Not synchronized with the system clock. An event is asynchronous if its occurrence depends on signals other than the system clock.

Block Diagram: A drawing, consisting primarily of labeled boxes, showing the logical structure of part of a computer.

Bus Controller: The logic on a microprocessor chip controlling the protocol at the chip's pins and providing synchronized signals to the state sequencer.

Cache: A smaller, faster memory that sits between the CPU and the main memory to give the main memory the appearance of higher speed. CPUs are faster than memory, and so caches can serve a speed-matching function.

Central Processing Unit: See CPU.

Condition Code: An attribute of the result of an ALU operation. The following are fairly standard condition codes: c = carry out (this is borrow—not for subtraction; n = sign of the result (1 is negative, 0 is positive); v = overflow (significant result bits lost); z = zero result.

CPU (Central Processing Unit): The engine controlling the operation of a computer system. It contains the arithmetic and logic units, program counter, register file, and state sequencer.

Decoder: Logic that maps an input bit pattern to some new output bit pattern.

IC: Integrated circuit; a chip; literally, a "circuit in one place."

Implementation: The logical representation of a functional specification (e.g., the logic design implements the specification in the user's manual).

Interrupt: A signal that disrupts the normal sequence of events.

Microcode: The contents of a control store. Microcode is the collective term for the bit patterns controlling the actions of the execution unit and the state sequencer.

Microprocessor: A single integrated circuit containing at least the functions of a CPU in a computer system.

Multiplexer: Logic that selects a single output from several inputs.

Operand: A quantity on which an operation is to be performed.

Register: Separately addressed (from memory) storage location, at least conceptually, in the CPU.

Semiconductor: A material whose electrical conductivity is between that of an insulator and that of a conductor.

Synchronous (Hardware): Synchronized with the system clock. An event is synchronous if its occurrence depends only on a clock.

Works Cited

Bell, C. G., Mudge, J. C., McNamara, J. E. (1978), *Computer Engineering*, Bedford, MA: Digital, Chap. 1.

Davis, H., Wharton, M. (1983), *Inside the Chip*, London: Usborne.

Diefendorff, K., Allen, M. (1992), "Organization of the Motorola 88110 Superscalar RISC Microprocessor," *IEEE Micro* **12** (2), 40–63.

Faggin, F. (1992), "The Birth of the Microprocessor," *Byte* **17** (3), 145–150.

Hennessy, J. L., Jouppi, N. P., Gill, J., Baskett, F., Strong, A., Gross, T., Rowen, C., Leonard, J. (1982), in: *The MIPS Machine*, Proceedings of COMPCON, pp. 2–7, New York: IEEE.

Hennessy, J. L., Jouppi, N. P. (1991), "Computer Technology and Architecture: An Evolving Interaction," *Computer* **24** (9), 18–29.

McWilliams, G. (1992), "The Reward of Many RISCs," *Business Week* No. 3258, p. 84.

Money, S. A. (1990), *Microprocessor Data Book,* 2d ed., San Diego: Academic, Chap. 1.

Patterson, D. A., Séquin, C. H. (1980), "Design Considerations for Single-Chip Computers of the Future," *IEEE Trans. Comput.* **C-29,** 108–116.

Rao, G. V. (1978), *Microprocessors and Microcomputer Systems,* New York: Van Nostrand Reinhold, Chap. 4.

Slater, M. (1992), "Wave of High-End Processors Due," *Microproc. Rep.* **6** (2), 9–15.

Smith, B. R. (1981), *Computers,* London: Usborne.

Smith, R. (1992), "Intel to delay rollout of next-generation microprocessor," *San Jose Mercury News* 23 July, pp. 1F, 6F.

Stallings, W. (1990), *Computer Organization and Architecture: Principles of Structure and Function,* 2d ed., New York: Macmillan, Chap. 1.

Texas Instruments (1973), *The TTL Data Book for Design Engineers,* Dallas: Texas Instruments.

Tredennick, N. (1991), "MIPs and Sunset," *Microproc. Rep.* **5** (12), 12–17.

Further Reading

Anceau, F. (1986), *The Architecture of Microprocessors,* Workingham, England: Addison-Wesley.

Comerford, R. (1992), "How DEC developed Alpha," *IEEE Spectrum* **29** (7), 26–31.

Glasser, L. A., Dobberpuhl, D. W. (1985), *The Design and Analysis of VLSI Circuits,* Reading, MA: Addison-Wesley.

Gupta, A. (1987), *Advanced Microprocessors, II,* New York: IEEE.

Linderholm, O., Malloy, R., Reinhardt, A., Sheldon, J. M. (1991), "A Talk with Intel," *Byte* **16** (6), 131–140.

Mirapuri, S., Woodacre, M., Vasseghi, N. (1992), "The MIPS R4000 Processor," *IEEE Micro.* **12** (4), 10–22.

Money, S. A. (1990), *Microprocessor Data Book,* 2d ed., San Diego: Academic.

Myers, W. (1991), "The Drive to the Year 2000," *IEEE Micro.* **11** (1), 10–13, 68–74.

Oehler, R. R., Blasgen, M. W. (1991), "IBM RISC System/6000: Architecture and Performance," *IEEE Micro.* **11** (3), 14–17, 56–62.

Patterson, D. A., Hennessy, J. L. (1990), *Computer Architecture: A Quantitative Approach,* San Mateo, CA: Morgan Kaufmann.

Sachs, H. G., McGhan, H., Hanson, L. F., Brookwood, N. A. (1991), "Design and Implementation Tradeoffs in the Clipper C400 Architecture," *IEEE Micro.* **11** (3), 18–21, 74–80.

Slater, M. (1987), *Microprocessor-Based Design: A Comprehensive Guide to Effective Hardware Design,* Mountain View, CA: Mayfield.

Stallings, W. (1990), *Computer Organization and Architecture: Principles of Structure and Function,* 2nd ed., New York: Macmillan.

Stockley, C., Watts, L. (1983), *Usborne Guide to Computer Jargon,* London: Usborne.

Tatchell, J., Bennett, B. (1982), *Understanding the Micro,* London: Usborne.

Tredennick, N. (1987), *Microprocessor Logic Design,* Bedford, MA: Digital.

Williams, T. (1992), "Superscalar Sparc chips offer performance gains, compatibility," *Comput. Design,* July, 32–36.

MICROSCOPY, ACOUSTICAL

See SCANNING ACOUSTICAL MICROSCOPY

MICROSCOPY, ELECTRON

See ELECTRON MICROSCOPY

MICROSCOPY, FIELD-ION

See FIELD-ION MICROSCOPY

MICROSCOPY, OPTICAL

See OPTICAL MICROSCOPY

MICROSENSORS

SIMON MIDDELHOEK, *Department of Electrical Engineering, Delft University of Technology, Delft, The Netherlands*

3-527-28132-0/94/$5.00 + .50

INTRODUCTION

The rapid development of microelectronics over the last two decades has been celebrated in many places and on many occasions. There is general agreement that the phenomenal improvement in the performance/price ratio of microelectronic components has given impetus to what is commonly referred to as the "information society," in which not only materials and energy but also information and information-processing systems play a major role. A further penetration of microelectronic components into electronic and currently nonelectronic information processing systems can be foreseen. However, this penetration is being impeded by the lack of input transducers or sensors for the measurement of physical and chemical parameters whose performance/price ratio is comparable to that of microelectronic components (Middelhoek and Audet, 1989). One solution to this problem might be to apply the batch-fabrication technology used for the production of integrated circuits to the fabrication of sensors. The resulting sensors are called *microsensors*, since the technology by which they are made allows them to be quite small.

We begin by defining the role of sensors in measurement systems and then present some microsensors for a variety of physical and chemical parameters. We conclude with a discussion of the special processing sequences used to fabricate sensors and electronic circuits which interconnect sensors and microprocessors.

1. GENERAL PRINCIPLES

1.1 Measurement and Control Systems

The term "information-processing systems" encompasses a wide spectrum ranging from computers and word processors to oscilloscopes, door locks, clinical thermometers, satellites, automatic vending machines, and so on. All these systems process information in one way or another. An important subgroup is that of measurement and control systems. Like all information-processing systems, these systems can be represented by a sequence of three elements (Fig. 1). The first of these is the input transducer, more often called the *sensor*. In the input transducer, a physical or chemical quantity, such as light intensity, displacement, temperature, magnetic field, or *p*H value, is converted into an electrical or electronic signal. A great number of physical and chemical effects are available for this purpose. A sensor is distinguished by the fact that the energy carrying the signal is converted from a nonelectrical form into an electrical form.

In the second element—the signal processor—the electronic signal is, as the name implies, processed in some way. The processor can be as simple as an amplifier, but it can also consist of a microprocessor or even of the central processing unit of a large-scale computer. In contrast to sensors, in a processor the form of the energy carrying the signal is not converted into another form.

Last, in the output transducer the electrical signal is once again converted into a nonelectrical signal, which can be detected by at least one of the human senses. For example, in a display device the signal is carried by radiant energy and can be perceived by our eyes.

1.2 The Six Signal Domains

From a theoretical point of view, the process of information transduction in a sensor can be viewed as a coupling between the states of two systems (Duyn and Middelhoek, 1992). If a change of state in one system affects the state in the other system, we say that the two systems are coupled and that this coupling enables information to be exchanged between the systems. To devise a useful classification of sensors, it is necessary to examine the state variables used to describe a

FIG. 1. Functional block diagram of a measurement system.

system. Two types of state variables exist: extensive state variables, such as mass, charge, strain, entropy, and polarization, and intensive state variables, such as electric potential, stress, temperature, and electric and magnetic fields. If we consider the products of the matching intensive and extensive state variables, we can define six energy domains and, since signals are carried by energy, also six signal domains: radiant, mechanical, thermal, electrical, magnetic, and chemical.

Using the six energy domains, we can depict the signal transduction in a sensor by the general diagram shown in Fig. 2. Each circle represents a signal or an energy domain. A sensor is represented by an arrow from a nonelectrical signal domain to the electrical domain. For instance, a photodetector (vertical arrow) will convert a radiant signal into an electrical signal.

When the various physical and chemical effects in sensors are reviewed, it appears that some effects can be used to make sensors that do not require auxiliary energy sources for the signal conversion, while others only yield useful sensors when some form of energy is supplied. The first group, referred to as *self-generating sensors,* includes, for example, the solar cell based on the photovoltaic effect and the thermocouple based on the Seebeck effect. The second group, referred to as *modulating sensors,* includes the photoconductor based on the photoelectric effect and the pressure sensor based on the piezoresistive effect.

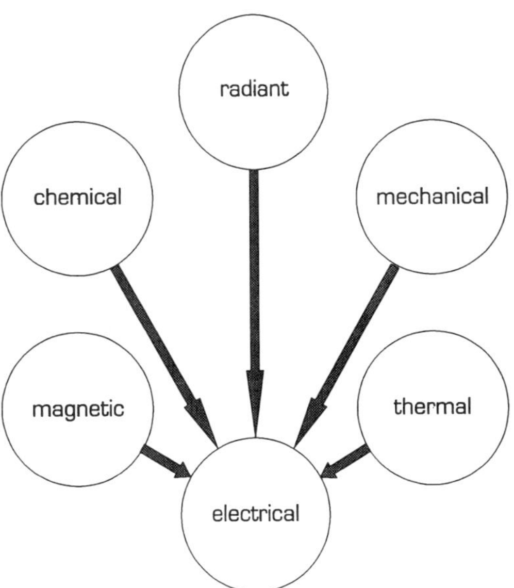

FIG. 2. Diagram indicating the five possible signal conversions in sensors.

1.3 Sensor Technologies

As already stated, the introduction of microelectronic components into traditionally nonelectrical information-processing systems could be more successful if more sensors with an acceptable performance/price ratio were available. Therefore, a global search for new and improved sensor technologies was initiated. These days, the main technologies for fabricating sensors are based on piezoelectric materials, such as quartz, ceramics, and polymers, on gas-sensitive metal oxides, on III-V and II-VI semiconductors, such as GaAs, GaP, AlSb, InSb, CdS, CdSe, ZnO, and ZnS, on thick- and thin-film materials, on optical glass fibers, and last but not least, on monocrystalline silicon. Progress in silicon planar technology has exceeded the most daring predictions. This has led to an increasing number of very sophisticated and yet cost-effective very large-scale integrated (VLSI) components, such as microprocessors and memories. Therefore, it made sense to apply this technology to sensors as well. As a result, the first silicon photodetectors were produced as early as 1960. Ten years later, silicon microsensors for the measurement of magnetic fields, pressure, and *p*H were presented. Scientific literature currently abounds with descriptions of a wide variety of silicon microsensors, many of which are commercially available. The attractiveness of using silicon technology for sensors lies in the fact that a large number of wafers, each containing hundreds of highly similar sensors, can be simultaneously processed as one batch, markedly improving the performance/price ratio. Moreover, advanced photolithographic techniques permit the fabrication of very fine patterns on a silicon substrate so that the size of silicon sensors can be very small; hence the name microsensors. Their small size makes it possible to use them for many applications, particularly in the medical field, where small size is crucial.

Silicon has a wide variety of physical and chemical effects that are suitable for the construction of sensors for many measurands. In many sensors, the silicon must be able to be machined into shapes such as thin membranes, cantilevers, and resonators. Several chemical, electrochemical, and dry plasma etching techniques have been developed to achieve this in silicon. These so-called micromachining techniques are discussed in Sec. 7. Fabricating sensors and electronic circuits with almost similar processing steps makes it possible to combine the sensors and circuits on one chip. Such sensors are called *smart sensors*, and they are discussed in the last section (Sec. 8).

2. MICROSENSORS FOR RADIANT SIGNALS

2.1 The Electromagnetic Spectrum

The radiant-signal domain involves the electromagnetic radiation of all frequencies as well as nuclear-particle radiation. Electromagnetic radiation includes microwaves, far- and near-infrared light, visible light, ultraviolet light, soft x rays, x rays, and gamma rays. The difference between the various types of electromagnetic radiation lies in their characteristic frequencies. Nuclear-particle radiation includes alpha particles, beta particles, neutrons, and minimum-ionizing particles, resulting from targets bombarded with charged particles in particle accelerators.

Thanks to research efforts dating back to the 1960s, numerous types of microsensors have been developed for both application fields. In this section, we briefly discuss the basic principles and the most important microsensors for radiant signals.

2.2 Basic Physical Principles

2.2.1 Low Photon Energies A wide range of interactions occurs between photons and silicon. Some interactions result in the conversion of the photons into an electrical charge (the photoelectric effect), while other interactions convert impinging photons into entities that have no net electric charge (phonons, excitons, and other photons). The only effects of interest here will be those involving the absorption of photons in silicon. During this absorption process, electron-hole pairs are freed. The pairs are then collected, in the process of which radiant energy is transformed into electrical energy. When the energy of an incident photon exceeds the band gap E_g of a semiconductor, the photon can be absorbed by exciting a valence electron into the conduction band. This band-to-band absorption results in the creation of an electron-hole pair. The free electron and hole then become excess carriers and are free to contribute to the conductivity of the material.

If the semiconductor is a direct band-gap material such as gallium arsenide, an incident-light photon can excite an electron directly from the valence band into the conduction band because the electron does not need to change its momentum. Thus, once the incident-photon energy equals the band-gap energy, the absorption coefficient stays virtually the same, even when the photon energy is increased further.

In order for an incident-light photon to excite a valence electron into the conduction band of an indirect band-gap semiconductor such as silicon or germanium, the electron must change its momentum. The probability of such an interaction taking place depends on the presence of a phonon with the required momentum. Because the phonon density increases as the momentum decreases, and as a relatively smaller change in momentum is required to generate electron-hole pairs of higher-energy photons, absorption increases with increasing photon energy. This effect can be put to great use in a color microsensor (Wolffenbuttel, 1987).

2.2.2 High Photon Energies Interactions of incident electromagnetic radiation at high photon energies, i.e. within the x-ray and gamma-ray regions, occur with atomic electrons. X rays and gamma rays are absorbed by three main processes, each characterized by different absorption coefficients: the photoelectric effect, the Compton effect, and pair production. The total absorption coefficient is the sum of these three. The process that takes place depends on the photon energy (Fig. 3). The photoelectric effect dominates at lower energies, from visible light up to about 0.1 MeV. If a photon interacts in a photoelectric process, it will transfer all of its energy to the interacting electron and disappear. The highest probability that this process will occur is with a *K*-shell electron. The excited electron, falling down to the conduction-band edge, will behave like a charged particle by creating other electron-hole pairs.

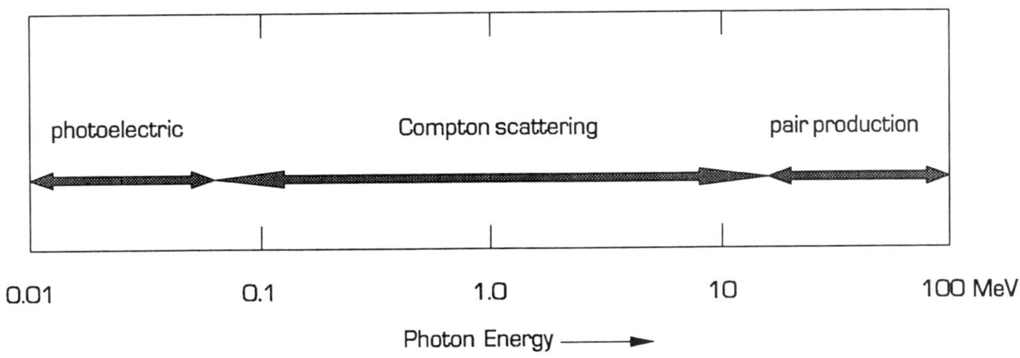

FIG. 3. Interaction processes in silicon with photons of high energy. The photoelectric effect dominates at lower energies, from visible light up to about 0.1 MeV. From 0.1 MeV up to 15 MeV the Compton process dominates, and above 15 MeV production of electron-positron pairs will dominate.

From 0.1 MeV up to 15 MeV, the Compton process dominates. In this type of interaction, only part of the photon energy is transferred to the electron. The original photon is deflected and decreased in energy. The deflected photon can then be reabsorbed by the photoelectric or Compton processes, or it can escape from the silicon.

Electron-positron pairs will start to be produced at 1.022 MeV and will dominate at energies above 15 MeV. In this interaction, the original photon will disappear while the excess energy will be imparted to the electron-positron pair. The electron and positron lose their energy through collisions and eventually come to rest. The positron and an electron from the crystal then annihilate. This interaction produces two gamma rays, each having an energy of 511 keV, which are emitted in opposite directions. The gamma rays can then be reabsorbed by the photoelectric or Compton processes, or they can escape from the silicon.

2.2.3 Interaction of Nuclear Particles with Silicon When alpha particles and other heavy charged particles enter a silicon crystal, they lose their energy through a succession of Coulomb interactions that occur between the particle's electromagnetic fields and the crystal's electrons. In each electronic interaction, a bound electron is freed and brought into the conduction band. The average ionization energy, i.e., the average energy necessary to create one electron-hole pair in silicon at room temperature, is 3.6 eV and is much smaller than the 30 eV needed by gaseous detectors. The ionization energy in silicon is constant for different types of radiation and for different energies. As the free electron in the conduction band will have an energy of only a few keV, it will lose this energy by creating secondary electron-hole pairs. The resulting ionization is distributed as a track, centered on the path of the particle. For heavy charged particles, the ionization track is comparatively straight, with only minor deflections from the original path direction, as these particles do not become deflected from their track by collisions with crystal electrons (Fig. 4).

The interactions of beta particles and other light charged particles are similar to those previously mentioned for heavy charged particles. Despite their physical differences, all high-energy nuclear particles tend for the most part to interact with silicon through Coulomb interactions, producing comparatively straight ionization tracks. They are therefore known as minimum-ionizing particles and produce approximately 85 electron-hole pairs per micron in silicon. Several silicon-based detectors for nuclear particles have been constructed. The best known of these are the silicon-microstrip detector and the silicon drift chamber. These detectors can be used to determine both the energy and the position of incident nuclear particles (Ten Kate, 1986).

Neutron detectors cannot be made with silicon as the interaction of neutrons with silicon is nil.

2.3 Microsensors for Visible Light

2.3.1 Photoconductors Photoconductors or photoresistors can be made in silicon by creating *n*-type or *p*-type areas in the usual

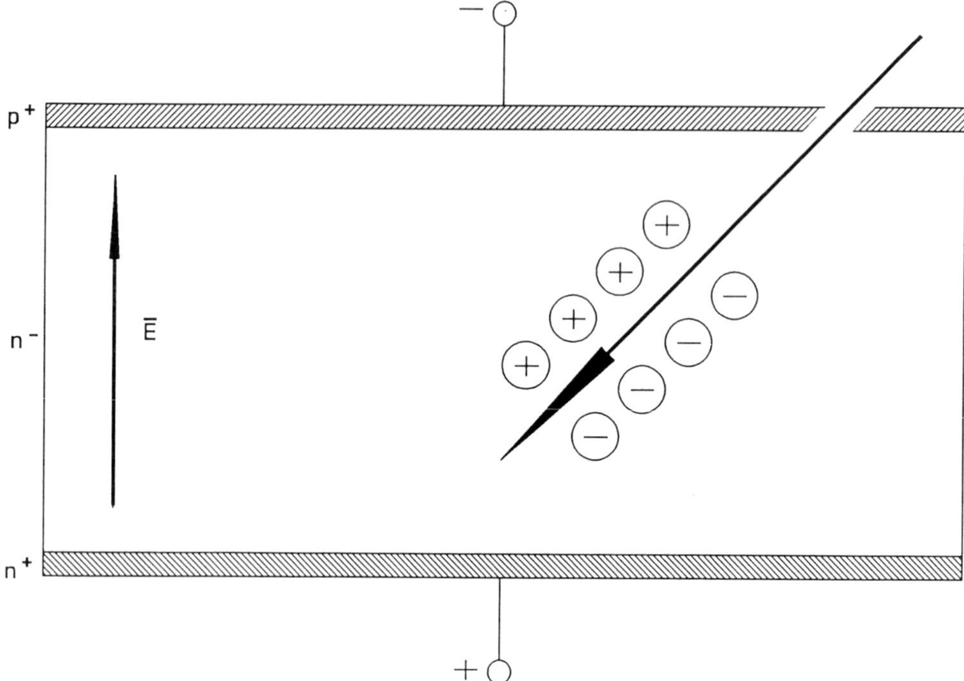

FIG. 4. Schematic of a silicon nuclear-particle detector.

way. Incident radiation creates electron-hole pairs, causing an increase in conductivity. To obtain a high sensitivity and a low switching time, the lifetime of the charge carriers must be large with respect to the transit time, i.e., the time required for the charge carriers to move between the electrodes of the photoconductor.

2.3.2 Photodiodes Silicon photodiodes can be used either without a bias voltage or with a reverse bias voltage. When a reverse bias voltage is applied, incident light creates electron-hole pairs in and around the depletion region, and these pairs are swept across the junction (Fig. 5). This creates a photocurrent, which is a measure of the light intensity. Because of the wavelength dependence of the absorption coefficient in silicon, radiation with a short wavelength (blue) is absorbed relatively close to the silicon surface, while radiation with a longer wavelength (red) penetrates deeper into the silicon. In order to collect the electrons and holes generated by the red light, the reverse bias voltage must be taken larger than for blue light. Measuring the photocurrents at three or more different bias voltages allows us to extract information about the spectrum and color of the incident light (Wolffenbuttel, 1987).

A photodiode can also make use of the photovoltaic effect. Here as well, the electric field in the depletion layer will, without a bias voltage, separate the charge carriers generated by the incident light. The n-type layer will become negative with respect to the p-type layer, and the voltage will be a measure of the incident light.

2.3.3 Position-Sensitive Photodetectors (PSDs) A special kind of photodiode is the position-sensitive photodetector. If the illumination of a p-n junction is nonuniform, i.e., produced by a light spot much smaller than the photodiode surface, a lateral photovoltage is established. By placing a pair of Ohmic contacts on opposite edges of an n-type layer of the junction, we can fabricate a device and with it measure the lateral photovoltage $V(x)$ as a function of the location of the light spot. $V(x)$ will vary in magnitude and polarity as the light spot is moved between the two contacts. The x coordinate of the light spot can be obtained from this voltage. If two additional contacts in the y direction are made to

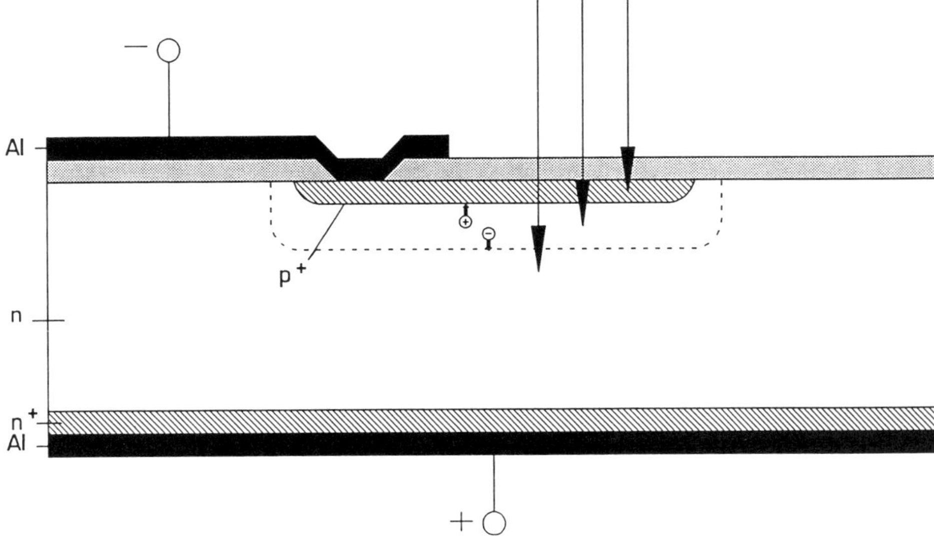

FIG. 5. Cross section of a photodiode. The depletion region is indicated with the dotted line.

the *p*-type layer, both the *x* and *y* coordinates can be obtained from voltages $V(x)$ and $V(y)$. Devices with a resolution of better than 1 μm have been fabricated.

2.3.4 Charge-Coupled Devices (CCDs) A very interesting and commercially valuable microsensor is the charge-coupled device (*q.v.*). CCDs are currently being applied in solid-state TV cameras. When a MOS (metal–silicon dioxide–silicon) capacitor is exposed to light, charge carriers are generated and are stored on the capacitor. Placing many such MOS capacitors side by side makes it possible to transfer this charge between the capacitors. The basic structure of a CCD consists of an array of electrodes placed on the silicon-dioxide layer. Clock voltages can be applied to these interconnected electrodes to allow for the transfer of stored charges. The CCD structures require a three-phase clock system to ensure one-directional charge movement. Three MOS capacitors are thus required for each elemental cell. A schematic of a basic three-phase CCD is shown in Fig. 6. Different configurations have been introduced to form two-dimensional CCD arrays. One approach uses a two-dimensional CCD array with an image and a storage section. During a certain time period, the image section of the CCD is exposed to the light, and charge is formed in this section. These charge packets are then transferred to the storage section and are read out one line at a time (Collet, 1986). The fabrication of CCDs is extremely difficult, as an acceptable image sensor has to contain at least 1 million functioning pixels.

3. MICROSENSORS FOR MECHANICAL SIGNALS

3.1 Basic Physical Principles

Microsensors for mechanical signals are used to measure

1. motion-related measurands, such as position, displacement, surface roughness, velocity, flow, and speed of rotation; and
2. force-related measurands, such as weight, pressure, acceleration, torque, strain, attitude, and vibration.

Most solid-state sensors for mechanical signals are based on the piezoresistive effect. The change in resistance of a metallic conductor when subjected to a mechanical strain was first reported by Lord Kelvin in 1856. As the change is caused by changes in the geometry of the resistors, it is referred to as the *geometric piezoresistive effect*. This effect has been used to produce strain gauges that are widely used in industry today. Since the piezoresistive effect in metals is rather small,

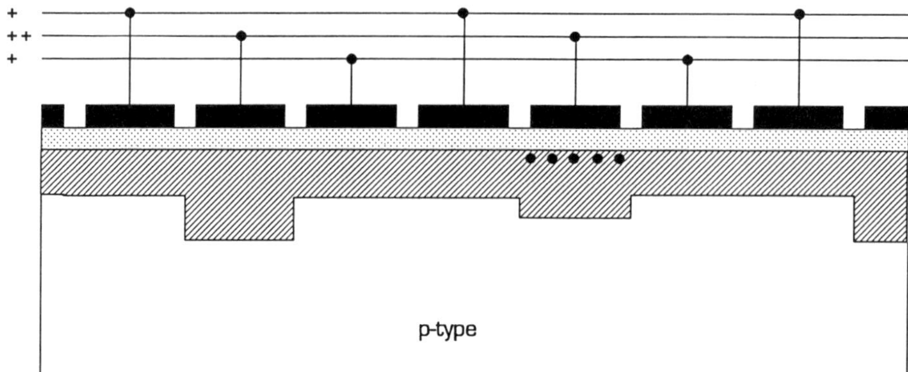

FIG. 6. Schematic of a three-phase MOS *n*-channel CCD.

researchers turned their attention to semiconductors.

Smith (1954) reported experiments indicating that the effect in silicon was about a hundred times larger than in metallic conductors. Smith showed that *p*-type silicon has a very large gauge factor of up to 175 and that *n*-type silicon has a large negative gauge factor of −135. The gauge factor of a resistance strain gauge is defined as the relative change of the resistance divided by the relative change of the length of the resistance.

To understand the piezoresistive effect in *n*-type silicon, we have to study how the band structure is affected by the application of a stress. The forbidden energy band in silicon is the result of the interaction between moving electrons and the periodic lattice. When an anisotropic stress is applied, the lattice spacing increases in one direction, whereas it decreases in the directions perpendicular to this direction. As we might expect, the interaction between the electrons and the lattice is also affected: The mobility of electrons moving in the [100] direction decreases when the crystal is compressed in this direction. The conductivity in this direction decreases as well. The relative change in the resistivity can be expressed by

$$d = dR/R = d\rho/\rho = \pi_{\parallel}\sigma_{\parallel} + \pi_{\perp}\sigma_{\perp} \quad (1)$$

in which $\sigma_{\parallel}$ is the stress parallel to and $\sigma_{\perp}$ the stress perpendicular to the current direction. Further, $\pi_{\parallel}$ and $\pi_{\perp}$ are the parallel and perpendicular piezoresistive coefficients. These coefficients depend very much on the orientation of the elongated resistor with respect to the crystal axes and on the type of silicon. For (100) planes, $\pi_{\parallel}$ and $\pi_{\perp}$ are virtually identical and are highly angle dependent. The (111) crystal plane is sometimes favored because in this plane the coefficients for longitudinal as well as transverse stresses do not depend on the crystal direction.

3.2 Diffused-Diaphragm Pressure Microsensors

Gieles (1969), working at the Philips Research Laboratories in The Netherlands, was the first to describe a miniature silicon diffused-diaphragm pressure sensor based on the piezoresistive effect. Many research laboratories around the world then set to work on this novel sensing device, drawn by the unique features offered by silicon pressure microsensors. The general structure of a pressure microsensor is shown in Fig. 7. It consists of a housing with one or two pressure ports. At the heart of the pressure sensor is a silicon chip etched to form a thin diaphragm. The diaphragms can be fabricated by spark erosion or chemical etching techniques. The piezoresistors are diffused or ion-implanted on the upper surface of the diaphragm. When the pressures on the two sides of the diaphragm are unequal, the diaphragm will bend. This will cause a change in the piezoresistors which can easily be detected.

A wide variety of configurations and positions of the resistors on the diaphragm can be found in the literature. There are four different types of sensors: absolute, sealed, gauge, and differential pressure sensors. Absolute pressure sensors feature the right pressure

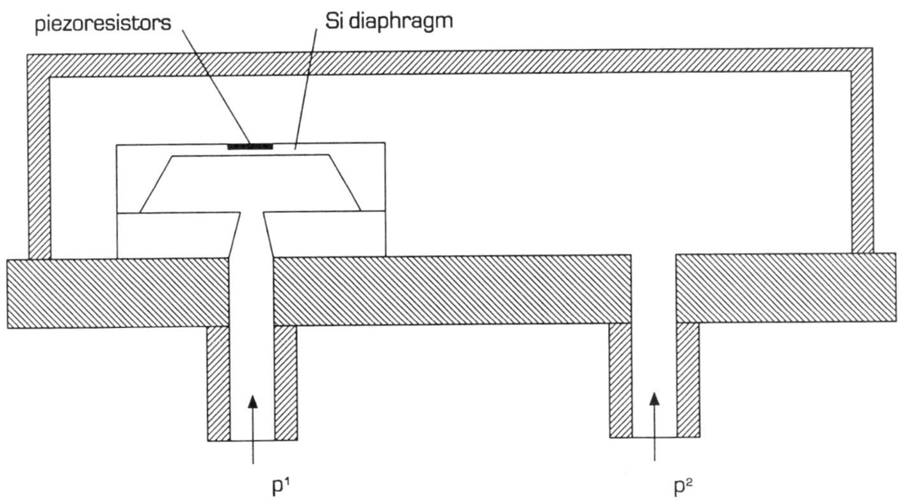

FIG. 7. Schematic drawing of a silicon-diffused diaphragm pressure microsensor.

port only, which connects to the upper surface. The sensor cavity is sealed in a vacuum and is the reference at the bottom of the diaphragm. The upper surface of the diaphragm might be covered by a silicone gel or other material to protect the surface from humidity or corrosive fluids. Differential pressure sensors have two pressure inlet ports. Pressure sensors with all kinds of operating pressure ranges can be obtained by varying the diameter and thickness of the diaphragms. While very sensitive pressure sensors with a range from 0 to 10 kPa exist, sensors with a range from 0 to 200 MPa are also available. Bridge voltages usually lie between 5 and 10 V, and the sensitivity of silicon pressure sensors varies from 10 mV/kPa for low-pressure devices to 0.001 mV/kPa for high-pressure devices.

Today, silicon pressure sensors are used in a large number of applications. They are especially successful in the medical and automotive markets. Silicon pressure sensors can be found in altimeters, blood-pressure gauges, tire-leak detectors, pushbuttons, microphones, and level detectors.

3.3 Capacitive Pressure Sensors

The pressure sensors presented in the preceding section are based on the piezoresistivity effect in silicon. Piezoresistive pressure sensors use a deformable diaphragm in which the strain-sensitive components are diffused or implanted. However, a metallized silicon diaphragm can also form an electrical capacitance with, for instance, an overlying metallized glass plate (Ko, 1986). When the diaphragm bends, because of a pressure change, this capacitance will also change. This capacitance change can be measured. The measurements show that capacitive pressure sensors are much more sensitive than piezoresistive sensors. However, in capacitive sensors the leads connecting the pressure sensor to the outside tend to show stray capacitances that are of the same order as the sensor capacitance. This makes it necessary to include electronic circuitry on the sensor chip in order to convert the capacitance change into a reliable electrical signal.

3.4 Accelerometers

With the help of micromachining, it is also possible to make silicon accelerometers. Roylance and Angell (1979) were the first to present such a device. Their accelerometer, which still forms the basis for many present-day devices, consists of a very thin 15-μm cantilever containing a diffused piezoresistor. A 200-μm-thick silicon seismic mass is formed at the end of the cantilever, and this serves as the inertial reference (Fig. 8). A 200-μm-thick silicon supporting rim surrounds the beam and the mass. When the device is accelerated, the seismic mass will change its position with respect to the surroundings and

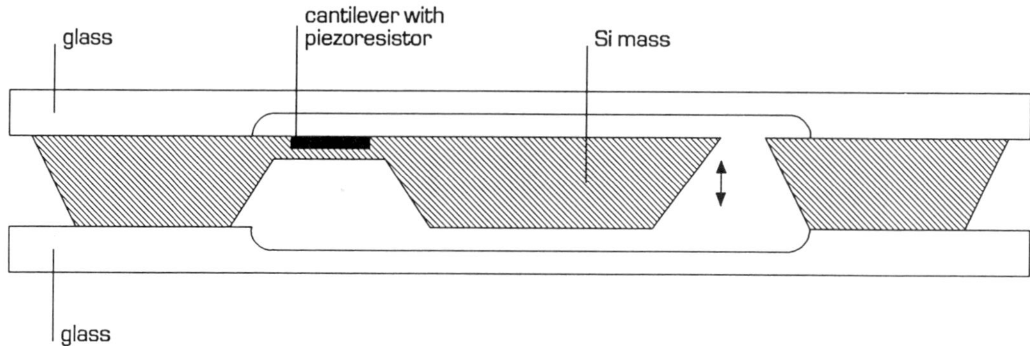

FIG. 8. Cross section of miniature piezoresistive accelerometer.

cause the cantilever to bend. The piezoresistor in the cantilever will change its value by an amount that can easily be detected. A device of this sort is capable of measuring accelerations from 0.01*g* to 100*g*.

4. MICROSENSORS FOR THERMAL SIGNALS

4.1 Basic Physical Principles

Since in most applications of conventional electronic circuits the temperature sensitivity of the components, such as diodes and transistors, is undesirable, this sensitivity has been subject to extensive research. Not unexpectedly, this accumulated knowledge has also been useful for the design of silicon-based temperature sensors, where temperature sensitivity is a desirable property. As discussed in the Introduction, sensors can be divided into self-generating and modulating sensors. The self-generating sensors require no source of power other than the signal being measured. Temperature sensors based on thermocouples belong to this category. Modulating sensors are temperature sensors based on the thermal modulation of electric currents or voltages supplied by auxiliary energy sources, for example resistors, diodes, and transistors. In the following sections, we first discuss self-generating temperature sensors based on the thermoelectric effect and then modulating sensors based on resistors, diodes, and transistors.

4.2 Thermocouple-Based Microsensors

Three thermoelectric effects are known to exist: the Seebeck effect, the Peltier effect, and the Thomson effect. The Seebeck effect is the only one of the three that can be used to design microsensors for the measurement of temperature. The Seebeck effect describes the experimental observation that when two conductors made of different materials are joined together at one point and a temperature difference is maintained between the joined and the nonjoined parts, an open-circuit voltage will develop between the nonjoined parts. For small temperature differences, the voltage is proportional to the temperature difference and is strongly dependent on the materials that are used to form the junctions. When the temperature difference between the hot junction (T_2) and the cold junction (T_1) is given by ΔT, the voltage ΔV can be expressed by

$$\Delta V = \alpha_{ab} \Delta T, \qquad (2)$$

where α_{ab} is defined as the relative Seebeck coefficient between materials *A* and *B*. Many phenomena contribute to the Seebeck effect in silicon, such as the change in the Fermi level, band-gap changes, charge-carrier-concentration gradient changes, diffusion-coefficient changes, thermodiffusion, and phonon drag. The Seebeck effect in silicon still requires further theoretical and experimental study in order to explain all observed findings.

The Seebeck coefficients can be measured by using structures as shown in Fig. 9. *p*-type strips are made by ion implantation or dif-

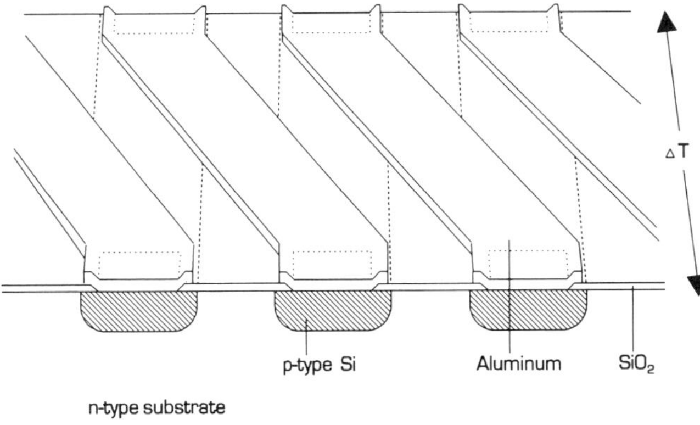

FIG. 9. Schematic drawing of a silicon/aluminum thermopile.

fusion in an n-type epitaxial layer. Many strips can be fabricated simultaneously and connected by means of evaporated aluminum strips, forming a thermopile. Such thermopiles have been shown to have a very high sensitivity, and Seebeck coefficients of 1 mV/K per couple can be achieved. The largest Seebeck coefficients are obtained for small impurity concentrations and thus large resistivities. Silicon thermopiles can be used for the design of quite a number of tandem sensors, such as thermal gas flow, vacuum, infrared, and acceleration sensors (van Herwaarden and Sarro, 1986). By way of example, Fig. 10 depicts a thermopile-based infrared microsensor. An infrared sensor consists of a thermopile located in a thin etched silicon beam. An infrared absorbing layer is placed at the end of the thermopile. The absorbed infrared radiation generates heat, which flows through the beam to the rim where the beam connects to the rest of the silicon chip. The output voltage produced by the thermopile is a direct measure of the infrared radiation. Infrared sensors with a responsivity of 10 V/W have been made.

4.3 Silicon Thermoresistors

The dependence of the resistivity of metals on temperature is fairly predictable, so that the temperature can be determined by measuring the resistance. Microsensors for the measurement of temperature can easily be obtained by depositing metal resistors on top of an oxide-covered silicon substrate. In general, the resistivity of a metal increases as the temperature increases. The temperature coefficient of the resistivity in silicon itself displays a more complex pattern than in

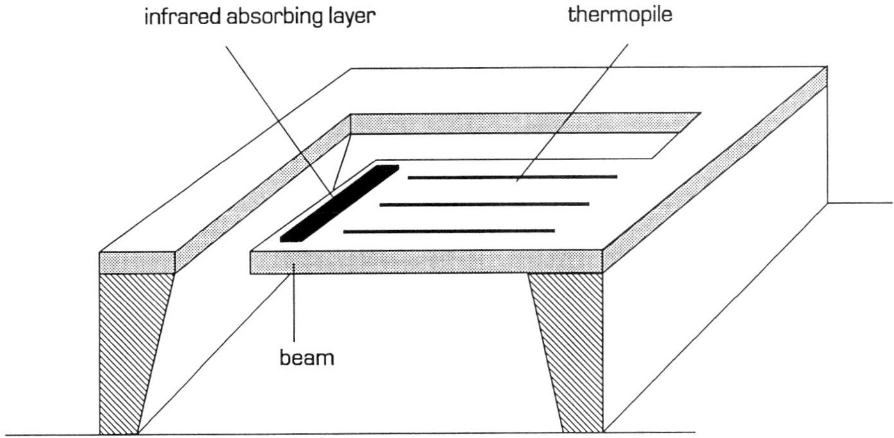

FIG. 10. Cross section of a thermopile-based infrared microsensor.

metals and depends on both the impurity concentration and the temperature. Around room temperature, the temperature coefficient of a silicon resistor is positive for low-doped silicon and constant for high-doped silicon. For temperatures above 200 °C, silicon shows a negative temperature coefficient.

Silicon thermoresistors can be made by diffusing meander-shaped *n*-type resistors in a *p*-type substrate. The *p-n* junction isolates the resistor from the substrate. In thermoresistors processed in this way, there is a large spread in nominal value and temperature coefficient, because both the *p-n* junction and the leakage current are also temperature-dependent. To obtain better silicon resistors, researchers set out to find silicon structures that are not dependent on *p-n* junctions for isolation. The so-called spreading resistance temperature sensor was the result of their efforts. A common technique for measuring the resistivity of a semiconductor wafer is to press a needle on the surface of the wafer. When the diameter of the needle is much smaller than the thickness of the wafer, the resistance R appears to be

$$R = \rho/2d, \tag{3}$$

in which d is the diameter of the needle and ρ is the resistivity of the material. Thus the needle-silicon surface combination provides us with a resistor that depends only on the resistivity of the silicon and the diameter of the needle and not on the other dimensions of the silicon crystal or on a *p-n* junction. This makes it feasible to design a thermoresistor based on this structure and on the temperature sensitivity of the resistivity. In such a thermoresistor, the needle contact is replaced by a small diffused contact. In Fig. 11, a device of this type consists of a small crystal (0.5×0.5 mm^2) with one large flat Ohmic contact on the bottom and a point contact with a diameter of 20 μm on the upper surface. For better contact, n^+ layers are diffused at both the bottom and the point contact so that an n^+nn^+ structure results. A spreading resistance contact can be applied to obtain an even more interesting device that can measure temperatures up to 350 °C.

In order to explain this, it is worth while to consider the product of hole and electron densities in a semiconductor, which can be written as

$$n_n n_p = n_i^2, \tag{4}$$

in which n_n is the electron density, n_p is the hole density, and n_i is the intrinsic carrier density. The variable n_i depends only on the temperature. This means that almost no holes will occur in the layers. When the temperature is increased beyond 200 °C, electrons from the valence band will be thermally excited to the conduction band, and the number of charge carriers will increase markedly so that the resistivity will start to decrease rapidly. When the large bottom contact is made pos-

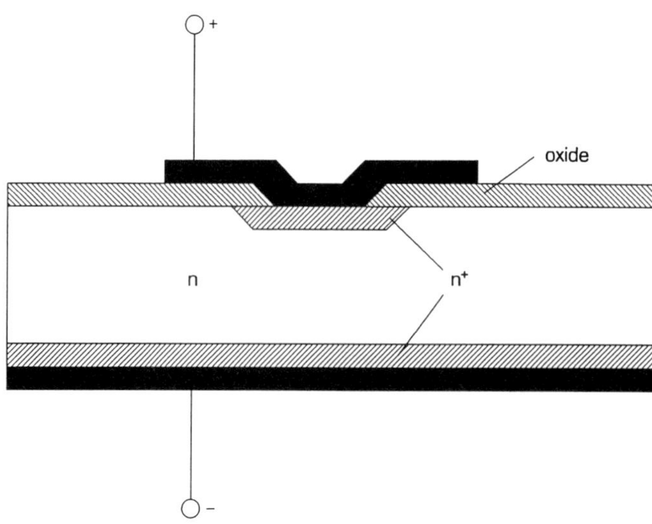

FIG. 11. Spreading resistance thermoresistor.

itive, this is indeed what is experimentally observed for any magnitude of current. However, when the spreading contact is made positive very different and interesting things start to happen: The resistivity continues to increase at higher temperatures, and the silicon shows extrinsic behavior. This unexpected feature of the spreading resistors has to do with the decreasing ability of the n^+ layer at the tiny spreading contact to supply sufficient holes to support the conduction process in the n substrate. This interesting effect has been used to good advantage to produce thermoresistors and other microsensors for temperatures up to 350 °C (Middelhoek and Audet, 1989).

4.4 Silicon Diodes and Transistors as Thermal Sensors

Diodes were already proposed many years ago as simple and low-cost devices for the measurement of temperature in the range between −50 °C and +150 °C. The characteristics of the temperature behavior of p-n junctions are well understood, and diodes are quite often used for simple temperature measurements. When a silicon diode is used as a thermal microsensor, the forward voltage is usually measured as a function of the temperature for a constant small forward current. In practice, measurements on many different diodes show temperature coefficients between 1 and 3 mV/K in the forward direction.

Not unexpectedly, the base-emitter junction of a transistor also shows a temperature sensitivity, which can be employed for temperature measurement. Transistors are generally much better temperature sensors than diodes because their bases are very thin and because they are three-port devices, which makes it much more convenient to incorporate them in electronic circuits.

The use of a transistor as a temperature sensor is usually based on the I_c-V_{be} characteristics. The emitter current I_e consists of a diffusion, a surface leakage, and a recombination component. Because of the thin base, in a transistor I_c consists mainly of the diffusion component, whereas the two other components of I_e are drained away by the base current. The expression for the dependence of V_{be} on the collector current I_c and the temperature T is

$$V_{be} = V_g(0) + (kT/q)\ln[I_c - \ln(KT^r/\eta)]. \qquad (5)$$

The values of $V_g(0)$ and the structure-dependent parameters K and r can be measured for each transistor. The factor η is the ionization factor, which is 1 above 20 K. It has been found that $V_g(0)$ varies between 1.12 and 1.19 eV and r between 3 and 5 for commercially available devices. According to Eq. (5), a transistor can be used as an electrical thermometer. However, in most cases a more sophisticated method is employed. By making use of only one transistor, the collector current can be varied between a fairly high value I_{c1} and a fairly low value I_{c2}, with corresponding base voltages V_{be1} and V_{be2} (Fig. 12). With the aid of Eq. (5) we then find for the voltage difference that

$$\Delta V_{be} = V_{be1} - V_{be2} = (kT/q)\ln(I_{c1}/I_{c2}). \qquad (6)$$

Differentiating this voltage difference with respect to temperature gives

$$\alpha = d(\Delta V_{be})/dT = (k/q)\ln(I_{c1}/I_{c2}). \qquad (7)$$

The result is interesting: The temperature coefficient α is a constant, and ΔV_{be} is exactly proportional to the absolute temperature T (Meijer, 1986). Further, both ΔV_{be} and α are

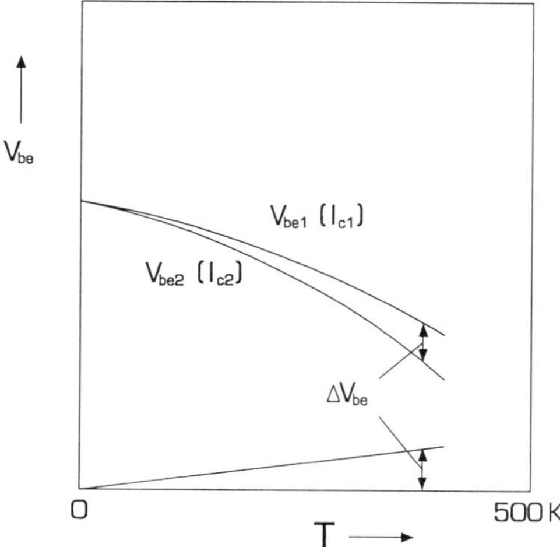

FIG. 12. Base-emitter voltage V_{be} as a function of the absolute temperature for two collector currents. The difference between the base-emitter voltages ΔV_{be} is exactly proportional to the absolute temperature.

independent of material or transistor parameters. Such sensors are known as PTAT (Proportional To Absolute Temperature) sensors. In PTAT sensors, only the current ratio I_{c1}/I_{c2} has to be known. Instead of using one transistor and two currents, it is also possible to use two neighboring transistors with unequal emitter surfaces at equal collector currents. To obtain a well-defined ratio between the emitter surfaces, it is advisable to design the large transistor in such a way that it consists of a multiple of N small transistors in parallel, where N is the desired ratio.

5. MICROSENSORS FOR MAGNETIC SIGNALS

5.1 Basic Physical Principles

In 1879 Edwin Hall, then a graduate student in physics at Johns Hopkins University, discovered, when working with gold foil, an effect that later would bear his name: the Hall effect. Hall found that when a transverse magnetic field is applied to a current-carrying conductor, an electric field results that is perpendicular to the current as well as to the magnetic field. The Hall voltage arises because, as was later shown, a charge carrier traveling in a magnetic field is subject to the so-called Lorentz force.

Since the Hall effect in semiconductors is rather high, semiconductors such as silicon, GaAs, and InSb have been studied extensively. The Hall effect can be employed for the construction of all kinds of practical devices, such as keyboards, magnetometers, magnetic recording heads, brushless electromotors, and displacement sensors.

In addition to the Hall effect, the so-called magnetoresistive effect can also be observed in metals, semiconductors, and magnetic materials. The resistivity of these materials changes when a magnetic field is applied.

As we wish to combine the microsensors for magnetic signals and electronic circuits on one chip, silicon, in particular, is of great interest as a semiconductor. Two kinds of silicon devices have been amply studied:

1. Hall plates, which are sensitive to a magnetic field perpendicular to the silicon surface; and
2. magnetotransistors, which are sensitive to a field parallel to the silicon surface.

Since the magnetoresistive effect in silicon is very small, practical silicon devices using this effect cannot be produced. In contrast, the magnetoresistive effect in magnetic layers is very large. Magnetic recording heads, consisting of silicon substrates on which magnetoresistive structures are fabricated by means of modern photolithographic techniques, are very useful for high-density recording equipment and are briefly discussed in the last section.

5.1.1 Hall Effect in Metals and Semiconductors The Hall effect in metals is very small, and only by using a very thin sample was Hall able to detect the Hall voltage. The Hall effect in semiconductors is much larger, because as we shall see later on, the Hall constant is inversely proportional to the number of charge carriers.

A Hall plate is typically a small, thin square of a metal or a semiconductor with length l, width w, and thickness t. When a current I passing through the plate is subjected to a magnetic flux density B_z perpendicular to the plane of the plate, a Hall voltage V_H occurs at the contacts, as shown in Fig. 13.

The Hall voltage V_H is proportional to the current and the magnetic field and can be expressed by (Popovic, 1991)

$$V_H = R_H I_x B_z / t, \tag{8}$$

in which R_H is the so-called Hall constant. The Hall voltage appears to be inversely proportional to the thickness of the plate t. For n-type silicon, the Hall constant is

$$R_H(\text{electrons}) = -1/nq. \tag{9}$$

The Hall constant is inversely proportional to the carrier density. This explains why the Hall effect is so much more pronounced in semiconductors than in metals. Holes also take part in the electrical conduction process in semiconductors. For p-type silicon, we find for the Hall constant

$$R_H(\text{holes}) = 1/pq \tag{10}$$

When the experimentally determined and theoretically calculated values of the Hall constant are compared, it appears that the theoretical values are a factor r smaller than the experimental values. This factor varies

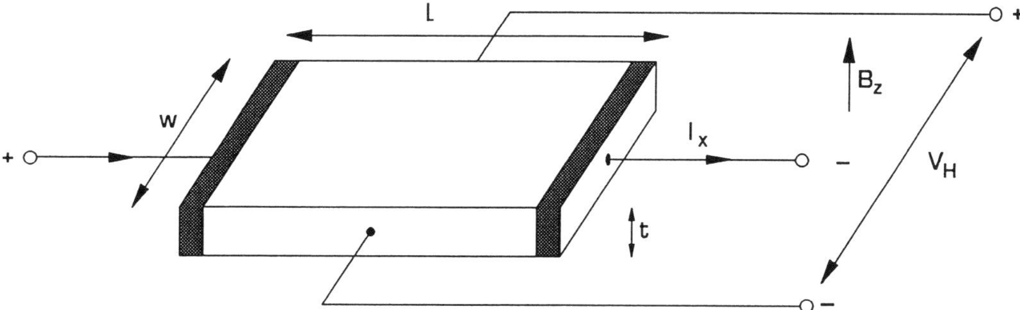

FIG. 13. Schematic representation of a Hall element.

between 1 and 2 and depends on the kind of semiconductor and the temperature. For the Hall constant, we obtain

$$R_H(\text{electrons}) = -r/nq \qquad (11)$$

and

$$R_H(\text{holes}) = +r/nq. \qquad (12)$$

The correction factor is necessary because in a semiconductor the charge carriers show a certain velocity distribution, and so the Lorentz force is not equal for all charge carriers.

5.1.2 Magnetoresistance Effects

5.1.2.1 Physical Magnetoresistance Effect. Not all charge carriers in a semiconductor have the same velocity. In the stationary state, this will cause some charge carriers to deflect upward and others to deflect downward. Consequently, the path of these charge carriers in crossing the Hall plate will be somewhat longer, which will slightly increase the resistance. This effect, the physical magnetoresistance effect, can be expressed as follows

$$\sigma(B) = \sigma(0)(1 - r^2\mu^2 B_z^2). \qquad (13)$$

The mobility μ in silicon is relatively low, and because the square of the mobility appears in the above equation, the physical magnetoresistive effect in silicon can be disregarded.

5.1.2.2 Geometrical Magnetoresistance Effect. A standard Hall plate consists of a rectangular semiconductor platelet with current contacts usually extending along the edges of the platelet in the y direction and pointlike Hall contacts halfway between the current contacts. When the length is much smaller than the width, the current conductors form a short-current path for the Hall voltage. In this case, the Hall voltage is reduced and does not fully balance the Lorentz force. As a result, the average path of the charge carriers deviates from the x direction, making the path longer and the resistance larger. This increase in resistance caused by the geometry of the Hall platelet is generally known as the geometrical resistance effect. For the resistance change, we get

$$R = R(0)(1 + \mu^2 B_z^2). \qquad (14)$$

In the geometrical magnetoresistive effect, the mobility again plays a very important role. Therefore, this effect is also very small in silicon and cannot be used for practical devices. The geometrical resistance effect is very large in certain high-mobility materials such as InSb. InSb might be of interest once it becomes possible to deposit this material on a silicon substrate. This would enable us to construct smart sensors consisting of magnetically sensitive InSb layers and electronic circuits on the same silicon substrate.

5.1.2.3 Magnetoresistivity in Magnetic Layers. In magnetic materials, the magnetoresistive effect is based on the interaction between the charge carriers and the magnetization of the material. This interaction is much stronger than the Lorentz force in silicon. When a small magnetic field is applied, the magnetization in suitable magnetic layers will rotate. This rotation of the magnetization affects the path of the charge carriers through the magnetic layer. Magnetic layers are characterized by a so-called uniaxial anisotropy. In the absence of a magnetic field, this anisotropy causes the magnetization to lie along the so-called easy axis. When a magnetic field is applied, the magnetization rotates in the plane of the film. When a current flows through the layer, an external magnetic field can thus rather easily change the angle between the current and the magnetization of the film, causing a change in the film's resistance.

The resistance of the magnetic film can be expressed by

$$R(\Theta) = R(0) - [R(0) - R(90)] \sin^2\Theta, \qquad (15)$$

in which $R(0)$ is the resistance when the current and the magnetization are parallel, $R(90)$ is the resistance when the magnetization is perpendicular to the current direction, and Θ is the angle between current and magnetization. In most cases, NiFe or NiCo films are used for the magnetic layers, as these films can be made with a small magnetostriction and a small crystal anisotropy. For fields of the order of several mT, relative resistance changes of a few percent can be achieved by using these materials.

5.2 Silicon Hall Plates

Smart magnetic microsensors can be constructed by integrating Hall plates and readout electronic circuits on one silicon substrate. Standard silicon bipolar technology can be used to fabricate the Hall devices. Though the characteristics of Hall plates can be improved by carefully optimizing all the processing steps and impurity concentrations, this is rarely done. To obtain low-cost devices, it is much better to use the standard processing steps used for the production of integrated circuits, omitting only those steps that are not needed for Hall plates. This is illustrated in Fig. 14, showing the cross section of a standard bipolar *npn* transistor with a buried layer and a Hall plate. The Hall plate is formed by an *n*-type epitaxial layer (e.g., 10 μm, 2 Ω cm), which constitutes the collector area in an *npn* transistor. This layer is deposited on top of a *p*-type substrate. The high-concentration *n*-type diffusion that is used for the emitter and collector contacts is used to make the current and Hall contacts in the Hall plate. The *n*-type buried layer normally used to reduce collector resistance in an *npn* transistor is not needed for Hall plates. The same applies to the *p*-type base diffusion. The collector layer is quite suitable for Hall plates because it offers a low carrier concentration and because, being *n*-type, it offers the mobility of electrons, which is higher than that of holes. The reverse-biased *pn* junction isolates the Hall plate from the neighboring devices and the substrate.

5.3 Magnetotransistors

The path of the charge carriers in a transistor can be influenced by a magnetic field. This fact is used in the construction of mag-

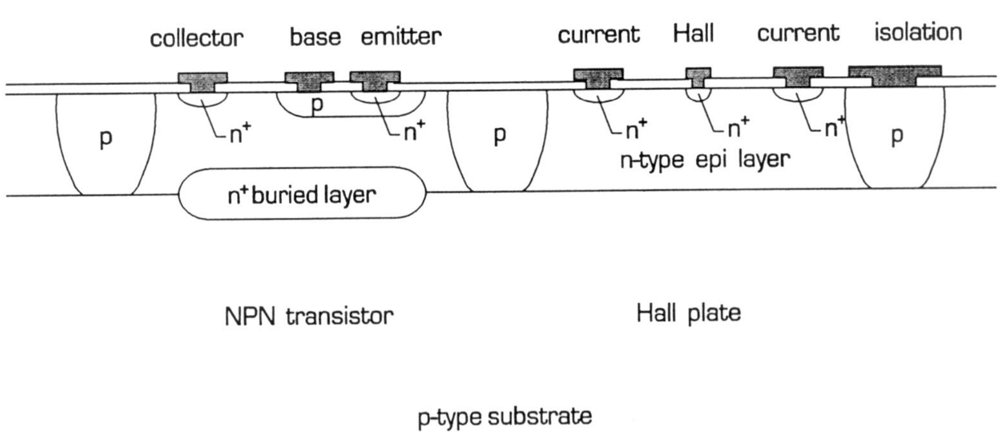

FIG. 14. Cross section of a bipolar *npn* transistor and a Hall plate.

netotransistors (Zieren and Middelhoek, 1982). Most devices consist of a single emitter and more than one collector contact to pick up the current. Figure 15 depicts a so-called vertical dual collector magnetotransistor. It consists of two *pnp* transistors with a common emitter and base. The collector area reaches beneath the emitter base structure by means of buried layers. There has to be a gap between the buried layers to avoid short-circuiting the collector contacts. Because of the rather thin base and the large surface of the base-emitter junction, the current is injected mainly in a direction normal to the substrate. A magnetic field parallel to the elongated transistor causes the electrons to deflect toward one of the collectors. The collector-current difference is proportional to the magnitude of the magnetic field. When the field is not parallel to the transistor, the current difference is proportional to the component of the field that is parallel to the transistor. For reasons of symmetry, an in-plane magnetic field perpendicular to the transistor does not cause a collector-current difference. Magnetotransistor-based devices have been developed that can simultaneously measure the magnitude and the direction in space of a magnetic field.

5.4 Magnetic Recording Heads

Over the last decade, the bit density of disk memories has increased remarkably thanks in large measure to improved read/write heads. The areal density has, at present, reached about 10^6 bits/cm^2. This increase was made possible when the inductive sensing of magnetic transitions gave way to direct field measurements by means of reading heads based on magnetoresistivity in Ni-Fe films. When a magnetoresistive head is brought into proximity to a magnetic disk or tape, the external magnetic field of the magnetic surface rotates the magnetization in the magnetic layer in the head. The resulting change in the resistance is measured and becomes an image of the stored signal on the tape or disk. The S/N (signal-to-noise) ratio of magnetoresistive heads has been shown to be superior to the S/N ratio of silicon Hall plates, and they also allow the required track and bit densities. Another advantage magnetoresistive heads have over inductive heads is that they are sensitive to a magnetic field and not to its time derivative, as would be the case in conventional inductive recording. Therefore, magnetoresistive heads are also suitable for the readout of magnetic tracks on credit cards and the like. Magnetoresistive heads consist of a combination of magnetic Ni-Fe layers and thin-film coils and are made by delicate deposition processes borrowed from semiconductor technology.

6. MICROSENSORS FOR CHEMICAL SIGNALS

Since time immemorial, mankind has taken a vivid interest in the composition of the materials used to make utensils or pre-

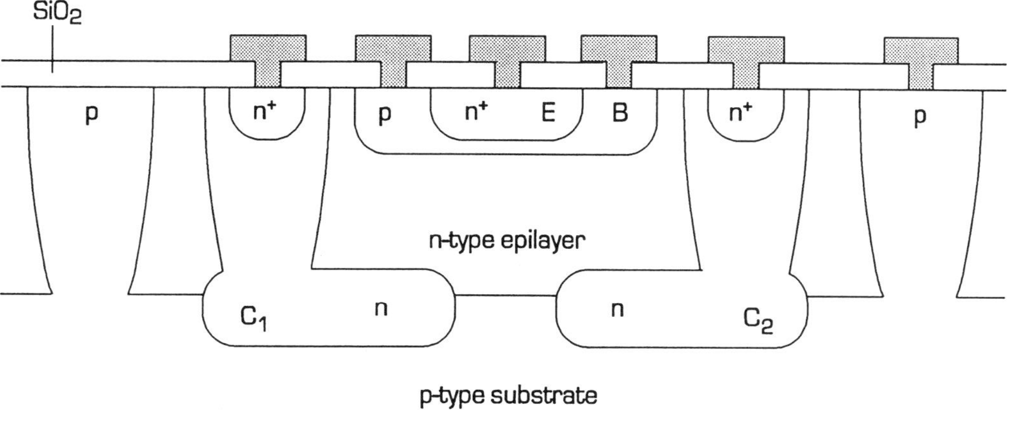

FIG. 15. Cross section of a dual-collector magnetotransistor.

pare food. Over the years, this need developed into the discipline of analytical chemistry. Today, thanks to a multitude of modern, often very expensive, analytical equipment and devices, very minute traces of practically any atom, molecule, or ion can be detected. However, analytical equipment is often bulky and can only be operated by trained personnel in suitable laboratory environments. Moreover, most measurements are time consuming. Given our increasing interest in our environment, our need for the on-line control of gas mixtures in industrial systems, our desire to monitor, for instance, oxygen in the blood during surgery, our need for reliable fire alarms, and our desire to optimize the performance of car engines, many scientists have been motivated to investigate the extent to which it is possible to make microsensors for chemical parameters. In the last few decades, many sensors for the detection of atoms, molecules, and ions have been presented in the literature. These sensors are based on materials such as metal oxides, organometallic compounds, ceramics, conducting polymers, and catalysts and are based on devices such as glass fibers, membranes, thermocouples, and ion-selective electrodes. In more recent years, silicon has also been proposed as a material for chemical microsensors.

6.1 Chemoresistors and Chemocapacitors

6.1.1 Chemoresistors Many chemical sensors depend on the measurement of the change in the conductivity or the dielectric constant of a chemical layer when subjected to a gas or an electrolyte. A very well-known metal oxide such as tin dioxide, which is an *n*-type semiconductor, changes its conductivity when heated and subjected to natural or town gas (methane). Standard metal-oxide gas sensors consist of a sintered block of the oxide that is heated to a few hundred degrees centigrade. Surrounding gases react with the oxygen on the hot surface, causing changes in the resistivity of the material. Properly doped gas sensors based on this principle and employing oxides of tin, zinc, iron, zirconium, or titanium have been shown to be sensitive to many gases. Metal-oxide gas sensors can be fabricated not only in thin- and thick-film technology but also in silicon technology (Demarne and Grisel, 1988). As is schematically shown in Fig. 16, this type of sensor consists of a silicon substrate on which a SiO_2 layer is grown thermally. A thin gold layer is deposited on top of this layer. This gold layer is patterned by the usual photolithographic methods in such a way that a meander-shaped resistance heater is obtained. Another SiO_2 layer is sputtered on top of the gold layer. This is followed by the sputtering and patterning of a tin-oxide layer to obtain a CO sensor. Last, to reduce the heat capacity of the structure, all silicon is removed from the back of the sensor by chemical etching. The resulting device is known as a chemoresistor.

6.1.2 Chemocapacitors The so-called chemocapacitor is based on the change that occurs in the dielectric constants of many organic materials when they are exposed to a gas. Interdigitated structures can be used to form a planar capacitor on top of a silicon substrate. By measuring the capacity, we can see whether the structure has been exposed to a gas. An accurate integrated dew-point humidity sensor was designed by using this type of chemocapacitor (Regtien, 1981). In this case, silicon technology makes batch production possible. Moreover, temperature-sensing elements and heating elements can easily be integrated on the same substrate. The real challenge is to find materials that are sensitive to only one chemical parameter and whose sensitivity does not change with time.

6.2 MOSFET-Based Chemical Microsensors

Microelectronics is dominated by two technologies: the bipolar technology and the metal-oxide-semiconductor field-effect transistor (MOSFET) technology. The MOSFET is based on the creation of conductive chan-

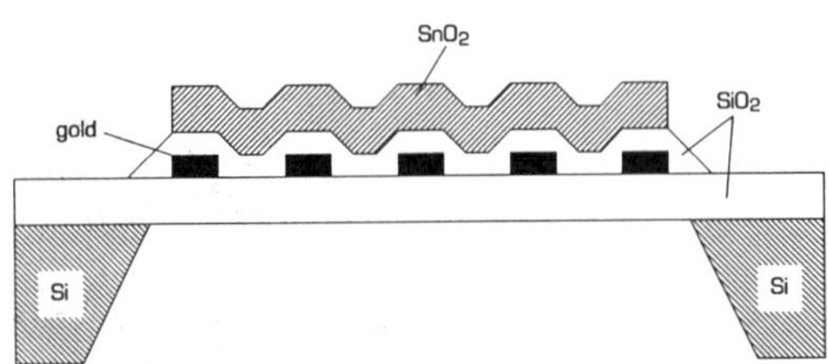

FIG. 16. Schematic representation of a silicon-based tin-oxide CO gas sensor.

nels along the surface of a silicon substrate and the modulation of the conductivity by properly applied electrical fields. MOSFET structures can also be used for the construction of chemical microsensors. A few decades ago, Bergveld (1970) proposed his so-called ion-sensitive field-effect transistor or ISFET. This is a MOSFET without a gate contact. When the ISFET is immersed in a solution containing ions, the drain current depends on the ion concentration. A few years later, Lundstrom *et al.* (1975) proposed a MOSFET with palladium as the gate metal for the detection of hydrogen. Since then, these devices have been studied extensively all over the world, with the result that many new microsensors have been found for a multitude of chemical parameters.

6.2.1 Ion-Sensitive Field-Effect Transistors (ISFETs) An ISFET, as is shown in Fig. 17, consists of a MOSFET in which the oxide layer is replaced by a combination of an oxide layer and a solution containing ions. The gate of an ISFET is positioned at a certain distance from the silicon substrate. The *p*H of an ionic solution can be determined by an ISFET. As standard *p*H-measurement cells are rather large and fragile, the invention of the ISFET in 1970 was greeted with avid interest, as it promised to allow the construction of elegant, miniature, and mass-produced *p*H-measurement cells. The first ISFETs used a thermally grown silicon oxide as the "ion-selective membrane." For calibration, the ISFET is brought into an electrolyte in which the *p*H can be varied. A frequently used electrolyte is a solution of NaCl in water. The *p*H

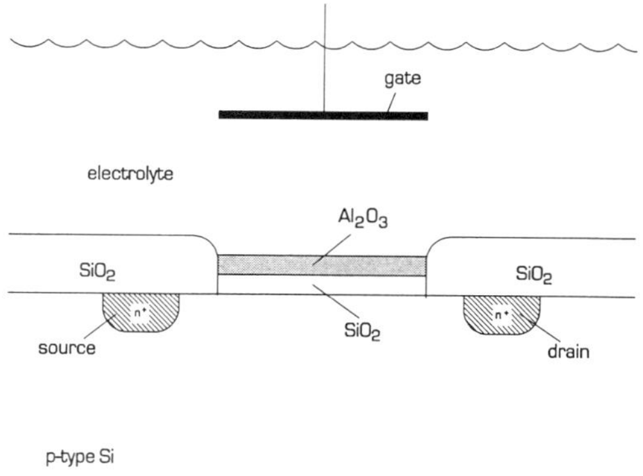

FIG. 17. Schematic cross section of an ISFET.

is increased by adding NaOH or decreased by adding HCl. The threshold-voltage shift is measured as a function of the *p*H for a constant drain current and voltage.

In the first decade following Bergveld's invention of the ISFET, the working mechanism of ISFETs was the focus of much discussion. It was initially assumed that the operation of ISFETs would be similar to that of a standard glass membrane cell and that hydrogen ions from the solution would diffuse through the thin SiO_2 layer to the oxide-silicon interface. In the case of the ISFET, they would change the interface charge and thus the threshold voltage of the transistor. Later on, it was shown that hydrogen diffusion is much too slow (seconds) to explain the experimentally observed quick response (milliseconds) of ISFETs. A new model, known as the *site binding model* in electrochemistry, was then applied to the ISFET. Today, almost all the features of ISFETs and related devices can be explained on the basis of this model. In the site-binding model, it is assumed that the *p*H response of an ISFET arises from surface reactions between the electrolyte and the oxide and not from hydrogen diffusion through the oxide. A gate oxide of SiO_2 can easily be obtained by thermal oxidation. However, other insulators such as Al_2O_3, Ta_3O_5, TiO_2, and Si_3N_5 have also been studied. It appears that Al_2O_3 is a much better material than SiO_2.

In most present-day experiments with ISFETs, SiO_2 has been replaced by a sandwich of SiO_2-Al_2O_3 layers. A disadvantage is that the technology of Al_2O_3 is much more complex, as these layers have to be made by chemical vapor deposition. ISFETs are mainly *p*H-sensitive devices. However, when the insulator is covered by a layer that allows some ion exchange to take place, the device can also be made sensitive to other ions. Such devices are known as CHEMFETs. When the gate area is covered with an enzyme, the resulting devices are known as *biosensors*.

6.2.2 Biosensors Many biological reactions must take place in the human body to keep the organism functioning. These reactions occur at a moderate temperature of 37 °C. They would be much too slow except that, fortunately, certain materials, known as *catalysts*, can increase the reaction speed. Biochemical catalysts, called *enzymes*, have complex molecular structures. Many thousands of enzymes have been identified to date. Enzymes also play an important role in the production of food, medicine, washing powder, etc. As living organisms depend on enzymes to survive, great interest has been shown in the possibility of controlling the enzyme concentration, especially when illness causes a deficiency. For example, insulin is an enzyme that plays an important role in diabetes. Diabetic patients have an excess of glucose, and this can be corrected by administering insulin. Of course, they would greatly benefit from a system that is capable of monitoring the blood glucose level and of automatically injecting the proper amounts of insulin and that is, moreover, small enough so that it can easily be carried around by the patient. Silicon-based chemical microsensors can be made very small. In addition, much of the electronic control circuitry can be placed on the sensor chip. As a result, these microsensors are understandably exciting great interest in the biomedical field.

A typical example of a biosensor is one that can measure the urea concentration in human blood, which gives an indication of the functioning of the kidneys. The urease enzyme catalyzes the reaction

$$(NH_2)_2CO + 2H_2O + H^+ \rightarrow 2NH_4^+ + HCO_3^-.$$

Since the pH of the solution is affected by the reaction, an ISFET can be used to measure the pH. Figure 18 depicts a schematic cross section of an ENFET (ENzyme FET). An ENFET is fabricated by depositing a SiO_2 and a Si_3N_4 layer on top of the n-channel structure of an ISFET by means of thermal oxidation and chemical vapor deposition (CVD). Next,

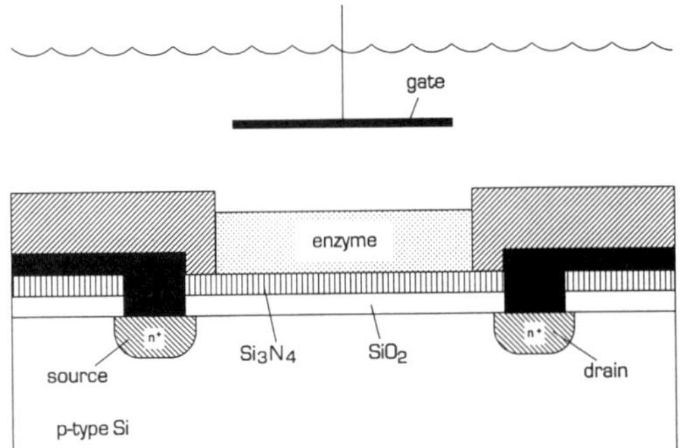

FIG. 18. Cross section of a biosensor.

in an elaborate process, a gel containing the urease enzyme is applied to the top of the structure. Then the ENFET is immersed in a solution that allows the urea content to be changed. The output signal of the ENFET is compared to that of a reference ISFET that lacks the enzyme cover. The result is a measure of the urea concentration (Miyahara *et al.*, 1983).

6.2.3 Pd-Gate MOSFET Gas Microsensors

When the aluminum gate of a MOSFET is replaced by a palladium gate, the device's I-V characteristics are hydrogen dependent. Lundstrom and co-workers (Lundstrom *et al.*, 1975) were the first to report on these types of devices. The metal palladium has unique properties with respect to hydrogen. Molecular hydrogen dissociates to atomic hydrogen on the surface of the palladium gate. The atomic hydrogen diffuses through the palladium layer and is adsorbed at the palladium–silicon dioxide interface in the form of dipoles. The adsorbed hydrogen atoms thus give rise to a change in the working mechanism of the palladium. This in turn changes the threshold voltage of the MOSFET. The change can be expected to depend on the hydrogen coverage of the interface, which in turn depends on the hydrogen pressure. The shift in the threshold voltage is equal to

$$\Delta V_t = \Delta V_{\max}/[1 + 1/C(P_{H_2})^{1/2}], \tag{16}$$

where $\Delta V_{\max}$ is the maximum observable voltage shift (about 0.5 V), C is a constant, and P_{H_2} is the partial pressure of hydrogen. In the palladium gate layer, at a certain hydrogen pressure an equilibrium occurs between the hydrogen molecules being dissociated into atoms and the process in which the atoms are recombined into hydrogen molecules. The actual dipole coverage of the interface layer is the result of the balance between these two processes. The most sensitive measurements are obtained when hydrogen is detected in an inert-gas atmosphere. Hydrogen concentrations down to 0.03 ppm can be detected.

Pd-gate devices are normally operated at temperatures from 60–150 °C, as the response time then assumes acceptable values of the order of some seconds. In a practical device, the Pd-gate MOSFET, a heating resistor, and a temperature sensor can be integrated on one silicon chip. There is still a lot of work that has to be done on the reproducibility, stability, poisoning, etc. of Pd-gate devices before a breakthrough in Pd-gate and related devices can be expected. Nevertheless, several possible applications have been studied, such as a simple hydrogen leak detector and a smoke detector.

7. PROCESSING OF MICROSENSORS

7.1 Silicon Planar Technology

Sensors can be fabricated in many technologies, including glass-fiber, metal-oxide, polymer, thin-film, thick-film, piezoelectric-material, and III-V and II-VI semiconductor technologies. However, only silicon-based sensor technologies have uniquely been able to progress by directly reaping the benefits of the expertise and skill gained in the development of the silicon integrated circuit and its fabrication technology. Silicon-based sensor technologies are of particular significance because they make it possible to produce smart sensors, in which the sensor and the signal-processing circuits are integrated on the same chip.

7.2 Micromachining Technologies

Micromachining pertains to the use of specialized fabrication techniques for the controlled, selective etching of silicon and films used in silicon-planar processing in order to obtain beams, diaphragms, valves, miniature electromotors, and many other structures. Early etching methods employed wet-chemical isotropic etchants. Anisotropic-etching technologies were later introduced to overcome the problems associated with precision, sensitivity, and temperature dependence.

There are two approaches to chemical etching: bulk micromachining and surface micromachining. In bulk micromachining, the silicon wafers are etched on both sides to achieve the desired structures. However, the most important advantage of silicon planar technology for the production of integrated circuits is that the circuits can be fully fabricated by processing only one side of a wafer. It therefore made sense to come up with methods in which the sensor and actuator structures could be obtained by processing only one surface of the wafer. The general term presently used for these methods is surface micromachining. In surface micromachining, selected materials are sequentially deposited and removed by selective etching procedures. Silicon etching has developed into an art with many followers. The achievements are very impressive and occasionally even artistic. Miniature electromotors having a diameter of less than 120 μm have attracted a great deal of attention and have shown how sophisticated structures can be made with present-day techniques (Muller, 1990).

7.2.1 Etching Solutions and Gases In wet-chemical isotropic etching solutions, usually consisting of a mixture of hydrofluoric, nitric, and acetic acids, the etch rate does not show a preference for any crystallographic orientation. Problems with isotropic etchants arise in the areas of etch control, selectability, and precision. The etch rate is agitation sensitive and temperature dependent and is limited by the transport of the etchant to the reacting surface. Narrow windows take longer to etch than wide windows because their access to fresh etchant is restricted. Anisotropic wet-chemical etchants differ from isotropic etchants in that they are orientation dependent. They are known to etch selectively the [100] and the [110] crystal orientations, leaving the [111] orientation relatively free from attack. Representative etch rates are typically 50 times slower in the [111] direction than in either the [100] or [110] direction. The use of anisotropic etchants with [110] silicon substrates results in openings with vertical sidewalls, while sidewalls set at an angle of 54.7° with respect to the surface are produced in [100] silicon. Lateral geometries and etch profiles can be controlled more precisely with anisotropic wet-chemical etchants than with isotropic ones.

The most commonly used wet-chemical anisotropic etchants include potassium hydroxide (KOH) and ethylenediamine–pyrocatechol–water (EDP). These etch solutions must be warmed to 85–115 °C to obtain reasonable etch rates, and the temperature must be kept very stable, since these etchants, like all other chemical etchants, are sensitive to temperature.

The dry etching in radio-frequency–generated plasmas of silicon, polysilicon, silicon dioxide, silicon nitride, resist, aluminum, and other films used in silicon planar technologies has become a well-accepted alternative to conventional wet-chemical techniques. Dry-etching techniques allow greater control over the etching procedure. Dry-plasma etching can be isotropic or anisotropic (i.e., respectively orientation-independent or -dependent) in nature. The directionality of the etching, the absolute etch rates, the etch-rate ratios, and the degree of radiation damage are all determined by procedural and instrumental parameters such as the composition, temperature, and flow rates of the reactant gases, the instrumental operating frequency, the pressure and the power density of the plasma, the voltage between the substrates and the plasma, and the wafer temperature.

Dry etching occurs by different mechanisms, which can mainly be broken down into ion-etching techniques and reactive-etching techniques. In ion-etching techniques, etching occurs primarily by physical means, such as ion bombardment. In reactive-etching techniques, a radio-frequency–generated plasma produces neutral atoms, neutral molecules, and radicals that react with the films to produce volatile compounds. Common gases used for silicon etching include Freons such as CF_4 and chlorine gases such as CCl_4 and Cl_2. Atomic fluorine and chlorine are produced in their respective plasmas, which react to produce the volatile compounds SiF_4 and $SiCl_4$, respectively. Silicon dioxide and silicon nitride are also etched in these plasmas.

7.2.2 Etching Procedures and Techniques

To fabricate precisely dimensioned micromechanical structures such as cantilever beams, diaphragms, and bridges, we have to be able to control the etching procedure so that it can be stopped completely at a predesignated point, thus providing dimension control on the order of 1 μm. The most common method of termination is predetermination of the etch rate followed by the timing of the etch procedure, but because of such factors as the sensitivity of the etch rate to agitation and temperature, as well as variations in the processing parameter and substrate thickness, this method of fabricating microstructures is unsatisfactory. Satisfactory etch-rate reduction methods that are currently employed include the boron etch-stop procedure and electrochemical etch (ECE) stops. In a number of applications, the boron etch-stop technique can be used when the anisotropic etchants EDP or KOH are used. When these anisotropic etchants are applied to the n regions of structures with p^+n junctions, the etch rate is significantly reduced at the interface between the n region and the heavily doped p^+ region (where the boron impurity is greater than approximately 5×10^{19} cm^{-3}). This procedure cannot be used in all applications. Considerable mechanical strain is introduced when the impurity level is so high, making it difficult to grow a high-quality epilayer on top of this p^+ layer. In addition, the high boron-doping level also prohibits the direct fabrication of microelectronics within this layer. Although anisotropic ECE etch-stop procedures are more difficult to use, they offer significant advantages in this area. The etching will terminate at the epitaxy/substrate interface of an n-type epitaxial layer grown on an n^+ or a p substrate with the standard doping level used in microelectronics fabrication.

The microelectronic or sensor devices can be fabricated in the epilayer by standard processing techniques followed by the use of the ECE etching procedure to define the microstructures. Figure 19 shows an experimental setup used for the electrochemical etching of silicon (Sarro and van Herwaarden, 1986). The hot-water control and the magnetic stirrer maintain the etchant at a selected temperature. An anode contact is made on the n-type epilayer, while a platinum cathode is installed within the solution. The etching of the p substrate begins as soon as contact is established within the etch solution between the anode and the n-type epilayer. The etching ends when the n epilayer is reached. The etching procedure is monitored by recording the cell current versus time. As the p substrate is being etched, the cell resistance will decrease. The cell current will simultaneously increase because the power supplied is kept constant. The termination of the etching procedure is then seen as a peak in the cell current.

The first well-established plasma-etching techniques were isotropic and nonuniform in nature, occurring primarily through the formation of volatile compounds and practiced in low-voltage barrel-type reactors. These in-

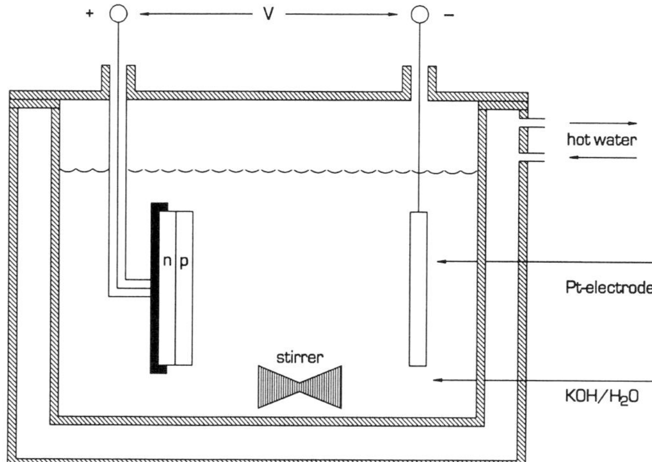

FIG. 19. Experimental setup for the electrochemical etching of silicon.

struments are still widely in use for applications not requiring a high resolution to etch the numerous films used in silicon planar technology (i.e., photoresist) with line widths on the order of a few microns. Etch ratios of 20:1 for $Si:SiO_2$ and 10:1 for $Si_3N_4:SiO_2$ are typical. Highly directional etching can be performed by reactive-ion etching in parallel-plate reactors. These systems are also primarily etched through the formation of volatile compounds, but ion bombardment also plays an important role. The reactors are similar to sputter-etching instruments. Vertical-etch rates that greatly exceed lateral-etch rates result with reactive-ion etching procedures, but the etch selectivity is reported to be poor because of the nonselective sputtering.

8. INTERFACES, SMART SENSORS, AND BUS SYSTEMS

In the past hundred years, myriads of sensors (as many as 50 000 according to a rough estimate) have been developed for the measurement of approximately 100 physical and chemical parameters. It might seem as if using silicon as the sensor material might merely add a few more sensors to the already existing mass of sensing devices. However, silicon offers something that other materials do not, namely, the possibility to integrate the sensor and part of the electronic processing circuit on one mass-produced silicon chip. These types of sensors are called *integrated smart sensors*. The on-chip integration may be the simple realization of a temperature-correction circuit, or in the future, a complete microprocessor might even be added to a sensing element. As we will elucidate below, smart sensors have an important role to play in future measurement and control systems.

Most present-day sensors are based on elements in which one of the parameters (resistivity, dielectric constant, resonance frequency, threshold voltage, etc.) shows a minute change in response to one or more measurands. This change is amplified and converted into a variety of electrical formats such as voltage, current, resistance, capacitance, frequency, and pulse duration. Moreover, the output signal will show a number of unwanted deficiencies such as offset, drift, nonlinearity, cross sensitivities, and poor interchangeability, to name a few. The multitude of output formats and the errors in the output signal make it very hard for system engineers to apply sensors unless they have an intimate knowledge of the workings of the sensing element itself.

The situation with respect to electronic components is even more favorable. Many standard components with well-defined input and output characteristics are available. For a trained system engineer, designing large systems based on these advanced components is not too difficult a task. As yet, this does not apply to sensors. Therefore, integrated smart sensors, which have a standard output and in which most of the deficiencies have been corrected by electronic circuits on the sensor chip, will have a significant impact on the development of measurement systems. In this last section, we delve into some of the issues concerning smart sensors.

8.1 Analog Signal Conversion

In most measurement and control systems, large numbers of sensors are connected to a central computer system. On the basis of the incoming information, the computer calculates relevant data about the process and, where necessary, sends data back along the bus system to actuators, such as valves or heaters, that adjust the process. One important feature in that process is the type of format the information should be in when it is sent along the bus data lines. Most physical measurands such as temperature, pressure, displacement, and so on, come to us in analog form. A simple measurement system may be based on locally amplifying the sensor signal to make it more resistant to interference and then sending it along the data lines in the form of an analog signal.

Various methods for analog signal conversion exist (Huijsing, 1986). In the simplest case, an analog signal from a sensor is electronically amplified into a voltage or a current representing the measurand. An analog output signal from a sensor can also be converted into a quasidigital signal, such as a frequency, a pulse count, a phase shift, or a pulse duration, since this type of signal is much less prone to noise and interference and can be recovered by filtering, phase-lock-loop techniques, or synchronous detection. Many sensors themselves generate a frequency output. In this case, while electronic circuitry might be needed for amplification or impedance matching, it will not be needed for the frequency-conversion step itself. One group of such sensors is based on resonators. Micromachining techniques can be applied to create suitable resonant structures, such as membranes or tuning forks, and the device can be activated electrically by means of a piezoelectric film or thermally.

Another group of sensors is based on integrating periodic geometrical structures in silicon. An example of this group is the CCD (charge-coupled device) (*q.v.*). CCDs can be used for one- and two-dimensional displacement measurements. With CCDs, displacement or position can be converted directly into number of pulses. The resolution of such devices depends almost solely on the size of the photosensitive elements of the CCDs.

Yet other sensors are based on electronic oscillators. One way to obtain a frequency output is to use ring oscillators in which the sensor element itself is the frequency-determining element. Pressure sensors based on two nine-stage MOS ring oscillators have already been achieved. A strain-sensitive ring oscillator can be obtained by making use of the fact that the drain current of a MOSFET is proportional to the mobility in the channel, which happens to be sensitive to strain. A temperature coefficient of the output signal as small as 0.008%/°C can be achieved when two ring oscillators are located in a silicon pressure diaphragm such that in one ring oscillator the stress is parallel and in the other the stress is perpendicular to the current in the channels, and the ratio of the oscillator frequencies is taken (Neumeister *et al.*, 1985).

8.2 Binary-Encoded Signals and Bus Systems

In the preceding section, we dealt with signal-conversion methods in which the resulting frequency, number of pulses, or pulse duration is a linear function of the measurand. Signal conversion can also be performed in such a way that the output contains an encoded form of the amplitude of the parameter being measured. The encoding of output signals is highly advantageous when the signal has to be transmitted or processed in some way. Binary encoding is the most frequently applied type of encoding: Since a bit can assume only two values (0 or 1), the output signal is virtually free of noise and interference.

In large measurement and control systems, communication between the central computer and the sensors and sensor subsystems scattered throughout a factory site is difficult. In all likelihood, the sensors will have been produced by several different companies. To make matters worse, the central computer will probably have been made by yet another manufacturer. To ensure that the data transfer is organized in an orderly and reliable fashion, bus systems, or networks, were designed. In the early days, bus systems were only implemented in spatially extended systems such as oil refineries, steel factories,

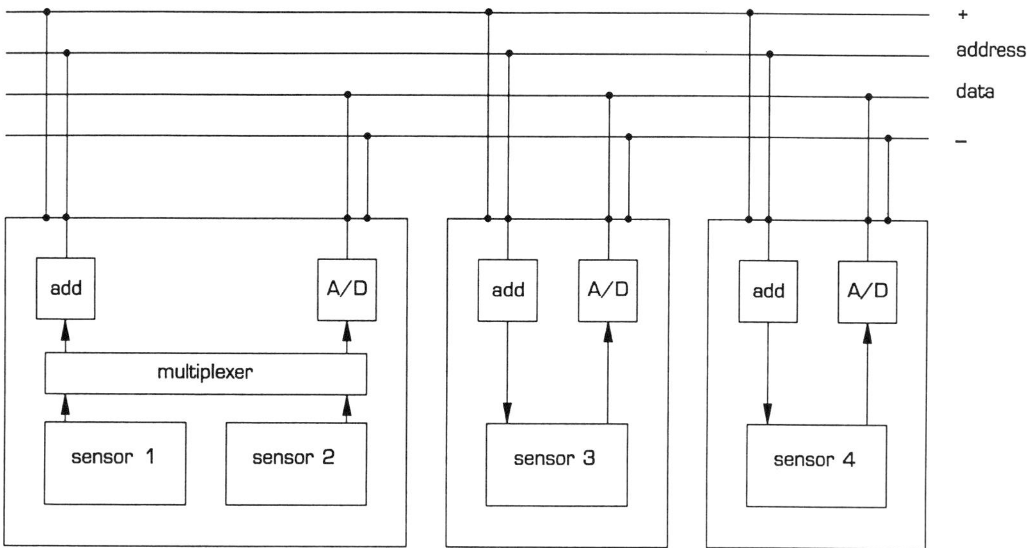

FIG. 20. Sensor bus system.

and large airplanes, but thanks to the introduction of low-cost microelectronic components, these days bus systems are being applied in increasingly smaller systems, even in consumer products. Moreover, the rise in the number of personal computers and workstations coupled to central computers has spurred the development of these data transfer systems and boosted the demand for cost-effective sensor systems.

A bus system generally consists of a central computer connected by several parallel wires to a large number of sensor systems. All sensor systems are switched in parallel, and electronic switches are required to switch the desired sensor to the data transfer lines. Many types of networks are feasible: ring structures, star networks, and a variety of trees. Figure 20 depicts a sensor subsystem with four bus lines: two for the voltage supply, one for sending addressing information, and one for the data line. When a sensor needs to be activated to send information, the address of the sensor is selected in the central computer, and that particular sensor is switched to the digital data line. The sensor subsystem may consist of a simple sensor, a signal conditioner, an analog-to-digital (A/D) converter, and an address identifier, although multiplexing techniques may also be applied so that the A/D converter and the addressing circuit are shared by several different sensors. Similarly, the data sent along the data lines may consist of simple data, such as the temperature or the pressure encoded in binary form, or other more complex information may be sent between the central computer and the sensors.

The central computer can also initiate all kinds of periodic tests and can order recalibration procedures. In addition, processing circuitry may be added to ensure that information about the fabrication date, the last calibration date, the range, or the average signal will periodically be sent to the central computer (Brignell, 1986). In the coming decade, silicon smart sensors will undoubtedly lead to many innovations in measurement and control systems. To meet the demands, it will be necessary to design new communication protocols and to process sensor signals at an even greater local level.

GLOSSARY

Anisotropic Etching: Etching method in which the use of certain etching solutions makes the etch rate depend on the crystal direction.

Band Gap: The energy states between the conduction and the valence band in a semiconductor or insulator that cannot be occupied by electrons.

Batch Fabrication: Manufacture of a number of products at the same time, so that they have very similar properties.

Biosensor: A sensor that can measure the biological reactions in a living organism.

Buried Layer: A layer that is obtained by diffusion or ion implantation prior to the growth of an epitaxial layer on the substrate.

Conduction Band: Band of allowed energy states in a semiconductor or insulator in which the lowest energy states are occupied by electrons.

Cross-Sensitivity: The sensitivity of a sensor to more than one parameter.

Dew Point: The temperature at which water vapor starts to condense.

Enzyme: A biochemical catalyst that increases the speed of biological reactions.

Exciton: A bound electron-hole pair with an energy in the band gap of a semiconductor.

***K* Shell:** Shell nearest the nucleus of an atom, containing two electrons.

Lorentz Force: The force exerted on a moving charge in a magnetic and an electric field.

Phase-Lock Loop: A system in which the frequency of a voltage-controlled oscillator can be made exactly equal (locked) to the frequency of an incoming periodic signal.

***pn* Junction:** The region between a *p*-type and an *n*-type semiconductor.

Valence Band: Band of allowed energy states in a semiconductor or insulator in which the highest energy states are not occupied by electrons.

Works Cited

Bergveld, P. (1970), *IEEE Trans. Biomed. Eng.* **19**, 70–71.

Brignell, J. (1986), in: S. Middelhoek, J. Van der Spiegel (Eds.), *Sensors and Actuators 10, State of the Art of Sensor Research and Development*, Lausanne: Elsevier Sequoia, pp. 249–261.

Collet, M. G. (1986), in: S. Middelhoek, J. Van der Spiegel (Eds.), *Sensors and Actuators 10, State of the Art of Sensor Research and Development*, Lausanne: Elsevier Sequoia, pp. 287–302.

Demarne, V., Grisel, A. (1988), *Sensors Actuators* **13**, 301–314.

Gieles, A. C. M. (1969), *Digest IEEE International Solid-State Circuits Conference*, New York: IEEE, pp. 108–109.

Huijsing, J. H. (1986), in: S. Middelhoek, J. Van der Spiegel (Eds.), *Sensors and Actuators 10, State of the Art of Sensor Research and Development*, Lausanne: Elsevier Sequoia, pp. 219–237.

Ko, W. H. (1986), in: S. Middelhoek, J. Van der Spiegel (Eds.), *Sensors and Actuators 10, State of the Art of Sensor Research and Development*, Lausanne: Elsevier Sequoia, pp. 303–320.

Lundstrom, I., Shivaraman, S., Svensson, C., Lundkvist, L. (1975), *Appl. Phys. Lett.* **26**, 55–57.

Meijer, G. C. M. (1986), in: S. Middelhoek, J. Van der Spiegel (Eds.), *Sensors and Actuators 10, State of the Art of Sensor Research and Development*, Lausanne: Elsevier Sequoia, pp. 103–125.

Middelhoek, S., Audet, S. A. (1989), *Silicon Sensors*, London: Academic Press.

Miyahara, Y., Matsu, F., Moriizumi, T., Matsuoka, H., Karube, I., Susuki, S. (1983), in: *Proceedings of the International Meeting on Chemical Sensors, Fukuoka, Japan*, Tokyo: Kodansha Ltd., pp. 501–506.

Muller, R. S. (1990), *Sensors Actuators* **A21–A23**, 1–8.

Neumeister, J., Schuster, G., von Münch, W. (1985), *Sensors Actuators* **7**, 167–176.

Popovic, R. S. (1991), *Hall Effect Devices*, Bristol: Adam Hilger.

Regtien, P. P. L. (1981), *Sensors Actuators* **2**, 85–89.

Roylance, L., Angell, J. B. (1979), *IEEE Trans. Electron Dev.* **ED-26**, 1911–1917.

Sarro, P. M., van Herwaarden, A. W. (1986), *J. Electrochem. Soc.* **133**, 1724–1729.

Smith, C. S. (1954), *Phys. Rev.* **94**, 42–49.

Ten Kate, W. R. Th. (1986), in: S. Middelhoek, J. Van der Spiegel (Eds.), *Sensors and Actuators, State of the Art of Sensor Research and Development*, Lausanne: Elsevier Sequoia, pp. 83–101.

van Duyn, D. C., Middelhoek, S. (1992), in: P. Ciureanu, S. Middelhoek (Eds.), *Thin Film Resistive Sensors*, Bristol: Adam Hilger, Chap. 1.

van Herwaarden, A. W., Sarro, P. M. (1986), in: S. Middelhoek, J. Van der Spiegel (Eds.), *Sensors and Actuators, State of the Art of Sensor Research and Development*, Lausanne: Elsevier Sequoia, pp. 321–346.

Wolffenbuttel, R.F. (1987), *IEEE J. Solid-State Circuits*, **22**, 350–356.

Zieren, V., Middelhoek, S. (1982), *Sensors and Actuators* **2**, 251–262.

Further Reading

Middelhoek, S., Audet, S. A. (1989), *Silicon Sensors*, Academic Press: London.

Göpel, W., Hesse, J., Zemel, J. N. (Eds.) (1991–1993), *Sensors—A Comprehensive Survey*, 8 vols., Weinheim: VCH.

Vol. 1, Grandke, T., Ko, W. H. (Eds.), *Fundamentals and General Aspects*.

Vol. 2, Göpel, W., Jones, T. A., Kleitz, M., Lundström, J., Seiyama, T. (Eds.), *Chemical and Biochemical Sensors Part I.*

Vol. 3, Göpel, W., Jones, T. A., Kleitz, M., Lundström, J., Seiyama, T. (Eds.), *Chemical and Biochemical Sensors Part II.*

Vol. 4, Ricolfi, T., Scholz, J. (Eds.), *Thermal Sensors.*

Vol. 5, Boll, R., Overshott, K. J. (Eds.), *Magnetic Sensors.*

Vol. 6, Wagner, E., Dändliker, R., Spenner, K. (Eds.), *Optical Sensors.*

Vol. 7, De Rooij, N. F., Kloeck, B., Bau, H. H. (Eds.), *Mechanical Sensors.*

Vol. 8, *Cumulative Index and Selected Topics.*

Details on current developments can be found in the proceedings of the biennial international TRANSDUCERS conferences (1981 Boston, 1983 Delft, 1985 Philadelphia, 1987 Tokyo, 1989 Montreux, 1991 San Francisco, 1993 Yokohama, 1995 Stockholm) published in *Sensors and Actuators* or as a separate proceedings. In Europe, the proceedings of the annual EUROSENSOR conferences are published in *Sensors and Actuators*. In the U.S.A., the proceedings of the American biennial sensor conferences are published by the IEEE. In Japan, English-language proceedings of the annual conferences are published by the IEE of Japan. Many topical conferences and workshops focusing on such topics as biosensors, micromachining, chemical sensors, and the automotive applications of sensors are held every year.

MICROWAVE CIRCUITS

PETER STAECKER, *M/A-COM, Inc., Corporate Research and Development Center, Lowell, Massachusetts, U.S.A.*

INTRODUCTION

Electronic circuit elements (resistors, capacitors, and inductors are elementary examples) and their terminal electrical characteristics measured in voltage and current are approximate realizations of the laws of electromagnetism; *microwave circuits* provide the framework in which these approximations are validated, modified, or replaced at high frequencies. Historically, the basis for microwaves dates to a series of papers written by Maxwell between 1855 and 1865 which established the theory of electromagnetic wave propagation, and Hertz, who in 1886 experimentally verified Maxwell's theory. Contemporary applications of microwaves developed rapidly during World War II with the formation of the Royal Radar Establishment in England, research activities in Germany and Japan, and the MIT Radiation Laboratory in the United States.

In general, electromagnetic field descriptions of circuit behavior become necessary when the wavelength (speed of light in the medium divided by the frequency) approaches the size of the circuit or component. For this discussion, the microwave region is bounded from above by the frequency of 100–500 GHz, where quasi-optical techniques apply, and given this definition, includes the millimeter-wave frequency region, which extends from 30 to 300 GHz. A lower frequency boundary is set by the somewhat fuzzy interface between analog and digital circuits, presently in the few-GHz range, but as the speed of digital logic increases, microwave circuit techniques will embrace this technology also. The following provides an overview of microwave electronics, from its basis in electromagnetic theory (Sec. 1), to propagation media used in microwave circuits (Sec. 2), to single-function components (Sec. 3), to circuit integration techniques (Sec. 4). Contemporary practices as well as projections of future activity are noted within these

3-527-28132-0/94/$5.00 + .50

sections as well as in the summary (Sec. 5). Applications of microwave circuits include radar, navigation, communications, medical applications and other interaction with materials (heating), remote sensing, and energy transmission. Radar (RAdio Detection And Ranging) is so commonplace it rarely is given uppercase acronym status. In 1940, on the eve of the formation of the MIT Radiation Laboratory when detection research was in its infancy, the focus was on detection of aircraft by UHF waves, but curious scientists were attracted to other moving targets. The success with which the observers detected violations of the highway speed limit of 50 miles per hour led one scientist to remark, "Don't let the cops hear about this." Collision-avoidance and object-detection systems for automobiles are current topics of intense interest in commercial radar applications.

Operating at 1.7 GHz, the Global Positioning Satellite (GPS) network broadcasts information allowing users with commercial receiving equipment to locate themselves anywhere on earth to a precision of 20–40 m. Palm-size GPS receivers are available now for about $1000 and will appear in automobiles as navigation aids in the mid-1990s. (1993 U.S. dollars are used throughout this article.)

For communications applications, bandwidth (Hz) is key to the rate of information transfer (bits/s). Communications systems are being extended to light-wave frequencies for Tbit/s network requirements of supercomputer interconnections, full-motion color graphics, and medical imaging, which may require high-resolution images to be transmitted uncompressed. Microwave signal transmission is used for terrestrial radio telephone communications. Satellite communication networks service television, telephone, and other commercial communication network needs. Cellular telephones operating at 900 MHz and 1.8 and 2.4 GHz presently serve local areas with terrestrial transmitters. Military communications users favor new frequencies in the millimeter-wave region, some as high as 60 GHz, to take advantage of an oxygen absorption band in the earth's atmosphere, which ensures privacy [Fig. 1(a)]. Commercial wireless applications (automotive communications) planned for the same frequency band will achieve greater signal densities (channels/km^2) through the same characteristic. Wireless markets have exploded in the past five years and will probably overtake military microwave applications near the turn of the century [Fig. 1(b)].

Medical uses of microwaves represent more contemporary applications of microwave technology. Microwave hyperthermia uses the heating effect of microwaves in human tissue to destroy malignant tumors. Radiometric techniques developed at the MIT Radiation Laboratory are today in use to detect temperature differentials characteristic of these same tumors. Microwaves are used to heat blood instantly in trauma cases where it is crucial to elevate body temperature to normal levels in a short time.

The microwave oven was a direct spinoff of MIT Radiation Laboratory activity, using a magnetron as the power source operating at 2450 MHz where absorption of electromagnetic energy by water is extremely efficient. The same frequency region has been proposed for use in beaming power collected by orbiting solar arrays to antenna farms on earth.

The markets supporting these applications are currently strongly military but by the year 2000 should be exceeded by the wireless data and telephony markets.

1. CONCEPTS AND DEFINITIONS

Central to the discussion of microwave circuits is the understanding of electromagnetic wave propagation. Indeed, the concepts of distributed circuits and dispersion find their basis in the wave equation, which in turn is derived from the time-dependent Maxwell's equations. Equally important notions of loss and lumped equivalent circuit elements result from the static approximations to these same equations. Maxwell's equations in differential form, with steady-state sinusoidal time dependence, are

$$\nabla \times \mathbf{E} = -j\omega\mathbf{B}, \tag{1}$$

$$\nabla \times \mathbf{H} = j\omega\mathbf{D} + \mathbf{J}, \tag{2}$$

$$\nabla \cdot \mathbf{D} = \rho, \tag{3}$$

$$\nabla \cdot \mathbf{B} = 0, \tag{4}$$

where **E** and **D** are the electric field and electric displacement intensities, **H** and **B** are the

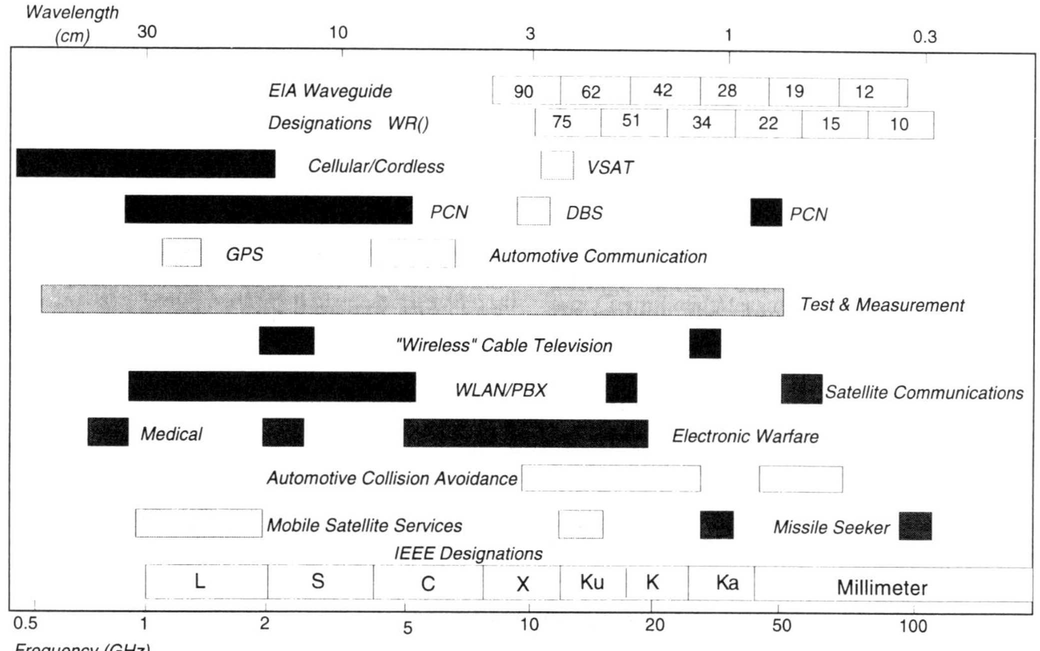

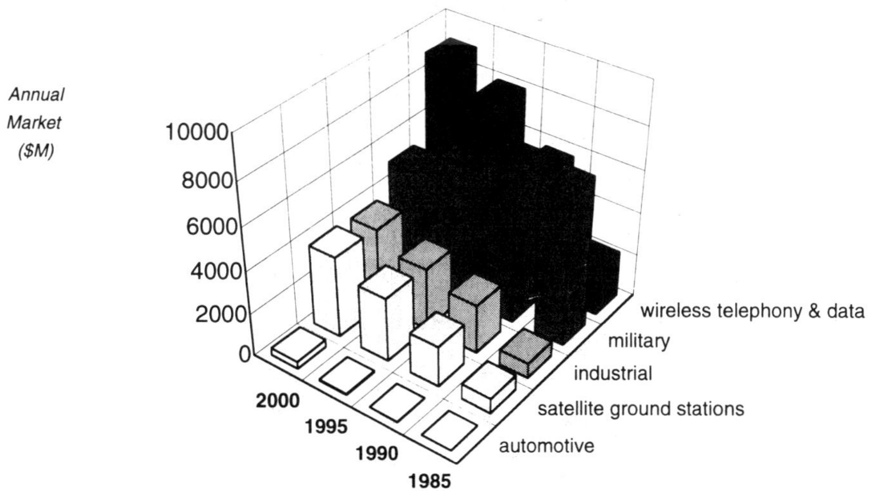

FIG. 1. (a) Microwave frequency spectrum, band designations, and applications. In this figure, the following acronyms are used: EIA: Electronics Industry Association. DBS: Direct Broadcast Satellite. GPS: Global Positioning Satellite, a system of satellites that broadcast time-coded messages to Earth, which, when received and decoded, allow latitude and longitude information to an accuracy of 30 m (commercial) and 2 m (military). IEEE: Institute of Electrical and Electronics Engineers. PBX: Private Branch Exchange, a private telecommunications exchange that includes access to a public telecommunications exchange. PCS: Personal Communications Services, a family of mobile or portable radio communications services providing services to individuals and businesses and through a variety of competing networks. PCN: Personal Communications Network, a form of PCS that offers metropolitan-area portable radio telephony competitive with cellular and public telephone networks. VSAT: Very small-aperture terminal. WLAN: Wireless Local Area Network. (b) Microwave world-wide market segments: history and projections. (Courtesy of Strategies Unlimited, Mountain View, CA.)

magnetic field and magnetic flux intensities, **J** is the conduction current, and ρ is the electric charge density. All fields are harmonic in time ($e^{j\omega t}$) and are vector quantities. The associated constitutive relations are

$$\mathbf{D} = \epsilon\mathbf{E} = \epsilon_0\epsilon_r\mathbf{E}, \tag{5}$$

$$\mathbf{B} = \mu\mathbf{H} = \mu_0\mu_r\mathbf{H}, \tag{6}$$

$$\mathbf{J} = \sigma\mathbf{E}, \tag{7}$$

where ϵ is the electrical permittivity, μ is the magnetic permeability, and σ is the electrical conductivity of the medium (all of which are, in general, tensor quantities), ϵ_0 is the permittivity of free space ($\epsilon_0 = 8.854 \times 10^{-12}$ F/m), and ϵ_r is the relative permittivity of the dielectric medium. Likewise, μ_0 is the permeability of free space ($\mu_0 = 4\pi \times 10^{-7}$ H/m), and μ_r is the relative permeability of the medium. The properties of materials are embedded in their permittivities and permeabilities; the imaginary part of ϵ introduces dielectric loss, an important consideration at high frequencies. The anisotropic nature of μ allows operation of magnetic circulators described below.

1.1 Boundary Conditions

From the integral formulation of Eq. (1), it can be shown that the tangential component of the electric field is continuous across any boundary in the medium. The same manipulation of Eq. (2) reveals that the tangential magnetic field discontinuities across any boundary in the medium require the presence of surface current on the boundary. In a similar manner, integral formulations of Eqs. (3) and (4) yield boundary conditions on field components normal to surfaces: The normal component of magnetic flux must be continuous across media interfaces, whereas discontinuities in the normal component of electric displacement are due to the presence of surface charge ρ_s at the interface. The special case of electrical conductor–vacuum interface (electrical wall) leads to the relations

$$E_t = 0, \qquad H_t = K, \tag{8}$$

$$D_n = p_s, \qquad B_n = 0, \tag{9}$$

at the metal wall (where K is surface current in A/m). Equations (8) and (9) are used extensively in the solution of wave propagation in transmission media with conducting boundaries. For solutions to problems in fiber optics, the appropriate boundary conditions are applied at dielectric interfaces, and the field solutions are more complicated, but the basic electromagnetic framework is the same.

1.2 Dissipative Media (Loss)

Solving the Helmholtz (or wave) equation [derived from Eqs. (1) and (2)]

$$\nabla^2\mathbf{E} - \gamma^2\mathbf{E} = 0 \tag{10}$$

for fields harmonic in time and space [$e^{(j\omega t - \gamma z)}$] leads to an expression for the complex propagation constant $\gamma = \alpha + j\beta$ in terms of the complex permittivity $\epsilon = \epsilon' - j\epsilon''$ and the conductivity as

$$\gamma^2 = j\omega\mu(\sigma + j\omega\epsilon) = j\omega\mu([\sigma + \omega\epsilon''] + j\omega\epsilon'). \tag{11}$$

The dielectric damping term $\omega\epsilon''$ can be combined with the conductivity σ to give the total effective electrical conductivity. The loss tangent (sometimes called dissipation factor) of a dielectric medium is defined as the ratio of the sum of the conduction current ($\sigma E + \omega\epsilon'' E$) to the displacement current $j\omega\epsilon' E$:

$$\tan\delta = \frac{\sigma + \omega\epsilon''}{\omega\epsilon'}. \tag{12}$$

Dielectric damping accounts for the absorption of microwaves in nonconductive materials but for transmission media is usually ignored over most of the microwave frequency region, i.e., $\epsilon \approx \epsilon'$. For situations in which the conduction current σE is much less than the displacement current $\omega\epsilon' E$, as might be the case in a slightly imperfect dielectric, propagation is described by

$$\gamma \approx \omega\sqrt{\mu\epsilon}\left(\frac{\tan\delta}{2} + j\right), \quad \tan\delta \ll 1; \tag{13}$$

whereas when conduction currents greatly exceed displacement currents, as would be the case in a slightly imperfect conductor, for example, the real and imaginary parts of the propagation constant are equal and are given by

$$\gamma = \sqrt{\frac{\omega\mu\sigma}{2}}(1 + j), \quad \tan\delta \gg 1. \tag{14}$$

Under these conditions, field amplitudes decay exponentially in the direction of propagation, and give meaning to the concept of *skin depth*, δ_s, the inverse of the attenuation constant α. Surface resistance ($R_s = 1/\sigma\delta_s$ ohms/square) is used to describe properties of conductive films.

The ratio of electric to magnetic field is defined as the *wave impedance*, which has dimensions of ohms [a subscript zero (η_0) is used when the propagation medium is vacuum].

$$\eta \equiv |\mathbf{E}|/|\mathbf{H}| = \sqrt{\mu/\epsilon} \tag{15}$$

and is the basis for and related to circuit-theoretical impedance concepts discussed below.

In resonant circuits, the ratio of average energy stored in the resonant circuit to the resonator energy loss per cycle is the (unloaded) quality factor Q_u. [Two additional quality measures which consider energy storage/loss in the external circuit (Q_E), or energy storage/loss in the resonator *and* external circuit (loaded Q: Q_L), are used frequently in resonator design.] This convenient descriptor of losses in both distributed and lumped circuit elements can be defined for transmission-line resonators with resonant wavelength λ_r:

$$Q_u = 2\pi \frac{\text{average stored energy}}{\text{energy loss/cycle}} = \frac{\beta}{2\alpha} = \frac{\pi}{\alpha\lambda_r}. \tag{16}$$

The quality factor is a function of frequency as well as geometry, having typical values (at 10 GHz) of 10–100 for lumped element reactances and ~1000 for distributed resonators, and can reach values greater than 10 000 for cavity resonators, and greater than 100 000 for cavities with superconducting walls. This figure of merit is very useful in the design of frequency-selective structures used in microwave filters and oscillators. When size is a consideration, low-loss dielectrics (whose dielectric constant can reduce λ_r by as much as a factor of 5) or acoustic resonators (with resonant wavelengths a factor of 10 000 less than air-filled cavities) can be used.

1.3 Guided Waves

The electromagnetic boundaries of the guiding structure define the nature of field solutions for wave propagation in the structure under consideration. The wave equation in a lossless medium may be written as

$$\frac{\partial^2 \mathbf{H}}{\partial x^2} + \frac{\partial^2 \mathbf{H}}{\partial y^2} + (\omega^2\mu\epsilon - \beta^2)\mathbf{H} = 0. \tag{17}$$

Transverse electromagnetic (TEM) fields by definition have no longitudinal component (z component, in the notation of this article). These fields exist in free space (no guiding medium) and in certain guiding structures described below. Solutions to Eq. (17) are impossible unless $\omega^2\mu\epsilon - \beta^2 = 0$ (TEM solutions therefore exist to zero frequency), which reduces the form of Eq. (17) to Laplace's equation in the transverse plane, familiar from electrostatics. As a direct consequence, the voltage between conductors defining the guiding medium can be found from the integral form of Eq. (3) as

$$\mathcal{V}_{12} = \int_1^2 \mathbf{E} \cdot d\mathbf{l}. \tag{18}$$

Likewise, from the integral form of Eq. (2), the current on the conductors can be found from

$$I = \oint_c \mathbf{H} \cdot d\mathbf{l}, \tag{19}$$

where c is the cross-sectional contour of the conductor surface. Because of the uniqueness of solutions to Laplace's equation, the voltage and current, and therefore the ratio $\mathcal{V}/I$, are unambiguous and useful for circuit representations in TEM structures.

For regions enclosed by a single conductor, solutions to Eq. (17) depend upon the exact geometry of the surrounding region. For a conducting boundary of infinite extent in the z direction, Eq. (17) can be written as three scalar equations. For the z component,

$$\frac{\partial^2 H_z}{\partial x^2} + \frac{\partial^2 H_z}{\partial y^2} + k_c^2 H_z = 0. \tag{20}$$

If the cross section of the conducting boundary is rectangular, solutions to Eq. (20) exist of the form

$$H_z = [A\cos(k_x x) + B\sin(k_x x)] \times [C\cos(k_y y) + D\sin(k_y y)], \quad (21)$$

where k_x and k_y are both multivalued:

$$k_x^2 + k_y^2 = k_c^2 \equiv \omega^2\mu\epsilon - \beta^2 \equiv k^2 - \beta^2. \quad (22)$$

For frequencies greater than a cutoff frequency f_c $(=k_c/2\pi\sqrt{\mu\epsilon})$, propagating modes ($\beta$ real) exist, whereas for frequencies below f_c, β is imaginary, and waves are evanescent (attenuated exponentially with distance, not propagating). The phase velocity, which is greater than the speed of light in the same medium, is given by

$$v_p \equiv \frac{\omega}{\beta} > \frac{\omega}{k} = \frac{1}{\sqrt{\mu\epsilon}}. \quad (23)$$

This treatment defines transverse electric (TE) waves, i.e., where $H_z \neq 0$ and $E_z = 0$. The same boundary conditions allow transverse magnetic (TM) field solutions, where $E_z \neq 0$ and $H_z = 0$. In general, discrete modes of propagation [solutions to Eq. (20)] exist for each possible value of k_c allowed by the boundary conditions.

Some additional characteristics of TE (TM) waves may now be noted:

1. Because there is a longitudinal component of field, unambiguous definitions of voltage and current are not rigorously correct. These concepts are used only under restricted conditions as will be mentioned below.
2. Because $k_c \neq 0$, TE (TM) wave solutions do not exist to zero frequency but propagate for frequencies above a cutoff frequency f_c that is determined by the cross-section dimensions. In addition, the phase velocity of TE (TM) waves varies with frequency, causing individual frequency components of a wave form to change their original phase relationships with respect to each other during propagation through the guiding structure, resulting in wave form distortion. This property, called *dispersion*, is of extreme importance in pulse propagation over long distances.

1.4 Lumped and Distributed Circuits, Characteristic Impedance

Energy storage in the magnetic and electric fields of a transmission line is associated with a series inductance and shunt capacitance per unit length. Likewise, power loss in the conductors and dielectric can be modeled by a series resistance and shunt conductance per unit length. The equivalent circuit of a transmission line of differential length dz can thus be represented as an equivalent circuit whose element values are related to the material properties of the transmission medium [Fig. 2(a)]. Trading heuristic arguments for rigorous computer-based electromagnetic field solvers, equivalent circuit models can be generated for more complex structures such as spiral inductors [Fig. 2(b)].

Applying Kirchoff's voltage and current laws to the model of Fig. 2, and assuming voltage wave propagation of the form $V = V^+e^{j\omega t-\gamma z} + V^-e^{j\omega t+\gamma z}$ (and similar solutions for current waves), the lumped-circuit version of the propagation characteristics can be expressed as

$$\gamma = \sqrt{(R + j\omega L)(G + j\omega C)} = \omega\sqrt{LC}\left\{\left[\frac{R}{L} + \frac{G}{C}\right] + j\right\}, \quad (24)$$

$$Z_0 \equiv \frac{V^+}{I^+} = \frac{V^-}{I^-} = \sqrt{\frac{R + j\omega L}{G + j\omega C}} \approx \sqrt{\frac{L}{C}}, \quad (25)$$

where the approximations apply for small losses ($R \ll \omega L$ and $G \ll \omega C$). If the transmission line is terminated in its characteristic impedance Z_0, there will be no reflected wave.

Because the impedance concept embodies the phase and amplitude characteristics of electromagnetic wave propagation in a network description, it is an extremely important tool in all aspects of microwave circuit design. Specific applications include maximizing power-transfer or power-generating characteristics, or minimizing additive thermal noise of active devices.

1.5 Reflection Coefficient and the Smith Chart

The reflection coefficient Γ is defined as the ratio of the reflected wave to the incident wave:

$$\Gamma(z) \equiv \frac{V^-(z)}{V^+(z)} = \frac{V^-(0)}{V^+(0)}e^{-2\gamma z} \equiv \Gamma_R e^{-2\gamma z}, \quad (26)$$

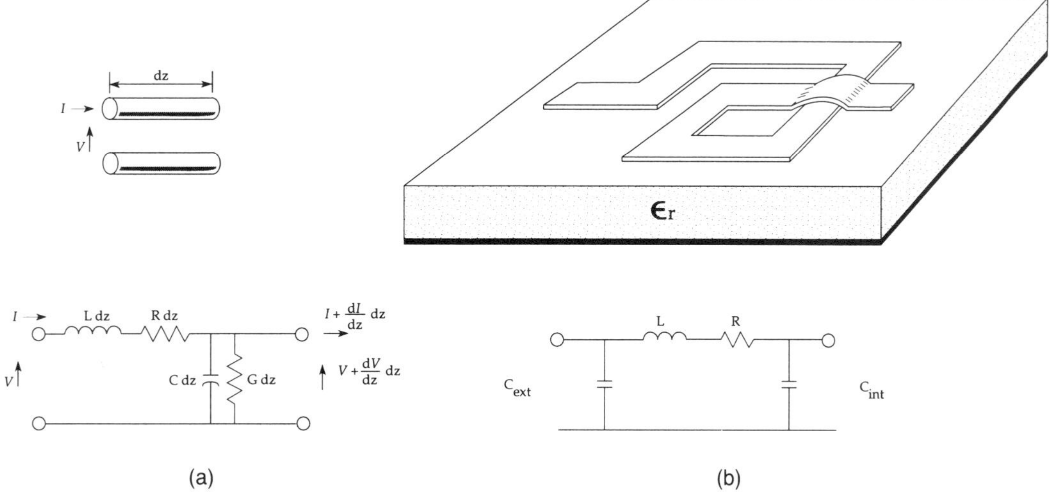

FIG. 2. (a) Transmission-line equivalent circuit; and (b) printed spiral inductor and its equivalent circuit (for frequencies below resonance).

where $z = 0$ denotes the wave amplitude reference location. (These reference locations, where amplitude and phase are specified, are called reference planes and define the phase datum for each network port, since phase is a relative quantity. These locations can be coincident with the planes of electromagnetic access to the network, such as a waveguide flange or a coaxial connector, and can be analytically relocated within the network using the propagation characteristics of its transmission lines.) In this notation, a perfect open (short) circuit at $z = 0$ has a reflection coefficient of $+1$ (-1), and a transmission line terminated in its characteristic impedance has a reflection coefficient of zero.

Reflection coefficient is related to the normalized impedance Z_n by the bilateral transformation

$$\Gamma_R = (Z_n - 1)/(Z_n + 1) \tag{27}$$

(where the normalized impedance Z_n is equal to the reference impedance Z_R divided by the characteristic impedance Z_0), which maps the infinite half-plane of impedance space into the unit circle (Fig. 3). This unit circle representation is called the Smith chart (Wheeler, 1984), in recognition of Philip Smith, its inventor, and has been widely used by microwave circuit designers since its disclosure. Circuit synthesis with the Smith chart uses simple graphical constructs: From a given reference impedance, lengths of transmission line are rotations in electrical length about the center of the chart; inductive or capacitive reactance is added by moving on constant-resistance circles. Impedance to admittance conversions are executed by rotating halfway around the chart on a constant radius. Having suffered the same fate as its computational counterpart, the slide rule, the Smith chart today is rarely used as a design aid but still serves as the universal template for displaying measured and computed data as a function of impedance and frequency.

The ratio of maximum voltage magnitude to minimum voltage magnitude is called the voltage standing-wave ratio (VSWR) of the line, and is defined as

$$\text{VSWR} \equiv \frac{1 + |\Gamma_R|}{1 - |\Gamma_R|} = \frac{|V(z)|_{\max}}{|V(z)|_{\min}}. \tag{28}$$

Loci of constant VSWR on the Smith chart are circles with centers at the origin of the unit circle, making infinite VSWR locus the edge of the chart, and perfect match (unity VSWR) the center of the chart.

2. PROPAGATION MEDIA

Electromagnetic wave propagation structures used for microwave and millimeter-wave components and circuitry fit into a broad

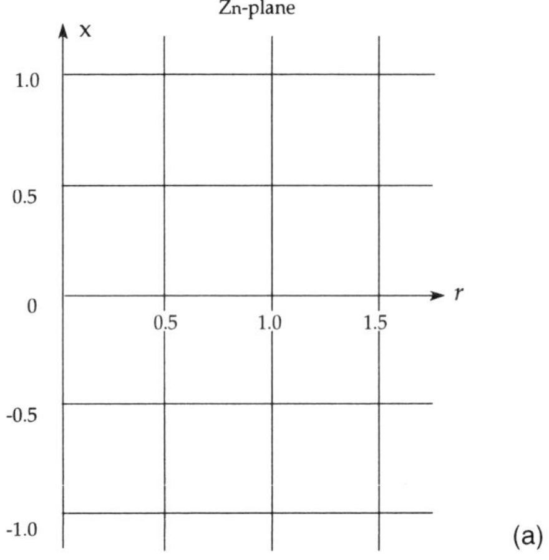

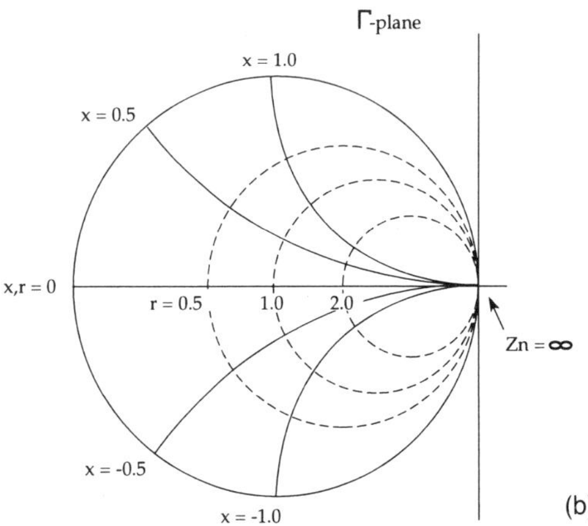

FIG. 3. (a) Mathematical representation of the reflection coefficient Γ in rectangular coordinates; (b) transformation of $\Gamma < 1$ to the unit circle (Smith chart).

collection of methods of confining fields and guiding power. The simplest physical method is to use electrically conductive pipes or tubes; indeed, metallic waveguide development prior to World War II depended upon unlikely sources of supply. The extrusions available at that time defined rectangular waveguide cross sections available to microwave practitioners and have survived in many instances to today. An example of particular note is WR 137 waveguide, whose outside dimensions are $1\frac{1}{2}$ in. by $\frac{3}{4}$ in., and which graces architectural designs as exterior stair railings and metal storm doors in the northeastern United States. Dielectrics in an air surround can also be used to confine energy (dielectric waveguide), the best example of which is single-mode optical fiber cable, with an installed base of greater than 4 Gm in the United States alone. At the circuit level, inexpensive planar guiding structures are easily defined with precise photolithographic processes on a wide range of dielectric substrates. Discrete active and passive devices can be integrated on the surface of these dielectrics, or can be monolithically integrated with waveguiding structures on semiconducting materials that serve both as low-loss substrate and as host material for active devices. Selection criteria for propagation media include producibility and cost, size, mechanical interfaces to devices and other structures, impedance, power handling, electrical loss, and finally accuracy of analytical circuit models.

2.1 TEM Structures

Transverse electromagnetic fields, by definition, have no field component in the direction of propagation. The cutoff wave number is zero (TEM waves are dispersionless), and field solutions are found from Laplace's equation in the transverse plane, familiar from electrostatics. As a direct consequence, the voltage between and current on the conductors defining the guiding medium (and therefore the impedance V/I) are unambiguous and directly applicable as circuit representations for TEM structures. Where defined, these impedance levels are functionally related to the dielectric constant and the ratio of characteristic transverse dimensions of the transmission line.

The simplest guiding medium supporting TEM waves is coaxial transmission line [Fig. 4(a)]. The frequency dependence of attenuation per unit length resulting from conductor losses depends on R_s, whereas the dielectric loss depends directly on frequency. This frequency dependence is universally valid for TEM structures and is an accurate low-frequency approximation for quasi-TEM structures. At very high frequencies (in particular, when the wavelength in the medium approaches the median circumference of the coaxial structure), higher-order (non-TEM) modes are supported by the coaxial structure. Cost and single-mode propagation at high frequencies favor smaller diameters, at the expense of conductor loss. These compromises are general and apply to many other guiding structures.

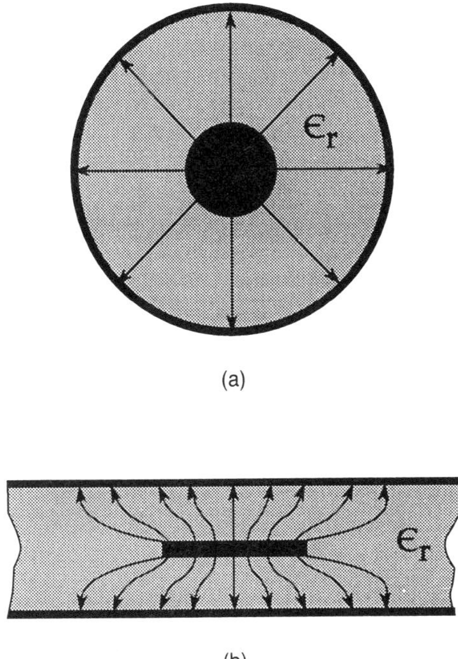

FIG. 4. TEM structures: (a) coaxial transmission line; (b) symmetric stripline.

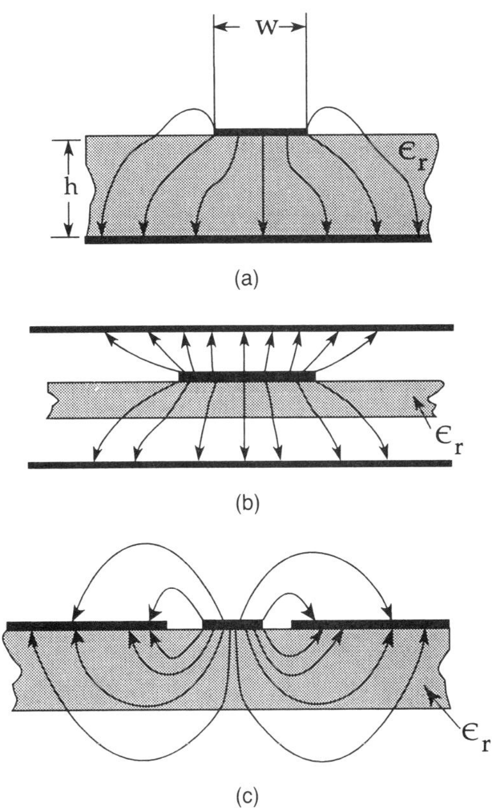

FIG. 5. Quasi-TEM structures: (a) microstrip; (b) suspended substrate stripline; (c) coplanar waveguide.

In symmetric stripline, a conductor is embedded in a dielectric medium with ground planes located symmetrically about it [Fig. 4(b)]. Commonly used materials for stripline construction employ polytetrafluoroethylene (PTFE) in a woven glass matrix, a so-called "soft" substrate with a dielectric constant slightly larger than 2.

2.2 Quasi-TEM Structures

Quasi-TEM lines are comprised of two separate conductor electrodes embedded in an inhomogeneous dielectric. Longitudinal (small with respect to transverse) components of fields exist and lead to ambiguous definitions of impedance. An effective dielectric constant valid at nonzero frequency can be defined and computed using a number of methods. These procedures give results that converge to pure TEM field solutions as frequency approaches zero (Hoffman, 1987).

Microstrip consists of a single conductor of width w on a dielectric medium (substrate) of thickness h with a ground plane on its opposite face [Fig. 5(a)]. The electrical ground plane is also an effective system thermal ground, accommodating devices that generate or dissipate heat. Because of its ease of fabrication, microstrip is the workhorse of contemporary microwave circuit realizations. Microstrip geometry allows propagation of higher-order TE or TM modes as surface waves on a grounded dielectric slab, which exhibit exponential decay in regions outside the dielectric. These modes are minimized when the substrate thickness is chosen to be less than a quarter wavelength at the highest operating frequency.

Removing the microstrip ground plane and suspending the resulting structure in the H plane of a rectangular conducting channel results in a low-loss quasi-TEM transmission medium called suspended stripline [Fig. 5(b)]. This configuration is useful as a transition from planar media to waveguide and has low losses, thus offering a planar alternative to waveguide at millimeter-wave frequencies.

Coplanar waveguide (CPW) consists of a single conductor centered in a gap between conducting ground strips on the top surface of a dielectric medium of semi-infinite extent [Fig. 5(c)]. Because both conductor electrodes are on the top surface of the substrate, CPW can be used where metallization of both sides of the substrate is difficult. In addition, the relative ease of integrating shunt and series devices in CPW offers a complementary alternative to microstrip integration techniques. Advances in the technology of microwave measurements have incorporated CPW in a probe/transition configuration in order to extend the nondestructive vector (amplitude and phase) measurement capability on planar circuits to 100 GHz (Fig. 6).

2.3 Non-TEM Structures

For regions enclosed by a single conductor, solutions to the wave equation depend upon the exact geometry of the surrounding region. For a conducting boundary of rectangular cross section and of infinite extent in the propagation direction, TE and TM waves can exist. In general, discrete modes of propagation exist for each possible value of k_c allowed by the boundary conditions. Unambiguous definitions of voltage and current do not exist, and therefore circuit definitions of impedance apply only under very restrictive conditions.

Metallic waveguide confines the power while retaining most of the low-loss properties of air. The TE and TM solutions each exist in an infinite set. Dimensions of standard rectangular guides are such that fundamental-mode operating bandwidths are less than an octave. A number of variations of rectangular guide have been devised to take advantage of its low-loss characteristics. Metallic guiding structures in the E plane of rectangular guide, usually supported by dielectric and called finline, allow a smooth transition from waveguide to printed substrates. To increase bandwidth while minimizing loss, ridge guide (Hopfer, 1955), a modification of rectangular guide, is used. Circular waveguide (which can be envisioned as a deformation of square waveguide) also has an infinite set of TE and TM modes.

Signal transmission using optical fibers offers a low-loss, low-cost alternative to circular waveguide for long-distance communications applications and has been the subject of intensive research and development since the development of the semiconductor laser in 1964. Losses as low as 0.2 dB/km have been measured at a wavelength of 1.55 μm, the wavelength of one available type of semiconductor laser frequency source. This frequency (194 THz!) hardly fits the microwave boundaries mentioned above, but since fiber-optic transmission lines are displacing present microwave transmission media and because modulation and demodulation techniques will continue to require microwave circuit methods, the distinction between microwave and light-wave techniques and vocations will continue to blur.

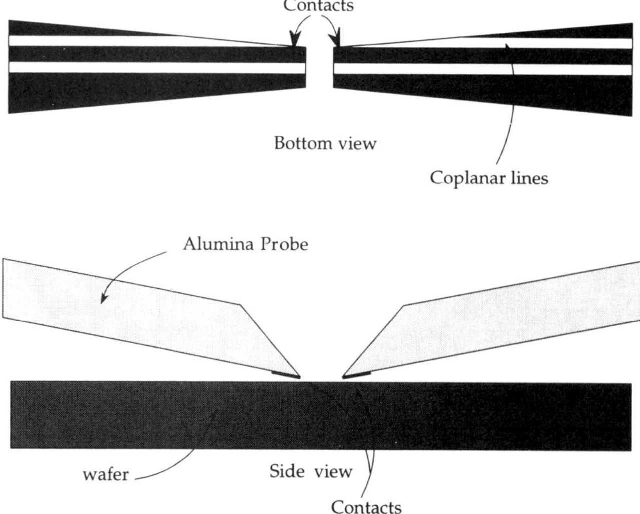

FIG. 6. Coplanar waveguide microwave probing geometry, allowing in-circuit (or on-wafer) measurements at frequencies as high as 100 GHz.

2.4 Enabling Technology: High-Temperature Microwave Superconductivity

A major driver in the development of transmission media for microwave circuits has been the desire to minimize both dielectric and conductor losses. Recent spectacular results in superconducting oxides show great promise for microwave applications. Superconductivity was first observed in 1911 by H. Kamerlingh Onnes, the first person to liquefy helium (Gorter, 1964), but applications of the effect were not realized until the early 1960s when superconducting properties of Nb_3Sn wire, and the invention of the Josephson junction, initiated the superconducting magnet and Josephson electronics. The recent intense interest in superconductivity dates to the 1986 announcement of a 30-K transition temperature in a superconducting oxide by Bednorz and Müller. In a two-year burst of activity, thin-film superconducting oxides with transition temperatures to 125 K were produced, giving rise to the description high-temperature superconductivity (HTSC). Recent activity in the mercury-based layered copper oxides has raised this temperature to 133 K (Schilling *et al.*, 1993).

Although the frequency dependence of superconductor surface resistance varies as f^2 compared to the more gradual $f^{1/2}$ variation in normal conductors, much lower absolute values of resistance can be attained by superconductors at frequencies in the microwave region (Fig. 7). At 77 K, the most-studied HTSC oxide $YBa_2Cu_3O_7$ (or YBCO) has a surface resistance of 0.2–0.6 mΩ at 10 GHz, compared to 12 mΩ for copper. This advantage increases markedly below 10 GHz and disappears at frequencies above 100 GHz. As the technology of the oxide superconductors matures, the liquid-nitrogen loss of YBCO should approach that of niobium at 4.2 K, and the liquid-hydrogen loss of YBCO should approach that of Nb_3Sn at 4.2 K.

The physics of growth techniques, physical properties, and low-temperature (4 K) performance of the HTSC oxide films are not well understood; nevertheless, performance at liquid nitrogen temperatures may be approaching practical limits. A particular item of interest is film stability; patterning of YBCO films causes increases in film resistance due to the heavy concentration of current at the edges of the conductor. This issue is of vital significance to successful realization in media like microstrip, and for devices requiring narrow line structures. Substrates for these oxides necessarily match the crystal lattice structure of the as-deposited films. The compounds $LaAlO_3$, MgO, and α-Al_2O_3 (sapphire) are all acceptable substrate materials but except for sapphire are dominated by dielectric losses in the microwave region at 77 K. Development of low-loss substrate materials, therefore, will continue to be an important part of HTSC work for future microwave applications.

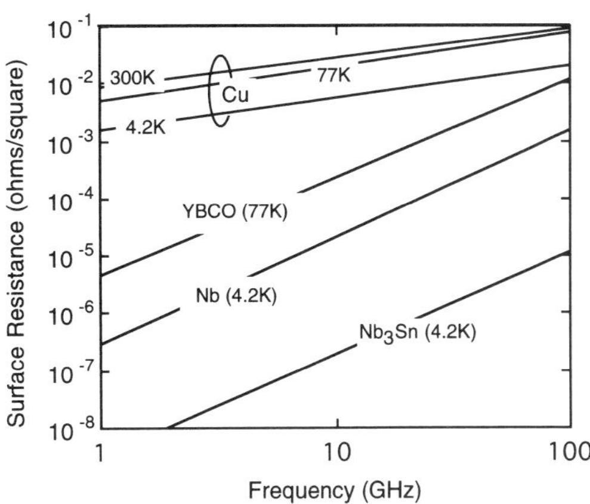

FIG. 7. Surface resistance vs frequency for selected normal conductors and superconductors.

Electromagnetic coupling between adjacent conductors in planar transmission media depends on the ratio of conductor spacing to dielectric thickness, and can be precisely computed for applications such as filters and directional couplers. Unintentional coupling (crosstalk) is reduced to minimal effect when the conductor spacing is more than a few dielectric thicknesses. Circuit density limitations due to crosstalk can be decreased while maintaining constant impedance by reducing the thickness of the substrate and the line width, but at the cost of increased loss. While all conductors experience increased losses with reduced conductor widths, superconductors, for the same loss and impedance level as normal conductors, allow increases of greater than two orders of magnitude in circuit density considering coupling effects.

Applications of HTSC thin films in the microwave frequency region exist where the system cost of providing the cooling necessary to keep the film cooler than the critical temperature is justified by the reduction of loss. Immediate uses of this technology exist in spacecraft signal-processing systems, where passive radiation-cooling techniques can achieve component temperatures below 100 K, and simple single-stage closed-cycle refrigeration systems can reach temperatures below 60 K.

2.5 Future Developments

Attenuation characteristics vary widely among different propagation media (Fig. 8) and therefore serve different needs of signal transmission. Although the most recent transmission-line configurations discussed above are nearly 20 years old, new life for transmission media in those applications where cooling to 77 K can be accommodated is promised by HTSC technology.

Development in substrate materials and packaging technology has helped to maintain a steady growth in this area, which has a great deal of overlap with circuit integration. Thermoset microwave materials, for example, combine the dimensional stability attributes of hard substrates with the machinable and low-cost characteristics of soft substrates. Typical costs for soft dielectric substrates are \$0.15/cm^2, compared to ceramic substrate costs of greater than \$1/cm^2, and semiconductor substrate costs of \$10/cm^2.

In the area of semiconductor materials, work will continue on integration of lightwave components such as light-emitting diodes, semiconductor lasers, and waveguide with microwave and millimeter-wave components used for modulation and frequency conversion purposes. Multilayer propagation media address the need for increased packaging density by stacking layers of signal and ground planes sandwiched between dielectrics. A popular version of this technique is low-temperature co-fired ceramic (LTCC). Thick-film conductive paste defines signal and ground traces on unfired ceramic tapes, which are stacked and fired to final configuration. Low-temperature co-fired ceramic finds use to frequencies as high as 20 GHz but is lossier and more expensive than thin-film techniques on single-layer ceramic. Organic polymers such as benzocyclobutane, an organic dielectric with low dielectric constant and low loss tangent, offer an alternative method of

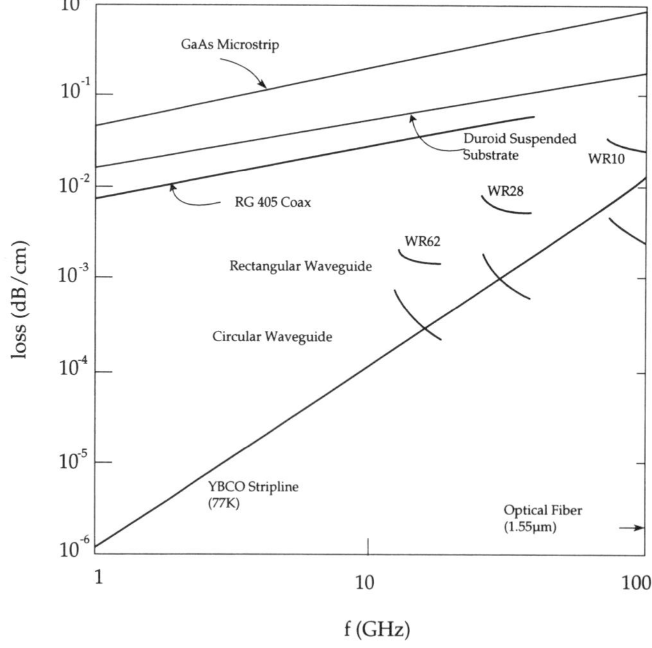

FIG. 8. Attenuation vs frequency of selected microwave and optical propagation media. Waveguides are copper. Circular guide diameters are chosen to match cutoff frequency of corresponding rectangular guide.

constructing three-dimensional propagation media using thin-film metallization techniques.

3. COMPONENTS

Single-function circuit elements used for signal control, frequency conversion, and amplification employ lumped and distributed circuits and propagation media described above with active elements such as transistors and diodes. Control components modify the amplitude or phase of the input signal, providing methods of routing the signal through a microwave subassembly. Signal generation and amplification components, classified by signal level, play a key role in both receiver and transmitter sections of microwave subsystems. Frequency conversion components include mixers and frequency multipliers. Both are used extensively in subsystems where (increasingly digital) signal *processing* is most efficiently implemented at low (baseband) frequencies, whereas signal *transmission* is at higher microwave or millimeter-wave frequencies. At these frequencies, large information content (or equivalently, bandwidth) is a small percentage of the carrier frequency, allowing manageable component design.

3.1 Passive and Signal Control Functions

Attenuators are used to control the amplitude of a microwave signal in microwave systems. One-port attenuators (terminations) eliminate reflections from transmission lines and are realized in either lumped-element (smaller, less expensive) or distributed-circuit (higher-frequency, lower-reflection coefficient) form. Mechanically variable attenuators are realized in waveguide with the controlled insertion of lossy E-plane card into the center of the guide. Variable attenuators of this type are used in precise measurements of power or control of signal level in instrumentation applications. Input- and output-matched attenuators of the pi or tee (Fig. 9) construction provide losses of less than 1 dB to greater than 20 dB, and represent a lumped-element approach to the fixed attenuator. Thin-film resistors on planar substrates or shunt discs and series rods for coaxial use are common. For waveguide, a tapered resistive card located in the E plane of the guide will provide fixed loss with very small return loss. Using *p-i-n* diodes or metal-semiconductor field-effect transistors (MESFETs) as current- or voltage-controlled (respectively) variable resistors, and the matched-tee approach as described above, inexpensive ($0.60/chip) low- to medium-power variable attenuators exhibiting extremely large bandwidths can be fabricated. Analog dc voltages control the drain–source resistance of the field-effect transistor (FET) to achieve the required attenuation.

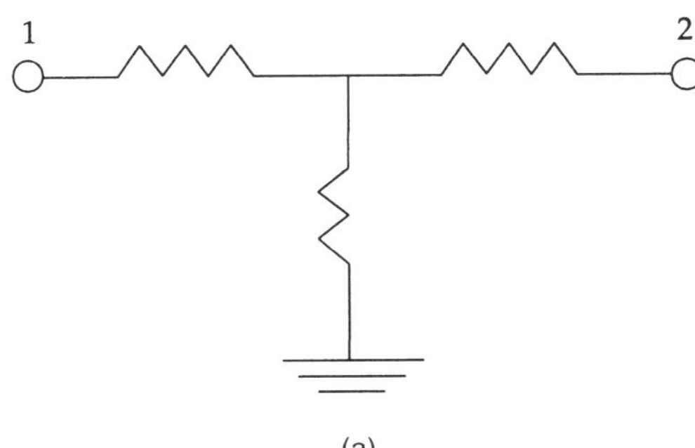

(a)

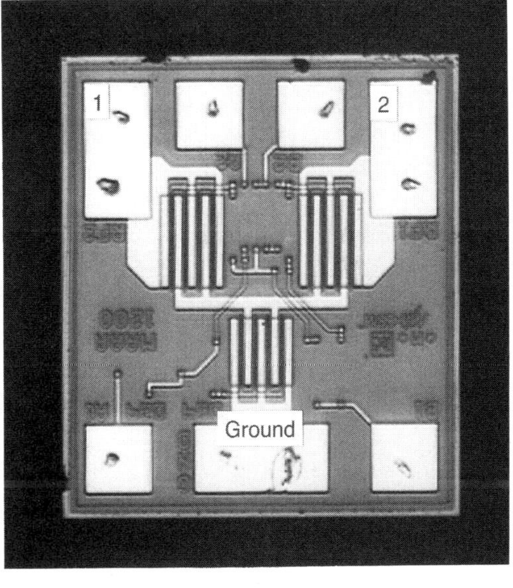

(b)

FIG. 9. (a) Matched-tee variable attenuator (schematic). (b) Matched-tee attenuator in MESFET MMIC technology; chip size: 600 × 600 μm. (Courtesy of M/A-COM, Inc., Lowell, MA.)

Microwave switches can be of mechanical construction, or can be realized with semiconductor devices (*p-i-n* diodes or FETs), or with ferrite devices. Mechanical waveguide switches have extremely low mismatches (VSWR < 1.05) and are used in instrumentation applications where switching speed, size, and cost are of secondary importance. Switching with *p-i-n* diodes or FETs offers advantages of fastest switching speeds (<10 nsec), low (*p-i-n*) or extremely low (FET) drive-current requirements, and the possibility of monolithic integration to realize higher functional integration. Both devices have a high-impedance state and a low-impedance state that depend upon bias, and can be used in the shunt or series configuration to increase on/off ratio (isolation). Although *p-i-n* diodes have been successfully used in applications to 100 GHz, and offer lower losses than FETs, low-cost monolithic integration techniques have made FET-based switches at lower microwave frequencies one of the early successes of gallium arsenide (GaAs) commercial foundry operations (Fig. 10). "Drivers" for switches transform digital-logic voltage levels to proper current waveforms for *p-i-n* diodes or proper voltage levels for FETs. These circuits are evolving from a hybrid form (using discrete components on a separate printed circuit board) to a monolithic form, either as a separate application-specific integrated-circuit (ASIC) chip or as part of the microwave monolithic circuit fabricated to support both digital and analog functions. Ferrite switches are three-port devices that are inherently single-pole, double-throw. These devices offer lower loss than the semiconductor switches and higher power handling, with the disadvantages of microsecond switching speeds, larger size, and more complicated driver circuitry. These devices require no bias in the steady state; i.e., they are latching switches. Issues of size, cost, speed, integration ability, and driver complexity are causing semiconductors to displace ferrites in all but a few custom applications.

Because of the importance of phase in any signal-processing system, phase shifters serve important roles as calibration tools, as balance or trim elements in multichannel systems, or to realize efficient power combining. They are, of course, key elements in phased-array radars, where beam steering is achieved electronically by controlling the phase relationship among thousands of individual transmit/receive modules. Continuous phase shift can be realized mechanically with "trombone"-like coaxial lines up to 20 GHz, and with waveguide configurations similar to the variable attenuator described above, where the resistive card is replaced by a low-loss dielectric card. Discrete values of phase shift can be electronically realized using FETs or *p-i-n* diodes as electronic switches, to select from one of two phase-shift states ("bits"). Analog techniques using the variable (with dc bias) reactance properties of low-loss varactor diodes are used to achieve continuous phase shift with application of dc bias. Here, too, ferrites offer alternatives to semiconductors for phase-shift applications but show ad-

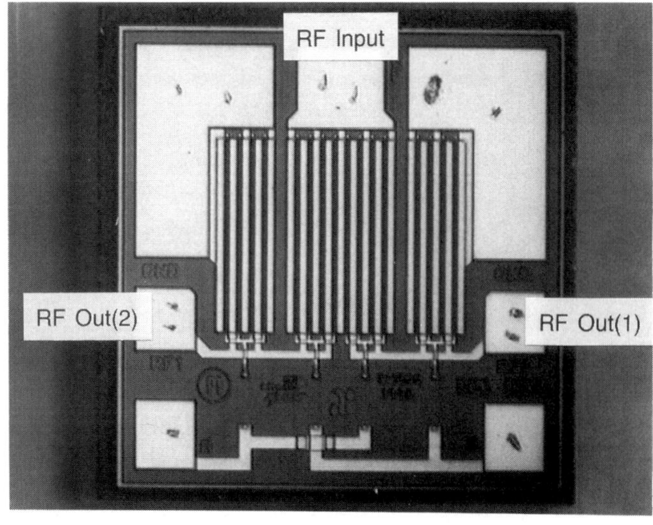

FIG. 10. Single-pole, double-throw MMIC switch in MESFET MMIC technology; chip size: 600 × 600 μm. (Courtesy of M/A-COM, Inc., Lowell, MA.)

vantage only where highest powers (tens of kilowatts) are required. Monolithic realizations of semiconductor phase shifters will operate to higher frequencies, switch faster, but handle less power than hybrid realizations.

The anisotropic permeability in ferrites allows the direction of wave propagation to be modified by the application of a dc magnetic field. This property has been exploited in the development of nonreciprocal devices, two examples of which are circulators and isolators. A junction, or Y, circulator is a three-port network in which signals (waves) may "circulate" from port 1 to 2, 2 to 3, and 3 to 1, but not in the opposite direction, while all ports are matched at their inputs. These junctions are commonly implemented in stripline (Fig. 11). With the application of an axial dc magnetic field, an electric field null is placed at port 3, and "circulation" of power from port 1 to port 2 is allowed. The direction of circulation is controlled by the polarity of the dc magnetic field. The conditions for circulation (as well as impedance matching and loss) depend on operating frequency, magnetic properties of the material, and physical dimensions of the ferrite disc. By terminating the isolated port, three-port junction circulators become two-port isolators. Z-directed magnetic fields resulting from a current-carrying wire placed within the ferrite allow operation without permanent magnets. By reversing the direction of the current, the circulation pattern is reversed, and switching behavior is achieved. Because of magnetic field hysteresis within the ferrite material, no sustaining current is required to maintain a given state. Switching speeds in the microsecond range are attained with these ferrite latching devices.

Used to define the frequency band of interest at the component or subsystem level, microwave filters are categorized as low-, band-, or high-pass, or band-reject, depending on the desired frequency selectivity. Realizations are primarily passive at microwave frequencies, although active filters, including switched-capacitor topologies with high-gain amplifiers as key elements, are used extensively at frequencies below 100 MHz, where inexpensive monolithic chips provide gain block functions. The theory of direct coupled-cavity filters has been advanced continuously since work initiated at the MIT Radiation Laboratory (Levy and Cohn, 1984). Distributed circuit theory in filters traces its origin to Richards, who established a simple relationship between lumped and distributed circuits in 1948, allowing the existing lumped-element filter theory to be applied to distributed realizations. In the intervening time, synthesis techniques have been developed for increasingly compact designs. Higher-order mode cavities and dielectric resonators have allowed further size reduction. Today, filter theory provides a valuable base for integrated designs requiring large

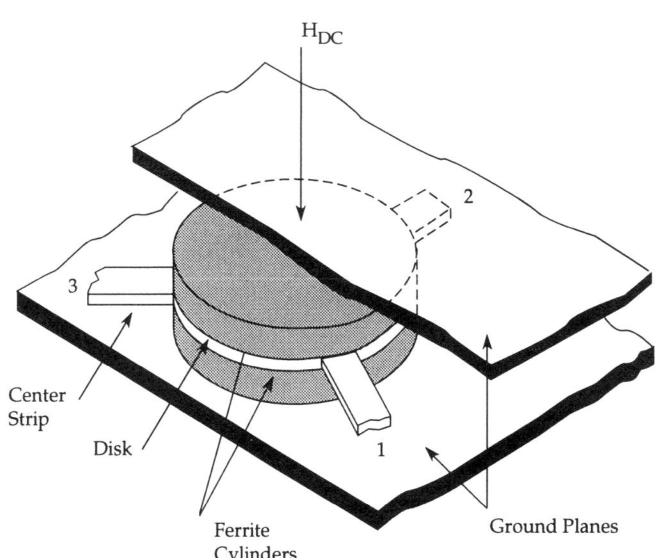

FIG. 11. Stripline junction circulator.

percentage bandwidths. Conventional component requirements for filters include limiting the noise input power to a sensitive receiver, reducing spurious signals generated by active or nonlinear components in a system, separating input and output signals in nonlinear components such as mixers and multipliers, and segmenting wide-bandwidth (octave or greater) receivers into frequency bands that can be processed by components with modest bandwidths. Microwave spectrum analyzers are instrument-grade versions of wide-band receivers, which use fixed and tunable filters to identify frequency of input signals.

Four-port junctions have special properties which have made them very useful in microwave circuits, including applications of power combining, signal duplexing, and power coupling. Important elements of these junctions were discovered by Schwinger and Dicke in the early 1940s at the MIT Radiation Laboratory using mathematical symmetry arguments (Montgomery *et al.*, 1948). These plus later synthesis techniques in planar transmission media are major milestones in the history of passive microwave circuit development. Hybrids, a special class of four-ports, are impedance matched at all ports simultaneously. [This special use of the word "hybrid" should not be confused with its other use in microwave circuits (see Sec. 4, below) to describe a mixture of integrated circuits and discrete devices on a common substrate.] A signal incident on any port is coupled to two others and isolated from the fourth. Hybrids with equal power coupling are called 3-dB hybrids or *magic tees* and have been realized in waveguide and distributed- and lumped-element forms. The magic tee in waveguide consists of an H-plane tee junction with the fourth port forming an E-plane tee [Fig. 12(a)]. When a TE_{10} mode is incident on port 1, the electric field pattern distribution is such that port 4 is completely isolated, and the power split is equal and in phase between ports 2 and 3. Likewise, for TE_{10} excitation at port 4, port 1 is isolated, and power is equally split with phase difference of 180° between ports 2 and 3. Lumped-element and planar versions of the magic tee also exist. Directional couplers are four-port hybrids in which portions of the forward and reverse traveling waves are separately coupled to two additional ports. Realizations of these components usually involve transmission lines whose relative proximity determines the coupling, but can be implemented with lumped elements as well [Fig. 12(b)]. Synthesis procedures in many media have been developed (Wadell, 1991). Special constructions have been invented for specific media. In microstrip, for example, couplings tighter than 6 dB require extremely narrow line spacing or interdigital construction techniques. Waveguide directional couplers exhibit extremely low loss and directivities as high as 50 dB, which find extensive use in instrumentation or measurement applications where amplitude and phase information of a traveling wave is desired.

3.2 Signal Generation and Amplification

Techniques for microwave signal generation address a variety of application-specific requirements. For communication, radar, and instrumentation purposes where efficient use of allocated bandwidths is important, stability (frequency change with time) and spectral purity (spectral line width and spurious signals) of frequency sources are key requirements. For industrial applications such as microwave heating, these requirements are much less demanding but are replaced by the need for high power and dc/rf power conversion efficiency. Microwave frequency generation at the fundamental frequency uses active devices with gain in a circuit which provides feedback appropriate to the conditions for oscillation. Oscillators can be tuned mechanically or electronically by perturbations of the resonant circuit. The quality factor of the resonator will determine the tuning range and stability of the oscillator, with high Q's resulting in narrow-band, low–phase-noise components. [The figure of merit used for oscillator comparison, phase noise, is caused by phase fluctuations within the oscillator, which broaden the frequency spectrum of the source. The term $\mathscr{L}_m(f_m)$ describes the rf power spectrum and is defined as the ratio of the single-sideband phase noise in a 1-Hz bandwidth f_m Hz away from the carrier frequency to the total signal power.] Quartz crystal resonators at frequencies of 5–100 MHz, with unloaded Q's greater than 10^6 at 5 MHz, followed by frequency multiplication to the microwave region, or with dielectric or cavity resonators used as the fre-

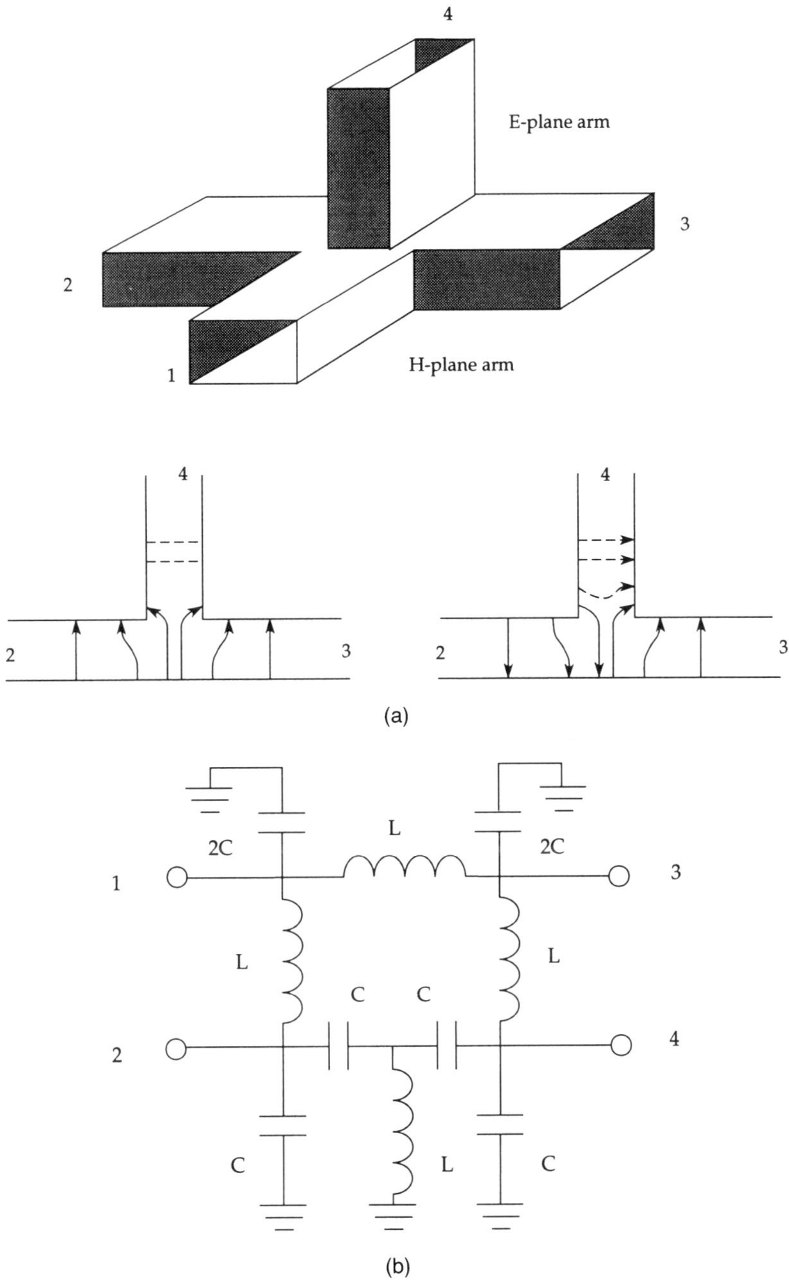

FIG. 12. (a) Waveguide magic tee; and (b) lumped element magic tee.

quency-determining element in fundamental frequency oscillators are commonly used to provide low–phase-noise sources. The active element used in the oscillator also affects the phase noise: Oscillators using Si bipolar junction transistors have better phase noise performance than FETs in otherwise identical oscillator environments (Fig. 13).

Microwave amplifiers are designed for various applications, including a low-noise region, where very weak signals at receiver inputs are amplified with a minimum of internally generated noise and spurious signals, and a high-power region, where power output and conversion efficiency are critical characteristics. Low-noise amplifiers (LNAs)

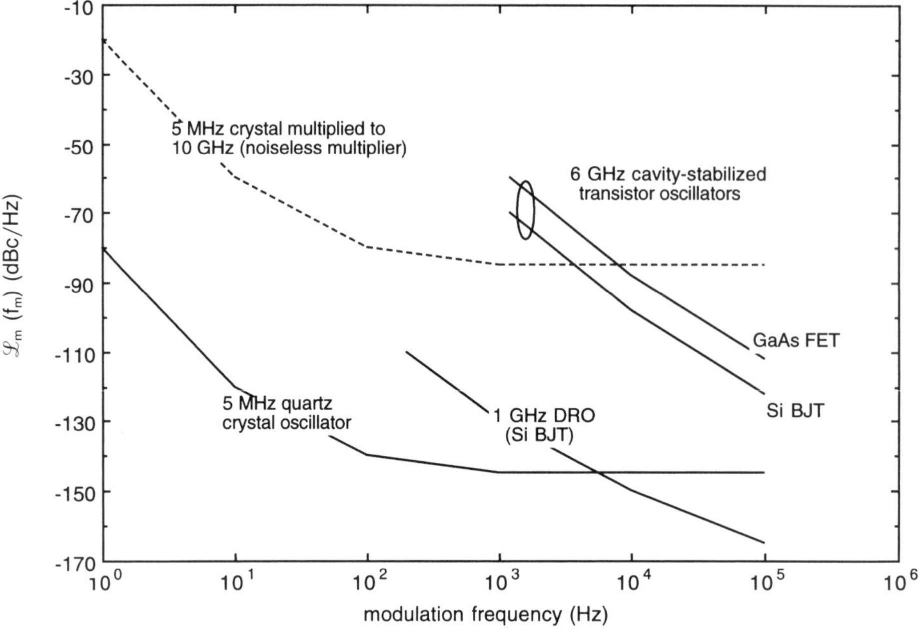

FIG. 13. Oscillator phase noise vs modulation frequency (normalized to operation at 10 GHz).

used at the input of sensitive receivers amplify signals that in some applications can be lower than background thermal noise power of −174 dBm (measured in a 1-Hz bandwidth at room temperature). An ideal (noiseless) component will add no noise to the thermal noise available at its input. Noise figure F is the ratio of output noise power to the product of amplifier gain by input noise, whose minimum value of 1 is being approached by a number of components in the microwave region (Fig. 14). The maser provides the ultimate in low-noise amplification, but its expense and need for cryogenic operating temperatures limit its use to the most demanding of applications such as radio astronomy. Cooled three-terminal solid-state devices offer lower cost and size advantage with a slight degradation in performance. Parametric amplifiers, developed during the early 1960s, use variable reactance (*varactor*) diodes to accomplish frequency conversion with gain and little additive noise, but are being replaced by cooled high–electron-mobility transistor (HEMT) amplifiers, which show excellent low-noise results to frequencies as high as 100 GHz. The 3-K galactic noise temperature, measured at 6 GHz by Penzias and Wilson in 1964, is one of the most profound examples of microwave remote sensing. Measurements on the galaxy made with an extremely low-noise antenna/receiver at Bell Laboratories repeatedly yielded noise powers higher than expected. Later verified as accurate readings, these noise measurements were used to support the Big-Bang theory of the birth of the universe.

Methods of amplifying large signals include vacuum-tube techniques, which are capable of megawatts of power but are expensive and large, and solid-state approaches, which are suited for lower power applications but offer the cost advantages of high-volume semiconductor manufacturing technology (Fig. 15). Vacuum-tube techniques include traveling-wave tubes, magnetrons, crossed-field amplifiers, gyrotrons, and klystrons (see VACUUM AND GASEOUS ELECTRON TUBES). Developments in this field date from the mid-1930s. These components use structures whose fabrication requires a great deal of mechanical precision (high cost), whose operation requires high-voltage power supplies (>1 kV), and whose lifetime is limited by tube "wearout" and power-supply failure. In contrast, solid-state methods of generating power in the microwave and millimeter-wave frequency region include two-terminal

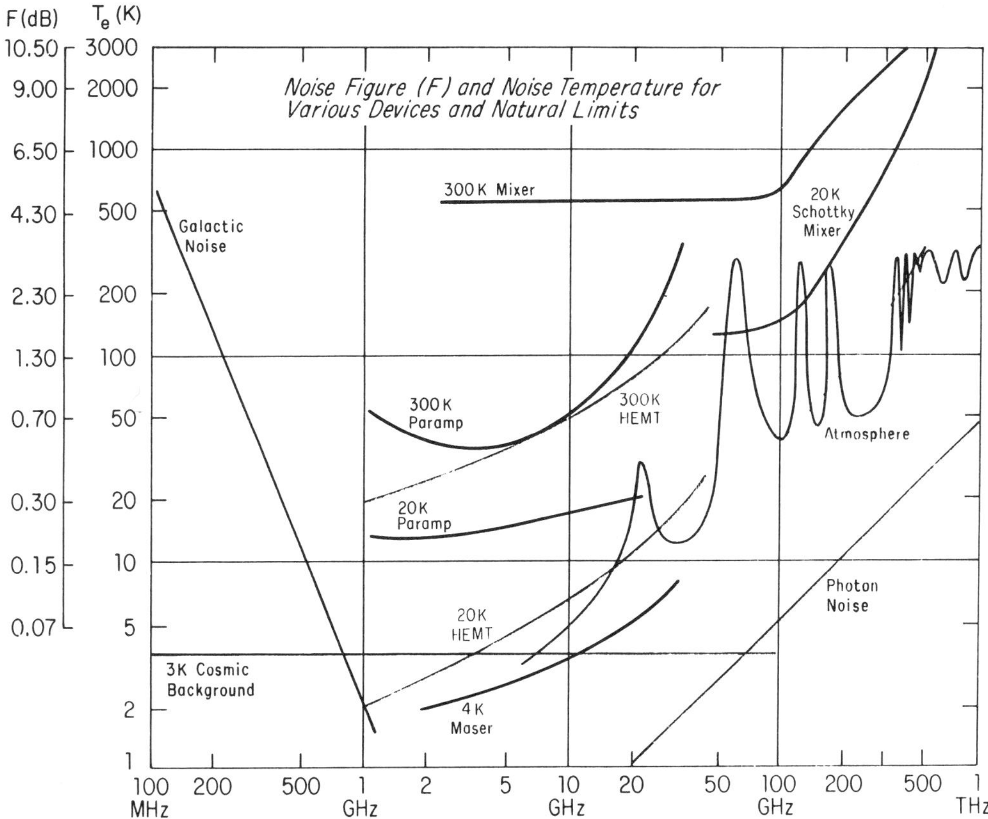

FIG. 14. Noise performance (in equivalent measures of noise figure and noise temperature) of microwave receiver components. Indicated temperatures are physical temperatures of devices or components. (Courtesy of National Radio Astronomy Observatory, Charlottesville, VA.)

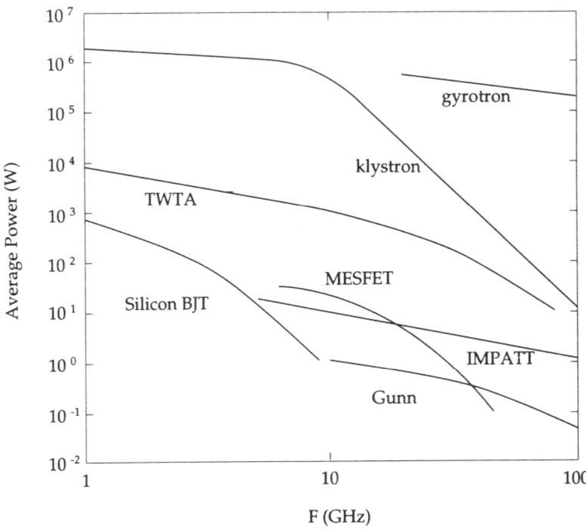

FIG. 15. Output power vs frequency for solid-state and vacuum-tube amplifiers.

devices such as IMPATT and Gunn diodes, and three-terminal devices such as MESFETs, heterojunction bipolar transistors (HBTs), and HEMTs. General features of solid-state power generation techniques include use of low-voltage power supplies (<40 V) and device lifetime predictions in the millions of hours. In the three-terminal device area, much effort is being applied in the GaAs homojunction and heterojunction areas to produce watt-level modules at frequencies as high as 35 GHz. Power densities twenty times higher than those achieved in FETs have been observed in HBTs, which also show extremely linear operation and high power-added efficiency ideal for cellular radio handset applications. In the portable telephone marketplace, "talk-time," or battery life, is crucially dependent on the efficiency of the microwave power amplifier. Advances in three-terminal solid-state power capability in the millimeter-wave frequency region during the next decade will be driven by automotive applications which are targeted for the 60–80 GHz range; already FETs, HEMTs and HBTs exhibit small-signal oscillation frequencies which satisfy these requirements.

3.3 Frequency Conversion Components: Mixers and Frequency Multipliers

The frequency mixer, a three-port device, converts an input frequency signal (f_{in}) to an output signal using a third signal, called the local oscillator (f_{LO}). Based upon device nonlinearities within the mixer, the output signal spectrum consists of all the multiplicative single-tone intermodulation products, namely $\pm mf_{in} \pm nf_{LO}$ (m and n are the positive integers, including zero). It is the task of the mixer or circuit designer to select only the output frequency or interest (usually the $m = n = 1$ product). Other intermodulation (IM) products are suppressed using filtering external to the mixer or by symmetry within the mixer. The higher-order responses (high values of m and n) are not efficiently converted within the mixer. Devices used in mixer design are chosen for low loss, low noise, and nonlinearity; Schottky diodes, FETs, and, more recently, HBTs are most popular. Silicon diodes exhibit lower spectral noise than GaAs diodes but have greater conversion loss. Mixers based on transistors as nonlinear elements are lower in performance than the best diode mixers but offer the possibility of monolithic integration with other transistor-based receiver functions. In receivers, mixers can contribute a large part of the IM distortion, which sets the large-signal limit on dynamic range, the range of signals that allow spurious-free or linear operation. Reducing the gain in front of the mixer stage to reduce the mixer IM distortion contribution will increase the system noise figure and increase the small-signal limit on dynamic range. Receiver design is a compromise between dynamic range and sensitivity, which requires careful choice and design of the individual components as well as detailed attention to the frequency plan to avoid unwanted mixing products and other spurious signals. The simplest mixer uses a single semiconductor diode. (Single) balanced mixers use two diodes and either a 90° or 180° hybrid to provide LO-rf isolation or even-order harmonic rejection. Double-balanced mixers using four diodes provide isolation among all three ports while suppressing even-order harmonics (Fig. 16). Additional balancing techniques employing increased circuit complexity can be used to reduce intermodulation terms further. Conversion losses of 3–8 dB are typical, although careful reactive termination of unwanted harmonic outputs can reduce the loss below 3 dB.

Analytically, frequency multipliers are degenerate cases of mixers (no local oscillator), with output frequency spectrum given by mf_{in}. Multipliers offer methods of converting low-frequency, ultralow-noise signal sources (produced by crystal oscillators in the 5–100 MHz frequency range) to microwave and millimeter-wave frequencies. Conversion efficiencies vary from a few percent (typical of multipliers that use resistive nonlinearities) to nearly 100% for multipliers that use reactive nonlinearities. These latter devices have been developed in high-power versions for providing watt-level power in the millimeter-wave frequency range. Spurious responses are reduced using circuit balance and filter techniques, as for mixer circuits.

3.4 Future Developments

Control functions such as switches and attenuators will increase in complexity with the addition of digital logic and digital-to-analog conversion to the analog semiconductor functions noted above. Switch and amplifier topologies using active elements, and designed as filter topologies, will allow efficient space-conservative broadband designs. Improvements in materials and electromagnetic design tools will allow continuous reductions in the size of filter functions while maintaining performance. At the same time, digital techniques will invade the lower-GHz frequencies and establish equivalent functions that can be integrated into the rest of the digital circuit functions. Thin-film acoustic resonators (Lakin *et al.*, 1993) will compete with high–dielectric-constant materials for filter applications where small size and integration with other functions are important. Ferrite circulators, isolators, and

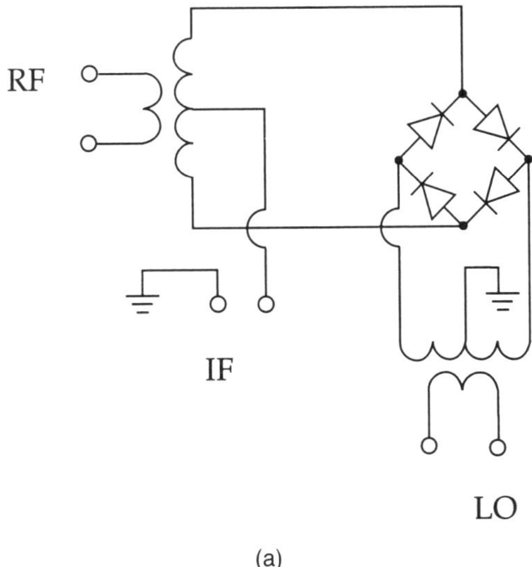

(a)

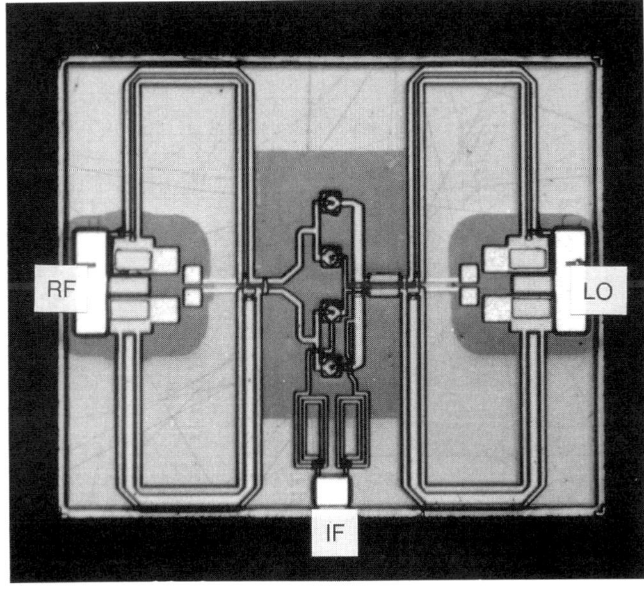

(b)

FIG. 16. (a) Double-balanced mixer schematic; (b) realization using Si monolithic Schottky diodes and glass circuit integration; chip size, 1.9 mm × 2.3 mm. (Courtesy of M/A-COM, Inc., Lowell, MA.)

switches will be replaced by semiconductor equivalents as needs for small size and low cost force alternative approaches.

With increases in the power capability of solid-state devices, most likely to occur within the three-terminal category, the area of overlap with vacuum techniques will gradually shift to higher power–frequency products. Advances in multiple-channel HEMTs and pseudomorphic HEMTs for power, extension of the HBT to millimeter-wave frequencies, and development of FET material properties allowing higher power density will allow transistors to compete with tubes at higher power levels. Quantum effects will be used to push operation to higher frequencies, and diamond will be used as a semiconductor to achieve reliable operation at extremely high temperatures (Trew *et al.*, 1991). Combining the advantages of vacuum microelectronics with modern solid-state fabrication techniques, research in field-emitter arrays (FEAs) has been recently rejuvenated. The structure of the FEA allows significant current densities from micron-sized conical semiconductor mesas with modest applied voltages, giving projections of performance competitive with solid-state devices. Projected improvements in printed antenna distribution networks, and the possibility of integration of HTSC films with the active components of monolithic microwave integrated circuit (MMIC) technology such as the gate structure of FETs or HEMTs will improve receiver noise characteristics.

4. CIRCUIT INTEGRATION

Markets for single-function components include instrumentation and moderate-quantity microwave subassemblies, where performance is more critical than size and cost. With automated assembly techniques, discrete devices are used to manufacture multifunction assemblies in large volume. As the microwave integrated circuits (MICs) and monolithic integration of multiple circuit functions improve in performance and become less expensive to prototype, these methods of integration will serve component customers who need increasingly complex functionality, over broader frequency bandwidths, in less space, and at a lower price.

4.1 Microwave Integrated Circuits

MIC technology uses planar dielectric substrates that carry printed transmission lines and other planar passive circuitry, together with discrete passive and active devices to realize complex circuit functions. The earliest attempts to construct planar transmission lines on dielectric substrates used conductive pressure-sensitive tape to define the transmission lines. By plating both sides of a soft dielectric substrate with copper and skillfully using a sharp blade, similar results could be achieved, with resolution dependent upon the expertise of the individual, but limited to tenths of millimeters. With the development of photoresist and metal thin-film technology together with the availability of smooth dielectric substrates, however, the definition and resolution of patterns have moved to the range of microns, and batch processing has become a low-cost method of reproducing circuits. Microwave integrated circuits are used where timely empirical design verification, modifications, and repairs are important. Discrete devices can be selected before integration to assure high performance, and relatively quick design realizations can be achieved (a week or so from finished photomask to circuit).

Both soft and hard substrates are popular for use. Soft substrates made of materials such as PTFE or PTFE-based material are easy to machine into intricate shapes and are available in relative dielectric constant range from 2 to 10 (Table 1). Dielectric constants at the higher end of this range are obtained by using ceramic filler in the PTFE base. Pattern resolution is of the order of 20 μm. Hard substrates can be ground or polished to surface finishes (<1 μm) that result in better pattern resolution but require diamond machining tools or laser cutting techniques. Relative dielectric constants of commonly used hard substrates vary from 4 to 40, although dielectrics with $\epsilon_r > 100$ are available. Glass offers an attractive substrate choice because of its low microwave loss, smooth surface finish, and compatability with semiconductor processing (Fig. 17). Both substrate types allow multilayer structures consisting of alternating layers of conductive transmission lines or ground planes separated by insulating dielectric. Conductive interconnects between transmission-line layers or to electrical ground planes are formed by metallic plugs or drilled and plated "via holes" through the insulating dielectric layers.

Mounting of active power devices or other dissipative elements requires attention to the thermal path from the device to thermal ground. A number of methods have been devised to handle this problem in the MIC environment. Because of its high thermal conductivity and relatively low loss tangent, and in spite of its toxicity, beryllium oxide (BeO) has been used as microwave substrate material for microwave power amplifiers as well as for device packages. Aluminum nitride (AlN), an excellent thermal conductor and good dielectric, is an environmentally acceptable alternative to BeO. Epitaxial diamond, combining the highest thermal conductivity with excellent insulator properties, is commercially available in thicknesses of 1 mm and areas of 50 cm^2, allowing substantial improvements in power handling on MICs.

Table 1. Electrical and thermal properties of materials used in microwave circuits.

Material	Bulk conductivity (S/m)	Relative dielectric constant	Loss tangent (at 10 GHz)	Thermal conductivity (W/cm-°C)	Thermal expansion coefficient (ppm/°C)
Soft dielectrics					
PTFE glass random fiber	5×10^{-12}	2.2–2.4	1×10^{-3}	1.7×10^{-4}	100
Benzocyclobutane (BCB)	1×10^{-18}	2.65	8×10^{-4}	0.007	35–60
Cross-linked polystyrene	1×10^{-14}	2.55	5×10^{-4}	0.015	70
Hard dielectrics					
Fused silica	2×10^{-17}	3.8	1×10^{-4}	0.01	0.55
7070 glass	$<1 \times 10^{-15}$	4	2×10^{-3}	0.01	3.2
CVD diamond	$<1 \times 10^{-14}$	5.6	5×10^{-4}	13–15	2.6
BeO	$<1 \times 10^{-16}$	6.6	6×10^{-3}	2.6	7.8
Low-temperature co-fired ceramic	$<1 \times 10^{-16}$	7.1	1×10^{-3}	0.2	7.9
Si_3N_4		7	3×10^{-3}	0.3	2.3
AlN	$<1 \times 10^{-15}$	8.8	4×10^{-3}	1.7	4.4–4.8
(99.5%) Al_2O_3	$<1 \times 10^{-16}$	9.6	2×10^{-4}	0.19	6.7
Titanium dioxide	1×10^{-14}	40–100	4×10^{-3}	0.06	7.5
High-temperature superconductivity materials (properties at T = 77 K)					
YBCO	$7 \times 10^{+9}$				10
TCCBO	$7 \times 10^{+9}$				10
α-Al_2O_3 (sapphire)	1×10^{-16}	11.5	1×10^{-6}	0.42	6
MgO	1×10^{-16}	9.9	5×10^{-4}	6	0.8
$LaAlO_3$	1×10^{-16}	24.2	1×10^{-4}	5	10
Semiconductors (properties for semi-insulating material)					
GaAs	1×10^{-7}	12.9	2×10^{-3}	0.46	6.9
Si	1×10^{-3}	11.7	1.5×10^{-2}	1.5	4.2
InP	1×10^{-7}	12.6	1×10^{-3}	0.68	4.8
Metals/alloys					
Cu	$6 \times 10^{+7}$			4	16
Al	$3 \times 10^{+7}$			2.2	24
Cu/W (20%/80%)	$3 \times 10^{+7}$			2.5	8.5
Pb/Sn (eutectic)	$7 \times 10^{+6}$			0.5	25
Ni/Co/Fe (Kovar™)	$(1–2) \times 10^{+6}$			0.17	5.5

Prices today are nearly \$200/cm^2 (about as expensive as a completely processed GaAs monolithic integrated circuit) but are projected to drop to less than \$1/cm^2 by 1997. Even with the present price structure, overall circuit assembly costs may favor diamond substrates over more complicated heat-spreading techniques for some applications.

In normal MIC assembly practice, dielectric substrates are attached to metal carriers for integration into higher-level subassemblies. Thermal coefficients of expansion (TCE) of substrate and carrier must be well matched to avoid fracture of the substrate during thermal cycling of the subassembly. While alloys of Ni/Co/Fe provide good TCE match to Si and GaAs and are relatively inexpensive, the thermal properties are relatively poor. Copper/tungsten alloys achieve good thermal match to both alumina and GaAs while maintaining excellent thermal properties.

4.2 Monolithic Microwave Integrated Circuits

Properties of semiconductors such as Si and GaAs allow the addition of active devices such as transistors and diodes to a library of process-defined passive components. The key to success of passive functional integration is the low electrical conductivity of the substrate; in this regard, GaAs significantly outperforms Si. The monolithic integration of device and circuit functions on a single substrate holds the promise of reducing or eliminating expensive skilled manual labor used for discrete die attach and wirebonding operations, and has received support of in-

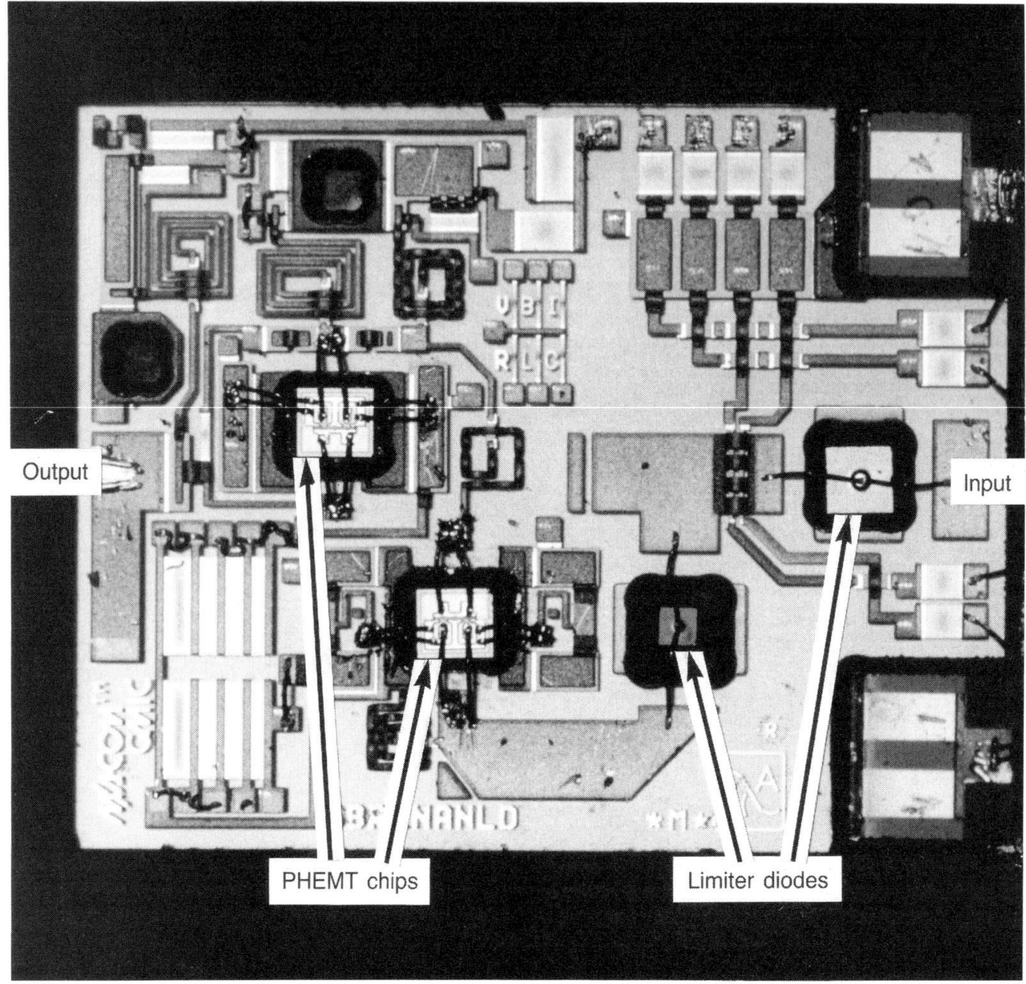

FIG. 17. 9–10 GHz limiter/low-noise amplifier MIC; component size, 3 mm × 4 mm.

dustry and government funding since the early 1980s.

The use of MMIC technology is prescribed when high volume and low unit cost are paramount, or in certain instances where circuit performance can be achieved only with monolithic fabrication. Large-scale integration and multifunction capability in a very small area are possible. The higher density of circuit elements minimizes undesirable interconnect lengths and results in higher operating bandwidths than MICs. Fewer wirebonds result in higher reliability.

Surface finish of semiconductor substrates is held to much closer tolerances than that of MIC substrates. Because active devices with feature sizes (FET gate finger "lengths") of as little as 0.1 μm must be made with high yield, surfaces must be optically smooth and free from scratches and particles. The resulting surface finish allows planar passive components such as resistors and capacitors to be defined as part of the processing of the substrate. These processing steps require more complex substrate preparation and processing but result in significant savings in assembly time, yielding cost savings in high-volume applications. The *airbridge* process photolithographically defines transmission lines (typically gold) to connect conductors separated by an air gap and an obstacle such as another transmission line. The airbridge replaces the wirebond with a precise, repeatable connection, and, because it is part of a batch manufacturing process, many subsequent labor-intensive wirebonding assembly procedures are eliminated.

Monolithic microwave integrated circuit processes are categorized according to the active device required for the bulk of the functional realization. These processes can be diode-based or transistor-based, and use either Si, GaAs, or more advanced semiconductor compounds. A process in use today for inexpensive low-noise as well as power MMIC functions is the GaAs ion-implanted process (Fig. 18). The starting (substrate) material is semi-insulating GaAs of extremely high quality, available as 3-in.- and 4-in.-diam wafers. The active channel layer impurity distribution is defined using a 100–500 keV accelerator. Subsequent wafer processing steps include the deposition and patterning of one or more thin films. Processing of the back side of the wafer can be as simple as mechanically thinning the wafer, or, in the case of a process which involves power devices, can include the formation of back-side metal vias to lower the parasitic inductance to electrical ground, as well as thermal resistance of power devices.

Monolithic device/circuit fabrication has resurrected the technique of distributed amplification (Ginzton *et al.*, 1948), developed to increase the operating bandwidth of vacuum-tube amplifiers. In this idea, transistors, each with finite input and output reactances, are absorbed into separate input and output transmission-line sections, and the resulting structure can be designed to have constant gain from very low frequencies right up to the maximum operation frequency of the individual active devices. Today's transistor technology yields gains to frequencies of 200 GHz, and monolithic distributed amplifier results have been quoted to 100 GHz. Distributed amplifier designs are used heavily in the 6–18-GHz electronic warfare applications (Fig. 19).

Substantial initial fabrication and design costs are required for MMIC development. Material-growth and processing-technology capital costs can be in the \$10 M range for even a modest operation. Approximately 20 GaAs foundries exist in the United States and Europe, offering production capabilities in MESFETs, HEMTs, and HBTs. A new MMIC realization with via holes and 0.5-μm feature size costs between \$50 K and \$100 K and requires three to six months per design iteration. Because the circuit cannot be tuned once fabricated, and because of the high cost of a single foundry run, stability and accuracy of every circuit fabrication step are imperative.

The growth of semiconductor foundries has been paralleled by the increase in speed and complexity of computer-aided design (CAD) programs, providing linear and nonlinear

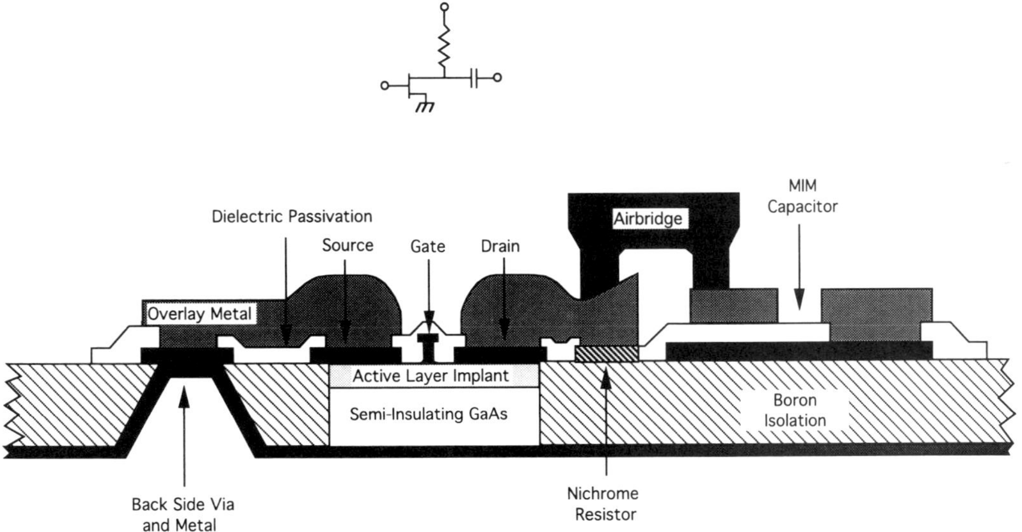

FIG. 18. Gallium arsenide MMIC ion-implanted process (cross-sectional view of processed wafer).

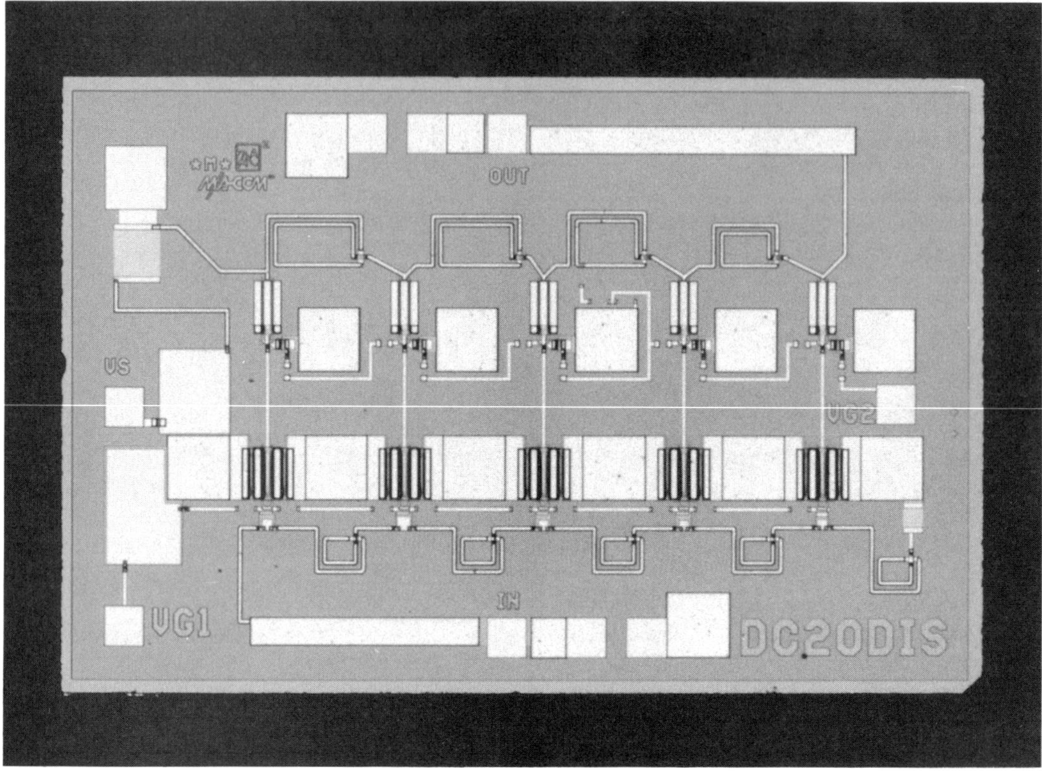

FIG. 19. 2–20 GHz MMIC distributed amplifier; chip size, 1.0 × 1.5 mm.^{2} (Courtesy of M/A-COM, Inc., Lowell, MA.)

analysis, synthesis, and layout of the MMIC circuitry (see COMPUTER-AIDED DESIGN IN ELECTRONICS). A typical "mid-range" [40 million floating point operations per second (Mflops)] workstation with these design capabilities costs approximately \$50 K, although prices on hardware and software are dropping. A "high-end" personal computer (1.5 Mflops) with linear analysis/optimization software can be obtained for around 1/20 the cost of a full workstation. These improvements in processing speed allow detailed electromagnetic field simulations of complex two- and three-dimensional geometries such as transitions between different transmission media, multiline coupling, and package cavities, which have no models in present circuit libraries. As these tools grow in accuracy, circuit realization will consist of design-simulate-analyze iterations using CAD tools, followed by a single circuit-fabrication operation, instead of design-build-measure iterations, the present costly and time-consuming practice.

4.3 Integrated Subassemblies: Selecting from the Menu of Technologies

For high-volume commercial applications, MMIC technology provides a method for producing quantities of millions per year at prices in the \$1–10 range. A complete direct broadcast satellite (DBS) receiver downconverter, for example, consisting of a rf amplifier, oscillator, mixer, and i.f. amplifier, occupies approximately 1 mm^2 of area, allowing inexpensive production of this circuit function.

For applications where quantities are lower (1000–20 000 parts) but performance is extremely important, and where circuit repeatability is a performance driver, higher-performance, generic single-function MMICs may be profitably used as building blocks in a larger, more labor-intensive subassembly. Multichannel wideband receivers where phase and amplitude of signals over wide bandwidths must be identical among channels are examples of multifunction assemblies where

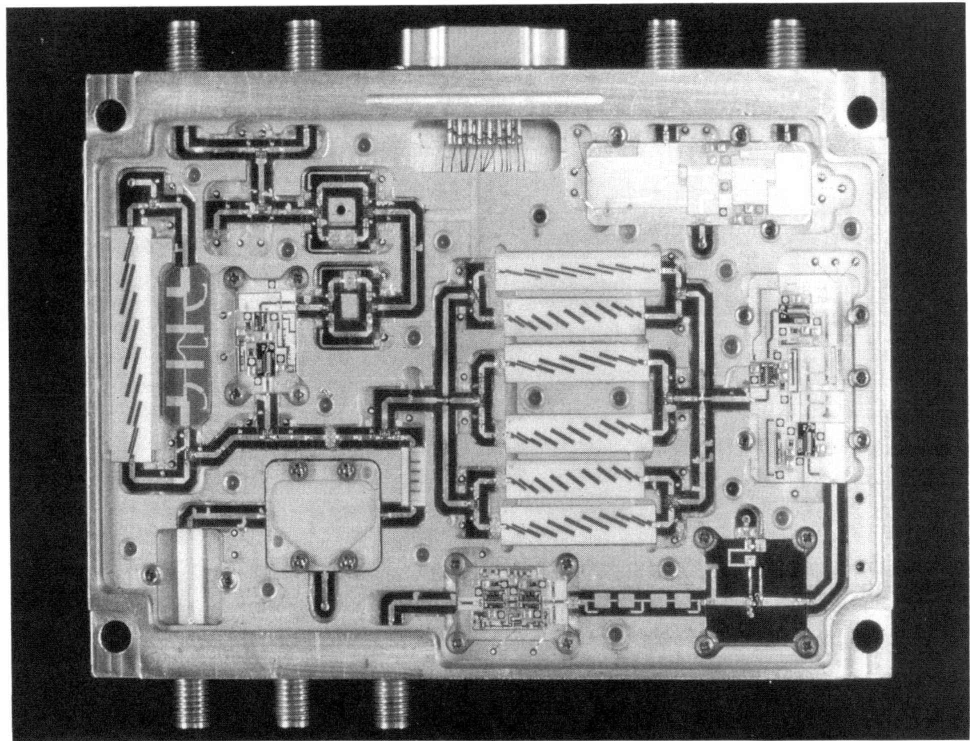

(a)

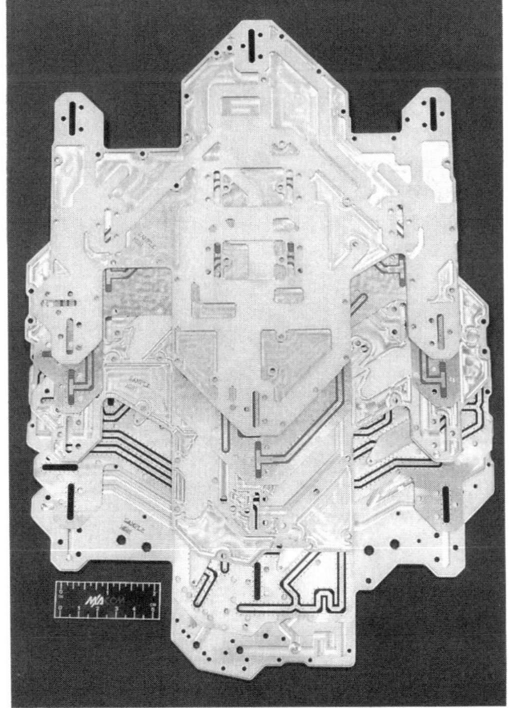

(b)

FIG. 20. Examples of microwave subassembly integration techniques: (a) Multichannel microwave receiver using combination of MIC and MMIC techniques; module size, 15 × 20 × 2.5 cm. (Courtesy of M/A-COM, Inc., Lowell, MA.) (b) X-band antenna element signal-distribution network using waveguide and suspended-stripline techniques; module size, 25 × 35 × 1 cm^3. (Courtesy of M/A-COM, Inc., Lowell, MA.)

circuit tuning is still required to achieve performance specifications; MMIC here is used for low-cost single-function components [Fig. 20(a)].

For other low-quantity applications with extremely low loss requirements, an integration method is used consisting of milled aluminum plates that combine waveguide, suspended stripline, microstrip, and other transmission media. Machining is done with fully automatic milling equipment to lower manufacturing costs [Fig. 20(b)].

4.4 Future Developments

For commercial applications, cost reduction will drive technology development for circuit integration. Monolithic microwave integrated circuit technology will require circuit size reduction, process automation, and manufacturing yield improvements to compete with mature MIC techniques. The cost requirements of a functionally complicated commercial assembly like a cellular telephone rf section will not be met by a single multifunction monolithic circuit. Microwave integrated circuit techniques, which have the cost advantage of inexpensive and low-loss substrates over MMIC technology, are easily adaptable to automated assembly techniques to chip sizes as small as 400 μm square, and thus allow selection of semiconductor chips or MMICs from many processes. This multichip module (MCM) integration philosophy has the advantage that device and circuit innovation can be transferred to production very quickly. Size reduction at the MCM level will use multilayer techniques to increase circuit density and allow process-oriented, chip-connect, multilayer hybrid techniques. Silicon will continue to compete with GaAs, with Si holding the edge on process costs (12-in. Si wafers are routinely being processed for digital memory applications, compared to 6-in. GaAs fabrication lines still in development.). Gallium arsenide, on the other hand, claims the advantage on performance, an assertion that becomes clear at frequencies above 5 GHz.

5. SUMMARY

Microwave circuits comprise concepts such as distributed effects and device–circuit interactions that are constant with time; because the circuit functions and component names are also time invariant, one might incorrectly conclude that the discipline is static. In fact, growing commercial markets for microwave functions and the need for new communications spectra at millimeter-wave frequencies are driving forces behind the remarkable changes in *implementation* and *synthesis* in microwave circuits. These changes are increasingly focused toward semiconductor device and process development, which includes the two-dimensional world of MMIC, but will be extended to the third dimension with multilayer techniques as the requirements for increased functional density accelerate. Methods of defining new semiconductor structures with (vertical) growth control at the molecular level will continue to change the manner in which performance limits in power, noise, and cost are defined. Improvements in design tools represent a technology overlay on the future of microwave circuits, as do developments in enabling technologies such as superconductivity, and the related fields of digital electronics and optoelectronics.

GLOSSARY

Characteristic Impedance: That impedance which, when terminating a transmission line, causes no reflection to result from an incoming wave.

E (H) Plane: That plane in guided-wave media, usually rectangular waveguide, which is parallel to the electric (magnetic) field lines.

FET: See MESFET.

Gate: The control electrode in a field-effect transistor. Voltage applied to the gate controls the current from source to drain electrodes.

Gunn Diode: Device operating on principle of negative bulk resistivity resulting from transfer of electrons in a high-energy (from applied electric field), low-mass (high-velocity) state to low-energy, high-mass (low-velocity) state. With proper circuit matching can oscillate or amplify to frequencies exceeding 100 GHz. Also called TEDs (transferred electron devices).

HBT: Heterojunction bipolar transistor. Extends bipolar (homo)junction transistor (BJT) operation to microwave and millimeter-wave frequencies by using an extremely low-loss base region with material whose energy band structure is different from that of emitter and collector, but whose crystal structure is similar.

HEMT: High–electron-mobility transistor. A field-effect device where the conduction path (channel) between source and drain has no impurities and therefore produces very low loss. The supply of electrons is in an adjacent material layer whose ability to contribute electrons to current is controlled by the gate voltage. Operation frequency is as high as 150 GHz. Also called MODFET (modulation-doped FET), TEGFET (two-dimensional electron gas FET).

Heterojunction: The interface between two materials of different band structure, but similar crystal structure.

Homojunction: The interface between regions of the same material with different impurity concentrations.

Hybrid Junction: A waveguide or transmission-line arrangement with four ports, which, when the ports have reflectionless terminations, has the property that energy entering any one port is transferred to two of the remaining three.

IMPATT Diode: IMPAct-ionization Transit-Time diodes use the high electric fields of reverse breakdown in semiconductor junctions to create electron or hole multiplication. As these free carriers pass through a drift region, the resulting phase shift in current causes negative resistance at the terminals of the diode. Appropriate circuit termination produces oscillation or amplification to frequencies greater than 200 GHz.

LTCC: Low-temperature co-fired ceramic. A ceramic-based manufacturing method of achieving multiple (as many as 20) layers of signal propagation in a single fabrication process.

Maser: Microwave Amplification by Stimulated Emission of Radiation, which defines the operation of a device for amplifying or generating microwave radiation by induced transitions of electrons, atoms, molecules, or ions between two energy levels.

MESFET: Metal Epitaxial Semiconductor Field-Effect Transistors (commonly denoted as FETs in this article) have gate control electrodes that are deposited directly on the semiconducting channel region, distinguished from junction FETs (JFETs) where the gate is formed by a semiconducting junction on the channel, or a MOSFET (metal-oxide field-effect transistor) where the gate metal is formed on an oxide grown on the channel. These devices are used in many microwave signal control and amplification applications.

(M)MIC: (Monolithic) Microwave Integrated Circuit.

Ohmic Contact: A region between two materials where voltage is linearly proportional to current.

On/Off Ratio: Used to describe the signal attenuation ratio in a switch from the transmission or "on" state to the "off" state.

Phase Velocity: The speed at which a constant-phase point travels in a propagating wave.

Quadrature Hybrid: A hybrid junction having the property that a wave leaving one output port is in phase quadrature (90° shifted in phase) with respect to the wave leaving the other output port.

Spectrum Analyzer: A broadband receiver that is capable of measuring amplitude and frequency of multiple signals.

$\boldsymbol{T_e}$**:** (Effective) noise temperature. A figure of merit used to describe extremely low-noise operation. When normalized to 290 K, T_e is equal to the noise figure minus unity.

Transition Temperature: That temperature (also called critical temperature) which separates normal conducting regime from superconducting regime in a superconductor. Similar descriptions apply to current and magnetic field.

Varactor Diode: VARiable reACtance diodes, semiconductor junction diodes which are designed to give large (as high as 10:1) capacitance changes with an applied voltage, while maintaining low loss.

Works Cited

Ginzton, E. L., Hewlett, W. R., Jasberg, J. H., Noe, J. D. (1948), *Proc. IRE* **36**, 956–969.

Gorter, C. J. (1964), *Rev. Mod. Phys.* **36**, 1.

Hoffmann, R. K. (1987), *Handbook of Microwave Integrated Circuits*, Norwood, MA: Artech, Chap. 2.

Hopfer, S. (1955), *IRE Trans. Microwave Theory Techniques* **MTT-2**, 20–29.

Lakin, K. M., Kline, G. R., McCarron, K. T. (1993), *1993 IEEE MTT-S International Microwave Symposium Digest*, 1517–1520.

Levy, R., Cohn, S. B. (1984), *IEEE Trans. Microwave Theory Techniques* **MTT-32**, 1055–1067.

Montgomery, C. G., Dicke, R. H., Purcell, E. M. (1948), *Principles of Microwave Circuits*, MIT Rad. Lab. Series, Vol. 8, New York: McGraw-Hill, Chaps. 5, 12.

Trew, R. J., Yan, J.-B., Mook, P. M. (1991), *Proc. IEEE* **79**, 598–620.

Schilling, A., Cantoni, M., Guo, J. D. Ott, H. R. (1993) *Nature*, **363**, 56–58.

Wadell, B. C. (1991), *Transmission Line Design Handbook*, Norwood, MA: Artech. Chap. 4.

Wheeler, H. A. (1984), *IEEE Trans. Microwave Theory Techniques* **MTT-32**, 1008–1021.

Further Reading

Collin, R. E. (1992), *Foundations for Microwave Engineering*. New York: McGraw Hill. This recently revised and expanded text covers a broad scope of optical and microwave topics.

Guerlac, H. E. (1987), *Radar in World War II*, New York: American Institute of Physics/Tomash Publishers. An excellent and comprehensive review of the research activity during World War II in laboratories in England and the United States that served as the foundation for contemporary microwave theory and applications. Henry Guerlac, retained by the MIT Radiation Laboratory to document the events of that organization, was prevented from publishing the work during his lifetime.

Maas, S. A. (1993), *Microwave Mixers*, Norwood, MA: Artech.

Marcuvitz, N. (1951), *Waveguide Handbook*, MIT Radiation Laboratory Series Vol. 10, New York: McGraw-Hill. Many of the 28-volume set of books written at the close of the MIT Radiation Laboratory are classics of microwave theory. Two (Montgomery *et al.*, 1948, and Ragan *et al.*, 1948) are mentioned in Works Cited; Marcuvitz's text is still widely used as the definitive reference on waveguide models.

Matthaei, G. L., Young, L., Jones, E. M. T. (1964), *Microwave Filters, Impedance-Matching Networks and Coupling Structures*. New York: McGraw-Hill, pp. 775–842.

Rizzi, P. A. (1988), *Microwave Engineering*, Englewood Cliffs, NJ: Prentice-Hall. A good general reference for passive components, which includes electromagnetic theory while giving insight into the practical applications.

Rodrigue, G. P. (1988), *Proc. IEEE* **76**, 121–137. This invited paper summarizes the development and status of microwave ferrites and their applications.

Saad, T., special editor (1984), *IEEE Trans. Microwave Theory Techniques* **MTT-32**, 955–1263. This special (centennial) issue of the *IEEE Transactions on Microwave Theory and Techniques* describes in detail the technical contemporary milestones of microwave theory and applications.

White, J. F. (1982), *Microwave Semiconductor Engineering*, New York: Van Nostrand Reinhold. This text covers semiconductor control devices thoroughly.

MICROWAVE TRANSISTORS

See TRANSISTORS, MICROWAVE

MISSILES

See SPACE MISSILES

MODULATORS AND DEMODULATORS, ELECTRICAL

Kohji Hohkawa, *Faculty of Engineering, Kanagawa Institute of Technology, Atsugi, Kanagawa, Japan*

INTRODUCTION

Today's highly developed information society is supported by a wide range of communication systems, such as telephones, television (TV) and radio broadcasting systems, citizens band radio, computer networks, and special security systems. These systems have been introduced all over the world, and use a variety of transmission media, such as coaxial cables, optical-fiber transmission lines, mobile, satellite, and microwave radio transmission links, and many special-purpose private networks.

Modulation and demodulation are among the main functions of communication systems for transmitting a large amount of information signals economically in real time. When we need to transmit and receive message signals, such as speech, video signals, or digital data, the information must be modified into a form suitable for transmission over the medium. This modification is achieved by means of a process known as *modulation*, which involves varying some parameters of a *carrier wave* (e.g., the amplitude, frequency, or phase of a sinusoidal wave). After transmission, the modulated signal is distorted by several degradation sources, such as noise, interference, and imperfection of frequency characteristics of the transmission medium. The *demodulation* process, which is the reverse of the modulation process, is to create original message signals from the distorted signal at a receiver. These are the primary applications of modulators and demodulators, although they can also be applied to other fields—for example, measurement equipment, sensing systems, and various kinds of consumer electronics.

Electrical modulation and demodulation processes can now be achieved using the high-performance signal-processing capabilities of advanced semiconductor technology. Operating frequencies of tens of gigahertz are currently available, and up to 100 GHz should be available in the future. Modulation and demodulation of signals with higher frequency can be done by means of optical technology. In most optical modulation and de-

3-527-28132-0/94/$5.00 + .50

modulation systems, almost all the basic modulation processes, such as forming the modulation wave, extracting baseband signals, coherent signal generation, and precise timing control, are carried out electrically, except for the electrical-to-optical and the optical-to-electrical conversion processes.

This article gives an overview of the most widely used electrical modulation methods, giving examples of actual modulator and demodulator circuits and devices. Some examples of applications in communication systems are also reviewed.

1. ELECTRICAL MODULATION TECHNIQUES

1.1 Amplitude Modulation (AM)

The wave forms produced by a variety of constant-wave modulation methods are illustrated in Fig. 1. The frequency spectrum of an AM signal consists of the carrier signal at the center and two sidebands located symmetrically on both sides of the carrier. AM techniques are categorized in terms of how this spectrum is modified before transmission. DSB (double sideband), in which both sidebands are sent in addition to the carrier wave, has the advantage that it can be demodulated by simple circuitry without the need for carrier recovery, but this technique is inefficient in terms of power dissipation. This method is used for commercial AM radio broadcasts, which have one source and numerous receivers. DSBSC (double sideband, suppressed carrier), in which both sidebands are sent without a carrier wave, is suitable for point-to-point communication involving one transmitter and one receiver. SSB (single sideband), in which only one sideband is sent, has many advantages over the others in terms of power consumption, frequency bandwidth, and noise characteristics against fading. This method is suitable for long-distance transmission of voices over wire links, and is also used in marine radiotelephony. VSB (vestigial sideband) involves sending not just one sideband, but also the low-frequency components of the other sideband near the carrier wave. VSB is suitable for transmitting, for example, TV signals, which contain significant low-frequency components.

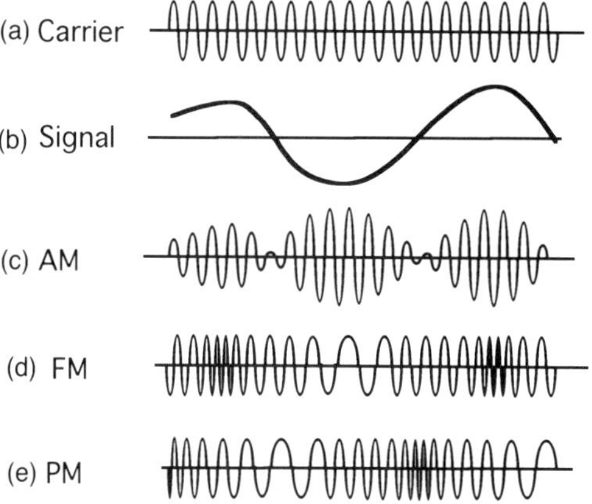

FIG. 1. Comparison of various constant-wave modulation methods. (a) Carrier signal. (b) Baseband information signal. (c) AM, in which the amplitude of a sinusoidal carrier signal is varied in accordance with the baseband information signal. (d) FM, in which the frequency of a carrier wave is modulated continuously in accordance with the baseband signal. (e) PM, in which the phase of a carrier signal is modulated. As shown in the figure, the instantaneous frequency of the FM wave varies in proportion to the baseband signal level, so it has the highest value when the baseband signal is at maximum level. On the other hand, a PM wave has the highest instantaneous frequency when the transition of the baseband signal level is at maximum because its instantaneous frequency varies in proportion to the deviation of the baseband signal levels.

1.2 Angle Modulation (Frequency Modulation, FM; Phase Modulation, PM)

In FM, the frequency of the carrier wave is modulated continuously by the baseband signal, while in PM, the phase of the carrier wave is modulated. In both methods, the amplitude of the carrier wave remains constant. The spectra of the modulated waves have the carrier frequency at the center, with an infinite number of side spectra extending from both sides of the carrier. An important feature of these techniques is that they can provide better discrimination against noise and interference than AM. However, this improvement in performance is achieved at the expense of increased transmission bandwidth. That is, angle modulation provides us with a practical means of exchanging transmission bandwidth for improved noise performance. Such a tradeoff is not possible with AM. Angle modulation is extensively used in public communication systems and is also used in special-purpose communication systems, such as VHF-FM marine radio.

1.3 Pulse Modulation

1.3.1 Analog Pulse Modulation In this system, a periodic pulse train is used as the carrier wave, and some characteristic of each pulse is varied continuously according to the sampled value of the baseband signal. Figure 2 illustrates various pulse modulation schemes. Among them, pulse-amplitude modulation (PAM) is the most efficient in terms of power and utilization, and pulse-position modulation (PPM) is more efficient than pulse-width modulation (PWM) because the latter uses long pulses that expend considerable power while bearing no additional information. Most pulse systems require synchronization of the receiver to the transmitter. A simple synchronization method used in many analog pulse modulation systems is the so-called start-stop method, in which a time marker signal is inserted in every frame interval in addition to the message signal. The marker signal is identified by making its amplitude, width, and interval exceed those of PAM, PWM, and PPM signals, respectively.

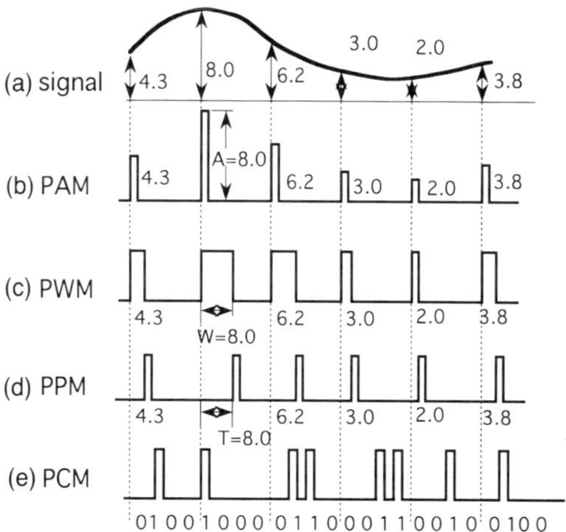

FIG. 2. Comparison of various pulse modulation methods. (a) Continuous-baseband message signal. (b) PAM, in which the amplitude of the regularly spaced rectangular pulses varies with the instantaneous sampled values of the baseband signal. (c) PWM, in which the width of pulses is varied. (d) PPM, in which the position of each pulse is varied according to the baseband signal. (e) PCM. For a comparison of digital pulse modulation technique, PCM is shown. In this technique, the baseband analog signal is converted into digital codes.

1.3.2 Digital Pulse Modulation In pulse-code modulation (PCM), the message is sampled, and the amplitude of each sample is quantized and coded into a series of binary signals, as shown in Fig. 2(e). In the quantization process, the amplitude is rounded off to the nearest integer within a finite set of allowable values, so that both time and amplitude are in discrete form. This process causes quantizing errors in the demodulated signals. Reduction of this error by increasing the number of levels also increases the bandwidth. In actual systems, this problem can be overcome by using nonuniform quantization, in which frequently used signal levels are quantized with more precision than those that are rarely used. For example, the range of voltages covered by voice signals, from the peaks of loud talk to the weak passages of weak talk, is on the order of 1000 to 1. By using a nonuniform quantizer with the feature that the step size increases as the separation from the origin of the input-output amplitude is increased, the large-end step of the quantizer can take care of possible excursions of voice signal into the large-amplitude ranges, which occur relatively frequently. In other words, the weak passages, which need more precision, are favored at the expense of loud passages. In this way, a nearly uniform percentage precision is achieved through the greater part of the amplitude range of the input signal, with the result that fewer steps are needed than would be the case if a uniform quantizer were used. PCM modulation systems are considerably more complex than analog pulse modulation systems and require increased transmission bandwidth. However, they offer the following advantages:

1. ruggedness to transmission noise;
2. efficient regeneration of the coded signal; and
3. the possibility of adopting a uniform format for different kinds of baseband signals (video, speech, and digital-computer data).

Generally, baseband signals, such as speech and video signals, have redundancy in that there exists a high correlation between adjacent samples. Differential PCM (DPCM) is an efficient coding method that reduces the bandwidth by transmitting information about the correlation between the adjacent samples, thereby removing the redundancy. Delta modulation is a one-bit version of DPCM. In this method, the baseband signal is oversampled (i.e., at a rate much higher than the Nyquist rate), and the difference between adjacent signals is quantized to only two levels, namely δ and $-\delta$, corresponding to positive and negative differences, respectively. Recently, the bandwidth required for transmitting video and speech signals by pulse coding has been reduced drastically by using complex signal-processing circuits consisting of advanced very large-scale integrated circuits (VLSIs). For example, in video signals such as TV and HDTV, bandwidth reduction by efficient coding is carried out by reducing redundancies not only between adjacent samples but also between lines, or even frames. In line with the recent development of high-speed semiconductor devices and large-scale integrated circuits, PCM has become the main technology.

1.3.3 Modulation of Bandpass Pulses For transmitting digital signals such as PCM waves or computer signals along a channel with limited bandwidth, the incoming pulses are modulated onto a sinusoidal carrier wave. The modulation process involves switching or "keying" the amplitude, frequency, or phase according to the incoming data. Based on these methods for bandpass pulses are amplitude-shift keying (ASK), frequency-shift keying (FSK), and phase-shift keying (PSK) methods, as illustrated in Fig. 3. Extending these methods, various improved modulation methods have been introduced for transmitting larger quantities of data without errors in a limited bandwidth. For example, in CPFSK (continuous-phase FSK), the duration time of the baseband pulses and the frequencies of carriers used for data 1 and 0 are chosen so as to make the modulated signal continuous at the data transition points. This is because abrupt phase changes in the modulated wave cause a large signal distortion at the transition points and spectral spreading when the wave passes through nonlinear elements. In quadrature-PSK (QPSK), the carrier takes on one of four possible phases relative to the carrier wave, such as $\pi/4$, $3\pi/4$, $5\pi/4$, or $7\pi/4$. The advantage of QPSK is that

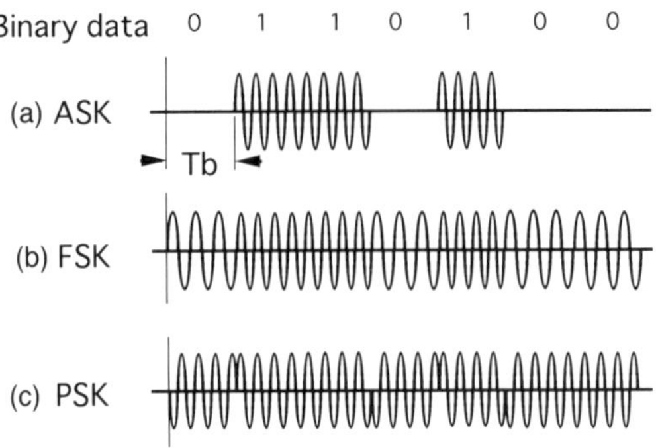

FIG. 3. Bandpass signal waveforms. (a) Amplitude-shift keying, in which the carrier signal's amplitude is modulated by the baseband pulses in the same way as that of DSBSC. (b) Frequency-shift keying, in which the frequency of the carrier is varied according to the baseband pulse. (c) Phase-shift keying, in which the phase of the carrier wave is varied according to the baseband pulse signals.

twice as much information compared to binary-PSK (BPSK) can be transmitted with the same bandwidth because the modulated signal of the QPSK is the superposition of two BPSK signals with the same bandwidth. Generally, FSK signals require a larger bandwidth than PSK signals. However, the frequency deviation between carrier signals can be reduced by using coherent detection (in which the reference carrier signal for detection has the same phase and the same frequency as the carrier signal of the modulated wave), achieved at the expense of increased receiver complexity. The so-called MSK (minimum-shift keying) technique makes it possible to transmit information with half the frequency deviation of that with conventional CPFSK (Pasupathy, 1979).

2. MODULATORS

2.1 Circuits for AM

Ordinary AM signals are modulated by means of a square-wave modulator or switching modulators. A DSBSC wave is generated by product modulators, such as balanced modulators, switching modulators, and ring modulators, as shown in Fig. 4. The balanced modulator uses two AM modulators arranged in a balanced configuration so as to suppress the carrier wave. Several types of product modulators have been implemented

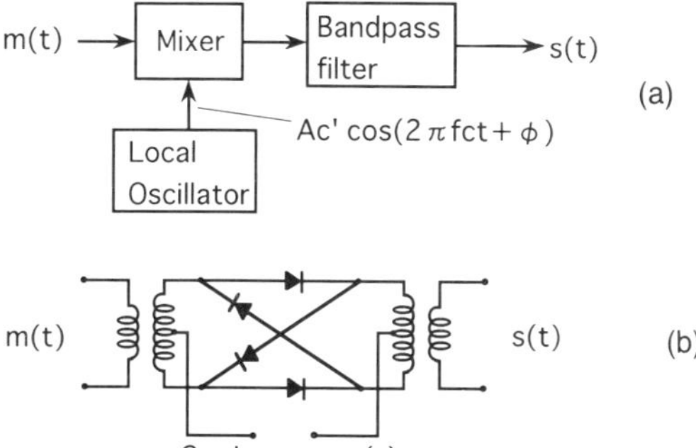

FIG. 4. AM wave modulators. (a) Switching modulator, which is one of the most useful product modulators suited for DSBSC waves. Modulation is done by multiplying the carrier signal with the baseband signal by means of a switching device, such as a diode, transistor, or mixer, and by extracting a bandpass signal with a bandpass filter. This circuit is also used for phase-shift keying technology, discussed in Sec. 2.3: If the baseband signal consists of a bipolar-pulse sequence having amplitude ± 1, the modulated signal forms a bandpass-pulse-sequence signal whose amplitude is almost constant and whose phase is switched into 0 and π according to the polarity of the baseband signal. (b) Ring modulator, in which the four diodes form a ring in which they all point the same way, as shown in the figure. These diodes are controlled by a square-wave carrier that is applied by means of two center-tapped transformers. When the carrier supply is positive, the outer diodes are forward biased and have zero impedance, so that the modulator multiplies the baseband signal by $+1$. When the carrier supply is negative, the situation is reversed, and the modulator multiplies the baseband signal by -1.

in integrated circuits and are now in practical use. SSB and VSB waves are both generated by use of a product modulator followed by a bandpass filter that extracts the desired signal spectrum from the DSBSC wave.

2.2 Circuits for Angle Modulation (FM and PM)

A frequency modulator and a phase modulator are basically the same circuit because an instantaneous frequency is equivalent to the time differentiation of a phase; if we connect an integrator in front of a phase modulator, it operates as a frequency modulator, and if we connect a differentiator in front of a frequency modulator, it operates as a phase modulator. An indirect frequency modulator generates FM waves by employing PM. One method for generating PM waves is a vector synthesis method, in which the PM signal is generated by the vector sum of a carrier signal and an AM signal that has the same frequency as the carrier but a different phase. In a direct frequency modulator, the instantaneous frequency of the carrier is varied directly by the baseband signal by means of a device known as a voltage-controlled oscillator (VCO). One way of implementing a VCO is to use a sinusoidal oscillator with a relatively high-Q frequency-determining network, and to control the oscillator by symmetrical incremental variation of the reactive components. A voltage-variable capacitor consisting of a reverse-biased p-n-junction diode is commonly used for the variable-reactance component. In order to generate wideband frequency modulation with a required frequency deviation, we may use a configuration that consists of a voltage-controlled oscillator followed by a series of frequency multipliers and mixers, as shown in Fig. 5. This permits the attainment of good oscillator stability, constant proportionality between output frequency change and input voltage change, and the necessary bandwidth for wideband FM.

2.3 Circuits for Pulse Modulation

PAM waves are generated by switching the baseband signal with short-duration pulses at the Nyquist interval. PWM waves are generated by using a sawtooth sweeper and a slicer; the sum of the baseband signal and the sweep wave is sliced at a suitable signal level by the slicer. PPM waves are generated from PWM by simply using a monostable multivibrator that is designed to trigger on the

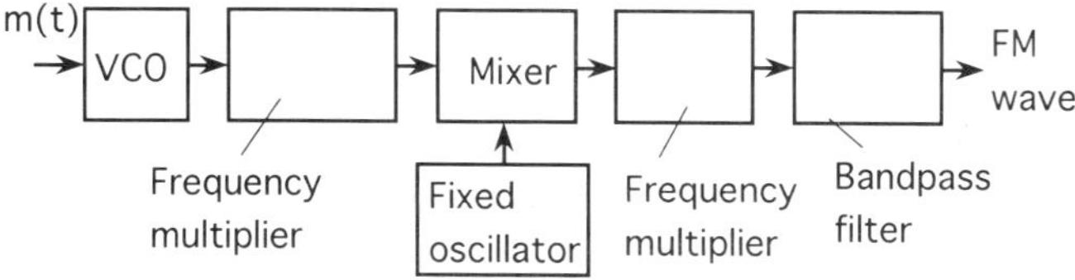

FIG. 5. Wideband FM modulator.

trailing edge of each width-modulated pulse (assuming that the time of occurrence of the trailing edge of each width-modulated pulse is varied in accordance with the message signal), as shown in Hakin (1983).

The process of PCM boils down to analog-to-digital (A/D) conversion. This involves quantizing the amplitude of a signal to a number of discrete levels by sampling the baseband analog signal at regular time intervals, and then generating a digital code consisting of 1's and 0's aligned to each time interval. Figure 6 illustrates an example of an A/D converter for PCM. Another A/D converter is obtained by replacing the sense amplifiers with parallel comparators and exclusive-OR gates. These circuits require a number of parallel amplifiers. Such circuits are complicated but have the greatest speed because all the existing outputs are capable of generating the required code simultaneously. Several types of A/D conversion circuits have been developed and are in practical use, as previously reported (Glasford, 1986). By using a counter-type converter, such as a counter-ramp A/D converter or a dual-slope counter-ramp converter, a simple circuit configuration can be achieved with one or two comparators, but such circuits require large

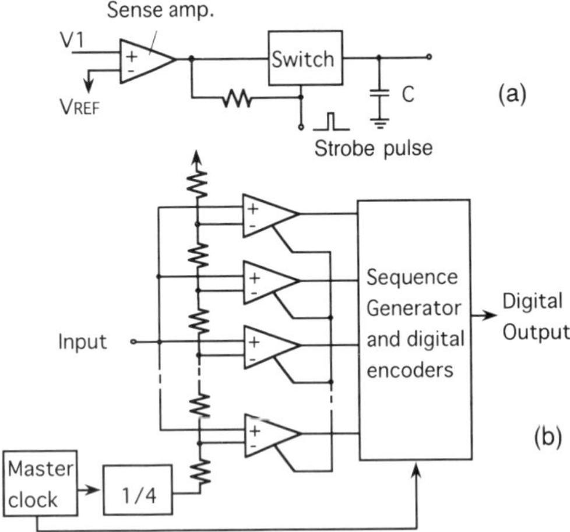

FIG. 6. PCM modulators. (a) A sample-and-hold sense amplifier. The sense amplifier outputs a high-level signal if the input signal is between the lower and upper threshold levels for a certain quantum window during the strobe pulse. (b) PCM wave modulator using it. The sense amplifier outputs a high level when the input signal is within the window of a quantized signal level. The sequence generator and encoder outputs PCM waves.

processing time. Another type of A/D conversion extensively used employs a D/A converter in the feedback loop.

Some shift-keying techniques bear a close resemblance to DSBSC modulation, except that the input baseband signal consists of binary digits. Using a product modulator, shown in Fig. 4(a), ASK or PSK is achieved by modulating the carrier with baseband pulses of amplitude 0 and 1, or 1 and −1, respectively. On the other hand, FSK signals are produced using two ASK modulators with different carrier frequencies. The data input to one modulator are inverted before being input to the other, and the two inputs are summed together. In QPSK, the modulation is done by using two BPSK modulators, as shown in Fig. 7.

3. DEMODULATORS

3.1 AM Demodulators

A simple yet highly effective device known as an envelope detector can be used for narrow-band AM signals in which the signal bandwidth is small compared to the carrier frequencies, and the modulation depth is less than 100%. Alternatively, AM waves can be detected by a square-law detector, which uses the square-law modulator for the purpose of demodulation.

The baseband signal can be uniquely recovered from a DSBSC wave by using coherent detection or synchronous detection, as shown in Fig. 8(a). In this method, the phase error between the reference local wave and the time-varying signal carrier wave generated in the modulation process causes distortion of the demodulated signal. Therefore, the receivers must include circuitry to maintain the local oscillator in perfect synchronization, as previously mentioned. One method is to use the Costas receiver (Costas, 1965) shown in Fig. 8(b). Another method is to use a squaring loop, in which a sinusoidal wave with twice the carrier frequency, $2f_c$, is generated by a phase-locked loop that follows the $2f_c$ component produced by squaring the original DSBSC. The local wave is synchronized to the carrier wave by dividing its frequency by 2. SSB signals are also demodulated by synchronous detection. However, errors in the frequency and phase of the

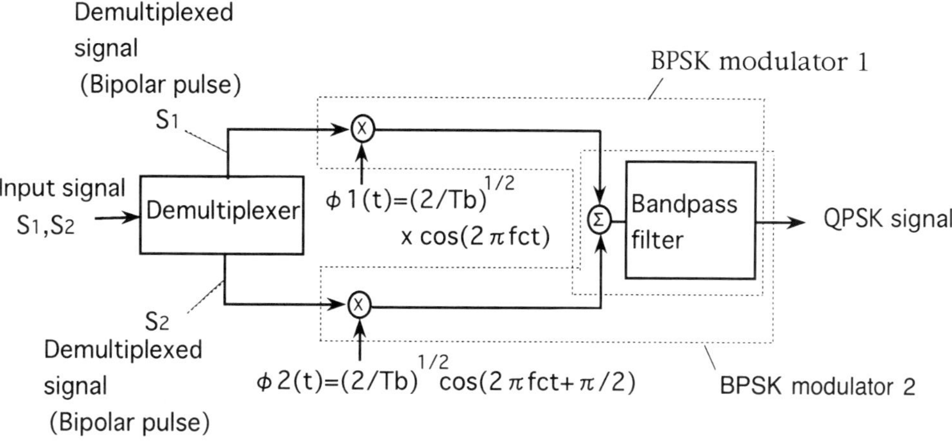

FIG. 7. QPSK modulator. The binary data are converted to two-bit bipolar symbols, which are then used to phase-modulate the carrier. The modulation is done by using two BPSK modulators with a common oscillator that produces two different carrier signals with a phase shift of $\pi/2$. The two bits of each symbol are fed to the modulators and phase-modulate the carrier waves. QPSK wave is formed by combining the two BPSK waves.

local carrier cause large distortion of the demodulated waves. Perfect synchronization can be achieved either by transmitting a pilot signal with the selected sideband or by using a highly stable local oscillator.

3.2 Circuits for Angle Demodulation

The demodulation of angle-modulated signals basically involves producing a voltage whose instantaneous amplitude is directly proportional to the instantaneous frequency offset of the input signal. One common technique is to use a frequency discrimination circuit, which consists of a differentiator circuit followed by an envelope detector. Another technique is to use a phase-locked loop, as shown in Fig. 9, which is a negative-feedback system consisting of three major components: a multiplier, a loop filter, and a voltage-controlled oscillator. The circuit is designed so that the VCO control signal corresponds to the baseband signal when the VCO oscillation frequency is tracked to the

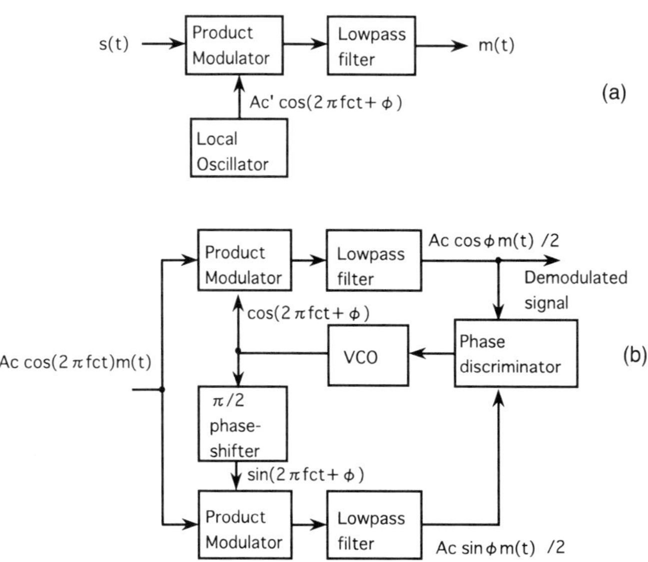

FIG. 8. AM demodulators. (a) Product modulator. (b) Costas receiver. This consists of two coherent detectors with individual local oscillators that are in phase quadrature to each other. These two detectors are coupled together to form a negative-feedback system designed in such a way as to synchronize the local oscillator with the carrier wave.

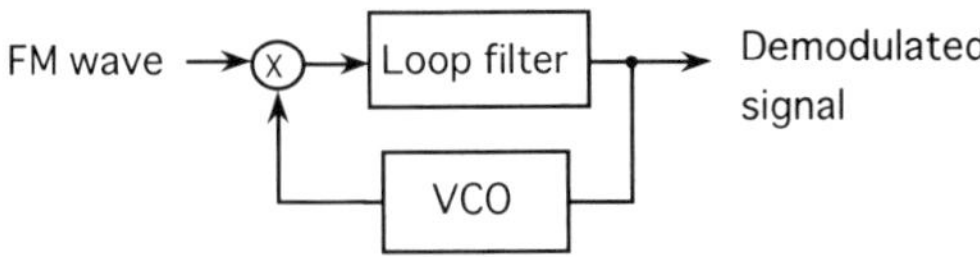

FIG. 9. FM demodulator.

incoming FM wave. As in the modulation process, a PM wave is demodulated using an FM demodulator by simply attaching an integration circuit.

3.3 Circuits for Pulse Demodulation

Since PAM and PWM signals include the baseband signal in their frequency spectrum, they can be demodulated by a simple low-pass filtering. PPM signals are demodulated by converting them to the PWM and applying a low-pass filter.

The demodulation of PCM signals involves a digital-to-analog (D/A) conversion. D/A conversion is inherently simpler and consumes less hardware than A/D conversion, and can be achieved using a resistive network, switches, and a high-gain operational amplifier used as a summing amplifier for improved stability. An example of a resistive-network D/A converter is shown in Fig. 10. In this configuration, a wide range of resistance values is required when there is a large number of bits to convert. This can cause problems in integrated implementations.

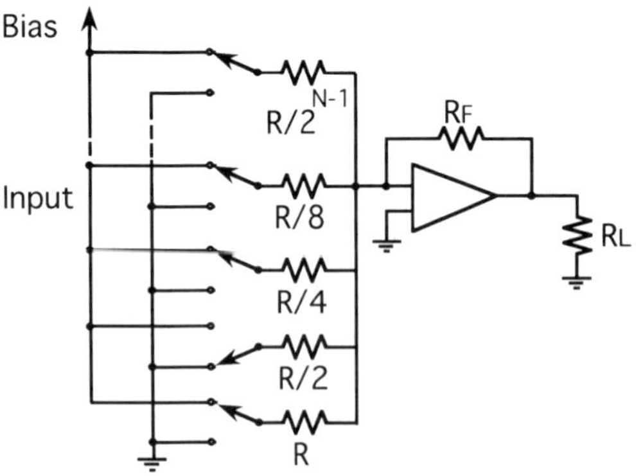

FIG. 10. D/A converter for PCM demodulation. As is shown, the resistors are weighted so as to generate signals with quantized levels. In this configuration, the signal is decoded by using the amplifier to sum up the weighted voltage levels corresponding to each bit in the digital code.

However, several kinds of resistor networks, free from this problem have been introduced so far. For example, the *R-2R* ladder network and the inverted-ladder network D/A converters are implemented by using only two kinds of resistance values, as previously reported (Glasford, 1986). Switching is implemented in a variety of ways using various forms of clamp circuit.

At the receiver, the input PSK signals are bandpass filtered to produce raised cosine pulses and limit the input noise. The signal is then demodulated in the same way as DSBSC signals. Most systems use coherent detection, as shown in Fig. 8(a), in which incoming shift-keyed signals are multiplied by a signal that is synchronized to the original carrier waves. The baseband binary signal is extracted by feeding the multiplied signal to a low-pass filter. A coherent BPSK uses a carrier recovery circuit, such as a squaring loop, in which there is an ambiguity of π in the phase of the reference carrier wave. To overcome this, a known binary signal is transmitted as part of the reference signal. By comparing the demodulated reference sequence with a stored version of the known sequence, the ambiguity can be resolved. The receiver extracts a clock signal of binary data, and this clock signal is used as a timing reference for recognizing the data in a decision circuit. As shown in Fig. 11, a QPSK demodulator consists of two BPSK demodulators.

Another demodulation method for PSK signals is differential detection, which detects pulses by comparing the incoming datum with that received one bit earlier. This method does not require coherent detection.

4. DEVICES FOR MODULATION AND DEMODULATION

The rapid progress of semiconductor technology over the last few decades has resulted in great advances in information-processing systems and communication systems. Communication systems are not only less expensive but also highly integrated, are capable of highly complex signal-processing capability, and operate at much higher frequencies. Even in high-frequency radio communication systems, vacuum tubes have been replaced by highly developed semiconductor elements or integrated circuits, except for

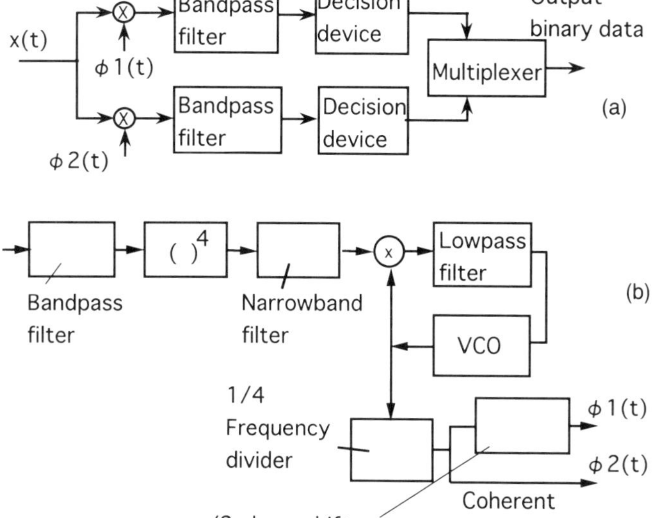

FIG. 11. QPSK demodulation circuit. (a) Demodulator. Data d_1 and d_2, which are demultiplexed in the modulator, are recovered by feeding the coherent reference carrier and $\pi/2$ phase-shifted signal to each BPSK demodulator. (b) Carrier extraction circuit. The reference carrier is synchronized using fourth-power harmonic generator, a phase-locked loop oscillator, and a 1/4 frequency divider, whose circuit structure and operation are similar to that of the square loop.

special-purpose high-frequency high-power amplifiers, such as traveling-wave tubes (TWTs) and klystrons.

As shown in various circuit examples, the main devices used for modulators and demodulators are nonlinear elements, amplifiers, filters, oscillators, delay elements, and logic devices. Many modulator and demodulator integrated circuits (ICs) have been developed and are in practical use. Examples include standard ICs, such as amplifiers, phase-locked loop oscillators, synthesizers, and switches, and many custom ICs for modulation and demodulation in radio and TV receivers, wireless telephones, and citizen's-band transceivers. For digital pulse signals, a variety of coders and decoders and A/D and D/A converters have been developed using monolithic ICs containing analog and digital devices. At microwave frequencies, various kinds of monolithic microwave ICs have been developed using GaAs technology.

Among devices required for modulation and demodulation, high-*Q* circuit elements, such as filters and stable oscillators, are the devices that are hard to implement with semiconductor technology. Filters with a high *Q* or a sharp cutoff are made with electromechanical devices, such as bulk-wave crystals and surface acoustic-wave (SAW) devices, as shown in Fig. 12. Crystal filters, such as energy-confined multiple-mode–coupled filters, are extensively used—for example, as channel filters and pilot signal filters in radio communication systems and SSB frequency-division multiplexing systems. However, their upper frequency is limited to 100 MHz. SAW filters, including resonators and delay-line type filters whose operation is the same as that of transversal filters,* have been extensively used in the frequency range from 10 MHz to several GHz. SAW filters have the advantage that they can be mass produced using advanced semiconductor fabrication technologies, enabling small, high-performance filters to be produced at a low cost. In spite of a relatively large insertion loss, SAW devices are suitable for analog processing of bandpass signals because of their transversal filter nature, in which both amplitude and phase characteristics are controlled (designed) independently. Various

*A transversal filter consists of a delay line with multiple weighted taps. The input signal is fed to the tapped delay line, and the output signal is obtained as a sum of the tapped signals. The delay time and weighting value of each tap are designed so as to correspond to the sampling time and sampled value, respectively, of the impulse response of the required filter. The filter can be designed directly in the time domain by simply estimating its impulse response characteristic. It presents a superior design flexibility; for example, we can design a filter with an arbitrary phase and amplitude characteristic. This filter is applied to various kinds of signal processing circuits, such as an equalizer, a matched filter, or a dispersive delay device.

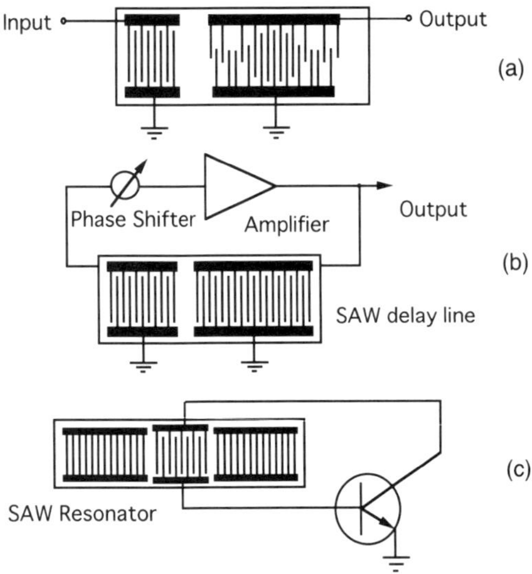

FIG. 12. SAW devices. Schematic diagram of (a) SAW transversal filter, (b) a delay line oscillator, and (c) a resonator oscillator.

types of functional circuits have been developed, including modulators and demodulators of AM, FM, and bandpass pulse signals. At the frequencies of microwaves and millimeter waves, dielectric resonators and microwave components such as microwave cavities, wave-guide filters, and strip-line filters are used.

An oscillator consists of an amplifier with a positive feedback loop containing reactance elements. Quartz oscillators, which use a steep reactance variation with frequency of the crystal resonator, are extensively used in modulators and demodulators as a source of carrier waves and reference waves to stabilize other oscillators such as VCOs by means of phase-locked loops. Quartz oscillators are stable against temperature variation and aging. For high-frequency applications, harmonics of their outputs are used. A SAW oscillator also has a relatively high performance and can be used at high frequencies. There are two types of oscillator in SAW devices, as shown in Figs. 12(b) and 12(c): resonator oscillators and delay-line oscillators. The former uses a steep reactance variation in the same way as quartz oscillators. The latter consists of a loop with an amplifier and a delay line. The delay line has a bandpass characteristic for which only one component in multiples of $1/T$ is passed with a small loss and fed back to the amplifier, where T is the total delay time of the loop. The delay-line oscillator has the advantage that the oscillation frequency can be varied over a wide range by changing the loop delay.

In the microwave frequency range, Gunn-effect and IMPATT diodes, as described in Liao (1985), with reactance elements have been used for high-power signal sources. Since these oscillators are not inherently stable, phase-locking technologies using harmonics of a stable oscillation source, such as a quartz oscillator, are required for frequency stabilization. Recently, oscillators using high-frequency GaAs transistors have also been applied to this frequency range. Extraordinarily stable oscillators, such as rubidium and cesium beam oscillators, which use the inherent transition frequency of the atom, are employed for the standard carrier signal source in various systems.

5. APPLICATIONS OF MODULATORS AND DEMODULATORS

For public communication applications, such as the telephone network, the bandwidth of the transmission medium is far greater than that required by a single user. A large number of baseband signals are multiplexed before transmission, and demodulation is carried out after demultiplexing. Frequency-division multiplexing (FDM) and time-division multiplexing (TDM) technologies, shown in Fig. 13, are used for analog waves and pulse waves, respectively. For both multiplexing methods, hierarchies are introduced to facilitate signal handling and to simplify instrumentation.

FDM has been used in many conventional communication systems—for example, coaxial cable transmission systems, microwave radio transmission systems, and personal, mobile, and satellite communication systems. The multiplexed signal is either directly transmitted after passing through bandpass filters to reject spurious signals outside the signal band, or amplified for long-distance transmission before filtering. For long-distance transmission, repeaters are used for compensating transmission loss. Two types of repeaters are available. In direct repeaters, amplification is carried out at the high frequency of the transmission band. In indi-

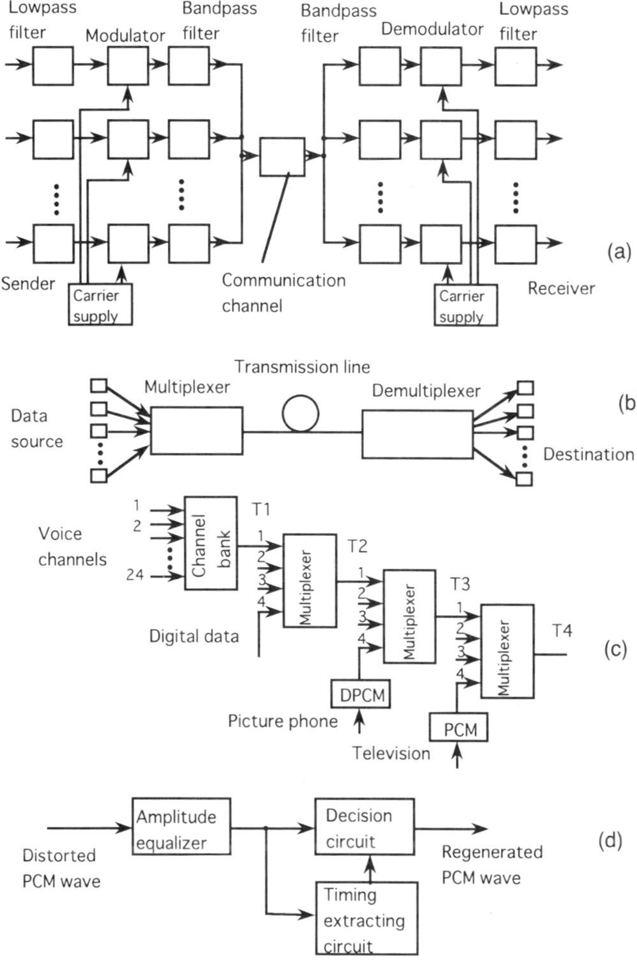

FIG. 13. Block diagram of multiplexing and demultiplexing systems. (a) FDM, in which the modulated signals are arranged onto the frequency domain with suitable separation so as to reject interchannel interference. The baseband signal of each channel is directly modulated by cw analog modulation methods, such as FM and SSB, onto a predetermined frequency band with a corresponding carrier frequency. Alternatively, each baseband signal is modulated onto a fixed intermediate frequency, and each modulated signal is modulated onto the predetermined different frequency by a product modulator with a corresponding carrier frequency. In an SSB system, a pilot signal is inserted for synchronization in the multiplexed waves. The demultiplexing process is the reverse of the multiplexing process. (b) TDM, in which modulated digital pulses of each channel are arranged on the time domain. (c) An example of hierarchical structure, in which multiplexing is carried out by classifying in several groups and higher-group signals are composed by multiplexing a number of multiplexed signals in lower groups. (d) Regenerative repeater.

rect repeaters, a heterodyne amplifier is used to convert the received signal to an intermediate frequency, at which low-noise, high-performance amplifiers and filters can be used to compensate phase or delay distortions caused in the transmission lines. The amplified intermediate-frequency signals are then converted back to the transmission band and retransmitted. In microwave systems, each repeater retransmits the signals at a different frequency from that of the received signal, so as to avoid interference.

TDM technology is used for the integrated-service digital network (ISDN) and will be used for the coming broadband ISDN (BISDN), which provides various new services and functions and is gradually replacing the conventional analog system. The reliability, flexibility, and ease of implementation provided by large-scale integrated (LSI) technology are not obtainable in analog systems. In most coaxial cable systems, the data is formed into bipolar pulses, such as the alternate mark inversion (AMI) pulse format. In intensity-modulation direct-detection optical-fiber transmission systems, data are formed into nonreturn-to-zero (NRZ) pulses. In radio communication systems, such as personal, mobile, and satellite communication systems operating at frequencies from UHF to the millimeter band, and in coherent optical-fiber systems, bandpass pulse waves are used. In all systems, the data are inserted in a fixed-format container called a frame and then multiplexed. After multiplexing them into a predetermined frame format, the data are transmitted over the line. At repeaters, the distorted and attenuated data are reformed, using a clock extracted from the data, and retransmitted as shown in Fig. 13(d).

In digital systems, such as PCM systems, we need to extract a timing clock and to identify the exact data position from the sequence in order to demodulate the data correctly. After amplifying received waves to a certain level by means of an equalizing amplifier, a timing signal is extracted by feeding the signals to a timing extraction circuit consisting of, for example, a high-Q amplifier and a limiting amplifier. If the input data contain a long sequences of marks or spaces, the data lose their timing signal. In addition, the equalizer amplifier is capacitive coupled and does not pass the component. As a result, the data cannot maintain their exact signal amplitude; for example, a long sequence of high-level pulses in AMI and NRZ signals drifts into the low levels. This effect causes incorrect operation in the decision process. To overcome this problem, bit-sequence-independent (BSI) line-coding and data-scrambling technologies are introduced. BSI line coding is done by feeding extra bits into the data to increase the data transitions between marks and spaces. For example, nMIC inserts one complementary signal to the last bit of every n bits in the data sequence. Data scrambling is done by modulating the data with a pseudorandom-noise binary sequence (PRBS) with a mark-to-space ratio of 1/2 that is generated by a feedback shift register. Modulation is done by obtaining the exclusive-OR of the signal data and the PRBS data. The output signal has a mark-to-space ratio of 1/2. Demodulation is done by the same process: The data are reproduced by obtaining an exclusive-OR of the input data and the same PRBS codes.

The data position is identified by inserting a predetermined sequence of extra pulses into the transmitting data. Usually, the data are inserted into a frame with a predetermined format to each hierarchy, with a frame pulse and other service bits that control the operation of the network. Recently, several systems have been introduced with synchronous digital hierarchy network-node interface (SDH-NNI) and are now in commercial use. SDH-NNI is a worldwide standard for network interfaces and should permit efficient construction of digital networks, including BISDN, and effective network operation, administration, and maintenance, as previously reported (Asatani, 1988). The technology can be applied to coaxial cable networks, to optical-fiber networks, and even to the radio communication systems whose bandwidth is limited.

Electrical modulation and demodulation technologies are also applied to optical-fiber transmission systems. In most optical systems, almost all the basic modulation processes, such as forming the modulation wave, extracting baseband signals, coherent signal generation, and precise timing control, are carried out electrically, except for the electrical-to-optical and the optical-to-electrical conversion processes. The electrically modulated signals are converted to optical signals before transmission, and at the receiver, the received optical signal is converted back to the electrical signal.

6. CONCLUSIONS

The rapid development of semiconductor technologies, especially digital technology, has made drastic changes in signal processing and communication systems. Further development is expected in semiconductor technologies. Recent deep submicron silicon processes make it possible to fabricate ultra-large-scale integrated circuits (ULSIs) with several million gates. A small-scale Si bipolar IC operating at more than 20 Gb/s already has been reported, and circuits operating at several GHz are commercially available. Compound semiconductor devices, such as GaAs MESFETs, HEMTs, and HBTs, operate at higher frequencies, and devices operating around 10 GHz are now available. The operating frequency of GaAs devices is likely to reach 40 GHz in small-scale IC devices. Higher-frequency operation up to 100 GHz should be possible using InGaAs. These developments should increase the operating frequency range of various modulators and demodulators. These device technologies will provide high-frequency communication systems with small size and a low cost. In addition, the increasing sophistication of LSI technologies should make the modulation and demodulation of digital signals more complicated but also more efficient.

ABBREVIATIONS

A/D: Analog to digital.
AM: Amplitude modulation.
AMI: Alternate mark inversion.

ASK: Amplitude-shift keying.
BISDN: Broadband integrated-service digital network.
BPSK: Binary phase-shift keying.
BSI: Bit-sequence-independent.
CPFSK: Continuous-phase frequency-shift keying.
D/A: Digital to analog.
DPCM: Differential pulse-code modulation.
DSB: Double sideband.
DSBSC: Double-sideband suppressed-carrier.
FDM: Frequency division multiplexing.
FM: Frequency modulation.
FSK: Frequency-shift keying.
HBT: Hetero-bipolar transistor.
HDTV: High-definition television.
HEMT: High–electron-mobility transistor.
IC: Integrated circuit.
IMPATT: Impact-ionization avalanche transit time.
ISDN: Integrated-service digital network.
MESFET: Metal-semiconductor gate field-effect transistor.
NNI: Network-node interface.
NRZ: Nonreturn to zero.
PAM: Pulse-amplitude modulation.
PCM: Pulse-code modulation.
PM: Phase modulation.
PPM: Pulse-position modulation.
PRBS: Pseudorandom binary sequence.
PSK: Phase-shift keying.
PWM: Pulse-width modulation.
OPSK: Quadrature phase-shift keying.
SAW: Surface acoustic wave.
SDH: Synchronous digital hierarchy.
SSB: Single sideband.
TDM: Time-division multiplexing.
TWT: Traveling-wave tube.
ULSI: Ultralarge-scale integrated circuit.
VCO: Voltage-controlled oscillator.
VHF: Very high-frequency.
VLSI: Very large-scale integrated circuit.
VSB: Vestigial sideband.

Works Cited

Asatani, K. (1988), in: *Digital Technology Spanning the Universe*, Proceedings of the International Conference on Communication, New York: IEEE, pp. 118–124.

Costas, J. P. (1965), *Proc. IRE* **44**, 1713–1718.

Glasford, G. M. (1986) *Analog Electronic Circuits*, Englewood Cliffs, NJ: Prentice-Hall, chap. 10.

Haykin, S. (1983), *Communication Systems*, New York: Wiley, Sec. 7.7.

Liao, S. Y. (1985), *Microwave Devices and Circuits*, 2nd Ed., Englewood Cliffs, NJ: Prentice-Hall, chaps. 9 and 10.

Psupathy, S. (1979), *IEEE Commun. Soc. Mag.*, **17**, 14–22.

Further Reading

Glasford, G. M. (1986), *Analog Electronic Circuits*, Englewood Cliffs, NJ: Prentice-Hall.

Korn, I. (1985), *Digital Communications*, New York: Van Nostrand Reinhold Electrical/Computer Science and Engineering Series.

Roddy, D. (1989), *Satellite Communications*, Englewood Cliffs, NJ: Prentice-Hall.

Sinnema, M., McGovern, T. (1986), *Digital, Analog and Data Communication*, A Reston Book, Englewood Cliffs, NJ: Prentice-Hall.

MODULATORS AND DEMODULATORS, OPTICAL

S. K. YAO, *Optech Laboratory, Santa Ana, California, U.S.A.*

E. H. YOUNG, *NEOS Technologies, Inc., Melbourne, Florida, U.S.A.*

INTRODUCTION

With the invention of lasers, there has been a major expansion in the utility of modulated light for communications, manufacturing, test equipment, signal processing, medical applications, entertainment, etc. Most laser applications require some sort of modulation. Often the development of new laser applications depends on successful development of a suitable optical modulation process.

With coherent laser light, which provides predictable optical wavelength, frequency, amplitude, and phase-front parameters, tremendous advances in the art of the light

3-527-28132-0/94/$5.00 + .50

modulation have been achieved. Modulation speed has been increased from a few hertz to microwave frequencies. Modulation format includes all possible optical wave parameters, such as amplitude, phase, intensity, polarization, carrier frequency, and wavelength.

Modulator devices have experienced steady improvements in efficiency and power consumption, with constantly shrinking geometry. Integrated optical (waveguide) devices are very small in size compared to bulk devices and have many applications in communications.

In this article, optical modulation and demodulation techniques, as well as associated device principles and examples, will be described.

1. DEFINITIONS

Modulation: A process that combines a signal containing the desired information with a much higher frequency "carrier wave" so that the information or a group of information can be delivered through certain transmission media over distance. In the modulation process, one or more parameters of the carrier wave are modified in proportion to the signal information strength, making the modified parameter a function of time. In order to achieve clean and noiseless modulation, the high-frequency carrier wave to be modulated must be stable and constant before the modulation process.

The use of a much higher frequency carrier wave is necessary to take advantage of certain desirable wave transmission properties in the specific media. Very often, the information to be transmitted occurs at rather low frequencies, which are extremely difficult to send over great distances. In addition, at low frequencies, similar signals transmitted from other sources at the same time would interfere with each other. Usually, the size of signal emitter needed is proportional to the wavelength of signal and, therefore, inversely proportional to the frequency of the signal. Use of a high-frequency carrier wave reduces the size of signal emitter as well as associated equipment, allows proper separation of carrier waves from different sources because of the wider frequency span at higher frequencies, and provides the possibility of

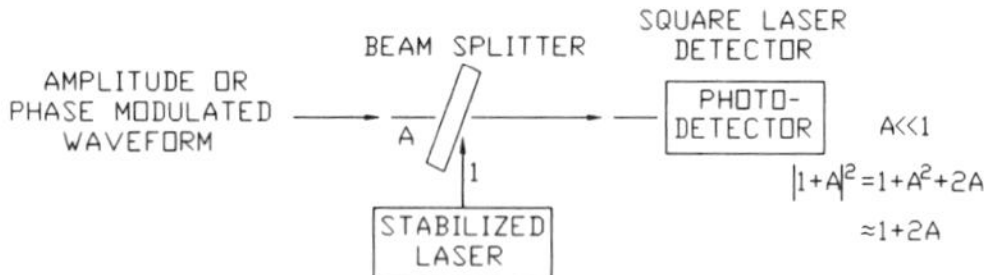

FIG. 1. Detection of amplitude or phase-modulated optical wave forms.

simultaneous transmission of a large group of signals through the same transmission medium at the same time.

There are many forms of modulation characterized by each of the wave parameters. For instance, there are amplitude modulation, phase modulation, frequency modulation, polarization modulation, and several forms of pulse modulation.

Demodulation: Demodulation is a process that recovers the signal information from the appropriate wave parameter of a modulated carrier wave. In a noiseless modulated carrier wave, the demodulation process simply extracts the time-variable content of the appropriate parameter of the carrier wave. Therefore, for each type of modulation, there is one or more corresponding types of demodulation. Complementary demodulation of a modulated information transmission channel is necessary at the receiving end. If this is not implemented, the receiving end will only notice the existence of the carrier wave, while the information will be lost.

Two forms of optical modulation and demodulation are shown in Figs. 1 and 2.

Optical Modulation and Optical Modulator: Light waves are a form of electromagnetic waves at a frequency about 10^5 times higher than radio waves. Light waves can also be used as a carrier wave. The process that modifies one or more parameters of light according to the time-variable content of an information signal is called optical modulation. The device that facilitates optical modulation is called an optical modulator.

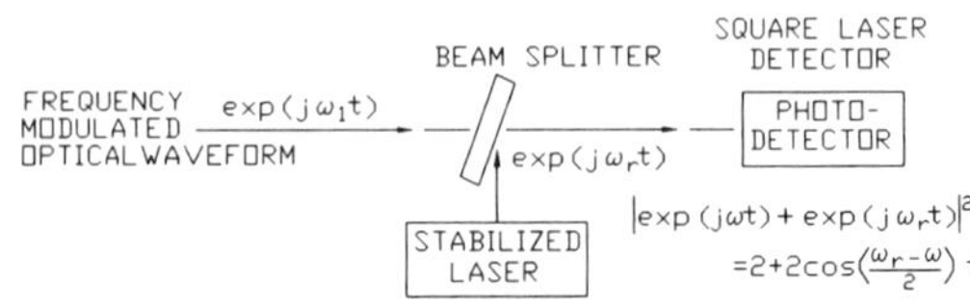

FIG. 2. Detection of frequency-modulated optical wave form.

Conventional light waves are incoherent, consisting of a collection of a large number of independent oscillators or sources, each having its own wavelength (frequency), phase, polarization, direction of propagation, etc. However, in order to qualify as a carrier wave the light wave (or the selected parameters of the light wave) must be stable and constant with respect to the signal information. Light with well-defined frequency and phase parameters, temporal and/or spatial, is called coherent light. Conventional incoherent light can approach the well-defined frequency and phase parameters of coherent light with the use of polarization, wavelength, and spatial filters at the expense of significantly reduced optical power and intensity. The invention of the laser provides high-quality, high-intensity coherent light sources well suited for optical modulation and communication applications.

2. METHODS IN THE MODULATION OF LIGHT WAVES

Optical modulations are unique because of the physical properties of laser light and light detection. The greatest difference between radio-wave amplitude modulation techniques and optical modulation techniques is that the output of radio receivers is usually proportional to the amplitude of the incoming signal, while the output of optical receivers is usually proportional to the intensity of the incoming signal, because of the use of square-law photodetectors. The amplitude of a signal contains both magnitude and phase information, while the intensity of a signal is the square of its magnitude only.

2.1 Intensity Modulation

When the information signal is used to alter the intensity of light waves (laser beam), it is called intensity modulation. The intensity-modulated wave form appears similar to an amplitude-modulated signal, except that the amplitude variation is proportional to the square root of the signal and that the phase bears no information. The advantage of intensity modulation is its compatibility with simple demodulation using photodetectors. Another advantage is that it does not require a high degree of frequency and phase stability of the optical carrier, as long as the carrier intensity is constant. This is very important since long-term frequency and phase stability of optical sources including lasers is very difficult to achieve. Therefore, intensity modulation is the most popular form of optical modulation in use to date.

Since light-wave frequencies are much higher than radio-wave frequencies, modulated radio waves in all forms may be used for intensity modulation of a laser beam. In particular, intensity modulation is well suited for digital pulse information. Intensity modulation can also be used to carry amplitude-modulated or frequency-modulated radio frequencies (rf) through an optical communication channel. However, because the intensity of light is always positive, while most analog signal wave forms are bipolar with zero mean, a bias level is often built into the optical intensity-modulation process for bipolar analog signal transmission. The dc bias level is easily removed by an electrical dc-blocking circuit after the photodetector.

2.2 Pulse Modulation

In pulse modulation, the parameters of a pulse train are disturbed according to the information signal. Often, the information signal is digital, in the form of a binary data stream comprising two levels usually designated 0 or 1 (OFF or ON). The pulse train parameters that can be disturbed for pulse modulation include the pulse amplitude (amplitude-shift keying or ASK), the center frequency of each pulse burst in the pulse train (frequency-shift keying or FSK), or the phase angle of each pulse burst in the pulse train (phase-shift keying or PSK), as well as the individual pulse widths, and relative positions of each pulse. Most of these pulse modulations are similar to pulse modulation of rf carriers.

However, in optical systems, pulse-amplitude modulation is implemented as optical pulse-intensity modulation.

Optical implementation of frequency-shift keying and phase-shift keying can be either in the form of an intensity-modulated optical system or in the form of a so-called coherent optical communication system. The term "coherent optical communication" specifies that long-term frequency- and phase-stable laser sources are employed and that the phase parameter of the optical carrier wave is used for the modulation process. Note that, despite the common use of "coherent" laser sources for intensity-modulated optical systems, the temporal coherence properties of the laser source are often not a key part of the modulation process because conventional lasers are not temporally coherent enough to provide sufficient long-term frequency and phase stability. Instead, the use of laser sources for most intensity-modulated optical systems is due to the spatial coherence property of lasers, which allows high directivity for space communication, tight focus capability for optical instruments, and high intensity-coupling efficiency through optical fibers.

Implemented in the form of intensity modulation, electrical FSK or PSK wave forms are simply intensity modulated on an optical beam. The frequency or phase information is carried by the rf carrier, which is intensity modulated on top of the optical carrier.

In the form of a coherent optical communication system, FSK or PSK data are used to shift the frequency or phase of the optical wave form of the laser beam. In order to detect the FSK or PSK signal in a coherent optical communication system, a long-term stabilized source laser and a local oscillator laser are employed as frequency or phase reference in a heterodyne arrangement. Although such a system is much more difficult to build, it can provide significantly greater signal bandwidth and sensitivity and is highly desirable for wide-band long-haul communication systems.

2.3 Polarization Modulation

In a polarization modulation system, the polarization of a laser beam is modulated or switched (digital system) between two orthogonal states. Demodulation of a polarization-modulated signal into an intensity-modulated signal is easily accomplished by means of an optical polarizer or an optical polarization beam splitter. Like intensity modulation, such a system does not require sophisticated stabilized lasers. Since the optical intensity of a polarization-modulated signal is about twice as high as an intensity-modulated signal, it provides better receiver signal-to-noise ratio.

2.4 Amplitude Modulation

In amplitude modulation, the modulating signal appears as a change in the amplitude of the crests of the carrier wave. The greater the change in the amplitude of the carrier-wave crests, the greater is the degree of the modulation. The degree of modulation is expressed as a percentage of the unmodulated carrier amplitude.

Since photodetectors are square-law devices, direct detection of an amplitude-modulated optical signal results in significant distortion of the original signal. Heterodyne detection with a constant-amplitude (magnitude and phase) optical source in an interferometric arrangement is necessary, requiring a long-term stabilized laser source. Amplitude modulation is seldom used, except recently in the development of coherent optical communication systems.

2.5 Phase Modulation

In phase modulation, the phase of a highly stable carrier wave is modulated by an analog signal or switched between two or more states according to a digital signal. In optical systems, frequency- and phase-stabilized lasers are required as the source for the carrier wave and as a local oscillator for reference. In optical communication, using visible and near-infrared lasers, phase modulation is seldom used, except recently in the development of coherent optical communication systems using stabilized diode lasers.

At far-infrared wavelengths, light sources such as CO_2 lasers can be stabilized relatively easily, and phase modulation is more common in laser communication systems and in cw-type laser radars.

2.6 Frequency Modulation

In frequency modulation, the modulating signal changes the frequency of the carrier wave, while the amplitude (or intensity) of the carrier remains constant (unchanged). Thus, the instantaneous center frequency of the modulated signal swings about the average center frequency according to the information signal. As the information signal approaches its positive peak, the instantaneous center frequency of the modulated signal decreases toward its maximum lower deviation; as the information signal approaches its negative peak, the instantaneous center frequency of the modulated signal increases toward its maximum upper deviation from the average center frequency.

Optical implementation of frequency modulation requires frequency-stabilized lasers as a source and as a reference. It is seldom used for visible laser systems because of the lack of stabilized lasers. In the infrared, stabilized CO_2 lasers are available, and fm systems are often used in laser range-finder and cw laser radar applications.

2.7 Wavelength Modulation

Wavelength modulation differs from frequency modulation by the amount of frequency deviation in the modulation process. In frequency modulation, the amount of frequency deviation is generally less than a few parts per million or a few GHz. The wavelength change in a frequency-modulated signal is extremely difficult to detect by optical spectrometers, and must be detected by means of interferometric comparison with a stabilized reference optical wave form. On the other hand, the frequency deviation in a wavelength-modulated signal is generally on the order of parts per thousand, and the associated wavelength change can be detected by spectrometers.

Wavelength modulation does not require the use of particularly stabilized lasers since the wavelength deviation is large compared with the drift of many lasers. Wavelength hopping, in which the optical wavelength of a laser beam hops among several predetermined values, is also useful for digital communications. Wavelength modulation is not widely employed in most practical applications because of the relative difficulty in constructing an efficient and fast device. A few experimental examples include a proposed electro-optical analog-to-digital conversion system that consists of a real-time spectrogram of a wavelength-modulated optical signal from a laser, a gray-code optical mask, and a detector array.

2.8 Deflection Modulation

Laser and optical beams can be formed with high directivity because of their short wavelengths. The pointing direction of a highly directive laser beam can be modulated by beam deflection devices with many useful applications such as scanning, pointing and tracking, plotting, and pattern generation. When intercepted with optical aperture through properly designed optics, it can be converted into intensity modulation.

2.9 Modulation of Incoherent Light

Here, incoherent light means light with a low degree of spatial coherence—for instance, light from nonlaser sources. Modulation of incoherent light is generally limited to intensity modulation, polarization modulation, and intensity-based pulse modulation. Because of the lack of spatial coherency, incoherent light tends to possess large beam divergence angles, requiring the use of large–numerical-aperture devices for modulation. Effective modulation means include mechanical shutters, electrical drive-current modulation, certain types of electro-optic shutters, and liquid-crystal light valves.

Examples of modulated incoherent light include drive-current–modulated light-emitting diodes (LED) for display, fiber-optical communications, and printing instruments, and liquid-crystal light valves (LCD) for display and instrumentation applications. Generally, the speed in modulation of incoherent light tends to be limited to under 10 to 100 ns rise time.

2.10 Modulation of Laser Light

Laser light can be modulated by any one of the optical modulation methods. Use of stabilized lasers exhibiting long-term frequency and phase stability can allow for sophisticated coherent light modulation and demodulation techniques that result in unparalleled sensitivity and bandwidth.

Laser light is characterized by a high degree of spatial coherency and, therefore, high beam directivity. This gives the ability to focus a laser beam into a small spot diameter with relatively large depth of focus. Thus, in addition to the modulation means mentioned in the section on modulation of incoherent light (Sec. 2.9), it can be modulated using long–interaction-length and narrow–numerical-aperture electro-optic, acousto-optic, or guided-wave optical devices specifically constructed for high modulation speed and low drive voltage. Modulation speeds of greater than 10 to 100 GHz with rise times in the range of picoseconds have been demonstrated.

3. OPTICAL MODULATION, DEVICES AND MATERIALS

3.1 Mechanical Modulation

Simple mechanical shuttering action produced by a mechanical shutter or a rotating chopper wheel has been among the oldest and most widely used methods of optical modulation. Although slow, it can be extremely cost effective.

3.2 Electrical Modulation

Most light sources are driven by electrical current and can be modulated by controlling the electrical drive current. The modulator can be relatively simple and low cost. However, the modulation tends to be rather nonlinear and is more suitable for pulse modulation or switches. Gas lasers and solid-state lasers can be modulated or switched at up to a fraction of a megahertz speed. Light-emitting diodes can be switched to as fast as 100 MHz. Injection laser diodes have been switched from a few gigahertz to about 20 GHz.

3.3 Electro-optical Modulation

Optical modulation can be achieved either by electrically induced absorption in semiconductor materials or by perturbing the optical index of refraction in an optical or semiconductive medium.

For *electro-absorptive modulation*, the application of an electrical field in bulk semiconductive materials will cause a perturbation or shift in the band-gap structure, the Franz–Keldysh effect, which can result in large changes in the optical transmission properties at certain wavelengths. The required large electrical field can be obtained at reverse biased junctions. Modulation of optical transmission can be achieved by varying the reverse bias voltage at semiconductor junctions.

Recently, significant research interest has been devoted to electro-optic effects in quantum wells (QWs) and superlattice (SL) materials. These are new classes of material systems in which layers of extremely thin (on the order of several angstroms) semiconductive material of different compositions have been grown on compound semiconductor substrates, altering the electronic energy levels, wave functions, and optical properties. Large third-order susceptibility, proportional to the square of the field, in such modified material systems has been observed resulting from conduction-band nonparabolicity, exciton line saturation, and the quantum confined Stark effect (QCSE). For instance, the excitonic absorption edge can be red-shifted by an applied field producing a positive optical refractive-index change at wavelengths below the band edge proportional to the square of the applied field. In contrast to the Franz–Keldysh effect in bulk semiconductors, the QCSE can simultaneously yield large values of index change and absorption change. Practical devices using QCSE, the so-called self-electro-optic effect–devices or SEEDs, have been successfully demonstrated in the form of large two-dimensional arrays.

In order to provide second-order effects, such as the linear electro-optic effect, in such material systems, the inversion symmetry of the QW or SL can be removed by growing structures with a built-in electric field. Practical means for the fabrication of asymmetric QW structures include strained layers, doping, or graded-band-gap techniques. Large Pockels coefficients at optical wavelengths near the semiconductor band gap have been predicted. With sophisticated epitaxial growth techniques, QW or SL structures have been fabricated as part of injection laser diodes, photo-detectors, and optical waveguides. This is still an area of intensive current research activities.

For conventional electro-optic modulation, the application of an electrical field in bulk electro-optic materials will result in a redistribution of the bond charges and possibly a slight deformation of the ion lattice of the optical medium, causing a change in the optical index of refraction. This is known as the electro-optic effect.

In general, the change in optical index of refraction can be described from the change in the impermeability tensor as

$$\Delta N_{ij} = r_{ijk}E_k + S_{ijkl}E_kE_l, \qquad (1)$$

where E is the applied electrical field. The constants r_{ijk} are the linear (Pockels) electro-optic coefficients, and s_{ijkl} are the quadratic (Kerr) electro-optical coefficients. The higher-order terms are much smaller and are neglected here. In crystalline materials without central symmetry, the linear electro-optic coefficient is the dominant effect. Thus, with the application of electrical field on an electro-optic material, the index of refraction of the crystal can be changed slightly, usually less than 10^{-4}. Such change in the optical index of refraction, although small, can cause the accumulated optical path length of a passing laser beam to vary up to a few microns, comparable to the optical wavelength. Electro-optical modulators are devices constructed to perturb the optical wave parameters by perturbation of the optical index of refraction.

3.3.1 Electro-optical Phase Modulator With the electro-optic effect, the optical path length of an electro-optic medium can be perturbed according to an applied electrical field. When an electrical signal voltage is applied to an electro-optic material, the signal will appear as phase changes in the optical wave form. This makes an electro-optical phase modulator the simplest form of an electro-optic modulator. The voltage required to cause a phase change of π, which corresponds to an optical path-length change of half wavelength, is called the "half-wave voltage" of the modulator.

Electro-optic phase modulators suffer from phase drift in other parts of the optical path of transmission, as well as phase drift in the phase-modulator crystal itself due to temperature-induced optical-index drift. Also, a stable reference optical beam is required to detect the phase change at the receiver end. Therefore, it is seldom employed in its simplest form, except in a few special applications. However, the principle of phase modulation is very important since it forms the basis for other types of electro-optical modulators.

In a "longitudinal electro-optic device" (see Fig. 3), the voltage is applied parallel to the optical beam path. Since the field-induced optical path-length change is the product of the electrical field and the physical thickness of the material, electro-optical modulation in such a device arrangement is therefore independent of the thickness of the electro-optic material. Usually, several thousand volts is required for achieving a phase shift of π. Longitudinal electro-optic modulator devices are often made with crystal slabs of moderate thickness providing relatively large optical acceptance angles (numerical aperture).

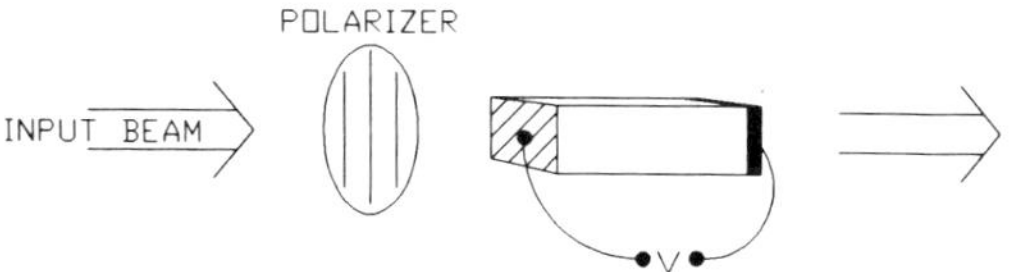

FIG. 3. Longitudinal electro-optic devices.

In a "transverse electro-optic device," as shown in Fig. 4, the voltage is applied transverse to the optical beam path in the crystal. The field-induced optical path-length change is proportional to the voltage applied, proportional to the physical length of the modulator crystal, and inversely proportional to the height of the electro-optic crystal. Increasing the modulator aspect ratio, i.e., long length and short height, can significantly reduce the required drive voltage. As a result, electro-optic modulators with drive voltages as low as 10 to 100 V have been demonstrated with the optical acceptance angle (numerical aperture) as a tradeoff.

3.3.2 Electro-optical Frequency Modulator The frequency of a wave function is the rate of change in phase, $\omega = df/dt$. When a linear voltage ramp is applied to an electro-optic phase modulator, the optical frequency will be shifted by a constant proportional to the slope of the voltage ramp. When a sinusoidal voltage is applied, there will be a sinusoidal frequency modulation in quadrature to the applied voltage. Thus, when an amplitude-modulated electrical wave form including its carrier with sufficient magnitude is applied to an electro-optical phase modulator, the optical wave form becomes

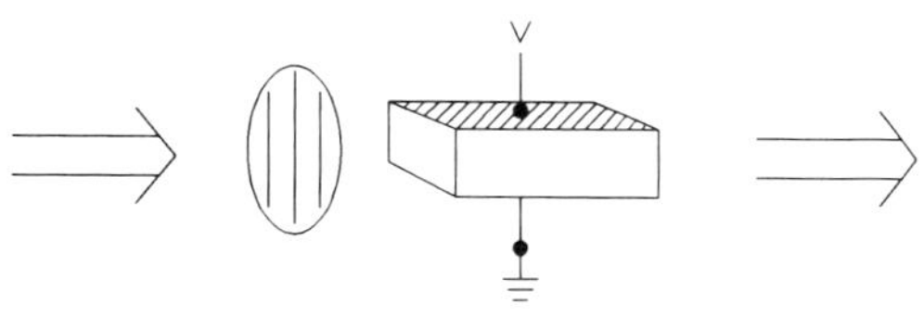

FIG. 4. Transverse electro-optic devices.

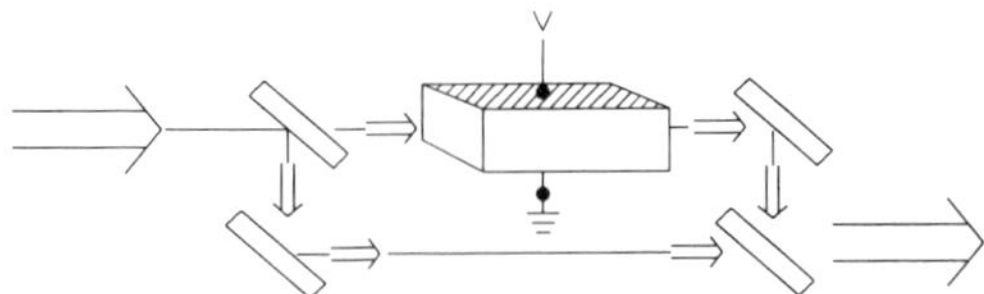

FIG. 5. Intensity modulator by means of a phase modulator inside an interferometer.

frequency modulated. Frequency modulation is much less sensitive to slow phase drifts in the optical transmission path since it is proportional to the rate of change in optical index of refraction.

3.3.3 Electro-optical Intensity Modulator To alleviate the unknown phase drifts in a phase-modulated optical system, a reference optical beam may be sent along with the modulated signal beam. When designed properly, all of the uncontrolled phase drifts in the system can be canceled as common mode between the signal beam and the reference beam. Figure 5 shows an interferometric intensity modulator in which a phase modulator is incorporated in one arm of an optical interferometer to provide intensity modulation. When the induced phase of the modulated arm becomes 180°, the interferometer output becomes zero.

In general, the output intensity follows Fig. 6, which shows an electro-optical phased-array intensity modulator using a linear array of electro-optical phase modulators. In this case, the reference beams are provided through the spacing among the electro-optical phase-modulator electrodes. With an applied voltage, the optical path length through the electroded areas is perturbed. The optical spatial phase front becomes nonuniform, and the laser beam with such a broken phase front can no longer be focused into a tight spot. When an optical aperture is placed at the focal point, the light intensity passing the optical aperture is therefore modulated by the application of an electrical field to the phase-modulator array.

3.3.4 Electro-optical Polarization Modulator In a polarization modulator, the reference beam is introduced in the form of another polarization orthogonal to the polarization of the signal beam. In most electro-optic materials, the optical index of refraction and the associated electro-optic coefficients are anisotropic and are polarization dependent. Light propagation through such a medium can be analyzed by decomposing the polarization state into two orthogonal components, each parallel to a certain crystalline axis and each capable of traveling through the thickness of the crystal without changing its polarization direction. The polarization state of the light output from such anisotropic material is the vector summation (interference) of the two exiting orthogonal polarization components. Since the electro-optic coefficients for the two orthogonal polarizations are different, a differential phase retardation can be introduced between the two polarization components by applying a voltage to the electro-optic crystal. As a result, the polarization state of the output light can be modulated by an applied voltage.

In the configuration shown in Fig. 7, the input optical polarization is oriented 45° to the crystalline axis of the electro-optic material so that the two orthogonal polarizations have equal magnitude. For example, when the relative phases of the two orthogonal polarization components are retarded by π as a result of an introduced optical path-length difference of half a wavelength, the output optical polarization becomes rotated by 90°.

Using an optical polarizer on the polarization-modulated optical beam converts polarization modulation into intensity modulation.

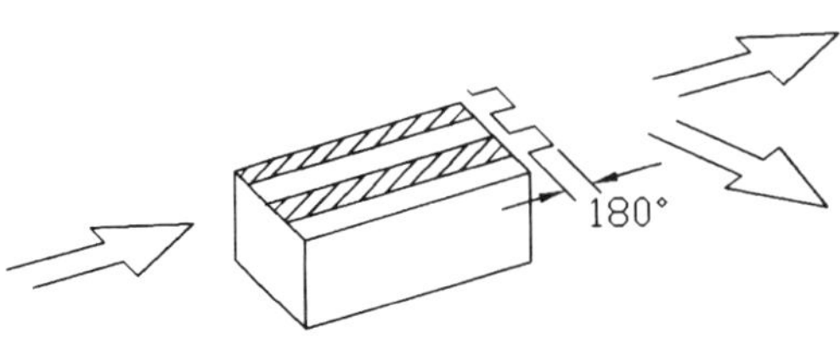

FIG. 6. Phase array intensity module.

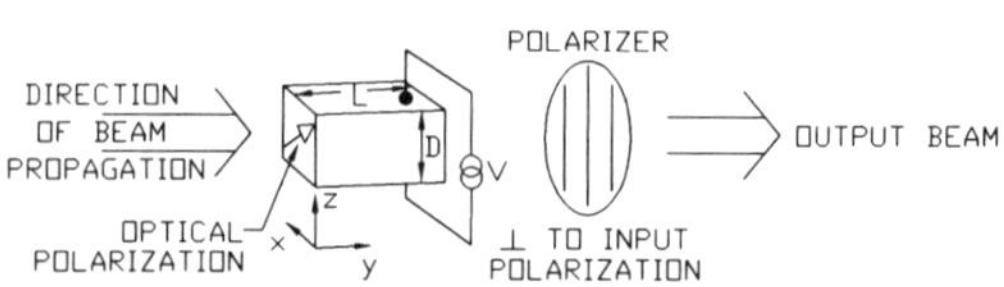

FIG. 7. Electro-optic intensity modulator by means of polarization switching.

The driving voltage required for the accumulated optical path-length difference to reach half an optical wavelength is called the "half wave voltage," which is often used as an indicator for the effectiveness of electro-optic materials or modulators. The drive power required of electro-optic modulators is determined by the driving voltage, the capacitance of the electrodes, and the load resistor value.

3.3.5 Electro-optic Crystals The electro-optic effect results from the redistribution of charges due to the application of an electric field. The redistribution is small since the applied field is typically many orders of magnitude smaller than the intra-atomic electric field binding the electrons and ions. However, the effect of such a redistribution of charges is noticeable in single-crystal materials because of their well-organized atomic structure. It can be shown that in crystals with a centrosymmetric lattice structure, the linear electro-optic effect vanishes, and the quadratic effect dominates. Also, since the applied field is small compared to the intra-atomic field, the quadratic electro-optic effect is expected to be small compared to the linear electro-optic effect and is often neglected when the linear effect is present. Electro-optic crystals that do not have a center of symmetry typically exhibit a large linear electro-optic effect, and are the most commonly used materials in the construction of electro-optic modulators. Examples of these electro-optic crystals include single crystals of $LiNbO_3$, $LiTaO_3$, potassium dihydrogen phosphate (KDP), ammonium dihydrogen phosphate (ADP), potassium trihydrogen phosphate (KTP), CdTe, GaAs, etc.

Most electro-optic crystals are associated with the photorefractive effect. With minute optical absorption, charge carriers are generated by light and are diffused and trapped to form a space-charge pattern inside the crystals. Coupled with the electro-optical effect, the space-charge pattern results in a nonuniform pattern of optical index of refraction. This effect is related to so-called "optical damage" of electro-optical modulators and can cause the optical quality of the laser beam passing through the electro-optic modulator to deteriorate as a function of time. Optical damage often limits the usefulness of many electro-optical modulators for the visible and UV wavelength ranges. For instance, lithium niobate, which has its transmission band edge at 0.4 μm, exhibits significant optical damage at blue and green wavelengths. The optical damage threshold can be improved with the use of carefully prepared high-purity crystals or special doping. On the other hand, ADP, which has its transmission band edge at 0.13 μm, does not exhibit photorefractive optical damage in the visible wavelength range.

However, the photorefractive effect can be used to alter the optical property of a material with light. Information can be written into an electro-optic crystal by controlled laser beams. Recent developments in photorefractivity have found potential new applications in optical signal processing, optical memory, etc.

3.4 Liquid Crystals

Liquid crystals represent an intermediate phase of materials that displays some of the properties of both liquids and crystals. In typical solid-state crystals, the molecules maintain both rotational and translational orders. During the melting transition of solids into liquids, kinetic energy is given to the molecules, and both rotational and translational orders are lost. However, in the situation where the molecules are very asymmetrical or oblong, the translational and rotational motions may be sufficiently decoupled, resulting in the separation of the melting transition into two stages. At first, the rod-shaped molecules will undergo a melting phase transition that involves only the translational degree of freedom. In this resulting liquid, molecules will have the freedom to move about, but because of their rod-shaped structure, their rotational motion about an axis perpendicular to the long rod directions will be largely inhibited. Such a liquid that maintains some of its rotational order is called a liquid crystal.

Liquid crystals are highly useful electro-optic materials as a result of the extremely large anisotropy in dielectric constant as well as optical index of refraction due to their highly asymmetric molecular shape, and as a result of the relative ease in perturbing the molecular orientations with the application of external electrical, thermal, or other disturbances.

There are three main classes of liquid crystals useful for the development of electro-optic devices; the smectics, the cholesterics, and the nematics. Figure 8 shows the schematic structure of each of these classes. The smectic class possesses not only an orientation order but also a certain degree of translational order associated with its layered structure. The nematic class is completely disordered translationally other than some imperfect alignment of the rod-shaped molecules parallel to their long dimensions. The addition of chiral molecules to the nematic phase will result in the formation of a helical structure referred to as the cholesteric class of liquid crystals. A similar structure of helical shape can also be obtained by imposing different alignment directions on the cell surfaces of a nematic liquid crystal, resulting in twisted nematic liquid crystal cells. The twisted nematic liquid crystals are the most commonly used configurations of liquid crystals to date.

The design of electro-optical modulators using liquid crystals may use any one of the classes of liquid crystals in various configurations. In general, a liquid crystal responds to an applied electrical field by tilting its higher dielectric constant direction toward the applied electric field.

Physical movement of such highly optical anisotropic molecules results in large changes in the observed optical index of refraction along a certain direction. The change in optical index of refraction can be used either to cause large scattering or to rotate the polarization of incident light similar to electro-optic modulators.

3.5 Acousto-optical Modulator

The optical index of refraction of all solid materials can be perturbed by the application of stress through the photo-elastic effect.

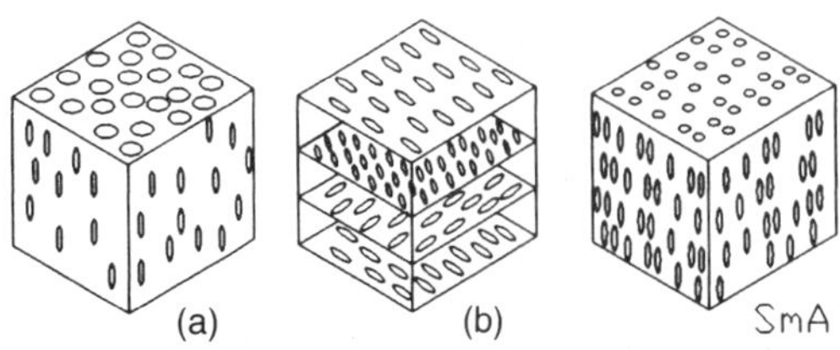

FIG. 8. Three types of liquid crystal materials.

With the presence of an acoustic wave (a stress wave), periodical localized stress regions cause a periodical optical index profile to appear in the acoustic medium, traveling at the speed of sound. With sufficiently small periodicity, the periodic optical index can be viewed as a dynamic optical phase grating with adjustable wavelength and amplitude created by the introduction of acoustic waves to photoelastic materials. Acousto-optical modulators are based on modulation of light by means of the acoustically generated dynamic phase grating.

Passing through a phase grating, a light beam will be diffracted into many diffraction orders, each containing part of the incident optical intensity. Most acousto-optical modulators operate in the Bragg diffraction regime in which the phase grating thickness is much larger than Λ^2/λ, where λ is optical wavelength and Λ is acoustic wavelength. With Bragg diffraction, high-efficiency devices can be developed since the majority of light intensity is concentrated in only two diffraction orders. With sufficient grating thickness and with well-collimated light incident at the Bragg angle condition, the diffraction efficiency of Bragg gratings can be extremely high, approaching 100%. The diffracted light is deflected by an angle of approximately λ/Λ, which is proportional to the acoustic frequency, from the incident direction. Using Bragg-diffracted light, acousto-optical modulators of high efficiency and high extinction ratio become available, making acousto-optical modulators, which are sometimes called Bragg cells, the most widely used type of laser modulator.

The diffracted light from a grating assumes the phase information of the grating, i.e., the diffracted light experiences a phase shift as the grating is shifted perpendicular to the grating lines. An acoustic grating traveling at the speed of sound causes the diffracted light to acquire a frequency shift equal to the frequency of the acoustic wave. Thus, it is possible to use an acousto-optical modulator for amplitude modulation, intensity modulation, phase modulation, and frequency modulation. Polarization switching can be accomplished by means of an anisotropic acousto-optical modulator design in which the polarization of the diffracted light is orthogonal to the polarization of the incident light. Changing the frequency of the acoustic wave can cause the diffracted light to change its deflection angle rapidly over a relatively wide range.

Typical construction of acousto-optical modulators is given by Fig. 9. A thin piezoelectric layer, such as $LiNbO_3$ or ZnO, is sandwiched between conductive electrodes, is fabricated to resonate at the desired center frequency of the acoustic bandwidth, and is attached to the acousto-optical interaction medium. The metal electrode layers, typically made of indium, gold, or chromium, are carefully designed to enhance the passage of broad-band acoustic waves from the piezoelectric layer to the acousto-optical medium. An electrical impedance-matching circuit facilitates efficient and broad-band transfer of electrical power to the acoustic transducer. The far end of the acousto-optical medium is acoustically terminated to prevent undesirable acousto-optical interaction due to reflected acoustic waves. Commonly used acousto-optical media are $PbMoO_4$, $LiNbO_3$, TeO_2, GaP, fused quartz, and various flint glasses. Acousto-optical modulators can also be made of surface acoustic waves when the propagation of light is confined to the surface of acousto-optical materials.

3.6 Magneto-optical Modulator

Magneto-optical modulators are based on the rotation of optical polarization as light propagates along the magnetic field in a material, by the Faraday effect. The Faraday effect originates from the effect of the static magnetic field on the motion of electrons in the presence of light and the Lorentz force. Because of the polarity of the magnetic field parallel to the light beam, Faraday rotation is nonreciprocal and is highly useful for optical isolators that separate backward reflection light from forward light. The measure of the Faraday effect is the Verdet constant, which is defined as the specific rotation (i.e., rotation per unit length) divided by the magnetic field B. Large Verdet constants are found in many glasses and in garnet-based material systems.

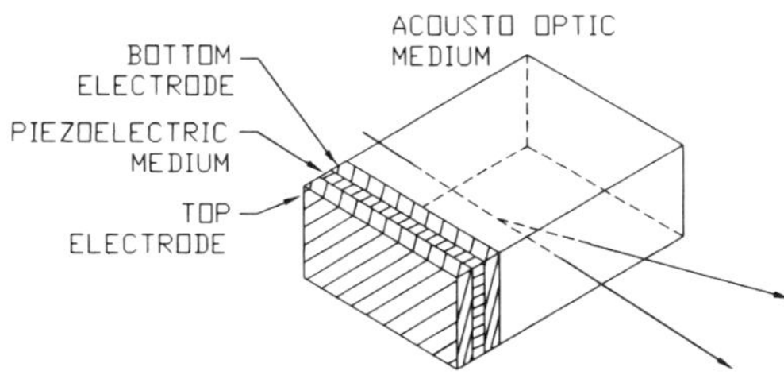

FIG. 9. Construction of acousto-optic modulator.

Magneto-optical modulators can be made of a Faraday material placed inside an electrical coil with modulating electrical current. By use of high-frequency magnetic spin waves, optical phase gratings can be set up in magneto-optical materials for modulation of light in a manner similar to that in acousto-optical modulators.

3.7 Waveguide Optical Modulator

Recently, a low-loss optical waveguide has been made in the form of thin-film planar waveguides or channel waveguides on electro-optic crystals such as $LiNbO_3$, GaAs, InP, etc., by means of solid-state diffusion or epitaxy. In an optical waveguide, light is trapped inside a region of higher optical index of refraction than its surroundings by means of total internal reflection at the boundaries. High-index waveguide films on various substrates can be prepared by vapor-phase deposition. Single-crystalline high-index layers with electro-optic properties can be prepared by epitaxial growth on single-crystal substrates. Solid-state diffusion is another means of making waveguides with electro-optic properties on an electro-optic crystal substrate. The number of waveguide modes supported by an optical waveguide is proportional to waveguide thickness. Most high-performance waveguide optical modulators use single-mode optical waveguides, because each waveguide mode may behave differently in electro-optic devices.

Since typical single-mode optical waveguides are only a few microns in thickness, light is confined within the high–optical-refractive-index layer or channel on the substrate's surface. Use of a guided-wave light beam overcomes the trade-off between a tight focal spot and limited depth of focus in conventional focusing optics. With an optical waveguide, high-intensity light can be maintained over a long propagation length, allowing the construction of electro-optic modulators with an extremely large aspect ratio. Electro-optical modulators with extremely low drive voltage, in the range of a few volts, become possible.

Figure 10 illustrates a channel waveguide electro-optical phase modulator. The single-mode channel waveguide has a width of about 3 to 5 μm, which nearly equals the waveguide depth. The top electrode is a fin-line structure on the top surface of the substrate. The electrical fringing field of the top electrode penetrates the substrate for a depth approximately equaling the gap width and thus covers the channel waveguide cross section. In such an arrangement, the waveguide index is perturbed by the applied electrical field, causing phase modulation of the guided light mode.

There are many other types of waveguide optical modulators. In a Mach–Zehnder waveguide optical modulator, one optical channel waveguide is split into two channel waveguides in parallel, with each one phase modulated with a signal voltage of opposite polarity. These two phase-modulated waveguide channels are merged together into one channel waveguide similar to an optical interferometer. Only when the two merging waveguide modes are in phase does strong coupling into the output channel waveguide occur. The result is an intensity modulator.

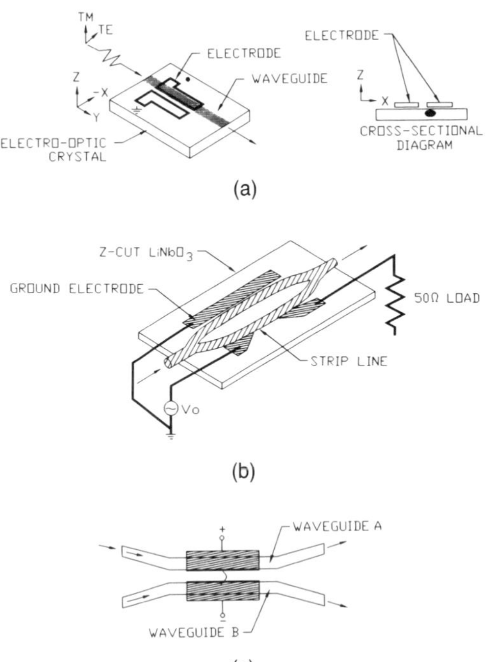

FIG. 10. Waveguide optical modulator: (a) phase modulator, (b) interferometric waveguide modulator, (c) switched-coupler type of waveguide modulator ($\Delta\beta$ waveguide modulator).

In the $\Delta\beta$ waveguide modulator, two single-channel waveguides are brought to close proximity in a region influenced by electro-optic fields. The coupling coefficient of the pair of closely located channel waveguides is related to the guided-mode propagation constant, β, of the two channels. Since the guided-mode propagation constants in the electro-optic waveguides are adjustable by the application of an electric field, the coupling coefficient can be controlled by the applied electrical signal. With strong coupling, optical power can switch from one channel to another channel after a short distance of propagation through the coupling region.

Waveguide acousto-optic modulators can also be constructed using surface acoustic waves to interact with light in planar optical waveguides.

4. OPTICAL MODULATOR PERFORMANCE AND EXAMPLES

4.1 Frequency Response of Modulation

When electro-optic modulators are properly designed, their frequency response can cover more than 20 GHz of bandwidth. The frequency response of lumped-element electro-optical modulators is limited to a few gigahertz by the capacitance of the electrodes and the load resistor as a result of the *RC* time constant. To extend the frequency response beyond the *RC* time-constant limit, the electrodes of the electro-optical modulator can be made into a broad-band microwave transmission line—for example, a microstrip line or a fin-line structure. Often, waveguide electro-optical modulator configurations are employed to reduce the electrical drive voltage as well as the drive power consumption.

The ultimate limit of broad-band electro-optical modulators is the velocity mismatch between the electrical signal in the microwave transmission line and the light wave in the optical waveguide. Both electrical signal and light wave travel at the speed of light. However, the speed of the electrical signal is reduced by the square root of the material dielectric constant, while the speed of light is reduced by the optical index of refraction.

For instance, in $LiNbO_3$, the speed of a light beam is more than twice the speed of the electrical signal since the dielectric constant is about 30, while the optical index of refraction is about 2.2. Various means have been developed to match better the velocities of light and electrical signal: choice of material, changing the phase of the electrical signal, lengthening the optical path, etc.

The frequency response of acousto-optic modulators is limited by the acoustic transit time across the finite width of the light beam, the acoustic propagation loss in the acousto-optic medium, and the frequency response of the acoustic transducer. With proper choice of materials and transducer fabrication techniques, the limitations in frequency response of practical acousto-optic modulators are usually due to the acoustic transit time. Acoustic-wave velocities in most solid materials are in the range of 3 to 10 mm μs^{-1} or $\mu m\ ns^{-1}$. A sharply focused laser beam is necessary to reduce the acoustic transit time. However, a sharply focused beam has a large cone angle so that only part of the focused beam may truly satisfy the Bragg-angle condition for efficient Bragg diffraction of light. The state of the art of wide-band acousto-optic modulator optimizes these conflicting requirements and to date results in bandwidth tradeoffs of up to a few hundred megahertz of bandwidth.

4.2 Efficiency of Modulation

The efficiency of optical modulators is the ratio of modulated optical output power divided by the input optical power. Since laser light is expensive, it is very important to keep the modulator efficiency as high as possible. The efficiency of electro-optical modulators is limited by the optical transmissivity of optical components, including the crystal itself, the optical polarizer, and beam-shaping optics. The efficiency of acousto-optic modulators is limited by the Bragg diffraction efficiency and the beam-shaping optics. With proper design, efficiencies greater than 70% can be achieved.

4.3 Multispectral Modulation

In certain applications, such as color separation and laser entertainment, it is important to modulate several colors of laser light independently and simultaneously. Instead of using multiple modulators in physically separated channels, it is possible to do it with just one acousto-optic modulator if the modulation bandwidth for an individual channel is sufficiently small. Figure 11 illustrates such a device. Each color component of the multispectral light beam requires a different acoustic wavelength in the acousto-optic modulator to satisfy the Bragg condition. Thus, the acoustic signal at a certain acoustic frequency in the acousto-optic modulator intended for one color will not interact with other color components of the multispectral input light. Simultaneous delivery of several signals at different acoustic frequencies to the multispectral acousto-optic modulator can provide independent modulation of the multispectral light.

4.4 Examples of Modulation of Lasers—Laser Diodes

Most laser applications require some kind of modulation. For example, in fiber-optics communications, the laser or LED output needs to be modulated to bear the data signal. The laser beams in laser printers, copier machines, or laser entertainment applications need to be turned on and off, i.e., modulated, while scanning across the printing media surface or the display surface. In laser radar, the laser beam needs to be modulated with pulses sometimes themselves further modulated to contain codes for counter-counter-measures.

4.5 Modulation of Lasers—Intracavity

When a laser modulator is placed inside a laser cavity, special effects can be generated out of the laser. With intracavity modula-

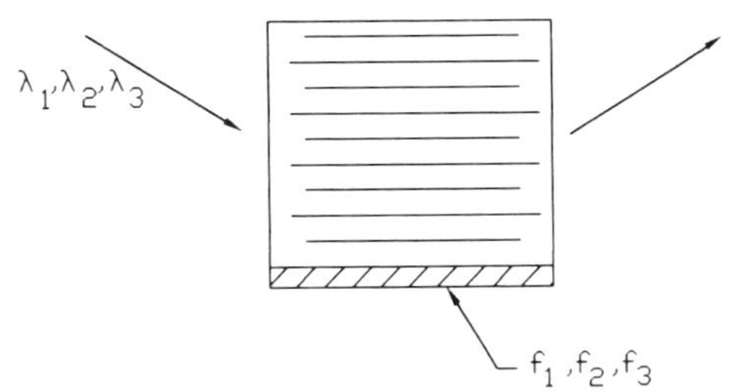

FIG. 11. Multispectral acousto-optic modulator.

tion, the laser output can be configured to several different types of pulse trains.

In the case of the Q-switching type of intracavity laser modulation, the modulator introduces sufficient optical loss in the laser cavity to prohibit laser oscillation during most of the time. Without oscillation, photon energy is stored in the laser rod as in a charged capacitor. The moment the intracavity laser modulator is turned to low-loss state, the laser starts to oscillate, draining the stored photon energy into laser output. Because the stored energy can reach a higher level when laser oscillation is quenched than when there is laser oscillation during the charging cycle, the sudden oscillation enacted by the Q switch can provide a giant output laser pulse with much higher intensity than a continuously oscillating laser. Since it takes several round-trip passes of photon travel for the oscillation to start, the length of Q-switch pulses is on the order of nanoseconds to a few hundred nanoseconds. Typical Q switches are made of either $LiNbO_3$ or ADP electro-optic modulators or fused-quartz acousto-optic modulators.

In the case of cavity dumping, which is usually used with cw-type gas lasers, the laser is made to oscillate continuously in a closed laser cavity with no output window. The laser light that is continuously supplied by the discharging gas builds up in the closed cavity until suddenly an output window is switched open by a fast–rise-time intracavity acousto-optic modulator. The light output is typically shorter than one round trip of the laser cavity, on the order of nanoseconds. Cavity dumping is often used in conjunction with laser mode locking to generate a giant pulse of less than a 1-ns-length transit time.

In the case of laser mode locking, the laser intracavity modulator is driven by a periodic signal synchronized to the round-trip photon traveling time in the laser cavity. Although light is continuously generated by spontaneous emission from the laser medium, only that part of the light that arrives at the intracavity modulator when the modulator loss is minimum will experience a low loss and therefore will grow as it accumulates more and more passes in the laser cavity. The result is a giant circulating optical pulse within the laser cavity in synchronization with the periodic intracavity modulator. Each time the intracavity pulse hits the output mirror of the laser cavity, a portion of the pulse is coupled out of the laser. Thus, the output of a mode-locked laser is a continuous pulse train with a pulse separation equaling the round-trip traveling time of the laser cavity. Use of a cavity-dumping modulator can switch this giant circulating intracavity optical pulse to the outside, resulting in a single giant output pulse of duration much shorter than the laser cavity length.

5. DEMODULATION AND DETECTION

Demodulation and detection of the optical signal is retrieving the signal content from the modulated light. Since photodetectors are sensitive only to the intensity of light, key elements of optical demodulation are the conversion of specific optical modulation into intensity-modulated optical wave forms and the use of photodetectors of proper working wavelength range, speed, and sensitivity. Thus, demodulation means the conversion from various forms of optical modulation into intensity-modulated light, while detection means the conversion of intensity-modulated light into electrical wave forms.

5.1 Demodulation and Detection of Intensity-Modulated Light

Demodulation or detection of intensity-modulated light, including pulse-modulated light, is the straightforward use of proper photodetectors with suitable wavelength response, speed, and sensitivity. This is the base for all other optical demodulations.

In a case where the modulation rate is faster than the speed of available photodetectors, optical demodulation may include the use of fast electro-optical switches or gates for sampling certain portions of the modulated light in a systematic manner. In case the intensity of modulated light is too weak for simple detection with conventional high-speed detectors, special schemes may be used for the enhancement of optical detection, such as optical preamplification with a laserlike optical gain medium, optical heterodyning using a strong local-oscillator laser, or the use of photomultiplier tubes, avalanche photodiodes, and/or optical intensifiers (night vision devices).

5.2 Demodulation and Detection of Polarization-Modulated Light

Polarization-modulated light can be converted to intensity-modulated light by means of an optical polarizer or a polarization optical beam splitter of proper type and orientation. With a polarization beam splitter, there will be two intensity-modulated optical outputs of complementary polarity. Detection of both can improve sensitivity of the system.

5.3 Demodulation and Detection of Deflection-Modulated Light

Deflection-modulated light can be demodulated into intensity-modulated light with the use of optical apertures, detectors with limited field, or a detector array. For instance, a triangular optical aperture provides linear conversion of deflection position to intensity variation.

5.4 Demodulation and Detection of Wavelength-Modulated Light

Wavelength-modulated light can be demodulated into intensity-modulated light with the use of wavelength-sensitive optical elements such as a single-layer or multilayer thin film, detector band edges, color-filter absorption edges, etc. It can also be converted to deflection-modulated light with the use of optical dispersive elements such as a prism, a grating, or a wedge Fabry–Pérot interferometer of proper spectral resolution.

5.5 Interferometric Demodulation and Detection

Information carried in optical phase or frequency modulations can be recovered by means of optical interferometry, which compares two optical wave fronts. A simple optical beam splitter can be employed to combine the modulated optical beam with a reference or local-oscillator optical beam; see Fig. 12. When properly aligned, the optical intensity output from one of the output ports will be

$$|A(t)e^{j(\omega t+f)} + Re^{j\omega_0 t}|^2 = |A(t)|^2 + |R|^2 + 2A(t)R\cos[(\omega - \omega_0)t + f], \qquad (2)$$

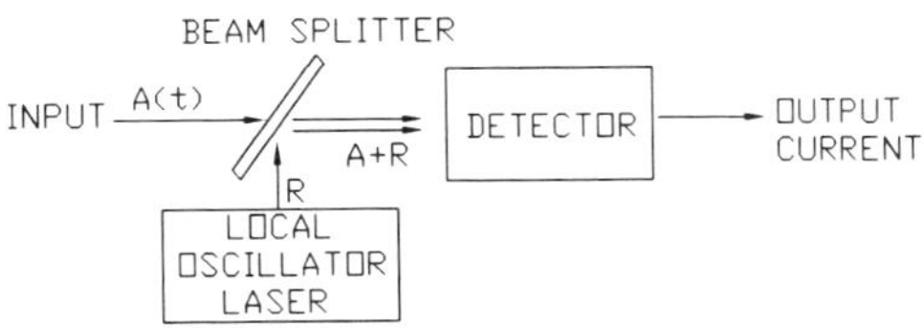

FIG. 12. Heterodyne optical receiver.

where $A(t)$ is the information signal, R is a constant denoting the magnitude of the reference beam or local oscillator, and f is phase noise due to the laser carrier and/or the reference laser. Recovery of the signal $A(t)$ can be made by separating the last term from the first two terms using either a band-pass filter or dc-blocking filters. Often, the reference R is made much stronger than the signal $A(t)$, $R \gg A(t)$, so that the time-variable term $|A(t)|^2$ can be ignored, and the useful term, $2A(t)R$, includes signal amplification.

A special case is the use of a delayed copy of the signal light as the reference beam so that the signal is compared to a delayed section of itself, as illustrated by Fig. 13 in the form of an uneven-arms Michelson interferometer. The optical output of this interferometer is an intensity-modulated signal proportional to the differential phase of the input, $A(t)^2 \cos[f(D) - f(0)]$. The optimal length of delay, D, is customized to specific modulation parameters.

5.6 Demodulation and Detection of Amplitude- or Phase-Modulated Light

Interferometric demodulation is required for amplitude- or phase-modulated light. Although the information is carried in the envelope of the light wave in amplitude-modulated light, direct detection of amplitude-modulated light will produce a highly

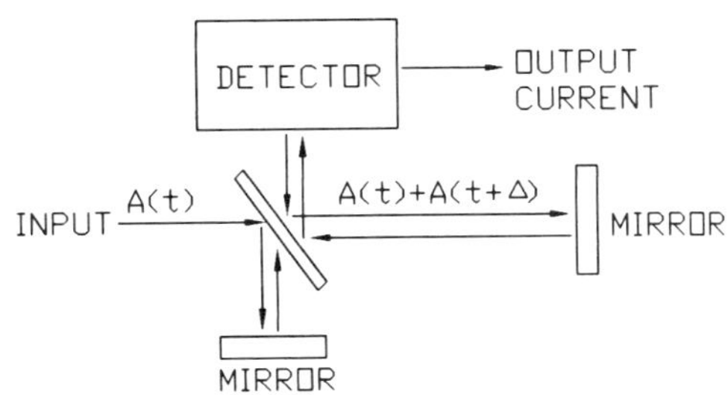

Fig. 13. Homodyne optical interferometric receiver.

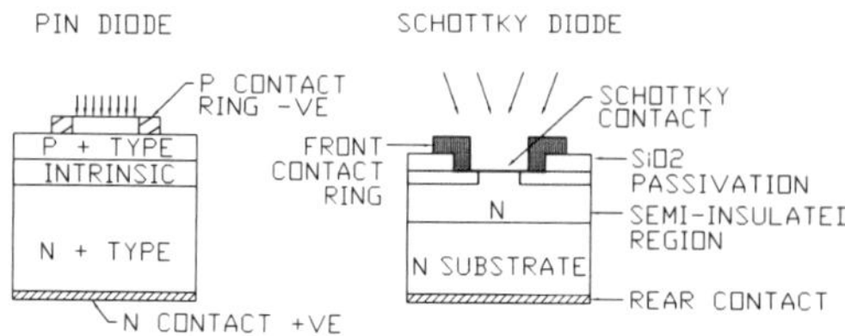

FIG. 14. *p-i-n* and Schottky diode configuration.

distorted electrical wave form proportional to the square of signal. As shown by the formula for the heterodyne interferometer output, the amplitude of the optical input can be obtained by the intensity cross term, $2A(t)R \cos[(\omega - \omega_0)t + f]$. The envelope of the cross term, $2A(t)R$, can be produced with a rectifier. In such a case, stability in the phase and frequency of the reference laser is not necessary.

For demodulation of phase-modulated light, $A(t) = A\, e^{jg(t)}$ in which A is constant, the interferometer output becomes $A^2 + R^2 + 2AR \times \cos[(\omega - \omega_0)t + g(t) + f]$. The phase signal goes into the intensity cross term. The filtered output of the photodetector is the phase signal on the electrical carrier of frequency $\omega - \omega_0$, with phase noise f. Stability of the source and reference (local oscillator) lasers is necessary. For the case of homodyne interferometric demodulation using a delayed signal, the output is simply the differential phase signal given by $2A^2\{1 + \cos[g(t + D) - g(t)]\}$, if we assume that short-term laser noises are negligible.

5.7 Laser Light Detection

Light is converted back to electrical energy through the use of an optical detector. Vacuum tubes and solid-state devices are typical detectors available. In vacuum tubes, photomultipliers have very good sensitivity due to their very high gain. In solid-state detectors, there are many versions, different doping arrangements of the semiconductor, and various physical geometries. Two common types for high-speed applications are the *p-i-n* diode and the Schottky diode whose geometries are shown in Fig. 14. The discussions below center on these two types of detectors.

When light hits the photodiode, photon energy is absorbed in the depletion layer to generate hole and electron carriers, and is then collected as current at the electrodes before recombination. A high quantum efficiency results from maximizing the absorption and recombination time of the carriers.

When a diode operates in the open-circuit condition to a high-impedance load, it is called the photovoltaic mode. The dark current from the diode is the lowest in this mode. When the diode is reversed biased, working into a low impedance, the bandwidth and the quantum efficiency are improved. For optical-communication applications, the detector operates in the reverse biased mode.

The common detector materials are Si, InGaAs, and Ge. In the visible wavelength, Si has a good responsitivity R (A/W), i.e., output current in amperes per input optical power in watts. In the near-IR and the 1.3- and 1.55-μm laser diode wavelength regions, InGaAs has a good responsitivity. The 1.3- and 1.55-μm regions are particularly important in optical communications because of low fiber losses and the availability of optical amplifiers at the 1.55-μm wavelength. Figure 15 shows the responsitivity for the three materials. The quantum efficiency of the detector

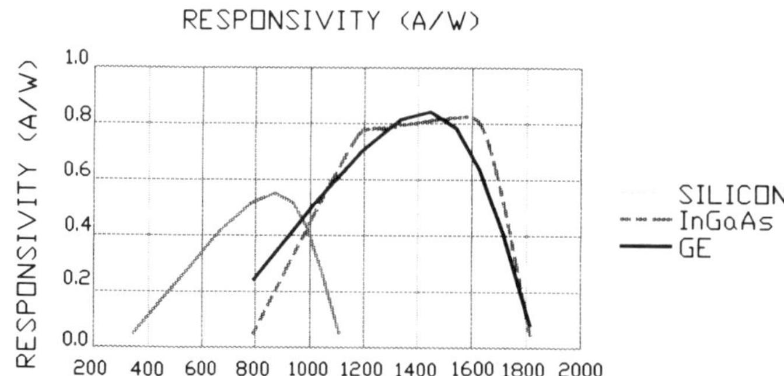

FIG. 15. Responsitivity for SI, InGaAs, and Ge.

material is calculated from the formula

$$\eta = (1.24R/\lambda) \times 100\%. \quad (3)$$

Typical quantum efficiency values range from 50% to 90%.

The detector performance is characterized by bandwidth, efficiency, noise equivalent power, and linearity. The bandwidth of the device is inversely proportional to the size of the detector, i.e., the larger the size, the slower the response. Another key factor is the device design in making the carriers move at a high velocity, thereby resulting in a fast response time. In high-speed applications, Schottky diodes are used for the visible wavelength, while heterostructure photodiodes are used in the longer-wavelength applications.

For a 1-GHz device, the detector size is about 100 μm in diameter. At 20-GHz bandwidth, the detector diameter drops to 25 μm of active area.

Noise in the detector is an important consideration. In a silicon photodetector, the noise sources are thermal motion of particles in the resistor and shot noise in the photoelectron generation.

Noise equivalent power (NEP) is the incident light power such that the signal-to-noise ratio is equal to 1. NEP is defined as

$$\mathrm{NEP} = i_n/B^{1/2}R \quad \text{in} \quad \mathrm{W/Hz^{1/2}}, \quad (4)$$

where i_n is the rms noise current. Some typical values are 10–15 $\mathrm{W/Hz^{1/2}}$ for small-size, high-speed diodes. Another parameter its detectivity, which is equal to

$$D = A^{1/2}/\mathrm{NEP}, \quad (5)$$

where A is the active detector area.

As it turns out, the noise contribution from detectors is usually less significant than noise from post-amplifier electronics. Therefore, the responsitivity is the key deciding parameter in critical high-speed applications.

For high speed and better linearity, the detectors are back biased with voltages from several volts to about 100 V. The maximum linear input for high-speed detectors is one to a few milliwatts. In addition to biasing the detector, mechanical packaging is also very important for high-speed performance, and the techniques are similar to microwave component packaging (see PACKAGING TECHNOLOGY).

GLOSSARY

Band Gap: The difference in energy between the highest state in the valence band of an insulator or semiconductor and the lowest state of the conduction band.

Carrier Wave: The high-frequency wave to be modulated by the signal.

Frequency-Shift Keying (FSK): The coding of a digital signal by shifting the frequency of the carrier wave to one of several preset values.

Injection Laser Diode: A laser device made of a *pn* junction in a direct–band-gap semiconductor, such as GaAs.

Interferometer: An optical instrument that brings two optical beams together and emits the sum of the amplitudes of the two beams as output.

Phase-Shift Keying (PSK): The coding of a digital signal by shifting the phase of the carrier wave by one of several preset amounts.

Photodetector: Electronic device that responds to the arrival of an optical photon.

Polarization: Orientation of the electrical field direction of an electromagnetic wave.

Quantum Well: A system in which at least one of the three degrees of translational freedom of the electrons in a solid is restricted, or even suppressed, by a spatial variation of the properties of the solid, as in a thin layer of one material embedded within a different material; this gives rise to quantization of movement in the direction affected.

Waveguide Optical Modulator: Optical modulator made for use with light waves traveling in optical waveguides.

Further Reading

Chaimowicz, J. C. A. (1989), *Lightwave Technology: An Introduction*, London: Butterworths.

Chang, K. (1991), *Handbook of Microwave and Optical Components*, Vol. 4, New York: Wiley.

Huang, H., Koch, S. W. (1990), *Semiconductors*, Singapore: World Scientific.

Kaminow, I. (1974), *An Introduction to Electro-Optic Devices*, New York: Academic.

Karim, M. A. (1990), *Electro-Optic Devices and Systems*, Boston: PWS-KENT Publishing Co.

Narasimhamurty, T. S. (1981), *Photoelasticity and*

Electro-Optic Properties of Crystals, New York: Plenum.

Nishihara, H., Hruna, M., Suhara, T. (1989), *Optical Integrated Circuit*, Optical and Electro-Optical Engineering Series, New York: McGraw-Hill.

Marsland, R. A. (1993), "Selection and Application of High Speed Photodetectors," in *Photonics Design and Applications Handbook*, No. 3, *The Photonics Directory*, Pittsfield, MA: Laurin Publishing Co.

Watson, J. (1988), *Optoelectronics*, London: Van Nostrand Reinhold.

Xu, J., Stroud, R. (1992), *Acousto-Optic Devices: Principles, Designs and Applications*, New York: Wiley.

Yariv, A. (1971), *Introduction to Optoelectronics*, New York: Holt, Rinehart and Winston.

Yariv, A., Yeh, P. (1984), *Optical Waves in Crystals*, New York: Wiley.

MOLECULAR AND ATOMIC CLUSTERS

JOANNA M. HUNTER AND MARTIN F. JARROLD, *Department of Chemistry, Northwestern University, Evanston, Illinois, U.S.A.*

INTRODUCTION

The diverse field of molecular and atomic clusters lies at the interface of physics, chemistry, and materials science. A cluster is an aggregate of atoms or molecules, the number of which lies somewhere between the limits of a single atom and bulk matter. The physical and chemical properties of these microscopic aggregates evolve as a function of size. Clusters can be considered the building blocks of bulk matter, and a better understanding of bulk properties may be gained through their study. In addition, clusters have unique characteristics that are distinct from those of the bulk. The physical and chemical properties of clusters often do not vary smoothly with increasing size, and in some cases there are "special" clusters (containing a particular number of atoms) that have exceptional properties—enhanced stability, for example. The main topics of this article are effects due to the finite size of clusters, which can be structural, electronic, thermodynamic, or quantum in nature.

Two main categories of clusters arise from a classification of the bonding interaction between the individual components of the aggregate. We define molecular clusters as aggregates that are generally bound by less than about 0.5 eV/atom, such as by weak van der

3-527-28132-0/94/$5.00 + .50

Waals interactions, hydrogen bonding, or ion–molecule interactions. Atomic clusters are typically more strongly bound, having ionic or covalent chemical bonds between component atoms. Although the distinction between these two types of cluster sometimes becomes blurred, it is nevertheless useful to divide the vast amount of information in this way.

Weakly bound molecular clusters are generally not stable at room temperature and pressure. Therefore, special techniques, usually employing molecular beams under high-vacuum conditions, are required to generate and study these species. These aggregates can only be studied during their transient lifetime in the vacuum chamber. Chemically bound atomic clusters are, in principle, more stable, but, because of their extreme reactivity, naked (unprotected) atomic clusters are also usually studied in an interaction-free beam environment. An alternative approach involves isolation of the clusters in an inert matrix such as solid argon at cryogenic temperatures. Passivation of the cluster surface with ligands is another way of circumventing the reactivity problem. Although attaching ligands clearly modifies the electronic and geometric structure, these passivated clusters can be generated in macroscopic quantities and manipulated in air. Both gas-phase (naked) and condensed-phase (ligated) clusters will be discussed in this article, spanning the size range from three to several thousand atoms.

1. SYNTHESIS OF CLUSTERS

There is a plethora of methods for fabricating clusters. These techniques are constantly being improved and tailored to suit the requirements of investigating a specific property. Here we will merely outline the function of the most common types of cluster beam sources, as well as some general methods for condensed-phase cluster synthesis. For further detailed descriptions of the techniques, the reader is referred to the articles cited at the end of the text.

1.1 Molecular and Atomic Cluster Beams

1.1.1 Supersonic Expansions Weakly bound molecular clusters can be generated by the adiabatic expansion of a gas from a high-pressure source through a small orifice into a region of low pressure. Cluster growth is promoted by the cooling that occurs in the expansion and takes place through low-energy monomer–cluster and cluster–cluster collisions. The nascent clusters can reach vibrational and rotational temperatures of as low as a few kelvins, and have their translational energy directed along the beam axis. Cluster-size distributions, internal temperatures, and terminal velocities are dependent on parameters such as pressure, temperature, and nozzle shape. Some experimental control over cluster-size distribution is possible, although this approach always generates a broad distribution of cluster sizes. Because of the high source-pressure requirement, pulsed valves are often used to generate large, weakly bound molecular clusters. Short pulses of gas, 10 μs to 1 ms long, eliminate the need for a large pumping capacity required to maintain low pressures in the instrument (Gentry, 1988).

1.1.2 Oven Sources While simple supersonic expansions can be used to generate many types of molecular clusters, for atomic clusters it is necessary first to volatilize the element. The most straightforward method of volatilizing elements is known as a Knudsen cell, which involves heating bulk material in an oven or crucible and then allowing the vapor to effuse into a region of low pressure. Clusters generated in this way have a broad distribution of velocities and sizes and are internally hotter than those formed by supersonic expansion. Colder clusters can be formed by coexpanding the vapor with an inert carrier gas that collisionally cools the clusters.

1.1.3 Laser Vaporization At the high temperatures necessary to vaporize many bulk materials, oven sources often encounter severe materials limitations. Refractory materials such as tungsten and carbon, for example, are difficult to vaporize by direct heating. A versatile approach that avoids the use of high temperatures is pulsed-laser vaporization, which was developed by Smalley and co-workers (Powers *et al.*, 1982). Focusing a pulsed laser onto a sample transiently heats the material to extremely high temperatures. The resulting plasma is cooled by collisions with a carrier gas and expanded through a nozzle to form an intense pulsed cluster beam containing both neutral and ionic species. Although continuous carrier-gas flows produce more reproducible beams, a pulsed carrier-gas flow is often used to reduce the pumping requirements.

1.1.4 Sputtering Direct sputtering with high-energy ions or neutral atoms is another useful method of producing atomic clusters. The impact of high–kinetic-energy (10–20 keV) particles on a solid surface removes neutral and ionic clusters directly. The intensity of the resulting clusters decreases exponentially with increasing cluster size, and, thus, this approach is generally not suitable for making large clusters. Furthermore, the clusters are internally hot and must be thermalized before subsequent study (Parent and Anderson, 1992).

1.2 Condensed-Phase Clusters

1.2.1 Colloidal Dispersions Small metal particles, of gold in particular, have been prepared in colloidal solutions since the middle of the last century (Faraday, 1857). Typically, colloidal particles can range in size from 4 to 400 nm in diameter. The most general synthesis of colloidal metal particles involves reduction of a metal salt in solution. The size and shape of the resulting colloids depend on the reducing agent, *p*H, and reagent concentration. Thus, it is possible to synthesize selectively particles of a desired average size and shape (Milligan and Morriss, 1964). Protective polymers are sometimes added to prevent coalescence, but because of their inherent instability, colloidal particles have not been isolated from solution. Incorporation of the colloidal particles into solid matrices such as polymers and glasses solves this problem. As it is clearly desirable to prepare a narrow particle size distribution, chromatographic methods such as capillary-zone electrophoresis and size-exclusion chromatography have been applied to reduce the broad size distributions found in colloids.

1.2.2 Capped Clusters An alternative to colloidal particles, which have relatively broad size distributions, is the direct chemical synthesis of ligand-stabilized (also called capped) clusters with well-defined structures. The size range of these compounds extends from nanometers down to several atoms. For example, Schmid has been particularly successful at developing methods to generate ligated metal clusters of specific size (Schmid, 1992). A series of clusters with the general formula $M_{55}L_{12}Cl_x$ (M = Rh, Ru, Pt, Au; L = PR_3, AsR_3; x = 6, 20, where R = phenyl, *t*-butyl) are formed by reduction of metal salts with B_2H_6 in organic solvents. Numerous other ligated metal clusters have been crystallized, as described in recent reviews (Mingos and Wales, 1990; Schmid, 1992; Lewis, 1993).

Considerable effort has been expended to synthesize semiconductor clusters in which to study quantum confinement effects. The size control attainable here is generally not as great as in the ligand-stabilized metal clusters. One method used to generate capped II-VI semiconductor clusters (e.g., CdSe) is the inverse micelle preparation (Steigerwald *et al.*, 1988). In this procedure, surfactant molecules encapsulate and stabilize growing CdSe clusters to form micelles. When all of the reactant Cd^{2+} and Se^{2-} have been depleted, cluster growth is terminated, thus allowing some control over the cluster size. Treating the clusters with organoselenides, such as $C_6H_5Se^-$, covalently links organic caps to the Cd^{2+}-rich cluster surface. Because this chemically modified surface is hydrophobic, the clusters precipitate as stable entities, which can be collected and subsequently redispersed. Other semiconductor cluster synthesis methods also involve terminating the cluster surface with stabilizing caps. For example, $Cd_{10}S_4$ has been capped with thiophenolate groups, and these pyramidal $Cd_{10}S_4(SPh)_{16}{}^{4-}$ precursors can then be fused to form stable clusters containing 55 Cd and S atoms (Wang and Herron, 1991). The occurrence of surface defects in clusters prepared by these methods can be reduced by surface passivation with, for example, ammonia, hydroxide ions, amines, and ZnS. Well-defined CdS cluster arrays have also been generated in templates such as zeolites that dictate the cluster size and shape (Herron *et al.*, 1989).

Several recently developed methods of generating well-defined silicon nanocrystals are noteworthy. Efforts in this area are partially driven by the desire to understand luminescence from porous silicon. Thermolysis of SiH_4 in a gas-phase flow tube, followed by oxidation to passivate the cluster surface, generates Si clusters of controlled size (Littau *et al.*, 1993). Also, Heath (1992) has produced nanocrystalline silicon in a high-pressure and -temperature bomb by reducing $SiCl_4$ or $RSiCl_3$ (R = H or octyl) with sodium metal.

1.2.3 Nanocrystalline Materials Consolidation of small particles produces materials consisting of regions, 1–100 nm in diameter, of crystalline structure separated by grain boundaries. Bulk quantities of these nanocrystalline materials can be produced by evaporating the component materials into a vacuum chamber containing an inert gas. The small particles that grow under these conditions are collected on a liquid-nitrogen–cooled surface and then compacted at high pressures to form a nanophase material (Birringer *et al.*, 1984). A more detailed discussion can be found under NANOPHASE MATERIALS.

2. MOLECULAR CLUSTERS

Molecular clusters are weakly bound aggregates of stable molecules. (We include rare-gas atomic clusters here because of the weak interactions that exist between closed-shell atoms.) A variety of different interactions are responsible for binding molecular clusters. These include van der Waals interactions [e.g., Ar_n, $(N_2)_n$], dipole-dipole interactions [e.g., $(CO)_n$, $(NO)_n$], hydrogen bonding [e.g., $(H_2O)_n$, $(NH_3)_n$], and ion-induced dipole interactions [e.g., $(CO_2)_n^+$, Ar_n^+]. The properties of the individual (monomer) components in these systems are only slightly perturbed, and, consequently, the properties of molecular clusters can be considered a composite of the slightly perturbed properties of the monomer and the collective properties of the weakly bound aggregate. There are limited opportunities for exploiting molecular clusters in practical applications. The impetus for studying molecular clusters instead lies in probing the nature of weak interactions between molecules and atoms. These interactions are important in, for example, solvation phenomena and nucleation processes.

2.1 Spectroscopic Methods

A comparison of theoretical calculations and the results of high-resolution spectroscopic studies of cold molecular clusters permits assignment of the ground-state (minimum potential energy) structure and yields valuable information about the role of intermolecular forces. Tracking the evolution of spectroscopic observables as a function of cluster size, together with computer simulations, has led to a deeper understanding of processes such as the structural and energetic changes that occur with stepwise solvation. Most high-resolution spectroscopic techniques couple lasers with supersonic beams and mass spectroscopic detection.

2.1.1 Electronic Spectroscopy Resonance-enhanced multiphoton ionization (REMPI) is an extremely sensitive and versatile method for probing the excited states of molecules and clusters (Brutschy, 1992). Resonant absorption of the first photon to an intermediate electronic state is followed by absorption of a second photon that raises the energy of the molecule to above the ionization threshold. Scanning the excitation laser frequency and monitoring the ion signal results in a resonance-enhanced two-photon ionization (R2PI) spectrum, which is an indirect measurement of the absorption spectrum of the neutral chromophore. Both rotational and vibrational resolution are possible, providing a spectroscopic fingerprint indicative of the molecule and its environment. If the experimental conditions are optimized such that ionization does not lead to evaporation of monomers, mass analysis of the resulting photoions enables these measurements to be performed as a function of cluster size.

A variety of doped molecular clusters, $M \cdot A_n$ (A = rare-gas atom), have been thoroughly studied by R2PI spectroscopy (Whetten and Hahn, 1990). The dopant molecule is predominantly an aromatic chromophore such as benzene, carbazole, or dichloroanthracene (DCA). Solvation of a molecule, M, generally lowers the energy of the first excited electronic (S_1) state with respect to the ground (S_0) state, primarily through stabilizing dispersive interactions. As a result, the $S_0 \rightarrow S_1$ transition shifts to the red. Size-dependent spectral shifts and linewidths of the $S_0 \rightarrow S_1$ transition reflect the extent of solvation and have been interpreted by comparison to molecular-dynamics simulations of cluster structure and spectral features (Jortner, 1992). For small clusters, the spectra exhibit sharp features due to intermolecular van der Waals modes, and dramatic changes occur with each addition of a rare-gas atom. As the size increases, these sharp features give way to broad features that evolve more slowly with size. The point at which this change occurs is attributed to the completion of the first close-packed solvation shell. The number of atoms in the first solvent shell is dependent on the system. A filled shell occurs at $n = 17$ for benzene, $n = 30$ for carbazole, and $n = 36$ for DCA. With increasing cluster size, the spectral shifts and linewidths converge smoothly toward those of the molecule in the bulk matrix.

2.1.2 Vibrational Spectroscopy Vibrational spectra of neutral molecular clusters in supersonic beams can be recorded by a technique developed by Gough *et al.* (1985). The cluster beam is irradiated with photons from an infrared laser, and, if a cluster undergoes vibrational transitions that are in resonance with the laser frequency, then it will absorb energy. The associated increase in temperature is detected when the neutral clusters and dissociation fragments strike a bolometer. This approach has been used to study doped molecular clusters and provides a probe of both the structure and properties of the "host" cluster and the effects of solvation on the "guest" chromophore. Because the species involved are in their ground electronic states, interaction potentials are well understood, and accurate modeling of the system is possible. Studies of SF_6 embedded in rare-gas clusters show that the shift in the frequency of the fundamental ν_3 band of SF_6 can be correlated to the packing density of rare-gas atoms around the SF_6 (Goyal *et al.*, 1993). The shift of this band in small argon clusters is greater than that in solid argon matrices, indicating that the argon atom density around the SF_6 is higher than in the bulk. One possible explanation for this is that the rare-gas atoms are packed in an icosahedral arrangement, where the atoms are more densely packed than in fcc packing. As the cluster size increases, peaks begin to appear at frequencies identifiable as those of bulk argon. Eventually, the fcc features dominate over the icosahedral features, and this crossover is taken to be the structural phase transition. As this method involves no size selection, the spectra necessarily pertain only to an average cluster size. The transition is observed to take place gradually over a broad cluster-size range, and, in clusters of both argon and krypton, it occurs at around 1800 atoms.

Fundamental aspects of ion solvation can be probed by vibrational predissociation spectroscopy of size-selected solvated cluster ions. The interaction between Na^+ and methanol molecules has been probed by monitoring the methanol C–O ν_4 stretching mode in $Na^+(CH_3OH)_n$, $n = 3$–25 (Selegue *et al.*, 1992). Monte Carlo simulations present a picture of the structure of the solvent shell, as well as the kinetics of dissociation. Analysis of the measured spectra indicates that the first solvation shell comprises six methanol molecules. No hydrogen bonding occurs between methanol molecules in the first shell, and the structure is dominated by interactions of the Na^+ with methanol oxygen atoms. The C–O stretching frequency shifts toward that of bulk methanol for $n > 6$, indicating that the ion has less influence on additional methanol molecules, and hydrogen bonding interactions are observed. Figure 1 illustrates the geometry determined for the $n = 10$ cluster by Monte Carlo calculations. In clusters containing more than 15 methanol molecules, hydrogen bonding interactions dominate as in bulk methanol.

2.1.3 Photoelectron Spectroscopy The existence of hydrated electrons in the condensed phase has prompted investigations of electron localization and solvation in gas-phase polar molecular clusters (Barnett *et al.*, 1990). The photodetachment energy (vertical binding energy) of a solvated anion or electron may be determined by photoelectron spectroscopy (PES). Of interest is the minimum cluster size that is necessary to stabilize a bound electron. Theoretical modeling of the $(H_2O)_n^-$ system predicts that, for $11 \leq n \leq 64$, the electron is in a relatively strongly bound surface state. A gradual transition to internal solvation occurs for $32 \leq n \leq 64$. The electron occupies a diffuse, weakly bound surface state in clusters containing fewer than ten water molecules. In agreement with these predictions, measured vertical binding energies suggest that 11 water molecules are required to fully "solvate" an excess electron. In cluster anions containing fewer than 11 molecules, the electron is weakly bound. For

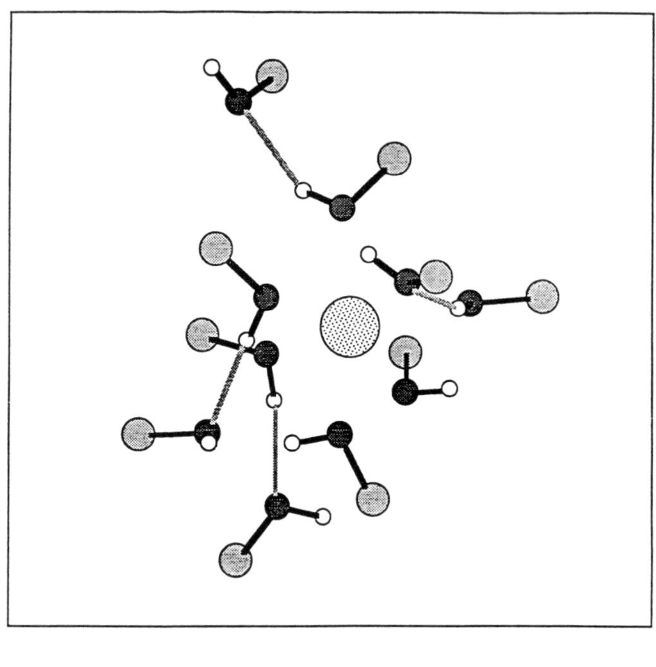

FIG. 1. Monte Carlo simulation of the configuration of $Na^+(CH_3OH)_{10}$ (Selegue *et al.*, 1992). The six methanol molecules that comprise the first solvent shell are arranged about the ion with octahedral symmetry, and hydrogen bonding to the remaining four methanols is indicated by the shaded lines.

$n \geq 11$, the electron is more strongly bound, and the binding energy increases slowly with cluster size.

The electrostatic stabilization of a solvated anion is defined as the difference between photodetachment energy and the electron affinity of the bare ion. A drop in the incremental stabilization energy with the addition of a solvent molecule suggests the existence of a complete solvation layer. Figure 2 plots the evolution of the stabilization energy of I^- as a function of the number of surrounding H_2O molecules (Markovich *et al.*, 1993). Six water molecules form the first solvation layer, as indicated by the large incremental increases in stabilization energy for the first six water molecules and the sudden drop with the addition of the seventh water molecule. The magnitude of the binding energies suggests that the I^- is located near the center of the cluster. As the cluster increases in size to $n = 34$–40, the appearance of small peaks at lower binding energies in the photoelectron spectrum has been interpreted as evidence that there is also a probability that the anion will be located near the surface of the cluster rather than in the interior.

2.2 Chemical Reaction Dynamics in Clusters

In addition to the static investigations discussed above, the fairly recent advances in ultrafast laser techniques are invaluable for probing the detailed dynamics of chemical processes occurring in solvated molecules. Following photoexcitation of the solute mol-

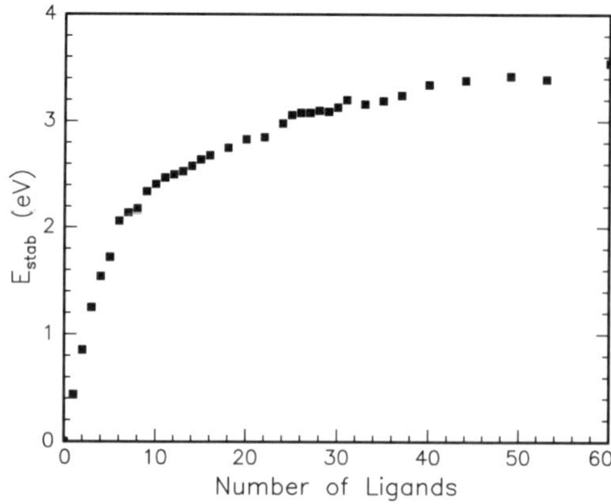

FIG. 2. Evolution of the stabilization energy of I^- in $(H_2O)_n$ as a function of the number of H_2O molecules (Markovich *et al.*, 1993).

ecule, solvent reorganization, proton or electron transfer, or photodissociation of molecules in the solvent cage can occur. These dynamical events can now be interrogated by time-resolved pump-probe spectroscopic methods.

A central issue in understanding charge-transfer processes is the number of solvent molecules that are needed to stabilize an excited-state ion pair. Excited-state proton transfer in phenol–ammonia clusters, $C_6H_5OH \cdot (NH_3)_n$, has been followed by picosecond pump-probe ionization (Syage, 1992). A potential energy diagram illustrating this process for $n = 5$ is given in Fig. 3. A pump pulse excites the phenol (PhOH), which can then undergo proton transfer to the basic "solvent" cluster (B_n) to form the excited-state ion-pair complex $PhO^{*-} \cdots H^+B_n$. If the system is probed before proton transfer occurs, then $PhOH^+B_n$ is detected, whereas ionization of the ion-pair complex results in dissociation to PhO and H^+B_n. Thus, the time evolution of the proton transfer process can be elucidated. Successive addition of NH_3 molecules lowers the energy of the ion-pair state. When this charge-transfer stabilization energy falls below the energy of the S_1 state, excited-state proton transfer will occur. In the phenol–ammonia system, the threshold for excited-state proton transfer occurs at $n = 5$.

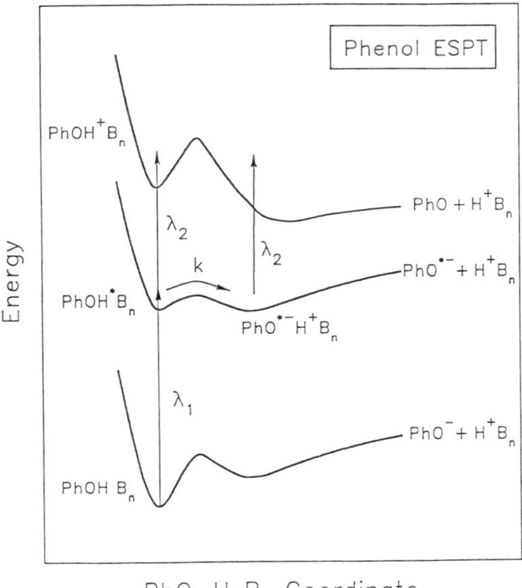

FIG. 3. Potential energy diagram corresponding to $PhOH \cdot (NH_3)_5$, where $B = NH_3$ (Syage, 1992).

Picosecond pump-probe photodissociation spectroscopy of size-selected cluster ions has been used to isolate the dynamics of the solvent cage effect at the molecular level. This is accomplished by photodissociating I_2^-, for example, and then determining the rate of I-I$^-$ recombination within a $(CO_2)_n$ solvent cage (Papanikolas *et al.*, 1993). By appropriate tuning of the probe laser frequency, these experiments are capable of providing a detailed picture of coherent I···I$^-$ internuclear motion following photoexcitation. Wave-packet motion on a dissociative excited-state potential energy surface has been observed for $n = 14$–22. In this cluster-size range, the mechanism for recombination involves "reflection" of the wave packet off the surrounding solvent cage. Recombination dynamics are qualitatively different for $n \leq 12$, where attractive forces between the CO_2 molecules and the dissociating diatom dominate the recombination process.

2.3 Rare-Gas Clusters

2.3.1 Structure From a theoretical perspective, investigation of bonding interactions and geometric structure in rare-gas clusters is appealing because of the simplicity of the interatomic potential. The rare gases have completely filled valence-electronic shells and spherically symmetric electron density. Hoare and Pal performed a comprehensive search for minimum-energy structures in noble-gas clusters using Lennard-Jones pair potentials (Hoare, 1979). In small clusters, structures based on tetrahedral, pentagonal, and icosahedral symmetry are found to be more stable than those with face-centered-cubic (fcc) symmetry. The icosahedral packing scheme, first introduced by Mackay, appears to be an important packing geometry in small particles (Mackay, 1962). The higher atomic coordination numbers in these densely packed geometries make them energetically favorable over the less-compact fcc packing. The first five complete icosahedral shells contain 13, 55, 147, 309, and 561 atoms, as illustrated in Fig. 4. The fivefold symmetry prohibits extended icosahedral packing without introducing additional strain energy, and, as the cluster size increases, this noncrystal-

FIG. 4. The first five Mackay icosahedra contain 13, 55, 147, 309, and 561 atoms, respectively.

lographic symmetry gives way to more energetically favorable bulk fcc packing. The number of atoms required for this structural transition to occur is an important quantity because it indicates the emergence of bulk properties. Calculating the cluster size where this transition occurs is fraught with difficulty because the precise number of atoms is sensitive to the interatomic potential, and it varies from 2000 to 10 000, depending on the method of calculation.

Experimentally, electron diffraction is the only method of directly determining the structure of molecular clusters larger than the limit of direct spectroscopic methods. Although a broad cluster-size distribution is sampled, comparison of simulated and experimental diffraction patterns makes possible a structural assignment. The electron-scattering technique has been used extensively by Farges and co-workers (1986), who find that, for clusters of argon, the transition from icosahedral to bulk fcc structure lies at around 800–1000 atoms. Another structural transition occurs in smaller clusters, which adopt less symmetric "polyicosahedral" structures, clusters with interpenetrating 13-atom icosahedra. This type of structure is observed in the size range $30 < n < 60$ atoms. Further evidence for icosahedral structure in rare-gas clusters is found in electron-impact–ionization mass spectra (Miehle *et al.*, 1989). Clusters of Ar, Kr, and Xe with up to 1000 atoms exhibit peaks of local stability at n = 147, 309, and 561, all of which are complete icosahedral shells. Intermediate abundance maxima that correspond to partially filled shells are also observed. As discussed in Sec. 2.1.2, addition of atomic or molecular dopants to rare-gas clusters provides another indirect structural probe, although the dopant may perturb the geometry of the cluster.

2.3.2 Phase Transitions Another type of transition, that from rigid (solidlike) to nonrigid (liquidlike), has been extensively investigated. For very small clusters, the definition of "melting temperature" as applied to macroscopic systems no longer applies, and "melting" can be considered the rapid interconversion between many isomers with similar energy. Molecular-dynamics simulations indicate that this type of transition is not always abrupt, and "phase coexistence" has been observed, where both fluid and rigid phases may be in equilibrium (Berry, 1993). This is characterized by an S-shaped bend (van der Waals loop) in the microcanonical caloric curve (T vs E). Although the thermodynamic transitions that occur in small clusters do not necessarily have a solid-state analog, this phenomenon is attributed to a rigid–nonrigid transition, similar to a bulk first-order phase transition. Molecular van der Waals clusters, such as $(SF_6)_n$, $(CH_3OH)_n$, and $(H_2O)_n$, as well as rare-gas clusters, can exhibit this type of behavior. Spectroscopic evidence for a possible nonrigid–rigid phase transition has recently been observed in $(\text{benzene})_{13}$ clusters (Easter *et al.*, 1993). The environment of C_6H_6 in $(C_6H_6)(C_6D_6)_{12}$ has been probed by recording has been probed by recording R2PI spectra as a function of the distance from the nozzle of a supersonic expansion source. In this way, the time evolution of the cluster structure can be obtained. Initially, broad features in the R2PI spectrum suggest a nonrigid state. Several microseconds later, sharp spectroscopic features emerge, suggesting that the cluster has solidified.

2.3.3 Quantum Clusters Helium (and also H_2 and D_2) clusters are appropriate systems in which to investigate quantum effects. Because of the very weak interatomic interactions and relatively large zero-point energy, bulk helium (^4He and ^3He) is a quantum fluid and remains liquid even at 0 K. Consequently, clusters of helium are also expected to be liquid and thus cannot be adequately described using the classical potentials that are used to describe heavier noble-gas clusters. Differences between clusters of ^4He and ^3He are of interest because ^4He is a boson and ^3He is a fermion. Calculations predict that because of these statistical effects, all clusters of ^4He are bound, whereas, in the ^3He system, only clusters with more than about 30 atoms are bound (Stringari, 1990). Experimental information is relatively scarce, a result in part of the difficulty of the experiment and also interpretation of the results. However, efforts have been directed at unraveling their unique properties (Toennies, 1990). Because of the weak binding energies, extensive fragmentation occurs upon ionization of He clusters, making mass spectra difficult to interpret. Although He_2^+ has been observed in mass spectra, it is unclear whether the origin of this ion signal is from fragmentation of larger clusters or from ionization of He_2. However, recent experiments have verified the existence of $^4He_2$ (Luo *et al.*, 1993). There is an extremely long-range interaction between the He atoms, and the bond energy of roughly 1 mK makes this the weakest chemical bond that has been observed.

3. ATOMIC CLUSTERS

Unlike molecular clusters, atomic clusters are strongly bound entities. The chemical bonding between atoms can be either covalent, metallic, or ionic in nature. Because of this strong bonding, addition of a single atom can dramatically alter the properties of the cluster. The organization of this section is roughly along the lines of the periodic table. We begin by considering clusters of the Group I metals, which have been extensively studied because they provide simple model systems in which to examine the emergence of metallic properties.

3.1 Alkali-Metal Clusters

3.1.1 Electronic Shell Model Mass-spectral studies of alkali-metal clusters reveal the presence of particularly abundant clusters containing 8, 20, 40, 58, 92, . . . atoms. The existence of these "magic" numbers can be rationalized by the electronic shell model. According to this model, closed–electronic shell clusters are unusually stable (de Heer *et al.*, 1987). Manifestations of electronic shell structure also appear in size-dependent measurements of the ionization potential, electron affinity, binding energy, static electric polarizability, and other properties (Cohen and Knight, 1990). In the electronic shell model, the jellium approximation (see COLLECTIVE PHENOMENA IN SOLIDS) is used to replace the lattice with a uniformly charged positive background, and the valence electrons are treated as a quantized Fermi gas confined by the cluster surface. The effective one-electron eigenfunctions of a spherically symmetric potential are characterized by a main quantum number, n, and an angular momentum quantum number, l. The electronic energy levels (subshells) are $2(2l + 1)$-fold degenerate. A closed shell occurs when subshells are filled with valence electrons and there is a large energy gap to the next empty level. The ordering of the levels (and, thus, the location of the shell closings) is dependent on the potential. Periodic oscillations resulting from the filling of angular momentum shells are a common physical phenomenon. The filling of electronic angular momentum shells in atoms results in highly stable closed-shell atoms, such as the rare-gas elements, and the nuclear shell model (developed in the 1940s) accounts for particularly stable nuclei.

When the restriction of spherical symmetry is relaxed, spheroidal or ellipsoidal distortions of the jellium background potential partially lift the energy-level degeneracies (Clemenger, 1985). Some levels increase in energy and some decrease, depending on the nature of the distortion. Distortion raises the total energy of clusters with filled subshells, and so these clusters retain spherical symmetry. However, ellipsoidal or spheroidal distortions can lower the total energy of clusters with partially filled subshells, and this gives rise to additional shell structure.

A low-frequency oscillation envelops the higher-frequency shell oscillations, resulting in a "supershell" beating pattern in the electronic level density and total electronic binding energies (Nishioka *et al.*, 1990). Classically, this periodicity results from a superposition of the amplitudes associated with triangular and square valence-electronic orbits. Calculated periodic oscillations in valence-electron energy are indicative of shell and supershell structure. This is plotted in Fig. 5 for the case of spherically symmetric sodium clusters. Electronic shell closings are indicated, and the first supershell node occurs at about 1000 atoms. Agreement of the shell-model predictions with experimentally observed magic numbers provides evidence for electron delocalization. Multiphoton ionization mass spectra of Na cluster beams containing up to 22 000 atoms exhibit abundance maxima at the predicted shell closings (Martin *et al.*, 1991). In addition to electronic structure, abundance distributions in the mass spectrum in Fig. 6 suggest that a transition from filled electronic shells to atomic shells occurs at approximately 1500 atoms. At this point, geometric constraints become dominant over electronic requirements, such that the cluster stability is dictated by closing of atomic shells rather than electronic shells.

3.1.2 Collective Electronic Effects There has been extensive theoretical and experimental work on the collective electronic behavior of alkali-metal cluster systems (Kresin, 1992). The classical description of the interaction of small metallic spheres with electromagnetic waves was derived by Mie, and extended to prolate and oblate ellipsoidal particles by Gans (Bohren and Huffman, 1983). The response of bulk free-electron metals to electromagnetic radiation is attributed to excitation of collective electronic motion (plasmon oscillations). Both surface and

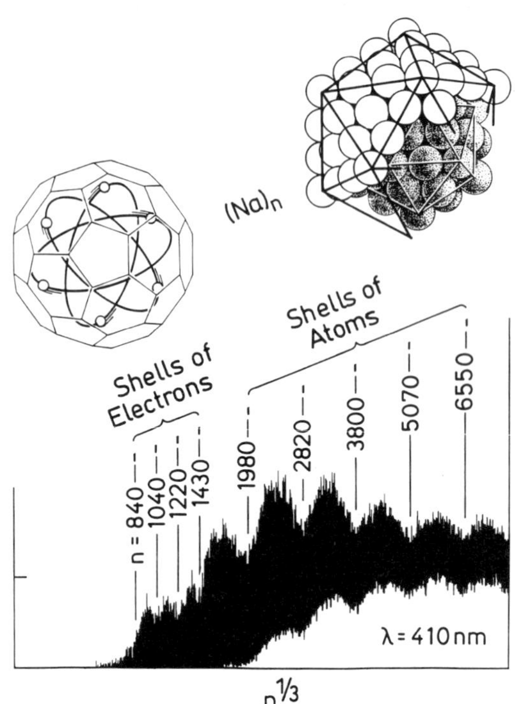

FIG. 6. Photoionization mass spectrum of sodium clusters (Martin *et al.*, 1991). Shells of electrons are observed up to n = 1430, and shells of atoms are observed in larger clusters.

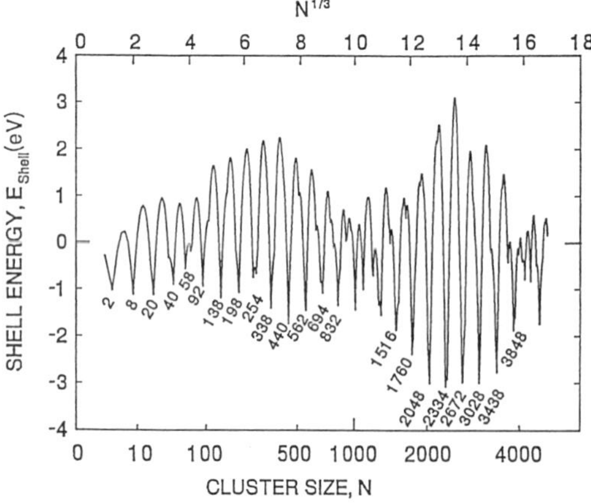

FIG. 5. Calculated valence-electron contribution to the binding energy of spherically symmetric sodium clusters (Nishioka *et al.*, 1990). Magic numbers occur at low energies. Supershell structure is also observed.

volume plasmons can be observed in small clusters, although surface effects dominate. Generally, the electronic properties of large clusters are well described by the Mie–Gans theory. However, in the smaller clusters, the finite boundaries cause a modification of the electronic oscillation that further red-shifts the resonance frequency with respect to the classical prediction. Deformations from spherical symmetry can cause the single collective electronic absorption to be split into multiple components. For example, open-shell clusters such as Na_{11}^{+} (and isoelectronic Na_{10}) exhibit a two-component absorption, which agrees with that predicted for an ellipsoidally distorted cluster.

Experimentally, the collective electronic properties of free alkali-metal clusters have been determined by resonant optical absorption and cross-section measurements (by photodepletion spectroscopy). Because of the fast relaxation of the intermediate electronic excited state in metal clusters, R2PI techniques using nanosecond lasers do not lead to ionization. Under these circumstances, photodepletion provides a method for measuring absorption spectra (Morse *et al.*, 1983). Cluster-ion signals are recorded as a function of the frequency of the excitation laser. Absorption of photons results in photofragmentation and is detected as a decrease in the ion signal. If all of the photoexcited clusters dissociate before reaching the detector, then the photofragmentation cross section is equivalent to the photoabsorption cross section. Photoabsorption spectra for several Na_n are shown in Fig. 7. The single resonance in the spectrum of closed-shell Na_8 is predicted by Mie theory. Other size-dependent resonances are discussed further below.

3.1.3 Geometric Structure For small clusters, quantum-mechanical calculations that explicitly consider the positions of the nuclei give more meaningful descriptions of geometric structure than the structureless

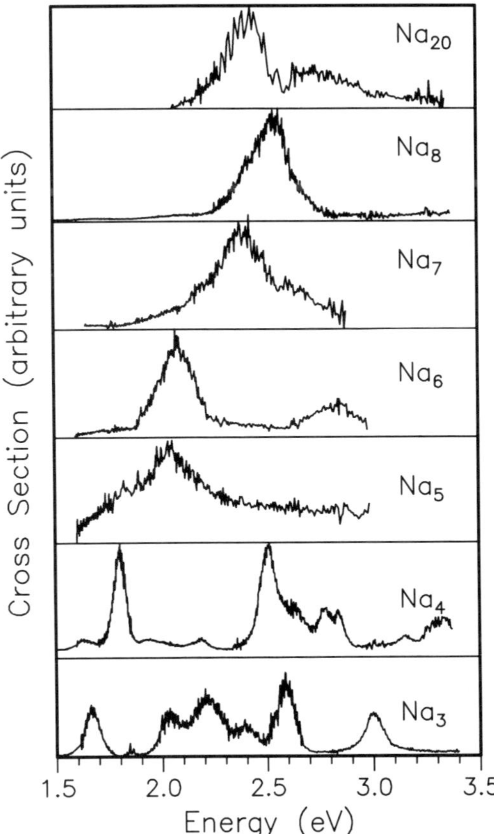

FIG. 7. Photoabsorption spectra for Na_n (n = 3–8, 20) obtained from photodepletion measurements (Wang *et al.*, 1992).

jellium approximation. Most *ab initio* quantum-chemical structure calculations have been done on lithium, the simplest metal, but most experimental work has centered on sodium (Bonačić-Koutecký *et al.*, 1991). A combination of theoretical modeling and photodepletion measurements has led to an understanding of the bonding in small alkali-metal clusters. Figure 8 depicts Hartree–Fock optimized geometries and calculated binding energies for neutral and cationic Na_n. Electronic transitions calculated for minimum-energy geometries are correlated to features in absorption spectra such as those of small Na_n in Fig. 7. In this manner, the minimum-energy ground-state geometry of Na_4 has been determined to be a planar rhombus, and Na_5 has a planar C_{2v} structure. The cluster geometries undergo a transition from planar to three dimensional at the hexamer, which has several conformers that are close in energy. In agreement with the jellium-model prediction, clusters with 8 and 20 atoms are particularly stable, close-packed structures. High-level calculations have been done on clusters with up to 14 atoms, above which *ab initio* theory becomes intractable. Semiempirical methods predict an odd-even alternation in stability that persists to greater than about 40 atoms. This effect is observed experimentally and is attributed to electron-pairing effects. Even-numbered (paired electrons) systems are more stable than those with unpaired electrons.

3.1.4 Alkali-Halide Clusters Alkali-halide (MX) cluster systems can be considered microscopic analogs of ionic crystals and surfaces. Simple electrostatic interaction potentials allow accurate calculations of ground-state geometries and binding energies. Of particular interest are alkali-halide cluster anions containing extra electrons that are not associated with a halogen ion. These are of the form $(MX)_n^-$, $M(MX)_n^-$, and $M_2(MX)_n^-$. Molecular-dynamics calculations predict that the excess electron could be localized either near an M^+ or in an empty X^- lattice site (as in a bulk F-center defect), or it could occupy a delocalized surface state. Ultraviolet photoelectron spectroscopy has been used to determine the location of the extra electron in $(KI)_n^-$ systems (Yang *et al.*, 1992). Electron binding energies for n = 3, 4, 6, and 12 indicate that the excess electron is loosely

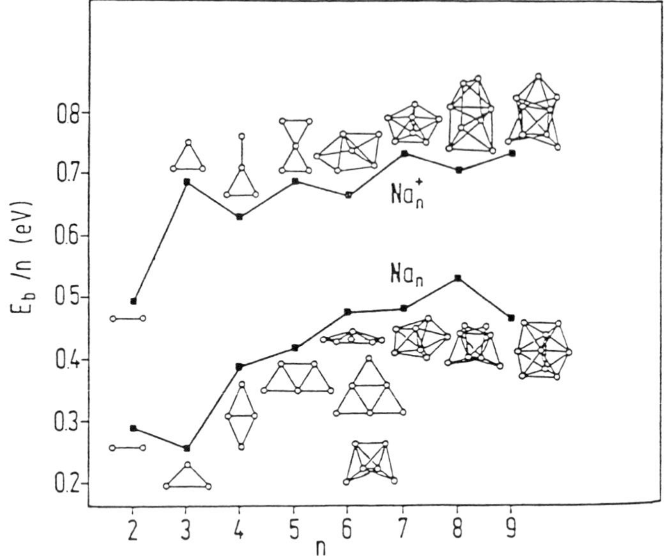

FIG. 8. Binding energy per atom calculated using configuration interaction and the corresponding Hartree–Fock optimized geometries for Na_n and Na_n^+, n = 2–9 (Bonačić-Koutecký *et al.*, 1991).

bound. Because the neutral cluster has a low electron affinity, the extra electron does not significantly change the neutral ground-state geometry and is delocalized over all of the K^+ ions. A relatively high binding energy is observed for n = 13. For this cluster, it has been suggested that the electron is localized at the missing corner I^- site of the 3 x 3 x 3 atom cube.

Investigations of the reactivity of alkali-halide clusters such as $(Na_nF_{n-1})^+$ with H_2O and NH_3 confirm that these clusters have structures that are chunks of the bulk lattice (Homer *et al.*, 1993). Defect structures occur in clusters with one less NaF unit than a complete cuboidal cluster. These clusters, such as n = 13 (3 x 3 x 3), 22 (3 x 3 x 5), 31 (3 x 3 x 7), and 37 (3 x 5 x 5), are significantly more reactive, and calculations indicate that NH_3 fits readily into the hole created by the missing atoms.

3.2 Group II Elements

In contrast to the Group I elements, little is known about clusters of the divalent Group II elements. To a first approximation, the n s orbitals of an n-atom cluster result in $n/2$ bonding and $n/2$ antibonding orbitals, which are both filled by the $2n$ valence electrons. Because the resulting bond order is zero, the cluster should only be bound by weak van der Waals interactions. In contrast, chemical bonding in the bulk results from overlap of the filled s band and the empty p band. An insulator-to-metal transition is thus expected with increasing cluster size. Nonmetal–metal transitions in Group II elements are a subject of intense study, and this issue is addressed further for mercury clusters in Sec. 3.3.5.

High level *ab initio* quantum-chemical calculations of the ground-state structures of Group II clusters have proved to be somewhat difficult because of the involvement of nearly degenerate s and p levels (Bonačić-Koutecký *et al.*, 1991). The same trends are observed for both Be_n and Mg_n. For n = 3–7, calculations predict that the ground-state geometries are equilateral triangle, tetrahedron, trigonal bipyramid, octahedron, and pentagonal bipyramid. *Ab initio* molecular-dynamics simulations suggest that, for Be clusters with $7 < n < 20$, the growth pattern leads to structures that resemble hexagonal-close-packed (hcp) lattice fragments (Kawai and Weare, 1990). Experimentally, evidence for icosahedral packing has been found in mass-spectral abundance patterns of Mg clusters containing from 147 to 2869 atoms and Ba clusters containing 13 to 35 atoms (Martin *et al.*, 1991).

3.3 Transition Metals

The unfilled d shells in isolated transition-metal atoms result in a large number of low-lying excited electronic states. Consequently, transition-metal clusters, even small ones, have exceedingly high electronic state densities. This makes spectroscopic studies extremely difficult to interpret. The complex electronic structure of these systems results in their rich and varied chemical properties.

3.3.1 Chemical Reactivity Studies of the variations in the reactivity of transition-metal clusters as a function of size can reveal information about their electronic and geometric structure. Most heterogeneous catalysts consist of supported transition-metal clusters, so that understanding the chemical properties of isolated transition-metal clusters may have important practical implications. A large amount of data on the chemistry of transition-metal clusters is presently available, but only a few relevant examples will be presented below. Further references to reviews providing additional information are given at the end of this article.

The fast-flow reactor has been widely used to study the reactivity of atomic clusters. The clusters are typically formed by pulsed-laser vaporization (see Sec. 1.1.3) and are carried by the buffer gas along a channel into the flow tube, where a reagent gas is added. After a fixed reaction time (determined by the flow velocity), the clusters, reagent, and buffer gas expand into vacuum. The clusters are then photoionized and analyzed by time-of-flight mass spectrometry. Thermodynamic properties such as reaction enthalpy and entropy can be determined from equilibrium constants recorded as a function of temperature. A potential source of uncertainty is fragmentation of the products, induced by either the reaction exothermicity or the ionization process.

Many systems display a reactivity that is a strong function of cluster size. For example, in reactions of Fe_n with H_2, the rate constants for $n < 23$ vary by several orders of magnitude. For $n > 23$, only slight variations occur, and the reaction rate is close to that of bulk iron (Riley and Parks, 1987). There appears to be an approximate correlation between the chemical reactivity and the ionization potential (IP) of Fe_n. As shown in Fig. 9, Fe_n with low IPs are generally more reactive toward H_2. This has been rationalized in terms of the ease of electron transfer from the Fe_n to H_2. H_2 adsorption is dissociative, and transfer of an electron from the cluster to the lowest unoccupied molecular orbital on H_2, which is an antibonding orbital, will weaken the H–H bond, decrease the activation energy, and increase the reaction rate. The origin of these variations in reactivity is not a completely resolved issue, however. Other studies suggest that variations in both reactivity and IP may result from simultaneous changes in geometric structure. It is clearly difficult to isolate effects due to electronic structure from those due to geometric structure.

Chemical reactivity has been used as an indirect probe of the geometric structure of neutral transition-metal clusters (Parks *et al.*, 1993). The maximum number of weakly bound adsorbates, such as H_2, N_2, NH_3, or H_2O, that can physisorb onto a cluster is dependent on the number of available surface sites. Saturation coverage is therefore deter-

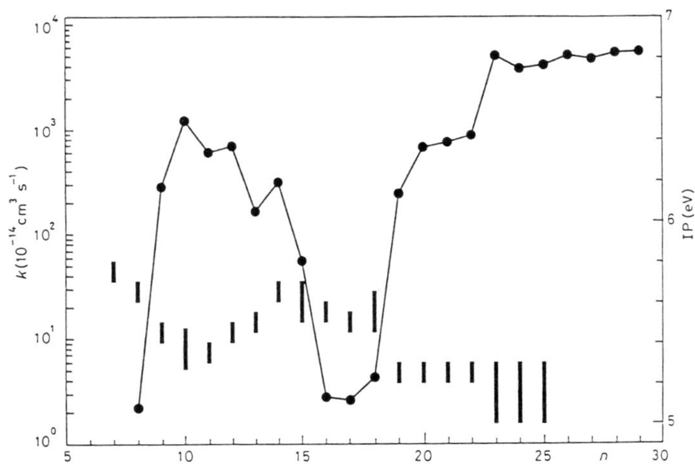

FIG. 9. Correlation of reactivity (left scale, ●) of $Fe_n + H_2$ and ionization potentials (right scale, ■) of Fe_n (Riley and Parks, 1987).

mined by the underlying geometric structure. These studies suggest polyicosahedral geometries for clusters of Fe, Co, and Ni containing 19 to 34 atoms, and icosahedral structure for Cu clusters containing more than 70 atoms. Evidence for the existence of multiple structural isomers appears in many of these systems. Because of the possibility of adsorbate-induced structural rearrangements, these geometries correspond to those of the saturated cluster, and the geometry of the bare metal cluster may be different.

To avoid complications associated with product fragmentation, investigations of cluster reactivity are ideally performed on one cluster size at a time. Mass selection necessitates a study of the cluster ion rather than the neutral species. Probes of size-selected metal-cluster ions offer a controlled method of studying reactivity (Parent and Anderson, 1992). Cluster ions of a specific mass may be selected by quadrupole mass filters, magnetic sectors, or Wien filters. The selected ions are then focused into a low-energy ion beam. Reaction cross sections, as functions of kinetic energy, can be recorded by passing the ion beam through a gas cell or ion guide containing a low pressure of reagent gas and mass analyzing the products. Reaction-rate constants, under thermal conditions, can be recorded as functions of temperature by injecting the ions into a drift tube containing a high pressure of buffer gas (Jarrold, 1991). The mobility of an ion in the drift tube is related to its geometry (von Helden *et al.*, 1991), and increasing the injection energy to transiently heat (anneal) the clusters provides a way to study structural transitions.

Fourier-transform ion cyclotron resonance (FT-ICR) is an alternative approach to probing cluster-ion chemistry. Here, the ions are trapped in a cell by magnetic and electrostatic fields, and a mass spectrum is recorded by coherently exciting the cyclotron motion of the ions and recording the decay of an image signal. Fourier transforming the resulting time-domain signal yields the mass spectrum. The power of this method lies in the ability to store ions selectively and to follow multiple reaction steps. Sequential reaction pathways have been elucidated in the reaction of Fe_4^+ with ethene (C_2H_4) (Irion and Schnabel, 1992). Up to four ethene units can react with the Fe_4^+, and each addition results in elimination of H_2. The reaction products are then probed by collision-induced dissociation. In the case of the $Fe_4(C_2H_2)_3^+$ product, the lowest-energy fragmention pathway is the one-step elimination of $(C_2H_2)_3$ (presumably benzene). The Fe_4^+ is then available to restart the reaction cycle. Thus, Fe_4^+ can effectively catalyze the formation of benzene from three ethene units. This remarkable behavior is specific to the Fe tetramer.

Naked transition-metal cluster ions bind and then dehydrogenate many hydrocarbons. Dehydrogenation occurs to a varying degree, depending on the metal and the cluster size. An example is the reaction of transition-metal clusters with benzene. For niobium cluster cations containing up to 13 atoms, the extent of dehydrogenation increases with size. Complete dehydrogenation to $Nb_nC_6^+$ can occur in the reaction of clusters containing more than four atoms, although partial dehydrogenation also occurs (Zakin *et al.*, 1989). Similar reactivity trends are observed for neutral niobium clusters, and this suggests that geometric structure is an important factor in determining reactivity. Cationic clusters of another Group V element, vanadium, are found to be less reactive than niobium. Although complete dehydrogenation is observed for $n = 5$–7, partial dehydrogenation is the major reaction channel. The difference in reactivity between Nb and V may be interpreted in terms of electronic configuration. Niobium may be more reactive because it is a d^4s^1 metal, while V has a d^3s^2 electronic configuration.

3.3.2 Magnetic Deflection Although the ferromagnetic properties of bulk transition metals are reasonably well understood, the size-dependent magnetic behavior of transition-metal clusters is still under investigation. The magnetic properties of metal clusters can be determined by Stern–Gerlach magnetic deflection experiments. The cluster magnetic moment μ_n will align with the magnetic field B, to an extent that is dependent on the counteracting effects of thermal motion. From the magnitude of the deflection of the cluster in the magnetic field, the magnetic moment and magnetization can be determined. In the case where $\mu_n B/kT \ll 1$, the magnetization M can be derived as

$$M = \mu_n^2 B/3\,kT. \qquad (1)$$

This expression predicts simple thermodynamic behavior known as superparamagnetism. Magnetic deflection experiments have recently been performed for mass-selected iron clusters containing up to 700 atoms (Billas *et al.*, 1993). The deflection of small Fe clusters is always in the high-field direction, which suggests that rapid spin relaxation occurs via coupling of spin and rotational angular momentum. The magnetic moment per atom decreases from about $3\mu_B$ for $25 < n < 130$ to near the bulk limit of about $2.2\mu_B$ for clusters containing more than about 500 atoms. The rotational temperature of iron clusters has a significant effect on the measured magnetization. When the precession frequency of the spin is similar to the rotational frequency, resonant spin–rotation coupling can occur. In this case, the approximations of Eq. (1) are no longer valid, and the magnetization of rotationally cold clusters is greatly reduced and exhibits a nonlinear dependence on the magnetic field.

3.3.3 Photoionization Trends in metal-cluster ionization potentials (IPs) may be reproduced by a classical electrostatic image-charge model (Wood, 1981). The ionization potential for a spherical metal droplet is expressed in terms of the bulk work function (WF) and the radius of the cluster, R:

$$\mathrm{IP} = \mathrm{WF} + \tfrac{3}{8} e^2/R. \tag{2}$$

Although the constant term $\frac{3}{8}$ is mathematically correct in the electrostatic limit, if a more realistic cutoff to the image potential is considered, the constant term becomes $\frac{1}{2}$ (Makov *et al.*, 1988). The limitations of this description of metal-cluster IPs stem from the fact that clusters do not have uniform valence-electron density, and, in many cases, the assumption of spherical shape is not valid. Inclusion of a term for spillout of the electron density beyond the edge of the cluster gives the semiclassical expression

$$\mathrm{IP} = \mathrm{WF} + \tfrac{1}{2} e^2/(R + a), \tag{3}$$

where a is the element-specific extent of electron spillout. Figure 10 compares measured IPs for silver clusters to the predictions of Eqs. (2) and (3) (Alameddin *et al.*, 1992). The best agreement is obtained for clusters with nearly spherical geometries, which correspond to clusters with closed electronic shells within the electronic shell-model description. The agreement also improves for larger cluster sizes. For small clusters, large size-dependent oscillations in IP result from quantum size effects, and deviations are significant. Although this model fits the IPs of simple metal clusters such as the alkalis and noble metals, other transition-metal clusters do not conform to the classical description. Clusters of transition metals such as Ni, Nb, Fe, and Co exhibit similar trends in their IPs. There is a rapid decrease in the IP for up to approximately $n = 25$, and then, for $25 < n < 100$, the IP levels off to a value that is about 0.5 eV above the bulk work function. Disagreement with the classical model suggests that the d band has a significant influence on the electronic structure of these clusters.

A delayed ionization phenomenon, attrib-

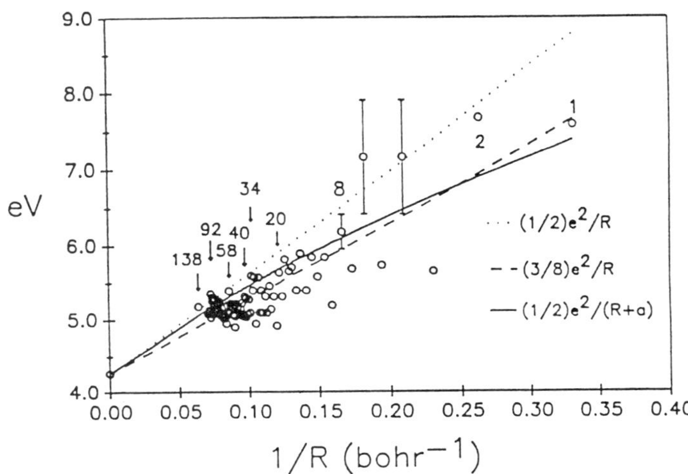

FIG. 10. Silver-cluster ionization potentials compared to the classical predictions (Alameddin *et al.*, 1992).

uted to thermionic emission, has been observed in clusters of W, Nb, and Ta (Amrein *et al.*, 1991; Leisner *et al.*, 1991). Multiphoton excitation (with photon energies less than the ionization energy) may not lead to direct ionization of metal clusters. Rapid internal conversion results in a highly vibrationally excited (hot) cluster that can cool either by evaporation of atoms or by emission of an electron. For strongly bound transition-metal clusters, the rate of dissociation of the excited cluster is slow compared to that of ionization, and delayed ionization may be observed. A statistical model has been developed by Klots (1991) to predict rate constants for electron emission from small neutral clusters. Taking into account the Coulombic attraction between the separating particles, the rate constant is

$$k(T) = \left(\frac{2k_bT}{h}\right)\left(\frac{Q_{\mathrm{vib}}}{Q^0_{\mathrm{vib}}}\right)\exp\left(\frac{-IP}{k_bT}\right) \times \left[\frac{2b}{a_0} + 2\left(\frac{Q_s\pi}{4}\right)^{1/2} + Q_s\right], \quad (4)$$

where

$$Q_s = 8\pi^2\mu b^2 k_b T/h^2.$$

The terms Q^0_{vib} and Q_{vib} are the vibrational partition functions before and after electron emission, μ is the reduced mass (essentially the mass of an electron), b is the hard-sphere collision radius, and a_0 is the Bohr radius. In the limit of large n, this expression converges to the bulk description of thermionic emission given by the Richardson–Dushman equation. Verification of this model requires that the ionization rate be measured as a function of cluster temperature, which has not yet been performed.

3.3.4 Collision-Induced Dissociation Dissociation energies of atomic clusters provide information about cluster stability. Collision-induced dissociation (CID) is a widely used method of determining adiabatic bond dissociation energies. These experiments are performed by colliding a beam of internally cold ions with a collision gas (such as Xe), and recording the CID cross section as a function of kinetic energy. After the finite lifetime of the collisionally excited cluster is taken into account by means of a statistical RRKM (Rice, Ramsperger, Kassel, and Marcus) model, the bond dissociation energy (BDE) of the cluster ion can be extracted from the CID threshold. Armentrout and co-workers (Su *et al.*, 1993) have measured dissociation energies for a series of transition-metal cluster ions. The CID of V_n^+, $n = 2$–20, with Xe will be discussed as an example. All V_n^+ dissociate by sequentially losing atoms except for $n = 4$, which fragments by losing the dimer. Fragmentation by loss of individual atoms is characteristic of nearly all metal-cluster ions and suggests that the cohesive energies increase relatively smoothly with cluster size. Bond energies for neutral V clusters are derived from the BDEs of the ion and IPs of the neutral. An even-odd oscillation in the BDE of both neutral and cationic clusters, $4 \leq n \leq 10$, correlates with trends in the reactivity of V_n with D_2 and N_2. Even-numbered clusters are more strongly bound and less reactive, suggesting that these are closed-shell clusters.

Cluster cohesive energies are the sum of the BDEs divided by the number of atoms. It is informative to compare the cohesive energies to the predictions of the liquid-drop model (Miedema, 1978). In this model, an increase in the surface energy (per atom) with decreasing cluster size results in a decrease in the cohesive energy. The cohesive energy of a cluster with n atoms is expressed as

$$E_c(n) = \Delta H_{\mathrm{vap}}{}^0 - (36\pi/n)^{1/3}\gamma^0 V_a^{2/3}, \quad (5)$$

where γ^0 is the surface energy, $\Delta H_{\mathrm{vap}}{}^0$ is the bulk heat of vaporization, and V_a is the atomic volume. A plot of E_c against $n^{1/3}$ in Fig. 11 shows that the measured cohesive energies approach the spherical-drop prediction with increasing cluster size and are in remarkably good agreement for clusters containing more than approximately 13 atoms. Similar results have been obtained for other transition metals.

3.3.5 Nonmetal-to-Metal Transitions An unresolved issue in studying clusters of metal atoms is the minimum number of atoms that is necessary for the cluster to exhibit properties of the bulk metal. Here the exact definition of "metallic" behavior is critical. Intrinsically metallic properties appear when the valence electrons are delocalized and the gap between highest occupied and lowest un-

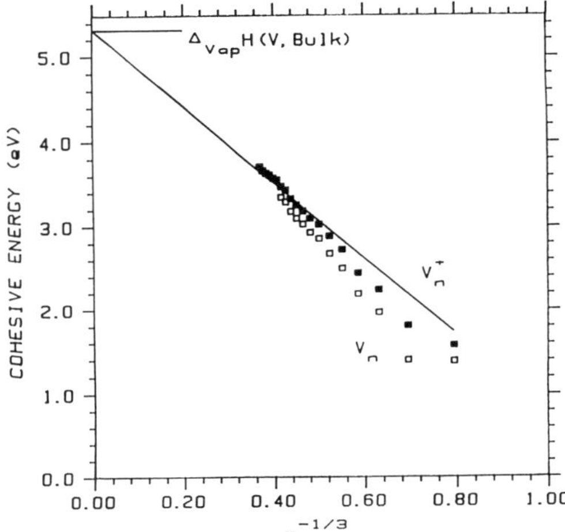

FIG. 11. Cohesive energies for neutral (□) and cationic (■) vanadium clusters (Su *et al.*, 1993). The line corresponds to cohesive energies calculated using the spherical-drop model.

occupied electronic energy levels is small relative to kT. Nonmetal–metal transitions in Group II clusters, briefly discussed in Sec. 3.2, have been systematically studied in mercury clusters. Because of the closed-shell atomic configuration ($5d^{10}6s^2$), small mercury clusters are expected to be bound by weak van der Waals interactions. With increasing cluster size, the 6*s* and 6*p* shells broaden until the full 6*s* shell eventually overlaps with the empty 6*p* level, delocalizing the electrons into metal-like levels. Cohesive energies and ionization energies have been calculated using a tight-binding electronic theory, and these suggest that covalent bonding begins to occur in Hg clusters containing 13–20 atoms. Calculations of the electronic band gap and density of states suggest that clusters containing approximately 20–80 atoms are semiconducting, whereas those containing more than 80 atoms are metallic. The transition from van der Waals to metallic is therefore expected to be gradual, and experimental evidence supports this prediction. Electron binding energies for Hg_n, $n < 110$, have been measured by vacuum-ultraviolet photoelectron spectroscopy (Kaiser and Rademann, 1992). In small Hg_n clusters, $n < 18$, the IP deviates significantly from that predicted by the spherical-droplet model, providing evidence for the dominance of van der Waals bonding. For clusters containing about 18–70 atoms, the IP begins to evolve gradually toward the spherical-droplet prediction. In addition, the photon-energy dependence of the photoemission spectra indicates that Hg_n, $n > 60$, already have a high density of *p* states near the Fermi energy, providing further evidence of the emergence of metallic properties in this size range.

3.3.6 Noble Metals Because of their $d^{10}s^1$ valence-electronic structure, a description that is conceptually similar to that of the alkali metals can be applied to the noble metals, Cu, Ag, and Au. Although the metal–metal bonding in clusters of these metals is dominated by the *s-s* interaction, contributions of the underlying *p* and *d* electrons significantly influence the properties. Therefore, *d-s-p* hybridization and *d-s* correlation energy must be taken into account in theoretical studies (Bonačić-Koutecký *et al.*, 1991). Because of the large number of electrons in these systems, *ab initio* calculations have thus far been limited to $n < 9$. Calculated ground-state geometries and size-dependent variations in the binding energies are similar to those of the alkali elements.

There have been several experimental investigations of the evolution of electronic properties in noble-metal clusters (both ionic and neutral). Vertical electron affinities of the anions of copper, silver, and gold clusters have been measured by ultraviolet photoelectron spectroscopy (Taylor *et al.*, 1992). For up to 70 atoms, variations in electron affinity are found to agree with the predictions of the electronic shell model for ellipsoidally distorted clusters. (Clusters with closed electronic shells have high electron affinities.) In clusters containing only a few atoms, the *d*-*s* band separation is fixed at approximately that of the bulk, and, with increasing cluster size, the *d* band evolves smoothly toward that of the bulk. Photofragmentation studies of size-selected silver-cluster cations provide verification of the near-spherical geometric structure of closed electronic-shell clusters (Tiggesbäumker *et al.*, 1992). The resulting optical spectra show collective electronic resonances that are in qualitative agreement with the predictions of Mie theory. Also in agreement with the predictions of the electronic shell model are the ionization potentials of Cu_n, $n < 150$, and Ag_n, $n < 100$ (Knickelbein, 1992; Alameddin *et al.*, 1992).

3.4 Group III Elements

The elements of Group III (B, Al, Ga, In, and Tl) have an s^2p^1 valence configuration. The properties vary dramatically down the group. Bulk boron is a large–band-gap semiconductor, whereas bulk aluminum is a nearly ideal free-electron metal. Gallium, indium, and thallium are also metals. There is a considerable amount of information available for clusters of boron and aluminum, but relatively little is known about the clusters of the other elements in this group.

3.4.1 Boron In the solid state, many boron compounds consist of three-dimensional networks of octahedral B_6 and cuboctahedral B_{12} clusters. However, there is no evidence to suggest that these species are important building blocks of the isolated atomic clusters. Mass spectra of boron clusters generated by laser vaporization show particularly abundant clusters at 5, 10, 11, and 13 atoms. Collision-induced dissociation measurements of the bond dissociation energies indicate that B_{13}^+ is particularly stable. The bond dissociation energies rapidly approach the bulk cohesive energy, suggesting that these clusters have close-packed, highly coordinated geometries. This is consistent with quantum-chemical calculations, which suggest that small boron clusters adopt three-dimensional geometries. According to these calculations, B_5 is a trigonal bipyramid, and B_6 is a capped pentagon. Molecular-dynamics calculations for B_{13}^+ suggest that the plausible icosahedron geometry is not even a local minimum. The lowest-energy geometry of B_{13}^+ appears to be a relatively low-symmetry structure consisting of three hexagonal pyramids. The enhanced stability of this cluster seems to be related to a closed-shell electronic structure.

3.4.2 Aluminum Since bulk aluminum is almost an ideal free-electron metal, the electronic shell model might be expected to provide a reasonable description for the properties of aluminum clusters. Assuming that aluminum is trivalent, the shell closings should occur with approximately one-third of the atoms required for shell closings with the s^1 alkali metals. Evidence for shell structure of aluminum clusters has been found in dissociation-energy measurements, ionization-energy measurements, and studies of chemical reactivity. The expected shell closings are observed for clusters with 20 and 40 valence electrons (Al_7^+ and Al_{13}^-). However, for clusters with >20 atoms, there are some significant discrepancies between the predictions of the shell model and experimental results. The origin of these discrepancies has not been resolved. However, perturbations provided by the highly charged positive cores (3+) and issues related to the overlap of the s and p bands may be important. In bulk aluminum, the $3s$ and $3p$ bands are broad and overlap. These bands are well separated in the atom, and the cluster-size regime where they begin to overlap is an issue of considerable interest. Photoelectron spectra of Al_n^- suggest that the $3p$ band begins to overlap with the $3s$ band for clusters with $n > 20$.

Quantum-chemical calculations for small aluminum clusters suggest that they adopt relatively open, covalently bound geometries, where the bonding arises mainly from the p orbitals. Closed-shell Al_7^+ (20 valence electrons) has a distorted capped octahedron geometry. The significance of this geometry is that all of the atoms lie approximately the same distance from the center of mass, as if distributed on the surface of a sphere. (Spherical geometries are expected for closed–electronic-shell clusters.) The electronic structure of this cluster can be related to the shell-model prediction, even though there is a substantial gap between the s and p levels. Al_{13}^- is the next largest cluster to have to have a closed shell. Quantum-chemical calculations for this cluster suggest that the gap between the s and p levels has disappeared. The geometry is icosahedral, and the bonding is more metallic. For large, cool (room temperature) aluminum clusters, a geometric shell structure consistent with octahedral symmetry has been observed in mass spectra. However, for warmer (possibly liquid) aluminum clusters, the geometric shell structure vanishes and electronic shell structure appears to emerge (Baguenard *et al.*, 1993). Gallium and indium clusters display a similar electronic shell structure.

3.5 Group IV Elements

3.5.1 Carbon Carbon clusters have been extensively studied. Most of the interest in carbon clusters has recently been focused on the fullerenes, which are discussed in the article FULLERENES. Fullerenes [spheroidal-shell carbon clusters consisting of 12 pentagons and $(n - 20)/2$ hexagons] exist for carbon clusters containing more than around 30 atoms. Bulk quantities of C_{60} and a few other larger fullerenes have been isolated. Mobility measurements for gas-phase carbon-cluster ions have revealed the existence of a large number of different structural isomers, including a range of roughly planar monocyclic and bicyclic rings (von Helden *et al.*, 1991). The formation of these unusual geometries reflects carbon's tendency to form multiple bonds. A number of mixed metal–carbon clusters have also been studied. These include the metallofullerenes (Bethune *et al.*, 1993) and metallocarbohedrenes (Wei *et al.*, 1992). The latter are a class of M_8C_{12} clusters (M = Ti, V, Zr, Hf) that appear to be relatively stable, but they have not yet been isolated in bulk quantities.

3.5.2 Silicon and Germanium Below carbon in the periodic table are silicon and germanium, which are also covalently bound materials at room temperature and pressure. However, unlike carbon, these elements show little tendency to form double bonds. Thus, there is no analog of the graphite geometry for these elements in their bulk states, and there is no evidence for any fullerene analogs in the clusters. Silicon, in particular, is an important material in the semiconductor industry. With the continuous push to smaller and smaller features in semiconductor devices, there is now considerable interest in how the properties of silicon change as dimensions approach atomic scales. Although much less is known about germanium clusters than silicon clusters, the preliminary indications are that, like the bulk materials, silicon and germanium clusters have similar properties (Jarrold, 1993).

Unlike bulk Si and Ge, which evaporate mainly individual atoms when heated, the atomic clusters dissociate by preferential loss of fragments such as Si_6, Si_{10}, and Ge_{10}. These fragments are particularly strongly bound (with cohesive energies close to that of the bulk), and they appear as "magic numbers" in mass spectra of the clusters generated by pulsed laser vaporization. The high stability of Si_6 and Si_{10} has been reproduced in *ab initio* quantum-chemical calculations (Raghavachari, 1990). These calculations suggest that small silicon clusters have geometries that cannot be considered as fragments of the bulk diamondlike lattice. The geometry of Si_6 is a distorted octahedron, and Si_{10} is probably a tetracapped biprism. (A tetracapped octahedron geometry appears to be slightly higher in energy.) The bulk fragments (the cyclohexane and adamantane skeletons) are considerably higher in energy because of the large number of dangling bonds. Reconstruction to the more compact and highly coordinated geometries mentioned above is driven by these dangling bonds, in much the same way that bulk silicon surfaces reconstruct. The size regime where silicon clusters adopt a bulklike geometric structure is not yet known.

The chemical properties of silicon clusters have also been extensively studied (Jarrold *et al.*, 1991). Surprisingly, even quite large clusters, such as Si_{70}, appear to be substantially less reactive than bulk silicon surfaces. The low reactivity probably reflects the low density of dangling bonds on the cluster's surface. Whether or not there exist some clusters (such as Si_{33}, Si_{39}, and Si_{45}) with particularly low reactivity is an unresolved controversy. The chemical properties of silicon clusters can be modified by annealing, which demonstrates that cluster growth may not lead to the lowest-energy geometric isomer. The relatively large number of structural isomers suggested by these studies is probably a consequence of the directed nature of the chemical bonding in these clusters.

Direct experimental information on the geometries of silicon clusters has been very difficult to obtain. Spectroscopic information has been obtained for silicon clusters containing 2–7 atoms using photoelectron spectroscopy and surface-enhanced Raman spectroscopy of matrix-isolated, size-selected clusters (Honea *et al.*, 1993). These experimental studies generally confirm the main features of calculated predictions for these smaller clusters. For clusters with more than ten atoms, there is little direct experimental information about the geometries. Mobility measurements shown in Fig. 12 suggest that clusters with 10 to ~27 atoms follow a one-dimensional growth sequence to give prolate geometries (Jarrold and Bower, 1992). At ~27 atoms, a structural transition occurs, and the clusters reconstruct to a more spherical ge-

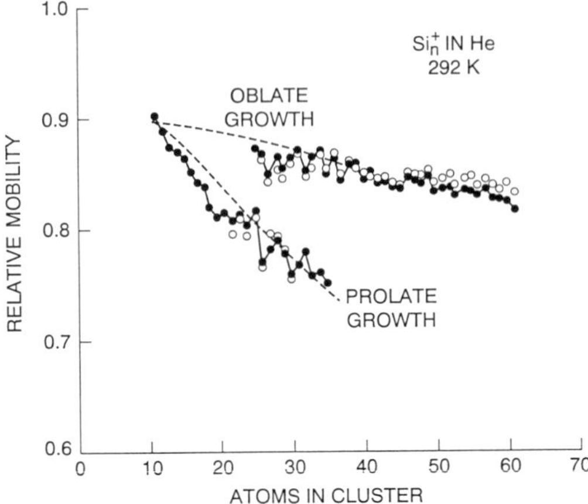

FIG. 12. The relative mobility of silicon clusters as a function of cluster size (Jarrold and Bower, 1992). Dashed lines indicate predicted oblate and prolate growth sequences.

ometry. For these larger clusters, the theoretical calculations become less reliable (and more controversial). Results obtained from molecular-dynamics simulations using the local-density approximation suggest that clusters with 30–50 atoms consist of two strongly interconnected and buckled shells (Röthlisberger *et al.*, 1994). The buckling arises because of the tendency of silicon to form sp^3 hybrids. For Si_{45}, the core consists of 7–9 atoms, and the rest of the atoms are in the surrounding shell.

4. CONDENSED-PHASE CLUSTERS

4.1 Deposited Clusters

While a well-defined cluster beam provides the best environment in which to probe the physical and chemical properties of atomic clusters, exploitation of these properties requires that the clusters be collected either on a surface or in a matrix. Deposited atomic clusters can be studied using a variety of techniques that cannot be applied to gas-phase clusters. These include x-ray photoelectron spectroscopy (XPS), extended x-ray absorption fine structure (EXAFS), electron energy-loss spectroscopy (EELS), and scanning tunneling microscopy (STM). A critical consideration is the nature of the cluster–surface and cluster–cluster interactions. The importance of cluster–surface interactions is well known in studies of the catalytic properties of supported transition-metal clusters because these interactions can modify both the electronic and geometric structure of the cluster (Poppa, 1984; Lewis, 1993). If the cluster–substrate interaction is strong, then the cluster will be strongly perturbed and anchored to the surface. Conversely, weak cluster–substrate interactions mean greater mobility of the cluster on the surface. Coalescence may occur, even at low surface coverages, when cluster–cluster interactions are stronger than cluster–substrate interactions. Many of the issues raised above remain largely unexplored, as there have been relatively few controlled experimental studies of deposited, monodisperse clusters.

There have been some efforts to generate high-quality thin films by ionized cluster beam deposition. If the kinetic energy of the cluster is raised to the point that it is a substantial fraction of the cohesive energy, the violence of the cluster–surface impact will either cause the cluster to shatter or embed it into the surface, depending on the nature of the cluster and the surface. If the cluster sticks to the surface, then the kinetic energy is initially deposited in a relatively small area, leading to locally high temperatures and possibly even local melting (Landman and Luedtke, 1992). The energy provided by the cluster impact increases surface mobility and leads to highly crystalline thin films at low substrate temperatures.

The large surface-to-volume ratio of clusters gives them unique thermodynamic properties. One example is the reduction of the melting point with decreasing particle size. Several models have been developed to account for this effect. These approaches describe the modification of the cluster chemical potential by the surface free energy. Experimentally, the melting temperature range for a distribution of small gold particle sizes has been measured by observing the loss of sharp rings in the electron diffraction pattern (Buffat and Borel, 1976). From the good agreement between the experimental data and the model (Fig. 13), it appears that the thermodynamic model provides a reasonable description of the size dependence of the melting temperature.

4.2 High-Resolution Imaging

Of critical importance in the study of metal clusters is the experimental determination of cluster geometry. Direct spectroscopic deter-

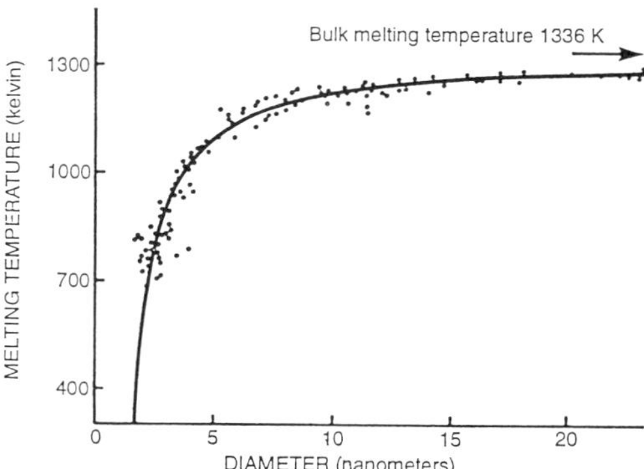

FIG. 13. Size dependence of the melting temperature of Au particles as a function of particle diameter (Buffat and Borel, 1976). The line is calculated using a thermodynamic model discussed in the text.

mination of the structure of small, unsupported metal clusters has so far only been possible for clusters with fewer than four atoms. Atomic-resolution imaging techniques such as scanning tunneling microscopy (STM), atomic force microscopy (AFM), and high-resolution (transmission) electron microscopy (HREM) have provided some insight into the geometric structure of larger clusters, even though the exact relationship between gas-phase structure and surface structure is unknown. On crystalline substrates, HREM is capable, in theory, of imaging clusters containing as few as several atoms (Spence, 1988). However, STM and AFM are better techniques for imaging the smallest clusters (down to single atoms), and additional information about the electronic states of deposited clusters may be obtained from the voltage dependence of the tunneling current (Chen, 1993).

There have been extensive studies of the structure of small metal particles by transmission electron microscopy since the discovery that fcc metals often form noncrystallographic multiply twinned particles (MTPs) (Ajayan and Marks, 1990). Decahedral, or pentagonal bipyramidal, MTPs contain five twin-related tetrahedral units, while icosahedral MTPs contain twenty such units. In HREM images of gold particles, single-crystalline and singly twinned structures are observed to predominate in smaller particles (1–2 nm diameter), while icosahedral and decahedral MTPs, as well as other highly twinned structures, are observed in larger particles. A variety of theoretical models also indicate that clusters containing fewer than a few thousand atoms do not pack as chunks of the bulk lattice. Molecular-dynamics simulations, using many-body potentials, have been performed for clusters of a number of transition metals. The relative energies of optimized decahedral, icosahedral, and fcc cuboctahedral clusters have been compared, and an element-specific cluster size for the icosahedral to fcc cuboctahedral transition is found (Uppenbrink and Wales, 1992). Although electron-microscope images do not allow determination of the exact number of atoms for this transition, the results of this model agree qualitatively with the electron-microscope observations.

Under the influence of the energetic electron beam in the electron microscope, small metal particles are often observed to undergo rapid ($<1/30$ s) structural transitions. This structural instability is an intriguing phenomenon because of the possibility of phase coexistence. A possible mechanism for these structural fluctuations is quasimelting, in which rapid isomerization between energetically similar geometries occurs at temperatures lower than the cluster thermodynamic melting point. Molecular-dynamics simulations suggest the coexistence of a variety of different structures with similar stabilities, which may be a reason for structural fluctuations (Uppenbrink and Wales, 1993). However, recent estimates of the temperature of gold particles in an electron microscope suggest that a mechanism based on rapid recrystallization through a highly excited, possibly molten, intermediate state (created by the high-energy electron beam) is more plausible (Ugarte, 1993).

4.3 Quantum Dots

Semiconductor clusters in the nanometer size range exhibit unique quantum-confinement effects. In the bulk semiconductor, because electrons are strongly delocalized and the effective mass is small, the electron–hole pair that results from an electronic transition from the valence band to the conduction band has an effective diameter that is relatively large, on the order of 1–20 nm (depending on the material). When the dimension of the cluster is of the same magnitude or less, this exciton becomes spatially confined (a quantum dot). This quantum confinement increases the band gap as the cluster size decreases (Brus, 1991).

The size-dependent band gap can be qualitatively described by a simple model that reduces the problem to a two-body interaction between the electron in the bottom of the conduction band and the hole that it left in the top of the valence band (Steigerwald and Brus, 1990). The electron and the hole are considered particles with effective masses m_e and m_h, respectively. The allowed electronic states are given by the solutions to the Schrödinger equation for two particles confined to a sphere of radius R:

$$E(R) = E_g + (h^2\pi^2/2R^2)(1/m_e + 1/m_h) - 1.8e^2/\epsilon R, \tag{6}$$

where ϵ is the bulk dielectric constant. The bulk band gap, E_g, is modified by the positive kinetic energy and by the negative Coulomb attraction between the electron and the hole. More accurate electronic-structure calculations employ an empirical psuedopotential method (Ramakrishna and Friesner, 1993). Diagonalization of the Hamiltonian matrix at a set of points in the Brillouin zone results in the band structure. A comparison of experimental and calculated band gaps is shown in Fig. 14. Although the simple description of Eq. (6) is a good qualitative prediction of the increase in the energy of the lowest electronic transition with decreasing cluster size, it fails for small clusters. On the other hand, the psuedopotential model agrees surprisingly well with the measured energies, given that cluster surface effects are not taken into account.

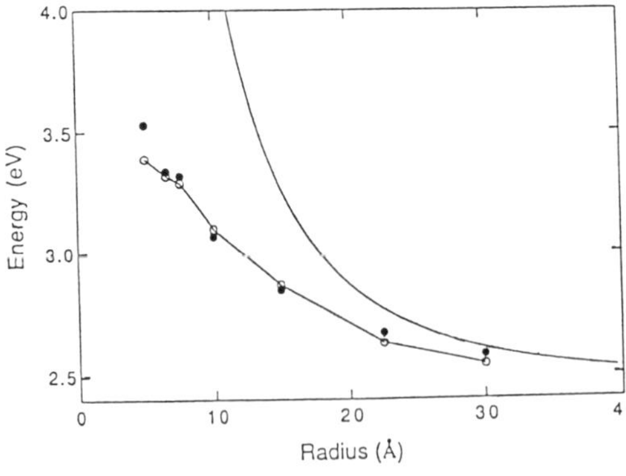

FIG. 14. Size-dependent band gaps of CdS clusters. The solid line is the prediction of the simple particle-in-a-box model given in Eq. (6), the experimental points are the filled circles, and the open circles (connected by lines) are calculated using the empirical pseudopotential method (Ramakrishna and Friesner, 1993).

Efforts to understand luminescence from porous silicon have prompted the investigation of this property in nanocrystalline silicon (Littau *et al.*, 1993). Bulk silicon is an indirect-gap semiconductor. However, relaxation of translational symmetry in finite-size silicon crystals causes the electronic transition to the lowest excited state to become weakly dipole allowed. Emission occurs at approximately 0.5 eV above the bulk Si band gap in 2-nm-diam Si particles, in agreement with the predicted quantum size effect.

4.4 Inorganic Clusters

Although the perturbing influence of the ligands prohibits direct comparison to free metal clusters of the same size, ligated clusters provide an alternative route to the investigation of monosized clusters (Schmid, 1992). One of the best-characterized ligated metal clusters is $Au_{55}(P(C_6H_5)_3)_{12}Cl_6$. EXAFS measurements show that this complex has a slightly distorted closed-shell cuboctahedral rather than an icosahedral core. The fourth and fifth closed cuboctahedral shells, containing 309 and 561 atoms, have also been investigated, for Pt and Pd complexes. As expected, the electronic states of the ligated surface atoms are distinctly different from those of the core atoms. Mössbauer data and optical-absorption spectra for the Au_{55} complex suggest that the inner Au_{13} atomic core is partly metallic in character. In ligated platinum systems, large Knight shifts, indicative of delocalized electrons, in ^{195}Pt NMR studies provide evidence that the core of the Pt_{309} complex manifests bulk behavior.

A mathematical scheme called the Hume–Rothery rule has been extended and developed to predict metal-packing arrangements in ligated "high nuclearity" metal clusters by counting electrons and atoms (Teo and Zhang, 1990). If the total nuclearity of a cluster, m, is the sum of bulk (core) and surface atoms, $m_b + m_s$, then the total number of electrons (both metal valence and ligand donated) is given by

$$N = m_b n_b + m_s n_s = (m_b + r m_s) n_b, \tag{7}$$

where n_b and n_s are the number of electrons in bulk and surface atoms, respectively, and r is the ratio of n_s to n_b, empirically determined to be $18/16 = 1.125$. The predicted electronic requirements correlate well with the number of electrons in experimentally observed stable clusters. Two general classifications arise from a systematic analysis of cluster structures. The first is called "supraclusters," which are vertex-, edge-, or face-sharing clusters of primary clusters (tetrahedral or icosahedral units, for example). The other class is termed "polyhedral" clusters, which grow from primary units in a layer-by-layer fashion. Recently, these electron- and atom-counting rules have been applied to enumerate all possible stereoisomers of binary (two metals) icosahedral clusters (Teo *et al.*, 1993).

The structures of a wide variety of ligated clusters have been determined by x-ray diffraction (Mingos and Wales, 1990; Schmid, 1992). An intriguing indium-containing fullerene cage analog deserves mention (Sevov and Corbett, 1993). Double-hexagonal close-packed arrays of pentagon-sharing In_{74} fullerenes are stuffed with $In_{10}Ni$ clusters and Na_{39} polyhedra. The electronic structure of these $Na_{96}In_{97}Z_2$ (Z = Ni, Pd, or Pt) arrays apparently parallels that of carbon fullerenes.

4.4 Nanophase Materials

A logical extension of isolated cluster research is the assembly of nanophase structures by selective manipulation of physical and chemical properties of nanoscale aggregates. The effects of spatial confinement of electrons that results in unique cluster properties are also responsible for novel properties in nanoscale materials. This subject is mentioned only briefly here. More detailed information may be found in the article NANOPHASE MATERIALS and in several recent reviews (Siegel, 1993; Andres *et al.*, 1989).

GLOSSARY

Capped Cluster: Condensed-phase cluster that is stabilized by chemical modification (with ligands) of the cluster surface.

Doped Cluster: A pure cluster containing a foreign (guest) atom or molecule.

Jellium Approximation: An approximation in which the nuclei are smeared out into a uniform positively charged background surrounded by the nearly free valence electrons.

Molecular Beam: A stream of clusters (of atoms or molecules) in a collision-free vacuum environment.

Photodepletion Spectroscopy: A form of absorption spectroscopy obtained by monitoring a decrease (depletion) in the cluster-ion signal when the excitation laser is tuned to an electronic resonance.

Quantum Dot: A cluster in which electronic excitation is spatially confined in three dimensions by the cluster boundaries.

R2PI: Resonance-enhanced two-photon ionization, where the first photon is absorbed into an intermediate electronic state, and absorption of a second photon results in ionization.

Vertical Binding Energy: The diabatic energy difference between initial and final states.

Works Cited

Ajayan, P. M., Marks, L. D. (1990), *Phase Transitions* **24–26**, 229–258.

Alameddin, G., Hunter, J., Cameron, D., Kappes, M. M. (1992), *Chem. Phys. Lett.* **192**, 122–128.

Amrein, A., Simpson, R., Hackett, P. (1991), *J. Chem. Phys.* **95**, 1781–1800.

Andres, R. P., Averback, R. S., Brown, W. L., Brus, L. E., Goddard, W. A., III, Kaldor, A., Louie, S. G., Moscovits, M., Peercy, P. S., Riley, S. J., Siegel, R. W., Spaepen, F., Wang, Y. (1989), *J. Mater. Res.* **4**, 704–736.

Baguenard, B., Pellarin, M., Lermé, J., Vialle, J. L., Broyer, M. (1993), *J. Chem. Phys.* **100**, 754–755.

Barnett, R. N., Landman, U., Rajagopal, G., Nitzan, A. (1990), *Israel J. Chem.* **30**, 85–105.

Berry, S. D. (1993), *Chem. Rev.* **93**, 2379–2394.

Bethune, D. S., Johnson, R. D., Salem, J. R., de Vries, M. S., Yannoni, C. S. (1993), *Nature* **366**, 123–128.

Billas, I. M. L., Becker, J. A., Châtelain, A., de Heer, W. A. (1993), *Phys. Rev. Lett.* **71**, 4067–4070.

Birringer, R., Gleiter, H., Klein, H.-P., Marquardt, P. (1984), *Phys. Lett. A* **102**, 365–369.

Bohren, C. F., Huffman, D. R. (1983), *Absorption and Scattering of Light by Small Particles*, New York: Wiley.

Bonačić-Koutecký, V., Fantucci, P., Koutecký, J. (1991), *Chem. Rev.* **91**, 1035–1108.

Brus, L. E. (1991), *Appl. Phys. A* **53**, 465–474.

Brutschy, B. (1992), *Chem. Rev.* 92, 1567–1587.

Buffat, Ph., Borel, J.-P. (1976), *Phys. Rev. A* **13**, 2287–2298.

Chen, C. J. (1993), *Introduction to Scanning Tunneling Microscopy*, New York: Oxford University Press.

Clemenger, K. (1985), *Phys. Rev. B* **32**, 1359–1362.

Cohen, M. L., Knight, W. D. (1990), *Phys. Today* **43** (12), 42–50.

de Heer, W. A., Knight, W. D., Chou, M. Y., Cohen, M. L. (1987), in: H. Ehrenreich, D. Turnbull (Eds.), *Solid State Physics*, Vol. 40, New York: Academic.

Easter, D. C., Baronavski, A. P., Hawley, M. (1993), *Chem. Phys. Lett.* **206**, 329–336.

Faraday, R. (1857), *Philos. Trans. R. Soc. London* **141**, 145.

Farges, J., deFeraudy, M. F., Raoult, B., Torchet, G. (1986), *J. Chem. Phys.* **84**, 3491–3501.

Gentry, W. R. (1988), in: G. Scoles (Ed.), *Atomic and Molecular Beam Methods*, Vol 1, New York: Oxford University Press, p. 54.

Gough, T. E., Mengel, M., Rowntree, P. A., Scoles, G. (1985), *J. Chem. Phys.* **83**, 4958–4961.

Goyal, S., Schutt, D. L., Scoles, G. (1993), *Acc. Chem. Res.* **26**, 123–130.

Heath, J. R. (1992), *Science* **258**, 1131–1133.

Herron, N., Wang, Y., Eddy, M., Stucky, G., Cox, D. E., Moller, K., Bein, T. (1989), *J. Am. Chem. Soc.* **111**, 530–540.

Hoare, M. R. (1979), *Adv. Chem. Phys.* **40**, 49–135.

Homer, M. L., Livingston, F. E., Whetten, R. L. (1993), *Z. Phys. D* **26**, 201–203.

Honea, E. C., Ogura, A., Murray, C. A., Raghavachari, K., Sprenger, W. O., Jarrold, M. F., Brown, W. L. (1993), *Nature* **366**, 42–44.

Irion, M. P., Schnabel, P. (1992), *Ber. Bunsenges. Phys. Chem.* **96**, 1091–1100.

Jarrold, M. F. (1991), *J. Cluster Sci.* **2**, 137–181.

Jarrold, M. F., Ijiri, Y., Ray, U. (1991), *J. Chem. Phys.* **94**, 3607–3618.

Jarrold, M. F., Bower, J. E. (1992), *J. Chem. Phys.* **96**, 9180–9190.

Jarrold, M. F. (1993), in: C.-Y. Ng, T. Baer, I. Powis (Eds.), *Cluster Ions*, Chichester: Wiley.

Jortner, J. (1992), *Z. Phys. D* **24**, 247–275.

Kaiser, B., Rademann, K. (1992), *Phys. Rev. Lett.* **69**, 3204–3207.

Kawai, R., Weare, J. H. (1990), *Phys. Rev. Lett.* **65**, 80–83.

Klots, C. E. (1991), *Chem. Phys. Lett.* **186**, 73–76.

Knickelbein, M. B. (1992), *Chem. Phys. Lett.* **192**, 129–134.

Kresin, V. V. (1992), *Phys. Rep.* **220**, 1–52.

Landman, U., Luedtke, W. D. (1992), *Appl. Surf. Sci.* **60/61**, 1–12

Leisner, T., Athanassenas, K., Echt, O., Kandler, O., Kreisle, D., Recknagel, E. (1991), *Z. Phys. D* **20**, 127–129.

Lewis, L. N. (1993), *Chem. Rev.* **93**, 2693–2730.

Littau, K. A., Szajowski, P. J., Muller, A. J., Kortan, A. R., Brus, L. E. (1993), *J. Phys. Chem.* **97**, 1224–1230.

Luo, F., McBane, G. C., Kim, G., Giese, C. F., Gentry, W. R. (1993), *J. Chem. Phys.* **98**, 3564–3567.

Mackay, A. L. (1962), *Acta. Crystallogr.* **15**, 916–918.

Makov, G., Nitzan, A., Brus, L. E. (1988), *J. Chem. Phys.* **88**, 5076–5085.

Markovich, G., Pollack, S., Giniger, R., Cheshnovsky, O. (1993), *Z. Phys. D* **26**, 98–100.

Martin, T. P., Bergmann, T., Göhlich, H., Lange, T. (1991), *J. Phys. Chem.* **95**, 6421–6429.

Miedema, A. R. (1978), *Z. Metall.* **69**, 287–292.

Miehle, W., Kandler, O., Leisner, T., Echt, O. (1989), *J. Chem. Phys.* **91**, 5940–5952.

Milligan, W. O., Morriss, R. H. (1964), *J. Am. Chem. Soc.* **86**, 3461–3467.

Mingos, D. M. P., Wales, D. J. (1990), *Introduction to Cluster Chemistry*, Englewood Cliffs, NJ: Prentice-Hall.

Morse, M. D., Hopkins, J., Langridge-Smith, P., Smalley, R. (1983), *J. Chem. Phys.* **79**, 5316–5328.

Nishioka, H., Hansen, K., Mottelson, B. R. (1990), *Phys. Rev. B* **42**, 9377–9386.

Papanikolas, J. M., Vorsa, V., Nadal, M. E., Campagnola, P. J., Buchenau, H. K., Lineberger, W. C. (1993), *J. Chem. Phys.* **99**, 8733–8750.

Parent, D. C., Anderson, S. L. (1992), *Chem. Rev.* **92**, 1541–1565.

Parks, E. K., Zhu, L., Ho, J., Riley, S. J. (1993), *Z. Phys. D* **26**, 41–45.

Poppa, H. (1984), *Vacuum* **34**, 1081–1095.

Powers, D. E., Hansen, S. G., Geusic, M. E., Puiu, A. C., Hopkins, J. B., Dietz, T. G., Duncan, M. A., Langridge-Smith, P. R. R., Smalley, R. E. (1982), *J. Phys. Chem.* **86**, 2556–2560.

Raghavachari, K. (1990), *Phase Transitions* **24–26**, 61–90.

Ramakrishna, M. V., Friesner, R. A. (1993), *Israel J. Chem.* **33**, 3–8.

Riley, S. J., Parks, E. K. (1987), in: P. Jena, B. K. Rao, S. N. Khanna (Eds.), *Physics and Chemistry of Small Clusters*, New York: Plenum, p. 727.

Röthlisberger, U., Andreoni, W., Parrinello, M. (1994), *Phys. Rev. Lett.* **72**, 665–668.

Schmid, G. (1992), *Chem. Rev.* **92**, 1709–1727.

Selegue, T. J., Moe, N., Draves, J. A., Lisy, J. M. (1992), *J. Chem. Phys.* **96**, 7268–7278.

Sevov, S. C., Corbett, J. D. (1993), *Science* **262**, 880–883.

Siegel, R. W. (1993), *Physics Today* **46** (10), 64–68.

Spence, J. C. H. (1988), *Experimental High-Res-*

olution Electron Microscopy, Oxford: Oxford University Press, chap. 6.

Steigerwald, M. L., Alivisatos, A. P., Gibson, J. M., Harris, T. D., Kortan, R., Muller, A. J., Thayer, A. M., Duncan, T. M., Douglass, D. C., Brus, L. E. (1988), *J. Am. Chem. Soc.* **110**, 3046–3050.

Steigerwald, M. L., Brus, L. E. (1990), *Acc. Chem. Res.* **23**, 183–188.

Stringari, S. (1990), in: G. Scoles (Ed.), *The Chemical Physics of Atomic and Molecular Clusters*, Amsterdam: North-Holland, p. 199.

Su, C.-X., Hales, D. A., Armentrout, P. B. (1993), *J. Chem. Phys.* **99**, 6613–6623.

Syage, J. A. (1992), in: J. D. Simon (Ed.), *Ultrafast Spectroscopy in Chemical Systems*, Boston: Kluwer Academic.

Taylor, K. J., Pettiette-Hall, C. L., Cheshnovsky, O., Smalley, R. E. (1992), *J. Chem. Phys.* **96**, 3319–3329.

Teo, B. K., Zhang, H. (1990), *J. Cluster Sci.* **1**, 155–187.

Teo, B. K., Zhang, H., Kean, Y., Dang, H., Shi, X. (1993), *J. Chem. Phys.* **99**, 2929–2941.

Tiggesbäumker, J., Köller, L., Lutz, H. O., Meiwes-Broer, K. H. (1992), *Chem. Phys. Lett.* **190**, 42–47.

Toennies, J. P. (1990), in: G. Scoles (Ed.), *The Chemical Physics of Atomic and Molecular Clusters*, Amsterdam: North-Holland, p. 597.

Ugarte, D. (1993), *Z. Phys. D* **28**, 177–181.

Uppenbrink, J., Wales, D. J. (1992), *J. Chem. Phys.* **96**, 8520–8534.

Uppenbrink, J., Wales, D. J. (1993), *Z. Phys. D* **26**, 258–260.

von Helden, G., Hsu, M.-T., Kemper, P. R., Bowers, M. T. (1991), *J. Chem. Phys.* **95**, 3835–3837.

Wang, C. R. C., Pollack, S., Dahlseid, T. A., Koretsky, G. M., Kappes, M. M. (1992), *J. Chem. Phys.* **96**, 7931–7937.

Wang, Y., Herron, N. (1991), *J. Phys. Chem.* **95**, 525–532.

Wei, S., Guo, B. C., Purnell, J., Buzza, S., Castleman, A. W., Jr. (1992), *Science* **256**, 818–820.

Whetten, R. L., Hahn, M. Y. (1990), in: E. R. Bernstein (Ed.), *Atomic and Molecular Clusters*, Amsterdam: Elsevier, chap. 8.

Wood, D. M. (1981), *Phys. Rev. Lett.* **46**, 749.

Yang, Y. A., Bloomfield, L. A., Jin, C., Wang, L. S., Smalley, R. E. (1992), *J. Chem. Phys.* **96**, 2453–2459.

Zakin, M. R., Cox, D. M., Brickman, R. O., Kaldor, A. (1989), *J. Phys. Chem.* **93**, 6823–6827.

Further Reading

Haberland, H. (Ed.) (1994), *Clusters of Atoms and Molecules*, Berlin: Springer-Verlag.

Schmidt, R., Lutz, H. O., Dreizler, R. (Eds.) (1992), *Nuclear Physics Concepts in the Study of Atomic Cluster Physics*, Berlin: Springer-Verlag.

Scoles, G. (Ed.) (1990), *The Chemical Physics of Atomic and Molecular Clusters*, Amsterdam: North-Holland.

Scoles, G. (Ed.) (1988), *Atomic and Molecular Beam Methods*, Vol. 1, New York: Oxford University Press.

Sugano, S. (1991), *Microcluster Physics*, Berlin: Springer-Verlag.

MOLECULAR AND ATOMIC COLLISION PROCESSES WITH ION BEAMS

REINHARD BRUCH, ELIZABETH RAUSCHER, AND STEPHAN FÜLLING, *Department of Physics, University of Nevada, Reno, Nevada, U.S.A.*

DIETER SCHNEIDER, *Lawrence Livermore National Laboratory, Livermore, California, U.S.A.*

SVEN MANNERVIK AND MATS LARSSON, *Manne Siegbahn Institute of Physics, Stockholm, Sweden*

3-527-28132-0/94/$5.00 + .50

INTRODUCTION

Research in the field of collision processes of molecules and atoms has developed at an explosive rate since the early 1960s (Rudd and Macek, 1974; Kessel and Fastrup, 1974; Bashkin, 1976; Crasemann, 1975; Sellin, 1978; Mott and Massey, 1987; Fano and Rau, 1988; Alonso *et al.*, 1989). The knowledge of such collision processes has been of great value in numerous applications from pure to applied physics, in particular plasma diagnostics, condensed matter physics, gas laser design, x-ray lasers, physics and technology of ion sources and accelerators, optics of charged particle beams, surface, thin film, and materials analysis, ion implantation, atmospheric collision physics and chemistry, radiation physics and biology, and astrophysics. In order to understand the physics of planetary and stellar atmospheres, including those of the Earth and the sun, detailed knowledge is required of a number of molecular and atomic collision processes. In fusion energy research, molecular and atomic collision physics (De Heer, 1982) plays a significant role in the heating, cooling, and loss mechanisms, as well as in diagnostics. Consequently, a fundamental knowledge of molecular and atomic collision processes (CIAMDA, 1987) is necessary for modeling of high-temperature plasmas (Vainshtein and Shevelko, 1986). Such collision processes occur when molecules and atoms are ionized and accelerated and allowed to interact with atomic, molecular, liquid, surface, or solid targets. The target can also comprise electrons, ions, or photons.

In any ion-atom or ion-molecule interaction, if no excitation of the target occurs, the process is called an elastic collision. Elastic atomic collisions are of primary importance at very low energies (Scoles, 1988). Inelastic collisions may take place if the collision partners come close enough together for atomic or molecular excitation or ionization to occur. As the projectile velocity is increased, first the outer-shell electrons and then the inner-shell electrons are affected and participate in the dynamical process. Inelastic collision events are characterized by the emission of interaction products, such as photons, electrons, scattered ions, or molecular fragments. The use of ion beams for the study of collision processes of molecules and atoms involves many different reaction channels. For readers not familiar with the field of ion interactions and ion-beam techniques, these topics are discussed in general terms in Sec. 1. A more detailed description of individual high-, medium-, and low-energy techniques is provided in Sec. 2, and in Sec. 3, we present briefly some basic atomic and molecular theoretical models.

1. AN OVERVIEW OF COLLISION PROCESSES OF MOLECULAR AND ATOMIC ION BEAMS

Ion-molecule and ion-atom collision processes are controlled by many parameters including ion velocity or energy, projectile and target size, composition, charge state, and mass, interaction time, etc. Useful parameter estimates and simple equations relevant for collision physics are defined in Table 1. A schematic illustration of a typical collisional process characterized by different emission products and reaction channels is shown in Fig. 1. When the projectile beam interacts with a gas, vapor, foil, or solid target, a number of observable elementary reactions occur. These processes involve, for example, photon and electron emission, recoil ions, scattered projectiles, and fragments. Typical detection systems used are optical monochromators, electron spectrometers, time-of-flight analyzers, translational energy spectrometers, lasers for molecular beam fragmentation, precision atomic spectroscopy, photodetachment of negative ions, polarization and angular distribution studies, and multiparameter coincidence techniques. One of the major goals of these collision studies is to understand the many-body aspect of the interaction dynamics, which supersedes independent-particle models. Generally, the most important forces are the Coulomb forces. In few-electron heavy ions at relativistic velocities, quantum electrodynamic (QED) effects must be taken into account. Also, relativistic effects occupy an important role in the collision dynamics and spectroscopy of electrons and photons at low velocities, if the projectile is heavy and highly charged.

It is convenient to classify collisional processes according to the projectile velocity v and the velocity v_e of the participating elec-

Table 1. Some characteristic properties used in ion-atom collision experiments.

Quantity	Symbol	Formula	Units	Typical values
Beam energy	E		MeV	10^{-2}–10^{3-}
Reduced beam energy[a,b]	ϵ	$545E/M$	eV	10^{-2}–10^{5}
		$20E/M$	a.u.	10^{-3}–10^{4}
Velocity[a,c]	v	$\sqrt{40E/M}$	a.u.	10^{-2}–10^{2}
		$13.8\sqrt{E/M}$	mm/ns	10^{-1}–10^{2}
Time of flight	t	$72\sqrt{M/E}$	ps/mm	10^{-11}–10^{-6}
Time resolution	Δt	$\Delta z/v$	s	10^{-12}–10^{-9}
Collision time[d]	T	$2b/v$	s	
Ion-beam current[e]	I		μA	10^{-6}–10^{2}
Particle rate	$\Delta N/\Delta t$	I/g	Ions/s	10^{7}–10^{15}
Gas-target density[f]	n	$10^{17}p/3$	Particles/cm^3	10^{14}–10^{17}
Mean free path length[f]	l	$3 \times 10^{-17}/(p\sigma)$	cm	
Cross section for single electron capture[g]	$\sigma_{q,q-1}$	πR_C^2	cm^2	10^{-14}–10^{-26}
Single-collision condition[f]		$n\sigma l \ll 1$		

[a]Beam energy E (MeV), projectile mass M (amu), nonrelativistic velocities.
[b]1 a.u. = 2 Ry = 27.2 eV.
[c]Bohr velocity v_0 = 2.18765 10^8 cm/s.
[d]Impact parameter b.
[e]Ion current for singly charged ions.
[f]Gas pressure p (Torr), cross section σ (cm^2).
[g]R_C = critical distance (1 Å = 10^{-8} cm), only valid for slow collisions.

trons. The low-energy regime is characterized by $v < v_e$. In this region, the semiempirical approximation is appropriate, which treats the heavy particles classically and the electron motion quantum mechanically. Excitation cross sections increase with v and reach a maximum at about $v/v_e \approx 1$. At higher velocities, $v > v_e$, perturbation theory is used to describe scattering amplitudes. In this case, the interaction between electrons and nuclei

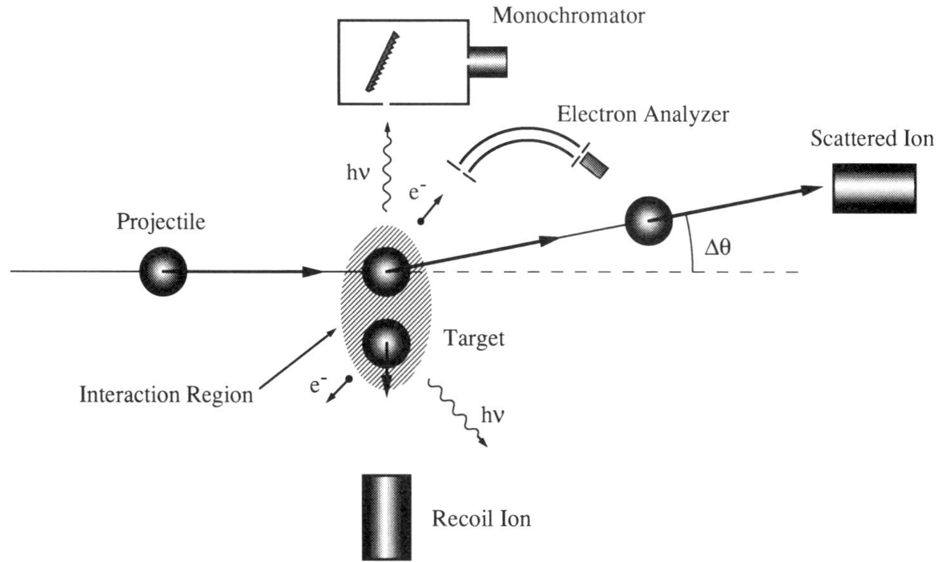

FIG. 1. View of the target region with characteristic ion-molecule and ion-atom interaction processes and reaction fragments (electrons, photons, scattered projectiles, and recoil ions) and corresponding detection systems.

Table 2. Overview of collision processes.

Collision process	Particles detected	Experimental technique
Beam-foil interaction: $A^{q+} + \mathrm{C\ foil} \rightarrow A^{m+*}$ Molecular fragmentation: $AB^{+} + \mathrm{C\ foil} \rightarrow A^{+} + B$	Photons (visible, UV, EUV, x rays), electrons (autoionization, Auger effect, continuum electrons), scattered ions, molecular fragments.	*Multiple collision process* (formation of highly excited states): Beam-foil spectroscopy, lifetime measurement, quantum beats, polarization, field ionization, charge state measurement, Auger-electron spectroscopy, molecular dissociation, spectroscopy of cusp electrons.
Beam-laser interaction: $A^{q+} + h\nu \rightarrow A^{q+*}$ Photodetachment: $A^{-} + h\nu \rightarrow A + e^{-}$ Photofragmentation: $AB^{+} + h\nu \rightarrow A^{+} + B$ (or $A + B^{+}$)	Fluorescence studies (infrared, visible, UV photons), electrons, atoms, ions, molecular fragments.	*Laser excitation*: Laser spectroscopy of fast beams, Doppler-free methods, laser cooling, collinear laser ion-beam method, photodetachment, photoionization, photofragmentation spectroscopy, half collisions.
Beam surface interaction: $A^{q} + \mathrm{surface} \rightarrow A^{m+*}$ Beam-solid interaction: $A^{q} + \mathrm{solid} \rightarrow A^{m+*}$	Photons (visible, UV, EUV, x rays), Auger electrons, scattered ions.	*Ion-surface collision* (angle of ion impact from normal to grazing incidence): X-ray and Auger-electron spectrometry, ion-beam crystallography, ion implantation, Rutherford backscattering, particle-induced x-ray emissions, channeling.
Electron capture: $A^{q+} + B \rightarrow A^{(q-1)+*} + B^{+}$ Ionization: $A^{q+} + B \rightarrow A^{q+} + B^{+} + e^{-}$ Transfer ionization: $A^{q+} + B \rightarrow A^{(q-1)+} + B^{2+} + e^{-}$ General case: $A^{q+} + B \rightarrow A^{(q-s)+} + B^{r+} + (r-s)e^{-} + \Delta E$	Photons (visible, UV, EUV, x rays), electrons (Auger decay, autoionization, continuum electrons, cusps), and scattered ions.	*Single collision process*: Charge-state, energy, and angle selection of projectile and target ions, impact-parameter–dependent coincidence measurements, photon and electron emission studies, translational spectroscopy, accel-decel method, recoil-ion spectroscopy, zero-degree electron spectroscopy, EUV and x-ray studies, dissociative charge transfer.
Few (mostly two) electron transitions: Double ionization: $A^{q+} + B \rightarrow A^{q+} + B^{2+} + 2e^{-}$ Double excitation: $A^{q+} + B \rightarrow A^{q+} + B^{**}$ Ionization plus excitation: $A^{q+} + B \rightarrow A^{q+} + B^{+*} + e^{-}$	Projectile and target photons, electrons and ions.	*Single collision process* (medium to high energy): Charge state, photon and electron emission cross section measurements, photon and electron angular distribution.

Information obtained	Theoretical model	Review articles
Multielectron effects, energy levels of highly excited states, lifetimes, casacade effects, spin-forbidden transitions, fine and hyperfine structure, Lamb shift, magnetic substate population, coherence parameters, charge-state distribution, energy-loss mechanisms, molecular properties, photon yields, population of Rydberg levels.	Many-body theories, atomic structure calculations, relativistic and QED effects, time-dependent phenomena, Auger decay, ion-solid and ion-surface theory, molecular dissociation.	Andrä (1979), Bashkin (1976), Berry (1977), Betz (1983), Cowan (1981), Gemmel (1980), Lindgren and Morrison (1982), Mohr (1989), Sellin (1978).
Accurate measurements of energy levels, fine and hyperfine splitting, Lamb shift, lifetimes, resonance widths, photodetachment and molecular fragmentation channels and energies, formation of ordered ion structures.	Precision calculations of atomic and molecular properties, excitation and decay dynamics of selectively excited states, quantum electrodynamics, quantum optics, time-dependent phenomena.	Andrä (1979), Hotop and Lineberger (1985), Letokov (1989), Müller-Dethlefs and Schlag (1991), Neugart (1987), Walther (1992).
Photon and electron spectra, ion scattering, charge state of ions, secondary-ion yield, surface modification, short-wavelength lasers, stopping power, ion-beam modification, surface composition and structure.	Deexcitation and neutralization processes near surfaces, ion scattering, secondary-electron, ion, and photon production, cascades, optical polarization, charge-exchage mechanisms, surface modifications, radiation damage.	Andrä (1989), Datz (1983), Elton (1990), Hasselkamp *et al.* (1992), Marrus (1989), Tully and White (1977).
Total and differential capture or excitation cross sections (inner- and outer-shell contributions), state-selective photon- and electron-emission cross sections, polarization, angular distributions, multiple-electron transitions, threshold effects, atomic and molecular data for fusion, astrophysics, x-ray lasers.	*Low energy*: Overbarrier model, Landau-Zener theory, post-collisional interaction, molecular orbital picture. *Intermediate and high energy*: Perturbative methods: Born, eikonal, Glauber, distorted-wave, higher-order contributions. Non-perturbative methods: Close coupling, R- and T-matrix approaches. Statistical methods. Hartree-Fock methods.	*Low energy*: Marrus (1989), Niehaus (1990), Ivanov (1988), Salzborn *et al.* (1991). *Medium to high energy*: Briggs *et al.* (1988), Cocke and Olson (1991), Dubé (1986), Fritsch and Lin (1991), Rudd and Macek (1974), Stolterfoht (1987).
Multielectron effects and electron correlation, kinematics in several-body scattering, ionization of many electrons, multiple excitation, ionization plus excitation, multiple-electron capture, transfer, and ionization, transfer and excitation, few-electron transitions, differential cross sections, Fano shape profiles.	Independent-electron approximation, expansion in projectile charge (Z_p) and velocity (v), 1st- and 2nd-order terms, higher-order terms, many-body perturbation expansion, sudden approximation, forced impulse method, coupled-channel calculations, time-dependent Hartree-Fock, binomial distributions, multiple-center scattering.	Datz (1990), McGuire (1992).

occurs during an extremely short time interval (for 1-MeV H^+, $\Delta t \approx 10^{-17}$ s), and high fields are obtained leading to inner-shell excitation and ionization. Currently, ions with energies of GeV/u (1 u = atomic mass unit) can be employed using relativistic velocities and the ability to produce few-electron or bare ions throughout the periodic table. These time-dependent, high-energy, few-body scattering problems provide a challenging theoretical endeavor.

Even though multiply charged ions dominate over neutral species in the universe, accurate experimental data and theoretical models on the structure and dynamics of multiply charged ions in low-, medium-, and high-energy collisions are limited. In particular, few-electron ions are ideal for testing relativistic many-body theories and QED effects (Wilson *et al.*, 1991). Therefore, experiments with highly charged ions are currently of great interest.

Collision processes between a wide range of atomic and molecular species have been investigated using a variety of experimental techniques, and some theoretical models have been developed. Representative collision processes that elucidate the field of molecular and atomic collisions are summarized in Table 2 along with additional information concerning the type of particles detected, experimental techniques used, experimental parameters obtained, and theoretical methods applied. Also listed in Table 2 are some review articles and books covering specific topics and categories of ion-molecule and ion-atom collisions. For more information, the reader is referred to the conference proceedings on the physics of electronic and atomic collisions (ICPEAC; see, for example, Datz, 1990) and the Proceedings of the NATO Advanced Study Institutes on Fundamental Processes of Atomic Dynamics (Briggs *et al.*, 1988). The physics of highly charged ions has recently been reviewed by Marrus (1989).

2. EXPERIMENTAL METHODS USING MOLECULAR AND ATOMIC ION BEAMS

2.1 Typical Ion Accelerator Setup

Tunability in energy, mass, charge state, current, repetition rate, etc. makes ion accelerators powerful tools for the exploration of atomic and molecular structure and collision dynamics. Several informative reviews and conference proceedings exist that cover practical aspects of ion accelerators, e.g., optics of charged particles (Wollnik, 1987), principles of charged-particle acceleration (Humphries, 1986), the physics and technology of ion sources (Brown, 1989), ion beams for material analysis (Bird and Williams, 1989), and the excellent Denton Conference proceedings on "Applications of Accelerators in Research and Industry" published periodically in *Nuclear Instruments and Methods*. A typical ion-beam collision apparatus consists of a positive- or negative-ion source and a means to accelerate the ions such as a Van de Graaff (2 or 3 MeV) or any other type of electrostatic accelerator (see Fig. 2). The ion beam is usually focused, mass analyzed, and collimated before entering the target chamber, which may contain a foil, gas, vapor, jet, surface, or solid target. Particle detectors may be employed to analyze target or projectile fragments, or a spectrometer may be used to observe the photons, electrons, and ions emitted by excited target or projectile states.

Atomic and molecular collisions at low to medium energy are conducted mostly with electrostatic machines. Recently, radio-frequency boosters have been designed, extending the range to higher energies and charge states and heavier ions. Relatively large-scale accelerators, such as the Kansas State, Oak Ridge, Brookhaven, Berkeley, and Argonne accelerators, and ion storage rings (Aarhus, GSI Darmstadt, Heidelberg, and Stockholm) as well as other large-scale accelerator facilities like ATLAS, BEVALAC, RIKEN, GANIL, UNILAC, etc. are frequently used for atomic and molecular collision studies.

2.2 Beam-Foil Excitation

The beam-foil technique was independently proposed by Kay and Bashkin in the early 1960s. The basic concept is to produce an ion beam in an accelerator and to let the ion beam pass through a thin foil. The ions in the beam will be excited and can be observed by spectroscopic methods. In particular, since the emitted photons stem from fast-moving particles, time-of-flight techniques can be used to measure decay curves and time-delayed spectra. Many review articles and books about this method have been pub-

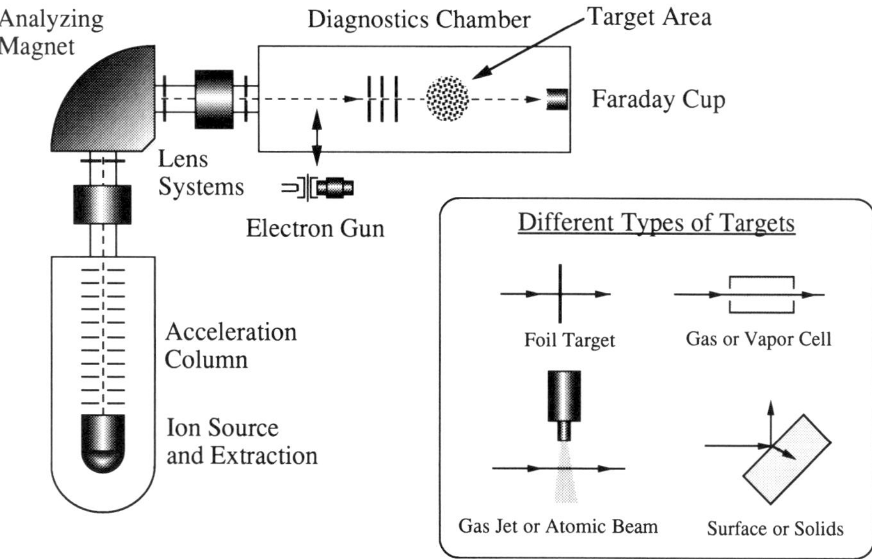

FIG. 2. Schematic layout of a low- to medium-energy ion accelerator with target chamber and different types of possible targets.

lished; see, e.g., contributions by Bashkin (1976), Berry (1977), and Andrä (1979).

2.2.1 Experimental Techniques Early beam-foil experiments were performed mainly at Van de Graaff machines or electrostatic isotope separators. These accelerators deliver typically beam energies between 50 keV and 3 MeV with beam currents of the order of 1 μA. Such beam energies correspond to velocities of a few mm/ns (see Table 1). As more powerful accelerators have been constructed, beam-foil experiments have more recently been performed at much higher energies (many GeV). Two illustrative examples are given here. For example, a 250-keV He beam will mainly leave the foil in charge state +1, while a 150-MeV U beam will have a mean charge of about +32. The excitation process is generally nonselective. It has been found that the population of the different excited Rydberg states usually follows approximately a $1/(n^*)^3$ dependence where n^* is the effective principal quantum number (Stebbings and Dunning, 1983). When the ion beam traverses the foil, it interacts with many atoms, causing energy loss and energy spread. These kinematic effects often have to be taken into account as spectroscopic measurements are evaluated. As an example, the stopping power dE/dx for 100-keV H^+ ions interacting with C foils is about 0.7 keV/(μg cm^{-2}).

In a typical experimental setup for beam-foil measurements, either the photon or electron emission spectrum is observed close to the foil, or the intensity decay of a specific spectral line is measured by increasing the distance between the excitation (foil) and the observation region (usually by moving the foil upstream in the vacuum chamber). The distance is converted to time through the simple relation $t = x/v$, where x and v are the distance and beam velocity after foil excitation, respectively. Measurements are normalized to a specific particle flux by monitoring either the total light emitted at the foil or the ion-beam current in a Faraday cup. An optical spatial resolution of the order of 0.1 mm is usually sufficient to measure atomic lifetimes. The high velocity of the ion beam generally causes large Doppler broadening and shifts. Even for a perpendicular observation, which eliminates the Doppler shift in first order, there will be a large Doppler broadening as a result of the angular spread of the monochromator. To a large extent, this can be compensated for by moving the exit slit of the monochromator ("refocusing"). This method works for photon as well as electron spectroscopy.

2.2.2 Lifetime Measurements There is a great need for experimental measurements of atomic transition probabilities. Such measurements are important for astrophysics, plasma diagnostics studies, and tests of many-body theories. The atomic lifetime (τ) associated with a specific state is closely related to the transition probabilities (A_{ki}) by

$$\tau_i = 1/\Sigma A_{ik}, \tag{1}$$

where the sum involves all lower states k to which the state i can decay. Generally, beam-foil lifetime curves are composed of more than one exponential profile as a result of cascades from higher-lying states. The differential equation for the decay process is

$$\frac{dN_i}{dt} = -\frac{N_i}{\tau} + \Sigma N_j A_{ji}, \tag{2}$$

where the last term describes the repopulation of the state i by direct cascades. When the lifetimes of the cascading transitions are very different from the primary lifetime to be measured, it is generally possible to fit the decay curve by a multiexponential function. However, if this is not the case a more careful cascade analysis is needed. The most frequently used method is the so-called ANDC method (arbitrarily normalized decay curves). In this method, the decay curves of all direct cascades are experimentally measured. Another problem to handle in the extraction of lifetimes from a decay curve is to determine the beam velocity. The beam energy and the energy spread after the passage of the foil can be measured by an energy analyzer. In favorable cases, accuracies for lifetimes of the order of 1% have been achieved by the beam-foil method.

2.2.3 Spectroscopy of Ions The beam-foil light source has additional advantages compared to other, more conventional, light sources. First, the ion beams used are isotopically pure. Second, a specific beam energy can be chosen to optimize a specific charge state. In principle, all charge states of any element are presently available with accelerators in operation today. Third, the time-resolved character of the light source can also facilitate line identification. Spectroscopic studies using the beam-foil method have been performed all over the periodic table (Bashkin and Stoner, 1975, 1978, 1981). In particular, access to heavy-ion accelerators has recently made it possible to study relativistic interactions and QED effects in H-like and He-like uranium.

It was also recognized that the beam-foil interaction process can efficiently produce exotic multiply excited states that lie in the continuum of the atomic system (i.e., above the lowest ionization limit) and may decay by autoionization. However, there exist some multiply excited states that are not allowed to decay via autoionization because of specific selection rules. These states instead decay predominantly to lower states by radiative transitions. Much experimental and theoretical work has been devoted to the study of such exotic systems (see Mannervik, 1989). Some of the most spectacular findings include the optical emission lines from multiply excited states in the negative ions of lithium and beryllium (see also Sec. 2.6). The beam-foil method is also ideally suited to the study of spin- or angular-momentum–forbidden transitions—for instance, intercombination lines. "Forbidden" means in this context that the transition is much less likely to occur. For the higher members of an isoelectronic sequence, relativistic interactions become more pronounced, and hence spin-forbidden transitions become more likely. Thus, by observing time-delayed photon or electron spectra most prompt-decaying states will be suppressed. Using the beam-foil method, such time-delayed spectra are easily obtained by measuring the photon or electron emission from the projectile beam some distance behind the foil so that a certain time has elapsed after excitation (time-of-flight method).

2.2.4 Quantum Beats Unresolved multiplet transitions can in some cases be studied via quantum beat measurements. Since the excitation process is coherent, a superposition of fine-structure or hyperfine states is produced. The decay curve for such a superposition of states can exhibit interference patterns similar to that from a double slit where the beat frequencies correspond directly to the splitting of the levels. For some light elements, it has been possible to determine fine- and hyperfine-structure splittings with an accuracy even below 1%.

2.2.5 Molecular Effects Furthermore, experiments have been performed with primary beams consisting of molecular ions. As the ions pass the foil, electrons will be stripped off, and the ion fragments will repel each other by Coulomb interaction ("Coulomb explosion"). For example, energies and angular distributions of these fragments can be measured. The electrostatic potential energy stored in such molecular ions is of the order of eV, and because of the high velocity of the ion beam transforms to higher energies in the laboratory system (typically keV). These energies can be measured with high accuracy to study internuclear distances in molecular ions. In these types of experiments, the response of the solid also has been seen; in particular, a phenomenon called the "wake effect" has been observed (Hasselkamp *et al.*, 1992).

2.3 Laser Excitation

In Sec. 2.2, excitation of fast-ion beams by thin foils was discussed. Problems with that technique concern the nonselective excitation mechanism and also substantial energy loss and the energy and angular spread resulting from multiple collisions in the foil. A method to circumvent these problems is to excite the beam with laser photons directly. With laser excitation, much higher spectral resolution can be obtained (typically 10^{-3} Å) and for lifetime measurements, because of the selective excitation process, a cascade-free decay is observed. Besides the spectroscopic applications, laser excitation of fast-ion beams can be used for other purposes. For example, excitation of an ion beam with a laser is an excellent tool to determine accurately beam velocities and velocity distributions of the ions. As a laser photon is absorbed by an ion, momentum is transferred to the projectile. If this process is repeated many times, the velocity distribution of ions can be changed (laser cooling). This method has been used to narrow velocity distributions in fast- and slow-ion beams and ion storage rings (Sec. 2.10.5). Review articles about beam-laser excitation and applications have been published by, e.g., Andrä (1979) and Neugart (1987). Laser photoionization spectroscopy and lasers in atomic, molecular, and nuclear physics are treated in the book by Letokov (1989).

2.3.1 Atomic and Molecular Spectroscopy

For excitation with a narrow-bandwidth single-mode laser, the spectroscopic resolution is limited by the ion-beam properties and geometry. The linewidth $\Delta\nu$ is determined by the relation (including first-order terms)

$$\Delta\nu = \frac{v}{c}\,\nu(\Delta\theta + \Delta\phi)\sin\theta + \frac{\Delta v}{c}\,\nu\cos\theta + \frac{v}{\pi\Delta x}\sin\theta, \tag{3}$$

where v is the ion velocity and c is the speed of light. The angle θ and the angular divergences $\Delta\theta$ and $\Delta\phi$ are illustrated in Fig. 3. From Eq. (3), it is seen that for the collinear geometry ($\theta = 180°$) all terms except the second will become zero. For a fast beam, there is an effect called "velocity compression" that reduces Δv. Thus, for a certain energy spread ΔE in the ion source, and corresponding velocity spread Δv_i, the resulting velocity spread Δv_a after acceleration is reduced, i.e., $\Delta v_a \ll \Delta v_i$. Therefore, in the longitudinal direction an accelerated beam will appear as a very cold ion ensemble. For the collinear geometry, the resulting Doppler shift

$$\nu_0 - \nu = \frac{v}{c}\,\nu_0\cos\theta \tag{4}$$

will reach its maximum. Such shifts can, however, be corrected for if the ion velocity is accurately known. If the laser resonance frequency is measured both for parallel (ν_+) and antiparallel (ν_-) laser beams, an accurate measurement of the velocity can be per-

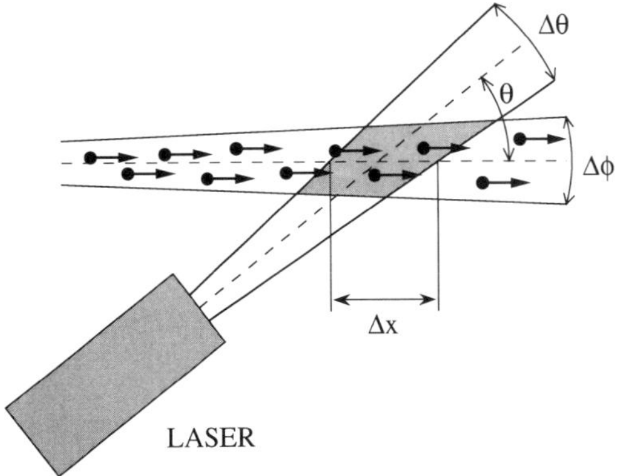

FIG. 3. Schematic illustration of typical ion beam–laser beam interaction region. $\Delta\phi$, $\Delta\theta$, and Δx are the ion-beam divergence, laser-beam divergence, and interaction length, respectively.

formed according to

$$v = c(\nu_+ - \nu_-)/(\nu_+ + \nu_-). \tag{5}$$

Specifically, the collinear geometry has been used to perform precision measurements of the Lamb shift for one- and two-electron systems, of hyperfine structure splitting of atoms and ions, and of vibrational–rotational structure of molecular species (Sec. 2.4). A typical experimental setup is shown in Fig. 4, where the excitation region can be shifted through a local change of the ion velocity to maximize the fluorescence signal at the detector, which usually consists of a broad-band monochromator or an interference filter in conjunction with a photomultiplier.

2.3.2 Lifetime Measurements If the excitation region is made very narrow when compared to the mean decay length of a specific projectile state, cascade-free lifetime measurements can be performed. This is usually accomplished in a crossed-beam geometry. To enhance the laser power and interaction time, intracavity excitation is often used; i.e., the ion beam crosses the laser beam many times inside the laser cavity. By rapid Doppler shifting, it is also possible to restrict the excitation region to a narrow region even for a collinear geometry.

2.3.3 Photofragment Spectroscopy Laser spectroscopy of fast molecular ions is also an important field. A very attractive method is to excite the molecular ion to a dissociative state. The fragments can then be collected with a very high efficiency, since they travel in the forward direction and can be separated by an electrostatic analyzer and collected by a particle detector (see Fig. 4). A fast beam will act as an amplifier and transform a small released energy to an energy that is relatively easy to measure. For a diatomic molecular ion, the fragment will, in the forward direction, have a kinetic energy

$$E_f = \tfrac{1}{2}E_0 \pm \sqrt{E_0 W} + \tfrac{1}{2}W, \tag{6}$$

where E_0 is the initial energy and W is the energy release in the dissociation process. For excitation in a collinear geometry, the effect of velocity compression discussed above (Sec. 2.3.1) corresponds to spectroscopy of very cold

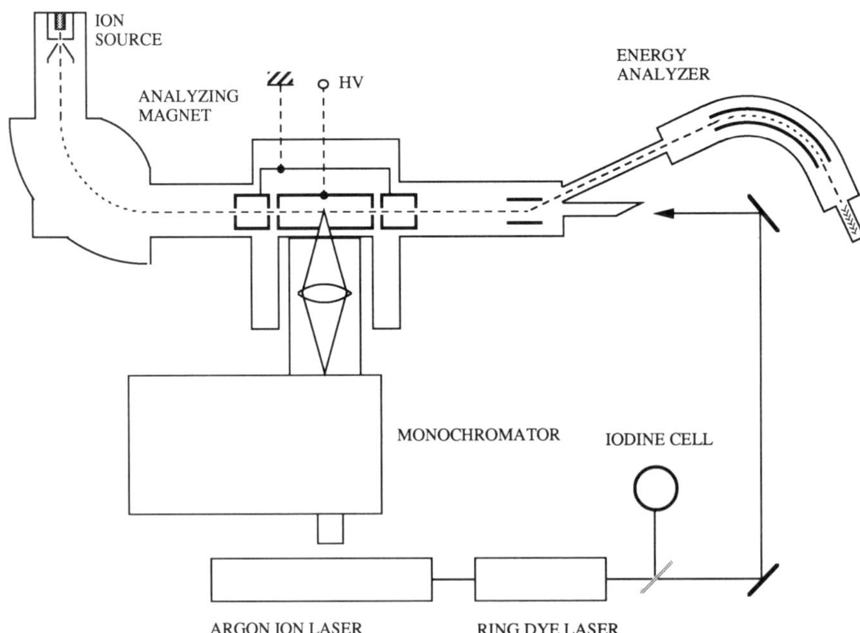

FIG. 4. Experimental setup for collinear ion beam–laser spectroscopy, lifetime measurements, and molecular-ion photofragment spectroscopy. Photoproduced fragment ions are deflected into an electrostatic energy analyzer. Moreover, fluorescence photons can be detected by an optical monochromator, equipped with a photomultiplier.

ions with small Doppler width (sub-Doppler spectroscopy). In Sec. 2.4, the application of this technique is discussed.

2.4 Spectroscopy and Dynamics of Molecular Ions

Molecular ions are major chemical agents in many crucial processes that occur in physics, chemistry, astronomy, and biology. Despite this fundamental importance, the structure and properties of neutral and charged molecules in the gas phase were for a long time poorly understood, and the rich chemistry and physics of molecular ions were the almost exclusive domain of mass spectroscopists. While mass spectroscopy can give very useful information about the existence and stability of molecular ions, it provides little information about their structures. In recent years, a number of more sensitive techniques to examine molecular structure have been developed and applied to these species. Most of these techniques involve the absorption or emission of photons or the emission of electrons and fall into the general area of spectroscopy. There has also been a rapid development in *ab initio* quantum mechanical methods during the last decades, which, in combination with the enormous enhancement in computer power, has led to a close interplay between experiment and theory. Several books and reviews that give in-depth coverage of the general area of spectroscopy and dynamics of molecular ions have been published (Miller and Bondybey, 1983; Gudeman and Saykally, 1984; Maier, 1989; Müller-Dethlefs and Schlag, 1991).

2.4.1 Experimental Techniques Molecular spectroscopy is traditionally and naturally divided into electronic, vibrational, and rotational spectroscopy. Vibrational and rotational spectroscopy are the more direct routes to molecular structure, while electronic spectroscopy also reveals information about excited electronic states. Most simple stable molecules have been studied by means of these spectroscopies. Two difficulties arise when these methods are applied to molecular ions. First, ions are generally, because of their high reactivity, present in very low concentrations. This means that highly sensitive techniques are required to detect them. Second, neutral molecules are usually present among the ions in much larger abundance. Hence, neutral absorption or emission spectra usually overwhelm those from the charged molecules.

The first obstacle discussed above can be overcome by using lasers as light sources (see Sec. 2.3) in absorption experiments. In the infrared region (1–100 μm), a variety of systems are in use, such as difference-frequency laser systems, color-center lasers, diode lasers, and CO_2 lasers. Spectroscopic transitions occurring in the infrared region are generally those between vibrational levels within a given electronic state (vibrational spectroscopy). In the visible and UV regions, dye lasers have dominated for a long time but have in recent years been replaced by solid-state lasers in some wavelength regions (most notably titanium sapphire for the 700–1100-nm region). Spectroscopic transitions occurring in the visible and UV regions are generally those between different electronic states in a molecule (electronic spectroscopy). Spectroscopy in the microwave region (~4–600 GHz), which is not laser based, has advanced by the availability of monochromatic sources. Spectroscopic transitions occurring in the microwave region are generally those between rotational levels of a given vibronic state of a molecule. Microwave spectroscopy in general fits into the conceptual framework of a classic absorption experiment, with a source, a cell, and a detector. Spectroscopy in the other wavelength regions has most often employed indirect detection schemes in order to increase the sensitivity (see below).

The second obstacle (i.e., that of higher neutral abundance) can be overcome by using ion-beam techniques (Miller and Bondybey, 1983). Figure 4 shows an example of an experimental setup for ion-beam laser spectroscopy. The ions are generally produced by electron-impact ionization/fragmentation of a neutral molecule, extracted by applying a 1–200-kV (in some cases, even as low as 10–230 eV) potential to the ion source, and are collimated into a narrow beam by a series of electrostatic lenses. The advantages with the resultant beam are that unambiguous identification of the ions in the beam can be made by the introduction of a mass analyzer (magnetic or quadrupole) in the beam line, an efficient interaction with the laser beam can be obtained by a collinear arrangement, frequency scanning may be achieved with a fixed-frequency laser line by sweeping the source potential, and, finally, sub-Doppler resolution is obtainable as a result of an effect known as kinematic compression or velocity bunching (see Sec. 2.3). The disadvantage with a fast-ion beam is that ion densities are low, typically 10^4 to 10^6 ions/cm^3. This means that the direct observation of the attenuation of the laser beam by the ion beam is very difficult, although it has been demonstrated in a few cases in the infrared region. In general, however, it is necessary to use indirect methods to detect the absorption of radiation. One such method, which is based on the occurrence of fragment ions as a result of photon absorption, will be discussed in Sec. 2.5; it suffices to mention here that other indirect schemes involve detection of charge transfer that is vibrational-state dependent, and detection of laser-induced fluorescence perpendicular to the ion beam.

Another way to enhance spectra of ions at the expense of neutrals is to pass a laser beam through a cell in which an alternating-current plasma is maintained. The molecular ions are accelerated back and forth by the alternating electric field, thus frequency modulating the absorption line through the Doppler effect. Neutral molecules are essentially unaffected by the electric field and do not exhibit observable Doppler shifts in their absorption spectra. A further advantage with this so-called velocity modulation technique is that ion densities in the plasma are much higher than in ion beams, of the order of 10^{13} ions/cm^3. The technique was developed at Berkeley around 1982–83 (Gudeman and Saykally, 1984) and has yielded information on more than 60 molecular ions (both positively and negatively charged), mainly through studies in the infrared region.

A technique that involves absorption of radiation by the neutral precursor while still revealing information on the molecular ion is photoelectron spectroscopy. This technique was developed already in the 1950s and must now be considered to be a mature field as far as stable neutral molecule precursors are concerned. Briefly, radiation in the VUV or x-ray regions is used to ionize a neutral molecule. The kinetic energy (E_k) of the electron is measured, and from the relation $E_k = h\nu - I$, where $h\nu$ is the photon energy and I is the binding energy of the electron, information on I is obtained. Compared to the techniques described previously, which allow the rotational structure to be resolved, photoelectron spectroscopy in its traditional use is a low-resolution technique (10 meV, or about 100 cm^{-1}). Recently, a technique has been developed (Müller-Dethlefs and Schlag, 1991) that allows a three–order-of-magnitude increase in resolution. The new method is based on the idea that ionization is carried out so that electrons with $E_k = 0$ are created in a field-free region. After a time delay of some μs, the photoelectrons are extracted, and the zero–kinetic-energy (ZEKE) electrons are separated from near-ZEKE electrons by a time-of-flight device. Lasers are used for the initial ionization step.

2.4.2 Structure of Molecular Ions

Structural information derived from vibrational and rotational spectroscopy involves equilibrium rotational constants and corresponding equilibrium moments of inertia, bond angles, vibrational frequencies, etc., which may in some cases lead to a construction of the ground-state potential energy surface. Electronic spectroscopy also contributes to information on the ground state; in addition, it gives information on excited electronic states.

Around 1970, about 30 diatomic ions and 6 polyatomic ions had been studied by high-resolution spectroscopic methods, where high resolution means that the rotational structure is resolved. With the exception of H_2^+, these studies had been carried out with absorption or emission techniques in the UV–visible wavelength region. Today, the number of molecular ions studied in the infrared and microwave regions exceeds 60 and 20, respectively. A large number of ions have also been studied in the UV–visible region during the last 20 years. It is not possible to discuss all these molecules here, and only some characteristic examples will be given.

The infrared absorption spectrum of the simplest triatomic positive ion, H_3^+, was discovered in 1980 by Oka (see Miller and Bondybey, 1983). The difficulties involved in the search for spectra of molecular ions are well illustrated by the fact that the existence of H_3^+ was reported already in 1912. The molecular ion H_3^+ is, because of its relative simplicity, suitable for theoretical treatments. It plays a key role in the chemistry that takes place in the interstellar molecular clouds.

One of the most fundamental processes in chemistry and biology is the proton transfer of liquid water to form a hydronium ion (H_3O^+) and a hydroxide ion (OH^-). The characterization of these ions in the gas phase was therefore a goal for high-resolution spectroscopists for a long time. It was, however, not until the development of the velocity modulation technique that high-resolution spectra of H_3O^+ and OH^- were observed.

The C_2^+ fragment ion has no chemically stable neutral precursor, is the smallest carbon cluster, and is believed to occupy an important role in many terrestrial and extraterrestrial environments. Fragment ions in general have mainly been identified by mass spectrometry. A spectrum of C_2^+ in the visible-wavelength region was recently observed and analyzed (Maier, 1989) by a combination of low-resolution absorption spectroscopy of C_2^+ frozen into a Ne-crystal matrix and high-resolution laser excitation spectroscopy.

2.4.3 Dynamics of Molecular Ions The subject of the dynamics of molecular ions is a vast field because their reaction and scattering processes can be very complex. The close examination of the individual steps of gas-phase reaction processes, i.e., elementary reactions, has resulted in the detection of a large number of short-lived molecules as reaction intermediates. Identification of these species is a prerequisite for understanding the reaction mechanism. The high-resolution spectra of reaction intermediates are very valuable in examining, in real time, distributions over quantum states of molecules taking part in reactions. A typical environment where molecular ions act as reaction intermediates is a laboratory plasma. The ions are initially produced by ionization from electron bombardment and are destroyed by recombination. If the ions are produced in a pulsed discharge, one can measure the infrared absorption signal as a function of time and thereby monitor the population of particular vibrational-rotational states of the ion. If a molecular ion recombines with an electron, a common scenario is $AB^+ + e \rightarrow A + B$, i.e., dissociative recombination. If the laser is tuned to a specific transition in AB^+, the signal will then decrease as a function of time and reflect the rate of destruction of AB^+. It should be mentioned that there are several other techniques suited for studies of dissociative recombination—for example, merged ion-electron beam and flowing afterglow techniques.

Photoionization of a neutral molecule can be regarded as a half collision between an electron and the molecular ion core. The specific features of the ZEKE method make it particularly suited to study the short-range interaction between the photoejected electron and the ion core, the angular distribution of the ejected electron, and coupling of electronic and nuclear motions. There is a clear analogy between photoionization and photodissociation; the latter process is the subject of the next section.

2.5 Photodissociation and Half Collisions of Molecular Ions

Photodissociation of neutral molecules is an important subfield of molecular reaction dynamics (i.e., the study of the molecular mechanism of elementary physical and chemical rate processes). A typical question posed in studies of the dynamics of photodissociation is: What is the fate of a molecule after it has absorbed a photon whose energy exceeds the threshold for dissociation of one or more bonds? The experimental techniques employed for these studies in general involve a neutral molecular beam, a pulsed laser for photodissociation, and a second pulsed laser for probing the internal-state distribution of one of the photofragments.

Photodissociation of molecular ions differs from that of neutral molecules both when it comes to experimental techniques and in terms of information extracted from the experiments (Miller and Bondybey, 1983; Maier, 1989). The technique used to detect photodissociation of molecular ions has already been discussed in Sec. 2.4.1 and is known as ion photofragment spectroscopy (IPS). Figure 5 shows two examples of how photon absorption may induce a bond-breaking process in a molecule. The potential curves, which are the potentials in which the nuclei move, are for a fictive diatomic molecule, but the arguments can easily be generalized to polyatomic molecules. In Fig. 5(a), it is shown how the molecule is promoted by a photon from its electronic ground state to a repulsive state. Since this repulsive state does not support any bound vibrational levels, the molecule dissociates on a time scale of 10^{-13} s. If the cross

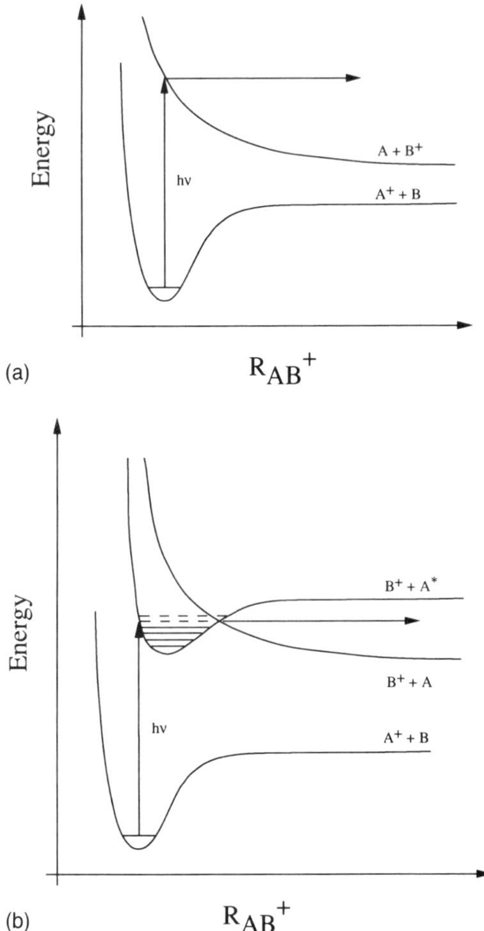

FIG. 5. Schematic potential curves for a diatomic molecule. (a) The process of direct photodissociation where the molecule is promoted by a photon from its ground to a repulsive state, which leads to a breaking of the molecular bond. (b) An indirect electronic predissociation process. The molecule is excited to a bound state that interacts with a repulsive state. This leads to a nonradiative transition to the repulsive state, and the molecule dissociates.

section for photodissociation is measured as a function of photon wavelength, there will be only a slow variation and no structure. The wavelength range over which the molecule photodissociates will depend on the initial state distribution. In Fig. 5(b), the molecule is excited into a bound state that interacts with a repulsive state. This interaction leads to a rearrangement of electronic energy from the bound state to the repulsive state, and the molecule dissociates. The rate-determining step is in this case the nonradiative transition from the bound to the repulsive state (predissociation). Photodissociation only occurs when the laser is tuned into resonance with one of the discrete lines of the transition between the two bound states. The photodissociation cross section as a function of laser wavelength will be structured. Thus, for the molecular spectroscopist IPS serves as a highly sensitive tool to study absorption spectra of molecular ions, often at a spectral resolution of the natural width of the excited state. For those interested in slow-ion–atom collisions, IPS allows the study of half collisions in a well-defined system of reference, exploring the dynamic paths that connect a precisely defined molecular system with an equally precisely defined continuum of separated states.

A photofragment spectroscopy experiment can briefly be described as follows. A highly collimated, mass-selected, fast-ion beam is irradiated coaxially by a single-frequency laser. A transition from a molecular level populated in the ion beam to a level lying above the dissociation limit is induced by tuning the laser (or Doppler tuning the ion beam) into resonance. Provided that the excited level interacts with a continuum state, dissociation will occur, and the photoinduced ionic or neutral fragments can be detected, often with nearly unity detection efficiency. The laser wavelength provides information about the transition frequency, the measured linewidth reveals the upper-level lifetime toward predissociation, and measurement of the released kinetic energy imparted to the fragments reveals the location of the excited level with respect to the dissociation limit. If the transition takes place directly to a repulsive state, measurement of the angular distribution of fragment ions gives information on what type of electronic transition is involved and the product-state distributions. In this latter case, it is necessary to irradiate the ion beam perpendicular, so that the plane of polarization of the laser beam can be varied. An impressive number of molecular ions have been investigated by means of IPS.

2.6 Photodetachment

The process whereby a negative ion loses an electron through the absorption of a photon is known as photodetachment (see Table

2). The process may be symbolized as

$$A^- + h\nu \rightarrow A + e^-, \tag{7}$$

where the products are an atom, which may be in an excited state, and an electron with a kinetic energy sufficient to satisfy energy conservation. The interaction of a photon and a negative ion has been used to determine the binding energy of the attached electron and to study the initial and final states of the photodetachment process as well as the dynamics of the emission process.

The pioneering work in photodetachment was performed in the 1950s and early 1960s (McDaniel *et al.*, 1993). A crossed-beams apparatus, using an arc lamp as the light source, was used to measure absolute photodetachment cross sections. The advent of the laser as a light source for photodetachment studies (see Sec. 2.3), with the inherent increase in photon flux, accelerated the experimental investigation of negative ion properties. In particular, Lineberger's group at the Joint Institute for Laboratory Astrophysics investigated a large fraction of the negative-ion systems during the 1970s using single-line and tunable lasers. Most of these experiments were designed to measure electron affinities and determine relative cross sections for photodetachment. Hotop and Lineberger (1985) discuss electron affinities for elements up to $Z = 85$. The recommended electron affinities of the noble gases and group II*A* elements have negative values; i.e., stable negative ions are not formed in these groups. The formation of a stable negative ion in these groups would involve the attachment of an electron to the core of the parent atom. Calculations had supported the general belief that under such circumstances, the enhanced screening of the nuclear charge by the atomic electrons would prohibit stable negative-ion formation (Massey, 1976; Smirnov, 1982). Recently, a crossed-beams laser-photodetachment electron spectroscopy technique (LPES) was used by Pegg and coworkers (1987) to determine that Ca^- is a stable negative ion.

Photodetachment of the simplest negative ion, H^-, has been studied in detail by Halka *et al.* (1991) using a unique experimental technique. In these experiments, a relativistic H^- beam from the Los Alamos Meson Physics Facility is intersected by a fixed-frequency laser. The Doppler effect allows a systematic variation (factor of 10) of the center-of-mass photon energy, simply by adjusting the laboratory intersection angle between the two beams. This technique has been used to study doubly excited resonances of H^-. Relative cross sections for multiphoton detachment above the one-electron detachment threshold have also been measured.

2.7 Delta-Electron and Other Secondary Emission Processes in Fast Heavy-Ion–Atom Collisions

2.7.1 Introduction When an ion of high velocity interacts with a target atom, one of the possible interaction processes is the ejection of a free electron. Secondary-electron emission has been studied for a large variety of ion-atom collision systems. Here we summarize measurements involving projectiles at energies larger than 3 MeV/u and incident ions of high charge state up to uranium incident primarily on He and Ar gas targets. These recent experiments have been performed at the Lawrence Berkeley Laboratory SUPERHILAC accelerator.

For fast light-ion impact, the electron emission is primarily determined by the potential of the target atom, and the ionization is normally well described by the first Born approximation (see Sec. 3). For fast heavy-ion impact, however, it has been demonstrated that the doubly differential electron emission cross sections (DDCS) are not generally in good agreement with this approximation. Therefore, investigations of the electron emission probabilities for heavy-ion projectiles provide a sensitive measure of the dynamics of the ionization process during the collision. It has been shown that in fast collisions using highly charged projectiles, particularly at high electron emission energies, at forward and backward angles, higher-order scattering effects are important where both target and projectile fields contribute to the ionization process. In such cases, the "single-center emission" concept fails and has to be replaced by a "two-center emission" picture. In other words, the high-energy electrons are affected strongly by the target as well as the projectile potentials. Such studies and the determination of reliable absolute cross sections are relevant to plasma physics, fusion research, astrophysics, biophysical tissues (energy deposition in tissues), and basic collision theory.

2.7.2 Experimental Procedure The experimental method of electron-emission cross-section measurements has been pioneered by Rudd, Stolterfoht, and Toburen (Rudd and Macek, 1974; Briggs and Macek, 1991). For a more recent discussion, the reader is referred to the review by Toburen (1991). A characteristic setup for DDCS measurements for secondary-electron observation between a few eV and about 8 keV energies is shown in Fig. 6. In these experiments, a tightly collimated ion beam interacts with a gas target jet in the center of a magnetically shielded scattering chamber. The secondary-electron emission is observed at different observation angles using an electrostatic electron analyzer that is rotated in a plane around the scattering center.

Absolute electron-emission cross sections can be determined for the observation angles ranging from approximately 27° to 160°. The target density in the gas jet is typically of the order of 10^{14} atoms/cm^3 over a target length of about 3 mm. This low density ensures single-collision conditions. The analyzer typically has an energy resolution of a few percent [full width at half maximum (FWHM)] and a solid angle of about 4×10^{-3} steradians. The accurate target density is obtained by comparison of electron-yield measurements with the gas target jet in place and with the jet removed (homogeneous target). Furthermore, the electron detector is calibrated with electron beams at different electron energies.

2.7.3 Double-Differential Cross Sections A comparison of theoretical and experimental DDCS for 3.5-MeV/u Fe^{22+} + He is shown in Fig. 7 for $\theta = 27°$ and 155°. In this figure, the DDCS are multiplied by the electron energy for forward angles, which makes the specific differences in the cross-section behavior more visible. For example, the DDCS for Fe^{22+} + He indicate a maximum below an electron energy of 10 eV and fall off by several orders of magnitude toward the high-energy limit at electron energies of about 8 keV. At high electron energies and at forward angles, a broad characteristic peaked structure, the "binary encounter peak" (BEP), resulting from electron emission following collisions with maximum momentum transfer, dominates the cross section. The BEP energy is strongly dependent on the electron observation angle θ and is given by

$$E(\text{BEP}) = 4t_p \cos^2\theta, \qquad (8)$$

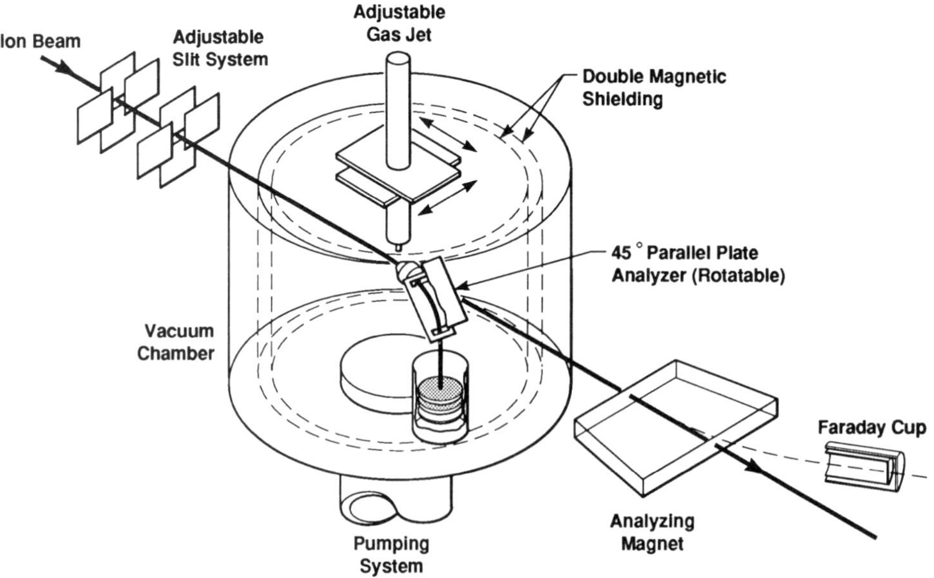

FIG. 6. Experimental setup for double-differential electron-emission cross-section measurements, consisting of ion-beam collimation system, target chamber with gas jet and electron spectrometer, and Faraday cup for charge normalization.

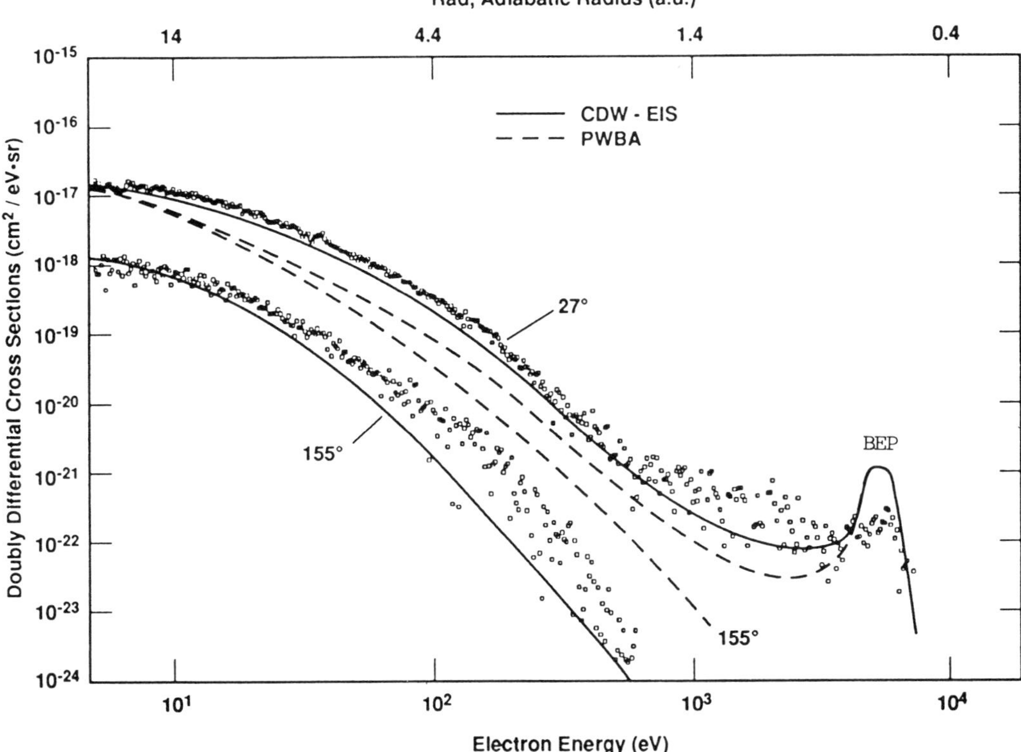

FIG. 7. Comparison of calculated double-differential electron-emission cross sections (PWBA and CDW-EIS) with experimental data for 3.5 MeV/u Fe^{22+} on He collisions. The upper scale represents the approximate impact parameter dependence. PWBA represents the plane-wave Born approximation and CDW-EIS the continuum distorted-wave eikonal initial-state calculation (Schneider *et al.*, 1992).

with

$$t_p = (m_e/M_p)E_p \tag{9}$$

the reduced projectile energy. Here m_e, M_p, and E_p are the electron mass, projectile mass, and energy, respectively. Another broader electron structure, which may occur at the reduced beam energy t_p, is the "electron loss peak" (ELP). This structure, which is not a predominant peak in the present case, is independent of the electron observation angle. As can be seen from Fig. 7, the intensity of the BEP contributions decreases with increasing observation angles. Continuum electrons centered at the projectile (ELP) and the target (BEP) may overlap depending on the projectile, the angle of observation, the projectile charge state, and target atom used. Measurements with heavier projectiles like U^{q+} show interference effects in the BEP, which have been related to the elastic differential cross section for scattering of target electrons from the impinging ion, similar to the phenomenon of rainbow scattering.

2.8 Auger-Electron Spectroscopy Following Ion-Atom and Ion-Molecule Collisions

Projectile ion beams colliding with multielectron target atoms, molecules, thin foils, or surfaces also give rise to the emission of Auger or autoionization electrons. These discrete line structures are superimposed on the continuous electron background. Auger-electron spectroscopy (*q.v.*) has become a well-established spectroscopic method (Sevier, 1972) for studying the structure and collision dynamics of atoms, molecules, and ions (Mehlhorn, 1985).

2.8.1 Forward-Angle Projectile Auger-Electron Spectroscopy The availability of intense, highly charged ion beams makes possible the production of inner-shell excitation and/or ionization at high velocities and multielectron capture processes at low velocities. Such highly excited states subsequently decay by two competing processes, radiative and radiationless via Auger decay or autoionization to one or more adjacent continuum states. The later Auger processes result in the ejection of free electrons with well-defined kinetic energies depending on the excitation energies of the initial and final ionic states. In the limit of nonrelativistic theory, the Auger decay is caused by the electron-electron interaction, whereas in the relativistic regime, spin-induced transitions also may occur. The identification and classification of such highly excited states has been one of the challenges in atomic physics over the last decade. These states are of fundamental importance for the understanding of high-temperature laboratory plasmas and electron correlation effects. They also provide important test cases to probe relativistic interactions. If Auger electrons are emitted from a moving emitter frame (projectile), they exhibit kinematic effects; i.e., Auger lines are modified in energy position, width, and intensity (Rødbro *et al.*, 1979; Stolterfoht, 1987). For example, the Auger-line energy position E in the laboratory frame is transformed to the center-of-mass (source particle) energy E' by

$$E' = E + t_p - 2\sqrt{t_p E}\cos\theta, \qquad (10)$$

where θ is the observation angle in the laboratory frame. When the ion velocity increases, the projectile electrons are strongly affected by kinematic broadening effects originating mainly from the variation of the observation angle θ and the finite angular spread $\Delta\theta$ of the electron analyzer. Such broadening effects can be reduced when the electrons are observed at forward angles, especially zero degrees.

Recently, the zero-degree electron spectroscopy method (Stolterfoht, 1987) has been used to study high-resolution Auger spectra of a variety of multiply charged ions. In these measurements, electrons emitted at 0° and 180° can be observed with energies E' as low as 0.1 eV. These low-energy line structures are centered around the electron cusp peak, which has a maximum at t_p [see Eq. (9)].

2.8.2 Auger-Electron Angular Distributions The measurement of magnetic substate specific alignment and orientation parameters in ion-atom and ion-molecule collisions can reveal more detailed information on dynamical effects that is not directly accessible via total cross-section measurements (Kleinpoppen and Williams, 1980; Andersen *et al.*, 1988). Thus, close collisions between fast-moving ions and target atoms or molecules may result in anisotropic "aligned" population of excited projectile states with inner-shell vacancies. One technique to study such anisotropies is by means of Auger-electron angular distributions (Bisgaard *et al.*, 1980). An interesting case is represented by the projectile Auger electron spectra for Be^+ + He and Be^+ + CH_4 collisions. The Be^+ ions were obtained from the 600-keV ion accelerator of the University of Aarhus, Denmark, and directed through a He or CH_4 gas target. The ejected projectile Auger electrons were analyzed by an electrostatic parallel-plate spectrometer, which was continuously rotated between 0° and 150°. The Be^+ $(1s2s^2)^2S \rightarrow (1s^2\epsilon s)^2S$ Auger transition is isotropic and was used for normalization purposes. In order to obtain the cross sections in the projectile emitter frame, it was necessary to correct for the Doppler anisotropy. Representative results of Be^+ $1s(2s2p\ ^3P^\circ)^2P^\circ$ and $1s(2s2p\ ^1P^\circ)^2P^\circ$ angular distributions are shown in Fig. 8. The anisotropy coefficient A_2 was found by fitting

$$I(\theta_0) = C(1 + A_2 P_2 \cos\theta_0) \qquad (11)$$

to the angular distribution where θ_0 is the emission angle in the projectile emitter frame. The large positive alignment parameter A_2 corresponds to an overpopulation of $M_L = 0$ magnetic substates, when compared to $M_L = \pm 1$ states. Large anisotropies have also been observed recently in $C^{5+}(1s)$ + $He(1s^2) \rightarrow C^{3+}(1s2l2l)^2L$ + He^+ double electron-capture processes at low collision velocities by Prior and co-workers at the Lawrence Berkeley Laboratory.

2.9 Photon Emission Processes in Fast Projectile Collisions

2.9.1 Multielectron Transitions Many-electron processes like double ionization, double excitation, and ionization-excitation,

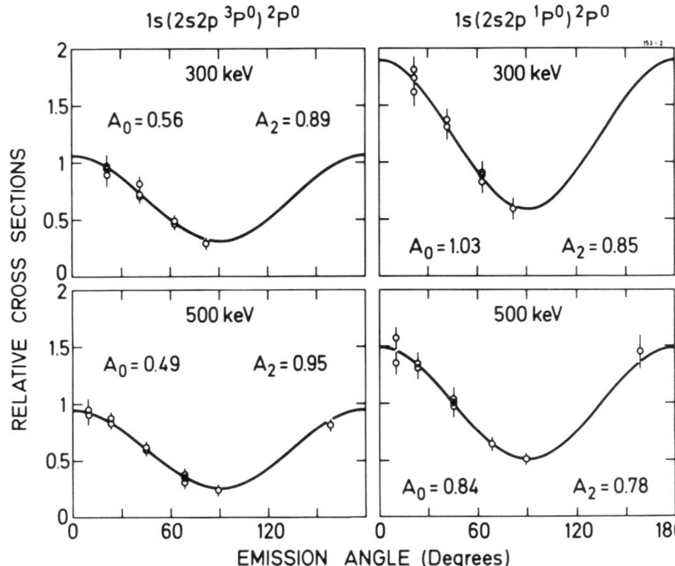

FIG. 8. Angular distributions of ejected projectile Auger electrons from Be^+ $1s(2s2p\ ^3P^{\circ})\ ^2P^{\circ}$ and $1s(2s2p\ ^1P^{\circ})\ ^2P^{\circ}$ core-excited states excited in $Be^+ \rightarrow$ He ion-atom collisions.

which occur in collisions of fast charged particles with few-electron targets, have received much attention recently (McGuire, 1992). Such two-electron processes are of fundamental importance to an understanding of the dynamics of few-body atomic collisions. Specifically, the measurements of absolute extreme-ultraviolet (EUV) photon and electron emission cross sections may allow development of a more comprehensive formulation of few-body atomic processes. Some important applications in which those few-electron transitions occur are plasma dynamics and astrophysical processes including solar atmospheric physics. For example, a challenging problem in astrophysics is the transformation of stored electromagnetic energy and particle emission in solar flares, which is still not well understood. Also, electron correlation occupies a fundamental role in inner-shell photoionization of atoms and molecules (Amusia, 1990). Furthermore, these many-body effects are relevant to plasma diagnostics, x-ray laser research, chemical physics, and quantum biology of simple and complex systems.

2.9.2 EUV Spectroscopy State-selective cross-section measurements from excited target atoms can be studied by using the method of high-resolution EUV spectroscopy. In these measurements, a positively or negatively charged or neutral particle beam enters a differentially pumped gas target or intercepts a gas jet. After exciting the target gas, the beam is collected in a Faraday cup for the purpose of charge normalization. A schematic view of a gas target and EUV monochromator is shown in Fig. 9. The photons emitted by the target cell are analyzed by a grazing-incidence monochromator and counted by a Channeltron electron detector. All measurements are performed under single-collision conditions. The gas pressure in the cell is accurately monitored with a capacitance manometer and kept constant with a feedback control system. Typical ion-beam currents in the target area are 1–20 μA in the energy range of 50 keV to 2 MeV. Identical excitation and detection geometries are used for electron and ion impact. The gas pressure used in these experiments typically ranges from about 0.1 to 100 mTorr.

Characteristic spectra of EUV Lyman series of He^+ are shown in Fig. 10 for both electron and proton impact at equal velocities. The Lyman series is clearly resolved up to $n = 5$. To derive excitation cross sections from measured line intensities, the following formula can be used:

$$\sigma = \frac{4\pi}{\omega}\frac{qe}{Q}\frac{I(\lambda)}{k(\lambda)BNL}, \tag{12}$$

where e is the electron charge, q the charge

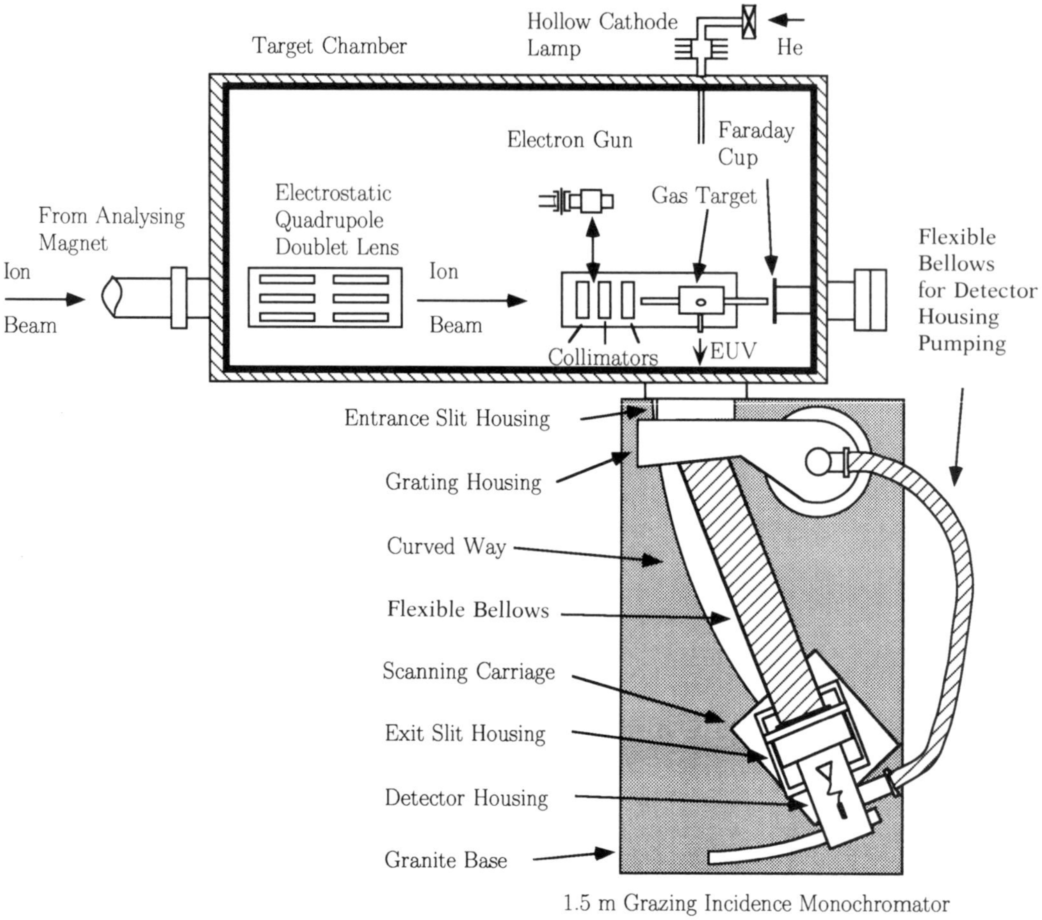

Target Chamber
Hollow Cathode Lamp
He
Electron Gun
Faraday Cup
Electrostatic Quadrupole Doublet Lens
Gas Target
From Analysing Magnet
Ion Beam
Ion Beam
Flexible Bellows for Detector Housing Pumping
Collimators
EUV
Entrance Slit Housing
Grating Housing
Curved Way
Flexible Bellows
Scanning Carriage
Exit Slit Housing
Detector Housing
Granite Base
1.5 m Grazing Incidence Monochromator

FIG. 9. Target chamber for EUV emission spectroscopy with gas cell, electron gun, and grazing-incidence monochromator.

163 eV e⁻, 50 mTorr, 100 μC
(1s−∞p)
(1s−5p)
(1s−4p)
(1s−3p)
(1s−2p)
0.3 MeV H⁺, 50 mTorr, 500 μC
Relative Intensity
1000
100
10
1
22
23
24
25
26
27
28
29
30
31
Wavelength [nm]

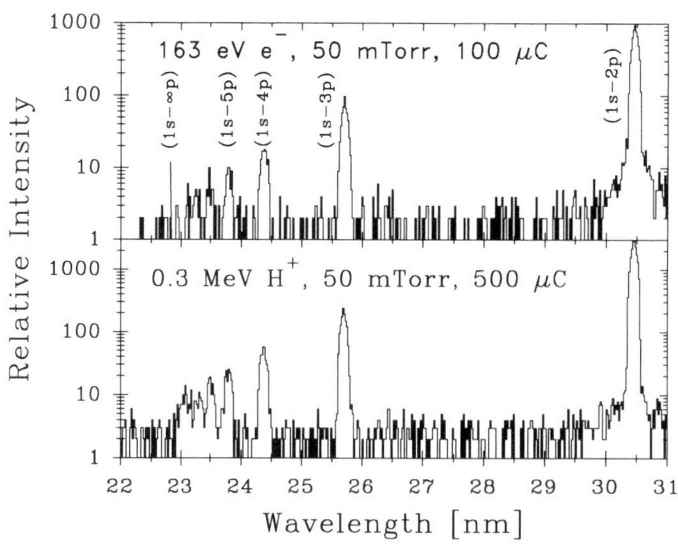

FIG. 10. EUV emission spectrum of the Lyman series of He^+ $np \rightarrow 1s$ for (top) 163-eV electron impact on He and (bottom) 300-keV proton impact on He at equal projectile velocities. The data were generated at the University of Nevada, Reno, 2-MV Van de Graaff accelerator facility.

state of the projectile, Q the integrated beam charge or charge normalization, $I(\lambda)$ the measured line intensity, $K(\lambda)$ the sensitivity of the detection system, B the branching ratio associated with a specific transition, N the target number density, ω the solid angle of acceptance for the monochromator, and L the observation length (the reaction region visible to the monochromator).

2.9.3 Ionization-Excitation of He In particular, ionization plus excitation is a fundamental process relevant to the testing of basic scattering theories. Helium is the simplest many-electron atom and therefore ideally suited for achieving a better understanding of two-electron processes. According to McGuire (1992), cross terms between first-order and second-order perturbation-theory contributions may be responsible for the observed large differences in ionization-excitation cross sections at equal velocities for protons and electrons (see Fig. 11). In this figure, absolute $He^+(2p)$ cross sections are plotted for H^+ + He and e^- + He as a function of the projectile velocity. Higher np states have also been measured (Fuelling *et al.*, 1992). The following observations can be made from the plotted cross section data: For velocities v/v_0 above 4, the e^- + He cross sections lie consistently above the corresponding proton results. Two basic reaction mechanisms contribute to the observed cross sections, namely few-electron processes where the projectile interacts only once with the target, and a second Born term where the projectile interacts twice with the target in a single collision event. Such processes can be distinguished by their Z_p dependence, where Z_p is the projectile charge. Furthermore, if these first- and second-order scattering amplitudes add coherently, a Z_p^3-dependent term appears in the cross section, which varies as the square of the total amplitude. Hence, cross-section differences for electrons ($Z_p = -1$) and protons ($Z_p = +1$) at equal velocities may provide direct evidence of first- and second-order effects. Such Z_p^3 cross terms occupy also a significant role in double ionization and double excitation of He. These two-electron transitions have been studied experimentally and theoretically for double ionization of He by antiprotons ($\bar{p}$) and protons (p) as well as electron (e^-) and positron (e^+) impact on He (Andersen, 1988). Moreover, measurements have been made for double excitation of He as a function of the projectile charge, mass, and velocity (Giese *et al.*, 1990; Wang *et al.*, 1993).

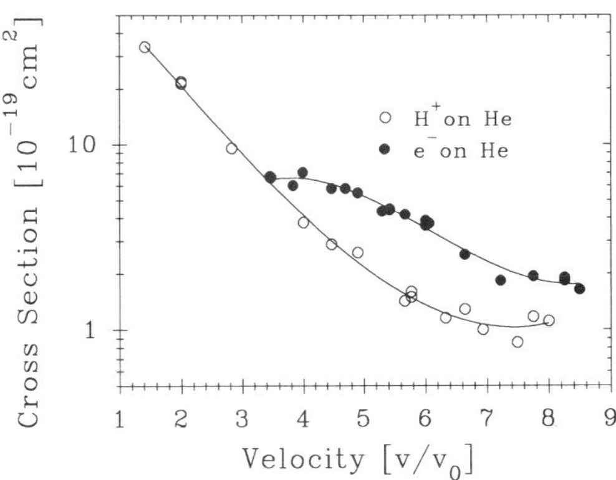

FIG. 11. Absolute ionization-excitation cross sections for electron and proton impact on He for $He^+(2p)$ states, as a function of the projectile velocity in units of v_0 (Bohr velocity).

2.9.4 Excitation of Molecular Gases Laboratory studies of ion–molecule collisions are an important aspect of ionospheric research and can add to the understanding of the processes affecting auroral phenomena. Describing the aurora borealis presents a challenging problem, the solution to which depends on the knowledge of photon emission cross sections for the excitation and ionization-excitation of atmospheric gases by the impact of energetic electrons, protons, and other positively and negatively charged species. The experimental method is identical with the one described in Sec. 2.9.2 except that the EUV grazing-incidence spectrometer is replaced, for example, by a 1-m air Czerny–Turner optical monochromator. Recently observed nitrogen and oxygen spectra in the visible range following H^+, H_2^+, and H_3^+ impact on N_2 and O_2 have revealed some spectral lines which arise from dissociation of molecules and ionization and excitation of the reaction fragments. The dominant spectral features arise from neutral and singly charged molecules. High spectral resolution in the visible or ultraviolet (UV) is needed to resolve the electronic transitions, each accompanied by additional vibrational structures and clusters of rotational components. Because of the many degrees of freedom, it is very difficult to determine accurate state-specific molecular cross sections (Bashkin *et al.*, 1991). Studies of these molecular collision processes in the laboratory can also yield important knowledge relevant to a better understanding of extra-terrestrial plasmas.

2.9.5 Electron Capture to Rydberg States The mechanism of electron capture at higher velocities, $v > v_e$, has received much attention recently. In this case, the first Born approximation is inadequate for electron capture by charged particles, and higher-order contributions are required to explain experimental cross-section data. As a prototype example, the following capture reaction is discussed here:

$$C^{4+}(1s^2) + A \rightarrow C^{3+}(1s^2nl) + A^+$$
$$\hookrightarrow C^{3+}(1s^2n'l') + h\nu, \qquad (13)$$

where the target A is either H_2 or He (Bruch *et al.*, 1982; Dubé, 1986). High-resolution EUV spectroscopy of the subsequent photon emission is used to extract selective information about the final-state (nl) population produced during the capture process. The impact energies ($v \approx 2.6$ to 4.1 a.u.) were selected such that the relative velocities were large enough for perturbative charge transfer theories to be applicable but small enough for the cross sections to remain significant. A common feature of all experiments performed at those velocities is that, in contrast to similar investigations at lower velocities (Janev and Winter, 1985), where the electron is preferentially captured into one or two well-defined principal shells, several prominent lines can be identified as transitions from Rydberg states (nl) with $n \leq 10$. The cross-section measurements on s, p, and d series obtained in C^{3+} give evidence of the inadequacy of the single-scattering theory and provide a direct measure of the importance of multiple-scattering effects—for example, theories based on a distorted-wave formalism like the continuum distorted-wave approximation (CDW) (see Sec. 3). Other experimental studies (Chetioui *et al.*, 1988) have reached the same conclusions and confirm the superiority of multiple-scattering descriptions for single-electron capture mechanism over the first Born approximation.

2.10 Collisions with Slow Multiply Charged Ions

2.10.1 Introduction The prospect of new exciting physics that can be studied with highly charged heavy ions in various areas of physics has stimulated new developments of advanced types of ion sources (Marrus, 1989), new accelerator facilities (Stöckli *et al.*, 1992), ion storage rings (Datz, 1987; Schuch, 1989), and ion traps (Levine *et al.*, 1988; Church, 1988). Low-velocity, highly charged ions strongly attract electrons in their vicinity and therefore represent the most reactive species in the universe (Schneider, 1991). For example, close interaction of one such ion with a molecule can result in the Coulomb explosion of this molecule.

The possible impact regarding basic physics as well as applied physics related to highly charged ions is expressed in recent reviews and international workshops and conferences (Ivanov, 1988; Alonso *et al.*, 1989; Salzborn *et al.*, 1991; Poth, 1990). Nuclear, atomic, solid-state, and surface physics (Andrä, 1989; Tully and White, 1977), as well as x-ray laser research and space science, are realizing the potential of new-generation facilities including advances in instrumentation, detection, and laser technology. The physics of highly ionized atoms is equally important to those questions at the boundary of pure and applied science (e.g., input for plasma modeling) and to those questions having a far-reaching impact on the understanding of the fundamental laws of nature, i.e., strong relativistic interactions (Wilson *et al.*, 1991) and multielectron (QED) effects (Mohr, 1989). The most recently developed facilities that are able to produce highly charged high-Z ions (HCHZI) throughout the periodic table are electron cyclotron resonance (ECR) ion sources (Meyer and Kirkpatrick, 1991), electron-beam ion sources (EBIS) or traps (EBIT) (Schneider *et al.*, 1991), and heavy-ion storage rings (HISR).

The EBIT, in particular in conjunction with additional ion traps, can compensate partly for the lack of storage rings. A wealth of experiments have been suggested for storage rings and ion traps for HCHZI, and some of the work has been in progress over the last few years. In this context, it should be noted that this new scope of atomic and molecular physics involves rather sophisticated, highly efficient x-ray and extreme-ultraviolet (EUV) optics, and high-resolution spectrometers, including absolute intensity standards and detection systems. The resulting highly accurate measurements may serve as benchmark data both for basic theoretical calculations of atomic processes and as input for high-temperature plasma modeling codes and in other areas of applied and basic physics. In ion traps and storage rings, a major experimental effort is, for instance, to produce ion crystals in the fast beam (Schuch, 1989).

2.10.2 Charge-Transfer Reactions Electron-capture processes following low- to medium-energy ($E \leq 25$ keV/amu) ion-atom and ion-molecule collisions can be characterized by

$$A^{q+} + B \rightarrow A_{n,l}^{(q-s)+} + B^{r+} + (r-s)e, \quad (14a)$$

$$A^{q+} + B + h\nu_L \rightarrow A_{n',l'}^{(q-s)+} + B^{r+} + (r-s)e, \quad (14b)$$

where A^{q+} is a multiply charged ion of charge state q, B describes a target atom (molecule or surface), and $h\nu_L$ characterizes a possible additional laser field. Process (14a) represents a typical charge transfer plus target ionization reaction, whereas mechanism (14b) can be viewed as the "photon-assisted" counterpart of reaction (14a). If the laser frequency is tuned into resonance of the target system, reaction (14b) is considered electron capture from an excited state of the target atom (molecule). The motivation for these experiments is

1. to determine relative and absolute cross-section data relevant for plasma diagnostics, fusion research, and laser gain in the EUV and soft x-ray region, and
2. to extend knowledge and understanding of the charge-transfer process to the case of excited target atoms and quantum optics of photon-assisted interactions.

For example, x-ray laser pumping by virtue of charge-transfer processes has a great potential for multiply charged ions because of the extremely large electron-capture cross sections ($\sigma \geq 10^{-15}$ cm^2) and the state-selective nature of the transfer process.

In the past few years, single and double electron-capture processes of multiply charged ions with atoms and molecules have provided new insight into few-electron dynamical phenomena that are dominated by electrostatic interactions. For the more highly charged ions, relativistic effects need to be considered for theoretical models describing these systems. We discuss here the formation and Auger decay of metastable Fe^{15+} $(2p^5 3s3p)^4D_{7/2}$ states by means of the zero-degree electron spectroscopy technique (Sec. 2.8.1). Assuming single-collision conditions, such quartet states can be formed in two-electron–capture processes as follows:

$$Fe^{17+}(2p^5) + He(1s^2) \rightarrow Fe^{15+}(2p^5 3snl)^4L + He^{2+}. \quad (15)$$

In Fig. 12, typical high-resolution projectile-electron spectra produced in 170-keV Fe^{17+} + He collisions are displayed. The most intense

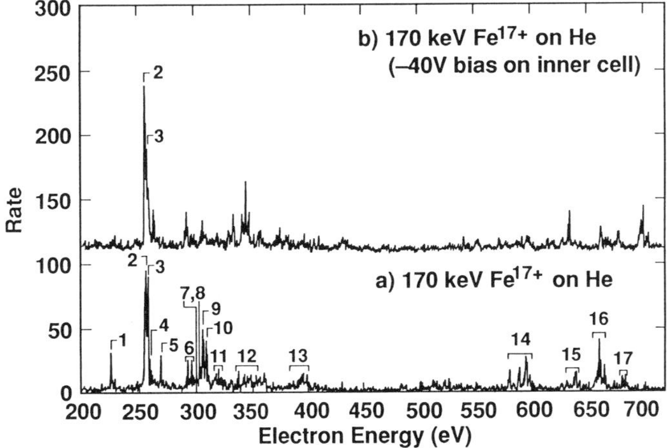

FIG. 12. (a) Auger-electron spectrum of sodium-like $Fe^{15+}(2p^5 3lnl')$ states following 170-keV Fe^{17+} + He collisions in the energy range from 200 to 700 eV. The electron energies are given in the projectile emitter frame. (b) Separation of long-lived metastable autoionizing states from prompt Coulomb autoionizing states. The delayed lines 2 and 3 are mainly due to the $(2p^5 3s3p)^4D_{7/2} \rightarrow (2p^6 \epsilon g)^2G_{7/2}$ and $(2p^5 3s3p)^4P_{5/2} \rightarrow (2p^6 \epsilon d)^2D_{5/2}$ transitions.

peaks arise from initial states associated with the $2p^53s3p$, $2p^53p3p$, $2p^53s3d$, and $2p^53p3d$ configurations. In particular, the prominent lines in the spectrum, labeled 2 and 3, are due to spin-induced Auger decay of the $(2p^53s3p)^4D_J$ and $(2p^53s3p)^4P_J$ levels. "Prompt" and "time-delayed" Auger transitions have been isolated. The direct excitation of the $^4D_{7/2}$ state involves a spin flip in Fe^{17+} + He single collisions.

2.10.3 Electron-Beam Ion Traps: New Facilities for Collision Experiments with Heavy, Highly Charged Ions A schematic diagram of EBIT, including the ion extraction system, is shown in Fig. 13. A 3-*T* axial magnetic field from two superconducting Helmholtz coils compresses an electron beam to current densities of up to 6000 A/cm^2 in the drift-tube region. Typical operating electron-beam currents are of the order of 100 mA. The axial magnetic field varies by less than 0.2% over the 2-cm length of the ion trap. The drift-tube assembly accelerates the beam electrons to their full interaction energy and provides an axial electrostatic trap for the positive ions. At present, EBIT is capable of producing electron energies of up to 30 keV. An upgrade of EBIT allows use of electron energies up to 150 keV (Super EBIT). Neutral atoms can be injected through one of the side ports of the trap. In order to inject ions, EBIT is also equipped with a MEVVA (metallic vapor vacuum arc) source that produces the initially low-charged ions up to 4, e.g., U ions. Only a fraction of these ions are trapped in the central drift-tube region, where the ions are confined axially by a potential well provided by two end drift tubes. The axial potential-well depth is typically around 100–300 eV. Radial confinement is provided by a combination of the space-charge field of the electron beam and the applied axial magnetic field. The ions undergo successive stages of ionization in the electron beam. The highest ionization stage present in the trap is determined by the electron-beam energy. The number of ions available depends on the ionization, recombination, and charge-exchange cross sections. In order to inject the ions into the trap, confine them for a given time and ionize them, and extract them, the potentials on the trap have to be varied in a fast switching mode. In the extraction phase, the ions are electrostatically deflected and focused through an einzel lens in the path between the MEVVA and the trap. The same lens provides focusing of the initially injected ions from the MEVVA into the trap. After passing the deflector, the ions are refocused via a second einzel lens and are momentum analyzed in a 90° magnet. A recent example of charge-state measurements of extracted uranium ions illustrates the capability of EBIT (see Fig. 14). These measured charge-state distributions allow detailed modeling calculations including ion cooling which is essential for heavy-ion trapping. Using such highly charged ions, new experiments in different areas of physics can be performed. Spectroscopy experiments focus on precision (one per million) measurements on highly charged few-electron ions. Photon interaction experiments will look at photoionization, photoexcitation, and lifetime measurements with lasers. Ion-ion and ion-atom collisional processes can be used to study resonant transfer and excitation, charge exchange, recoil ion production, and ion-ion collisional cross sections. Furthermore, electron-ion interactions, such as dielectronic recombination and electron-impact excitation and ionization, can be studied. All these processes are needed to model inertial confinement and other relevant plasmas, such as astrophysical sources (Burke and Eissner, 1983).

2.10.4 Ion–Surface Interaction Much effort is presently directed toward research on ion–surface interactions using very highly charged ions (Andrä, 1989). In particular, the recently developed technique (Schneider *et al.*, 1991) of extracting highly charged ions from EBIT allows one to use ions up to Th^{80+}, e.g., for ion-surface interaction studies. The goal of these studies is to provide an understanding of the neutralization dynamics of very highly charged ions as they approach the surface and penetrate into the solid. Such highly charged projectile ions carry up to several hundred keV potential energy, and x-ray emission studies are particularly suitable to illuminate the different interaction processes that lead to the transfer of this energy to the surface. The satellite intensities and energy positions of the emitted x rays and Auger electrons provide important information on the history of the projectile ions interacting with the surface and the bulk material. As illustrated in Fig. 15, highly charged slow ions capture electrons efficiently into

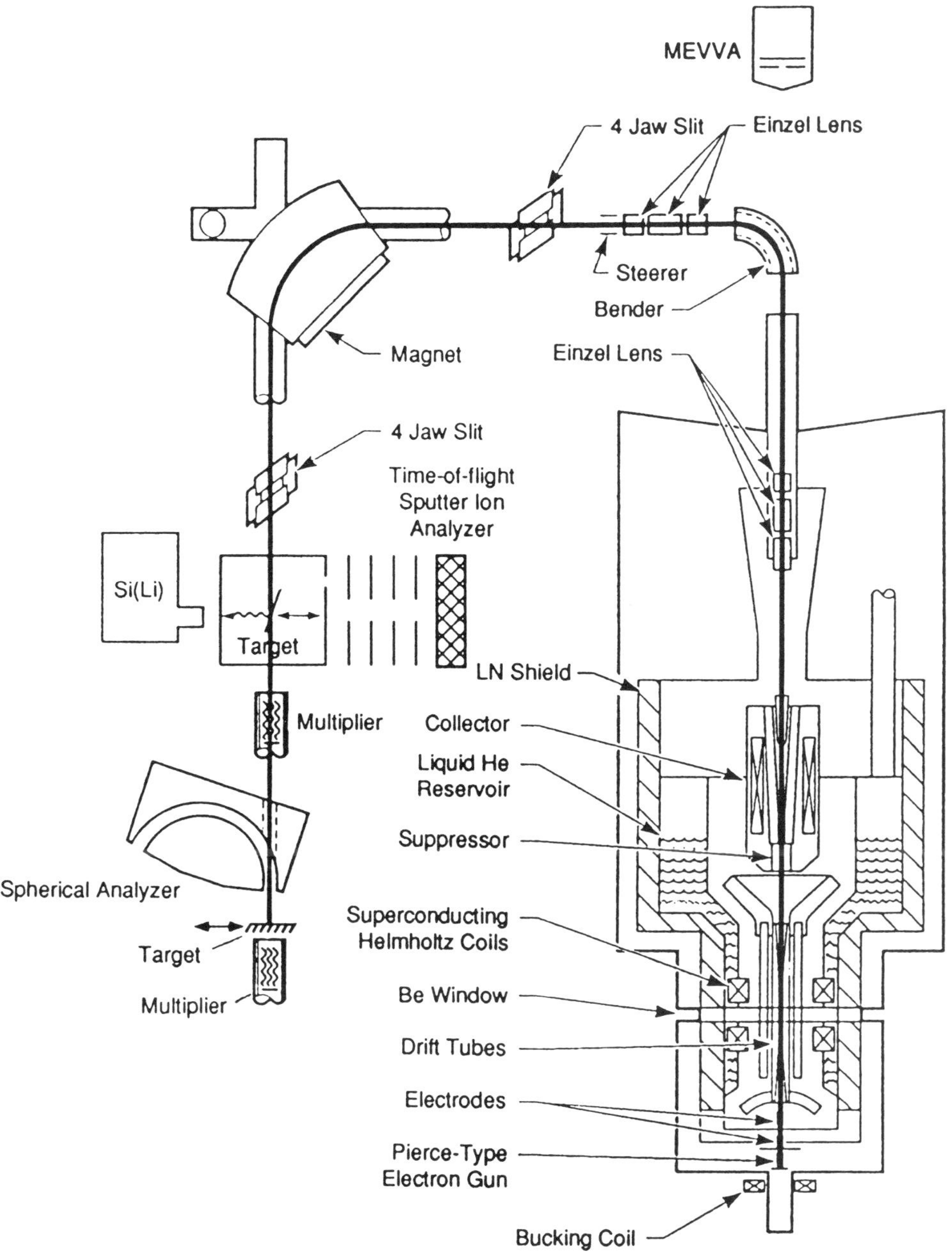

FIG. 13. Electron-beam ion trap, including electron gun and extraction system with focusing elements, analyzing magnet, and target arrangements. Also shown is the experimental setup for highly charged ion-surface studies, consisting of a Si(Li) x-ray detector, time-of-flight sputter ion analyzer, and spherical electron spectrometer.

high-*nl* states at relatively large distances depending on the ionic charge (Tully and White, 1977). Thus, the ion is promoted into a multiply excited state as it approaches the surface with electrons occupying high-*nl* levels while the core is virtually empty (hollow atoms) (Briand *et al.*, 1990).

Following the classical overbarrier model (Folkerts, 1992), the neutralization process starts at a critical distance $R_c \sim 4.6q$ (Å) where q is the projectile charge (assuming a metal work function $W_\phi = 4.5$ eV). From these measurements, the occurrence of "hollow atoms" during the neutralization process has been

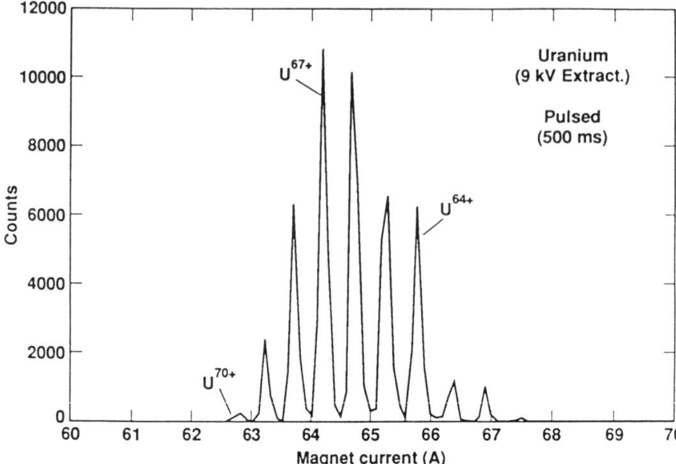

FIG. 14. Spectrum of charge analyzed uranium ions following 9-kV electron impact.

inferred. The created multiple inner-shell vacancy states provide time windows (atomic clocks) that allow study of stepwise deexcitation and evaporation of the excitation energy. In these studies, the following physical processes and parameters are of major importance (Clark *et al.*, 1993):

1. The initial electronic configuration of the approaching ion and the surface potential of the solid determine the transfer of electrons from the surface into states with high n quantum numbers.
2. The approach velocity and the autoionization and radiative decay rates determine the transfer of electrons to lower states or back to the solid, together with the distribution and time evolution of the population of different quantum states during neutralization.
3. Transfer of electrons into states with low quantum numbers occurs when the ion enters into the surface.

The ion–bulk interaction is more pronounced at higher approach velocities. In this case, the inner-shell vacancies have a larger probability of interpenetrating the first surface layer, and the interaction with the bulk causes a fast filling of shells with lower n quantum numbers with a simultaneous "peeling off" of electrons in shells with higher n manifolds.

Recently, doubly differential electron-emission yields following the impact of fast (3.95×10^7 cm/s) highly charged ions on Cu and Au targets have been measured (McDonald *et al.*, 1992). The electron emission is dominated by low-energy electrons (<20 eV). Specifically, it is found that the total yield, which rises to about 200 electrons per ion, is a nonlinear function of the total potential energy of the incident ion (up to about 200 keV). Ion–surface interaction studies with 1–3-keV/amu ions up to Th^{75+} have also been reported for conductor, semiconductor, and insulator materials. Microscopic surface analysis using an "atomic force microscope" revealed interesting nanoscale defects caused by single highly charged ion bombardment.

2.10.5 Ion Storage Rings: A New Tool for Atomic and Molecular Collision Physics

Heavy-ion storage rings make possible experiments with highly ionized and cooled ion beams (Datz, 1987; Schuch *et al.*, 1989). The main motivation for the design and construction of such heavy-ion storage rings is the study of fundamental dynamical processes of ions interacting with ions, atoms, molecules, electrons, and photons. One major advantage achieved in cooled storage-ring systems is the gain in luminosity and emittance as well as interaction time when compared to conventional ion experiments (see Sec. 2.1), leading to a substantial increase in sensitivity and precision in collisions with heavy, highly charged ions. These new facilities are also ideally suited for accurate measurements of atomic and molecular properties such as binding energies and relativistic and QED effects over a wide range of nuclear charges.

A simplified schematic view of the Stockholm ion storage ring is presented in Fig. 16.

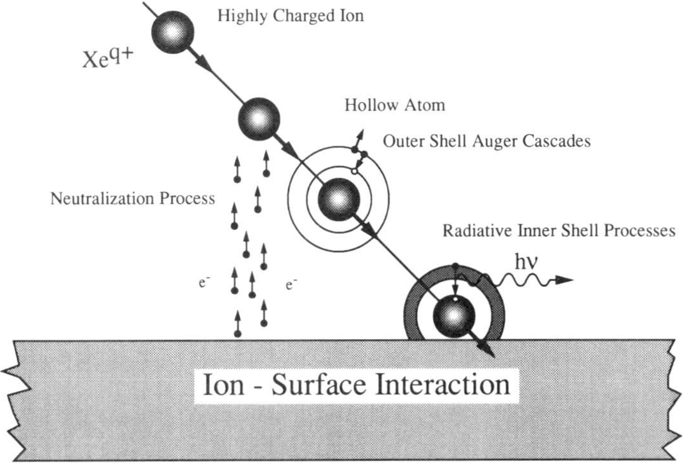

FIG. 15. Schematic diagram illustrating the neutralization process of highly charged Xe^{q+} (q = 44 to 48) ions close to a metal surface. The figure shows the creation of a highly excited hollow atom by multiple electron capture. This superexcited state then decays via outer-shell Auger cascades and inner-shell radiative processes.

This facility is uniquely suitable for studies of the fundamental interaction of very highly charged heavy ions with other ions, electrons, or photons. Investigations also involve ground-state molecular ions. The ring consists basically of a periodic structure of magnet components, i.e., dipole magnets for bending, focusing, and defocusing the beam. Also inserted in the ring are beam-cooling and synchrotron acceleration devices. The highly charged ions are created from a cryogenic electron-beam ion source (CRYSIS) and are injected into the ring after preacceleration. Fast acceleration is necessary (~100 ms) to minimize electron-capture beam losses.

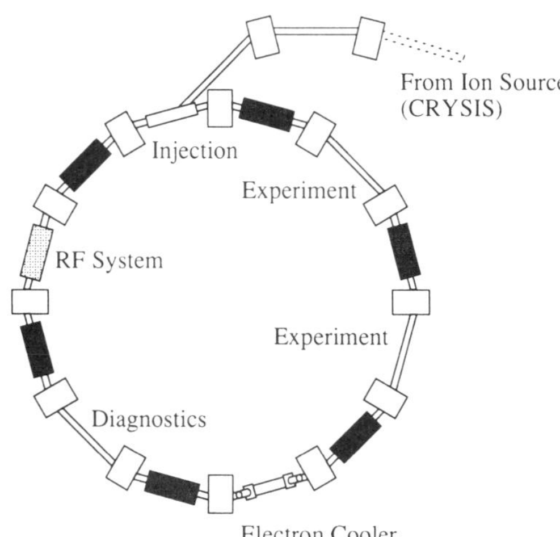

FIG. 16. Schematic layout of the Stockholm heavy-ion storage ring with rf acceleration system, main magnets, dipoles, focusing elements, electron cooler, straight experimental sections, and injection system.

The ions circulating in the storage ring have slightly different velocities. This velocity spread can be regarded as "thermal" motion. The reduction of the ion temperature in the ring is important for precision experiments. At present, there are three methods available for cooling hot ions, i.e., electron cooling, stochastic cooling, and laser cooling. Electron cooling is accomplished by the exchange of momentum through repeated Coulomb scattering between the ion beam in the ring and a low-temperature electron beam. The two beams are merged together over a distance of about 1 m with the cold electrons having the same average velocity as the ions. Through electron-ion interaction, heat is transferred from the ions to the electrons. Cooling times for 10-MeV/u particles are approximately 0.02 s for Xe^{44+} ions. The electron cooler can also be used as a target in experiments with cooled ion beams. Typical electron densities in the cooler are of the order 10^8 cm^{-3}. Important electron-ion recombination processes that can be studied include radiative recombination, laser-induced recombination, dielectronic recombination, and three-body recombination.

For singly charged ions such as Li^+, Be^+, Mg^+, etc., laser cooling also can be used in the ring to study ion-ion plasmas and crystals. Recently, Walther and co-workers have demonstrated crystallization of Mg^+ ions in a tiny table-top ring structure (Waki *et al.*, 1992). One of the ultimate goals is to observe phase transitions from ion clouds to liquids or crystals. These measurements can reveal information on the Coulomb coupling in one-dimensional (linear string, necklace) or three-dimensional helical structures. A recent simulation of 5-MeV/u Ar^{18+} ions has shown that a crystallized beam would retain its order after about four revolutions in a large storage ring. Similar results have been predicted at GSI in Darmstadt. Ion-beam crystallization of highly charged ions can be monitored by laser-induced fluorescence, Bragg scattering of laser photons, position-sensitive detection as a function of a varying deflection field, or measurement of the Schottky noise.

These processes of transforming ionic (plasma) beams to aligned crystalline phases are unique and will yield new insight into the nature of phase transitions and mechanisms of rotational and vibrational states in highly charged ionic solids. Other challenging experiments are concerned with precision measurements of atomic energy levels, fine-structure and hyperfine-structure splitting, QED effects of highly charged ions, and the masses of radioactive nuclei. Accurate mass measurements are of interest to chemists for the analysis of structure and modeling of complex molecules. These investigations are performed with ion traps (Church, 1988), in which ions can be stored by electromagnetic fields. At heavy-ion ring facilities, using ions with masses up to 100 and charge states up to 50, such experiments become feasible for the first time. Using this technique, it is planned to measure masses of heavy ions with a relative precision of $1:10^8$ using ion traps.

3. THEORETICAL MODELS FOR ATOMIC AND MOLECULAR COLLISION PROCESSES

Phenomenological and *ab initio* theoretical models have been applied to atomic and molecular systems undergoing collision, reaction, emission, and decay processes. The relative simplicity and extreme success of the Bohr atomic model for the two-body hydrogen system does not extend to three- or more-body systems. Hence, a number of theoretical models have been applied to few- and many-body systems to elucidate some of the single-particle and multiparticle collision and photon reaction mechanisms. Theoretical many-body treatments have met with some limited success. But the field of theoretical atomic and molecular physics is a dynamic one with many exciting discoveries to be made.

Although there are a number of theoretical treatments, these methods can be divided into several different classes of few- and many-body approaches. In atomic and molecular as well as nuclear systems, these approaches may involve semiempirical formulations for reaction-product parameters and relative and absolute cross-section predictions as well as fits to other useful experimental parameters. All theories involve such basic input parameters as projectile and target potentials, atomic and molecular radii or sizes, projectile charge and mass, projectile velocity, particle masses, and single-particle or multiparticle wave functions. Specifics of potential strength and form, reaction size such as channel radius, and wave functions depend on the specific system and relative velocity of the collision partners.

The usual approach is to define an operator that acts on appropriate wave functions, which are defined in terms of the mechanism of interaction such as scattering, reaction, or transition processes within the system. In Table 3, we list and summarize some basic mathematical forms of the major categories of theoretical models. In any system containing more than two bodies, we are immediately confronted with the need to use a much more complex description compared to two-body hydrogenic form.

The basic tenet of atomic and molecular processes is that they are describable by the Schrödinger wave equation. Most of these processes are nonrelativistic so that we do not require the Dirac wave equation, which is applied to more energetic processes. Some reactions are taken in the time-independent picture, and in some cases time dependence is used. Primarily, we have dealt with molecular and atomic processes such as pho-

Table 3. Theoretical models applicable to many-body physics.

Theory	Structure
1. The Born approximation in powers of the interaction potential and Green's function G.	$T = V + VG_0^+ + VG_0^+V + \ldots$ for $T^{(1)} = \langle \Psi_f \vert V \vert \Psi_i \rangle$ $T^{(2)} = \left\langle \Psi_f \middle\vert V \frac{1}{E - H + i\delta} V \middle\vert \Psi_i \right\rangle$ $V(r)G = G_0^+$ free particle, $G = \frac{1}{E - H}$ or $G = \frac{1}{4\pi r} e^{-qr}$
2. Coupled radial channels with channel radius r_c.	$\left(\frac{d^2}{dr^2} - \frac{l(l+1)}{r^2} + V_0^2\right) u_0(r) = u_{l'}(r)$ $\left(\frac{d^2}{dr^2} - \frac{l'(l'+1)}{r^2} + V_{l'}^2\right) u_{l'} = u_0(r)$
3. R matrix or reaction matrix.	$R = R_0 + R'$ for $R' = \sum \frac{\gamma_\lambda \times \gamma_\lambda}{E_\lambda - E}$ Sum over singularities $\gamma_\lambda = \left(\frac{\hbar^2}{2\mu r_c}\right)^{1/2} u_\lambda$ for $u_\lambda = G_\lambda + iF_\lambda$ $\gamma_\lambda = \frac{2\Gamma_\rho^2}{\hbar}\left(\frac{\rho}{(F_l^2 + G_l^2)r_c}\right)$ where γ_λ^2 is termed the reduced width.
4. T matrix or transition matrix.	$T = V + V \frac{1}{E - H - i\delta} V$ $R = V + R \frac{\mathrm{P}}{E - H} V$ for P = Cauchy principle value $T = R - i\pi R\delta(E - H)T$
5. S matrix or scattering matrix, interaction domain in complex plane.	$S_{if} = \lim_{t\to\infty} \langle \Psi_f^- \vert \Psi_i^+ \rangle$ $S_{if} = \delta_{if} - 2\pi i\delta(E_f - E_i)(H - E_f)\langle \Psi_f^- \vert \Psi_i^+ \rangle$ $S_{if} = \delta_{if} - 2\pi i\delta(E_f - E_i)T_{fi}$

toexcitation and photoemission, atomic and molecular collision, scattering processes, and reactions. Most of these processes involve collision energies from a few keV to MeV energies, and below 1 eV molecular rotation-vibration excitations are involved. In this energy region, plasma collective phonon and plasmon states can occur in an ionized medium (Rauscher, 1968). At the upper end of this energy range, we have K-shell electron ionization and excitation and electron capture including L and M electron processes (Briggs and Macek, 1991).

Details on the Schrödinger or Lippmann-Schwinger equation and other quantum mechanical terminology, such as Green's functions, Born-Oppenheimer approximation, diagonalization, and transition probabilities, can be found in any good standard atomic physics textbooks such as Friedrich (1990) and Yoachain (1975).

The Born approximation summarized in Table 3 has found widespread use. For example, an expansion of the cross section for ion-atom collisions can be made in terms of Z_p^n, where Z_p represents the projectile charge (Sec. 2.9.3). The transition probabilities for a given process are determined from the transition operator or T matrix, which acts on an initial incoming unperturbed plane-wave state to produce a perturbed outgoing state. If several possible "open channel" states mix, coupling can exist between these various states. The Schrödinger equation contains channel-dependent potentials with coupled channel terms dependent on the various possible states, such as angular momentum variables, of the system beyond the channel radius, which can be taken as approximately the atomic radius. One method of solving complex equations, which can be anywhere from two to more channels depending on the possible final states of the system, is to use the Green's function method. Channel wave functions are expressed in terms of the Green's function, which is the solution to the hypergeometric equation in which the time-dependent part is taken as a delta function between the possible states. The basis wave function is the solution to corresponding homogeneous equations, and a perturbative expansion is made in terms of the appropriate Green's functions and coupling potential between the various relevant states.

The number of terms in the perturbative expansion and the identification of the specific order of Born terms with specific processes depend on the problem under consideration. This technique is the basic approach in atomic and molecular collision physics. There are dominating terms that usually occur and are identified with the major features of the interactive processes. Currently of interest is the consideration of time ordering of the perturbative expansion terms and the use of time-ordering operators. It is often useful to treat scattering problems using an equivalent integral equation in place of the Schrödinger equation,

$$[-(\hbar^2/2\mu)\nabla^2 + V(\mathbf{r})]\Psi(\mathbf{r}) = E\Psi(\mathbf{r}), \tag{16a}$$

rewritten as

$$[E + (\hbar^2/2\mu)\nabla^2]\Psi(\mathbf{r}) = V(\mathbf{r})\Psi(\mathbf{r}), \tag{16b}$$

where μ is the reduced mass. We can solve this equation using the three-particle Green's function

$$G(\mathbf{r},\mathbf{r}') = \frac{-\mu}{2\pi\hbar^2}\frac{e^{i\mathbf{k}\cdot(\mathbf{r}-\mathbf{r}')}}{|\mathbf{r}-\mathbf{r}'|}; \tag{17}$$

using the equation

$$[E + (\hbar^2/2\mu)\nabla^2]G(\mathbf{r},\mathbf{r}') = \delta(\mathbf{r}-\mathbf{r}'), \tag{18}$$

the wave function

$$\Psi(\mathbf{r}) = e^{ikz} + \int G(\mathbf{r},\mathbf{r}')V(\mathbf{r}')\Psi(\mathbf{r}')d^3r', \tag{19}$$

solves the above Schrödinger equation with the $V(\mathbf{r})\Psi(\mathbf{r})$ right-hand term. To transform from space-time dependence of the variables to equivalent momentum-energy dependence, we can use the Lippmann-Schwinger equation. Its solutions automatically fulfill the free-particle boundary conditions at large distances from the scattering event.

Using the T matrix, defined as the transition operator between initial and final states, we can describe the scattering process in an analogous manner to the time-dependent perturbation theory as a transition from the incoming plane wave to the outgoing plane waves. With the incoming "unperturbed" wave function $\Psi_i(\mathbf{r}') = e^{ikz'}$, we can define the T matrix as $T_{fi} = \langle\Psi_f|T_{op}|\Psi_i\rangle$. This assumption leads directly to the Born approximation. Once in this form, we can express the transition process, $i \rightarrow f$, in terms of an angular dependent scattering amplitude.

The scattering amplitude in the Born approximation is derived by a Fourier transformation of the potential. If the potential $V = V(\mathbf{r})$ is radially symmetric, then the time-independent Schrödinger equation can be reduced to the radial Schrödinger equation. Since the basic forces in atomic and molecular systems are Coulombic, we can employ the Born-Oppenheimer approximation where we solve the electron problem quantum mechanically and treat the nucleus classically.

In general, we treat the problem of an incoming plane wave which undergoes scattering under the influence of an appropriate potential to an outgoing spherical wave. But, for example for electron scattering, the potential corresponds to a pure Coulomb potential at large separations only, and short-range modifications resulting from electron screening need to be used. Then the incoming plane wave is taken to be a distorted wave in the distorted-wave Born approximation, DWBA, leading to an additional term in the scattering amplitude.

Other forms of wave functions such as Hartree-Fock, harmonic oscillator, or rotational wave functions can be used for various molecular systems with internal-state quantum numbers of vibration, rotation, excitation, and ionization. The cross section can be expressed in terms of the T matrix and scattering amplitude, as an expansion which yields the T- or S-matrix elements (see Table 3).

Symbolically, we represent the total cross section in terms of the T matrix as $\sigma = \tilde{T}^*T$ where $\tilde{T}$ is the transpose of T and T^* is the complex conjugate of T. The notation $T^\dagger$ is used for $\tilde{T}^*$. Furthermore, the R-matrix or reaction method is extremely useful for describing isolated states and is expressed in terms of the regular and irregular Coulomb wave functions, F_l and G_l, respectively. Both functions solve the free equation, but the irregular wave function does not vanish at $r = 0$. Their general form can be expressed asymptotically for $r \rightarrow \infty$ in terms of sine and cosine, respectively, with arguments that depend on momentum k, radius r, and Coulomb phase shift δ_l. For smaller radial dependence at lower energies, the regular and irregular Coulomb-wave functions are expressed in terms of ordinary Bessel functions. Because of the form of the dependence of the R matrix on the potential, this approach is more suitable for describing higher-energy processes. Observed experimental data for the energy and angular dependence of secondary electron emission following fast ion-atom collisions (see Sec. 2.7) can be compared with the BEA (binary encounter approximation; Garcia *et al.*, 1973), PWBA (plane wave Born approximation; Merzbacher and Lewis, 1958; Fano, 1963), and CDW-EIS (continuum distorted wave eikonal initial state) calculations. The BEA and PWBA calculations account only for the interaction between the static Coulomb fields of the projectile and target electrons. A representative illustration of the PWBA and CDW-EIS approximation is shown in Fig. 7 from the work of Schneider and co-workers. The collision system is 3.5-MeV/u Fe^{22+} on He. A number of statistical methods applied to molecular and atomic collision processes have been developed over the last two decades. One of the most successful is the n-body classical trajectory Monte Carlo method (nCTMC), which has been used extensively by Olson and co-workers (Cocke and Olson, 1991) to describe molecular and atomic collision processes.

Various semiempirical expansions can be represented by Feynman graphical techniques (Mattuck, 1992), which are frequently used in many-body collision physics. The approaches briefly described in Table 3 and other theories currently in use can assist us in analyzing the vast amount of diverse experimental collision data. By analyzing these models, we will be able to design new experiments and better understand atomic and molecular interactions. This fundamental knowledge is also of importance for applications to industrial problems.

4. SUMMARY

We have given some characteristic examples of molecular and atomic collision processes in three broad categories, namely, fragmentation, spectroscopy, and few-body collision dynamics. With each improvement in experimental methods, ion-beam facilities, and computational physics, the understanding of complex yet fundamental collision processes has advanced dramatically in the last few years. In fact, the richness of recent molecular and atomic collision research has yielded valuable insight into few-body dynamics. Nonetheless, a detailed quantum mechanical description of many-body collision processes has still to be developed. In particular, few-body problems still require extensive investigations to elucidate the interplay of electron-correlation, relativistic, and quantum electrodynamic effects in collective particle interactions as a function of the velocity, charge state, atomic number, participating electrons, and atomic and nuclear spins. Also, phenomena like multipole transitions, or hyperfine-induced mixing in multiply charged heavy ions, require precision atomic many-body calculations. The availability of new highly charged ion sources, ion-storage rings, and ultrashort-duration high-power laser systems has opened completely new avenues of research in atomic and molecular collision physics that are of vital interest for pure and applied research in the academic and industrial environment. The advances at the forefront of science and technology are closely related to the fundamental nature of matter and collisional phenomena involving atoms, molecules, and ions. Collisional processes are, for example, crucial to the understanding of auroral, magnetic, and ionospheric phenomena and also contribute to the understanding of plasmas, x-ray lasers, astrophysics, ion implantation, surface diagnostics and modification, nanostructure and subnanostructure technologies, and exotic new electron-electron, electron-ion, ion-ion, and electron-nuclear interactions.

ABBREVIATIONS

BEA: Binary encounter approximation.
BEP: Binary encounter peak.
CDW: Continuum distorted-wave approximation.
CRYSIS: Cryogenic electron-beam ion source.
CSW-EIC: Continuum distorted-wave eikonal initial-state calculations.
DDCS: Double-differential electron cross section.
EBIT: Electron-beam ion trap.
EBIS: Electron-beam ion source.
ECR: Electron cyclotron resonance.
ELP: Electron loss peak.
EUV: Extreme ultraviolet.
HCHZI: Highly charged high-Z ions.
HISR: Heavy-ion storage ring.
LPES: Laser photodetachment electron spectroscopy.
nCTMC: n-body classical trajectory Monte Carlo method.
PWBA: Plane-wave Born approximation.

GLOSSARY

Auger Effect and Autoionization: A radiationless decay of a highly excited molecular or atomic state where one electron undergoes a transition to a lower-lying state and another electron is transferred to the adjacent continuum, resulting in a free electron and the formation of a residual ion.

Born Approximation: High-energy limit of the scattering amplitude in ion-atom collisions in which the Born series converges at sufficiently high projectile velocities.

Cross Section: The cross section (σ) represents an effective area seen by the incoming projectile and has the units cm^2 per atom.

Works Cited

Alonso, J. R., Dietrich, D. D., Prior, M. H., Schneider, D. H. G. (Eds.) (1989), "Proceedings of a Workshop on Highly Charged Ions: New Physics and Advanced Techniques," *Nucl. Instrum. Methods Phys. Res.* **B43**, 265–470.

Amusia, M. Ya (1990), *Atomic Photoeffect*, New York: Plenum Press.

Andersen, L. H. (1988), in: H. B. Gilbody, W. R. Newell, F. H. Read, A. C. H. Smith (Eds.), *Electronic and Atomic Collisions*, Amsterdam: North-Holland, p. 451.

Andersen, N., Gallagher, J., Hertel, I. V. (1988), *Phys. Rep.* **165**, 1–188.

Andrä, H. J. (1979), in: W. Hanle, H. Kleinpoppen (Eds.), *Progress in Atomic Spectroscopy*, New York: Plenum Press, p. 829.

Andrä, H. J. (1989), in: Marrus, R. (Ed.), *Physics of Highly Ionized Atoms*, NATO Advanced Study Institutes, Series B, Physics, Vol 201, New York: Plenum.

Bashkin, S. (Ed.) (1976), *Beam-Foil Spectroscopy*, Topics in Current Physics No. 1, Berlin: Springer.

Bashkin, S., Träbert, E., Thiede, D. A., Sercel, P. C., Lin, P.-C., Li, M.-L., Jenkins, D. G., Shemansky, D. E., Wells, K. Bruch, R., Fülling, S., DeWitt, D. (1991), *Phys. Rev. A*43, 3553–3562.

Bashkin, S., Stoner, J. O., Jr. (1975), *Atomic Energy Levels and Grotrian Diagrams*, Vol. 1, Amsterdam: North-Holland; (1978), Vol. 2; (1981), Vol. 3.

Berry, H. G. (1977), *Rep. Prog. Phys.* **40**, 155–217.

Betz, D. (1983), in: S. Datz (Ed.), *Applied Atomic Collision Physics*, Vol. 4, Orlando, FL: Academic, pp. 1–42.

Bird, J. R., Williams, J. S. (1989), *Ion Beams for Materials Analysis*, Sydney: Academic.

Bisgaard, P., *et al.* (1980), in: H. Kleinpoppen, J. F. Williams (Eds.), *Coherence and Correlation in Atomic Collisions*, New York: Plenum.

Briand, J. P., De Billy, L., Charles, P., Essabaa, S., Briand, P., Geller, R., Desclaux, J. P., Bliman, S., Ristori, C. (1990), *Phys. Rev. Lett.* **65**, 159–162.

Briggs, J. S., Kleinpoppen, H. J., Lutz, H. O. (1988), *Fundamental Processes of Atomic Dynamics*, NATO Advanced Study Institutes, Series B, Physics, New York: Plenum.

Briggs, J. S., Macek, J. H. (1991), in: Sir David Bates, Benjamin Bederson (Eds.), *The Theory of Fast Ion-Atom Collisions*, Advances in Atomic, Molecular, and Optical Physics Vol. 28, Boston: Academic.

Brown, I. G. (Ed.) (1989), *The Physics of Ion Sources*, New York: Wiley.

Bruch, R., Dubé, L. J., Träbert, E., Heckmann, P. H., Raith, B., Brand, K. (1982), *J. Phys. B* **15**, L857–L862.

Burke, P. A., Eissner, W. (Eds.), (1983), *Atoms in Astrophysics*, New York: Plenum.

Chetioui, A., Politis, M. F., Wohrer, K., Rozet, J. P., Blumenfeld, L., Touati, A., Vernhet, D., Roncin, P., Gaboriaud, M. N., Barat, M., Laurent, H. (1988), in: H. B. Gilbody, W. R. Newell, F. H. Read, A. C. H. Smith (Eds.), *Electronic and Atomic Collisions*, Amsterdam: North-Holland, p. 309.

Church, D. A. (1988), *Physica Scr.* **T22**, 164–170.

CIAMDA (1987), *An Index to the Literature on Atomic and Molecular Collision Data Relevant to Fusion Research*, Vienna: International Atomic Energy Agency.

Clark, M. W., Schneider, D., Dewitt, D., McDonald, J. W., Bruch, R. Safronova, U. I., Tolstikhina, I. Yu., Schuch, R. (1993), *Phys. Rev. A* **47**, 3983–3997.

Cocke, C. L., Olson, R. E. (1991), *Phys. Rep.* **205**, 153–219.

Cowan, R. D. (1981), *Theory of Atomic Structure and Spectra*, Berkeley, CA: Univ. of California Press.

Crasemann, B. (Ed.) (1975), *Atomic Inner-Shell Processes*, Vols. 1 and 2, New York: Academic.

Datz, S. (Ed.) (1983), *Applied Atomic Collision Physics*, Vol. 4, *Condensed Matter*, Orlando, FL: Academic.

Datz, S. (1987), *Nucl. Instrum. Methods Phys. Res.* **B24/25**, 3–10.

Datz, S. (1990), in: A. Dalgarno, R. S. Freund, M. S. Lubell, T. B. Lucatorto (Eds.), *The Physics of Electronic and Atomic Collisions*, AIP Conference Proceedings No. 205, New York: American Institute of Physics, p. 2.

De Heer, F. J. (1982), in: C. J. Joachain, D. E. Post (Eds.), *Atomic and Molecular Physics of Controlled Thermonuclear Fusion*, New York: Plenum, pp. 269–312.

Dubé, L. (1986), in: D. C. Lorents, W. E. Meyerhof, J. R. Petersen (Eds.), *Electronic and Atomic Collisions*, Amsterdam: North-Holland, pp. 345–364.

Elton, R. C. (1990), *X-Ray Lasers*, Boston: Academic.

Fano, U. (1963), *Annu. Rev. Nucl. Sci.* **13**, 1–66.

Fano, U., Rau, A. R. P. (1988), *Atomic Collisions and Spectra*, New York: Academic.

Folkerts, L. (1992), "Charge Exchange of Multiply Charged Ions with Metal Surfaces," Ph.D. Thesis, University of Groningen.

Friedrich, H. (1990), *Theoretical Atomic Physics*, Berlin: Springer.

Fritsch, W., Lin, C. D. (1991), *Phys. Rep.* **202**, 1–97.

Fülling, S., Bruch, R., Rauscher, E. A., Weill, P. A., Träbert, E., Heckmann, P. H., McGuire, J. H. (1992), *Phys. Rev. Lett.* **68**, 3152–3155.

Garcia, J. D., Fortner, R. J., Kavanagh, T. M. (1973), *Rev. Mod. Phys.* **45**, 111–177.

Gemmel, D. S. (1980), *Chem. Rev.* **80**, 301–311.

Giese, J. P., Schulz, M., Swenson, J. H., Schöne, H., Benhenni, M., Varghese, S. L., Vane, C. R., Dittner, P. F., Shafroth, S. M., Datz, S. (1990), *Phys. Rev. A* **42**, 1231–1244.

Gudeman, C. S., Saykally, R. J. (1984), *Annu. Rev. Phys. Chem.* **35**, 387–418.

Halka, M., Bryant, H. C., Mackerrow, E. P., Miller, W., Mohagheghi, A. H., Tang, C. Y., Cohen, S., Donahue, J. B., Hsu, A., Quick, C. R., Tiee, J., Rozsa, K. (1991), *Phys. Rev. A* **44**, 6127–6129.

Hasselkamp, D., Varga, P., Groenveld, K. O., Winter, H., Kemmler, J., Rothard, H. (1992), *Particle-Induced-Electron Emission II*, Springer Tracts in Modern Physics, Vol. 123, Berlin: Springer.

Hotop, H., Lineberger, W. C. (1985), *J. Chem. Ref. Data* **14**, 731–750.

Humphries, S., Jr. (1986), *Principles of Charged Particle Acceleration*, New York: Wiley.

Ivanov, L. N. (1988), *Phys. Rep.* **164**, 315–375.

Janev, R. K., Winter, H. P. (1985), *Phys. Rep.* **117**, 265–387.

Joachain, C. J. (1975), *Quantum Collision Theory*, Amsterdam: North-Holland.

Kessel, Q. C., Fastrup, B. (1974), in: E. W. McDaniel, M. R. C. McDowell (Eds.), *Case Studies in Atomic Physics III*, Amsterdam: North-Holland, pp. 139–213.

Kleinpoppen, H., Williams, J. F. (Eds.) (1980), *Coherence and Correlation in Atomic Collisions*, New York: Plenum.

Letokov, V. S. (1989), *Lasers in Atomic, Molecular, Nuclear Physics*, Singapore: World Scientific.

Levine, M. A., Marrs, R. E., Hendersen, J. P., Knapp, D. A., Schneider, M. B. (1988), *Phys. Scr.* **T22**, 157–163.

Lindgren, I., Morrison, J. (1982), *Atomic Many-Body Theory*, Berlin: Springer.

Maier, J. D. (1989), *Ion and Cluster Ion Spectroscopy and Structure*: Amsterdam: North-Holland.

Mannervik, S. (1989), *Phys. Scr.* **40**, 28–52.

Marrus, R. (Ed.) (1989), *Physics of Highly Ionized Atoms*, NATO Advanced Study Institutes, Series B, Physics, Vol. 201, New York: Plenum.

Massey, H. (1976), *Negative Ions*, Cambridge, U.K.: Cambridge Univ. Press.

Mattuck, R. D. (1992), *A Guide to Feynman Diagrams in the Many-Body Problem*, 2nd ed., New York: Dover.

McDaniel, E. W., Mitchell, J. B. A., Rudd, M. E. (1993), *Atomic Collisions: Heavy Particle Projectiles*, New York: Wiley.

McDonald, J. W., Schneider, D., Clark, M. W., Dewitt, D. (1992), *Phys. Rev. Lett.* **68**, 2297–2300.

McGuire, J. H. (1992), *Adv. At. Mol. Opt. Phys.* **29**, 217–323.

Mehlhorn, W. (1985), in: B. Crasemann (Ed.), *Atomic Inner-Shell Physics*, New York: Plenum.

Merzbacher, E., Lewis, H. (1958), *Encyclopedia of Physics*, Vol. 34, Berlin: Springer.

Meyer, F. W., Kirkpatrick, M. I. (1991), *Proceedings of the 10th International Workshop on ECR Ion Sources*, ORNL Conf-9011136.

Miller, T. A., Bondybey, V. E. (1983), *Molecular Ions: Spectroscopy, Structure and Chemistry*, Amsterdam: North-Holland.

Mohr, P. J. (1989), in: Marrus (Ed.), *Physics of Highly Ionized Atoms*, NATO Advanced Study Institutes, Series B, Physics, Vol. 201, New York: Plenum.

Mott, N. F., Massey, H. S. (1987), *The Theory of Atomic Collisions*, 3rd ed., Vols. 1 and 2, Oxford: Clarendon.

Müller-Dethlefs, K., Schlag, E. W. (1991), *Annu. Rev. Phys. Chem.* **42**, 109–136.

Neugart, R. (1987), in: H. J. Beyer, H. Kleinpoppen (Eds.), *Progress in Atomic Spectroscopy*, Part D, New York: Plenum, p. 75.

Niehaus, A. (1990), *Phys. Rep*, **186**, 149–214.

Pegg, D. J., Thomson, J. S., Compton, R. N., Alton, G. D. (1987), *Phys. Rev. Lett.* **59**, 2267–2270.

Poth, H. (1990), *Nature* **345**, 399–405.

Rauscher, E. A. (1968), *J. Plasma Phys.* **2**, 517–541.

Rødbro, M., Bruch, R., Bisgaard, P. (1979), *J. Phys. B.* **12**, 2413–2447.

Rudd, M. E., Macek, J. H. (1974), in: E. W. McDaniel, M. R. C. McDowell (Eds.), *Case Studies in Atomic Physics III*, Amsterdam: North-Holland.

Salzborn, E., Mueller, A., Mokler, P. H. (1991), *Atomic Physics of Highly Charged Ions*, Berlin: Springer.

Scoles, G. (Ed.) (1988), *Atomic and Molecular Beam Methods*, Vol. 1, New York: Oxford Univ. Press.

Schneider, D., Clark, M. W., Penetrante, B. M., McDonald, J., DeWitt, D., Bardsley, J. N. (1991), *Phys. Rev. A* **44**, 3119–3124.

Schneider, D. H., DeWitt, D. R., Bauer, R. W., Mowat, J. R., Graham, W. G., Schlachter, A. S., Skogvall, B., Fainstein, P., Rivarola, R. D. (1992), *Phys. Rev. A* **46**, 1296–1302.

Schuch, R., Bárány, A., Danored, H., Elander, N., Mannervik, S. (1989), *Nucl. Instrum. Methods Phys. Res.* **B43**, 411–424.

Sellin, I. A. (Ed.) (1978), *Structure and Collisions of Ions and Atoms*, Topics in Current Physics, Vol. 5, New York: Springer.

Sevier, K. D. (1972), *Low Energy Electron Spectroscopy*, New York: Wiley.

Smirnov, B. M. (1982), *Negative Ions*, New York: McGraw-Hill.

Stebbings, R. F., Dunning, F. B. (1983), *Rydberg States of Atoms and Molecules*, Cambridge, U.K.: Cambridge Univ. Press.

Stöckli, M. P., Ali, R. M., Cocke, C. L., Raphaelian, M. L. A., Richard P., Tipping, T. N. (1992), *Rev. Sci. Instrum.* **63**, 2822–2824.

Stolterfoht, N. (1978), in: I. A. Sellin (Ed.), *Structure and Collisions of Ions and Atoms*, Topics in Current Physics Vol. 5, New York: Springer.

Stolterfoht, N. (1987), *Phys. Rep.* **146**, 315–424.

Toburen, L. H. (1991), *Atomic and Molecular Physics in the Gas Phase*, Physical and Chemical Mechanisms in Molecular Radiation Biology, New York: Plenum.

Tully, J. C., White, C. W. (1977), *Inelastic Ion Surface Collisions*, New York: Academic.

Vainshtein, L. A., Shevelko, V. (1986), *Structure and Properties of Ions in Hot Plasmas*, Moscow: Nauka.

Waki, I., Kassner, S., Birkl, G., Walther, H. (1992), *Phys. Rev. Lett.* **68**, 2007–2010.

Wang, H., Bruch, R., Hao, F., Fuelling, S., Xu, Z., Wang, Z., Rauscher, E. (1993), *Nucl. Instrum. Methods Phys. Res.* **B79**, 114–116.

Wilson, S., Grant, I. T., Gyorffy, B. L. (1991), *The Effects of Relativity in Atoms, Molecules, and The Solid State*, New York: Plenum.

Wollnik, H. (1987), *Optics of Charged Particle Beams*, Orlando, FL: Academic.

Further Reading

Bransden, B. H., Joachain, C. J. (1986), *Physics of Atoms and Molecules*, London: Longman.

Lin, C. D. (1993), *Review of Fundamental Processes and Applications of Atoms and Ions*, Singapore: World Scientific.

McDaniel, E. W. (1989), *Atomic Collisions*, New York: Wiley.

McDaniel, E. W., Mitchell, J. B. A., Rudd, M. E. (1993), *Atomic Collisions: Heavy Particle Projectiles*, New York: Wiley.

Svanberg, J., (1992), *Atomic and Molecular Spectroscopy*, Berlin: Springer.

MOLECULAR BEAM EPITAXY

R. J. FISCHER AND A. Y. CHO, *AT&T Bell Laboratories, Murray Hill, New Jersey, U.S.A.*

INTRODUCTION

The term *molecular beam epitaxy* (MBE) encompasses a variety of specific techniques for deposition of thin-film electronic materials. These thin films are the basis of many electronic and photonic devices. Cost and reliability factors place stringent requirements on the purity, number of defects, and controllability with which these films are deposited. The molecular beam epitaxy process offers a high degree of controllability and reproducibility. It is for this reason that many companies in the compound-semiconductor industry are using MBE in their processes. For example, at the time of this writing, more than half of the lasers used in compact-disk players worldwide are produced by MBE.

In the conventional form, MBE is a highly controlled evaporation process. Elemental sources are heated in an ultrahigh-vacuum environment, and thus the source material is transported to a heated substrate. Epitaxy, which is a process by which a deposited film takes on a crystalline structure that is related to that of the substrate, occurs on the substrate in this process. The reason why this occurs is that the substrate, being heated, provides the impinging molecules with enough energy to be mobile on the substrate, such that these atoms or molecules can find appropriate lattice sites on which to condense.

The typical rates used for deposition are about one atomic layer per second. This view of the process allows one to see why there is such controllability—the process is building up a thin film one atomic layer at a time. This feature allows many phenomena of quantum physics to be investigated. In what follows, a more detailed description of MBE will be given, and many variants of the process will be described, and material and device examples will be given.

1. HISTORY AND GENERAL PRINCIPLES

The modern-day MBE apparatus has evolved from its earlier form, which was an experimental system designed to measure adsorption and desorption kinetics on GaAs surfaces. From these studies, a method was discovered for epitaxially depositing GaAs. In contrast, today's multiwafer MBE systems are true production tools in the semiconductor industry. Figure 1 shows the first MBE effusion-cell assembly, while Fig. 2 shows a modern production system. In this section, the

3-527-28132-0/94/$5.00 + .50

FIG. 1. Photograph of early MBE system.

hardware of MBE will be described, highlighting many of the innovations in design through the generations that have brought the technique to its present maturity.

1.1 The MBE Apparatus

Figure 3 shows a cutaway view of a modern MBE system and highlights most of the important features. MBE is an ultrahigh-vacuum technique with system base pressures of typically 10^{-11} Torr (a hundred-trillionth of our atmospheric pressure). These vacuum levels are required in order to maintain purity of the deposited film. To achieve such a vacuum, liquid-nitrogen-cooled panels are fitted to surround the entire deposition chamber. This liquid-nitrogen cooling substantially reduces the amount of water vapor present. Another feature that aids the vacuum is the use of a sample-exchange load lock. In this way, the deposition chamber need not be exposed to atmospheric pressure each time a new sample is introduced. Sample transfer mechanisms are required for this feature and on modern systems have become completely automated.

With several important differences, MBE is in essence a specialized form of vacuum evaporation. The so-called effusion cells that are aimed at the substrate (e.g., a GaAs wafer) are the sources in the technique and form the molecular beams. Figure 4 is a photograph of such an effusion cell. The cell has a crucible (usually made of pyrolytic boron nitride) in which a charge of the appropriate element is contained. For growth of GaAs, three such effusion-cell sources would usually be required; one for Ga, one for As, and one for a dopant. The crucible with its elemental charge is heated in the cell by resistive heater windings to a temperature that will produce the molecular beams by evaporation (sublimation, in the case of solid charges). In front of each effusion cell a shutter serves to block the beam when that particular element is not wanted. These shutters, which swing in and out of the beam path, allow composition and doping profiles that are atomically abrupt. A typical MBE system

FIG. 2. Modern production MBE system. This system is fully automated and can deposit films on several wafers simultaneously. (Photo courtesy of ISA-Riber.)

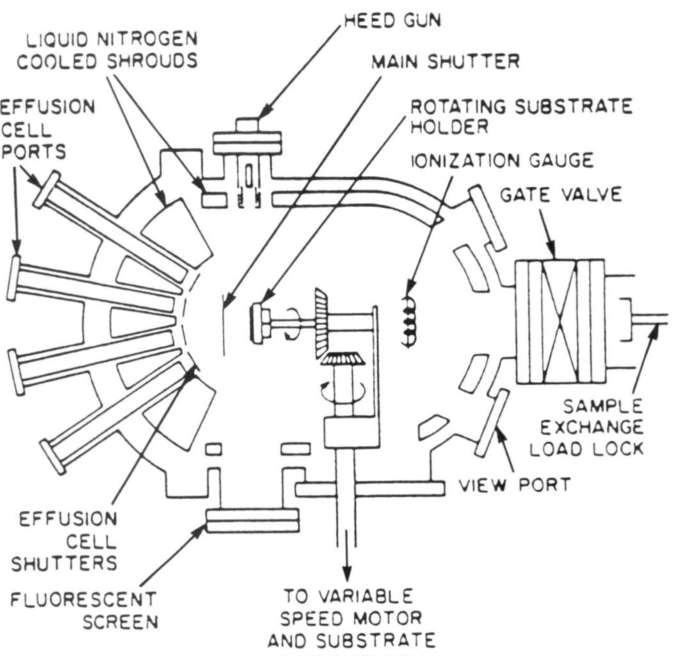

FIG. 3. Cutaway view of a modern MBE system viewed from the top. (After Cho and Cheng, 1981.)

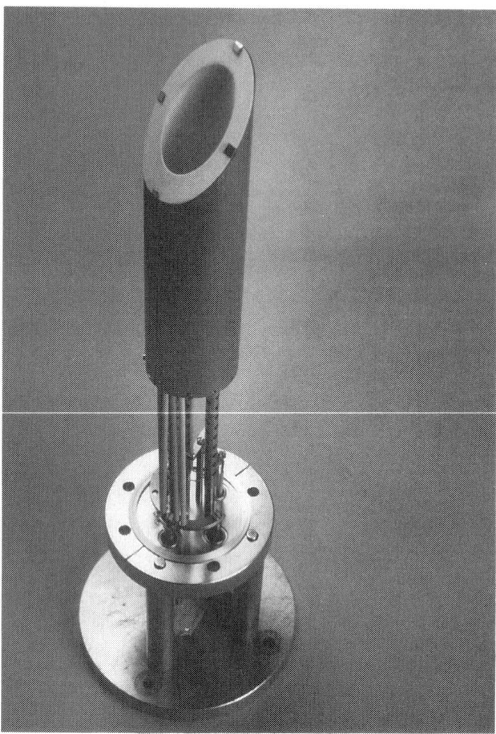

FIG. 4. Photograph of an effusion-cell source. (Photo courtesy of ISA-Riber.)

has eight effusion cells with eight shutters, and a main shutter in front of the substrate.

As outlined above, the effusion cells generate molecular beams that are directed toward the substrate. A heated substrate is required, because epitaxy is a process of depositing a crystalline thin film on a crystalline substrate such that the film has a well-defined crystalline relationship to the substrate. When the molecular beams impinge on the substrate they are absorbed (in other words, they stick to the surface but are not incorporated into the film). Since the substrate is heated, these atoms have the ability to migrate on the surface until they encounter an appropriate site to incorporate. If the substrate were not heated, epitaxy would not generally occur because the impinging atoms would not have this mobility, resulting in an amorphous film.

One of the most significant advantages of MBE is the uniformity and reproducibility that can be achieved. This surface atomic motion mentioned above happens on the scale of atomic dimensions. The rate of deposition is controlled by the intensity of the molecular beam. Therefore, the uniformity of the film depends on the uniformity of the beams. For many reasons, it is impractical to design a source that by itself produces a highly uniform beam. Therefore, a rotating sample (substrate) holder is used to average out beam nonuniformities. The design of these rotating mechanisms is again not a simple problem, and these designs have evolved over many generations. However, it is this innovation that has allowed the degree of uniformity necessary for production of optical and electronic devices (Cho and Cheng, 1981).

Since MBE was a development prompted by investigations of surface physics, many of the surface analytical methods can be used to monitor the MBE process. Such analyses are termed *in situ* diagnostics. One of the most used techniques is that of reflection high-energy electron diffraction (RHEED). An electron gun is mounted in the deposition chamber such that a beam of high-energy (approximately 10 keV) electrons hits the sample at a grazing angle. This technique can monitor the condition of the surface of a growing film and will be described in more detail in a later section. Another diagnostic commonly used in MBE is beam flux measurement. In this technique, an ionization gauge, which measures pressure, is mounted in such a way that it can be positioned in the path of the molecular beams. The pressure reading from this gauge can be converted into a molecular flux (number of molecules passing a unit area per unit time), which is directly related to the deposition rate. In this way, the deposition rate is measured without actual deposition. The pressures measured in this way are called beam equivalent pressures. The relative flux values can be measured to four decimal places if a digital voltmeter is used in conjunction with the commercial gauge readout.

1.2 Stoichiometry and the Surface Phase Diagram

In general, for a compound semiconductor such as GaAs, one of the constituent species will evaporate more readily than the other or, in other words, have a higher vapor pressure. For example, when GaAs is heated, As, which has a higher vapor pressure than Ga, will preferentially evaporate. Therefore it is not desirable in general to use a III-V compound as a source to deposit the same III-V material, because the ratio of the group-V to group III element varies as a function of the effusion-cell temperature. For high growth rates, i.e., high effusion-cell temperatures, the arsenic component in the GaAs source will quickly be depleted.

Historically, several schemes were used to achieve stoichiometric GaAs films. The first method was the "flash evaporation" technique (Richards *et al.* 1963), in which small particles of the source material were completely vaporized on a hot filament. Since the temperatures were very high, essentially all of the source compound was evaporated simultaneously, thus producing a stoichiometric beam flux. A second method was to use two crucibles situated in a way that caused large variations in beam flux across a substrate. In this way the optimum beam ratio would be achieved at some point on the wafer (Richards *et al.*, 1963). The third method, the "three-temperature technique" (Gunther, 1958), approximate that used for MBE. Again, two separate crucibles, one for Ga and one for As, were used and were held each at a different temperature. In this experiment, stoichiometry was achieved in the film because any excess As was reevaporated from the substrate. Note that this technique did not produce single-crystal epitaxial films, because of the vacuum conditions and the amorphous glass substrate. For a more complete account, the reader is referred to Cho and Arthur (1975).

As the development of MBE proceeded, pulsed-beam experiments revealed new insights into the behavior and interaction of Ga and As beams on the GaAs surface. It was found that adsorption and desorption of Ga on the (111) GaAs face obeyed a simple first-order rate equation with temperature-dependent time constants (Arthur, 1968; Cho and Arthur, 1975). In contrast, the behavior of As was found to depend on the surface coverage of Ga. Therefore, stoichiometry could be achieved by providing a significantly higher flux of As such that all Ga reacts and excess As would not stick to the surface. This insight formed the basis for epitaxial growth of III-V compounds by MBE.

While the above-mentioned conclusion is a useful concept, it is somewhat oversimplified. Changes in surface structure occur as the As-to-Ga ratio and substrate temperature are varied. In RHEED studies, various surface structures have been identified and compiled to form the so-called surface phase diagram. Figure 5 shows such a diagram for the GaAs(100) surface (Cho, 1971b, 1983). GaAs is typically grown under As-stabilized conditions because stoichiometry is more readily obtained there.

1.3 Flux Control of Molecular Beams

As already outlined, effusion-cell sources are used to generate the molecular beams used for epitaxy. Each cell contains a thermo-couple for temperature measurement and regulation. Knowing the geometry of the apparatus and the temperature, the beam flux can be calculated as follows:

$$\Gamma = p(T)AN/(2\pi MRT)^{1/2}, \tag{1}$$

where Γ is the number of molecules leaving the effusion-cell aperture, of area A, per second (flux), $p(T)$ is the vapor pressure of the element at the cell temperature T (in kelvins), N is Avogadro's number, M is the mass, and R is the gas constant. If the substrate is positioned at a distance d from the aperture,

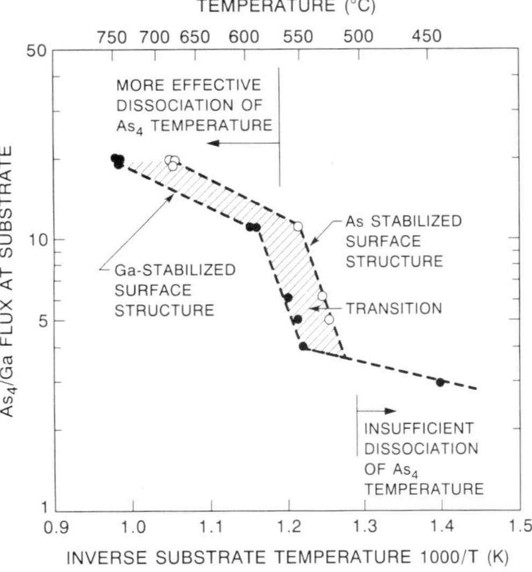

FIG. 5. Surface phase diagram for GaAs on GaAs(100). The plot shows the surface condition at a given temperature and As-to-Ga flux ratio (After Cho, 1983.)

the expression for the number of molecules striking the substrate per unit area is

$$G = 1.118 \times 10^{22} p(T)A/d^2 \times (MT)^{1/2} \text{ molecules cm}^{-2}\text{ s}^{-1}, \quad (2)$$

where $p(T)$ is expressed in Torr.

Once the beam flux for the group-III elements is established, there is a one-to-one correspondence to the deposition rate. In III-V epitaxy, it is usually the case that alloys (or mixtures) of different III-V compounds are deposited. One important example would be the aluminum-gallium-arsenide ($Al_xGa_{1-x}As$) system. Compositional control of such ternary compounds also follows from establishment of beam flux. For example, the amount of Al or the mole fraction x in $Al_xGa_{1-x}As$ can be found as follows:

$$x = G(\text{AlAs})/[G(\text{GaAs}) + G(\text{AlAs})], \quad (3)$$

where G(AlAs) and G(GaAs) are the respective growth rates for AlAs and GaAs and further are directly related to the beam fluxes of Al and Ga. Note that the deposition rates and composition are independent of As flux.

The flux equation above is valid for any effusion source, including dopants. However, the beam flux of a dopant source is usually many orders of magnitude lower than a constituent and therefore is usually not measurable directly. Therefore the beam flux is inferred by growing thin films of, say, GaAs with dopant sources set at various temperatures. Dopant concentrations are then measured *ex situ*. Figure 6 shows a typical plot of dopant concentration versus temperature for many dopants used for GaAs with a nominal growth rate of 1 μm/h. Si and Be are the most commonly used.

2. III-V COMPOUNDS

Molecular beam epitaxy, as outlined in the preceding section, was developed as a growth method for III-V compounds. While many of the original techniques for MBE growth are still in use today, there have also been numerous advances in the technology. In this section, the growth techniques for III-V compounds will be described.

2.1 Conventional Techniques

Given the understanding of the growth process outlined in Sec. 1, the most crucial element of epitaxial growth is that of obtaining a clean and defect-free starting surface. As outlined earlier, epitaxy is a process of depositing a crystalline material on a substrate, also crystalline, such that the structure of the deposited film has a well-defined relationship to that of the substrate. If the starting surface has defects or is contaminated, the overlayer will be inhibited from replicating the crystal surface of the substrate. Under these conditions the deposited film will be of little use.

The technique for substrate surface preparation developed early in MBE technology is still predominantly in use today. The method consists of first polishing, to achieve a nearly atomically smooth surface, and then a chemical etching step to remove surface damage. After this etching step, a thin oxide layer is formed on the surface by immersion in deionized water. This oxide layer protects the surface of the wafer until it can be loaded into the deposition system. Once ready to initiate growth, the wafer is heated to about 580 °C where the oxide layer is desorbed from the surface. When samples prepared and heated in this way are analyzed, there are no detectable surface contaminants. Figure 7 demonstrates this (Cho, 1983), showing Auger-electron spectra both before and after the heating step. Auger-electron spectroscopy (*q.v.*) is an analytical technique that can detect and is especially sensitive to elemental species on a surface.

Once the high-quality surface is obtained in this way, epitaxial growth is initiated. It is beyond the scope of this article to go into detail about specific procedures and detailed recipes used for growth of the various compounds. The power of MBE is the ability to make highly abrupt heterostructures between various compound or elemental materials. Growth of these structures requires optimization of growth parameters.

The growth of these heterostructure materials can be quite complex. Figure 8 shows a cross-sectional view of an innovative laser structure called the vertical-cavity surface-emitting laser. In this structure very precise control of layer thickness and composition is required. As can be seen, the interfaces are

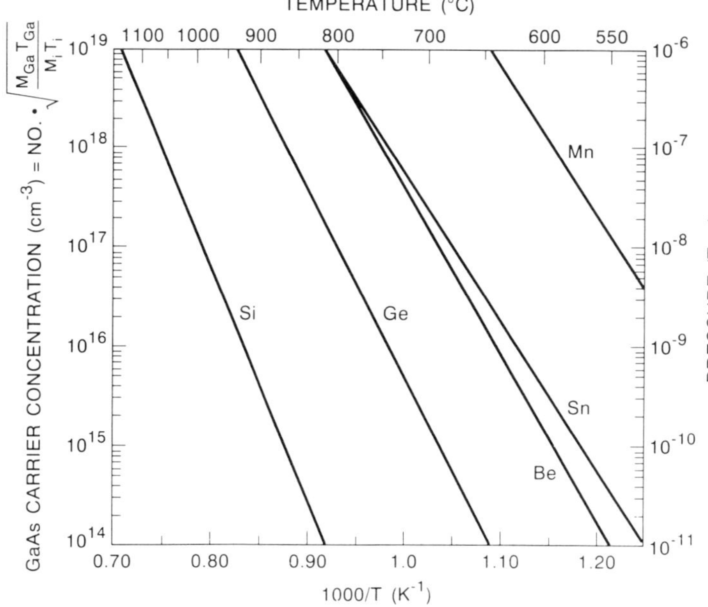

FIG. 6. Dopants used in MBE growth of GaAs. Concentration is plotted as a function of effusion-cell temperature. (After Cho, 1983.)

quite abrupt. In fact, the control offered by MBE allows precise control to the monolayer level (Cho, 1971a). The type of structure shown in Fig. 8 is called a compositional superlattice. It is also possible to modulate doping on this dimensional scale, producing what would be called a doping superlattice.

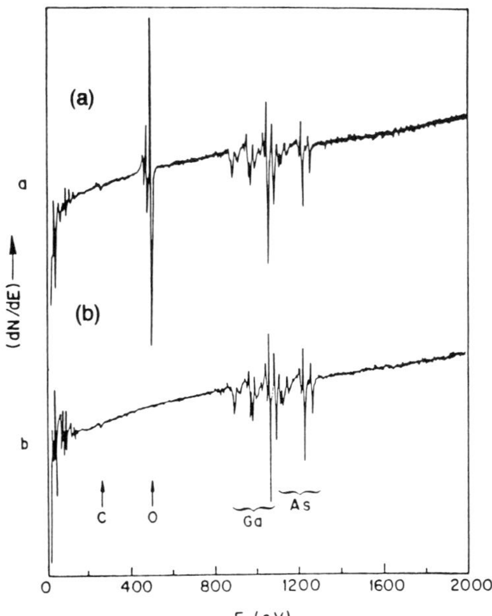

FIG. 7. Auger-electron spectra of GaAs (curve *a*) before and (curve *b*) after heating to desorb the oxide. (After Cho, 1983.)

2.2 *In Situ* Diagnostics and RHEED Oscillations

The above discussion was centered on the importance of clean surfaces. The process of MBE, as pointed out earlier, is a clean-surface technique. In this section we will discuss the methods used to examine surface cleanliness and structure. Because MBE is done in an ultrahigh-vacuum environment, many surface analytical techniques may be used. The important point is that these techniques may be used during the deposition—so called *in situ* analysis.

One of the most important tools of MBE growth is reflection high-energy electron diffraction (RHEED), which was introduced in Sec. 1.1. Figure 9 shows the geometry of the RHEED apparatus. A collimated beam of high-energy electrons is directed toward the sample at a grazing angle (usually 1° to 3°). Because the de Broglie wavelength of an electron at this energy is comparable to the atomic spacings on the surface, a diffraction pattern is formed on a fluorescent screen mounted opposite the electron source. Prior to the use of this geometry, diffraction experiments were carried out by diffracting

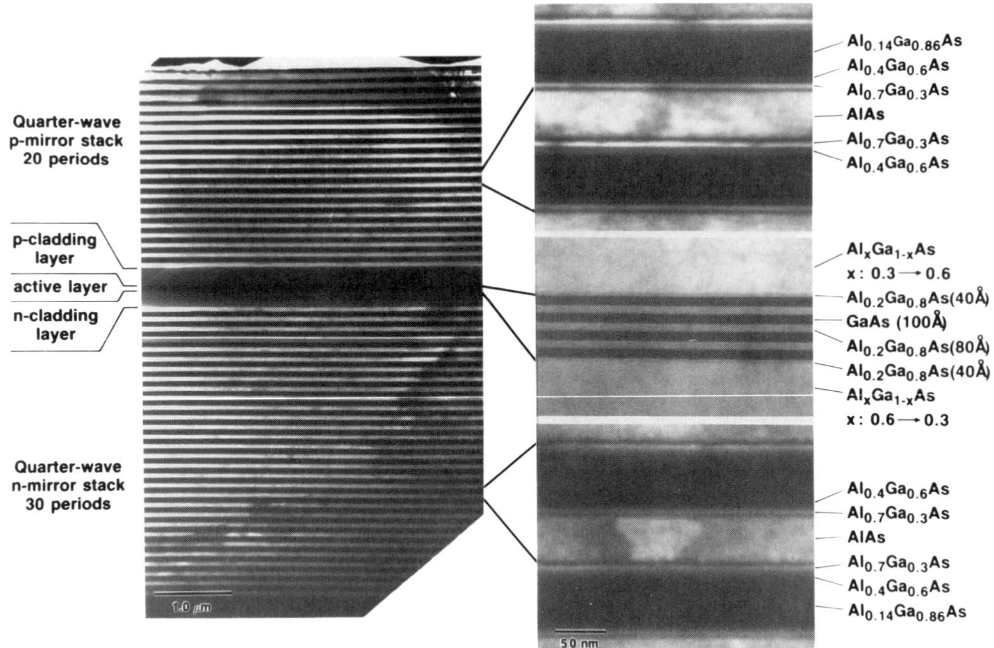

FIG. 8. Cross-sectional view of an MBE-grown vertical-cavity surface-emitting laser. The complexity and precision required in this structure are readily achieved by MBE.

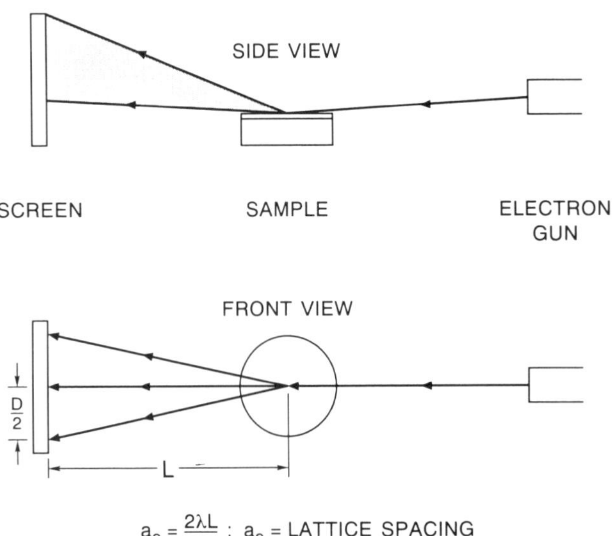

FIG. 9. Geometry of RHEED. This configuration permits observation of diffraction patterns during deposition.

electrons to a screen situated behind the substrate [low-energy electron diffraction (LEED)]. This geometry could not be used during deposition because the sample could not face the effusion cells, and the fluorescent screens would get coated very quickly by the sources.

The diffraction pattern formed by RHEED geometry can be used to distinguish the surface condition (Cho, 1971b), whether primarily As covered or primarily Ga covered. Crystalline surfaces have unsatisfied bonds. Therefore the surface atoms rearrange themselves so as to satisfy these so-called dangling bonds. When GaAs is grown under As-stabilized conditions (surface is primarily As covered) on the ⟨100⟩ orientation, the surface atoms rearrange themselves with a periodicity (or surface structure) twice that of the underlying bulk crystal in one direction and four times that of the bulk crystal in the other. This pattern is termed the (2×4) reconstruction.

The RHEED technique is useful to determine the surface cleanliness. If the surface is covered by an oxide layer, the diffraction pattern either will not appear at all or will not be clear, depending on the thickness of that oxide. In practice, this is used to determine when the oxide layer is completely removed.

The ideal diffraction pattern from a perfectly flat surface consists of lines spaced in accordance with the surface atomic arrangement. In the presence of roughness, these lines become broken or spotty in appearance. This effect is shown in Fig. 10, where growth is initiated on a slightly rough surface. As growth proceeds, this roughness tends to smooth out, and the diffraction lines become uniform streaks.

Another useful feature of RHEED in MBE growth is the RHEED intensity oscillations (Wood, 1981). MBE growth proceeds in a layer-by-layer mode. When the details of how each layer is formed are examined more closely, it is found that the individual layer starts at random places on the surface (called nucleation sites) and starts to form monolayer "islands." These islands grow in size until they run into one another and coalesce. Per the preceding discussion, such island formation is a height variation and can cause interference in the diffracted beam. The intensity would be maximum at the completion of a monolayer and minimum at some intermediate (half-monolayer) point. These oscillations, then, correspond exactly with the deposition of a monolayer. This fact may be used for calibration of growth rates, as the deposition of an individual monolayer can be accurately determined (Van Hove, 1983).

2.3 Gas-Phase Sources

The preceding discussions have centered on the use of evaporation sources. A more recent development in MBE technology has been the use of gas-phase sources. These sources are similar to those used in chemical vapor deposition [see CVD (CHEMICAL VAPOR DEPOSITION)]. The important distinction is that use of these sources in MBE is done at much lower pressures and at much lower flow rates. Since pressures are kept very low, individual gas molecules do not interact, and therefore the beam nature of MBE is preserved.

There are several variations of technique for gas-phase MBE. The sources for group-III elements are typically organometallic (such as trimethyl gallium), and the group V are typically group V hybrides (such as arsine or phosphine) (Tsang, 1985). In principle, it is possible to use combinations of solid and gas-phase sources, but the two most common configurations are using either all gas-phase sources (organometallic group III and hydride group V) or standard evaporative sources for group III and hydride group V (Panish, 1980).

The initial motivation for using gas-phase sources for MBE was to grow phorphorus-containing compounds. It is difficult to use solid phosphorus evaporation sources, because the evaporation of phosphorus is not easy to control and reproduce. While recent innovations in valved cracker-cell design hold promise for obtaining this control and reproducibility in solid-evaporation P sources, it is quite simple to control the flow of a gas.

The design of a gas-phase source includes a source, such as a bubbler for organometallics or a gas cylinder for hydrides, a flow-control device, such as an electronic mass-flow controller, and an injector. The metal atom in an organometallic compound is usually not strongly bound to the organic radicals, and therefore the heat of the substrate is more than adequate to crack the molecule. Thus, the design of a organometallic injector is only re-

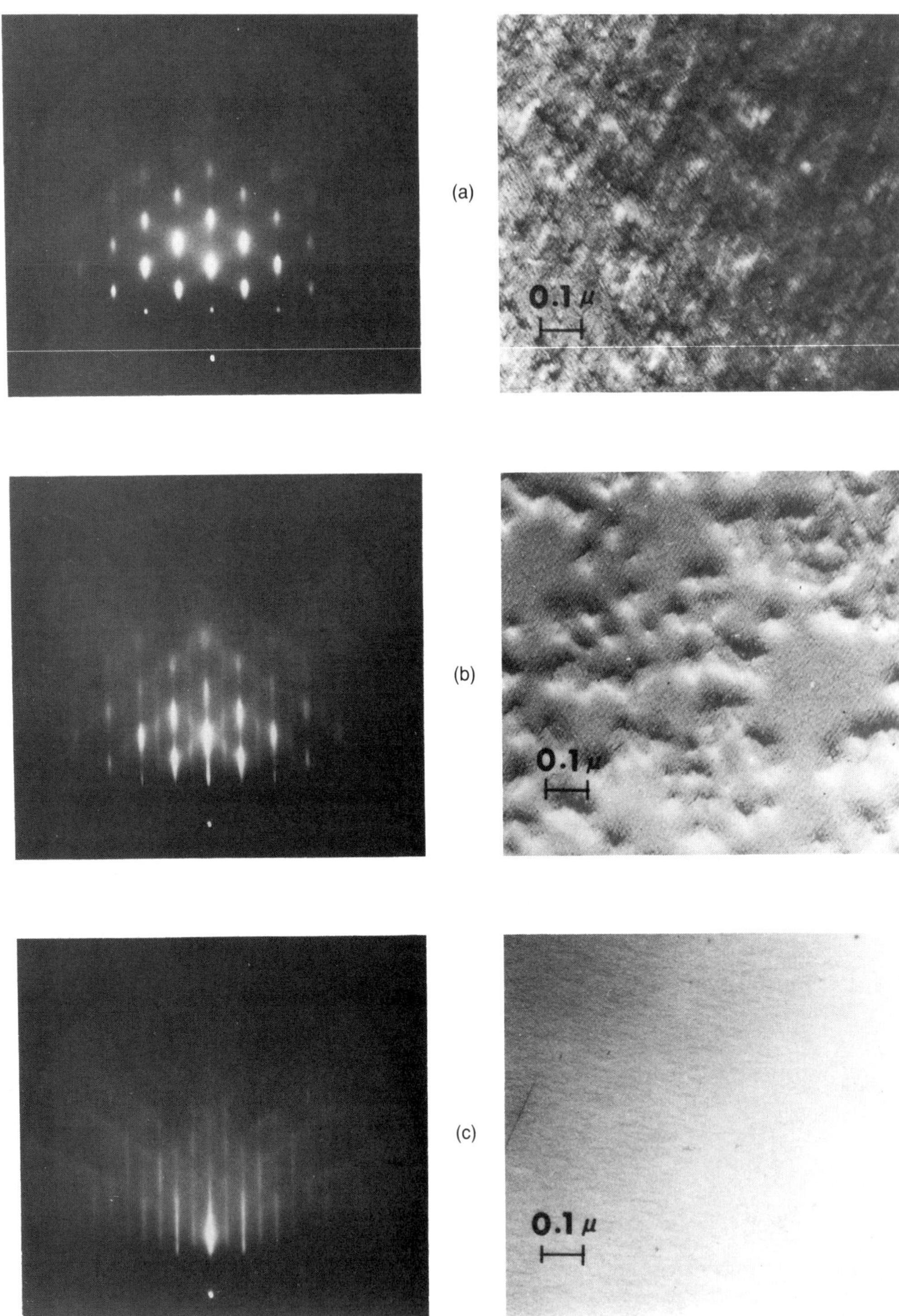

FIG. 10. RHEED pattern of a typical sequence of growth. (a) The pattern after the oxide is desorbed, (b) after deposition of 150 Å, and (c) after 1 μm of deposition. Corresponding electron micrographs are PtC replica of the surface. (After Cho 1983.)

quired to provide a uniform flux of gas molecules at the substrate. In contrast, the hydride injector must crack (thermally decompose) the hydride molecule.

While there are some differences in behavior between hydrides and solid group-V sources, aside from the controllability issue, there is little difference in use of these sources. On the other hand, metalorganic sources have some important differences from solid group-III sources. The distinction is that there is a chemical reaction that takes place on the substrate surface. This means that the species present just before condensation into the solid phase is quite different. There is evidence that the organometallic sources give much larger surface diffusion lengths, give different nucleation behavior, and, because of the different chemical kinetics, show a dependence of growth rate on substrate temperature. These gas-phase techniques have allowed MBE crystal growers to achieve high-quality InP/InGaAsP optical devices, and visible lasers using AlInP/GaInP.

2.4 Device Examples

As pointed out already, MBE has matured to the point of being a production tool in the semiconductor industry. In this section, we will discuss some examples of device structures that are in production using MBE material. While there are many new and interesting device structures made possible by MBE (the field of band-gap engineering) we will limit our discussion to only two examples.

The first example is the modulation-doped field-effect transistor (MODFET, sometimes called HEMT for high–electron-mobility transistor). This device has found a great deal of use in low-noise microwave receivers. Its low noise and high speed are attributable to the superior electron transport of the structure (Fischer and Morkoc, 1986). In a normal field-effect transistor, electrons travel through doped parts of the crystal. When a dopant is introduced into a crystal, it leaves a charged ion in the crystal lattice. This charged ion affects the electrons passing by, deflecting them through the electrostatic interaction. This effect of disturbing the electron motion results in inferior transport. The MODFET uses a heterostructure, that is, a junction between two dissimilar materials, to isolate the electrons from these ionized dopants.

Figure 11 shows an electron energy diagram of a MODFET. Doping is introduced into the material with higher band-gap energy (AlGaAs), and the free electrons transfer into the lower band-gap energy material (GaAs) to minimize their total energy. As can be seen, the electrons now travel in a part of the crystal where there is no doping and are essentially unaffected by the ionized impurities.

A metal electrode is placed on the surface to control the number of electrons available to carry current. In this way the applied voltage controls the output current of the device. From a materials point of view, it is important that the thickness and doping of the AlGaAs be highly reproducible. If there is variation in either one of these parameters, the devices will not switch on and off at the same voltage, making integration difficult, if not impossible. The other important aspect is the quality of the interface between the AlGaAs and GaAs. Any roughness or irregularity will degrade the electron transport properties. MBE has demonstrated its ability to produce highly uniform layers, in composition, thickness, and doping. The layer-by-layer nature of MBE growth also ensures a high-quality interface.

The second example of MBE as a production tool is in the AlGaAs lasers for compact-disk players. The key issues for this technology are the reproducibility in composition and thickness, and the purity of the material. The

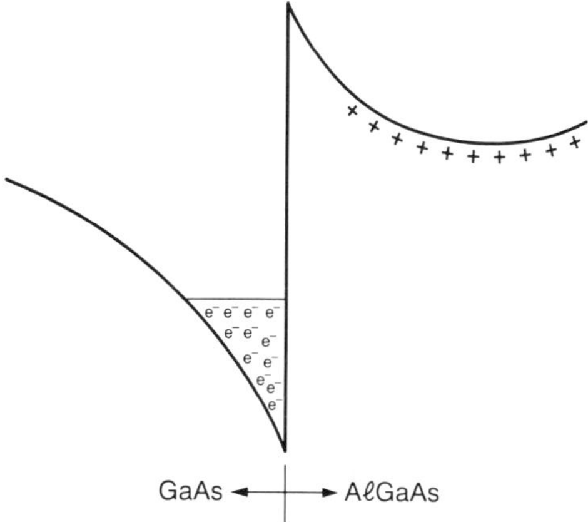

FIG. 11. Band-edge diagram of a typical MODFET. Electrons transfer from the AlGaAs into the GaAs and are separated from the ionized donors.

injection laser uses a *p-n* junction to inject electrons and holes into a semiconductor slab such that it will have optical gain. If certain types of defects are present, the electrons and holes will recombine without emitting light. If the thickness of the layer structure is not well controlled, the threshold current and emission wavelength will vary. Fortunately, MBE again provides excellent control over both these parameters.

The leading producer of compact-disk lasers, Rohm Co., Ltd., of Japan, uses a two-step MBE growth process for production of their lasers (Tanaka and Mushiage, 1991). This two-step process automatically produces a stripe geometry, which controls the optical (transverse and longitudinal) modes of the laser. Since MBE provides a high degree of reproducibility, the yields are very high in this process, which results in high-quality and low-cost lasers for many applications.

3. GROUP-IV MATERIALS

In this section, the growth of group-IV materials by MBE will be outlined. There are many important distinctions between the group-IV semiconductors and the III-V compounds. MBE as it has been described thus far relates to the III-V compound growth. Since the group-IV semiconductors are quite different, the MBE apparatus will also be quite different.

3.1 Apparatus and Growth Technique

The most technologically important group-IV semiconductor, is, of course, silicon. One of the primary differences is that the temperature required to produce an adequate flux of Si is far too high for an effusion cell of the conventional type. Therefore, electron-beam sources must be used. These sources use a beam of high-energy electrons at a relatively high current to melt a source material contained in a hearth. The walls of the hearth are cooled so that only the center portion of the source material is hot enough to evaporate, and thus serves as its own crucible. The use of electron-beam evaporation is not as stable as an effusion cell, and real-time flux monitoring must be used to control the power supplied to evaporate the source.

Another difference is that almost all evaporated doping elements surface segregate on the growing Si surface (Bean, 1981). Segregation is an effect where the surface concentration is larger than that incorporated in the film. In other words, there is a buildup and tailing off of doping atoms on the surface when the doping is switched on and off, respectively. Obviously, this is an undesirable effect for devices that require sharp doping profiles. This effect also limits the maximum concentration of doping attainable, and in some cases can cause degradation of crystal quality. To get around this problem, molecular doping sources and low-energy implants are used.

The molecular species used for doping are boron oxide, B_2O_3, and boric acid, HBO_3. Effusion cells can produce a wide range of concentrations with these sources, and with good control. The use of B_2O_3, however, results in high concentrations of oxygen in the films, and only at high temperature (so high that the doping profile cannot be maintained) does the oxygen content become negligible. The use of HBO_3, on the other hand, seems to work much better, with temperatures of only about 650 °C required to eliminate oxygen. At this temperature, there is no evidence of boron movement or surface segregation.

The other method of doping is to use simultaneous low-energy ion implantation. In this process, the dopant species are ionized and accelerated to energies of 200–1000 eV. In this way, the dopant atom is buried within the film so that the surface segregation effect is reduced. Figure 12 shows the behavior of In ions co-implanted into Si at various energies as a function of temperature (Hasan, 1989). The drawback of this method is that there is the possibility of residual radiation damage. Many studies have been done, and it was found that when growth temperature is reduced from 650 to 500 °C, some features were observed that indicate incomplete annealing of the radiation damage. Other studies showed that a 2-keV implant grown at 550 °C produced little or no degradation in minority-carrier lifetime.

Growth of Si films by MBE proceeds in much the same way as for III-V MBE discussed earlier. The surface cleaning method is again a critical step. The Si oxide is a much more stable compound than the oxide on a III-V compound semiconductor. Therefore unless a special chemical process is used to

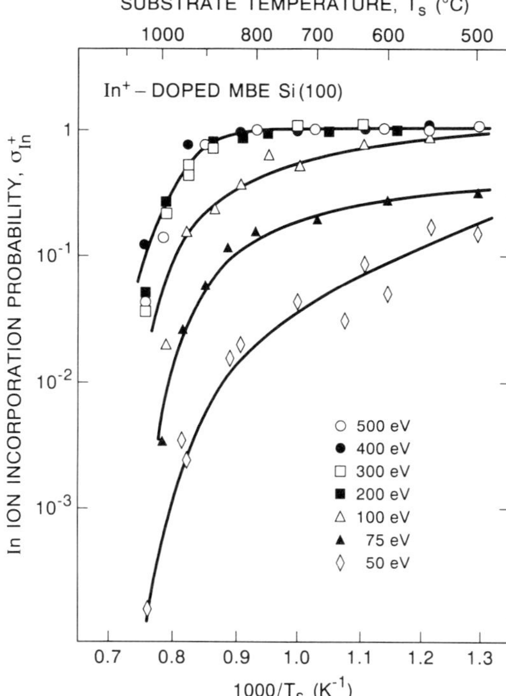

FIG. 12. Incorporation probability of In co-implanted into Si during MBE as a function of substrate temperature. The curves each correspond to a different ion energy. (After Hasan *et al.*, 1989; used by permission.)

produce a very thin oxide, it is impossible to remove the oxide thermally once loaded into the MBE chamber. This so-called "Shiraki clean" is typically used and results in an extremely thin and volatile oxide, which is sufficient to protect the substrate surface during the loading operation. With the wafer prepared in this way, it is possible to remove the oxide thermally.

Silicon technology is much more mature, and therefore particulate levels are of great concern, because of the scale of integration. Material deposited on the walls of a Si MBE chamber can be released as fine dust, and therefore many precautions are taken to reduce this effect. For example, in Si MBE, water cooling of the shrouding can be used in place of liquid nitrogen. This reduced thermal cycling greatly reduces the particle generation. Another approach is to steer particulates away from the substrate surface by means of electrostatic fields.

As was the case with III-V MBE, there are certain advantages to using gas-phase sources in Si MBE. While this concept was investigated at quite an early stage, the experiments used pumping systems that resulted in formation of SiC. More recently, improvements in pumping technology have eliminated these problems. The most obvious source for growth by Si gas-source MBE is silane, SiH_4. However, growth temperatures of greater than 700 °C are required to achieve reasonable growth rates. Furthermore, with silane as a source, the growth rate is quite sensitive to growth temperature. Most recently, the use of disilane, Si_2H_6, has increased low-temperature growth rates and is much less temperature sensitive.

3.2 Growth of SiGe

One of the greatest strengths of MBE is its ability to make highly abrupt compositional profiles. In this section, we will discuss the growth and material properties of a group-IV heterostructure system, SiGe/Si.

Germanium, while it is group-IV material as is Si, does not lattice match to Si. In other words, the atomic spacing of Ge is about 4% larger than that of Si. This mismatch makes growth of these structures difficult, because if thickness or temperatures are not within certain limits, the films will have many dislocations, which are defects that degrade the electronic performance.

One of the most sensitive tests of a material is its performance in a bipolar transistor. Optimization of SiGe growth parameters by MBE has allowed extremely high current gains of 750 to 2000 (depending on the structure), and operating frequencies of 20 GHz. This performance demonstrates the quality of SiGe films produced by MBE (Bean, 1990).

As discussed in the previous section, MODFETs, which use the properties of a heterojunction to improve electron transport, can also be done with SiGe/Si. This concept has been successfully applied in the MBE-grown SiGe/Si system. Hole mobilities as high as 18 000 cm^2/V s and electron mobilities as high as 180 000 cm^2/V s (Xie, *et al.* 1993) have been obtained in MBE-grown structures at 4.2 K.

4. II-VI COMPOUNDS

Another class of compound semiconductors are the II-VIs, which are made up from an element from column II and an element from column VI of the periodic table. The most important examples of II-VI compounds are HgTe, CdTe, ZnTe, ZnS, and ZnSe. The interest in HgTe and CdTe has been for long-wavelength infrared detectors, while in contrast, ZnTe, ZnS, and ZnSe, having wide band gaps, are interesting for visible-light emitting devices.

4.1 Narrow-Bandgap Materials

The growth of narrow-band-gap II-VI materials by MBE can be done using sources that are in principle the same as conventional ones. Usually for the growth of HgTe, a source of Hg and a source of Te are used. For the growth of CdTe, a CdTe compound source can be used, since Te is more volatile and therefore produces the excess flux required for growth of this compound. The growth of Hg-containing compounds is quite difficult in practice since at the typical growth temperatures, the sticking coefficient of Hg can be as low as 1 part in 1000. Therefore, a high flux of Hg is required to grow HgTe. This high-Hg flux depletes the source rapidly and therefore source designs which allow replenishment of Hg are required.

The first experiments with growth of HgTe/CdTe were done on ⟨111⟩-oriented CdTe substrates. It was later found that although there is a very large mismatch (14.5%) between CdTe and GaAs, HgCdTe grows well on GaAs substrates. This is a great advantage since the relatively well-developed technology of GaAs substrates can be used. It is interesting that in the growth of this material system on GaAs, the temperature used to desorb the oxide (preheating temperature) has a profound effect on the crystal orientation of the epitaxial layer. If the substrate is preheated to 480 °C or less, then CdTe grows in the ⟨100⟩ orientation. If preheated to 580 °C, then the CdTe grows in the ⟨111⟩ orientation. When the ⟨111⟩ orientation is grown, about four times less Hg flux is required compared to the ⟨100⟩ orientation. The concerns in growing this material system are interdiffusion of the HgTe/CdTe interface, and the exact electronic nature of that interface (Faurie, 1986).

4.2 Wide-Band-Gap Materials

As pointed out earlier, the wide-band-gap II-VI materials are of interest because of their application in visible-light emitters. While ZnTe emits in the green portion of the spectrum, ZnSe is in the blue. Only recently have semiconductor lasers been fabricated in this system (Haase *et al.*, 1991; Jeon *et al.*, 1991) that emit in the blue-green. While there are still many technological problems to be solved, MBE growth has been instrumental in the demonstration of these devices.

The single most problematic issue in the growth of these materials is doping. The II-VI compounds have a tendency to resist substitutional doping—the tellurides tend to be *p* type and the selenides *n* type. When these crystals are grown, introduction of impurities tends to generate compensating defects. The reason MBE has been successful in achieving this doping is that it is a nonequilibrium process.

The sources used for this system can be compound or elemental. In the growth of laser material mentioned above, elemental Zn, Se, and Cd were used, while a compound source, ZnS, was used for sulfur-containing materials. The elemental sources in II-VI growth tend to have very high vapor pressure at relatively low source temperature, making control of flux difficult. To make matters worse, very fine control of the group-II to group-VI ratio must be maintained to avoid generation of defects. Here as in earlier discussions, RHEED is an invaluable tool for determining the surface stoichiometry (whether a group-II-rich or a group-VI-rich surface). For example, a C(2 × 2) reconstruction pattern is observed for a Zn-stabilized surface, while the (2 × 1) reconstruction appears for a Se-stabilized surface. For a more complete discussion of II-VI compound growth the reader is referred to Gunshor *et al.* (1991).

The *p*-type doping of ZnSe has been historically a very difficult problem. In order to enhance the doping properties, a photoassisted technique was developed (Bicknell *et al.*, 1986) for In doping in CdTe. While dramatic improvements were made, *p*-type doping of ZnSe could not be adequately controlled. Most recently, use of nitrogen plasma sources have been successful to introduce *p*-type doping into ZnSe grown by MBE. This, coupled with the ability of MBE to produce high-quality heterostructures, has allowed the successful demonstration of blue-green semiconductor lasers in this system. Figure 13 shows a light output versus current input characteristic for this laser device.

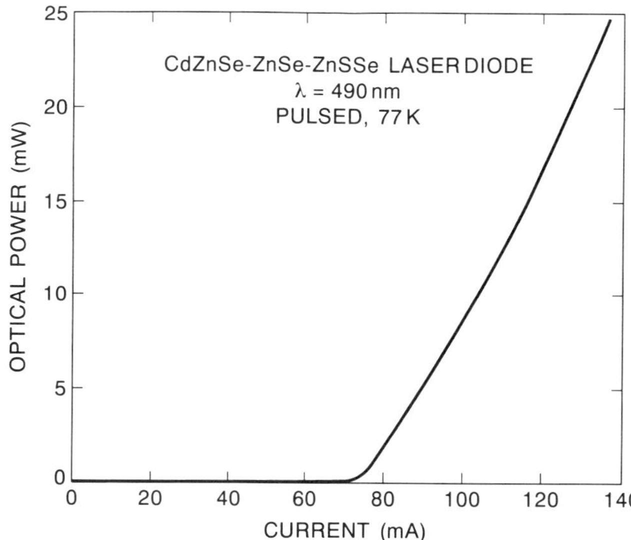

FIG. 13. Light output versus current input characteristic for a blue-green laser diode. (After Haase *et al.*, 1991; used by permission.)

5. OTHER MATERIAL SYSTEMS

In this section we will discuss some of the other material systems that have been grown using MBE or variations of MBE. Examples include epitaxial metal films, silicides, and high-temperature superconducting materials. The wide range of materials produced by MBE shows the diversity of the technique.

5.1 Epitaxial Metal Films

Metals as they occur in nature are polycrystalline in form; that is, there are small crystalline domains that are randomly oriented with respect to one another. If thin metal films are deposited under carefully controlled conditions, the metal films can be epitaxial. These epitaxial films, which uniformly have the same crystal structure and orientation, have useful properties. Because of the high degree of perfection for an epitaxial metal film interface, Schottky barriers could potentially be more reproducible and reliable, and there is the possibility of making metallic quantum wells and devices using buried metal films. Recent efforts have also included thin magnetic films for data storage applications.

Sources for metal deposition usually include both conventional effusion cells for metals with high vapor pressures, and electron-beam sources for those with low vapor pressures. While not used exclusively, semiconductor substrates are normally used, since many of the applications for these films are related to semiconductor devices.

The earliest work on depositing epitaxial metal films was Al on GaAs (Cho and Dernier, 1978) and Fe (Prinz and Krebs, 1981). More recent examples of metal films grown by MBE include NiAl on GaAs (Harbison *et al.*, 1988), single-crystal Dy (Yang *et al.*, 1988), Gd/Y superlattices (Kwo *et al.*, 1986), Dy/Y superlattices (Salamon *et al.*, 1986), and $Fe_3(Al,Si)$ on GaAs (Hong *et al.*, 1991). Figure 14 shows a sequence of RHEED patterns from deposition of $Fe_3(Al,Si)$ on GaAs. As can be seen through the quality of the patterns, highly ordered and atomically flat films were obtained.

5.2 Epitaxial Silicides

For electronic device fabrication, highly perfect interfaces between heterolayers is crucial. Higher quality and higher purity at these interfaces nearly always leads to superior electronic performance. Single-crystal epitaxial silicides on Si substrates are among the most perfect metal-semiconductor interfaces. The applications for such structures range from contacts to high-speed devices to the possibility for three-dimensional integration. Aside from the technological importance, the degree of perfection provides a model system to study the intrinsic properties of metal-semiconductor interfaces. Ex-

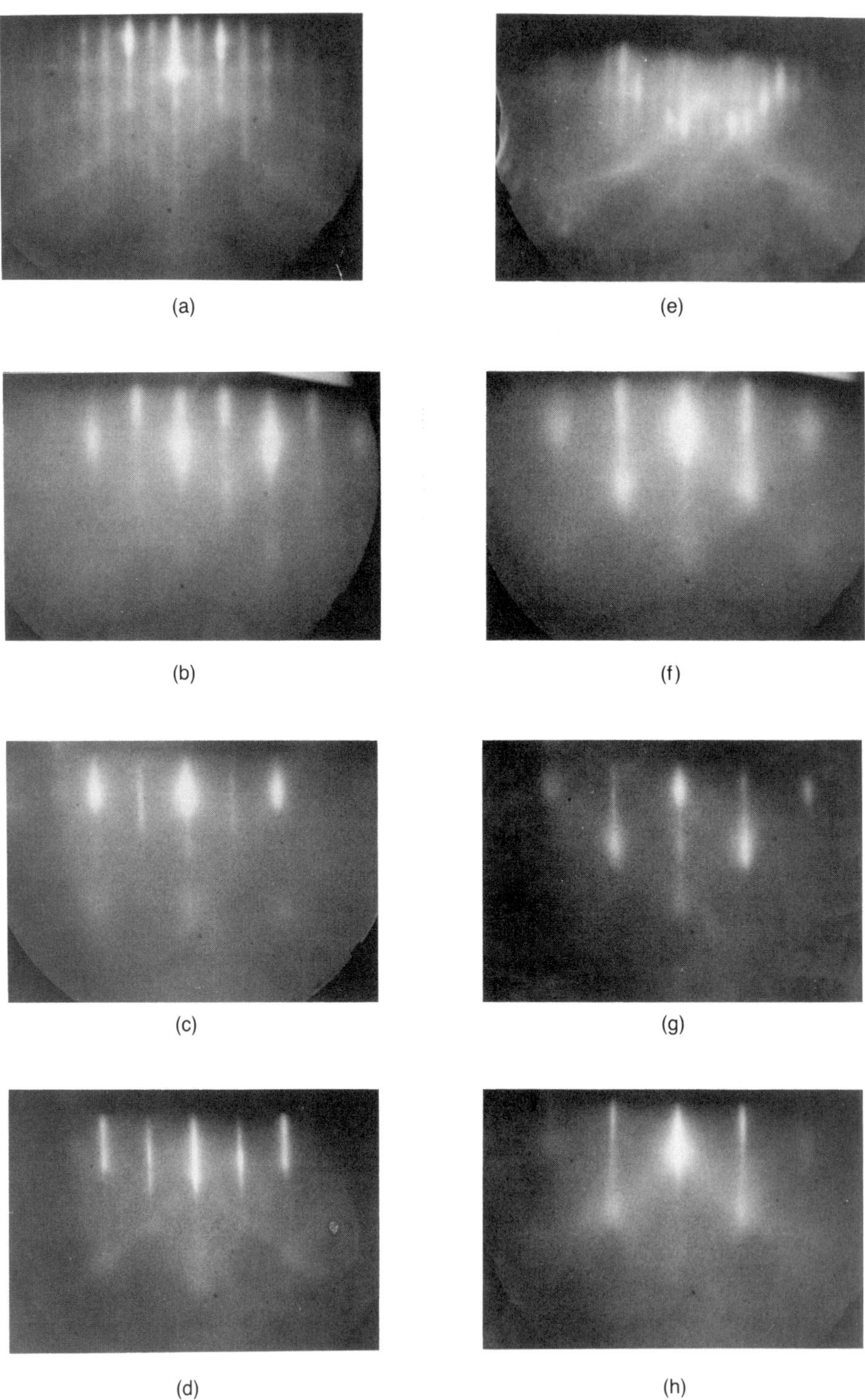

FIG. 14. RHEED patterns from the deposition of Fe(Al,Si). Patterns (a)–(d) are taken along the [011] direction, while (e)–(h) are along $[0\bar{1}1]$. (a), (e) The starting GaAs surface; (b), (f) after deposition of one monolayer of Fe_3Al; (c), (g) after growth of three monolayers; (d), (h) after growth of 12 monolayers. (After Hong *et al.*, 1991; used by permission.)

amples of epitaxial silicides would be $NiSi_2$ and $CoSi_2$.

Epitaxial silicides on Si have been grown by several methods, but the ability of MBE to produce clean interfaces and to maintain high purity in deposition allow it to produce the highest-quality films. Sources for this type of growth are typically electron-beam guns, because the elements have very low vapor pressures. In order to control stoichiometry, a coevaporation technique is used; that is, there are two electron-beam sources, one for the metal and the other for Si. The fluxes are adjusted so that a stoichiometric ratio is obtained.

The problems of silicide growth in MBE are in the deposition of the first monolayers. Whenever monolayers or submonolayers of silicides are deposited on Si, there is a tendency for the Si to deposit as crystalline Si on the substrate. As can be seen in Fig. 15, there is a distinct structural difference between crystalline Si and $NiSi_2$. To suppress this tendency, a method was developed called the template technique (Tung, 1992). This method is a two-step growth process in which the initial monolayers are deposited under radically different conditions than the remainder of the film. In silicides, this first nucleation process is done at very low temperatures, which promotes the formation of the silicide as opposed to the Si structure. Once the first few monolayers are deposited, there is no longer this tendency, because the Si atoms no longer see the Si substrate. At this point the temperature can be raised to the optimum for silicide growth.

Because of the difference in structure between Si and the silicide, noted above, it turns out that there are in fact two different ways that the epitaxial layer can orient itself with respect to the substrate. These two orientations are termed type A, which has the same orientation as the substrate, and type B, which is rotated from the axis of the Si. The details of the nucleation process determine which structure will predominate.

5.3 Superconducting Materials

Over the past several years, a new field of materials science and physics has emerged, that being high-temperature superconductivity. This has been brought about by the discovery of the so-called layered perovskite oxides. An example of such a material is $Y_1Ba_2Cu_3O_{7-x}$ (yttrium barium copper oxide). While the initial discovery of this material was not made by MBE, its structure is a layered one, and because of the layer-by-layer nature of MBE growth, much can be learned about the properties of these materials through the MBE growth with its associated *in situ* analysis. Furthermore, it allows one to tailor the structure on an atomic level.

Figure 16 shows the configuration of an MBE apparatus used to grow these compounds. Again a combination of electron-beam sources and effusion-cell sources is used. To provide oxygen, a separate injection tube is located in close proximity to the sample. In fact, in order to reduce the amount of oxygen required for growth, an activated oxygen source is used. The substrates used for these compounds vary and include MgO, $SrTiO_3$, and $LaAlO_3$. Because of the highly oxidizing environment, many changes in the materials of construction for the system must be made.

Use of the MBE technique has yielded very

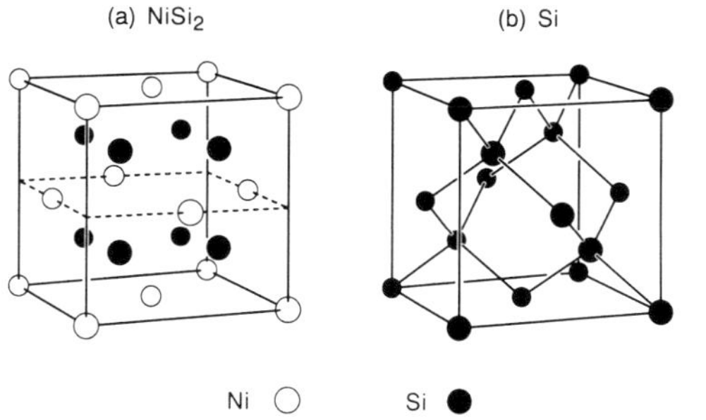

(a) $NiSi_2$ (b) Si

Ni ○ Si ●

FIG. 15. Structure of NiSi and of Si. (After Tung, 1992; used by permission.)

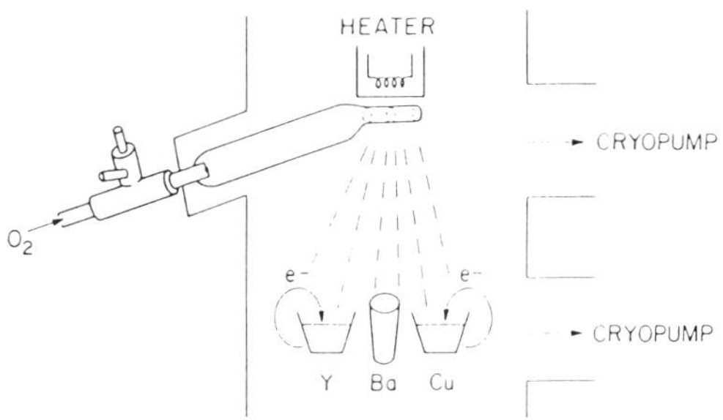

FIG. 16. Configuration of an MBE apparatus used for superconductor films. (After Kwo, 1991; used by permission).

high-quality superconducting films. Because of the controllability of the MBE process, some progress has been made in achieving tunnel junctions in this material. Josephson current has been observed (Kwo, 1991) between a high-temperature superconducting oxide and a conventional superconductor. This field has now been termed "molecular engineering of oxides."

GLOSSARY

Adsorption: Process by which an atom or molecule is bound to a surface, but is not incorporated into the underlying material.

Compound Semiconductor: A semiconductor that is composed of two or more elements, e.g., GaAs, AlGaAs.

Desorption: Process by which atoms or molecules bound to the surface are released into the vapor phase.

Dopant: An impurity added to a semiconductor in order to change its conductivity.

Epitaxy: The process of depositing one crystalline material on another such that the deposited material has a definite crystalline relationship to the substrate.

Heterostructure: A metallurgical junction between two dissimilar materials.

Mismatch: The difference, measured in percentage, between the atomic spacing in the substrate compared to that of the material being deposited.

Segregation: The tendency of some dopant species to stay at the surface when present during epitaxy.

Stoichiometry: The precise chemical ratio of one element to another in a compound semiconductor.

Substrate: A crystalline material that is usually flat and polished (in the form of a wafer), and that is used as a starting material for epitaxy.

Vapor Pressure: The tendency of a species to evaporate into the vapor form, measured in units of pressure at a given temperature.

Works Cited

Arthur, J. R. (1968), *J. Appl. Phys.* **39**, 4032–4034.

Bean, J. C. (1981), in: F. F. Y. Wang (Ed.), *Growth of Doped Silicon by Molecular Beam Epitaxy*, Amsterdam: North-Holland, Chap. 4, p. 177.

Bean, J. C. (1990), *J. Electron. Mater.* **19**, 1055–1059.

Bicknell, R. N., Giles, N. C., Schetzina, J. F. (1986), *Appl. Phys. Lett.* **49**, 1095–1097.

Cho, A. Y. (1971a), *Appl. Phys. Lett.* **19**, 467–468.

Cho, A. Y. (1971b), *J. Appl. Phys.* **42**, 2074–2081.

Cho, A. Y. (1983), *Thin Solid Films* **100**, 291–317.

Cho, A. Y., Arthur, J. R. (1975), in: J. McCaldin, G. Somorjai (Eds.), *Progress in Solid State Chemistry*, Vol. 10, Oxford: Pergamon Press, pp. 157–191.

Cho, A. Y., Cheng, K. Y. (1981), *Appl. Phys. Lett.* **38**, 360–362.

Cho, A. Y., Dernier, P. D. (1978), *J. Appl. Phys.* **49**, 3328–3332.

Faurie, J. P. (1986), *IEEE. J. Quantum Electron.* **QE-22**, 1656–1665.

Fischer, R., Morkoc, H. (1986), *IEEE Circuits Devices Mag.* **1**, 35–38.

Gunshor, R. L., Kolodziejski, L. A., Nurmikko, A. V., Otsuka, N. (1991), *Semiconductors and Semimetals*, Vol. 33, Chap. 6, p. 337.

Gunther, Von K. G. (1958), *Z. Naturforsch.* **13a**, 1081–1089.

Haase, M. A., Qui, J., DePuydt, J., Cheng, H. (1991), *Appl. Phys. Lett.* **59**, 1272–1274.

Harbison, J. P., Sands, T., Tabatabaie, N., Chan, W. K., Florez, L. T., Kerimidas, V. G. (1988), *Appl. Phys. Lett.* **53**, 1717–1719.

Hasan, M. A., Knall, J., Barnett, S. A., Sundgren, J. E., Markert, L. C., Rockett, A., J. E. Greene (1989), *J. Appl. Phys.* **65**, 172–179.

Hong, M., Chen, H. S., Kwo, J., Kortan, A. R., Mannaerts, J. P., Weir, B. E., Feldman, L. C. (1991), *J. Crystal Growth* **111**, 984–988.

Jeon, H., Ding, J., Patterson, W., Nurmikko, A. V., Xie, W., Grillo, D. C., Dobayashi, M., Gunshor, R. L. (1991), *Appl. Phys. Lett.* **59**, 3619–3621.

Kwo, J. (1991), *J. Crystal Growth* **111**, 965–972.

Kwo, J., Gyorgy, E. M., Disalvo, F. J., Horg, M., Yafet, Y., McWhan, D. B. (1986), *J. Magn. Magn. Mater.* **54–57**, 771–772.

Panish, M. B. (1980), *J. Electrochem. Soc.* **127**, 2729–2733.

Prinz, G. A., Krebs, J. J. (1981), *Appl. Phys. Lett.* **39**, 397–399.

Richards, J. L., Hart, P. B., Gallone, L. M. (1963), *J. Appl. Phys.* **34**, 3418–3420.

Salamon, M. B., Sinha, S., Rhyne, J. J., Cunningham, J. E., Erwin, R. W., Borchers, J., Flynn, C. P. (1986), *Phys. Rev. Lett.* **56**, 259–262.

Tanaka, H., Mushiage, M. (1991), *J. Crystal Growth* **111**, 1043–1046.

Tsang, W. T. (1985), *Appl. Phys. Lett.* **46**, 1086–1088.

Tung, R. (1992), in: S. Mahajan (Ed.), *Handbook on Semiconductors*, 2nd ed., Vol. 3, Amsterdam: Elsevier.

Van Hove, J. M., Lent, C. S., Pukite, P. R., Cohen, P. I. (1983), *J. Vac. Sci. Technol.* B **1**, 741–746.

Wood, C. E. C. (1981), *Surf. Sci.* **108**, L441–L443.

Xie, Y. H., Fitzgerald, E. A., Monroe, D. P., Silverman, P. J., Kortan, A. R. (1993) (unpublished).

Yang, K. Y., Homma, H., Schuller, I. K. (1988), *J. Appl. Phys.* **63**, 4066–4068.

Further Reading

For further reading there are two books devoted to various topics of molecular beam epitaxy. One is L. L. Chang and K. Ploog (Eds.), *Molecular Beam Epitaxy*, Dordrecht: Martinus Nijhoff (1985), and the other is E. H. C. Parker (Ed.), *Physics and Technology of Molecular Beam Epitaxy*, New York: Plenum (1985). Scientific articles relating to MBE work can be found in many journals, but primarily in *Journal of Applied Physics* and *Applied Physics Letters*. The proceedings of the MBE Workshop are published in the *Journal of Vacuum Science and Technology B* and the *Journal of Crystal Growth*. We, the authors, have attempted to cite review articles that cover the topics we have discussed in more depth.

MOLECULAR DYNAMICS

See SIMULATION BY MOLECULAR DYNAMICS

MOLECULAR INTERACTIONS WITH PHOTONS

See PHOTON INTERACTIONS WITH MOLECULES

MOLECULAR SPECTROSCOPY

D. A. Ramsay, *Herzberg Institute of Astrophysics, National Research Council of Canada, Ottawa, Ontario, Canada*

INTRODUCTION

Molecular spectroscopy is primarily the study of the emission or absorption of electromagnetic radiation by molecules. The subject covers the whole range of electromagnetic radiation from radio waves to microwaves and millimeter waves, through the infrared to the visible and ultraviolet, and on to the X-ray region. The various regions overlap to some extent but may be characterized by the *wavelength* or *frequency* of the electromagnetic radiation involved (see Table 1).

Molecules exhibit various types of motion which give rise to series of *energy levels*, denoted by E_i. Transitions from one level to another may be accompanied by the emission or absorption of electromagnetic radiation whose frequency ν is given by

$$E' - E'' = h\nu \qquad (1)$$

where h is Planck's constant (6.626075×10^{-34} J s) and E' and E'' are the energies of the upper and lower levels, respectively. Frequencies of spectral lines are measured in units of hertz (cycles/s) or, more frequently, kilohertz (kHz, 10^3 s^{-1}), megahertz (MHz, 10^6 s^{-1}),

3-527-28132-0/94/$5.00 + .50

Table 1. Regions of the electromagnetic spectrum.

Region	Wavelength (cm)	Frequency (GHz)	Wave number (cm^{-1})
Radio frequency	3×10^4–3×10^1	0.001–1	
Microwave	3×10^1–3×10^{-1}	1–100	
Millimeter	3×10^{-1}–3×10^{-2}	100–1000	3–30
Far infrared	3×10^{-2}–2.5×10^{-3}		30–400
Mid infrared	2.5×10^{-3}–2.5×10^{-4}		400–4000
Near infrared	2.5×10^{-4}–8×10^{-5}		4000–12 500
Visible	8×10^{-5}–4×10^{-5}		12 500–25 000
Near ultraviolet	4×10^{-5}–2×10^{-5}		25 000–50 000
Vacuum ultraviolet	2×10^{-5}–2×10^{-6}		50 000–500 000
X ray	2×10^{-6}–1×10^{-8}		

gigahertz (GHz, 10^9 s^{-1}), or terahertz (THz, 10^{12} s^{-1}). In the infrared, visible, and ultraviolet, it is customary to use wave number $\tilde{\nu}$, which represents the number of waves per unit distance and is the inverse of the wavelength. The unit of wave number most commonly used is cm^{-1}, which represents the number of waves per centimeter. The relation between frequency and wave-number units is

$$\nu = c\tilde{\nu}, \qquad (2)$$

where c is the velocity of light.

The energy levels that molecules possess are governed by quantum theory and are characterized by a series of quantum numbers. It is not always possible to go from one energy level to another under the influence of electromagnetic radiation. The changes in quantum number that accompany an electromagnetic transition are governed by restrictions known as *selection rules*.

We will now consider the various types of energy level that molecules can occupy. Molecules in the gas phase rotate freely and occupy one of a number of discrete rotational energy levels. Transitions to neighboring levels give rise to emission or absorption of electromagnetic radiation, depending on whether the transition goes to a lower or higher energy level. The resulting radiation usually lies in the radio-frequency, microwave, or millimeter-wave region.

Molecules with N atoms have $3N - 6$ fundamental vibrational degrees of freedom ($3N - 5$ for a linear molecule; see Sec. 2.2). Transitions involving the vibrational energy levels usually lie in the infrared region of the spectrum.

Molecules are composed of nuclei and electrons. In their lowest energy state, the ground state, the electrons occupy certain orbits. Excitation of an electron to an orbit of higher energy produces an excited state of the molecule. Transitions between electronic states give rise to radiation that usually lies in the near-infrared, visible, or ultraviolet region of the spectrum.

Some molecular species, notably free radicals, have a resultant electron spin of $\frac{1}{2}h/2\pi$. Some nuclei, e.g., ^1H, ^{13}C, ^{19}F, ^{31}P, have a nuclear spin of $\frac{1}{2}h/2\pi$. In a magnetic field, the electron or nuclear spins can be aligned either parallel or antiparallel to the field. This produces two states of slightly different energies, the separation being proportional to the field strength. For typical laboratory fields, transitions between the two states arising from electron spin normally lie in the radio-frequency or microwave region, and the type of spectroscopy that studies these transitions is called electron spin resonance (ESR) or electron paramagnetic resonance (EPR) (see ELECTRON PARAMAGNETIC RESONANCE). The corresponding study of nuclear spins is known as nuclear magnetic resonance (NMR) and has widespread applications in chemistry and medicine (see MAGNETIC RESONANCE, NUCLEAR; MAGNETIC RESONANCE IMAGING).

The measurement of spectra involves instruments that measure the frequencies or wavelengths of transitions and the intensity of the radiation being emitted or absorbed. A variety of instruments is used, depending on the spectral region.

Spectroscopy is carried out on molecules in the gas, liquid, and solid phases. The subject is too vast for coverage in a single article; hence, the emphasis of this contribution will be on the spectroscopy of molecules in the gas phase. References will be given to related work in other areas.

1. ROTATIONAL SPECTRA

1.1 Diatomic Molecules

Let us consider a molecule composed of two atoms with masses m_1 and m_2 separated by a distance r (Fig. 1). Quantum mechanics shows that the rotational energy E_r of the molecule can have only certain discrete values given by the equation (Townes and Schawlow, 1955, Chap. 1)

$$E_r = (h^2/8\pi^2 I)J(J + 1), \tag{3}$$

where J is the rotational quantum number, which can take values 0, 1, 2, ..., and I is the moment of inertia of the molecule about its center of mass:

$$I = \mu r^2, \tag{4}$$

where μ is called the reduced mass and is given by

$$1/\mu = 1/m_1 + 1/m_2. \tag{5}$$

For radiative transitions to occur, the molecule must posses an electric *dipole moment*, and the transitions are restricted by the selection rule

$$\Delta J = J' - J'' = +1. \tag{6}$$

The frequency of the transition ν is given by

$$h\nu_{J+1-J} = E_{J+1} - E_J. \tag{7}$$

Hence,

$$\begin{aligned}\nu_{J+1-J} &= (h/8\pi^2 I)\,[(J + 1)(J + 2) - J(J + 1)] \\ &= 2B(J + 1),\end{aligned} \tag{8}$$

where B is called the *rotational constant* and is given by

$$B = h/8\pi^2 I. \tag{9}$$

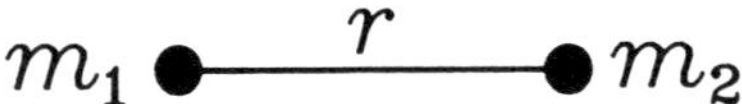

FIG. 1. Molecular parameters for a diatomic molecule.

The rotational spectrum of a diatomic molecule therefore consists of a series of equally spaced lines at frequencies $2B$, $4B$, $6B$, and so on. A diagram showing the rotational energy levels and transitions is given in Fig. 2. It is customary to plot the energy levels in units of E/hc, which are known as *term values*. The rotational term values $F(J)$ for a diatomic molecule are given by

$$F(J) = BJ(J + 1) \tag{10}$$

and take the values 0, $2B$, $6B$, $12B$, $20B$ and so on. In these units,

$$B = h/8\pi^2 cI. \tag{10a}$$

Rotational spectra usually lie in the radiofrequency, microwave, or millimeter-wave regions, but, for higher values of J or for light molecules, the transitions may lie in the far-infrared region. As an example, the 1–0, 2–1, and 3–2 transitions of the carbon monoxide molecule CO lie at 115.271, 230.538, and 345.796 GHz, respectively. Transitions of this molecule have been observed up to 38–37 at 4340.138 GHz. It is found that the frequencies do not obey exactly the linear relation-

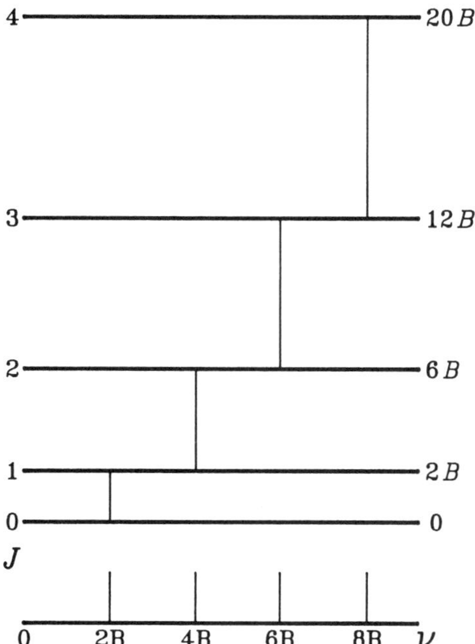

FIG. 2. Rotational term values and allowed transitions for a diatomic molecule.

ship given by Eq. (8). The reason is that the molecule stretches slightly with increasing rotation, thereby increasing its effective moment of inertia or decreasing its effective B value. Equation (10) is modified by the addition of a small correction term:

$$F(J) = BJ(J+1) - DJ^2(J+1)^2, \tag{11}$$

where D is known as the centrifugal distortion constant (see Sec. 2.1.1).

The frequencies in the rotational spectrum are then given by

$$\nu_{J+1-J} = 2B(J+1) - 4D(J+1)^3, \tag{12}$$

where D is very small compared with B. For example, for CO, B = 57 636 MHz, but D is only 183.5 kHz.

Rotational constants for diatomic molecules can easily be determined with an accuracy of 1 part in 10^7. By the use of molecular beam techniques, this accuracy can be improved by another factor of 10. Provided that the atomic masses are known, accurate values for internuclear distances r can be obtained from Eqs. (4), (5), and (9). However, Planck's constant h is known only to slightly better than 1 part in 10^6.

If one or both of the atoms exist in isotopic modifications, then rotational spectroscopy can be used to obtain mass ratios. For example, by measuring the spectra of $^{127}I^{35}Cl$ and $^{127}I^{37}Cl$, the ratio of the masses $^{35}Cl/^{37}Cl$ can be determined with an accuracy of about 1 part in 10^7. Such a determination does not depend on the accuracy with which Planck's constant is known. However, the B values must be corrected for the effects of vibration (see Sec. 2.1.1).

1.2 Polyatomic Molecules

All molecules rotate freely in space about their center of mass, G. There is always an axis through G about which the moment of inertia of the molecule is a minimum. This axis is called the a axis and the least moment of inertia I_A. At right angles to this axis is another axis, c, about which the moment of inertia I_C is a maximum. A third axis, b, is chosen at right angles to the a and c axes, and the moment of inertia about the b axis is I_B. By definition,

$$I_A \leqslant I_B \leqslant I_C. \tag{13}$$

Rotational constants A, B, and C are defined so that

$$A = h/8\pi^2 cI_A, \quad B = h/8\pi^2 cI_B, \\ C = h/8\pi^2 cI_C. \tag{14}$$

Again, by definition,

$$A \geqslant B \geqslant C. \tag{15}$$

There are now four types of molecule to consider:

1. Spherical-top molecules for which $A = B = C$,
2. Symmetric-top molecules,
   a. Prolate tops with $A \neq B = C$
   b. Oblate tops with $A = B \neq C$,
3. Linear molecules with $B = C$ and $I_A = 0$,
4. Asymmetric-top molecules with $A \neq B \neq C$.

The rotational term values for spherical-top and linear molecules are the same as for diatomic molecules [Eq. (11)]. However, spherical-top molecules—e.g., CH_4—have no dipole moment and, hence, no rotational spectrum (except a very weak spectrum produced by certain vibrations or by centrifugal distortion). Linear molecules with dipole moments produce rotational spectra that obey the same formula as for diatomic molecules [see Eq. (12)].

We will now consider the other two types of rotor.

1.2.1 Symmetric-Top Molecules The rotational term values for symmetric-top molecules depend on two quantum numbers, J and K. The total angular momentum of the molecule is quantized and equals $\sqrt{J(J+1)}h/2\pi$. Similarly, the component of the total angular momentum along the direction of the symmetric-top axis is quantized and equals $Kh/2\pi$, where K is an integer ($K \leqslant J$). The rotational term values for a prolate symmetric top are given by

$$F(J,K) = (A-B)K^2 + BJ(J+1), \tag{16}$$

and, for an oblate symmetric top,

$$F(J,K) = (C-B)K^2 + BJ(J+1). \tag{17}$$

Energy-level diagrams for the two types of symmetric top are given in Fig. 3. Examples of prolate tops are the methyl halides; benzene provides an example of an oblate top. It should be noted that since the component of J in the direction of the symmetric-top axis can be either positive or negative, all states with $K > 0$ are *doubly degenerate*.

Since the dipole moment of the molecule always lies along the symmetric-top axis, the selection rules are

$$\Delta K = 0, \qquad \Delta J = +1. \tag{18}$$

In the absence of centrifugal distortion, the rotational spectrum is similar to that of a diatomic molecule. However, when centrifugal terms are added, the rotational term values are given by

$$F(J, K) = (A - B)K^2 + BJ(J + 1) - D_K K^4 - D_{JK}K^2 J(J + 1) - D_J J^2(J + 1)^2 \tag{19}$$

for a prolate top. For an oblate top, the rotational constant A should be replaced by C. The frequency of a rotational transition $(J + 1, K)$—(J, K) is

$$\nu = 2(J + 1)(B - D_{JK}K^2) - 4D_J(J + 1)^3. \tag{20}$$

The series for given J values with different values of K will be separated if the value of D_{JK} is sufficiently large.

The analysis of rotational spectra of symmetric-top molecules yields only the moments of inertia about the center of mass in a direction perpendicular to the symmetric-top axis. No information is obtained regarding the moment of inertia about the symmetric-top axis. Partial structural information, however, can be obtained if more than one isotopic species is studied.

1.2.2 Asymmetric-Top Molecules There is no simple formula for the rotational term values of an asymmetric-top molecule. For each value of J, the total angular momentum quantum number, there are $2J + 1$ rotational energy levels. The term values are obtained by diagonalization of matrices. Each level is designated by J_{K_a,K_c}, where K_a and K_c are the K quantum numbers of the levels with which the asymmetric-top level correlates in the

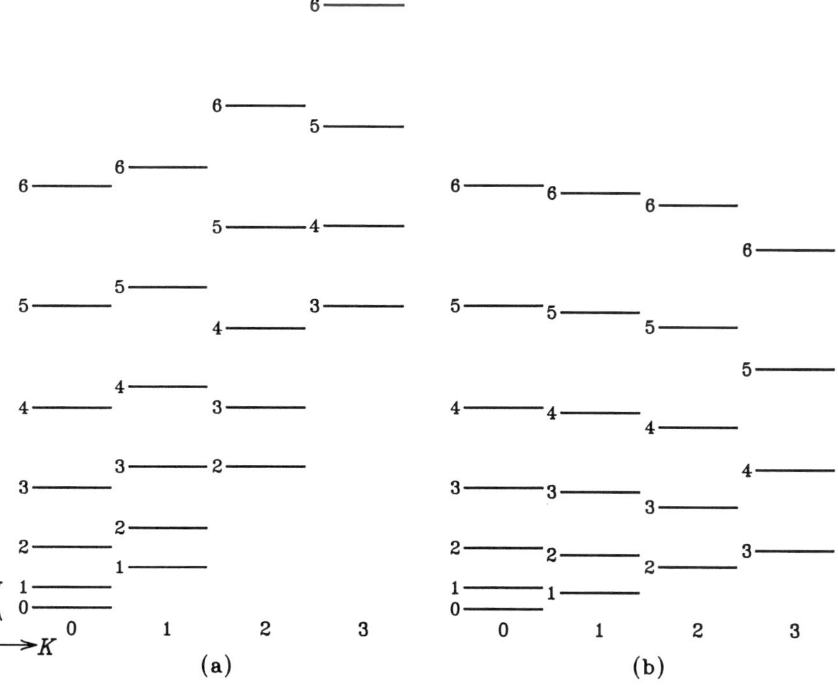

FIG. 3. Rotational term values for symmetric-top molecules: (a) prolate top with $A = 3B$, (b) oblate top with $C = \frac{1}{2}B$.

prolate and oblate symmetric-top limits, respectively. A correlation diagram is shown in Fig. 4 and is based on the principle that no levels with the same value of J can cross (the *non-crossing rule*).

Selection rules depend on the direction of the dipole moment of the molecule with respect to the inertial axes. For all transitions,

$$\Delta J = J' - J'' = 0, \pm 1. \qquad (21)$$

If the dipole moment lies in the a direction,

$$\Delta K_a = 0, \pm 2, \ldots; \quad \Delta K_c = \pm 1, \pm 3, \ldots. \qquad (22)$$

If the dipole moment lies in the b direction,

$$\Delta K_a = \pm 1, \pm 3, \ldots; \quad \Delta K_c = \pm 1, \pm 3, \ldots. \qquad (23)$$

If the dipole moment lies in the c direction,

$$\Delta K_a = \pm 1, \pm 3, \ldots; \quad \Delta K_c = 0, \pm 2, \ldots. \qquad (24)$$

If the dipole moment does not lie in the direction of an inertial axis, then more than one type of transition will be allowed, depending on the components of the dipole moment in the directions of the inertial axes.

An example of an asymmetric top is provided by the water molecule H_2O (Fig. 5). The a and b inertial axes lie in the plane of the molecule, while the c axis lies perpendicular to the plane. The dipole moment lies in the b direction; hence, the selection rules obey Eqs. (21) and (23). One of the transitions that produces absorption in the Earth's atmosphere and interfered with early radar is the 6_{16}—5_{23} transition at 22.235 GHz.

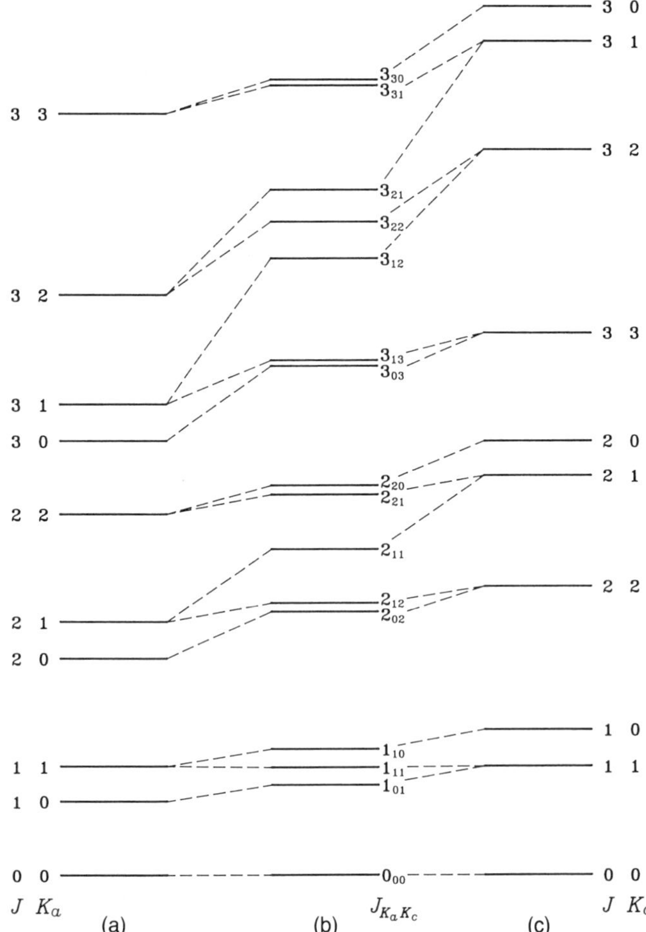

FIG. 4. Correlation diagram connecting the rotational energy levels of (a) a prolate symmetric top with $A = 2$, $B = C = 1$, (b) an asymmetric top with $A = 2$, $B = 1.5$, $C = 1$, (c) an oblate top with $A = B = 2$, $C = 1$.

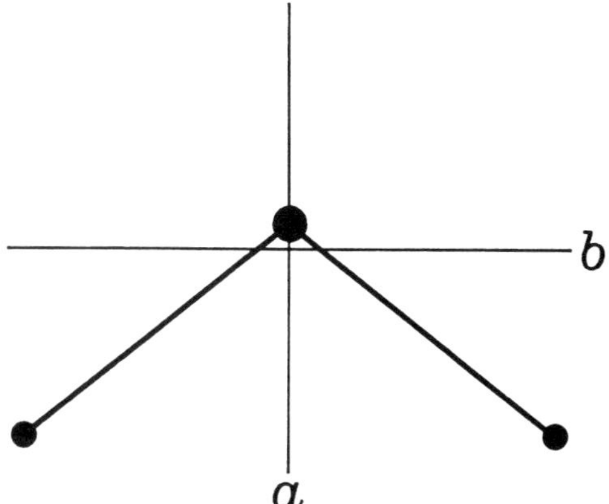

FIG. 5. Principal axes of inertia for the H_2O molecule. The c axis is perpendicular to the plane.

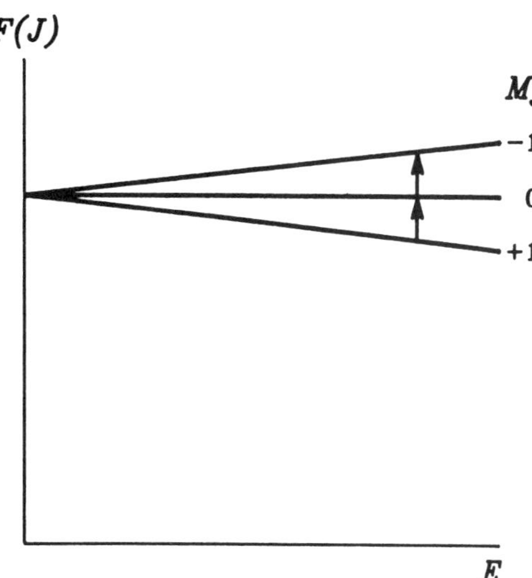

FIG. 6. Stark splittings for a $J = 1$, $K = 1$ level of a symmetric-top molecule. In perpendicular polarization, the splittings between the $M_J = 0$ level and the $M_J = \pm 1$ levels give a measure of the dipole moment of the molecule. (The M_J quantum numbers are given for $K = +1$; for the $K = -1$ component, the M_J quantum numbers should be reversed.)

1.3 Determination of Dipole Moments

The energy levels of a symmetric-top molecule, except those with $K = 0$, split into $2J + 1$ components in the presence of an electric field, a phenomenon known as the Stark effect. The components are labeled by a quantum number M_J, which represents the component of J in the direction of the electric field and takes values $J, J - 1, J - 2, \ldots, 0 \ldots, -(J - 1), -J$. The splittings ΔW are given by the formula

$$\Delta W = -\mu E M_J K / J(J + 1), \tag{25}$$

where μ is the dipole moment of the molecule and E is the electric-field strength.

An energy-level diagram showing the effect of an electric field on a $J = 1$, $K = 1$ level is given in Fig. 6. The level splits into three components with $M_J = -1$, 0, and +1.

Selection rules for transitions depend on the direction of the electric vector of the electromagnetic radiation relative to the direction of the electric field. If the two are parallel, then

$$\Delta M_J = 0. \tag{26}$$

If the two are perpendicular, then

$$\Delta M_J = \pm 1. \tag{27}$$

In Fig. 6, the splittings between the level with $M_J = 0$ and the levels with $M_J = \pm 1$ give a measure of the dipole moment of the molecule.

For diatomic and linear polyatomic molecules, the splittings are much smaller and depend on the square of the electric field strength. Furthermore, levels with $+M_J$ and $-M_J$ remain degenerate; i.e., they are not split. Consequently, only $J + 1$ components are observed in an electric field. Asymmetric-top molecules behave in a similar manner except when the K-type doubling is very small; i.e., levels with the same K_a (or K_c) but different K_c (or K_a) show a very small splitting. In the latter case, the levels behave like those of a symmetric top and show a first-order Stark effect.

Dipole moments of many molecules have been measured, some with accuracies of the order of 1 part in 10^6.

1.4 Hyperfine Structure

Spectra of molecules containing atoms with nuclear spin show very small splittings called hyperfine structure. If all the nuclei in a molecule have zero nuclear spin except one with a nuclear spin I, then, in a rotational state J, the level splits into components with a resultant quantum number F which takes values $J + I, J + I - 1, \ldots, J - I$. Selection rules for transitions between hyperfine components are

$$\Delta J = 0, \pm 1; \quad \Delta F = 0, \pm 1; \quad \Delta I = 0. \tag{28}$$

The largest hyperfine splittings in molecular spectra occur if one (or more) of the atoms has an electric quadrupole moment. Of necessity, this is possible only for atoms with $I = 1$ or more.

For a linear molecule, the quadrupole energy W_Q is

$$W_Q = -eq_m Q f(I, J, F), \tag{29}$$

where e is the electronic charge, q_m is the electric-field gradient at the nucleus in the direction of the linear axis, Q is the quadrupole moment of the nucleus, and $f(I, J, F)$ is Casimir's function

$$f(I, J, F) = \frac{\frac{3}{4}C(C+1) - I(I+1)J(J+1)}{2I(2I-1)(2J-1)(2J+3)}, \tag{30}$$

where

$$C = F(F+1) - I(I+1) - J(J+1). \tag{31}$$

Since q_m is rarely known, the quantity determined experimentally is the product eqQ, which is known as the quadrupole coupling constant. Values are normally of the order of 100 MHz and may be positive or negative, but, for the ^{127}I nucleus in ICN, the large value of $eqQ = -2420$ MHz is found. An example of quadrupole hyperfine structure in the $J = 1$—0 transition of $^{16}O^{12}C^{33}S$ is shown in Fig. 7. The ^{33}S nucleus has a spin of $\frac{3}{2}$, and $eqQ = -29.2$ MHz.

The quadrupole energy for a nucleus on the axis of a symmetric-top molecule is similar to Eq. (29), but, for asymmetric-top molecules, the expressions are more complicated.

Hyperfine structure may also be observed resulting from the magnetic dipole moments of nuclei. The splittings are very small in most molecules but may become appreciable if a molecule has a resultant electron spin as in a free radical.

An important application of hyperfine structure lies in nuclear quadrupole resonance (NQR) spectroscopy. Transitions between hyperfine levels are measured directly using a radio-frequency spectrometer that usually operates over the range 2–1000 MHz. The applications are generally confined to solids (Semin *et al.*, 1975).

1.5 Applications to Astronomy

In 1968, only four molecules were known to exist in the interstellar medium. Three of these, CH, CH^+, and CN, had been found by optical spectroscopy between 1937 and 1941. The fourth, OH, was found by radioastron-

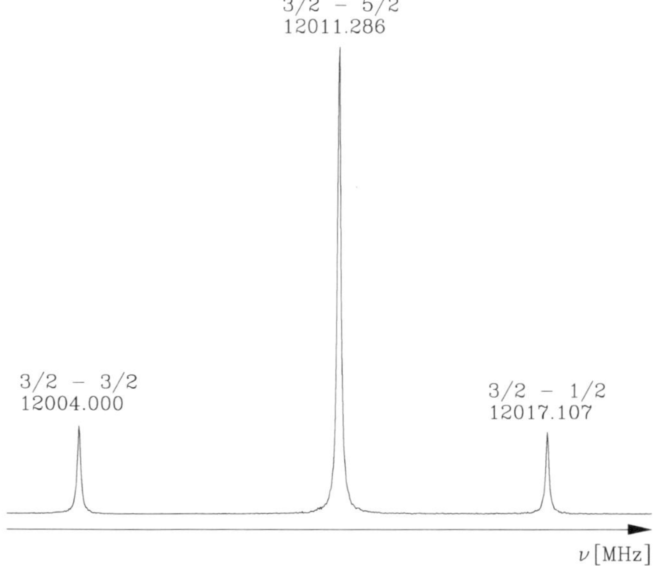

FIG. 7. Quadrupole hyperfine structure observed in the $J = 1$–0 transition of $OC^{33}S$, by courtesy of H. Dreizler, Universität Kiel, Germany.

omy in 1963. In 1968, interstellar NH_3 was detected by means of its microwave spectrum; this was followed in quick succession by the detection of interstellar H_2O (1968), H_2CO (1969), CO, HCN, HCO^+, HC_3N, CH_3OH, HCOOH (1970), NH_2CHO, SiO, OCS, CS, CH_3CN, HNCO, CH_3C_2H, $HCOCH_3$, and H_2CS (1971) (Snyder, 1972). Today, approximately 100 molecules have been identified in the interstellar medium by means of radioastronomy, and some of these have also been seen in the atmospheres of stars. Most of the spectra have been observed in emission, but some have been seen in absorption against the 2.7-K background radiation.

Some species, e.g., HCO^+, were found in the interstellar medium before they were detected in the laboratory. Another group of molecules that has been observed in the interstellar medium is the straight-chain cyanopolyacetylenes $H—(C≡C—)_nC≡N$, where $n = 1–5$. The lower members ($n = 1, 2, 3$) had been synthesized and measured in the laboratory, but the higher members with n equal to 4 and 5 were first seen in the interstellar medium.

The conditions of low temperature and exceedingly low pressures in the interstellar medium are favorable for the observation of very sharp lines. Frequencies can be measured very accurately and usually differ slightly from laboratory frequencies on account of motions within the interstellar medium, i.e., Doppler shifts. Furthermore, microwaves, on account of their long wavelengths, can penetrate dark molecular clouds, which are opaque to visible light. These clouds are of particular interest since they are regions where new stars are formed. Consequently, microwave spectroscopy of the interstellar medium provides an important technique for probing the dynamics and structure of regions of star formation in the universe.

1.6 Rotational Raman Spectra

Molecules with a zero dipole moment, e.g., H_2, N_2, C_6H_6, have no microwave absorption. Nevertheless, a rotational Raman spectrum can be observed. This spectrum is produced by irradiating the molecules with a strong monochromatic source and recording the lines that appear on both sides of the exciting line. In the early work, a mercury lamp was usually used, but nowadays excitation is achieved by the use of lasers. The selection rules for a diatomic or linear polyatomic molecule are

$$\Delta J = 0, \pm 2 \tag{32}$$

and, for a symmetric top,

$$\Delta J = 0, \pm 1, \pm 2, \tag{33}$$

except that Eq. (32) applies if $K = 0$.

As an example, the rotational Raman spectrum of N_2 is shown in Fig. 8. A series of equidistant lines (called S branch since $\Delta J = +2$) is observed on either side of the exciting line, the series going to lower frequencies being called the Stokes branch and the series going to higher frequencies the anti-Stokes branch. The frequency shifts, neglecting centrifugal distortion, are given by

$$|\Delta\nu| = 4B(J + \tfrac{3}{2}); \tag{33a}$$

hence, the spacing between neighboring lines in the branches is $4B$, and the separation between the exciting line and the first line in each branch is equal to $6B$. The alternation in the intensities of the lines is caused by the spin statistics of the N nuclei (see Sec. 2.2.1).

The rotational Raman spectrum of nitrogen is of particular interest, since, when it was first reported, the alternation of intensities was opposite to that observed for hydrogen and led to the deduction that nuclei could not

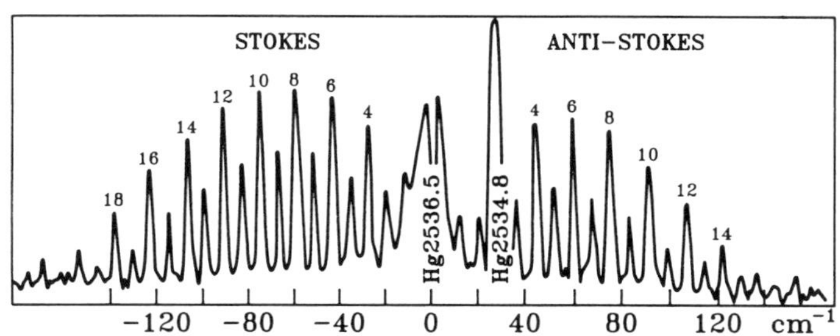

FIG. 8. Rotational Raman spectrum of N_2 (Stoicheff, 1950). The spectrum was excited by the 2536.5 = Å Hg line and shows the Stokes and anti-Stokes S branches with a 2:1 intensity alternation. There is some pressure broadening of the lines, resulting from the high pressure used (10 atm.).

be composed of protons and electrons (Heitler and Herzberg, 1929). This result preceded the discovery of the neutron.

2. VIBRATIONAL SPECTRA

2.1 Diatomic Molecules

According to quantum mechanics, the energy levels of a simple harmonic oscillator are given by (Pauling and Wilson, 1935)

$$E_v = h\nu(v + \tfrac{1}{2}), \tag{34}$$

where v is the vibrational quantum number, which takes values 0, 1, 2, ... In term values,

$$G_v = E_v/hc = \omega(v + \tfrac{1}{2}), \tag{35}$$

where $G(v)$ is the vibrational term value and ω is the vibrational frequency expressed in cm^{-1} units. The lowest energy level that the molecule can occupy has an energy $\frac{1}{2}hc\omega$, which is known as the zero-point energy and exists even at the absolute zero of temperature.

The vibrational frequency is related to the *force constant* k for the vibration by the relation

$$k = 4\pi^2\mu c^2\omega^2, \tag{36}$$

where $k(r - r_e)$ is the restoring force when the molecule is displaced from its equilibrium position and μ is the reduced mass [Eq. (5)].

In reality, diatomic molecules behave like anharmonic oscillators and dissociate into atoms when the vibrational energy is sufficiently large. A potential energy function and vibrational energy levels for an *anharmonic* oscillator are shown in Fig. 9. Equation (35) is modified by the addition of terms involving a series of anharmonic constants $\omega_e x_e$, $\omega_e y_e$, $\cdots$:

$$G(v) = \omega_e(v + \tfrac{1}{2}) - \omega_e x_e(v + \tfrac{1}{2})^2 + \omega_e y_e(v + \tfrac{1}{2})^3 + \cdots. \tag{37}$$

This form of equation does not satisfactorily explain the behavior when the molecule approaches the dissociation limit. Hence, another potential function is frequently used,

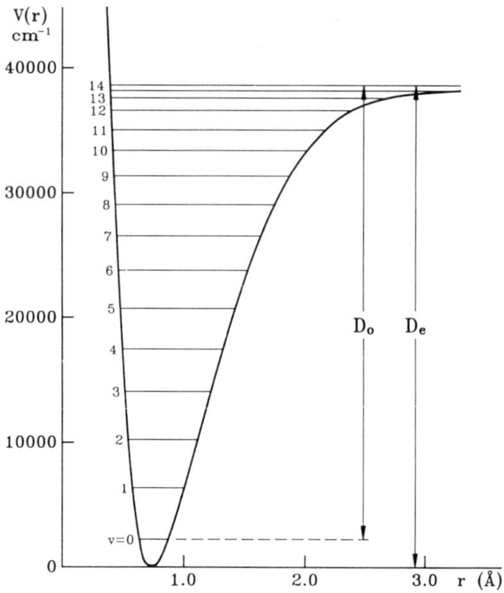

FIG. 9. Potential curve and energy levels of an anharmonic oscillator.

the *Morse potential*, given by

$$V(r) = D_e\{1 - \exp[-\beta(r - r_e)]\}^2, \tag{38}$$

where D_e is the dissociation energy—that is, the energy difference between the minimum of the potential curve and the dissociation limit. If D_e is in cm^{-1}, then

$$\beta = \sqrt{\frac{2\pi^2 c\mu}{D_e h}}\,\omega_e. \tag{39}$$

This function has the merit that $V = 0$ when $r = r_e$ and $V = D_e$ when r becomes infinite. Note, however, that the quantity that is measured experimentally is D_0, which is the energy difference between the zeroth vibrational level and the dissociation limit.

The selection rule for the vibrational quantum number of a simple harmonic oscillator is

$$\Delta v = \pm 1. \tag{40}$$

However, molecules are not exactly harmonic oscillators; hence, transitions with $\Delta v > 1$ can also appear, though weakly. Transitions with $\Delta v = 1$ lie in the infrared region of the spectrum. Overtone spectra with $\Delta v =$

2, 3, 4, . . . are much weaker, the decrease in intensity being roughly an order of magnitude or more for each higher value of v. Nevertheless, overtone spectra may be observed in absorption using progressively longer path lengths or intracavity techniques using lasers. For example, the 6—0 band of HD has been observed in absorption in the visible region using a path length of 3 km and a pressure of 1 atm.

Vibrational transitions may also be observed in emission. The intensity in emission is proportional to the fourth power of the frequency, as compared to the first power in absorption. Consequently, $\Delta v = 2$ transitions may be observed in emission, with intensities comparable to those of $\Delta v = 1$ transitions.

The intensity of an infrared transition depends on the derivative of the dipole moment with respect to the internuclear distance. Hence, homonuclear molecules such as H_2, O_2, and N_2 do not have an infrared spectrum, although a vibrational Raman spectrum can be observed (Long, 1977). In the case of H_2, an extremely weak electric quadrupole spectrum has been observed and is common in astrophysical sources.

2.1.1 Rotational Fine Structure When a molecule vibrates, the value of r changes during the vibration. For each vibrational level, we can define a rotational constant B_v such that

$$B_v = \frac{h}{8\pi^2 c\mu}\left[\frac{1}{r^2}\right]_{\text{av}}. \tag{41}$$

The quantity B_v can be expressed as a power series in $v + \frac{1}{2}$:

$$B_v = B_e - \alpha_e(v + \tfrac{1}{2}) + \cdots \tag{42}$$

and

$$B_e = h/8\pi^2 c\mu r_e^2. \tag{43}$$

α_e is usually small compared with B_e, e.g., ~1%. The quantity B_e is the equilibrium rotational constant, i.e., the rotational constant at the minimum of the potential curve. Since r_e is independent of the atomic masses, B_e is the quantity that must be used in the determination of the ratio of isotopic masses (see Sec. 1.1).

During a vibrational transition, the rotational quantum numbers may also change. The selection rule $\Delta J = J' - J'' = \pm 1$ gives rise to an R branch ($\Delta J = +1$) and a P branch ($\Delta J = -1$). An energy-level diagram is given in Fig. 10. Note that all the transitions are numbered according to the lower-state J value.

For the R branch,

$$\begin{aligned} R(J) &= \nu_0 + F'(J + 1) - F''(J) \\ &= \nu_0 + B'(J + 1)(J + 2) - B''J(J + 1) \\ &= \nu_0 + 2B' + (3B' - B'')J \\ &\quad + (B' - B'')J^2, \end{aligned} \tag{44}$$

where ν_0 is the frequency of the band origin and $J = 0, 1, 2, \ldots$.

For the P branch,

$$\begin{aligned} P(J) &= \nu_0 + F'(J - 1) - F''(J) \\ &= \nu_0 + B'(J - 1)J - B''J(J + 1) \\ &= \nu_0 - (B' + B'')J + (B' - B'')J^2, \end{aligned} \tag{45}$$

where $J = 1, 2, 3, \ldots$.

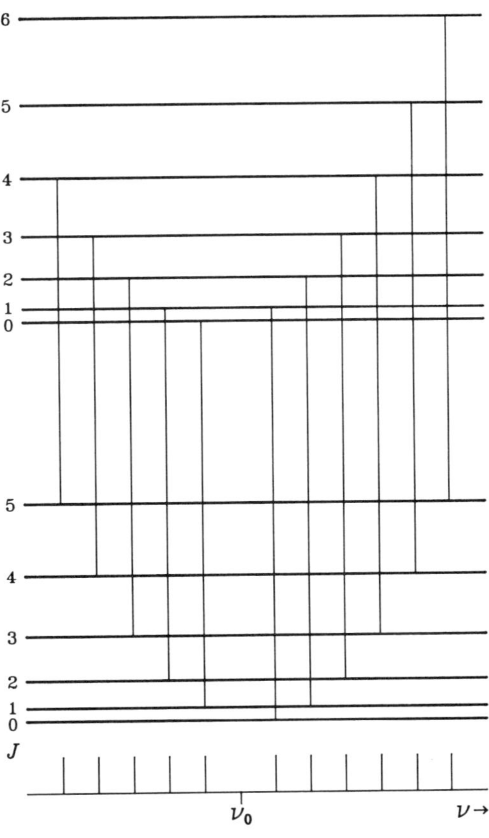

FIG. 10. Energy-level diagram and transitions for the P and R branches of a simple diatomic molecule.

Both branches can be fitted by a single equation

$$\nu = \nu_0 + (B' + B'')m + (B' - B'')m^2, \qquad (46)$$

where $m = J + 1$ for the R branch and $m = -J$ for the P branch. This equation is known as the *Fortrat parabola*. A more complete expression, including the effects of centrifugal distortion, is

$$\nu = \nu_0 + (B' + B'')m + (B' - B'' - D' + D'')m^2 - 2(D' + D'')m^3 - (D' - D'')m^4, \qquad (47)$$

where

$$D \sim 4B^3/\omega^2. \qquad (48)$$

An example is provided by the fundamental vibration-rotation band of carbon monoxide, which is shown in Fig. 11.

2.2 Polyatomic Molecules

The positions of the nuclei of a molecule with N atoms can be described by $3N$ coordinates. Three of these coordinates are required to specify the position of the center of mass and three more angular coordinates to specify the orientation of the molecule in space (only two angular coordinates are required for a linear molecule since there is no rotation of the nuclei about the internuclear axis). Consequently, $3N - 6$ coordinates are needed to specify the internal modes of vibration of the molecule ($3N - 5$ for a linear molecule).

In the simplest approximation, the vibrational motion may be considered to be the sum of a series of independent simple harmonic oscillators, and the energy levels are given by

$$G(v_1, v_2, \ldots) = \sum_{i=1}^{3N-6} \omega_i(v_i + \tfrac{1}{2}d_i), \qquad (49)$$

where G is the vibrational term value, the ω_i are the zero-order frequencies of the various modes of vibration, and d_i is the degeneracy of the ith vibration (normally 1, but may be 2 or higher).

A more complete expression that takes account of the interactions between the vibrations and of their anharmonicities is

$$G(v_1, v_2, \ldots) = \sum_i \omega_i(v_i + \tfrac{1}{2}d_i) + \sum_i \sum_{k \geq i} x_{ik}(v_i + \tfrac{1}{2}d_i)(v_k + \tfrac{1}{2}d_k) + \sum_i \sum_{k \geq i} g_{ik} l_i l_k + \cdots. \qquad (50)$$

The x_{ik}'s are the anharmonic constants, and

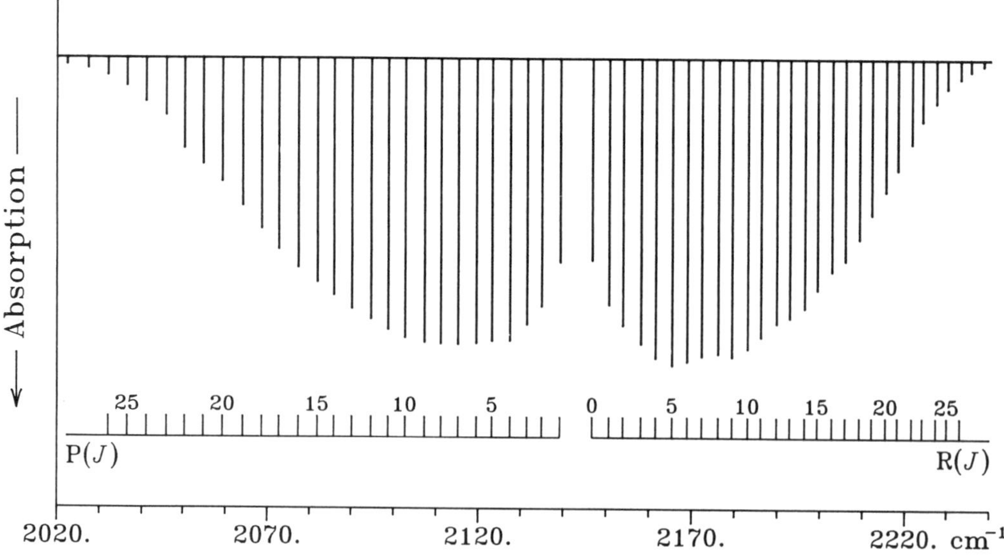

FIG. 11. The fundamental vibration–rotation band of CO in absorption (courtesy of J. W. C. Johns, National Research Council of Canada).

the last term is a small correction term that must be applied when one or more degenerate vibrations are excited. Note that the anharmonic terms are preceded by + signs, in contrast to the formula for a diatomic molecule for which − signs are customary [Eq. (37)].

As an example, let us consider the CO_2 molecule. The normal modes of vibration are ν_1, the symmetric CO stretching vibration; ν_2, the doubly degenerate bending vibration; and ν_3, the antisymmetric CO stretching vibration (Fig. 12). The two components of ν_2 can generate an angular momentum $l_2 h/2\pi$ about the linear axis depending on their relative phases. If v_2 quanta of this vibration are excited, then l_2 can take the values $v_2, v_2 - 2, \ldots, 1$ or 0 depending on whether v_2 is odd or even.

The vibrational term values are given by

$$\begin{aligned}G(v_1, v_2, v_3) = {} & \omega_1(v_1 + \tfrac{1}{2}) \\ & + \omega_2(v_2 + 1) + \omega_3(v_3 + \tfrac{1}{2}) \\ & + x_{11}(v_1 + \tfrac{1}{2})^2 + x_{12}(v_1 + \tfrac{1}{2}) \\ & \times (v_2 + 1) + x_{13}(v_1 + \tfrac{1}{2})(v_3 + \tfrac{1}{2}) \\ & + x_{22}(v_2 + 1)^2 + x_{23}(v_2 + 1) \\ & \times (v_3 + \tfrac{1}{2}) + x_{33}(v_3 + \tfrac{1}{2})^2 \\ & + g_{22}l_2^2.\end{aligned} \tag{51}$$

The zero-point energy is

$$\begin{aligned}G(0, 0, 0) = {} & \tfrac{1}{2}\omega_1 + \omega_2 + \tfrac{1}{2}\omega_3 + \tfrac{1}{4}x_{11} + \tfrac{1}{2}x_{12} \\ & + \tfrac{1}{4}x_{13} + x_{22} + \tfrac{1}{2}x_{23} + \tfrac{1}{4}x_{33}.\end{aligned} \tag{52}$$

The energy levels can also be expressed by a formula based on the zero-point level, viz.

$$\begin{aligned}G(v_1, v_2, v_3) = {} & \omega_1^0 v_1 + \omega_2^0 v_2 + \omega_3^0 v_3 + x_{11}^0 v_1^2 \\ & + x_{12}^0 v_1 v_2 + x_{13}^0 v_1 v_3 + x_{22}^0 v_2^2 \\ & + x_{23}^0 v_2 v_3 + x_{33}^0 v_3^2 + g_{22}l_2^2,\end{aligned} \tag{53}$$

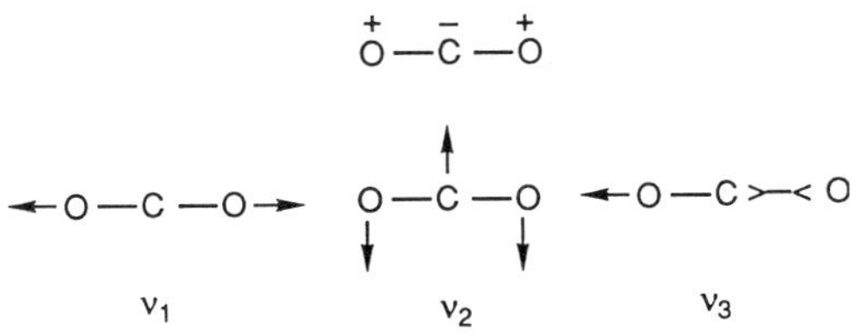

FIG. 12. Normal modes of vibration of the CO_2 molecule. The ν_2 mode is doubly degenerate.

where

$$\omega_1^0 = \omega_1 + x_{11} + x_{12} + \tfrac{1}{2}x_{13}, \tag{54}$$

$$\omega_2^0 = \omega_2 + 2x_{22} + \tfrac{1}{2}x_{12} + \tfrac{1}{2}x_{23}, \tag{55}$$

$$\omega_3^0 = \omega_3 + x_{33} + \tfrac{1}{2}x_{13} + x_{23}, \tag{56}$$

$$x_{ik}^0 = x_{ik}. \tag{57}$$

In general, for any polyatomic molecule,

$$\begin{aligned}G_0(v_1, v_2, \ldots) = {} & \sum_i \omega_i^0 v_i + \sum_i \sum_{k \geq i} x_{ik}^0 v_i v_k \\ & + \sum_i \sum_{k \geq i} g_{ik} l_i l_k + \cdots,\end{aligned} \tag{58}$$

where

$$\omega_i^0 = \omega_i + x_{ii}d_i + \tfrac{1}{2}\sum_{k \neq i} x_{ik}d_k. \tag{59}$$

The quantity that is measured experimentally and is called the fundamental vibration frequency ν is the interval between the levels $v = 1$ and $v = 0$. For the CO_2 molecule, we see from Eqs. (51) and (54) that

$$\nu_1 = \omega_1 + 2x_{11} + x_{12} + \tfrac{1}{2}x_{13} \tag{60}$$

$$= \omega_1^0 + x_{11}. \tag{61}$$

Similar expressions may be derived for ν_2 and ν_3.

2.2.1 Rotational Fine Structure

2.2.1.1 Linear Molecules. The general expression for the rotational term values of a linear molecule in a vibrational level v is

$$\begin{aligned}F_v(J) = {} & B_v[J(J + 1) - l^2] \\ & - D_v[J(J + 1) - l^2]^2 + \cdots,\end{aligned} \tag{62}$$

where l is the vibrational angular momentum about the internuclear axis, J takes the values $l, l + 1, l + 2, \ldots$ and

$$B_v = B_e - \sum_i \alpha_i(v_i + \tfrac{1}{2}d_i); \tag{63}$$

d_i is the degeneracy of the level and equals 1 if $l = 0$ and 2 if $l \neq 0$. Levels with $l = 0, 1, 2, \ldots$ are referred to as $\Sigma, \Pi, \Delta, \ldots$ levels.

We now need to consider some symmetry properties of rotational levels. The parity of a level is *positive* or *negative* depending on whether the total wave function remains unchanged or changes sign when the coordinates of all the particles (electrons and nuclei), relative to the center of mass and measured in a space-fixed coordinate system, are replaced by their negatives. The total wave function is a product of an electronic, a vibrational, and a rotational contribution, but, for the present, we will consider the electronic contribution to be totally symmetric. For a Σ^+ state, the rotational levels with $J = 0, 1, 2, 3, \ldots$ are $+, -, +, -, \ldots$, respectively. States with $l > 0$ are doubly degenerate, and, for each value of J, there is a + and a − level.

For molecules with identical nuclei, there is an additional symmetry property. Levels are called *symmetric* (s) if the total wave function is unchanged when the identical nuclei are exchanged. Levels are called *antisymmetric* (a) if the total wave function changes sign when the identical nuclei are exchanged. For a Σ_g^+ state, the positive (+) levels are symmetric (s), and the negative levels are antisymmetric (a) (see Sec. 3.5 for further examples). If the identical nuclei have zero nuclear spin ($I = 0$), then only the s levels occur. An example is provided by the CO_2 molecule.

If $I \neq 0$, then both s and a levels occur but with different statistical weights. For a molecule with only one pair of equivalent nuclei, the ratio of the statistical weights for the s and a levels is $(I + 1)/I$ or $I/(I + 1)$ according as I is integral or half-integral, respectively.

For electric dipole transitions, *positive* levels only combine with *negative* levels, i.e.,

$$+ \leftrightarrow -, \qquad + \not\leftrightarrow +, \qquad - \not\leftrightarrow -. \tag{64}$$

Furthermore, *symmetric* levels only combine with *symmetric* levels and *antisymmetric* levels with *antisymmetric* levels. Intercombinations between *symmetric* and *antisymmetric* levels are strictly forbidden, i.e.,

$$s \leftrightarrow s, \qquad a \leftrightarrow a, \qquad s \not\leftrightarrow a. \tag{65}$$

Further selection rules for transitions are

$$\Delta l = 0, \qquad \Delta J = \pm 1 \quad \text{if } l = 0; \tag{66}$$

$$\Delta l = 0, \qquad \Delta J = 0, \pm 1 \quad \text{if } l \neq 0; \tag{67}$$

$$\Delta l = \pm 1, \qquad \Delta J = 0, \pm 1. \tag{68}$$

Σ–Σ transitions have only P and R branches, which can be expressed by Eq. (47). Π–Π and Δ–Δ transitions have, in addition, weak Q branches ($\Delta J = 0$), which are given by

$$\begin{aligned} Q(J) = \nu_0 &+ (B' - B'')[J(J + 1) - l^2] \\ &- (D' - D'')[J(J + 1) - l^2]^2. \end{aligned} \tag{69}$$

Σ–Π, Π–Σ, Π–Δ, and Δ–Π transitions have P and R branches and strong Q branches given by Eqs. (47) and (69). It should be noted that when $l = 1$, there is a small splitting of the levels, called l-type doubling, and additional terms should be added to Eqs. (47) and (69).

2.2.1.2 Symmetric-Top Molecules. The rotational term values for a symmetric-top molecule in a nondegenerate state are given by Eq. (19). If a degenerate vibration is excited, then additional Coriolis terms need to be added.

As with linear molecules, the rotational levels are either *positive* or *negative*, depending on whether the total wave function remains unchanged or changes sign when the coordinates of all the particles are replaced by their negatives. This symmetry property is important for planar molecules. For nonplanar molecules, each rotational level is doubly degenerate and consists of a + and a − component. The splitting is not resolved, except for a few molecules such as NH_3, which can tunnel fairly freely through the mean molecular plane.

For a molecule with an n-fold axis of symmetry, there are additional symmetry properties analogous to the *symmetric* and *antisymmetric* properties for linear molecules. If we consider a molecule with a threefold axis of symmetry, e.g., CX_3Y, then rotation of the molecule by 120° about the symmetry axis exchanges identical nuclei. In the ground state, levels with $K = 0, 3, 6, 9, \ldots$ have a larger statistical weight than levels with $K = 1, 2, 4, 5, 7, 8, \ldots$. The ratio of the statistical weights depends on the nuclear spin of X. There is a selection rule, analogous to Eq. (65), that only levels with the same overall symmetry can combine with each other.

If we consider only transitions between nondegenerate vibrational levels, two types of band structure are possible, depending on the direction of the change of dipole moment, known as the *transition moment*. If the transition moment is parallel to the symmetric-top axis, then the selection rules are

$$\Delta K = 0, \qquad \Delta J = \pm 1 \quad \text{if } K = 0; \tag{70}$$

$$\Delta K = 0, \qquad \Delta J = 0, \pm 1, \quad \text{if } K \neq 0. \tag{71}$$

A parallel band therefore is composed of a series of sub-bands similar to Σ–Σ, Π–Π, Δ–Δ, . . . bands of a linear molecule. The origins of the sub-bands are given by

$$\nu_0^{\text{sub}} = \nu_0 + [(A' - A'') - (B' - B'')]K^2. \tag{72}$$

A diagram showing the structure of a parallel band is given in Fig. 13.

If the transition moment is perpendicular to the symmetric-top axis, then

$$\Delta K = \pm 1, \qquad \Delta J = 0, \pm 1. \tag{73}$$

The band consists of a superposition of sub-bands similar to the Σ–Π, Π–Σ, Π–Δ, Δ–Π, . . . bands of a linear molecule. The origins of the sub-bands are given by

$$\nu_0^{\text{sub}} = \nu_0 + (A' - B') \pm 2(A' - B')K + [(A' - B') - (A'' - B'')]K^2, \tag{74}$$

where the + sign refers to $\Delta K = +1$ (with $K = 0, 1, 2, \ldots$) and the − sign to $\Delta K = -1$ (with $K = 1, 2, 3 \ldots$). A typical band structure is given in Fig. 14.

The intensity I of a line in an absorption band is given by

$$I(J, K) = C\nu S_{JK} g_{JK} \exp\left(-\frac{hcF''(J, K)}{kT}\right), \tag{75}$$

where C is a constant, S_{JK} is a Hönl–London line-strength factor (see Table 2), g_{JK} is the statistical weight of the lower state, and the last term is the Boltzmann factor for the lower state. The statistical weight factor equals $2J + 1$ if $K = 0$ and $2(2J + 1)$ if $K \neq 0$. The table also applies to linear molecules if K is replaced by l.

The line strength S_{JK} is not highly dependent on J, the statistical weight factor increases linearly with J, while the Boltzmann factor decreases exponentially with increasing J. Consequently, the intensity in a branch usually increases with J for lower J values and decreases for higher J values (see Figs. 13 and 14). It may be noted that, when $K = 0$, the intensity of the Q branch is zero.

In emission, the intensity distribution is given by

$$I(J, K) = C\nu^4 S_{JK} g_{JK} \exp\left(-\frac{hcF'(J, K)}{kT}\right) \tag{76}$$

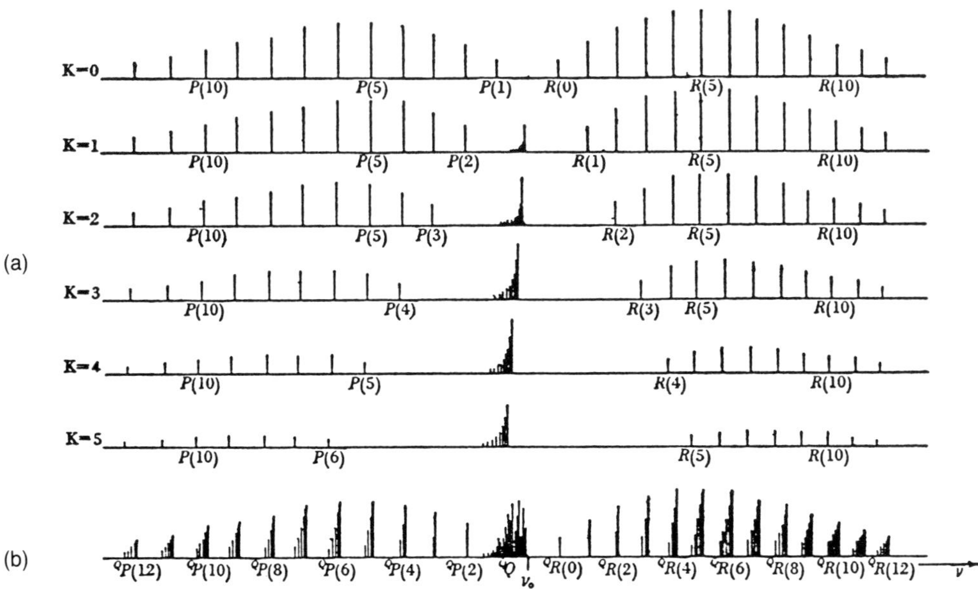

FIG. 13. Parallel band of a symmetric-top molecule: (a) the sub-band structure; and (b) the summation (from Herzberg, 1945, by courtesy of the author and publishers).

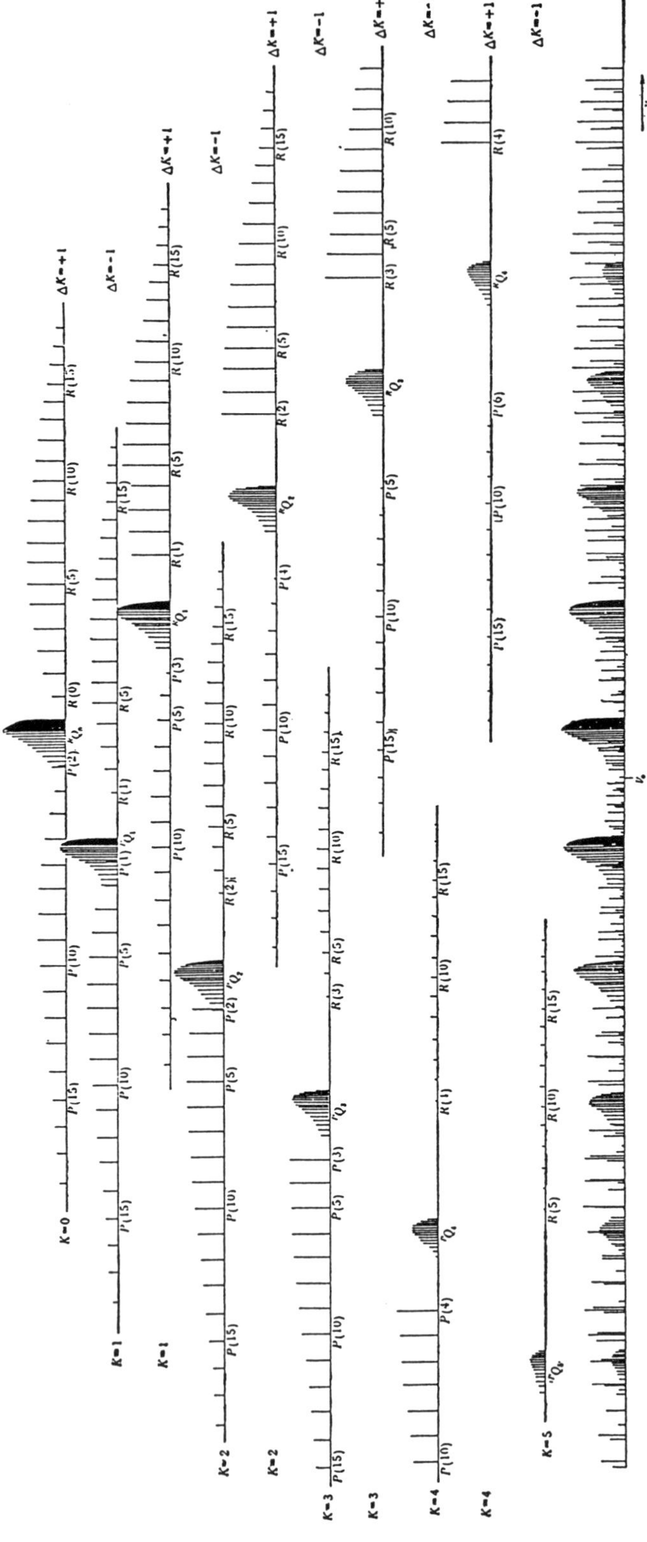

FIG. 14. Perpendicular band of a symmetric-top molecule. The sub-band structure is shown, with the summation in the bottom strip (from Herzberg, 1945, by courtesy of the author and the publishers).

Table 2. Hönl–London line strength factors. If $K = 0$ and $\Delta K = +1$, the values given have to be multiplied by 2.

	$\Delta K = +1$	$\Delta K = 0$	$\Delta K = -1$
R	$\dfrac{(J+2+K)(J+1+K)}{(J+1)(2J+1)}$	$\dfrac{(J+1)^2 - K^2}{(J+1)(2J+1)}$	$\dfrac{(J+2-K)(J+1-K)}{(J+1)(2J+1)}$
Q	$\dfrac{(J+1+K)(J-K)}{J(J+1)}$	$\dfrac{K^2}{J(J+1)}$	$\dfrac{(J+1-K)(J+K)}{J(J+1)}$
P	$\dfrac{(J-1-K)(J-K)}{J(2J+1)}$	$\dfrac{J^2 - K^2}{J(2J+1)}$	$\dfrac{(J-1+K)(J+K)}{J(2J+1)}$

and depends on the fourth power of the frequency and the Boltzmann factor for the upper state. The product $S_{JK}g_{JK}$ is the same as for absorption.

2.2.1.3 Asymmetric-Top Molecules. The rotational structures are very similar to those for pure rotational spectra, except that the band type depends on the direction of the change of dipole moment rather than on the dipole moment itself. There are three types of band: *a* type, *b* type, and *c* type, depending on whether the transition moment is along the axis of least, intermediate, or largest moment of inertia, respectively.

The selection rules are the same as those given in Eqs. (22), (23), and (24). If the direction of the transition moment is not along one of the inertial axes, then hybrid bands are observed with more than one band type.

The rotational structures of asymmetric-top molecules are not, in general, simple and will not be discussed further.

2.2.1.4 Spherical-Top Molecules. Spherical top molecules, e.g., CH_4, SiH_4, pose special problems on account of the high symmetry of some of the levels and splittings caused by Coriolis interactions. Since very few molecules belong to this class, the rotational fine structures will not be considered further.

2.3 Infrared Techniques

Most of the early infrared spectra were taken with prism or grating spectrometers. Nowadays, most infrared spectra are recorded with commercial Fourier-transform spectrometers. The resolution can be varied from a few cm^{-1} for low-resolution scans to 0.001 cm^{-1} for high-resolution scans. An example of a high-resolution recording is shown in Fig. 15, which illustrates the Q branch of the 588.8-cm^{-1} band of N_2O.

A technique that has been used for the detection of free-radical species is laser magnetic resonance (LMR). The species are flowed through a cell placed within the cavity of a fixed-frequency laser. The CO_2 and N_2O lasers are used in the region 900 to 1100 cm^{-1}, while CO is used at higher frequencies (1250 to 2000 cm^{-1}). The laser is tuned to one of the frequencies in the vibration-rotation band. Since these frequencies rarely coincide with a frequency of a free-radical species, the latter is tuned to resonance by the application of a magnetic field. An example of an LMR spectrum of CH_2 is shown in Fig. 16. This technique has been used for the study of many free-radical species, e.g., FO, CH_2, NH_2, HCO, and FeH. One of the advantages of this technique is that it provides a sensitive means of detection for paramagnetic species but does not detect stable diamagnetic species such as the parent and product molecules. The technique was originally developed for use in the far-infrared region of the spectrum (Evenson *et al.*, 1980).

A similar technique uses electric instead of magnetic fields and is known as laser electric resonance or laser Stark spectroscopy. The technique is sensitive to all species possessing a dipole moment and is thus less selective than laser magnetic resonance.

Other types of laser in common use are infrared diode lasers, difference-frequency lasers, and color-center lasers. Infrared diode lasers are very widely used and available for most regions of the infrared spectrum (3–30 μm). They can be tuned over limited frequency ranges, normally ± 50 cm^{-1}, but the tuning characteristics frequently display dis-

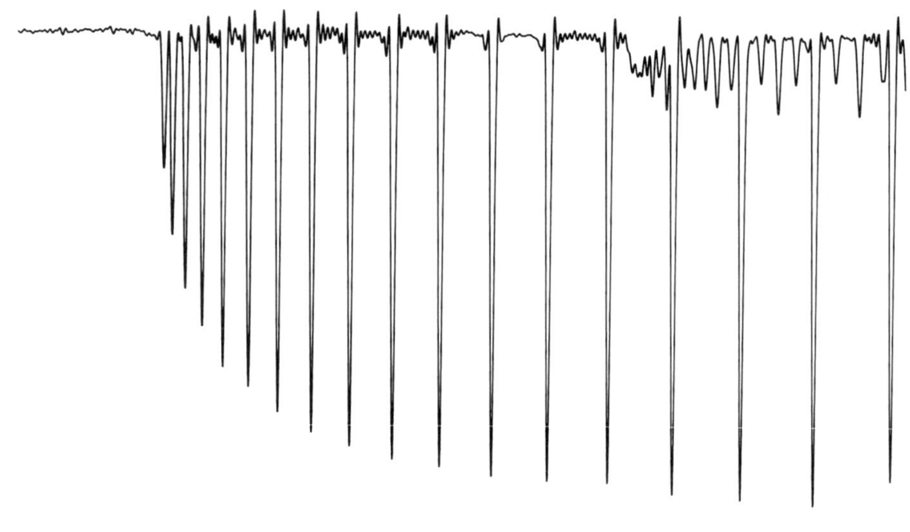

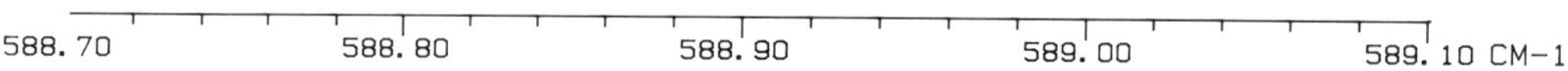

FIG. 15. *Q* branch of the 588-cm^{-1} band of N_2O recorded with a Bomem Fourier-transform spectrometer with a resolution of 0.0014 cm^{-1} (by courtesy of J. W. C. Johns, National Research Council of Canada).

continuities due to what are called "mode hops."

Difference-frequency lasers involve the mixing of two visible lasers, one fixed and the other tunable, in a nonlinear crystal such as lithium niobate. Tunable infrared radiation in the region 2000 to 4500 cm^{-1} is obtained with powers of about 10 μW. The laser has been used for the detection of molecular ions and other transient species. The first molecular ion to be detected by this method was H_3^+. An ion may be produced in an ac electric discharge so that its velocity is modulated, first in one direction and then in the other. The change in frequency caused by the Doppler effect is then sufficient so that the modulated frequency can be detected by a lock-in amplifier. Both positive and negative ions have been detected using this technique.

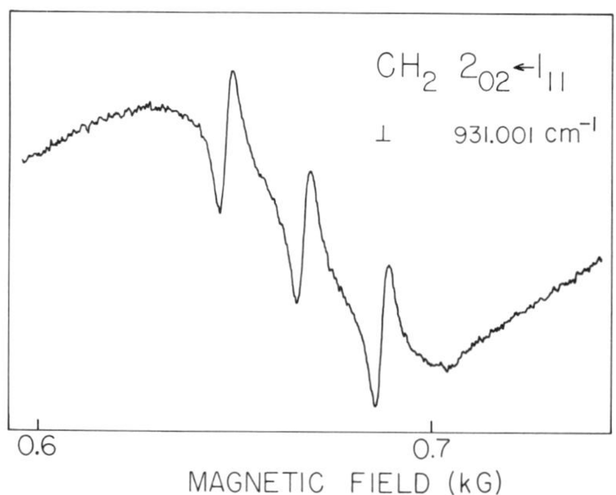

FIG. 16. Laser magnetic resonance (LMR) spectrum of CH_2 (unpublished figure from Sears *et al.*, 1982). One magnetic component only is shown. The triplet structure arises from proton hyperfine interactions.

Color-center lasers are produced by pumping alkali halide crystals with a Kr^+ laser. The lasers operate over the frequency range 4000 to 3000 cm^{-1} and have been used mainly for the detection of transient species in this region. The most notable example at present is the C_2H radical.

2.4 Applications of Infrared Spectroscopy

Infrared spectroscopy has widespread applications in many branches of chemistry. It was stated as early as the 1940s that "the in-

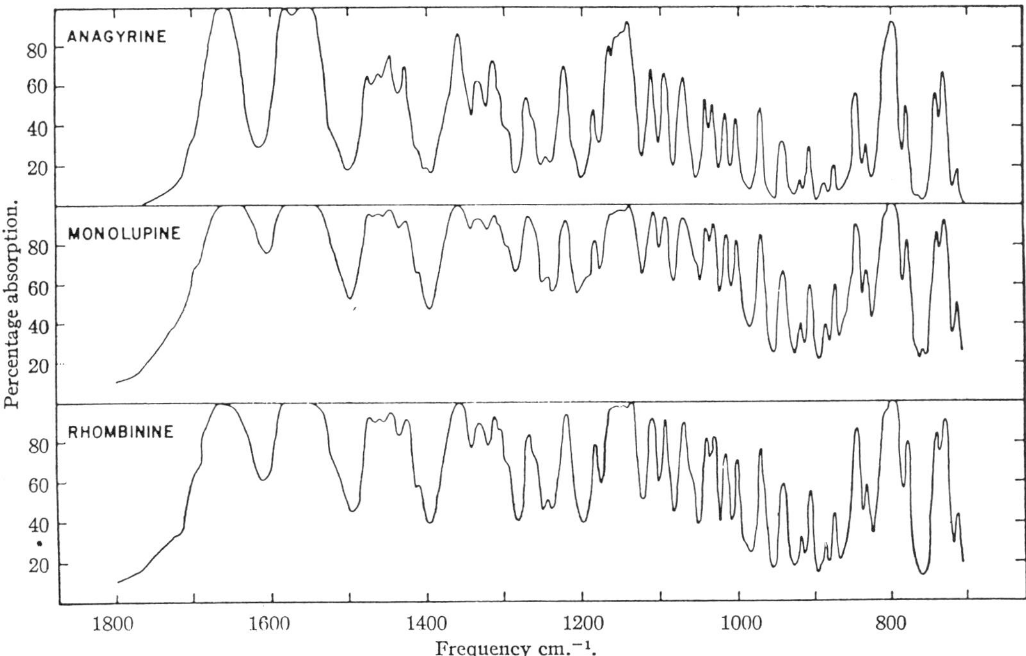

FIG. 17. Infrared spectra of anagyrine, monolupine, and rhombinine (from Marion and Ouellet, 1948, by permission of the journal).

fra-red spectrum of a chemical compound is probably the most characteristic physical property of that compound" (Sutherland and Thompson, 1945). Organic molecules containing large numbers of atoms have many fundamental bands spanning the range from about 10 to 3500 cm^{-1}, as well as overtone and combination bands. Since these bands may have widely different intensities, it is not difficult to understand why the infrared spectrum provides a unique fingerprint of the compound.

An early example involves three alkaloids, anagyrine, monolupine, and rhombinine, which were initially thought to be different compounds. Classical studies provided strong evidence that the compounds were identical, and this was convincingly confirmed by the identity of their infrared spectra (Fig. 17) (Marion and Ouellet, 1948).

Infrared spectra are frequently used for the analysis of mixtures of compounds or the detection of impurities. These analyses may use substances in the gas phase or in condensed phases. An example is provided by the detection of small amounts of impurities, e.g., SO_2 or chlorofluorocarbons in the Earth's atmosphere.

Infrared techniques are also used for the detection of molecules in astronomical sources. The presence of CO_2, NH_3, and CH_4 in planetary atmospheres has been known for a long time, and recently a spectrum of the H_3^+ ion has been observed in emission from Jupiter (Fig. 18). Molecules such as CO, CH_4, and C_2H_2 have also been found in infrared spectra of regions of star formation and circumstellar envelopes.

3. ELECTRONIC SPECTRA

3.1 Classification of Electronic States

The various electronic states of a molecule are classified according to the resultant spin $Sh/2\pi$ of all the electrons and the symmetry species of the electronic wavefunction. The *multiplicity* of an electronic state is defined as $2S + 1$. Molecules with an even number of electrons give rise to states with $S = 0$ (singlet), $S = 1$ (triplet), $S = 2$ (quintet), and so on. Species with an odd number of electrons give rise to states with $S = \frac{1}{2}$ (doublet), $S = \frac{3}{2}$ (quartet), $S = \frac{5}{2}$ (sextet), and so on.

For diatomic and linear polyatomic mol-

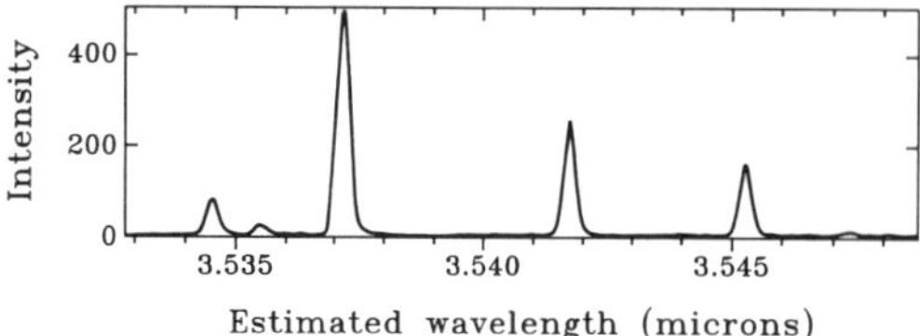

FIG. 18. Emission lines of H_3^+ observed in the spectrum of Jupiter (courtesy of T. Oka, University of Chicago).

ecules, the symmetry species of the electronic wave function depends on the component $\Lambda h/2\pi$ of the orbital angular momentum of the electrons along the axis of the molecule. For $\Lambda = 0, 1, 2, \ldots$, the states are designated as $\Sigma, \Pi, \Delta, \ldots$ (compare the usage of $s, p, d, \ldots$ in atoms).

For nonlinear molecules, the symmetry species of the electronic wave function depends on the symmetry group (point group) of the molecule and will belong to one of its symmetry types, e.g., A_1, B_2, B_{1u}, E_{2g}, ... An example will be given below.

3.2 Intensities of Transitions

The probability P of an electric dipole transition between two electronic states with wave functions ψ_e' and ψ_e'' is proportional to

$$P \propto |\int \psi_e'^* \mu \psi_e'' d\tau_e|^2, \tag{77}$$

which is often written in matrix notation as

$$P \propto |\langle \psi_e' | \mu | \psi_e'' \rangle|^2, \tag{78}$$

where the asterisk denotes the complex conjugate and μ is the dipole-moment operator with components μ_x, μ_y, μ_z such that $\mu_x = \Sigma_i e_i x_i$, $\mu_y = \Sigma_i e_i y_i$, and $\mu_z = \Sigma_i e_i z_i$. The product of the symmetry species of ψ_e', μ, and ψ_e'' must be totally symmetric; otherwise, the integral (matrix element) will vanish, in which case the transition is forbidden.

Let us consider a symmetrical AH_2 molecule that belongs to the point group C_{2v}. The axis system is given in Fig. 19 with the x axis perpendicular to the plane of the molecule. The group table is given in Table 3. I denotes the identity operator, $C_2(z)$ a rotation by 180° about the z axis, $\sigma(xz)$ a reflection in a plane perpendicular to the plane of the molecule, and $\sigma(yz)$ a reflection in the plane of the molecule. A +1 denotes that the operation leaves the wave function unchanged, while a −1 indicates that the operation changes the sign of the wave function.

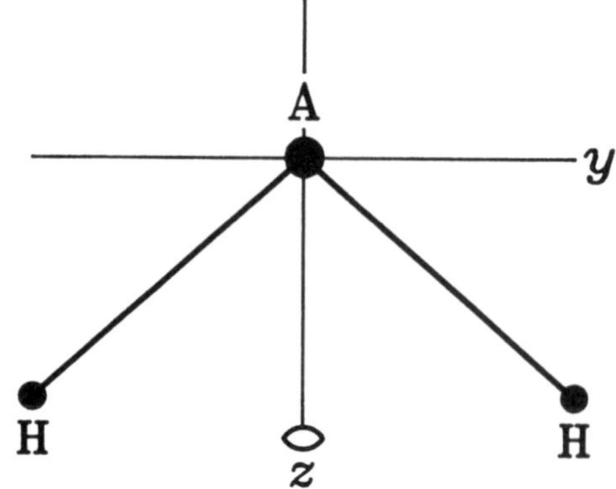

FIG. 19. Axes for an AH_2 molecule. The x axis lies perpendicular to the molecular plane.

Let us consider the symmetry species of the dipole-moment operators μ_x, μ_y, and μ_z. The term μ_x changes sign when rotated by 180° about the z axis, remains unchanged on reflection in the xz plane, and changes sign when reflected in the yz plane. It therefore belongs to the symmetry type B_1. Similarly, μ_y and μ_z have species B_2 and A_1, respectively. There is no component of the dipole-moment operator with symmetry type A_2.

Consider now a transition between two electronic states: The product of the symmetry species of ψ_e', μ, and ψ_e'' must be totally symmetric (A_1); otherwise, the integral (matrix element) will vanish. The low-lying electronic states of the NH_2 free radical have species $^2B_1$, $^2A_1$, and $^2B_2$ (see Fig. 20). The transition between the ground state ($^2B_1$) and the first excited state ($^2A_1$) is allowed with a transition moment in the x direction since $A_1 \times B_1 \times B_1 = A_1$. Similarly, the transition between the two excited states is allowed with a transition moment in the y direction since $B_2 \times B_2 \times A_1 = A_1$. The transition between the ground state and the second excited state

Table 3. Group table for a C_{2v} molecule.

C_{2v}	I	$C_2(z)$	$\sigma(xz)$	$\sigma(yz)$
A_1	+1	+1	+1	+1
A_2	+1	+1	−1	−1
B_1	+1	−1	+1	−1
B_2	+1	−1	−1	+1

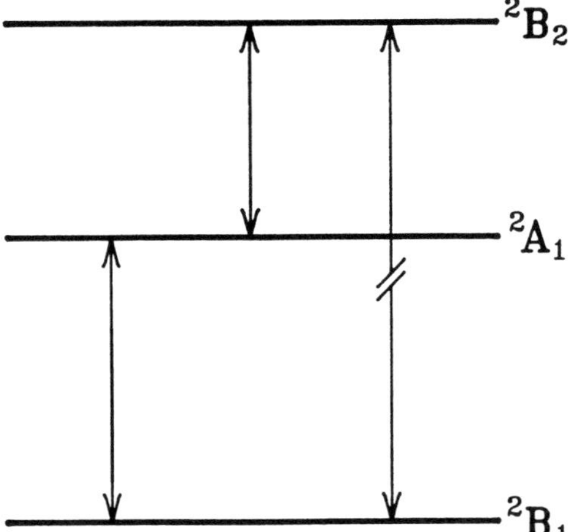

FIG. 20. Low-lying electronic states of NH_2.

is forbidden since the product of the wave functions $B_2 \times B_1 = A_2$, and there is no component of the dipole-moment operator with symmetry A_2.

3.3 Franck–Condon Principle

According to the classical model proposed by Franck, the positions and velocities of the nuclei remain virtually unchanged during an electronic transition, and the molecule then executes vibrational motion about the new equilibrium configuration. This concept was treated on a quantum-mechanical basis by Condon, who showed that the probability P of a transition between two states with total wave functions ψ' and ψ'' is given by

$$P \propto |\int \psi'^* \mu \psi'' d\tau|^2. \tag{79}$$

If the electronic and vibrational motions are completely separable, then

$$\psi = \psi_e \psi_v, \tag{80}$$

and the intensity I of a transition is given by

$$I \propto |\int \psi_e'^* \, \psi_v'^* \mu \psi_e'' \psi_v'' \, d\tau_e d\tau_v|^2. \tag{81}$$

If we separate the dipole-moment operator μ into one part that depends on the electronic coordinates and another part that depends on the nuclear coordinates, then

$$\mu = \mu_e + \mu_n, \tag{82}$$

and

$$I \propto |\int \psi_e'^* \mu_e \psi_e'' \, d\tau_e \int \psi_v'^* \psi_v'' \, d\tau_v + \int \psi_v'^* \mu_n \psi_v'' \, d\tau_v \int \psi_e'^* \psi_e'' \, d\tau_e|^2. \tag{83}$$

But since the wave functions for different electronic states are orthogonal to each other,

$$\int \psi_e'^* \psi_e'' \, d\tau_e = 0. \tag{84}$$

Hence,

$$I \propto |\int \psi_e'^* \mu_e \psi_e'' \, d\tau_e \int \psi_v'^* \psi_v'' \, d\tau_v|^2, \tag{85}$$

or, in matrix notation,

$$I \propto |\langle \psi_e' | \mu | \psi_e'' \rangle \langle \psi_v' \psi_v'' \rangle|^2. \tag{86}$$

The first term in Eq. (86) (unsquared) is called the electronic transition moment, and the second term is the vibrational overlap integral.

It is readily seen from Eq. (86) that, for the intensity of a transition to be nonvanishing, the product of the symmetry species of ψ_v' and ψ_v'' must be totally symmetric. This condition is automatically satisfied for a diatomic molecule since ψ_v is always totally symmetric, but for a polyatomic molecule, it gives rise to selection rules originally stated by Herzberg and Teller, viz., for totally symmetric vibrations,

$$\Delta v = 0, 1, 2, \ldots; \tag{87}$$

for non-totally symmetric vibrations,

$$\Delta v = 0, 2, 4, \ldots. \tag{88}$$

A series of bands in which one vibrational quantum number changes is known as a *progression*. For totally symmetric vibrations, the length of a progression depends on the change in geometric structure accompanying the electronic transition (see HCN below). For a v'–0 transition involving a non-totally symmetric vibration,

$$\frac{I(0—0)}{\Sigma_{v'} I(v'—0)} = \frac{(\nu' \nu'')^{1/2}}{\frac{1}{2}(\nu' + \nu'')}, \tag{89}$$

where I denotes band intensity. If $\nu' = \nu''$, this ratio equals 1, and all the intensity in the progression lies in the 0—0 band. If $\nu' = \frac{1}{2}\nu''$, the above ratio equals 0.943; hence, 94.3% of

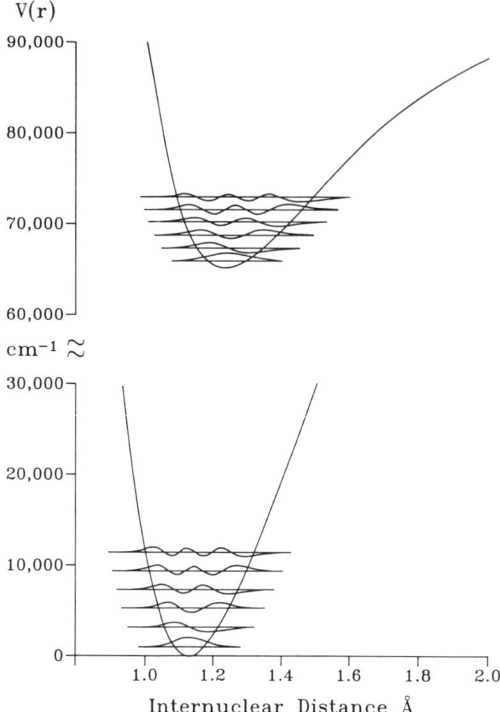

FIG. 21. Potential curves, vibrational levels, and wave functions for the $A\ ^1\Pi$–$X\ ^1\Sigma^+$ transition of the CO molecule (courtesy of R. W. Nicholls, York University, Toronto).

the intensity of the progression lies in the 0—0 band and only 5.7% in the 2—0, 4—0, . . . bands. Thus non-totally symmetric vibrations do not form strong progressions in molecular spectra. Such vibrations may give rise to *sequence* bands, i.e., bands of the type 0—0, 1—1, 2—2, . . ., particularly if the vibrational frequency is small.

An example of the vibrational levels and wave functions for two electronic states of a diatomic molecule is given in Fig. 21 for the $A\ ^1\Pi$–$X\ ^1\Sigma^+$ band system of CO. It is interesting to note that the $v = 0$ wave functions have a single maximum at the midpoint of the vibration, in contrast to classical mechanics, where the greatest probabilities are at the turning points. Furthermore, the wave function is finite, though small, outside the range of the classical turning points. The numbers of nodes in the wave functions are equal to v. A table of Franck–Condon factors $\langle\psi'_{v}\psi''_{v}\rangle^2$ is given in Table 4. In absorption, the $v'' = 0$ progression shows a single maximum at $v' = 2$; in emission, the $v' = 0$ progression also shows a single maximum at $v'' = 2$. The $v'' = 1$ progression shows two intensity maxima at $v' = 0$ and 5, while the $v' = 1$ progression shows maxima at $v'' = 0$ and 4, consistent with the fact that there are two maxima in a $v = 1$ wave function.

For a polyatomic molecule, a large change in equilibrium geometry for different electronic states is often found, especially for smaller species. For example, HCN is linear in its ground state, and the CN bond length is 1.156 Å. In the first excited state, the molecule is bent with an equilibrium angle of 125° and a CN bond length of 1.297 Å. As a result, the absorption spectrum of HCN in the ultraviolet consists of a long progression of the bending vibration in the excited state, with values of the quantum number v'_2 ranging from 0 to 10 and a maximum of intensity in the region of $v'_2 = 7$. Similar progressions are observed in which one and two quanta of the CN stretching vibration are also excited (see Fig. 22).

3.4 Vibronic Interactions

In the event that the electronic and vibrational motions are not completely separable, the system can be described by a vibronic wave function ψ_{ev} whose symmetry species is

Table 4. Franck–Condon factors for the $A^1\Pi$-$X^1\Sigma$ band system of CO (from Nicholls, 1962).

v' \ v''	0	1	2	3	4	5	6
0	1.13×10^{-1}	2.61×10^{-1}	2.85×10^{-1}	1.96×10^{-1}	9.60×10^{-2}	3.55×10^{-2}	1.03×10^{-2}
1	2.16×10^{-1}	1.55×10^{-1}	3.05×10^{-3}	7.64×10^{-2}	1.93×10^{-1}	1.86×10^{-1}	1.08×10^{-1}
2	2.30×10^{-1}	1.22×10^{-2}	9.01×10^{-2}	1.16×10^{-1}	5.08×10^{-3}	5.72×10^{-2}	1.65×10^{-1}
3	1.81×10^{-1}	2.05×10^{-2}	1.17×10^{-1}	6.45×10^{-4}	8.96×10^{-2}	8.42×10^{-2}	4.71×10^{-4}
4	1.19×10^{-1}	8.73×10^{-2}	3.44×10^{-2}	5.76×10^{-2}	6.66×10^{-2}	6.17×10^{-3}	9.80×10^{-2}
5	6.88×10^{-2}	1.23×10^{-1}	3.24×10^{-4}	9.13×10^{-2}	1.38×10^{-5}	8.24×10^{-2}	2.27×10^{-2}
6	3.67×10^{-2}	1.16×10^{-1}	3.21×10^{-2}	4.22×10^{-2}	4.29×10^{-2}	4.23×10^{-2}	2.18×10^{-2}

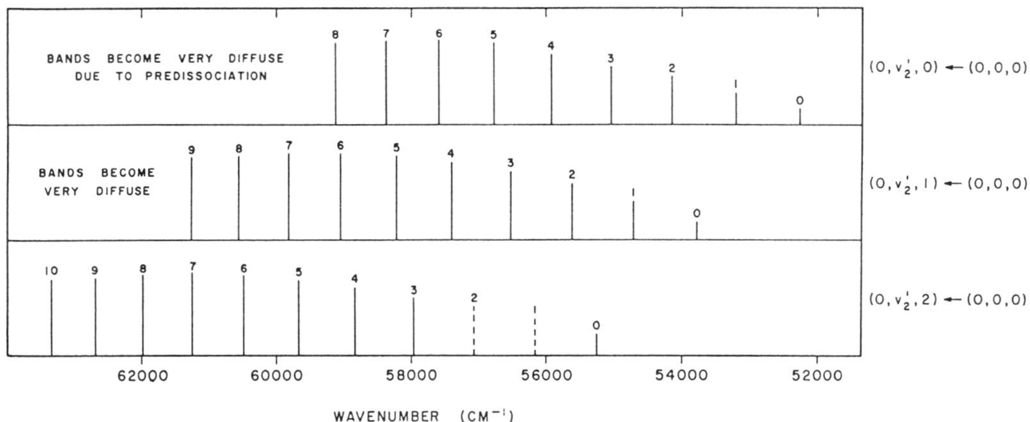

FIG. 22. Relative intensities of the bands in the ultraviolet spectrum of HCN (from Herzberg and Innes, 1957). The intensities are plotted on a logarithmic scale. The bands are plotted as separate progressions (0, v_2', v_3')–(0, 0, 0) with v_3' = 0, 1, and 2.

the direct product of the electronic and vibrational species.

Suppose we have an electronically forbidden transition, so that

$$\langle \psi_e' | \mu | \psi_e'' \rangle = 0. \tag{90}$$

There may be non-totally symmetric vibrations of appropriate symmetry so that

$$\langle \psi_{ev}' | \mu | \psi_{ev}'' \rangle \neq 0. \tag{91}$$

Weak bands may then be observed and are said to be vibronically induced. Classical examples are found in the spectra of formaldehyde and benzene of which the former will be considered here.

Formaldehyde, H_2CO, belongs to the point group C_{2v} and has $3 \times 4 - 6 = 6$ normal modes of vibration (see Fig. 23). Three are totally symmetric (a_1) and may be described as a symmetric C—H stretching vibration ν_1, a CO stretching vibration ν_2, and a symmetric H—C—H bending vibration ν_3. There is one vibration of the type b_1, which is the out-of-plane C—H wagging motion ν_4. The remaining two vibrations belong to the type b_2 and are the antisymmetric C—H stretching vibration ν_5 and the CH_2 rocking vibration ν_6. There are no fundamental vibrations of the type a_2. The frequencies of these vibrations in the ground ($\tilde{X}\ ^1A_1$) and first excited singlet state ($\tilde{A}\ ^1A_2$) are given in Table 5. The CO stretching frequency ν_2 changes considerably from the ground state to the excited state, corresponding to a lengthening of the CO bond from 1.21 Å to 1.32 Å. The most dramatic change involves ν_4. In the ground state, the molecule is planar, and the out-of-plane wagging frequency is 1167.3 cm^{-1}. In the excited state, the molecule is slightly nonplanar, and the out-of-plane frequencies are very anharmonic. Successive levels are 124.5 cm^{-1} (b_1), 542.3 cm^{-1} (a_1), 947.9 cm^{-1} (b_1), 1429.3 (a_1), . . ., where the symmetry of the vibrational level is b_1 for odd values of v_4' and a_1 for even values of v_4'.

An energy-level diagram showing some of the bands observed in the near-ultraviolet spectrum of formaldehyde is given in Fig. 24. The 0–0 band is forbidden by the electric dipole selection rules. The first band in the ab-

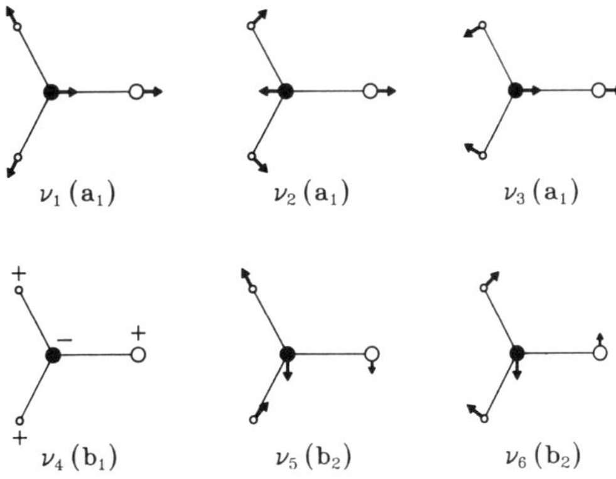

FIG. 23. Normal modes of vibration for formaldehyde, H_2CO.

Table 5. Vibrational frequencies for formaldehyde in its ground ($\tilde{X}\ ^1A_1$) and excited ($\tilde{A}\ ^1A_2$) states in cm^{-1} (from Clouthier and Ramsay 1983).

Vibration	$\tilde{X}\ ^1A_1$	$\tilde{A}\ ^1A_2$
$\nu_1(a_1)$	2782.5	2846
$\nu_2(a_1)$	1746.0	1183
$\nu_3(a_1)$	1500.2	1293.1
$\nu_4(b_1)$	1167.3	[a]
$\nu_5(b_2)$	2843.3	2968.3
$\nu_6(b_2)$	1249.1	904

[a]These levels are very anharmonic and take successive values of 124.5 (b_1), 542.3 (a_1), 947.9 (b_1), 1429.3 (a_1), ···.

sorption spectrum from the ground state is called the A band and terminates on the level with $v_4' = 1$. The vibronic symmetry of the latter is $^eA_2 \times {}^vB_1 = {}^{ev}B_2$. The transition is therefore vibronically B_2—A_1, and the transition moment lies in the plane of the molecule in a direction at right angles to the CO bond. The rotational structure is found to be b type, which is consistent with the above interpretation. The A band is the first band of a progression involving the stretching vibration of the CO bond ν_2' in the excited state. Another band similar in type to the A band is the B' band, which terminates on the level with $v_4' = 3$ in the excited state. The B' band is the first member of another progression involving ν_2'.

In emission, the first band from the zero level in the excited state terminates on the level with $v_4'' = 1$ in the ground state and is called the α band. This band is the first member of a progression involving the CO stretching vibration ν_2'' in the ground state. The first band in emission from the $\nu_4 = 1$ level of the excited state is the A band, which is weak on account of Franck–Condon considerations. The next band terminates on $v_4'' = 2$ and is called the β band. This band is also the first member of a progression involving ν_2''. It should be mentioned that the levels involving ν_4 in the ground state are coupled to levels involving ν_6 by a mechanism known as Coriolis interaction. The b-type bands in emission that terminate on levels involving ν_4 are accompanied by weaker c-type bands at lower frequencies terminating on the levels involving ν_6.

3.5 Rotational Fine Structure

Electronic bands of diatomic and simple polyatomic molecules frequently show rotational fine structure that can be analyzed to give some information on the geometrical structure of the molecule in the excited state.

For linear molecules, the expression for the

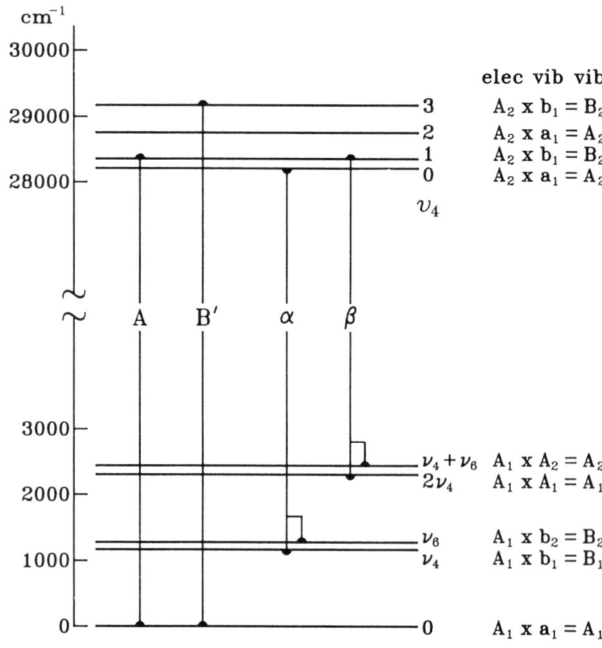

FIG. 24. Energy-level diagram showing some of the principal bands in the near-ultraviolet spectrum of formaldehyde.

rotational term values is the same as in Eq. (62), except that l is replaced by Λ for a diatomic molecule and K for a polyatomic molecule. In the latter case, there are two possible contributions to the resultant angular momentum about the internuclear axis. The first is the electronic angular momentum $\Lambda h/2\pi$, and the other is the vibrational angular momentum $lh/2\pi$. The resultant is $Kh/2\pi$, where

$$K = |\Lambda \pm l|. \tag{92}$$

We now need to consider two additional symmetry properties. If the molecule has a center of symmetry, then the *electronic* wave function either remains unchanged or changes sign when the coordinates of all the electrons are replaced by their negatives. In the former case, the state is labeled g (*gerade*) and in the latter u (*ungerade*). For a molecule in a Σ electronic state, the *electronic* wave function either remains unchanged (Σ^+ state) or changes sign (Σ^- state) when reflected in any plane through the linear axis. The symmetry properties of the rotational levels for various types of electronic state are given in Fig. 25.

The selection rules for transitions involving singlet states are the same as those for infrared bands, replacing l by Λ or K [see Eqs. (64) to (68)]. The intensity factors are the same as those given in Table 2, with K replaced by Λ or l as required. A further selection rule is

$$\Delta S = 0, \tag{93}$$

although many weak systems are known, e.g., triplet-singlet, which violate this rule.

The B values for the molecule in the ground and excited states are frequently quite different, and so a "head" may form in either the P or the R branch. It is seen that the Fortrat parabola [Eq. (46)] forms a head when $d\nu/dm = 0$ and

$$m = -(B + B'')/2(B' - B''). \tag{94}$$

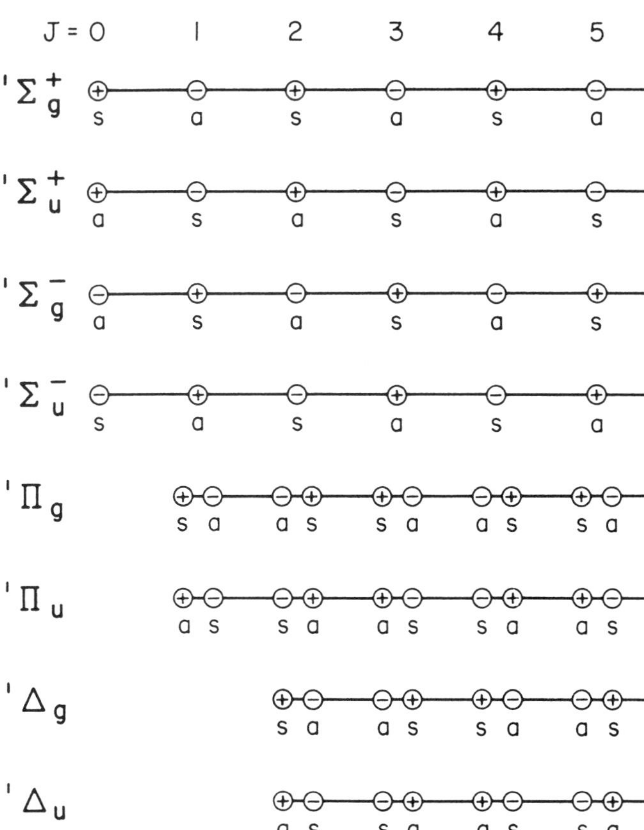

FIG. 25. Symmetry properties for the rotational levels of linear molecules. The group symbols refer to the vibronic species of the molecule. (+) denotes positive and (−) negative rotational levels; *s* stands for symmetric and *a* for antisymmetric rotational levels. For molecules which do not possess a center of symmetry, the *g* and *u*, and the *s* and *a* symbols may be deleted.

The separation between the band head and the band origin is given by

$$\nu_{\text{Head}} - \nu_0 = -(B' + B'')^2/4(B' - B''). \quad (95)$$

If $B' < B''$, m is positive, and the head is formed in the R branch. If $B' \geq B''$, m is negative, and the head is formed in the P branch. R heads are observed more frequently than P heads since bond lengths are usually larger in the excited state and $B' < B''$.

The nature of an electronic transition can frequently be deduced from the observation of the first lines in the branches. Remembering that J must always be equal to or larger than Λ or K, it is easy to deduce that the first lines in the P, Q, and R branches are $P(2)$, $Q(1)$, and $R(0)$ for a $^1\Pi$—$^1\Sigma$ transition, but $P(1)$, $Q(1)$, and $R(1)$ for a $^1\Sigma$—$^1\Pi$ transition.

For linear molecules in states other than singlet states, the nature of the coupling between the resultant electron spin and the orbital angular momentum needs to be considered. Various coupling cases were treated originally by Hund and expressions for rotational term values deduced. Additional selection rules also apply.

For symmetric-top and asymmetric-top molecules, the analysis of bands is similar to that which has been described for infrared spectra.

3.6 Magnetic Rotation Spectra

In this procedure, radiation from a continuous source, e.g., a high-pressure xenon arc, is passed through a polarizer and an absorption tube that is placed inside a solenoid. On emerging, it encounters a second polarizer, which is crossed with respect to the first one. No radiation emerges unless a molecule that produces a rotation of the plane of polarization in the presence of a magnetic field is present in the absorption tube. Weak triplet–singlet transitions can be detected in this way, whereas stronger singlet–singlet transitions are unaffected. An example of a magnetic rotation spectrum is shown in Fig. 26.

Plane-polarized light can be resolved into two circularly polarized components, one left-handed and the other right-handed. If the magnetic field introduces any optical activity, then the two components travel with different velocities when passing through the cell and, when recombined on emerging, give plane-polarized light in which the plane of polarization has been rotated. This process is called magnetic circular birefringence (MCB). If the absorption coefficients for left- and right-handed circularly polarized light are different, then the two components will emerge from the cell with different amplitudes and, on recombination, will produce elliptically polarized light. This process is called magnetic circular dichroism (MCD). The combination of MCB and MCD produces the magnetic rotation spectrum.

For theoretical interpretation, it is preferable to measure either the MCB or the MCD. The latter is experimentally simpler. Before entering the cell, the plane-polarized light is passed through a quarter-wave plate with its principal axis arranged at +45° to the plane of polarization. This produces left- or right-handed circularly polarized light. By means of a photoelastic modulator, the polarization can be modulated from left- to right-handed circular polarization at frequencies of ~50 kHz. Any magnetic circular dichroism is then detected as a modulation in the intensity of

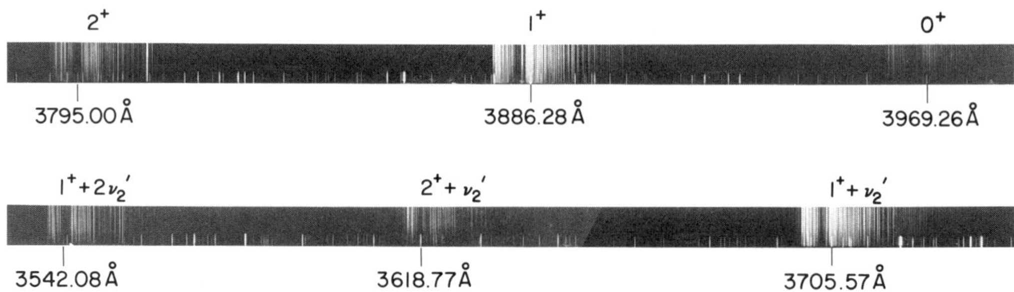

FIG. 26. Magnetic rotation spectrum of the $\tilde{a}\ ^3A_2$–$\tilde{X}\ ^1A_1$ system of formaldehyde (from Brown *et al.*, 1976, by courtesy of the authors and the journal).

the beam emerging from the cell using a detector and a lock-in amplifier.

3.7 Dissociation and Predissociation

For some molecules, vibration–rotation structure has been observed right up to a dissociation limit. If the states of the atoms at the limit are known, then an accurate value for the dissociation energy of the molecule is obtained. As an example, the $B\ {}^3\Sigma_u^- \rightarrow X\ {}^3\Sigma_g^-$ bands of oxygen, known as the Schumann–Runge bands, have been observed right up to a limit at 1750 Å. The oxygen atoms produced are in the atomic states $O({}^3P)$ + $O({}^1D)$. Subtracting the excitation energy of the $O({}^1D)$ atom, a value for the dissociation energy of the molecule in the ground state is obtained, viz., 41 260 ± 15 cm^{-1}.

More frequently, a phenomenon known as predissociation is observed in a spectrum. In absorption, the lines become broad or diffuse; in emission, a series of lines may suddenly terminate at a certain J value. The reason is that the potential curve for the excited state is crossed by a repulsive potential curve, and the molecule may dissociate along this curve rather than continue to vibrate and rotate in the initial excited state. The energy at which this phenomenon occurs gives an upper limit for the dissociation energy of the molecule.

3.8 Linewidths and Lifetimes

The relation between the width of a line and the lifetime of the excited state is given by the uncertainty relation,

$$\Delta E \Delta t \geqslant h/2\pi. \tag{96}$$

Since $\Delta E = hc\Delta\nu$, where $\Delta\nu$ is the width of a level in cm^{-1}, then

$$\Delta\nu \geqslant 1/2\pi c \Delta t. \tag{97}$$

Hence, the greater the mean life of a state, the smaller is its width. For strongly allowed electronic transitions, the lifetime of the excited state is of the order of 10–20 ns, and the corresponding broadening is of the order of 0.001 cm^{-1}. This is considerably less than the widths of spectral lines caused by the motions of the molecules (Doppler widths), which in the ultraviolet may be of the order of 0.05 cm^{-1} at room temperature. When an excited state is crossed by a repulsive curve, predissociation may occur during the period of one vibration, i.e., in a time of the order of 10^{-13}–10^{-14} s, for which the corresponding broadening is of the order of 50–500 cm^{-1}. Not all predissociation processes proceed this rapidly, but, even if predissociation occurs in a time of 10^{-10} s, the contribution to the line width will be 0.05 cm^{-1}, which is comparable with the Doppler width.

In larger molecules, the lifetime of an excited state may not be determined by predissociation but by rapid conversion from an excited-state level to a high vibrational level of the ground state (*internal conversion*), or to a level of a triplet state (*intersystem crossover*). Azulene provides a classic example of internal conversion, a process that has been studied using picosecond spectroscopy (Rentzepis and Struve, 1976). Formaldehyde provides another example for which fluorescence quantum yields have been measured and found to be less than unity. However, in this case, there is competition between internal conversion and predissociation (Moore, 1983).

3.9 Molecular Orbital Theory

Molecules are composed of atoms in which the electrons can occupy various orbits, 1*s*, 2*s*, 2*p*, 3*s*, 3*p*, 3*d*, 4*s*, and so on. In molecules, the electrons also occupy various orbits, called orbitals, which in simple molecular orbital theory are taken as linear combinations of the atomic orbitals. By solving the Schrödinger wave equation using computers, the energy levels of molecules are obtained (see ELECTRONIC STRUCTURE OF ATOMS AND MOLECULES). It is found that the electrons occupy certain shells as in the case of atoms. The outermost of these shells are sometimes called valence shells, and the electrons occupying them valence electrons.

The electronic structure of the ground state of a molecule is obtained by filling the orbitals with the available electrons, starting with the orbital of lowest energy. Frequently, a valence shell will not be completely filled in this manner, and excited states can be generated by promoting one or more of the valence electrons to vacancies in the valence shell. Such excited states lie usually in the range 0–5 eV, and spectra involving these states lie in the near-infrared, visible, or near-ultraviolet region. If the ground state of a molecule involves a complete valence shell, then the first excited state of the molecule will involve the excitation of an electron to a shell with a higher value of the principal quantum number n. Such states are called Rydberg states and usually occur in the vacuum-ultraviolet region of the spectrum below 2000 Å.

3.10 Rydberg States and Ionization Potentials

The simplest example of Rydberg states is provided by the hydrogen atom. The term values are given by the formula

$$T_n = I - R/n^2, \tag{98}$$

where R is the Rydberg constant (109 677.58 cm^{-1}) and is equal to the ionization potential (I = 13.59840 eV); n is the principal quantum number.

Molecules also have Rydberg states, and the term values can be represented to a good approximation by a formula

$$T_n = I - R/(n - \mu)^2, \tag{99}$$

where I is the ionization potential of the molecule and μ is known as the quantum defect. The value of μ gives some information on the nature of the Rydberg electron.

Rydberg states may be studied in absorption, and the transitions lie mainly in the vacuum ultraviolet, in the region between 2000 and 500 Å. Frequently, series are observed to the first and higher ionization limits and give values for the ionization potentials. Alternatively, transitions between Rydberg states may be studied in emission at longer wavelengths. Rydberg spectra frequently display vibrational and sometimes rotational structure; hence, to obtain accurate values for ionization potentials, it is necessary to cool the molecules in a supersonic jet.

An example of a Rydberg series is shown in Fig. 27. Members can be seen up to n = 30, and the series converges to an ionization limit, which can be determined to better than 0.1 cm^{-1}.

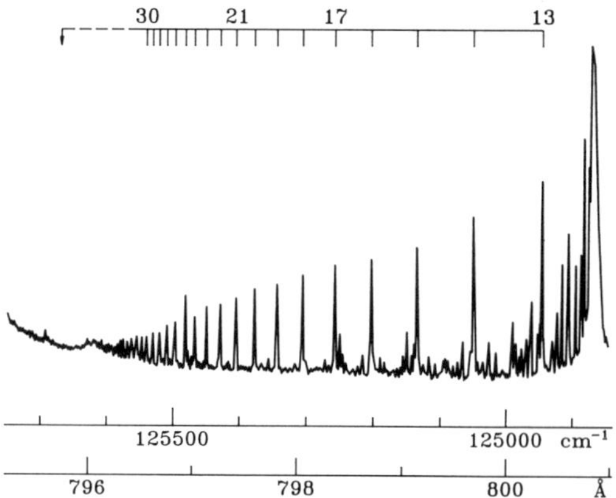

FIG. 27. Rydberg series for N_2 (courtesy of K. P. Huber, National Research Council of Canada). The sample was cooled in a supersonic jet to a temperature of ~20 K. The principal quantum numbers n are indicated.

3.11 Photoelectron Spectra and ZEKE Spectra

An alternative method of studying ionization limits of molecules is by the use of photoelectron spectroscopy. Radiation from a helium lamp at a wavelength of 584 Å has sufficient energy (21.2168 eV) to ionize all molecules, producing ions in their ground states and in excited states. The energy spectrum of the electrons emitted is measured and gives information complementary to that obtained from an optical spectrum and also gives a determination of the various ionization limits (Turner *et al.*, 1970).

A photoelectron spectrum of nitrogen is shown in Fig. 28. The peak labeled 15.57(9) corresponds to the ionization limit shown in Fig. 27. The peaks labeled 16.69(1) and 18.75(9) correspond to the second and third ionization limits, respectively. Vibrational structure is seen in all the transitions and is most pronounced for transitions to the second ionization limit. One of the limitations of this type of spectroscopy lies in the accuracy with which the electron energies can be measured, which is ~0.01 eV (~80 cm^{-1}).

A more recent development uses tunable lasers and measures only electrons with zero kinetic energy; hence the acronym ZEKE (zero electron kinetic energy) spectroscopy. The experimental technique uses zero-field ionization and delayed pulsed-field extraction. However, care must be taken to correct for

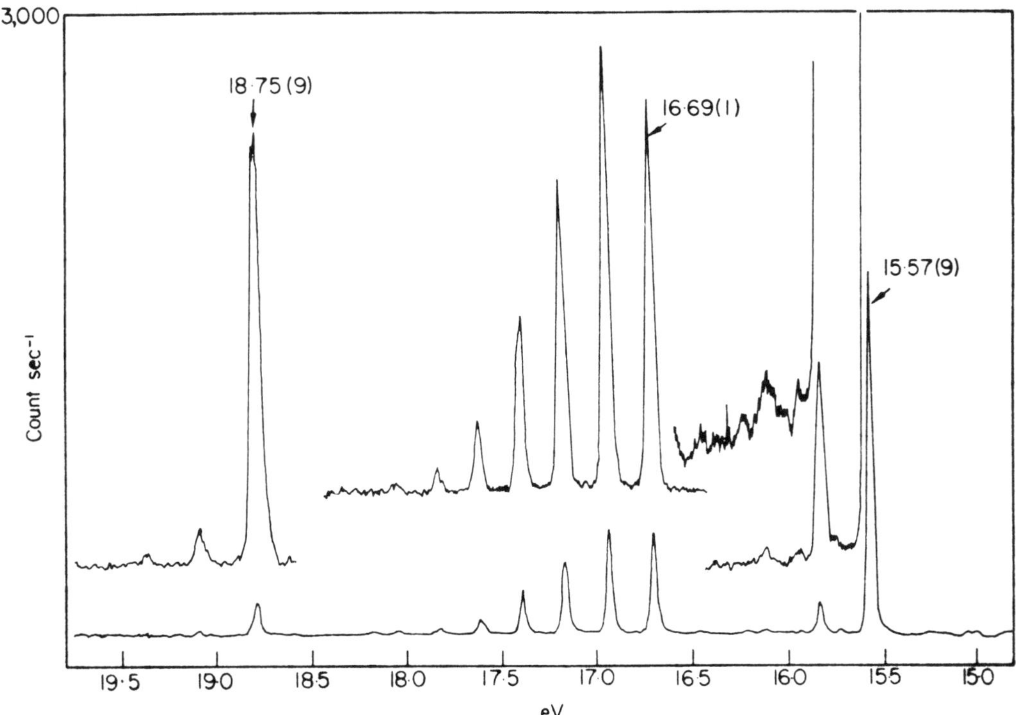

FIG. 28. Photoelectron spectrum of N_2 (from Turner *et al.*, 1970, by courtesy of the authors and publisher).

field ionization of high Rydberg states. The technique employs two-photon and multi-photon excitation, often involving near-resonant intermediate states (see Sec. 3.14.3). The technique uses supersonic jets to produce molecules at very low temperatures and is capable of giving ionization limits with an accuracy of $\sim 0.5\ cm^{-1}$ or better. Vibrational structure is readily seen, and, for some molecules, rotational structure has also been observed (Müller-Dethlefs and Schlag, 1991).

3.12 X-ray Spectroscopy (ESCA)

The most tightly bound orbitals in a molecule are those derived from 1*s* electrons. It is usually assumed that these are almost pure atomic orbitals. However, they are influenced slightly by molecule formation and are studied by means of x-ray spectroscopy. The shifts of lines from their positions in atoms give some information on the electronic structure of the molecule.

Most of the work has been carried out using the Mg $K\alpha$ line as the source of excitation. The energy spectrum of the electrons emitted is measured, and linewidths of ~0.80 eV have been achieved. This corresponds almost entirely to the inherent width of the exciting radiation. The technique is very similar to photoelectron spectroscopy and was initially called x-ray photoelectron spectroscopy (XPS). It is now usually referred to as ESCA (electron spectroscopy for chemical analysis; Siegbahn *et al.*, 1969). More recently, the technique has been extended to studying molecules adsorbed on surfaces (Stohr, 1992).

3.13 Negative-Ion Spectroscopy

A technique that measures electron affinities rather than ionization potentials is negative-ion spectroscopy. A beam of negative ions is crossed with a tunable laser (see Sec. 3.14), and the beam current is monitored. When the frequency of the laser reaches the limit at which the ion dissociates into a neutral molecule and an electron, there is a sharp decrease in the beam current. Many electron affinities have been measured in this way, and, from the structure observed in the spectrum,

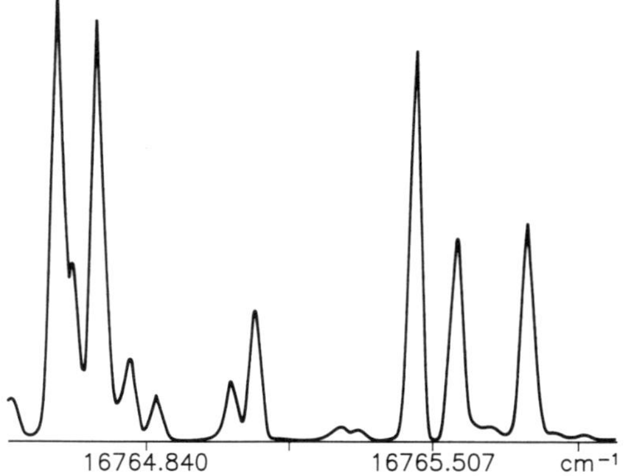

FIG. 29. Laser-excited fluorescence spectrum of thioformaldehyde, H_2CS. The spectrum covers a region of approximately 1 cm^{-1} at 16 765 cm^{-1} (courtesy of M. Barnett, National Research Council of Canada).

some deductions can be made concerning the geometry of the negative ion (Ervin and Lineberger, 1992).

3.14 Laser Spectroscopy

Many types of laser are now available. Fixed-frequency lasers, such as the argon-ion and krypton-ion lasers, can deliver powers up to 20 W at specific wavelengths. Tunable dye lasers can be operated continuously from 4000 to 8600 Å by a suitable choice of dye and optics and can yield linewidths of a few megahertz (approx. 0.0001 cm^{-1}) with powers up to 1 W in the case of rhodamine dyes. Pulsed dye lasers can be used with similar dyes giving pulses with a duration of a few nanoseconds and pulse energies of several millijoules. Frequency-doubling crystals can be used with both cw and pulsed dye lasers to produce radiation at shorter wavelengths. Other multiplication methods exist for producing tunable radiation in the vacuum ultraviolet down as far as 500 Å.

3.14.1 Laser-Excited Fluorescence Spectra One of the common applications of tunable dye lasers is to record fluorescence spectra of molecules and especially species that may be present in only small concentrations, e.g., free radicals. Since the linewidths of the laser are small compared to the Doppler widths of the lines being studied, there is no difficulty in recording high-resolution spectra. The relative intensities of the emission lines, however, may not accurately reflect the intensities of the lines in absorption, since fluorescence efficiencies vary from level to level in the excited state. The spectra obtained in this way are called laser-induced fluorescence (LIF) spectra or laser-excited fluorescence (LEF) spectra. A typical laser-excited fluorescence spectrum is shown in Fig. 29.

3.14.2 Saturated Absorption Spectra One of the great advantages of lasers lies in the narrow linewidths, providing an opportunity for carrying out sub-Doppler spectroscopy. The increase in resolution compared with Doppler-limited spectroscopy may be as high as 1000. One technique, known as saturated absorption spectroscopy, may be studied using the apparatus shown in Fig. 30. Radiation from a laser is divided into two

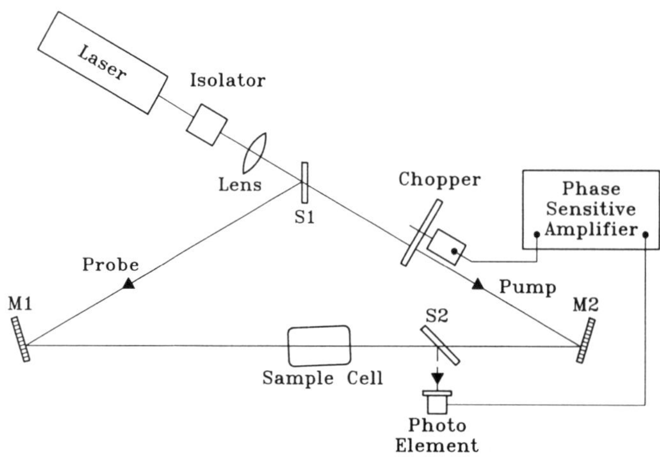

FIG. 30. Experimental arrangement for a saturated absorption experiment. The pump beam saturates the transition, and the probe beam detects the dip seen at the center of the absorption profile. The isolator prevents feedback into the laser.

beams, a stronger one known as the pump beam and a weaker one called the probe beam. The two beams pass through a sample cell in opposite directions. The pump beam excites molecules with a certain velocity v with respect to the beam, which corresponds to excitation on one side of the Doppler profile. The probe beam monitors molecules with the same velocity but in the reverse direction, which corresponds to absorption on the other side of the Doppler profile. When the frequency of the laser corresponds to the center frequency of the line, it interacts with molecules having a zero velocity with respect to the laser beam. On account of the high power of the laser, some of these molecules are transferred to an excited state by the pump beam, and the probe beam, finding a reduced population, records a reduced absorption. If the laser is scanned over the whole profile of the line, a Doppler line shape is obtained with a sharp reduction of the absorption at the line center. The latter is called an *inverse Lamb dip*, and gives an accurate measure of the line center. This technique was first used to observe the hyperfine structure of some of the iodine lines (Fig. 31). A similar technique, which monitors emission rather than absorption, is called *intermodulated fluorescence* (Sorem and Schawlow, 1972).

3.14.3 Two-Photon and Multiphoton Spectra With high-powered pulsed lasers, it is possible to observe transitions that involve the simultaneous absorption of two or more photons and probe excited states of molecules that cannot be accessed directly by one-photon spectroscopy.

The intensity I_{ik} for a two-photon transition between an initial state i and a final state k is given by

$$I_{ik} \propto \left| \sum_j \frac{\langle i|\mu|j\rangle\langle j|\mu|k\rangle}{\nu - \nu_{ij}} \right|^2, \tag{100}$$

where ν is the frequency of the photon and j represents an intermediate state. The intensity of absorption is greatly enhanced if the frequency of the photon is resonant or nearly resonant with a single-photon transition, and the latter is strongly allowed.

General two-photon selection rules are of the type

$$\Delta J = 0, \pm 1, \pm 2; \tag{101}$$

$$+ \leftrightarrow +, \qquad - \leftrightarrow -, \qquad + \not\leftrightarrow -; \tag{102}$$

$$g \leftrightarrow g, \qquad u \leftrightarrow u, \qquad g \not\leftrightarrow u. \tag{103}$$

In principle, it should be possible to access g electronically excited states from ground states with g symmetry, but, in practice, it is found that the transitions observed usually involve u vibrational levels of u electronically excited states, i.e., levels with g vibronic symmetry. Nevertheless, the spectroscopy is complementary to one-photon spectroscopy, which permits the study primarily of the totally symmetric levels in the excited state [see Eqs. (87) and (88)].

One method of detection of two-photon and

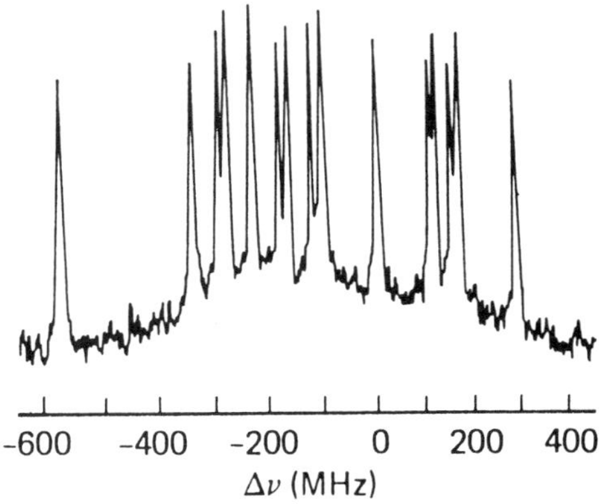

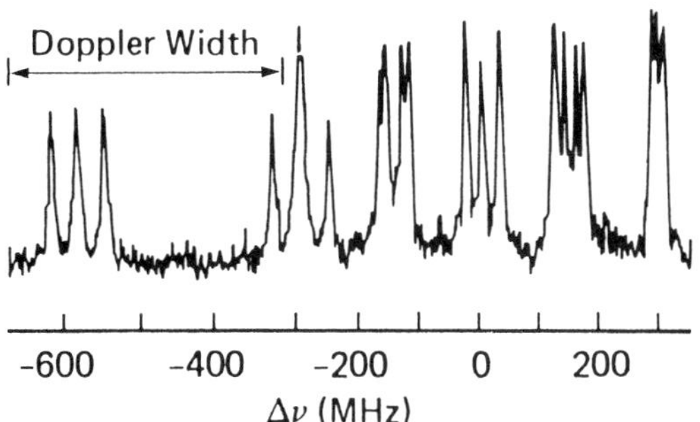

FIG. 31. Hyperfine structure of iodine, I_2, obtained by saturated absorption (from Levenson and Kano, 1988, by permission of the authors and publishers). The two lines show, respectively, 15 and 21 hyperfine components.

multiphoton spectra involves the measurement of fluorescence from the excited state. If the latter does not fluoresce, then an alternative and sensitive method involves the ionization of the molecule and the measurement of the ions. From energy considerations, this method is limited primarily to multiphoton spectra. The method may use one or, more frequently, two dye lasers. In the latter case, the process is described by the number of photons required to reach an excited state and the number of additional photons required to ionize the molecule, e.g., 3 + 1, 2 + 2, ... As with two-photon spectra, the probability of absorption is enhanced if the intermediate steps lie close to single-photon transitions, a process known as *resonance-enhanced multiphoton ionization* (REMPI).

Sub-Doppler two-photon spectra may be obtained if the experiment is carried out in an optical cavity. If a molecule is traveling with a velocity v with respect to the laser beam in one direction, it is traveling with a velocity $-v$ with respect to the reflected laser beam. The Doppler shifts cancel for two-photon absorption, and all molecules absorb two photons corresponding to the center frequency, not just those with zero velocity with respect to the laser beam. This results in greatly enhanced sensitivity (Levenson and Kano, 1988).

3.14.4 Nonlinear Laser Spectroscopy With high-powered lasers, the strength of the electric vector E is sufficient to induce electric moments that depend not only on the first power of E but on higher powers E^2, E^3, ... These higher powers of E give rise to nonlinear optical phenomena (see OPTICS, NONLINEAR). Some of the processes that have been discussed earlier, e.g., difference-frequency laser, saturated absorption, and two-photon and multiphoton processes, are nonlinear phenomena. In addition, there are many other nonlinear optical phenomena, e.g., coherent anti-Stokes Raman spectroscopy (CARS), stimulated Raman gain spectroscopy (SRGS), Raman-induced Kerr-effect spectroscopy (RIKES), and hyper-Raman spectroscopy (Levenson and Kano, 1988).

As an example, let us consider CARS. If two lasers with frequencies ν_1 and ν_2 are incident on a sample such that the difference $\nu_1 - \nu_2$ is equal to the frequency of a Raman-active vibration of the molecule, then the scattered light contains a frequency equal to $\nu_1 + (\nu_1 - \nu_2) = 2\nu_1 - \nu_2$. By scanning one of the lasers, information on the Raman spectrum of the molecule is obtained.

3.15 Applications of Electronic Spectroscopy

The applications of electronic spectra in the laboratory are less widespread than those of infrared or other forms of spectroscopy such as NMR and ESR. Nevertheless, spectra in the near-ultraviolet region are frequently used for monitoring compounds with "chromophores," i.e., compounds with double or triple bonds such as carbonyl compounds, benzene and its derivatives, and heterocyclic compounds.

An important application of electronic spectroscopy lies in astronomy. The spectra of numerous stars have been recorded and the molecules present identified. Most of the species are diatomic, e.g., TiO and MgH, but a few polyatomic species have been found, e.g., SiC_2, C_3. The spectra of comets reveal the presence of several simple species such as CN, C_2, NH_2, and H_2O^+ that are produced by photodissociation of parent molecules under the influence of the Sun's radiation. Finally, processes in the Earth's atmosphere can be monitored by spectra observed in absorption or emission, e.g., the light of the night sky or aurorae.

4. DOUBLE-RESONANCE SPECTROSCOPY

Double-resonance spectroscopy involves the use of two sources, which may be selected from any region of the electromagnetic spectrum. The studies use klystrons or traveling-wave-tube oscillators in the microwave region and tunable lasers in the infrared or optical region. One of the aims of these studies is to assist the assignment of spectra. For example, in optical-optical double-resonance studies, one laser is used to pump a molecule to a level of an excited state for which all the quantum numbers are known—electronic, vibrational, and rotational. This level is usually monitored by means of the fluorescence it emits. The second laser is then scanned and resonances detected by a change in the fluorescence emitted. The second laser may excite the molecule to a higher excited state or stimulate emission to a lower state. In each case, the assignment of rotational quantum numbers is relatively straightforward.

Another application involves the study of collisional processes. Extensive microwave-microwave double-resonance studies showed that, when a molecule collides with another molecule, the change in the rotational quantum numbers is not random but shows a preference for the changes permitted by electric dipole selection rules (Oka, 1973). Similar results have been obtained from microwave–optical double-resonance studies (Petersen and Ramsay, 1985).

Works Cited

Brown, J. M., Buckingham, A. D., Ramsay, D. A. (1976), *Can. J. Phys.* **54**, 895–906.

Clouthier, D. J., Ramsay, D. A. (1983), *Annu. Rev. Phys. Chem.* **34**, 31–38.

Ervin, K. M., Lineberger, W. C. (1992), *Advances in Gas Phase Ion Chemistry*, Vol. 1, Greenwich, CT: JAI Press, pp. 121–166.

Evenson, K. M., Saykally, R. J., Jennings, D. A., Curl, R. F., Jr., Brown, J. M. (1980), *Chem. Biochem. Appl. Lasers* **5**, 95–138.

Heitler, W., Herzberg, G. (1929), *Naturwissenschaften* **17**, 673.

Herzberg, G. (1945), *Infrared and Raman Spectra of Polyatomic Molecules*, Princeton, NJ: Van Nostrand.

Herzberg, G., Innes, K. K. (1957), *Can. J. Phys.* **35**, 842–879.

Levenson, M. D., Kano, S. S. (1988), *Introduction to Nonlinear Laser Spectroscopy*, San Diego, CA: Academic Press.

Long, D. A. (1977), *Raman Spectroscopy*, New York: McGraw-Hill.

Marion, L., Ouellet, J. (1948), *J. Am. Chem. Soc.* **70**, 3076–3078.

Moore, C. B. (1983), *Annu. Rev. Phys. Chem.* **34**, 525–555.

Müller-Dethlefs, K., Schlag, E. W. (1991), *Annu. Rev. Phys. Chem.* **42**, 109–136.

Nicholls, R. W. (1962), *J. Quant. Spectrosc. Radiat. Transfer* **2**, 433–449.

Oka, T. (1973), *Adv. At. Mol. Phys.* **9**, 127–206.

Pauling, L., Wilson, E. B. (1935), *Introduction to Quantum Mechanics*, New York: McGraw-Hill.

Petersen, J. C., Ramsay, D. A. (1985), *Chem. Phys. Lett.* **118**, 34–37.

Rentzepis, P. M., Struve, W. S. (1976), in: D. A. Ramsay (Ed.), *MTP International Review of Science*, Physical Chemistry, Series 2, London: Butterworths, Vol. 3, pp. 263–300.

Sears, T. J., Bunker, P. R., McKellar, A. R. W. (1982), *J. Chem. Phys.* **77**, 5363–5369.

Semin, G. K., Babushkina, T. A., Yakobson, G. G. (1975), *Nuclear Quadrupole Resonance in Chemistry*, New York: Wiley.

Siegbahn, K., Nordling, C., Johansson, G., Hedman, J., Hedén, P. F., Hamrin, K., Gelius, U., Bergmark, T., Werme, L. O., Manne, R., Baer, Y. (1969), *ESCA Applied to Free Molecules*, Amsterdam: North-Holland.

Snyder, L. E. (1972), in: D. A. Ramsay (Ed.), *MTP International Review of Science*, Physical Chemistry, Series 1, London: Butterworths, Vol. 3, pp. 193–240.

Sorem, M. S., Schawlow, A. L. (1972), *Opt. Commun.* **5**, 148–151.

Stohr, J. (1992), *NEXAFS Spectroscopy*, Berlin: Springer-Verlag.

Stoicheff, B. P. (1950), thesis, University of Toronto.

Sutherland, G. B. B. M., Thompson, H. W. (1945), *Trans. Faraday Soc.* **41**, 174–179.

Townes, C. H., Schawlow, A. L. (1955), *Microwave Spectroscopy*, New York: McGraw-Hill.

Turner, D. W., Baker, C., Baker, A. D., Brundle, C. R. (1970), *Molecular Photoelectron Spectroscopy*, New York: Wiley-Interscience.

Further Reading

Bunker, P. R. (1979), *Molecular Symmetry and Spectroscopy*, New York: Academic.

Demtroder, W. (1981), *Laser Spectroscopy: Basic Concepts and Instrumentation*, Berlin: Springer-Verlag.

Gordy, W., Cook, R. L. (1984), *Microwave Molecular Spectra*, 3rd ed., New York: Wiley-Interscience.

Herzberg, G. (1945), *Infrared and Raman Spectra of Polyatomic Molecules*, Princeton, NJ: Van Nostrand.

Herzberg, G. (1950), *Spectra of Diatomic Molecules*, Princeton, NJ: Van Nostrand.

Herzberg, G. (1967), *Electronic Spectra of Polyatomic Molecules*, Princeton, NJ: Van Nostrand.

Herzberg, G. (1971), *The Spectra and Structures of Simple Free Radicals*, Ithaca, NY: Cornell University Press.

Hollas, J. M. (1982), *High Resolution Spectroscopy*, London: Butterworths.

Kroto, H. W. (1975), *Molecular Rotation Spectra*, New York: Wiley.

Papousek, D., Aliev, M. R. (1982), *Molecular Vibrational/Rotational Spectra*, Prague: Academia.

Wilson, E. B., Jr., Decius, J. C., Cross, P. C. (1955), *Molecular Vibrations*, New York: McGraw-Hill.

MOLECULES

PAUL ENGELKING, *Department of Chemistry, University of Oregon, Eugene, Oregon, U.S.A.*

INTRODUCTION

The study of molecules has been primarily the focus of chemistry. The subject of molecules is so broad, however, that chemistry needs considerable assistance from related areas such as materials science, solid state physics, and molecular biology.

Today, the concept of molecule adapts to encompass new structures, many only recently encountered in the laboratory. Polymer chemists now make "dynamic polymers" with unique physical properties that depend upon long chains of atoms continually forming, breaking, and re-forming networks of bonds. Physicists and physical chemists now produce "clusters" of atoms in free jet expansions that blur the distinctions between physical and chemical bonding forces. Electronic engineers align "liquid crystals," partially oriented fluids, in electric fields to produce changing, flat-panel video displays. Materials scientists construct "superlattices," containing repeating units of "nanostructure" aggregates of atoms, to give materials new physical properties. Laser spectroscopists study molecular "free radicals" and ions, so reactive that they are destroyed on contact with any container and can only be confined and manipulated by electromagnetic fields. Astronomers discover "interstellar molecules" in space, even before these molecules can be identified and studied in laboratories on earth. Biochemists manipulate and "splice" the molecules of genes, definable by replicating, even self-repairing, sequences of repeating nucleic acid units, to make new pharmaceuticals and even new organisms.

The modern concept of molecule entered

3-527-28132-0/94/$5.00 + .50

science in two ways: Chemists introduced the molecular concept to explain the definite combining proportions of elements in compounds; physicists introduced the molecular concept to explain the physical properties of gases. By the beginning of the 20th century, the two concepts had converged into the single unifying idea at the foundation of chemistry and the basis of our understanding of how atoms combine to form ordinary matter.

At the start of the 19th century, chemists such as Dalton began to infer the existence of elemental atoms and compound molecules from the definite combining ratios of constituent elements in chemical compounds. The concept of molecules originally explained the constant combining ratios of sodium and chlorine in table salt (NaCl), or of carbon and oxygen in carbon dioxide (CO_2): In the first case, each atom of sodium was imagined paired (incorrectly, as we now understand) with one atom of chlorine in individual molecules of sodium chloride; in the second case, two atoms of oxygen are attached to one atom of carbon in individual molecules of carbon dioxide. Today we have revised that image of common salt to explain the physical and chemical properties of sodium chloride, an ionic lattice solid, but the existence of individual triatomic molecules of carbon dioxide still dominates our understanding of this gas.

In the middle of the 19th century, physicists such as Maxwell introduced molecules into the kinetic theory of gases. Assemblages of hard, elastic, nearly pointlike molecules of matter explained many of the physical properties of gases such as carbon dioxide. Today a researcher using high-pressure carbon dioxide in a supercritical extraction would not explain the solvent properties of high-pressure carbon dioxide by considering carbon dioxide molecules as discrete, hard, pointlike particles, but the molecular concept underlies our understanding of even condensed phases such as liquids and solids.

Even while we currently understand that a crystalline material such as sodium chloride is more properly considered to be a lattice of Na^+ and Cl^- ions, and a solution of carbon dioxide and water contains solvated H^+, HCO_3^-, $CO_3^=$, and OH^- ions and H_2CO_3 molecules, the concept of molecule still would be at the base of that more complex explanation and understanding. There is no more firmly established and useful concept of the chemist or materials scientist than that of molecule.

Nevertheless, the definition given the term molecule has changed and continues to change.

The strong influence of men such as Ostwald (1900, 1922), a firm believer in a continuum theory of matter, led chemists to define molecules traditionally as a limit, in a way reminiscent of taking derivatives in calculus: A "molecule" is the smallest amount of a substance that has all the properties of that substance. Generations of students have found that definition of molecule inadequate, even as they have learned it.

We will simply define *molecules* broadly as *collections of atoms held together by bonding forces for a sufficient time that the collection can be considered as having a fixed composition*. Even so defined, the line between strong *intra*molecular chemical bonding forces and weak *inter*molecular physical interaction forces is a matter of convention and usage; the requisite time is a matter of context.

This definition of molecule is not unique, and several important variations occur. For example, while strictly speaking a molecule that becomes charged is changed to an *ion*, it is possible to think of degree of charge as an attribute and widen the definition of molecule to include charged molecules. For example, Herzberg's classic books (1945, 1950, 1966) on the spectroscopy of molecules treat ions on the same footing as molecules. Likewise, if one were to quibble with a molecular biologist, insisting that the field should more properly be named "ionic and molecular" biology because most of the molecules that are important to this field are actually ions, one would be deserving of a stern look. However, just as often, the term molecule is used specifically to mean "neutral molecule" as in the term "molecular crystal," denoting an array of neutral molecules (such as in "dry ice," solid CO_2), as opposed to "ionic crystal," denoting an array of charged ions (such as "salt," solid NaCl).

Molecule sometimes is restricted to denote more than one atom; other times, the definition is extended, following Maxwell, to include single atoms, such as the "molecules" of monatomic, inert, rare-gas elements He, Ne, Ar, Kr, Xe, Rn.

Different spatial arrangements of atoms are sometimes considered different molecules when their contrasts are emphasized; different spatial arrangements of atoms are sometimes considered different conformations of the same molecule when their similarities are emphasized.

Finally, while the word molecule derives from the diminutive, *molecula*, of the Latin term for mass, there is no hesitancy in using the term *macromolecule* to indicate a large number of atoms, as in a polymer or a protein. Nor do authors consider molecules as necessarily being invisibly small: Many make statements that often whole crystals, such as those of diamond containing atoms of carbon chemically bonded in a rigid network, are properly macromolecules.

As history has shown, attempts to define "molecule" uniquely must be taken with a grain of salt! In what follows, we will concentrate on a concept that derives its strength from its plasticity, its permanence from its adaptability.

1. NOMENCLATURE

The language of chemistry can confound physicists, biologists, engineers, and chemists. Knowledge of a few common names of familiar molecules and an acquaintance with the conventions of naming new molecules and compounds will help negotiating in the molecular realm, just as foreign travelers are aided by knowledge of a few indigenous words or phrases.

1.1 Names

Every molecule has at least one systematic name and often an array of common names or acronyms. It also has a sequence of chemical formulas, each of increasing detail in specifying the spatial relationships of atoms to one another.

For example, the organic molecule (formula CH_3COOH) with the *proper* IUPAC (International Union of Pure and Applied Chemists) name "ethanoic acid" has the *common* name "acetic acid." Dilute solutions of this compound with water (the common name for the hydrogen oxide molecule with the formula H_2O) are commonly known to us as "vinegar." The chemical formula may be abbreviated by an organic chemist by an *acronym* "HAc." Any, or all, of these names, formulas, or abbreviations may occur in one paper. As molecules become more exotic or more uncommon—with longer proper names—they are less likely to have "common" names and more prone to designation by acronyms.

There are two major systematic naming systems: organic and inorganic. The nomenclature of both fields is established and continuously revised by committees of the International Union of Pure and Applied Chemists. The inorganic rules are published in *Nomenclature of Organic Chemistry* (IUPAC, 1979), and the inorganic rules have been recently revised in *Nomenclature of Inorganic Chemistry, Recommendations 1990* (IUPAC, 1990). Synopses of these naming rules are contained in the *Handbook of Chemistry and Physics* (Lide, 1993).

In addition to the IUPAC nomenclature, all compounds are given a CAS name and number by the Chemical Abstracting Service. Knowing the name under which a particular compound is indexed by the editors of *Chemical Abstracts* helps locate information on that compound elsewhere. For example, Material Safety Data Sheets (MSDS's) of commercial compounds are indexed under their CAS numbers and names.

Two major compilations of data about compounds exist in the chemical encyclopedias known by their original editor's names Beilstein and Gmelin. Information on virtually any organic compound is contained in Beilstein (1918, *et seq.*) A peculiar numbering and indexing system is used here, and the novice is advised to enlist the aid of a chemical librarian when approaching this collection. A slightly more friendly indexing system is used for inorganic and organometallic compounds by Gmelin (1922, *et seq.*). These two encyclopedias offer invaluable compilations of the literature of most compounds, especially the older literature, which is not covered by the newer computerized data bases.

The easiest way to learn a new language is perhaps a crash conversational course. In addition to reading standard organic and inorganic texts, a student may make use of programed texts on nomenclature (Benfry, 1966; Banks, 1967; Tranham, 1980; Block, 1990).

Chemical etiquette would suggest that authors and speakers should not assume that their audiences know the names or formulas of any novel compounds they are discussing.

Chemical formulas, or structural drawings, should accompany the first mention of any novel compound and should be made available whenever an unfamiliar or novel compound is referred to by name.

1.2 Formulas

Molecules are given three types of chemical formulas: *empirical, molecular*, and *structural*.

An *empirical formula* lists the symbols of elements present; subscripts on each indicate the proportions in which each element's atoms are present in the molecule, usually expressed as smallest integers. For example, benzene, which consists of a hexagonal arrangement of six carbon and six hydrogen atoms, may be given the empirical formula CH (by convention the subscript "1" is dropped), while ethene, a compound of two carbon atoms and four hydrogen atoms, could be given the empirical formula CH_2.

A *molecular formula* indicates the exact numbers of atoms within each molecule by subscripts. The molecular formula of benzene is C_6H_6 to represent molecules each containing six carbon atoms and six hydrogen atoms. Ethene is represented as C_2H_4 in a molecular formula.

A *structural formula* indicates to greater or lesser degree the spatial relationship of atoms. Bonds are usually indicated. For example, benzene may be represented in one of the following structural formulas:

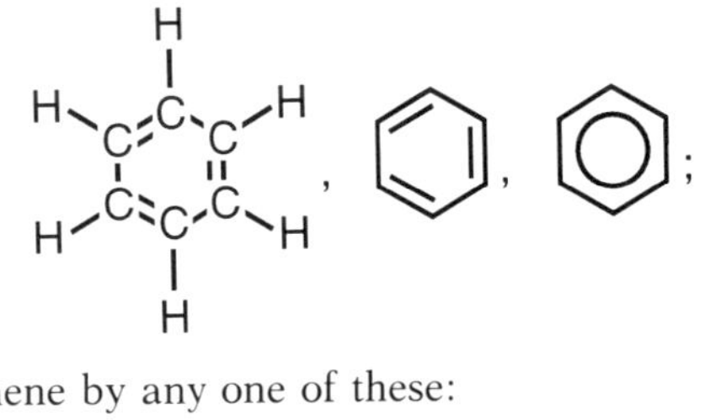

ethene by any one of these:

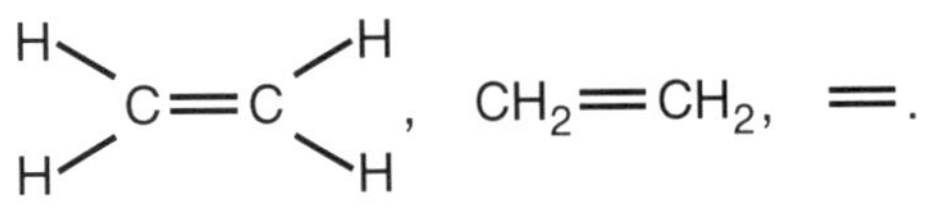

$CH_2{=}CH_2$, $=$.

In the last form in each case, only the bonds between carbon atoms—the carbon *skeletal structure*—are shown. Hydrogen atoms are implied, and their number at each carbon can be inferred from the rule that each carbon has four bonds.

Groups of atoms within molecules are termed *radicals*. Separated from stable molecules, "free" radicals possess high reactivity and readiness to recombine.

If the molecule has a net charge, it is formally an *ion*, and the charge in multiples of the fundamental charge e is denoted by a signed integer, usually in the upper right-hand corner of its empirical or molecular formula, as in NO_3^-, SO_4^{-2}, and NH_4^+.

Sometimes it is necessary to indicate the arrangement of atoms three dimensionally with either specialized structural notation or a structural drawing. For example, while $CH_2{=}CH_2$ suffices to indicate that two hydrogens are attached to each of two carbon atoms in ethene, two structural drawings are needed to distinguish between the different *geometrical isomers cis*-1-2-dichloroethene and *trans*-1-2-dichloroethene:

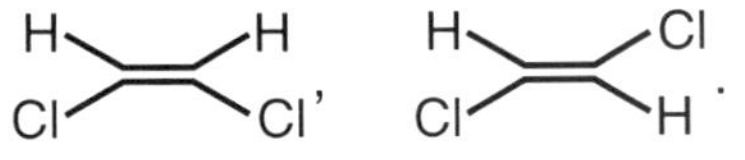

2. STRUCTURE

The structure of molecules is determined by the electrostatic force between charged electrons and nuclei moving quantum mechanically. The distribution of the electrons results from a delicate interplay of kinetic and potential effects: If the electrons are too distant from the nuclei, the potential energy is too large, while if the electrons are confined too tightly to the nuclei, the kinetic energy is excessive.

The overall energy operator for molecules is given by (Kotani, *et al.*, 1961; Parr 1964; Platt, 1961)

$$H = T_n + T_e + V_{nn} + V_{ne} + V_{ee} + H_{\text{rel+mag}}. \tag{1}$$

Here T_e and T_n represent the kinetic energies of the electrons and nuclei, while V_{ee} and V_{nn} represent electrostatic repulsive potential energies between electrons and nuclei, respectively, and V_{ne} represents the electrostatic attractive potential energies between electrons and nuclei. Details of these terms may be found in ELECTRONIC STRUCTURE OF ATOMS AND MOLECULES.

The last term $H_{\text{rel+mag}}$ represents relativistic corrections and magnetic interactions. This term is usually too small to determine the bonding within molecules made of elements from the top half of the periodic table and is often considered as a correction. (Richards *et al.*, 1981) For example, the interaction of electron spin and orbital motion in ground state OH is only about 8.7 meV, which is much smaller than the net bonding energy between the hydrogen and oxygen atoms of 4.392 eV (Huber and Herzberg, 1979).

Born and Oppenheimer (1927) originally removed nuclear motion from electronic motion in a perturbation expansion for the energy that begins with nuclear and electronic coordinates completely separated. That is, they separated the wave function into the product

$$\psi_N(R_N) \cdot \psi_e(r_e), \tag{2}$$

where R_N and r_e represent nuclear and electronic coordinates. In this treatment, corrections to the wave function enter as powers of

$$(m_e/m_N)^{1/4} \approx 0.1, \tag{3}$$

where m_e/m_N is the ratio of electron to nuclear mass. The relatively rapid convergence of this perturbation series is taken as justification of approximate separability of electronic and nuclear motion; hence, all subsequent separations of nuclear and electronic motions are termed *Born-Oppenheimer approximations*.

Our usual understanding of bonding is based upon one of these Born-Oppenheimer approximations, the *adiabatic approximation* (Born, 1931; Born and Huang, 1951). Here nuclear motion is first neglected while solutions to the quantum mechanical energy problem for only the electrons are obtained (or assumed) *at each, fixed, nuclear configuration*. This approximation starts with a parametrically separated wave function

$$\psi_N(R_N) \cdot \psi_e(r_e;R_N), \tag{4}$$

in which the electronic part $\psi_e(r_e;R_N)$ depends upon the nuclear coordinates only as slowly varying parameters. Only T_e, V_{ee}, and V_{ne} for each fixed configuration of nuclei are involved in the quantum mechanical calculation on the electrons. When the solution $\psi_N(R_N)$ to the motion of nuclei is subsequently obtained, the kinetic energy of the nuclei T_n is usually small (on the order of 10–100 meV per atom near the equilibrium geometry of typical molecules). Wave-function corrections appear in the adiabatic approximation on the order of $(m_e/m_N)^{1/2}$, an order of magnitude smaller than in the original Born-Oppenheimer scheme.

The nuclear-nuclear electrostatic interaction V_{nn} may be considered as just a classical addition to the electronic energy obtained in the adiabatic approximation. This *adiabatic potential* energy can be represented as a function of the nuclear positions; this is the function often represented in figures of molecular energy vs bond length—for example, Fig. 1. Minima on an effective, adiabatic potential energy surface are expected to correspond to experimental molecular structures.

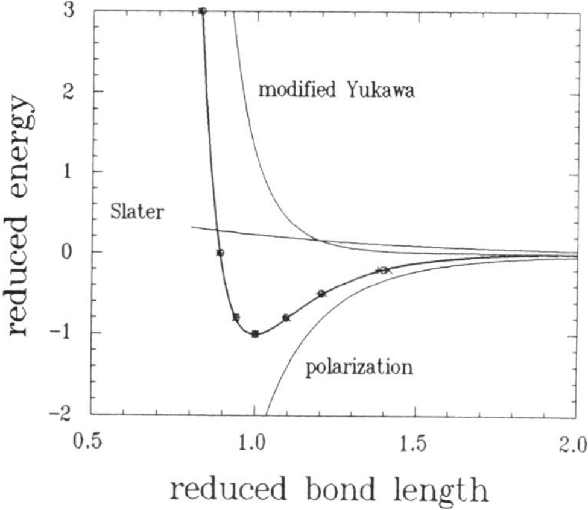

FIG. 1. Adiabatic potential energy for rare-gas atom pairs. The shape is typical of atomic interactions. Reduced units scale bond lengths and energies for each pair: Bond lengths are divided by the bond length at the minimum of the potential; energies are divided by the value of the bond energy at the minimum of the potential. For each pair, dissociation occurs at 0 energy; the minimum of the potential occurs at (1,−1) in reduced units. The curve that fits argon data includes a short-range, modified Yukawa-type repulsion term, two Slater-type terms expressing overlap of electronic functions (one attractive and the other repulsive), and a long-range r^{-6} polarization interaction; these are plotted to show their individual contributions. In addition to the experimental points for argon (o), values for helium (x) and neon (+) are plotted in their own reduced units. Except in the outer wall region, all rare-gas potentials are similar to within 1%. The potential is deep enough to form diatomic "molecules" for all but the lightest rare gas, helium. (Data compiled by Scoles, 1980.)

2.1 Bonding

Bond energy is the difference between the adiabatic potential energy corresponding to separated atoms and the adiabatic potential energy of these same atoms when combined and in their minimum energy conformation. Bond energy can be roughly factored into derived bonding "forces" that dominate at different internuclear distances.

2.1.1 Short-Range Forces The overlap of atomic electronic wave functions on different atoms gives rise to short-range forces, which can be either attractive or repulsive.

The electronic density in atoms decreases approximately exponentially with distance from the nucleus. For the outer electrons at large distances r from a charged center, the decreasing electron density becomes dominated by Slater-type terms of the form $r^x \times \exp(-ar)$ (a is a range parameter, x usually a positive integer), such as are prevalent in hydrogen-atom wave functions. Again, near the nuclei where nuclear attraction dominates the mutual electronic repulsion, the electronic density is also dominated by Slater-type terms (see ATOMS).

To lowest order, if the electronic distributions of the individual atoms are not permitted to relax, adjusting in the presence of nearby atoms, the short-range force is mostly repulsive, dominated by modified Yukawa-type effective potential terms $r^{-y} \exp(-ar)$, resulting from the r^{-1} Coulomb repulsion between nuclei as they penetrate the Slater-type $r^x \exp(-ar)$ charge shells of adjacent atoms.

When the electron distributions are allowed to relax under the influence of adjacent nuclei and electrons, the interaction may be either attractive or repulsive.

Emphasis on the diffuseness of the electronic wave function leads to the picture of an electron fluid under forces. A collection of charges in equilibrium possesses a total kinetic energy which is half of the total (net attractive) potential energy, according to the quantum virial theorem. This balance between kinetic and potential energy is not uniform throughout molecules, however. Comparison of regions of space inside molecules shows a relatively greater amount of kinetic energy for electrons near the nuclei; this condition reverses itself for distant electrons. As a result, the electrons near the nuclei act as a fluid in compression, while the electrons distant from the nuclei, near the "surface" of the molecule, act as a fluid in tension, with "surface tension" holding the molecule together.

The forces between all atoms become repulsive at very short internuclear distances, when the region between nuclei is too small for electrons to accumulate there without acquiring enormous kinetic energies; at very short range, the forces between nuclei correspond to their mutual Coulomb repulsion. This repulsion produces an inner "wall" on the adiabatic potential that is often modeled by a steep r^{-9} or r^{-12} function that diverges at the origin, or by a modified Yukawa-type $r^{-y} \exp(-ar)$ leading term, as in the case of unrelaxed, atomic electron distributions.

If the electron density increases in a region "between" nuclei, attraction results (Hirshfelder *et al.* 1954; Deb, 1973; Bamzai and Deb, 1981). Figure 2 shows the region between pairs of nuclei in which electron buildup leads to bonding.

In order for the atomic electron distributions to relax toward the intranuclear bonding region, the atomic wave functions must have sufficient flexibility and must not act as rigid shells, which only leads to repulsion. This flexibility is present when "open shell" atoms, having degenerate (or nearly degenerate) electronic wave functions, approach one another.

Emphasis on the discreteness of electrons leads to the chemist's picture of bonds. Chemical bonds are associated with pairs of electrons possessing similar spatial states, but of opposite spins. Because electrons can have only two spin states, only two electrons in the same spatial state can take advantage of the attractive region of low potential energy between nuclear charges without violating the Pauli exclusion principle. Additional electrons are excluded from having similar space and momentum coordinates; these additional electrons will anti-correlate their motion with that of a bonding pair.

Chemists have discovered empirical rules for determining the likelihood of the formation of a bond between two atoms. Their most important tool, *Lewis structures*, relies upon the observation that the rare-gas elements possess very little tendency to react chemically; these elements exist commonly as monatomic molecules, with closed s- and p-orbital shells having eight electrons. (Helium

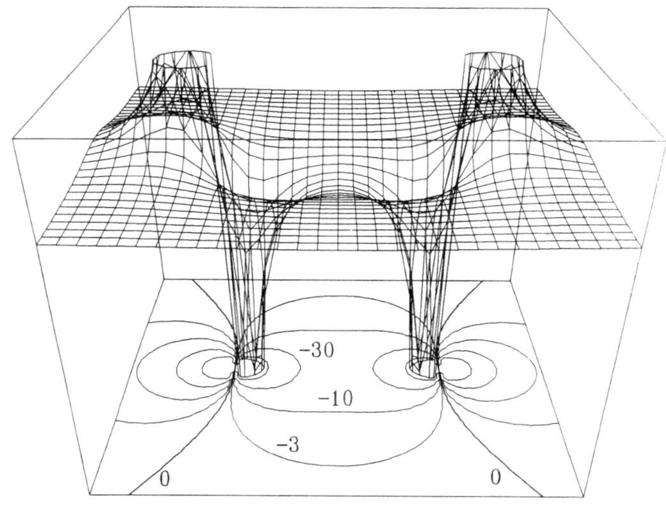

(a)

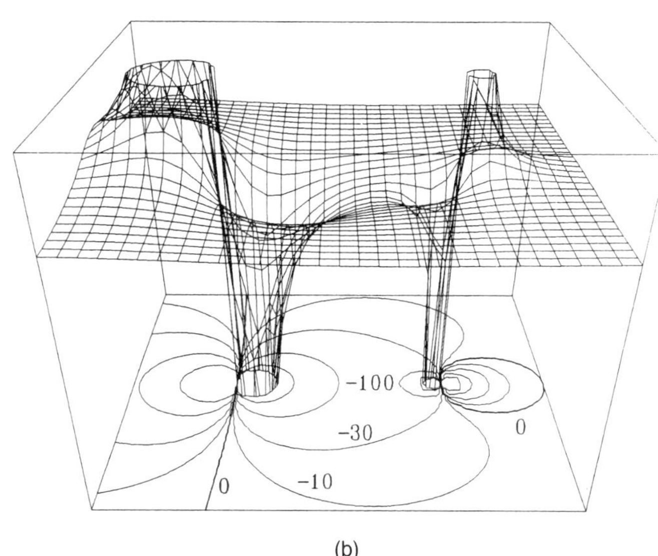

(b)

FIG. 2. Regions of bonding electron density between two atoms, shown in both contour and surface-grid plot representations. The component of bonding force between two nuclei, which is produced by an element of negative charge located variously in a plane containing the axis between nuclei, is plotted in arbitrary units for two cases: (a) homonuclear charge case showing the attractive (negative) region between two hyperbolas passing through the nuclei; (b) heteronuclear charge case corresponding to a 1:9 charge ratio, such as in HF. In the last case, the repulsive (positive) region behind the smaller charge has shrunk to a small oval, while the attractive, bonding region has expanded to nearly anywhere to the right of the larger charge. The surface plots are cut off at arbitrary positive and negative force values to aid representation; the actual force magnitudes diverge at the nuclei.

is the exception, with only two *s*-type electrons, without any *p*-type electrons.) Lewis's octet rule just generalizes this observation, predicting that stable molecules generally will have each atom surrounded by eight electrons that simulate the closure of the *s* and *p* electron shells in the rare gas atoms. To keep track of the *s*- and *p*-type electrons, Lewis invented the "electron dot" diagrams that receive his name. Basically, a dot is assigned to each *s* or *p* electron in the atoms to be combined, and then atoms are allowed to share pairs of dots between them until each atom is surrounded by eight electrons. (The major exception, hydrogen, only requires two electrons to resemble the rare gas helium, however.) For example, the Lewis structures for H, C, and O atoms are

H·, ·Ċ·, :Ö .

The resulting Lewis structures for hydrogen, oxygen, water, and carbon dioxide molecules satisfy the octet rule by sharing electrons between atoms:

H:H, Ö::Ö, H:Ö:H, Ö::C::Ö .

Atoms possessing partially filled *d* orbitals, such as the transition metals, or atoms possessing low-energy, empty *d* orbitals, such as sulfur or phosphorus, often violate the octet rule, but chemists have found ways of extending simple rules for electron counting to predict the likelihood of bond formation.

Atoms generally can aggregate until the resulting molecules become "closed shell," no longer electronically degenerate, at which point additional atoms no longer readily add. The exceptions to this rule usually involve atoms with partially filled *d* or *f* orbitals. Many compounds containing transition-metal, lanthanide, and actinide elements have closed *s*- and *p*-orbital shells, while the deeply buried *d*- and *f*-type orbitals only marginally participate in the bonding and remain partially open.

If the increase in bonding electron density between two nuclei comes from both atoms, and no net charge transfer occurs as evidenced by a lack of dipole moment, then the bond is termed *covalent*. If the increase in bonding electron density between two nuclei is accomplished by transfer of electrons from one atom to another, the bond is termed *ionic*. A third type, the *coordinate covalent bond* (or "Lewis acid-base" or "dative" bond), results when one atom contributes a pair of electrons to the internuclear region. Most bonds between dissimilar atoms are mixtures of these pure cases, but tending to one extreme or another. The amount of ionic character of a bond significantly influences chemical and physical properties. Covalently bonded molecules tend to remain whole when they dissolve in solutions and when they form crystals. Ionically bonded compounds typically dissociate as ions when they dissolve, and form large lattices of alternating ions when they crystalize. Typically, covalent bonding dominates organic chemistry, ionic bonding dominates inorganic chemistry.

Chemists empirically deduced the existence of multiple bonds between pairs of atoms long before multiple bonds were explained quantum mechanically. The Pauli principle requires that each bonding pair of electrons in a molecule must differ spatially from all other electron pairs. Electrons distinguish their orbitals spatially by adding extra nodes. Since a node perpendicular to the internuclear axis would exclude electrons from a large part of the attractive region between nuclei, just where electron density is needed to increase bonding, the most bond-favorable way of including this node is along the internuclear axis. Chemists denote the existence of these extra nodal surfaces along internuclear axes by the greek letters σ, π, δ, ϕ, . . ., reminiscent of the atomic series *s*, *p*, *d*, *f*, . . . to denote 0, 1, 2, 3, . . . angular nodal surfaces in atoms (see ATOMS). This notation is fitting, since it is unlikely for nodes along a bond axis to arise unless the bonding atoms originally have atomic orbitals with the correct number of nodes. Thus, atoms with unpaired electrons in *s* orbitals may form σ bonds, atoms with partially filled *p* orbitals may form either σ bonds or π bonds, while electrons in *d* atomic orbitals may form σ, π, or δ bonds, as shown in Fig. 3. The requirement that nodes along bond axes arise

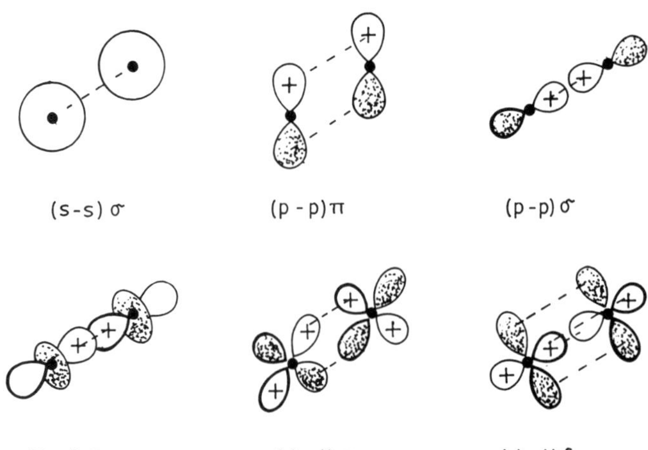

FIG. 3. The σ, π, and δ bond orbitals formed from *s*, *p*, and *d* atomic orbitals.

from the nodes originally present in the bonding atoms explains why hydrogen, with only a partially filled *s* orbital, typically bonds with single σ bonds, while oxygen, with two vacancies in a *p*-orbital shell, may form double bonds consisting of one σ and one π bond. It also explains the rarity of "quadruple" bonds between main group elements not having access to partially occupied *d*-electron shells.

For the most part, the picture of electron pairs localized in bonds between atoms gives a qualitatively correct picture of bonding. Nevertheless, electrons, being described by quantum mechanics that does not restrict electrons to certain regions of space, will "delocalize" over the whole molecule, if for no other reason than to satisfy the Pauli principle requiring that the electronic wave function be antisymmetric under exchange of electrons. The tendency of electrons to delocalize can be accounted for by the use of *molecular orbitals* extending over the whole molecule, in analogy to atomic orbitals extending over the space around atoms. This is shown schematically in Fig. 4.

The lowest-energy electronic wave function composed of molecular orbitals, each containing a maximum of two electrons and having the minimum correlation of electron motion necessary to be consistent with the Pauli principle, represents the "Hartree-Fock limit." As in atoms, this orbital picture accounts for most of the electron bonding energy in molecules. It is also fortunate that this same Hartree-Fock orbital picture can account for bonding between atoms, at least qualitatively.

Attempts to improve bonding calculations quantitatively beyond the Hartree-Fock limit quickly leave the simple molecular orbital picture behind. The farthest outpost that still associates electrons with individual electron orbitals is the Geshkenbein-Ioffe (GI) type of wave function, in which molecular orbitals are each singly occupied, and an elaborate antisymmetrization procedure insures that the overall wave function obeys the Pauli principle. The molecular orbital picture is so compelling, however, that even in cases where electronic correlation of motion is explicitly taken into account, the wave function usually is constructed by superimposing many (10^3–10^6) "configurations" composed of molecular orbital functions.

That the concept of atomic structure dominates molecular electronic structure is both an advantage and a disadvantage in understanding molecular bonding. On one hand, molecular orbitals can be made up of overlapping atomic orbital functions, and the qualitative electronic structure of a molecule often can be guessed to surprising accuracy by drawing a few atomic orbitals, estimating the shapes and relative energies of the resulting molecular orbitals, and then just

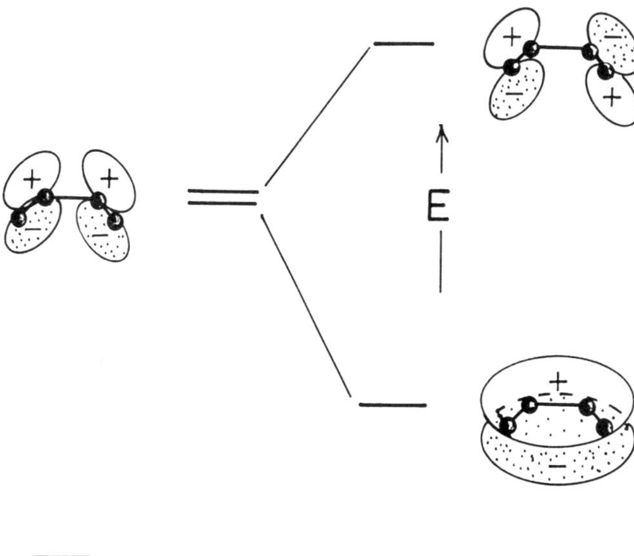

FIG. 4. Schematic of delocalization of π-bonding orbitals in *cis*-butadiene, C_4H_6. A pair of degenerate π bonds shown on the left become two delocalized molecular orbitals of different energies on the right. Delocalization occurs when it causes the overall electronic energy to decrease. One effect of delocalization is to make the otherwise single carbon-carbon bond in the center of the molecule torsionally rigid.

counting out electrons to occupy the lowest-energy orbitals. On the other hand, making quantitative predictions of bond energies is complicated by the facts that 95–99% of the electronic energy in molecules can be ascribed to the binding of electrons in atoms and that the chemical properties of molecules are predicted by perhaps the last 0.1% of the total electron binding energy. Whereas atomic orbitals provide a convenient basis for describing a molecular wave function, many linear combinations of atomic orbitals are required for the necessary precision, and 10^5–10^6 configurations of electrons occupying these orbitals is not unusual for a molecular electron wave function that gives chemically reliable adiabatic potential energies (see ELECTRONIC STRUCTURES OF ATOMS AND MOLECULES).

2.1.2 Long-Range Forces Long-range forces are dominated by multipole Coulomb effects, combined with anisotropic polarization of the electron shells of atoms. A number of these interactions are summarized in Table 1. For example, ionic crystals can be modeled to good accuracy as collections of hard, polarizable, charged ions (Stoneham and Harding, 1986). Here the forces consist of, first, the Coulomb interactions between the ions, which are attractive for arrays in which charges alternate; second, the polarizations of the atoms; and third, a short-range repulsion of the ions produced by the partial interpenetration of their electron distributions.

Between neutral atoms or molecules, an almost universal r^{-6} attraction results from the polarization of an atom by the fluctuating, instantaneous electric field of the electrons on another atom. At very long distances for which the rate at which atomic fields fluctuate exceeds the time it takes the fields to reach the other atoms, relativity modifies the form of this attraction to r^{-7}. (The wavelength of the lowest electronic transition in the isolated atoms gives a rough idea of the distance at which this occurs: 10 to 1000 picometers.) These forces induced by polarization effects appear under the names "Landau," "van der Waals," or "dispersion."

A particularly interesting and extremely important bonding force between closed shell molecules is the hydrogen bond. Partway between a strong chemical bond and a weak intermolecular interaction, the hydrogen bond is responsible for the structure of ice and much of the physical and solvating characteristics of water. It has been observed that molecules in which hydrogen is bonded to oxygen, nitrogen, or fluorine atoms have especially strong intramolecular interactions, approximately 5–10 times greater than caused by dispersion forces alone. Furthermore, these interactions are directional, whereby a hydrogen atom of one molecule attracts an O, N, or F atom of another molecule. The reason why a hydrogen bond forms between the relatively positively charged atoms of hydrogen in one molecule and the relatively negatively charged atoms of nitrogen, oxygen, or fluorine in another compound has been the subject of controversy. At first thought to be primarily an electrostatic phenomenon, it was observed that a purely multipole expansion

Table 1. Effective interatomic forces. Subscript indicates atom 1 or 2. Z_i = charge; μ_i = dipole moment; Q_i = quadrupole moment; α_i = polarizability; θ_i = angle between the axis of moment and the internuclear axis; ϕ = dihedral angle of moments; r = internuclear distance; a, x, y = parameters.

Interaction	Potential model form
	Orientation-independent potentials
Ion-ion	Z_1Z_2/r
Screened nuclear repulsion	
Modified Yukawa	$r^{-y}\exp(-ar)$
Inverse power	r^{-9}, r^{-12}
Covalent bonding	$-r^x\exp(-ar)$
Pauli repulsion	$+r^x\exp(-ar)$
Dispersion (van der Waals)	r^{-6}
Charge-induced dipole	$-Z_1^2\alpha_2/2r^4$
	Orientation-dependent potentials
Charge dipole	$Z_1\mu_2\cos\theta/r^2$
Dipole-dipole	$\mu_1\mu_2[2\cos\theta_1\cos\theta_2 - \sin\theta_1\sin\theta_2\cos\phi]/r^3$
Dipole-quadrupole	$3\mu_1Q_2[\cos\theta_1(3\cos^2\theta_2 - 1) - 2\sin\theta_1\sin\theta_2\cos\theta_2\cos\phi]/(4r^4)$

about molecular centers gave a poor quantitative treatment of the effect. It was also noticed that the hydrogen would locate itself in particular sites on the surface of the negative atoms. This led to speculation that quantum mechanical calculations on whole, intermolecular complexes would be needed to explain hydrogen bonds. Presently, the original idea of a particularly strong electrostatic effect appears sufficient to explain the hydrogen bond. Expansions of the electrostatic potentials about atoms, rather than about molecular centers, explain most cases of the hydrogen bond. For example, the hydrogen bond between molecules of water can be understood as the interaction of the charge distribution on oxygen with that on the hydrogen. Expansions of the atomic electrostatic potentials up to quadrupole terms on the oxygen and up to dipole terms on the hydrogen give quantitative agreement with the presently available data (Buckingham *et al.*, 1988).

The overall effective forces between neutral atoms are often represented by model potentials. Some of these forms are given in Table 2. The justification given for using any particular single form is usually mathematical convenience. Very few interatomic molecular potentials are known accurately.

More on long-range interactions and their effects can be found under CRITICAL PHENOMENA, EQUATIONS OF STATE, PHASE TRANSITIONS, PHASE EQUILIBRIA, RARE GASES, and THERMAL PROPERTIES OF GASES.

Table 2. Examples of model adiabatic potential energy functions. ϵ = |potential minimum|; $\sigma = 2^{1/16} r_e$; r_e = location of potential minimum; r = internuclear distance; α, a, b, c, d = parameters.

Lennard-Jones (6–12):

$$V(r) = 4\epsilon[(\sigma/r)^{12} - (\sigma/r)^6]$$

Morse:

$$V(r) = \epsilon\{\exp[-2a(r - r_e)] - 2\exp[-a(r - r_e)]\}$$

Yukawa-6:

$$V(r) = \epsilon \frac{1 + \alpha r_e}{\alpha r_e - 5}\left\{\frac{6 r_e}{(1 + \alpha r_e) r} \exp[\alpha(r_e - r) - r] - \frac{r_e}{r^6}\right\}$$

Buckingham:

$$V(r) = b\exp(-ar) - cr^{-6} - dr^{-8}$$

Modified Buckingham:

$$V(r) = \frac{\epsilon}{1 - 6/a}\left\{\frac{6}{a}\exp\left[a\left(1 - \frac{r}{r_e}\right)\right] - \left(\frac{r_e}{r}\right)^6\right\} \text{ if } r \geq r'$$
$$= \infty \quad \text{if } r < r'$$

r' given by root of

$$\left(\frac{r'}{r_e}\right)^7 \exp\left[a\left(1 - \frac{r'}{r}\right)\right] = 1$$

2.2 Bond Angles and Lengths

Three circumstances simplify our understanding of bonding in most molecules. First, the almost hard-sphere, repulsive, short-range interaction between atoms allows us to assign almost consistent interatomic bond lengths between pairs of atoms. Second, bonding between atoms can be represented to a great degree as pairwise interactions between two atom centers. Third, bond angles at atomic centers are relatively constant for a series of molecules possessing similar elements. This leads to the "ball and stick" model that serves so well in much of chemistry. Table 3 summarizes much of this model for bonding among main-group elements.

Bond lengths and bond angles have to be adjusted to account for multiple bonds between adjacent atoms; this is readily included in ball and stick models. What is more difficult to include are the subtle variations of bond angles in response to bonding situations. A purely empirical model to describe bond-angle variations has been introduced under the name Valence Shell Electron Pair Repulsion (VSEPR), which is now taught to almost every chemistry student. This model extends the ideas of electron pairs prevalent in Lewis structures: Each atom is imagined as surrounded by pairs of electrons, both bonding and nonbonding, which "repel" one another. It gives qualitative predictions for bond angles about each atomic center in a molecule, which are surprisingly effective, considering the difficulty in justifying this model *ab initio*.

One should be careful about extending a set of bond lengths and bond angles derived for closed-shell molecules to open-shell radicals. In the latter, the electron distributions present around a bonding atom will differ considerably from those present under the

Table 3. Representative interatomic bond lengths (pm). — single, = double, ≡ triple bonds. Data from Kuchitsu *et al.* (1992) and Vilkov *et al.* (1983).

	H	C	N	O	F	Si	P	S	Cl
H	75								
C	109	—154 =134 ≡120							
N	101	—148 =127 ≡116	—145 =125 ≡110						
O	96	—142 =121	—145 =114	—145 =121					
F	92	—136	—137	—157	—142				
Si	148	—187	—172	—163 =(150)	—(156)	—233			
P	142	—185 ≡154	—165 =154 ≡149	—160 =(150)	—(156)		—222 ≡189		
S	134	—182 =(161–171)		=147	—(159)		—211	—206 =189	
Cl	128	177	176	—169	163	205	204	201	199

conditions for which these parameters are derived, and serious errors may result. For example, the methyl radical CH_3 derived from methane by removal of one hydrogen would be thought to be a carbon on a tripod of three hydrogen atoms having bond angles of approximately 109° based upon the methane (CH_4) tetrahedral geometry. In actuality, the methyl radical is planar in its ground state, with all three hydrogen atoms forming an equilateral triangle with a carbon atom in the center; the bond angles are 120°. However, the closed-shell methide anion, CH_3^-, is again nonplanar. It is a dangerous extrapolation to treat an unsatisfied, "dangling bond" on an atom as just another substituent atom or group.

For some compounds, the concept of the two-centered bond breaks down. Boron and hydrogen form networks of three-centered bonds; the expected compound borane dimerizes into the structure diborane,

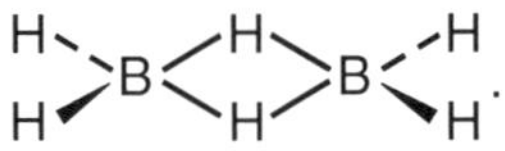

The two-electron, three-center, B—H—B bond is one of the few cases in which hydrogen is situated symmetrically between two bonding partners.

Transition-metal compounds also provide numerous examples of extended, multicenter bonds. For example, the sandwich compounds the metallocenes, of which ferrocene, $Fe(C_5H_5)_2$, is most famous, attach the two planar cyclopentadienyl rings to the central iron by a system of diffuse, eleven-center bonds.

2.3 Stereochemistry

The relative permanence of the spatial configuration of atoms within molecules allows the isolation of different molecules having the same numbers of atoms of each element (i.e., the same molecular formula), but arranged differently in space.

2.3.1 Isomers and Conformers A relatively stable arrangement of atoms is termed a *conformer*; different conformers that are stable long enough to allow physical separation from one another are termed *isomers*. Rapidly interconverting conformers are termed *tautomers*. For example, the carbon skeleton of cyclohexane (C_6H_{12}) has two types of conformers, "chair" and "boat" forms, but these readily interconvert at room temperatures and cannot be isolated:

The carbon skeleton of butadiene (C_4H_6), however, is kept planar by the delocalized interaction of the electrons in the double bonds. Two *geometric isomers, cis-* and *trans*-butadiene, may be isolated, differing only in the spatial arrangements of atoms, but having the same bond topology:

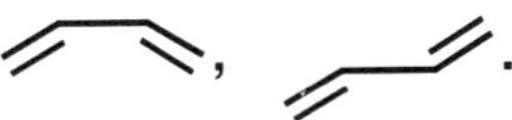

Topological isomers (or constitutional isomers) differ even more strongly from one another by having topologically different networks of bonds. Thus, bicylobutane C_4H_6 is a topological isomer of butadiene, as can be seen from the bonding topology of its carbon skeleton compared to that of butadiene:

The various classes of isomers are related as shown in Fig. 5.

2.3.2 Chirality A spatial configuration of four atoms introduces the possibility of mirror-image conformers. For example, hydrogen peroxide H_2O_2 has two interconverting conformers, one left handed and the other right handed:

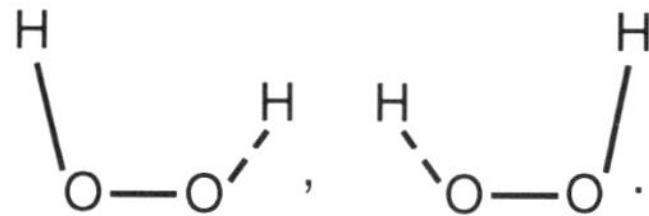

The relative rigidity of bonds about atoms such as carbon allows the isolation of mirror-image isomers termed *enantiomers* or optical isomers, based upon their ability to rotate the plane of polarization of light in opposite directions. Whenever an isomer is not rotationally superimposable upon its mirror image, it acquires the property of "handedness," and is termed *chiral*. Organic chemists are specially alert to this possibility whenever a carbon atom is bonded to four different atoms or groups, thus making it chiral. An example of this is lactic acid, which has two enantiomeric forms. Here a stereochemical convention indicates bonds receding from the plane as dotted; bonds approaching from the plane are emphasized:

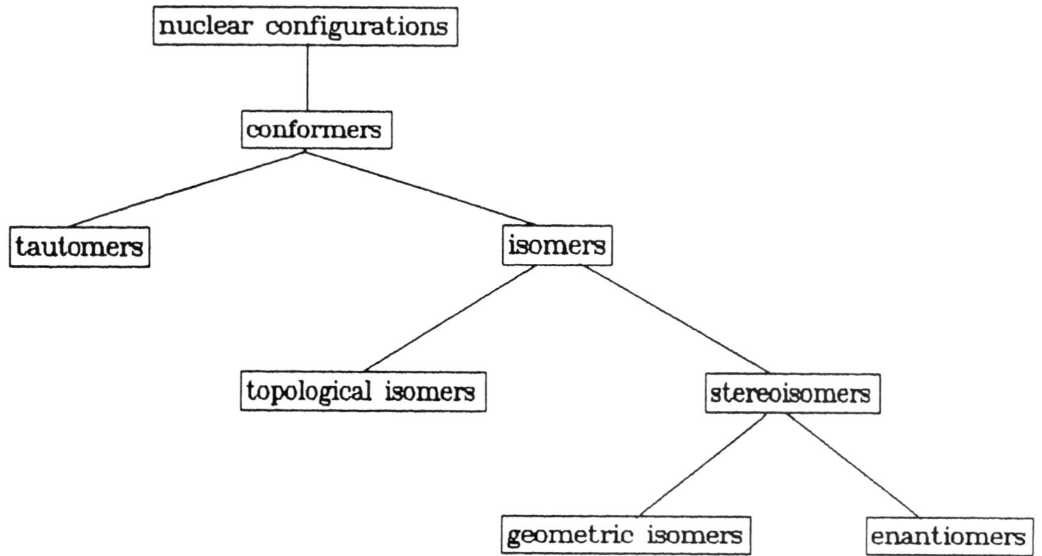

FIG. 5. Classification of nuclear conformers.

D (−) L (+)

Prepared by ordinary laboratory synthesis, a *racemic* mixture of both enantiomers is produced; prepared by bacterial fermentation of sucrose with bacillus *acidi levolactii*, predominantly the left-handed L-lactic acid is produced.

Racemic mixtures may be separated into enantiomers by physical, chemical, or biological techniques. Pasteur first used the fact that under the cool conditions of a 19th-century European laboratory, D- and L-tartaric acid would crystallize separately—it was just a matter of picking through crystals by hand! A seed crystal of one or the other enantiomer can be used to promote the crystallization of that form from a mixture.

Alternatively, reactions may separate enantiomers. If a chiral molecule is added to a racemic mixture, one enantiomer of a mixture may react more readily, and the reaction products will be enriched in one or another enantiomer.

An elegant physicochemical technique resolves enantiomers with a chiral chromatography column (see CHROMATOGRAPHY): A racemic mixture is made to diffuse along a chiral material such as a polymer.

One method of separating racemic mixtures of biological molecules is to find a biological species that preferentially consumes one enantiomer. Numerous optically active centers may be present in a large molecule. Life processes depend upon many molecules that possess chiral carbon centers, such as sugars and amino acids, and often preferentially consume or produce only specific enantiomers.

More information on the topic of stereochemistry will be found in CONFORMATIONAL ANALYSIS.

2.3.3 Symmetry The molecular Hamiltonian is invariant under exchange of identical particles. For molecules possessing identical nuclei, this invariance leads to classification of molecular states according to the symmetry group representations for exchange of nuclei (see GROUP THEORY).

The full spatial symmetry group of the Hamiltonian of a molecule consists of the operations of translation, rotation, electron permutation, nuclear permutation, and inversion. The first three correspond to classification of molecular states according to total linear momentum, overall angular momentum, and the Pauli principle for electrons. The last two make up a *permutation-inversion group*, according to which all molecular states may be further classified. This is an exact symmetry group of the molecule and does not depend upon a particular conformation of nuclei. It is useful for fluxional molecules or for molecules undergoing reactions. Examples of typical permutation-inversion operations are shown in Fig. 6.

If a molecule is assumed to have a fixed conformation, molecular states may be classified according to another group, the geometrical *point group* of the molecule. This group consists of all proper and improper rotations, inversions, and mirror reflections that transform the electronic wave function but which leave the positions of the nuclei invariant; examples of these operations are

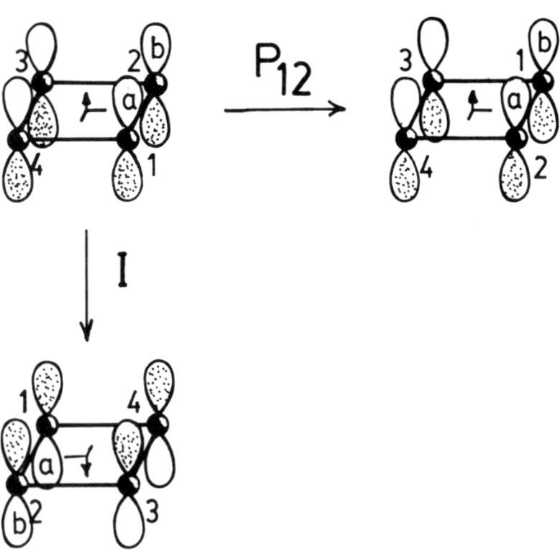

FIG. 6. Nuclear permutation-inversion group operations illustrated on four equivalent nuclei. Nuclei are numbered, while letters indicate wave functions. The whole group is generated by the operations of permuting pairs of nuclei (P_{ij}) and inverting the coordinate system (I). For four nuclei, the group contains $4! \times 2 = 48$ operations. Further description is given by Bunker (1978).

shown in Fig. 7. The molecular point group assumes that the molecule is rigid or, at least, that the vibrations of the nuclei take them only small distances from their equilibrium positions (Wilson, *et al.*, 1955).

Molecular point groups are normally used to classify molecular states in spectroscopy. Any subgroup of the permutation-inversion group may be used to classify molecular states, and in spite of being founded upon the approximation of molecular rigidity, a molecular point group acquires dynamical significance by being isomorphic to a subgroup of the larger, permutation-inversion group, which has exact symmetry. (The point group *is* a subgroup of the product of two invariance groups of the molecular Hamiltonian: the permutation-inversion group and the group of overall rotations of the molecule.) Molecular point groups have the advantages of an extensive literature, readily available character tables, and representations amenable to three-dimensional visualization. They have the disadvantage of being dynamically incomplete, not accounting for the complete permutation symmetry of the nuclei.

For fluxional molecules, interconverting between conformations, one may use (Bunker, 1978) the *molecular symmetry group*, also known as the *feasible group*, which incorporates all operations allowed under the point group, along with the "feasible" nuclear permutations accompanying interconversions between conformations.

2.4 Extended Bonding Structures

2.4.1 Clusters Extensive research currently proceeds on clusters of atoms. The bonding forces holding clusters together vary from strong electronic bonding as in clusters

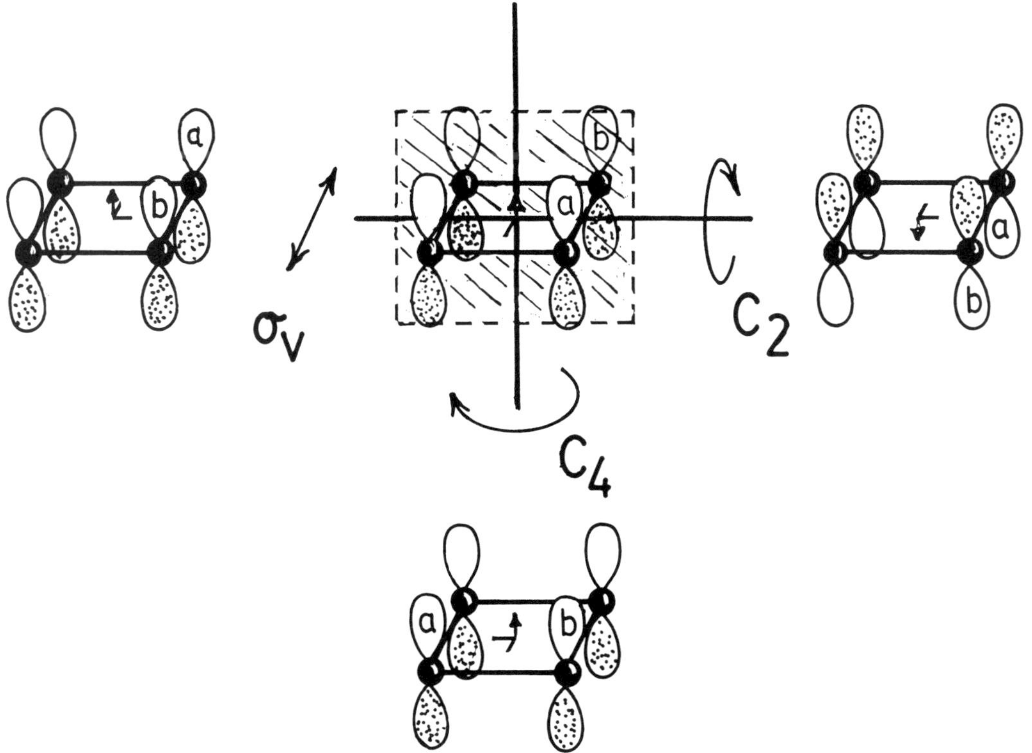

FIG. 7. Molecular point-group operations illustrated on a square arrangement of four identical atoms. All operations transform wave functions, leaving the fixed nuclear positions invariant. The whole group (in this case D_{4h}) can be generated by C_n operations (rotations about n-fold symmetry axes) and σ operations (reflections about mirror planes). For D_{4h}, there are 16 group operations, considerably fewer than for the corresponding permutation-inversion group. Further description and applications of molecular symmetry are given by Herzberg (1966).

of atoms of normally metallic elements, to weak, van der Waals bonding, as in clusters of noble-gas atoms. Work on clusters is made easier by inclusion of a charge, which permits the determination of mass with a mass spectrometer. Types of questions being explored with clusters include "How big a cluster is required to have 'metallic' behavior, as evidenced by ionization energy, electron affinity, and band structure?" Other questions focus on the statistical mechanical properties of clusters, such as "How big does a cluster have to be to resemble bulk material?" "Is a cluster crystalline or liquid?" "How tightly are the components held within clusters?" For small clusters, another question is "What is the structure of a particular cluster?"

Cluster researchers discovered that there are certain "magic numbers" in the size-distribution spectra of the clusters that they produce, generally by condensation from a cold molecular jet. Argon seems to prefer to form clusters consistent with formation of icosahedral shells. Carbon jets give relatively large amounts of C_{60}. There is good evidence that what researchers found was a strong propensity for carbon to form spherical shells of alternating hexagonal and pentangular bonding patterns, given the name "fullerene" because they resembled the geodesic domes of Buckminster Fuller (see Fig. 8). Gram quantities of this material can now be produced and treated as a chemist would treat an ordinary chemical substance.

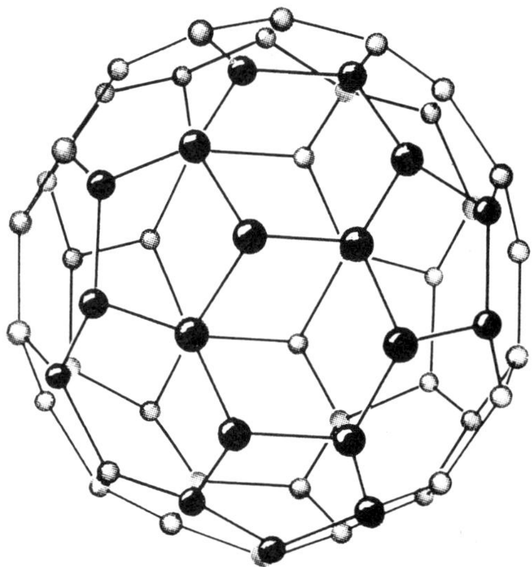

FIG. 8. Fullerene, a highly symmetrical form of carbon C_{60}.

There is some interest in producing monodisperse size distributions of clusters for the purpose of creating nanostructure materials. These could find use, for example, as catalysts in the chemical industry or as "quantum dot" structures in the electronics industry (Bawendi *et al.*, 1990).

More on this topic can be found in MOLECULAR AND ATOMIC CLUSTERS.

2.4.2 Excimers Closed-shell molecules, when excited, often increase their interactions with other molecules. For example, excited metastable rare-gas atoms form respectable chemical bonds, with binding energies on the order of an electron volt, between atoms of the same type:

$$Ar^* + Ar \rightarrow Ar_2^*$$

This behavior of excited rare-gas atoms contrasts with their notoriously weak interatomic interactions in the ground state. Since the bonding depends upon the electronic excitation, molecules that form from otherwise distinct molecules only when one is excited, and which dissociate when the excitation relaxes, are termed *excimers*.

The increased interaction between excited molecules and other ground-state molecules is responsible for increased rates of energy transfer, increased collisional cross sections, reduced diffusion coefficients, and clustering. Excimers can serve as incipient nucleation sites. For example, a saturated or nearly saturated vapor can often be induced to form droplets by illumination with a wavelength of light that excites a constituent molecule to an energetic electronic state. Or, they serve as lasing systems in the vacuum ultraviolet, whose ground states conveniently disappear as rapidly as they are formed by stimulated emission! (See LASERS, GAS.)

2.4.3 Polymers Some molecules, designated as *monomers*, can react with many similar monomer units to form long-chain or network *polymers*. For example, ethene molecules may serve as monomers in the additive polymerization of the linear polymer polyethylene,

$$-(CH_2-CH_2)_n-.$$

When they react, two or more different types of monomers may copolymerize. For example, the condensation polymerization of formaldehyde and resorcinol, with the elimination of water, forms a hard resin used to cement wood:

$$\mathrm{C_6H_4(OH)_2} + \mathrm{O{=}CH_2} \longrightarrow \left[-\mathrm{CH_2}-\mathrm{C_6H_2(OH)_2}-\right]_n + \mathrm{H_2O}$$

Most polymerization reactions give *polydisperse* mixtures of polymer molecules of varying size, shape, and molecular weight. The distribution of polymer mass and the frequency of branched chains strongly affect mechanical properties. Industrial production of plastics and resins attempts to control these carefully.

Some polymer samples are *monodisperse*, having all molecules the same mass (or, at least, within a narrow range of masses) and of similar structure. Monodisperse samples of common polymers are especially valuable to researchers investigating the thermodynamic and transport properties of polymeric materials. Measurements of the properties of pure monodisperse samples, and carefully prepared mixtures of these, provide the empirical parameters needed by theoretical models to predict viscosity, melting, diffusion, and other thermodynamic and transport properties of polydisperse mixtures.

Recent interest in dendritic polymers stems from the desire both to generate monodisperse polymer samples and to control chain branching. Reactions forming these carefully tailored large molecules double the molecular mass at each step by combining two polymers, rather than incrementing it by a constant amount by adding small monomers to a large, growing polymer. Because at each step the product polymer is twice the size of the starting polymers, it can be easily separated and purified before being combined with itself again in a molecular size-doubling reaction.

Many biological molecules are formally polymers composed of mixed monomer units. Proteins are chains of amino acid monomers. The structural proteins of cells contain repeating units of similar amino acids. Other cell parts, such as enzymes, consist of proteins containing careful sequences of different amino acids determined by the genetic material of the cell. The genetic material itself, DNA and RNA, consists of repeating units which encode the genetic information.

A major unsolved problem at this time is understanding how a specific sequence of amino acids affects the folding of the molecular chain in a protein, and hence determines the protein shape, and ultimately its chemical and biological function. A very delicate interplay of intermolecular and intramolecular forces, including van der Waals attraction and hydrogen bonding, is at work in the molecule and the solvent. Numerous local minima occur on the adiabatic potential surface for proteins. Direct simulations of protein folding face serious difficulties with proteins larger than about one hundred amino acids. In reality, a protein in a distorted conformation may take seconds, or minutes, to relax to its active conformation; this is a much longer time than achievable in a molecular dynamics calculation, in which time steps are often sub-femtosecond, and a nanosecond is computationally a long time! (See Molecular Dynamics of Biological Molecules; Monte-Carlo Methods.)

Three very large fields of active work are introduced in Polymers, Structure of; Proteins and Enzymes; Protein Dynamics; and Nucleic Acids.

2.4.4 Crystals Crystals of chemically bonded atoms may be considered to be the largest molecules. For example, the covalent bonding network of carbon atoms in a crystal of diamond meets all the criteria of a molecule. The ionic bonding network of ordinary sodium chloride table salt, NaCl, holds sodium and chloride ions in place with electrostatic bonding forces. Other crystals consist of molecules held together by weaker, intramolecular forces. More on crystals can be found in Crystalline State.

2.5 Structure Determination

2.5.1 Theoretical As noted under the topic of bonding, the stability of molecules is determined primarily by the quantum mechanical distribution of the electrons; the general theoretical background of this subject is contained in Quantum Mechanics; recent advances have been summarized by Payne *et al.* (1992); specific aspects of this topic are contained in ELECTRONIC STRUCTURE OF ATOMS AND MOLECULES.

2.5.2 Experimental

2.5.2.1 Mass Spectrometry Mass spectrometry makes use of the mass-to-charge ratio of the ions produced from molecules introduced into an ionization source of a mass spectrometer (see SPECTROMETERS, MASS). If molecules are singly ionized in the source without fragmentation, they are termed *parent ions* and appear in the mass spectrum at the molecular mass, which may serve to establish the molecular weight. *Fragment* or *daughter ions* may also be formed during ionization, and the masses of these ions typically correspond to stable structural groups within the parent molecule, or to stable rearrangement products. The mass spectrum of these daughter ions may help establish the structure of unknown molecules. For example, the presence of a benzyl group $C_6H_5CH_2\cdot$ in an organic molecule can be signaled by the appearance of mass 91 in the mass spectrum.

2.5.2.2 Diffraction The most detailed information on large-molecule spatial structure comes from diffraction data on crystalline samples. Atomic spacings in molecules are on the order of 10^2 picometers, while crystal unit-cell dimensions are of the order of 10^3–10^4 picometers; the difference in the wave vectors between incoming and scattered radiation must have Fourier components in this range to satisfy the Bragg condition for coherent diffraction from multiple centers. The Bragg condition requires x rays in the 10-keV range, neutrons (or protons) in the 0.1-eV range, or electrons in the 100-eV range (possibly higher to emphasize forward scattering). The large inelastic scattering cross sections for charged particles limit electron or proton scattering to thin samples, whereas the inconveniently low energy for protons severely limits their use in diffraction. Monochromatic thermal neutrons are of limited availability; their use generally complements x-ray diffraction, the primary method of crystallographic structure determination.

X rays scatter off the electrons, providing data that strongly emphasize heavy atoms with many electrons; for large molecules, the positions of hydrogen atoms may be very difficult to determine. Neutrons scatter off nuclei; hydrogen and, even more so, deuterium scatter neutrons strongly, and thus neutron scattering provides information about the lighter atoms that is often difficult to obtain by x-ray diffraction (see X-RAY DIFFRACTION; NEUTRON DIFFRACTION).

Electron diffraction finds use in the determination of structures of small gas molecules, thin (<500 μm thick) solid samples, or even molecules on surfaces. Interatomic distances in small molecules have been determined in the gas phase by electron diffraction, and questions of conformations of moderately large molecules of a dozen atoms have been solved with electron diffraction on gases (see ELECTRON DIFFRACTION; ELECTRON SCATTERING BY ATOMS AND MOLECULES).

2.5.2.3 Rotational Spectroscopy The highest resolution techniques for determination of small-molecule structures involve measuring the moments of inertia for molecular rotation. Molecular moments of inertia may be obtained directly from the microwave spectra of molecules or indirectly from electronic, infrared, or Raman spectra of molecules (see MOLECULAR SPECTROSCOPY).

The basic equation useful here relates the observed rotational energy levels to the molecular moments of inertia. For a linear rigid molecule of moment of inertia I, possessing only rotational angular momentum J, the energy levels E_J are given by

$$E_J = BJ(J + 1), \qquad (5)$$

$$B = h^2/8\pi^2 I. \qquad (6)$$

Other expressions for nonlinear, vibrating, and open-shell molecules can be found elsewhere (Herzberg, 1945, 1950, 1966; Townes and Schawlow, 1955; Hirota, 1985).

In rigid molecules, at most three moments of inertia may be determined spectroscopically; for molecules in which the number of internal atomic coordinates exceeds three, individual atomic masses may be varied by isotopic substitution to change both the center of mass and the relative moments of inertia, thus providing needed additional information to determine molecular structure. In the best cases, each isotopic substitution may provide three more independent constraints on the atomic coordinates. Difficulties occur if

1. the necessary isotopes are unavailable;
2. not all moments of inertia may be measured in an experiment;

3. the system of equations that results becomes ill conditioned; or
4. the molecule is insufficiently rigid.

The last problem often occurs in molecules containing hydrogens, since ground-state vibrational motion of a light proton is roughly 10% of its bond length. On the whole, bond lengths may be determined to within a picometer for heavy atoms and five picometers for hydrogen atoms in usual cases. Vilkov *et al.* (1983) compare the different practical methods, while Kuchitsu (1992) has supervised a recent compilation of results obtained on actual molecules with these methods, combined with those from electron diffraction.

2.5.2.4 Vibrational Spectroscopy Certain groups of atoms vibrate at, or near, characteristic frequencies in molecules (Wilson *et al.*, 1955). This property can be used to identify the presence or absence of particular structural groups, such as represented in Table 4. Infrared, Raman, uv/visible, and electron spectroscopies, and even inelastic neutron scattering, have been applied to determination of molecular vibration frequencies. Summaries of data on specific molecules can be found in the compilations of Shimanouchi (1972, 1977, 1978, 1980) for stable molecules and Jacox (1984, 1988, 1990) for free radicals and ions.

2.5.2.5 Magnetic Resonance Modern organic chemistry could not exist without nuclear magnetic resonance (NMR) to provide information needed to understand the bond topology of the molecules it produces. The chemical usefulness of NMR depends upon two major physical effects: the interaction of nuclear spins to identify adjacent nuclei and the diamagnetic shielding shifts, which indicate the relative electron density near nuclei. Before NMR, only a "complete" synthesis of a molecule by firmly understood chemical reactions served to confirm molecular structure; now modern spectroscopic techniques, principally NMR, establish molecular structure by physical methods in absence of a complete chemical synthesis.

To a lesser degree, where applicable, electron paramagnetic resonance has been used to determine molecular structure (see MAGNETIC RESONANCES, DETERMINATION OF STRUCTURES BY; MAGNETIC RESONANCE AND QUADRUPOLE RESONANCE, NUCLEAR; ELECTRON PARAMAGNETIC RESONANCE).

Table 4. Characteristic vibrational frequencies.

Group or motion	Wave number (cm^{-1})
C—H stretch	2900–3400
C—H bend	1300–1500
C—O—C	1050–1300
C—OH	1000–1300
=C=O	1000–1350
N—H	3300–3600
—C≡N	2200–2400
—S—H	2600
=P—H	2400
≡Si—H	2100–2300
=S=O	1000–1150
$=SO_2$	1150–1250
—NO	1320–1220
—O—NO	1600–1700
	1500–1600
$—NO_2$	1300–1350
—C—F	1000–1350
—C—Cl	650–800
—C—Br	200–600

3. PHYSICAL PROPERTIES

3.1 Molecular Mass

The fundamental unit of molecular mass is the atomic mass unit u, now defined as $\frac{1}{12}$th the mass of the carbon-12 atom.

For macroscopic use, molecules are counted out in units of *moles* (the "chemist's dozen"!), which contain N_A molecular particles. Avogadro's constant, $N_A \approx 6.022 \times 10^{23}$, gives the ratio of the mass of 1 g to that of 1 u. Deslattes (1980) has reviewed how Avogadro's constant is determined. Numerically, the number of grams per mole is equivalent to the number of atomic mass units per molecule of a substance.

One of the most useful tools for identification of an unknown compound is the mass spectrometer. A revolution in the detection and identification of small amounts of chemicals has occurred with the coupling of the mass spectrometer with the capillary bore chromatography column. As mentioned earlier, one useful product of mass spectrometry is the direct determination of the molecular mass from the mass of a parent ion.

For large molecules, such as polymers or biopolymers, mass spectrometry has been less useful, since production of an unfragmented parent ion of a large, nonvolatile molecule by traditional means is problematic. The technique of "electrospray" ionization shows promise here, in which droplets of a solution of molecules are sprayed into a vacuum. Evaporation leaves the molecule of interest surrounded by only a few of the solvent molecules and solution ions; this process produces an aggregate ion having a mass only slightly greater than that of the molecule whose mass is to be determined.

Traditional methods for determining molecular mass involve one form or another of an equation of state; for a dilute gas, one may use the perfect gas equation

$$PV = NkT, \tag{7}$$

where P, V, and T are the pressure, volume, and temperature variables for N molecules (k is Boltzmann's constant). For molecules that can form a dilute gas, the molecular mass can be easily determined from the implication of this equation that 1.00 mole of gas at 1.00 atm and 0°C will occupy 22.4 L (see MASS AND DENSITY, MEASUREMENT OF; EQUATIONS OF STATE).

In cases where molecules cannot be conveniently made into gases but can be dissolved in dilute liquid solutions, van 't Hoff showed they again behave much like a perfect gas. For example, the osmotic pressure Π generated by a solution in contact with its pure solvent through a semipermeable membrane takes the perfect-gas form in the limit of low concentration:

$$\Pi V = NkT, \tag{8}$$

where V is now the volume of the solution and N is the number of molecules dissolved in the solvent.

Determining molecular weight by osmosis (*q.v.*), using the van't Hoff equation (8), or its Huggins-Flory modification (see POLYMERS, MOLECULAR STRUCTURE), can be very sensitive, even for large molecular masses (30 000 u). In fact, it can be too sensitive for small masses, and large molecules with which it is used must be free of small ions and molecules. The sensitivity of this method can be seen by comparing osmotic pressure to the vapor-pressure difference of solvent and solution in the dynamic equilibrium of the apparatus shown in Fig. 9. For equilibrium to occur, the vapor pressure of the pure solvent

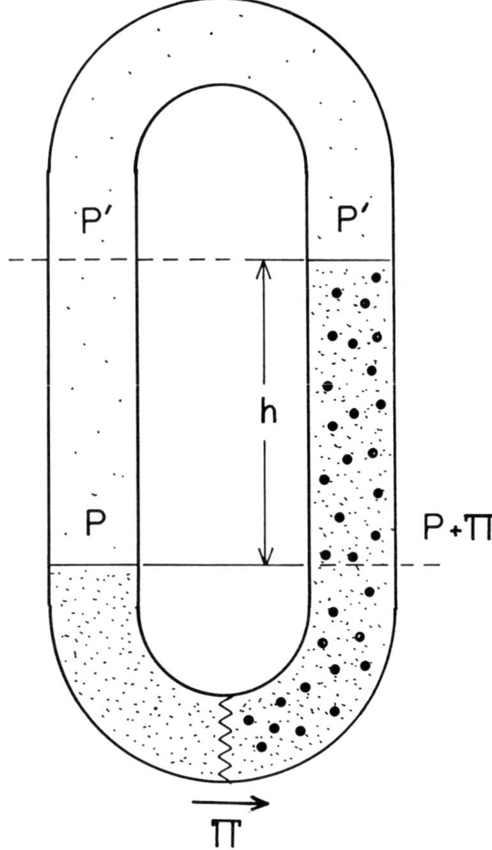

FIG. 9. Thermodynamic equivalence of a large osmotic pressure and a small vapor-pressure lowering produced by addition of a small amount of solute molecules into a pure solvent. The vapor-pressure difference P-P' is given by the hydrostatic head of a column of vapor equivalent in height to a column of solution held elevated by the osmotic pressure difference Π.

at the bottom of a column of vapor must equal the vapor pressure over the solution at the top of the column, plus the small pressure height of the vapor resulting from its finite density. The osmotic pressure, on the other hand, is given by the pressure height difference of a column of solvent vs a column of its vapor!

Although much smaller than osmotic pressure differences, vapor-pressure lowering of a solvent upon addition of solute molecules can be useful for determining molecular mass of small or moderate-size molecules. Thermodynamics applied to Fig. 9 gives the quantitative effect

$$\Delta P = \frac{N}{N_s} \frac{kT}{(v' - v)}. \quad (9)$$

Here a molecule of solvent occupies volumes v' and v in the vapor and liquid states, respectively; N is the number of molecules dissolved in N_s molecules of solvent.

It is often more convenient to measure the change of temperature at which a state change occurs. From Eq. (9) one obtains

$$\Delta T = \frac{N}{N_s} \frac{kT}{(s' - s)}. \quad (10)$$

Here s' and s refer to the entropy per molecule of the solvent in the two states, respectively. When additional solute molecules are present only in the liquid phase, the boiling point is elevated, while the freezing point is depressed. The changes in the freezing and boiling points of water are relatively small because of large entropy changes that occur on boiling or freezing, but even so water has a −1.85 °C freezing-point depression for 1-mol/L solutions (probably the concentration limit for applying a theory developed for "dilute" solutions).

Light scattering can be used to determine the molecular mass of dissolved molecules if their index of refraction differs from that of the solvent. Rayleigh showed that the incoherent scattering of unpolarized light from small inhomogeneities is proportional to their number density and inversely proportional to the fourth power of the wavelength λ. In the limit of low concentration c and long wavelength (much larger than the molecules of interest),

$$m = R_\theta / Kc, \quad (11)$$

where

$$K = 2\pi^2 n_s^2 (dn_s/dc)^2 \lambda^{-4}, \quad (12)$$

$$R_\theta = \frac{r^2 I(\theta)}{I_0(1 + \cos^2\theta)}, \quad (13)$$

and n_s is the index of refraction of the solution, r is the observation distance from the scattering region, and $I(\theta)$ and I_0 are the scattering intensity observed at angle θ and the incident intensity, respectively. This relationship, and others like it, find applications in polymer physics and chemistry.

Two other methods used routinely in biochemical research are electrophoresis and ultracentrifugation. Both depend upon the resistance to motion of a large molecule in a solvent and can be size and shape dependent (Noolandi, 1992).

3.2 Energy of Formation

The energies (or enthalpies) of formation of many molecules have been tabulated; tables include those of Wagman *et al.* (1982) and the JANAF tables (Chase *et al.*, 1985); also, a useful compilation is available in recent editions of the *Handbook of Chemistry and Physics* (Lide, 1993). For more specific information on methods, see THERMOCHEMISTRY.

A compilation of typical bond energies, along with actual determinations for specific molecules, is given by J. A. Kerr in the *Handbook of Chemistry and Physics* (Lide, 1993).

A useful summary of thermodynamic data of molecular ions has been compiled by Lias *et al.* (1988). Recent advances have been made by the increased precision available through electron spectroscopy (see SPECTROMETERS, ELECTRON AND ION).

The *ionization energy* E_i represents the minimum energy necessary to remove an electron from a molecule, typically in the 5–15-eV range. A related quantity, the *electron affinity* E_a, represents the energy released when an electron is attached to a molecule. These range from negative values (associated with temporary negative ions) to positive 5 eV. A recent compilation of electron affinities, the binding energies of electrons to neutral molecules, is provided by Miller in recent editions of the *Handbook of Chemistry and Physics* (Lide, 1993). Many earlier compilations of numbers obtained before the prevalence of laser photodetachment techniques contain numerous erroneous entries (see PHOTOEMISSION AND PHOTOELECTRON SPECTRA; ELECTRON SCATTERING BY ATOMS AND MOLECULES.)

The tendency of an atom within a molecule to attract electrons is referred to as *electronegativity*. Mulliken defined electronegativity of an atom as the average of its ionization energy and electron affinity. This definition of electronegativity for atoms has been extended to radicals; it has been generalized by making it a function of the degree of charge the atom or radical possesses in the molecule, or the type of atomic orbital involved. A number of other scales of electronegativity have been developed; one by Pauling (1960) uses the energy released in the formation of heteronuclear diatomics from the parent homonuclear diatomics,

$$X_2 + Y_2 \rightarrow 2XY.$$

The Pauling numerical scale approximates $(E_i + E_a)/(5.4 \text{ eV})$.

Electronegativity differences are useful in predicting the amount of ionic character in a chemical bond. Electronegativity differences of 2 units (Pauling scale) between bond-forming groups produce essentially ionic bonds (Gordy *et al.*, 1953).

3.3 Electric and Magnetic Constants

Electric dipole moments are most accurately determined by the Stark effect of free molecules; magnetic dipole moments are most accurately determined by the Zeeman effect (see ZEEMAN AND STARK EFFECTS). The determination of the electric dipole moment of molecules is complicated by the fact that rotations average away dipole moments: Molecular rotational states must be parity eigenstates, which have no dipole moments inherently. For most molecules, without degenerate rotational levels, the shift of rotational energy levels in an electric field is a second-order effect. In contrast, molecules having unpaired nuclear or electronic spins, or electronic orbital angular momentum, have magnetic degeneracies of their rotational energy levels, which are split by a magnetic field in a first-order effect (Townes and Schalow, 1955; Hirota, 1985). Thus, the Zeeman effect can be large, and it has been useful recently in investigating magnetic molecules by laser magnetic resonance (LMR) in the far infrared with only fixed-frequency lasers: The molecules can be "tuned" into resonance with the laser frequency by scanning a magnetic field. For molecules in condensed phases, the electric dipole moment can be extracted from the bulk dielectric constant, if some assumptions are made about the inherent polarizability of the molecules and the statistical mechanics of the bulk. Magnetic resonance is preferred for studying paramagnetism in the bulk.

Molecular dipole moments have been extensively tabulated—for example, by Nelson *et al.* (1967).

Higher-order electric moments and polarizabilities have been discussed theoretically by Bishop (1990), who points out that except for cases where they vanish by symmetry, higher moments of both atoms and molecules can be expected to be generated by charge asymmetries on the order of electronic charges displaced by atomic radii and thus are non-negligible on the atomic scale. Except for polarizabilities, many tabulated by T. M. Miller in the *Handbook of Chemistry and Physics* (Lide, 1993), few measurements of higher electronic moments and polarizabilities have been made. Although nuclear quadrupole coupling coefficients measure the local inhomogeneous second derivative of the electric potential at the nucleus, the overall molecular quadrupole moment is relatively difficult to determine. The tabulated values of Gordy *et al.* (1953), obtained indirectly from microwave collisional line broadening, have been widely quoted. This situation is unfortunate, since, as implied in the discussion on bonding above, covalent bond formation is accompanied by the development of a quadrupole moment.

Bulk magnetic susceptibilities have been readily studied. The weak diamagnetism of closed-shell molecules is a quantum mechanical effect. That the energy associated with electronic motion increases in a magnetic field might best be understood in this picture: The magnetic field B can be locally transformed away by moving to a rotating frame and solving the electronic motion there; transforming back to the physical fixed frame imparts to the electronic motion a small amount of additional, rotational kinetic energy. Because the molecular diamagnetic susceptibility is so small, to a good approximation it can be considered to be additive up to the high concentrations associated with bulk liquids or solids. Tabulations include those of R. R. Gupta in the *Handbook of Chemistry and Physics* (Lide, 1993).

Fine and hyperfine structure constants are discussed in MOLECULAR SPECTROSCOPY.

4. CHEMICAL DYNAMICS

4.1 Separation of Nuclear and Electronic Motion

The dynamics of molecules are heavily influenced by the adiabatic potential surfaces on which the nuclei move; in most cases, the motions of nuclei can be adequately modeled with semi-classical, or even classical, equations of motion on the potential energy surfaces corresponding to the ground electronic state (Schatz, 1988). The adiabatic approximation neglects effects of the nuclear kinetic energy operator T_{nn} in calculation of the purely electronic energy. The operator T_{nn} contains derivatives with respect to nuclear coordinates; in addition to accounting for nuclear momenta, these derivatives also detect changes in the electronic wave function that depend upon nuclear coordinates. When these terms are included, the nuclei acquire additional inertia as the electrons follow the nuclei; couplings between adiabatic potential energy surfaces also result. The most serious of these "nonadiabatic" couplings between adiabatic potential energy surfaces are of "velocity" type: They introduce an interaction between adiabatic surfaces that is proportional to the nuclear velocity. To first order, the adiabatic approximation holds only if

$$E_k(R_N) - E_j(R_N) \gg \hbar \dot{R}_N \langle \psi_k(r_e,R_N) | \partial / \partial R_N | \psi_j(r_e,R_N) \rangle, \quad (14)$$

where $E_k - E_j$ is the energy separation between adiabatic surfaces having electronic wave functions ψ_k and ψ_j that depend upon the nuclear and electronic coordinates R_N and r_e; $\dot{R}_N$ is the nuclear velocity.

Whenever adiabatic surfaces approach closely, or nuclear motions are rapid, the molecular system may "hop" from one adiabatic surface to another.

4.2 Conservation Rules

During a chemical reaction, mass (or, more properly, mass-energy) and charge are conserved. Additionally, ordinary chemical, as opposed to nuclear, reactions conserve atomic (or, more properly, nuclear) identities. A statement of a reaction that conserves these basic quantities is said to be "balanced."

Additional quantities may be conserved, or partially conserved, during a reaction. If the reaction occurs without outside fields or collisions to exert torque, angular momentum will be conserved. Electronic and spin quantum numbers are partially conserved separately during reactions, giving rise to the Wigner-Witmer propensity rules for atomic and diatomic reactions: Electronic spin is conserved, while the projection of electronic angular momentum along the axis of bond formation or destruction is conserved. For example, the energetically allowed reaction of CH with N_2 in flames to form the very stable HCN molecule plus atomic nitrogen N is relatively slow. The radical CH is commonly in its spin-$\frac{1}{2}$ ground state, both N_2 and HCN are spin 0, and the reaction must produce N atoms in a spin-$\frac{3}{2}$ state. It may be fortuitous, but this reaction requires a thermal activation energy almost exactly the amount necessary to promote CH to its lowest spin-$\frac{3}{2}$ state.

Also during a reaction, nuclear permutation symmetry must be preserved. Diatomic hydrogen, for example, consists of almost independent populations of even rotational angular momentum *para* states and odd rotational angular momentum *ortho* states, having opposite nuclear permutation symmetry. If diatomic hydrogen is eliminated from a molecule, the nuclei retain the permutation symmetry they originally had within the molecule. These statistics can modify the rates at which reactions occur, such as exhibited in the intensity alternation of the vibrational bands occurring in the photodissociation of cold ammonia. Especially severe for spin-0 nuclei, for which some rotational states may be wholly absent, this effect of nuclear permutation symmetry may account for isotopic enrichment, such as naturally occurs in ozone (O_3) in the stratosphere.

Less restrictive, but still effective, are the reaction propensity rules that result from the partial conservation of molecular orbital symmetry elements, such as node planes. For example, the Woodward-Hoffmann rules for organic reactions explain, among other things, why the Diels-Alder reaction involving six π-bonded carbons is facile, while the analogous reactions forming four and eight carbon rings are difficult:

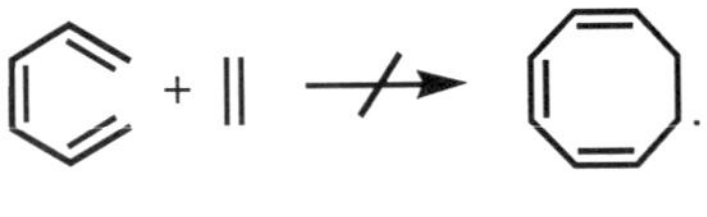

In the first case, where bonds form, the phase of the highest occupied molecular orbital of butadiene matches the phase of the lowest unoccupied molecular orbital on ethylene.

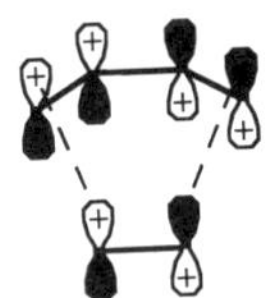

In the latter cases, where the reaction does not occur, the molecular orbital wave functions meet out of phase, inhibiting bond formation.

4.3 Theories of Reaction Rates

Reaction rate theory is discussed in CHEMICAL KINETICS. Bimolecular processes are the subject of MOLECULAR AND ATOMIC COLLISION PROCESSES. Aspects of dynamical modeling appear in SIMULATION BY MOLECULAR DYNAMICS, while relevant mathematical topics include MONTE-CARLO METHODS and CHAOTIC PHENOMENA.

Except when the adiabatic approximation fails or quantum mechanical tunneling occurs, the dynamics of reactions are dominated by classical motion on adiabatic surfaces. Over the years, modeling of chemical reactions has revealed what properties of adiabatic surfaces influence reactions. Energy barriers are most critical, acting as walls, requiring activation of the molecules undergoing reactions. Placement of these barriers is also important: An energy barrier requires energy stored in the nuclear coordinates normal to it in order to surmount it.

Integration of the equations of motion over many nuclear coordinates is a formidable task; even when done, what is required for comparison with experiments is often just a thermally averaged rate constant. Usually, the rate of a reaction step can be expressed according to

$$\text{Rate} = k(T)[X]^a[Y]^b[Z]^c \ldots, \qquad (15)$$

where factors of the form $[X]$ represent the concentrations of molecules of type X entering the reaction step. The thermal rate constant is usually parameterized in an Arrhenius form,

$$k(T) = AT^q \exp(E^*/kT), \qquad (16)$$

where A, q, and E^* are "constants." Roughly, E^* corresponds to the "height" of an energy barrier, while A is an "entropic" factor.

The Arrhenius form receives partial justification from statistical rate theories. Almost all of these assume that those molecular configurations corresponding to the minimum activation energy barrier are in statistical equilibrium with all the other reactant configurations. The fraction of molecules at the energy barrier is determined by statistical mechanics, or thermodynamics. These "activated" molecules in the *transition state* then proceed to leave this configuration, some passing to products, at a rate given by a "frequency factor." If statistical mechanical densities of states are used to calculate the fraction of activated molecules, one arrives at theories such as the Rice-Ramsperger-Kassels (RRK) and Rice-Ramsperger-Kassels-Marcus (RRKM) theories for gas-phase reactions. If one uses thermodynamic functions such as free energy in place of energy, including possibly solvation effects, one arrives at activated-complex theory that is useful in condensed phases. If reactions are considered in extremely viscous fluids, the frequency factor can be modified to take into account diffusion, as in Kassel's theory of diffusion-controlled reactions. All of these exhibit the basic Arrhenius form for reaction at a barrier.

Reactions that proceed without a barrier can be influenced greatly by long-range interactions and often do not show the typical Arrhenius form (Clary, 1990).

4.4 Reaction Steps

An overall molecular reaction mechanism may proceed as a sequence, or network, of fundamental reaction steps. For each step, the number of participating molecules determines its *molecularity*. *Unimolecular* reaction steps involve only one reactant molecule, in which nuclei or groups of nuclei either rearrange (*isomerize*) or separate (*dissociate*). *Bimolecular, termolecular,* and higher-order reaction steps involve first the association of two, three, or more molecules in a fundamental reaction step. More details can be found under CHEMICAL REACTIONS.

4.5 Mechanisms

4.5.1 Elimination Elimination splits a molecule into two or more parts. An example of elimination is the formation of highly reactive methylene radicals by the elimination of CO from ketene,

$$CH_2CO \rightarrow CH_2 + CO.$$

4.5.2 Addition The opposite of elimination, addition combines two or more molecules into a third. An example is the addition of a methylene radical into the double bond of an ethylene molecule,

$$\| + CH_2 \rightarrow \triangle$$

4.5.3 Substitution An elimination may occur after, or in concert with, an addition. Thus, the replacement of bromide at an optically active carbon by hydroxide may proceed through an addition of hydroxide first, then elimination of the bromide. In this case, the optically active carbon "inverts," but retains its chirality.

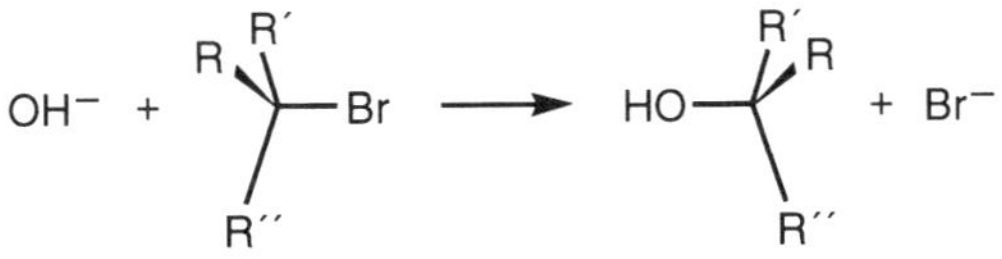

OH⁻ + (R, R′, R″)C–Br → HO–C(R, R′, R″) + Br⁻

An elimination may also precede an addition in an overall substitution. Most ionic reactions in solution occur this way. In the above example of bromine replacement at a carbon, if the elimination of bromine instead occurs first, the carbon center loses its optical activity, since the hydroxide can now add to either side of the carbon, and a racemic mixture results:

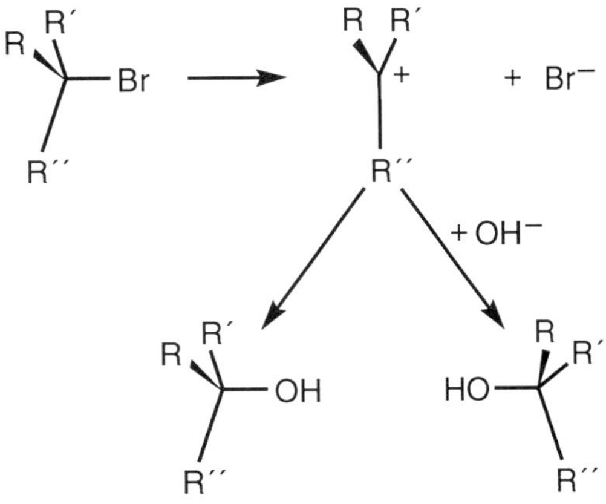

(R, R′, R″)C–Br → (R, R′, R″)C⁺ + Br⁻; + OH⁻ → (R, R′, R″)C–OH and HO–C(R, R′, R″)

Clues like this help chemists piece together the order of steps in a reaction mechanism.

4.5.4 Rearrangement The most exotic reactions involve often surprising rearrangements of atoms from one isomer to another. Most of the "name reactions" of chemistry involve rearrangements. These are a joy to some chemists and a bane to everyone else, since they prevent chemical synthesis from being simply a matter of adding, removing, or substituting atoms or groups. An example of rearrangement is the hydrolysis of neopentylbromide (1-bromo-2,2-dimethylpropane). Here the neopentane carbon skeleton rearranges to one of isopentane:

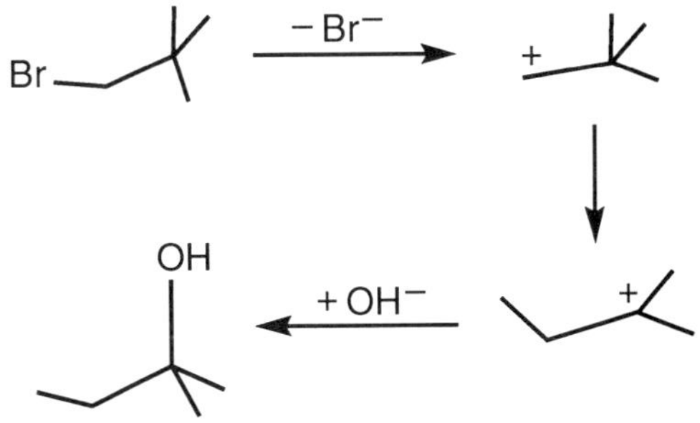

Br– (neopentyl) —(− Br⁻)→ + (neopentyl cation) → + (tert-amyl cation) —(+ OH⁻)→ OH (2-methyl-2-butanol)

4.5.5 Catalysis A catalyst has the property that it speeds up a reaction mechanism, without itself being consumed in the overall reaction. However, a catalyst must actually participate in the fundamental reaction steps. A catalytic molecule first reacts but is subsequently regenerated. An important example of this process is the destruction of stratospheric ozone by the catalytic action of chlorine atoms in the presence of light. This reaction also illustrates the possibility that catalytic intermediates may exist, and may even be isolated, in some catalytic mechanisms:

$$O_3 + Cl\cdot \rightarrow O_2 + \cdot OCl$$
$$\cdot OCl + h\nu \rightarrow O: + Cl\cdot$$
$$O_3 + O: \rightarrow 2O_2$$
$$\overline{2O_3 + Cl\cdot + h\nu \rightarrow 3O_2 + Cl\cdot}$$

The major effect of a catalyst on a reaction is a lowering of the activation energy necessary to cause the reaction. Because the activation energy appears in the exponent of an Arrhenius expression for a single reaction step, reduction in activation energy may greatly increase a reaction rate, making up for the additional complexity that catalysis brings to the reaction mechanism.

Most of the critical steps of biological reactions are catalyzed by enzymes. Further information may be found in CATALYSIS and PROTEINS AND ENZYMES.

4.6 Processes

More information on a number of molecular processes can be found in COMBUSTION, EXPLOSIVES, PHOTOCHEMISTRY, RADIOCHEMISTRY, LASER PHOTOCHEMISTRY, ELECTROCHEMISTRY, POLYMERS MOLECULAR STRUCTURE OF and POLYMER REACTIONS. Relevant topics include CVD (CHEMICAL VAPOR DEPOSITION) and PLASMA ETCHING.

GLOSSARY

Adiabatic Approximation: A specific method of separating nuclear and electronic motions in which the electronic motion is first solved for each set of nuclear positions; nuclear positions enter as parameters during the solution of the electronic motion, becoming dynamical variables during the subsequent solution of the nuclear motions.

Adiabatic Potential: The energy obtained by solving the electronic motion at fixed nuclear positions and adding to the electronic energy the classical nuclear repulsion energy. ("Adiabatic" does not mean "without heat" in this context; it means moving the nuclei "slowly," without inducing transitions between different resulting electronic states.)

Bimolecular: Involving two molecules at once.

Born-Oppenheimer Approximation: Separation of nuclear and electronic motion, now usually in the sense of adiabatic, rather than complete, separation of motion.

Chemical Bond: An attractive interaction between two nuclei, generally requiring more than one electron volt to break, and involving an electron pair.

Chiral: Possessing "handedness"; not superimposable on its mirror image.

Conformer: Relatively stable configuration of atoms of a specified molecular formula. Possibly transient.

Coordinate Covalent Bond: A covalent bond formed with a pair of electrons contributed by one of the two bonding atoms.

Covalent Bond: A bond formed by sharing pairs of electrons between two atoms.

Crystal: A spatially repeating array of atoms or ions.

Electronegativity: The tendency of an atom, or group of atoms, to attract electrons within a compound.

Empirical Formula: A formula denoting the relative proportions of elements in a compound.

Enantiomer: Either of a pair of isomers whose structures are mirror images of each other; a chiral molecule. Differing enantiomers are typically designated by the letters D- or L- (*dextro* or *levo*) according to their effect on the plane of polarization of light or, if known, by the actual spatial orientations of their nuclei.

Feasible Group: A subgroup of the permutation-inversion group of a molecule that includes transformations between specified "feasible" conformers and the point groups of those conformers.

Geometric Isomer: A stereoisomer that differs from another by not being superimposable on that stereoisomer or its mirror image.

Ionic Bond: Electrostatic attraction of two ions.

Isomers: Different, isolatable molecules corresponding to the same molecular formula.

Lewis Structure: A depiction of bonding

in which electrons are represented by dots surrounding elemental atoms.

Mole: The amount of matter that contains the number of particles (atoms, molecules, . . .) equal to the number of atoms in 1 gram of carbon-12; an Avogadro's number $N_A \approx 6.022 \times 10^{23}$ of particles.

Molecular Formula: A formula denoting the types and numbers of atoms in a molecule.

Molecularity: The number of molecules involved at once in a reaction.

Molecular Orbital: An electron orbital extending over one or more atoms.

Monomer: A molecule that can combine with others to form a polymer.

Optical Isomer: Enantiomer. (Named for the ability to rotate the plane of polarization of light.)

Permutation-Inversion Group: The symmetry group of a molecule involving all permutations of identical nuclei and inversion.

Point Group: The symmetry group of a molecular conformer involving proper and improper rigid rotations that transform the wave function but leave the reference positions of the nuclei fixed.

Polymer: A large chain or network molecule made by combining many similar, smaller molecules (monomers).

Racemic Mixture: A mixture of pairs of enantiomers in equal proportions.

Radical: A relatively stable group of atoms, either as part of a molecule or free. A radical generally has one or more unpaired electrons.

Skeletal Structure (Diagram): A diagram showing the bonds between atoms. Names of some atoms, such as carbon, are often omitted, and other atoms, such as hydrogen, are often implied.

Structural Formula: A molecular formula indicating the relative positions of atoms.

Tautomer: A rapidly interchanging conformer.

Termolecular: Pertaining to three molecules at once.

Topological Isomer: An isomer that differs from another by having a different bond topology.

Stereoisomer: An isomer that differs from another by having the same bond topology but a different spatial arrangement of atoms. *Cis* and *trans* geometric isomers are one example; D and L enantiomers are another example.

Works Cited

Bamzai, A. S., Deb, B. M. (1981), *Rev. Mod. Phys.* **53**, 95–126.

Banks, J. (1967), *Naming Organic Compounds, a Programmed Introduction to Organic Chemistry*, Philadelphia: Saunders.

Bawendi, M. G., Steigerwald, M. L., Brus, L. E. (1990), *Annu. Rev. Phys. Chem.* **41**, 477–496.

Beilstein, F. (1908, *et seq.*) [Beilstein Institute, Frankfurt], *Beilstein's Handbuch der Organischen Chemie*, 4th ed., Berlin: Springer-Verlag [Basic Series 1908 through Fifth Suppl. 1984 and continuing].

Benfry, O. T. (1966), *The Names and Structures of Organic Compounds*, Malabar, FL: Krieger.

Bishop, D. M. (1990), *Rev. Mod. Phys.* **62**, 343–374.

Block, B. P. (1990), *Inorganic Chemical Nomenclature*, Washington D.C.: American Chemical Society.

Born, M. (1931), *Gött. Nachr. Math Phys. Kl.* 1.

Born, M., Huang, K. (1951), *Dynamical Theory of Crystal Lattices*, Oxford: Clarendon.

Born, M., Oppenheimer, R. (1927), *Ann. Phys. (Leipzig)* **84**, 457–484.

Buckingham, A. D., Fowler, P. W., Hutson, J. M. (1988), *Chem. Rev.* **88**, 963–988.

Bunker, P. R. (1978), *Molecular Symmetry and Spectroscopy*, New York: Academic Press.

Chase, M. W., Jr., Davies, C. A., Downey, J. R., Jr., Frurip, D. J., McDonald, R. A., Syverrud, A. N. (1985), *J. Phys. Chem. Ref. Data* **14**, Suppl. 1.

Clary, D. C. (1990), *Annu. Rev. Phys. Chem.* **41**, 61–90.

Deb, B. M. (1973), *Rev. Mod. Phys.* **45**, 22–43.

Deslattes, R. D. (1980), *Annu. Rev. Phys. Chem.* **31**, 435–461.

Gmelin, L. (1922, *et seq.*), *Gmelin Handbook of Inorganic Chemistry*, 8th ed., Berlin: Springer-Verlag.

Gordy, W., Smith, W. V., Trambarulo, R. F. (1953), *Microwave Spectroscopy*, New York: Wiley.

Herzberg, G. (1945), *Infrared and Raman Spectra of Polyatomic Molecules*, New York: Van Nostrand.

Herzberg, G. (1950), *Spectra of Diatomic Molecules*, New York: Van Nostrand.

Herzberg, G. (1966), *Electronic Spectra and Electronic Structure of Polyatomic Molecules*, New York: Van Nostrand.

Hirota, E. (1985), *High-Resolution Spectroscopy of Transient Molecules*, Berlin: Springer-Verlag.

Hirshfelder, J. O., Curtis, C. F., Bird, R. B. (1954), *Molecular Theory of Gases and Liquids*, New York: Wiley.

Huber, K. P., Herzberg, G. (1979), *Constants of Diatomic Molecules*, New York: Van Nostrand Reinhold.

IUPAC, (1979), *Nomenclature of Organic Chemistry*, New York: Pergamon.

IUPAC (1990), *Nomenclature of Inorganic Chemistry, Recommendations 1990*, Oxford: Blackwell Scientific Publications.

Jacox, M. E. (1984), *J. Phys. Chem. Ref. Data* **13**, 945–1068.

Jacox, M. E. (1988), *J. Phys. Chem. Ref. Data* **17**, 269–511.

Jacox, M. E. (1990), *J. Phys. Chem. Ref. Data* **19**, 1387–1546.

Kotani, M., Ohno, K., Kayama, K. (1961), "Quantum Mechanics of Electronic Structure of Simple Molecules," in S. Flügge (Ed.), *Handbuch der Physik*, Vol. XXXVII/2, Berlin: Springer-Verlag, pp. 1–172.

Kuchitsu, K. (Ed.) (1992), *Structure Data of Free Polyatomic Molecules; Landolt-Börnstein New Series II/21*, Berlin: Springer-Verlag.

Lias, S. G., Bartmess, J. E., Liebman, J. F., Holmes, J. L., Levin, R. D., Mallard, W. G. (1988), *J. Chem. Phys. Ref. Data* **17**, Suppl. 1.

Lide, D. R. (Ed.) (1993), *Handbook of Chemistry and Physics*, 74th ed., Boca Raton: CRC Press.

Nelson, R. D., Jr., Lide, D. R., Jr., Maryott, A. A. (1967), *Selected Values of Electric Dipole Moments for Molecules in the Gas Phase*, National Standard Reference Data Series, National Bureau of Standards, No. 10, Washington, DC: U.S. G.P.O.

Noolandi, J. (1992), *Annu. Rev. Phys. Chem.* **43**, 237–236.

Ostwald, W. (1900, 1922), *Grundlinien der anorganishchen Chemie* [1900; 5th ed. 1922].

Parr, R. G. (1964), *Quantum Theory of Molecular Electronic Structure*, New York: Benjamin.

Pauling, L. (1960), *The Nature of the Chemical Bond*, Ithaca, N.Y.: Cornell University Press.

Payne, M. C., Teter, M. P., Allan, M. C., Arias, T. A., Joannopoulos, J. D. (1992), *Rev. Mod. Phys.* **64**, 1045–1097.

Platt, J. R. (1961), "The Chemical Bond and the Distribution of Electrons in Molecules," in: S. Flügge (Ed.), *Handbuch der Physik*, Vol. XXXVII/2, Berlin: Springer-Verlag., pp. 173–281.

Richards, W. G., Trivedi, H. P., Cooper, D. L. (1981), *Spin-Orbit Coupling in Molecules*, Oxford: Clarendon Press.

Schatz, G. C. (1988), *Annu. Rev. Phys. Chem.* **39**, 317–340.

Scoles, G. (1980), *Annu. Rev. Phys. Chem.* **31**, 81–96.

Shimanouchi, T. (1972), *Tables of Molecular Vibrational Frequencies. Consolidated Volume 1*, National Standard Reference Data Series, National Bureau of Standards, No. 39, Washington, DC: U.S. G.P.O.

Shimanouchi, T. (1977), *J. Chem. Phys. Ref. Data* **6**, 993–1102.

Shimanouchi, T. (1978), *J. Chem. Phys. Ref. Data* **7**, 1323–1443.

Shimanouchi, T. (1980), *J. Chem. Phys. Ref. Data* **9**, 1149–1254.

Stoneham, A. M., Harding, J. H. (1986), *Annu. Rev. Phys. Chem.* **37**, 53–80.

Townes, C. H., Schawlow, A. L. (1955), *Microwave Spectroscopy*, New York: McGraw-Hill.

Tranham, J. G. (1980), *Organic Nomenclature, A Programed Introduction*, Englewood Cliffs: Prentice-Hall.

Vilkov, L. V., Mastryukov, V. S., Sadova, N. I. (1983), *Determination of the Geometrical Structure of Free Molecules*, Moscow: Mir.

Wagman, D. D., Evans, W. H., Parker, V. B., Schumm, R. H., Halow, I., Bailey, S. M., Churney, K. L., Nuttall, R. L. (1982), *J. Phys. Chem. Ref. Data* **11**, Suppl. 2.

Wilson, E. B., Jr., Decius, J. C., Cross, P. C. (1955), *Molecular Vibrations*, New York: McGraw-Hill.

Further Reading

Atkins, P. W. (1987), *Molecules*, New York: Freeman.

Bunker, P. R. (1978), *Molecular Symmetry and Spectroscopy*, New York: Academic Press.

Burdett, J. K. (1980), *Molecular Shapes*, New York: Wiley.

Cotton, F. A., Wilkinson, G. (1980), *Advanced Inorganic Chemistry*, New York: Wiley-Interscience.

Deb, B. M. (1973), *Rev. Mod. Phys.* **45**, 22–43.

Flygare, W. H. (1978), *Molecular Structure and Dynamics*, Englewood Cliffs, NJ: Prentice-Hall.

Gray, H. B., Ballhausen, C. J. (1964), *Molecular Orbital Theory*, New York: Benjamin.

Gardener, W. C., Jr. (Ed.) (1984), *Combustion Chemistry*, Berlin: Springer-Verlag.

Hirshfelder, J. O., Curtis, C. F., Bird, R. B. (1954), *Molecular Theory of Gases and Liquids*, New York: Wiley.

Johnston, H. S. (1966), *Gas Phase Reaction Rate Theory*, New York: Ronald Press.

Oxtoby, D. W., Nachtrieb, N. H., Freeman, W. A. (1990), *Chemistry: Science of Change*, Philadelphia: Saunders.

Setser, D. W. (1979), *Reactive Intermediates in the Gas Phase*, New York: Academic Press.

Steinfeld, J. I., Fransisco, J. S., Hase, W. L. (1989), *Chemical Kinetics and Dynamics*, Englewood Cliffs, NJ: Prentice-Hall.

Stryer, L. (1981), *Biochemistry*, 2nd ed., San Francisco: Freeman.

Vilkov, L. V., Mastryukov, V. S., Sadova, N. I. (1983), *Determination of the Geometrical Structure of Free Molecules*, Moscow: Mir.

Wilson, E. B., Jr., Decius, J. C., Cross, P. C. (1955), *Molecular Vibrations*, New York: McGraw-Hill.

MONOCHROMATORS

GIORGIO MARGARITONDO AND JEAN-LOUIS STAEHLI, *École Polytechnique Fédérale de Lausanne, Lausanne, Switzerland*

INTRODUCTION

Spectroscopy and other experimental techniques are based on the use of monochromatized photon beams. Most of the photon sources are not monochromatic, and this requires the use of narrow-bandpass filters called monochromators. The technology of monochromators is markedly different for different spectral domains. This is because photons of different wavelengths or energies interact in different ways with the materials used to build monochromators. For example, the transmission of soft x rays through all solids is very inefficient; therefore, most optical components for soft x rays, including monochromators, are reflection devices rather than transmission devices (Margaritondo, 1988; Koch, 1983).

The differences between spectral domains notwithstanding, there are some broad concepts that are valid for most types of monochromators, discussed in the next section. The following sections will analyze the main classes of monochromators, those based on gratings, prisms, the Fabry-Pérot type instruments, and crystal instruments. For each class, some of the most commonly used specific monochromators are presented.

1. GENERAL CONCEPTS

The quality of a monochromator is defined by parameters that are more or less the same for all spectral ranges. The most important of them are the *resolving power*, the *photon throughput* for a given resolving power, and the *spectral purity*. Other parameters define the monochromator's performance in specialized applications; typical examples are the degree of linear or circular polarization of the monochromatized radiation, and the angular acceptance of the instrument.

The resolving power is defined as

$$P = h\nu/\Delta h\nu, \quad (1)$$

where $h\nu$ is the photon energy and $\Delta h\nu$ is the minimum distance in energy of two spectral features that can be separated by the monochromator. This last quantity, commonly called the monochromator's *resolution*, can be defined, for example, with the standard

3-527-28132-0/94/$5.00 + .50

Rayleigh criterion. Note that in modern spectroscopy this definition is somewhat restrictive, since advanced data processing can easily separate features much closer to each other than required by the Rayleigh criterion.

The resolving power can be modified by changing some of the monochromator's parameters, and its value can be chosen according to the needs of each experiment. In general, an increase in the monochromator's resolving power causes a decrease in the photon throughput, and therefore in the signal level. Many monochromators, for example, use dispersion devices and entrance and/or exit slits to stop all photons except those traveling in a given direction; narrower slits mean higher resolving power, but also smaller photon throughput. In practice, it is preferable not to use excessive resolving power, but to set the monochromator's parameters for the minimum resolving power required by the experiment, with the maximum possible throughput.

The *spectral purity* defines the amount of light leaking through the monochromator, outside the nominal wavelength bandpass interval. For visible and infrared monochromators, it is normally measured using a narrow laser line, as the ratio of the photon throughput at some wavelength other than that of the laser line to the maximum throughput at the laser wavelength. The leakage is caused by secondary maxima in the diffraction pattern, imperfections of the dispersing device, and the presence of light of unwanted wavelength inside the monochromator (stray light). The spectral purity increases with the resolution and depends on the wavelength and the geometry of the input beam. In particular, the stray light caused by a laser beam becomes stronger if the monochromator is set to a wavelength close to the laser wavelength. Stray light can be effectively reduced by spatial filter devices, and by using two or three monochromators in series; for example, in Raman spectroscopy double or triple grating monochromators are commonly used.

Resolving power, photon throughput, and spectral purity give only a partial view of the monochromator's performance. The most advanced approach is to measure the instrument's *spectral response* as a function of the slit width and of other parameters that determine the resolving power. The spectral response $R(h\nu)$ is defined as the spectral power distribution produced by the monochromator when the input is a spectral δ function (which can be approximated by a narrow laser line for visible and infrared instruments) whose integral corresponds to unit power.

In practical data processing, this last point is often neglected and the response function is arbitrarily normalized neglecting the problem of photon throughput. Also, in lieu of the theoretical δ function, $R(h\nu)$ is determined by sending into the monochromator a photon beam of known spectral distribution, $I(h\nu)$, and measuring the spectral distribution produced by the instrument, $O(h\nu)$. The function $O(h\nu)$ is the convolution of the functions R and I:

$$O(h\nu) = I(h\nu) \otimes R(h\nu), \tag{2}$$

which, in terms of the Fourier transforms o, i, and r of the same functions, corresponds to

$$o = ir. \tag{3}$$

This last equation appears deceivingly simple in solving the problem of relating spectral input and spectral output. The problem is that Fourier transforms and antitransforms cannot be calculated for the entire spectral range of photon energies, up to infinity. An arbitrary high–photon-energy cutoff must be adopted. The cutoff can be displaced to higher values by using powerful computer hardware and software. However, even a high-frequency cutoff can introduce significant artifacts in the calculated spectral distribution, and can quite often simulate spectral features that do not exist.

Several factors influence the resolving power of a monochromator besides the slit widths. One of them affects instruments that use energy-filtering periodic arrays, such as reflection and transmission gratings, crystals, and multilayers. The resolving power of these devices is proportional to the total number N_l of the periods in the array that are illuminated by the unmonochromatized photon beam. This property has a number of practical consequences. In a grating monochromator, for example, the resolving power depends on the size of the grating area that is reached by the input photon beam, and on the number of lines per unit length. Hence the need to manufacture large-size gratings.

Several components in an optical system can work as photon-energy bandpass devices, and therefore complement or replace the monochromator. For example, the source can be *per se* a bandpass emitter: such is the case of undulators for soft x rays or of lasers. The multilayer coatings that enhance or suppress reflections (Cerrina *et al.*, 1989) can also work as bandpass elements. The optical system may also include nonbandpass devices whose effectiveness depends nevertheless on the photon energy, and which therefore contribute to the energy filtering; a typical example is the photon-energy low-pass function of x-ray mirrors (Margaritondo, 1988; Koch, 1983).

2. PRISM MONOCHROMATORS

These instruments are used for visible and infrared radiation, but their applications are decreasing because of the competition of grating instruments. Dispersion of light by a prism is based on Snell's (or Descartes's) law of refraction, and on the dependence on wavelength of the refractive index n. Figure 1 shows the principles of the layout of a prism monochromator. The chromatic resolving power of a prism monochromator is limited by diffraction at the edges of the prism, and it is given by

$$\lambda/\Delta\lambda = b\frac{dn}{d\lambda}, \tag{4}$$

where b is the base of the prism. Thus, the larger the base and the stronger the wavelength dependence of the refractive index, the better the resolution.

As an example, let us consider the resolution of a fully illuminated flint glass prism with a 5-cm base, at a wavelength of 550 nm: $dn/d\lambda$ (550 nm) $\approx 10^3$ cm^{-1}; thus $\Delta\lambda \approx 0.1$ nm. This is about 10 times worse than the typical resolution of a grating monochromator (see below). Because of their relatively modest resolution, prism monochromators are often used as prefilters in series to a grating or Fabry-Pérot device to remove the light having wavelengths outside the desired free spectral range. The main advantages of prism monochromators are the absence of overlapping orders and the very high spectral purity.

3. FABRY-PÉROT TYPE MONOCHROMATORS

Instruments based on parallel-plate filtering play an important role for visible and infrared applications that require high energy resolution. Consider a parallel transparent plate of refractive index n and thickness d, and suppose that a polychromatic light beam of parallel plane waves falls onto it with an incidence angle θ (Fig. 2). Because of multiple-beam interference, there will be maximum transmission (minimum reflection) for light with wavelength λ if (Born and Wolf, 1980)

$$m\lambda = 2nd\cos\Theta', \tag{5}$$

where the integer m is the order of interfer-

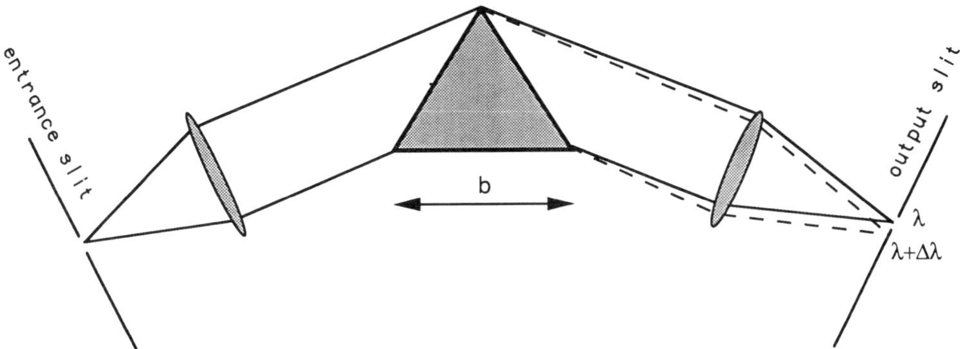

FIG. 1. Scheme of a prism monochromator.

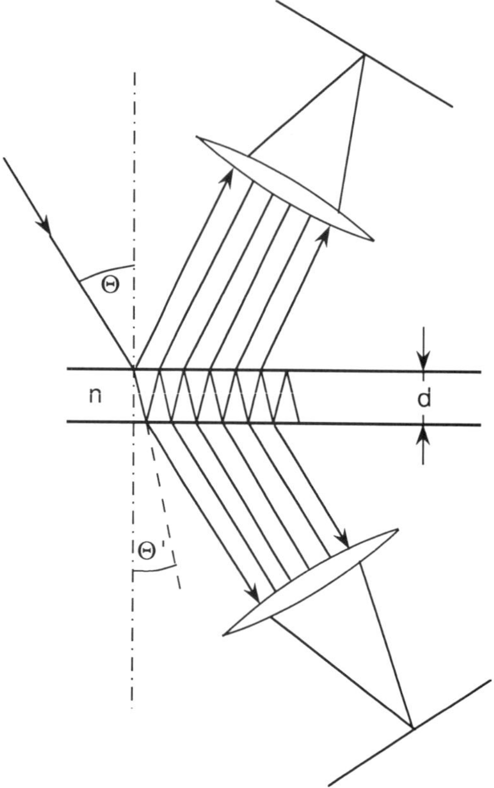

FIG. 2. Formation of multiple-beam fringes of inclination θ on a plane parallel plate.

ence, and Θ' is the angle of inclination inside the plate, which is related to Θ through Snell's law of refraction. The spectral shape of the interference fringes, outlined in Fig. 3, is characterized by a parameter called *finesse,*

$$F = D\lambda/2\Delta\lambda = \pi r^{1/2}/(1 - r), \tag{6}$$

where the *free spectral range*

$$D\lambda = \lambda/m = (2nd \cos \Theta')/m^2 \tag{7}$$

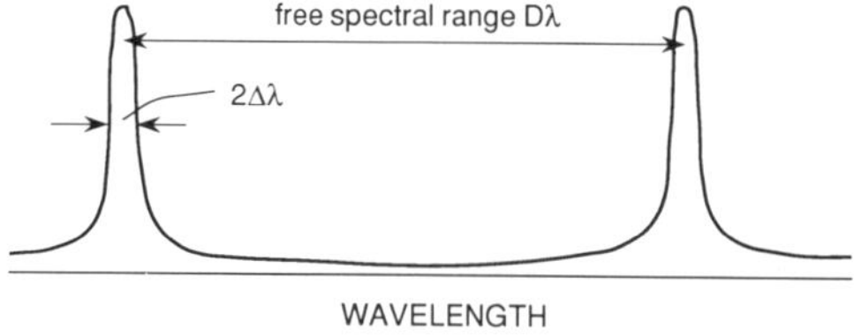

FIG. 3. Outline of the spectral distribution of the interference fringes of a Fabry-Pérot monochromator, illustrating *free spectral range* and *resolving power.* The relationship between these two parameters is given by Eqs. (6)–(8).

is the spectral distance between two interference maxima, $2\Delta\lambda$ is the full half width of one fringe, and r is the reflection coefficient for intensity of the surfaces of the plate. The finesse, which is directly proportional to the resolution power,

$$\Delta\lambda/\lambda = mF, \tag{8}$$

increases with increasing r (because the number of interfering beams also grows with r), and therefore the surfaces of the plate are often metallized. The measured finesse includes also the effects of the imperfections of the two surfaces. And finally, the ratio of the maximum to the minimum transmitted intensities of the interference fringes, the *contrast,* is given by

$$C \equiv I_{\max}/I_{\min} = 1 + 4F^2/\pi^2, \tag{9}$$

and it corresponds roughly to the inverse of the spectral purity, defined previously.

3.1 The Fabry-Pérot Etalon as a Monochromator

A practical Fabry-Pérot monochromator is shown in Fig. 4. The light beam is at normal incidence onto the two parallel and metallized surfaces separated by air ($\Theta = 0$ and $n \approx 1$). The wavelength is set by adjusting the width of the gap between the two plates, today using piezoelectric devices. The spatial filter consisting of a lens focusing the outgoing light onto an exactly axially positioned pinhole cuts away the stray light mainly generated at the front and back surfaces of the two plates.

The overall finesse of a simple etalon, having surfaces with a reflection coefficient of 99% or more, is typically 100–1000, includ-

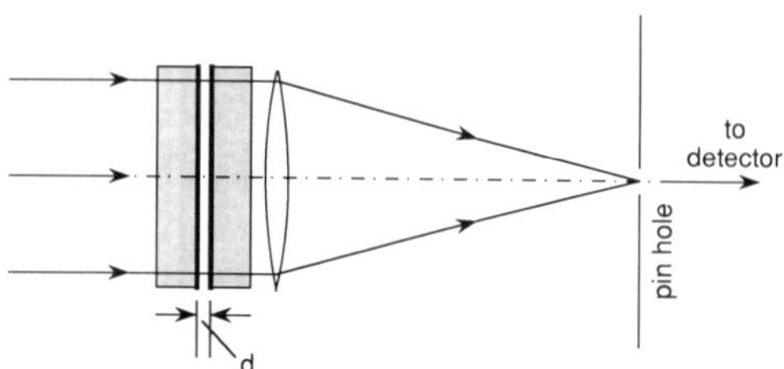

FIG. 4. The basic arrangements of a Fabry-Pérot etalon.

ing the effects of diffraction and of the pinhole. These finesse values correspond to contrasts of $\approx 4 \times 10^3$–4×10^5. The achievable resolution is proportional to m, the order of interference, and thus to the distance d between the two surfaces; with $d = 50$ μm and $\lambda = 5000$ Å, we have $m = 100$, and a free spectral range of 50 Å can be scanned with a resolution of 0.5–0.05 Å (corresponding to the above finesse range); for $d = 10$ mm at the same wavelength the free spectral range reduces to 0.25 Å, and the resolution becomes 0.0025–0.00025 Å.

The main advantages of a Fabry-Pérot etalon are the high resolution that can be achieved with a compact device (the etalon including the piezoelectric mounts occupies a volume that can be as small as 100–200 cm^3), and the very high photon throughput (typically 90%). They are used, for example, in optical spectroscopy of molecules, and inside laser cavities to obtain laser emission of very narrow spectral width.

Problems are caused by the high precision and stability that are needed in the alignment of the two parallel plates. Alignments with a precision of 10 Å are possible today using piezoelectric mounts. The alignment becomes simpler if instead of plane parallel surfaces, the so-called confocal systems are used (the surfaces are concave with a curvature radius equal to the spacing d); however, such systems often have lower throughput, and for each spacing a separate mirror set is required. Finally, note that because of the high interference order m that is normally used, the free spectral range is strongly limited, and an additional monochromator with a low resolution (a prism or a grating device) must often be used in series with the Fabry-Pérot interferometer, to reject the light that could be transmitted in another order.

3.2 Interference Filters

Interference filters consist of a series of thin dielectric layers deposited on a substrate, designed to work like a Fabry-Pérot monochromator at normal incidence. The transmitted band of wavelengths is fixed, and wavelengths outside the interesting range are absorbed by color glass filters added in series to the Fabry-Pérot cavity. Through an appropriate choice of the sequence of the thin dielectric layers, the spectral shape of the transmitted band can be designed as desired. A particularly interesting shape is close to rectangular, with very steep edges.

These filters are quite commonly used in laser spectrometry, for which their wavelength is set to a laser line, and thus the unwanted emissions of the laser cavity can be cut away.

4. GRATING MONOCHROMATORS

The basic elements of most grating monochromators are the grating itself and the entrance and exit slits, which determine the effective source size and separate photons of different energies, diffracted in different directions. These basic elements are often complemented by other components inside and outside the monochromator; several specialized instruments have been developed in this class, a development that has been accelerated by the growth of synchrotron radiation (Margaritondo, 1988; Koch, 1983) in the early 1970s.

4.1 General Properties of Gratings

Most of the gratings used in monochromators are reflection gratings, obtained by engraving a periodic series of lines on a reflecting metal or metal-coated surface, plane or non-plane. The most important characteristics (Hutley, 1982) of a reflection grating are

1. the period d, which coincides with the distance between two adjacent lines; the *pitch* or number of lines per millimeter is inversely related to d;
2. the total number of active lines in the grating, N_l, where a line is "active" if it is reached by the photon beam that must be monochromatized;
3. the profile of the grating's surface; referring to Fig. 5, when the profile is a sawtooth, it is characterized by the blaze angle θ_b;
4. the geometrical shape of the grating surface, which can be plane, concave spherical, or concave aspherical.

The period defines the diffraction conditions; for the mth order diffracted beam, again referring to Fig. 5 for the definition of the in-

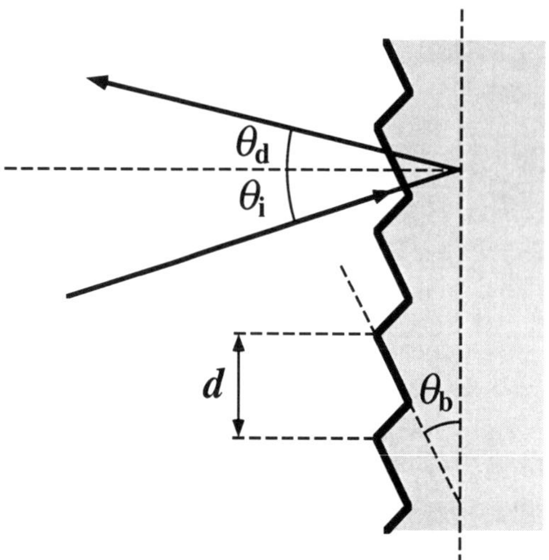

FIG. 5. Scheme of a reflection diffraction grating defining the incidence and diffraction angles, θ_i and θ_d. The grating surface shown here has a sawtooth profile, defined by the blaze angle θ_b.

cidence and diffraction angles, we have

$$d(\sin\theta_i + \sin\theta_d) = m\lambda, \tag{10}$$

and the effectiveness of the grating in separating different wavelengths can be expressed in terms of the angular dispersion, $d\theta_d/d\lambda = (m/d)\cos\theta_d$. The total number of active lines determines the resolving power:

$$h\nu/\Delta h\nu = (mN_l)(\sin\theta_i + \sin\theta_d). \tag{11}$$

The profile of the grating surface affects the photon throughput. The facets of a sawtooth profile reflect the photons in a preferential direction; the throughput is enhanced when the diffraction angle coincides with the direction of reflection from the facets. The corresponding geometrical condition (see again Fig. 5) is $\theta_d = 2\theta_b - \theta_i$, which together with Eq. (10) gives

$$m\lambda_b = 2d\sin\theta_b\cos(\theta_i - \theta_b); \tag{12}$$

this equation defines the wavelength λ_b for which the condition of optimized throughput is valid—the so-called *blaze wavelength.*

These mathematical conditions become quite simple when the grating is mounted with the facets perpendicular to the incident rays, the *Littrow configuration* with $\theta_i = \theta_b$. The corresponding Littrow blaze wavelength is defined as the blaze wavelength in the first order for the Littrow configuration, and is derived from Eq. (12):

$$\lambda_b^{\mathrm{L}} = 2d\sin\theta_b. \tag{13}$$

The Littrow blaze wavelength is often used instead of the blaze angle to define the sawtooth profile of the grating surface. Inserting Eq. (13) in Eq. (12), we find the relation between the (maximum efficiency) blaze wavelength and the Littrow blaze wavelength for any grating geometry and any order:

$$m\lambda_b = \lambda_b^{\mathrm{L}}\cos(\theta_i - \theta_b). \tag{14}$$

As far as the overall grating surface shape is concerned, the simplest plane gratings do not produce any beam focusing, and require additional focusing mirrors. Since part of the beam intensity disappears for each reflection, it would be preferable to combine the dispersing and focusing functions in the same optical component. The simplest solution is to use a spherical grating, since the grating's lines can be accurately produced on a spherical surface with mechanical methods. For both plane and spherical gratings, a substantial reduction in cost is possible (Hutley, 1982) by first producing an original and then using it to manufacture replicas with techniques that were developed in the 1940s.

4.2 Grating Monochromators for Visible and Infrared Radiation

Each reflection causes a loss of intensity, and in some spectral ranges this requires combining dispersion and focusing capabilities in the same component, for example by using spherical gratings. This is not necessary for visible and infrared radiation, since metal coatings produce highly reflecting surfaces. Mostly plane reflection gratings working in first (or in very low) order, in combination with focusing concave mirrors, are used. Concave gratings are employed only in special cases, for example to save space.

The most recent major progress in the development of grating monochromators for the visible and infrared spectral regions was the introduction of interference gratings in the late 1960s. Such gratings have a much lower number of defects than ruled gratings and therefore much less stray light and almost no ghosts (Hutley, 1982).

The most commonly used design for grating monochromators is the Czerny–Turner mounting (also known as Ebert mounting with two separate mirrors), shown in Fig. 6. The advantage of this mounting lies in the fact that aberrations (mainly astigmatism) introduced by each of the two concave mirrors are mutually compensated to a large extent. For very good resolution and high throughput, a long curved output slit can be used (James and Sternberg, 1969).

4.3 Grating Monochromators for Ultraviolet and Soft X Rays

The main problem in constructing a monochromator for this spectral domain is that except for its lowest photon energies the transmission through materials is extremely poor, and the reflection inefficient (Samson, 1980; Henke, 1972; Margaritondo, 1988; Koch, 1983). Specifically, for a metal or metal-coated surface the reflection drops dramatically at grazing angles larger than the critical value defined by the Lorentz-Drude equation,

$$\theta_c = \sin^{-1}[\lambda(e^2D/m_0c^2\pi)^{1/2}], \qquad (15)$$

where D is the number of free electrons per unit volume of the metal, and e and m_0 are the electron charge and mass. In practice, the reflectivity is low when the grazing angle in degrees is larger than one-tenth the wavelength in angströms. For vacuum ultraviolet photons of 20 eV, this means approximately 60°, but for a soft–x-ray beam of 1000 eV, it becomes of the order of 1°.

Therefore, most optical components for ultraviolet and soft x rays, including the monochromators, work at grazing reflection—except for the lowest-energy portion of this spectral range. The filtering elements in monochromators are typically reflection gratings. There are two notable exceptions to this rule. First, it is difficult but possible to manufacture transmission gratings consisting of very thin periodic films over random-structured supporting substrates. Second, and most important, it is possible (Vladimirsky *et al.*, 1989; Rarback *et al.*, 1988) to utilize Fresnel zone plates (see Fig. 7). These are arrays of concentric photon-absorbing rings over a transmitting substrate. The primary objective of a Fresnel zone plate is to focus a photon beam of wavelength λ onto a focal point at a distance f from the device. This requires constructive interference of the photons passing through the transmitting zones, which in turn implies, in first approximation, a relation $\lambda f \approx$ const. Thus, photons of different wavelength are focused at different distances from the zone plate, making it possible to use the plate as a monochromator.

It is not easy, however, to manufacture Fresnel zone plates suitable for soft x rays. The total power transmitted by the plate increases with the plate's width. In turn, this width is determined by the radius of the narrowest external zone, which is limited by the capability to fabricate very narrow features. In practice, to obtain plates of reasonable size it is necessary to obtain (with electron-beam lithography) external zones of the order of one-

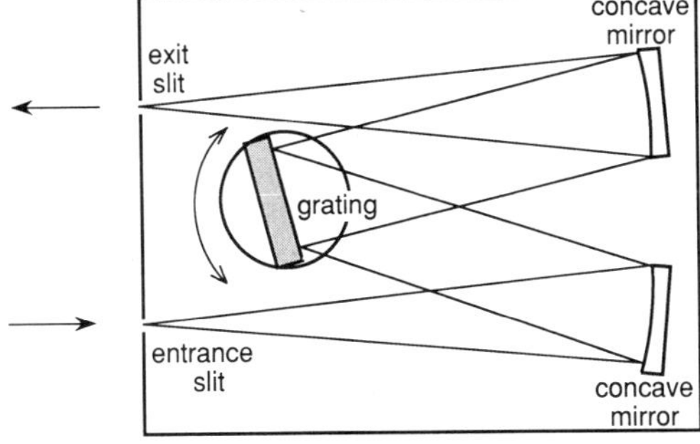

FIG. 6. Outline of the Czerny–Turner mounting for a plane-grating monochromator. Two spherical mirrors are used, and the slits are placed on either side of the grating.

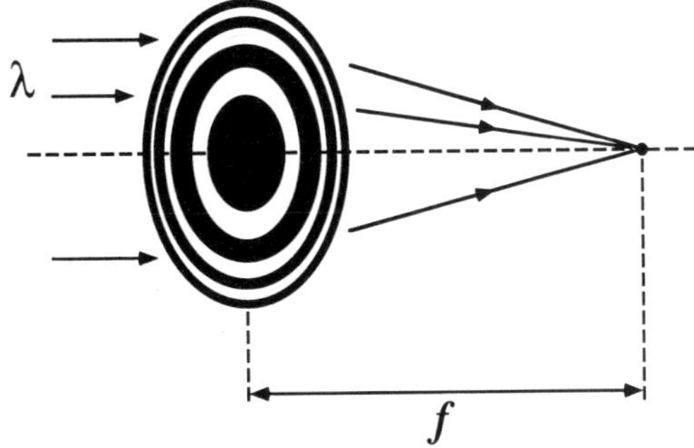

FIG. 7. Scheme of a transmission Fresnel zone plate (Vladimirsky *et al.*, 1988). Photons of different wavelengths λ are focused at different distances, f, hence the possibility to monochromatize the beam.

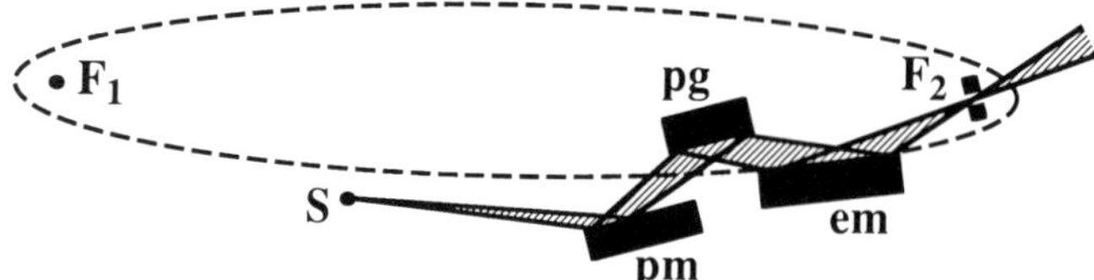

FIG. 8. An advanced type of PGM: the SX-700 instrument (Kaindl *et al.*, 1992). The scheme includes no entrance slit. The photon energy is scanned by simultaneously rotating both the entrance plane mirror (pm) and the plane grating (pg); for all positions of these components, the source (S) is transformed into a virtual source at one of the foci (F_1) of an ellipsoid; the focusing ellipsoidal mirror (em) brings the radiation into the exit slit at the other focus F_2.

tenth of a micron or less (Vladimirsky *et al.*, 1989; Rarback *et al.*, 1988).

In order to reduce the number of reflections and the resulting losses, focusing gratings are often used in these spectral domains, for example spherical gratings. The focusing properties of a spherical surface are far from ideal for soft x rays (Margaritondo, 1988; Koch, 1983). The aberrations become quite severe at grazing incidence; the focusing occurs in two different points for the direction in the plane of incidence and for that perpendicular to it. Some of these problems are removed (Schmahl and Rudolph, 1976) by using toroidal reflecting surfaces instead of spherical surfaces. Toroidal gratings can be manufactured with sophisticated holographic methods.

4.3.1 Plane-Grating Monochromators (PGM's) The PGM's are typically used for synchrotron radiation beam lines in a grazing-incidence geometry. They offer the advantage of using simpler and less expensive gratings. In principle, the wavelength selection could be accomplished with a simple rotation of the grating and fixed slits, but the actual PGM's use more complicated schemes to enhance their performances (Eberhardt *et al.*, 1978).

One of the most advanced types of PGM is the SX-700 monochromator, schematically shown in Fig. 8. Tests at the BESSY synchrotron radiation laboratory have demonstrated the excellent resolving power of this instrument. Thanks to its grazing-incidence geometry, the spectral range extends from 35 to 1000 eV (Kaindl *et al.*, 1992).

4.3.2 Spherical-Grating Monochromators (SGM's) The optimal condition for an SGM is that the beam emerging from the entrance slit be focused onto the exit slit. Referring to Fig. 9 for the definition of the symbols (note that the grating grooves are perpendicular to the plane of incidence), it can be demonstrated (Hutley, 1982) from this condition and from Eq. (10) that

$$(\cos^2\theta_i)/r_i - (\cos\theta_i + \cos\theta_d)/R_c + (\cos^2\theta_d)/r_d = 0. \qquad (16)$$

The most important solution of this equation

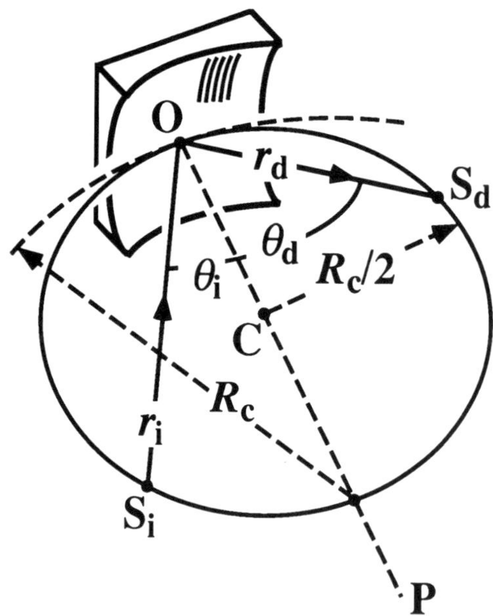

FIG. 9. Geometrical scheme of the Rowland-circle solution [Eq. (17)] for the focusing-diffraction conditions [Eq. (16)] valid for spherical focusing gratings. R_c and $R_c/2$ are the radius of the grating and that of the Rowland circle. Both the entrance slit S_i and the exit slit S_d are on the circle, at distances r_i and r_d from the grating's center, O. P is the direction perpendicular to the grating, C is the center of the Rowland circle, θ_i is the angle of incidence, and θ_d that of diffraction.

is

$$r_i = R_c \cos\theta_i, \quad r_d = R_c \cos\theta_d, \tag{17}$$

which defines a circle of radius $R_c/2$ in the plane of incidence, shown in Fig. 9: the *Rowland circle*. An alternative but less important solution for Eq. (16) is provided by the Wadsworth mounting, in which the entrance slit is at an infinite distance from the grating (Margaritondo, 1988; Koch, 1983).

In the Rowland-circle geometry, the simplest way to separate photons of different energies is to keep the entrance slit and the grating fixed, and move the exit slit along the circle. This approach is all but abandoned today, because it requires moving the vacuum experimental chamber together with the slit; many other types of Rowland-circle monochromators with fixed exit slit have been invented and implemented.

The Rowland-circle geometry is not affected by low-order aberrations, but it does produce spherical aberrations. The focusing occurs at different points for the plane of the Rowland circle and for the direction perpendicular to it, except in the trivial case of normal incidence and normal dispersion.

Rowland-circle monochromators can in fact work with a normal or near-normal incidence geometry, of course within a spectral range limited to low photon energies. Some of these normal-incidence monochromators (NIM's) reach very high energy resolution (Hansen *et al.,* 1989) by having a long distance between slits and grating. The Seya-Namioka monochromator illustrated in Fig. 10 stresses the photon flux instead of the resolution, which is moderate. This optimization determines the "magic angle" between incident and diffracted beams. Seya-Namioka monochromators played a crucial role in the early years of synchrotron radiation, and still are excellent general-purpose instruments for the photon energy range up to 30 eV.

Higher energies require grazing-incidence instruments. For example, Fig. 11 shows another widely used synchrotron-radiation instrument: the "grasshopper" monochromator (Brown *et al.,* 1974). The rather complex mechanism makes it possible to achieve high resolution in a wide spectral range (up to over 1 keV), with a fixed exit slit. The operation is based on a coordinated motion of three components, the first focusing mirror, the second mirror at the entrance slit, and the grating. This effectively produces a rotation of the Rowland circle around the exit slit, without changing its radius.

A recent type of grazing-incidence SGM has reached some of the best performances in this spectral domain, in particular for energy resolution: the "Dragon" monochromator (Chen and Sette, 1989). Some of the most crucial

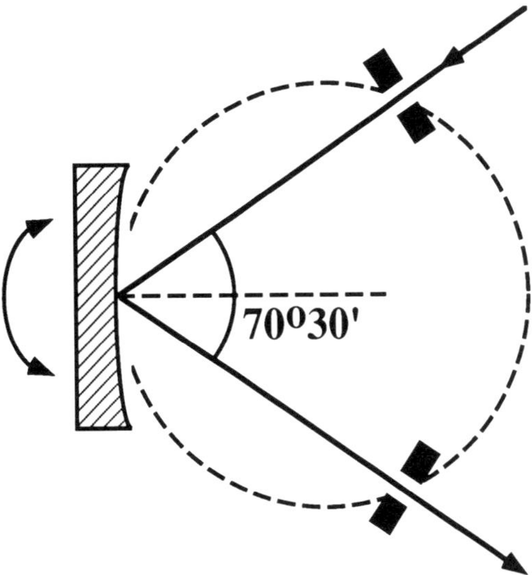

FIG. 10. Scheme of a vacuum-ultraviolet Seya-Namioka monochromator, with its simple grating motion and a "magic angle" of 70°30′.

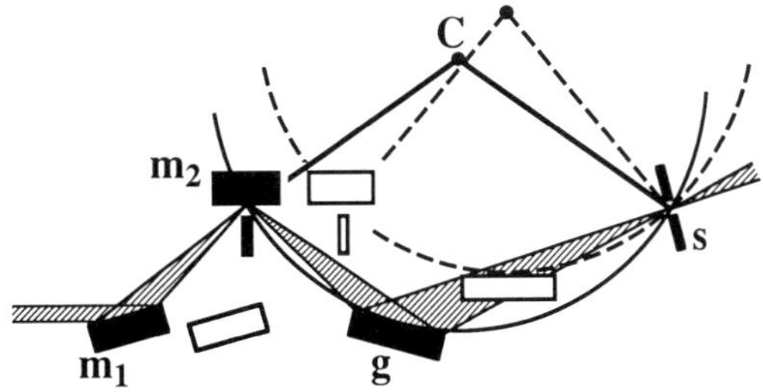

FIG. 11. The "grasshopper," a popular spherical-grating monochromator for soft x rays (Brown *et al.,* 1974). The photon energy scanning is accomplished without changing the position of the exit slit (S), with a coordinated motion of three different components: the first mirror m, the second mirror m_2 (at the entrance slit), and the grating (g). This complex coordinated motion can be visualized by comparing the two different sets of positions of the three components: note that the center of the Rowland circle has been shifted by a rotation of the circle around S.

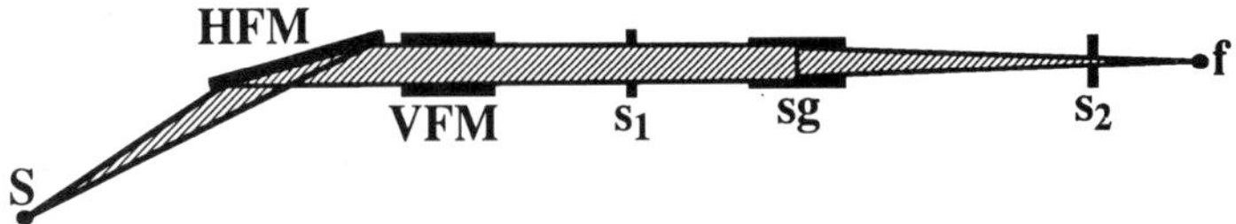

FIG. 12. Top-view scheme of the "Dragon" monochromator (Chen and Sette, 1989), with its source *S*, the decoupled-function entrance mirrors HFM (horizontal focusing mirror) and VFM (vertical focusing mirror), the entrance and exit slits s_1 and s_2, the holographic spherical grating (sg), and focalization onto the sample at the point *f*.

features of Dragon, which is schematically shown in Fig. 12, are the decoupling of the optical functions of the two mirrors HFM and VFM, and the possibility to move the exit slit, thereby enhancing the resolving power.

4.3.3 Toroidal-Grating Monochromators (TGM's) Figure 13 shows the typical geometry of a TGM; the photon-energy scanning is accomplished with a rotation of the grating. The complete name of each TGM instrument identifies the distances between the slits and the grating, which affect the resolving power—for example, "5–10 m TGM."

A typical TGM works with grazing but not extreme grazing incidence; therefore its spectral range is not as extended to high photon energies as for a grasshopper monochromator. Energies above a few hundred electron volts would require TGM instruments of very large and often unrealistic size. Furthermore, TGM's are affected by higher-order contamination of the monochromatized beam, and by the high cost of the gratings. Except for these problems, they are excellent instruments, widely used for synchrotron radiation activities.

5. HARD–X-RAY MONOCHROMATORS

The upper photon-energy limit of the x-ray domain served by grating monochromators is determined by the technical difficulties in manufacturing gratings for hard x-rays. Beyond 1–2 keV, crystal monochromators dominate. These instruments exploit the diffraction properties of the periodic arrays of atoms in crystals.

Such properties are schematized in Fig. 14, where we see the *Bragg reflection* of a photon beam of wavelength λ by a family of crystal planes (Margaritondo, 1988; Koch, 1983). The Bragg condition for constructive interference is

$$2d_c \sin \theta_d = \lambda. \tag{18}$$

This equation means that each family of crystal planes (identified by its Miller indices) produces a Bragg reflection of a monochromatic beam in a given direction. Conversely, a given family of crystal planes diffracts photons of different wavelengths in different directions, according to Eq. (18). This makes it possible to separate photons of different energies, with a resolving power that is derived from Eq. (18):

$$h\nu/\Delta h\nu = \lambda/\Delta\lambda = (\Delta\theta_d \cot \theta_d)^{-1}; \tag{19}$$

this equation shows that the resolution is affected by the "angular spread," $\Delta\theta_d$. In turn, $\Delta\theta_d$ is influenced by the angular divergence of the x-ray source, and by the quality of the diffracting crystal. The first problem is solved by using a highly collimated synchrotron radiation source, either a bending magnet or a

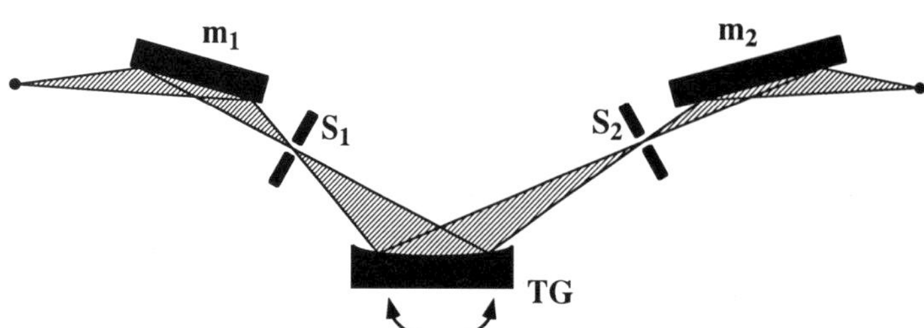

FIG. 13. Scheme of a toroidal-grating monochromator for soft x-rays, with the toroidal grating (TG) (Schmahl and Rudolph, 1976) and its simple motion, the focusing and refocusing mirrors m_1 and m_2, and the entrance and exit slits S_1 and S_2.

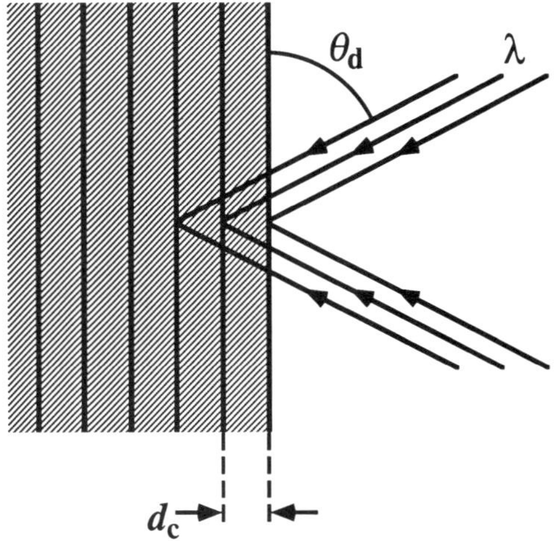

FIG. 14. Bragg reflection (diffraction) of a photon beam of wavelength λ by a family of crystal planes of periodicity d_c.

"wiggler" (Margaritondo, 1988; Koch, 1983). The second problem requires high-quality single crystals such as silicon, quartz, and a few others offered by the present crystal growth technology.

The crystal quality effect on the angular spread can be assessed by measuring the so-called *rocking curve,* which is the plot of the Bragg-reflected intensity as a function of the diffraction angle [referred to the ideal Bragg angle of Eq. (18)]. The width of the rocking curve is affected by the angular spread produced by the crystal quality.

The rocking curve also contains other types of valuable information; for example, its amplitude shows what portion of the incident beam intensity is Bragg-reflected and therefore still available for experiments. Note that this is true only when the input beam has an angular spread much smaller than the width of the rocking curve; in the opposite limit, the fraction of the incident beam intensity that is Bragg-reflected corresponds to the rocking curve's integral.

It should be emphasized that the simple Bragg condition of Eq. (18) is valid only for single scattering. This hypothesis is correct for mosaic crystals, but may not apply to large-size, high-quality single crystals typically used in x-ray monochromators. For these, a more rigorous approach based on dynamic calculations must include multiple scattering.

5.1 Geometries for Crystal Monochromators

The simplest type of crystal monochromator uses a single crystal mounted on a goniometer for high-accuracy positioning. In many cases, however, more complicated geometries are adopted. Many of these are double-crystal geometries like those shown in Fig. 15. Such geometries are based on two subsequent Bragg reflections by similar crystals.

The "parallel" or (+, −) geometry of Fig. 15 keeps the exit beam direction constant while scanning the photon energy, except for a small lateral displacement. This characteristic matches the requirements of EXAFS (extended x-ray-absorption fine structure) studies (Margaritondo, 1988; Koch, 1983), which constitute a large fraction of the pres-

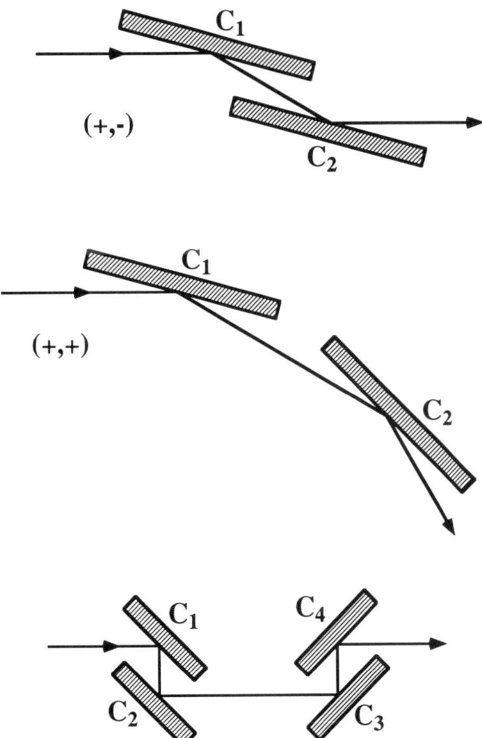

FIG. 15. Schemes of three different multiple-crystal monochromator geometries. From top to bottom: the (+, −) geometry, in which the two crystals C_1 and C_2 are often in a single block, obtained by channel-cutting a large-size crystal; the (+, +) geometry that implies a substantial displacement of the output beam when the photon energy is scanned; and a geometry including two channel-cut crystals, with a total of four Bragg-reflecting crystals (C_1, C_2, C_3, and C_4).

ent x-ray experiments. The (+, −) geometry is often implemented with the two reflecting crystals in a single block, obtained by channel-cutting a large crystal of high quality, typically silicon.

EXAFS experiments do not require very high energy resolution, and the (+, −) geometry is suitable for them. For higher resolution, corresponding to higher dispersion, it is preferable to use the (+, +) geometry of Fig. 15. Note that the exit beam direction changes while scanning the photon energy. This complicates the experiment's mechanism, and in particular requires moving the photon detector. Alternative geometries are possible to achieve high resolution without large displacements of the exit beam during scanning. An example, based on two channel-cut double crystals, is shown in the lower part of Fig. 15.

5.2 Curved Crystals

The crystal-monochromator geometries discussed in the previous section are often complemented by other optical elements such as focusing mirrors, in particular when using noncollimated sources of x rays. As in the case of grating monochromators, it would be preferable to avoid using too many reflecting devices, because each reflection causes a loss of intensity. It is possible, for example, to combine the dispersion and focusing functions by bending the monochromator's crystal(s).

Focusing both in the plane of incidence and in the direction perpendicular to this plane requires a double bending of the crystal, which is technically difficult. In many cases, multiple single-bent crystals are used as a compromise. The theory of bent-crystal monochromators is similar to that of focusing-grating monochromators (Hutley, 1982). For example, the Rowland-circle and Wadsworth geometries are still solutions for the general focusing-monochromatization problem. The Rowland-circle geometry, however, creates difficulties because the source must be quite close to the crystal: for crystal monochromators at high-energy synchrotron radiation sources, this is often made impossible by the radiation shielding.

Even without reaching the (Wadsworth) limit of infinite source-grating distance, in many crystal monochromators the source is well outside the Rowland circle. The image is demagnified and falls inside the Rowland circle; there is also a loss of resolution with respect to the Rowland-circle geometry. This is often acceptable, since high resolution is not required for many hard–x-ray experiments (Margaritondo, 1988; Koch, 1983), and the demagnification may in fact be desirable.

GLOSSARY

Blaze Angle: The angle (Fig. 5) that defines a sawtooth profile of a reflection grating.

Blaze Wavelength: The wavelength for which the throughput of a given reflection grating in a given geometrical configuration is optimized; see Eq. (12) and the corresponding discussion.

Bragg Reflection: The diffraction of an x-ray beam by a family of crystal planes, whose conditions are expressed by Eq. (18).

Contrast: Ratio of the maximum to the minimum transmitted intensities of interference fringes; it corresponds roughly to the reciprocal of the spectral purity.

Critical Angle: The grazing angle of an incident photon beam beyond which the reflection becomes very inefficient. See Eq. (15).

Crystal: A high-quality single crystal or mosaic, used to monochromatize x rays.

Dispersion Device: An optical component that disperses in different directions photons of different energies; for example, a grating.

Double Crystal: A pair of crystals for a parallel-geometry crystal monochromator, often obtained in a single block by channel-cutting a large-size crystal.

Dragon: An advanced spherical-grating grazing-incidence monochromator with very high resolving power; see Fig. 12.

Finesse: Ratio between free spectral range and spectral width of the interference fringes; corresponds to the number of resolvable elements in one free spectral range.

Free Spectral Range: Spectral distance between two neighboring interference maxima; corresponds to the wavelength (or energy) interval that can be scanned without changing the order of interference.

Fresnel Zone Plate: A regular array of concentric circular lines that focuses and/or monochromatizes photon beams. Inexpensive Fresnel zone plates are widely available for visible radiation, but similar devices for soft x rays are extremely difficult to fabricate.

Ghosts: Spurious spectral features appearing in a grating monochromator, caused by periodic errors in the grooves of the grating. Such periodic imperfections are typical for ruled gratings.

Grasshopper: A widely used grazing-incidence spherical-grating monochromator (see Fig. 11).

Grating: An artificially fabricated periodic array of lines, used as a dispersion device. It can be a reflection grating or a transmission grating.

Holographic Gratings: Special type of interference reflection gratings with a nonplanar surface, fabricated with photographic methods based on holography. In this way the spacing and the shape of the grooves can be controlled so as to achieve correction of aberrations and other desired functions.

Interference Gratings: Gratings made using interference fringes produced with coherent light, in combination with photolithography methods.

Littrow Blaze Wavelength: The blaze wavelength for the Littrow mounting geometry. See Eqs. (13) and (14) and the corresponding discussion.

Littrow Mounting: The geometry for a reflection grating with sawtooth profile in which the incident beam is perpendicular to the grating surface's facets.

Multilayer: A periodic array of thin films, used to enhance or suppress reflections, that also acts as a bandpass filter and therefore as a monochromator.

NIM: Normal-incidence monochromator, typically a reflection spherical-grating instrument for the vacuum ultraviolet.

Original: A master grating used to manufacture low-cost replicas.

Parallel Geometry: A crystal-monochromator configuration, also called "(+, −) geometry," that minimizes the deflection of the output beam during the photon energy scanning. Also see **Double Crystal,** and Fig. 15.

Period: The parameter that defines the periodicity of the elements in a dispersion device such as a grating or a crystal.

PGM: Plane-grating monochromator for ultraviolet photons and soft x rays.

Pitch: See **Period:** the pitch is inversely related to the period.

Rayleigh Criterion: Two features are considered as just resolved when the central maximum in the diffraction pattern caused by the first one falls onto the first minimum in the diffraction pattern of the other.

Replica: A low-cost grating manufactured from an original.

Resolving Power: By definition, the photon energy divided by the energy resolution.

Rocking Curve: Reflected intensity as a function of the angular position of the crystal in a crystal monochromator; it is used to assess the monochromator's performance.

Rowland Circle: The circle that contains the entrance and exit slits for optimized focusing and resolution of a spherical reflection grating.

Ruled Grating: A grating that has been ruled on metal, typically a thin layer of metal (for example, aluminum or gold) deposited on a glassy substrate.

Seya-Namioka: A popular near-normal-incidence reflection spherical-grating monochromator for synchrotron radiation (see Fig. 10).

SGM: Spherical-grating monochromator.

Spectral Purity: The relative throughput for the light of a narrow laser line having a wavelength outside the nominal bandpass interval of the monochromator. It remains finite due to the presence of secondary maxima in the diffraction pattern and of stray light.

Spectral Response: The function of the photon energy that defines the response of the monochromator to different photon energies, and in particular its resolution and throughput. It is defined in the text.

Stray Light: Light traveling randomly inside a monochromator. It is due to defects of the dispersing device, and to incomplete absorption of light having wavelengths outside the nominal bandpass interval of the monochromator.

SX-700: One of the most advanced PGM's, with high resolving power (see Fig. 8).

TGM: Toroidal-grating monochromator.

Throughput: The output beam power within the photon-energy band filtered by the monochromator. The throughput depends on the input beam intensity, and can be defined normalized to it; it depends also on the resolving power.

Toroidal Grating: A reflection grating for soft x rays, which focuses in both directions (in the plane of incidence and perpendicular to it) with limited aberrations. Also see **Holographic grating.**

Wadsworth Mounting: A reflection spherical-grating geometry that can be used instead of the Rowland-circle geometry, placing the entrance slit at infinite distance from the grating.

X Rays, Soft and Hard: The definition of the boundary between these two photon-energy domains changes from author to author; some authors place it at 3–4 keV, others at lower energies such as 1–2 keV.

Works Cited

Born, M., Wolf, E. (1980), *Principles of Optics,* Oxford: Pergamon.

Brown, F. C., Bachrach, R. Z., Hagström, S. B. M., Lien, N., Pruett, C. H. (1974), in *Vacuum Ultraviolet Radiation Physics,* E. E. Koch, R. Haensel and C. Kunz, eds. Braunschweig: Pergamon-Vieweg, pp. 785–787.

Cerrina, F., Lai, B., Gong, C., Ray Chaudhuri, A., Margaritondo, G., Green, M. A., Hochst, H., Cole, R., Crossley, D., Collier, S., Underwood, J., Brillson, L. J., Franciosi, A., DeLuca, P. M., Jr., Gould, M. N. (1989), *Rev. Sci. Instrum.* **60**, 2249–2240.

Chen, C. T., Sette, F. (1989), *Rev. Sci. Instrum.* **60**, 1616–1621.

Kaindl, G., Domke, M., Laubschat, C., Weschke, E., Xue, C. (1992), *Rev. Sci. Instrum.* **63**, 1234–1240.

Eberhardt, W., Kalkoffen, G., Kunz, C. (1978), *Nucl. Instrum. Methods* **152**, 81–85.

Hansen, R. W. C., Pruett, C. H., Salehzadeh, A., Wallace, D. J. (1989), *Rev. Sci. Instrum.* **60**, 2113–2115.

Henke, B. (1972), *Phys. Rev.* **A6**, 94–104.

Hutley, M. C. (1982), *Diffraction Gratings,* New York: Academic.

James, J. F., Sternberg, R. S. (1969), *The Design of Optical Spectrometers,* London: Chapman and Hall.

Koch, E.-E. (1983), *Handbook of Synchrotron Radiation,* Amsterdam: North-Holland.

Margaritondo, G. (1988), *Introduction to Synchrotron Radiation,* New York: Oxford University.

Rarback, H., Shu, D., Feng, S. C., Ade, H., Kirz, J., McNulty, I., Kern, D. P., Chang, T. H. P., Vladimirsky, Y., Iskander, N., Attwood, D., McQuaid, K., Rothman, S. (1988), *Rev. Sci. Instrum.* **59**, 52–59.

Schmahl, G., Rudolph, D. (1976), in: E. Wolf (Ed.), *Progress in Optics,* Vol. 14, Amsterdam: North-Holland, pp. 195–244.

Samson, J. A. R. (1980), *Techniques of Vacuum Ultraviolet Spectroscopy,* Lincoln, Nebraska: Pied.

Vladimirsky, Y., Kern, D., Meyer-Ilse, W., Attwood, D. (1989), *Appl. Phys. Lett.* **54**, 286–288.

Further Reading

There are many books where more information on monochromators can be obtained, and we just mention a few of them. The book by Jenkins, F. A., White, E. (1957), *Fundamentals of Optics,* New York: McGraw-Hill, treats many physical principles. The work by Born, M., Wolf, E. (1980): *Principles of Optics,* Oxford: Pergamon, published for the first time in 1959, remains an important reference book for today's engineers and scientists. The books by Hutley, M. C. (1982), *Diffraction Gratings,* New York: Academic, and by James, J. F., Sternberg, R. S. (1969), *The Design of Optical Spectrometers,* London: Chapman and Hill, are more application oriented.

In the ultraviolet or x-ray domains, most of the progress in monochromators and optical components in general has been stimulated by the advent and expansion of synchrotron radiation. For an elementary review of synchrotron radiation, we suggest Margaritondo, G. (1988), *Introduction to Synchrotron Radiation,* New York: Oxford University Press. For a more advanced presentation, the reader is referred to Volumes 1–4 of North-Holland's *Handbook of Synchrotron Radiation* (see in particular Koch, 1983).

The progress in synchrotron radiation instrumentation has been consistently reported at specialized conferences, whose regular proceedings are excellent sources of information. The most important examples are the *Synchrotron Radiation Instrumentation Conference,* whose proceedings have been published by several different companies; the *Vacuum-Ultraviolet Research Conference,* some of whose proceedings are landmarks in this field, notably the one from the Hamburg meeting in 1974. We also suggest the *Proceedings of the Annual Symposium of the American Vacuum Society,* published as a regular issue of the *Journal of Vacuum Science and Technology,* and those of the triennial meeting of the International Vacuum Society.

We also suggest that the reader consult the annual reports of the major synchrotron radiation laboratories, and in particular those of the National Synchrotron Light Source in Brookhaven, New York, of the Stanford Synchrotron Radiation Laboratory, Stanford, California, of the LURE in Orsay, France, of HASYLAB in Hamburg, Germany, of the Daresbury Laboratory, Daresbury, England, of BESSY in Berlin, of the Photon Factory in Tsukuba, and the abstract book of the annual users' meeting of the Wisconsin Synchrotron Radiation Center, Stoughton, Wisconsin.

MONTE-CARLO METHODS

K. BINDER, *Institut für Physik, Johannes-Gutenberg-Universität Mainz, Mainz, Germany*

3-527-28132-0/94/$5.00 + .50

INTRODUCTION AND OVERVIEW

Many problems in science are very complex: e.g., statistical thermodynamics considers thermal properties of matter resulting from the interplay of a large number of elementary particles. A deterministic description in terms of the equation of motion of all these particles would make no sense, and a probabilistic description is required. A probabilistic description may even be intrinsically implied by the quantum-mechanical nature of the basic processes (e.g., emission of neutrons in radioactive decay) or because the problem is incompletely characterized, only some degrees of freedom being considered explicitly while the others act as a kind of background causing random noise. While thus the concept of probability distributions is ubiquitous in physics, often it is not possible to compute these probability distribution functions analytically in explicit form, because of the complexity of the problem. For example, interactions between atoms in a fluid produce strong and nontrivial correlations between atomic positions, and, hence, it is not possible to calculate these correlations analytically.

Monte-Carlo methods now aim at a numerical estimation of probability distributions (as well as of averages, that can be calculated from them), making use of (pseudo) random numbers. By "pseudorandom numbers" one means a sequence of numbers produced on a computer with a deterministic procedure from a suitable "seed." This sequence, hence, is not truly random—see Sec. 1 for a discussion of this problem.

The outline of the present article is as follows. Since all Monte-Carlo methods heavily rely on the use of random numbers, we briefly review random-number generation in Sec. 1. In Sec. 2, we then elaborate on the discussion of "simple sampling," i.e., problems where a straightforward generation of probability distributions using random numbers is possible. Section 3 briefly mentions some applications to transport problems, such as radiation shielding, and growth phenomena, such as "diffusion-limited aggregation" (DLA).

Section 4 then considers the importance sampling methods of statistical thermodynamics, including the use of different thermodynamic ensembles. This article emphasizes applications in statistical mechanics of condensed matter, since this is the field where most activity with Monte-Carlo methods occurs.

Some more practical aspects important for the implementation of algorithms and the judgment of the tractability of simulation approaches to physical problems are then considered in Sec. 5: effects resulting from the finite size of simulation boxes, effects of choosing various boundary conditions, dynamic correlation of errors, and the application to studies of the dynamics of thermal fluctuations.

The extension to quantum-mechanical problems is mentioned in Sec. 6 and the application to elementary particle theory (lattice gauge theory) in Sec. 7. Section 8 then illustrates some of the general concepts with a variety of applications taken from condensed matter physics, while Sec. 9 contains concluding remarks.

We do not discuss problems of applied mathematics such as applications to the solution of linear operator equations (Fredholm integral equations, the Dirichlet boundary-value problem, eigenvalue problems, etc.); for a concise discussion of such problems, see, e.g., Hammersley and Handscomb (1964). Nor do we discuss simulations of chemical kinetics, such as polymerization processes (see, e.g., Bruns et al. 1981).

1. RANDOM-NUMBER GENERATION

1.1 General Introduction

The precise definition of "randomness" is a problem in itself (see, e.g., Compagner, 1991) and is outside the scope of the present article. Truly random numbers are unpredictable in advance and must be produced by an appropriate physical process such as radioactive decay. Series of such numbers have been documented but would be very inconvenient to use for Monte-Carlo simulations, and usually their total number is too limited anyhow. Thus, we do not discuss them here any further.

Pseudorandom numbers are produced in the computer by one of several simple algorithms, some of which will be discussed below, and thus are predictable, as their sequence is exactly reproducible. (This reproducibility, of course, is a desirable property, as it allows detailed checks of Monte-Carlo simulation programs.) They are thus not truly random, but they have statistical properties (nearly uniform distribution, nearly vanishing correlation coefficients, etc.) that are very similar to the statistical properties of truly random numbers. Thus, a given sequence of (pseudo)random numbers appears "random" for many practical purposes. In the following, the prefix "pseudo" will be omitted throughout.

1.2 Properties That a Random-Number Generator Should Have

What one needs are numbers that are uniformly distributed in the interval [0,1] and that are uncorrelated. By "uncorrelated" we not only mean vanishing pair correlations for arbitrary distances along the random-number sequence, but also vanishing triplet and higher correlations. No algorithm exists that satisfies these desirable requirements fully, of course; the extent to which the remaining correlations between the generated random numbers lead to erroneous results of Monte-Carlo simulations has been a matter of long-standing concern (Knuth, 1969; James, 1990); even random-number generators that have passed all common statistical tests and have been used successfully for years may fail for a new application, in particular if it involves a new type of Monte-Carlo algorithm (see, e.g., Ferrenberg *et al.*, 1992, for a recent example). Therefore, the testing of random-number generators is a field of research in itself (see, e.g., Marsaglia, 1985; Compagner and Hoogland, 1987).

A limitation due to the finite word length of computers is the *finite period*: Every generator begins after a long but finite period to produce exactly the same sequence again. For example, simple generators for 32-bit computers have a maximum period of 2^{30} ($\approx 10^9$) numbers only. This is not enough for recent high-quality applications! Of course, one can get around this problem (Knuth, 1969; James, 1990), but, at the same time, one likes the code representing the random-number generator to be "portable" (i.e., in a high-level programing language like FORTRAN usable for computers from different manufactures) and "efficient" (i.e., extremely fast so it does not unduly slow down the simulation program as a whole). Thus, inventing new random-number generators that are in certain respects a better compromise between these partially conflicting requirements is still of interest (e.g., Marsaglia *et al.*, 1990).

1.3 Comments about a Few Frequently Used Generators

Best known is the linear multiplicative algorithm (Lehmer, 1951), which produces random integers X_i recursively from the formula

$$X_i = aX_{i-1} + c \text{ (modulo } m), \tag{1}$$

which means that m is added when the results otherwise were negative. For 32-bit computers, $m = 2^{31} - 1$ (the largest integer that can be used for that computer). The integer constants a, c, and X_0 (the starting value of the recursion, the so-called "seed") need to be appropriately chosen {e.g., $a = 16\,807$, $c = 0$, X_0 odd}. Obviously, the "randomness" of the X_i results since, after a few multiplications with a, the result would exceed m and hence is truncated, and so the leading digits of X_i are more or less random. But there are severe correlations: If d-tuples of such numbers are used to represent points in d-dimensional space having a lattice structure, they lie on a certain number of hyperplanes (Marsaglia, 1968).

Equation (1) produces random numbers between 0 and m. Converting them into real numbers and dividing by m yields random numbers in the interval [0,1], as desired.

More popular now are shift-register generators (Tausworthe, 1965; Kirkpatrick and Stoll, 1981), based on the formula

$$X_i = X_{i-p} \cdot \mathrm{XOR} \cdot X_{i-q}, \tag{2}$$

where $\cdot \mathrm{XOR} \cdot$ is the bitwise "exclusive or" operation, and the "lags" p, q have to be properly chosen [the popular "R250" (Kirkpatrick and Stoll, 1981) uses $p = 109$, $q = 250$ and thus needs 250 initializing integers]. "Good" generators based on Eq. (2) have fewer correlations between random numbers than those resulting from Eq. (1), and much larger period.

A third type of generators is based on Fibonacci series and also recommended in the literature (Knuth, 1969; Ahrens and Dieter, 1979; James, 1990). But a general recommendation is that every user of random numbers should not rely on their quality blindly, and should perform his own tests in the context of his application.

2. SIMPLE SAMPLING OF PROBABILITY DISTRIBUTIONS USING RANDOM NUMBERS

In this section, we give a few nearly trivial examples of the use of Monte-Carlo methods, which will be useful for the understanding of later sections. More material on this subject can be found in standard textbooks like Koonin (1981) and Gould and Tobochnik (1988).

2.1 Numerical Estimation of Known Probability Distributions

A known probability distribution p_i that a (discrete) state i occurs with $1 \leq i \leq n$, with $\Sigma_{i=1}^{n} p_i = 1$, is numerically realized using random numbers uniformly distributed in the interval from zero to unity: defining $P_i = \Sigma_{j=1}^{i} p_i$, we choose a state i if the random number ζ satisfies $P_{i-1} \leq \zeta < P_i$, with $P_0 = 0$. In the limit of a large number (M) of trials, the generated distribution approximates p_i, with errors of order $1/\sqrt{M}$.

Monte-Carlo methods in statistical mechanics can be viewed as an extension of this simple concept to the probability that a point $\mathbf{X}$ in phase space occurs,

$$P_{\text{eq}}(\mathbf{X}) = (1/Z)\exp\{-\mathcal{H}(\mathbf{X})/k_{\text{B}}T\},$$

k_{B} being Boltzmann's constant, T the absolute temperature, and $Z = \Sigma_{\mathbf{X}} \exp\{-\mathcal{H}(\mathbf{X})/k_{\text{B}}T\}$ the partition function, although in general neither Z nor $P_{\text{eq}}(\mathbf{X})$ can be written explicitly (as function of the variables of interest, such as T, particle number N, volume V, etc.). The term $\mathcal{H}(\mathbf{X})$ denotes the Hamiltonian of the (classical) system.

2.2 "Importance Sampling" versus "Simple Sampling"

The sampling of the Boltzmann probability $P_{\text{eq}}(\mathbf{X})$ by Monte-Carlo methods is not completely straightforward: One must not choose the points $\mathbf{X}$ in phase space completely at random, since $P_{\text{eq}}(\mathbf{X})$ is extremely sharply peaked. Thus, one needs "importance sampling" methods which generate points $\mathbf{X}$ preferably from the "important" region of space where this narrow peak occurs.

Before we treat this problem of statistical mechanics in more detail, we emphasize the more straightforward applications of "simple sampling" techniques. In the following, we list a few problems for which simple sampling is useful. Suppose one wishes to generate a configuration of a randomly mixed crystal of a given lattice structure, e.g., a binary mixture of composition A_xB_{1-x}. Again, one uses pseudorandom numbers ζ uniformly distributed in [0,1] to choose the occupancy of lattice sites $\{j\}$: If $\zeta_j < x$, the site j is taken by an A atom, and else by a B atom. Such configurations now can be used as starting point for a numerical study of the dynamical matrix, if one is interested in the phonon spectrum of mixed crystals. One can study the distribution of sizes of "clusters" formed by neighboring A atoms if one is interested in the "site percolation problem" (Stauffer, 1985), etc.

If one is interested in simulating transport processes such as diffusion, a basic approach is the generation of simple random walks. Such random walks, resulting from addition of vectors whose orientation is random, can be generated both on lattices and in the continuum. Such simulations are desirable if one wishes to consider complicated geometries or boundary conditions of the medium where the diffusion takes place. Also, it is straightforward to include competing processes (e.g., in a reactor, diffusion of neutrons in the moderator competes with loss of neutrons due to nuclear reactions, radiation going to the outside, etc., or gain due to fission events). Actually, this problem of reactor criticality (and related problems for nuclear weapons!) was the starting point for the first large-scale applications of Monte-Carlo methods by Fermi, von Neumann, Ulam, and their co-workers (Hammersley and Handscomb, 1964).

2.3 Monte Carlo as a Method of Integration

Many Monte-Carlo computations may be viewed as attempts to estimate the value of a (multiple) integral. To give the flavor of this idea, let us discuss the one-dimensional integral

$$I = \int_0^1 f(x)dx$$
$$\equiv \int_0^1 \int_0^1 g(x,y)dxdy, \qquad \text{with}$$

$$g(x,y) = \begin{cases} 0 & \text{if } f(x) < y, \\ 1 & \text{if } f(x) \geq y, \end{cases} \tag{3}$$

as an example (suppose, for simplicity, that also $0 \leq f(x) \leq 1$ for $0 \leq x \leq 1$). Then I simply is interpreted as the fraction of the unit square $0 \leq x,y \leq 1$ lying underneath the curve $y = f(x)$. Now a straightforward (though often not very efficient) Monte-Carlo estimation of Eq. (3) is the "hit or miss" method: We take n points (ζ_x,ζ_y) uniformly distributed in the unit square $0 \leq \zeta_x \leq 1$, $0 \leq \zeta_y \leq 1$. Then I is estimated by

$$\bar{g} = \frac{1}{n}\sum_{i=1}^{n} g(\zeta_{xi},\zeta_{yi}) = \frac{n^*}{n}, \tag{4}$$

n^* being the number of points for which $f(\zeta_{xi}) \geq \zeta_{yi}$. Thus, we count the fraction of points that lie underneath the curve $y = f(x)$. Of course, such Monte-Carlo integration methods are inferior to many other techniques of numerical integration, if the integration space is low dimensional, but the situation is worse for high-dimensional integration spaces: For any method using a regular grid of points for which the integrand needs to be evaluated, the number of points sampled along each coordinate is $M^{1/d}$ in d dimensions, which is small for any reasonable sample size M if d is very large.

2.1 Infinite Integration Space

Not always is the integration space limited to a bounded interval in space. For example, the ϕ^4 model of field theory considers a field variable $\phi(\mathbf{x})$, where $\mathbf{x}$ is drawn from a d-dimensional space and $\phi(\mathbf{x})$ is a real variable with distribution

$$P(\phi) \propto \exp[-\alpha(-\tfrac{1}{2}\phi^2 + \tfrac{1}{4}\phi^4)]; \qquad \alpha > 0. \tag{5a}$$

While $-\infty < \phi < +\infty$, the distribution $P'(y)$

$$P'(y) = \int_{-\infty}^{y} P(\phi)d\phi \Big/ \int_{-\infty}^{+\infty} P(\phi)d\phi \tag{5b}$$

varies in the unit interval, $0 \leq P' \leq 1$. Hence, defining $Y = Y(P')$ as the inverse function of $P'(y)$, we can choose a random number ζ uniformly distributed between zero and one, to obtain $\phi = Y(\zeta)$ distributed according to the chosen distribution $P(\phi)$. Of course, this method works not only for the example chosen in Eq. (5) but for any distribution of interest. Often it will not be possible to obtain $Y(P')$ analytically, but then one can compute numerically a table before the start of the sampling (Heermann, 1986).

2.5 Random Selection of Lattice Sites

A problem that occurs very frequently (e.g., in solid-state physics) is that one considers a large lattice (e.g., a model of a simple cubic crystal with $N = L_x \times L_y \times L_z$ sites), and one wishes to select a lattice site (n_x,n_y,n_z) at random. This is trivially done using the integer arithmetics of standard computers, converting a uniformly distributed random number ζ_x $(0 \leq \zeta_x < 1)$ to an integer n_x with $1 \leq n_x \leq L_x$ via the statement $n_x = \text{int}(\zeta_x L_x + 1)$. This is already an example where one must be careful, however, when three successive pseudorandom numbers drawn from a random-number generator (RNG) are used for this purpose: If one uses a RNG with bad statistical qualities, the frequency with which individual sites are visited may deviate distinctly from a truly random choice. In unfavorable cases, successive pseudorandom numbers are so strongly correlated that certain lattice sites would be never visited!

2.6 The Self-Avoiding Walk Problem

As an example of the straightforward use of simple sampling techniques, we now discuss the study of self-avoiding walks (SAWs) on lattices (which may be considered as a simple model for polymer chains in good solvents; see Kremer and Binder, 1988). Suppose one considers a square or simple cubic lattice with coordination number (number of nearest neighbors) z. Then, for a random walk (RW) with N steps, we would have $Z_{RW} = z^N$

configurations, but many of these random walks intersect themselves and thus would, not be self-avoiding. For SAWs, one only expects of the order of Z_{SAW} = const. × $N^{\gamma-1}z_{\mathrm{eff}}^N$ configurations, where $\gamma > 1$ is a characteristic exponent (which is not known exactly for $d = 3$ dimensions), and $z_{\mathrm{eff}} \leq z - 1$ is an effective coordination number, which also is not known exactly. But it is already obvious that an exact enumeration of all configurations would be possible for rather small N only, while most questions of interest refer to the behavior for large N; e.g., one wishes to study the end-to-end distance of the SAW,

$$\langle R^2 \rangle_{\mathrm{SAW}} = \frac{1}{Z_{\mathrm{SAW}}} \sum_{\mathbf{X}} [\mathbf{R}(\mathbf{X})]^2, \tag{6}$$

the sum being extended over all configurations of SAWs which we denote formally as points $\mathbf{X}$ in phase space. One expects that $\langle R^2 \rangle_{\mathrm{SAW}} \propto N^{2\nu}$, where ν is another characteristic exponent. A Monte-Carlo estimation of $\langle R^2 \rangle_{\mathrm{SAW}}$ now is based on generating a sample of only $M \ll Z_{\mathrm{SAW}}$ configurations $\mathbf{X}_l$, i.e.:

$$\overline{R^2} = \frac{1}{M} \sum_{l=1}^{M} [\mathbf{R}(\mathbf{X}_l)]^2 \approx \langle R^2 \rangle_{\mathrm{SAW}}. \tag{7}$$

If the M configurations are statistically independent, standard error analysis applies, and we expect that the relative error behaves as

$$\frac{\overline{(\delta R^2)^2}}{(\overline{R^2})^2} \approx \frac{1}{M-1} \left[\frac{\langle R^4 \rangle_{\mathrm{SAW}}}{\langle R^2 \rangle_{\mathrm{SAW}}^2} - 1 \right]. \tag{8}$$

While the law of large numbers then implies that $\overline{R^2}$ is Gaussian distributed around $\langle R^2 \rangle_{\mathrm{SAW}}$ with a variance determined by Eq. (8), one should note that the variance does not decrease with increasing N. Statistical mechanics tells us that fluctuations decrease with increasing number N of degrees of freedom; i.e., one equilibrium configuration differs in its energy $E(\mathbf{x})$ from the average $\langle E \rangle$ only by an amount of order $1/\sqrt{N}$. This property is called "self-averaging." Obviously, such a property is not true for $\langle R^2 \rangle_{\mathrm{SAW}}$. This "lack of self-averaging" is easy to show already for ordinary random walks (Binder and Heermann, 1988).

2.7 Simple Sampling versus Biased Sampling: the Example of SAWs Continued

Apart from this problem, that the accuracy of the estimation of R^2 does not increase with the number of steps of the walk, it is also not easy to generate a large sample of configurations of SAWs for large N. Suppose we do this at each step by choosing one of $z - 1$ neighbors at random (eliminating from the start immediate reversals, which would violate the SAW condition). Whenever the chosen lattice site is already taken, we also would violate the SAW condition, and the attempted walk is terminated. Now the fraction of walks that will continue successfully for N steps will only be of the order of $Z_{\mathrm{SAW}}/(z-1)^N \propto [z_{\mathrm{eff}}/(z-1)]^N N^{\gamma-1}$, which decreases to zero exponentially $\{\propto \exp(-N\mu)$ with $\mu = \ln[(z-1)/z_{\mathrm{eff}}]$ for large $N\}$; this failure of success in generating long SAWs is called the "attrition problem."

The obvious recipe, to select at each step only from among the lattice sites that do not violate the SAW restriction, does not give equal statistical weight for each configuration generated, of course, and so the average would not be the averaging that one needs in Eq. (6). One finds that this method would create a "bias" toward more compact configurations of the walk. But one can calculate the weights of configurations $w(\mathbf{X})$ that result in this so-called "inversely restricted sampling" (Rosenbluth and Rosenbluth, 1955), and in this way correct for the bias and estimate the SAW averages as

$$\overline{R^2} = \left\{ \sum_{l=1}^{M} [w(\mathbf{X}_l)]^{-1} \right\}^{-1} \sum_{l=1}^{M} [w(\mathbf{X}_l)]^{-1} [\mathbf{R}(\mathbf{X}_l)]^2. \tag{9}$$

However, error analysis of this biased sampling is rather delicate (Kremer and Binder, 1988).

A popular alternative to overcome the above attrition problem is the "enrichment technique," founded on the principle "Hold fast to that which is good." Namely, whenever a walk attains a length that is a multiple of s steps without intersecting itself, n independent attempts to continue it (rather than a single attempt) are made. The numbers n, s are fixed, and, if we choose $n \approx \exp(\mu s)$, the numbers of walks of various lengths generated will be approximately equal. Enrichment has the advantage over inversely restricted sampling that all walks of a given length have equal weights, while the weights in Eq. (9) vary over many orders of magnitude for large N. But the disadvantage is, on the other hand, that the linear dimensions of the walks are highly correlated, since some of them have many steps in common! For these reasons, simple sampling and its extensions are useful only for a small fraction of problems in polymer science, and now importance sampling (Sec. 4) is much more used. But we emphasize that related problems are encountered for the sampling of "random surfaces" (this problem arises in the field theory of quantum gravity), in path-integral Monte-Carlo treatments of quantum problems, and in many other contexts.

3. SURVEY OF APPLICATIONS TO SIMULATION OF TRANSPORT PROCESSES

The possibilities to simulate the random motions of particles are extremely widespread. Therefore, it is difficult to comment about such problems in general. Thus, we rather prefer again to proceed by briefly discussing a few examples that illustrate the spirit of the approach.

3.1 The "Shielding Problem"

A thick shield of absorbing material is exposed to γ radiation (energetic photons), of specified distribution of energy and angle of incidence. We want to know the intensity and energy distribution of the radiation that penetrates that shield.

The level of description is here that one may generate a lot of "histories" of those particles traveling through the medium. The paths of these γ particles between scattering events are straight lines, and different γ particles do not interact with each other. A particle with energy E, instantaneously at the point $\mathbf{r}$ and traveling in the direction of the unit vector $\mathbf{w}$, continues to travel in the same direction with the same energy, until a scattering event with an atom of the medium occurs. The standard assumption is that the atoms of the medium are distributed randomly in space. Then the total probability that the particle will collide with an atom while traveling a length δs of its path is $\sigma_c(E)\delta s$, $\sigma_c(E)$ being the cross section. In a region of space where $\sigma_c(E)$ is constant, the probability that a particle travels without collision a distance s is $F_c(s) = 1 - \exp[-\sigma_c(E)s]$. If a collision occurs, it may lead to absorption or scattering, and the cross sections for these types of events are assumed to be known.

A Monte-Carlo solution now simply involves the tracking of simulated particles from collision to collision, generating the distances s that the particles travel without collision from the exponential distribution quoted above. Particles leaving a collision point are sampled from the appropriate conditional probabilities as determined from the respective differential cross sections. For increasing sampling efficiency, many obvious tricks are known. For example, one may avoid losing particles by absorption events: If the absorption probability (i.e., the conditional probability that absorption occurs given that a collision has occurred) is α, one may replace $\sigma_c(E)$ by $\sigma_c(E)(1 - \alpha)$, and allow only scattering to take place with the appropriate relative probability. Special methods for the shielding problem have been extensively developed and already have been reviewed by Hammersley and Handscomb (1964).

3.2 Diffusion-Limited Aggregation (DLA)

Diffusion-limited aggregation is a model for the irreversible formation of random aggregates by diffusion of particles, which get stuck at random positions on the already formed object if they hit its surface in the course of their diffusion [sec Vicsek (1989), Meakin (1988), and Herrmann (1986, 1992) for detailed reviews of this problem and related phenomena]. Many problems (shapes of snowflakes, size distribution of asteroids, roughness of crack surfaces, etc.) can be understood as the end product of similar random dynamical and irreversible growth processes. Diffusion-limited aggregation is just one example of them. It may be simulated by iterating the following steps: From a randomly selected position on a spherical surface of radius R_m that encloses the aggregate (that has already been grown in the previous steps, its center of gravity being in the center of the sphere), a particle of unit mass is launched to start a simple random-walk trajectory. If it touches the aggregate, it sticks irreversibly on its surface. After the particle has either stuck or moved a distance R_f from the center of the aggregate such that it is unlikely that it will hit in the future, a new particle is launched. Ideally one would like to have $R_f \to \infty$, but, in practice, $R_f = 2R_m$ is sufficient. By this irreversible aggregation of particles, one forms fractal clusters. That means that the dimension d_f characterizing the relation between the mass of the grown object and its (gyration) radius R, $M \propto R^{d_f}$, is less than the dimension d of space in which the growth takes place. Again there are some tricks to make such simulations more efficient: For example, one may allow the particles to jump over larger steps when they travel in empty regions. From such studies, researchers have found that $d_f = 1.715 \pm 0.004$ for DLA in $d = 2$, while $d_f = 2.485 \pm 0.005$ in $d = 3$ (Tolman and Meakin, 1989). Such exponents as yet cannot be analytically predicted.

4. MONTE-CARLO METHODS IN STATISTICAL THERMODYNAMICS: IMPORTANCE SAMPLING

4.1 The General Idea of the Metropolis Importance Sampling Method

In the canonical ensemble (see STATISTICAL MECHANICS, CLASSICAL), the average of an observable $A(\mathbf{X})$ takes the form

$$\langle A \rangle = \frac{1}{Z} \int_{\Omega} d^k X A(\mathbf{X}) \exp\left[-\frac{\mathcal{H}(\mathbf{X})}{k_B T}\right], \qquad (10)$$

where Z is the partition function,

$$Z = \int_{\Omega} d^k X \exp\left[-\frac{\mathcal{H}(\mathbf{X})}{k_B T}\right], \qquad (11)$$

Ω denoting the (k-dimensional) volume of phase space $\{\mathbf{X}\}$ over which is integrated, $\mathcal{H}(\mathbf{X})$ being the (classical) Hamiltonian. For this problem, a simple sampling analog to Sec. 2 would not work: The probability distribution $p(\mathbf{X}) = (1/Z) \exp[-\mathcal{H}(\mathbf{X})/k_B T]$ has a very sharp peak in phase space where all extensive variables $A(\mathbf{X})$ are close to their average values $\langle A \rangle$. This peak may be approximated by a Gaussian centered at $\langle A \rangle$, with a relative half-width of order $1/\sqrt{N}$ only, if we consider a system of N particles. Hence, for a practically useful method, one cannot sample the phase space uniformly, but the points $\mathbf{X}_\nu$ must be chosen preferentially from the important region of phase space, i.e., the vicinity of the peak of this probability distribution. This goal is achieved by the importance sampling method (Metropolis *et al.*, 1953): Starting from some initial configuration $\mathbf{X}_1$, one constructs a sequence of configurations $\mathbf{X}_\nu$ defined in terms of a transition probability $W(\mathbf{X}_\nu \to \mathbf{X}'_\nu)$ that rules stochastic "moves" from an old state $\mathbf{X}_\nu$ to a new state $\mathbf{X}'_\nu$, and, hence, one creates a "random walk through phase space." The idea of that method is to choose $W(\mathbf{X} \to \mathbf{X}')$ such that the probability with which a point $\mathbf{X}$ is chosen in this process converges toward the canonical probability

$$P_{eq}(\mathbf{X}) = \left(\frac{1}{Z}\right) \exp\left[-\frac{\mathcal{H}(\mathbf{X})}{k_B T}\right]$$

in the limit where the number M of states $\mathbf{X}$ generated goes to infinity. A condition sufficient to ensure this convergence is the so-called principle of detailed balance,

$$P_{eq}(\mathbf{X}) W(\mathbf{X} \to \mathbf{X}') = P_{eq}(\mathbf{X}') W(\mathbf{X}' - \mathbf{X}). \qquad (12)$$

For a justification that Eq. (12) actually yields this desired convergence, we refer to Hammersley and Handscomb (1964), Binder (1976), Heermann (1986), and Kalos and Whitlock (1986). In this importance sampling technique, the average Eq. (10) then is estimated in terms of a simple arithmetic average,

$$\bar{A} = \frac{1}{M - M_0} \sum_{\nu=M_0+1}^{M} A(\mathbf{X}_\nu). \qquad (13)$$

Here it is anticipated that it is advantageous to eliminate the residual influence of the initial configuration $\mathbf{X}_1$ by eliminating a large enough number M_0 of states from the average. [The judgment of what is "large enough" is often difficult; see Binder (1976) and Sec. 5.3 below.] It should also be pointed out that this Metropolis method can be used for sampling any distribution $P(\mathbf{X})$: One simply must choose a transition probability $W(\mathbf{X} \to \mathbf{X}')$ that satisfies a detailed balance condition with $P(\mathbf{X})$ rather than with $P_{eq}(\mathbf{X})$.

4.2 Comments on the Formulation of a Monte-Carlo Algorithm

What is now meant in practice by the transition $\mathbf{X} \rightarrow \mathbf{X}'$? Again there is no general answer to this question; the choice of the process may depend both on the model under consideration and the purpose of the simulation. Since Eq. (12) implies that $W(\mathbf{X} \rightarrow \mathbf{X}')/W(\mathbf{X}' \rightarrow \mathbf{X}) = \exp(-\delta\mathcal{H}/k_BT)$, $\delta\mathcal{H}$ being the energy change caused by the move from $\mathbf{X} \rightarrow \mathbf{X}'$, typically it is necessary to consider small changes of the state $\mathbf{X}$ only. Otherwise the absolute value of the energy change $|\delta\mathcal{H}|$ would be rather large, and then either $W(\mathbf{X} \rightarrow \mathbf{X}')$ or $W(\mathbf{X}' \rightarrow \mathbf{X})$ would be very small. Then it would be almost always forbidden to carry out that move, and the procedure would be poorly convergent. For example, in the lattice gas model at constant particle number, a transition $\mathbf{X} \rightarrow \mathbf{X}'$ may consist of moving one particle to a randomly chosen neighboring site. In the lattice gas at constant chemical potential, one removes (or adds) just one particle at a time, which is isomorphic to single flips in the Ising model of anisotropic magnets.

Another arbitrariness concerns the order in which the particles are selected for considering a move. Often one chooses to select them in the order of their labels (in the simulation of a fluid or lattice gas at constant particle number) or to go through the lattice in a regular typewriter-type fashion (in the case of spin models, for instance). For lattice systems, it may be convenient to use sublattices (e.g., the "checkerboard algorithm," where the white and black sublattices are updated alternatively, for the sake of an efficient "vectorization" of the program; see Landau, 1992). An alternative is to choose the lattice sites (or particle numbers) randomly. The latter procedure is somewhat more time consuming, but it is a more faithful representation of a dynamic time evolution of the model described by a master equation (see below).

It is also helpful to realize that often the transition probability $W(\mathbf{X} \rightarrow \mathbf{X}')$ can be written as a product of an "attempt frequency" times an "acceptance frequency." By clever choice of the attempt frequency, it is possible sometimes to attempt large moves and still have a high acceptance, and thus make the computations more efficient.

For spin models on lattices, such as Ising or Potts models, *XY* and Heisenberg ferromagnets, etc., algorithms have been devised where one does not update single spins in the move $\mathbf{X} \rightarrow \mathbf{X}'$, but, rather, one updates specially constructed clusters of spins (see Swendsen *et al.*, 1992, for a review). These algorithms have the merit that they reduce critical slowing down, which hampers the efficiency of Monte-Carlo simulations near second-order phase transitions. "Critical slowing down" means a dramatic increase of relaxation times at the critical point of second-order phase transitions, and these relaxation times also control statistical errors in Monte-Carlo simulations, as we shall see in Sec. 5. Since these "cluster algorithms" work for rather special models only, they will not be discussed further here. But further development of such algorithms is an important area of current research (e.g., Barkema and Marko, 1993).

There is also some arbitrariness in the choice of the transition probability $W(\mathbf{X} \rightarrow \mathbf{X}')$ itself. The original choice of Metropolis *et al.* (1953) is

$$W(\mathbf{X} \rightarrow \mathbf{X}') = \begin{cases} \exp(-\delta\mathcal{H}/k_BT) & \text{if } \delta\mathcal{H} > 0, \\ 1 & \text{otherwise.} \end{cases} \tag{14}$$

An alternative choice is the so-called "heat-bath method." There one assigns the new value α_i' of the ith local degree of freedom in the move from $\mathbf{X}$ to $\mathbf{X}'$ irrespective of what the old value α_i' was. One thereby considers the local energy $\mathcal{H}_i(\alpha_i')$ and chooses the state α_i' with probability

$$\exp[-\mathcal{H}_i(\alpha_i')/k_BT] \Big/ \sum_{\{\alpha_i''\}} \exp[-\mathcal{H}_i(\alpha_i'')/k_BT].$$

We now outline the realization of the sequence of states $\mathbf{X}$ with chosen transition probability W. At each step of the procedure, one performs a *trial move* $\alpha_i \rightarrow \alpha_i'$, computes $W(\mathbf{X} \rightarrow \mathbf{X}')$ for this trial move, and compares it with a random number η, uniformly distributed in the interval $0 < \eta < 1$. If $W < \eta$, the trial move is rejected, and the old state (with α_i) is counted once more in the average, Eq. (13). Then another trial is made. If $W > \eta$, on the other hand, the trial move is accepted, and the new configuration thus generated is taken into account in the average, Eq. (13). It serves then also as a starting point of the next step.

Since subsequent states $\mathbf{X}_\nu$ in this Markov chain differ by the coordinate α_i of one particle only (if they differ at all), they are highly correlated. Therefore, it is not straightforward to estimate the error of the average, Eq. (13). Let us assume for the moment that, after n steps, these correlations have died out. Then we may estimate the statistical error δA of the estimate $\bar{A}$ from the standard formula,

$$\overline{(\delta A)^2} = \frac{1}{m(m-1)} \sum_{\mu=\mu_0}^{m+\mu_0-1} [A(\mathbf{X}_\mu) - \bar{A}]^2, \quad m \gg 1, \tag{15}$$

where the integers μ_0, μ, m are defined by $m = (M - M_0)/n$, μ_0 labels the state $\nu = M_0 + 1$, $\mu = \mu_0 + 1$ labels the state $\nu = M_0 + 1 + n$, etc. Then also $\bar{A}$ for consistency should be taken as

$$\bar{A} = \frac{1}{m} \sum_{\mu=\mu_0}^{m+\mu_0-1} A(\mathbf{X}_\mu). \tag{16}$$

If the computational effort of carrying out the "measurement" of $A(\mathbf{X}_\mu)$ in the simulation is rather small, it is advantageous to keep taking measurements every Monte-Carlo step per degree of freedom but to construct block averages over n successive measurements, varying n until uncorrelated block averages are obtained.

4.3 The Dynamic Interpretation of the Monte-Carlo Method

It is not always easy to estimate the appropriate number of configurations M_0 after which the correlations to the initial state $\mathbf{X}_1$, which typically is a state far from equilibrium, have died out, nor is it easy to estimate the number n between steps after which correlations in equilibrium have died out. A formal answer to this problem, in terms of relaxation times of the associated master equation describing the Monte-Carlo process, is discussed in the next section. This interpretation of Monte-Carlo sampling in terms of master equations is also the basis for Monte-Carlo studies of the dynamics of fluctuations near thermal equilibrium, and is discussed now. One introduces the probability $P(\mathbf{X},t)$ that a state $\mathbf{X}$ occurs at time t. This probability then decreases by all moves $\mathbf{X} \to \mathbf{X}'$, where the system reaches a neighboring state $\mathbf{X}'$; on the other hand, inverse processes $\mathbf{X}' \to \mathbf{X}$ lead to a gain of probability. Thus, one can write down a rate equation, similar to chemical kinetics, considering the balance of all gain and loss processes:

$$\frac{d}{dt} P(\mathbf{X},t) = -\sum_{\mathbf{X}'} W(\mathbf{X} \to \mathbf{X}')P(\mathbf{X},t) + \sum_{\mathbf{X}'} W(\mathbf{X}' \to \mathbf{X})P(\mathbf{X}',t). \tag{17}$$

The Monte-Carlo sampling (i.e., the sequence of generated states $\mathbf{X}_1 \to \mathbf{X}_2 \to \ldots \to \mathbf{X}_\nu \to \ldots$) can hence be interpreted as a numerical realization of the master equation, Eq. (17), and then a "time" t is associated with the index ν of subsequent configurations. In a system with N particles, we may normalize the "time unit" such that N single-particle moves are attempted per unit time. This is often called a "sweep" or "1 Monte-Carlo step (MCS)."

For the thermal equilibrium distribution $P_{eq}(\mathbf{X})$, because of the detailed balance principle, Eq. (12), there is no change of probability with time, $dP(\mathbf{X},t)/dt = 0$; thus, thermal equilibrium arises as the stationary solution of the master equation, Eq. (17). Thus, it is also plausible that Markov processes described by Eq. (17) describe a relaxation that always leads toward thermal equilibrium, as desired.

Now, for a physical system (whose trajectory in phase space, according to classical statistical mechanics, follows from Newton's laws of motion), it is clear that the stochastic trajectory through phase space that is described by Eq. (17) in general has nothing to do with the actual dynamics. For example, Eq. (17) never describes any propagating waves (such as spin waves in a magnet, or sound waves in a crystal or fluid, etc.).

In spite of this observation, the dynamics of the Monte-Carlo "trajectory" described by Eq. (17) sometimes does have physical significance. In many situations, one does not wish to consider the full set of dynamical variables of the system, but rather a subset only: For instance, in an interstitial alloy where one is interested in modeling the diffusion processes, the diffusion of the interstitials may be modeled by a stochastic hopping between the available lattice sites. Since the mean time between two successive jumps is orders of magnitude larger than the time scale of atomic vibrations in the solid, the phonons can be reasonably well approximated as a heat bath, as far as the diffusion is concerned.

There are many examples where such a separation of time scales for different degrees of freedom occurs: For example, for a description of the Brownian motion of polymer chains in polymer melts, the fast bond-angle and bond-length vibrations may be treated as heat bath, etc. As a rule of thumb, any very slow relaxation phenomena (kinetics of nucleation, decay of remanent magnetization in spin glasses, growth of ordered domains in adsorbed monolayers at surfaces, etc.) can be modeled by Monte-Carlo methods. Of course, one must pay attention to building in relevant conservation laws into the model properly (e.g., in an interstitial alloy, the overall concentration of interstitials is conserved; in a spin glass, the magnetization is not conserved) and to choosing microscopically reasonable elementary steps representing the move $\mathbf{X} \rightarrow \mathbf{X}'$. The great flexibility of the Monte-Carlo method, where one can choose the level of the modeling appropriately for the model at hand and identify the degrees of freedom that one wishes to consider, as well as the type and nature of transitions between them, is a great advantage and thus allows complementary applications to more atomistically realistic simulation approaches such as the molecular dynamics (*q.v.*) method where one numerically integrates Newton's equations of motion (Heermann, 1986; Ciccotti and Hoover, 1986; Hockney and Eastwood, 1988). By a clever combination with cluster-flip algorithms, one sometimes can construct very efficient algorithms and hence span a very broad range of time scales (Barkema and Marko, 1993).

4.4 Monte-Carlo Study of the Dynamics of Fluctuations near Equilibrium and of the Approach toward Equilibrium

Accepting Eq. (17), the average in Eq. (13) then is simply interpreted as a time average along the stochastic trajectory in phase space,

$$\bar{A} = \frac{1}{t_M - t_{M_0}} \int_{t_{M_0}}^{t_M} A(t)dt,$$
$$t_M = M/N, \qquad t_{M_0} = M_0/N. \tag{18}$$

It is thus no surprise that, for the importance-sampling Monte-Carlo method, one needs to consider carefully the problem of ergodicity: Time averages need not agree with ensemble averages. For example, near first-order phase transitions there may be long-lived metastable states. Sometimes the considered moves do not allow one to reach all configurations (e.g., in dynamic Monte-Carlo methods for self-avoiding walks; see Kremer and Binder, 1988).

One can also define time-displaced correlation functions: $\langle A(t)B(0)\rangle$, where A, B stand symbolically for any physical observables, is estimated by

$$\overline{A(t)B(0)} = \frac{1}{t_M - t - t_{M_0}} \int_{t_{M_0}}^{t_M - t} A(t+t')B(t')dt',$$
$$t_M - t > t_{M_0}. \tag{19}$$

Equation (19) refers to a situation where t_{M_0} is chosen large enough such that the system has relaxed toward equilibrium during the time t_{M_0}; then the pair correlation depends on the difference t and not the two individual times t', $t' + t$ separately.

However, it is also interesting to study the nonequilibrium relaxation process by which equilibrium is approached. In this region, $A(t) - \bar{A}$ is systematically dependent on the observation time t, and an ensemble average $\langle A(t)\rangle_T - \langle A(\infty)\rangle_T$ [$\lim_{t\to\infty} \bar{A} = \langle A\rangle_T \equiv \langle A(\infty)\rangle_T$ if the system is ergodic] is nonzero. One may define

$$\langle A(t)\rangle_T = \sum_{\{\mathbf{X}\}} P(\mathbf{X},t)A(\mathbf{X}) = \sum_{\{\mathbf{X}\}} P(\mathbf{X},0)A(\mathbf{X}(t)). \tag{20}$$

In the second step of this equation, the fact was used that the ensemble average involved is actually an average weighted by $P(\mathbf{X},0)$ over an ensemble of initial states $\mathbf{X}(t=0)$, which then evolve as described by the master equation, Eq. (17). In practice, Eq. (20) means an average over a large number $n_{\text{run}} \gg 1$ statistically independent runs,

$$[\bar{A}(t)]_{av} = \frac{1}{n_{\text{run}}} \sum_{l=1}^{n_{\text{run}}} A(t,l), \tag{21}$$

where $A(t,l)$ is the observable A observed at time t in the lth run of this nonequilibrium Monte-Carlo averaging.

Many concepts of nonequilibrium statistical mechanics can immediately be introduced in such simulations. For instance, one can introduce arbitrary "fields" that can be switched off to study the dynamic response functions, for both linear and nonlinear response (Binder, 1984).

4.5 The Choice of Statistical Ensembles

While so far the discussion has been (implicitly) restricted to the case of the canonical ensemble (*NVT* ensemble, for the case of a fluid), it is sometimes useful to use other statistical ensembles. Particularly useful is the grand canonical ensemble (μVT), where the chemical potential μ rather than the particle number N is fixed. In addition to moves where the configuration of particles in the box relaxes, one has moves where one attempts to add or remove a particle from the box.

In the case of binary (*AB*) mixtures, a useful variation is the "semi–grand canonical" ensemble, where $\Delta_\mu = \mu_A - \mu_B$ is held fixed and moves where an A particle is converted into a B particle (or vice versa) are considered, in an otherwise identical system configuration.

The isothermal-isobaric (NpT) ensemble, on the other hand, fixes the pressure, and then volume changes $V \to V' = V + \Delta V$ need to be considered (rescaling properly the positions of the particles).

It also is possible to define artificial ensembles that are not in the textbooks on statistical mechanics. An example is the so-called Gaussian ensemble (which interpolates between the canonical and microcanonical ensembles and is useful for the study of first-order phase transitions, as the so-called "multicanonical ensemble"). Particularly useful is the "Gibbs ensemble," where one considers the equilibrium between two simulation boxes (one containing liquid, the other gas), which can exchange both volume ΔV and particles (ΔN), while the total volume and total particle number contained in the two boxes are held fixed. The Gibbs ensemble is widely used for the simulation of gas-fluid coexistence, avoiding interfaces (Panagiotopoulos, 1992; Levesque and Weis, 1992).

A simulation at a given state point (*NVT*) contains information not only on averages at that state point but also on neighboring state points (*NVT'*), via suitable reweighting of the energy distribution $P_N(E)$ with a factor $\exp(E/k_BT)\exp(-E/k_BT')$. Such "histogram methods" are particularly useful near critical points (Swendsen *et al.*, 1992).

5. ACCURACY PROBLEMS: FINITE-SIZE PROBLEMS, DYNAMIC CORRELATION OF ERRORS, BOUNDARY CONDITIONS

5.1 Finite-Size–Induced Rounding and Shifting of Phase Transitions

A prominent application of Monte-Carlo simulation in statistical thermodynamics and lattice theory is the study of phase transitions. Now it is well known in statistical physics that sharp phase transitions can occur in the thermodynamic limit only, $N \to \infty$. Of course, this is no practical problem in everyday life—even a small water droplet freezing into a snowflake contains about $N = 10^{18}$ H_2O molecules, and, thus, the rounding and shifting of the freezing are on a relative scale of $1/\sqrt{N} = 10^{-9}$ and thus completely negligible. But the situation is different for simulations, which often consider extremely small systems (e.g., a hypercubic d-dimensional box with linear dimensions L, $V = L^d$, and periodic boundary conditions), where only $N \sim 10^2$ to 10^4 particles are involved.

In such small systems, phase transitions are strongly rounded and shifted (Barber, 1983; Binder, 1987, 1992a; Privman, 1990). Thus, care needs to be applied when simulated systems indicate phase changes. It turns out, however, that these finite-size effects can be used as a valuable tool to infer properties of the infinite system from the finite-size behavior. As a typical example, we discuss the phase transition of an Ising ferromagnet (Fig. 1), which has a second-order phase transition at a critical temperature T_c. For $L \to \infty$, the spontaneous magnetization M_{spont} vanishes according to a power law, $M_{\text{spont}} = B(1 - T/T_c)^\beta$, B being a critical amplitude and β a critical exponent (Stanley, 1971), and the susceptibility χ and correlation length ξ diverge,

$$\chi \propto |1 - T/T_c|^{-\gamma}, \qquad \xi \propto |1 - T/T_c|^{-\nu}, \tag{22}$$

where γ, ν are the associated critical exponents. In a finite system, ξ cannot exceed L, and, hence, these singularities are smeared out.

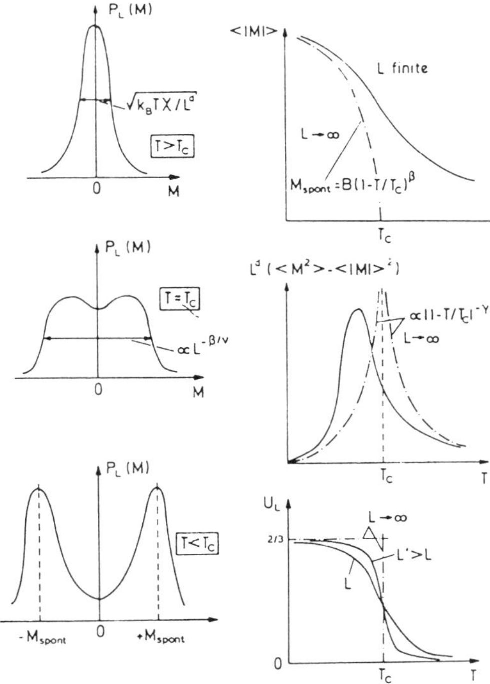

FIG. 1. Schematic evolution of the order-parameter probability distribution $P_L(M)$ from $T > T_c$ to $T < T_c$ (from above to below, left part), for an Ising ferromagnet (where M is the magnetization) in a box of volume $V = L^d$. The right part shows the corresponding temperature dependence of the mean order parameter $\langle|M|\rangle$, the susceptibility $k_BT\chi' = L^d(\langle M^2\rangle - \langle|M|\rangle^2)$, and the reduced fourth-order cumulant $U_L = 1 - \langle M^4\rangle/[3\langle M^2\rangle^2]$. Dash-dotted curves indicate the singular variation that results in the thermodynamic limit, $L \to \infty$.

Now finite-size scaling theory (Barber, 1983; Privman, 1990) implies that basically these finite-size effects can be understood from the principle that "L scales with ξ"; i.e., the order-parameter probability distribution $P_L(M)$ can be written (Binder, 1987, 1992a)

$$P_L = L^{\beta/\nu}\tilde{P}(L/\xi, ML^{\beta/\nu}), \tag{23}$$

where $\tilde{P}$ is a "scaling function." From Eq. (23), one immediately obtains the finite-size scaling relations for order parameter $\langle|M|\rangle$ and the susceptibility (defined from a fluctuation relation) by taking the moments of the distribution P_L:

$$\langle|M|\rangle = L^{-\beta/\nu}\tilde{M}(L/\xi), \tag{24}$$

$$k_BT\chi' = L^d(\langle M^2\rangle - \langle|M|\rangle^2) = L^{\gamma/\nu}\tilde{X}(L/\xi), \tag{25}$$

where $\tilde{M}$, $\tilde{\chi}$ are scaling functions that follow from $\tilde{P}$ in Eq. (23). At T_c where $\xi \to \infty$, we thus have $\chi' \propto L^{\gamma/\nu}$; from the variation of the peak height of χ' with system size, hence, the exponent γ/ν can be extracted.

The fourth-order cumulant U_L is a function of L/ξ only,

$$U_L \equiv 1 - \langle M^4\rangle/(3\langle M^2\rangle^2) = \tilde{U}(L/\xi). \tag{26}$$

Here $U_L \to 0$ for a Gaussian centered at M = 0, i.e., for $T > T_c$; $U_L \to \frac{2}{3}$ for the double-Gaussian distribution, i.e., for $T < T_c$; while $U_L = \tilde{U}(0)$ is a universal nontrivial constant for $T = T_c$. Cumulants for different system sizes hence intersect at T_c, and this can be used to locate T_c precisely (Binder, 1987, 1992a).

A simple discussion of finite-size effects at first-order transitions is similarly possible. There one describes the various phases that coexist at the first-order transition in terms of Gaussians if $L \gg \xi$ (note that ξ stays finite at the first-order transition). In a finite system, these phases can coexist not only right at the transition but over a finite parameter region. The weights of the respective peaks are given in terms of the free-energy difference of the various phases. From this phenomenological description, energy and order-parameter distributions and their moments can be worked out. Of course, this description applies only for long enough runs where the system jumps from one phase to the other many times, while for short runs where the systems stay in a single phase, one would observe hysteresis.

5.2 Different Boundary Conditions: Simulation of Surfaces and Interfaces

We now briefly mention the effect of various boundary conditions. Typically one uses periodic boundary conditions to study bulk properties of systems not obscured by surface effects. However, it also is possible to choose different boundary conditions to study surface effects deliberately; e.g., one may simulate thin films in a $L \times L \times D$ geometry with two free $L \times L$ surfaces and periodic boundary conditions otherwise. If the film thickness D is large enough, the two surfaces do not influence each other, and one can infer the properties of a semi-infinite system. One may choose special interactions near the free surfaces, apply surface "fields" (even if they cannot be applied in the laboratory, it may nevertheless be useful to study the response to them in the simulation), etc.

Sometimes the boundary conditions may stabilize interfaces in the system (e.g., in an Ising model for $T < T_c$ a domain wall between phases with opposite magnetization will be present, if we apply strong enough surface fields of opposite sign). Such interfaces also are often the object of study in simulation. It may be desirable to simulate interfaces without having the systems disturbed by free surfaces. In an Ising system, this may simply be done by choosing antiperiodic boundary conditions. Combining antiperiodic and staggered periodic boundary conditions, even tilted interfaces may be stabilized in the system. In all such simulations of systems containing interfaces one must keep in mind, however, that because of capillary-wave excitations, interfaces usually are very slowly relaxing objects, and often a major effort in computing time is needed to equilibrate them. A further difficulty (when one is interested in interfacial profiles) is the fact that the center of the interface is typically delocalized.

5.3 Estimation of Statistical Errors

We now return to the problem of judging the time needed for having reasonably small errors in Monte-Carlo sampling. If the subsequent configurations used were uncorrelated, we simply could use Eq. (15), but in the case of correlations we have rather

$$\langle(\delta A)^2\rangle = \left\langle\left[\frac{1}{n}\sum_{\mu=1}^{n} A_\mu - \langle A\rangle\right]^2\right\rangle = \frac{1}{n}\left[[\langle A^2\rangle - \langle A\rangle^2 + 2\sum_{\mu=1}^{n}\left(1-\frac{\mu}{n}\right) \times (\langle A_0 A_\mu\rangle - \langle A\rangle^2)\right]. \tag{27}$$

Now we remember that a time $t_\mu = \mu\delta t$ is associated with the Monte-Carlo process, δt being the time interval between two successive observations A_μ, $A_{\mu+1}$. Transforming the summation to a time integration yields

$$\langle(\delta A)^2\rangle = \frac{1}{n}(\langle A^2\rangle - \langle A\rangle^2) \times \left[1 + \frac{2}{\delta t}\int_0^{t_n}\left(1-\frac{t}{t_n}\right)\phi_A(t)dt\right], \tag{28}$$

where

$$\phi_A(t) \equiv \frac{\langle A(0)A(t)\rangle - \langle A\rangle^2}{\langle A^2\rangle - \langle A\rangle^2}.$$

Defining a relaxation time $\tau_A = \int_0^\infty dt\phi_A(t)$, one obtains for $\tau_A \ll n\delta t = \tau_{\text{obs}}$ (the observation time)

$$\langle(\delta A)^2\rangle = \frac{1}{n}[\langle A^2\rangle - \langle A\rangle^2]\left(1 + \frac{2\tau_A}{\delta t}\right) \approx 2\left(\frac{\tau_A}{\tau_{\text{obs}}}\right)[\langle A^2\rangle - \langle A\rangle^2]. \tag{29}$$

In comparison with Eq. (15), the dynamic correlations inherent in a Monte-Carlo sampling as described by the master equation, Eq. (17), lead to an enhancement of the expected statistical error $\langle(\delta A)^2\rangle$ by a "dynamic factor" $1 + 2\tau_A/\delta t$ (sometimes also called the "statistical inefficiency").

This dynamic factor is particularly cumbersome near second-order phase transitions (τ_A diverges: critical slowing down) and near first-order phase transitions (τ_A diverges at phase coexistence, because of the large lifetime of metastable states). Thus, even if one is interested only in static quantities in a Monte-Carlo simulation, understanding the dynamics may be useful for estimating errors. Also the question of how many configurations (M_0) must be omitted at the start of the averaging for the sake of equilibrium [Eq. (18)] can be formally answered in terms of a nonlinear relaxation function

$$\phi^{(nl)}(t) = \frac{\langle A(t)\rangle_T - \langle A(\infty)\rangle_T}{\langle A(0)\rangle_T - \langle A(\infty)\rangle_T}$$

and its associated time $\tau_A^{(nl)} = \int_0^\infty \phi_A^{(nl)}(t)dt$ by the condition $t_{M_0} \gg \tau_A^{(nl)}$.

6. QUANTUM MONTE-CARLO TECHNIQUES

6.1 General Remarks

Development of Monte-Carlo techniques to study ground-state and finite-temperature properties of interacting quantum many-body systems is an active area of research (for reviews see Ceperley and Kalos, 1979; Schmidt and Kalos, 1984; Kalos, 1984; Berne and Thirumalai, 1986; Suzuki, 1986; Schmidt and Ceperley, 1992; De Raedt and von der Linden, 1992). These methods are of interest for problems such as the structure of nuclei (Carlson, 1988) and elementary particles (De Grand, 1992), superfluidity of ^3He and ^4He (Schmidt and Ceperley, 1992), high-T_c superconductivity (e.g., Frick *et al.*, 1990), magnetism (e.g., Reger and Young, 1988), surface physics (Marx *et al.*, 1993), etc. Despite this widespread interest, much of this research has the character of "work in progress" and hence cannot feature more prominently in the present article. Besides, there is not just one quantum Monte-Carlo method, but many variants exist: variational Monte Carlo (VMC), Green's-function Monte Carlo (GFMC), projector Monte Carlo (PMC), path-integral Monte Carlo (PIMC), grand canonical quantum Monte Carlo (GCMC), world-line quantum Monte Carlo (WLQMC), etc. Some of these (like VMC, GFMC) address ground-state properties, others (like PIMC) finite temperatures. Here only the PIMC technique will be briefly sketched, following Gillan and Christodoulos (1993).

6.2 Path-Integral Monte-Carlo Methods

We wish to calculate thermal averages for a quantum system and thus rewrite Eqs. (10) and (11) appropriately,

$$\langle A\rangle = \frac{1}{Z}\,\mathrm{Tr}\exp\left(-\frac{\hat{\mathcal{H}}}{k_BT}\right)\hat{A},$$
$$Z = \mathrm{Tr}\exp\left(-\frac{\hat{\mathcal{H}}}{k_BT}\right), \tag{30}$$

using a notation that emphasizes the operator character of the Hamiltonian $\hat{\mathcal{H}}$ and of the quantity $\hat{A}$ associated with the variable A that we consider. For simplicity, we consider first a system of a single particle in one dimension acted on by a potential $V(x)$. Its Hamiltonian is

$$\hat{\mathcal{H}} = -\frac{\hbar^2}{2m}\frac{d^2}{dx^2} + V(x). \tag{31}$$

Expressing the trace in the position representation, the partition function becomes

$$Z = \int dx\langle x|\exp(-\hat{\mathcal{H}}/k_BT)|x\rangle, \tag{32}$$

where $|x\rangle$ is an eigenvector of the position operator. Writing $\exp(-\hat{\mathcal{H}}/k_BT)$ formally as $[\exp(-\hat{\mathcal{H}}/k_BTP)]^P$, where P is a positive integer, we can insert a complete set of states between the factors:

$$Z = \int dx_1 \ldots \int dx_P\langle x_1|\exp(-\hat{\mathcal{H}}/k_BTP)|x_2\rangle \langle x_2| \ldots |x_P\rangle\langle x_P|\exp(-\hat{\mathcal{H}}/k_BTP)|x_1\rangle. \tag{33}$$

For large P, it is a good approximation to ignore the fact that kinetic and potential energy do not commute. Hence, one gets

$$\langle x|\exp(-\hat{\mathcal{H}}/k_BTP)|x'\rangle \approx \left(\frac{k_BTmP}{2\pi\hbar^2}\right)^{1/2}\exp\left[\frac{-k_BTmP}{2\pi\hbar^2}(x-x')^2\right] \times \exp\left\{\frac{-1}{2k_BTP}[V(x)+V(x')]\right\}, \tag{34}$$

and

$$Z \approx \left(\frac{k_BTmP}{2\pi\hbar^2}\right)^{P/2}\int dx_1 \ldots \int dx_P \exp\left\{-\frac{1}{k_BT} \times \left[\tfrac{1}{2}\sum_{s=1}^{P}\kappa(x_s-x_{s+1})^2 + P^{-1}\sum_{s=1}^{P}V(x_s)\right]\right\}, \tag{35}$$

where

$$\kappa \equiv (k_BT/\hbar)^2 mP \tag{36}$$

In the limit $P \to \infty$, Eq. (35) becomes exact. Apart from the prefactor, Eq. (35) is precisely the configurational partition function of a classical system of a ring polymer, consisting of P beads coupled by harmonic springs with spring constant κ. Each bead is under the action of a potential $V(x)/P$.

This approach is straightforwardly generated to a system of N interacting quantum particles—one ends up with a system of N classical cyclic "polymer" chains. As a result of this isomorphism, the Monte-Carlo method developed for simulating classical systems can be carried over to such quantum-mechanical problems, too. It is also easy to see that the system always behaves classically at high temperatures—κ gets very large, and then the cyclic chains contract essentially to a point, while at low temperatures they are spread out, representing zero-point motion. However, PIMC becomes increasingly difficult at low temperatures, since P has to be the larger the lower T: If σ is a characteristic distance over which the potential $V(x)$ changes, one must have $\hbar^2/m\sigma^2 \ll k_B TP$ in order that two neighbors along the "polymer chain" are at a distance much smaller than σ. In PIMC simulations, one empirically determines and uses that P beyond which the thermodynamic properties do not effectively change.

This approach can be generalized immediately to the density matrix $\rho(x - x') = \langle x|\exp(-\hat{\mathcal{H}}/k_B T)|x'\rangle$, while there are problems with the calculation of time-displaced correlation functions $\langle A(t)B(0)\rangle$, where t is now the true time (associated with the time evolution of states following from the Schrödinger equation, rather than the "time" of Sec. 4.3 related to the master equation).

The step leading to Eq. (34) can be viewed as a special case of the Suzuki–Trotter formula (Suzuki, 1986)

$$\exp(\hat{A} + \hat{B}) = \lim_{P\to\infty} [\exp(\hat{A}/P) \exp(\hat{B}/P)]^P, \tag{37}$$

which is also used for mapping d-dimensional quantum problems on lattices to equivalent classical problems (in $d + 1$ dimensions, because of the additional "Trotter direction" corresponding to the imaginary time direction of the path-integral).

6.3 An Application Example: the Momentum Distribution of Fluid ^{4}He

We now consider the dynamic structure factor $S(\mathbf{k},\omega)$, which is the Fourier transform of a time-displaced pair correlation function of the density at a point $\mathbf{r}_1$ at time t_1 and the density at point $\mathbf{r}_2$ at time t_2 [$\hbar\mathbf{k}$ being the momentum transfer and $\hbar\omega$ the energy transfer of an inelastic scattering experiment by which one can measure $S(\mathbf{k},\omega)$]. In the "impulse approximation," the dynamic structure factor $S(\mathbf{k},\omega)$ can be related to the Fourier transform of the single-particle density matrix $\rho_1(\mathbf{r})$, which for ^{4}He can be written in terms of the wave function $\psi(\mathbf{r})$ as $\rho_1(\mathbf{r}) = \langle\psi^+(\mathbf{r}' + \mathbf{r})\psi(\mathbf{r})\rangle$. This relation is

$$S(k,\omega) \propto J(Y) = \frac{1}{\pi}\int_0^\infty \rho_1(r)\cos(Yr)dr,$$

where $Y \equiv m(\omega - k^2/2m)/k$ (West, 1975). Since $\mathbf{S}(k,\omega)$ has been measured via neutron scattering (Sokol *et al.*, 1989), a comparison between experiment and simulation can be performed without adjustable parameters (Fig. 2). Thus, the PIMC method yields accurate data in good agreement with experiment.

The studies of ^{4}He have also yielded qualitative evidence for superfluidity (Ceperley and Pollock, 1987). For a quantitative analysis of the λ transition, a careful assessment of finite-size effects (Fig. 1) is needed since one works with very small particle numbers (of the order of 10^2 ^{4}He atoms only). This has not been possible so far.

6.4 A Few Qualitative Comments on Fermion Problems

Particles obeying Fermi–Dirac statistics (such as electrons or ^{3}He, for instance) pose particular challenges to Monte-Carlo simulation. If one tries to solve the Schrödinger equation of a many-body system by Monte-Carlo methods, one exploits its analogy with a diffusion equation (Ceperley and Kalos, 1979; Kalos, 1984). As mentioned at various places in this article, diffusion processes correspond to random walks and are hence accessible to Monte-Carlo simulation. However, while the diffusion equation (for one particle) considers the probability $P(\mathbf{r},t)$ that

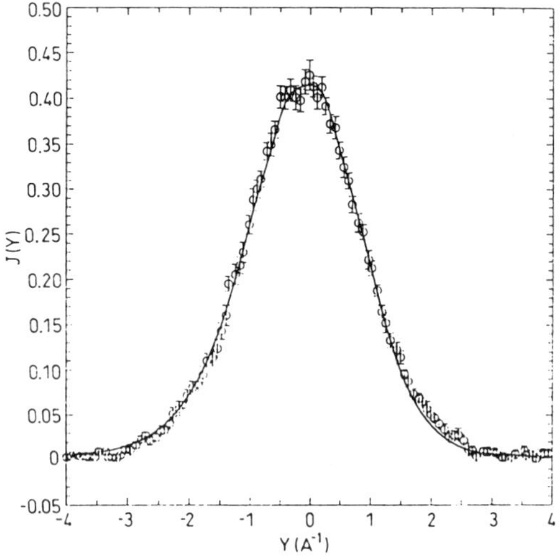

FIG. 2. The measured momentum distribution $J(Y)$ of ^{4}He at $T = 3.3$ K (circles, from Sokol *et al.*, 1989) compared with the PIMC result of Ceperley and Pollock (1987) (solid line). From Schmidt and Ceperley (1992).

a particle starting at time $t = 0$ at the origin has reached the position $\mathbf{r}$ at time t, the wave function ψ in the Schrödinger equation is not positive definite. This fact creates severe problems for wave functions of many-fermion systems, since these wave functions must be antisymmetric, and the "nodal surface" in configuration space (where ψ changes sign) is unknown.

Formally the difficulty of applying importance-sampling techniques to distributions $\rho(\mathbf{r})$ that are not always positive can be overcome by splitting $\rho(\mathbf{r})$ into its sign, $s = \text{sign}(\rho)$, and its absolute value, $\rho = s|\rho|$, and one can use $\tilde{\rho}(\mathbf{r}) = |\rho(\mathbf{r})|/\int|\rho(\mathbf{r})|d^3r$ as a probability density for importance sampling, and absorb the sign of $\rho(\mathbf{r})$ in the quantity to be measured (e.g., De Raedt and von der Linden, 1992). Symbolically, the average of an observable A is obtained as

$$\langle A \rangle = \frac{\displaystyle\int d^3r A(\mathbf{r}) s(\mathbf{r}) \tilde{\rho}(\mathbf{r})}{\displaystyle\int s(\mathbf{r}) \tilde{\rho}(\mathbf{r}) d^3r} = \frac{\langle As \rangle_{\tilde{\rho}}}{\langle s \rangle_{\tilde{\rho}}},$$

where $\langle \ldots \rangle_{\tilde{\rho}}$ means averaging with $\tilde{\rho}$ as weight function. However, as is not unexpected, using $\tilde{\rho}$ one predominantly samples unimportant regions in phase space; therefore, in sampling the sign $\langle s \rangle_{\tilde{\rho}}$, one has large cancellations from regions where the sign is negative, and, for N degrees of freedom, one gets $\langle s \rangle_{\tilde{\rho}} \propto \exp(-\text{const.} \times N)$. This difficulty is known as the "minus-sign problem" and still hampers applications to fermion problems significantly!

Sometimes it is possible to start with a trial wave function where nodes are a reasonable first approximation to the actual nodes, and, starting with the population of random walks from this fixed-node approximation given by the trial function, one now admits walks that cross this nodal surface and sample the sign as indicated above. In this way, it has been possible to estimate the exchange-correlation energy of the homogeneous electron gas (Ceperley and Alder, 1980) over a wide range of densities very well.

7. LATTICE GAUGE THEORY

Monte-Carlo simulation has become the primary tool for nonperturbative quantum chromodynamics, the field theory of quarks and hadrons and other elementary particles (e.g., Rebbi, 1984; De Grand, 1992). In this section, we first stress the basic problem, to make the analogy with the calculations of statistical mechanics clear. Then we very briefly highlight some of the results that have been obtained so far.

7.1 Some Basic Ideas of Lattice Gauge Theory

The theory of elementary particles is a field theory of gauge fields and matter fields. Choice of a lattice is useful to provide a cutoff that removes the ultraviolet divergences that would otherwise occur in these quantum field theories. The first step, hence, is the appropriate translation from the four-dimensional continuum (3 space + 1 time dimensions) to the lattice.

The generating functional (analogous to the partition function in statistical mechanics) is

$$Z = \int DAD\bar{\psi}D\psi \exp[-S_g(A,\bar{\psi},\psi)], \tag{38}$$

where A represents the gauge fields, $\bar{\psi}$ and ψ represent the (fermionic) matter field, S_g is the action of the theory (containing a coupling constant g, which corresponds to inverse temperature in statistical mechanics as const.$/g^2 \to 1/k_BT$), and the symbols $\int D$ stand for functional integration. The action of the gauge field itself is, using the summation convention that indices that appear twice are summed over,

$$S_G = \tfrac{1}{4} \int d^4x F^{\alpha}_{\mu\nu}(x) F^{\mu\nu}_{\alpha}(x), \tag{39}$$

$F^{\alpha}_{\mu\nu}$ being the fields that derive from the vector potential $A^{\alpha}_{\mu}(x)$. These are

$$F^{\alpha}_{\mu\nu}(x) = \partial_\mu A^{\alpha}_{\nu}(x) - \partial_\nu A^{\alpha}_{\mu}(x) + g f^{\alpha}_{\beta\gamma} A^{\beta}_{\mu}(x) A^{\gamma}_{\nu}(x), \tag{40}$$

$f^{\alpha}_{\beta\gamma}$ being the structure constants of the gauge group, and g a coupling constant.

The fundamental variables that one then introduces are elements $U_\mu(x)$ of the gauge group G, which are associated with the links of the four-dimensional lattice, connecting x and a nearest neighbor point $x + \mu$:

$$U_\mu(x) = \exp[igaT^{\alpha} A^{\alpha}_{\mu}(x)],$$
$$[U_m(x + m)]^{\dagger} = U_m(x), \tag{41}$$

where a is the lattice spacing and T^{α} a group generator. Here $U^{\dagger}$ denotes the Hermitean conjugate of U. Wilson (1974) invented a lattice action that reduces in the continuum limit to Eq. (39), namely

$$\frac{S_U}{k_B T} = \frac{1}{g^2} \sum_n \sum_{\mu > \nu} \mathrm{Re\,Tr} U_\mu(n) \times U_\nu(n+\mu) U_\mu^\dagger(n+\nu) U_\nu^\dagger(n), \tag{42}$$

where the links in Eq. (42) form a closed contour along an elementary plaquette of the lattice.

Using Eq. (42) in Eq. (38), which amounts to the study of a "pure" gauge theory (no matter fields), we recognize that the problem is equivalent to a statistical mechanics problem (such as spins on a lattice), the difference being that now the dynamical variables are the gauge group elements $U_\mu(n)$. Thus importance-sampling Monte-Carlo algorithms can be put to work, just as in statistical mechanics.

In order to include also matter fields, one starts from a partition function of the form

$$Z = \int DU D\bar{\psi} D\psi \exp\left\{ -\frac{S_U}{k_B T} + \sum_{i=1}^{n_f} \bar{\psi} \underline{M} \psi \right\} = \int DU (\det \underline{M})^{n_f} \exp\left(-\frac{S_U}{k_B T} \right), \tag{43}$$

where we have assumed fermions with n_f degenerate "flavors." It has also been indicated that the fermion fields can be integrated out analytically, but the price is that one has to deal with the "fermion determinant" of the matrix $\underline{M}$. In principle, for any change of the U's this determinant needs to be recomputed; together with the fact that one needs to work on rather large lattices in four dimensions, in order to reproduce the continuum limit, this problem is responsible for the huge requirement of computing resources in this field.

It is clear that lattice gauge theory cannot be explained in depth on two pages—we only intend to give a vague idea of what these calculations are about to a reader who is not familiar with this subject.

7.2 A Recent Application

Among the many Monte-Carlo studies of various problems (which include problems in cosmology, like the phase transition from the quark-gluon plasma to hadronic matter in the early universe), we focus here on the problem of predicting the masses of elementary particles. Butler *et al.* (1993) have used a new massively parallel supercomputer with 480 processors ("GF11") exclusively for one year to run lattice sizes ranging from $8^3 \times 32$ to $24^3 \times 32$, $24^3 \times 36$, and $30 \times 32^2 \times 40$. Their program executes at a speed of more than 5 Gflops (Giga floating point operations per second), and the rather good statistics reached allowed a meaningful elimination of finite-size effects by an extrapolation to the infinite-volume limit. This problem is important, since the wave function of a hadron is spread out over many lattice sites.

Even with this impressive effort, several approximations are necessary:

1. The fermion determinant mentioned above is neglected (this is called "quenched approximation").
2. One cannot work at the (physically relevant) very small quark mass m_q, but rather has to take data on the various hadron masses for a range of quark masses and extrapolate these data to zero (Fig. 3).

After a double extrapolation ($m_q \to 0$, lattice spacing at fixed volume $\to 0$), one obtains mass ratios that are in very satisfactory agreement with experiment. For example, for the nucleon the mass ratio for the finite volume is $m_N/m_\rho = 1.285 \pm 0.070$,, extrapolated to infinite volume 1.219 ± 0.105, the experimen-

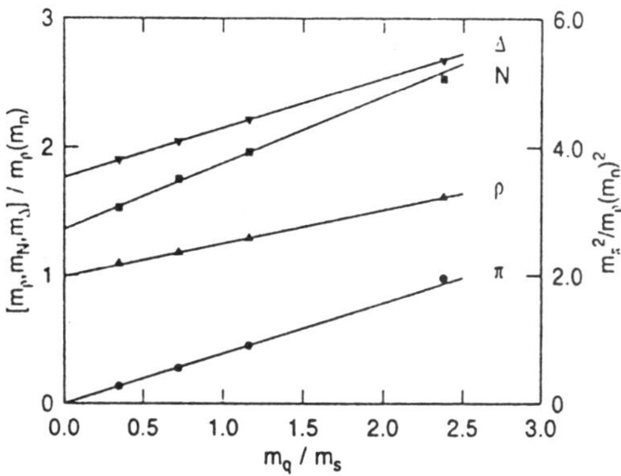

FIG. 3. For a $30 \times 32^2 \times 40$ lattice at $(k_B T)^{-1} = 6.17$, m_π^2, m_ρ, m_N, and m_Δ in units of the physical rho meson mass $m_\rho(m_n)$, as functions of the quark mass m_q in units of the strange quark mass m_s. Particles studied are pion, rho meson, nucleon, and delta baryon, respectively. From Butler *et al.* (1993).

tal value being 1.222 (all masses in units of the mass m_ρ of a rho meson).

8. SELECTED COMMENTS ON APPLICATIONS IN CLASSICAL STATISTICAL MECHANICS OF CONDENSED-MATTER SYSTEMS

In this section, we mention a few applications very briefly, just in order to give the flavor of the type of work that is done and the kind of questions that are asked and answered by Monte-Carlo simulations. More extensive reviews can be found in the literature (Binder, 1976, 1979, 1984, 1992b).

8.1 Metallurgy and Materials Science

A widespread application of Monte-Carlo simulation in this area is the study of order-disorder phenomena in alloys: One tests analytical approximations to calculate phase diagrams, such as the cluster variation (CV) method, and one tests to what extent a simple model can describe the properties of complicated materials.

An example (Fig. 4) shows the order parameter for long-range order (LRO) and short-range order (SRO, for nearest neighbors) as function of temperature for a model of Cu_3Au alloys on the fcc lattice. Here an Ising model with antiferromagnetic interactions between nearest neighbors only is studied, and the Monte-Carlo data (filled symbols and symbols with crosses) are compared to CV calculations (broken curve) and other analytical calculations (dash-dotted curve) and to experiment (open circles). The simulation shows that the analytical approximations describe the ordering of the model only qualitatively. Of course, there is no perfect agreement with the experimental data either; this is to be expected, of course, since in real alloys the interaction range is considerably larger than just extending to nearest neighbors only.

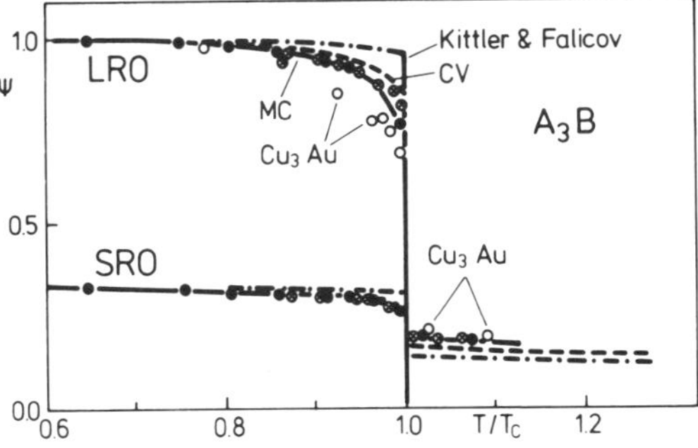

FIG. 4. Long-range order parameter (LRO) and absolute value of nearest-neighbor short-range order parameter (SRO) plotted versus temperature T (in units of the temperature T_c where the first-order transition occurs) for a nearest-neighbor model of binary alloys on the face-centered cubic lattice with A_3B structure. Open circles: experimental data for Cu_3Au; broken and dash-dotted curves: results of analytical theories. Full dots: Monte-Carlo results obtained from a simulation in the semi-grand canonical ensemble (chemical potential difference between A and B atoms treated as independent variable); circles with crosses: values obtained from a canonical ensemble simulation (concentration of B atoms fixed at 25%). From Binder *et al.* (1981).

8.2 Polymer Science

One can study phase transitions not only for models of magnets or alloys, of course, but also for complex systems such as mixtures of flexible polymers. A question heavily debated in the recent literature is the dependence of the critical temperature of unmixing of a symmetric polymer mixture (both constituents have the same degree of polymerization $N_A = N_B = N$) on chain length N. The classical Flory–Huggins theory predicted $T_c \propto N$, while a recent integral equation theory predicted $T_c \propto \sqrt{N}$ (Schweizer and Curro, 1990). This law would lead in the plot of Fig. 5 to a straight line through the origin. Obviously, the data seem to rule out this behavior, and are rather qualitatively consistent with Flory–Huggins theory (though the latter significantly overestimates the prefactor in the relation $T_c \propto N$).

Polymer physics provides examples for many scientific questions where simulations could contribute significantly to provide a better understanding. Figure 6 provides one more example (Paul *et al.*, 1991a). The problem is to provide truly microscopic evidence for the reptation concept (Doi and Edwards, 1986). This concept implies that, as a result of "entanglements" between chains in a dense melt, each chain moves snakelike along its own contour. This behavior leads to a special behavior of mean square displacements: After a characteristic time τ_e, one should see a crossover from a law $g_1(t) \equiv \langle[\mathbf{r}_i(t) - \mathbf{r}_i(0)]^2\rangle \propto t^{1/2}$ (Rouse model) to a law $g_1(t) \propto t^{1/4}$, and, at a still later time (τ_R), one should see an-

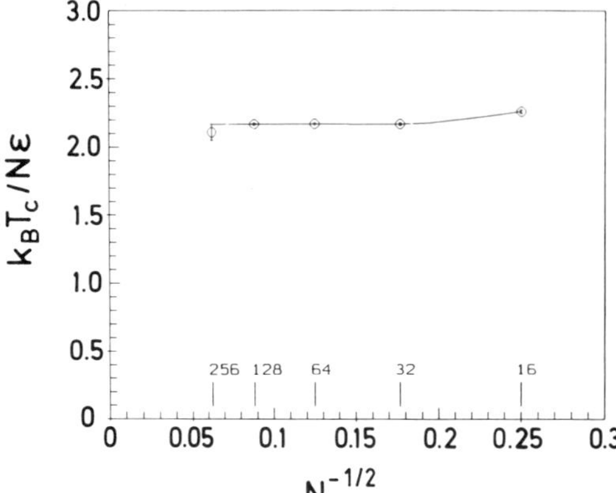

FIG. 5. Normalized critical temperature $k_BT/N\epsilon$ of a symmetric polymer mixture (N = chain length, ϵ = energy parameter describing the repulsive interaction between A-B pairs of monomers) plotted versus $N^{-1/2}$. Data are results of simulations for the bond-fluctuation model, using N in the range from N = 16 to N = 256, as indicated. The data are consistent with an asymptotic extrapolation $k_BT_c/\epsilon \approx 2.15N$, while Flory–Huggins theory (in the present units) would yield $k_BT_c/\epsilon \approx 7N$, and the integral-equation theory $k_BT_c/\epsilon \propto \sqrt{N}$. From Deutsch and Binder (1992).

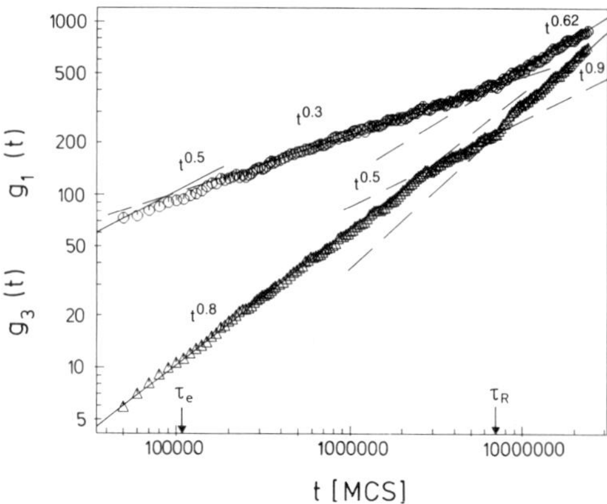

FIG. 6. Log-log plot of the mean square displacements of inner monomers [$g_1(t)$] and of the center of gravity of the chain [$g_3(t)$] versus time t (measured in units of Monte-Carlo steps, while lengths are measured in units of the lattice spacing). Straight lines show various power laws as indicated; various characteristic times are indicated by arrows (see text). Data refer to the bond-fluctuation model on the simple cubic lattice, for an athermal model of a polymer melt with chain length N = 200 and a volume fraction ϕ = 0.5 of occupied lattice sites. From Paul *et al.* (1991a).

other crossover to $g_1(t) \propto t^{1/2}$ again. At the same time, the center of gravity displacement should also show an intermediate regime of anomalously slow diffusion, $g_3(t) \propto t^{1/2}$. Figure 6 provides qualitative evidence for these predictions—although the effective exponents indicated do not quite have the expected values. A challenge for further theoretical explanation is the anomalous center-of-gravity diffusion in the initial Rouse regime, $g_3(t) \propto t^{0.8}$.

While this is an example where dynamic Monte-Carlo simulations are used to check theories—and pose further theoretical questions—one can also compare to experiment if one uses data in suitably normalized form. In Figure 7, the diffusion constant D of the chains is normalized by its value in the Rouse regime (limit for small N) and plotted versus N/N_e where the characteristic "entanglement chain length" N_e is extracted from τ_e shown in Figure 6 (see Paul *et al.*, 1991a, b, for details). The Monte-Carlo data presented in this scaled form agree with results from both a molecular dynamics (MD) simulation (Kremer and Grest, 1990) and experiment on polyethylene (PE) (Pearson *et al.*, 1987).

This example also shows that, for slow dif-

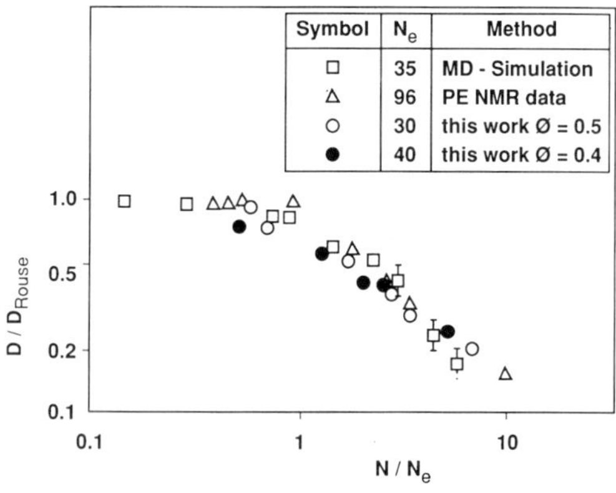

FIG. 7. Log-log plot of the self-diffusion constant D of polymer chains, normalized by the Rouse diffusivity, versus N/N_e (N_e = entanglement chain length, estimated independently and indicated in the inset). Circles: from Monte-Carlo simulations of Paul *et al.* (1991b); squares: from molecular dynamics simulations (Kremer and Grest, 1990); triangles: experimental data (Pearson *et al.*, 1987). From Paul *et al.* (1991b).

fusive motions, Monte-Carlo simulation is competitive to molecular dynamics, although it does not describe the fast atomic motions realistically.

8.3 Surface Physics

Our last example considers phenomena far from thermal equilibrium. Studying the ordering behavior of ordered superstructures, we treat the problem where initially the adatoms are adsorbed at random, and one gradually follows the formation of ordered domains out of initially disordered configurations. In a scattering experiment, one expects to see this by the gradual growth of a peak at the Bragg position $\mathbf{q}_B$. Figure 8 shows a simulation for the case of the (2×1) structure on the square lattice, where the Bragg position is at the Brillouin-zone boundary ($q_B = \pi$ if lengths are measured in units of the lattice spacing). Here a lattice gas with repulsive interactions between nearest and next-nearest neighbors (of equal strength) was used (Sadiq and Binder, 1984), using a single–spin-flip kinetics (if the lattice gas is translated to an Ising spin model), as is appropriate for a description of a monolayer in equilibrium with surrounding gas (the random "spin flips" then correspond to random evaporation-condensation events). Figure 9 presents evidence

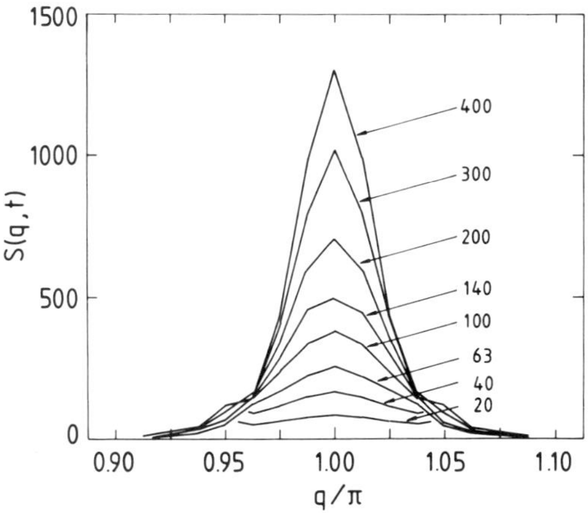

FIG. 8. Structure factor $S(q,t)$ versus wavevector q at various times t (in units MCS per lattice site), after the system was started in a completely random configuration. Temperature (measured in units of the nearest-neighbor repulsion W_{nn}) is 1.33 ($T_c \approx 2.07$ in these units), and coverage $\theta = \frac{1}{2}$. From Sadiq and Binder (1984).

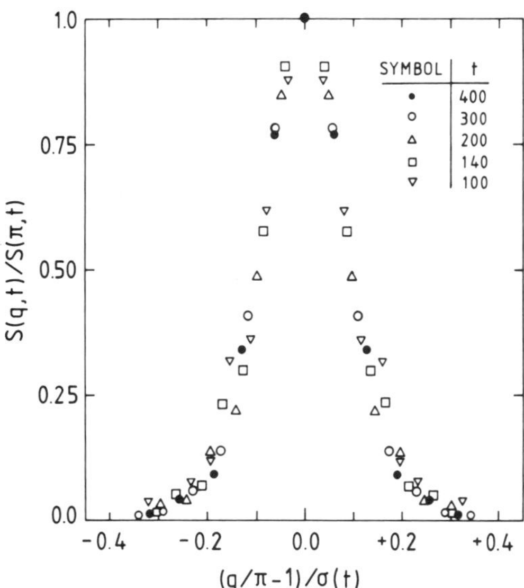

FIG. 9. Structure factor of Fig. 8 plotted in scaled form, normalizing $S(q,t)$ by its peak value $S(\pi,t)$ and normalizing $q/\pi - 1$ by the halfwidth $\sigma(t) = L^{-1}(t)$, where $L(t)$ thus defined is the characteristic domain size. From Sadiq and Binder (1984).

that these data on the kinetics of ordering satisfy a scaling hypothesis, namely

$$S(q,t) = [L(t)]^2 \tilde{S}(|\mathbf{q} - \mathbf{q}_B| L(t)), \qquad (44)$$

where $\tilde{S}$ is a scaling function. This hypothesis, Eq. (44), was first proposed on the basis of simulations, and later it was established to describe experimental data as well.

Of course, surface physics provides many more examples where simulations have been useful; see Binder and Landau (1989) for a review.

9. CONCLUDING REMARKS

In this article, the basic features of the most widely used numerical techniques that fall into the category of Monte-Carlo calculations were described. There is a vast literature on the subject—the author estimates the number of papers using Monte-Carlo methods in condensed-matter physics of the order of 10^4, in lattice gauge theory of the order of 10^3! Thus many important variants of algorithms could not be treated here, and interesting applications (e.g., the study of neural-network models) were completely omitted.

There also exist other techniques for the numerical simulation of complex systems, which sometimes are an alternative approach to Monte-Carlo simulation. The molecular dynamics (MD) method (numerical integration of Newton's equations) has already been mentioned in the text, and there exist combinations of both methods ("hybrid Monte Carlo," "Brownian dynamics," etc.). A combination of MD with the local-density approximation of quantum mechanics is the basis of the Car–Parrinello method.

Problems like that shown in Figs. 8 and 9 can also be formulated in terms of numerically solving appropriate differential equations, which may in turn even be discretized to cellular automata. When planning a Monte Carlo simulation, hence, some thought to the question "When which method?" should be given.

GLOSSARY

Critical Slowing Down: Divergence of the relaxation time of dynamic models of statistical mechanics at a second-order phase transition (critical point).

Detailed Balance Principle: Relation linking the transition probability for a move and the transition probability for the inverse move to the ratio of the probability for the occurrence of these states in thermal equilibrium. This condition is sufficient for a Markov process to tend toward thermal equilibrium.

Ergodicity: Property that ensures equality of statistical ensemble averages (such as the "canonic ensemble" of statistical mechanics) and time averages along the trajectory of the system through phase space.

Finite-Size Scaling: Theory that describes the finite-size–induced rounding of singularities that would occur at phase transitions in the thermodynamic limit.

Heat-Bath Method: Choice of transition probability where the probability to "draw" a trial value for a degree of freedom does not depend on its previous value.

Importance Sampling: Monte-Carlo method that chooses the states that are generated according to the desired probability distribution. For example, for statistical mechanics applications, states are chosen with weights proportional to the Boltzmann factor.

Lattice Gauge Theory: Field theory of quarks and gluons in which space and time are discretized into a four-dimensional lattice, gauge field variables being associated to the links of the lattice.

Master Equation: Rate equation describing the "time" evolution of the probability that a state occurs as a function of a "time" coordinate labeling the sequence of states (in the context of importance-sampling Monte-Carlo methods).

Molecular-Dynamics Method: Simulation method for interacting many-body systems based on numerical integration of the Newtonian equations of motion.

Monte-Carlo Step: Unit of (pseudo) time in (dynamically interpreted) importance sampling where, on the average, each degree of freedom in the system gets one chance to be changed (or "updated").

Random Number Generator (RNG): Computer subroutine to produce pseudorandom numbers that are approximately not correlated with each other and approximately uniformly distributed in the interval from zero to one. RNGs typically are strictly periodic, but the period is large enough that, for practical applications, this periodicity does not matter.

Simple Sampling: Monte-Carlo method that chooses states uniformly and at random from the available phase space.

Transition Probability: Probability that controls the move from one state to the next one in a Monte-Carlo process.

Works Cited

Ahrens, J. H., Dieter, U. (1979), *Pseudo Random Numbers*, New York: Wiley.

Barber, M. N. (1983), in: C. Domb, J. L. Lebowitz (Eds.), *Phase Transitions and Critical Phenomena*, Vol. 8 New York: Academic, Chap. 2.

Barkema, G. T., Marko, J. F. (1993), *Phys. Rev. Lett.* **71**, 2070–2073.

Berne, B. J., Thirumalai, D. (1986), *Annu. Rev. Phys. Chem.* **37**, 401.

Binder, K. (1976), in: C. Domb, M.S. Green (Eds.), *Phase Transitions and Critical Phenomena*, Vol. 5b, New York: Academic, p. 1.

Binder, K. (1979) (Ed.), *Monte Carlo Methods in Statistical Physics*, Berlin: Springer.

Binder, K. (1984) (Ed.), *Applications of the Monte Carlo Method in Statistical Physics*, Berlin: Springer.

Binder, K. (1987), *Ferroelectrics* **73**, 43–67.

Binder, K. (1992a), *Annu. Rev. Phys. Chem.* **43**, 33–59.

Binder, K. (1992b) (Ed.), *The Monte Carlo Method in Condensed Matter Physics*, Berlin: Springer.

Binder, K., Heermann, D. W. (1988), *Monte Carlo Simulation in Statistical Physics: An Introduction*, Berlin: Springer.

Binder, K., Landau, D. P. (1989), in: K. P. Lawley (Ed.), *Molecule-Surface Interaction*, New York: Wiley, pp. 91–152.

Binder, K., Lebowitz, J. L., Phani, M. K., Kalos, M. H. (1981), *Acta Metall.* **29**, 1655–1665.

Bruns, W., Motoc, I., O'Driscoll, K. F. (1981), *Monte Carlo Applications in Polymer Science*, Berlin: Springer.

Butler, F., Chen, H., Sexton, J., Vaccarino, A., Weingarten, D. (1993), *Phys. Rev. Lett.* **70**, 2849–2852.

Carlson, J. (1988), *Phys. Rev. C* **38**, 1879–1885.

Ceperley, D. M., Alder, B. J. (1980), *Phys. Rev. Lett.* **45**, 566–569.

Ceperley, D. M., Kalos, M. H. (1979), in: K. Binder (Ed.), *Monte Carlo Methods in Statistical Physics*, Berlin: Springer, pp. 145–194.

Ceperley, D. M., Pollock, E. L. (1987), *Can. J. Phys.* **65**, 1416–1420.

Ciccotti, G., Hoover, W. G. (1986) (Eds.), *Molecular Dynamics Simulation of Statistical Mechanical Systems*, Amsterdam: North-Holland.

Compagner, A. (1991), *Am. J. Phys.* **59**, 700–705.

Compagner, A., Hoogland, A. (1987), *J. Comput. Phys.* **71**, 391–428.

De Grand, T. (1992), in: H. Gausterer, C. B. Lang (Eds.), *Computational Methods in Field Theory*, Berlin: Springer, pp. 159–203.

De Raedt, H., von der Linden, W. (1992), in: K. Binder (Ed.), *The Monte Carlo Method in Condensed Matter Physics*, Berlin: Springer, pp. 249–284.

Deutsch, H. P., Binder, K. (1992), *Macromolecules* **25**, 6214–6230.

Doi, M., Edwards, S. F. (1986), *Theory of Polymer Dynamics*, Oxford: Clarendon Press.

Ferrenberg, A. M., Landau, D. P., Wong, Y. J. (1992), *Phys. Rev. Lett.* **69**, 3382–3384.

Frick, M., Pattnaik, P. C., Morgenstern, I., Newns, D. M., von der Linden, W. (1990), *Phys. Rev. B* **42**, 2665–2668.

Gillan, M. J., Christodoulos, F. (1993), *Int. J. Mod. Phys.* **C4**, 287–297.

Gould, H., Tobochnik, J. (1988), *An Introduction to Computer Simulation Methods/Applications to Physical Systems, Parts 1 and 2*, Reading, MA: Addison-Wesley.

Hammersley, J. M., Handscomb, D. C. (1964), *Monte Carlo Methods*, London: Chapman and Hall.

Heermann, D. W. (1986), *Computer Simulation Methods in Theoretical Physics*, Berlin: Springer.

Herrmann, H. J. (1986), *Phys. Rep.* **136**, 153–227.

Herrmann, H. J. (1992), in: K. Binder (Ed.), *The Monte Carlo Method in Condensed Matter Physics*, Berlin: Springer, Chap. 5.

Hockney, R. W., Eastwood, J. W. (1988), *Computer Simulation using Particles*, Bristol: Adam Hilger.

James, F. (1990), *Comput. Phys. Commun.* **60**, 329–344.

Kalos, M. H. (1984) (Ed.), *Monte Carlo Methods in Quantum Problems*, Dordrecht: Reidel.

Kalos, M. H., Whitlock, P. A. (1986), *Monte Carlo Methods*, Vol. 1, New York: Wiley.

Kirkpatrick, S., Stoll, E. (1981), *J. Comput. Phys.* **40**, 517–526.

Knuth, D. (1969), *The Art of Computer Programming*, Vol. 2, Reading, MA: Addison-Wesley.

Koonin, S. E. (1981), *Computational Physics*, Reading, MA: Benjamin.

Kremer, K., Binder, K. (1988), *Comput. Phys. Rep.* **7**, 259–310.

Kremer, K., Grest, G. S. (1990), *J. Chem. Phys.* **92**, 5057–5086.

Landau, D. P. (1992), in: K. Binder (Ed.), *The Monte Carlo Method in Condensed Matter Physics*, Berlin: Springer, Chap. 2.

Lehmer, D. H. (1951), in: *Proceedings of the 2nd Symposium on Large-Scale Digital Computing Machinery*, Harvard University, Cambridge, 1951, pp. 142–145.

Levesque, D., Weis, J. J. (1992), in: K. Binder (Ed.), *The Monte Carlo Method in Condensed Matter Physics*, Berlin: Springer, Chap. 6.

Marsaglia, G. A. (1985), in: L. Billard (Ed.), *Computer Science and Statistics: The Interface*, Amsterdam: Elsevier, Chap. 1.

Marsaglia, G. A. (1986), *Proc. Natl. Acad. Sci. U.S.A.* **61**, 25–28.

Marsaglia, G. A., Narasumhan, B., Zaman, A. (1990), *Comput. Phys. Comm.* **60**, 345–349.

Marx, D., Opitz, O., Nielaba, P., Binder, K. (1993), *Phys. Rev. Lett.* **70**, 2908–2911.

Meakin, P. (1988), in: C. Domb, J. L. Lebowitz (Eds.), *Phase Transitions and Critical Phenomena*, Vol. 12, New York: Academic, pp. 336–489.

Metropolis, N., Rosenbluth, A. W., Rosenbluth, M. N., Teller, A. M., Teller, E. (1953), *J. Chem. Phys.* **21**, 1087–1092.

Panagiotopoulos, A. Z. (1992), *Molec. Simul.* **9**, 1–23.

Paul, W., Binder, K., Heermann, D. W., Kremer, K. (1991a), *J. Chem. Phys.* **95**, 7726–7740.

Paul, W., Binder, K., Heermann, D. W., Kremer, K. (1991b), *J. Phys. (Paris)* **111**, 37–60.

Pearson, D. S., Verstrate, G., von Meerwall, E., Schilling, F. C. (1987), *Macromolecules* **20**, 113–1139.

Privman, V. (1990) (Ed.), *Finite Size Scaling and Numerical Simulation of Statistical Systems*, Singapore: World Scientific.

Rebbi, C. (1984), in: K. Binder (Ed.), *Application*

of the Monte Carlo Method in Statistical Physics, Berlin: Springer, p. 277.

Reger, J. D., Young, A. P. (1988), *Phys. Rev. B* **37**, 5978–5981.

Rosenbluth, M. N., Rosenbluth, A. W. (1955), *J. Chem. Phys.* **23**, 356–362.

Sadiq, A., Binder, K. (1984), *J. Stat. Phys.* **35**, 517–585.

Schmidt, K. E., Ceperley, D. M. (1992), in: K. Binder (Ed.), *The Monte Carlo Method in Condensed Matter Physics*, Berlin: Springer, pp. 203–248.

Schmidt, K. E., Kalos, M. H. (1984), in: K. Binder (Ed.), *Applications of the Monte Carlo Method in Statistical Physics*, Berlin: Springer, pp. 125–149.

Schweizer, K. S., Curro, J. G. (1990), *Chem. Phys.* **149**, 105–127.

Sokol, P. E., Sosnick, T. R., Snow, W. M. (1989), in: R. E. Silver, P. E. Sokol (Eds.), *Momentum Distributions*, New York: Plenum Press.

Stanley, H. E. (1971), *An Introduction to Phase Transitions and Critical Phenomena*, Oxford: Oxford University Press.

Stauffer, D. (1985), *An Introduction to Percolation Theory*, London: Taylor & Francis.

Suzuki, M. (1986) (Ed.), *Quantum Monte Carlo Methods*, Berlin: Springer.

Swendsen, R. H., Wang, J. S., Ferrenberg, A. M. (1992), in: K. Binder (Ed.), *The Monte Carlo Method in Condensed Matter Physics*, Berlin: Springer, Chap. 4.

Tausworthe, R. C. (1965), *Math. Comput.* **19**, 201–208.

Tolman, S., Meakin, P. (1989), *Phys. Rev. A* **40**, 428–437.

Vicsek, T. (1989), *Fractal Growth Phenomena*, Singapore: World Scientific.

West, G. B. (1975), *Phys. Rep.* **18C**, 263–323.

Wilson, K. (1974), *Phys. Rev. D* **10**, 2445–2453.

Further Reading

A textbook describing for the beginner how to learn to write Monte-Carlo programs and to analyze the output generated by them has been written by Binder, K., Heermann, D. W. (1988), *Monte Carlo Simulation in Statistical Physics: An Introduction*, Berlin: Springer. This book emphasizes applications of statistical mechanics such as random walks, percolation, and the Ising model.

A useful book that gives much weight to applications outside of statistical mechanics is Kalos, M. H., Whitlock, P. A. (1986), *Monte Carlo Methods, Vol. 1*, New York: Wiley.

A more general but pedagogic introduction to computer simulation is presented in Gould, H., Tobochnik, J. (1988), *An Introduction to Computer Simulation Methods/Applications to Physical Systems, Parts 1 and 2*, Reading, MA: Addison-Wesley.

A rather systematic collection of applications of Monte-Carlo studies in statistical mechanics and condensed matter physics has been compiled in a series of books edited by the author of this article: Binder, K. (1979) (Ed.), *Monte Carlo Methods in Statistical Physics*, Berlin: Springer; Binder, K. (1984) (Ed.), *Applications of the Monte Carlo Method in Statistical Physics*, Berlin: Springer; and Binder, K. (1992) (Ed.), *The Monte Carlo Method in Condensed Matter Physics*, Berlin: Springer.

Finally we draw attention to two important areas which are only briefly covered in the present article: For quantum problems, see Suzuki, M. H. (1987) (Ed.), *Quantum Monte Carlo Methods*, Berlin: Springer; Doll, J. D., Gubernaitis, J. E. (1990) (Eds.), *Quantum Simulations*, Singapore: World Scientific. For lattice gauge theory, see Bunk, B., Mütter, K. H., Schilling, K. (1986) (Eds.), *Lattice Gauge Theory, A Challenge in Large-Scale Computing*, New York: Plenum; De Grand, T. (1992), in: H. Gausterer, C. B. Lang (Eds.), *Computational Methods in Field Theory*, Berlin: Springer, pp. 159–203; Smit, J., van Baal, P. (1993) (Eds.), *Lattice 92*, Amsterdam: North Holland.

LIST OF RECOMMENDED UNITS AND SYMBOLS

RECOMMENDED UNITS AND CONVERSION FACTORS

The SI system provides six basic units: meter m, kilogram kg, second s, ampere A, kelvin K, candela cd, and mole mol.

Some important derived units are also allowed and bear special names, e.g.:

1 N (newton) = 1 kg m s^{-2}
1 J (joule) = 1 N m = 1 kg m^2 s^{-2}
1 W (watt) = 1 J s^{-1} = 1 kg m^2 s^{-3}
1 Pa (pascal) = 1 N m^{-2} = 1 kg m^{-1} s^{-2}

For mass, the gram g, or the metric ton t which equals 1000 kg, may be used instead of kilogram kg.

The so-called "long ton" (UK) and "short ton" (US) have been abandoned and will not be used in the *Encyclopedia of Applied Physics*.

For pressure, the bar (name and symbol alike) may be used, and for temperature, the degree celsius °C.

From all of these, decimal multiples or fractions can be derived:

Power of ten	Prefix	Symbol	Power of ten	Prefix	Symbol
10	deca	da	10^{-1}	deci	d
10^2	hecto	h	10^{-2}	centi	c
10^3	kilo	k	10^{-3}	milli	m
10^6	mega	M	10^{-6}	micro	μ
10^9	giga	G	10^{-9}	nano	n
10^{12}	tera	T	10^{-12}	pico	p
10^{15}	peta	P	10^{-15}	femto	f
10^{18}	exa	E	10^{-18}	atto	a

For mass, multiples or fractions are derived from g, not from kg (since the latter already contains a prefix), e.g., mg.

Units with the prefixes are considered as one entity and can, therefore, be raised to any power, e.g., cm^3.

The liter is now considered as synonymous with dm^3 (which does not hold in the older literature!). Please use the capital letter L as unit symbol (following a recent IUPAC recommendation). The use of the Ångström unit is discouraged; it should be replaced by fractional meters (1 Å = 100 pm = 0.1 nm).

SELECTED QUANTITIES, UNITS, AND SYMBOLS

Name of quantity[a]	Symbol[b]	SI unit[c]	Name of unit	Other units
Space and time				
length*	l	m	meter[d]	
breadth, width	b	m		
height	h	m		
radius	r	m		
thickness	d	m		
area	A,S	m^2		a (are) h (hectare)
volume	V	m^3		L, l(liter)[e]
plane angle	$\alpha,\beta,\gamma,$ ϑ,φ	1, rad	radian	° (degree) ′ (minute) ″ (second)
solid angle	ω,Ω	1, sr	steradian	
wavelength	λ	m		
wave number	σ,ν	m^{-1}		
time*	t	s	second	min (minute) h (hour) d (day)
frequency	ν, f	s^{-1}		Hz (hertz)[f]
relaxation time	τ	s		
velocity	u,v	$m\ s^{-1}$		km/h
acceleration	a	$m\ s^{-2}$		
Mechanics				
mass*	m	kg	kilogram	g (gram) t (tonne)[g]
(mass) density	ρ	$kg\ m^{-3}$		g/cm^3
momentum	$\mathbf{p}$	$kg\ m\ s^{-1}$		
angular momentum	$\mathbf{L}$	$kg\ m^2\ s^{-1}$		
force	$\mathbf{F}$	N	newton[f]	
moment of force	$\mathbf{M}$	N m		
weight	G, W	N		
pressure	p	Pa	pascal	bar (bar)[h]
energy	E, W	J	joule	W h (watt hour)[i] eV (electron volt)[j]
work	W,A	J		
power	P	W	watt	J/s, V A[k]
Molecular Physics and Thermodynamics				
thermodynamic temperature*	T	K	kelvin	°C (degrees Celsius)
Celsius temperature	ϑ,t			°C
number of entities	N			
Avogadro constant[l]	N_A,L	mol^{-1}	(particles) per mole	
Boltzmann constant[m]	k,k_B	$J\ K^{-1}$		
Planck constant[n]	h	J s		
(molar) gas constant[o]	R	$J\ mol^{-1}\ K^{-1}$		
(quantity of) heat	Q	J		
entropy[p]	S	$J\ K^{-1}$		
internal energy[p]	U	J		
Helmholtz function,[p] (Helmholtz) free energy, Helmholtz energy	F,A	J		
enthalpy[p]	H	J		
Gibbs function,[p] (Gibbs) free energy, Gibbs energy	G	J		
heat capacity[p]	C_p,C_V	$J\ K^{-1}$		

Name of quantity[a]	Symbol[b]	SI unit[c]	Name of unit	Other units
Chemical Physics				
amount of substance*	n	mol	mole	
relative atomic mass	A_r	1		
relative molecular mass	M_r	1		
atomic mass constant	m_u	kg		u (atomic mass unit)[q]
mass of a portion (of substance B)	$m_B, m(B)$	kg		g (gram)
molar mass (of substance B)	$M_B, M(B)$	kg mol^{-1}		
concentration (of substance B)	$c_B, c(B)$	mol m^{-3}		mol/L
mole fraction[r] (of substance B)	$\kappa_B, \kappa(B)$	1		
mass fraction (of substance B)	$\omega_B, \omega(B)$	1		%,‰,ppm,ppb
volume fraction (of substance B)	$\varphi_B, \varphi(B)$ $\phi_B, \phi(B)$	1		%,‰,ppm,ppb
mass concentration	ρ	kg m^{-3}		g/L
molality		mol kg^{-1}		mmol/kg
volume concentration[s]	σ	1		
molar volume	V_m	m^3 mol^{-1}		L/mol
molar heat capacity	C_m	J mol^{-1} K^{-1}		
molar conductivity	Λ_m	S m^2 mol^{-1}	(S: siemens)	
Faraday constant[t]	F	C mol^{-1}		
Electricity and Magnetism				
quantity of electricity[u]	Q	C	coulomb	
charge density	ρ	C m^{-3}		
electric potential	ϕ, V	V	volt	
electric potential difference, voltage	$U, \Delta\phi, \Delta V$	V		
electric dipole moment	$\mathbf{p}, \mathbf{p}_c$	C m		
electric current*	I	A	ampere	
electric current density	j	A m^{-2}		
electric field strength	$\mathbf{E}$	V m^{-1}		
electric displacement	$\mathbf{D}$	C m^{-2}		
capacitance	C	F	farad	
permittivity	ε	F m^{-1}		
relative permittivity	ε	1		
dielectric polarization	$\mathbf{P}$	C m^{-2}		
electric susceptibility	χ_e	1		
polarization (of a particle)	α	m^2 C V^{-1}		
magnetic flux	Φ	Wb	weber	
magnetic flux density	$\mathbf{B}$	T	tesla	
magnetic field strength	$\mathbf{H}$	A m^{-1}		
permeability	μ	H m^{-1}, N A^{-2}	(H: henry)	
relative permeability	μ_r	1		
magnetization	M	A m^{-1}		
magnetic susceptibility	χ	1		
molar magnetic susceptibility	χ_m	m^3 mol^{-1}		
(electrical) resistance	R	Ω	ohm	
(electrical) conductance	G	S	siemens	
(electrical) resistivity	ρ	Ω m		
(electrical) conductivity	κ, σ	S m^{-1}		
self-inductance	L	H	henry	
Radiation				
radiant energy	Q, W, Q_e	J		
luminous intensity*	I	cd	candela	
radiant intensity	I_e	W sr^{-1}, W		
emissivity, emittance	ε	1		

Name of quantity[a]	Symbol[b]	SI unit[c]	Name of unit	Other units
absorptance	α	1		
reflectance	ρ, R	1		
transmittance	τ	1		
absorption coefficient:				
linear (decadic)	a	m^{-1}		
molar (decadic)	ε	$m^2\ mol^{-1}$		
refractive index	n	1		
molar refraction	R_m	$m^3\ mol^{-1}$		
angle of optical rotation	α	1, rad		
Transport Properties				
flux of quantity X	J_X, J	(varies)		
mass flow rate	$q_m, \dot{m}$	$kg\ s^{-1}$		
volume flow rate	$q_V, \dot{V}$	$m^3\ s^{-1}$		
heat flow rate	Φ	W		
thermal conductivity	κ, k, λ	$W\ m^{-1}\ K^{-1}$		
coefficient of heat transfer	h	$W\ m^{-2}\ K^{-1}$		
thermal diffusivity	a	$m^2\ s^{-1}$		
diffusion coefficient	D	$m^2\ s^{-1}$		
thermal diffusion coefficient	D_T	$m^2\ s^{-1}$		
viscosity	η, μ	Pa s		
kinematic viscosity	ν	$m^2\ s^{-1}$		

[a]SI base quantities are marked by asterisks (*)
[b]Recommended by IUPAC
[c]SI base units as well as derived and supplementary units are listed; all are to be used with prefixes as needed
[d]do not use "metre"
[e]do not use "litre"; $1\ L = 10^{-3}\ m^3$
[f]$1\ Hz = 1\ s^{-1}$; $1\ N = 1\ kg\ m\ s^{-2}$
[g]Formerly metric ton; $1\ t = 10^3\ kg$
[h]$1\ bar = 10^5\ Pa$
[i]$1\ W\ h = 3.6 \times 10^3\ J$
[j]$1\ eV = 1.602\ 189 \times 10^{-19}\ J$
[k]$1\ W = 1\ J/s = 1\ VA$
[l]$N_A = 6.022\ 136\ 7 \times 10^{23}\ mol^{-1}$
[m]$k = 1.380\ 658 \times 10^{-23}\ J\ K^{-1}$
[n]$h = 6.626\ 075\ 5 \times 10^{-34}\ J\ s$
[o]$R = 8.314\ 510\ J\ mol^{-1}\ K^{-1}$
[p]Molar quantities can be distinguished from the quantity of a system by adding the subscript m; e.g., molar internal energy U_m, in $J\ mol^{-1}$
[q]$1\ u = 1.660\ 565\ 5 \times 10^{-27}\ kg$
[r]A more accurate, but rather uncommon, name is "amount-of-substance fraction"
[s]σ refers to the total volume of a mixture, whereas the volume fraction φ relates the volume of a substance to the volume of several components before mixing
[t]$F = 9.648\ 530\ 9 \times 10^4\ C\ mol^{-1}$
[u]Also called electric charge